# ELECTROMECHANICAL DESIGN HANDBOOK

## Other Handbooks of Interest from McGraw-Hill

*Avallone and Baumeister* • MARK'S STANDARD HANDBOOK FOR MECHANICAL ENGINEERS
*Bhushan and Gupta* • HANDBOOK OF TRIBOLOGY
*Brady and Clauser* • MATERIALS HANDBOOK
*Bralla* • HANDBOOK OF PRODUCT DESIGN FOR MANUFACTURING
*Brink* • HANDBOOK OF FLUID SEALING
*Brunner* • HANDBOOK OF INCINERATION SYSTEMS
*Corbitt* • STANDARD HANDBOOK OF ENVIRONMENTAL ENGINEERING
*Ehrich* • HANDBOOK OF ROTORDYNAMICS
*Elliot* • STANDARD HANDBOOK OF POWERPLANT ENGINEERING
*Freeman* • STANDARD HANDBOOK OF HAZARDOUS WASTE TREATMENT AND DISPOSAL
*Ganíc and Hicks* • THE McGRAW-HILL HANDBOOK OF ESSENTIAL ENGINEERING INFORMATION AND DATA
*Gieck* • ENGINEERING FORMULAS
*Grimm and Rosaler* • HANDBOOK OF HVAC DESIGN
*Harris* • HANDBOOK OF NOISE CONTROL
*Harris and Crede* • SHOCK AND VIBRATION HANDBOOK
*Hicks* • STANDARD HANDBOOK OF ENGINEERING CALCULATIONS
*Hodson* • MAYNARD'S INDUSTRIAL ENGINEERING HANDBOOK
*Jones* • DIESEL PLANT OPERATIONS HANDBOOK
*Juran and Gryna* • JURAN'S QUALITY CONTROL HANDBOOK
*Karassik et al.* • PUMP HANDBOOK
*Kurtz* • HANDBOOK OF APPLIED MATHEMATICS FOR ENGINEERS AND SCIENTISTS
*Mason* • SWITCH ENGINEERING HANDBOOK
*Nayyar* • PIPING HANDBOOK
*Parmley* • STANDARD HANDBOOK OF FASTENING AND JOINING
*Rosaler and Rice* • STANDARD HANDBOOK OF PLANT ENGINEERING
*Rothbart* • MECHANICAL DESIGN AND SYSTEMS HANDBOOK
*Schwartz* • COMPOSITE MATERIALS HANDBOOK
*Schwartz* • HANDBOOK OF STRUCTURAL CERAMICS
*Shigley and Mischke* • STANDARD HANDBOOK OF MACHINE DESIGN
*Townsend* • DUDLEY'S GEAR HANDBOOK
*Tuma* • ENGINEERING MATHEMATICS HANDBOOK
*Tuma* • HANDBOOK OF NUMERICAL CALCULATIONS IN ENGINEERING
*Wadsworth* • HANDBOOK OF STATISTICAL METHODS FOR ENGINEERS AND SCIENTISTS
*Walsh* • McGRAW-HILL MACHINING AND METALWORKING HANDBOOK
*Wang* • HANDBOOK OF AIR CONDITIONING AND REFRIGERATION
*Woodruff and Lammers* • STEAM-PLANT OPERATION
*Young* • ROARK'S FORMULAS FOR STRESS AND STRAIN

# ELECTROMECHANICAL DESIGN HANDBOOK

**Ronald A. Walsh**
*Manager, Research and Development*
*Powercon Corporation*
*Severn, Maryland*

**Second Edition**

**McGRAW-HILL, INC.**
New York San Francisco Washington, D.C. Auckland Bogotá
Caracas Lisbon London Madrid Mexico City Milan
Montreal New Delhi San Juan Singapore
Sydney Tokyo Toronto

**Library of Congress Cataloging-in-Publication Data**

Walsh, Ronald A.
Electromechanical design handbook / Ronald A. Walsh. — 2nd ed.
p. cm.
Includes bibliographical references and index.
ISBN 0-07-068035-3
1. Electromechanical devices—Design and construction—Handbooks, manuals, etc. I. Title.
TJ163.W35 1995
621—dc20 94-12956
CIP



1 2 3 4 5 6 7 8 9 0 DOC/DOC 9 0 9 8 7 6 5 4

ISBN 0-07-068035-3

*The sponsoring editor for this book was Robert W. Hauserman, the editing supervisor was Caroline R. Levine, and the production supervisor was Pamela A. Pelton. It was set in Times Roman by North Market Street Graphics.*

*Printed and bound by R. R. Donnelley & Sons Company.*

This book is printed on acid-free paper.

# CONTENTS

## ABOUT THE AUTHOR

Ronald A. Walsh is director of research and development at the Powercon Corporation in Severn, Maryland. An industrial product designer for almost 40 years with such companies as Bendix Radio, American Machine and Foundry, and Martin Marietta, he holds three U.S. patents for electronic packaging systems and mechanical devices, and five copyrights. Ronald Walsh worked at Cape Canaveral, Florida, as a liaison engineer on the Titan and Gemini aerospace vehicles and participated in the design of the Titan, Gemini, and Apollo aerospace vehicles and numerous manned military and commercial aircraft and military guided missiles. He has worked with some of the pioneers of advanced aircraft and aerospace vehicle design, atomic energy processing equipment, and military and commercial electrical and electronic equipment. He is the author of the *McGraw-Hill Machining and Metalworking Handbook.*

# PREFACE

This second edition of the *Electromechanical Design Handbook* is presented as a useful working tool for product designers, design engineers, and others who are involved directly in the product design disciplines and those in the design support groups throughout industry. The data in this handbook cover a broad range of subjects because of the nature of electromechanical design engineering practice.

Product designers who possess general and specific practical engineering knowledge are valuable to their particular company or organization. Those who have a broad range of practical electromechanical design knowledge not only are more secure in their jobs but can acquire the ability to find new jobs more easily and quickly if the need arises.

I have tried to consolidate enough basic and application-specific design data and procedures to allow the electromechanical designer to initiate many of the design calculations and procedures required to begin the development of mechanical, electrical, and electronic parts, mechanisms, assemblies, and processes. Most of the equations and techniques presented in this handbook have been previously used and accepted as standard practice throughout industry. At the same time, the experienced designer and student should understand that mathematical solutions to complex industrial design engineering problems are approximations at best in many cases. The mathematical solutions to problems in some areas of design are exact, while the solutions in others are unattainable with standard analytical procedures. There are no problem solution substitutes more effective than the prototype and *definitive testing* in proving the adequacy of a particular design or product. Some products cannot have a prototype stage, and in these cases we rely upon accepted standards and procedures formulated by the American standards organizations, and calculations and experience. The prototype and testing stages of product design are of prime importance to the industrial electromechanical product designer.

The intent of this handbook is to show practical, working design data and procedures; little emphasis is placed on theory. Successful industrial design engineers do not dwell on theory; rather, they use their practical engineering training, their intelligence, and testing and experience to solve actual industrial design problems quickly and effectively, with any means at their disposal. This handbook does not attempt to cover all subjects relevant to design engineering, only those that are used in most common and some specialized electromechanical design applications.

This second edition contains an expanded amount of information and design data that will prove even more useful than the first edition, which was so widely accepted in the design engineering community.

For those designers who are interested in or must have information concerning the manufacturing practices and processes directly related to electromechanical design engineering practices and also modern manufacturing procedures, the author recommends his *McGraw-Hill Machining and Metalworking Handbook* (1994), which, together with the *McGraw-Hill Electromechanical Design Handbook,* gives the product designer or design engineer an unprecedented amount and source of practical and effective design procedures and data, as well as manufacturing procedures and industrial processes.

American industry today, more than ever, needs people in the engineering professions who are trained in the multiple integrated disciplines of mechanical, electrical, and electronic engineering design practices (called *electromechanical design engineering,* not *mechatronics*). See the introduction to Chap. 6 and its contents for how to effectively implement these disciplines. America needs more highly trained electromechanical design engineers and fewer specialists in order to more effectively compete in the national and international markets. This has been the author's purpose and goal in writing this volume and its companion, the *McGraw-Hill Machining and Metalworking Handbook.*

*Ronald A. Walsh*

# ACKNOWLEDGMENTS

In the course of compiling data and procedures for inclusion in this handbook, I selected data, design information, and design procedures from many sources. I wish to express my gratitude particularly to the following individuals, companies, and organizations who granted permission to use their copyrighted design data or otherwise contributed to the production of this handbook: Ralph Siegel, president, Powercon Corporation (Severn, Maryland 21144); Alex Feygelman, design engineer; Katrina T. Walsh; Michelle E. Walsh; Exide Corporation; THE TIMKEN COMPANY; Boston Gear Division of IMO Industries; ALCOA (Aluminum Company of America). I am grateful also to the following societies, institutes, and specification authorities: AISC; SMI; ANSI; ASME; AGMA; NEMA; AISI; AFBMA; IFI; ASTM; SAE; Newark Electronics; Allied Electronics; Tandy Corporation; Bryant Electric—Division of Westinghouse; Thermometrics, Inc.; *Machine Design,* a Penton publication; The Chemical Rubber Publishing Company; Omega Engineering; T. B. Wood's Sons, Inc.; WATLOW, Inc.; Eastman Kodak Company; Waldes Truarc; Tinnerman Co.; Ruland Co.; PEM Inc.; Robert W. Hauserman, Senior Editor, McGraw-Hill, Inc.; and Larry S. Hager, Senior Editor, McGraw-Hill, Inc.

# INTRODUCTION

If you are involved in the design of industrial, military, or consumer products and perform mechanical, electrical, and electronic design functions, this handbook will be of value in your work.

The handbook is not intended for specialists in the various engineering disciplines, but rather for those designers who create the majority of general products, parts, mechanisms, and assemblies used throughout industry.

The handbook does not attempt to show an individual how to be a designer, but does present the basic reference data and techniques used to perform the various design functions.

The term *electromechanical designer* is appropriate today because many designers are involved with components and assemblies that contain elements that are mechanical, electrical, and electronic in nature, all dependently combined to form the finished product.

In many small companies, the electromechanical or product designer is responsible for the entire product—all aspects included. It is to the designer's advantage to be proficient in or to have reference data pertaining to as many of the disciplines involved in product design as possible.

Aside from containing a formidable amount of design data, the references listed in the bibliographies at the ends of chapters will be of assistance to many designers in carrying out their work, as will the listings of specification and standards authorities shown in Chap. 15. As stated in the handbook, the accumulation of accurate data is one of the basic design functions.

It was not my intent to show as many aspects of product design as possible, but to show those methods and data that are used again and again in the course of product-design engineering work. I have attempted to cover those subjects and areas which are of prime importance to the product designer, with emphasis on procuring additional information from various expert sources, which have been listed.

The handbook also will be of value to students of engineering attending technical teaching facilities, as it provides them with a good view of what is involved in modern product design and how to approach many basic design problems.

Electromechanical designers today have many technological advantages over their predecessors. In modern manufacturing facilities, the designer has access to accurate measuring and testing equipment, computer-aided design stations, and high-speed and accurate manufacturing equipment, such as computer-controlled machining centers and multistation punch presses, and highly accurate sheet-metal cutting and bending machines. Facilities such as these allow parts to be made consistently more accurately and at a higher rate of production than in the past.

Modern test and measuring equipment and devices, such as computers, digital oscilloscopes, digital multimeters and counters, pressure transducers, load cells, strain gauges, thermocouples, accelerometers, and high-speed cameras, allow the modern designer to acquire accurate answers to many design problems that would have been difficult or impossible to solve mathematically in the past, and that are still beyond the scope of engineering design manuals.

This equipment allows designers today to design and develop a particular product quickly, send it onto its prototype stage, and run definitive tests to prove the adequacy and safety of the design. Cost analyses can then be made and implemented, the prototype again tested, and finally, the product set up for production after the engineering documentation has been completed.

Using this handbook will enable you to perform many preliminary design studies and calculations that are necessary to define the basic requirements of the various parts, mechanisms, and assemblies that represent or make up the designer's concepts of the anticipated product.

On complex products, you will consult with specialists in the various engineering and scientific disciplines who will assist you in your efforts. It is good design practice to also consult the various manufacturers of purchased parts and the outside vendor producers of your designed parts or processes, such as springs, castings, molded-plastic parts, gears, bearings, plating, special finishes, and fabricating materials. These manufacturers are all willing and able to assist the designer in their particular specialties.

As a conclusion to this introduction, it must be said that the majority of the credit for this handbook is not due the author, but the thousands of individuals and companies who originated and developed the data and equations that fill the pages of this handbook. A consolidation of all their efforts can only be beneficial to the practicing product designer and to American industry.

# ELECTROMECHANICAL DESIGN HANDBOOK

# CHAPTER 1

# MATHEMATICS REFERENCE AND MEASUREMENT SYSTEM CONVERSIONS

This chapter reviews the standard mathematics references and measurement system conversions that are useful for many applications in the engineering disciplines. Not included are tables of logarithms, trigonometric functions, powers of numbers, involute functions, and other standard tables, the values of which are readily obtainable with the use of the pocket calculator in six or more decimal places. Also, the basic algebraic operations are not reviewed because it is assumed that the reader is thoroughly familiar with these operations and procedures.

Textbooks covering the different branches of mathematics are listed in the bibliography at the end of this chapter. Some of these texts are written in a style which facilitates self-teaching and provide a comprehensive review of difficult subject matter.

The examples contained in the sections on differential and integral calculus, including the partial derivative shown in Sec. 1.10.7, are designed to show in simplified form the computational power afforded by the use of these branches of mathematics. Most of the design equations listed in other chapters of the handbook were derived by mathematicians using the calculus and other branches and procedures of advanced mathematics.

The section on measurement system conversions includes all units that are most common to general engineering applications, taken from both the U.S. Customary and SI (Systeme International d'Unites) metric measuring systems.

The sections devoted to engineering applications of percentage operations and compound-interest procedures will be useful and practical in many applications. Applications of logarithmic procedures to solve otherwise difficult problems are also included.

In the quest to simplify the calculation procedures and increase accuracy in solving mathematics problems from all branches of industry and commerce, various methods have been used. For many years, the cumbersome and tedious numerical methods of logarithmic procedures prevailed. Engineering and scientific slide rules of various forms, including straight, circular, and cylindrical slide rules, were also extensively used until the introduction of the handheld electronic calculator in the late 1960s. Figure 1.1 shows a typical obsolete Keuffel-and-Esser (K&E) standard engineering slide rule together with its typical modern replacement; an equation programmable electronic calculator. The slide rule is shown for historical reference only.

Figure 1.2 shows a highly advanced equation graphing calculator, the Texas Instruments 81 (TI-81), which is programmable and capable of solving many complex mathematics problems. Handheld electronic calculators have made the modern design engineer's job easier, quicker, and more accurate than previously possible.

The advent of the personal computer (PC), has ushered in the introduction of mathematics programs of great computational power such as Mathematica, MathCad, and MuMath.

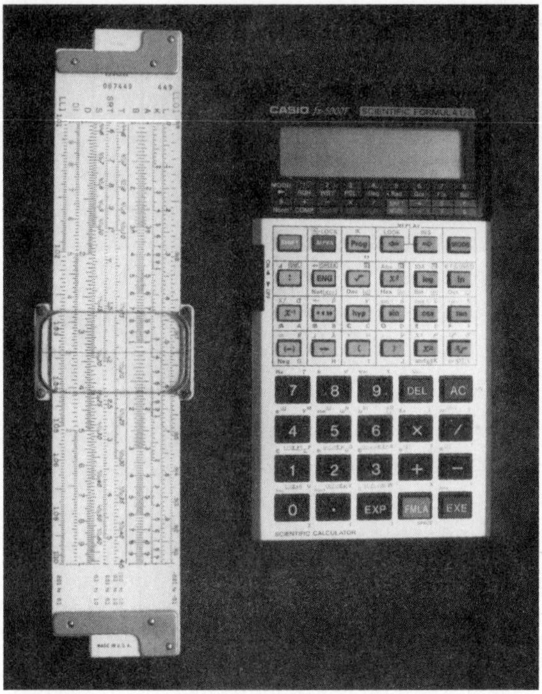


**FIGURE 1.1** Modern equation processing calculator and engineer's slide rule.

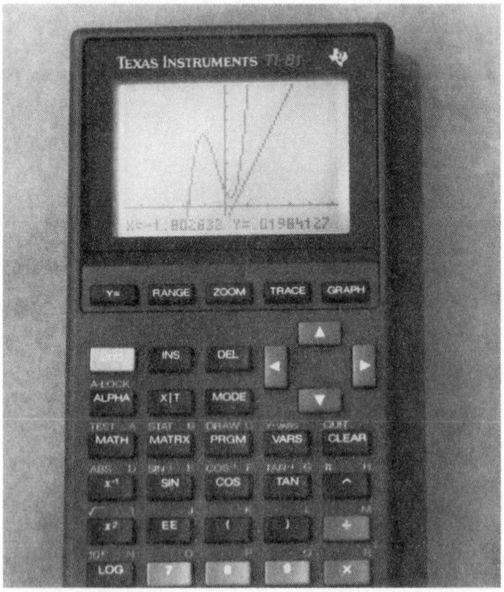


**FIGURE 1.2** A complex programmable and graphing calculator, TI-81.

Figure 1.3 shows a laser printout of a typical engineering problem as solved and presented using the MathCad 2.5 program. In this program, the equation is written as "WYSIWYG" (What You See Is What You Get) on the computer screen and the problem or equation is automatically solved. The entire screen may then be printed as it appears and kept for recording purposes. If a set of range variables is put into the problem, the MathCad program will

**RADIOACTIVE DECAY**

**This document shows the results of radioactive decay. Element A decays with a half-life of thalfA into element B. Element B decays with a half-life of thalfB.**

gm ≡ 1M          yr ≡ 1T

**Initial conditions:**

Initial amount of element A:          A0 := 100·gm

Half-lives:          thalfA := 10·yr
thalfB := 100·yr

Ratio: mass of 1 atom of A to 1 atom of B:          α := 1.1

**Compute decay constants:**

$$kA := \frac{\ln(2)}{thalfA} \qquad kB := \frac{\ln(2)}{thalfB}$$

**Decay functions:**

$$A(t) := A0 \cdot e^{-(kA \cdot t)}$$

$$B(t) := -\alpha \cdot A0 \cdot e^{-(kB \cdot t)} \left[ e^{-(kA \cdot t)} - 1 \right]$$

**Graph the mass of A and B over time:**

t := 0·yr, 2·yr ..200·yr          t2 := 0·yr, 10·yr ..200·yr

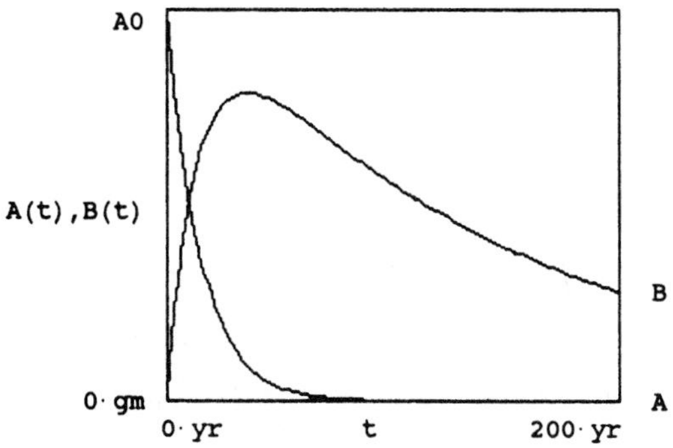


| t2 / yr | A(t2) / gm | B(t2) / gm |
|---|---|---|
| 0 | 100 | 0 |
| 10 | 50 | 51.3 |
| 20 | 25 | 71.8 |
| 30 | 12.5 | 78.2 |
| 40 | 6.3 | 78.2 |
| 50 | 3.1 | 75.4 |
| 60 | 1.6 | 71.4 |
| 70 | 0.8 | 67.2 |
| 80 | 0.4 | 62.9 |
| 90 | 0.2 | 58.8 |
| 100 | 0.1 | 54.9 |
| 110 | 0 | 51.3 |
| 120 | 0 | 47.9 |
| 130 | 0 | 44.7 |
| 140 | 0 | 41.7 |
| 150 | 0 | 38.9 |
| 160 | 0 | 36.3 |
| 170 | 0 | 33.9 |

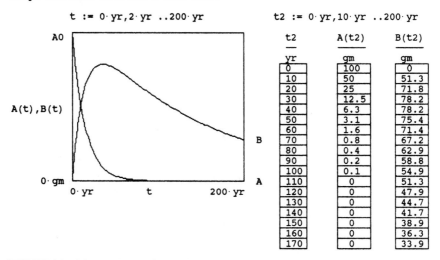

**FIGURE 1.3** Printout of a complex problem solved using MathCad 2.5.

solve for many answers, according to the extent of the range variables. Other MathCad problem printouts are shown in Chaps. 6 and 7.

## 1.1 PLANE GEOMETRY

In any triangle, $\angle A + \angle B + \angle C = 180°$, and $\angle A = 180° - (\angle B + \angle C)$, $\angle B = 180° - (\angle A + \angle C)$, and $\angle C = 180° - (\angle A + \angle B)$. See Fig. 1.4.

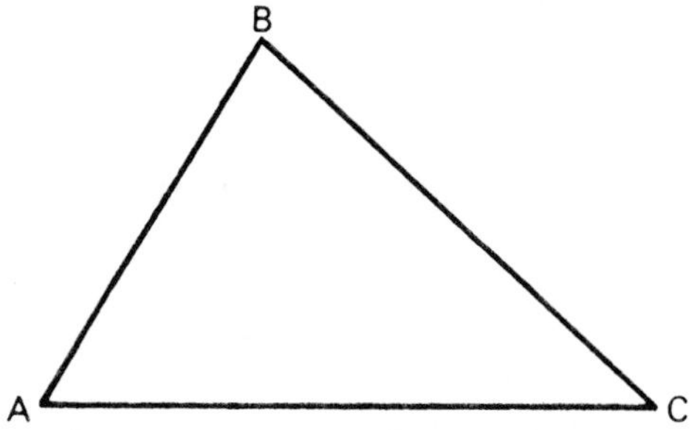


**FIGURE 1.4** Triangle.

If three sides of one triangle are proportional to the corresponding sides of another triangle, the triangles are similar. Also, if $a : b : c = a' : b' : c'$, then $\angle A = \angle A'$, $\angle B = \angle B'$, $\angle C = \angle C'$ and $a/a' = b/b' = c/c'$. Conversely, if the angles of one triangle are equal to the respective angles of another triangle, the triangles are similar and their sides are proportional; thus, if $\angle A = \angle A'$, $\angle B = \angle B'$, and $\angle C = \angle C'$, then $a : b : c = a' : b' : c'$ and $a/a' = b/b' = c/c'$. See Fig. 1.5.

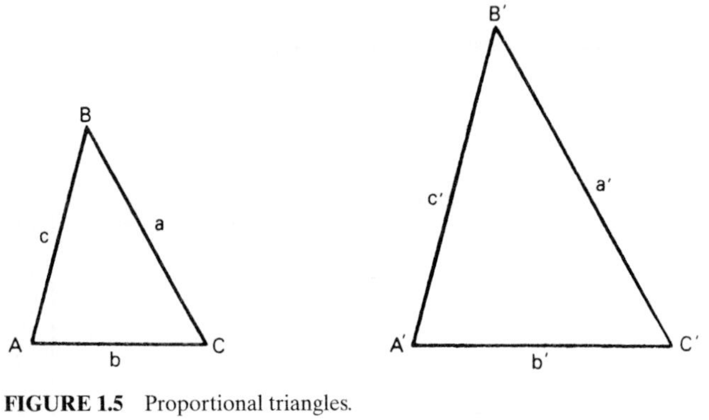


**FIGURE 1.5** Proportional triangles.

In isosceles triangles, if side $c$ = side $b$, then $\angle C = \angle B$. See Fig. 1.6.

In equilateral triangles, sides $a = b = c$ and angles $A$, $B$, and $C$ are equal (60°). See Fig. 1.7.

In right triangles, $c^2 = a^2 + b^2$ when $\angle C = 90°$:

$$c = \sqrt{a^2 + b^2} \qquad b = \sqrt{c^2 - a^2} \qquad a = \sqrt{c^2 - b^2}$$

These relations in a right triangle are called the *Pythagorean theorem.* See Fig. 1.8.

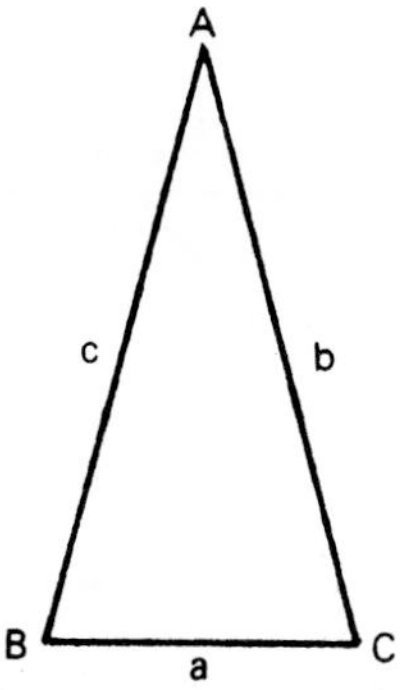


**FIGURE 1.6** Isosceles triangle.

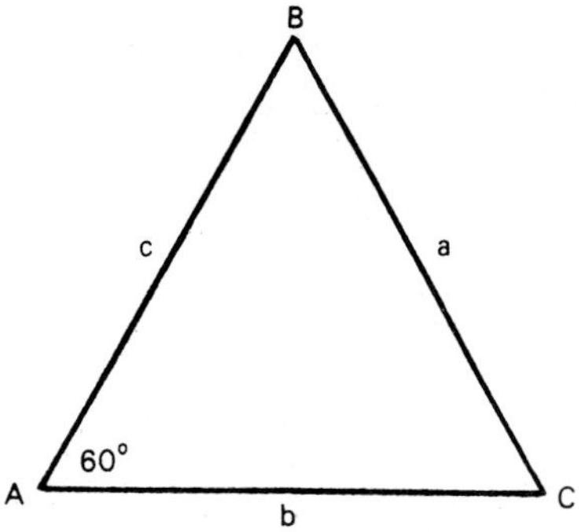


**FIGURE 1.7** Equilateral triangle.

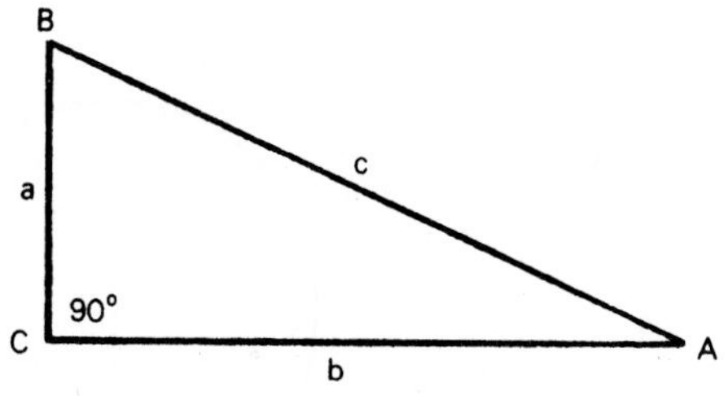


**FIGURE 1.8** A right triangle.

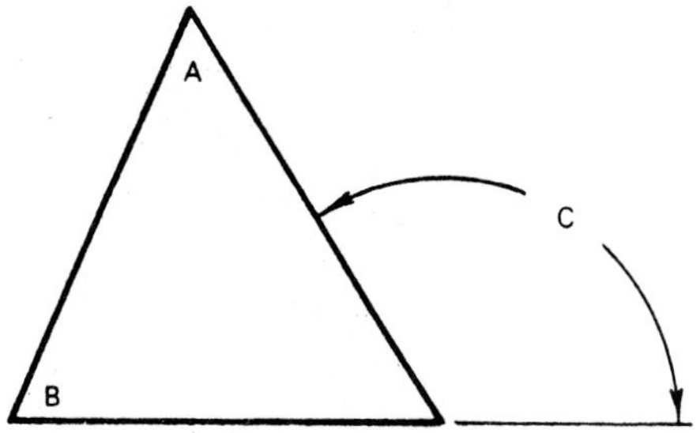


**FIGURE 1.9** External angle.

$\angle C = \angle A + \angle B$. See Fig. 1.9.

Intersecting straight lines: $\angle A = \angle A'$, $\angle B = \angle B'$. See Fig. 1.10.

Two parallel lines intersected by a straight line; alternate interior and exterior angles are equal: $\angle A = \angle A'$; $\angle B = \angle B'$. See Fig. 1.11.

In any four-sided figure, the sum of all interior angles equals 360°; $\angle A + \angle B + \angle C + \angle D = 360°$. See Fig. 1.12.

A line tangent to a point on a circle is at 90°, or *normal,* to a radial line drawn to the tangent point. See Fig. 1.13.

The common point of tangency of two circles is intersected by the line between their centers. See Fig. 1.14.

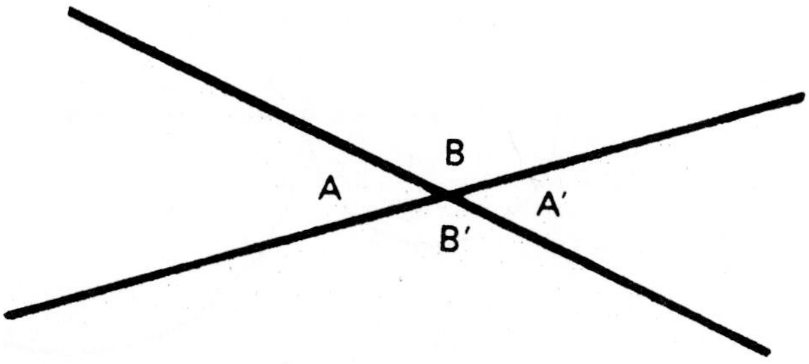


**FIGURE 1.10** Intersecting straight lines.

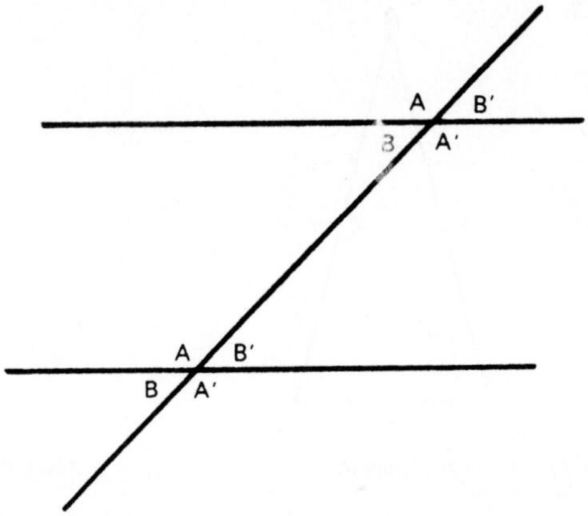


**FIGURE 1.11** Parallel lines intersected by straight line.

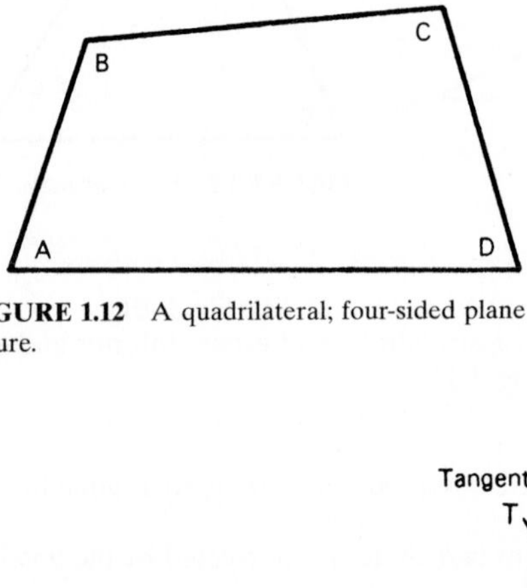


**FIGURE 1.12** A quadrilateral; four-sided plane figure.

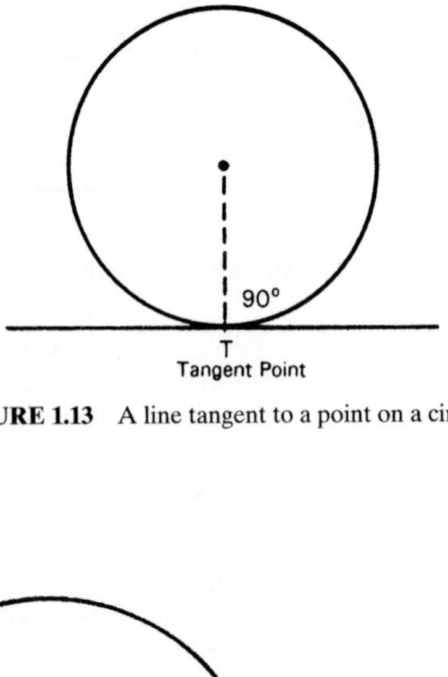


**FIGURE 1.13** A line tangent to a point on a circle.

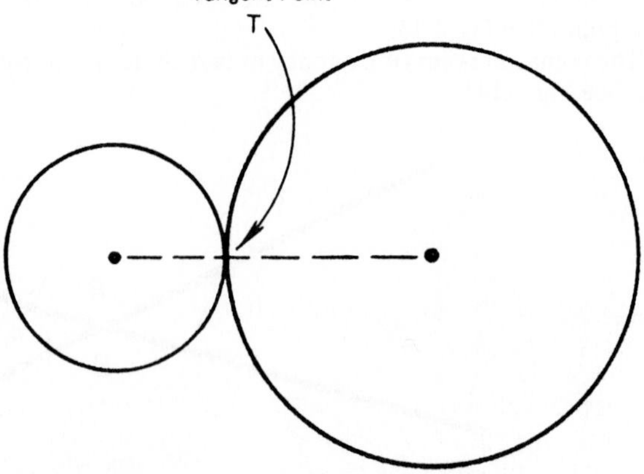


**FIGURE 1.14** Common point of tangency.

Side $a = a'$; $\angle A = \angle A'$. See Fig. 1.15.

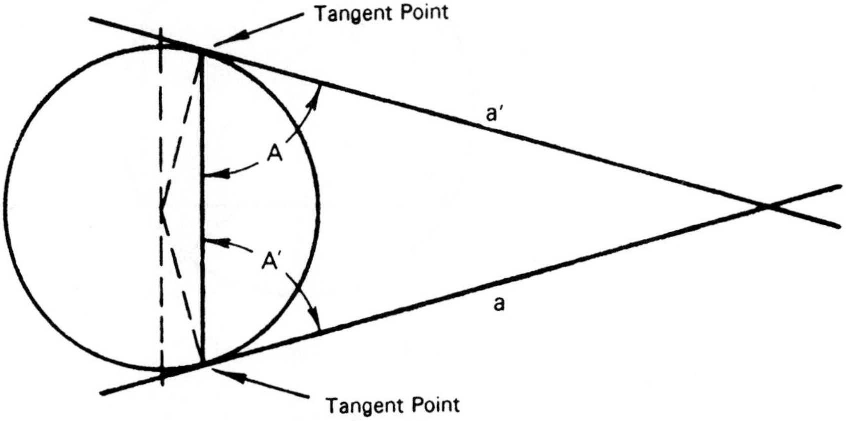


**FIGURE 1.15** Two lines tangent to a circle.

$\angle A = ½ \angle B$. See Fig. 1.16.

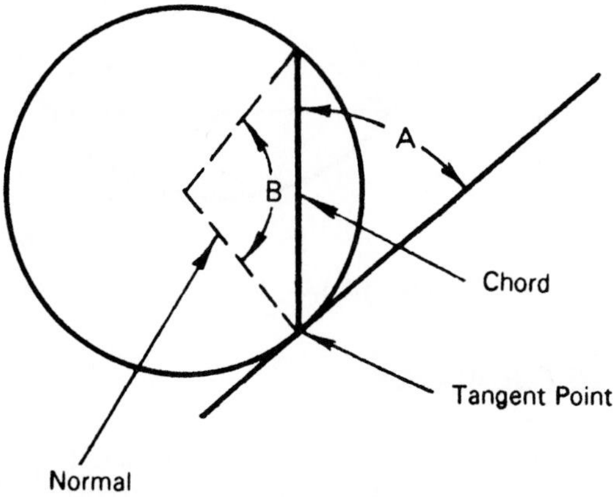


**FIGURE 1.16** Line, angle, and chord.

$\angle A = \angle B = \angle C$. All perimeter angles of a chord are equal. See Fig. 1.17.

$\angle B = ½ \angle A$. See Fig. 1.18.

$a^2 = bc$. See Fig. 1.19.

All perimeter angles in a circle, drawn from the diameter, are 90°. See Fig. 1.20.

Arcs are proportional to internal angles. $\angle A : \angle B = a : b$. See Fig. 1.21.

Circumferences are proportional to their respective radii. $C : C' = r : R$ and areas are proportional to the squares of the respective radii. See Fig. 1.22.

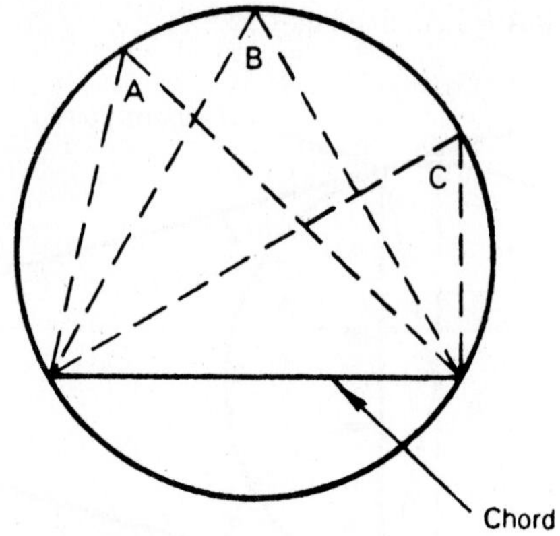


**FIGURE 1.17** Chordal angles.

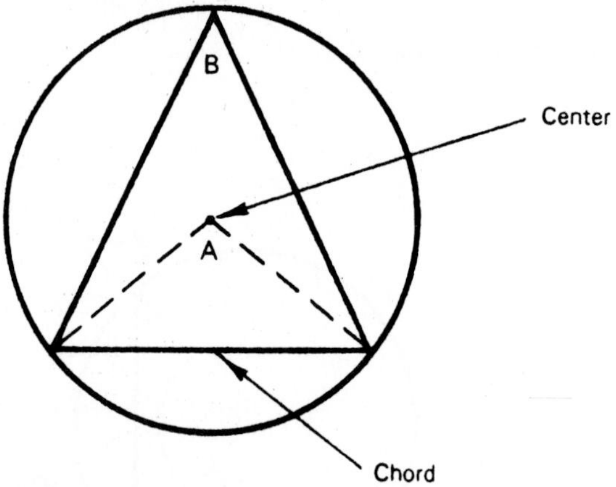


**FIGURE 1.18** Internal angles of a chord.

## 1.2 TRANSPOSING EQUATIONS

The members or terms of an equation with multiple variables may be transposed to solve for any variable where all variables are known except one. Thus, if

$$R = \frac{Gd^4}{8\,ND^3}$$

when written for $R$, then

$$\left.\begin{aligned} Gd^4 &= R\,8\,ND^3 \\ d^4 &= R\,8\,ND^3/G \\ d &= \sqrt[4]{R\,8\,ND^3/G} \end{aligned}\right\} \text{ solving for } d \qquad \left.\begin{aligned} Gd^4 &= R\,8\,ND^3 \\ D^3 &= Gd^4/R\,8\,N \\ D &= \sqrt[3]{Gd^4/R\,8\,N} \end{aligned}\right\} \text{ solving for D}$$

$$\left.\begin{aligned} Gd^4 &= R\,8\,ND^3 \\ N &= Gd^4/8\,RD^3 \end{aligned}\right\} \text{ solving for N} \qquad \left.\begin{aligned} Gd^4 &= R\,8\,ND^3 \\ G &= R\,8\,ND^3/d^4 \end{aligned}\right\} \text{ solving for G}$$

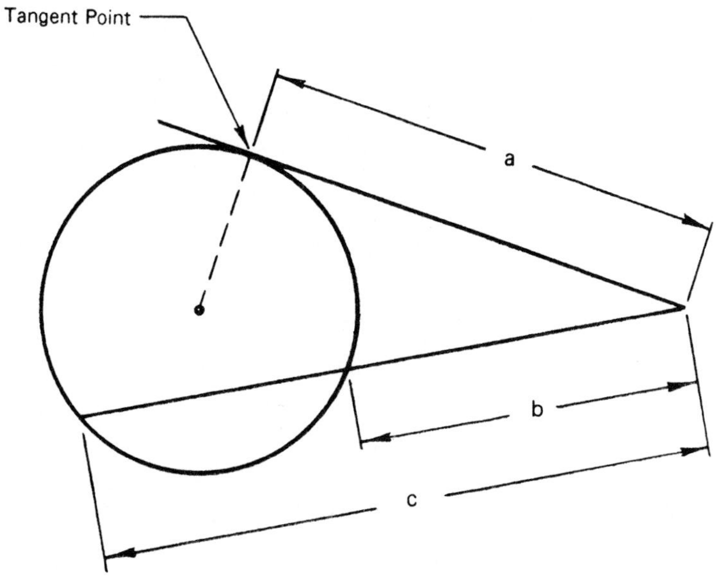


**FIGURE 1.19** Line-and-circle relationship.

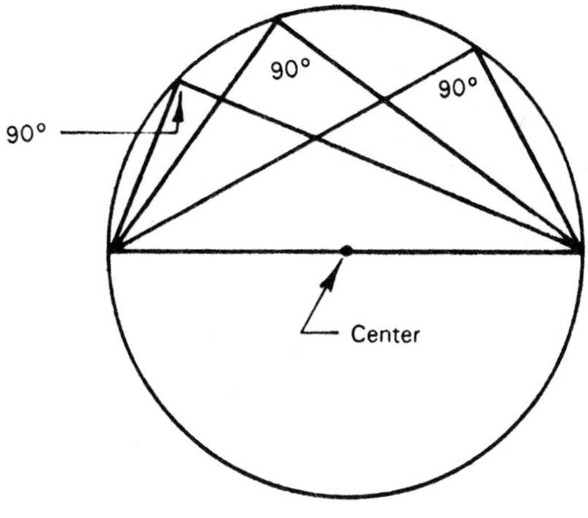


**FIGURE 1.20** The 90° chordal relationship.

## 1.3 ALGEBRAIC FACTORS AND EXPANSIONS

$$(a + b)^2 = a^2 + 2ab + b^2$$

$$(a - b)^2 = a^2 - 2ab + b^2$$

$$(a + b)^3 = a^3 + 3a^2b + 3ab^2 + b^3$$

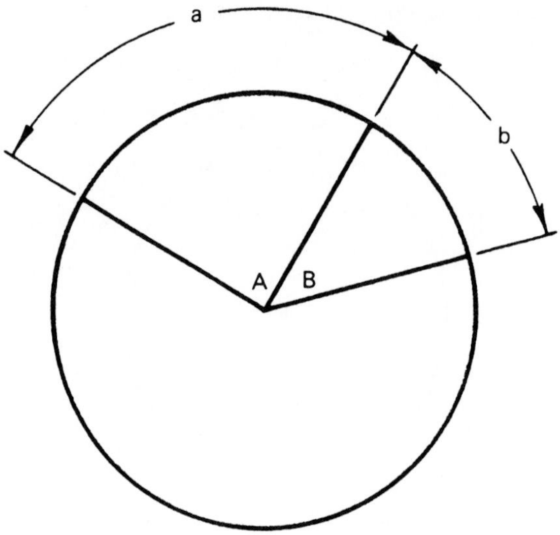


**FIGURE 1.21** Arcs and internal angles are similar.

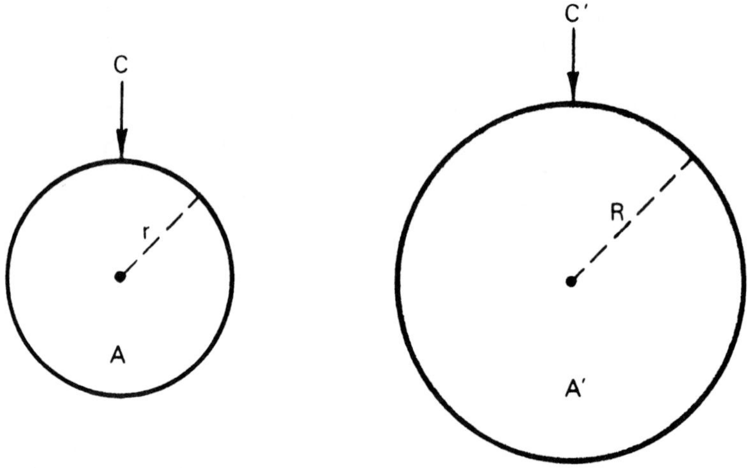


**FIGURE 1.22** Circumferences, areas, and radii.

$$(a - b)^3 = a^3 - 3a^2b + 3ab^2 - b^3$$

$$a^2 - b^2 = (a - b)(a + b)$$

$$a^3 - b^3 = (a - b)(a^2 + ab + b^2)$$

$$a^3 + b^3 = (a + b)(a^2 - ab + b^2)$$

$$a^4 + b^4 = (a^2 + ab\sqrt{2} + b^2)(a^2 - ab\sqrt{2} + b^2)$$

$$a^{\eta} - b^{\eta} = (a - b)(a^{\eta-1} + a^{\eta-2}b + \cdots + b^{\eta-1}) \text{ even } \eta$$

$$a^{\eta} - b^{\eta} = (a + b)(a^{\eta-1} - a^{\eta-2}b + \cdots - b^{\eta-1}) \text{ odd } \eta$$

$$a^{\eta} + b^{\eta} = (a + b)(a^{\eta-1} - a^{\eta-2}b + \cdots + b^{\eta-1})$$

$$(a + b + c)^2 = a^2 + b^2 + c^2 + 2ab + 2ac + 2bc$$

$$(a + b + c)^3 = a^3 + b^3 + c^3 + 3a^2(b + c) + 3b^2(a + c) + 3c^2(a + b) + 6abc$$

### 1.3.1 Imaginary Numbers

$\sqrt{-P}$ has the property $(\sqrt{-P})^2 = -P$ and $(-\sqrt{-P})^2 = -P$.
Thus, $(-P)$ has two imaginary numbers as square roots, $\pm\sqrt{-P}$.

### 1.3.2 Complex Numbers

Let $i$ represent $-1$ and have the property of $i^2 = -1$. If $P$ is positive, let $\sqrt{-P}$ represent $i\sqrt{P}$. Then $(-P)$ has two square roots: $\pm\sqrt{-P}$ or $\pm i\sqrt{P}$. For example

- $\sqrt{-36} = i\sqrt{36} = 6i$.
- If $x > 0$, then $\sqrt{-16x^2} = 4xi$.
- $n$ expressed as a complex number $= n + 0\,i$, where $n$ = any positive number.

### 1.3.3 Trigonometric Form of a Complex Number

The absolute value of any complex number is

$$|a + bi| = \sqrt{a^2 + b^2}$$

For example

$$|4 + 5i| = \sqrt{41} = 6.403.$$

Referring to Fig. 1.23, if we let $P$ represent $a + bi$, then

$$r = \sqrt{a^2 + b^2}$$

where $a = r\cos\theta$ and $b = r\sin\theta$. The expression $a + bi = r(\cos\theta + i\sin\theta)$ is called the *polar* form, and $\theta$ is labeled as amplitude. The value $r$ (positive length) is the modulus or absolute value.

To plot $r(\cos\theta + i\sin\theta)$, form angle $XOP = \theta$ where $OP = r$. Then $P$ represents the given complex number.

### 1.3.4 Radicals

$$(\sqrt[\eta]{a})^{\eta} = a$$

$$\sqrt[\eta]{a^{\eta}} = a$$

$$\sqrt[\eta]{ab} = \sqrt[\eta]{a}\sqrt[\eta]{b}$$

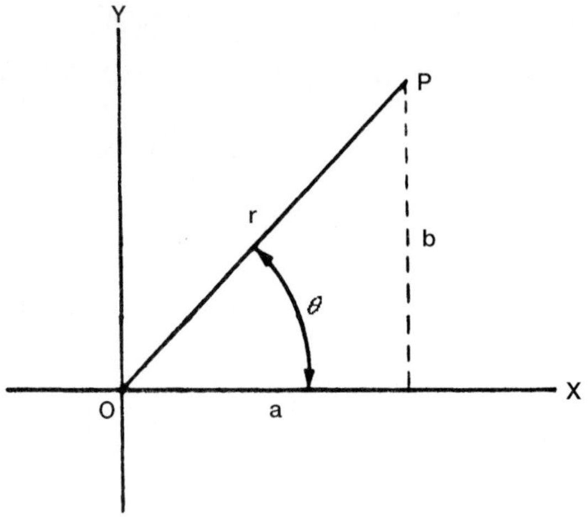


**FIGURE 1.23** *P* represents the complex number.

$$\sqrt[\eta]{\frac{a}{b}} = \sqrt[\eta]{a} \div \sqrt[\eta]{b}$$

$$\sqrt[\eta]{a^\chi} = a^{\chi/\eta} \qquad \therefore \qquad \sqrt[3]{6.1^2} = 6.1^{2/3}$$

$$\sqrt[\eta]{a} = a^{1/\eta} \qquad \therefore \qquad \sqrt{3} = 3^{1/2}$$

$$\sqrt[\eta]{a^\chi} = (\sqrt[\eta]{a})^\chi$$

$$a^{\chi/\eta} = (\sqrt[\eta]{a})^\chi$$

$$a^0 = 1$$

$$a^{-\eta} = \frac{1}{a^\eta}$$

$$\frac{a}{bx^\eta} = \frac{ax^{-\eta}}{b}$$

$$\sqrt[\eta]{a}\sqrt[\eta]{b} = \sqrt[\eta]{ab}$$

$$\frac{\sqrt[\eta]{a}}{\sqrt[\eta]{b}} = \sqrt[\eta]{\frac{a}{b}}$$

$$(a^{\chi/\eta})^\rho = a^{\chi\rho/\eta}$$

If $\chi$, $\eta$, $\rho$, and $\nu$ are positive, then

$$a^{\chi/\eta}a^{\rho/\nu} = a^{(\chi\nu + \eta\rho)/\eta\nu}$$

$$(a^{\chi/\eta}a^{\rho/\nu})^{\eta\nu} = a^{\chi\nu + \rho\eta}$$

$$(a^{\chi/\eta})^{\rho/\nu} = a^{\chi\rho/\eta\nu}$$

$$a^{-\eta}a^{-\chi} = a^{-\eta-\chi}$$

$$a^{\eta}a^{-\chi} = a^{\eta-\chi}$$

$$a^{-\eta}a^{-\chi} = \frac{1}{a^{\eta+\chi}} \text{ or } a^{-(\eta+\chi)}$$

$$(a^{\chi/\eta})^{\rho} = [(a^{1/\eta})^{\chi}]^{\rho} = a^{\chi\rho/\eta}$$

### 1.3.5 Factorials

5! is termed *5 factorial.* (The obsolete term was 5 inclusive.)

$$5! = 5 \times 4 \times 3 \times 2 \times 1 = 120$$

$$10! = 3{,}628{,}800$$

Uses: expansions and power-series equations.

### 1.3.6 Ratios and Proportions

If $$\frac{a}{b} = \frac{c}{d}$$

then $$\frac{a+b}{b} = \frac{c+d}{d}$$

$$\frac{a-b}{b} = \frac{c-d}{d}$$

and $$\frac{a-b}{a+b} = \frac{c-d}{c+d}$$

### 1.3.7 Summation

$$\sum (n) = 1 + 2 + 3 + 4 + \cdots + n = \frac{n(n-1)}{2}$$

$$\sum (n^2) = 1^2 + 2^2 + 3^2 + 4^2 + \cdots + n^2 = \frac{n(n+1)(2n+1)}{6}$$

$$\sum (n^3) = 1^3 + 2^3 + 3^3 + 4^3 + \cdots + n^3 = \frac{n^2(n+1)^2}{4}$$

### 1.3.8 Permutations, Combinations, and Progressions

If $P_1$ denotes the number of *permutations* of $n$ things taken $q$ at a time, then

$$P_1 = n\,(n-1)(n-2)\cdots(n-q+1)$$

If $P_2$ denotes the number of *combinations* of $n$ things taken $q$ at a time, then

$$P_2 = \frac{n(n-1)(n-2)\cdots(n-q+1)}{q!}$$

$$= \frac{n!}{q!\,(n-q)!}$$

***Arithmetic Progression.*** If $a$ is the first term; $l$, the last term; $d$, the common difference; $n$, the number of terms; and $s$, the sum of $n$ terms, then

$$l = a + (n-1)d \qquad \text{and} \qquad s = \frac{n}{2}(a+l) = \frac{n}{2}[2a + (n-1)d]$$

***Geometric Progression.*** If $a$ is the first term; $l$, the last term; $r$, the common ratio; $n$, the number of terms; and $s$, the sum of $n$ terms, then

$$l = ar^{n-1} \qquad \text{and} \qquad s = a\frac{(1-r^n)}{1-r} = \frac{lr-a}{r-1}$$

### 1.3.9 Quadratic Equations

Any quadratic equation may be reduced to the form

$$ax^2 + bx + c = 0$$

The two roots

$$x_{1,2} = \frac{-b \pm \sqrt{b^2-4ac}}{2a}$$

When $a$, $b$, and $c$ are real, if $b^2 - 4ac$ is positive, the roots are real and unequal. If $b^2 - 4ac$ is zero, the roots are real and equal. If $b^2 - 4ac$ is negative, the roots are imaginary and unequal.

### 1.3.10 Cubic Equations

A cubic equation $y^3 + py^2 + qy + r = 0$ may be reduced to the form

$$x^3 + ax + b = 0$$

by substituting for $y$ the value $x - (p/3)$, where

$$a = \frac{1}{3}(3q - p^2) \qquad \text{and} \qquad b = \frac{1}{27}(2p^3 - 9pq + 27r)$$

For solution, let

$$A = \sqrt[3]{-\frac{b}{2} + \left(\frac{b^2}{4} + \frac{a^3}{27}\right)^{1/2}}$$

and

$$B = \sqrt[3]{-\frac{b}{2} - \left(\frac{b^2}{4} + \frac{a^3}{27}\right)^{1/2}}$$

Then the values of $x$ will be

$$x_1 = A + B$$

$$x_2 = -\frac{A+B}{2} + \frac{A-B}{2}\sqrt{-3}$$

$$x_3 = -\frac{A+B}{2} - \frac{A-B}{2}\sqrt{-3}$$

If $b^2/4 + a^3/27 > 0$, there are one real root and two conjugate imaginary roots.

If $b^2/4 + a^3/27 = 0$, there are three real roots, with at least two being equal.

If $b^2/4 + a^3/27 < 0$, there are three real unequal roots.

In the last case, where $b^2 + (a^3/27)$ is $< 0$ (negative), the preceding cubic formulas for the roots are not in convenient form for numeric calculation; a trigonometric transformation reduces the roots to better form. If

$$\cos\phi = \sqrt{\frac{b^2}{4} + \left(-\frac{a^3}{27}\right)}$$

then the roots are as follows: (*Note:* In the following formulas the upper sign is used where $b > 0$ and the lower sign is used if $b < 0$.)

$$x_1 = \mp 2\sqrt{-\frac{a}{3}}\cos\frac{\phi}{3}$$

$$x_2 = \mp 2\sqrt{-\frac{a}{3}}\cos\left(\frac{\phi}{3} + 120°\right)$$

$$x_3 = \mp 2\sqrt{-\frac{a}{3}}\cos\left(\frac{\phi}{3} + 240°\right)$$

Solutions to general cubic equations can be found by the following methods:

- Make an estimate of the approximate root using parameters set by the problem. For example, you may know that the root cannot be negative or imaginary and that it will have a specific range. Trial and error can then establish the roots.
- Graph the equation to obtain approximate roots, then regraph in expanded steps to secure the desired accuracy.
- Use the form equations shown in this section to obtain the roots.
- Use Newton's method (see Sec. 2.3.28 of Chap. 2 on mechanics).
- Use an equation graphing calculator, such as the Texas Instrument TI-81, for approximate roots. See Fig. 1.2.

### 1.3.11 General Quartic Equations

The solution to the general quartic equation $x^4 + bx^3 + cx^2 + dx + e = 0$ was formulated by Lodorico Ferrari (1522–1560), a pupil of Cardano, who published the solution to the general cubic equation in 1545, after obtaining the method from Niccolò Tartaglia (1506–1557). Modern solutions to this type of equation are contained in the book *Brief College Algebra* (Hart 1947), listed in the bibliography at the end of this chapter.

Quartic equations may also be solved using Newton's method (see Sec. 2.3.28 in this handbook).

As in the case of cubic equations, an equation-graphing calculator may also be employed.

## 1.4 POWERS-OF-TEN NOTATION

Numbers with many digits may be expressed more conveniently in powers-of-ten notation, with the following:

$$0.0001759 = 1.759 \times 10^{-4}$$

$$3{,}756{,}314 = 3.756314 \times 10^{6}$$

You are actually counting the number of decimal places the decimal point is displaced from its original position to determine the power-of-ten multiplier. In the first example above, the decimal point was moved four places to the right; thus the multiplier is $10^{-4}$. Engineering calculations using powers-of-ten notation usually express the powers in multiples of 3, as, for example, in $17.59 \times 10^{-3}$, $6.345 \times 10^{6}$, or $16.366 \times 10^{9}$.

Multiplication and division using the powers-of-ten notation are handled easily.

$$1.246 \times 10^{4}\,(2.573 \times 10^{-4}) = 3.206 \times 10^{0} = 3.206$$

$$1.785 \times 10^{7} \div (1.039 \times 10^{-4}) = \frac{1.785}{1.039} \times 10^{7-(-4)} = 1.718 \times 10^{11}$$

$$(1.447 \times 10^{5})^{2} = (1.447)^{2} \times 10^{10} = 2.094 \times 10^{10}$$

$$\sqrt{1.391 \times 10^{8}} = 1.391^{1/2} \times 10^{8/2} = 1.179 \times 10^{4}$$

Powers-of-ten notation was particularly convenient when slide rules were used in engineering and other computational work. In the modern handheld calculator, such notation is automatically inserted and carried through by the machine. Even though this is the case, however, powers-of-ten notation is still used and indicated in all engineering and scientific work. It is an indispensable tool for these types of calculations.

## 1.5 LOGARITHMS

The logarithm of a number $N$ to the base $a$ is the exponent power to which $a$ must be raised to obtain $N$. Thus $N = a^{x}$ and $x = \log_a N$. Also, $\log_a 1 = 0$ and $\log_a a = 1$. Other relationships follow.

$$\log_a MN = \log_a M + \log_a N$$

$$\log_a \frac{M}{N} = \log_a M - \log_a N$$

$$\log_a N^{k} = k \log_a N$$

$$\log_a \sqrt[n]{N} = \frac{1}{n} \log_a N$$

$$\log_b a = \frac{1}{\log_a b} \qquad (\text{let } N = a)$$

Base 10 logarithms are referred to as *common logarithms,* or *Briggs logarithms,* after their inventor.

Base $e$ logarithms (where $e = 2.71828$) are designated as natural, hyperbolic, or naperian logarithms, where the last label refers to their inventor.

The base of the natural-logarithm system is defined by the infinite series

$$e = 1 + \frac{1}{1} + \frac{1}{2!} + \frac{1}{3!} + \frac{1}{4!} + \frac{1}{5!} + \cdots = \lim_n \rightarrow \infty \left(1 + \frac{1}{n}\right)^n$$

If $a$ and $b$ are any two bases, then

$$\log_a N = (\log_a b)(\log_b N), \qquad \text{or} \qquad \log_b N = \frac{\log_a N}{\log_a b}$$

$$\log_{10} N = \frac{\log_e N}{2.30261} = 0.43429 \log_e N$$

$$\log_e N = \frac{\log_{10} N}{0.43429} = 2.30261 \log_{10} N$$

Simply multiply the natural log by 0.43429 (a modulus) to obtain the equivalent common log.

Similarly, multiply the common log by the modulus 2.30261 to obtain the equivalent natural log. (Accuracy is to four decimal places for both cases.)

***Rules and Uses for Logarithms.*** The following rules of logarithms are useful in many applications.

If $\quad ab = X \quad$ then $\quad \log a + \log b = \log X$

$\quad a \div b = X \quad$ then $\quad \log a - \log b = \log X$

$\quad a^n = X \quad$ then $\quad n \log a = \log X$

If $\quad n\sqrt{a} = X \quad$ then $\quad \dfrac{\log a}{n} = \log X$

*Examples*

$$a^{1.34} = X \qquad \text{and} \qquad 1.34 \log a = \log X$$

If $a = 13.46$ and $1.34(1.129) = \log 1.513$, then antilog $1.513 = 32.584$.

*Proof.* $(13.46)^{1.34} = 32.58$.

$$1.34\sqrt{a} = X \qquad \text{and} \qquad \frac{\log a}{1.34} = \log X \quad \text{or} \quad 0.746 \log a = \log X$$

If $\quad a = 50; \ \dfrac{\log 50}{1.34} = \dfrac{1.699}{1.34} = \log 1.268$

$$\text{antilog } 1.268 = 18.535$$

*Proof.* $(50)^{1/1.34} = 18.53$ or $(50)^{0.7463} = 18.53$.

To show the practicality of problem solutions by the use of logarithms, the following sample problem is presented. The compound interest equation $P_n = P(1 + r)^n$ will be used as the basis for the following problem. If your company employs 250 people, in how many years will the number of employees triple if the employment rate is 15 percent per year?

*Solution.* The number 250 will become $3 \times 250 = 750$ in $n$ years at 15 percent increase per year. Let $P_n$ = new number of employees, $P$ = existing employees, and 15 percent = 0.15. Then

$$750 = 250\,(1 + 0.15)^n$$

$$3 = (1.15)^n$$

$$\log 3 = n\,(\log 1.15)$$

$$n = \frac{\log 3}{\log 1.15} = \frac{0.47712}{0.06070} = 7.86 \text{ years or 7 years 313.9 days}$$

Other problems such as population increases and other compound rate increases or decreases for engineering and design uses may be solved easily in the manner shown above, using logarithms.

## 1.6 BINOMIAL FORMULA

Any expression in the form $(a + b)^n$ may be expanded using the formula

$$(a+b)^n = a^n + na^{n-1}\,b + \frac{n(n-1)}{1 \cdot 2}a^{n-2}b^2 + \frac{n(n-1)(n-2)}{1 \cdot 2 \cdot 3}a^{n-3}b^3 + \cdots + b^n$$

Expanding $(x + 3)^4$, we get

$$x^4 + 4x^{4-1}(3) + \frac{4(4-1)}{1 \cdot 2}x^{4-2}(3)^2 + \frac{4(4-1)(4-2)}{1 \cdot 2 \cdot 3}x^{4-3}(3)^3 + (3)^4$$

$$= x^4 + 12x^3 + 54x^2 + 108x + 81$$

This expansion contains $(n + 1)$, or 5, terms.

The expression $(a + b)^n$ can also be rewritten in the form $a^n(1 + b/a)^n$, so that $\sqrt{50} = \sqrt{25 + 25} = (25 + 25)^{1/2}$ may be expanded as $(a + b)^n$.

## 1.7 DETERMINANTS

### 1.7.1 Second-Order Determinants

For the set of linear equations shown here (unknowns in $x$ and $y$)

$$a_1x + b_1y = c_1$$

$$a_2x + b_2y = c_2$$

solutions can be written as

$$x = \frac{c_1b_2 - c_2b_1}{a_1b_2 - a_2b_1} \quad \text{and} \quad y = \frac{a_1c_2 - a_2c_1}{a_1b_2 - a_2b_1}$$

Expressed as determinants, we define

$$\begin{vmatrix} a_1 & b_1 \\ a_2 & b_2 \end{vmatrix} \equiv a_1b_2 - a_2b_1 = D$$

$$\begin{vmatrix} c_1 & b_1 \\ c_2 & b_2 \end{vmatrix} \equiv c_1b_2 - c_2b_1 = D_x$$

$$\begin{vmatrix} a_1 & c_1 \\ a_2 & c_2 \end{vmatrix} \equiv a_1c_2 - a_2c_1 = D_y$$

$$\therefore x = \frac{D_x}{D} \qquad \text{and} \qquad y = \frac{D_y}{D}$$

The second-order determinant is read from the top left, down on a diagonal and from the bottom left, up on a diagonal. The center sign is always minus.

### 1.7.2 Third-Order Determinants

The solution of the system of equations

$$a_1x + b_1y + c_1z = d_1$$

$$a_2x + b_2y + c_2z = d_2$$

$$a_3x + b_3y + c_3z = d_3$$

is given by

$$\frac{\begin{vmatrix} d_1 & b_1 & c_1 \\ d_2 & b_2 & c_2 \\ d_3 & b_3 & c_3 \end{vmatrix}}{D} = x \qquad \frac{\begin{vmatrix} a_1 & d_1 & c_1 \\ a_2 & d_2 & c_2 \\ a_3 & d_3 & c_3 \end{vmatrix}}{D} = y$$

and

$$\frac{\begin{vmatrix} a_1 & b_1 & d_1 \\ a_2 & b_2 & d_2 \\ a_3 & b_3 & d_3 \end{vmatrix}}{D} = z \qquad \text{where} \qquad D = \begin{vmatrix} a_1 & b_1 & c_1 \\ a_2 & b_2 & c_2 \\ a_3 & b_3 & c_3 \end{vmatrix}$$

The fractional equivalents of this system of equations are

$$x = \frac{d_1b_2c_3 + d_3b_1c_2 + d_2b_3c_1 - d_3b_2c_1 - d_1b_3c_2 - d_2b_1c_3}{a_1b_2c_3 + a_3b_1c_2 + a_2b_3c_1 - a_3b_2c_1 - a_1b_3c_2 - a_2b_1c_3}$$

$$y = \frac{a_1d_2c_3 + \cdots}{a_1b_2c_3 + \cdots}$$

$$z = \frac{a_1b_2d_3 + \cdots}{a_1b_2c_3 + \cdots}$$

If $D = 0$, the equations are inconsistent or dependent.

## 1.8 TRIGONOMETRY

### 1.8.1 Functions of Angles in a Right Triangle

Referring to Fig. 1.24, we obtain

$$\sin A = \frac{a}{c} \text{ (sine)} \qquad \cot A = \frac{b}{a} \text{ (cotangent)}$$

$$\cos A = \frac{b}{c} \text{ (cosine)} \qquad \sec A = \frac{c}{b} \text{ (secant)}$$

$$\tan A = \frac{a}{b} \text{ (tangent)} \qquad \csc A = \frac{c}{a} \text{ (cosecant)}$$

The cotangent, secant, and cosecant functions are the reciprocals of the tangent, cosine, and sine, respectively. For example

$$\csc A = \frac{1}{\sin A} = \frac{1}{a/c} = \frac{c}{a}$$

Also

$$\text{vers } A = 1 - \cos A \text{ (versine)}$$
$$\text{covers } A = 1 - \sin A \text{ (coversine)}$$
$$\text{hav } A = \tfrac{1}{2} \text{ vers } A \text{ (haversine)}$$
$$\text{exsec } A = \sec A - 1 \text{ (exsecant)}$$

### 1.8.2 Signs and Limits of Trigonometric Functions

In a rectangular coordinate system, values for the quadrants are summarized as shown here.

| Quadrant II | Y | Quadrant I |
|---|---|---|
| (1 – 0) + sin | | sin + (0 – 1) |
| (0 – 1) – cos | | cos + (1 – 0) |
| (∞ – 0) – tan | | tan + (0 – ∞) |
| (0 – ∞) – cot | | cot + (∞ – 0) |
| (∞ – 1) – sec | | sec + (1 – ∞) |
| (1 – ∞) + csc | | csc + (∞ – 1) |
| X′ | 0 | X |
| **Quadrant III** | | **Quadrant IV** |
| (0 – 1) – sin | | sin – (1 – 0) |
| (1 – 0) – cos | | cos + (0 – 1) |
| (0 – ∞) + tan | | tan – (∞ – 0) |
| (∞ – 0) + cot | | cot – (0 – ∞) |
| (1 – ∞) – sec | | sec + (∞ – 1) |
| (∞ – 1) – csc | Y′ | csc – (1 – ∞) |

### 1.8.3 Equivalent Expressions (Identities)

$$\tan x = \frac{\sin x}{\cos x} \qquad \sin x = \cos (90^\circ - x)$$

$$\cot x = \frac{\cos x}{\sin x} \qquad \cos x = \sin (90^\circ - x)$$

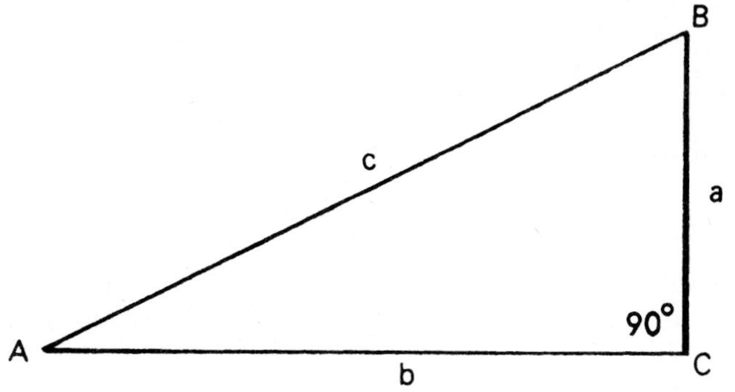


**FIGURE 1.24** Right-angle triangle.

$$\sin^2 x + \cos^2 x = 1 \qquad \tan x = \cot (90° - x)$$

$$\sin x = \pm \sqrt{1 - \cos^2 x} \qquad \cos x = \pm \sqrt{1 - \sin^2 x}$$

$$\tan x = \pm \sqrt{\sec^2 x - 1} \qquad \cot x = \pm \sqrt{\csc^2 x - 1}$$

$$\sec x = \pm \sqrt{\tan^2 x + 1} \qquad \csc x = \pm \sqrt{\cot^2 x + 1}$$

Note that in an expression such as $\cot^2 x + 1$, the cotangent value of $x$ is what is to be squared, and not the value of $x$. This is a common error that is sometimes made.

Additional identities include

$$\sin(x \pm y) = \sin x \cos y \pm \cos x \sin y$$

$$\cos (x \pm y) = \cos x \cos y \mp \sin x \sin y$$

$$\tan (x \pm y) = \frac{\tan x \pm \tan y}{1 \pm \tan x \tan y}$$

$$\sin 2x = 2\sin x \cos x$$

$$\cos 2x = \cos^2 x - \sin^2 x = 1 - 2 \sin^2 x$$

$$\sin 3x = 3 \sin x - 4\cos^3 x$$

$$\cos 3x = 4\cos^3 x - 3\cos x$$

$$\tan 2x = \frac{2\tan x}{1 - \tan^2 x}$$

$$\sin \frac{1}{2} x = \pm \sqrt{\frac{1 - \cos x}{2}}$$

$$\cos \frac{1}{2} x = \pm \sqrt{\frac{1 + \cos x}{2}}$$

$$\tan \frac{1}{2} x = \pm \sqrt{\frac{1 - \cos x}{1 + \cos x}} = \frac{\sin x}{1 + \cos x}$$

The ± in these three identities varies as the quadrant of $x$ varies.

$$\sin x \pm \sin y = 2\sin \frac{1}{2}(x \pm y) \cos \frac{1}{2}(x \mp y)$$

$$\cos x + \cos y = 2\cos \frac{1}{2}(x + y) \cos \frac{1}{2}(x - y)$$

$$\cos x - \cos y = -2\sin \frac{1}{2}(x + y) \sin \frac{1}{2}(x - y)$$

$$\tan x \pm \tan y = \frac{\sin (x \pm y)}{\cos x \cos y}$$

### 1.8.4 Solution of Triangles

Referring to Fig. 1.25 by the law of sines

$$\frac{a}{\sin A} = \frac{b}{\sin B} = \frac{c}{\sin C}$$

$$a : b : c :: A : B : C$$

Also $\frac{a}{b} = \frac{\sin A}{\sin B} \qquad \frac{b}{c} = \frac{\sin B}{\sin C} \qquad \frac{a}{c} = \frac{\sin A}{\sin C}$

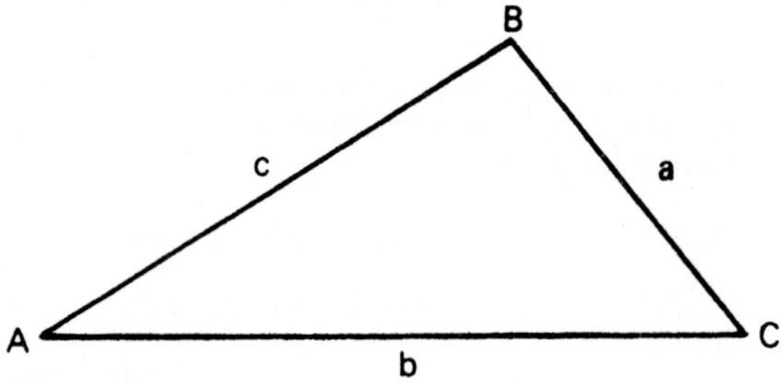


**FIGURE 1.25** Common triangle.

By the law of cosines

$$a^2 = b^2 + c^2 - 2bc \cos A$$

$$b^2 = a^2 + c^2 - 2ac \cos B$$

$$c^2 = a^2 + b^2 - 2ab \cos C$$

By the law of tangents

$$\frac{a + b}{a - b} = \frac{\tan[(A + B)/2]}{\tan [(A - B)/2]}$$

See Fig. 1.26, where

$$x = b \frac{\sin A \sin C}{\sin (A + C)} = \frac{b}{\cot A + \cot C}$$

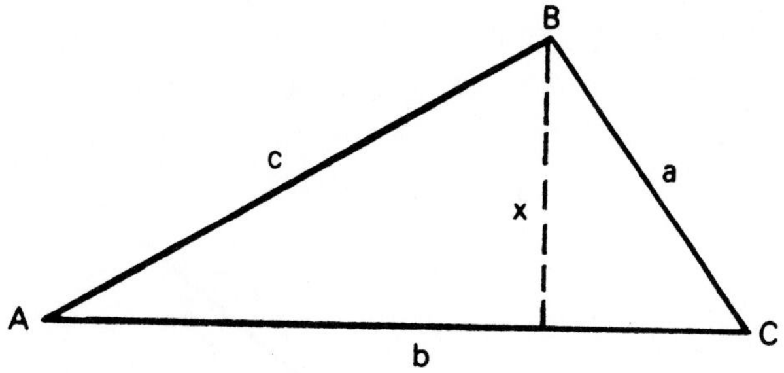


**FIGURE 1.26** Finding the height *x*.

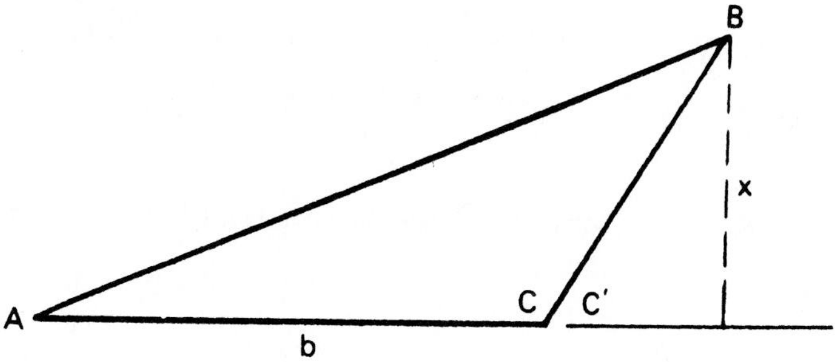


**FIGURE 1.27** Finding the external height *x*.

See Fig. 1.27, where

$$x = b\,\frac{\sin A \sin C}{\sin (C' - A)} = \frac{b}{\cot A - \cot C'}$$

Additional solutions include the following equations, where $s = ½(a + b + c)$ and $r = \sqrt{[(s-a)(s-b)(s-c)]/s}$

$$\sin\frac{A}{2} = \sqrt{\frac{(s-b)(s-c)}{bc}} \qquad \sin\frac{B}{2} = \sqrt{\frac{(s-c)(s-a)}{ca}} \qquad \text{and so forth for } \sin\frac{C}{2}$$

$$\cos\frac{A}{2} = \sqrt{\frac{s(s-a)}{bc}} \qquad \cos\frac{B}{2} = \sqrt{\frac{s(s-b)}{ac}} \qquad \text{and so forth for } \cos\frac{C}{2}$$

$$\tan\frac{A}{2} = \sqrt{\frac{(s-b)(s-c)}{s(s-a)}} = \frac{r}{s-a}$$

$$\tan\frac{B}{2} = \sqrt{\frac{(s-c)(s-a)}{s(s-b)}} \qquad \text{and so forth for } \tan\frac{C}{2}$$

$$\frac{a+b}{a-b} = \frac{\sin A + \sin B}{\sin A - \sin B} = \frac{\tan ½(A+B)}{\tan ½(A-B)}$$

To calculate the area of an oblique triangle, given only the lengths of the three sides *a, b,* and *c*, use

$$\text{Area} = \sqrt{s(s-a)(s-b)(s-c)} \qquad \text{where} \qquad s = \frac{a+b+c}{2}$$

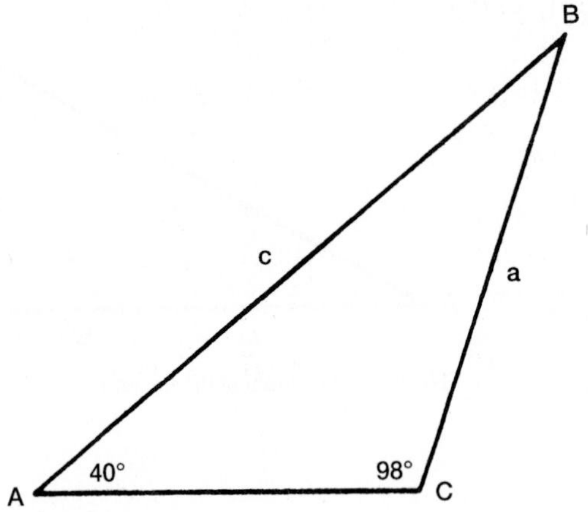


**FIGURE 1.28** Solve for $a$, $c$, and $B$.

*Example.* Solve the triangle shown in Fig. 1.28 for $a$, $c$, and $B$.

$$B = 180° - (40° + 98°) = 42°$$

By the law of sines

$$\frac{a}{\sin A} = \frac{b}{\sin B}$$

$$\frac{a}{\sin 40} = \frac{10}{\sin 42}$$

$$a = \frac{10 \sin 40}{\sin 42} = \frac{6.428}{0.669} = 9.608$$

Solving for side $c$ by the law of sines, $c = 14.801$.

In the solution of all triangles, it is wise to check calculated answers. This may be done with the use of the *Mollweide equation*, which involves all parts of the triangle.

$$\frac{a-b}{c} = \frac{\sin[(A-B)/2]}{\cos(C/2)} \quad \text{or} \quad \frac{a+b}{c} = \frac{\cos[(A-B)/2]}{\sin(C/2)}$$

Substituting the calculated values for the triangle of Fig. 1.28 into the Mollweide equation gives

$$\frac{9.608 - 10}{14.801} = \frac{\sin - 1°}{\cos 49°}$$

$$-0.392(\cos 49°) = 14.801(\sin - 1°)$$

$$-0.257 = -0.258$$

This is a very close check, indicating that the solution to the given triangle is correct.

## 1.9 ANALYTIC GEOMETRY

The distance $d$ between two points, $(x_1, y_1)$ and $(x_2, y_2)$, in a rectangular coordinate system is

$$d = \sqrt{(x_2 - x_1)^2 + (y_2 - y_1)^2}$$

In a polar coordinate system, between points $(r_1, \theta_1)$ and $(r_2, \theta_2)$, the distance is

$$d = \sqrt{r_1^2 + r_2^2 - 2r_1r_2 \cos(\theta_1 - \theta_2)}$$

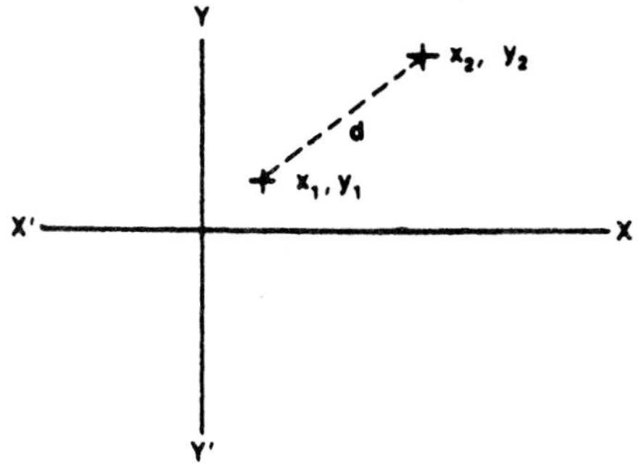


**FIGURE 1.29** Equation for a line: $y - y_1 = m(x - x_1)$.

### 1.9.1 Straight-Line Equations

Where $m$ is the tangent of inclination and $c$ is the distance of the intersection with axis $Y$ from the origin 0 (see Fig. 1.29)

$$y = mx + c$$

If a line (Fig. 1.6) with slope $m$ passes through point $(x_1, y_1)$, the equation is

$$y - y_1 = m(x - x_1)$$

### 1.9.2 Linear Functions of Lines

The point-slope form (Fig. 1.30) of the equation of a line is

$$\frac{y - y_1}{x - x_1} = m$$

$$y - y_1 = m(x - x_1)$$

The slope-intercept form (Fig. 1.31) of the equation is

$$y - b = m(x - 0) \qquad \text{or} \qquad y = mx + b$$

In the intercept form the slope is (see Fig. 1.32)

$$m = \frac{b - 0}{0 - a} = -\frac{b}{a}$$

Using point $(a,0)$ and the slope, we can rewrite the equation

$$y - 0 = -\frac{b}{a}(x - a) \qquad bx + ay = ab$$

Dividing both sides by $ab$ gives

$$\frac{x}{a} + \frac{y}{b} = 1$$

When the perpendicular length from the origin is $p$ and the inclination angle is $\phi$, the equation of the straight line is

$$x \cos \phi + y \cos \phi = p$$

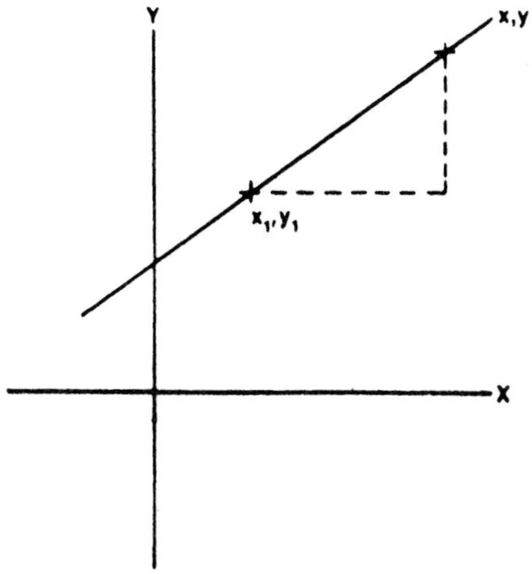


**FIGURE 1.30** Point-slope form.

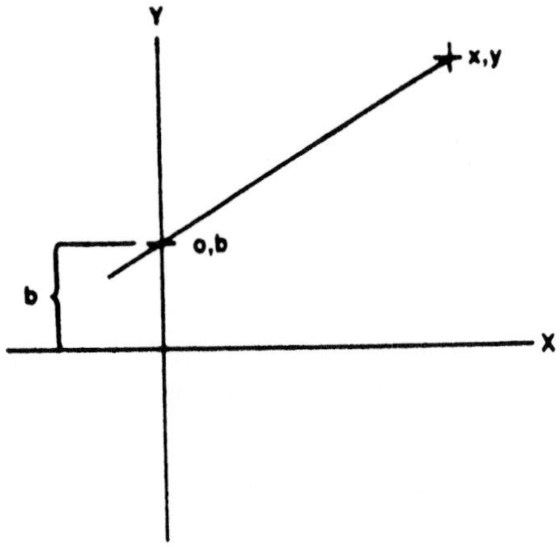


FIGURE 1.31 Slope-intercept form.

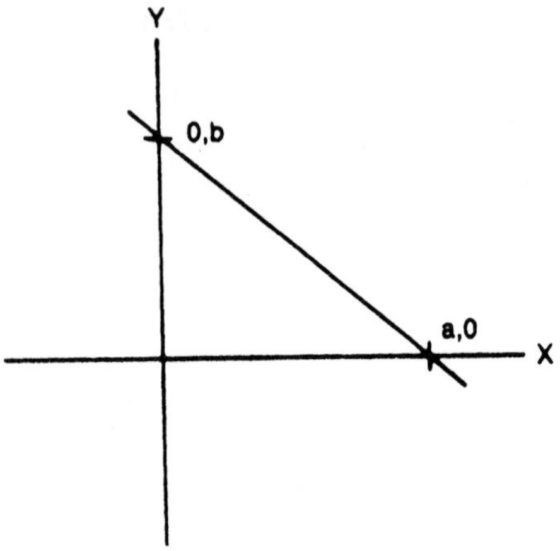


FIGURE 1.32 The intercept form.

### 1.9.3 Equations of the Common Geometric Forms

***Circle.*** With the center at $(a,b)$ and radius $r$,

$$(x-a)^2+(y-b)^2=r^2$$

With the center of the circle at the origin

$$x^2+y^2=r^2$$

***Ellipse.*** With the origin at the center of the ellipse and the major and minor axes of $a$ and $b$

$$\frac{x^2}{a^2}+\frac{y^2}{b^2}=1$$

***Parabola.*** With the origin at the vertex, and $a$ being the distance from focus to vertex

$$y^2=4ax.$$

***Hyperbola.*** With the origin at the center and major and minor axes of $a$ and $b$

$$\frac{x^2}{a^2}-\frac{y^2}{b^2}=1$$

### 1.9.4 Relating Rectangular and Polar Coordinates

$$x=r\cos\theta$$

$$y=r\sin\theta$$

$$r=\sqrt{x^2+y^2}$$

$$\theta=\tan^{-1}\left(\frac{y}{x}\right)$$

$$\sin \theta = \frac{y}{\sqrt{x^2 + y^2}}$$

$$\cos \theta = \frac{x}{\sqrt{x^2 + y^2}}$$

### 1.9.5 Other Useful Curves

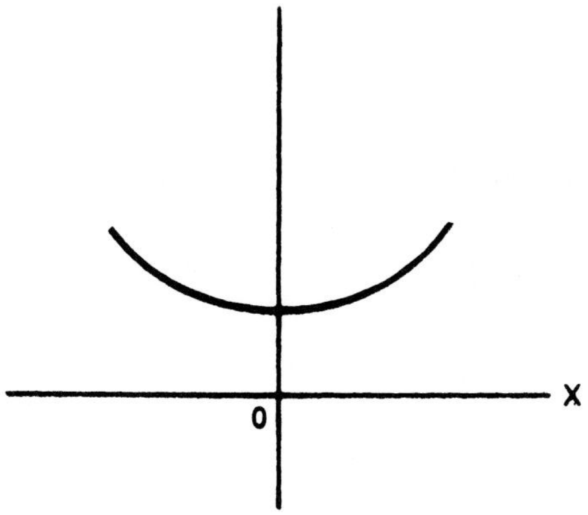

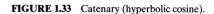

**FIGURE 1.33** Catenary (hyperbolic cosine).

***Catenary (Hyperbolic Cosine).*** (See Fig. 1.33.)

$$y = \frac{a}{2}\,(e^{x/a} + e^{-x/a})$$

$$y = a \cosh \frac{x}{a}$$

***Parabola.*** (See Fig. 1.34.)

$$\pm\sqrt{x} \pm \sqrt{y} = \sqrt{a}$$

$$x^2 - 2xy + y^2 - 2ax - 2ay + a^2 = 0$$

***Spiral of Archimedes.*** (See Fig. 1.35.)

$$(x^2 + y^2) = a^2\left[\tan^{-1}\left(\frac{y}{x}\right)\right]^2$$

$$p = a\,\theta$$

***Logarithmic Curve.*** (See Fig. 1.36.)

$$y = \log a^2$$

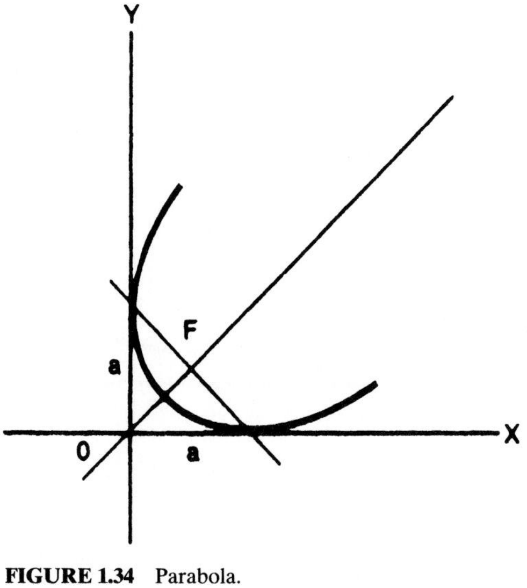


**FIGURE 1.34** Parabola.

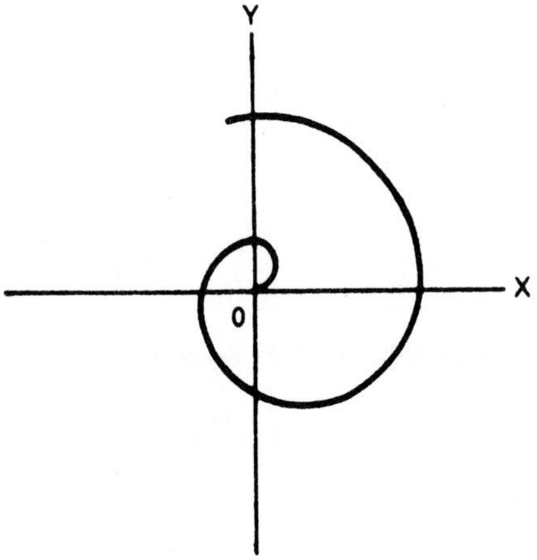


FIGURE 1.35 Spiral of Archimedes.

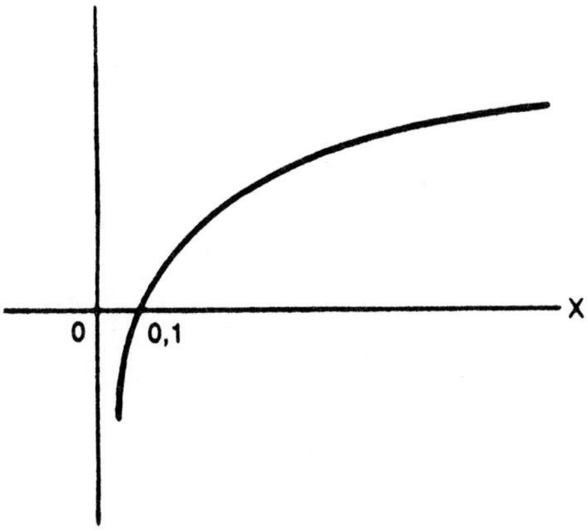


FIGURE 1.36 Logarithmic curve.

***Exponential Curve.*** (See Fig. 1.37.)

$$y = a^x$$

***Involute of a Circle.*** (See Fig. 1.38.)

$$x = r \cos \theta + r\theta \sin \theta$$

$$y = r \sin \theta - r\theta \cos \theta$$

***Logarithmic (Equiangular) Spiral.*** (Fig. 1.39.)

$$x^2 + y^2 = e^{2a\left[\tan^{-1}\left(\frac{y}{x}\right)\right]}$$

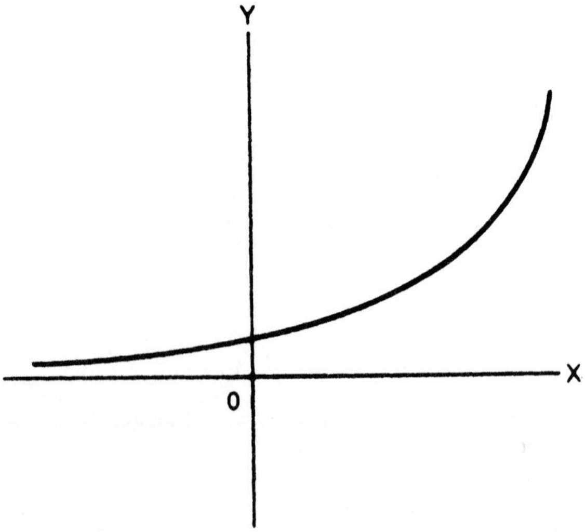


FIGURE 1.37 Exponential curve.

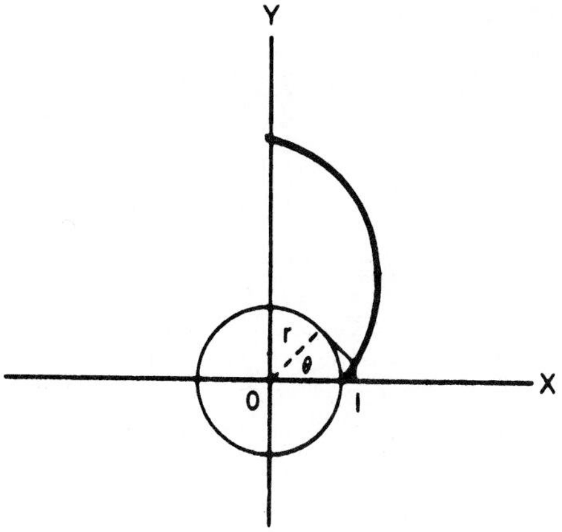


**FIGURE 1.38** Involute of a circle.

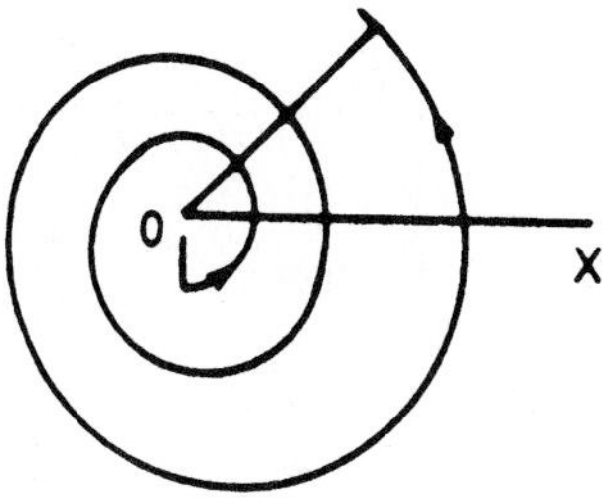


**FIGURE 1.39** Logarithmic spiral.

## 1.10 BASIC CALCULUS

### 1.10.1 Derivatives

The rate of change of a function is expressed by the derivative of the function.

Let $y = f(x)$ be a given function of one variable. Let $x_1$ be a chosen value of $x$, and let $\Delta x$ be an increment of $x$ to be added to $x_1$; then $\Delta y$ will be the corresponding increment of $y$:

$$\Delta y = f(x_1 + \Delta x) - f(x_1)$$

Now form the increment ratio:

$$\frac{\Delta y}{\Delta x} = \frac{f(x_1 + \Delta x) - f(x_1)}{\Delta x}$$

Let $\Delta x \rightarrow 0$; then $\Delta y/\Delta x$ will usually approach a limit that is called the *derivative* of $y$ with respect to $x$ at the value $x = x_1$. Expressed in another manner, the derivative of a function $f(x)$ is the limit that is approached by the increment of $f(x)$ to the increment of $x$, when the increment of $x$ approaches the limit 0.

The derivative of $y = f(x)$ with respect to $x$ is symbolized in various ways:

$$\frac{dy}{dx} \quad \text{or} \quad D_x y \quad \text{or} \quad f'(x) \quad \text{or} \quad y' \quad \text{(1st derivative)}$$

$$\frac{d^2y}{dx^2} \quad \text{or} \quad D_x^2 y \quad \text{or} \quad f''(x) \quad \text{or} \quad y'' \quad \text{(2d derivative)}$$

The derivative may be interpreted as

- *The slope of a curve:* If $x$ and $y$ are rectangular coordinates of a variable point, and if the function $y = f(x)$ represents a continuous curve, then the slope of the tangent line to the curve at a point $x = x_1$ is the value of the derivative of $f(x)$ at $x = x_1$.
- *The velocity of a moving particle:* If $s$ represents the distance traveled by a particle moving in a straight line over time $t$, then a function $s = f(t)$ represents the law of motion, and the derivative of $f(t)$ at $t = t_1$ represents the velocity of the particle at the instant $t_1$.

### 1.10.2 Differentiation Formulas for Algebraic Functions

Sum $$\frac{d}{dx}(u+v+w)=\frac{du}{dx}+\frac{dv}{dx}+\frac{dw}{dx}$$

Constant times a function, where $c$ = constant

$$\frac{d}{dx}(cu)=c\frac{du}{dx}$$

Product $$\frac{d}{dx}(uv)=u\frac{dv}{dx}+v\frac{du}{dx}$$

Constant $$\frac{dc}{dx}=0$$

Quotient of two functions

$$\frac{d}{dx}\left(\frac{u}{v}\right)=\frac{v(du/dx)-u(dv/dx)}{v^2}$$

Power $$\frac{d}{dx}(u^n)=nu^{n-1} \qquad \left[\text{e.g., } \frac{d}{dx}(5x^3)=15x^2\right]$$

Function of a function

$$\frac{dy}{dx}=\frac{dy}{dz}\frac{dz}{dx}$$

$$\frac{dy}{dx}=\frac{1}{dx/dy}$$

$$\frac{dy}{dx}=\frac{dy/du}{dx/du}$$

where $y$ = a function of $z$ and $z$ = a function of $x$. The differential of a function is the product of the derivative of the function with respect to its variable and the differential of the variable, whether the said variable is independent or is itself a function of another variable.

Listed here are some of the preceding formulas written in differential form.

$$d\,(u+v+w)=du+dv+dw$$

$$d\,(cu)=c\,du$$

$$d\,(c)=0$$

$$d\,(uv)=u\,dv+v\,du$$

$$d\left(\frac{u}{v}\right)=\frac{v\,du-u\,dv}{v^2}$$

$$d\,(u^n)=nu^{n-1}\,du$$

### 1.10.3 Differentiation Formulas for Transcendental Functions

Formulas for angles in radians

$$\frac{d}{dx}\sin u = \cos u \frac{du}{dx}$$

$$\frac{d}{dx}\cos u = -\sin u \frac{du}{dx}$$

$$\frac{d}{dx}\tan u = \sec^2 u \frac{du}{dx}$$

$$\frac{d}{dx}\log_a u = \frac{\log_a e}{u}\frac{du}{dx}$$

$$\frac{d}{dx}\ln u = \frac{1}{u}\frac{du}{dx}$$

$$\frac{d}{dx}a^u = a^u \ln a \frac{du}{dx}$$

$$\frac{d}{dx}e^u = e^u \frac{du}{dx}$$

In differential notation, these formulas take the forms:

$$d\sin u = \cos u \, du$$

$$d\cos u = -\sin u \, du$$

$$d\tan u = \sec^2 u \, du$$

and so forth.

### 1.10.4 Applications of the Derivative

***Maximum or Minimum.*** A container with planar sides and square ends is to be constructed from 50 square feet ($ft^2$) of material. What are the dimensions of the container (Fig. 1.40) that will allow a maximum volume to be achieved?

Now if $A_s$ = surface area and volume = $V = x^2 y$, then

$$A_s = x^2 + x^2 + 4xy$$

$$= 2x^2 + 4xy$$

$$50 = 2x^2 + 4xy$$

$$y = \frac{50 - 2x^2}{4x}$$

Restructuring the last equation, we get

$$\frac{50}{4x} - \frac{2x}{4}$$

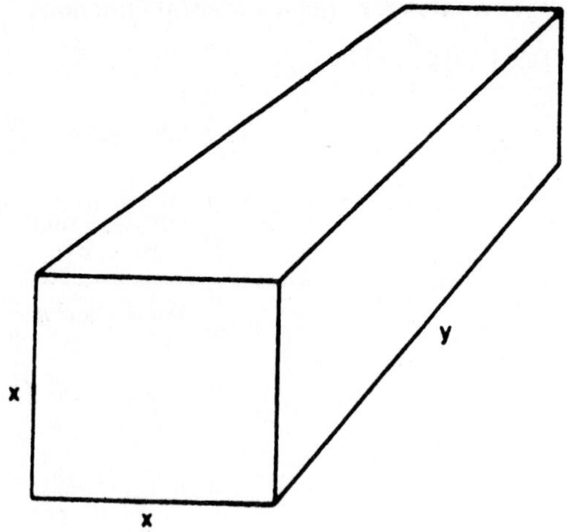


**FIGURE 1.40** Rectangular solid.

Next, list the volume equation:

$$V = x^2 y$$

Substitute the expression for $y$ back into the volume equation:

$$V = x^2\left(\frac{50}{4x} - \frac{2x}{4}\right) = \frac{50}{4} - \frac{2x^3}{4}$$

$$\frac{dv}{dx} = \frac{50}{4} - \frac{3x^2}{2}$$

Equate this expression to 0 and solve for $x$:

$$\frac{3x^2}{2} = \frac{50}{4} \qquad x = \sqrt{\frac{100}{12}} = 2.887$$

Substitute the value of $x$ into the $y$ equation above to solve for $y$:

$$y = \frac{50}{4x} - \frac{2x}{4} = \frac{50}{4\,(2.887)} - \frac{2\,(2.887)}{4} = 2.886$$

Substitute the calculated dimensions of $x$ and $y$ into the volume equation:

$$V = x^2 y = (2.887)^2 \times 2.886 = 24.054 \text{ ft}^3$$

It is interesting to note that this problem shows a cube to be the optimum six-sided solid for getting the most volume from a given amount of surface material. Had one side been left off the container, the dimensions would have been different than shown.

***Rectilinear Motion.*** The motion of an object $R$ along a straight line is described by an equation, $S = f(t)$. If $t \geq 0$ is time and $S$ the distance of the object from the fixed point in its line of motion, then

$$\text{Velocity of } R \text{ at time } t = v = \frac{ds}{dt}$$

If $v < 0$, $R$ is moving in the direction of decreasing distance $S$.
If $v > 0$, $R$ is moving in the direction of increasing distance $S$.
If $v = 0$, $R$ is instantaneously at rest.

The acceleration of $R$ at time $t$ is

$$a = \frac{dv}{dt} = \frac{d^2S}{dt^2}$$

If $a > 0$, $v$ is increasing.
If $a < 0$, $v$ is decreasing.
If $v$ and $a$ have the same sign, the speed of $R$ is increasing.
If $v$ and $a$ have opposite signs, the speed of $R$ is decreasing.

*Example.* An object moves in a straight line according to the equation

$$S = \frac{1}{2} t^3 - 3.5t$$

Determine its velocity and acceleration at the end of 5 sec.
Taking the first derivative, we obtain

$$\text{Velocity} = v = \frac{ds}{dt} = \frac{3}{2} t^2 - 3.5$$

When $t = 5$ sec

$$v = \frac{3}{2} (5)^2 - 3.5 = \frac{75}{2} - 3.5 = 34 \text{ ft/sec}$$

Taking the second derivative, we have

$$\text{Acceleration} = a = \frac{dv}{dt} = 3t = 3(5) = 15 \text{ ft/sec}^2$$

Note that the equation of motion in this example is for illustrative purposes only. In practice, the actual equation of motion must be determined or known before the problem can be solved.

The actual equation for motion of a body falling to Earth is

$$S(t) = kt^2$$

where $k$ is a constant due to Earth's gravity, and distance is a function of time. (We are neglecting air resistance.) In this case, $k = 16$ ft/sec$^2$, or in SI units, 4.9 m/sec$^2$.

Taking the first derivative of $S(t)$, we obtain

$$\frac{dS}{dt} = 2kt$$

Taking the second derivative, we have

$$\frac{dv}{dt} = 2k$$

With $k = 16$

$$\frac{dv}{dt} = 2k = 2\,(16) = 32 \text{ ft/sec}^2$$

which is the acceleration due to gravity, commonly denoted as $g$ or $G$. The actual value of the acceleration varies according to your location on Earth, but in engineering practice the commonly used figure is 32.16 ft/sec$^2$.

### 1.10.5 Integrals and Integration

The process of finding a quantity or function whose derivative is known is called *antidifferentiation* or *integration.* Integration is also a term used for the mathematical process of calculating areas and volumes.

If $f(t)$ is one antiderivative of $f'(t)$, then all antiderivatives are $f(t) + C$, where $C$ is any arbitrary constant. Some examples of antiderivatives are summarized here.

$$\text{Functions}\left\{\begin{array}{ll} nx^{n-1} & x^n + C \\ \cos x & \sin x + C \\ \cos 2x & \frac{1}{2}\sin x + C \\ \sin 4x & -\frac{1}{4}\sin x + C \\ e^{2x} & \frac{e^{2x}}{2} + C \end{array}\right\}\text{antiderivatives}$$

Some examples of indefinite integrals are:

$$\int x^n dx = \frac{1}{n+1}x^{n+1} + C$$

$$\int \cos x \, dx = \sin x + C$$

$$\int \sin x \, dx = -\cos x + C$$

Many mathematics references list tables of indefinite integrals that are actually antiderivatives of functions.

***Fundamental Theorem of Calculus.*** When $f$ is continuous on $(a,b)$, let $F(x) = \int f(x)\,dx$. Then

$$\int_a^b f(x)\,dx = F(b) - F(a)$$

To evaluate a definite integral by use of the fundamental theorem given above, an antiderivative is required.

*Example.* Calculate the area projected from the parabolic curve (Fig. 1.41) whose equation is $x = y^2 + 1$ to the $Y$ axis between $y = -2$ and $y = 2$.

$$A = \int_{-2}^{2} f(y^2 + 1)\,dy$$

$$= \left[\frac{y^3}{3} + y\right]_{-2}^{2} = \left[\frac{2^3}{3} + 2\right] - \left[\frac{(-2)^3}{3} + (-2)\right]$$

$$= \left(\frac{8}{3} + 2\right) - \left(\frac{-8}{3} - 2\right) = \frac{28}{3} = 9.333 \text{ in}^2$$

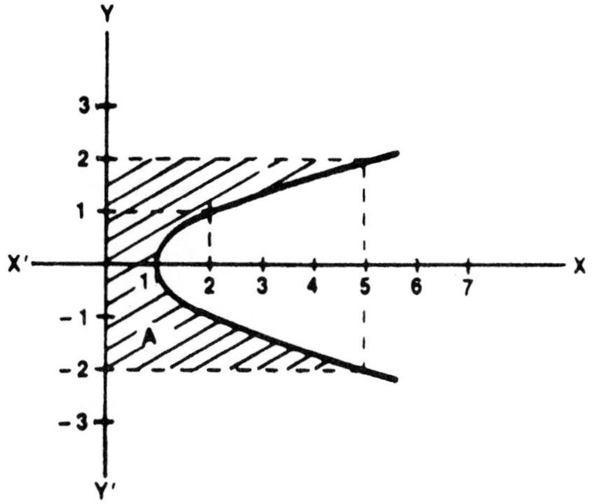


**FIGURE 1.41** Projected area of a parabolic curve.

The preceding examples of derivatives and integrals are shown in their most basic forms to introduce them to those not familiar with the basic principles of the calculus. Excellent texts on the calculus are listed in the bibliography at the end of this chapter for those who may wish to review or study the subject in depth. Complete lists of differentiation and integration formulas are too extensive to include in this handbook. Only the basic ones have been shown for reference.

### 1.10.6 Other Useful Integration Formulas

***Prismoidal Area Formula.*** For calculating approximate areas under curves (Fig. 1.42), we have

$$A = \frac{h}{3}\,(y_1 + 4y_2 + y_3)$$

where $A$ is exact only for third degree or lower polynomials $y = f(x)$, and approximate for other degree polynomial curves.

***Simpson's Rule (for Areas).*** Divide the required area (Fig. 1.43) by an even number of vertical strips (the $y$ ordinates). The more strips, the more accuracy. Then, if $y_1, y_2, y_3, \ldots, y_n$ = the values of the ordinates forming the boundaries of the strips, and $\Delta x$ = distance between strips and ordinates

$$A = \frac{\Delta x}{3}\,(y_1 + 4y_2 + 2y_3 + 4y_4 + 2y_5 + \cdots + 2y_{n-1} + 4y_n + y_{n+1})$$

The equation of the curve $y = f(x)$ does not have to be known if the $y$ ordinates are known or measured.

***Approximate-Volume Calculations.*** Referring to Fig. 1.44, if $A_1$, $A_2$, and $A_3$ equal areas of sections, then

$$A = \frac{h}{3}\,(A_1 + 4A_2 + A_3)$$

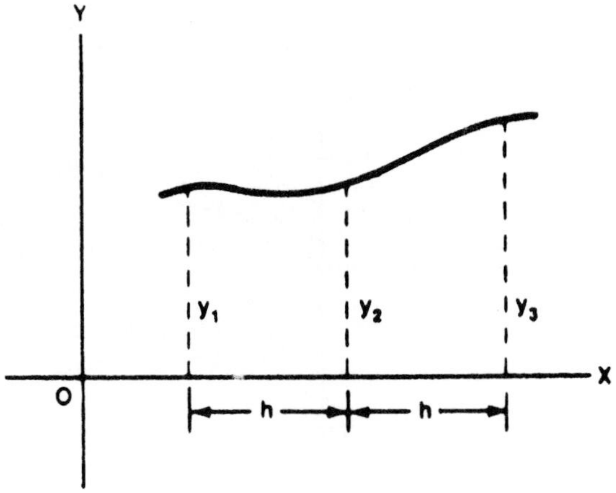


**FIGURE 1.42** Finding the area under the curve.

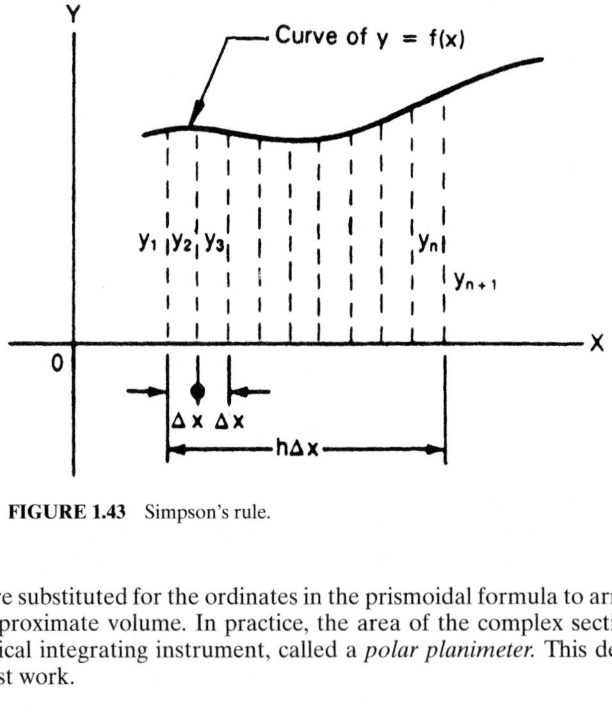


**FIGURE 1.43** Simpson's rule.

The areas are substituted for the ordinates in the prismoidal formula to arrive at the above formula for approximate volume. In practice, the area of the complex section is calculated with a mechanical integrating instrument, called a *polar planimeter.* This device is accurate enough for most work.

### 1.10.7 Partial Derivatives

When a function is expressed in terms of several variables rather than in terms of only one variable, the concept of the *partial derivative* is normally applicable. For example, if $z$ is a function of $x$ and $y$ (i.e., $f(x,y)$, then $\partial z/\partial x$ is the derivative of $z$ with respect to $x$, with $y$ treated as a constant, and $\partial z/\partial y$ is the derivative of $z$ with respect to $y$, with $x$ treated as a constant.

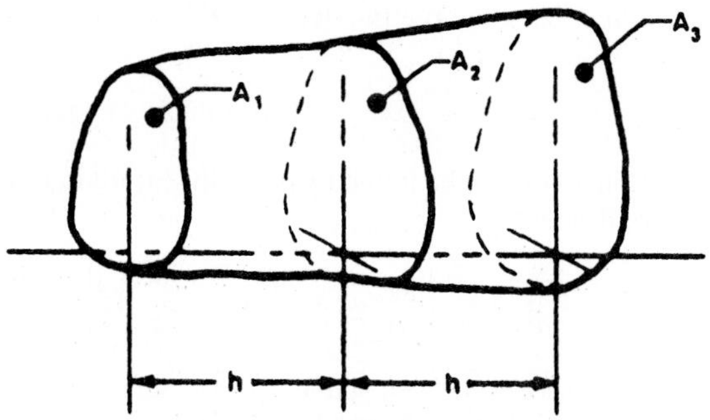


**FIGURE 1.44** Approximate volume.

Therefore, other than treating the variables one at a time, while the others are considered constant momentarily, there is no fundamental difference between the partial derivative and the total derivative.

When we consider the total differential, however, an additional concept is involved. Where for one variable we have

$$dy = \left(\frac{dy}{dx}\right) dx$$

For two variables, the differential must take into account both. Thus

$$dz = \frac{\partial z}{\partial x}\, dx + \frac{\partial z}{\partial y}\, dy$$

The total differential of $z$ is the sum of the partial derivatives, each multiplied by its proper differential. This concept can be extended from two variables to any number of variables.

*Example.* Given $z = 4x^3 - 3y^2 + 4xy$, find

$$\frac{\partial z}{\partial x} \quad \text{and} \quad \frac{\partial z}{\partial y}$$

*Solution.* When solving for the partial derivative $\partial z/\partial x$, we differentiate with respect to $x$, treating $y$ as a constant. We therefore obtain

$$\frac{\partial z}{\partial x} = 12x^2 + 4y$$

Solving for $\partial z/\partial y$, we differentiate with respect to $y$, treating $x$ as a constant to find

$$\frac{\partial z}{\partial y} = -6y + 4x \qquad \text{or} \qquad (4x - 6y)$$

*Note.* The first step in finding second partial derivatives is to find the first partial derivatives.

***Sample Problem Using Partial Derivatives.*** The length, depth, and width of a rectangular container are each increasing at a rate of 3 inches per minute (in/min). Find the rate at which the volume of the container is increasing at the instant when the length is 5 ft, the width is 3 ft, and the depth is 2 ft.

*Solution.* Let $x$ denote the depth, $w$ the width, and $l$ the length. The volume of the container is

$$v = xwl \qquad \text{(the volume of a rectangular solid)}$$

Its rate of increase can be found by partially differentiating the volume expression, and all the functions of time are

$$\frac{dv}{dt} = \frac{d}{dt}\left[\frac{\partial y}{\partial x}\,dx + \frac{\partial v}{\partial y}\,dy + \frac{\partial y}{\partial z}\,dz\right] = wl\,\frac{dx}{dt} + xl\,\frac{dw}{dt} + xw\,\frac{dl}{dt}$$

We know that $dx/dt = dw/dt = dl/dt = 3$ in/min = ¼ ft/min, and we wish to find $dv/dt$ at the instant that $x = 2$ ft, $w = 3$ ft, and $l = 5$ ft. Thus

$$\frac{dv}{dt} = \frac{1}{4}(3 \cdot 5 + 2 \cdot 5 + 2 \cdot 3) = \frac{1}{4}(31)$$

$$= 0.25 \cdot 31 = 7.75 \text{ ft}^3/\text{min}$$

### 1.10.8 Differentials

A differential is a change that the dependent variable ($y$) assumes when the independent variable ($x$) undergoes an infinitesimal change. The differential is also another way of interpreting the derivative. If

$$y = f(x) \qquad \text{(function)}$$

$$\frac{dy}{dx} = f'(x) \qquad \text{(derivative)}$$

$$dy = f'(x)\,dx \qquad \text{(differential form)}$$

This differential form is merely the derivative with the two sides multiplied by $dx$.

When we have more than two variables, the procedure is more complex. With three variables, $x$, $y$ and $z$, the differential of $z$ is

$$dz = \frac{\partial z}{\partial x}\,dx + \frac{\partial z}{\partial y}\,dy$$

Here, the symbol $\partial z/\partial x$ is the partial derivative of $z$ with respect to $x$, that is, the derivative of $z$ with respect to $x$ while considering $y$ as a constant. The partial derivative of $z$ with respect to $y$ is defined similarly, and the concept can be extended to any number of variables.

If we wish to find the change in a dependent variable, where the change in the independent variable is small but not *infinitesimal,* we can apply the same equation as an approximation:

$$\Delta z = \frac{\partial z}{\partial x}\,\Delta x + \frac{\partial z}{\partial y}\,\Delta y$$

where the $\Delta$ values represent small but not infinitesimal changes. This method of approximation of $dz$ by $\Delta z$ is very useful in the numerical calculation of changes in quantities. Typical differential problems are presented in the following paragraphs.

Given $y = 2x^3$, find $dy$ (differential $y$). By definition, for the differential $y = f'(x)\,dx = (dy/dx)\,dx$, we find $f'(x) = 6x^2$ (first derivative of $2x^3$), Therefore $dy = 6x^2\,dx$.

Find the differential of $(x^2 + 3)^3$:

$$\begin{aligned} d\,[(x^2+3)^3] &= 3(x^2+3)^{3-1}\cdot d(x^2+3) \\ &= 3(x^2+3)^2\cdot d(x^2) \\ &= 3(x^2+3)^2\cdot 2x\,dx \\ &= 6x(x^2+3)^2\,dx \end{aligned}$$

If $y = 2x^{3/2}$, what is the approximate change in $y$ when $x$ changes from 9 to 9.02?

*Solution.* The differential $dy$ is found from

$$\frac{dy}{dx} = \frac{3}{2}\,(2)\,x^{3/2-1} \qquad dy = 3x^{1/2}\,dx$$

The numerical value of $dy$ may be found by allowing $x = 9.00$ and $dx = 0.02$ in this equation:

$$dy = 3(9.00)^{1/2}\,(0.02) = 0.18$$

The exact change in $y$ may be found from

$$2(9.02)^{3/2} - 2(9.00)^{3/2} =$$

$$2(27.09005) - 2(27) = 0.1801$$

However, this is a more laborious method of finding $dy$.

By using differentials, find an approximation for the value of $2x^3 - 2x^2 + 3x - 1$, when $x = 2.995$.

*Solution.* We will consider the value 2.995 as the result of applying an increment of $-0.005$ to an original value of 3. Then, $x = 3$ and $\Delta x = -0.005 \approx dx$. Now, $dy/dx = 6x^2 - 4x + 3$ and $dy = (6x^2 - 4x + 3)\,dx$.

We now substitute $x = 3$ and $dx = -0.005$, obtaining

$$dy = [6(3)^2 - 4(3) + 3]\,(-0.005) = -0.225$$

which is approximately the change in $y$ caused by going from $x = 3$ to $x = 2.995$.

We must now find the value of $y$ when $x = 3$ and add this value to $dy$. When $x = 3$ in the original equation, we obtain

$$dy = 2(3)^3 - 2(3)^2 + 3(3) - 1 = 44$$

Therefore, $y$ at $x = 2.995 = y + dy = 44 + (-0.225) = 43.775$, which is the approximate value of the polynomial for $x = 2.995$.

To check, we substitute 2.995 in the original function $2x^3 - 2x^2 + 3x - 1$:

$$2(2.995)^3 - 2(2.995)^2 + 3(2.995) - 1 = 43.775$$

Using the differential in applications such as the preceding example provides very precise results.

### 1.10.9 Double or Iterated Integrals

The double integral is often used in complex problems involving areas, volumes, and moments. The basic form of the double integral is

$$\int_a^b \int_{g_1(x)}^{g_2(x)} f(x,y)\,dy\,dx$$

where the variable of the first differential ($y$) is integrated first. While this is done, $x$ is considered a constant. After integration with respect to $y$, the limits $g_2(x)$ and $g_1(x)$ are substituted. This results in an equation in $x$ and $dx$ alone, which is then integrated, and limits $a$ and $b$ are substituted for $x$.

The roles of the variables may be reversed, in which case the preceding integral can be written as

$$\int_c^d \int_{h_2(y)}^{h_1(y)} f(x,y)dx\, dy$$

In this case, integration is performed with respect to $x$ first and $h_1(y)$ and $h_2(y)$ are the limits. The constants $c$ and $d$ are the limits of $y$.

The end results are the same, and the choice of sequence depends on the complexity of the integration for either possibility. The choice of the order of integration is of no consequence, except that it may affect the amount of work required to solve the problem.

***Double-Integral References.*** Integrated areas in cartesian coordinates are

$$A = \int_R \int dy\, dx = \int_a^b dx \int_{f(x)}^{F(x)} dy$$

Integrated areas in polar coordinates are

$$A = \int\int p\, dp\, d\theta = \int_{\theta_1}^{\theta_2} d\theta \int_{f(\theta)}^{F(\theta)} p\, dp$$

## 1.11 PERCENTAGES AND COMPOUND-INTEREST CALCULATIONS

### 1.11.1 Percentages

Let us compare two arbitrary numbers, 22 and 37, as an illustration.

$$\frac{37-22}{22} = 0.6818$$

The number 37 is thus 68.18 percent larger than 22. We can also say that 22 increased by 68.18 percent is 37.

$$\frac{37-22}{37} = 0.4054$$

The number 37 minus 40.5 percent of itself is 22. We can also say that 22 is 40.54 percent less than 37.

$$\frac{22}{37} = 0.5946$$

The number 22 is 59.46 percent of 37.

*Example.* A spring is compressed to 417 lb of pressure, or load, and later decompressed to 400 lb. The percentage pressure drop is (417 – 400)/417 = 0.0408, or 4.08 percent.

The pressure is then increased to 515 lb. The percentage increase over 400 lb is therefore (515 – 400)/400 = 0.2875, or 28.75 percent.

Percentage problem errors are quite common, even though the calculations are simple. If you remember that the divisor is the number of which you want the percentage, either increasing or decreasing, the simple errors can be avoided.

### 1.11.2 Compound Interest

Following is a complete listing of compound-interest problems and formulas that may be used to your benefit in the handling of money.

***Compound Interest.*** Where $A_f$ = final value, $P_s$ = amount saved, $r$ = interest rate per year, $r/12$ = monthly rate, $r/365$ = daily rate, and $n$ = term (years with $r$, months with $r/12$, and days with $r/365$),

$$A_f = P_s(1 + r)^n$$

*Example.* A person places \$5000 in a savings account that pays 5¼ percent yearly interest. How much money will be in the account after 15 years?

$$A_f = 5000\ (1 + 0.055)^{15} = 5000\ (2.23248) = \$11{,}162.40$$

***Ammortization (Sinking Funds).*** Where $P_L$ = loan amount, $n$ = number of months to pay the loan, $r$ = yearly rate/12, and $R$ = monthly payment,

$$P_L = R\ \frac{1 - (1 + r)^{-n}}{r}$$

*Example.* A person buys an automobile and the total loan amount is \$7500, to be paid in 48 monthly payments. The interest rate is 17½ percent per year. What are the monthly payments?

$$7500 = R\ \frac{1 - (1 + 0.0145833)^{-48}}{0.0145833}$$

$$R = \frac{7500}{34.3474} = \$218.36 \text{ per month}$$

Home loans are calculated using this equation, as are other types of standard loans.

***Annuities.*** A sum of \$80,000 is in an account that pays 7½ percent annual interest. What annuity can be paid out of this sum for 15 years at monthly intervals before the sum is fully depleted?

In this case

$$A_m = \frac{P_s r(1 + r)^n}{(1 + r)^n - 1}$$

where $A_m$ = monthly payout, $P_s$ = \$80,000, $r$ expressed as a monthly rate = 0.075/12 = 0.00625, and $n = 15 \times 12 = 180$ months.

$$A_m = \frac{80{,}000\ (0.00625)(1.00625)^{180}}{(1.00625)^{180} - 1} = \$741.61 \text{ per month}$$

*Example.* If an annuity of $A$ per month is to be paid for $X$ years, what amount of money $P_a$ must be put into the account if interest is $R$ percent annually? Using $r = R/12$ and $n = X$ (12), we obtain

$$P_a = A\ \frac{(1 + r)^n - 1}{(1 + r)^n r}$$

***Individual Retirement Accounts (IRAs) and Long-Term Accounts.*** Setting aside a sum $A$ at the beginning of each year at $r\%$ yearly interest, the total value $P_a$ of the account in $n$ years will be

$$P_a = A\,\frac{(1+r)[(1+r)^n - 1]}{r}$$

In this equation, eliminating the first $1 + r$ term will yield the total amount $P_a$ when the sum $A$ is set aside at the end of each year.

If an account with $P$ amount is increased or decreased by a sum $A$ at the end of each year, then the account will be valued at $P_a$ after $n$ years.

$$P_a = P\,(1+r)^n \pm A\,\frac{(1+r)^n - 1}{r}$$

The second term in this equation is added or subtracted according to whether the sum $A$ is added to or subtracted from the account at the end of each year.

When using the listed equations for compound interest, be careful to use the correct interest rate, whether it is yearly, monthly, or daily. Note also that when monthly rates are used, $n$ must be the number of months; when yearly rates are used, $n$ must be the number of years. The preceding equations are mathematically exact. With these equations, loan payments, annuities, and savings can be calculated to the penny.

***Compound Interest.*** If we wish to know what interest rate is required to increase a sum by a certain amount after $n$ years

$$A_f = P_s\,(1+r)^n \qquad P_s = \frac{A_f}{(1+r)^n}$$

Then

$$(\text{Rate})\; r = \sqrt[n]{\frac{A_f}{P_s}} - 1$$

Also

$$n = \frac{\log A_f - \log p_s}{\log\,(1+r)} \qquad \text{(number of years lent)}$$

These equations apply not only to problems involving money also but to any compound-rate problems where these equations may be employed.

## 1.12 THE INVOLUTE FUNCTION

The involute function ($\text{inv}\,\phi = \tan\phi - \text{arc}\,\phi$) is widely used in gear calculations (see Chap. 8). The angle $\phi$ for which involute tables are tabulated is the slope of the involute with respect to a radius vector $R$ (see Fig. 1.45).

### 1.12.1 Involute Geometry

The *involute* of a circle is defined as the curve traced by a point on a straight line which rolls without slipping on the circle. It is also described as the curve generated by a point on a nonstretching string as it is unwound from a circle. The circle is called the *base circle* of the involute. A single involute curve has two branches of opposite hand, meeting at a point on the base circle, where the radius of curvature is zero. All involutes of the same base circle are congruent and parallel, while involutes of different base circles are geometrically similar.

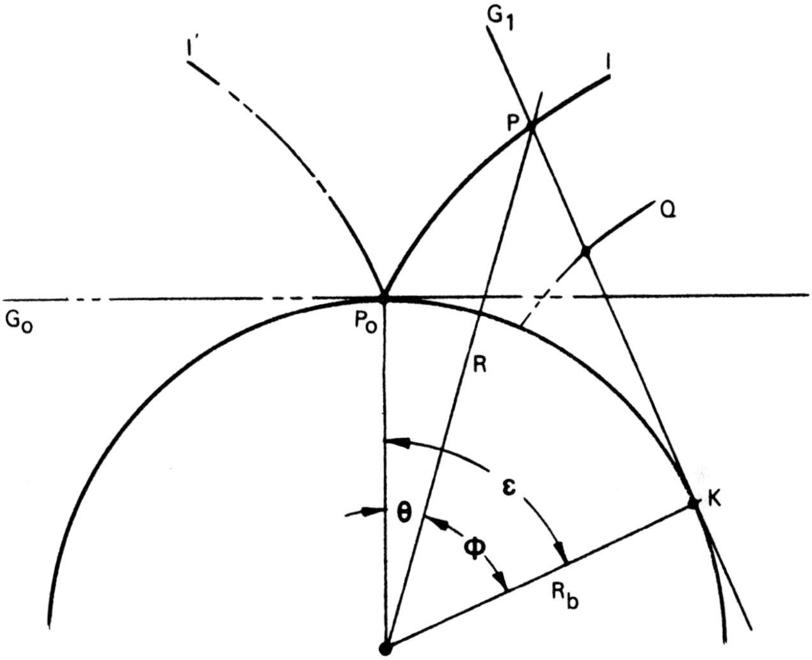


**FIGURE 1.45** Geometry of the involute to a circle.

Figure 1.45 shows the elements of involute geometry. The generating line was originally in position $G_0$, tangent to the base circle at $P_0$. The line then rolled about the base circle through the roll angle ε to position $G_1$, where it is tangent to the base circle at $K$. The point $P_0$ on the generating line has moved to $P$, generating the involute curve $I$. Another point on the generating line such as $Q$ generates another involute curve which is congruent and parallel to curve $I$.

Since the generating line is always normal to the involute, the angle $\phi$ is the slope of the involute with respect to the radius vector $R$. The polar angle $\theta$ together with $R$ constitutes the coordinates of the involute curve. The parametric polar equations of the involute are

$$R = R_b \sec \phi \qquad \theta = \tan \phi - \phi$$

The quantity $(\tan \phi - \phi)$ is called the *involute function* of $\phi$

*Note.* The roll angle ε in radians is equal to $\tan \phi$.

To calculate the involute function ($\text{inv}\ \phi = \tan \phi - \text{arc}\ \phi$), first find the involute function for 20.00°:

$$\text{inv}\ \phi = \tan \phi - \text{arc}\ \phi$$

where $\tan \phi$ = natural tangent of the given angle, arc $\phi$ = numerical value in radians of the given angle. Thus

Inv $\phi$ = tan 20° – 20° converted to radians.

Inv $\phi$ = 0.3639702 – (20 × 0.0174533). (*Note:* 1° = 0.0174533 radians.)

Inv $\phi$ = 0.3639702 – 0.3490659.

Inv 20° = 0.0.0149043.

The involute function for 20° = 0.0149043 (accurate to seven decimal places)

Using the procedure shown above, it becomes obvious that a table of involute functions is not required for gearing calculation procedures. It is also safer to calculate your own involute functions as the handbook tables may contain typographical errors.

### 1.12.2 Plotting the Involute Curve

The $x$ and $y$ coordinates of the points on an involute curve (see Fig. 1.38) may be calculated from

$$x = r\cos\theta + r\theta\sin\theta \qquad \text{and} \qquad y = r\sin\theta - r\theta\cos\theta$$

## 1.13 U.S. CUSTOMARY AND METRIC (SI) MEASURES AND CONVERSIONS

Refer to Table 1.1.

## 1.14 TEMPERATURE SYSTEMS

There are four common temperature systems used in engineering and design calculations: (°F) Fahrenheit, (°C) Celsius (formerly centigrade), (K) Kelvin, and (°R) Rankine.

The conversion equation for Celsius to Fahrenheit or Fahrenheit to Celsius is given here.

$$\frac{5}{9} = \frac{°\text{C}}{°\text{F} - 32}$$

This exact relational equation is all that you need to convert from either system. Enter the known temperature and solve the equation for the unknown value.

*Example.* You wish to convert 66°C to Fahrenheit.

$$\frac{5}{9} = \frac{66}{°\text{F} - 32}$$

$$5°\text{F} - 160 = 594$$

$$°\text{F} = 150.8$$

This method is much easier than trying to remember the two equivalent equations, which are

$$°\text{C} = \frac{5}{9}(°\text{F} - 32)$$

and

$$°\text{F} = \frac{9}{5}°\text{C} + 32$$

The other two systems, Kelvin and Rankine, are converted as described here.

The Kelvin and Celsius scales are related by

$$\text{K} = 273.18 + °\text{C}$$

Thus 0°C = 273.18 K. Absolute zero is equal to –273.18°C.

*Example.* A temperature of −75°C = 273.18 + (−75°C) = 198.18 K.
The Rankine and Fahrenheit scales are related by

$$°R = 459.69 + °F$$

Thus 0°F = 459.69°R. Absolute zero is equal to −459.69°F.
*Example.* A temperature of 75°F = 459.69 + (+75°F) = 534.69°R.

## 1.15 MATHEMATICAL SIGNS AND SYMBOLS

| | |
|---|---|
| $+$ | Plus, positive |
| $-$ | Minus, negative |
| $\times$ or $\cdot$ | Times, multiplied by |
| $\div$ or / | Divided by |
| $=$ | Is equal to |
| $\equiv$ | Is identical to |
| $\cong$ | Is congruent to or approximately equal to |
| $\sim$ | Is approximately equal to or is similar to |
| $<$ and $\nless$ | Is less than, is not less than |
| $>$ and $\ngtr$ | Is greater than, is not greater than |
| $\neq$ | Is not equal to* |

**TABLE 1.1** U.S. Customary and Metric (SI) Measures and Conversions

### Part A

**Conversions for Length, Pressure, Velocity, Volume, and Weight**

| To convert from: | to: | Multiply by: |
|---|---|---|
| | **Length** | |
| Centimeters | Inches | 0.3937 |
| Centimeters | Yards | 0.01094 |
| Feet | Inches | 12.0 |
| Feet | Meters | 0.30481 |
| Feet | Yards | 0.333 |
| Inches | Centimeters | 2.540 |
| Inches | Feet | 0.08333 |
| Inches | Meters | 0.02540 |
| Inches | Micrometers | 25,400. |
| Inches | Millimeters | 25.400 |
| Inches | Yards | 0.02778 |
| Kilometers | Feet | 3,281. |
| Kilometers | Miles (nautical) | 0.5336 |
| Kilometers | Miles (statute) | 0.6214 |
| Kilometers | Yards | 1,094. |

*List continued on p. 1.59.

**TABLE 1.1** U.S. Customary and Metric (SI) Measures and Conversions (*Continued*)

| Multiply: | By: | To obtain: |
|---|---|---|
| Meters | Feet | 3.2809 |
| Meters | Yards | 1.0936 |
| Micrometers | Inches | 0.0000394 |
| Micrometers | Meters | 0.000001 |
| Miles (statute) | Feet | 5,280. |
| Miles (statute) | Kilometers | 1.6093 |
| Miles (statute) | Meters | 1,609.34 |
| Miles (statute) | Yards | 1,760. |
| Miles (nautical) | Feet | 6,080.2 |
| Miles (nautical) | Kilometers | 1.8520 |
| Miles (nautical) | Meters | 1,852.0 |
| Millimeters | Inches | 0.03937 |
| Rods | Meters | 5.0292 |
| Yards | Centimeters | 91.44 |
| Yards | Feet | 3.0 |
| Yards | Inches | 36.0 |
| Yards | Meters | 0.9144 |
| | **Pressure** | |
| Dynes per cm$^2$ | Pascals | 0.1000 |
| Grams per cm$^3$ | Ounces per in$^3$ | 0.5780 |
| Kilograms per cm$^2$ | Pounds per in$^2$ | 14.223 |
| Kilograms per cm$^2$ | Pascals | 98,066.5 |
| Kilograms per sq. meter | Pascals | 9.8066 |
| Kilograms per sq. meter | Pounds per ft$^2$ | 0.2048 |
| Kilograms per sq. meter | Pounds per yd$^2$ | 1.8433 |
| Kilograms per cu. meter | Pounds per ft$^3$ | 0.06243 |
| Ounces per in$^3$ | Grams per cm$^3$ | 1.7300 |
| Pounds per ft$^3$ | Kilograms per m$^3$ | 16.019 |
| Pounds per ft$^2$ | Kilograms per m$^2$ | 4.8824 |
| Pounds per ft$^2$ | Pascals | 47.880 |
| Pounds per in$^2$ | Kilograms per cm$^2$ | 0.0703 |
| Pounds per in$^2$ | Pascals | 6,894.76 |
| Pounds per yd$^2$ | Kilograms per m$^2$ | 0.5425 |
| | **Velocity** | |
| Feet per minute | Meters per second | 0.00508 |
| Feet per second | Meters per second | 0.3048 |
| Inches per second | Meters per second | 0.0254 |
| Kilometers | Meters per second | 0.2778 |
| Knots | Meters per second | 0.5144 |

**TABLE 1.1** U.S. Customary and Metric (SI) Measures and Conversions (*Continued*)

| | | |
|---|---|---|
| Miles per hour | Meters per second | 0.4470 |
| Miles per minute | Meters per second | 26.8224 |
| | Volume | |
| Cubic centimeters | Cubic inches | 0.06102 |
| Cubic feet | Cubic inches | 1,728.0 |
| Cubic feet | Cubic meters | 0.0283 |
| Cubic feet | Cubic yards | 0.0370 |
| Cubic feet | Gallons | 7.481 |
| Cubic feet | Liters | 28.32 |
| Cubic feet | Quarts | 29.9222 |
| Cubic inches | Cubic centimeters | 16.39 |
| Cubic inches | Cubic feet | 0.0005787 |
| Cubic inches | Cubic meters | 0.00001639 |
| Cubic inches | Liters | 0.0164 |
| Cubic inches | Gallons | 0.004329 |
| Cubic inches | Quarts | 0.01732 |
| Cubic meters | Cubic feet | 35.31 |
| Cubic meters | Cubic inches | 61,023. |
| Cubic meters | Cubic yards | 1.3087 |
| Cubic yards | Cubic feet | 27.0 |
| Cubic yards | Cubic meters | 0.7641 |
| Gallons | Cubic feet | 0.1337 |
| Gallons | Cubic inches | 231.0 |
| Gallons | Cubic meters | 0.003785 |
| Gallons | Liters | 3.785 |
| Gallons | Quarts | 4.0 |
| Liters | Cubic feet | 0.03531 |
| Liters | Cubic inches | 61.017 |
| Liters | Gallons | 0.2642 |
| Liters | Pints | 2.1133 |
| Liters | Quarts | 1.057 |
| Liters | Cubic meters | 0.0010 |
| Pints | Cubic meters | 0.004732 |
| Pints | Liters | 0.4732 |
| Pints | Quarts | 0.50 |
| Quarts | Cubic feet | 0.03342 |
| Quarts | Cubic inches | 57.75 |
| Quarts | Cubic meters | 0.0009464 |
| Quarts | Gallons | 0.25 |
| Quarts | Liters | 0.9464 |
| Quarts | Pints | 2.0 |

**TABLE 1.1** U.S. Customary and Metric (SI) Measures and Conversions (*Continued*)

| To convert from: | to: | Multiply by: |
|---|---|---|
| | Weight | |
| Grams | Kilograms | 0.001 |
| Grams | Ounces | 0.03527 |
| Grams | Pounds | 0.002205 |
| Kilograms | Ounces | 35.274 |
| Kilograms | Pounds | 2.2046 |
| Ounces | Grams | 28.35 |
| Ounces | Kilograms | 0.02835 |
| Ounces | Pounds | 0.0625 |
| Pounds | Grams | 453.6 |
| Pounds | Kilograms | 0.4536 |
| Pounds | Ounces | 16.0 |

## Part B

### Standard Conversion Table: Measures Are Found from the Table

| Multiply: | By: | To obtain: |
|---|---|---|
| Acres | 43,560 | Square feet |
| Acres | 4047 | Square meters |
| Acres | $1.562 \times 10^{-3}$ | Square miles |
| Acres | 4840 | Square yards |
| Acre—feet | 43,560 | Cubic feet |
| Acre—feet | 325,851 | Gallons |
| Acre—feet | 1233.49 | Cubic meters |
| Atmospheres | 76.0 | Centimeters of mercury |
| Atmospheres | 29.92 | Inches of mercury |
| Atmospheres | 33.90 | Feet of water |
| Atmospheres | 10,333 | Kilograms per square meter |
| Atmospheres | 14.70 | Pounds per square inch |
| Atmospheres | 1.058 | Tons per square foot |
| Barrels—oil | 42 | Gallons—oil |
| Barrels—cement | 376 | Pounds—cement |
| Bags or sacks—cement | 94 | Pounds—cement |
| Board feet | $144\ in^2 \times 1\ in$ | Cubic inches |
| British thermal units (Btu) | 0.2520 | Kilogram-calories |
| British thermal units | 777.5 | Foot pounds |
| British thermal units | $3.927 \times 10^{-4}$ | Horsepower-hours |
| British thermal units | 107.5 | Kilogram-meters |
| British thermal units | $2.928 \times 10^{-4}$ | Kilowatt-hours |
| Btu/min | 12.96 | Foot pounds per second |

**TABLE 1.1** U.S. Customary and Metric (SI) Measures and Conversions (*Continued*)

| | | |
|---|---|---|
| Btu/min | 0.02356 | Horsepower |
| Btu/min | 0.01757 | Kilowatts |
| Btu/min | 17.57 | Watts |
| Centares (centiares) | 1 | Square meters |
| Centigrams | 0.01 | Grams |
| Centiliters | 0.01 | Liters |
| Centimeters | 0.3937 | Inches |
| Centimeters | 0.01 | Meters |
| Centimeters | 10 | Millimeters |
| Centimeters of mercury | 0.01316 | Atmospheres |
| Centimeters of mercury | 0.4461 | Feet of water |
| Centimeters of mercury | 136.0 | Kilograms per square meter |
| Centimeters of mercury | 27.85 | Pounds per square foot |
| Centimeters of mercury | 0.1934 | Pounds per square inch |
| Centimeters per second | 1.969 | Feet per minute |
| Centimeters per second | 0.03281 | Feet per second |
| Centimeters per second | 0.036 | Kilometers per hour |
| Centimeters per second | 0.6 | Meters per minute |
| Centimeters per second | 0.02237 | Miles per hour |
| Centimeters per second | $3.728 \times 10^{-4}$ | Miles per minute |
| Centimeters per second per second | 0.03281 | Feet per second per second |
| Cubic centimeters | $3.531 \times 10^{-5}$ | Cubic feet |
| Cubic centimeters | $6.102 \times 10^{-2}$ | Cubic inches |
| Cubic centimeters | $10^{-6}$ | Cubic meters |
| Cubic centimeters | $1.308 \times 10^{-6}$ | Cubic yards |
| Cubic centimeters | $2.642 \times 10^{-4}$ | Gallons |
| Cubic centimeters | $10^{-3}$ | Liters |
| Cubic centimeters | $2.113 \times 10^{-3}$ | Pints (liq.) |
| Cubic centimeters | $1.057 \times 10^{-3}$ | Quarts (liq.) |
| Cubic feet | $2.832 \times 10^{4}$ | Cubic cms. |
| Cubic feet | 1728 | Cubic inches |
| Cubic feet | 0.02832 | Cubic meters |
| Cubic feet | 0.03704 | Cubic yards |
| Cubic feet | 7.48052 | Gallons |
| Cubic feet | 28.32 | Liters |
| Cubic feet | 59.84 | Pints. (liq.) |
| Cubic feet | 29.92 | Quarts (liq.) |
| Cubic feet per minute | 472.0 | Cubic centimeters per second |
| Cubic feet per minute | 0.1247 | Gallons per second |
| Cubic feet per minute | 0.4720 | Liters per second |
| Cubic feet per minute | 62.43 | Pounds of water per minute |
| Cubic feet per second | 0.646317 | Millions gallons per day |
| Cubic feet per second | 448.831 | Gallons per minute |
| Cubic inches | 16.39 | Cubic centimeters |
| Cubic inches | $5.787 \times 10^{-4}$ | Cubic feet |

**TABLE 1.1** U.S. Customary and Metric (SI) Measures and Conversions (*Continued*)

| Multiply: | By: | To obtain: |
|---|---|---|
| Cubic inches | $1.639 \times 10^{-5}$ | Cubic yards |
| Cubic inches | $2.143 \times 10^{-5}$ | Cubic meters |
| Cubic inches | $4.329 \times 10^{-3}$ | Gallons |
| Cubic inches | $1.639 \times 10^{-2}$ | Liters |
| Cubic inches | 0.03463 | Pints (liq.) |
| Cubic inches | 0.01732 | Quarts (liq.) |
| Cubic meters | $10^{6}$ | Cubic centimeters |
| Cubic meters | 35.31 | Cubic feet |
| Cubic meters | $61.023 \times 10^{3}$ | Cubic inches |
| Cubic meters | 1.308 | Cubic yards |
| Cubic meters | 264.2 | Gallons |
| Cubic meters | $10^{3}$ | Liters |
| Cubic meters | 2113 | Pints (liq.) |
| Cubic meters | 1057 | Quarts (liq.) |
| Cubic yards | $7.646 \times 10^{5}$ | Cubic centimeters |
| Cubic yards | 27 | Cubic feet |
| Cubic yards | 46,656 | Cubic inches |
| Cubic yards | 0.7646 | Cubic meters |
| Cubic yards | 202.0 | Gallons |
| Cubic yards | 764.6 | Liters |
| Cubic yards | 1616 | Pints (liq.) |
| Cubic yards | 807.9 | Quarts (liq.) |
| Cubic yards per minute | 0.45 | Cubic feet per second |
| Cubic yards per minute | 3.367 | Gallons per second |
| Cubic yards per minute | 12.74 | Liters per second |
| Decigrams | 0.1 | Grams |
| Decilieters | 0.1 | Liters |
| Decimeters | 0.1 | Meters |
| Degrees (angle) | 60 | Minutes |
| Degrees (angle) | 0.01745 | Radians |
| Degrees (angle) | 3600 | Seconds |
| Degrees per second | 0.01745 | Radians per second |
| Degrees per second | 0.1667 | Revolutions per minute |
| Degrees per second | 0.002778 | Revolutions per second |
| Dekagrams | 10 | Grams |
| Dekaliters | 10 | Liters |
| Dekameters | 10 | Meters |
| Drams | 27.34375 | Grains |
| Drams | 0.0625 | Ounces |
| Drams | 1.771845 | Grams |
| Fathoms | 6 | Feet |
| Feet | 30.48 | Centimeters |
| Feet | 12 | Inches |
| Feet | 0.3048 | Meters |
| Feet | ⅓ | Yards |
| Feet of water | 0.02950 | Atmospheres |

**TABLE 1.1** U.S. Customary and Metric (SI) Measures and Conversions (*Continued*)

| | | |
|---|---|---|
| Feet of water | 0.8826 | Inches of mercury |
| Feet of water | 304.8 | Kilograms per square meter |
| Feet of water | 62.43 | Pounds per square foot |
| Feet of water | 0.4335 | Pounds per square inch |
| Feet per minute | 0.5080 | Centimeters per second |
| Feet per minute | 0.01667 | Feet per second |
| Feet per minute | 0.01829 | Kilometers per hour |
| Feet per minute | 0.3048 | Meters per minute |
| Feet per minute | 0.01136 | Miles per hour |
| Feet per second | 30.48 | Centimeters per second |
| Feet per second | 1.097 | Kilometers per hour |
| Feet per second | 0.5921 | Knots |
| Feet per second | 18.29 | Meters per minute |
| Feet per second | 0.6818 | Miles per hour |
| Feet per second | 0.01136 | Miles per minute |
| Feet per second per second | 30.48 | Centimeters per second per second |
| Feet per second per second | 0.3048 | Meters per second per second |
| Foot pounds | $1.286 \times 10^{-3}$ | British thermal units |
| Foot pounds | $5.050 \times 10^{-7}$ | Horsepower-hours |
| Foot pounds | $3.241 \times 10^{-4}$ | Kilogram-calories |
| Foot pounds | 0.1383 | Kilogram-meters |
| Foot pounds | $3.766 \times 10^{-7}$ | Kilowatt-hours |
| Foot pounds per minute | $1.286 \times 10^{-3}$ | Btu per minute |
| Foot pounds per minute | 0.01667 | Foot pounds per second |
| Foot pounds per minute | $3.030 \times 10^{-5}$ | Horsepower |
| Foot pounds per minute | $3.241 \times 10^{-4}$ | Kilogram-calories per minute |
| Foot pounds per minute | $2.260 \times 10^{-5}$ | Kilowatts |
| Foot pounds per second | $7.717 \times 10^{-2}$ | Btu per minute |
| Foot pounds per second | $1.818 \times 10^{-3}$ | Horsepower |
| Foot pounds per second | $1.945 \times 10^{-2}$ | Kilogram-calories per minute |
| Foot pounds per second | $1.356 \times 10^{-3}$ | Kilowatts |
| Gallons | 3785 | Cubic centimeters |
| Gallons | 0.1337 | Cubic feet |
| Gallons | 231 | Cubic inches |
| Gallons | $3.785 \times 10^{-3}$ | Cubic meters |
| Gallons | $4.95 \times 10^{-3}$ | Cubic yards |
| Gallons | 3.785 | Liters |
| Gallons | 8 | Pints (liq.) |
| Gallons | 4 | Quarts (liq.) |
| Gallons—Imperial | 1.20095 | U.S. gallons |
| Gallons—U.S. | 0.83267 | Imperial gallons |
| Gallons water | 8.3453 | Pounds of water |
| Gallons per minute | $2.228 \times 10^{-3}$ | Cubic feet per second |

**TABLE 1.1** U.S. Customary and Metric (SI) Measures and Conversions (*Continued*)

| Multiply: | By: | To obtain: |
|---|---|---|
| Gallons per minute | 0.06308 | Liters per second |
| Gallons per minute | 8.0208 | Cubic feet per hour |
| Gallons per minute | 8.0208 area (sq. ft.) | Overflow rate (ft/h) |
| Gallons water per minute | 6.0086 | Tons water per 24 hours |
| Grains (troy) | L | Grains (avour.) |
| Grains (troy) | 0.06480 | Grams |
| Grains (troy) | 0.04167 | Pennyweights (troy) |
| Grains (troy) | $2.0833 \times 10^{-3}$ | Ounces (troy) |
| Grains per U.S. gallon | 17.118 | Parts per million |
| Grains per U.S. gallon | 142.86 | Pounds per million gallons |
| Grains per Imperial gallon | 14.254 | Parts per million |
| Grams | 980.7 | Dynes |
| Grams | 15.43 | Grains |
| Grams | $10^{-3}$ | Kilograms |
| Grams | $10^{3}$ | Milligrams |
| Grams | 0.03527 | Ounces |
| Grams | 0.03215 | Ounces (troy) |
| Grams | $2.205 \times 10^{-3}$ | Pounds |
| Grams per centimeter | $5.600 \times 10^{-3}$ | Pounds per inch |
| Grams per cubic centimeter | 62.43 | Pounds per cubic foot |
| Grams per cubic centimeter | 0.03613 | Pounds per cubic inch |
| Grams per liter | 58.417 | Grains per gallon |
| Grams per liter | 8.345 | Pounds per 1000 gallons |
| Grams per liter | 0.062427 | Pounds per cubic foot |
| Grams per liter | 1000 | Parts per million |
| Hectares | 2.471 | Acres |
| Hectares | $1.076 \times 10^{5}$ | Square feet |
| Hectograms | 100 | Grams |
| Hectoliters | 100 | Liters |
| Hectometers | 100 | Meters |
| Hectowatts | 100 | Watts |
| Horsepower | 42.44 | Btu per minute |
| Horsepower | 33,000 | Foot pounds per minute |
| Horsepower | 550 | Foot pounds per second |
| Horsepower | 1.014 | Horsepower (metric) |
| Horsepower | 10.70 | Kilogram-calories per minute |
| Horsepower | 0.7457 | Kilowatts |
| Horsepower | 745.7 | Watts |

**TABLE 1.1** U.S. Customary and Metric (SI) Measures and Conversions (*Continued*)

| | | |
|---|---|---|
| Horsepower (boiler) | 33,479 | Btu per hour |
| Horsepower (boiler) | 9,803 | Kilowatts |
| Horsepower-hours | 2547 | British thermal units |
| Horsepower-hours | $1.98 \times 10^6$ | Foot pounds |
| Horsepower-hours | 641.7 | Kilogram-calories |
| Horsepower-hours | $2.737 \times 10^5$ | Kilogram-meters |
| Horsepower-hours | 0.7457 | Kilowatthours |
| Inches | 2.540 | Centimeters |
| Inches of mercury | 0.03342 | Atmospheres |
| Inches of mercury | 1.133 | Feet of water |
| Inches of mercury | 345.3 | Kilograms per square meter |
| Inches of mercury | 70.73 | Pounds per square foot |
| Inches of mercury | 0.4912 | Pounds per square inch |
| Inches of water | 0.002458 | Atmospheres |
| Inches of water | 0.07355 | Inches of mercury |
| Inches of water | 25.40 | Kilograms per square meter |
| Inches of water | 0.5781 | Ounces per square inch |
| Inches of water | 5.202 | Pounds per square foot |
| Inches of water | 0.03613 | Pounds per square inch |
| Kilograms | 980,665 | Dynes |
| Kilograms | 2.205 | Pounds |
| Kilograms | $1.102 \times 10^{-3}$ | Tons (short) |
| Kilograms | $10^3$ | Grams |
| Kilograms-calories | 3.968 | British thermal units |
| Kilograms-calories | 3086 | Foot pounds |
| Kilograms-calories | $1.558 \times 10^{-3}$ | Horsepower-hours |
| Kilograms-calories | $1.162 \times 10^{-3}$ | Kilowatthours |
| Kilogram-calories per minute | 51.43 | Foot pounds per second |
| Kilogram-calories per minute | 0.09351 | Horsepower |
| Kilogram-calories per minute | 0.06972 | Kilowatts |
| Kilograms per meter | 0.6720 | Pounds per foot |
| Kilograms per square meter | $9.678 \times 10^{-5}$ | Atmospheres |
| Kilograms per square meter | $3.281 \times 10^{-3}$ | Feet of water |
| Kilograms per square meter | $2.896 \times 10^{-3}$ | Inches of mercury |
| Kilograms per square meter | 0.2048 | Pounds per square foot |
| Kilograms per square meter | $1.422 \times 10^{-3}$ | Pounds per square inch |
| Kilograms per square millimeter | $10^6$ | Kilograms per square meter |

**TABLE 1.1** U.S. Customary and Metric (SI) Measures and Conversions (*Continued*)

| Multiply: | By: | To obtain: |
|---|---|---|
| Kiloliters | $10^3$ | Liters |
| Kilometers | $10^5$ | Centimeters |
| Kilometers | 3281 | Feet |
| Kilometers | $10^3$ | Meters |
| Kilometers | 0.6214 | Miles |
| Kilometers | 1094 | Yards |
| Kilometers per hour | 27.78 | Centimeters per second |
| Kilometers per hour | 54.68 | Feet per minute |
| Kilometers per hour | 0.9113 | Feet per second |
| Kilometers per hour | 0.5396 | Knots |
| Kilometers per hour | 16.67 | Meters per minute |
| Kilometers per hour | 0.6214 | Miles per hour |
| Kilograms per hour per second | 27.78 | Centimeters per second per second |
| Kilograms per hour per second | 0.9113 | Feet per second per second |
| Kilograms per hour per second | 0.2778 | Meters per second per second |
| Kilowatts | 56.92 | Btu per minute |
| Kilowatts | $4.425 \times 10^4$ | Foot pounds per minute |
| Kilowatts | 737.6 | Foot pounds per second |
| Kilowatts | 1.341 | Horsepower |
| Kilowatts | 14.34 | Kilogram-calories per minute |
| Kilowatts | $10^3$ | Watts |
| Kilowatthours | 3415 | British thermal units |
| Kilowatthours | $2.655 \times 10^6$ | Foot pounds |
| Kilowatthours | 1.341 | Horsepower-hours |
| Kilowatthours | 860.5 | Kilogram-calories |
| Kilowatthours | $3.671 \times 10^5$ | Kilogram-meters |
| Liters | $10^3$ | Cubic centimeters |
| Liters | 0.03531 | Cubic feet |
| Liters | 61.02 | Cubic inches |
| Liters | $10^3$ | Cubic meters |
| Liters | $1.308 \times 10^{-3}$ | Cubic yards |
| Liters | 0.2642 | Gallons |
| Liters | 2.113 | Pints (liq.) |
| Liters | 1.057 | Quarts (liq.) |
| Liters per minute | $5.886 \times 10^{-4}$ | Cubic feet per second |
| Liters per minute | $4.403 \times 10^{-3}$ | Gallons per second |
| $\frac{\text{Lumber width (in)} \times \text{thickness (in)}}{12}$ | Length (ft.) | Board feet |
| Meters | 100 | Centimeters |
| Meters | 3.281 | Feet |
| Meters | 39.37 | Inches |

**TABLE 1.1** U.S. Customary and Metric (SI) Measures and Conversions (*Continued*)

| | | |
|---|---|---|
| Meters | $10^{-3}$ | Kilometers |
| Meters | $10^{3}$ | Millimeters |
| Meters | 1.094 | Yards |
| Meters per minute | 1.667 | Centimeters per second |
| Meters per minute | 3.281 | Feet per minute |
| Meters per minute | 0.05468 | Feet per second |
| Meters per minute | 0.06 | Kilometers per hour |
| Meters per minute | 0.03728 | Miles per hour |
| Meters per second | 196.8 | Feet per minute |
| Meters per second | 3.281 | Feet per second |
| Meters per second | 3.6 | Kilometers per hour |
| Meters per second | 0.06 | Kilometers per minute |
| Meters per second | 2.237 | Miles per hour |
| Meters per second | 0.03728 | Miles per minute |
| Micrometers | $10^{-6}$ | Meters |
| Miles | $1.609 \times 10^{5}$ | Centimeters |
| Miles | 5280 | Feet |
| Miles | 1.609 | Kilometers |
| Miles | 1760 | Yards |
| Miles per hour | 44.70 | Centimeters per second |
| Miles per hour | 88 | Feet per minute |
| Miles per hour | 1.467 | Feet per second |
| Miles per hour | 1.609 | Kilometers per hour |
| Miles per hour | 0.8684 | Knots |
| Miles per hour | 26.82 | Meters per minute |
| Miles per minute | 2682 | Centimeters per second |
| Miles per minute | 88 | Feet per second |
| Miles per minute | 1.609 | Kilometers per minute |
| Miles per minute | 60 | Miles per hour |
| Milliers | $10^{3}$ | Kilograms |
| Milligrams | $10^{-3}$ | Grams |
| Milliliters | $10^{-3}$ | Liters |
| Millimeters | 0.1 | Centimeters |
| Millimeters | 0.03937 | Inches |
| Milligrams per liter | 1 | Parts per million |
| Million gallons per day | 1.54723 | Cubic feet per second |
| Miner's inches | 1.5 | Cubic feet per minute |
| Minutes (angle) | $2.909 \times 10^{-4}$ | Radians |
| Ounces | 16 | Drams |
| Ounces | 437.5 | Grains |
| Ounces | 0.0625 | Pounds |
| Ounces | 28.349527 | Grams |
| Ounces | 0.9115 | Ounces (troy) |
| Ounces | $2.790 \times 10^{-5}$ | Tons (long) |
| Ounces | $2.835 \times 10^{-5}$ | Tons (metric) |
| Ounces (troy) | 480 | Grains |
| Ounces (troy) | 20 | Pennyweights (troy) |
| Ounces (troy) | 0.08333 | Pounds (troy) |

**TABLE 1.1** U.S. Customary and Metric (SI) Measures and Conversions (*Continued*)

| Multiply: | By: | To obtain: |
|---|---|---|
| Ounces (troy) | 31.103481 | Grams |
| Ounces (troy) | 1.09714 | Ounces (avoir.) |
| Ounces (fluid) | 1.805 | Cubic inches |
| Ounces (fluid) | 0.02957 | Liters |
| Ounces per square inch | 0.0625 | Pounds per square inch |
| Overflow rate (ft/h) | 0.12468 × area ($ft^2$) | Gallons per minute |
| $\frac{1}{\text{Overflow rate (ft/h)}}$ | 8.0208 | Square feet per gallon per minute |
| Parts per million | 0.0584 | Grains per U.S. gallon |
| Parts per million | 0.07016 | Grains per Imperial gallon |
| Parts per million | 8.345 | Pounds per million gallons |
| Pennyweights (troy) | 24 | Grains |
| Pennyweights (troy) | 1.55517 | Grams |
| Pennyweights (troy) | 0.05 | Ounces (troy) |
| Pennyweights (troy) | $4.1667 \times 10^{-3}$ | Pounds (troy) |
| Pounds | 16 | Ounces |
| Pounds | 256 | Drams |
| Pounds | 7000 | Grains |
| Pounds | 0.0005 | Tons (short) |
| Pounds | 453.5924 | Grams |
| Pounds | 1.21528 | Pounds (troy) |
| Pounds | 14.5833 | Ounces (troy) |
| Pounds (troy) | 5760 | Grains |
| Pounds (troy) | 240 | Pennyweights (troy) |
| Pounds (troy) | 12 | Ounces (troy) |
| Pounds (troy) | 373.24177 | Grams |
| Pounds (troy) | 0.822857 | Pounds (avoir.) |
| Pounds (troy) | 13.1657 | Ounces (avoir.) |
| Pounds (troy) | $3.6735 \times 10^{-4}$ | Tons (long) |
| Pounds (troy) | $4.1143 \times 10^{-4}$ | Tons (short) |
| Pounds (troy) | $3.7324 \times 10^{-4}$ | Tons (metric) |
| Pounds of water | 0.01602 | Cubic feet |
| Pounds of water | 27.68 | Cubic inches |
| Pounds of water | 0.1198 | Gallons |
| Pounds of water per minute | $2.670 \times 10^{-4}$ | Cubic feet per second |
| Pounds per cubic foot | 0.01602 | Grams per cubic centimeter |
| Pounds per cubic foot | 16.02 | Kilograms per cubic meters |
| Pounds per cubic foot | $5.787 \times 10^{-4}$ | Pounds per cubic inch |
| Pounds per cubic inch | 27.68 | Grams per cubic centimeter |

**TABLE 1.1** U.S. Customary and Metric (SI) Measures and Conversions (*Continued*)

| | | |
|---|---|---|
| Pounds per cubic inch | $2.768 \times 10^4$ | Kilograms per cubic meter |
| Pounds per cubic inch | 1728 | Pounds per cubic foot |
| Pounds per foot | 1.488 | Kilograms per meter |
| Pounds per inch | 178.6 | Grams per centimeter |
| Pounds per square foot | 0.01602 | Feet of water |
| Pounds per square foot | 4.883 | Kilograms per square meter |
| Pounds per square foot | $6.945 \times 10^{-3}$ | Pounds per square inch |
| Pounds per square inch | 0.06804 | Atmospheres |
| Pounds per square inch | 2.307 | Feet of water |
| Pounds per square inch | 2.036 | Inches of mercury |
| Pounds per square inch | 703.1 | Kilograms per square meter |
| Quadrants (angle) | 90 | Degrees |
| Quadrants (angle) | 5400 | Minutes |
| Quadrants (angle) | 1.571 | Radians |
| Quarts (dry) | 67.20 | Cubic inches |
| Quarts (liq.) | 57.75 | Cubic inches |
| Quintal, Argentine | 101.28 | Pounds |
| Quintal, Brazil | 129.54 | Pounds |
| Quintal, Castile, Peru | 101.43 | Pounds |
| Quintal, Chile | 101.41 | Pounds |
| Quintal, Mexico | 101.47 | Pounds |
| Quintal, Metric | 220.46 | Pounds |
| Quires | 25 | Sheets |
| Radians | 57.30 | Degrees |
| Radians | 3438 | Minutes |
| Radians | 0.637 | Quadrants |
| Radians per second | 57.30 | Degrees per second |
| Radians | 0.1592 | Revolutions per second |
| Radians per second | 9.549 | Revolutions per minute |
| Radians per second per second | 573.0 | Revolutions per minute per minute |
| Radians per second per second | 0.1592 | Revolutions per second per second |
| Revolutions | 360 | Degrees |
| Revolutions | 4 | Quadrants |
| Revolutions | 6.283 | Radians |
| Revolutions per minute | 6 | Degrees per second |
| Revolutions per minute | 0.1047 | Radians per second |
| Revolutions per minute | 0.01667 | Revolutions per second |
| Revolutions per minute per minute | $1.745 \times 10^{-3}$ | Radians per second per second |
| Revolutions per minute per minute | $2.778 \times 10^{-4}$ | Revolutions per second per second |
| Revolutions per second | 360 | Degrees per second |
| Revolutions per second | 6.283 | Radians per second |

**TABLE 1.1** U.S. Customary and Metric (SI) Measures and Conversions (*Continued*)

| Multiply: | By: | To obtain: |
|---|---|---|
| Revolutions per second | 60 | Revolutions per minute |
| Revolutions per second per second | 6.283 | Radians per second per second |
| Revolutions per second per second | 3600 | Revolutions per minute per minute |
| Seconds (angle) | $4.848 \times 10^{-6}$ | Radians |
| Square centimeters | $1.076 \times 10^{-3}$ | Square feet |
| Square centimeters | 0.1550 | Square inches |
| Square centimeters | $10^{-4}$ | Square meters |
| Square centimeters | 100 | Square millimeters |
| Square feet | $2.296 \times 10^{-5}$ | Acres |
| Square feet | 929.0 | Square centimeters |
| Square feet | 144 | Square inches |
| Square feet | 0.09290 | Square meters |
| Square feet | $3.587 \times 10^{-8}$ | Square miles |
| Square feet | $\frac{1}{9}$ | Square yards |
| $\frac{1}{\text{Square feet per gallon per minute}}$ | 8.0208 | Overflow rate (ft/h) |
| Square inches | 6.452 | Square centimeters |
| Square inches | $6.944 \times 10^{-3}$ | Square feet |
| Square inches | 645.2 | Square millimeters |
| Square kilometers | 247.1 | Acres |
| Square kilometers | $10.76 \times 10^{6}$ | Square feet |
| Square kilometers | $10^{6}$ | Square meters |
| Square kilometers | 0.3861 | Square miles |
| Square kilometers | $1.196 \times 10^{6}$ | Square yards |
| Square meters | $2.471 \times 10^{-4}$ | Acres |
| Square meters | 10.76 | Square feet |
| Square meters | $3.861 \times 10^{-7}$ | Square miles |
| Square meters | 1.196 | Square yards |
| Square miles | 640 | Acres |
| Square miles | $27.88 \times 10^{6}$ | Square feet |
| Square miles | 2.590 | Square kilometers |
| Square miles | $3.098 \times 10^{6}$ | Square yards |
| Square millimeters | 0.01 | Square centimeters |
| Square millimeters | $1.550 \times 10^{-3}$ | Square inches |
| Square yards | $2.066 \times 10^{-4}$ | Acres |
| Square yards | 9 | Square feet |
| Square yards | 0.8361 | Square meters |
| Square yards | $3.228 \times 10^{-7}$ | Square miles |
| Temp. (°C.) + 273 | 1 | Abs. temp. (°C.) |
| Temp. (°C.) + 17.78 | 1.8 | Temp. (°F.) |
| Temp. (°F.) + 460 | 1 | Abs. temp. (°F.) |
| Temp. (°F.) − 32 | $\frac{5}{9}$ | Temp. (°C.) |
| Tons (long) | 1016 | Kilograms |

**TABLE 1.1** U.S. Customary and Metric (SI) Measures and Conversions (*Continued*)

| | | |
|---|---|---|
| Tons (long) | 2240 | Pounds |
| Tons (long) | 1.12000 | Tons (short) |
| Tons (metric) | $10^3$ | Kilograms |
| Tons (metric) | 2205 | Pounds |
| Tons (short) | 2000 | Pounds |
| Tons (short) | 32,000 | Ounces |
| Tons (short) | 907.18486 | Kilograms |
| Tons (short) | 2430.56 | Pounds (troy) |
| Tons (short) | 0.89287 | Tons (long) |
| Tons (short) | 29166.66 | Ounces (troy) |
| Tons (short) | 0.90718 | Tons (metric) |
| $\frac{1}{\text{Tons dry solids per 24 hours}}$ | Area ($ft^2$) | Square feet per ton per 24 hours |
| Tons of water per 24 hours | 83.333 | Pounds water per hour |
| Tons of water per 24 hours | 0.16643 | Gallons per minute |
| Tons of water per 24 hours | 1.3349 | Cubic feet per hour |
| Watts | 0.05692 | Btu per minute |
| Watts | 44.26 | Foot pounds per minute |
| Watts | 0.7376 | Foot pounds per second |
| Watts | $1.341 \times 10^{-3}$ | Horsepower |
| Watts | 0.01434 | Kilogram-calories per minute |
| Watts | $10^{-3}$ | Kilowatts |
| Watthours | 3.415 | British thermal units |
| Watthours | 2655 | Foot pounds |
| Watthours | $1.341 \times 10^{-3}$ | Horsepower-hours |
| Watthours | 0.8605 | Kilogram-calories |
| Watthours | 367.1 | Kilogram-meters |
| Watthours | $10^{-3}$ | Kilowatthours |
| Yards | 91.44 | Centimeters |
| Yards | 3 | Feet |
| Yards | 36 | Inches |
| Yards | 0.9144 | Meters |

| | |
|---|---|
| $\pm$ | Plus or minus, respectively |
| $\mp$ | Minus or plus, respectively |
| $\propto$ | Is proportional to |
| $\rightarrow$ | Approaches; e.g., as $x \rightarrow 0$ |
| $\leq, \leqq$ | Less than or equal to |
| $\geq, \geqq$ | More than or equal to |
| $\therefore$ | Therefore |
| : | Is to, is proportional to |
| Q.E.D. | Which was to be proved, end of proof |
| % | Percent |
| # | Number |
| @ | At |
| $\angle$ or $\measuredangle$ | Angle |

| | |
|---|---|
| ° ′ ″ | Degrees, minutes, seconds |
| ‖, // | Parallel to |
| ⊥ | Perpendicular to |
| e | Base of natural logs, 2.71828 . . . |
| π | Pi, 3.14159 . . . |
| ( ) | Parentheses |
| [ ] | Brackets |
| { } | Braces |
| ′ | Prime, $f'(x)$ |
| ″ | Double prime, $f''(x)$ |
| $\sqrt{\ }, \sqrt[n]{\ }$ | Square root, $n$th root |
| $\frac{1}{x}$ or $x^{-1}$ | Reciprocal of $x$ |
| ! | Factorial |
| ∞ | Infinity |
| Δ | Delta, increment of |
| ∂ | Curly "d," partial differentiation |
| Σ | Sigma, summation of terms |
| Π | The product of terms, product |
| arc | As in arcsine (the angle whose sine is) |
| $f$ | Function, as $f(x)$ |
| rms | Root mean square |
| $\lvert x \rvert$ | Absolute value of $x$ |
| $i$ | For $\sqrt{-1}$ |
| $j$ | Operator, equal to $\sqrt{-1}$ |
| $dx$ | Differential of $x$ |
| $\frac{\Delta x}{\Delta y}$ | Change in $x$ with respect to $y$ |
| $\frac{dx}{dy}, \frac{d}{dy}(x)$ | Derivative of $x$ with respect to $y$ |
| $\frac{dy}{dx}, \frac{d}{dx}(y)$ | Derivative of $y$ with respect to $x$ |
| $\int$ | Integral |

$\int_a^b$ Integral between limits $a$ and $b$

$\Big|_a^b$ Evaluated between limits $a$ and $b$

## 1.16 FACTORS AND PRIME NUMBERS

Refer to Table 1.2.

**TABLE 1.2** Factors and Prime Numbers

| *n* | 0 | 1 | 2 | 3 | 4 | 5 | 6 | 7 | 8 | 9 |
|---|---|---|---|---|---|---|---|---|---|---|
| 0 | | | | | $2^2$ | | 2·3 | | $2^3$ | $3^2$ |
| 1 | 2.5 | | $2^2\cdot3$ | | 2·7 | 3·5 | $2^4$ | | $2\cdot3^2$ | |
| 2 | $2^2\cdot5$ | 3·7 | 2·11 | | $2^3\cdot3$ | $5^2$ | 2·13 | $3^3$ | $2^2\cdot7$ | |
| 3 | 2·3·5 | | $2^5$ | 3·11 | 2·17 | 5·7 | $2^2\cdot3^2$ | | 2·19 | 3.13 |
| 4 | $2^3\cdot5$ | | 2·3·7 | | $2^2\cdot11$ | $3^2\cdot5$ | 2·23 | | $2^4\cdot3$ | $7^2$ |
| 5 | $2\cdot5^2$ | 3·17 | $2^2\cdot13$ | | $2\cdot3^3$ | 5·11 | $2^3\cdot7$ | 3·19 | 2·29 | |
| 6 | $2^2\cdot3\cdot5$ | | 2·31 | $3^2\cdot7$ | $2^6$ | 5·13 | 2·3·11 | | $2^2\cdot17$ | 3·23 |
| 7 | 2.5·7 | | $2^3\cdot3^2$ | | 2·37 | $3\cdot5^2$ | $2^2\cdot19$ | 7·11 | 2·3·13 | |
| 8 | $2^4\cdot5$ | $3^4$ | 2·41 | | $2^2\cdot3\cdot7$ | 5·17 | 2·43 | 3·29 | $2^3\cdot11$ | |
| 9 | $2\cdot3^2\cdot5$ | 7·13 | $2^2\cdot23$ | 3·31 | 2·47 | 5·19 | $2^5\cdot3$ | | $2\cdot7^2$ | $3^2\cdot11$ |
| 10 | $2^2\cdot5^2$ | | 2·3·17 | | $2^3\cdot13$ | 3·5·7 | 2·53 | | $2^2\cdot3^3$ | |
| 11 | 2·5·11 | 3·37 | $2^4\cdot7$ | | 2·3·19 | 5·23 | $2^2\cdot29$ | $3^2\cdot13$ | 2·59 | 7·17 |
| 12 | $2^3\cdot3\cdot5$ | $11^2$ | 2·61 | 3·41 | $2^2\cdot31$ | $5^3$ | $2\cdot3^2\cdot7$ | | $2^7$ | 3·43 |
| 13 | 2·5·13 | | $2^2\cdot3\cdot11$ | 7·19 | 2·67 | $3^3\cdot5$ | $2^3\cdot17$ | | 2·3·23 | |
| 14 | $2^2\cdot5\cdot7$ | 3·47 | 2·71 | 11·13 | $2^4\cdot3^2$ | 5·29 | 2·73 | 3·72 | $2^2\cdot37$ | |
| 15 | $2\cdot3\cdot5^2$ | | $2^3\cdot19$ | $3^2\cdot17$ | 2·7·11 | 5·31 | $2^2\cdot3\cdot13$ | | 2·79 | 3·53 |
| 16 | $2^5\cdot5$ | 7·23 | $2\cdot3^4$ | | $2^2\cdot41$ | 3·5·11 | 2·83 | | $2^3\cdot3\cdot7$ | $13^2$ |
| 17 | 2·5·17 | $3^2\cdot19$ | $2^2\cdot43$ | | 2·3·29 | $5^2\cdot7$ | $2^4\cdot11$ | 3·59 | 2·89 | |
| 18 | $2^2\cdot3^2\cdot5$ | | 2·7·13 | 3·61 | $2^3\cdot23$ | 5·37 | 2·3·31 | 11·17 | $2^2\cdot47$ | $3^3\cdot7$ |
| 19 | 2·5·19 | | $2^6\cdot3$ | | 2·97 | 3·5·13 | $2^2\cdot7^2$ | | $2\cdot3^2\cdot11$ | |
| 20 | $2^3\cdot5^2$ | 3·67 | 2·101 | 7·29 | $2^2\cdot3\cdot17$ | 5·41 | 2·103 | $3^2\cdot23$ | $2^4\cdot13$ | 11·19 |
| 21 | 2·3·5·7 | | $2^2\cdot53$ | 3·71 | 2·107 | 5·43 | $2^3\cdot3^3$ | 7·31 | 2·109 | 3·73 |
| 22 | $2^2\cdot5\cdot11$ | 13·17 | 2·3·37 | | $2^5\cdot7$ | $3^2\cdot5^2$ | 2·113 | | $2^2\cdot3\cdot19$ | |
| 23 | 2·5·23 | 3·7·11 | $2^3\cdot29$ | | $2\cdot3^2\cdot13$ | 5·47 | $2^2\cdot59$ | 3·79 | 2·7·17 | |
| 24 | $2^4\cdot3\cdot5$ | | $2\cdot11^2$ | $3^5$ | $2^2\cdot61$ | $5\cdot7^2$ | 2·3·41 | 13·19 | $2^3\cdot31$ | 3·83 |
| 25 | $2\cdot5^3$ | | $2^2\cdot3^2\cdot7$ | 11·23 | 2·127 | 3·5·17 | $2^8$ | | 2·3·43 | 7·37 |
| 26 | $2^2\cdot5\cdot13$ | $3^2\cdot29$ | 2·131 | | $2^3\cdot3\cdot11$ | 5·53 | 2·7·19 | 3·89 | $2^2\cdot67$ | |
| 27 | $2\cdot3^3\cdot5$ | | $2^4\cdot17$ | 3·7·13 | 2·137 | $5^2\cdot11$ | $2^2\cdot3\cdot23$ | | 2·139 | $3^2\cdot31$ |
| 28 | $2^3\cdot5\cdot7$ | | 2·3·47 | | $2^2\cdot71$ | 3·5·19 | 2·11·13 | 7·41 | $2^5\cdot3^2$ | $17^2$ |
| 29 | 2·5·29 | 3·97 | $2^2\cdot73$ | | $2\cdot3\cdot7^2$ | 5·59 | $2^3\cdot37$ | $3^3\cdot11$ | 2·149 | 13·23 |
| 30 | $2^2\cdot3\cdot5^2$ | 7·43 | 2·151 | 3·101 | $2^4\cdot19$ | 5·61 | $2\cdot3^2\cdot17$ | | $2^2\cdot7\cdot11$ | 3·103 |
| 31 | 2·5·31 | | $2^3\cdot3\cdot13$ | | 2·157 | $3^2\cdot5\cdot7$ | $2^2\cdot79$ | | 2·3·53 | 11·29 |
| 32 | $2^6\cdot5$ | 3·107 | 2·7·23 | 17·19 | $2^2\cdot3^4$ | $5^2\cdot13$ | 2·163 | 3·109 | $2^3\cdot41$ | 7·47 |
| 33 | 2·3·5·11 | | $2^2\cdot83$ | $3^2\cdot37$ | 2·167 | 5·67 | $2^4\cdot3\cdot7$ | | $2\cdot13^2$ | 3·113 |
| 34 | $2^2\cdot5\cdot17$ | 11·31 | $2\cdot3^2\cdot19$ | $7^3$ | $2^3\cdot43$ | 3·5·23 | 2·173 | | $2^2\cdot3\cdot29$ | |
| 35 | $2\cdot5^2\cdot7$ | $3^3\cdot13$ | $2^5\cdot11$ | | 2·3·59 | 5·71 | $2^2\cdot89$ | 3·7·17 | 2·179 | |
| 36 | $2^3\cdot3^2\cdot5$ | $19^2$ | 2·181 | $3\cdot11^2$ | $2^2\cdot7\cdot13$ | 5·73 | 2·3·61 | | $2^4\cdot23$ | $3^2\cdot41$ |
| 37 | 2·5·37 | 7·53 | $2^2\cdot3\cdot31$ | | 2·11·17 | $3\cdot5^3$ | $2^3\cdot47$ | 13·29 | $2\cdot3^3\cdot7$ | |
| 38 | $2^2\cdot5\cdot19$ | 3·127 | 2·191 | | $2^7\cdot3$ | 5·7·11 | 2·193 | $3^2\cdot43$ | $2^2\cdot97$ | |
| 39 | 2·3·5·13 | 17·23 | $2^3\cdot7^2$ | 3·131 | 2·197 | 5·79 | $2^2\cdot3^2\cdot11$ | | 2·199 | 3·7·19 |
| 40 | $2^4\cdot5^2$ | | 2·3·67 | 13·31 | $2^2\cdot101$ | $3^4\cdot5$ | 2·7·29 | 11·37 | $2^3\cdot3\cdot17$ | |
| 41 | 2·5·41 | 3·137 | $2^2\cdot103$ | 7·59 | $2\cdot3^2\cdot23$ | 5·83 | $2^5\cdot13$ | 3·139 | 2·11·19 | |
| 42 | $2^2\cdot3\cdot5\cdot7$ | | 2·211 | $3^2\cdot47$ | $2^3\cdot53$ | $5^2\cdot17$ | 2·3·71 | 7·61 | $2^2\cdot107$ | 3·11·13 |
| 43 | 2·5·43 | | $2^4\cdot3^3$ | | 2·7·31 | 3·5·29 | $2^2\cdot109$ | 19·23 | 2·3·73 | |
| 44 | $2^3\cdot5\cdot11$ | $3^2\cdot7^2$ | 2·13·17 | | $2^2\cdot3\cdot37$ | 5·89 | 2·223 | 3·149 | $2^6\cdot7$ | |
| 45 | $2\cdot3^2\cdot5^2$ | 11·41 | $2^2\cdot113$ | 3·151 | 2·227 | 5·7·13 | $2^3\cdot3\cdot19$ | | 2·229 | $3^3\cdot17$ |
| 46 | $2^2\cdot5\cdot23$ | | 2·3·7·11 | | $2^4\cdot29$ | 3·5·31 | 2·233 | | $2^2\cdot3^2\cdot13$ | 7·67 |
| 47 | 2·5·47 | 3·157 | $2^3\cdot50$ | 11·43 | 2·3·79 | $5^2\cdot19$ | $2^2\cdot7\cdot17$ | $3^2\cdot53$ | 2·239 | |
| 48 | $2^5\cdot3\cdot5$ | 13·37 | 2·241 | 3·7·23 | $2^2\cdot11^2$ | 5·97 | $2\cdot3^5$ | | $2^3\cdot61$ | 3·163 |
| 49 | $2\cdot5\cdot7^2$ | | $2^2\cdot3\cdot41$ | 17·29 | 2·13·19 | $3^2\cdot5\cdot11$ | $2^4\cdot31$ | 7·71 | 2·3·83 | |

**TABLE 1.2** Factors and Prime Numbers (*Continued*)

| | | | | | | | | | | |
|---|---|---|---|---|---|---|---|---|---|---|
| 50 | $2^2\cdot5^3$ | 3·167 | 2·251 | | $2^3\cdot3^2\cdot7$ | 5·101 | 2·11·23 | $3\cdot13^2$ | $2^2\cdot127$ | |
| 51 | 2·3·5·17 | 7·73 | $2^9$ | $3^3\cdot19$ | 2·257 | 5·103 | $2^2\cdot3\cdot43$ | 11·47 | 2·7·37 | 3·173 |
| 52 | $2^3\cdot5\cdot13$ | | $2\cdot3^2\cdot29$ | | $2^2\cdot131$ | $3\cdot5^2\cdot7$ | 2·263 | 17·31 | $2^4\cdot3\cdot11$ | $23^2$ |
| 53 | 2·5·53 | $3^2\cdot59$ | $2^2\cdot7\cdot19$ | 13·41 | 2·3·89 | 5·107 | $2^3\cdot67$ | 3·179 | 2·269 | $7^2\cdot11$ |
| 54 | $2^2\cdot3^3\cdot5$ | | 2·271 | 3·181 | $2^5\cdot17$ | 5·109 | 2·3·7·13 | | $2^2\cdot137$ | $3^2\cdot61$ |
| 55 | $2\cdot5^2\cdot11$ | 19·29 | $2^3\cdot3\cdot23$ | 7·79 | 2·277 | 3·5·37 | $2^2\cdot139$ | | $2\cdot3^2\cdot31$ | 13·43 |
| 56 | $2^4\cdot5\cdot7$ | 3·11·17 | 2·281 | | $2^2\cdot3\cdot47$ | 5·113 | 2·283 | $3^4\cdot7$ | $2^3\cdot71$ | |
| 57 | 2·3·5·19 | | $2^2\cdot11\cdot13$ | 3·191 | 2·7·41 | $5^2\cdot23$ | $2^6\cdot3^2$ | | $2\cdot17^2$ | 3·193 |
| 58 | $2^2\cdot5\cdot29$ | 7·83 | 2·3·97 | 11·53 | $2^3\cdot73$ | $3^2\cdot5\cdot13$ | 2·293 | | $2^2\cdot3\cdot7^2$ | 19·31 |
| 59 | 2·5·59 | 3·197 | $2^4\cdot37$ | | $2\cdot3^3\cdot11$ | 5·7·17 | $2^2\cdot149$ | 3·199 | 2·13·23 | |
| 60 | $2^3\cdot3\cdot5^2$ | | 2·7·43 | $3^2\cdot67$ | $2^2\cdot151$ | $5\cdot11^2$ | 2·3·101 | | $2^5\cdot19$ | 3·7·29 |
| 61 | 2·5·61 | 13·47 | $2^2\cdot3^2\cdot17$ | | 2·307 | 3·5·41 | $2^3\cdot7\cdot11$ | | 2·3·103 | |
| 62 | $2^2\cdot5\cdot31$ | $3^3\cdot23$ | 2·311 | 7·89 | $2^4\cdot3\cdot13$ | $5^4$ | 2·313 | 3·11·19 | $2^2\cdot157$ | 17·37 |
| 63 | $2\cdot3^2\cdot5\cdot7$ | | $2^3\cdot79$ | 3·211 | 2·317 | 5·127 | $2^2\cdot3\cdot53$ | $7^2\cdot13$ | 2·11·29 | $3^2\cdot71$ |
| 64 | $2^7\cdot5$ | | 2·3·107 | | $2^2\cdot7\cdot23$ | 3·5·43 | 2·17·19 | | $2^3\cdot3^4$ | 11·59 |
| 65 | $2\cdot5^2\cdot13$ | 3·7·31 | $2^2\cdot163$ | | 2·3·109 | 5·131 | $2^4\cdot41$ | $3^2\cdot73$ | 2·7·47 | |
| 66 | $2^2\cdot3\cdot5\cdot11$ | | 2·331 | 3·13·17 | $2^3\cdot83$ | 5·7·19 | $2\cdot3^2\cdot37$ | 23·29 | $2^2\cdot167$ | 3·223 |
| 67 | 2·5·67 | 11·61 | $2^5\cdot3\cdot7$ | | 2·337 | $3^3\cdot5^2$ | $2^2\cdot13^2$ | | 2·3·113 | 7·97 |
| 68 | $2^3\cdot5\cdot17$ | 3·227 | 2·11·31 | | $2^2\cdot3^2\cdot19$ | 5·137 | $2\cdot7^3$ | 3·229 | $2^4\cdot43$ | 13·53 |
| 69 | 2·3·5·23 | | $2^2\cdot173$ | $3^2\cdot7\cdot11$ | 2·347 | 5·139 | $2^3\cdot3\cdot29$ | 17·41 | 2·349 | 3·233 |
| 70 | $2^2\cdot5^2\cdot7$ | | $2\cdot3^3\cdot13$ | 19·37 | $2^6\cdot11$ | 3·5·47 | 2·353 | 7·101 | $2^2\cdot3\cdot59$ | |
| 71 | 2·5·71 | $3^2\cdot79$ | $2^3\cdot89$ | 23·31 | 2·3·7·17 | 5·11·13 | $2^2\cdot179$ | 3·239 | 2·359 | |
| 72 | $2^4\cdot3^2\cdot5$ | 7·103 | $2\cdot19^2$ | 3·241 | $2^2\cdot181$ | $5^2\cdot29$ | $2\cdot3\cdot11^2$ | | $2^3\cdot7\cdot13$ | $3^6$ |
| 73 | 2·5·73 | 17·43 | $2^2\cdot3\cdot61$ | | 2·367 | $3\cdot5\cdot7^2$ | $2^5\cdot23$ | 11·67 | $2\cdot3^2\cdot41$ | |
| 74 | $2^2\cdot5\cdot37$ | 3·13·19 | 2·7·53 | | $2^3\cdot3\cdot31$ | 5·149 | 2·373 | $3^2\cdot83$ | $2^2\cdot11\cdot17$ | 7·107 |
| 75 | $2\cdot3\cdot5^3$ | | $2^4\cdot47$ | 3·251 | 2·13·29 | 5·151 | $2^2\cdot3^3\cdot7$ | | 2·379 | 3·11·23 |
| 76 | $2^3\cdot5\cdot19$ | | 2·3·127 | 7·109 | $2^2\cdot191$ | $3^2\cdot5\cdot17$ | 2·383 | 13·59 | $2^8\cdot3$ | |
| 77 | 2·5·7·11 | 3·257 | $2^2\cdot193$ | | $2\cdot3^2\cdot43$ | $5^2\cdot31$ | $2^3\cdot97$ | 3·7·37 | 2·389 | 19·41 |
| 78 | $2^2\cdot3\cdot5\cdot13$ | 11·71 | 2·17·23 | $3^3\cdot29$ | $2^4\cdot7^2$ | 5·157 | 2·3·131 | | $2^2\cdot197$ | 3·263 |
| 79 | 2·5·79 | 7·113 | $2^3\cdot3^2\cdot11$ | 13·61 | 2·397 | 3·5·53 | $2^2\cdot199$ | | 2·3·7·19 | 17·47 |
| 80 | $2^5\cdot5^2$ | $3^2\cdot89$ | 2·401 | 11·73 | $2^2\cdot3\cdot67$ | 5·7·23 | 2·13·31 | 3·269 | $2^3\cdot101$ | |
| 81 | $2\cdot3^4\cdot5$ | | $2^2\cdot7\cdot29$ | 3·271 | 2·11·37 | 5·163 | $2^4\cdot3\cdot17$ | 19·43 | 2·409 | $3^2\cdot7\cdot13$ |
| 82 | $2^2\cdot5\cdot41$ | | 2·3·137 | | $2^3\cdot103$ | $3\cdot5^2\cdot11$ | 2·7·59 | | $2^2\cdot3^2\cdot23$ | |
| 83 | 2·5·83 | 3·277 | $2^6\cdot13$ | $7^2\cdot17$ | 2·3·139 | 5·167 | $2^2\cdot11\cdot19$ | $3^3\cdot31$ | 2·419 | |
| 84 | $2^3\cdot3\cdot5\cdot7$ | $29^2$ | 2·421 | 3·281 | $2^2\cdot211$ | $5\cdot13^2$ | $2\cdot3^2\cdot47$ | $7\cdot11^2$ | $2^4\cdot53$ | 3·283 |
| 85 | $2\cdot5^2\cdot17$ | 23·37 | $2^2\cdot3\cdot71$ | | 2·7·61 | $3^2\cdot5\cdot19$ | $2^3\cdot107$ | | 2·3·11·13 | |
| 86 | $2^2\cdot5\cdot43$ | 3·7·41 | 2·431 | | $2^5\cdot3^3$ | 5·173 | 2·433 | $3\cdot17^2$ | $2^2\cdot7\cdot31$ | 11·79 |
| 87 | 2·3·5·29 | 13·67 | $2^3\cdot109$ | $3^2\cdot97$ | 2·19·23 | $5^3\cdot7$ | $2^2\cdot3\cdot73$ | | 2·439 | 3·293 |
| 88 | $2^4\cdot5\cdot11$ | | $2\cdot3^2\cdot7^2$ | | $2^2\cdot13\cdot17$ | 3·5·59 | 2·443 | | $2^3\cdot3\cdot37$ | 7·127 |
| 89 | 2·5·89 | $3^4\cdot11$ | $2^2\cdot223$ | 19·47 | 2·3·149 | 5·179 | $2^7\cdot7$ | 3·13·23 | 2·449 | 29·31 |
| 90 | $2^2\cdot3^2\cdot5^2$ | 17·53 | 2·11·41 | 3·7·43 | $2^3\cdot113$ | 5·181 | 2·3·151 | | $2^2\cdot227$ | $3^2\cdot101$ |
| 91 | 2·5·7·13 | | $2^4\cdot3\cdot19$ | 11·83 | 2·457 | 3·5·61 | $2^2\cdot229$ | 7·131 | $2\cdot3^3\cdot17$ | |
| 92 | $2^3\cdot5\cdot23$ | 3·307 | 2·461 | 13·71 | $2^2\cdot3\cdot7\cdot11$ | $5^2\cdot37$ | 2·463 | $3^2\cdot103$ | $2^5\cdot29$ | |
| 93 | 2·3·5·31 | $7^2\cdot19$ | $2^2\cdot233$ | 3·311 | 2·467 | 5·11·17 | $2^3\cdot3^2\cdot13$ | | 2·7·67 | 3·313 |
| 94 | $2^2\cdot5\cdot47$ | | 2·3·157 | 23·41 | $2^4\cdot59$ | $3^3\cdot5\cdot7$ | 2·11·43 | | $2^2\cdot3\cdot79$ | 13·73 |
| 95 | $2\cdot5^2\cdot19$ | 3·317 | $2^3\cdot7\cdot17$ | | $2\cdot3^2\cdot53$ | 5·191 | $2^2\cdot239$ | 3·11·29 | 2·479 | 7·137 |
| 96 | $2^6\cdot3\cdot5$ | $31^2$ | 2·13·37 | $3^2\cdot107$ | $2^2\cdot241$ | 5·193 | 2·3·7·23 | | $2^3\cdot11^2$ | 3·17·19 |
| 97 | 2·5·97 | | $2^2\cdot3^5$ | 7·139 | 2·487 | $3\cdot5^2\cdot13$ | $2^4\cdot61$ | | 2·3·163 | 11·89 |
| 98 | $2^2\cdot5\cdot7^2$ | $3^2\cdot109$ | 2·491 | | $2^3\cdot3\cdot41$ | 5·197 | 2·17·29 | 3·7·47 | $2^2\cdot13\cdot19$ | 23·43 |
| 99 | $2\cdot3^2\cdot5\cdot11$ | | $2^5\cdot31$ | 3·331 | 2·7·71 | 5·199 | $2^2\cdot3\cdot83$ | | 2·499 | $3^3\cdot37$ |
| 100 | $2^3\cdot5^3$ | 7·11·13 | 2·3·167 | 17·59 | $2^2\cdot251$ | 3·5·67 | 2·503 | 19·53 | $2^4\cdot3^2\cdot7$ | |
| 101 | 2·5·101 | 3·337 | $2^2\cdot11\cdot23$ | | $2\cdot3\cdot13^2$ | 5·7·29 | $2^3\cdot127$ | $3^2\cdot113$ | 2·509 | |
| 102 | $2^2\cdot3\cdot5\cdot17$ | | 2·7·73 | 3·11·31 | $2^{10}$ | $5^2\cdot41$ | $2\cdot3^3\cdot19$ | 13·79 | $2^2\cdot257$ | $3\cdot7^3$ |
| 103 | 2·5·103 | | $2^3\cdot3\cdot43$ | | 2·11·47 | $3^2\cdot5\cdot23$ | $2^2\cdot7\cdot37$ | 17·61 | 2·3·173 | |

**TABLE 1.2** Factors and Prime Numbers (*Continued*)

| | | | | | | | | | | |
|---|---|---|---|---|---|---|---|---|---|---|
| 104 | $2^{4}\cdot5\cdot13$ | 3·347 | 2·521 | 7·149 | $2^{2}\cdot3^{2}\cdot29$ | 5·11·11 | 2·523 | 3·349 | $2^{3}\cdot131$ | |
| 105 | $2\cdot3\cdot5^{2}\cdot7$ | | $2^{2}\cdot263$ | $3^{4}\cdot13$ | 2·17·31 | 5·211 | $2^{5}\cdot3\cdot11$ | 7·151 | $2\cdot23^{2}$ | 3·353 |
| 106 | $2^{2}\cdot5\cdot53$ | | $2\cdot3^{2}\cdot59$ | | $2^{3}\cdot7\cdot19$ | 3·5·71 | 2·13·41 | 11·97 | $2^{2}\cdot3\cdot89$ | |
| 107 | 2·5·107 | $3^{2}\cdot7\cdot17$ | $2^{4}\cdot67$ | 29·37 | 2·3·179 | $5^{2}\cdot43$ | $2^{2}\cdot269$ | 3·359 | $2\cdot7^{2}\cdot11$ | 13·83 |
| 108 | $2^{3}\cdot3^{3}\cdot5$ | 23·47 | 2·541 | $3\cdot19^{2}$ | $2^{2}\cdot271$ | 5·7·31 | 2·3·181 | | $2^{6}\cdot17$ | $3^{2}\cdot11^{2}$ |
| 109 | 2·5·109 | | $2^{2}\cdot3\cdot7\cdot13$ | | 2·547 | 3·5·73 | $2^{3}\cdot137$ | | $2\cdot3^{2}\cdot61$ | 7·157 |
| 110 | $2^{2}\cdot5^{2}\cdot11$ | 3·367 | 2·19·29 | | $2^{4}\cdot3\cdot23$ | 5·13·17 | 2·7·79 | $3^{3}\cdot41$ | $2^{2}\cdot277$ | |
| 111 | 2·3·5·37 | 11·101 | $2^{3}\cdot139$ | 3·7·53 | 2·557 | 5·223 | $2^{2}\cdot3^{2}\cdot31$ | | 2·13·43 | 3·373 |
| 112 | $2^{5}\cdot5\cdot7$ | 19·59 | 2·3·11·17 | | $2^{2}\cdot281$ | $3^{2}\cdot5^{3}$ | 2·563 | $7^{2}\cdot23$ | $2^{3}\cdot3\cdot47$ | |
| 113 | 2·5·113 | 3·13·29 | $2^{2}\cdot283$ | 11·103 | $2\cdot3^{4}\cdot7$ | 5·227 | $2^{4}\cdot71$ | 3·379 | 2·569 | 17·67 |
| 114 | $2^{2}\cdot3\cdot5\cdot19$ | 7·163 | 2·571 | $3^{2}\cdot127$ | $2^{3}\cdot11\cdot13$ | 5·229 | 2·3·191 | 31·37 | $2^{2}\cdot7\cdot41$ | 3·383 |
| 115 | $2\cdot5^{2}\cdot23$ | | $2^{7}\cdot3^{2}$ | | 2·577 | 3·5·7·11 | $2^{2}\cdot17^{2}$ | 13·89 | 2·3·193 | 19·61 |
| 116 | $2^{3}\cdot5\cdot29$ | $3^{3}\cdot43$ | 2·7·83 | | $2^{2}\cdot3\cdot97$ | 5·233 | 2·11·53 | 3·389 | $2^{4}\cdot73$ | 7·167 |
| 117 | $2\cdot3^{2}\cdot5\cdot13$ | | $2^{2}\cdot293$ | 3·17·23 | 2·587 | $5^{2}\cdot47$ | $2^{3}\cdot3\cdot7^{2}$ | 11·107 | 2·19·31 | $3^{2}\cdot131$ |
| 118 | $2^{2}\cdot5\cdot59$ | | 2·3·197 | $7\cdot13^{2}$ | $2^{5}\cdot37$ | 3·5·79 | 2·593 | | $2^{2}\cdot3^{3}\cdot11$ | 29·41 |
| 119 | 2·5·7·17 | 3·397 | $2^{3}\cdot149$ | | 2·3·199 | 5·239 | $2^{2}\cdot13\cdot23$ | $3^{2}\cdot7\cdot19$ | 2·599 | 11·109 |
| 120 | $2^{4}\cdot3\cdot5^{2}$ | | 2·601 | 3·401 | $2^{2}\cdot7\cdot43$ | 5·241 | $2\cdot3^{2}\cdot67$ | 17·71 | $2^{3}\cdot151$ | 3·13·31 |
| 121 | $2\cdot5\cdot11^{2}$ | 7·173 | $2^{2}\cdot3\cdot101$ | | 2·607 | $3^{5}\cdot5$ | $2^{6}\cdot19$ | | 2·3·7·29 | 23·53 |
| 122 | $2^{2}\cdot5\cdot61$ | 3·11·37 | 2·13·47 | | $2^{3}\cdot3^{2}\cdot17$ | $5^{2}\cdot7^{2}$ | 2·613 | 3·409 | $2^{2}\cdot307$ | |
| 123 | 2·3·5·41 | | $2^{4}\cdot7\cdot11$ | $3^{2}\cdot137$ | 2·617 | 5·13·19 | $2^{2}\cdot3\cdot103$ | | 2·619 | 3·7·59 |
| 124 | $2^{3}\cdot5\cdot31$ | 17·73 | $2\cdot3^{3}\cdot23$ | 11·113 | $2^{2}\cdot311$ | 3·5·83 | 2·7·89 | 29·43 | $2^{5}\cdot3\cdot13$ | |
| 125 | $2\cdot5^{4}$ | $3^{2}\cdot139$ | $2^{2}\cdot313$ | 7·179 | 2·3·11·19 | 5·251 | $2^{3}\cdot157$ | 3·419 | 2·17·37 | |
| 126 | $2^{2}\cdot3^{2}\cdot5\cdot7$ | 13·97 | 2·631 | 3·421 | $2^{4}\cdot79$ | 5·11·23 | 2·3·211 | 7·181 | $2^{2}\cdot317$ | $3^{3}\cdot47$ |
| 127 | 2·5·127 | 31·41 | $2^{3}\cdot3\cdot53$ | 19·67 | $2\cdot7^{2}\cdot13$ | $3\cdot5^{2}\cdot17$ | $2^{2}\cdot11\cdot29$ | | $2\cdot3^{2}\cdot71$ | |
| 128 | $2^{8}\cdot5$ | 3·7·61 | 2·641 | | $2^{2}\cdot3\cdot107$ | 5·257 | 2·643 | $3^{2}\cdot11\cdot13$ | $2^{3}\cdot7\cdot23$ | |
| 129 | 2·3·5·43 | | $2^{2}\cdot17\cdot19$ | 3·431 | 2·647 | 5·7·37 | $2^{4}\cdot3^{4}$ | | 2·11·59 | 3·433 |
| 130 | $2^{2}\cdot5^{2}\cdot13$ | | 2·3·7·31 | | $2^{3}\cdot163$ | $3^{2}\cdot5\cdot29$ | 2·653 | | $2^{2}\cdot3\cdot109$ | 7·11·17 |
| 131 | 2·5·131 | 3·19·23 | $2^{5}\cdot41$ | 13·101 | $2\cdot3^{2}\cdot73$ | 5·263 | $2^{2}\cdot7\cdot47$ | 3·439 | 2·659 | |
| 132 | $2^{3}\cdot3\cdot5\cdot11$ | | 2·661 | $3^{3}\cdot7^{2}$ | $2^{2}\cdot331$ | $5^{2}\cdot53$ | 2·3·13·17 | | $2^{4}\cdot83$ | 3·443 |
| 133 | 2·5·7·19 | $11^{3}$ | $2^{2}\cdot3^{2}\cdot37$ | 31·43 | 2·23·29 | 3·5·89 | $2^{3}\cdot167$ | 7·191 | 2·3·223 | 13·103 |
| 134 | $2^{2}\cdot5\cdot67$ | $3^{2}\cdot149$ | 2·11·61 | 17·79 | $2^{6}\cdot3\cdot7$ | 5·269 | 2·673 | 3·449 | $2^{2}\cdot337$ | 19·71 |
| 135 | $2\cdot3^{3}\cdot5^{2}$ | 7·193 | $2^{3}\cdot13^{2}$ | 3·11·41 | 2·677 | 5·271 | $2^{2}\cdot3\cdot113$ | 23·59 | 2·7·97 | $3^{2}\cdot151$ |
| 136 | $2^{4}\cdot5\cdot17$ | | 2·3·227 | 29·47 | $2^{2}\cdot11\cdot31$ | 3·5·7·13 | 2·683 | | $2^{3}\cdot3^{2}\cdot19$ | $37^{2}$ |
| 137 | 2·5·137 | 3·457 | $2^{2}\cdot7^{3}$ | | 2·3·229 | $5^{3}\cdot11$ | $2^{5}\cdot43$ | $3^{4}\cdot17$ | 2·13·53 | 7·197 |
| 138 | $2^{2}\cdot3\cdot5\cdot23$ | | 2·691 | 3·461 | $2^{3}\cdot173$ | 5·277 | $2\cdot3^{2}\cdot7\cdot11$ | 19·73 | $2^{2}\cdot347$ | 3·463 |
| 139 | 2·5·139 | 13·107 | $2^{4}\cdot3\cdot29$ | 7·199 | 2·17·41 | $3^{2}\cdot5\cdot31$ | $2^{2}\cdot349$ | 11·127 | 2·3·233 | |
| 140 | $2^{3}\cdot5^{2}\cdot7$ | 3·467 | 2·701 | 23·61 | $2^{2}\cdot3^{3}\cdot13$ | 5·281 | 2·19·37 | 3·7·67 | $2^{7}\cdot11$ | |
| 141 | 2·3·5·47 | 17·83 | $2^{2}\cdot353$ | $3^{2}\cdot157$ | 2·7·101 | 5·283 | $2^{3}\cdot3\cdot59$ | 13·109 | 2·709 | 3·11·43 |
| 142 | $2^{2}\cdot5\cdot71$ | $7^{2}\cdot29$ | $2\cdot3^{2}\cdot79$ | | $2^{4}\cdot89$ | $3\cdot5^{2}\cdot19$ | 2·23·31 | | $2^{2}\cdot3\cdot7\cdot17$ | |
| 143 | 2·5·11·13 | $3^{3}\cdot53$ | $2^{3}\cdot179$ | | 2·3·239 | 5·7·41 | $2^{2}\cdot359$ | 3·479 | 2·719 | |
| 144 | $2^{5}\cdot3^{2}\cdot5$ | 11·131 | 2·7·103 | 3·13·37 | $2^{2}\cdot19^{2}$ | $5\cdot17^{2}$ | 2·3·241 | | $2^{3}\cdot181$ | $3^{2}\cdot7\cdot23$ |
| 145 | $2\cdot5^{2}\cdot29$ | | $2^{2}\cdot3\cdot11^{2}$ | | 2·727 | 3·5·97 | $2^{4}\cdot7\cdot13$ | 31·47 | $2\cdot3^{6}$ | |
| 146 | $2^{2}\cdot5\cdot73$ | 3·487 | 2·17·43 | 7·11·19 | $2^{3}\cdot3\cdot61$ | 5·293 | 2·733 | $3^{2}\cdot163$ | $2^{2}\cdot367$ | 13·113 |
| 147 | $2\cdot3\cdot5\cdot7^{2}$ | | $2^{6}\cdot23$ | 3·491 | 2·11·67 | $5^{2}\cdot59$ | $2^{2}\cdot3^{2}\cdot41$ | 7·211 | 2·739 | 3·17·29 |
| 148 | $2^{3}\cdot5\cdot37$ | | 2·3·13·19 | | $2^{2}\cdot7\cdot53$ | $3^{3}\cdot5\cdot11$ | 2·743 | | $2^{4}\cdot3\cdot31$ | |
| 149 | 2·5·149 | 3·7·71 | $2^{2}\cdot373$ | | $2\cdot3^{2}\cdot83$ | 5·13·23 | $2^{3}\cdot11\cdot17$ | 3·499 | 2·7·107 | |
| 150 | $2^{2}\cdot3\cdot5^{3}$ | 19·79 | 2·751 | $3^{2}\cdot167$ | $2^{5}\cdot47$ | 5·7·43 | 2·3·251 | 11·137 | $2^{2}\cdot13\cdot29$ | 3·503 |
| 151 | 2·5·151 | | $2^{3}\cdot3^{3}\cdot7$ | 17·89 | 2·757 | 3·5·101 | $2^{2}\cdot379$ | 37·41 | 2·3·11·23 | $7^{2}\cdot31$ |
| 152 | $2^{4}\cdot5\cdot19$ | $3^{2}\cdot13^{2}$ | 2·761 | | $2^{2}\cdot3\cdot127$ | $5^{2}\cdot61$ | 2·7·109 | 3·509 | $2^{3}\cdot191$ | 11·139 |
| 153 | $2\cdot3^{2}\cdot5\cdot17$ | | $2^{2}\cdot383$ | 3·7·73 | 2·13·59 | 5·307 | $2^{9}\cdot3$ | 29·53 | 2·769 | $3^{4}\cdot19$ |
| 154 | $2^{2}\cdot5\cdot7\cdot11$ | 23·67 | 2·3·257 | | $2^{3}\cdot193$ | 3·5·103 | 2·773 | 7·13·17 | $2^{2}\cdot3^{2}\cdot43$ | |
| 155 | $2\cdot5^{2}\cdot31$ | 3·11·47 | $2^{4}\cdot97$ | | 2·3·7·37 | 5·311 | $2^{2}\cdot389$ | $3^{2}\cdot173$ | 2·19·41 | |
| 156 | $2^{3}\cdot3\cdot5\cdot13$ | 7·223 | 2·11·71 | 3·521 | $2^{2}\cdot17\cdot23$ | 5·313 | $2\cdot3^{3}\cdot29$ | | $2^{5}\cdot7^{2}$ | 3·523 |
| 157 | 2·5·157 | | $2^{2}\cdot3\cdot131$ | $11^{2}\cdot13$ | 2·787 | $3^{2}\cdot5^{2}\cdot7$ | $2^{2}\cdot197$ | 19·83 | 2·3·263 | |

**TABLE 1.2** Factors and Prime Numbers (*Continued*)

| | | | | | | | | | | |
|---|---|---|---|---|---|---|---|---|---|---|
| 158 | $2^2\cdot 5\cdot 79$ | 3·17·31 | 2·7·113 | | $2^4\cdot 3^2\cdot 11$ | 5·317 | 2·13·61 | $3\cdot 23^2$ | $2^2\cdot 397$ | 7·227 |
| 159 | 2·3·5·53 | 37·43 | $2^3\cdot 199$ | $3^3\cdot 59$ | 2·797 | 5·11·29 | $2^2\cdot 3\cdot 7\cdot 19$ | | 2·17·47 | 3·13·41 |
| 160 | $2^6\cdot 5^2$ | | $2\cdot 3^2\cdot 89$ | 7·229 | $2^2\cdot 401$ | 3·5·107 | 2·11·73 | | $2^3\cdot 3\cdot 67$ | |
| 161 | 2·5·7·23 | $3^2\cdot 179$ | $2^2\cdot 13\cdot 31$ | | 2·3·269 | 5·17·19 | $2^4\cdot 101$ | $3\cdot 7^2\cdot 11$ | 2·809 | |
| 162 | $2^2\cdot 3^4\cdot 5$ | | 2·811 | 3·541 | $2^3\cdot 7\cdot 29$ | $5^3\cdot 13$ | 2·3·271 | | $2^2\cdot 11\cdot 37$ | $3^2\cdot 181$ |
| 163 | 2·5·163 | 7·233 | $2^5\cdot 3\cdot 17$ | 23·71 | 2·19·43 | 3·5·109 | $2^2\cdot 409$ | | $2\cdot 3^2\cdot 7\cdot 13$ | 11·149 |
| 164 | $2^3\cdot 5\cdot 41$ | 3·547 | 2·821 | 31·53 | $2^2\cdot 3\cdot 137$ | 5·7·47 | 2·823 | $3^3\cdot 61$ | $2^4\cdot 103$ | 17·97 |
| 165 | $2\cdot 3\cdot 5^2\cdot 11$ | 13·127 | $2^2\cdot 7\cdot 59$ | 3·19·29 | 2·827 | 5·331 | $2^3\cdot 3^2\cdot 23$ | | 2·829 | 3·7·79 |
| 166 | $2^2\cdot 5\cdot 83$ | 11·151 | 2·3·277 | | $2^7\cdot 13$ | $3^2\cdot 5\cdot 37$ | $2\cdot 7^2\cdot 17$ | | $2^2\cdot 3\cdot 139$ | |
| 167 | 2·5·167 | 3·557 | $2^3\cdot 11\cdot 19$ | 7·239 | $2\cdot 3^3\cdot 31$ | $5^2\cdot 67$ | $2^2\cdot 419$ | 3·13·43 | 2·839 | 23·73 |
| 168 | $2^4\cdot 3\cdot 5\cdot 7$ | $41^2$ | $2\cdot 29^2$ | $3^2\cdot 11\cdot 17$ | $2^2\cdot 421$ | 5·337 | 2·3·281 | 7·241 | $2^3\cdot 211$ | 3·563 |
| 169. | $2\cdot 5\cdot 13^2$ | 19·89 | $2^2\cdot 3^2\cdot 47$ | | $2\cdot 7\cdot 11^2$ | 3·5·113 | $2^5\cdot 53$ | | 2·3·283 | |
| 170 | $2^2\cdot 5^2\cdot 17$ | $3^5\cdot 7$ | 2·23·37 | 13·131 | $2^3\cdot 3\cdot 71$ | 5·11·31 | 2·853 | 3·569 | $2^2\cdot 7\cdot 61$ | |
| 171 | $2\cdot 3^2\cdot 5\cdot 19$ | 29·59 | $2^4\cdot 107$ | 3·571 | 2·857 | $5\cdot 7^3$ | $2^2\cdot 3\cdot 11\cdot 13$ | 17·101 | 2·859 | $3^2\cdot 191$ |
| 172 | $2^3\cdot 5\cdot 43$ | | 2·3·7·41 | | $2^2\cdot 431$ | $3\cdot 5^2\cdot 23$ | 2·863 | 11·157 | $2^6\cdot 3^3$ | 7·13·19 |
| 173 | 2·5·173 | 3·577 | $2^2\cdot 433$ | | $2\cdot 3\cdot 17^2$ | 5·347 | $2^3\cdot 7\cdot 31$ | $3^2\cdot 193$ | 2·11·79 | 37·47 |
| 174 | $2^2\cdot 3\cdot 5\cdot 29$ | | 2·13·67 | 3·7·83 | $2^4\cdot 109$ | 5·349 | $2\cdot 3^2\cdot 97$ | | $2^2\cdot 19\cdot 23$ | 3·11·53 |
| 175 | $2\cdot 5^3\cdot 7$ | 17·103 | $2^3\cdot 3\cdot 73$ | | 2·877 | $3^3\cdot 5\cdot 13$ | $2^2\cdot 439$ | 7·251 | 2·3·293 | |
| 176 | $2^5\cdot 5\cdot 11$ | 3·587 | 2·881 | 41·43 | $2^2\cdot 3^2\cdot 7^2$ | 5·353 | 2·883 | 3·19·31 | $2^3\cdot 13\cdot 17$ | 29·61 |
| 177 | 2·3·5·59 | 7·11·23 | $2^2\cdot 443$ | $3^2\cdot 197$ | 2·887 | $5^2\cdot 71$ | $2^4\cdot 3\cdot 37$ | | 2·7·127 | 3·593 |
| 178 | $2^2\cdot 5\cdot 89$ | 13·137 | $2\cdot 3^4\cdot 11$ | | $2^3\cdot 223$ | 3·5·7·17 | 2·19·47 | | $2^2\cdot 3\cdot 149$ | |
| 179 | 2·5·179 | $3^2\cdot 199$ | $2^8\cdot 7$ | 11·163 | 2·3·13·23 | 5·359 | $2^2\cdot 449$ | 3·599 | 2·29·31 | 7·257 |
| 180 | $2^3\cdot 3^2\cdot 5^2$ | | 2·17·53 | 3·601 | $2^2\cdot 11\cdot 41$ | $5\cdot 19^2$ | 2·3·7·43 | 13·139 | $2^4\cdot 113$ | $3^3\cdot 67$ |
| 181 | 2·5·181 | | $2^2\cdot 3\cdot 151$ | $7^2\cdot 37$ | 2·907 | $3\cdot 5\cdot 11^2$ | $2^3\cdot 227$ | 23·79 | $2\cdot 3^2\cdot 101$ | 17·107 |
| 182 | $2^2\cdot 5\cdot 7\cdot 13$ | 3·607 | 2·911 | | $2^5\cdot 3\cdot 19$ | $5^2\cdot 73$ | 2·11·83 | $3^2\cdot 7\cdot 29$ | $2^2\cdot 457$ | 31·59 |
| 183 | 2·3·5·61 | | $2^3\cdot 229$ | 3·13·47 | 2·7·131 | 5·367 | $2^2\cdot 3^3\cdot 17$ | 11·167 | 2·919 | 3·613 |
| 184 | $2^4\cdot 5\cdot 23$ | 7·263 | 2·3·307 | 19·97 | $2^2\cdot 461$ | $3^2\cdot 5\cdot 41$ | 2·13·71 | | $2^3\cdot 3\cdot 7\cdot 11$ | $43^2$ |
| 185 | $2\cdot 5^2\cdot 37$ | 3·617 | $2^2\cdot 463$ | 17·109 | $2\cdot 3^2\cdot 103$ | 5·7·53 | $2^6\cdot 29$ | 3·619 | 2·929 | $11\cdot 13^2$ |
| 186 | $2^2\cdot 3\cdot 5\cdot 31$ | | $2\cdot 7^2\cdot 19$ | $3^4\cdot 23$ | $2^3\cdot 233$ | 5·373 | 2·3·311 | | $2^2\cdot 467$ | 3·7·89 |
| 187 | 2·5·11·17 | | $2^4\cdot 3^2\cdot 13$ | | 2·937 | $3\cdot 5^4$ | $2^2\cdot 7\cdot 67$ | | 2·3·313 | |
| 188 | $2^3\cdot 5\cdot 47$ | $3^2\cdot 11\cdot 19$ | 2·941 | 7·269 | $2^2\cdot 3\cdot 157$ | 5·13·29 | 2·23·41 | 3·17·37 | $2^5\cdot 59$ | |
| 189 | $2\cdot 3^3\cdot 5\cdot 7$ | 31·61 | $2^2\cdot 11\cdot 43$ | 3·631 | 2·947 | 5·379 | $2^3\cdot 3\cdot 79$ | 7·271 | 2·13·73 | $3^2\cdot 211$ |
| 190 | $2^2\cdot 5^2\cdot 19$ | | 2·3·317 | 11·173 | $2^4\cdot 7\cdot 17$ | 3·5·127 | 2·953 | | $2^2\cdot 3^2\cdot 53$ | 23·83 |
| 191 | 2·5·191 | $3\cdot 7^2\cdot 13$ | $2^3\cdot 239$ | | 2·3·11·29 | 5·383 | $2^2\cdot 479$ | $3^3\cdot 71$ | 2·7·137 | 19·101 |
| 192 | $2^7\cdot 3\cdot 5$ | 17·113 | $2\cdot 31^2$ | 3·641 | $2^2\cdot 13\cdot 37$ | $5^2\cdot 7\cdot 11$ | $2\cdot 3^2\cdot 107$ | 41·47 | $2^3\cdot 241$ | 3·643 |
| 193 | 2·5·193 | | $2^2\cdot 3\cdot 7\cdot 23$ | | 2·967 | $3^2\cdot 5\cdot 43$ | $2^4\cdot 11^2$ | 13·149 | 2·3·17·19 | 7·277 |
| 194 | $2^2\cdot 5\cdot 97$ | 3·647 | 2·971 | 29·67 | $2^3\cdot 3^5$ | 5·389 | 2·7·139 | 3·11·59 | $2^2\cdot 487$ | |
| 195 | $2\cdot 3\cdot 5^2\cdot 13$ | | $2^5\cdot 61$ | $3^2\cdot 7\cdot 31$ | 2·977 | 5·17·23 | $2^2\cdot 3\cdot 163$ | 19·103 | 2·11·89 | 3·653 |
| 196 | $2^3\cdot 5\cdot 7^2$ | 37·53 | $2\cdot 3^2\cdot 109$ | 13·151 | $2^2\cdot 491$ | 3·5·131 | 2·983 | 7·281 | $2^4\cdot 3\cdot 41$ | 11·179 |
| 197 | 2·5·197 | $3^3\cdot 73$ | $2^2\cdot 17\cdot 29$ | | 2·3·7·47 | $5^2\cdot 79$ | $2^3\cdot 13\cdot 19$ | 3·659 | 2·23·43 | |
| 198 | $2^2\cdot 3^2\cdot 5\cdot 11$ | 7·283 | 2·991 | 3·661 | $2^6\cdot 31$ | 5·397 | 2·3·331 | | $2^2\cdot 7\cdot 71$ | $3^2\cdot 13\cdot 17$ |
| 199 | 2·5·199 | 11·181 | $2^3\cdot 3\cdot 83$ | | 2·997 | 3·5·7·19 | $2^2\cdot 499$ | | $2\cdot 3^3\cdot 37$ | |
| 200 | $2^4\cdot 5^3$ | 3·23·29 | 2·7·11·13 | | $2^2\cdot 3\cdot 167$ | 5·401 | 2·17·59 | $3^2\cdot 223$ | $2^3\cdot 251$ | $7^2\cdot 41$ |

## 1.17 NEWTON'S METHOD FOR SOLVING INTRACTABLE EQUATIONS

See Chap. 2, Sec. 2.3.28.

## 1.18 THE GREEK ALPHABET

| | |
|---|---|
| α, A alpha | ν, N nu |
| β, B beta | ξ, Ξ xi |
| γ, Γ gamma | ο, O omicron |
| δ, Δ delta | π, Π pi |
| ε, E epsilon | ρ, P rho |
| ζ, Z zeta | σ, Σ sigma |
| η, H eta | τ, T tau |
| θ, Θ theta | υ, Y upsilon |
| ι, I iota | φ, Φ phi |
| κ, K kappa | χ, X chi |
| λ, Λ lambda | ψ, Ψ psi |
| μ, M mu | ω, Ω omega |

## 1.19 DECIMAL CHART

See Table 1.3.

## 1.20 DEGREES AND RADIANS CHART

See Fig. 1.46.

## BIBLIOGRAPHY

Ayres, F., 1978: *Calculus,* 2d ed. Schaum's Outline Series. New York: McGraw-Hill.

Bronshtein, N., and K. A. Semendyayev, 1985: *Handbook of Mathematics.* New York: Van Nostrand Reinhold.

Downing, D., 1982: *Calculus the Easy Way.* New York: Barron's Educational Series.

Eves, H., 1969: *History of Mathematics,* 3rd ed. New York: Holt, Reinhart & Winston.

Fuller, G., 1979: *Analytic Geometry.* North Reading, Mass.: Addison-Wesley.

Gerolde, S., 1971: *Universal Conversion Factors.* Tulsa, Okla.: Petroleum Publishing Company.

Hart, W. L., 1947: *Brief College Algebra.* Boston: Heath.

Middlemiss, R. R., 1952: *College Algebra.* New York: McGraw-Hill.

Middlemiss, R. R., 1945: *Analytic Geometry,* 1st ed. New York: McGraw-Hill.

Mendelson, E., 1985: *Beginning Calculus,* Schaum's Outline Series. New York: McGraw-Hill.

Olenick, R. P., T. M. Apostol, and D. L. Goodstein, 1985: *The Mechanical Universe, Introduction to Mechanics and Heat,* 2 volumes. Cambridge, Mass.: Cambridge University Press.

Selby, S. M., R. C. Weast, R. S. Shankland, and C. D. Hodgman, eds., 1962: *Handbook of Mathematical Tables.* Cleveland, Ohio: Chemical Rubber Publishing Company.

Thompson, S. P., 1985: *Calculus Made Easy.* New York: St. Martin's Press.

Weir, M. D., 1982: *Calculus by Calculator.* Englewood Cliffs, N.J.: Prentice-Hall.

Weisbecker, H., 1986, *The Calculus Problem Solver,* New York: Research and Education Association.


**TABLE 1.3** Decimal Chart

| Fraction | DECIMALS | MILLIMETERS | Fraction | DECIMALS | MILLIMETERS |
|---|---|---|---|---|---|
| 1/64 | 0.015625 | 0.397 | 33/64 | 0.515625 | 13.097 |
| 1/32 | .03125 | 0.794 | 17/32 | .53125 | 13.494 |
| 3/64 | .046875 | 1.191 | 35/64 | .546875 | 13.891 |
| 1/16 | .0625 | 1.588 | 9/16 | .5625 | 14.288 |
| 5/64 | .078125 | 1.984 | 37/64 | .578125 | 14.684 |
| 3/32 | .09375 | 2.381 | 19/32 | .59375 | 15.081 |
| 7/64 | .109375 | 2.778 | 39/64 | .609375 | 15.478 |
| 1/8 | .1250 | 3.175 | 5/8 | .6250 | 15.875 |
| 9/64 | .140625 | 3.572 | 41/64 | .640625 | 16.272 |
| 5/32 | .15625 | 3.969 | 21/32 | .65625 | 16.669 |
| 11/64 | .171875 | 4.366 | 43/64 | .671875 | 17.066 |
| 3/16 | .1875 | 4.763 | 11/16 | .6875 | 17.463 |
| 13/64 | .203125 | 5.159 | 45/64 | .703125 | 17.859 |
| 7/32 | .21875 | 5.556 | 23/32 | .71875 | 18.256 |
| 15/64 | .234375 | 5.953 | 47/64 | .734375 | 18.653 |
| 1/4 | .2500 | 6.350 | 3/4 | .7500 | 19.050 |
| 17/64 | .265625 | 6.747 | 49/64 | .765625 | 19.447 |
| 9/32 | .28125 | 7.144 | 25/32 | .78125 | 19.844 |
| 19/64 | .296875 | 7.541 | 51/64 | .796875 | 20.241 |
| 5/16 | .3125 | 7.938 | 13/16 | .8125 | 20.638 |
| 21/64 | .328125 | 8.334 | 53/64 | .828125 | 21.034 |
| 11/32 | .34375 | 8.731 | 27/32 | .84375 | 21.431 |
| 23/64 | .359375 | 9.128 | 55/64 | .859375 | 21.828 |
| 3/8 | .3750 | 9.525 | 7/8 | .8750 | 22.225 |
| 25/64 | .390625 | 9.922 | 57/64 | .890625 | 22.622 |
| 13/32 | .40625 | 10.319 | 29/32 | .90625 | 23.019 |
| 27/64 | .421875 | 10.716 | 59/64 | .921875 | 23.416 |
| 7/16 | .4375 | 11.113 | 15/16 | .9375 | 23.813 |
| 29/64 | .453125 | 11.509 | 61/64 | .953125 | 24.209 |
| 15/32 | .46875 | 11.906 | 31/32 | .96875 | 24.606 |
| 31/64 | .484375 | 12.303 | 63/64 | .984375 | 25.003 |
| 1/2 | .5000 | 12.700 | 1 | 1.000 | 25.400 |

1 mm = .03937″

| MM | INCHES | MM | INCHES |
|---|---|---|---|
| .1 | .0039 | 46 | 1.8110 |
| .2 | .0079 | 47 | 1.8504 |
| .3 | .0118 | 48 | 1.8898 |
| .4 | .0157 | 49 | 1.9291 |
| .5 | .0197 | 50 | 1.9685 |
| .6 | .0236 | 51 | 2.0079 |
| .7 | .0276 | 52 | 2.0472 |
| .8 | .0315 | 53 | 2.0866 |
| .9 | .0354 | 54 | 2.1260 |
| 1 | .0394 | 55 | 2.1654 |
| 2 | .0787 | 56 | 2.2047 |
| 3 | .1181 | 57 | 2.2441 |
| 4 | .1575 | 58 | 2.2835 |
| 5 | .1969 | 59 | 2.3228 |
| 6 | .2362 | 60 | 2.3622 |
| 7 | .2756 | 61 | 2.4016 |
| 8 | .3150 | 62 | 2.4409 |
| 9 | .3543 | 63 | 2.4803 |
| 10 | .3937 | 64 | 2.5197 |
| 11 | .4331 | 65 | 2.5591 |
| 12 | .4724 | 66 | 2.5984 |
| 13 | .5118 | 67 | 2.6378 |
| 14 | .5512 | 68 | 2.6772 |
| 15 | .5906 | 69 | 2.7165 |
| 16 | .6299 | 70 | 2.7559 |
| 17 | .6693 | 71 | 2.7953 |
| 18 | .7087 | 72 | 2.8346 |
| 19 | .7480 | 73 | 2.8740 |
| 20 | .7874 | 74 | 2.9134 |
| 21 | .8268 | 75 | 2.9528 |
| 22 | .8661 | 76 | 2.9921 |
| 23 | .9055 | 77 | 3.0315 |
| 24 | .9449 | 78 | 3.0709 |
| 25 | .9843 | 79 | 3.1102 |
| 26 | 1.0236 | 80 | 3.1496 |
| 27 | 1.0630 | 81 | 3.1890 |
| 28 | 1.1024 | 82 | 3.2283 |
| 29 | 1.1417 | 83 | 3.2677 |
| 30 | 1.1811 | 84 | 3.3071 |
| 31 | 1.2205 | 85 | 3.3465 |
| 32 | 1.2598 | 86 | 3.3858 |
| 33 | 1.2992 | 87 | 3.4252 |
| 34 | 1.3386 | 88 | 3.4646 |
| 35 | 1.3780 | 89 | 3.5039 |
| 36 | 1.4173 | 90 | 3.5433 |
| 37 | 1.4567 | 91 | 3.5827 |
| 38 | 1.4961 | 92 | 3.6220 |
| 39 | 1.5354 | 93 | 3.6614 |
| 40 | 1.5748 | 94 | 3.7008 |
| 41 | 1.6142 | 95 | 3.7402 |
| 42 | 1.6535 | 96 | 3.7795 |
| 43 | 1.6929 | 97 | 3.8189 |
| 44 | 1.7323 | 98 | 3.8583 |
| 45 | 1.7717 | 99 | 3.8976 |
| | | 100 | 3.9370 |

.001″ = .0254 mm

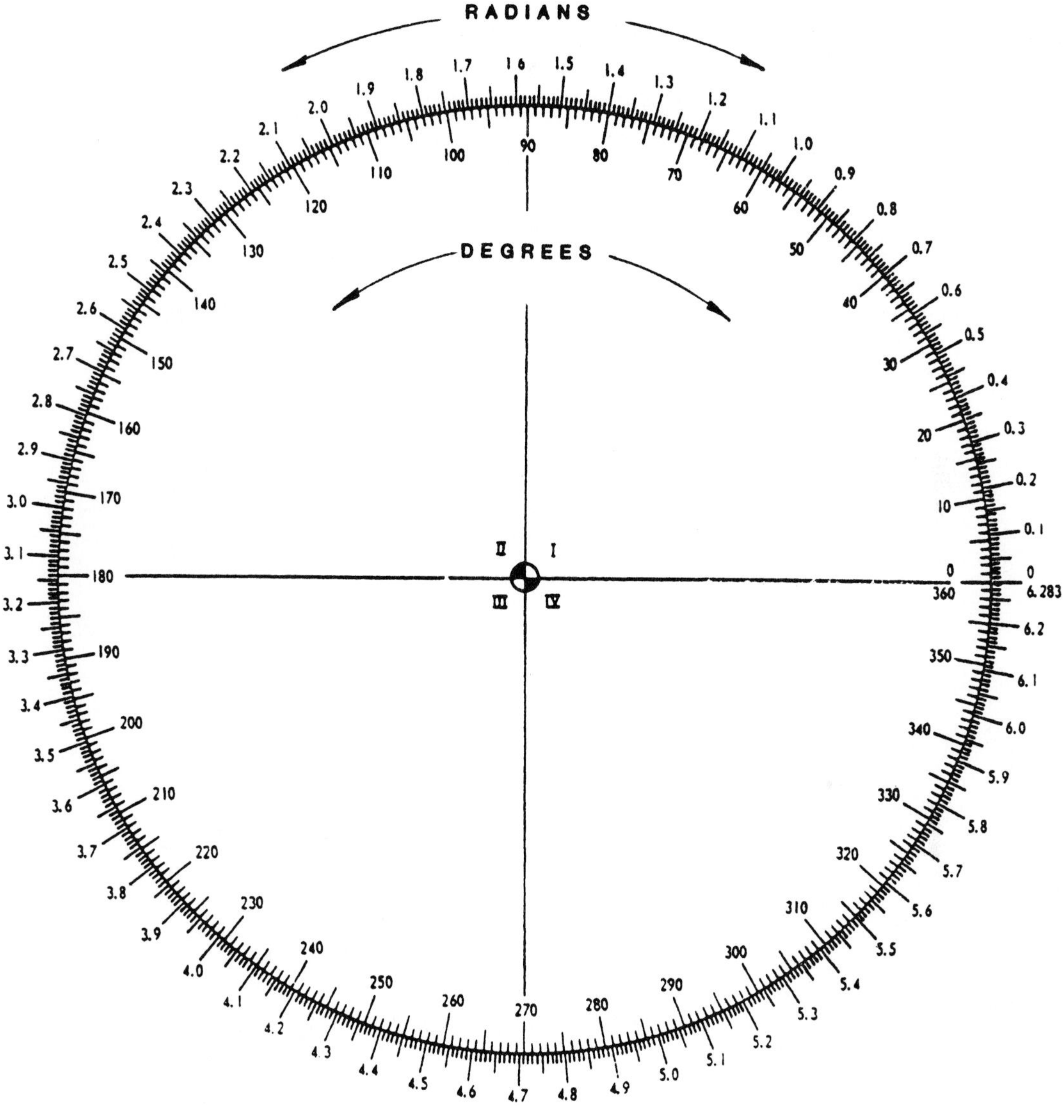


**FIGURE 1.46** Template for degrees and radians.

# CHAPTER 2

# PRACTICAL ENGINEERING MECHANICS

## 2.1 DEFINITIONS AND BASIC UNITS OF MECHANICS

In this section we will deal with two measurement systems: the U.S. customary gravitational system, and the SI absolute system, which uses metric units. Some commonly measured mechanical properties are defined here.

**Acceleration** Time rate of change of velocity. It is thus a time rate of change of a time rate of change.

**Components** The single individual forces that are the equivalent of the resultant.

**Concurrent** Forces whose lines of action or directions pass through a common point, or meet at a common point.

**Coplanar** Forces all in the same plane.

**Couple** A system of nonconcurrent forces tending to produce rotation about an axis.

**Dynamics** Concerns bodies not in equilibrium; i.e., not at rest or with nonuniform motion.

**Inertia** A property of matter that causes it to resist change in its motion or its state of rest.

**Kinematics** A branch of dynamics that deals only with motions of bodies, not considering forces. It can thus be described as the geometry of motion.

**Kinetics** A branch of dynamics that relates the actions of forces to the resulting motions of bodies.

**Mass** The measure of the inertia of a body. Mass also initiates gravitational attraction.

**Mechanics** The science that deals with forces and their effects on bodies at rest or in motion.

**Moment** A force with respect to a point, which is the product of the force and the perpendicular distance from the point to the direction of the force (line of force). The distance is termed the *moment arm.*

**Nonconcurrent** Forces whose lines of action do not meet at a common point.

**Noncoplanar** Forces that are not in the same plane.

**Power** Amount of work per unit of time.

**Resultant** In a system of forces, the resultant is the single-force equivalent of the entire system. When the resultant of a system of forces is zero, the system is in equilibrium.

**Statics** Concerns bodies in equilibrium; i.e., at rest or with uniform motion.

**Torque** The measurement of the tendency of a force to rotate a body, on which it acts, about an axis. Also called *moment.*

**Velocity** Time rate of change of distance.

**Work** The product of a force times the distance through which the force acts.

The basic units of the two measurement systems are given in Table 2.1 (see the end of this chapter for conversions among other units). Note that in the two systems shown, the U.S. customary system derives mass from force, whereas the SI system derives force from mass. The *slug* is the unit of mass in the U.S. customary system and is defined as $M = W/g$ or weight per 32.2 feet per second per second (ft/sec$^2$). The mass in the SI system is the kilogram (kg). The standard values for gravitational acceleration in the two systems are 32.2 ft/sec$^2$ and 9.81 meters per second per second (m/sec$^2$) at sea level.

**TABLE 2.1** Basic Units of Mechanics

| U.S. customary system | (SI) Système international |
|---|---|
| Length—foot | Length—meter |
| Time—second | Time—second |
| Force—pound-force | Force—newton, kg-m/sec$^2$ |
| Weight—pound | Weight—newton |
| Mass—slug, lb-sec$^2$/ft | Mass—kilogram |
| Work or energy—foot-pounds | Work or energy—newton-meters, kg-m$^2$/sec$^2$, joule |
| Power—foot-pounds per minute; 1 horsepower, 550 ft-lb/sec | Power—watt, joules per second; newton-meters per second, kg-m$^2$/sec$^3$ |
| Torque (moment)—pounds-feet | Torque (moment)—newton-meters |
| Velocity—feet per minute, feet per second | Velocity—meters per minute, meters per second |
| Acceleration—unit length per sec$^2$ | Acceleration—unit length per sec$^2$ |
| Stress—pounds per inch$^2$; 1 pound per inch$^2$ = 6894.757 pascals; 1 lb/in$^2$ = 6.894757 kPa | Stress—pascals; N/m$^2$; 1 pascal = 0.000145 lb/in$^2$; 1 kPa = 0.1450377 lb/in$^2$ |

***Definitions of Force.*** A *newton* (N) is that force that will impart an acceleration of 9.81 m/sec$^2$ to a mass of 1 kilogram. A *pound* (lb) is that force that will impart an acceleration of 1 ft/sec$^2$ to a 32.2-lb (1-slug) mass. This force is also termed *poundal.*

The variation of $g$ with altitude, determined by the gravitational law (Meriam and Kraige 1986), is

$$g = g_0 \frac{r^2}{(r+h)^2}$$

where $r$ = Earth's mean radius, m
$g_0$ = 9.81 m/sec$^2$ at sea level
$h$ = altitude, m

The mass and mean radius of Earth have been experimentally determined to be $5.976 \times 10^{24}$ kg and $6.371 \times 10^6$ m, respectively. The value of the gravitational constant has been determined at $6.673 \times 10^{-11}$ m$^3$/kg · sec$^2$.

Remember that in the U.S. customary system

$$W/g = \text{mass} = 1 \text{ slug} = 1 \text{ lb-sec}^2/\text{ft}$$

$$1 \text{ lb} = 1 \text{ slug-ft/sec}^2$$

In other words, the term *pound* is used as a label for both a unit of force and a unit of mass.

In the SI system the kilogram is normally used as a unit of mass. Where a mass is actually being used as weight, it must be stated as $Mg$. Thus, a mass of 150 kg used as a force becomes 150 (9.81) = 1471.5 N. Note that the load or force from the weight of a mass of $M$ kilograms is $Mg$ newtons, where $g = 9.81$ m/sec$^2$ (i.e., the load is the product of the mass in kilograms and the acceleration due to gravity).

***Scalars and Vectors in Mechanics.*** A *scalar* quantity has one specific property, such as time, volume, and density. A *vector* has two specific properties, such as force and direction, velocity and direction, and moment and direction. Put another way, a scalar has only magnitude, whereas a vector has both magnitude and direction.

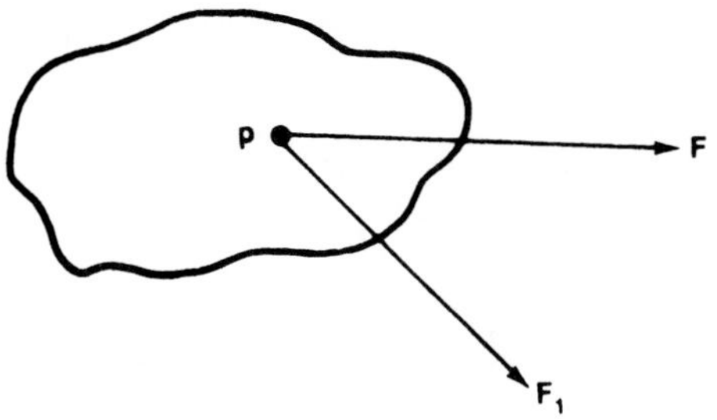

**FIGURE 2.1** Concurrent, coplanar forces.

## 2.2 STATICS

A force has three characteristics:

- Magnitude
- Point of application
- Direction

*Concurrent, coplanar forces* are represented in Fig. 2.1. The points of application of the forces are the same point *p*.

The resultant of any system of concurrent forces that are not in equilibrium is a single force.

*Nonconcurrent, coplanar forces* are represented in Fig. 2.2. The two are applied at separate points.

*Concurrent, noncoplanar forces* are represented in Fig. 2.3. *Nonconcurrent, noncoplanar forces* are represented in Fig. 2.4.

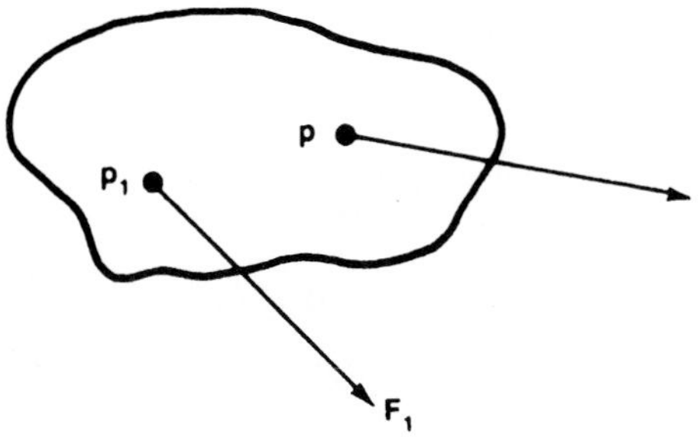

**FIGURE 2.2** Nonconcurrent, coplanar forces.

### 2.2.1 Resolution of Forces at 90°

The resolution of such forces that are coplanar and concurrent is illustrated in Fig. 2.5. $R$ denotes the resultant.

$$R_x = R\cos\phi \qquad R_y = R\sin\phi$$

$$R = \sqrt{R_x^2 + R_y^2}$$

$$\phi = \arctan\frac{R_y}{R_x} = \tan^{-1}\frac{R_y}{R_x}$$

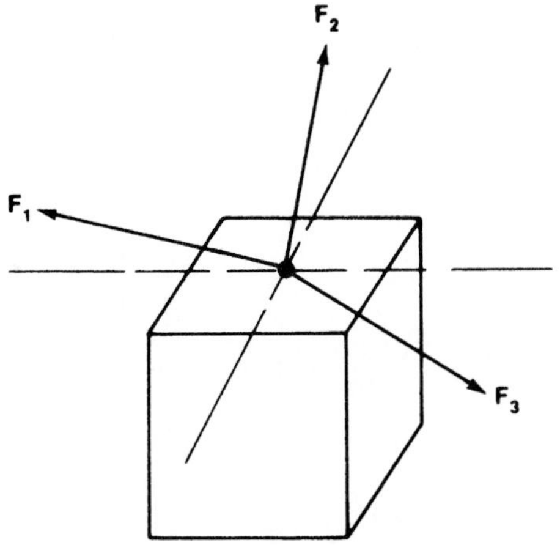

**FIGURE 2.3** Concurrent, noncoplanar forces.

### 2.2.2 Resolution of Forces Not at 90°

The case of coplanar, concurrent forces is illustrated in Fig. 2.6.

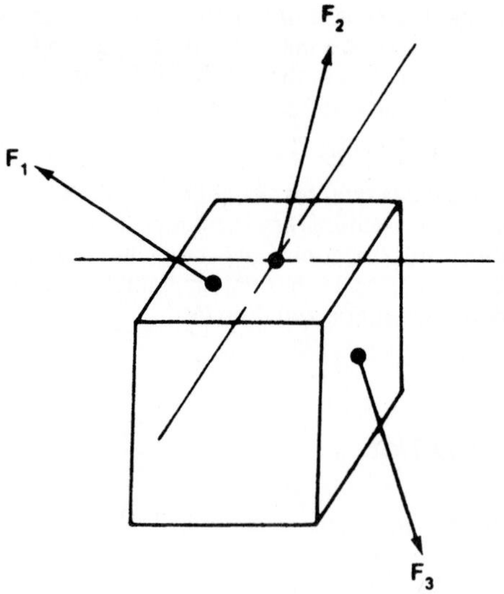


**FIGURE 2.4** Nonconcurrent, noncoplanar forces.

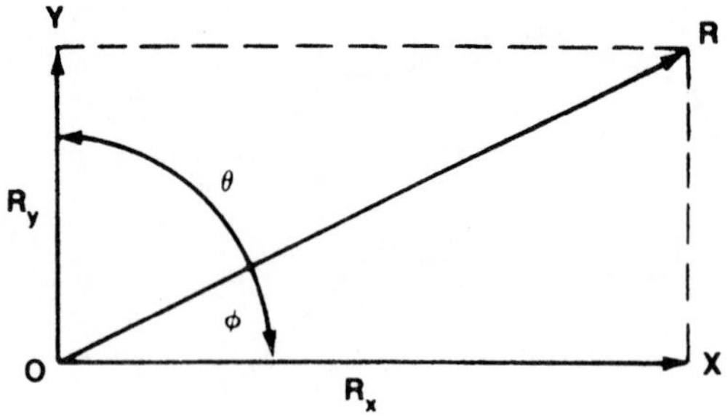


**FIGURE 2.5** Resolution of forces at 90°.

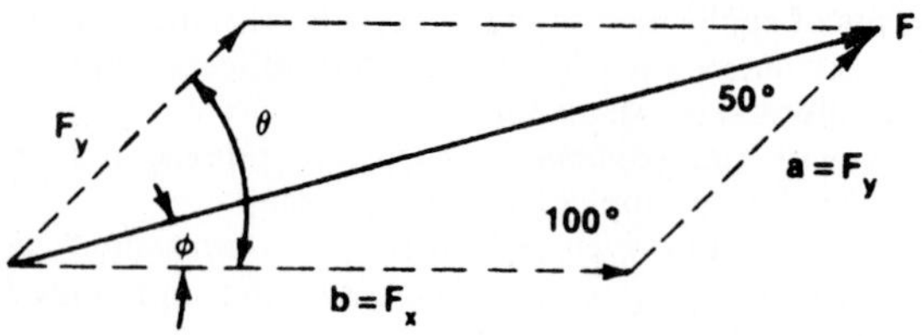


**FIGURE 2.6** Resolution of forces not at 90°.

Solving for $F_x$, if $\theta = 80$, $\phi = 30$, $b = F_x$, and $F =$ 150 lb, by the law of sines

$$\frac{\sin 50}{F_x} = \frac{\sin 100}{150}$$

$$0.9848\ F_x = 150\ (0.7660)$$

$$F_x = 116.673 \text{ lb}$$

Solving for $F_y$, if $a = F_y$, and $F = 150$ lb, then

$$\frac{\phi}{a} = \frac{100}{F}$$

and

$$\frac{\sin 30°}{F_y} = \frac{\sin 100°}{150}$$

by the law of sines

$$\frac{\sin 50°}{F_x} = \frac{\sin 100°}{150} \qquad \frac{0.5}{F_y} = \frac{0.9848}{150}$$

$$0.9848\ F_x = 150\ (0.7660) \qquad 0.9848\ F_y = 75$$

$$F_x = 116.673 \text{ lb} \qquad F_y = 76.158 \text{ lb}$$

### 2.2.3 Resolution of Noncoplanar Forces

The case in which component axes are at 90° and concurrent is shown in Fig. 2.7. The components are

$$F_x = F \cos \phi_x \qquad F_y = F \cos \phi_y \qquad F_z = F \cos \phi_z$$

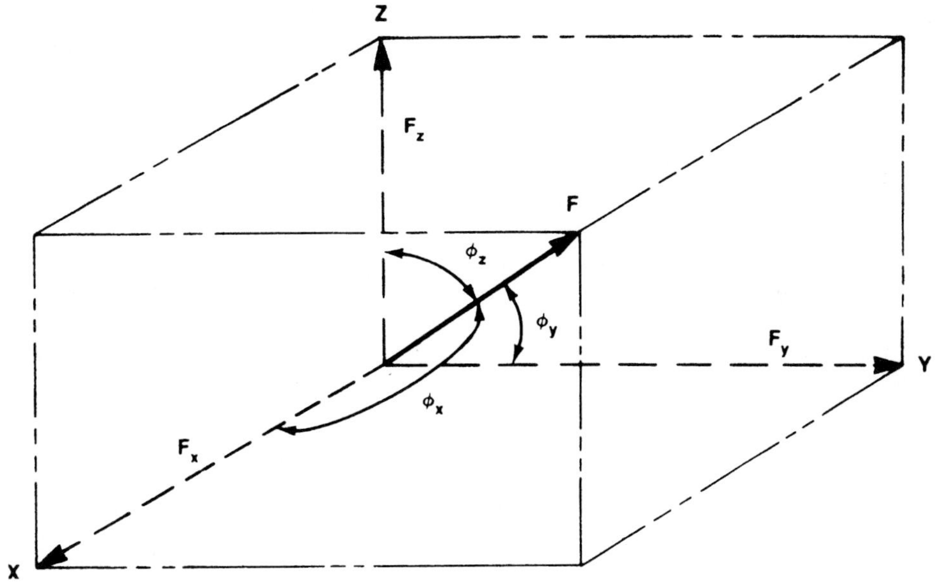


**FIGURE 2.7** Resolution of noncoplanar forces.

The resultant is

$$F = \sqrt{F_x^2 + F_y^2 + F_z^2}$$

Cos $\phi_x$, cos $\phi_y$, and cos $\phi_z$ are called the *direction cosines* of *F*.

When the component axes are not at 90°, standard trigonometric solutions are indicated.

### 2.2.4 Coplanar Force System

A nonconcurrent case is illustrated in Fig. 2.8.

*Example.* Find the resultants of the forces about the *X* and *Y* axes, and then find the resultant of the system.

$$R_x = \Sigma F_x = -20 - 40 \cos 30 + 45 \cos 20 + 40 = 27.65$$

$$R_y = \Sigma F_y = 40 \sin 30 + 25 + 45 \sin 20 = 60.39$$

$$R = \sqrt{R_x^2 + R_y^2} = \sqrt{(27.65)^2 + (60.39)^2} = 66.42$$

$$\phi = \arctan \frac{R_y}{R_x} = \arctan \frac{60.39}{27.65} = 65.4°$$

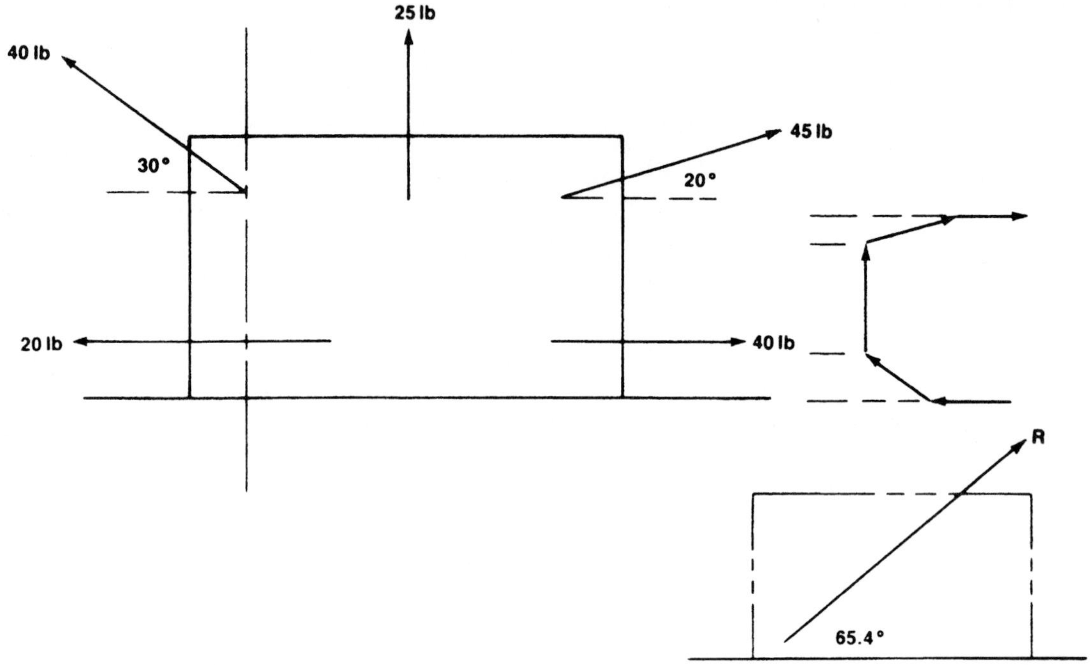


**FIGURE 2.8** Coplanar force system, nonconcurrent.

Obtain the position of the line of action from the principle of moments:

$$RD = \Sigma M_0$$

$$66.42D = \Sigma M_0$$

$$D = \frac{\Sigma M_0}{66.42}$$

which is the algebraic sum of moments. The resultant of a coplanar system of forces is either a force or a couple.

### 2.2.5 Noncoplanar, Concurrent Force System

Three-dimensional systems are best analyzed with the aid of graphical representation, as in Fig. 2.9.

The resultant vector $R$ is obtained by adding the forces $F_1, \ldots, F_n$ into a space polygon. The three *rectangular* components of the resultant vector $R$ are the algebraic sums of the corresponding components of the initial forces $F_1, \ldots, F_n$. So

$$R_x = \Sigma F_x \qquad R_y = \Sigma F_y \qquad R_z = \Sigma F_z$$

Then

$$R = \sqrt{(\Sigma F_z)^2 + (\Sigma F_y)^2 + (\Sigma F_z)^2}$$

With the angles associated with R equal to $\theta_x$, $\theta_y$, and $\theta_z$, the direction cosines of $R$ are

$$\cos \phi_x = \frac{\Sigma F_x}{R} \qquad \cos \phi_y = \frac{\Sigma F_y}{R} \qquad \cos \phi_z = \frac{\Sigma F_z}{R}$$

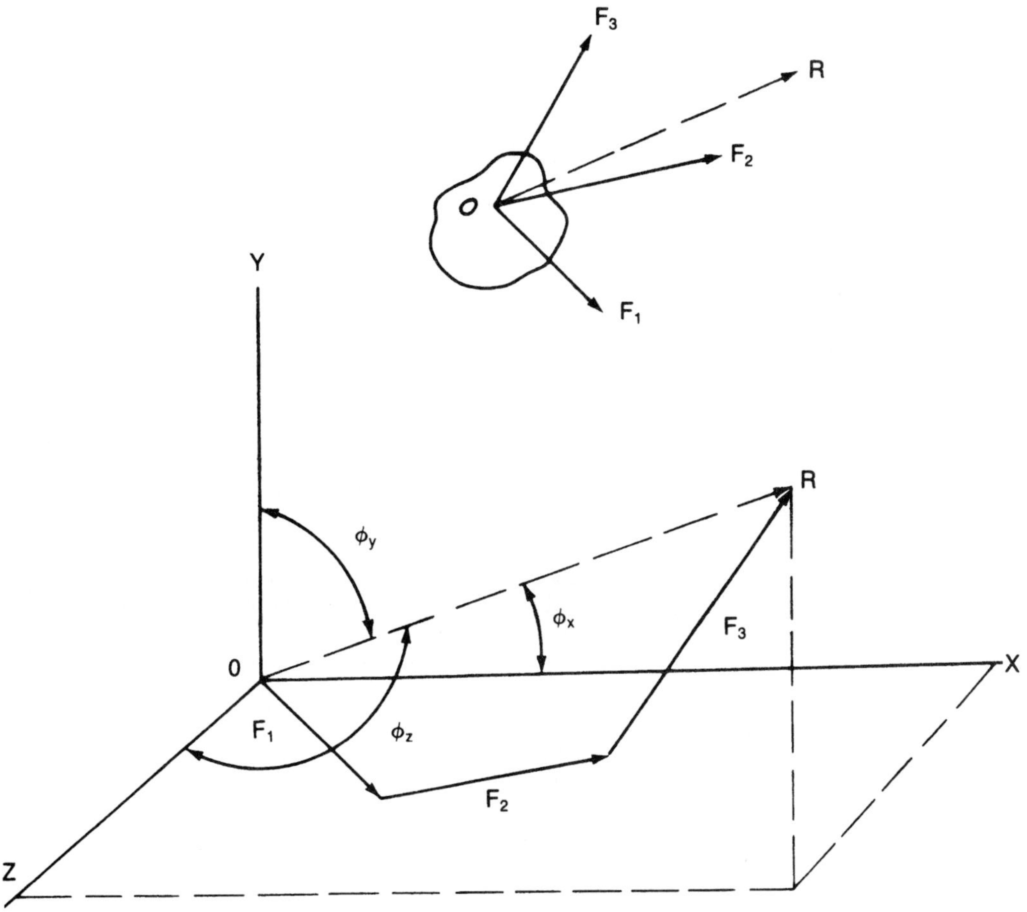


**FIGURE 2.9** Noncoplanar force system, concurrent.

The system *couples* are now determined from $M = Fd$, where $d$ is moment-arm length. The couple vectors $M_1, \ldots, M_n$ must be resolved into the three rectangular components $M_x$, $M_y$, and $M_z$. The resultant couple is

$$M = \sqrt{M_x^2 + M_y^2 + M_z^2}$$

The direction of the couples is then specified by the direction cosines: $\cos \phi_x = M_x/M$, etc.

### 2.2.6 Graphical Resolution of Coplanar Forces

*Coplanar forces not having the same application point:* These are diagrammed in Fig. 2.10. Extend the force lines $l$ and $m$ until they intersect; then form the force parallelogram with $A$ as the point of application. Measure the resultant with a scale to find its value.

*Coplanar polygon:* Consult Fig. 2.11. Starting from the end of $F_1$, transfer the force lines parallel to the original direction of each individual force until all forces are transferred end to end. The distance from the end of the last force to the beginning point is the resultant.

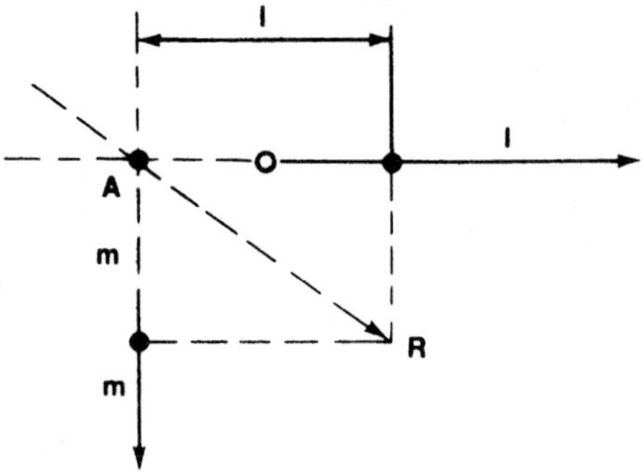

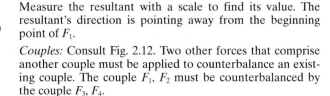

**FIGURE 2.10** Coplanar forces with different application points.

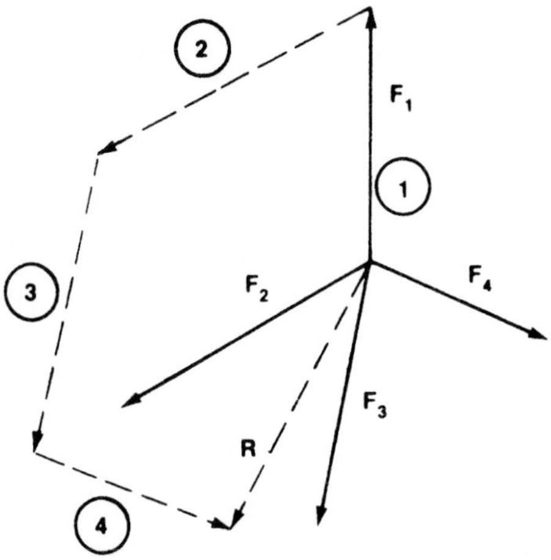


**FIGURE 2.11** Coplanar polygon of forces.

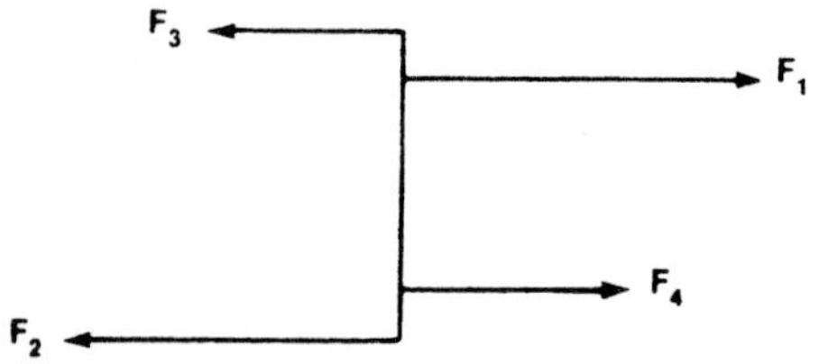


**FIGURE 2.12** A couple.

Measure the resultant with a scale to find its value. The resultant's direction is pointing away from the beginning point of $F_1$.

*Couples:* Consult Fig. 2.12. Two other forces that comprise another couple must be applied to counterbalance an existing couple. The couple $F_1$, $F_2$ must be counterbalanced by the couple $F_3$, $F_4$.

*Parallel forces:* Fig. 2.13. The resultant runs between the component forces and (5,6) = (1,2) + (3,4). The resultant, $F_3$, is equal to $F_2 + F_1$. Also

$$\frac{F_2}{F_1} = \frac{a}{b}$$

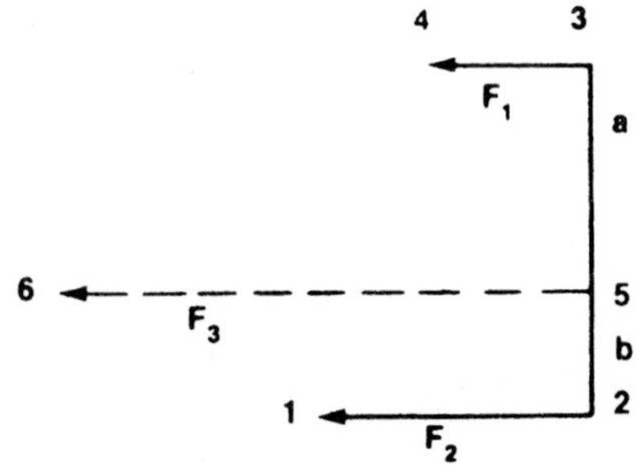


**FIGURE 2.13** Parallel forces.

*Parallel forces acting in opposite directions:* Consult Fig. 2.14. The resultant lies outside the two component forces and in the direction of the larger force: (5,6) = (1,2) – (3,4). The resultant, $F_3$, is equal to $F_2 - F_1$, and $F_2 > F_1$.

$$\frac{F_1}{F_2} = \frac{a}{a+b}$$

The *moment* of a couple is the measure of the tendency of a couple to produce rotation. It is the product of one of the forces in the couple times the distance between the forces. A couple may be represented by a vector in the direction of the axis about which it rotates. The magnitude of the couple and the direction of this vector can be compared with those of a right-hand screw advancing as it is being rotated by the couple.

When the axis is viewed as a point, a clockwise couple will advance the vector away from your line of sight (and into the plane of the drawing). A counterclockwise couple will advance the vector toward your line of sight (and out of the plane of the drawing). The point on which a couple acts is called the *origin,* or the center of moments.

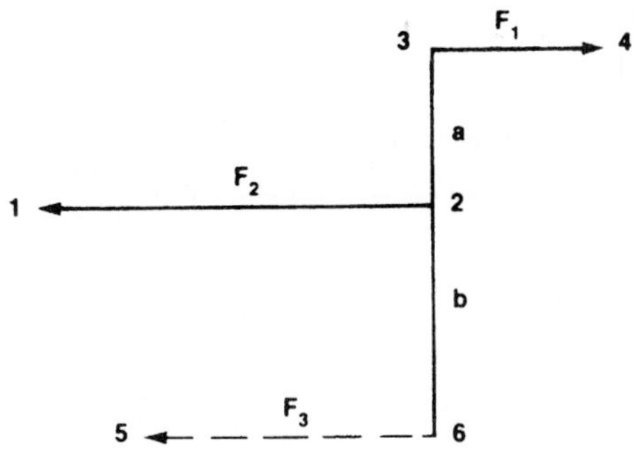


FIGURE 2.14 Forces in opposite directions.

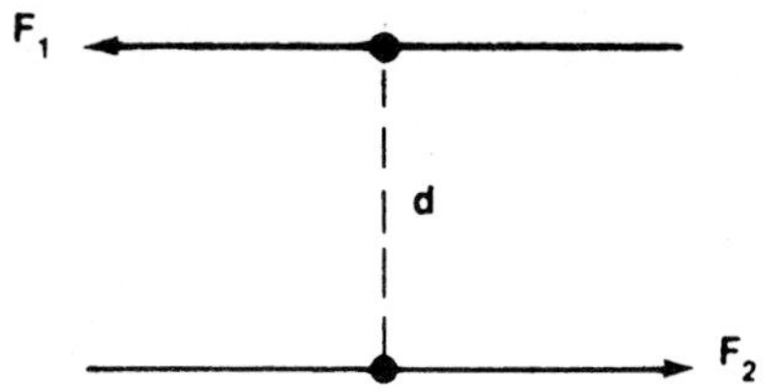


FIGURE 2.15 A couple.

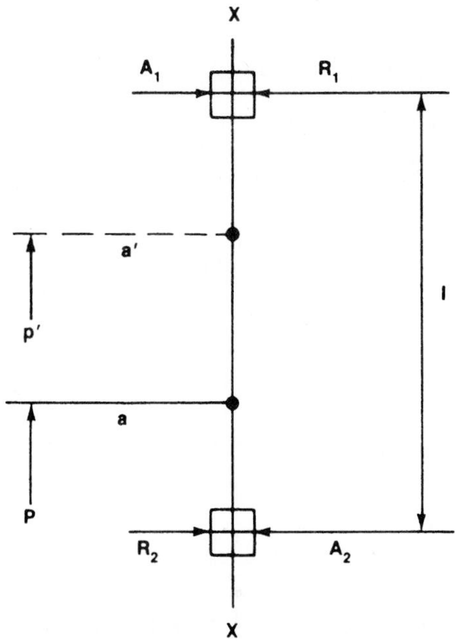


FIGURE 2.16 Force $p$ acting on arm $a$.

Another way to look at a couple is as two forces of equal magnitude, acting in parallel but opposite directions. The couple shown in Fig. 2.15 tends to produce rotation but has no resultant.

In Fig. 2.16, force $P$ produces equal pressures at $A_1$ and $A_2$, irrespective of the position of $P$ as it acts on arm $a$ along the axis $X$–$X$. (See, e.g., $P'$.) Therefore, $P \times a = R \times l$, or $R = (P \times a)/l$. The reactions of $A_1$, $A_2$ are $R_1$, $R_2$.

### 2.2.7 Systems in Equilibrium

*Equilibrium* is a condition where the resultant of all the forces acting on a body is equal to zero. It is a balance of both forces and moments.

If a body is statically *indeterminate,* the equilibrium equations are not valid. You must be able to recognize whenever redundant supports are contained in an equilibrium system. Removing the redundant support or supports will bring the system back to the point where it is statically determinate. Then you can do calculations using the equilibrium equations.

***Equilibrium of Parallel Forces.*** The algebraic sum of the moments of separate forces about any point must equal zero: $F_x = 0$, $F_y = 0$, and $M_0 = 0$.

Referring to Fig. 2.17, if we take moments about point $A$, we obtain

$$MA = 0 = (RA \times 0) + (2 \times 3.5) + (3 \times 14) + (25 \times 29) - (F \times 3.5)$$

$$0 = 0 + 7 + 42 + 725 - 3.5F$$

$$3.5F = 774$$

$$F = 221.14 \text{ lb}$$

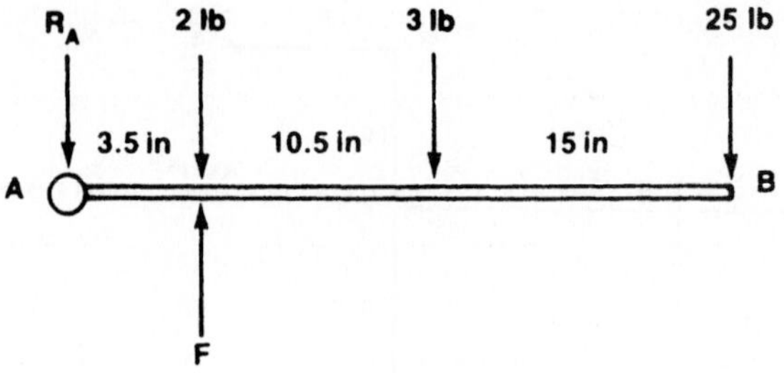


**FIGURE 2.17** A system in equilibrium.

Referring to Fig. 2.18, to find the reactions at point $A$ and point $B$ we proceed as follows:

$$F_x = 0$$

$$F_y = 0 = R_A + R_B - (40 + 60 + 100 + 50 + 90 + 80)$$

$$M_A = 0 = (6 \times 60) + (10 \times 100) + (14 \times 50) + (18 \times 90) + (20 \times 80) - 18R_B$$

$$18R_B = 5280$$

$$R_B = 293.33 \text{ lb}$$

$$M_B = 0$$

$$= (2 \times 80) - (50 \times 4) - (100 \times 8) - (60 \times 12) - (40 \times 18) + 18R_A - 18R_A$$

$$= -2280$$

$$R_A = 126.67 \text{ lb}$$

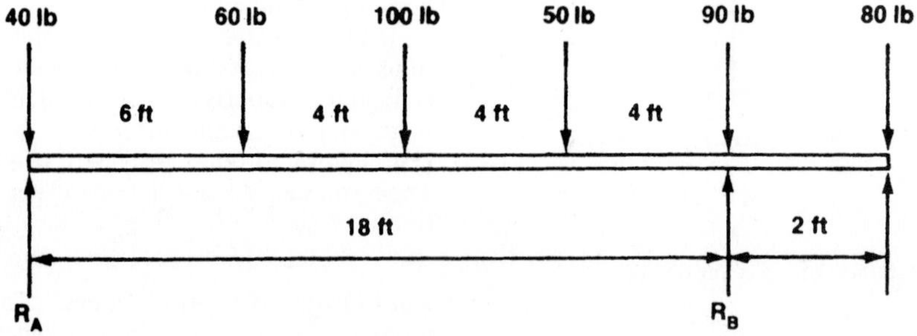


**FIGURE 2.18** Finding the reactions at $A$ and $B$.

Check

$$F_y = 126.67 + 293.33 - 420 = 0$$

***Equilibrium of Nonconcurrent Forces.*** Referring to Fig. 2.19, if tension $T$ is placed on a cable, let $HA$ equal horizontal reaction and $VA$ equal vertical reaction.

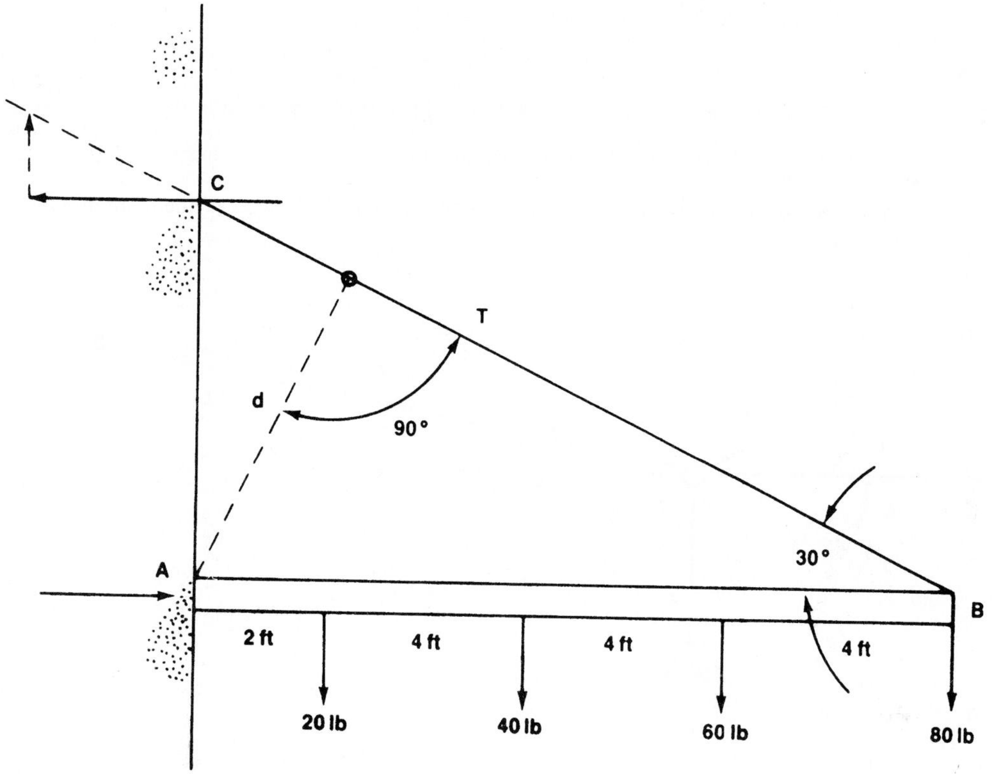


**FIGURE 2.19** Nonconcurrent force system.

$$\sin 30° = \frac{d}{14}$$

$$d = 14 \sin 30° = T$$

$$T = 14 \sin 30°$$

$$F_x = 0 = HA + 0 + 0 + 0 + 0 + 0 - T \cos 30°$$

Therefore $HA = T \cos 30°$.

$$F_y = 0 = 0 + VA - 20 - 40 - 60 - 80 + T \sin 30°$$

Therefore $VA = T \sin 30° - 200$.

$$M_A = 0$$

$$= (HA \times 0) + (VA \times 0) + (2 \times 20) + (6 \times 40) + (10 \times 60) + (14 \times 80) - (T \times 14 \sin 30°)$$

$$T \times 14 \sin 30° = 2000$$

$$T = 285.71 \text{ lb} \qquad \text{(tension in cable)}$$

The horizontal and vertical reactions are calculated as

$$HA = T \cos 30° = 285.71\ (0.866) = 247.42$$

$$VA = T \sin 30° - 200 = 285.71\ (0.5) - 200 = -57.15$$

The negative sign of $VA$ indicates an upward direction.

***Equilibrium of Concurrent Forces.*** The solution for Fig. 2.20 involves linear simultaneous equations.

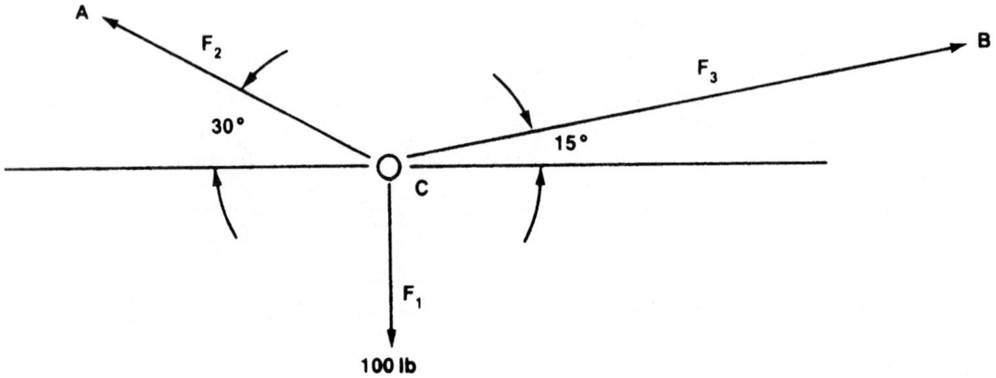


**FIGURE 2.20** Determining $F_2$ and $F_3$.

Find $F_2$ and $F_3$.

$$F_x = 0 = B(\cos 15°) - A(\cos 30°) + 0$$

$$F_y = 0 = B(\sin 15°) + A(\sin 30°) - 100$$

These relations yield linear simultaneous equations

$$0.966B - 0.866A = 0$$

$$0.259B + 0.500A = 100$$

Eliminate $A$ by multiplying the upper equation by 0.500 and the lower equation by 0.866:

$$0.483B - 0.433A = 0$$

$$\underline{0.224B + 0.433A = 86.6}$$

$$0.707B \qquad = 86.6$$

$$B = 122.49 \text{ lb}$$

Next, solve either of the original equations for $A$ in terms of the calculated $B$.

$$0.966B - 0.866A = 0$$

$$0.966(122.49) - 0.866A = 0$$

$$A = 136.64 \text{ lb}$$

You may also solve the set of linear equations for $A$ by eliminating $B$. Use whichever method is more convenient.

Simultaneous equations involving two or more unknown forces can be solved by means of determinants. (See Chap. 1 for more on this method.)

Note that in all the equations and examples shown in the preceding statics section, the meter can be substituted for the foot, and the newton substituted for pound-force. The answers, in the SI system units, are then rendered as newton-meters.

## 2.3 DYNAMICS

### 2.3.1 Moments of Inertia

These moments derive from the product of elementary areas or masses and the squares of their distances from a reference axis. The location of the reference axis determines the moment of inertia. The moment has a minimum value when the reference axis passes through the center of gravity of the area or mass.

Moments of inertia of plane areas are generally given as $I$. When areas are in square inches, $I$ is given in in$^4$, which is read as "inches squared, square inches," or inches to the fourth power. (In SI units, $I$ is given in mm$^4$, or m$^4$.)

### 2.3.2 Polar Moments of Inertia

Polar moments of inertia of plane areas $I_{pa}$ are defined by the axis being at right angles (or perpendicular) to the plane of the area. When areas are measured in square inches, $I_{pa}$ is given in in$^4$. The polar moment of inertia is derived from the plane area when $I$ is known for both the $X$–$X$ and $Y$–$Y$ axes. Summarizing:

$$I_{pa} = I_{xx} + I_{yy}$$

When SI units are used, equations for these moments remain valid for values given in millimeters or meters. Then $I$ is measured in mm$^4$ or m$^4$, and $I_{pa}$ in mm$^4$ and m$^4$.

### 2.3.3 Polar Moments of Inertia of Masses

Polar moments of inertia of masses, designated as $I_{pm}$, are used in dynamics equations involving rotary motion. In the U.S. customary system, the units of $I_{pm}$ are foot-pound-second$^2$ (squared), or slug-foot$^2$. (One slug is equal to one pound-second$^2$ per foot.) In the SI system, units are given in terms of kilogram-meters$^2$.

The $I_{pm}$ of a body may be calculated from

$$I_{pm} = \frac{dl}{g} I_{pa}$$

This equation applies to a part having a constant cross section, where $d$ is density in pounds per cubic foot, $l$ is the length of the part in feet, $g$ is 32.2 feet per square second, and $I_{pa}$ is the polar moment of inertia of the cross-sectional area in feet to the fourth power. (In the SI system, this equation becomes $I_{pm} = dlI_{pa}$, where $d$ is density in kilograms per cubic meter, $l$ is the length of part in meters, $I_{pa}$ is meters to the fourth power, and $I_{pm}$ is given in kilogram-meters$^2$.)

### 2.3.4 Moments of Complex Areas and Masses

Moments of inertia for the case illustrated in Fig. 2.21 may be derived using the following procedures. First

$$I_{pm} = I_{pm1} + 2(I_{pm2}) - I_{pm3}$$

(with reference to the $Z$–$Z$ axis).

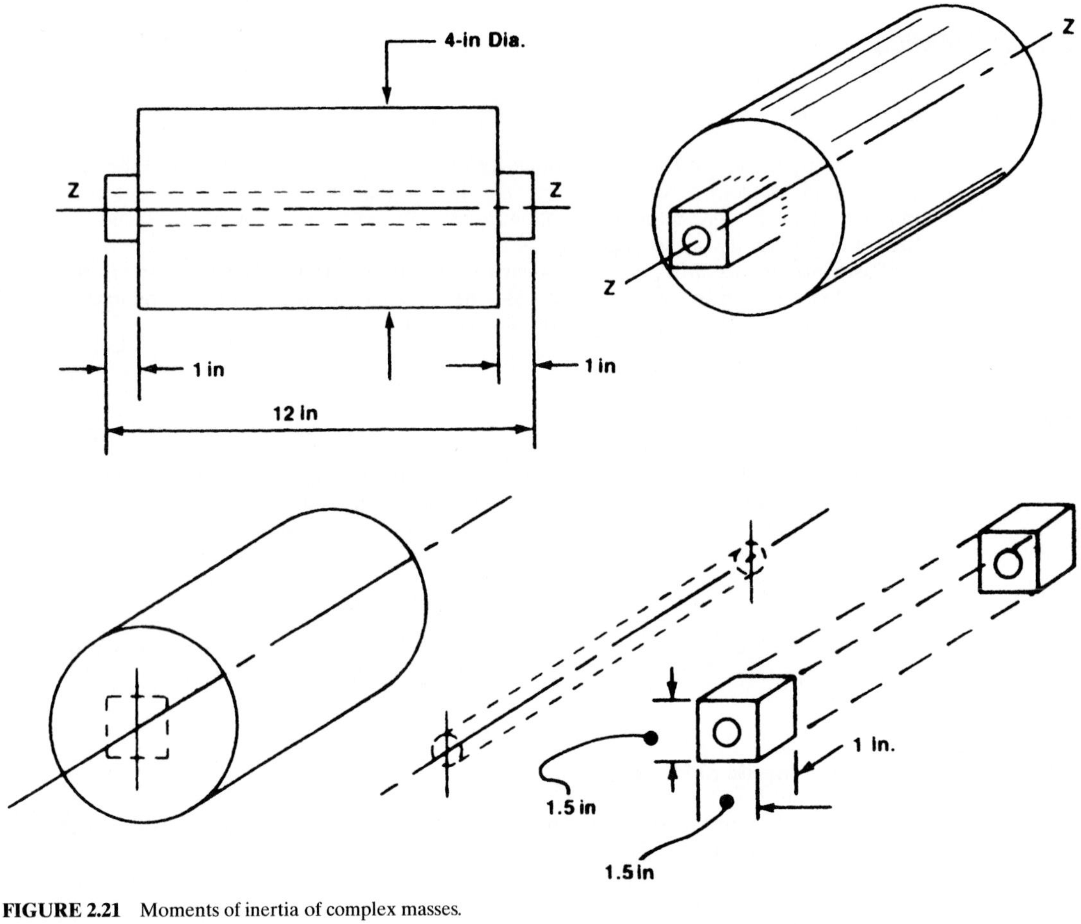


**FIGURE 2.21** Moments of inertia of complex masses.

Referring to the figure, for the cylinder

$$I_{pm1} = \frac{dl}{g} I_{pa1}$$

and for the two cubes

$$I_{pm2} = 2\left(\frac{dl}{g}\right) I_{pa2}$$

For a mass representing the hole

$$I_{pm3} = \frac{dl}{g} I_{pa3}$$

where $I_{pm3}$ and $I_{pa3}$ represent the polar moments of the masses and the polar moments of the plane areas, respectively.

### 2.3.5 Transfer of Axes

When the complex part has masses that are not on the principal, $Z$–$Z$, axis, the $I_{pm}$ is calculated for its shift off of the $Z$–$Z$ axis as follows (see Fig. 2.22).

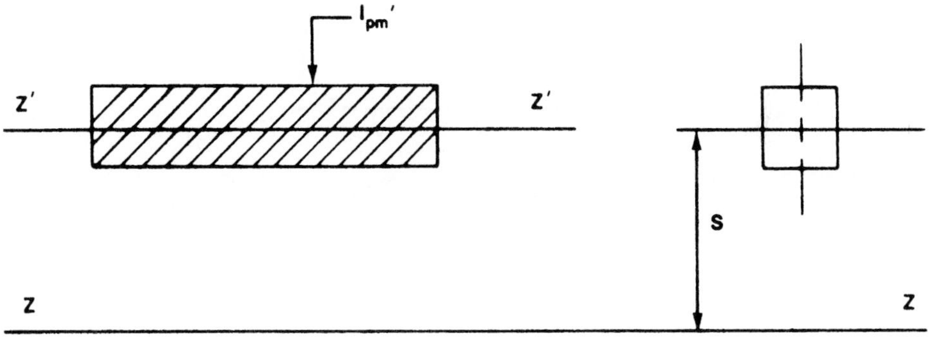


**FIGURE 2.22** Transferring the axis.

About $Z$–$Z$

$$I_{pm} = I'_{pm} + s^2 M$$

where $I_{pm}$ and $I'_{pm'}$ represent the final polar moment of the mass of the part that is shifted, and the polar moment of the offset mass, respectively.

Note that the $I'_{pm}$ can represent an additional member of the main part, in which case it is added to the $I_{pm}$, or it may represent a hole and be subtracted from the $I_{pm}$ of the main part. The $I_{pm'}$ is defined with respect to the center of gravity of the offset mass.

The same procedure is applied to respective calculations of $I$ and $I_{pa}$, the moment and polar moment of inertia of a plane area. That is

$$I_{pa} = I_{pa'} + s^2 A \text{ (polar)}$$

$$I = I' + s^2 A \text{ (plane)}$$

The $I_{pa'}$ and $I'$ are defined with respect to the center of gravity of $A$, the area of the section. The symbol $s$ denotes distance between $Z$–$Z$ and shifted axis $Z'$–$Z'$.

Additional equations for calculating the $I$ and $I_{pa}$ of standard plane areas are contained in Chap. 5.

### 2.3.6 Approximate Calculations for *I*

For the case shown in Fig. 2.23,

$$I_{xx} = (d_1^2 A_4) + (d_2^2 A_3) + (d_3^2 A_2) + (d_4^2 A_1) + (d_5^2 A_5) + (d_6^2 A_6) + (d_7^2 A_7)$$

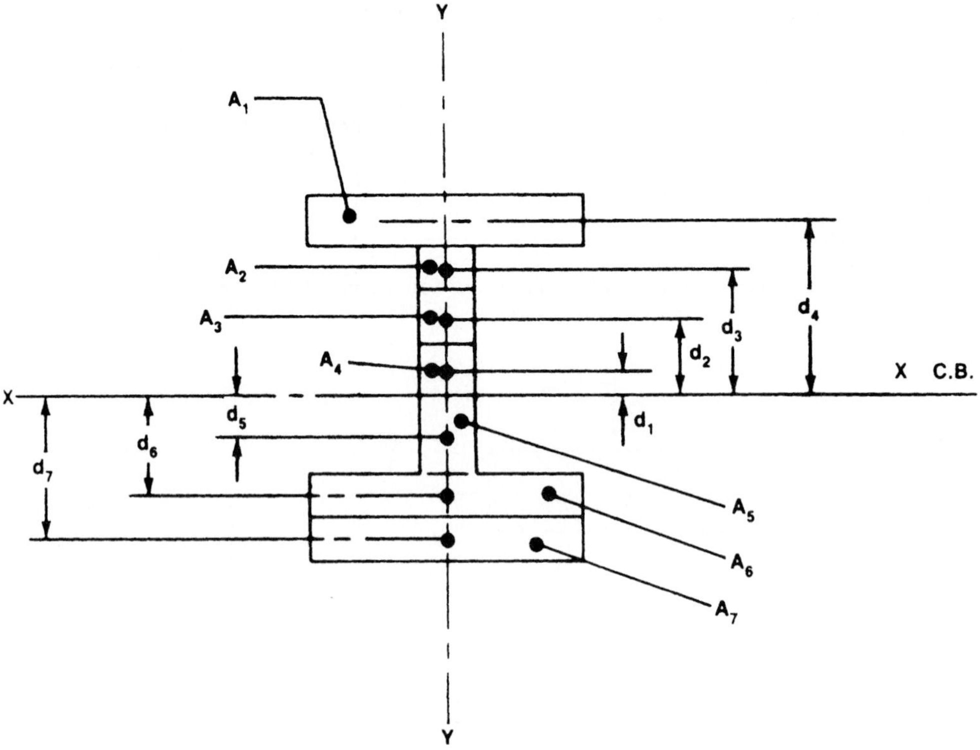


**FIGURE 2.23** A typical structural shape.

Select $A_1$ through $A_7$, elementary areas, according to the shape of the section.

The distances $d_1$ through $d_7$ are measured from the center of gravity of each elementary area to the $X$–$X$ axis, which must pass through the center of gravity of the whole area or section. These distances are squared.

The more elementary areas you divide the section into, the more accurate is the calculated moment of inertia.

To determine the polar moment of inertia of the area shown in Fig. 2.23, draw axis $Y$–$Y$ at right angles to $X$–$X$ and through the center of gravity. Proceed to take the moment of inertia about $Y$–$Y$ in a manner similar to the way it was done at $X$–$X$. Then, the polar moment of inertia of the area shown in Fig. 2.23 is

$$I_{pa} = I_{xx} + I_{yy}$$

### 2.3.7 Moments of Plane Areas by Integration

The example in Fig. 2.24 shows a rectangular area ($bd$), for which moments of inertia can be integrated as

$$I_{xx} = \int_{s=0}^{s=d} s^2 dA = \int_0^d s^2(b \cdot ds) = \frac{1}{3}\,(bs^3)_0^d = \frac{1}{3}\,bd^3$$

Referring to Fig. 2.25, by the fundamental theorem on definite integrals:

$$I_{xx} = \int_{s=a}^{s=b} s^2 dA$$

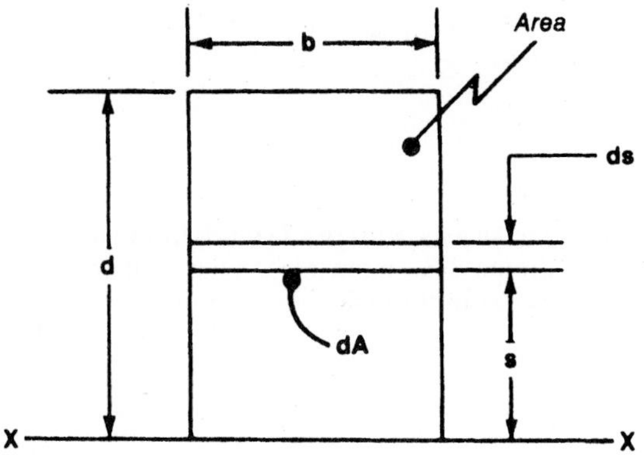


FIGURE 2.24 *I* by integration.

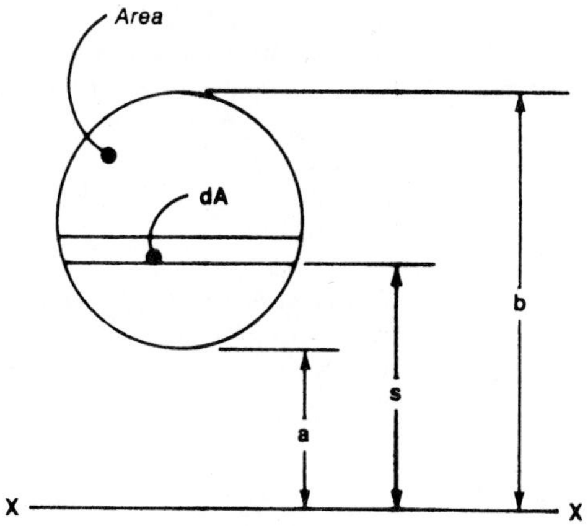


FIGURE 2.25 *I* by integration.

### 2.3.8 Section Modulus

The section modulus of an area is equal to the moment of inertia of the plane area, divided by the distance from the center of gravity of the area to the most remote "fiber" (or more remote point) (Fig. 2.26).

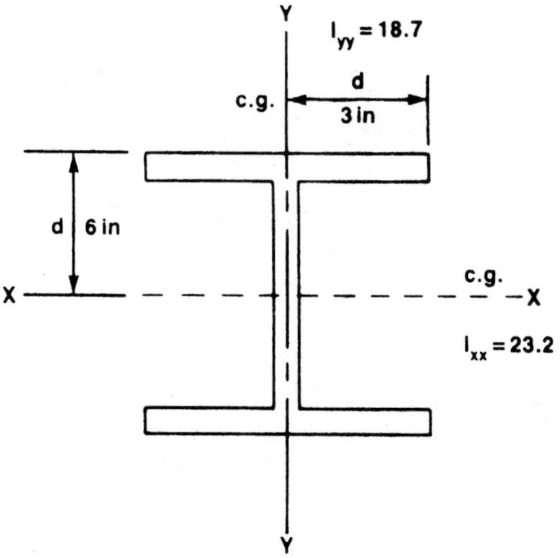


FIGURE 2.26 Finding *Z* or *S* of structural shapes.

In the example, the section modulus about axis *X–X*, $Z_{xx}$ and the section modulus about axis *Y–Y*, $Z_{yy}$ are

$$Z_{xx} = \frac{I_{xx}}{d} = \frac{23.2}{6} = 3.867 \text{ in}^3$$

$$Z_{yy} = \frac{I_{yy}}{d} = \frac{18.7}{3} = \frac{6.233 \text{ in}^3}{10.100 \text{ in}^3} \quad \text{(Sum is polar modulus.)}$$

***Polar Section Modulus.*** The polar section modulus of an area is equal to the sum (in this example, 10.100 $in^3$) of the section moduli about both axes, *X–X* and *Y–Y*:

$$Z_p = Z_{xx} + Z_{yy}$$

### 2.3.9 Radius of Gyration

***Radius of Gyration of a Mass.*** With reference to an axis, $R_m$ is the distance from the axis at which the mass may be considered as concentrated. The moment of inertia of the mass remains unchanged. It is minimum when the axis passes through the center of gravity of the mass.

$$R_m = \sqrt{\frac{I_{pm}g}{W}} \quad \text{and} \quad I_{pm} = \frac{WR_m^2}{g}$$

where $W$ = weight of the body (mass), lb
$I_{pm}$ = polar moment of inertia of the mass, ft · lb · $sec^2$
$R_m$ = radius of gyration, ft
$g$ = 32.2 ft/$sec^2$

In the SI system, $R_m$ is given in meters, when $I_{pm}$ = polar moment of mass, kg · $m^2$; and $M$ = mass, kg. The relationships become

$$R_m = \sqrt{\frac{I_{pm}}{M}} \quad \text{and} \quad I_{pm} = MR_m^2$$

***Radius of Gyration of an Area.*** To find $R_a$ in feet, where $A$ = area of section, $ft^2$; $I$ = moment of inertia of plane area, $ft^4$

$$R_a = \sqrt{\frac{I}{A}}$$

This may also be given in inches consistently. (In the SI system, $I$ is in $meters^4$, $A$ is in $meters^2$, and $R_a$ is in meters.)

***Transfer Axis of Radius of Gyration.*** When the axis of the radius of gyration passes through the center of gravity, the radius of gyration is minimum.

If $R_m$ is the radius of gyration of a body in which the axis passes through the center of gravity, then the radius of gyration $R'_m$ with respect to another parallel axis is given by

$$R'_m = \sqrt{R_m + d^2}$$

where $d$ = distance between the two axes, in inches or feet. (In the SI system, $R_m$ and $d$ are given in meters.)

### 2.3.10 Center of Percussion

The resultant of all forces acting on a body that rotates about a fixed axis passes through a point called the *center of percussion.*

When the radius of gyration of the mass and its center of gravity are known, the center of percussion is located on a line that passes through the center of rotation and the center of gravity. The distance from the pivot axis (center of rotation), to the center of percussion $P$ is given by

$$P = \frac{R_m^2}{r}$$

where $R_m$ is the radius of gyration of the mass with respect to the pivot axis and $r$ is the distance from the pivot axis to the center of gravity of the mass. (In the SI system, $P$ is given in meters.)

### 2.3.11 Center of Gravity

The center of gravity of a body is termed the *centroid* of the forces of gravitation acting on all the particles of the body.

To find the center of gravity of a plane or complex plane area, draw the area to scale on a stiff paper such as Bristol board. Cut out the area accurately and hang the figure from one extreme point. Suspend a plumb line through the point. Rehang the figure from another extreme point and suspend another plumb line through this point. The intersection of the two plumb lines is the approximate center of gravity of the figure (see Fig. 2.27). Calculations for locating the center of gravity are given for standard plane figures in Chap. 5.

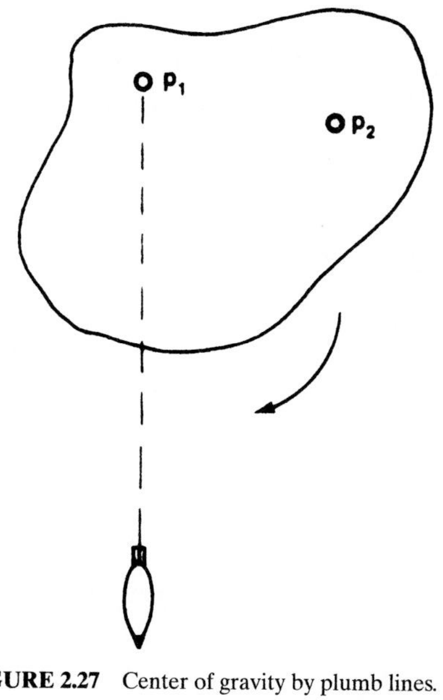


**FIGURE 2.27** Center of gravity by plumb lines.

### 2.3.12 Acceleration Resulting from Unbalanced Forces

For a body moving with pure translation, where $F_R$ is the resultant force of all forces acting on a mass

$$F_R = Ma = \frac{W}{g}a$$

In the U.S. customary system, $M$ = mass, or weight in pounds per 32.2 ft/sec$^2$; $g$ = 32.2 ft/sec$^2$; and $a$ = acceleration in ft/sec$^2$. (In the SI system the above equation is $F_R = Ma$, where $F_R$ = the resultant force, N; $M$ = mass, kg; and $a$ = acceleration of gravity, m/sec$^2$.) Note that the weight of a mass of $M$ kg is $Mg$ newtons, where $g$ is 9.81 m/sec$^2$, i.e., the weight or force of a 12-kg mass is 12(9.81) = 117.72 N.

*Example.* An elevator is being pulled up by its cable mechanism, and the tension in the cable is 12,000 N. The elevator weighs 900 kg. What is the upward acceleration of the elevator?

$$F = Ma$$

$$a = \frac{F_R}{M} = \frac{12{,}000 - 900(g)}{900} = \frac{12{,}000 - 8829}{900} = 3.52 \text{ m/sec}^2$$

### 2.3.13 Rectilinear Motion with Constant Velocity

*Linear, uniform velocity* is described by the following basic equation and its transposed versions:

$$v = \frac{s}{t} \qquad s = vt \qquad t = \frac{s}{v}$$

*Angular, uniform velocity* is described by another basic equation set:

$$\omega = \frac{\theta}{t} \qquad \theta = \omega t \qquad t = \frac{\theta}{\omega}$$

where $s$ = distance, ft
$t$ = time, sec
$\omega$ = angular velocity, rad/sec
$\theta$ = angle of rotation, rad
$v$ = velocity, ft/sec

### 2.3.14 Rectilinear Motion with Constant Acceleration

For uniform acceleration from rest ($v_0 = 0$), the basic equations are

$$s = \frac{1}{2}at^2 \qquad s = \frac{1}{2}v_f t \qquad s = \frac{v_f^2}{2a}$$

where $s$ = distance, ft
$t$ = time of acceleration, sec
$a$ = acceleration, ft/sec$^2$
$v_f$ = final velocity, ft/sec
$v_0$ = initial velocity, ft/sec

Transpose the basic equations to solve for $a$, $v_f$, and $t$.
For uniform acceleration from some initial velocity $v_0$, the basic equations are

$$s = v_0 t + \frac{1}{2}at^2 \qquad s = \frac{(v_f + v_0)t}{2}$$

$$s = \frac{(v_f^2 - v_0^2)}{2a} \qquad s = v_f t - \frac{1}{2}at^2$$

Transpose the basic equations to solve for $v_f$, $v_0$, $t$, and $a$.

***Derivatives and Integrals of Rectilinear Motion.*** *Instantaneous velocity* is given by

$$v = \frac{ds}{dt} = \text{limit}\left(\frac{\Delta s}{\Delta t}\right)$$

$$\Delta t \rightarrow 0$$

(Units are ft/sec, U.S. customary, and m/sec, SI system.)

With regard to acceleration, if $v_1$ = velocity at time $t_1$, and $v_2$ = velocity at time $t_2$, then $\Delta v = v_2 - v_1$ = change over velocity in interval $\Delta t = t_2 - t_1$, and the average acceleration = $\Delta v/\Delta t$.

*Instantaneous acceleration* is then

$$a = \frac{dv}{dt} = \frac{d^2s}{dt^2} = \text{limit}\left(\frac{\Delta v}{\Delta t}\right)$$

$$\Delta t \rightarrow 0$$

(Units are ft/sec$^2$, U.S. customary, and m/sec$^2$, SI system.)

Determine $a$, $v$, $s$, and $t$ when $s$ is given in terms of $t$ algebraically; then determine $v$ and $a$ in terms of $t$ by integration.

$$v = \frac{ds}{dt} \qquad a = \frac{dv}{dt} = \frac{d^2s}{dt^2} \qquad \frac{a}{v} = \frac{dv}{ds}$$

$$s_2 - s_1 = \int_{t_1}^{t_2} v\,dt \qquad v_2 - v_1 = \int_{t_1}^{t_2} a\,dt \qquad t_2 - t_1 = \int_{s_1}^{s_2} \frac{dv}{da}$$

$$v_2^2 - v_1^2 = 2\int_{s_1}^{s_2} a\,ds$$

### 2.3.15 Rotary Motion with Constant Acceleration

The basic equations for a body uniformly accelerated *from rest* are

$$\theta = \frac{1}{2}\,\alpha t^2 \qquad \theta = \frac{1}{2}\,\omega_f\,t^2 \qquad \theta = \frac{\omega_f^2}{2\alpha}$$

where $\omega_f$ = final angular velocity, ft/sec
$\omega_0$ = initial angular velocity, ft/sec
$\alpha$ = angular acceleration, rad/sec$^2$
$\theta$ = angular displacement of rotation, rad
$t$ = time, sec

Transpose the basic equations to solve for $\alpha$, $\omega_f$, and $t$.

The basic equations for a body uniformly accelerated *from an initial angular velocity* are

$$\theta = \omega_0 t + \frac{1}{2}\,\alpha t^2 = \frac{(\omega_f + \omega_0)t}{2}$$

$$= \frac{(\omega_f^2 - \omega_0^2)}{2\alpha} = \omega_f\,t - \frac{1}{2}\,\alpha t^2$$

Transpose the basic equations to solve for $\omega_f$, $\omega_0$, $\alpha$, and $t$.

***Derivatives of Curvilinear Motion.***

*Velocity:* When $s$ is the distance moved by a body along a curved path, the velocity at any instant is

$$v = \frac{ds}{dt}$$

and the direction of velocity is tangent to the path at the instantaneous position of the particle. The tangential component is $ds/dt$, and the normal component is $v^2/p$, the value $p$ being the radius of curvature at the point.

*Components of velocity and acceleration:* With coordinates $x$, $y$, and $z$ defining the position of a particle $P$, the axial components of velocity are

$$v_x = \frac{dx}{dt} \qquad v_y = \frac{dy}{dt} \qquad v_x = \frac{dz}{dt}$$

The resultant is

$$v = \sqrt{v_x^2 + v_y^2 + v_z^2}$$

and the direction cosines are

$$\cos \phi_x = v_x/v \qquad \cos \phi_y = v_y/v \qquad \cos \phi_z = v_z/v$$

The axial components of acceleration are

$$a_z = \frac{dv_x}{dt} = \frac{d^2x}{dt^2} \qquad a_y = \frac{dv_y}{dt} = \frac{d^2y}{dt^2} \qquad a_z = \frac{dv_z}{dt} = \frac{d^2z}{dt^2}$$

The resultant acceleration is then

$$a = \sqrt{a_x^2 + a_y^2 + a_z^2}$$

and the direction cosines are

$$\cos \phi_x = a_x/a \qquad \cos \phi_y = a_y/a \qquad \cos \phi_z = a_z/a$$

The tangential and normal components of acceleration are

$$A_t = \frac{dv}{dt} = \frac{d^2s}{dt^2} \qquad a_n = \frac{v^2}{p}$$

where $p$ = the radius of curvature. The resultant acceleration is then

$$a = \sqrt{a_t^2 + a_n^2} = \sqrt{a_x^2 + a_y^2 + a_z^2}$$

When the path is a plane curve, $v_z = 0$ and $a_z = 0$. Observe that velocities and accelerations, the same as forces, can be resolved according to the parallelogram and parallelopiped laws.

### 2.3.16 Torque and Angular Acceleration

Using Newton's Second Law of Motion as a basis, we may write

$$T_u = I_{pm}\alpha = MR_m^2\alpha = \frac{WR_m^2\alpha}{g}$$

where $T_u$ = unbalanced torque, lb · ft
$I_{pm}$ = polar moment of inertia of the mass about the axis of rotation, ft · lb · sec²
$R_m$ = radius of gyration of the mass with respect to the axis of rotation, ft
$\alpha$ = angular acceleration, rad/sec²
$M$ = mass of body, lb
$W$ = weight of body, lb

(In the SI system $T_u$ is in N · m, $I_{pm}$ is in kg · m², $R_m$ is in m, and $\alpha$ is in rad/sec².)

### 2.3.17 Kinetic Energy

The kinetic energy due to translation is

$$E_K^T = \frac{1}{2} Mv^2 = \frac{Wv^2}{2g}$$

where $M$ = mass ($W/g$), slugs; and $v$ = velocity of the center of gravity of the mass, ft/sec.
Kinetic energy due to rotation is

$$E_K^R = \frac{1}{2} I_{pm} \omega^2$$

and that due to translation plus rotation is

$$E_T = \frac{1}{2} Mv^2 + \frac{1}{2} I_{pm}\omega^2$$

where $T_K^T$, $E_K^R$, and $E_T$ are in ft · lb.
Also

$$E_T = \frac{Wv^2}{2g} + \frac{1}{2} I_{pm}\omega^2 = \frac{W}{2g} (v^2 + R_{mm}\omega^2)$$

In the preceding equation, $R_{mm}$ denotes radius of gyration with respect to the axis through the center of gravity (minimum $R_{mm}$), $E_T$ denotes total kinetic energy, and $I_{pm}$ denotes polar moment of inertia of the mass about its center-of-gravity (c.g.) axis.

[In SI, $E_K^T$, $E_K^R$, and $E_T$ are in joules (J), with $E_T = ½\, M\,(v^2 + R_{mm}{}^2\omega^2$.] The $M$ is mass, kg; $I_{pm}$ is moment of inertia of the mass, kg · m$^2$; $\omega$ is angular velocity, rad/sec.; $R_{mm}$ is minimum radius of gyration, with respect to an axis through the center of gravity; and $v$ is velocity, m/sec. Note that J = 1 N · m.

### 2.3.18 Potential Energy

The potential energy in foot-pounds of a weight $W$, elevated $s$ feet, is

$$E_P = Ws$$

(In SI the $E_P$ of a mass of $M$ kilograms raised to a height of $s$ meters is $Mgs$ joules, where $g$ = 9.81 m/sec$^2$.)

### 2.3.19 Force and Energy in a Force Field

If an electromagnetic force of 100 lb is displaced a distance of 1 in, then the energy expended is

$$Fd = E_T$$

$$100\,(1) = 100 \text{ in} \cdot \text{lb}$$

where $F$ = force, lb and $d$ = distance, in. With $d$ given in feet (1 in = 0.0833 ft), the energy is equal to 100 (0.0833) = 8.33 ft · lb.

(In SI, $F$ is in newtons, $d$ is in meters, and $E_T$ will be in newton-meters.)

### 2.3.20 Work and Power

To determine the work and power in terms of an applied force and the velocity at the point of application use these relationships:

$$s = \frac{Pt}{F} = \frac{J}{F} = \frac{550t(hp)}{F}$$

$$v = \frac{P}{F} = \frac{J}{Ft} = \frac{550(hp)}{F}$$

where $s$ = distance, ft
$v$ = constant or average velocity, ft/sec
$t$ = time, sec
$F$ = constant or average force, lb
$P$ = power, ft · lb/sec
$J$ = work, ft · lb
$hp$ = horsepower (550 ft · lb/sec = 1 hp)

[In SI, $J$ is in joules; $P$ is in watts (1 J/sec); $s$ is in meters; $v$ is in meters per second; $F$ is in newtons; and $t$ is in seconds.]

### 2.3.21 Centrifugal Force

If a body rotates about an axis other than the axis that passes through its center of gravity or mass, an outward radial force is developed from the axis that keeps it from moving tangentially.

$$C_F = \frac{Wv^2}{gd_p} = \frac{4\,Wd_p\,\pi^2 n^2}{3600\,g}$$

$$W = \frac{C_F d_p g}{v^2} \qquad d_p = \frac{Wv^2}{C_F g}$$

where $C_F$ = centrifugal force, lb
$W$ = weight of revolving mass, lb
$v$ = velocity at radius $d_p$ on the body, ft/sec
$n$ = rev/min (rpm)
$g$ = 32.2 ft/sec$^2$
$d_p$ = perpendicular distance from axis of rotation to the center of the rotating mass, ft

In SI, $W/g$ is replaced by $M$, mass in kilograms; $C_F$ is in newtons; $v$ is in meters per second; $n$ is in revolutions per minute; and $d_p$, the offset radius, is in meters, so that

$$C_F = \frac{M\,v^2}{p_d} = \frac{Mn^2\,(2\,\pi\,d_p)^2}{(60)^2\,d_p} = 0.01097\,Md_p n^2$$

In a *flywheel*, $d_p$ is given as the mean radius of the weighted rim, i.e., the distance of the center of mass from the center of rotation or the flywheel axis of rotation.

*Example.* A flywheel with a mean rim radius of 200 mm rotates at 3500 rpm. If the rim mass is 2 kg, what is the $C_F$?

$$C_F = \frac{Mn^2\,(2\pi d_p)^2}{(60)^2\,d_p} \qquad d_p = \frac{200}{1000} = 0.2\text{ m}$$

$$C_F = \frac{2\,(3500)^2[2(3.1416)(0.2)]^2}{3600\,(0.2)} = \frac{2(1.225\times 10^7)(1.579)}{720} = 53{,}730\text{ N} = 53.73\text{ kN}$$

### 2.3.22 Conversions

For conversions among units employed in this chapter, refer to Tables 2.2 and 2.3.

### 2.3.23 Mechanics in Product Design

Many of the preceding principles are applied in the following sample problem (Fig. 2.28). A high-power switch mechanism closes (travels from point $A$ to point $B$), in 4 Hz, as measured by an oscillograph. What is the velocity and energy of the three-pole blade assembly at point $B$? The known data: the blades are pure copper, whose density is 0.318 lb/in$^3$; they are 14 in long and the distance between the pivot point $C$ and point $A$ is 12 in.

First, calculate $I_{pa}$ from $I_{xx}$ and $I_{yy}$ as shown (see Fig. 2.29$a$).

$$I_{xx} = \frac{bd^3}{12} = \frac{0.188\,(13)^3}{12} = 34.4 \text{ in}^4$$

$$I_{yy} = \frac{b_1 d_1^3}{12} = \frac{13(0.188)^3}{12} = 0.0072 \text{ in}^4$$

Next, find the polar moment of the plane area of one blade.

$$I_{pa} = I_{xx} + I_{yy} = 34.4 + 0.0072 = 34.407 \text{ in}^4$$

Then find the polar moment of the blade mass about axis $X$–$X$ (see Fig. 2.29$a$).

$$I_{pm} = \frac{pl}{g} I_{pa} = \frac{0.318\,(13)}{386.4}(34.407) = 0.368 \text{ in} \cdot \text{lb/sec}^2$$

where $p = 0.318$ lb/in$^3$
$l = 13$ in
$g = 32.2$ ft/sec$^2$
$I_{pa} = 34.407$ in$^4$

Next, transfer the axis from $X$–$X$ to $Z$–$Z$, the pivot point (Fig. 2.29$b$). At axis $Z$–$Z$,

$$I_{pm} = I_{pm'} + s^2 M = 0.368 + 6^2(0.004023) = 0.513 \text{ in} \cdot \text{lb/sec}^2$$

where

$$M = \frac{\text{weight of one blade}}{386.4} = \frac{0.188 \times 2 \times 13 \times 0.318}{386.4} = 0.004023 \text{ lb} \cdot \text{sec}^2\text{/ft}$$

and $I_{pm'}$ = polar moment of inertia of mass of blade with respect to axis $X$–$X$, $s$ = 6-in offset distance of axis $Z$–$Z$, and $g = 386.4$ in/sec$^2$.

Since we do not know the equation of motion for the moving blades, we will make an assumption regarding the final angular velocity of the blades at point $B$ on Fig. 2.28$a$. If initial velocity = 0 (at rest), then average velocity = $(v_0 + v_1)/2$; 2(average velocity) = $v_1$.

We know the average angular velocity, since it was measured by an oscillograph. The average time is 4 Hz (0.067 sec), over which the area moves 52°. So

$$\frac{0.067}{52} = \frac{1}{x} \qquad x = 776°\text{/sec}$$

Convert to radians using $180/\pi = 1$ rad.

$$\frac{776}{180/\pi} = \frac{776}{57.296} = 13.54 \text{ radians/sec}$$

**TABLE 2.2** Conversions of Mechanics Units

| To convert from | To | Multiply by |
|---|---|---|
| Kilogram-force | Newton | 9.806650 |
| Newton | Kilogram-force | 0.1019716 |
| Newton | Dyne | 100,000 |
| Newton | Ounce force | 3.596942 |
| Newton | Pound force | 0.2248089 |
| Lbs-ft or pounds-foot | Newton-meter (N-m) | 14.59390 |
| Kilogram-meter | Newton-meter | 9.806650 |
| Ounce-inch | Newton-meter | 0.0070616 |
| Newton-meter | Pound-foot | 0.7375621 |
| Newton-meter | Ounce-inch | 141.6119 |
| Pound-foot | Newton-meter | 1.355818 |
| Moment of inertia, kg-$m^2$ | Pound-$foot^2$ | 23.73036 |
| Moment of inertia, kg-$m^2$ | Pound-$inch^2$ | 3,417.171 |
| Moment of section, $ft^4$ | $Meter^4$ ($m^4$) | 0.00863098 |
| Moment of section, $m^4$ | $Foot^4$ ($ft^4$) | 115.8618 |
| Section modulus, $ft^3$ | $Meter^3$ ($m^3$) | 0.02831685 |
| Section modulus, $m^3$ | $Foot^3$ ($ft^3$) | 35.31466 |
| Kilogram-meter per sec | Pound-foot per sec | 7.233011 |
| Pound foot per sec | Kilogram-meter per sec | 0.1382550 |
| Kilogram per $meter^2$ | Pascal | 9.806650 |
| Kilonewton/$meter^2$ | Pound per $inch^2$ | 0.1450377 |
| Newton/$meter^2$ | Bar | 0.00001 |
| Newton/$meter^2$ | Pascal | 1.000 |
| Newton/$meter^2$ | Pound per $inch^2$ | 0.00014504 |
| Pascal | Kilogram per $meter^2$ | 0.1019716 |
| Pascal | Newton per $meter^2$ | 1.000 |
| Pascal | Pound per $foot^2$ | 0.02088543 |
| Pascal | Pound per $inch^2$ | 0.00014504 |
| Pound per $foot^2$ | Pascal | 47.88026 |
| Pound per $inch^2$ | Pascal | 6,894.757 |
| Pound per $inch^2$ | Bar | 0.06894757 |
| Btu (mean) | Joule | 1,055.87 |
| Calorie (mean) | Joule | 4.19002 |
| Foot-pound | Joule | 1.355818 |
| Joule (J) | Foot-pound | 0.7375621 |
| Watt-hour | Joule | 3,600 |
| Btu per hour | Watt | 0.2930711 |
| Foot-pound per minute | Watt | 0.02259697 |
| Horsepower | Watt | 746 |
| Kilowatt (kW) | Horsepower | 1.341 |
| Watt | Foot-pound per minute | 44.2537 |
| Watt | Horsepower | 0.0013405 |
| Watt | Btu per hour | 3.41214 |
| Miles per hour | Feet per minute | 88 |

Symbols for SI (Système international) units: newton, N; watt, W; joule, J; pascal, Pa; kilogram, kg; gram, g; meter, m; millimeter, mm.

**TABLE 2.3** Equivalent Energy and Power Units

| | | |
|---|---|---|
| 1 Btu | = | 1,055 J |
| 1 Btu | = | 778.8 ft-lb |
| 1 W | = | 0.7376 ft-lb/sec |
| 1 W | = | 0.001341 hp |
| 1 kW-hr | = | 2,654,200 ft-lb |
| 1 kW-hr | = | 3,412 Btu |
| 1 kW-hr | = | 3,600,000 J |
| 1 hp | = | 0.746 kW |
| 1 hp | = | 33,000 ft-lb/min or 550 ft-lb/sec |
| 1 hp | = | 2545 Btu/h |

Therefore, the final velocity at point $B = 2 \times 13.54 = 27.1$ rad/sec. Now substitute this value for $\omega$ in the equation that follows, which expresses the kinetic energy of a blade rotating about the fixed axis $Z$–$Z$ as

$$E_K^R = \frac{1}{2} I_{pm}\, \omega^2 = \frac{1}{2}(0.512)(27.1)^2 = 188 \text{ in} \cdot \text{lb per blade}$$

Total energy at point $B = 6$ blades $\times 188 = 1128$ in · lb.

Next, we will compare this output to the input energy contained in the switch-spring mechanism that moves the blades.

Input energy to the spring mechanism is calculated by means of the potential-energy equation for springs

$$E_p = \frac{R_s S^2}{2}$$

where $R_s$ = spring rate, lb/in, and $S$ = distance the springs are compressed prior to discharge, in. The three power springs have rates of 88, 32, and 85 lb/in. Two are loaded to 4 in compression, and one to 2.25 in compression. Therefore

$$E_{p1} = \frac{88\,(4)^2}{2} = 704 \text{ in} \cdot \text{lb} \qquad E_{p2} = \frac{32\,(4)^2}{2} = 256 \text{ in} \cdot \text{lb} \qquad E_{p3} = \frac{85\,(2.25)^2}{2} = 215 \text{ in} \cdot \text{lb}$$

$$\text{Total input energy} = 1175 \text{ in} \cdot \text{lb}$$

The output derived from measurement and calculation is 1128 in · lb, a reduction from the input of $(1175 - 1128)/1175 = 0.04$, which is 4 percent.

We can assume that 4 percent of the input power is dissipated in the heat of friction during operation of the mechanism. This is a realistic conclusion based on measurement and calculation, and it shows in part the importance of mechanics and its applications to design problems.

The potential-energy equation for springs is derived by integration, as we will see in Chap. 7. As an alternate approach we could have evaluated the power produced by the closing-switch mechanism by using equations shown in the preceding sections.

(In SI, the spring energy equation remains valid if $R_s$ is in newtons per meter, $S$ is in meters, and $E_p$ is in joules.)

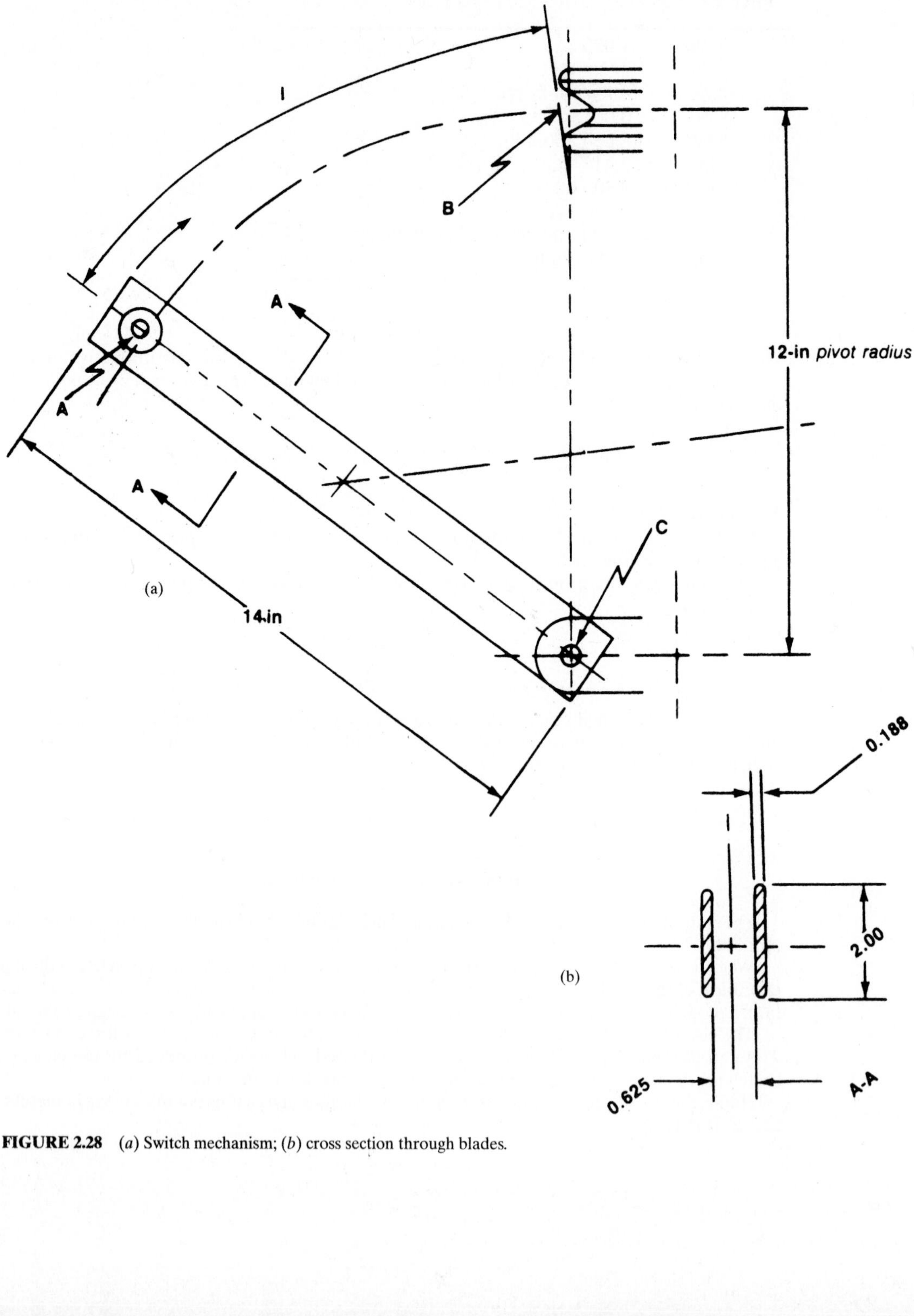


**FIGURE 2.28** (*a*) Switch mechanism; (*b*) cross section through blades.

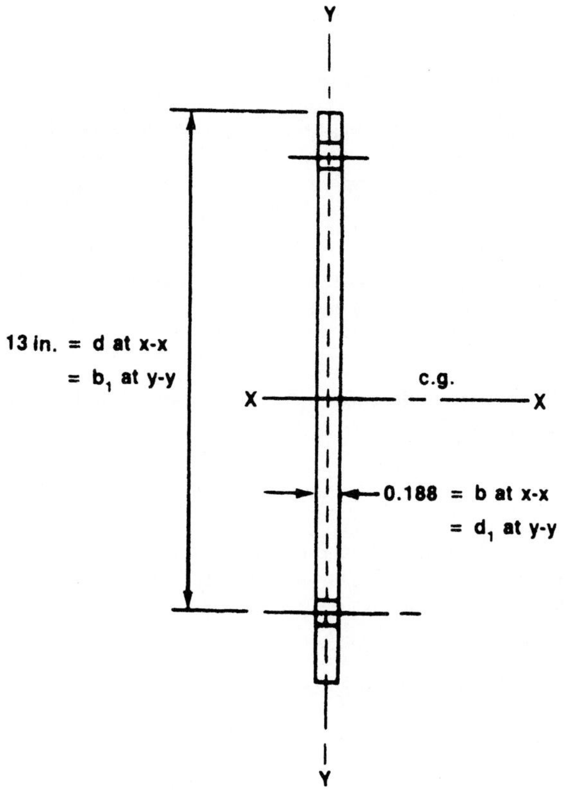


**FIGURE 2.29*a*** Calculating $I_{xx}$ and $I_{yy}$ of one blade.

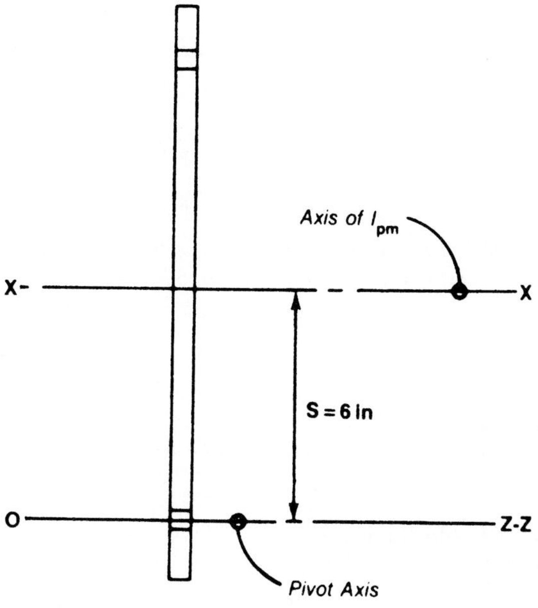


**FIGURE 2.29*b*** Transferring the axis.

### 2.3.24 Impulse and Linear Momentum

*Linear momentum* $L_m$ is the product of the mass and the velocity of the center of gravity of a body:

$$L_m = Mv = \frac{Wv}{g}$$

*Linear impulse* $L_i$ is the product of the resultant of forces acting on a body and the time over which the resultant acts:

$$L_i = Rt$$

The change in the momentum of a body is equal to the linear impulse causing the change in momentum:

$$L_i = \frac{W}{g} v_f - \frac{W}{g} v_0 = \frac{W}{g}(v_f - v_0)$$

where $v_f$ = final velocity and $v_0$ = initial velocity. (In SI, $L_m = Mv$ and $Rt = Mv_f - Mv_0$, where $M$ is mass, kg; $v$ is velocity, m/sec; $R$ is force, N; $t$ is time, sec; and $L_i$ is the linear impulse, which is equal to $Rt$ N · sec.)

### 2.3.25 Angular Impulse and Momentum

Calculate angular momentum from

$$A_m = I_{pm}\,\omega$$

Angular impulse is

$$A_i = M_0 t$$

where $I_{pm}$ = polar moment of mass, lb · ft · sec$^2$
$\omega$ = angular velocity, rad/sec
$M_0$ = torque, lb · ft

Angular impulse, which equals change in angular momentum, can also be written

$$M_0 t = I_{pm}\,\omega_f - I_{pm}\,\omega_0 = I_{pm}\,(\omega_f - \omega_0)$$

where $\omega_f$ = final angular velocity and $\omega_0$ = initial angular velocity.

(In SI, $I_{pm}$ is in kg · m$^2$; $\omega_0$, $\omega_f$ are in rad/sec; and $M_0$ is in N · m.)

The linear velocity of a point on a rotating body is

$$v = \omega r$$

where $\omega$ is angular velocity, rad/sec, and $r$ is the radius of the revolving point. When $r$ is in in, $v$ is in in/sec; when $r$ is in ft, $v$ is in ft/sec.

(In SI, $\omega$ is in rad/sec; $r$ is in m; and $v$ is in m/sec.)

### 2.3.26 Symbols Used in Mechanics

Refer to Table 2.4.

**TABLE 2.4** Symbols Used in Mechanics

| Symbol | Definition | Units | |
|---|---|---|---|
| | | U.S. customary | Systèmes International |
| $a$ | Acceleration | ft/sec$^2$ | m/sec$^2$ |
| $\alpha$ | Angular acceleration | rad/sec$^2$ | rad/sec$^2$ |
| $A_i$ | Angular impulse | ft-lb-sec | kg-m$^2$/sec |
| $A_m$ | Angular momentum | ft-lb-sec | kg-m$^2$/sec |
| $C_f$ | Centrifugal force | lb | N |
| $d$ | Distance | ft, in | m |
| $d$ | Density | lb/in$^3$ | kg/m$^3$ |
| $E$ | Energy | ft-lb | J |
| $F$ | Force | lbf | N-m |
| $g$ | Acceleration-gravity | ft/sec$^2$ | m/sec$^2$ |
| $I$ | Moment of inertia | in$^4$, ft$^4$ | m$^4$ |
| $I_{pa}$ | Moment of inertia (area) | in$^4$, ft$^4$ | m$^4$ |
| $I_{pm}$ | Moment of inertia (mass) | slug-ft$^2$ | kg-m$^2$ |
| $l$ | Length | ft, in | m |
| $L_i$ | Linear impulse | lb-sec | kg-m/sec |
| $L_m$ | Linear momentum | lb-sec | kg-m/sec |
| $M$ | Mass | slug | kg |
| $M_O$ | Moment of force | lb-ft | Nm |
| $n$ | Number of revolutions | | |
| $P$ | Power | ft-lb/sec | W |
| $P$ | Center of percussion | ft, in | m |
| $r$ | Radius | ft, in | m |
| rpm | Revolutions per minute | | |
| $R$ | Resultant | lb | N |
| $R_a$ | Radius of gyration (area) | ft, in | m |
| $R_m$ | Radius of gyration (mass) | ft, in | m |
| $R_s$ | Spring rate | lb/in | N/m |
| $S,s$ | Distance | ft, in | m |
| $T_u$ | Unbalanced torque | lb-ft | N-m |
| $\omega$ | Angular velocity | rad/sec | rad/sec |
| $t$ | Time | sec | sec |
| $\theta$ | Angle of rotation | rad | rad |
| $v$ | Velocity | ft/sec | m/sec |
| $W$ | Weight | lb | MgN |
| $Z$ | Section modulus | in$^3$, ft$^3$ | m$^3$ |
| $Z_p$ | Polar section modulus | in$^3$, ft$^3$ | m$^3$ |

### 2.3.27 Constants

A number of useful constants are displayed in Table 2.5.

**TABLE 2.5** Some Useful Constants

| | | |
|---|---|---|
| e | = | 2.7182818285 |
| $\pi$ | = | 3.1415926536 |
| R | = | 57.2957795131° |
| 1° | = | 0.0174532925 rad |
| 360° | = | $2\pi$ rad = 6.2831853072 rad, 1 rev |
| g | = | 32.2 ft/sec$^2$, U.S. customary, sea level |
| g | = | 9.81 m/sec$^2$, SI, sea level |
| 60 mph | = | 88 ft/sec |
| 180°/$\pi$ | = | 1 rad = 57.2957...deg (angular) |

### 2.3.28 Newton's Method for Solving Intractable Equations

This method for finding roots of difficult equations is accurate enough for most applications in engineering, in addition to being a powerful tool in mechanics.

The method is valid when there are no maxima and minima nor inflection points between the intersection of the line $y = f(x)$, and the $X$ axis and the $x_1$ point on the line. (For the theory of this method, see the bibliography at the end of Chap. 1.)

To solve an equation such as

$$2.97x^3 - 2.075x^2 + 4.75x - 3.875 = 0$$

proceed as follows. Make a rough graph to find the approximate roots, or obtain the approximate roots using one of the handheld calculators, such as the Casio Fx-7000G or comparable Hewlett-Packard model. For the first approximation, we have

$$x_2 = x_1 - \frac{f(x_1)}{f'(x_1)}$$

where $x_1$ = first approximate root, and $x_2$ = second approximate root. Solve for $x_2$ by substituting your first a¹proximate root, say, 0.79787, for $x$ in the original equation. Substitute 0.79787 for $x$ in the first derivative of the original equation.

$$x_2 = x_1 - \frac{f(x)}{f'(x)}$$

$$f(x) = 2.97\ (0.50792) - 2.075\ (0.63660) + 4.75\ (0.79787) - 3.875$$

$$f'(x) = 8.91\ (0.63660) - 4.15\ (0.79787) + 4.75$$

$$x_2 = 0.79787 - \left(\frac{0.10245}{7.11095}\right) = 0.78346 \text{ (2d approximation)}$$

Next, substitute 0.78346 back into the $f(x)$ and $f'(x)$ terms and solve for the third-root approximation $x_3$.

Continue the approximations until the desired accuracy is obtained. The general form of the newtons-method equation is

$$X_{n+1} = X_n - \left[\frac{f(x_n)}{f'(x_n)}\right]$$

For your information, the third-root approximation was 0.78331. As a check, we will substitute this root back into the $f(x)$ function, $2.97x^3 - 2.075x^2 + 4.75x - 3.875 = 0$. If $x = 0.78331$

$$1.42744 - 1.27317 + 3.72072 - 3.875 = 0$$

$$0.00001000 \approx 0$$

This root is accurate to four decimal places.

### 2.3.29 Mechanics Conversions

***Torque Conversion Units.*** It is often necessary to convert torque from one system of units to another. The following charts will be useful and time-saving and should also help prevent conversion errors. (See *Note* following these tables for explanation of asterisks.)

| To convert pounds-feet to | Multiply by |
|---|---|
| Gram-inches | 5443.1088 |
| Ounce-inches | 192.0 |
| Pound-inches | 12.0 |
| Kilogram-centimeters* | 13.8257 |
| Kilogram-meters* | 0.138257 |

| To convert pound-inches to | Multiply by |
|---|---|
| Gram-inches | 453.5924 |
| Ounce-inches | 16.0 |
| Pounds-feet | 0.08334 |
| Kilogram-centimeters* | 1.152 |
| Kilogram-meters* | $1.152 \times 10^{-2}$ |

| To convert ounce-inches to | Multiply by |
|---|---|
| Gram-inches | 28.3495 |
| Pound-inches | 0.0625 |
| Pounds-feet | $5.2087 \times 10^{-3}$ |
| Kilogram-centimeters* | $72.808 \times 10^{-3}$ |
| Kilogram-meters* | $728.08 \times 10^{-6}$ |

| To convert gram-inches to | Multiply by |
|---|---|
| Ounce-inches | 0.03527 |
| Pound-inches | $2.205 \times 10^{-3}$ |
| Pounds-feet | $1.8376 \times 10^{-4}$ |
| Kilogram-centimeters* | $2.54 \times 10^{-3}$ |
| Kilogram-meters* | $2.54 \times 10^{-5}$ |

| To convert kilogram-centimeters to | Multiply by |
|---|---|
| Gram-inches | 393.7 |
| Ounce-inches | 13.8858 |
| Pound-inches | $85.8108 \times 10^{-2}$ |
| Pounds-feet | $72.346 \times 10^{-3}$ |
| Kilogram-meters* | 0.01 |

| To convert kilogram-meters to | Multiply by |
|---|---|
| Gram-inches | 39370.0 |
| Ounce-inches | 1388.58 |
| Pound-inches | 85.8108 |
| Pounds-feet | 7.2346 |
| Kilogram-centimeters* | 100.0 |

[* *Note.* In the preceding conversions for the (SI) system, kilograms, when used as a weight or force, must be multiplied by 9.81 to arrive at the newton conversion. Many modern SI torque listings are given as newton-meters (N · m) instead of kilogram-meters (kg · m), or gram-centimeters (g · cm). This confusion was created when the kilogram was previously used as a force or weight. In modern SI systems, the kilogram is considered only as a unit of mass, while the newton is considered as a force. A 1-kg mass used as a force may be specified as $1 \times 9.81 = 9.81$ N. The kilogram is multiplied by the metric gravitational constant of 9.81 m/sec$^2$ to arrive at the newton measure of force. One newton is equal to 0.10197 kg-force.]

***Equivalent Units for Energy, Work, Power, Force, and Mass.*** Units of energy or work and heat energy are

| U.S. customary system | SI (metric) |
|---|---|
| Foot-pound | Newton-meter |
| British thermal unit (Btu) [the amount of heat energy required to raise the temperature of 1 lb of water ($H_2O$) at 72°F by 1°F] | Joule (1 J = 1 N-m) |
| Horsepower-hour | Watthour, kilowatthour (Wh, kWh) |
| | Gram-calorie (the amount of heat energy required to raise the temperature of 1 g of water at 22°C by 1°C) |

Units of power are

| U.S. customary system | SI (metric) |
|---|---|
| Horsepower | Watt, kilowatt (1 W = 1 J/sec = 1 N · m/sec) |
| Btu per second, minute, or hour | Joule per second, minute, or hour |
| Foot-pounds per second, minute, or hour | Newton-meters per second, minute, or hour |

Units of force are

| U.S. customary system | SI (metric) |
|---|---|
| Pound-force, lbf (force) | Newton (force) |
| Pound-foot, pound-inch (torque or turnimg moment) | Newton-meter (torque or turning moment) |
| | Kilogram (when used as weight or force must be specified as mg · N; i.e., 5 kg weight expressed as a force is equal to $5 \times 9.81 = 49.05$ N) |

Units of mass are

| U.S. customary system | SI (metric) |
|---|---|
| Pounds, slugs (lb · $sec^2$/ft) (see previous text) | Kilograms |

By definition, *energy* is a force acting through a distance or heat equivalent (work is the expenditure of energy) and *power* is the time rate of expending energy or heat equivalent. Power and energy are thus related by

$$\text{Power, W or Btu/h} = \frac{\text{energy, Wh or Btu}}{\text{time, h}}$$

Occasionally in the literature you may notice errors in specification of the correct units in either energy, power, force, mass, or torque (e.g., foot-pounds used for torque instead of pounds-feet, watts used as an energy unit instead of watthours, Btu used as a power unit instead of Btu/h).

### 2.3.30 Reference Axes in Mechanics Problems

Mechanics problems seldom include reference axes, so their selection is arbitrary or usually indicated by the geometry of each individual problem. Therefore, we *must* be able to determine the correct components of a force regardless of how the axes are oriented or how the angles are measured.

Figure 2.30*a*–2.30*d* is a graphic representation of how the component forces may be determined when the axes and angles are in various orientations.

| Fig. 2.30*a* | Fig. 2.30*b* |
|---|---|
| $F_x = F \sin \phi$ | $F_x = F \sin (180° - \phi)$ |
| $F_y = F \cos \phi$ | $F_y = -F \cos (180° - \phi)$ |

| Fig. 2.30*c* | Fig. 2.30*d* |
|---|---|
| $F_x = F \cos (\phi - \alpha)$ | $F_x = -F \cos \phi$ |
| $F_y = F \sin (\phi - \alpha)$ | $F_y = -F \sin \phi$ |

Typical working linkage problems and axis orientation examples are presented in Chap. 8.

### 2.3.31 Systematic Simplification of Complex Dynamics Problems in Industrial Applications

The author has been an industrial product design engineer for 40 years and is often required to find practical, quick answers to statics and dynamics problems related to actual product designs. As an example, Fig. 2.31 shows an AutoCad drawing of a complex high-power electrical switching device, often referred to as an *automatic load-break interrupter switch.* The main moving parts of this device are seen more clearly in Fig. 2.32, which is a photograph of a similar high-power switching device. In an industrial product such as this, great care must be exercised in the preliminary design phases, so that an effective prototype device can be manufactured for evaluation and testing, to prove the adequacy and practicality of the design.

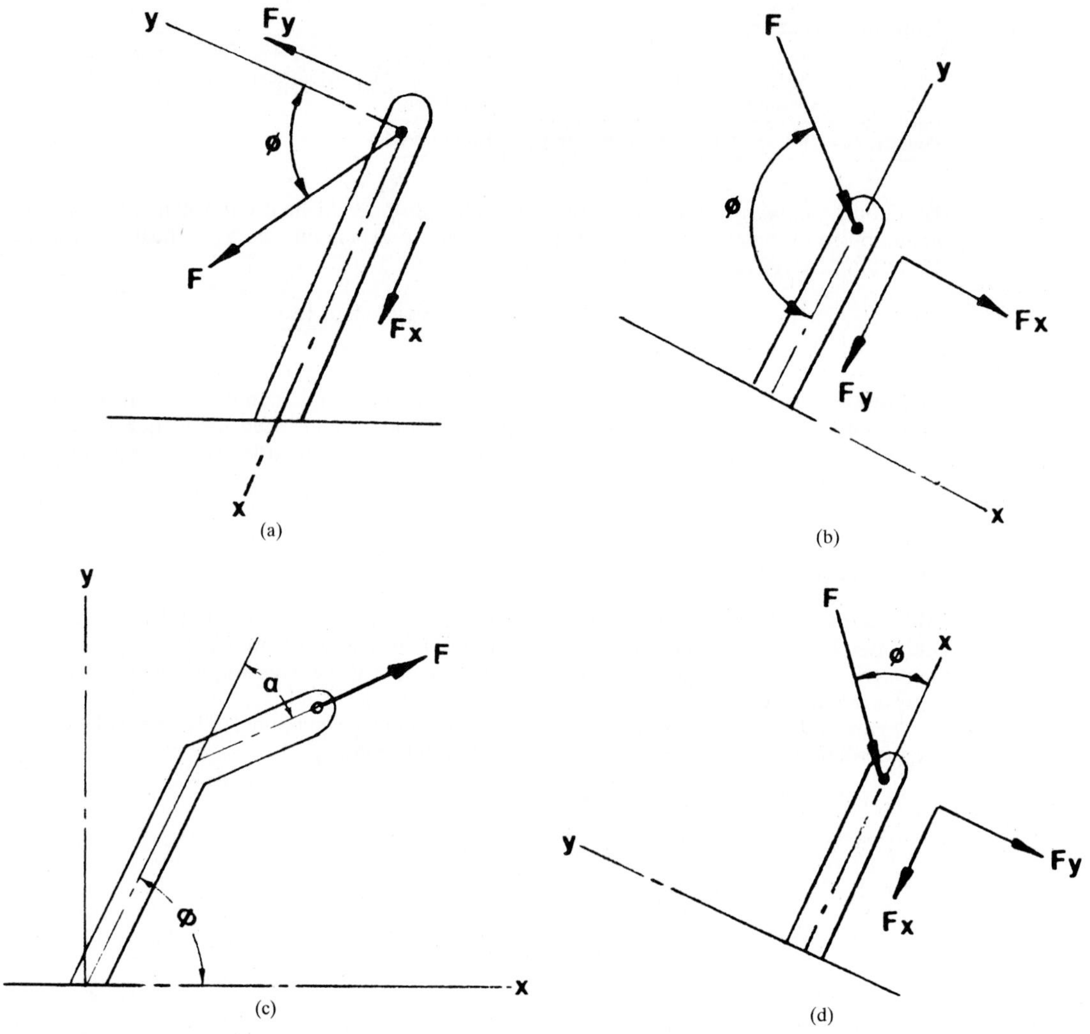


**FIGURE 2.30** (*a–d*) Orientation of axes.

The prime or most important aspect in the design of a new product and concept is your ability as a design engineer to render the prototype *functional.* Safety and producibility are next taken into account. Costing and tooling practices are then applied in order to make the product competitive. It is often detrimental to begin a design with costing as the prime goal, especially on complex products or where quality and safety are prime factors. The author has witnessed many design attempts in industrial applications, made with costing as the prime goal, with the inevitable result of partial or total product failure and wasted time, effort, and finances. In these instances, the designer usually failed to consider function as the first element in successful product design practice.

Refer to Figs. 2.31 and 2.32 during the following disclosures. In commercial, industrial, or comsumer product design practice, the design engineer seldom has the time required to run a detailed, totally mathematical analysis of an anticipated product's performance. The designer

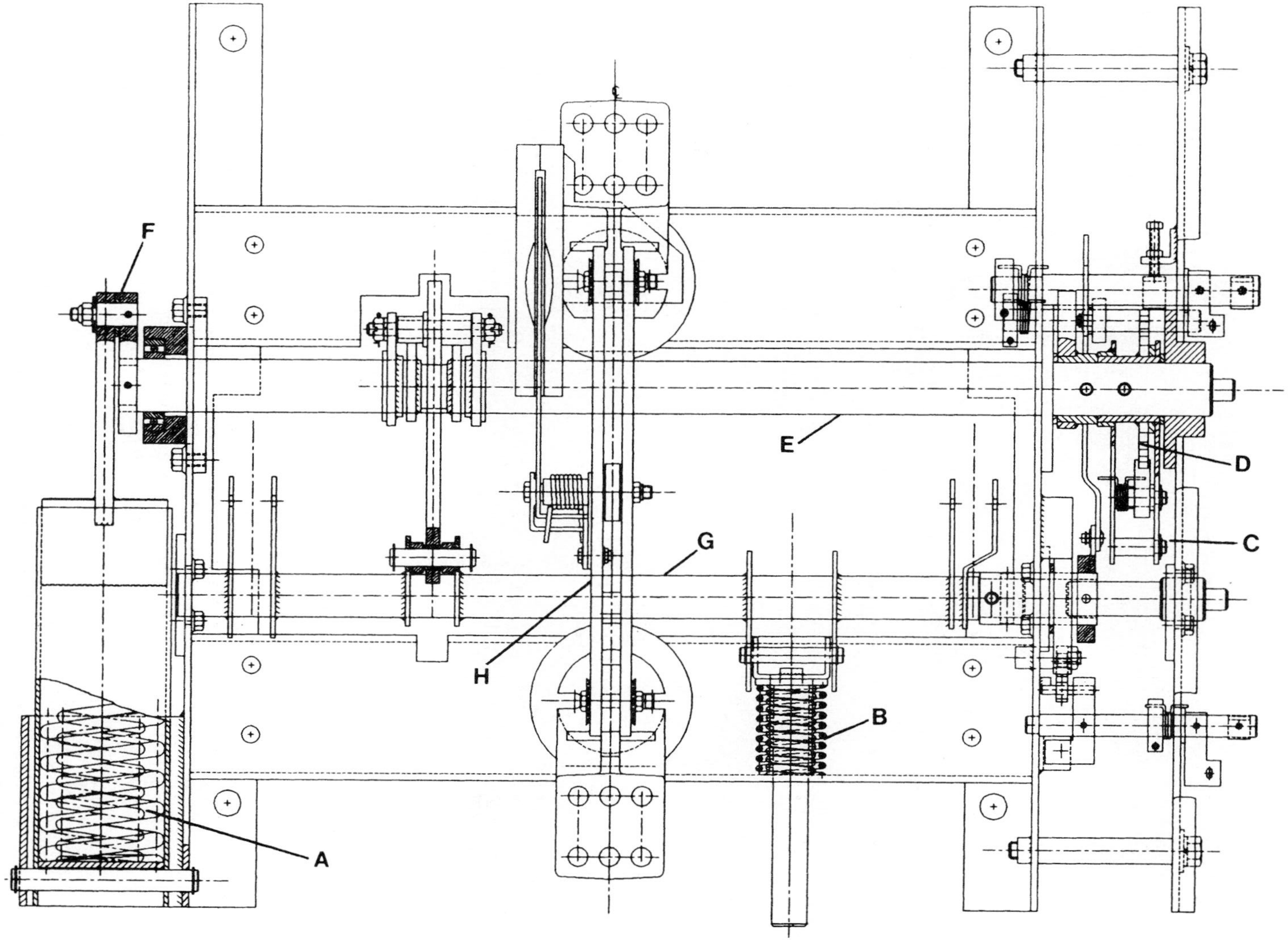

FIGURE 2.31 AutoCad design drawing of automatic interrupter switching device.

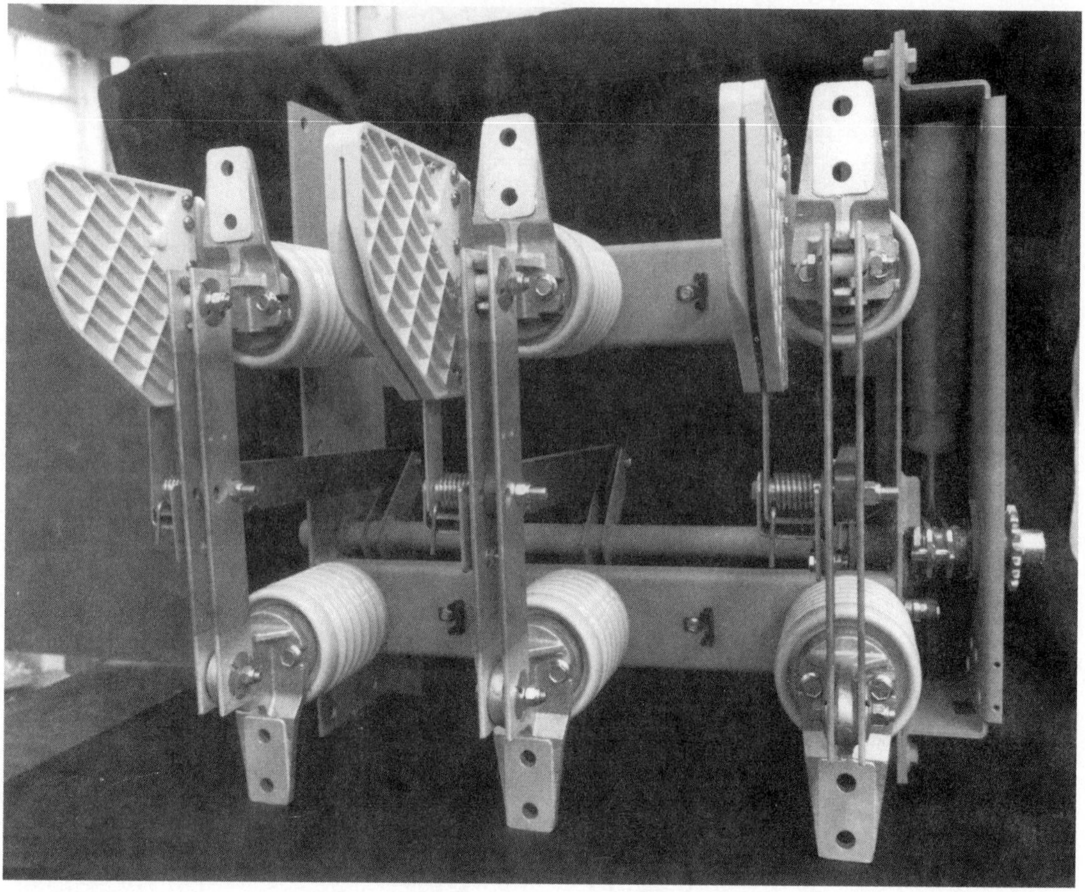

**FIGURE 2.32** A switching device similar to that shown in Fig. 2.31.

or design engineer may also not have the advanced mathematical ability to perform such a task. In fact, mathematical analyses of complex real-world problems are approximations in many cases. Rather, the experienced designer will often limit the analyses to the most critical elements in the design concept, using personal experience to design the least critical elements in their proper order. In Fig. 2.31, the critical elements of the design which require calculations are labeled *A, B,* and so on, and the analyses of these elements are performed according to the listing which follows.

*A.* Energy required in the main power springs, ft · lb.
*B.* Energy required in mechanism opening springs, ft · lb.
*C.* Draw-bar pull required to load the ratchet mechanism, lbf.
*D.* Forces imposed on the ratchet teeth, lb/in$^2$ (psi).
*E.* Torque and deflection of the main power shaft, lb · ft and in.
*F.* Forces on the crank arm and pin, lbf.
*G.* Torque and deflection on the blade operating shaft, lb · ft and in.
*H.* Angular velocity of main current-carrying blades, rad/sec (see Sec. 2.3.23 for a typical example).

In this example of actual design practice, no attempt is made to instruct the reader *how to design* a complex electromechanical product such as depicted. Rather, the basics of how to approach the procedures that are required to design such a product are outlined. In the total design analysis and procedures required of a product such as that shown in this example, the following elements had to be considered:

**1.** Electrical specifications
  *a.* Working voltages
  *b.* Continuous current-carrying requirements
  *c.* Voltage impulse levels
  *d.* Electromagnetic forces developed during electrical faults or short circuits
  *e.* Permissible heat rise temperatures
  *f.* Fault-closing ability of the main switch blades
  *g.* High current sustained ratings
**2.** Mechanical specifications
  *a.* Selection of proper materials of construction
  *b.* Selection of proper hardware items
  *c.* Shock loads
  *d.* Dynamic loads
  *e.* Mechanism movements (kinetics)
  *f.* Function
  *g.* Safety and liability
  *h.* Producibility (methods and processes of production)
  *i.* Costing
  *j.* Tooling

In the prototype stage of such a product, high-speed motion pictures are usually taken to show events during operation (electrical and mechanical) that cannot be accurately described by calculations or visualized by the design engineer during the preliminary design phases.

Most of the electrical and mechanical specifications listed previously are referenced throughout this handbook. For more engineering information concerning the manufacturing aspects encountered in product design, the author recommends the McGraw-Hill *Machining and Metalworking Handbook,* written by the author and first published in August 1993 by McGraw-Hill, Inc. When used in conjunction with this handbook, the design engineer has available a vast amount of practical technical data and information concerning product design and manufacturing.

## 2.4 NEWTON'S LAWS OF MOTION

Newton's three laws of motion have a special significance in dynamics and are stated here. In modern terminology, they are

**I.** A particle remains at rest or continues to move in a straight line with a constant velocity if there is no *unbalanced* force acting on it.

**II.** The acceleration of a particle is proportional to the resultant force acting on it and is in the direction of the force. (Another interpretation of this law states that the resultant force acting on a particle is proportional to the time rate of change of momentum of the particle and that this change is in the direction of the force.) Both interpretations are equally correct when applied to a particle of constant mass.

**III.** The forces of action and reaction between interacting bodies are equal in magnitude, opposite in direction, and collinear.

## FURTHER READING

Eshbach, O. W., and M. Souders, 1975: *Handbook of Engineering Fundamentals,* 3d ed. New York: Wiley.

Meriam, J. L., 1959: *Mechanics,* 2d ed.: Vol. 1, *Statics,* Vol. 2, *Dynamics.* New York: Wiley.

Meriam, J. L., and L. G. Kraige, 1986: *Engineering Mechanics,* 2d ed., English and SI version, Vol. 1, *Statics,* Vol. 2, *Dynamics.* New York: Wiley.

Oberg, E., F. Jones, and H. Horton, 1980: *Machinery's Handbook,* 21st ed. New York: Industrial Press.

Najder, K. W., 1958: *Machine Designer's Guide,* 4th ed. Ann Arbor, Mich.: Edwards Bros.

Niles, N. O., and G. E. Haborak, 1982: *Calculus with Analytic Geometry,* New York: Prentice-Hall.

# CHAPTER 3
# MENSURATION, DESCRIPTIVE GEOMETRY, AND OPTICS

## 3.1 MENSURATION

Plane areas, surface areas, and volumes of the regular geometric shapes and solids illustrated herein can be calculated by using the formulas which appear in parentheses next to the figure.

$$A = \frac{1}{2}\, bh \qquad \text{(Fig. 3.1)}$$

$$A = \frac{1}{2}\, ab \sin C$$

$$= \sqrt{s(s-a)(s-b)(s-c)} \qquad \left[\text{where } s = \frac{1}{2}\,(a+b+c)\right] \qquad \text{(Fig. 3.2)}$$

$$A = ab \qquad \text{(Fig. 3.3)}$$

$$A = bh \qquad \text{(Fig. 3.4)}$$

$$A = \frac{1}{2}\, cd \qquad \text{(Fig. 3.5)}$$

$$A = \frac{1}{2}\,(a+b)h \qquad \text{(Fig. 3.6}a\text{)}$$

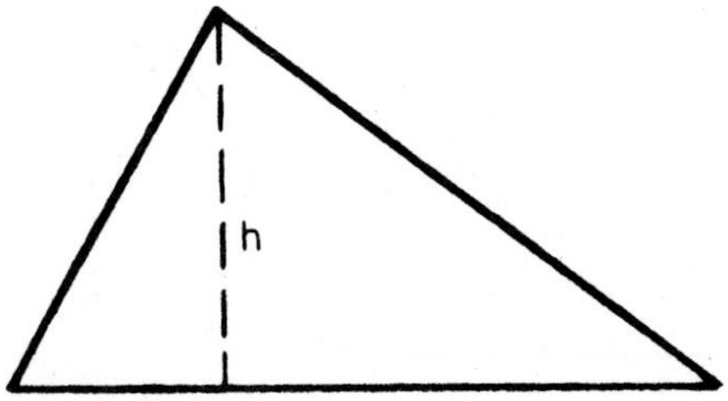


FIGURE 3.1 Triangle.

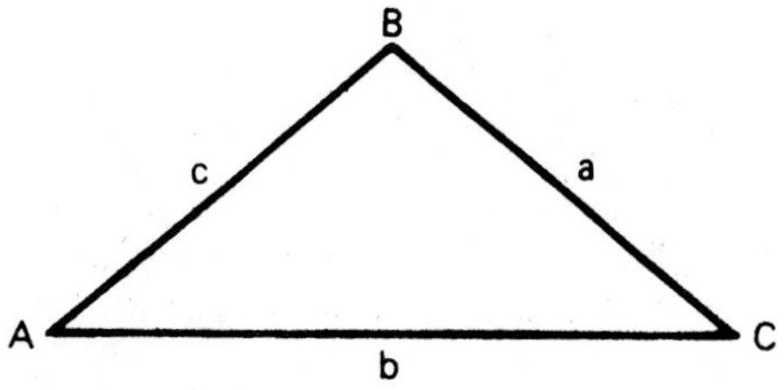


FIGURE 3.2 Triangle.

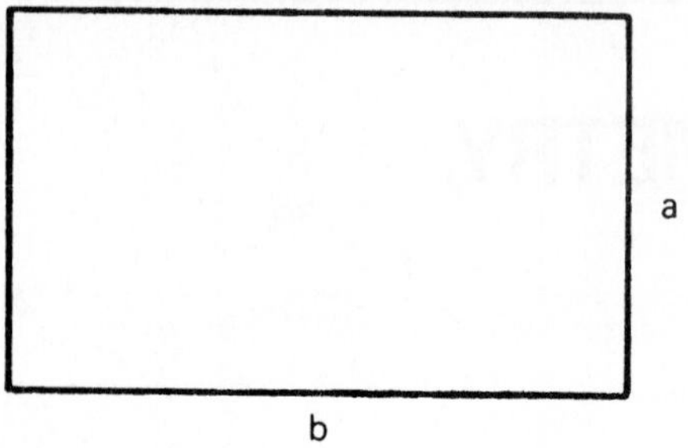


FIGURE 3.3 Rectangle.

$$A = \frac{(H+h)a + bh + cH}{2} \qquad \text{(Fig. 3.6}b\text{)}$$

$$A = \frac{1}{4}\, nL^2 \cot \frac{180}{n} \qquad \text{(Fig. 3.7)}$$

where $A$ = area
$V$ = volume
$R$ = radius
$S$ = surface area
$r$ = radius
$C$ = circumference
$n$ = number of sides
$L$ = length of a side

In a polygon of $n$ sides, each of which is $L$, the radius of the inscribed circle is

$$r = \frac{L}{2} \cot \frac{180}{n}$$

The radius of the circumscribed circle is

$$r_1 = \frac{L}{2} \operatorname{cosec} \frac{180}{n}$$

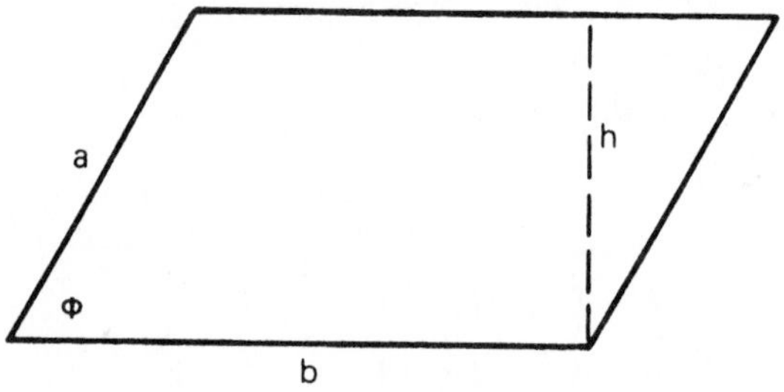


FIGURE 3.4 Parallelogram.

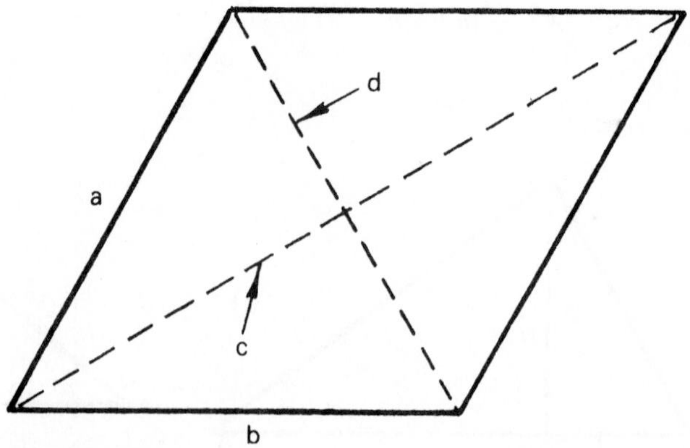


FIGURE 3.5 Rhombus.

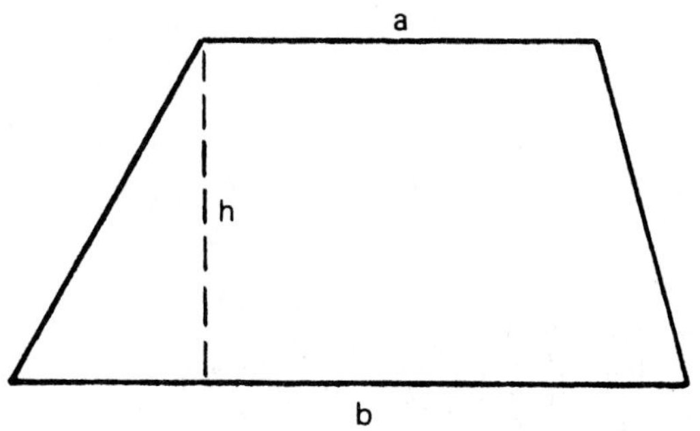


(a)

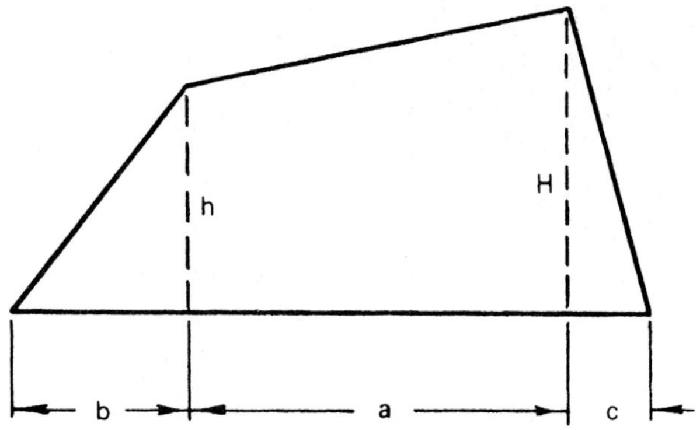


(b)

**FIGURE 3.6** (*a*) Trapezoid; (*b*) trapezium.

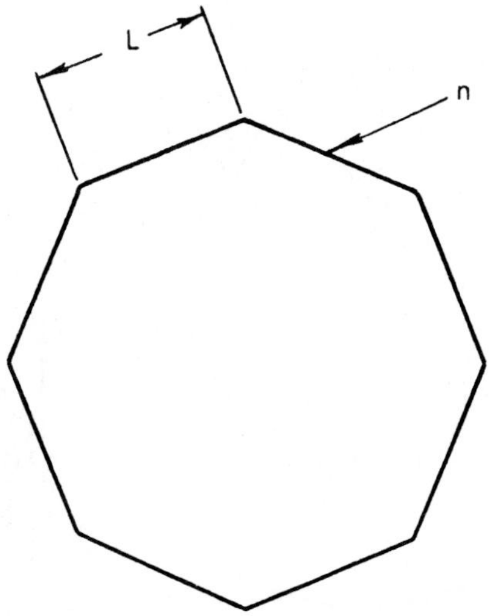


**FIGURE 3.7** Regular polygon.

The radius of a circle inscribed in any triangle whose sides are $a$, $b$, and $c$ is

$$r = \frac{\sqrt{s(s-a)(s-b)(s-c)}}{s} \qquad \text{(Fig. 3.8)}$$

where $s = (\frac{1}{2})(a + b + c)$. The radius of the circumscribed circle is

$$R = \frac{abc}{4\sqrt{s(s-a)(s-b)(s-c)}} \qquad \text{(Fig. 3.9)}$$

Area of an inscribed polygon is

$$A = \frac{1}{2} nr^2 \sin \frac{2\pi}{n} \qquad \text{(Fig. 3.10)}$$

where $r$ = radius of the circumscribed circle and $n$ = number of sides.

$$A = nR^2 \tan \frac{\pi}{n} \qquad \text{(Fig. 3.11)}$$

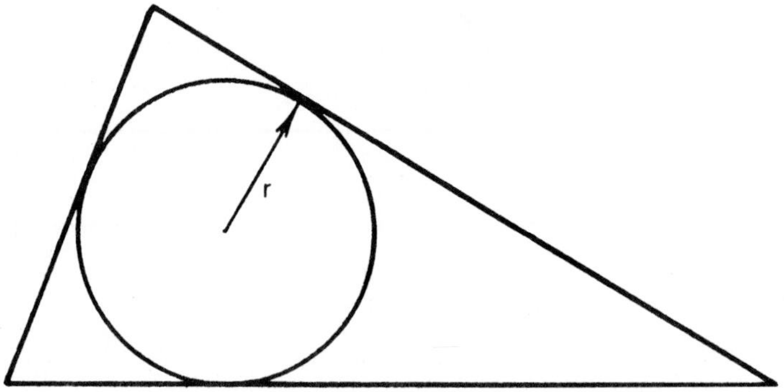


**FIGURE 3.8** Inscribed circle.

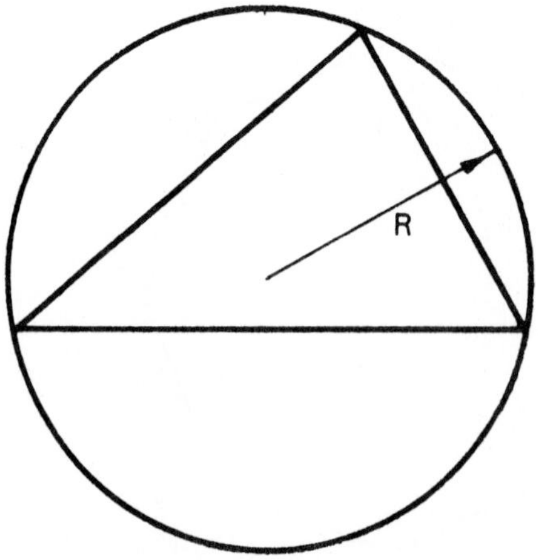


**FIGURE 3.9** Circumscribed circle.

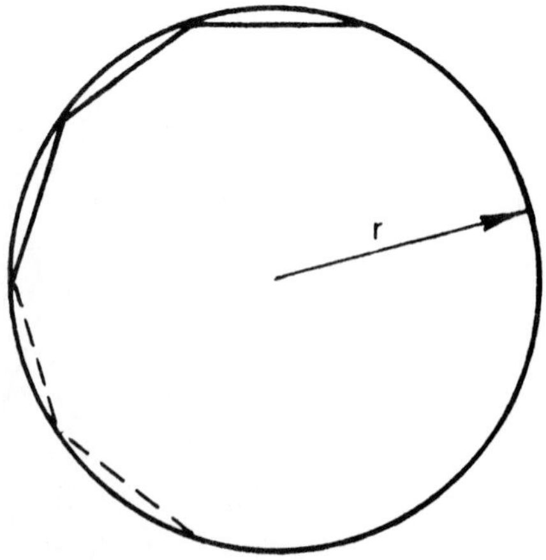


**FIGURE 3.10** Inscribed polygon.

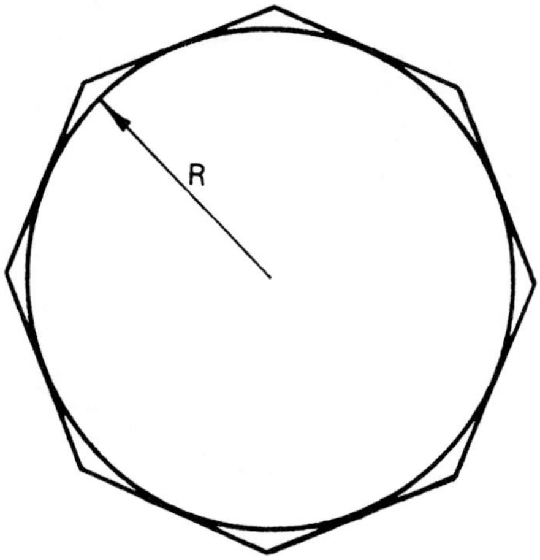


**FIGURE 3.11** Circumscribed polygon.

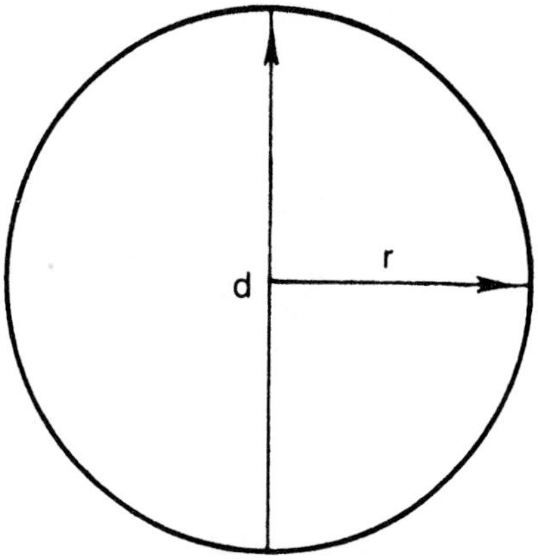


**FIGURE 3.12** Circle.

where $R$ = radius of the inscribed circle and $n$ = number of sides. Circumference of a circle:

$$C = 2\pi r = \pi d \qquad \text{(Fig. 3.12)}$$

Area of a circle:

$$A = \pi r^2 = \frac{1}{4}\pi d^2 \qquad \text{(Fig. 3.13)}$$

Length of arc $L$:

$$L = \frac{\pi r \phi}{180} \qquad \text{(Fig. 3.14)}$$

where $\phi$ is in radians and $L = r\phi$. Length of chord $AB$:

$$AB = 2r \sin \frac{1}{2}\phi$$

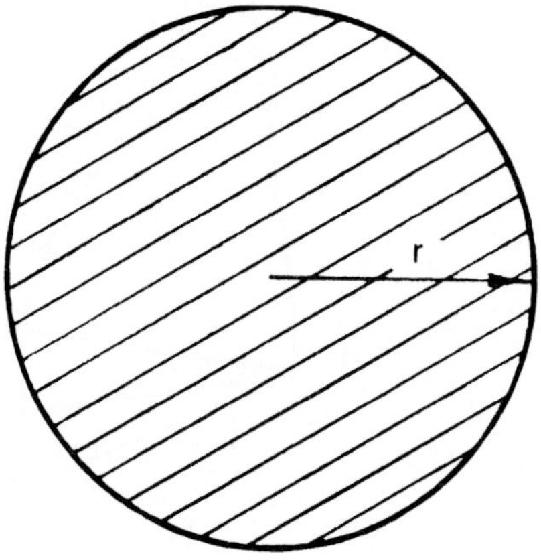


**FIGURE 3.13** Area.

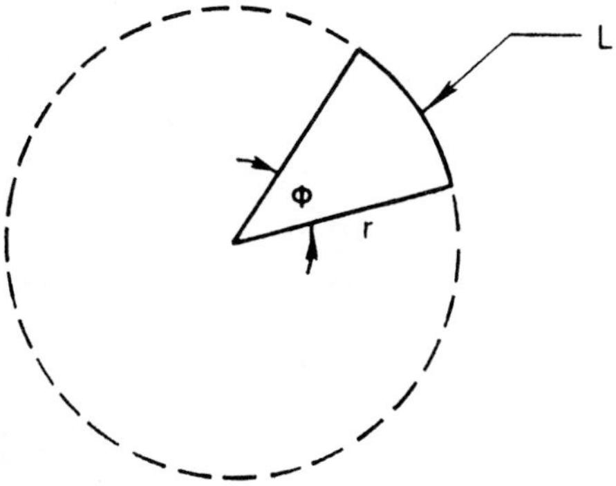


**FIGURE 3.14** Arc.

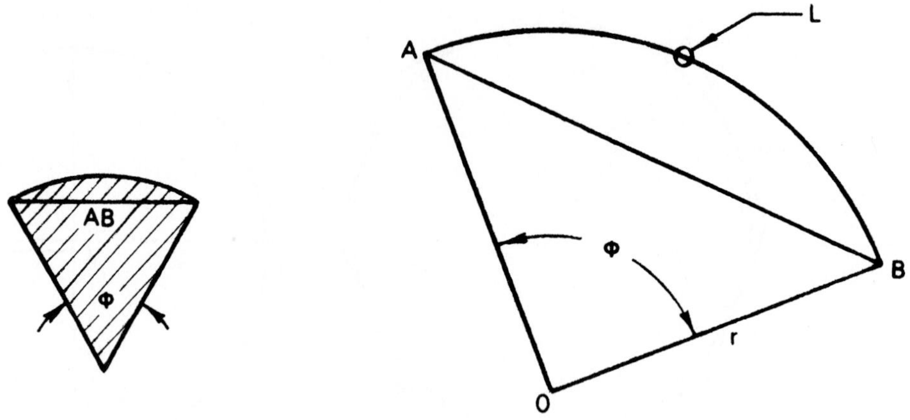


**FIGURE 3.15** Chord.

The area of the sector:

$$A = \frac{\pi r^2 \phi}{360} = \frac{rL}{2} \qquad \text{(Fig. 3.15)}$$

where $L$ is the length of the arc. Area of a segment of a circle:

$$A = \frac{\pi r^2 \phi}{360} - \frac{r^2 \sin \phi}{2} \qquad \text{(Fig. 3.16)}$$

where $\phi = 180° - 2 \arcsin x/r$ and $x$ = perpendicular distance, center to chord. If $\phi$ is in radians, $A = \frac{1}{2} r^2 (\phi - \sin \phi)$. Area of the ring between circles:

$$A = \pi(R + r)(R - r) \qquad \text{(Fig. 3.17)}$$

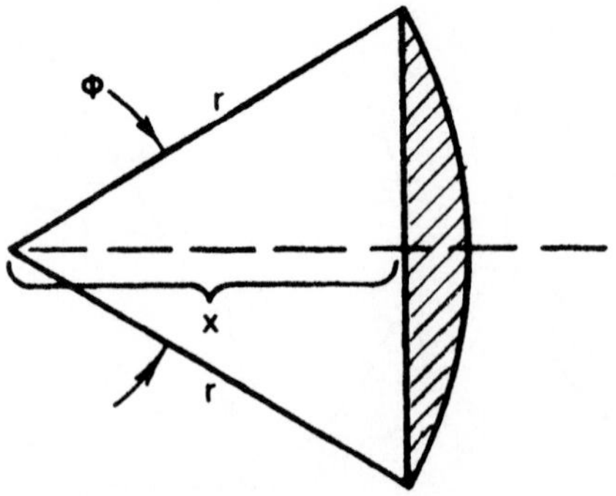

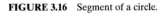

**FIGURE 3.16** Segment of a circle.

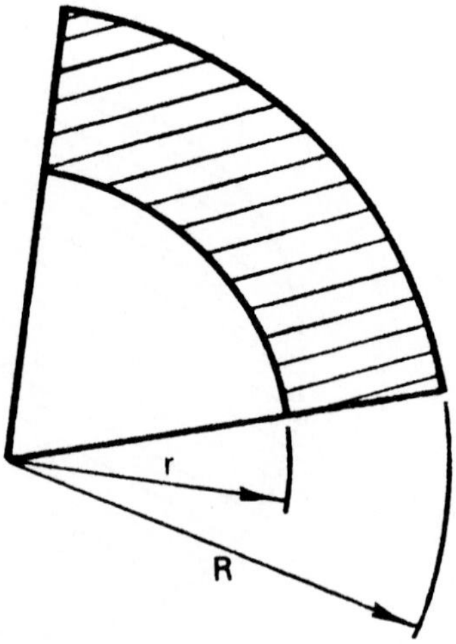


**FIGURE 3.17** Ring.

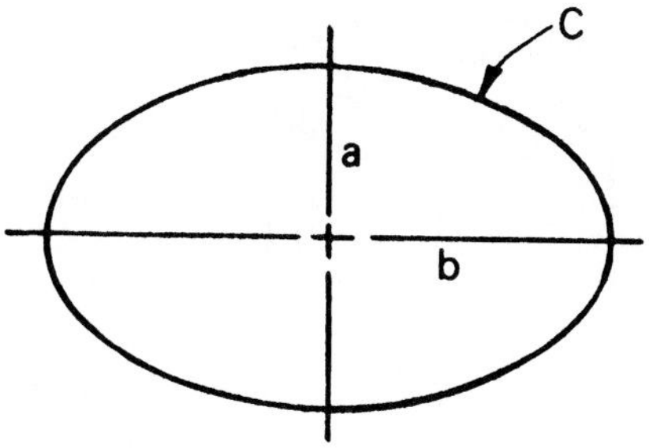


**FIGURE 3.18** Ellipse.

Note that the circles need not be concentric. Circumference of an ellipse:

$$C = 2\pi\sqrt{\frac{a^2 + b^2}{2}} \text{ (approximate)} \qquad A = \pi\, ab \qquad \text{(Fig. 3.18)}$$

$$V = \frac{1}{3} \text{ area of base} \times h \qquad \text{(Fig. 3.19)}$$

where $h$ is altitude. Surfaces and volumes of polyhedra (where $L$ = leg or edge)

| | Surface | Volume |
|---|---|---|
| Tetrahedron | $1.73205\ L^2$ | $0.11785\ L^3$ |
| Hexahedron (cube) | $6\ L^2$ | $1\ L^3$ |
| Octahedron | $3.46410\ L^2$ | $0.47140\ L^3$ |

Surface and volume of a sphere:

$$S = 4\pi\, r^2 = \pi\, d^2 \qquad V = \frac{4}{3}\,\pi r^3 = \frac{1}{6}\,\pi\, d^3 \qquad \text{(Fig. 3.20)}$$

Surface and volume of a cylinder:

$$S = 2\,\pi rh \qquad V = \pi\, r^2 h \qquad \text{(Fig. 3.21)}$$

Surface and volume of a cone:

$$S = \pi\, r\sqrt{r^2 + h^2} \qquad V = \frac{\pi}{3}\, r^2 h \qquad \text{(Fig. 3.22)}$$

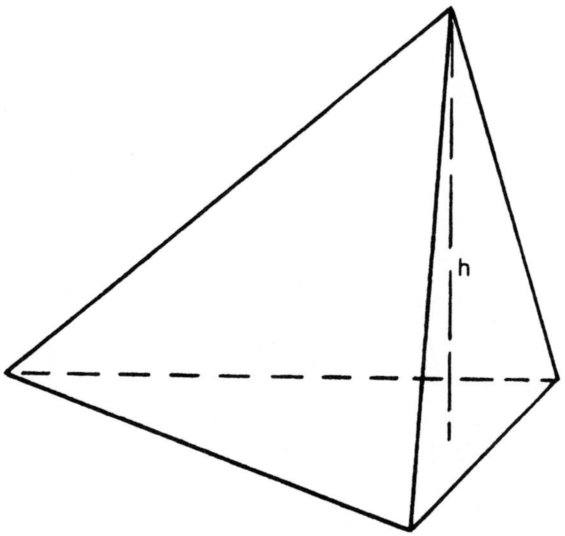


**FIGURE 3.19** Pyramid.

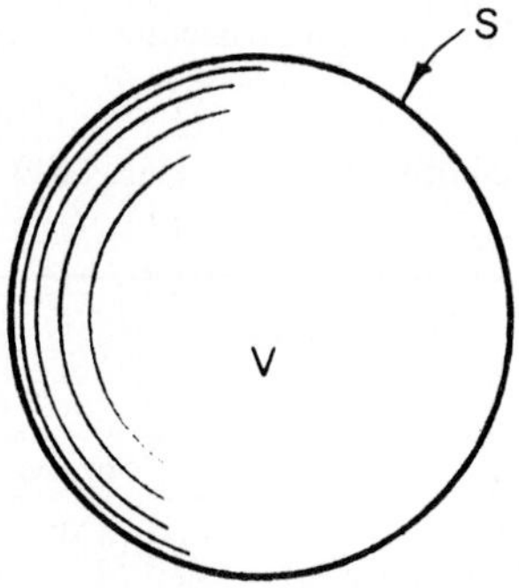


**FIGURE 3.20** Sphere.

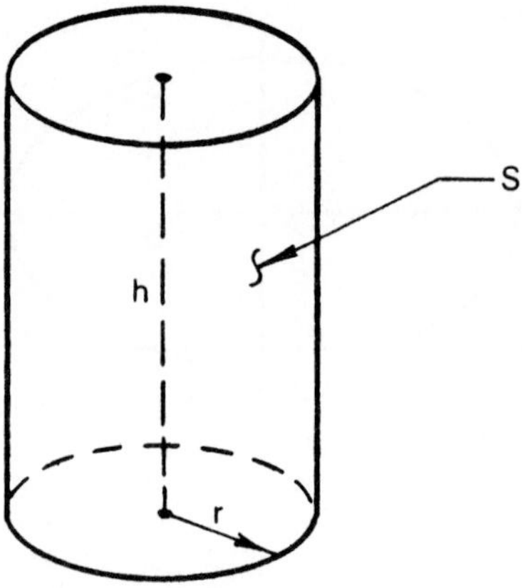


**FIGURE 3.21** Cylinder.

Area and volume of curved surface of spherical segment:

$$A = 2\pi rh \qquad V = \left(\frac{\pi}{3} h^2\right)(3r - h) \qquad \text{(Fig. 3.23)}$$

When $a$ is the radius of the base of the segment

$$V = \frac{\pi}{4} h (h^2 + 3a^2)$$

Surface area and volume of the frustum of a cone:

$$S = \pi(r_1 + r_2)\sqrt{h^2 + (r_1 - r_2)^2}$$

$$V = \frac{h}{3}(r_1^2 + r_1 r_2 + r_2^2)\pi \qquad \text{(Fig. 3.24)}$$

Area and volume of a truncated cylinder:

$$A = \pi\, r(h_1 + h_2) \qquad V = 1.5708r^2 (h_1 + h_2) \qquad \text{(Fig. 3.25)}$$

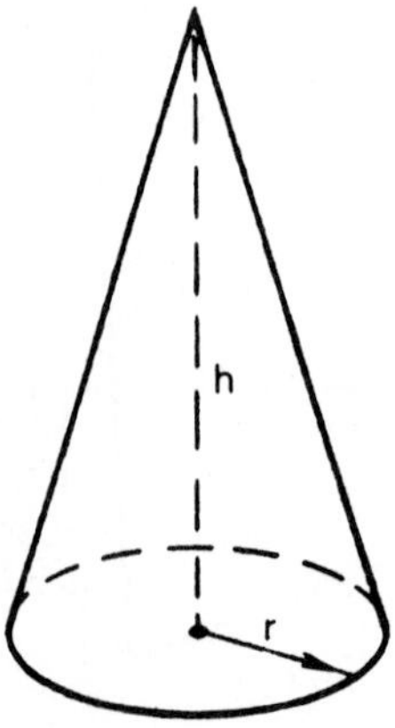


**FIGURE 3.22** Cone.

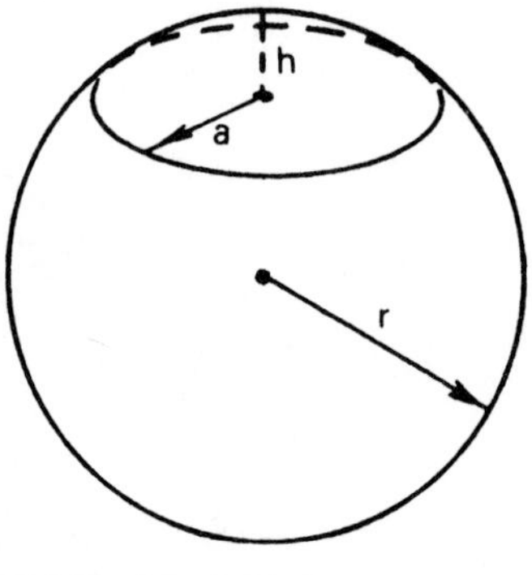


**FIGURE 3.23** Spherical segment.

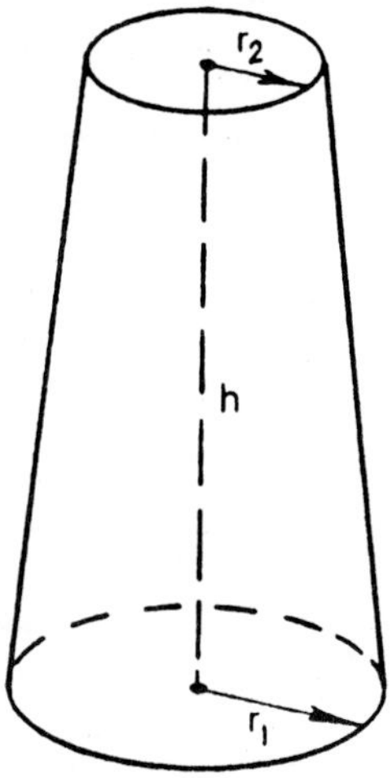


**FIGURE 3.24** Frustum of a cone.

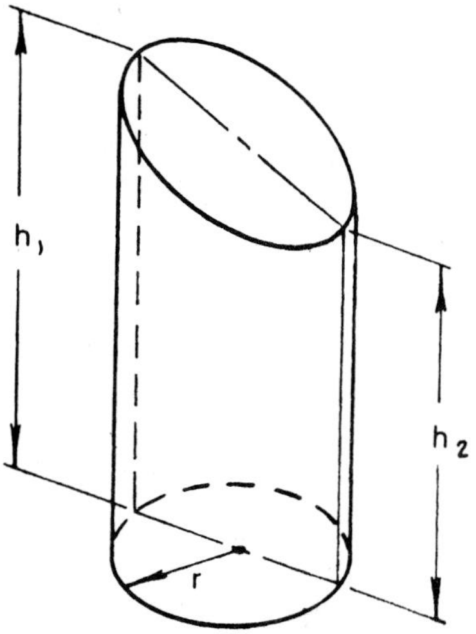


**FIGURE 3.25** Truncated cylinder.

Area and volume of a portion of a cylinder (base edge = diameter):

$$A = 2\,rh \qquad V = \frac{2}{3}\,r^2h \qquad \text{(Fig. 3.26)}$$

Area and volume of a portion of a cylinder (special cases):

$$A = \frac{h(ad \pm c \times \text{perimeter of base})}{r \pm c} \qquad V = \frac{h\,(\frac{2}{3}a^3 \pm cA)}{r \pm c} \qquad \text{(Fig. 3.27)}$$

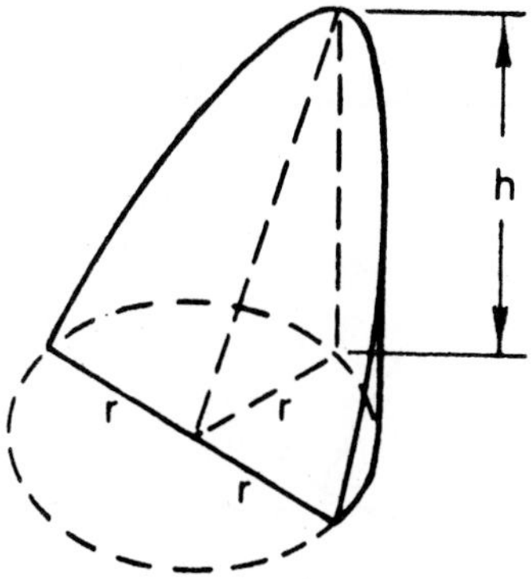


**FIGURE 3.26** Portion of a cylinder.

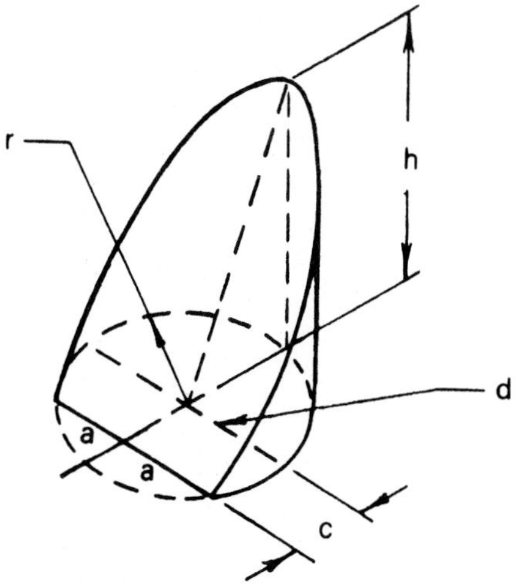


**FIGURE 3.27** Special case of a cylinder.

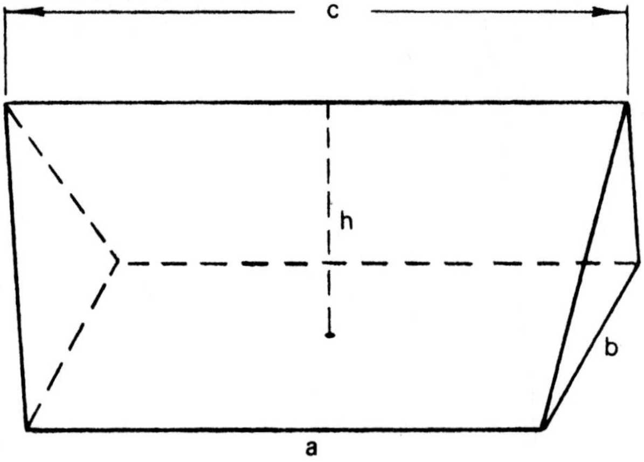


**FIGURE 3.28** Wedge.

where $d$ = diameter of base circle. Use $+c$ when base area is larger than half the base circle; use $-c$ when base area is smaller than half the base circle. Volume of a wedge:

$$V = \frac{(2b + c)\,ah}{6} \qquad \text{(Fig. 3.28)}$$

Area and volume of a spherical zone:

$$A = 2\,\pi\,rh \qquad V = 0.5236\,h\left(\frac{3c_1^2}{4} + \frac{3c_2^2}{4} + h^2\right) \qquad \text{(Fig. 3.29)}$$

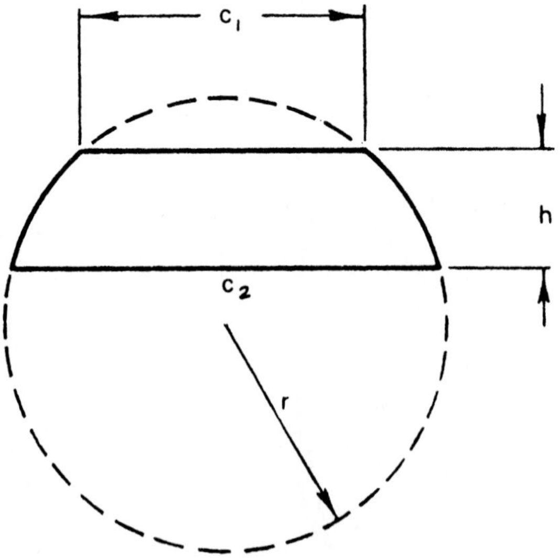


**FIGURE 3.29** Spherical zone.

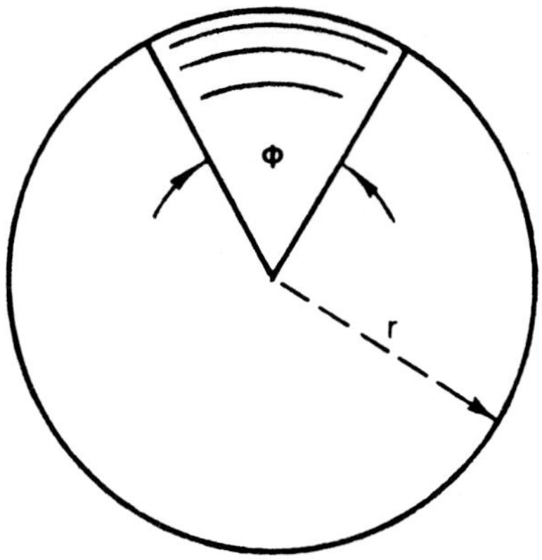


**FIGURE 3.30** Spherical wedge.

Area and volume of a spherical wedge:

$$A = \frac{\phi}{360} 4 \pi r^2 \qquad V = \frac{\phi}{360} \cdot \frac{4 \pi r^3}{3} \qquad \text{(Fig. 3.30)}$$

The volume of a paraboloid:

$$V = \frac{\pi r^2 h}{2} \qquad \text{(Fig. 3.31)}$$

Area and volume of a spherical sector:

$$A = \pi r \left(2h + \frac{c}{2}\right) \quad \text{yields total area}$$

$$V = \frac{2 \pi r^2 h}{3} \qquad c = 2 \sqrt{h (2r - h)} \qquad \text{(Fig. 3.32)}$$

Area and volume of a spherical segment:

$$A = 2 \pi r h \quad \text{(spherical surface)} = \pi \left(\frac{c^2}{4} + h^2\right)$$

$$c = 2 \sqrt{h (2r - h)}$$

$$r = \frac{c^2 + 4h^2}{8h} \qquad \text{(Fig. 3.33)}$$

$$V = \pi h^2 \left(r - \frac{h}{3}\right)$$

Area and volume of a torus:

$$A = 4 \pi^2 cr \quad \text{(total surface)} \qquad V = 2 \pi^2 cr^2 \quad \text{(total volume)} \qquad \text{(Fig. 3.34)}$$

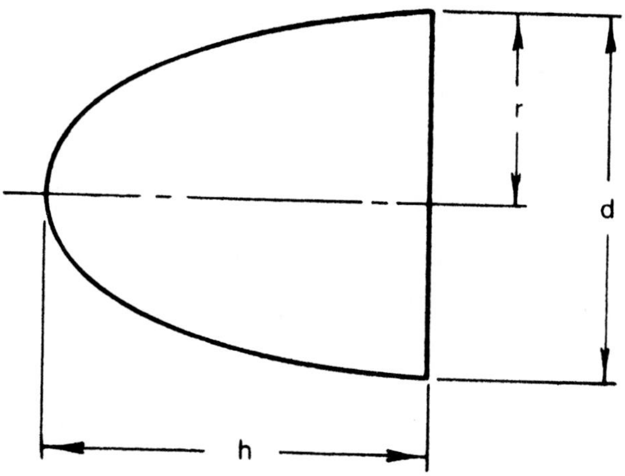


**FIGURE 3.31** Paraboloid.

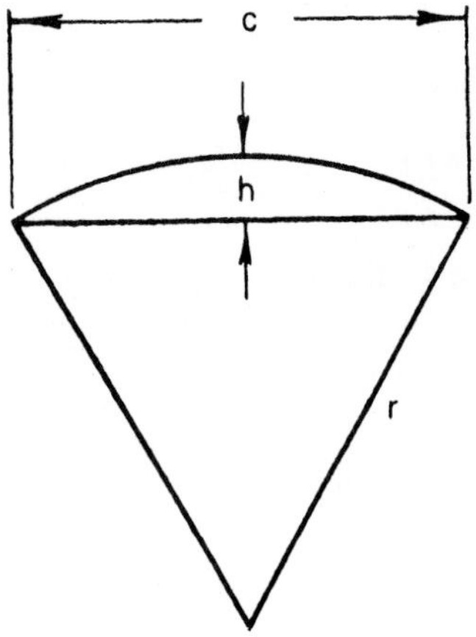


FIGURE 3.32 Spherical sector.

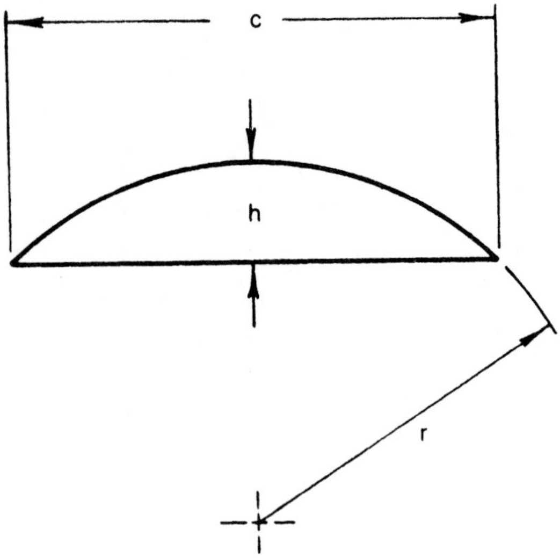


FIGURE 3.33 Spherical segment.

$$\text{Arc: } l = \frac{\pi\, r\, \theta°}{180}$$

$$\text{Angle: } \theta = \frac{180°\, l}{\pi\, r}$$

$$\text{Radius: } r = \frac{4b^2 + c^2}{8b};\ d = \frac{4b^2 + c^2}{4b} \qquad \text{(Fig. 3.35)}$$

$$\text{Chord: } c = 2\sqrt{2br - b^2} = 2\, r \sin\frac{\theta}{2} = d \sin\frac{\theta}{2}$$

$$\text{Rise: } b = r - \frac{1}{2}\sqrt{4r^2 - c^2} = \frac{c}{2}\tan\frac{\theta}{4} = 2\, r \sin^2\frac{\theta}{4}$$

$$\text{Rise: } b = r + y - \sqrt{r^2 - x^2}$$

where $y = b - r + \sqrt{r^2 - x^2}$ and $x = \sqrt{r^2 - (r + y - b)^2}$.

### 3.1.1 Area, Volume of Irregular Plane Areas, and Solid Masses

See Sec. 1.10.6 in Chap. 1 of this handbook for discussions of Simpson's rule and the prismoidal formula.

### 3.1.2 Properties of the Circle

Refer to Fig. 3.35, which depicts the geometric components to which the equations summarized here apply.

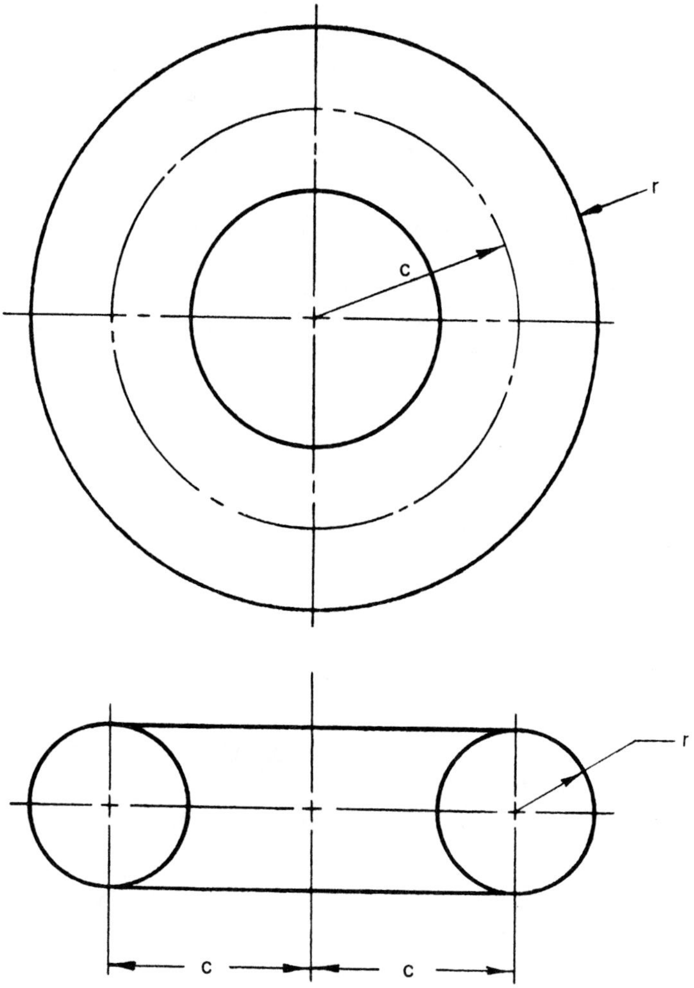


**FIGURE 3.34** Torus.

## 3.2 BASIC DESCRIPTIVE GEOMETRY

Descriptive geometry is a powerful tool in design engineering. It has many uses in determining true lengths of lines and planes, true distance between lines and planes, intersection of lines and planes, and resolution of both coplanar and noncoplanar forces in mechanics.

The basics included in this section may be used to solve a vast number of problems in design and mechanics. For a more detailed study of the entire subject, see the bibliography at the end of this chapter.

### 3.2.1 True Lengths of Lines

The true length of a line appears in a view where the line of sight for that view is perpendicular to the line: All frontal lines are true length in the front view. The reason for this property

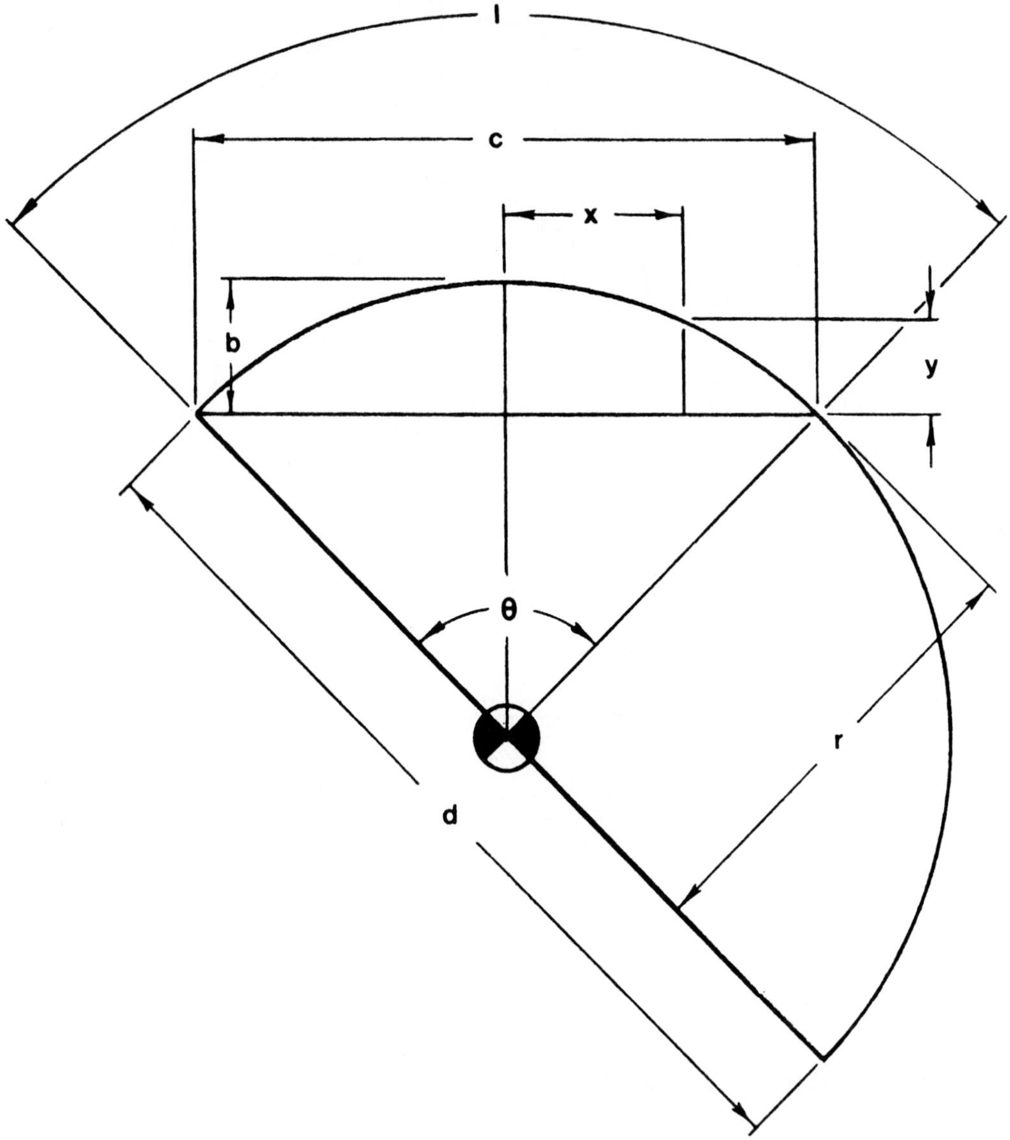


**FIGURE 3.35** Properties of the circle.

is that the line of sight for the front view is perpendicular to all frontal lines. In Fig. 3.36, the auxiliary plane is parallel to the line *AB*, and the line of sight for that auxiliary view is perpendicular to the line and the plane. So the line will appear true length in the auxiliary view.

The line *AB* in Fig. 3.37 is true length in the auxiliary view because the line of sight for that view is perpendicular to the line, as the top view indicates. The true length of *AB* could have been obtained in an auxiliary view projected off the front view, side view, or any auxiliary view of the line, provided the line of sight was made perpendicular to the line.

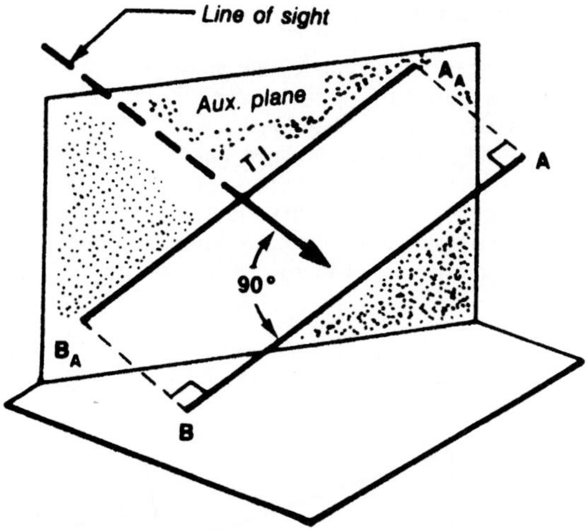


**FIGURE 3.36** True length of a line.

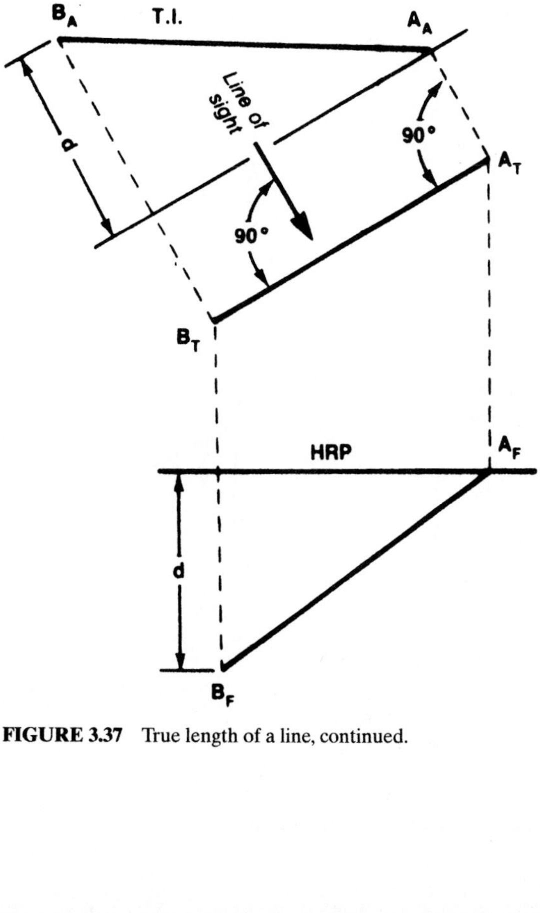


**FIGURE 3.37** True length of a line, continued.

### 3.2.2 Common Perpendicular to Two Skew Lines

The common perpendicular to two skew (nonparallel, nonperpendicular, nonintersecting) lines involves the double application of perpendicular lines, since the common perpendicular must be perpendicular to both lines.

The method of solution shown in Fig. 3.38*a* and 3.38*b* is based on finding the point view of either line. Line *CE* appears in its point view on the picture plane. In this view, the common perpendicular *PO* will appear true length, and since it is to be perpendicular in space to *AB*, it will appear perpendicular to *AB*. (When either of the two perpendicular lines appears in its point view, the other line will appear true length.)

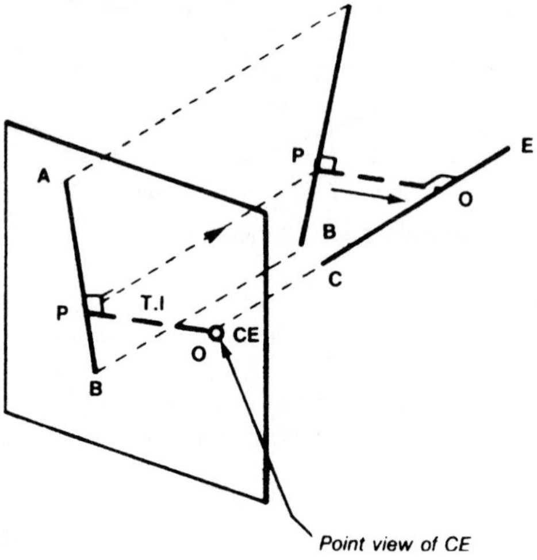


(a)

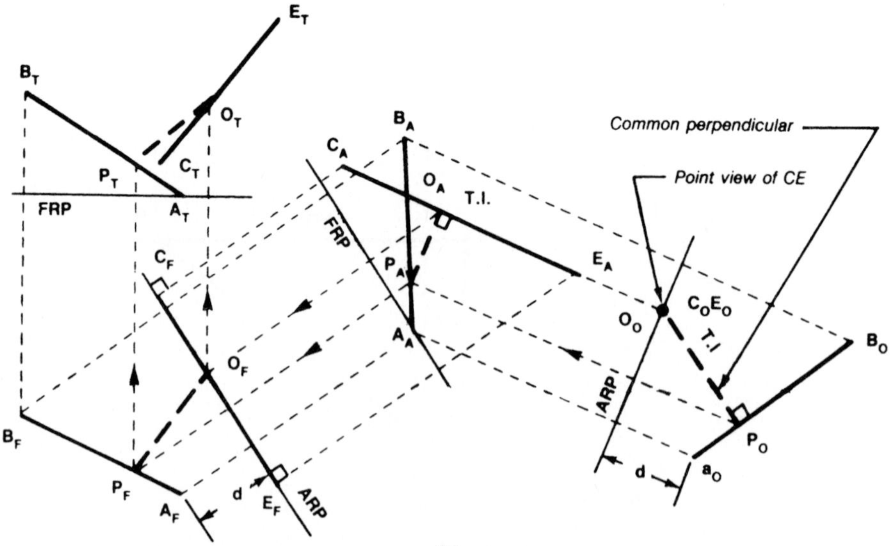


(b)

**FIGURE 3.38** (*a*) Common perpendicular to two skew lines; (*b*) common perpendicular to two skew lines, continued.

Drawing the common perpendicular through the point view of $CE$ and perpendicular to $AB$ will locate the point $P$, which is one end of the common perpendicular. In the true-length view of $CE$, the common perpendicular may be drawn perpendicular to the true-length view of $CE$ through the previously located point $P$. Thus, two views of the common perpendicular are found, and others can be obtained by regular projections.

### 3.2.3 True Size of a Plane

The true-size view of a plane is obtained by viewing (line of sight) perpendicularly at the edge view of the plane. The horizontal plane is true size in the top view, because the line of sight for that view is perpendicular in the front view to the edge view of that plane. An inclined plane is true size in a primary auxiliary view, because the plane is in its edge view in one of the regular views. The line of sight for the auxiliary view will appear perpendicular to the edge view of the inclined plane in the regular view.

An oblique plane must first be found in its edge view, as shown in Fig. 3.39. Since the profile line is true length in the side view, the line of sight for the edge view of the plane $ABC$ will be taken parallel to this line. The true size of the plane will show in the secondary auxiliary view when the line of sight for that view is perpendicular to the edge view of the plane.

### 3.2.4 The Angle between Any Two Planes

The angle between two planes may be measured by two lines drawn perpendicularly to the two planes through any point. In Fig. 3.40, point $O$ was chosen at random, line $OE$ was drawn perpendicularly to plane 1, and line $OK$ was drawn perpendicularly to plane 2.

By finding the true size of the plane $OKE$, the angle $\alpha$ may be measured. The angle between the planes is $180° - \alpha$.

In Fig. 3.41, point $O$ was chosen at a convenient place. Two lines were drawn through point $O$, with one line perpendicular to plane $ABC$ and the other perpendicular to plane 1-2-3. Note the use of horizontal and frontal lines to obtain the perpendicular lines. The horizontal line

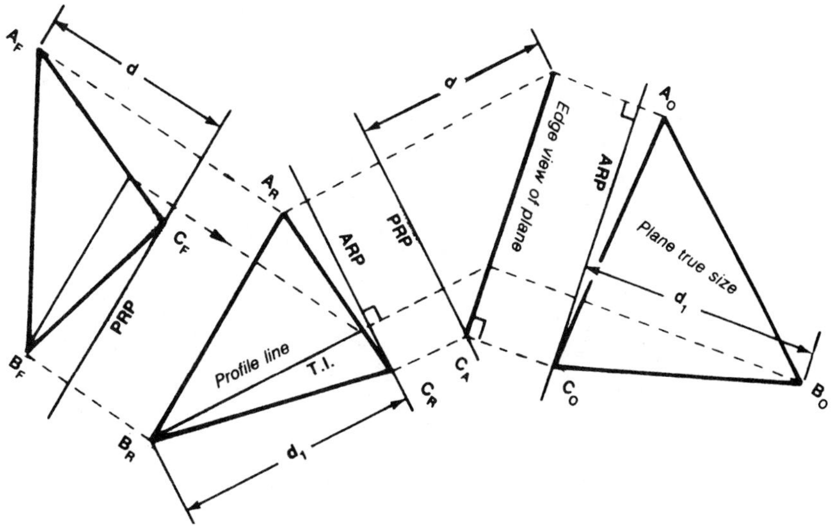


**FIGURE 3.39** True size of a plane.

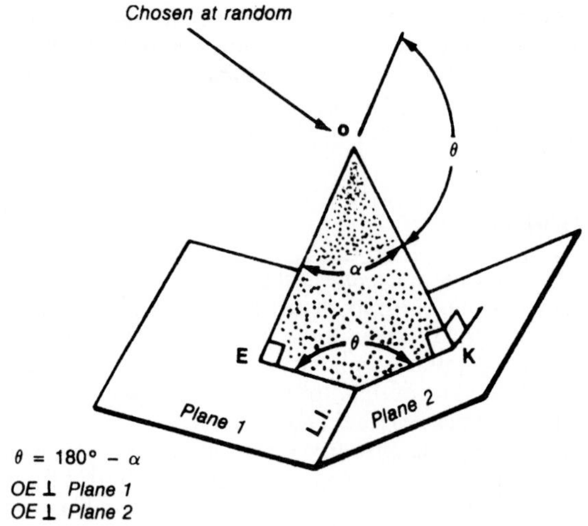


**FIGURE 3.40** Angle between two planes (perpendicular-lines method).

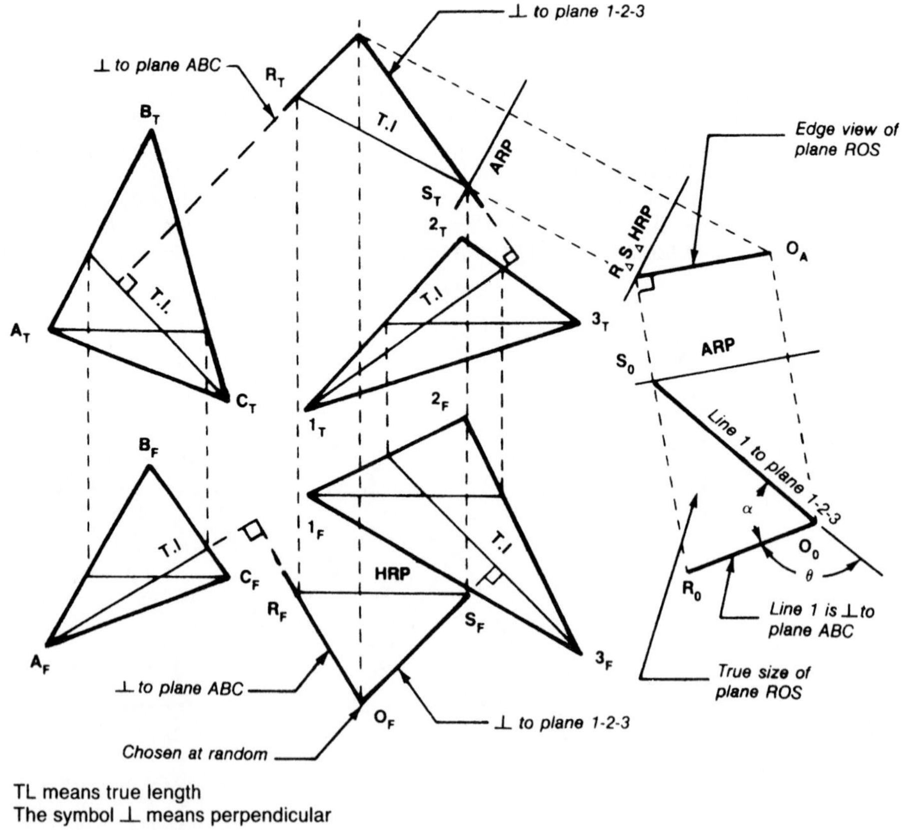


**FIGURE 3.41** Perpendicular lines method for angle between two planes.

*RS* has been added to the plane of the perpendicular lines, so that the edge and true-size views of this plane could be obtained. The first auxiliary view gives the edge view of plane *ROS*, and the secondary auxiliary view the true size of the plane and angles $\alpha$ and $\Theta$.

### 3.2.5 Resultant of Three Concurrent, Noncoplanar Forces

Finding the resultant in this case demonstrates one application of descriptive geometry to mechanics.

A parallelopiped may be formed by the loads in the three members shown in Fig. 3.42*a*. When these three members are placed in an orthographic drawing, some or all of the members may not be true length in the regular views. The loads may be scaled along members only in their true-length auxiliary views or their revolved true-length views.

After the loads have been properly scaled along the individual members using a scale suitable for all members, the parallelopiped can be drawn in each orthographic view, entirely independent of the foreshortening of the individual-member lengths. The fact that equal-length parallel lines foreshorten equally makes this possible. After the parallelopiped has been drawn, the diagonal through *O*, which is the resultant, can be revolved for its true length, or an auxiliary view can be drawn for the same purpose.

Fig. 3.42*b* shows a different method of determining the resultant of the same three forces. Force *OJ* is the resultant of the two forces *OG* and *OK*. It may be considered as replacing these two forces. This subresultant—it is not the total resultant—and force *OH*, when combined in a parallelogram, will produce the resultant *OF* of all three forces. The procedure will give the same answer arrived for Fig. 3.42*a*.

Figure 3.43 depicts the resultant of three noncoplanar forces.

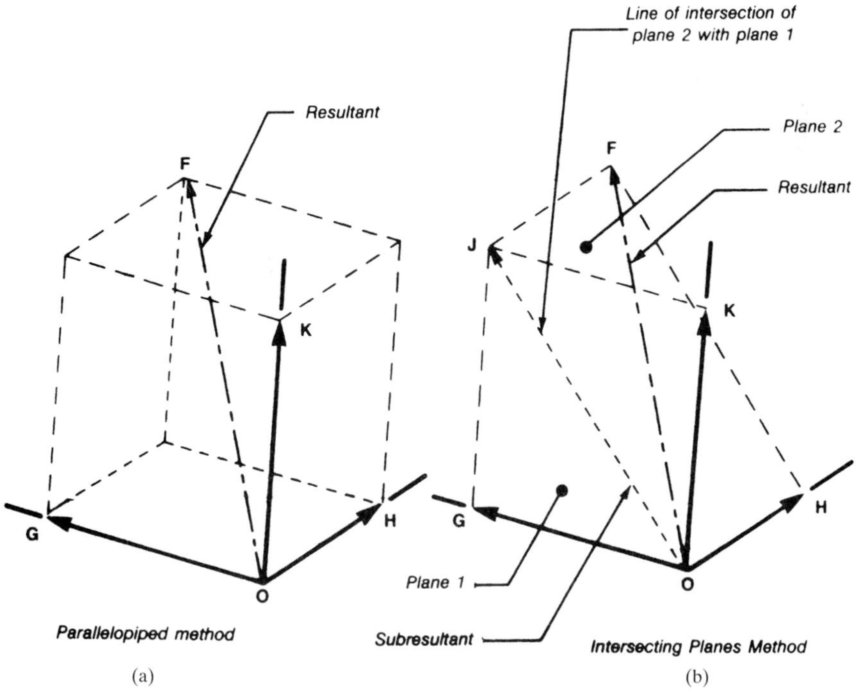


**FIGURE 3.42** (*a*) Parallelopiped method; (*b*) intersecting-planes method.

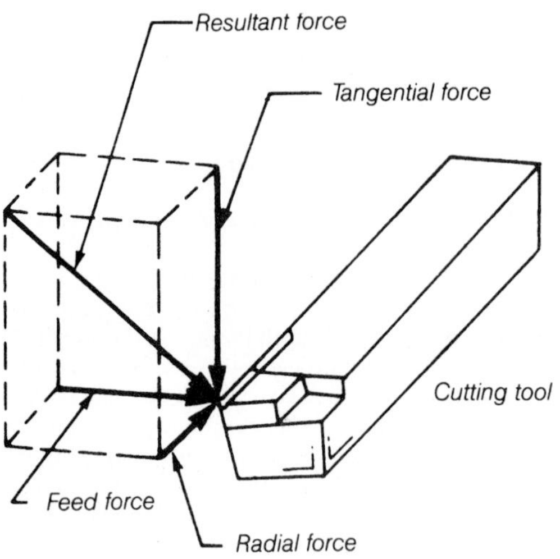


**FIGURE 3.43** Resultant of three forces.

Figure 3.44 depicts steps 1, 2, and 3 for the other method of obtaining the resultant of the forces.

## 3.3 BASIC OPTICS

Electromechanical designers are sometimes involved in the design of photographic and other equipment that contains optical elements. This section sets out many of the general working equations employed in designing basic optical equipment and systems.

Refer to Fig. 3.45, where *Ob* represents an object and *Im* its image, and where $FL$ = focal length, $U$ = distance from lens to object, $Y$ = distance from lens to image, $M$ = magnification, and $R$ = reduction. Solving for $FL$

$$\frac{1}{FL} = \frac{1}{U} + \frac{1}{V} \qquad FL = \frac{UM}{M+1} = \frac{U}{R+1}$$

$$FL = \frac{SR}{(R+1)^2} = \frac{UV}{S} = \frac{SM}{(M+1)^2}$$

Solving for $S$ (distance from object to image)

$$S = FL\left(M + \frac{1}{M} + 2\right) = FL\left(R + \frac{1}{R} + 2\right)$$

$$= \frac{FL(M+1)^2}{M} = \frac{FL(R+1)^2}{R}$$

Solving for $U$ (distance from lens to object)

$$U = \frac{V}{M} = \frac{S}{M+1}$$

$$= \frac{RS}{R+1} = \frac{FLV}{V-F}$$

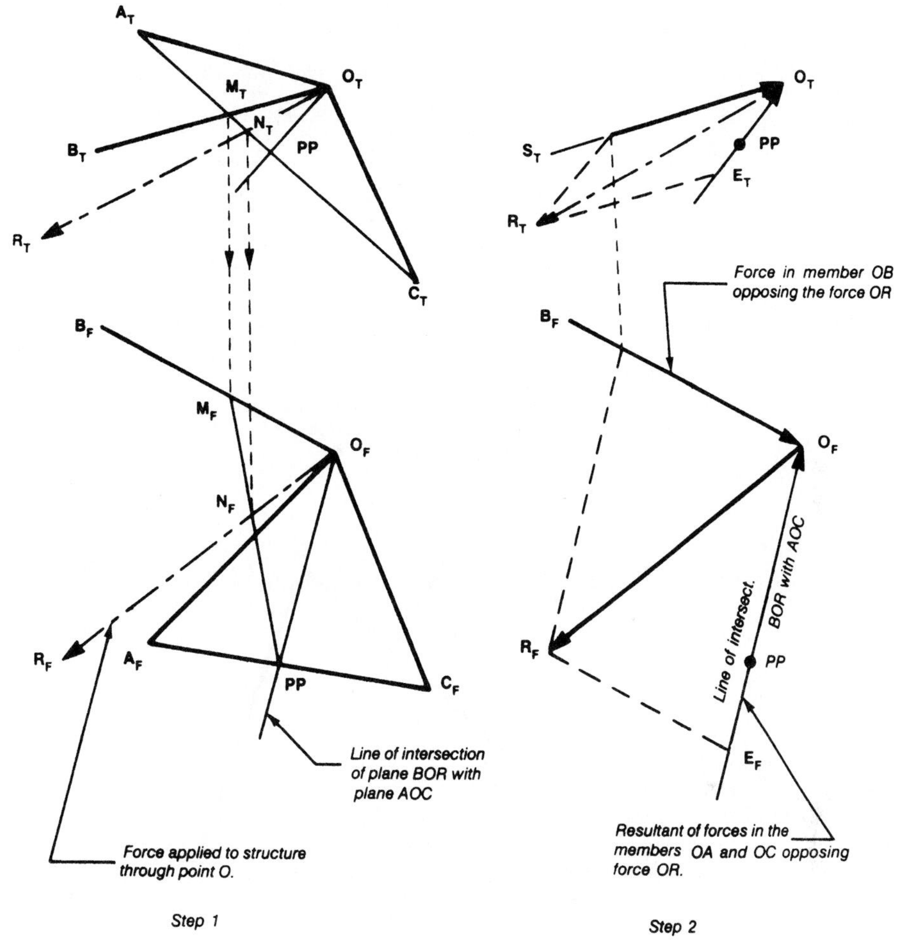


(a)

**FIGURE 3.44** Three steps in finding resultant of noncoplanar forces.

Solving for $V$ (distance from lens to image)

$$V = UM = \frac{F_L}{R} + F_L$$

$$= \frac{U}{R} = (F_L M) + F_L$$

Solving for $M$ (magnification)

$$M = \frac{V}{U} = \frac{F_L}{U - F_L} = \frac{V - F_L}{F_L}$$

Solving for $R$ (reduction)

$$R = \frac{U}{V} = \frac{U - F_L}{F_L} = \frac{F_L}{V - F_L}$$

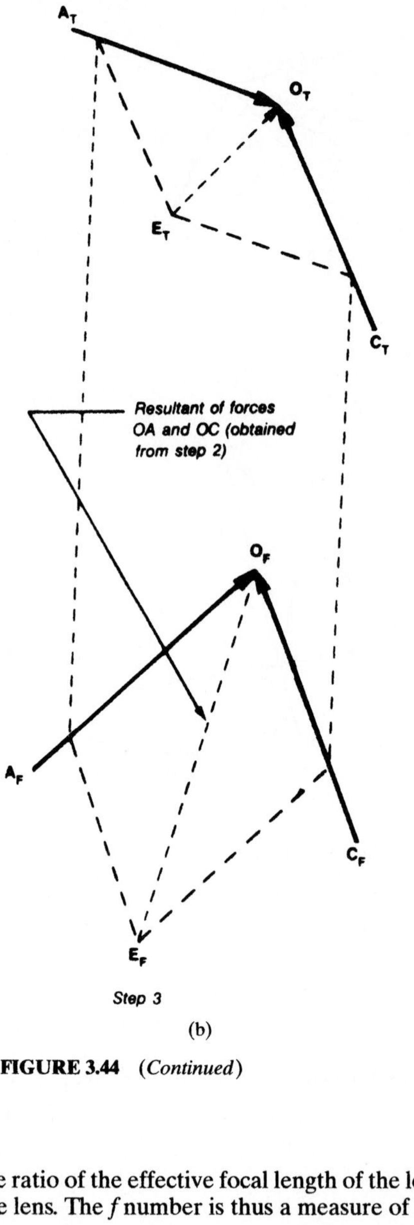


(b)

**FIGURE 3.44** (*Continued*)

### 3.3.1 Relative Aperture, or *f* Number

The *f number* of a lens is the ratio of the effective focal length of the lens divided by the diameter or clear aperture of the lens. The *f* number is thus a measure of the amount of light that can pass through the lens.

$$f\text{ number} = \frac{\text{effective focal length}}{\text{open diameter}}$$

Also

$$f\text{ number} = \frac{1}{2\,\text{NA}}$$

where NA = numerical aperture.

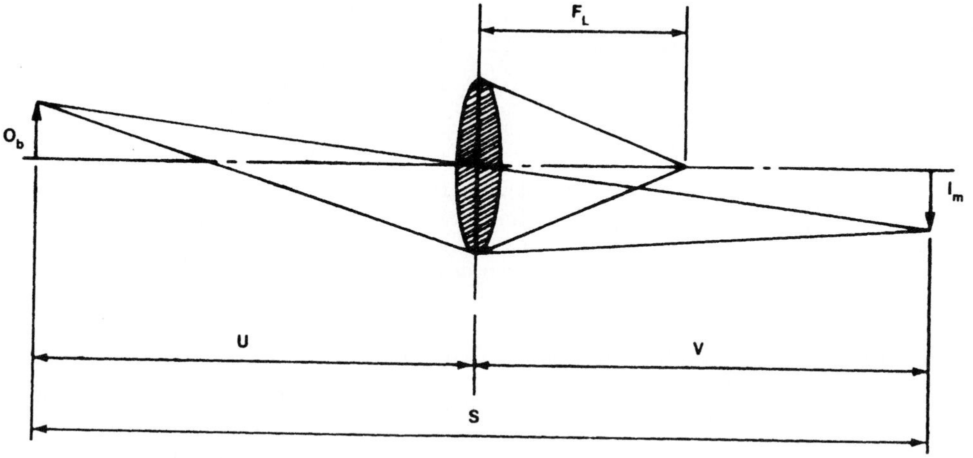


**FIGURE 3.45** Light rays through a lens.

### 3.3.2 Lensmaker's Equation

We write this equation

$$\phi = \frac{1}{f} = (n-1)\left(\frac{1}{R_1} - \frac{1}{R_2}\right)$$

where $\phi$ = power
$f$ = effective focal length (defined as the distance between the second principal point and the second focal point with parallel, incident light)
$R_1$ = first radius
$R_2$ = second radius
$n$ = index of glass

Taking lens thickness $t$ into account, this equation becomes

$$\phi = \frac{1}{f} = (n-1)\left[\frac{1}{R_1} - \frac{1}{R_2} + \frac{t(n-1)}{R_1 R_2 n}\right]$$

***Diopter.*** The reciprocal of the lens focal length given in meters is the diopter $D$.

$$D = \frac{1}{\text{effective focal length, m}}$$

Also

$$\text{Focal length, mm} = 1000 \div D$$

Thus the focal length of a 3 diopter lens = 1000 ÷ 3 = 333.3 mm.

Dioptric power is additive, so $D = D_1 + D_2$, where $D_1$ = dioptric power of first lens, and $D_2$ = dioptric power of second lens.

### 3.3.3 Lens Pairs

The following equations may be used to combine lenses in pairs to obtain selective focal lengths other than those of standard lenses. These equations are approximations, sufficiently accurate for most applications.

$$\frac{1}{F_L P} = \frac{1}{F_L 1} + \frac{1}{F_L 2} - \frac{d}{F_L 1 \cdot F_L 2} \qquad \text{or} \qquad F_L P = \frac{F_L 1 \cdot F_L 2}{F_L 1 + F_L 2 - d}$$

and

$$d = (F_L 1 + F_L 2) - \frac{F_L 1 \cdot F_L 2}{F_L P}$$

where $F_L P$ = focal length of lens pair
$F_L 1$ = focal length of first lens
$F_L 2$ = focal length of second lens
$d$ = distance between principal planes of the two lenses.

### 3.3.4 Magnification with Closeup Lenses (Cameras)

With the camera lens set at infinity focus, the magnification is calculated from

$$M = \frac{\text{focal length of camera lens, mm}}{\text{focal length of closeup lens, mm}}$$

The focal lengths of closeup lenses for various lens powers are listed:

| Lens power, diopters | Focal length, mm |
|---|---|
| 1 | 1000 |
| 2 | 500 |
| 3 | 333 |
| 1 + 3 | 250 |
| 2 + 3 | 200 |
| 1 + 2 + 3 | 167 |

If the camera lens is focused nearer than infinity, the magnification will increase and subject distance decrease. Therefore, the above equation for magnification gives the minimum attainable with a particular combination of closeup and camera lenses.

### 3.3.5 The Nature of Light

Visible light is one of many forms of radiant energy transmitted by waves. Thermal radiation and light waves are similar in nature and exhibit similar properties. Thermal radiation is not heat; it is energy in the form of wave motion.

The intensity of light falling on a nonluminous source is generally measured in footcandles (FC), or in terms of luminous flux radiated by the source per unit solid angle. The lumen (lm) is the amount of light which falls on an area one foot square at a distance of one foot from a standard light source of one candle. One foot candle is the intensity of a one candlepower light source falling on an object at a distance of one foot. The intensity of light radiating from a light source is inversely proportional to the square of the distance between the light source and the object. In photography, this fact is known as the *inverse-square law*.

The speed of light waves in a vacuum is 186,000 miles per second (mi/sec) or 300,000 km/sec. In any medium more dense than air, the speed of light is slower. The speed of light in various substances is summarized here.

| Substance | Speed of light, mi/sec |
|---|---|
| Quartz | 110,000 |
| Common crown glass | 122,691 |
| Rock salt | 110,000 |
| Borosilicate crown glass | 122,047 |
| Carbon disulfide | 114,000 |
| Medium flint glass | 114,320 |
| Ethyl alcohol | 137,000 |
| Water | 140,000 |
| Diamond | 77,000 |

The electromagnetic spectrum is shown in Fig. 3.46. From this, you will see the portion of the spectrum that is visible to the human eye. Figure 3.47*a* and 3.47*b* shows the colors created by mixing colored light and colored pigments.

### *Optical Quantities*

Luminous flux = $F$ = lumens

Intensity = $I$ = (lm)/steradians (sr) = $F/W$ ; $W$ = sr (solid angle)

$I$ = candlepower

Illumination = $E$ = lm/m$^2$ = meter-candle

Brightness or luminescence = $B$ = candle/m$^2$ = lm/($W \times$ m$^2$) = meter-lamberts

Luminous emittance = $L$ = lm/m$^2$

***Color Temperature.*** Color temperature is the method used for measuring the color of light in kelvins (degrees Celsius plus 273 = kelvins). Thus a light source with a temperature of 4500°C = 4500 + 273 = 4773 K, which is equivalent to a light source from a carbon arc (see Fig. 3.48).

***Reflection.*** A ray of light that strikes a reflecting surface is called the *incident ray* and the ray that bounces off the reflecting surface is called the *reflected ray.* The imaginary line that is perpendicular to the point where the incident ray strikes is called the *normal.*

The *law of reflection* is as follows. The angle of reflection equals the angle of incidence and lies on the opposite side of the normal. The incident ray, the reflected ray, and the normal all lie in the same plane. The incident ray and the reflected ray lie on opposite sides of the normal.

***Refraction.*** If a light wavefront strikes the first surface of a transparent medium (glass, water, etc.) at an angle, one edge of the first wavefront arrives at the surface an instant before the other edge, and the edge which arrives first is slowed down as it enters the denser medium before the second edge enters. This slowing down or retardation of one edge of the wavefront before the other edge slows down causes the wavefront to *pivot toward the normal.* The bending of the waves of light in this manner is called *refraction.* Figure 3.49*a* and 3.49*b* illustrates this principle or law of refraction.

The *laws of refraction* are as follows.

- When light travels from a medium of lesser density to a medium of greater density, the path of the light is bent toward the normal.
- When light travels from a medium of greater density to a medium of lesser density, the path of the light is bent away from the normal.

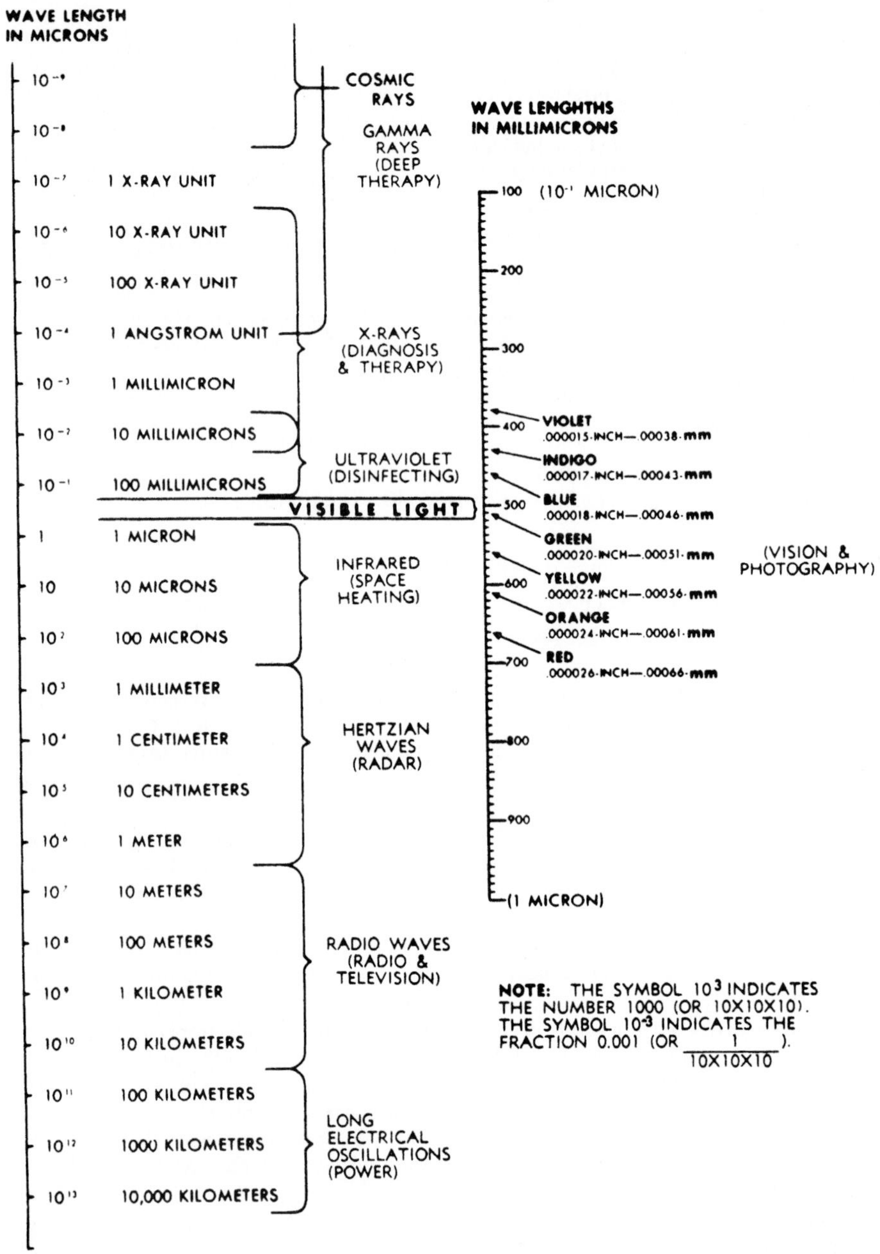


**FIGURE 3.46** The electromagnetic spectrum.

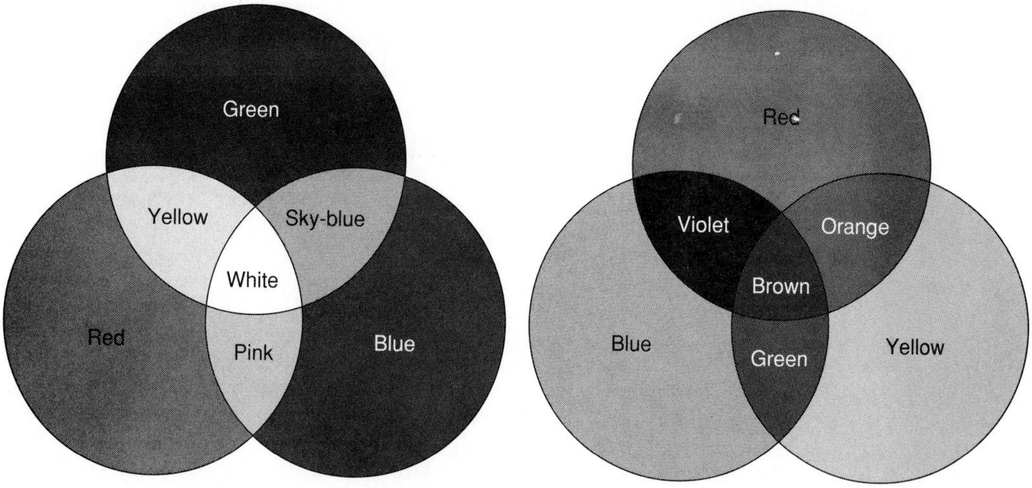


(a) (b)

**FIGURE 3.47** (*a*) Colors created by mixing beams of colored light; (*b*) colors produced by mixing paints.

- The incident ray, the normal and the refracted ray all lie in the same plane.
- The incident ray lies on the opposite side of the normal from the refracted ray.

The angle between the refracted ray of light and a straight extension of the incident ray of light through the optical medium is called the *angle of deviation.* This is the angle through which the refracted ray is bent from its original path by the optical density of the refracting medium. The amount of refraction is dependent on the angle at which light strikes an optical medium and the density of the new medium. The greater the angle of incidence and the denser the new medium, the greater the angle of refraction.

The ratio between the speed of light in a vacuum and the speed of light in an optical medium is known as the *index of refraction.*

$$\text{Index of refraction} = \frac{V_{\text{vacuum}}}{V_{\text{in medium}}}$$

You may determine the index of refraction of an optical medium by using *Snell's law.* (*Note:* For most calculations, the refractive index of air is considered to be 1.000. The exact figure is 1.000293.) The simple relationship between the angle of the incident ray with the surface normal ε and the angle of the refracted ray ε′ is shown by Snell's law to be

$$\frac{\sin \varepsilon}{\sin \varepsilon'} = \frac{n'}{n}$$

where $n$ and $n'$ are the indices of refraction of the two media. The indices of refraction of optical media are thus measured with the use of this fundamental law (see Fig. 3.50*a* and 3.50*b*). When a light ray travels from a dense medium to a less dense medium, according to Snell's law, the sine of the angle of refraction ε′ for large angles of incidence ε may be greater than 1. In this particular case, the light ray is reflected back into the medium from which it came (see Fig. 3.50*c* and 3.50*d*). Since sin ε′ cannot exceed 1, the *critical angle* is arcsin ($1/n$). Thus, if you know the indices of refraction of the optical media, you can calculate the critical angle. You can determine the index of refraction of a substance by using Snell's law by actually measur-

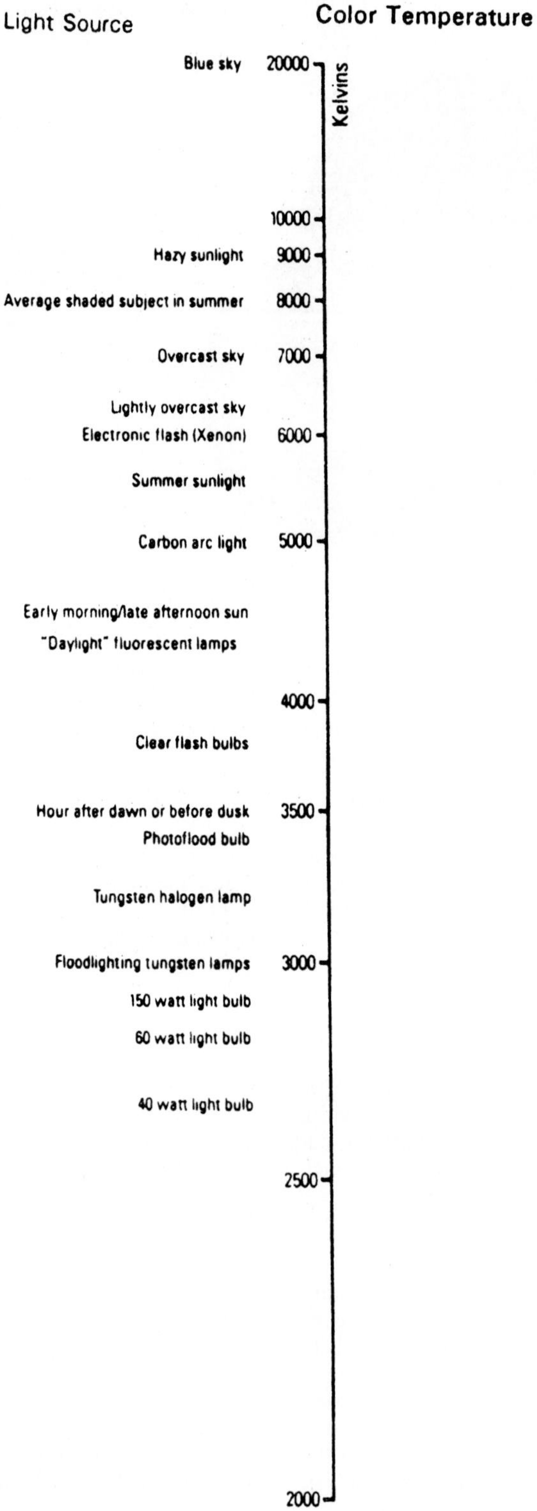


**FIGURE 3.48** Color temperature scale.

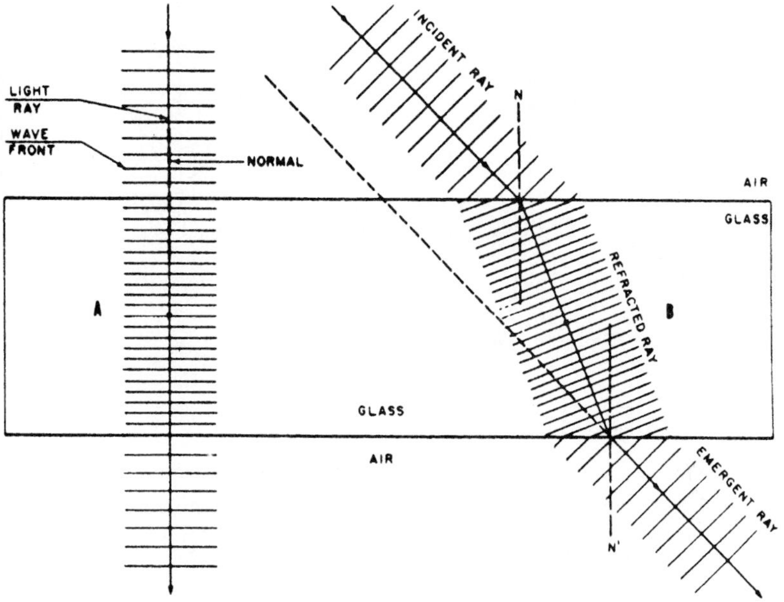


(a)

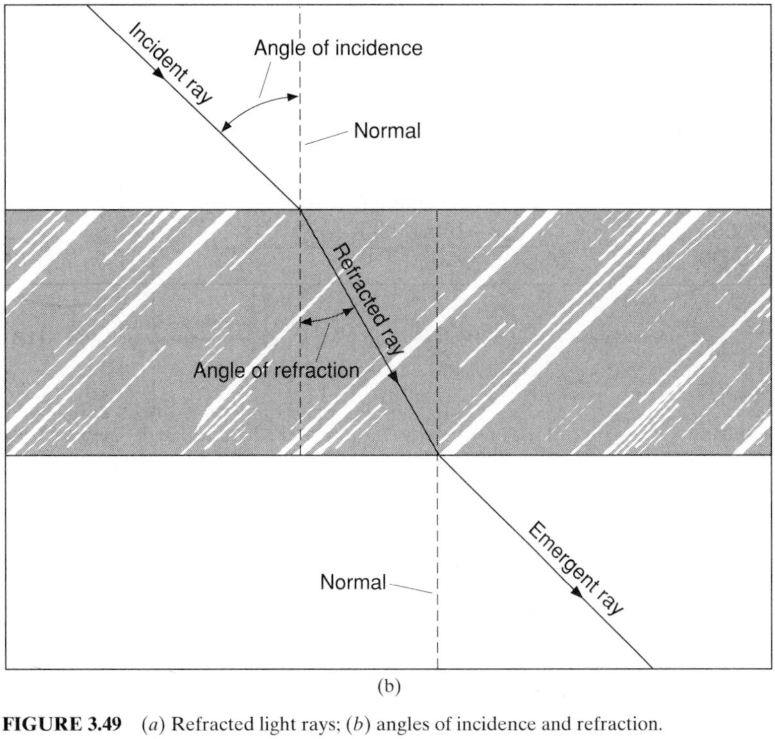


(b)

**FIGURE 3.49** (*a*) Refracted light rays; (*b*) angles of incidence and refraction.

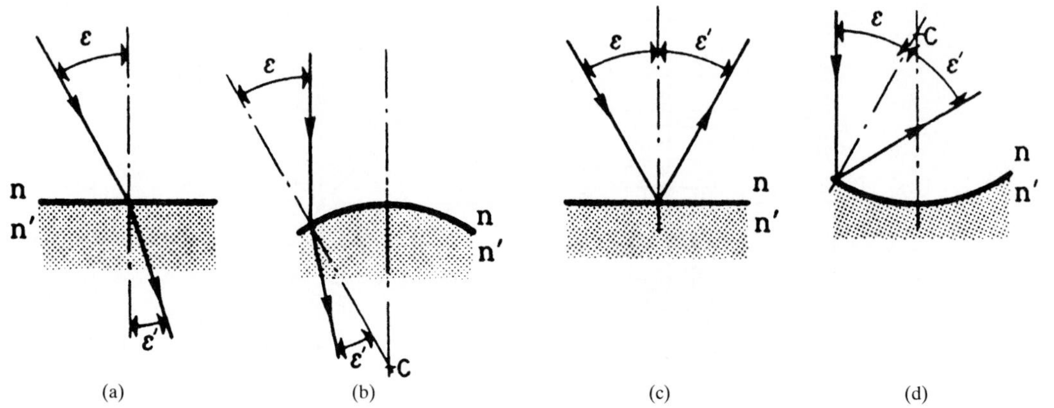


**FIGURE 3.50** (*a*) Refraction; (*b*) refraction; (*c*) reflection; (*d*) reflection.

ing the angles of incidence and refraction in a simple experiment, because air can be used instead of a vacuum for all practical purposes.

The following list shows the index of refraction for some substances:

| Substance | Index of refraction |
|---|---|
| Vacuum | 1.000000 |
| Air | 1.000293; normally taken as 1.000 |
| Water | 1.333 |
| Borosilicate crown glass | 1.517 |
| Thermosetting cement | 1.529 |
| Canada balsam | 1.530 |
| Gelatin | 1.530 |
| Light flint glass | 1.588 |
| Medium flint glass | 1.617 |
| Dense flint glass | 1.649 |
| Densest flint glass | 1.963 |
| Diamond | 2.416 |

The index of refraction of a transparent substance of high purity is a constant quantity of the physical properties of a substance. You can therefore determine the identity of a substance by measuring its index of refraction. The *refractometer* is an instrument that quickly and accurately measures the index of refraction of a substance.

### 3.3.6 Optical Lenses

The final design of optical lenses is beyond the scope of this book, but the various types of lenses and their characteristics are explained in this section. The lens equations shown in the beginning of the optics section, known as *first-order equations,* are used for the preliminary calculations required to describe or lay out optical systems prior to exacting design calculations. The books listed in the bibliography for this chapter are difinitive optical texts which thoroughly explain optical design procedures and the manufacture of optical elements.

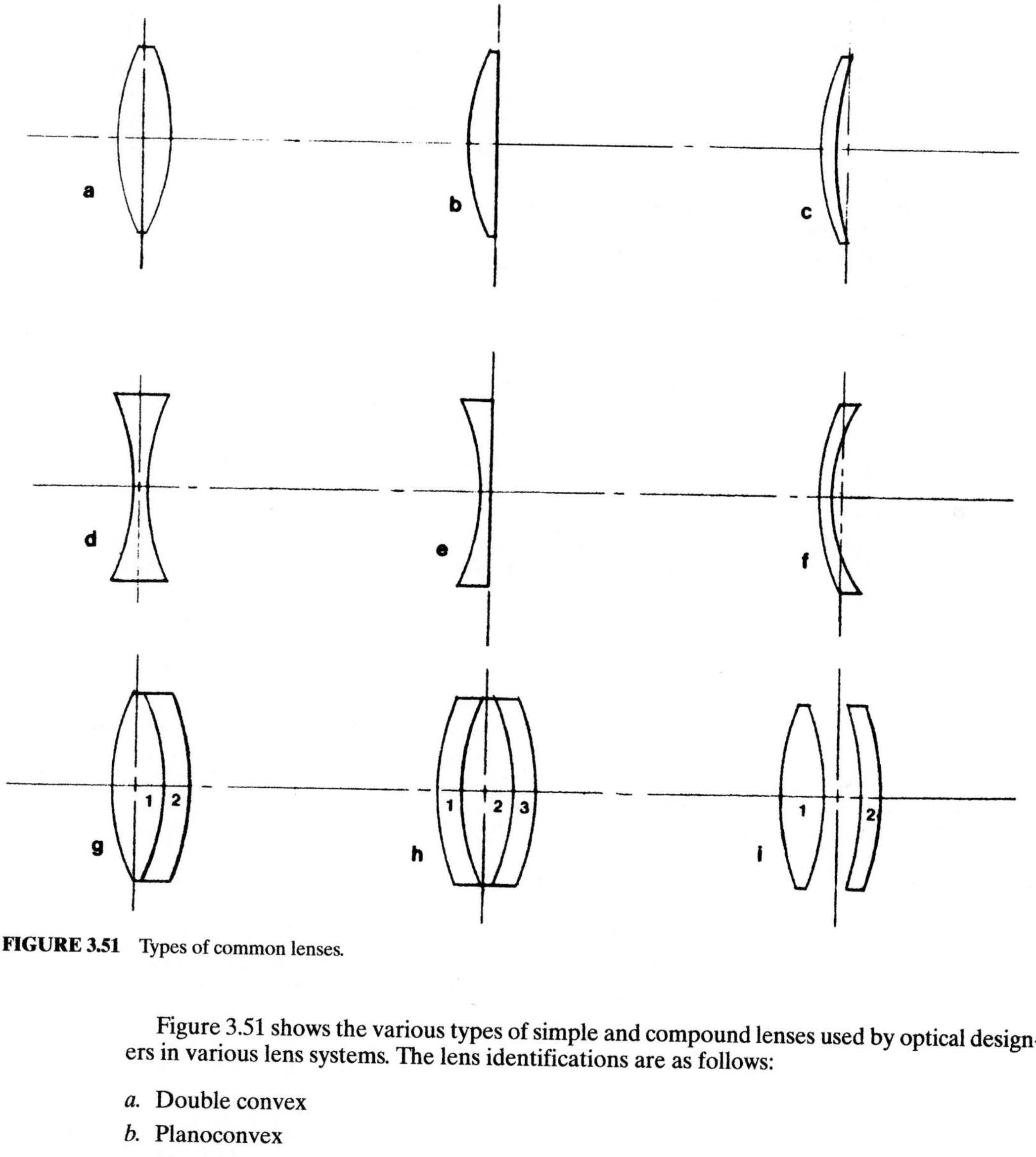


**FIGURE 3.51** Types of common lenses.

Figure 3.51 shows the various types of simple and compound lenses used by optical designers in various lens systems. The lens identifications are as follows:

*a.* Double convex
*b.* Planoconvex
*c.* Convexoconcave (meniscus converging)
*d.* Double concave
*e.* Planoconcave
*f.* Concavoconvex (meniscus diverging)
*g.* Doublet
*h.* Triplet } Compound lenses
*i.* Dialyte

***Camera Lenses.*** One of the most common applications of modern optics is in the various types of cameras produced worldwide. Optical design procedures and calculations are extremely complex and tedious and have progressed greatly since the age of the modern computer systems. The complex optical calculations are now handled by fast computers and have made it possible to design and produce lenses today that were impossible only 40 years ago. The development of improved optical glasses and new optical materials such as fluorite (synthetic monocrystalline calcium flourite) have made possible the excellent optical quality of many modern optical devices such as telescopes and camera lenses. The index of refraction of flourite is $n = 1.43$, which is low, and this material also has a very low dispersion, $V = 95.6$.

Camera lenses reached a high degree of perfection approximately 40 years ago, when the German optical companies such as Leitz (Wetzlar, Germany), Zeiss (Jena, Germany) and Schneider (Kreuznach, Germany) began to produce optical lenses and systems which are to this day considered to be of outstanding quality. Some of the early Leica lenses such as the Leitz Summicron 50mm, $f2$, are selling today for almost 400% of their original cost of 40 years ago. Figure 3.52 is the optical formula or layout for the Summicron 50mm, $f2$ camera lens. This design is a modification of an early Gauss symmetrical lens design that corrected many optical abberations present in other early lens designs.

Figure 3.53 is a more modern design of an extremely fast and highly corrected 50mm, $f1.2$ camera lens produced by the Asahi Optical Company in Japan, (Pentax). In both of the lens layouts shown, the front element faces left.

Interchangeable camera lenses are produced today in a wide range of focal lengths that begin at 15mm and go up to 2000mm. In all of these quality lens systems, distortion is kept to an absolute minimum, while color correction and contrast are excellent.

An actual camera lens system is shown disassembled in Fig. 3.54. This optical train is from a Canon 50mm, $f1.9$ Serenar, and is also of the Gauss symmetrical design. To show the complexity of a complete camera lens, Figure 3.55 depicts the Canon 50mm, $f1.9$ Serenar with all parts shown disassembled.

Figure 3.56 shows one of the early types of 35mm rangefinder cameras made by Leitz (Wetzlar). This camera is a Leica IIIf with black speed dial and is a totally mechanical type known for ruggedness, accuracy, and high-quality workmanship. Cameras of this type, although almost 40 years old, still take excellent photographs when fitted with lenses of the Summicron quality. It is doubtful whether any modern electronically, computer-controlled camera can photograph any better than these early, premium cameras.

Figure 3.57 shows a modern, mechanically controlled camera with integral exposure meter incorporated within the camera. This particular model is a Pentax Mx professional-type 35mm camera, and is one of the smallest 35mm reflex types ever manufactured. Although these cameras were produced almost 15 years ago, they still bring prices in excess of the price when the camera was new. The blackbody cameras also bring higher prices than do the

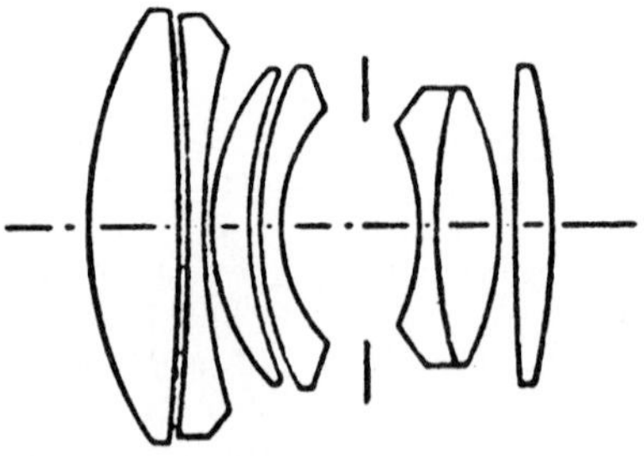

**FIGURE 3.52** The Leitz Summicron 50mm, $f2$ camera lens layout.

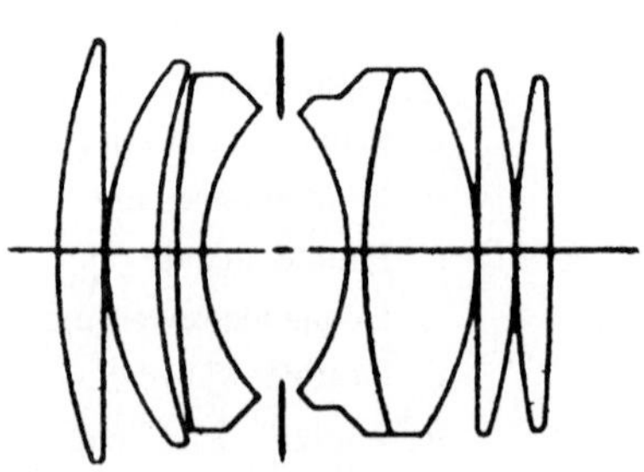

**FIGURE 3.53** The Asahi Optical Co. SMC Pentax 50mm, $f1.2$ camera lens layout.

**FIGURE 3.54** Canon Serenar $f$1.9 camera lens elements (Gauss design); circa 1952.

**FIGURE 3.55** Completely disassembled Serenar $f$1.9 camera lens showing complexity of parts.

chrome-finished ones. This particular model sold for approximately $170 when new, but will cost about $300 when in mint condition today.

Today's cameras are almost exclusively made of plastics and many contain a microprocessor or computer-control system. Many have complex and sophisticated exposure systems and operating modes to achieve optimum-quality transparencies and negatives. One must consider the fact that many of these modern features are perhaps unnecessary because the camera shutter can perform only one function: to open and close. Any camera with an accurate shutter timing mechanism and a good exposure or spot meter will produce the exact same results. Of course, modern cameras have a definite advantage because of their speed of operation, autofocusing capability, and other features which make them attractive to the general public. Professional cameras are usually of the mechanical type and are rugged and durable.

**FIGURE 3.56** A Leica IIIf rangefinder camera with Summicron, 50mm $f2$ lens (E. Leitz, Wetzlar, Germany).

**FIGURE 3.57** A Pentax SLR Mx camera with SMC Pentax M, 50mm, f1.7 lens (Asahi Optical Co., Japan).

## FURTHER READING

Bureau of Naval Personnel, 1969: *Basic Optics and Optical Instruments,* New York: Dover Publications.

De Vany, A. S., 1981: *Master Optical Techniques.* New York: Wiley.

Eshbach, O. W., and M. Souders, 1975: *Handbook of Engineering Fundamentals,* 3d ed. New York: Wiley.

Grant, H. E., 1952: *Practical Descriptive Geometry,* 1st ed. New York: McGraw-Hill.

Hecht, E., 1990: *Optics,* 2d ed. Reading, Mass.: Addison-Wesley.

Najder, K. W., 1958: *Machine Designer's Guide,* 4th ed. Ann Arbor, Mich.: Edwards Bros.

Oberg, E. V., 1980: *Machinery's Handbook,* 22d ed. New York: Industrial Press.

Rutten, H. G. J., and M. A. M. van Venrooij, 1988: *Telescope Optics, Evaluation and Design.* Richmond, Va.: Willmann-Bell.

Selby, S. M., R. C. Weast, R. S. Shankland, and C. D. Hodgman (ed.-in-chief), 1962: *Handbook of Mathematical Tables,* 1st ed. Cleveland, Ohio: Chemical Rubber Publishing Company.

# CHAPTER 4

# ENGINEERING MATERIALS, PROPERTIES, AND USES

The number of different materials used in engineering applications seems to be limitless. New materials and processes, as well as modifications to existing materials, are being developed constantly. This chapter contains reference data for only a small selection of the materials available for engineering uses; nonetheless it does include those materials that are most frequently encountered in product design applications.

The designer should be familiar with as many common materials as possible or at the least shall have access to materials reference sources. Included among such sources should be volume one of the Society of Automotive Engineers (SAE) handbook, on materials, the Copper Development Association handbook, Aluminum Institute handbook, and so forth. These references list not only the physical (mechanical), chemical, and electrical properties, but also the common uses and applications for the materials as an invaluable aid to the designer. Examples of data listed would be characteristics such as weldability, chemical resistance, and formability, and recommended product uses such as gears, brackets, and electrical contacts.

The collection of materials references is an important phase of design-engineering practice. Most of the data are available from manufacturers and societies free of charge. See Chap. 15 for a list of American societies, institutes, and specification authorities.

Further detailed technical data for engineering materials, including uses, heat treatment, and physical and chemical properties also may be obtained from the McGraw-Hill *Machining and Metalworking Handbook,* 1994. The material specifications for SAE and ASTM listed materials are covered in great detail in the previously mentioned handbook.

## *4.1 IRONS AND STEELS*

Included in this category are ferrous alloys and wrought products.

Alloys of iron or steel are the most common and useful materials for countless applications in industry. The accompanying tables of physical properties have been assembled for use in your design applications. All steels listed are wrought (hot-rolled, cold-drawn, etc.).

The acronym AISI designates the American Iron and Steel Institute, to whose identifying numbers the steels are referenced. Note that physical properties are given in ranges that represent the probable median expectancy values for each material listed.

Test values vary to a marked degree between specimens of the same material. For this reason, the values listed are not the minimum and maximum absolute values, but the probable ranges. Safety factors take into consideration these variations and must be applied during design. Safety factor values are determined by application and are covered in other sections of the handbook.

The safety factor approach is being employed somewhat less frequently today. Replacing it are the recommended or allowable stress levels for various materials classified by specification and standards authorities such as AISI, SAE, the American Society for Testing and Materials (ASTM), American National Standards Institute (ANSI), and American Institute of Steel Construction (AISC). See Chap. 15 for a listing of the American standards and specifications authorities.

### 4.1.1 Characteristics of Plain-Carbon and Common Alloy Steels

The following is a summary of the important characteristics of selected steels.

| AISI number | Characteristics |
|---|---|
| 1006–1015 | Low carbon, high formability, weldable, sheet-metal usage |
| 1030–1052 | Medium carbon, automotive applications, case- or through-harden, forgings, machinings, weldable with caution |
| 1055–1095 | Do not cold-form, flat stamping allowed, springs, usually heat-treated, excellent wear characteristics; welding not recommended |

***Common Low-Carbon or Case-Hardening Steels.*** These include AISI numbers 1018, 1020, 1025, 1117, 1215, 4615–4617, 4620, 8617, and 8620.

***Common Medium-Carbon or Direct-Hardening Steels.*** These include AISI numbers 1035, 1042, M1044, 1045, 1137, 1141, 1144, 4130, 4140, 4147–4150, 4340, 6150, and 8740.

***Common High-Carbon or Direct-Hardening Steels.*** These include AISI numbers 1055, 1060, 1065, 1070, 1080, 1090, 1095, 4150, 4140, and 4340.

***Structural-Quality Plates.*** ASTM A36 has yield strength 36,000 to 38,000 psi, tensile strength 56,000 to 82,000 psi, and elongation in 2 in of 22 to 24%.

ASTM A283, grade D, has yield strength 32,000 to 34,000 psi, tensile strength 58,000 to 74,000 psi, and elongation in 2 in of 22 to 24%. These steels are excellent choices for general structural work, buildings, and bridges, where a high-strength steel is not indicated.

***Pressure-Vessel-Quality Plates.*** ASTM A285 has a yield strength 30,000 to 32,000 psi and tensile strength 54,000 to 66,000 psi.

ASTM A515, grade 70, has yield strength 38,000 to 40,000 psi and tensile strength 69,000 to 86,000 psi and is selected for intermediate and high temperatures.

ASTM A516, grade 70, has yield strength 38,000 to 40,000 psi and tensile strength 69,000 to 86,000 psi and is selected for low temperatures.

***Common Alloy Steels.*** Low-carbon ones include 8620, 86L20, and 4617—case-hardened.

Medium-carbon ones include 4130, 4140, 4145, 4150, E6150, and 4340—direct-hardened (ASTM designations).

***Aircraft-Quality Alloy Steels.*** These include E4130 and E4340 (ASTM designations).

Note that when special conditions are encountered in the design of critical steel parts, a metallurgist should be consulted.

The steels listed in this section of the handbook are the most popular types of steels, which can be readily procured from a variety of sources and are usually stocked. Table 4.1 tabulates the mechanical properties of common-usage carbon, alloy, and stainless steels.

**TABLE 4.1** Typical Mechanical Properties of Common Carbon, Alloy, and Stainless Steels

| AISI no. | Form | Yield strength, psi | Tensile strength, psi | % elongation in 2 in | Brinell hardness |
|---|---|---|---|---|---|
| 1010 | HR | 29,000<br>31,000 | 46,000<br>48,000 | 29<br>31 | 90<br>92 |
| 1015 | HR | 44,000<br>46,000 | 60,000<br>62,000 | 38<br>40 | 125<br>127 |
| 1018 | HR | 39,000<br>41,000 | 68,000<br>70,000 | 37<br>39 | 142<br>144 |
| 1018 | CD | 69,000<br>71,000 | 81,000<br>83,000 | 19<br>21 | 162<br>164 |
| 1020 | HR | 39,000<br>41,000 | 68,000<br>70,000 | 37<br>39 | 142<br>144 |
| 1025 | HR | 44,000<br>46,000 | 66,000<br>68,000 | 35<br>37 | 142<br>144 |
| 1030 | HR | 49,000<br>51,000 | 79,000<br>81,000 | 31<br>33 | 178<br>180 |
| 1042 | CD | 88,000<br>90,000 | 101,000<br>103,000 | 15<br>17 | 206<br>208 |
| 1045 | HR | 58,000<br>60,000 | 97,000<br>99,000 | 23<br>25 | 211<br>213 |
| 1045 | WQ | 89,000<br>91,000 | 119,000<br>121,000 | 17<br>19 | 239<br>241 |
| 1060 | HR | 69,000<br>71,000 | 117,000<br>119,000 | 16<br>18 | 240<br>242 |
| 1080 | HR | 84,000<br>86,000 | 139,000<br>141,000 | 11<br>13 | 292<br>294 |
| 1095 | HR | 82,000<br>84,000 | 141,000<br>143,000 | 7<br>9 | 292<br>294 |
| 1095 | WQ | 137,000<br>139,000 | 200,000<br>202,000 | 11<br>13 | 387<br>389 |
| 1117 | HR | 41,000<br>43,000 | 70,000<br>72,000 | 32<br>34 | 142<br>144 |
| 1137 | CD | 89,000<br>91,000 | 104,000<br>106,000 | 14<br>16 | 206<br>208 |
| 1141 | CD | 94,000<br>96,000 | 111,000<br>113,000 | 15<br>17 | 222<br>224 |

**TABLE 4.1** Typical Mechanical Properties of Common Carbon, Alloy, and Stainless Steels (*Continued*)

| AISI no. | Form | Yield strength, psi | Tensile strength, psi | % elongation in 2 in | Brinell hardness |
|---|---|---|---|---|---|
| 1144 | CD | 96,000/98,000 | 113,000/115,000 | 13/15 | 234/236 |
| 4130 | N | 62,000/64,000 | 96,000/98,000 | 24/26 | 196/198 |
| 4140 | N | 94,000/96,000 | 147,000/149,000 | 16/18 | 301/303 |
| 4150 | N | 105,000/107,000 | 166,000/168,000 | 10/12 | 320/322 |
| 4340 | QR | 161,000/163,000 | 181,000/183,000 | 14/16 | 362/364 |
| 4620 | N | 52,000/54,000 | 82,000/84,000 | 28/30 | 173/175 |
| 8620 | N | 50,000/52,000 | 90,000/92,000 | 25/27 | 182/184 |
| 1141 | T | 175,000/177,000 | 236,000/238,000 | 5/7 | 460/462 |
| 1215 | CD | 67,000/69,000 | 75,000/77,000 | 14/16 | 238/240 |
| 1340 | T | 230,000/232,000 | 261,000/263,000 | 10/12 | 504/506 |
| **Chromium-nickel stainless steels (austenitic). Non-heat-treatable.** | | | | | |
| 201 | A | 54,000/56,000 | 114,000/116,000 | 54/56 | 209/211 |
| 301 | A | 39,000/41,000 | 109,000/111,000 | 49/51 | 179/181 |
| 303 | A | 34,000/36,000 | 89,000/91,000 | 49/51 | 160/162 |
| 304 | A | 34,000/36,000 | 84,000/86,000 | 54/56 | 179/181 |
| 309 | A | 39,000/41,000 | 94,000/96,000 | 44/46 | 200/202 |
| 310 | A | 44,000/46,000 | 94,000/96,000 | 49/51 | 180/182 |
| 316 | A | 34,000/36,000 | 84,000/86,000 | 59/61 | 198/200 |

**TABLE 4.1** Typical Mechanical Properties of Common Carbon, Alloy, and Stainless Steels (*Continued*)

| AISI no. | Form | Yield strength, psi | Tensile strength, psi | % elongation in 2 in | Brinell hardness |
|---|---|---|---|---|---|
| 321 | A | 34,000/36,000 | 84,000/86,000 | 54/56 | 198/200 |
| **Chromium stainless steels (martensitic). Heat treatable. (The values shown reflect heat treatment for maximum strength.)** | | | | | |
| 410 | HT | 179,000/181,000 | 199,000/201,000 | — | 400 |
| 416 | HT | 114,000/116,000 | 139,000/141,000 | — | 280 |
| 440C | HT | 274,000/276,000 | 284,000/286,000 | — | 600 |

FORM KEY:
HR = hot rolled, CD = cold drawn, N = normalized, WQ = water quenched, OQ = oil quenched, T = tempered, HT = heat treated, A = annealed.

*Note on stainless steels*: The 300 series is non-heat-treatable. Maximum strength may only be obtained through cold working. The steels may be ordered cold-worked in one-fourth, one-half, three-fourths, and fully hardened condition. In this case, maximum tensile strength for CD wire is 348,000/350,000 psi, and for strips 248,000/250,000 psi.

Austenitic stainless steels are nonmagnetic; martensitic and ferritic stainless steels are magnetic.

Steels may be purchased to particular specifications such that the chemical content, test data, and hardening requirements are *certified* on a "heat to heat" basis. A "heat of steel" is a batch quantity manufactured at one time.

For physical and chemical properties of steels not listed, the stock and data books of the large producers and suppliers are recommended. The data books may be obtained by letter request and should be kept in your engineering reference files.

For information on cast irons and steels, see the Chap. 12. section on metallic castings.

### 4.1.2 Identification of Stainless Steels

A method for determining whether a steel is stainless steel, carbon steel, or iron is as follows.

1. Make up a 7 to 8% aqueous solution of copper sulfate ($CuSO_4$). Use the proportion of 86 g of copper sulfate per liter of water; i.e., 11° Baumé (Bé), or a specific gravity of 1.08 g/ml.
2. Clean a small area of the metal to be tested. All grease and residue must be removed. Use a mild, nonchemical abrasive and flush with clean water.
3. Apply a few drops of the $CuSO_4$ solution to the cleaned metal on the test sample.
4. If the material is regular carbon steel or iron, a metallic copper film will form where the test solution contacts the sample. Stainless steel will show no copper deposit.

5. Apply a small magnet to the test sample to show the difference between austenitic and ferritic stainless steels; ferritic materials will be attracted to the magnet. Austenitic stainless steels are nonmagnetic.

## 4.2 BRINELL HARDNESS TESTING

To determine the hardness of metallic materials, a known load is applied to a steel ball of known diameter and the resulting impression in the material is measured. The hardness is then determined by the equation,

$$\text{Bhn} = \frac{P}{(\pi D/2)(D - \sqrt{D^2 - d^2})}$$

where Bhn = Brinell hardness number
$P$ = load applied to the ball, kg
$D$ = diameter of ball, mm
$d$ = measured diameter at the rim of the impression in the material, mm

The standard ball is 10 mm in diameter. Standard loads are 3000, 1500, and 500 kg. This test is not recommended when the material is above 630 Bhn.

## 4.3 AISI-SAE DESIGNATION SYSTEM FOR STEELS

The system adopted by these two organizations to designate carbon and alloy steels has been outlined in Table 4.2.

## 4.4 COMMON ALUMINUM ALLOYS

*Aluminum wrought products* are used to replace steel in many design applications. They are specified for bus bars and machined parts and structures and have many other uses in industry. Modern aluminum alloys are both lightweight and strong, weighing approximately 2.8 times less than steel.

Table 4.3 enumerates the mechanical properties of the common aluminum alloys.

***Popular Non-Heat-Treatable Alloys.*** Aluminum products in this category include numbers 1100,* 3003,* 5005,* 5052,* 5083,* 5086,* 5454,* and 5456.*

***Popular Heat-Treatable Strong Alloys.*** Aluminums in this group include numbers 2014, 2024, 6061,* 6063,* and 7075.

* Weldable alloys, i.e., suited for resistance and arc welding. Other alloys are not recommended for welding. For information on aluminum casting alloys, see the casting section of this handbook.

**TABLE 4.2** AISI-SAE Designations for Carbon and Alloy Steels

| AISI-SAE designation | Type of steel |
|---|---|
| 10xx<br>11xx<br>12xx<br>15xx | Carbon steels |
| 13xx | Manganese steels |
| 23xx<br>25xx | Nickel steels |
| 31xx<br>32xx<br>33xx<br>34xx | Nickel-chromium steels |
| 40xx<br>44xx | Molybdenum steels |
| 41xx | Chromium-molybdenum steels |
| 43xx 47xx<br>81xx 86xx<br>87xx 88xx<br>93xx 94xx<br>97xx 98xx | Nickel-chromium-molybdenum steels |
| 46xx<br>48xx | Nickel-molybdenum steels |
| 50xx<br>51xx<br>50xxx<br>51xxx<br>52xxx | Chromium steels |
| 61xx | Chromium-vanadium steels |
| 72xx | Tungsten-chromium steels |
| 92xx | Silicon-manganese steels |
| 9xx | High-strength, low-alloy steels |

*Note*: 13xx manganese steels and 300-series stainless steels are used in electrical equipment and switchgear when nonmagnetic properties are required, and a high strength material is indicated. Such applications are found in the vicinity of high-current-carrying bus systems, when the current exceeds 2,000 A. The use of nonmagnetic materials prevents the flow of induced currents that could cause heating problems in standard steels.

**TABLE 4.3** Mechanical Properties of Common Aluminum Alloys

| Alloy and temper | Yield strength, psi | Tensile strength, psi | % elongation in 2 in* | Brinell Hardness |
|---|---|---|---|---|
| 1100—O | 4,000/6,000 | 12,000/14,000 | 44/46 | 22/24 |
| 1100—H14 | 16,000/18,000 | 17,000/19,000 | 19/21 | 31/33 |
| 2014—O | 13,000/15,000 | 26,000/28,000 | 17/19 | 44/46 |
| 2014—T6,T651 | 59,000/61,000 | 69,000/71,000 | 12/14 | 134/136 |
| 2024—O | 10,000/12,000 | 26,000/28,000 | 21/23 | 46/48 |
| 2024—T4,T351 | 46,000/48,000 | 67,000/69,000 | 18/20 | 119/121 |
| 3003—O | 5,000/7,000 | 15,000/17,000 | 39/41 | 27/29 |
| 3003—H14 | 20,000/22,000 | 21,000/23,000 | 15/17 | 39/41 |
| 5005—H14 | 21,000/23,000 | 22,000/24,000 | 5/7 | — |
| 5052—O | 12,000/14,000 | 27,000/29,000 | 29/31 | 46/48 |
| 5052—H32 | 27,000/29,000 | 32,000/34,000 | 17/19 | 59/61 |
| 5083—H112 | 27,000/29,000 | 43,000/45,000 | 15/17 | 79/81 |
| 5054—O | 16,000/18,000 | 35,000/37,000 | 24/26 | 61/63 |
| 5054—H34 | 34,000/36,000 | 43,000/45,000 | 15/17 | 80/82 |
| 6061—O | 7,000/9,000 | 17,000/19,000 | 29/31 | 29/31 |
| 6061—T6,T651 | 39,000/41,000 | 44,000/46,000 | 16/18 | 94/96 |
| 6063—O | 6,000/8,000 | 12,000/14,000 | — | 24/26 |
| 6063—T6 | 30,000/32,000 | 34,000/36,000 | 17/19 | 72/74 |
| 7075—O | 14,000/16,000 | 32,000/34,000 | 15/17 | 59/61 |
| 7075—T6,T651 | 72,000/74,000 | 82,000/84,000 | 10/12 | 149/151 |
| 7178—T6,T651 | 77,000/79,000 | 87,000/89,000 | 10/12 | 159/161 |

*% elongation is for 0.50-in thick specimens.

## 4.5 COPPER AND ITS ALLOYS

Mechanical properties of wrought products have been compiled in Table 4.4. Explanations for the numbers associated with copper and copper-alloy designations are as follows:

| | |
|---|---|
| 102 | Oxygen-free |
| 110 | Electrolytic tough pitch |
| 210 | Gilding metal |
| 220 | Commercial bronze |
| 230 | Red brass |
| 260 | Cartridge brass |
| 268, 270 | Yellow brass |
| 342, 343 | High-leaded brass |
| 360 | Free-cutting brass |
| 464, 467 | Naval brass |
| 510 | Phosphor bronze |
| 614 | Aluminum bronze |
| 655 | Silicon bronze |
| 715 | Copper nickel |

For more detailed specifications and data on copper and copper alloys, see the *Standards Handbook—Copper and Copper Alloys,* published by the Copper Development Association and cited in the bibliography for this chapter.

For cast copper and its alloys, see the section on metallic castings.

## 4.6 OTHER COMMON METALS AND ALLOYS

Refer to Table 4.5 for property information on other common metals and alloys.

***Electrical Conductors.*** The best conductors of electric current are, in order:

1. *Silver:* Oxides and salts of silver are also conductors.
2. *Copper:* However, oxides of copper are not good conductors.
3. *Gold:* Gold does not normally oxidize.
4. *Aluminum:* However, oxides of aluminum are not good conductors.

## 4.7 COMMON PLASTICS

Listed here are some of the more prevalent materials:

Acetal (Delrin, Celcon)
Acetate (cellulose)
Acrylic (Lucite, Plexiglas)
Acrylonitrile-butadiene-styrene (ABS)

**TABLE 4.4** Mechanical Properties of Common Copper Alloys

| Copper or copper-alloy no. | Yield strength, psi | Tensile strength, psi | % elongation in 2 in* | Rockwell Hardness, B scale |
|---|---|---|---|---|
| 102 | 44,000 | 49,000 | 11 | 49 |
| hard | 46,000 | 51,000 | 13 | 51 |
| 110 | 43,000 | 47,000 | 15 | 46 |
| hard | 45,000 | 49,000 | 17 | 48 |
| 210 | 39,000 | 47,000 | 11 | 50 |
| 1/2 hard | 41,000 | 49,000 | 13 | 52 |
| 220 | 44,000 | 51,000 | 10 | 56 |
| 1/2 hard | 46,000 | 53,000 | 12 | 58 |
| 230 | 48,000 | 56,000 | 11 | 64 |
| 1/2 hard | 50,000 | 58,000 | 13 | 66 |
| 260 | 51,000 | 61,000 | 22 | 69 |
| 1/2 hard | 53,000 | 63,000 | 24 | 71 |
| 268,270 | 49,000 | 60,000 | 22 | 69 |
| 1/2 hard | 51,000 | 62,000 | 24 | 71 |
| 342, 343 | 49,000 | 60,000 | 19 | 69 |
| 1/2 hard | 51,000 | 62,000 | 21 | 71 |
| 360 | 44,000 | 57,000 | 24 | 77 |
| 1/2 hard | 46,000 | 59,000 | 26 | 79 |
| 464, 467 | 52,000 | 74,000 | 19 | 81 |
| 1/2 hard | 54,000 | 76,000 | 21 | 83 |
| 510 | 54,000 | 67,000 | 27 | 77 |
| 1/2 hard | 56,000 | 69,000 | 29 | 79 |
| 614 | 44,000 | 77,000 | 39 | 83 |
| hard | 46,000 | 79,000 | 41 | 85 |
| 655 | 44,000 | 77,000 | 16 | 86 |
| 1/2 hard | 46,000 | 79,000 | 18 | 88 |
| 715 | 69,000 | 74,000 | 14 | 80 |
| 1/2 hard | 71,000 | 76,000 | 16 | 82 |
| **Beryllium-copper* and beryllium-nickel (Δ) wrought products.** | | | | |
| C175* (10) | 80,000 | 100,000 | 10 | 92 |
| temper AT | 100,000 | 120,000 | 20 | 100 |
| C170* (165) | 130,000 | 150,000 | 4 | C33 |
| temper AT | 165,000 | 180,000 | 10 | C38 |
| Δ UNS #N033 | 230,000 | 270,000 | 7 | (83 |
| Ber.-nickel HT | 235,000 | 275,000 | 8 | 90) |
| | | | | 15N |

*Note*: Above are typical values for 0.50 in-thick specimens, $\frac{1}{2}$ hard and hard.

**TABLE 4.5** Other Common Metals and Alloys

| Metal | Density g/cm³ | Weight, lb/in³ | Young's modulus (E), tension psi | Torsional modulus (G), psi | Poisson's ratio | Electrical resistivity Ω/cm |
|---|---|---|---|---|---|---|
| Aluminum (pure) | 2.70 | 0.098 | $9 \times 10^6$ | — | 0.33 | $2.6 \times 10^{-6}$ |
| Aluminum alloy (high-strength) | 2.78 | 0.101 | $10–11 \times 10^6$ | — | 0.33 | $2.8 \times 10^{-6}$ |
| Antimony | 6.69 | 0.242 | $11.3 \times 10^6$ | $2.9 \times 10^6$ | — | $3.1 \times 10^{-6}$ |
| Beryllium-copper (C170) alloy | 8.4 | 0.303 | $19 \times 10^6$ | — | 0.29 | $3.1 \times 10^{-6}$ |
| Bismuth | 9.75 | 0.352 | — | — | — | $119 \times 10^{-6}$ |
| Brass (80Cu/20Zn) | 8.6 | 0.311 | $16 \times 10^6$ | $6 \times 10^6$ | 0.34 | $7 \times 10^{-6}$ |
| Cadmium | 8.65 | 0.312 | $10.1 \times 10^6$ | $3.5 \times 10^6$ | — | $7.5 \times -6$ |
| Cast iron (gray) | 7.2 | 0.260 | $14 \times 10^6$ | $5.6 \times 10^6$ | 0.21 | — |
| Chromium | 7.19 | 0.260 | — | — | — | $2.6 \times 10^{-6}$ |
| Copper | 8.96 | 0.324 | $17 \times 10^6$ | $6.4 \times 10^6$ | 0.34 | $1.72 \times 10^{-6}$ |
| Gold | 19.32 | 0.698 | $11.4 \times 10^6$ | — | — | $2.44 \times 10^{-6}$ |
| Iron | 7.87 | 0.284 | $28 \times 10^6$ | $11.2 \times 10^6$ | 0.30 | $10 \times 10^{-6}$ |
| Iron (malleable) | 7.85 | 0.284 | $25 \times 10^6$ | $11.5 \times 10^6$ | 0.17 | — |
| Lead | 11.35 | 0.410 | $2.4 \times 10^6$ | — | 0.43 | $22 \times 10^{-6}$ |
| Lithium | 0.53 | 0.019 | — | — | — | $9 \times 10^{-6}$ |
| Magnesium (cast) | 1.74 | 0.063 | $6.5 \times 10^6$ | $2.4 \times 10^6$ | 0.35 | $44 \times 10^{-6}$ |
| Mercury | 13.55 | 0.490 | — | — | — | $96 \times 10^{-6}$ |
| Molybdenum | 10.22 | 0.369 | — | — | — | $5.7 \times 10^{-6}$ |
| Nickel | 8.90 | 0.322 | $30 \times 10^6$ | $10.6 \times 10^6$ | 0.32 | $7.8 \times 10^{-6}$ |
| Palladium | 12.02 | 0.434 | — | — | — | $11 \times 10^{-6}$ |
| Phosphor bronze | 8.90 | 0.322 | $16 \times 10^6$ | $6 \times 10^6$ | 0.35 | $10 \times 10^{-6}$ |
| Platinum | 21.45 | 0.775 | $24.2 \times 10^6$ | $9.3 \times 10^6$ | — | $10 \times 10^{-6}$ |
| Silver | 10.50 | 0.379 | $11.2 \times 10^6$ | $3.8 \times 10^6$ | — | $1.6 \times 10^{-6}$ |
| Steel (medium-carbon) | 7.80 | 0.282 | $30 \times 10^6$ | $11.4 \times 10^6$ | 0.30 | $15 \times 10^{-6}$ |
| Steel (stainless 300-Type) | 8.03 | 0.290 | $28 \times 10^6$ | $11.4 \times 10^6$ | 0.28 | $30 \times 10^{-6}$ |
| Steel (stainless 400-Type) | 7.75 | 0.280 | $29 \times 10^6$ | $12.6 \times 10^6$ | 0.28 | $30 \times 10^{-6}$ |
| Tin | 7.31 | 0.264 | $7 \times 10^6$ | $2.4 \times 10^6$ | — | $11.5 \times 10^{-6}$ |
| Titanium | 4.54 | 0.164 | $16 \times 10^6$ | — | — | — |
| Tungsten | 19.30 | 0.697 | $51.5 \times 10^6$ | $21.5 \times 10^6$ | — | $5.51 \times 10^{-6}$ |
| Zinc | 7.13 | 0.258 | $14.5 \times 10^6$ | $5 \times 10^6$ | 0.11 | $6 \times 10^{-6}$ |

Benelex
Epoxy, epoxy glass
Diallyl phthalate, Melamine
Mylar (polyester film)
Nylon
Phenol formaldehyde
Phenolic laminates
Polycarbonate (Lexan)
Polyester glass
Polyethylene
Polyimide
Polypropylene
Polystyrene
Polysulfone
Polyurethane
Polyvinyl chloride (PVC)
Rtv (room-temperature vulcanizing) silicones
Styrofoam (polystyrene)
Teflon [polytetrafluoroethylene (PTFE)]
Urea formaldehyde

The following compilation of the properties and typical uses of common plastics will be useful in making preliminary selections for design purposes. You should secure a detailed specification sheet on your selected plastic from the manufacturer or supplier of the material. Samples of such sheets are shown in Figs. 4.1*a*–4.1*c*.

### Acetal (Delrin, Celcon)

*Properties:* High modulus of elasticity, low coefficient of friction, excellent abrasion and impact resistance, low moisture absorption, excellent machinability, ablative.

*Typical uses:* Bearings, gears, antifriction parts, electrical components, washers, seals, insulators, cams.

### Acetate (Cellulose)

*Properties:* Odorless, tasteless, nontoxic, grease resistance, high impact strength.

*Typical uses:* Badges, blister packaging, displays, optical covers, book covers.

### Acrylic (Lucite, Plexiglas)

*Properties:* Unusual optical clarity, high tensile strength, weatherability, good electrical properties, ablative.

*Typical uses:* Displays, signs, models, lenses, electrical and electronic parts.

# LEXAN® resin High Modulus 500 Grade

**LEXAN 500** resin is a member of the LEXAN family of polycarbonate resins particularly suited to applications requiring high rigidity combined with toughness and impact strength.

**LEXAN 500** resin has the highest combination of rigidity and impact strength available in any thermoplastic, plus excellent dimensional stability, low mold shrinkage, and colorability.

**LEXAN 500** resin also has these additional advantages:

**Mechanical:**
- Flexural modulus increased nearly 50% over standard LEXAN resins

**Flammability:**
- Listed 94 V-0* at 1/16″ (1.6 mm) per UL Std. 94
- Listed 94 5V* at 1/8″ (3.2 mm) per UL Std. 94

**Molding:**
- Reduced mold shrinkage allows for the production of more precise parts.

## Typical Property Values

English Units (SI Units)

| PROPERTY | ASTM TEST METHOD | LEXAN 500 resins** |
|---|---|---|
| **PHYSICAL** | | |
| Specific Gravity | D792 | 1.25 |
| Specific Volume, in³/lb(cm³/g) | — | 22.2(0.80) |
| Weight/volume, lbs/in³(g/cm³) | — | 0.045(1.25) |
| Water Absorption, % | D570 | |
| 24 hrs @ 73°F(23°C) | | 0.12 |
| Equilibrium, 73°F(23°C) | | 0.31 |
| Mold Shrinkage, in/in(mm/mm) | D955 | 0.002-0.004** |
| Transmittance, % | D1003 | Not applicable |
| Haze, % | D1003 | Not applicable |
| Refractive Index | — | Not applicable |
| **THERMAL** | | |
| Deflection Temperature, °F(°C) | D648 | |
| @ 66 psi(0.46 MPa) | | 295(146) |
| @ 264 psi(1.84 MPa) | | 288(142) |
| Specific Heat, btu/lb/°F(kj/kgK) | — | 0.29(1.20) |
| Thermal Conductivity, Btu-in/h-ft²-°F (W/km) | — | 1.41(0.20) |
| Coefficient of Thermal Expansion in/in/°F(m/m/°C) | D696 | 1.79x10⁻⁵(3.22x10⁻⁵) |
| Vicat Softening Temperature, °F(°C) | D1525 | 310(154) |
| Flammability Ratings | | |
| ASTM D635* | D635 | AEB <0.5″ |
| UL Standard 94* 1/16″(1.6 mm) | UL 94 | 94 V-0 |
| UL Standard 94* 1/8″(3.2 mm) | UL 94 | 94 V-0/94 5V |
| Oxygen Index | D2863 | 32.5 |

*This rating is not intended to reflect hazards of this or any other material under actual fire conditions.
** .002-.004 in/in for parts up to 8″ in length.
.0025-.0045 in/in for parts 8″-16″ in length.
.0035-.0055 in/in for parts over 16″ in length.
Property values may vary slightly for some colors.

| PROPERTY | ASTM TEST METHOD | LEXAN 500 resins** |
|---|---|---|
| **ELECTRICAL** | | |
| Dielectric Strength, volts/mi.(kV/mm) | D149 | |
| Short time, 125 mils(3.2 mm) | | 450(17.7) |
| Dielectric Constant | D150 | |
| 60H | | 3.10 |
| $10^6$H | | 3.05 |
| Power Factor | D150 | |
| 60H | | 0.0008 |
| $10^6$H | | 0.0075 |
| Volume Resistivity, ohm-cm | D257 | |
| @ 73°F, dry(23°C) | | >$10^{16}$ |
| Arc Resistance, sec | D495 | |
| Stainless Steel Electrodes | | 5-10 |
| Tungsten Electrodes | | 120 |
| **MECHANICAL** | | |
| Tensile Strength, psi(MPa) | D638 | |
| Yield | | 9,600(66) |
| Ultimate | | 8,000(55) |
| Elongation, % | D638 | |
| Yield | | 8-9 |
| Rupture | | 10-20 |
| Tensile Modulus, $10^5$ psi(MPa) | D638 | 4.50(3,100) |
| Flexural Strength, psi(MPa) | D790 | 15,000(100) |
| Flexural Modulus, $10^5$ psi(MPa) | D790 | 5.0(3,400) |
| Compressive Strength, psi(MPa) | D695 | 14,000(96) |
| Compressive Modulus, $10^5$ psi(MPa) | D695 | 5.20(3,600) |
| Shear Strength, psi(MPa) | D732 | |
| Yield | | 8,500(58) |
| Shear Modulus, $10^5$ psi(MPa) | — | 1.47(1,000) |
| Izod Impact Strength, ft-lbs/in(J/m) | D256 | |
| Notched, ⅛″ thick(3.2 mm) | | 2(100) |
| Unnotched, ⅛″ thick(3.2 mm) | | 40(2,000) |
| Tensile Impact Strength, ft-lbs/in²(kJ/m²) | D1822 | |
| S-type | | 75(158) |
| Falling Dart Impact Strength, ft-lbs, (J), ⅛″ thick(3.2 mm) | | 75(102) |
| Fatigue Strength, psi @ 2.5mm cycles(MPa) | D671 | 2,000(14.0) |
| Rockwell Hardness | D785 | |
| M | | 85 |
| R | | 124 |
| Deformation Under Load, % | D621 | |
| 4000 psi, @ 73°F(27 MPa @ 23°C) | | 0.2 |
| 4000 psi, @ 158°F(27 MPa @ 70°C) | | 0.4 |
| Taber Abrasion Resistance, mg weight loss/1000 cycles | D1044 | 11 |

**FIGURE 4.1*a*** Sample plastics specification sheet—Lexan 500 polycarbonate.

# POLYETHYLENE TEREPHTHALATE (PET)

## ERTALYTE®

Ertalyte® is an unreinforced, partly crystalline thermoplastic polyester based on polyethylene terephthalate (PET). It is characterized as having excellent wear resistance and a low coefficient of friction, together with high modulus, low creep and superior dimensional stability. Ertalyte's specific properties make it especially suitable for the manufacture of mechanical precision parts which are capable of sustaining high loads and wear conditions. Ertalyte's continuous service temperature is approximately 10% higher than acetals, and its melting point is almost 150°F higher.

In addition, Ertalyte® PET offers good strength combined with good chemical and abrasion resistance. Its low moisture absorption enables its mechanical and electrical properties to remain virtually unaffected by ambient moisture. These qualities, combined with FDA compliance, make Ertalyte® PET an excellent candidate for certain food contact applications.

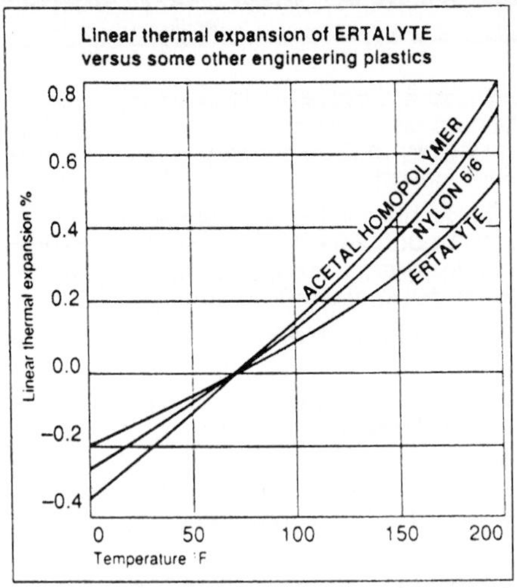


### Typical Applications

- Bearings
- Bushings
- Seals
- Spacers
- Thrust washers
- Rollers
- Guides
- Insulators
- Food contact parts
- Pump components
- Valve parts

### Availabilities

### Rod

Diameter: 1/2" to 6"
Length:
To 2 3/4" diameter –
8 ft. nominal
Over 2 3/4" diameter–
4 ft. nominal
Plate
Thickness: 1/2" to 2"
Size: 24" x 48" standard

### Key Benefits

- High strength and rigidity
- Low moisture absorption
- Superior chemical resistance compared to nylon and actal
- Good abrasion resistance
- FDA compliant
  (Regulation #177.1630)
- Good electrical properties
- Excellent dimensional stability
- Weather resistance
- Radiation resistance

| | Property | Ertalyte® PET |
|---|---|---|
| Product Description | | Semi-crystalline Thermoplastic Polyester |
| Mechanical | Specific Gravity | 1.39 |
| | Tensile Strength, 73°F | 12,400 |
| | Tensile Modulus of Elasticity, 73°F | 423,000 |
| | Elongation, 73°F | 20 |
| | Flexural Strength, 73°F | - |
| | Flexural Modulus of Elasticity, 73°F | - |
| | Shear Strength, 73°F | - |
| | Compressive Strength, 10% Def. | 15,000 |
| | Compressive Modulus of Elasticity, 73°F | - |
| | Coefficient of Friction (Dry vs. Steel) Dynamic ⑧ | - |
| | Hardness, Rockwell, 73°F | M93-100 |
| | Durometer, 73°F | D87 |
| | Tensile Impact | 50 |
| Thermal | Coefficient of Linear Thermal Expansion | $3.3 \times 10^{5}$ |
| | Deformation Under Load (122° F, 2,000 psi) | - |
| | Deflection Temperature 264 psi | 215 |
| | Tg-Glass transition (amorphous) | - |
| | Melting Point (crystalline) | 491 |
| | Continuous Service Temperature in Air (Max.) | 212 |
| Electrical | Dielectric Strength Short Time | 385 |
| | Volume Resistivity | $5.5 \times 10^{16}$ |
| | Dielectric Constant, 60Hz | - |
| | $10^3$Hz | - |
| | $10^6$Hz | - |
| | Dissipation Factor, 60Hz | - |
| | 10Hz | - |
| Chemical | Water Absorption Immersion 24 Hours | 0.07 |
| | Saturation | 0.50 |
| | Acids, Weak, 73°F | A |
| | Strong, 73°F | U |
| | Alkalies, Weak, 73°F | A |
| | Strong, 73°F | U |
| | Hydrocarbons-Aromatic, 73°F | - |
| | Hydrocarbons-Aliphatic, 73°F | U |
| | Ketones, 73°F | A |
| | Ethers, 73° | - |
| | Esters, 73°F | A |
| | Alcohols, 73°F | A |
| | Inorganic Salt Solutions, 73°F | - |
| | Continuous Sunlight, 73°F | L |

**FIGURE 4.1*b*** Sample plastics specification sheet—polyethylene terephthalate (PET).

# 800 Family Engineering Thermoplastics VALOX® resin

**800 Family.** A unique series of glass reinforced polyester alloys offering:

**Property Advantages**
• HDT's from 320° — 380° F (160° — 193° C) at 264 psi (1.82 MPa)
• Improved surface gloss • Mechanical strength and stiffness
• Good colorability • UL 94 V-0 and V-2 capability

**Processing Advantages**
• Superior flow (thin walls) • Good mold release

## Typical Property Values — English Units (SI Units)

| PROPERTY | UNITS | ASTM TEST METHOD | VALOX 815 resin 15% Glass Reinforced PBT Alloy | VALOX 830 resin 30% Glass Reinforced PBT Alloy | VALOX 850 resin 15% Glass Reinforced PBT Alloy UL 94 V-2 Recognized | VALOX 855 resin 15% Glass Reinforced PBT Alloy UL 94 V-0 Recognized | VALOX 865 resin 30% Glass Reinforced PBT Alloy UL 94 V-0 Recognized |
|---|---|---|---|---|---|---|---|
| **PHYSICAL** | | | | | | | |
| Specific Gravity | | D-792 | 1.43 | 1.54 | 1.51 | 1.54 | 1.66 |
| Specific Volume | in³/lb(cm³/kg) | D-892 | 19.5(701) | 18.6(669) | 18.3(659) | 18.0(649) | 16.8(607) |
| Water Absorption 24 hours | % | | .06 | .06 | .05 | .06 | .03 |
| Mold Shrinkage | in/in x $10^{-3}$ (mm/mmx$10^{-3}$) | | | | | | |
| Flow Direction | | | | | | | |
| 30-90 mil(0.76-2.3 mm) | | | 5-7 | 3-5 | 5-7 | 5-7 | 3-5 |
| 90-180 mil(2.3-4.6 mm) | | | 7-9 | 5-7 | 7-9 | 7-9 | 5-7 |
| Cross Flow Direction | | | | | | | |
| 30-90 mil(0.76-2.3 mm) | | | 7-8 | 6-7 | 7-8 | 7-8 | 6-7 |
| 90-180 mil(2.3-4.6 mm) | | | 8-11 | 7-9 | 8-11 | 8-11 | 7-9 |
| **MECHANICAL** | | | | | | | |
| Tensile Strength | psi(MPa) | D-638 | 12,000(83) | 15,000(100) | 13,000(90) | 14,000(96) | 16,000(110) |
| Elongation at Break | % | D-638# | 3 | 3 | 5 | 5 | — |
| Flexural Strength | psi(MPa) | D790 | 20,000(140) | 25,000(170) | 20,000(140) | 22,000(150) | 26,000(180) |
| Flexural Modulus | psi (MPa) | D790 | 650,000 (4,500) | 1,000,000 (6,900) | 750,000 (5,200) | 750,000 (5,200) | 1,100,000 (7,600) |
| Compressive Strength | psi(MPa) | D695 | 21,800(150) | 21,800(150) | 14,970(100) | 15,970(110) | 19,600(135) |
| Shear Strength | psi(MPa) | D732 | 8,900(61) | 8,900(61) | 7,510(52) | 7,390(51) | 9,000(60) |
| Izod Impact Strength | | D256 | | | | | |
| Notched, ⅛" thick(3.2 mm) | ft-lb/in (J/m) | | 0.7(37) | 1.5(80) | 0.8(43) | 1.0(53) | 1.4(75) |
| Unnotched, ⅛" thick(3.2 mm) | ft-lb/in (J/m) | | 5(270) | 12(640) | 5(270) | 6(320) | 12(640) |
| Rockwell Hardness R-scale | | D785 | 119 | 119 | 121 | 119 | 119 |
| **THERMAL** | | | | | | | |
| Heat Deflection Temp. | | D648 | | | | | |
| @ 66 psi(0.46 MPa) | °F(°C) | | 410(210) | 430(221) | 400(204) | 400(204) | 415(212) |
| @ 264 psi(1.82 MPa) | °F(°C) | | 320(160) | 380(193) | 370(187) | 370(187) | 380(193) |
| Coeff. of Thermal Expansion | in/in/°F (m/m/°C) | D696 | | | | | |
| Mold Direction x $10^{-5}$ | | | | | | | |
| Range: -40 → 100°F(-40 → 40°C) | | | 2.5(4.5) | 1.4(2.5) | 2.3(4.1) | 2.5(4.5) | 1.2(2.2) |
| Range: 140 → 280°F(60 → 140°C) | | | 3.0(5.4) | 1.4(2.5) | 2.3(4.1) | 3.4(6.1) | 1.2(2.2) |
| **ELECTRICAL** | | | | | | | |
| Dielectric Strength 1/16"(1.6 mm) | V/mil(kV/mm) | D149 | 600(24) | 630(25) | 580(23) | 570(22) | 590(23) |
| 1/8"(3.2 mm) | V/mil(kV/mm) | | 560(22) | 530(21) | 520(20) | 510(20) | 480(20) |
| Dielectric Constant 100 Hz | | D150 | 3.6 | 3.6 | 3.5 | 3.5 | 3.8 |
| $10^6$ Hz | | | 3.5 | 3.5 | 3.4 | 3.4 | 3.7 |
| Dissipation Factor 100 Hz | | D150 | .002 | .002 | .001 | .001 | .002 |
| $10^6$ Hz | | | .02 | .02 | .01 | .01 | .01 |
| Volume Resistivity | ohm-cm x $10^{16}$ (ohm-m x $10^{14}$) | D257 | 5.6 | 4.0 | 3.9 | 4.5 | 1.8 |
| **UL** | | | | | | | |
| Flammability† | | UL 94 | HB/.063" | HB/.063" | V-2/.062" | V-0/.062" | V-0/.062" 5V/.090" |
| Arc Resistance | sec | D-495 | 136/.125" | 84/.125" | 81/.125" | 68/.127" | 86/.124" |
| High Voltage Arc Tracking Rate | in/min | UL 746A | .8/.125" | 1.5/.125" | 6.9/.125" | 13.4/.127" | 11.4/.124" |
| High Ampere Arc Ignition | arcs | UL 746A | 44/.125" | 18/.125" | 14/.125" | 13/.127" | 16/.124" |
| Hot Wire Ignition | sec | UL 746A | 50/.125" | 139/.125" | 56/.125" | 33/.062" | 24/.062" |
| Comparative Track Index (CTI) | volts | UL 746A | 275/.125" | 260/.125" | 225/.125" | 205/.127" | 225/.124" |
| UL Temp. Index | | UL 746B | | | | | |
| Elec. Properties | °C | | 125/.063" | 120/.063" | 125/.062" | 125/.062" | 110/.062" |
| Mech. Properties with impact | °C | | 110/.063" | 75/.063" | 110/.062" | 110/.062" | 105/.062" |
| Mech. Properties without impact | °C | | 125/.063" | 120/.063" | 125/.062" | 125/.062" | 110/.062" |

† This rating is not intended to reflect hazards presented by this or any other material under actual fire conditions
# ASTM D638 type V @ 0.5"/min.

**FIGURE 4.1*c*** Sample plastics specification sheet—Valox reinforced polyester.

### Acrylonitrile-Butadiene-Styrene (ABS)

*Properties:* High rigidity and impact strength; excellent abrasion resistance; excellent electrical characteristics; resistance to many inorganic salts, alkalies, and many acids.

*Typical uses:* Containers, antifriction parts, electrical and electronic parts, machined parts.

### Benelex (Laminate)

*Properties:* High compressive strength; machinable; resists corrosion (alkalies or acids); good electrical insulation; high flexural, shear, and tensile strength.

*Typical uses:* Work surfaces, electrical panels and switchgear, bus braces (low voltage only), neutron shielding.

### Diallyl Phthalate, Melamine

*Properties:* High strength, chemical resistance, low water absorption, medium- to high-temperature use.

*Typical uses:* Terminal blocks and strips, dishware, automotive applications, aerospace applications.

### Epoxy Glass

*Properties:* High strength, high-temperature applications, flame retardant, low coefficient of thermal expansion, low water absorption.

*Typical uses:* High-quality printed-circuit boards, microwave stripline applications, very-high-frequency and ultra-high-frequency (VHF and UHF) applications, electrical insulation, services in temperature range −40 to 500°F.

### Mylar (Polyester Film; Polyethylene Terephthalate)

*Properties:* High dielectric strength, chemical resistance, high mechanical strength, moisture resistance, temperature range 70 to 105°C, does not embrittle with age.

*Typical uses:* Electrical and industrial applications, graphic arts applications.

### Nylon

*Properties:* Wear resistance, low friction, high tensile strength, excellent impact resistance, high fatigue resistance, easy machining, corrosion resistance, lightweight.

*Typical uses:* Bearings, bushings, valve seats, washers, seals, cams, gears, guides, wheels, insulators, wear parts.

### Phenol Formaldehyde (Bakelite)

*Properties:* Wear resistance, rigidity, moldability to precise dimensions, strength, excellent electrical properties, economical, will not support combustion.

*Typical uses:* Electrical and electronic parts, handles, housings, insulator parts, mechanism parts, parts that are to resist temperatures to 250°C.

## Phenolic Laminates

*Properties:* Immune to common solvents, lightweight, strong, easily machined.
*Typical uses:* Bearings, machined parts, insulation, gears, cams, sleeves, electrical and electronic parts.

## Polycarbonate (Lexan)

*Properties:* Virtually unbreakable, weather-resistant, optically clear, lightweight, self-extinguishing, thermoformable, machinable, solvent cementable.
*Typical Uses:* Thermoforming shapes, drill motor housings, glazing, electrical insulation, machined parts, thermoplastic injection mouldings, gears, bullet-proof glazing, plumbing parts.

## Polyester Glass

*Properties:* Extreme toughness, high dielectric strength, heat resistance, low water absorption; antitracking electrically, self-extinguishing, machinable.
*Typical uses:* Insulators and bus braces, switch phase barriers, general electrical insulation, mechanical insulated push rods for switches and breakers, contact blocks, terminal blocks.

## Polyethylene

*Properties:* Transparency in thin sheets, water resistance.
*Typical uses:* Bags for food storage, vapor barriers in construction, trays, rollers, gaskets, seals, radiation shielding.

## Polypropylene

*Properties:* Good tensile strength, low water absorption, excellent chemical resistance, stress crack resistance, electrical properties.
*Typical uses:* Tanks, ducts, exhaust systems, gaskets, laboratory and hospital ware, wire coating, sporting goods.

## Polystyrene

*Properties:* Outstanding electrical properties, excellent machinability, ease of fabrication, excellent chemical resistance, oil resistance, clarity, rigidity, hardness, dimensional stability.
*Typical uses:* Lighting panels, tote boxes, electronic components, door panels (refrigerators), drip pans, displays, furniture components.

## Polysulfone

*Properties:* Tough, rigid, high-strength, high-temperature thermoplastic, temperature range –150° to 300°F, excellent electrical characteristics, good chemical resistance, low creep and cold-flow properties, capable of being repeatedly autoclaved.

*Typical uses:* Food processing and medical industries, electrical and electronic, appliance, automotive, aircraft, and aerospace uses.

### Polyurethane

*Properties:* Elastomeric to rock hard forms available, high physical characteristics, toughness, durability, broad hardness range, withstands severe use, abrasion resistance, weather resistance, radiation resistance, temperature range –80 to 250°F, resistance to common solvents, available also in foam types.

*Typical uses:* Replaces a host of materials that are not performing well; extremely broad range of usage; replaces rubber parts, plastic parts, and some metallic parts.

### Polyvinyl Chloride (PVC)

*Properties:* Corrosion resistance, formability, excellent electrical properties, impact resistance, low water absorption; lightweight, cementable, machinable, weldable.

*Typical uses:* Machined parts, nuts, bolts, PVC pipe and fittings, valves, strainers.

### RTV Silicone Rubber

*Properties:* Resistance to temperature extremes (–75 to 400°F), excellent electrical characteristics, weather resistance, good chemical resistance, FDA-, USDA-, and UL-approved.

*Typical uses:* General-purpose high-quality sealant, gasket cement, food contact surfaces, electrical insulation, bonding agent, glass-tank construction, and countless other applications.

### Styrofoam

*Properties:* Low water absorption, floats, thermal insulator, extremely lightweight.

*Typical uses:* Insulation board for homes and buildings, cups, containers, thermos containers, shock-absorbing packaging, plates (food), flotation logs.

### Teflon (PTFE)

*Properties:* Unexcelled chemical resistance, cryogenic service, electrical insulation, very low friction, high dielectric strength, very low dissipation factor, very high resistivity, machinability.

*Typical uses:* Valve components, gaskets (with caution, due to cold flow), pump parts, seal rings, insulators (electrical), terminals, bearings, rollers, bushings, electrical tapes, plumbing tapes, machined parts, bondable with special etchant preparations.

### Urea Formaldehyde

*Properties:* Hard, strong, molds accurately, low water absorption, excellent electrical properties, ablative, economical, will not support combustion.

*Typical uses:* Electrical and electronic parts, insulators, small parts, housings.

## 4.8 COMMON PLASTICS—MECHANICAL PROPERTIES

Table 4.6 gives mechanical properties of plastics.

The property values listed in Table 4.6 are averaged minima and maxima and include values for glass-filled types. These data are for preliminary design considerations only. The specification sheets for a particular type of plastic should be obtained from the manufacturer of the material prior to initial design. Manufacturers such as Du Pont and General Electric will supply you with design data books and specification sheets on their materials on request. These material should be kept in your design data files.

### 4.8.1 Petrochemical Origin of the Plastics

The petrochemical origins of the main plastics families are shown in Fig. 4.2. The chart shows the chemical constituents that produce the plastic product. (*Source:* PMI, Plastics Manufacturing Co., Harrisburg, N.C.).

## 4.9 INSULATING MATERIALS

Table 4.7 summarizes properties of insulating materials.

## 4.10 ELASTOMERS

Properties of this class of materials are presented in Table 4.8.

## 4.11 COMMON WOODS

Table 4.9 draws together properties of common woods.

## 4.12 THERMAL EXPANSION OF MATERIALS

Table 4.10 presets coefficients of linear expansion for common materials.

## 4.13 SPECIAL-PURPOSE ALLOYS

Table 4.11 shows property data for selected special-purpose alloys.

***Magnesium Alloys.*** Wrought alloys include SAE numbers 51, 510, 511, 52, and 520. Casting alloys include SAE numbers 50 and 500.

***Nickel Alloys.*** These include Monel, R-Monel, K-Monel, S-Monel, Inconel, and Inconel X.

***Copper Alloys.*** These include tellurium copper 145, Hitenso 1622, chromium-copper 182, Everdur 637, 651, 655, 661, and HS commercial bronze 316.

**TABLE 4.6** Mechanical Properties of Plastics

| Plastic | Tensile strength, psi | Compressive strength, psi | Flexural yield, psi |
|---|---|---|---|
| ABS | 3,000/9,000 | 6,000/22,000 | 5,000/28,000 |
| Acetal | 6,000/12,000 | 10,000/19,000 | 12,000/29,000 |
| Acetate (cellulose) | 4,400/8,000 | — | 5,000/11,000 |
| Epoxy glass | 9,000/21,000 | 24,000/41,000 | 9,000/61,000 |
| Epoxy resin | 3,000/14,000 | 14,000/26,000 | 13,000/22,000 |
| Nylon | 7,000/26,000 | 12,000/30,000 | 6,000/42,000 |
| Phenolic resins | 4,000/10,000 | 10,000/35,000 | 4,000/18,000 |
| Polycarbonate | 7,000/26,000 | 8,000/22,000 | 13,000/33,000 |
| Polyester glass | 3,000/52,000 | 11,000/52,000 | 5,000/82,000 |
| Polyethylene | 500/5,600 | 1,500/5,600 | 1,500/7,200 |
| Polypropylene | 2,700/15,000 | 3,600/8,200 | 4,000/12,000 |
| Polystyrene | 1,500/22,000 | 3,000/23,000 | 2,500/27,000 |
| Polysulfone | 10,000/12,000 | 13,000/15,000 | 14,000/16,000 |
| Polyurethane | 4,000/12,000 | 19,000/21,000 | 700/20,000 |
| Polyvinyl chloride (pvc) | 5,000/8,000 | 7,000/14,000 | 9,000/17,000 |
| Rigid Teflon (tfe) | 4,000/7,000 | 4,500/7,500 | 7,000/10,000 |

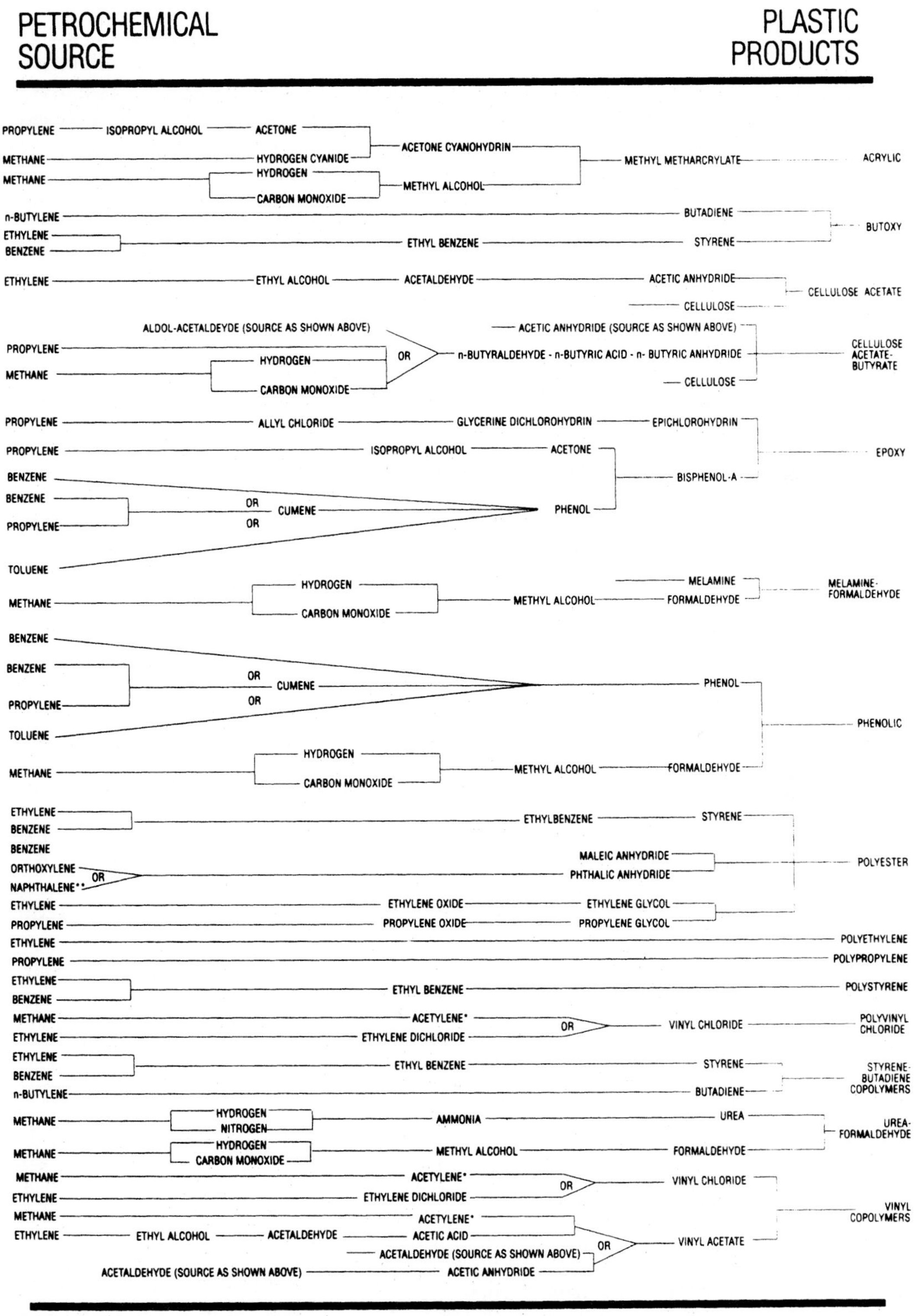


**FIGURE 4.2** Petrochemical source of the plastics. (*Source: PMI, Plastics Manufacturing Inc.*)

**TABLE 4.7** Properties of Insulating Materials

| Material | Specific gravity | Thermal resistivity, W/cm²/cm/°C | Dielectric constant | Electrical resistivity, ohm-cm |
|---|---|---|---|---|
| Air | 0.00129 | 4,000 | 1 | — |
| Asbestos paper | $\frac{2.0}{2.8}$ | 400 | — | $1.6 \times 10^{11}$ |
| Asphalt | $\frac{1.1}{1.5}$ | 140 | 2.7 | $6 \times 10^{14}$ |
| Bakelite resin | 1.25 | $\frac{300}{600}$ | $\frac{4.5}{7.5}$ | $2 \times 10^{16}$ |
| Buna S (RH–RW) | 0.94 | 520 | 2.9 | $10^{15}$ |
| Butyl | 0.91 | 520 | 2.4 | $10^{17}$ |
| Concrete | $\frac{1.8}{2.5}$ | $\frac{50}{100}$ | — | — |
| Cork | $\frac{0.22}{0.26}$ | 1,800 | — | — |
| Enamel (wire) | — | — | 5 | $10^{14}$ |
| Glass (common) | $\frac{2.4}{2.8}$ | $\frac{90}{100}$ | $\frac{5.5}{9.1}$ | $9 \times 10^{13}$ |
| Mica | $\frac{2.9}{3.2}$ | 280 | 5.7 | $10^{15}$ |
| Neoprene | $\frac{1.5}{1.24}$ | 520 | 9 | $10^{11}$ |
| Paper (dry) | $\frac{0.7}{1.15}$ | 800 | $\frac{1.7}{2.6}$ | $5 \times 10^{4}$ |
| Paper (in cable) | — | 700 | $\frac{3.4}{3.5}$ | $5 \times 10^{4}$ |
| Paraffin | 0.89 | $\frac{385}{400}$ | 2.1 | $10^{16}$ |
| Porcelain (wet process) | $\frac{2.3}{2.5}$ | 100 | 4.4 | $3 \times 10^{14}$ |
| PVC | $\frac{1.2}{1.7}$ | 600 | $\frac{6.5}{12}$ | $5 \times 10^{12}$ |
| Polyethylene | 0.92 | $\frac{300}{400}$ | 2.25 | $10^{16}$ |
| Silicone rubber | $\frac{1.4}{2.1}$ | $\frac{350}{450}$ | $\frac{3.2}{3.5}$ | $10^{14}$ |
| Teflon (tfe) | $\frac{2.1}{2.3}$ | 400 | 2 | $10^{16}$ |
| Varnished cloth | 1.25 | $\frac{600}{900}$ | 5 | $3 \times 10^{14}$ |
| Water | 1.00 | 170 | 80 | — |
| Wood (maple) | $\frac{0.62}{0.75}$ | 550 | 4.4 | $3 \times 10^{10}$ |

**TABLE 4.8** Elastomer Properties

| Base | Weight lb/in$^3$ | Tensile strength, psi | % elongation in 2 in | Durometer | Temperature range, °F |
|---|---|---|---|---|---|
| Acrylic | 0.039 / 0.041 | 1,900 / 2,100 | 400 / 500 | 40 / 90 | − 40 / + 350 |
| Butyl | 0.032 / 0.034 | 2,400 / 2,600 | 450 / 550 | 30 / 100 | − 65 / + 250 |
| Fluorocarbon | 0.045 / 0.075 | 2,400 / 2,600 | 250 / 350 | 60 / 95 | − 40 / + 500 |
| Natural rubber | 0.032 / 0.034 | 3,400 / 3,600 | 550 / 650 | 30 / 100 | − 65 / + 212 |
| Synthetic rubber | 0.032 / 0.034 | 3,400 / 3,600 | 550 / 650 | 40 / 80 | − 65 / + 212 |
| Neoprene | 0.043 / 0.045 | 2,900 / 3,100 | 550 / 650 | 40 / 100 | − 65 / + 212 |
| Nitrile | 0.035 / 0.037 | 2,900 / 3,100 | 550 / 650 | 20 / 90 | − 65 / + 300 |
| Silicone | 0.035 / 0.037 | 1,400 / 1,600 | 650 / 750 | 20 / 90 | − 120 / + 600 |
| Urethane | 0.038 / 0.040 | 4,400 / 4,600 | 600 / 700 | 50 / 100 | − 65 / + 212 |

***Beryllium Copper Alloys.*** These include numbers 25, 165, 10, 50, 20C, 275C, and 10C.

***Beryllium Nickel Alloys.*** UNS N03360 may be selected. Tensile strengths approaching 300,000 psi and yield strengths of 245,000 psi are attainable with beryllium nickel alloys.

*Caution:* Beryllium-bearing alloys can present a health hazard unless precautions are taken in handling and fabrication. Excessive inhalation of beryllium particles can cause serious chronic pulmonary illness. Although only one percent of people become sensitized to beryllium, adequate ventilation and other safety measures are required when working with these materials.

***Titanium Alloys.*** These include manganese-titanium and ferrochromium-titanium. As a general observation on titanium alloys, they are very strong for their relative weight, when compared to steel. The density of titanium is 4.54 g/cm$^3$, whereas that of iron is 7.87 g/cm$^3$. The yield strength of the common titanium alloys ranges from 80,000 to 165,000 psi, and the tensile strength from 100,000 to 175,000 psi. Elongation in 2 in is from 8 to 22%.

**TABLE 4.9** Properties of Common Woods

| Name | Density, g/cm$^3$ | Density, lb/ft$^3$ | Modulus of elasticity, psi × $10^6$ |
|---|---|---|---|
| Ash, white | 0.64 | 39.8 | 1.77 |
| Birch, paper | 0.60 | 37.5 | 1.59 |
| Cedar, eastern | 0.49 | 30.7 | 0.87 |
| Cherry, black | 0.53 | 33.3 | 1.49 |
| Cyprus | 0.48 | 30.1 | 1.44 |
| Ebony, African | 0.77 | 48.0 | 1.43 |
| Eucalyptus, Australian | 0.83 | 51.8 | 2.64 |
| Fir, Douglas | 0.51 | 32.0 | 1.93 |
| Gum, black | 0.55 | 34.5 | 1.19 |
| Hemlock, eastern | 0.43 | 27.0 | 1.20 |
| Hickory, shagbark | 0.84 | 52.2 | 2.17 |
| Ironwood, black | 1.01 | 67.3 | 2.99 |
| Locust, black | 0.71 | 44.2 | 2.06 |
| Mahogany, African | 0.67 | 41.7 | 1.53 |
| Maple, sugar | 0.68 | 42.2 | 1.83 |
| Oak, black | 0.67 | 41.8 | 1.64 |
| Oak, red | 0.66 | 41.1 | 1.81 |
| Oak, white | 0.71 | 44.3 | 1.78 |
| Pine, longleaf | 0.64 | 39.8 | 2.06 |
| Poplar, yellow | 0.43 | 26.7 | 1.50 |
| Redwood | 0.44 | 27.2 | 1.36 |
| Spruce, white | 0.43 | 27.0 | 1.42 |
| Teak, Indian | 0.58 | 36.3 | 1.70 |
| Walnut, black | 0.56 | 35.1 | 1.69 |

## 4.14 HARDENING PROCESSES FOR METALS

### 4.14.1 Hardening of Steels and Beryllium Copper

***Steels.*** Hardening of steel is accomplished by raising the temperature until the steel becomes austenitic in form and then quick-cooling it at a rate faster than the critical rate. This produces the martensitic structure of hardened steel. Carbon content determines the attainable hardness. The maximum temperature to which a steel is heated prior to quenching is termed the *hardening temperature.* The average hardening temperature for steels is 1375 to 1575°F. This range should be closely controlled to attain maximum hardness while inhibiting distortion.

Tempering reduces the brittleness of hardened steel and removes internal stresses caused by rapid cooling. Tempering temperatures range from 300 to 700°F.

The case hardening of steel leaves a skin on the surface of the part of extreme hardness, which prevents wear of faying or rubbing parts. Cases with hardnesses of Rockwell C40 through C60 have many general applications in mechanisms where parts work together. When a part is highly stressed and must also be hard, through-hardening is indicated. Case hardening can be done on carbon steels with carbon contents as low as 0.15 to 0.20%. This percentage also applies to low-carbon alloy steels.

**TABLE 4.10** Coefficients of Linear Expansion for Common Materials

| Metal, alloy, or other material | Linear expansion | |
|---|---|---|
| | in per 1°F | in per 1°C |
| Aluminum, wrought | 0.0000128 | 0.0000231 |
| Brass | 0.0000104 | 0.0000188 |
| Bronze | 0.0000101 | 0.0000181 |
| Copper | 0.0000093 | 0.0000168 |
| Cast iron, gray | 0.0000059 | 0.0000106 |
| Wrought iron | 0.0000067 | 0.0000120 |
| Lead | 0.0000159 | 0.0000286 |
| Magnesium alloy | 0.0000160 | 0.0000290 |
| Nickel | 0.0000070 | 0.0000126 |
| Cast steel | 0.0000061 | 0.0000110 |
| Hard steel | 0.0000073 | 0.0000132 |
| Medium steel | 0.0000067 | 0.0000120 |
| Soft steel | 0.0000061 | 0.0000110 |
| Stainless steel | 0.0000099 | 0.0000178 |
| Zinc, rolled | 0.0000173 | 0.0000263 |
| Concrete | 0.0000079 | 0.0000143 |
| Granite | 0.0000047 | 0.0000084 |
| Marble | 0.0000056 | 0.0000100 |
| Plaster | 0.0000092 | 0.0000166 |
| Slate | 0.0000058 | 0.0000104 |
| Fir | 0.0000021 | 0.0000037 |
| Maple | 0.0000036 | 0.0000064 |
| Oak | 0.0000027 | 0.0000049 |
| Pine | 0.0000030 | 0.0000054 |
| Plate glass | 0.0000050 | 0.0000089 |
| Hard rubber | 0.0000044 | 0.0000080 |
| Porcelain | 0.0000009 | 0.0000016 |
| Silver | 0.0000104 | 0.0000188 |
| Tin | 0.0000148 | 0.0000269 |
| Tungsten | 0.0000024 | 0.0000043 |

***Beryllium Copper.*** Hardening of beryllium copper reverses the process of hardening steels. The beryllium copper part is heated to a hardening temperature, which is held for an extended period, then allowed to slowly cool. The process employed to harden strips for spring use consists of the following steps.

1. Anneal the part by heating it to 1100 to 1150°F and then water quenching.
2. Raise the temperature of the part in an oven to 600 to 900°F. Hold this temperature for 3 h.
3. Air-cool the part to room temperature.

This process will harden beryllium copper alloy 25 (ASTM B194) to a tensile strength of 195,000 to 220,000 psi. Hardness range will be Rockwell C40 to C45.

**TABLE 4.11** Properties of Special-Purpose Alloys

| Alloy | Form | Yield strength, psi | Tensile strength, psi | % elongation in 2 in | Brinell hardness |
|---|---|---|---|---|---|
| **Wrought magnesium** | | | | | |
| SAE 51 | Hard | — | 31,000/33,000 | 3/5 | — |
| SAE 510 | Hard | 24,000/26,000 | 35,000/37,000 | 3/5 | — |
| SAE 511 | Hard | 27,000/29,000 | 38,000/40,000 | 2/4 | — |
| SAE 52 | Extruded | 24,000/26,000 | 36,000/38,000 | 10/12 | — |
| SAE 520 | Extruded | 39,000/41,000 | 25,000/27,000 | 10/12 | — |
| **Cast Magnesium** | | | | | |
| SAE 50 | As cast | 9,000/11,000 | 23,000/25,000 | 3/5 | — |
| SAE 500 | As cast | 9,000/11,000 | 19,000/21,000 | 1/2 | — |
| **Nickel alloys** | | | | | |
| Monel | CD | 79,000/81,000 | 98,000/102,000 | 24/26 | 185/195 |
| R-Monel | CD | 74,000/76,000 | 89,000/91,000 | 24/26 | 175/185 |
| K-Monel | CD | 84,000/86,000 | 113,000/115,000 | 24/26 | 208/210 |
| S-Monel | As cast | 99,000/101,000 | 128,000/130,000 | 1/2 | 318/322 |
| Inconel | CD | 89,000/91,000 | 114,000/116,000 | 18/20 | 198/202 |
| Inconel "X" | HR | 119,000/120,000 | 179,000/181,000 | 24/26 | 358/362 |
| **Special copper alloys** | | | | | |
| Tellurium copper 165 | Rod (H) | 38,000/40,000 | 46,000/48,000 | 10/12 | B 50* |
| Hitenso 1622 | Rod (H) | 48,000/50,000 | 56,000/58,000 | 13/15 | B 65* |
| Chromium copper 182 | Rod (HT) | 59,000/61,000 | 68,000/70,000 | 19/21 | B 85* |
| Bronze HS 316 | Rod (H) | 59,000/61,000 | 69,000/71,000 | 10/12 | B 78* |

*Note*: H = hard, HT = heat treated, HS = high strength, * = Rockwell-scale.

### 4.14.2 Testing Materials for Hardness

Material hardness is measured using various hardness systems such as Brinell, Rockwell, Vickers, Shore, and Durometer. Section 4.2 outlined the method used to calculate Brinell hardness. Figure 4.3 shows a typical hardness testing machine, wherein the applied load may be measured on the dial scale and the indentation measured using various methods. The Rockwell and Brinell systems are the most commonly used. Tables are available from the ASTM and SAE so that direct comparisons may be made when comparing one hardness measurement system with another. That is, a hardness of Brinell 355 may be translated to the Rockwell C scale system or the Vickers system and others as applicable.

**FIGURE 4.3** A typical hardness testing machine.

## 4.15 COLOR SCALE OF TEMPERATURE

The chart displayed in Table 4.12 can be used to determine the approximate temperatures of heated metals such as steel and copper.

**TABLE 4.12** Color Scale for Metal Temperatures

| Color | Temperature, °C | Temperature, °F |
|---|---|---|
| Dull red heat | 500–550 | 932–1,022 |
| Dark red heat | 650–750 | 1,202–1,382 |
| Bright red heat | 850–950 | 1,562–1,742 |
| Yellowish red heat | 1,050–1,150 | 1,922–2,102 |
| Dull white | 1,250–1,350 | 2,282–2,462 |
| White heat | 1,450–1,550 | 2,642–2,822 |

## 4.16 STANDARD PIPE DIMENSIONS

Table 4.13 tabulates standard pipe sizes for welded and seamless products (ASA denotes American Standards Association).

**TABLE 4.13** Standard Pipe Sizes

| Nominal pipe size | O.d., in | ASA pipe schedules (wall thickness, in) | | | | | |
|---|---|---|---|---|---|---|---|
| | | 40 | 60 | 80 | 100 | 120 | 160 |
| $\frac{1}{8}$ | 0.405 | 0.068 | | 0.095 | | | |
| $\frac{1}{4}$ | 0.540 | 0.088 | | 0.119 | | | |
| $\frac{3}{8}$ | 0.675 | 0.091 | | 0.126 | | | |
| $\frac{1}{2}$ | 0.840 | 0.109 | | 0.147 | | | |
| $\frac{3}{4}$ | 1.050 | 0.113 | | 0.154 | | | |
| 1 | 1.315 | 0.133 | | 0.179 | | | |
| $1\frac{1}{4}$ | 1.660 | 0.140 | | 0.191 | | | |
| $1\frac{1}{2}$ | 1.900 | 0.145 | | 0.200 | | | |
| 2 | 2.375 | 0.154 | | 0.218 | | | |
| $2\frac{1}{2}$ | 2.875 | 0.203 | | 0.276 | | | |
| 3 | 3.500 | 0.216 | | 0.300 | | | |
| $3\frac{1}{2}$ | 4.000 | 0.226 | | 0.318 | | | |
| 4 | 4.500 | 0.237 | | 0.337 | | 0.438 | 0.531 |
| 5 | 5.563 | 0.258 | | 0.375 | | 0.500 | 0.625 |
| 6 | 6.625 | 0.280 | | 0.432 | | 0.562 | 0.719 |
| 8 | 8.625 | 0.322 | 0.406 | 0.500 | 0.594 | 0.719 | 0.906 |
| 10 | 10.750 | 0.365 | 0.500 | 0.594 | 0.719 | 0.844 | 1.125 |
| 12 | 12.750 | 0.406 | 0.562 | 0.688 | 0.844 | 1.000 | 1.312 |

To determine the weight per linear foot of a pipe, use the relationship

$$W = \pi\left(\frac{\text{o.d.}}{2}\right)^2 - \pi\left(\frac{\text{i.d.}}{2}\right)^2 \times 12 \times \text{density}$$

When the outside diameter (o.d.) and inside diameter (i.d.) are given in inches, and the density is given in pounds per cubic inch, the weight will be calculated in pounds per linear foot.

## 4.17 COMMON AND FORMAL NAMES OF CHEMICALS

Table 4.14 displays information on chemical names.

## 4.18 CONDUCTIVE PROPERTIES OF VARIOUS METALS AND ALLOYS

Table 4.15 is a compendium of metals property data bearing on conductive behavior.

## 4.19 ELECTROMOTIVE SERIES

The following listings are the single potential differences (in volts, for 1.00-N solution), at 25°C, of certain elements in solutions of their ions with reference to hydrogen gas at 1 atmosphere (atm).

| *Element* | *Potential difference* |
|---|---|
| Magnesium | +2.40 |
| Aluminum | +1.70 |
| Beryllium | +1.69 |
| Manganese | +1.10 |
| Zinc | +0.76 |
| Chromium | +0.60 |
| Sulfur | +0.51 |
| Iron | +0.44 |
| Cadmium | +0.40 |
| Cobalt | +0.29 |
| Nickel | +0.22 |
| Tin | +0.13 |
| Lead | +0.12 |
| Bismuth | –0.20 |
| Copper | –0.34 to 0.51 |
| Mercury | –0.80 to 0.86 |
| Silver | –0.80 |
| Iodine | –0.54 |
| Bromine | –1.07 |
| Chlorine | –1.36 |

**TABLE 4.14** Common and Formal Names of Chemicals

| Common name | Formal name | Formula |
|---|---|---|
| Alum | Potassium aluminum sulfate | $K_2Al_2(SO_4)_4 \cdot 24H_2O$ |
| Alundum | Fused alumina | $Al_2O_3$ |
| Analine | Phenyl amine | — |
| Antichlor | Sodium thiosulfate | $Na_2S_2O_3 \cdot 5H_2O$ |
| Antimony black | Antimony trisulfide | $Sb_2S_3$ |
| Aqua fortis | Nitric acid | $HNO_3$ |
| Aqua regia | Nitric and hydrochloric acid | $HNO_3+3HCl$ |
| Bakelite | Resin from phenol and formaldehyde | — |
| Baking soda | Sodium bicarbonate | $NaHCO_3$ |
| Barium white | Barium sulfate | $BaSO_4$ |
| Bauxite | Hydrated alumina | $Al_2O_3 \cdot 2H_2O$ |
| Benzine | Gasoline | — |
| Blue vitriol | Copper sulfate | $CuSO_4 \cdot 5H_2O$ |
| Borax | Sodium tetraborate | $Na_2B_4O_7 \cdot 10H_2O$ |
| Burnt lime | Calcium oxide | $CaO$ |
| Calcite | Calcium carbonate | $CaCO_3$ |
| Carbolic acid | Phenol | $C_6H_5OH$ |
| Carbonic acid | Carbon dioxide | $CO_2$ |
| Carborundum | Silicon carbide | $SiC$ |
| Chalk | Calcium carbonate | $CaCO_3$ |
| China clay | Aluminum silicate | $Al_2O_3 \cdot 2SiO_2 \cdot 2H_2O$ |
| Chrome alum | Potassium chromium sulfate | $K_2Cr_2(SO_4)_4 \cdot 24H_2O$ |
| Common salt | Sodium chloride | $NaCl$ |
| Corundum | Aluminum oxide | $Al_2O_3$ |
| Emery powder | Aluminum oxide | $Al_2O_3$ |
| Epsom salts | Magnesium sulfate | $MgSO_4 \cdot 7H_2O$ |
| Formalin | 40% solution of formaldehyde and water | |
| Glycerin | Glycerol | — |
| Gypsum | Calcium sulfate | $CaSO_4 \cdot 2H_2O$ |
| Kaolin | Aluminum silicate | $Al_2O_3 \cdot 2SiO_4 \cdot 2H_2O$ |
| Lime | Calcium oxide | $CaO$ |
| Methanol | Methyl alcohol | $CH_3OH$ |
| Muriatic acid | Hydrochloric acid | $HCl$ |
| Oil of vitriol | Sulfuric acid | $H_2SO_4$ |
| Red lead | Lead tetroxide | $Pb_3O_4$ |
| Rouge | Ferric oxide | $Fe_2O_3$ |
| Sal ammoniac | Ammonium chloride | $NH_4Cl$ |
| Silica | Silicon dioxide | $SiO_2$ |
| Soda (washing) | Sodium carbonate | $Na_2CO_3 \cdot 10H_2O$ |
| Sugar of lead | Lead acetate | — |
| Talc | Hydrated magnesium silicate | $Mg_3Si_4O_{11} \cdot H_2O$ |
| Toluol | Toluene | $C_6H_6CH_3$ |
| White lead | Lead carbonate | $2PbCO_3+Pb(OH)$ |
| Whiting | Calcium carbonate | $CaCO_3$ |
| Zinc white | Zinc oxide | $ZnO$ |

**TABLE 4.15** Properties of Various Metals and Alloys

| Material | Relative resistivity* at 20°C | Density g/cm³ | Thermal conductivity at 20°C W/cm-°C | Thermal expansion × $10^{-6}$/°C, in | Melting °C |
|---|---|---|---|---|---|
| Aluminum | 1.54 | 2.70 | 2.22 | 23.6 | 660 |
| Beryllium | 2.3 | 1.85 | 1.46 | 11.6 | 1,277 |
| Bismuth | 67.0 | 9.80 | 0.084 | 13.3 | 271 |
| Brass, yellow | 3.7 | 8.47 | 1.17 | 20.3 | 930 |
| Cadmium | 4.3 | 8.65 | 0.92 | 29.8 | 321 |
| Carbon, graphite | 790 | 2.25 | 0.24 | 0.6–4.3 | Sublimes |
| Chromium | 7.4 | 7.19 | 0.67 | 6.2 | 1,875 |
| Cobalt | 3.6 | 8.85 | 0.69 | 13.8 | 1,495 |
| Columbium (see Niobium) | | | | | |
| Constantan | 29.0 | 8.9 | 0.21 | 14.9 | 1,290 |
| Copper, hard drawn | 1.03 | 8.94 | 3.91 | 16.8 | 1,083 |
| Gallium | 4.7 | 5.91 | 0.29 | 18.0 | 30 |
| Germanium | $2.7 \times 10^6$ | 5.33 | 0.59 | 5.75 | 937 |
| Gold | 1.36 | 19.32 | 2.96 | 14.2 | 1,063 |
| Inconel, 17–16–8 | 56.9 | 8.51 | 0.15 | 11.5 | 1,425 |
| Indium | 4.9 | 7.31 | 0.24 | 33.0 | 156 |
| Invar, 64–36 | 46.0 | 8.00 | 0.11 | 0–2 | 1,425 |
| Iron | 5.6 | 7.87 | 0.75 | 11.8 | 1,536 |
| Lead | 12.0 | 11.34 | 0.35 | 29.3 | 327 |
| Magnesium | 2.58 | 1.74 | 1.53 | 27.1 | 650 |
| Mercury | 55.6 | 13.55 | 0.082 | — | −38.9 |
| Molybdenum | 3.3 | 10.22 | 1.42 | 4.9 | 2,610 |
| Monel, 67–30 | 27.9 | 8.84 | 0.26 | 14.0 | 1,325 |
| Nichrome, 80–20 | 62.5 | 8.4 | 0.134 | 13.0 | 1,400 |
| Nickel | 5.5 | 8.89 | 0.61 | 13.3 | 1,440 |
| Niobium | 7.2 | 8.57 | 0.52 | 7.31 | 2,468 |
| Palladium | 6.3 | 12.02 | 0.70 | 11.8 | 1,552 |
| Phosphor-bronze 95–5 | 6.4 | 8.86 | 0.71 | 17.8 | 1,000 |
| Platinum, 99.9% | 6.16 | 21.45 | 0.69 | 8.9 | 1,769 |

**TABLE 4.15** Properties of Various Metals and Alloys (*Continued*)

| Material | Relative resistivity* at 20°C | Density g/cm³ | Thermal conductivity at 20°C W/cm-°C | Thermal expansion × $10^{-6}$/°C, in | Melting °C |
|---|---|---|---|---|---|
| Silicon | $10^{11}$ | 2.33 | 1.25 | 2.5 | 1,420 |
| Silver | 0.922 | 10.49 | 4.18 | 19.7 | 961 |
| Steel, .4–.5 C | 7–12 | 7.8 | 0.5 | 11.0 | 1,480 |
| Steel, stainless 304 | 42 | 7.9 | 0.16 | 17.0 | 1,430 |
| Steel, stainless 410 | 33 | 7.7 | 0.24 | 11 | 1,500 |
| Tantalum | 7.4 | 16.6 | 0.54 | 6.6 | 3,000 |
| Thorium | 8.1 | 11.6 | 0.37 | 12.5 | 1,750 |
| Tin | 7.0 | 7.30 | 0.63 | 23 | 232 |
| Titanium | 24.2 | 4.51 | 0.41 | 8.4 | 1,670 |
| Tungsten | 3.2 | 19.3 | 1.67 | 4.6 | 3,410 |
| Uranium | 17.5 | 18.7 | 0.3 | 6.8–14.1 | 1,132 |
| Zinc | 3.5 | 7.14 | 1.10 | 27 | 420 |
| Zirconium | 23 | 6.5 | 0.21 | 5.8 | 1,852 |

**Note*: Standard resistivity of 100% IACS copper at 20°C = 1.7241 × $10^{-6}$ ohm-cm.

Reading down the list, each element above loses electrons more readily than any element below. Accordingly, each element above has a higher reducing power than any element below. Reading up the list, each element below gains electrons more readily than any element above. Accordingly, each element below has a higher oxidizing power than any element above. Therefore, any "metal above" will displace, from the salt solution of any metal below, the metal of the latter from its own cation (–). Cathode is (–), anode is (+).

## 4.20 CHEMICAL SYMBOLS FOR METALS

Refer to Table 4.16 for a tabulation of these symbols.

## 4.21 THE ELEMENTS

Properties of the known elements are shown in Table 4.17. Elements that are unstable, have extremely short lives, and have no practical uses are not shown. These include numbers 104, 105, 106, 107, and 109.

## 4.22 LABORATORY ANALYSIS OF MATERIALS

Engineering materials suspected of not meeting the specifications under which they were purchased should be analyzed. This may be done in house if a materials testing laboratory is available, or sent to a certified materials test laboratory.

**TABLE 4.16** Chemical Symbols for Metals

| Metal | Symbol |
|---|---|
| Aluminum | Al |
| Antimony | Sb |
| Beryllium | Be |
| Bismuth | Bi |
| Boron | B |
| Cadmium | Cd |
| Carbon | C |
| Chromium | Cr |
| Cobalt | Co |
| Copper | Cu |
| Gold | Au |
| Iridium | Ir |
| Iron | Fe |
| Lead | Pb |
| Lithium | Li |
| Magnesium | Mg |
| Manganese | Mn |
| Mercury | Hg |
| Molybdenum | Mo |
| Nickel | Ni |
| Platinum | Pt |
| Selenium | Se |
| Silicon | Si |
| Silver | Ag |
| Tellurium | Te |
| Tin | Sn |
| Titanium | Ti |
| Tungsten | W |
| Vanadium | V |
| Zinc | Zn |
| Zirconium | Zr |

Many materials from foreign sources have not met their stipulated American standard specifications; some of these materials are intentionally counterfeited for profit. This counterfeiting practice is also being done with standard fastening devices such as bolts, screws, and nuts. Figure 4.4 is an actual test report as issued by a professional test laboratory, showing test results on aluminum alloy specimens.

## 4.23 OBSOLETE SAE STEELS

Former SAE standard steels and ex-steels are shown for reference in Tables 4.18*a* and 4.18*b*. These listed SAE numbers should not be specified on new engineering drawings or new specifications. Use an appropriate substitute that matches the properties of the obsolete steel.

## 4.24 IDENTIFICATION OF PLASTICS

Table 4.19 outlines the practical methods used to identify plastics.

**TABLE 4.17** The Elements—Properties

| Element Name | Symbol | Atomic Number | Atomic Weight | Melting Point Celcius | Electrical Resistivity Ohm-centimeters |
|---|---|---|---|---|---|
| Actinium | Ac | 89 | 227.028 | 1050 | ...... |
| Aluminum | Al | 13 | 26.9815 | 660.37 | $2.655 \times 10^{-6}$ |
| Americium | Am | 95 | (243)● | 994 ± 4 | ...... |
| Antimony | Sb | 51 | 121.75 | 630.74 | $39.0 \times 10^{-6}$ |
| Argon | A | 18 | 39.948 | - 189.2 | ...... |
| Arsenic | As | 33 | 74.9216 | 817■ | $35.0 \times 10^{-6}$ |
| Astatine | At | 85 | (210)● | 302 | ...... |
| Barium | Ba | 56 | 137.33 | 725 | ...... |
| Berkelium | Bk | 97 | (247)● | ...... | ...... |
| Beryllium | Be | 4 | 9.0122 | 1278 ± 5 | $4.0 \times 10^{-6}$ |
| Bismuth | Bi | 83 | 208.980 | 271.3 | $106.8 \times 10^{-6}$ |
| Boron | B | 5 | 10.81 | 2079 | $1.8 \times 10^{-6}$ |
| Bromine | Br | 35 | 79.904 | - 7.2 | ...... |
| Cadmium | Cd | 48 | 112.41 | 320.9 | $6.83 \times 10^{-6}$ |
| Calcium | Ca | 20 | 40.08 | 839 ± 2 | $3.91 \times 10^{-6}$ |
| Californium | Cf | 98 | (251)● | ...... | ...... |
| Carbon | C | 6 | 12.011 | 3652 ▲ | $1375 \times 10^{-6}$ |
| Cerium | Ce | 58 | 140.12 | 798 ± 2 | $75.0 \times 10^{-6}$ |
| Cesium | Cs | 55 | 132.9054 | 28.40 | $20.0 \times 10^{-6}$ |
| Chlorine | Cl | 17 | 35.453 | - 100.98 | ...... |
| Chromium | Cr | 24 | 51.996 | 1857 ± 20 | $129 \times 10^{-6}$ |
| Cobalt | Co | 27 | 51.9332 | 1495 | $6.24 \times 10^{-6}$ |
| Copper | Cu | 29 | 63.546 | 1083.4 | $1.673 \times 10^{-6}$ |
| Curium | Cm | 96 | (247)● | 1340 ± 40 | ...... |
| Dysprosium | Dy | 66 | 162.50 | 1409 | $90 \times 10^{-6}$ |
| Einsteinium | Es | 99 | (252)● | ...... | ...... |
| Erbium | Er | 68 | 167.26 | 1522 | $107.0 \times 10^{-6}$ |
| Europium | Eu | 63 | 151.96 | 822 ± 5 | $90.0 \times 10^{-6}$ |
| Fermium | Fm | 100 | (257)● | ...... | ...... |
| Fluorine | F | 9 | 18.9984 | - 219.62 | ...... |
| Francium | Fr | 87 | (223)● | 27 * | ...... |
| Gadolinium | Gd | 64 | 157.25 | 1311 ± 1 | $134 \times 10^{-6}$ |
| Gallium | Ga | 31 | 69.72 | 29.78 | $56.8 \times 10^{-6}$ |
| Germanium | Ge | 32 | 72.59 | 937.4 | $89.0 \times 10^{-6}$ |
| Gold | Au | 79 | 196.967 | 1064.4 | $2.44 \times 10^{-6}$ |

| | | | | | |
|---|---|---|---|---|---|
| Hafnium | Hf | 72 | 178.49 | 2227 ± 20 | 35.1 x $10^{-6}$ |
| Helium | He | 2 | 4.0026 | - 272.2 ♦ | ...... |
| Holmium | Ho | 67 | 164.93 | 1470 | 87.0 x $10^{-6}$ |
| Hydrogen | H | 1 | 1.0079 | - 259.14 | ...... |
| Indium | In | 49 | 114.82 | 156.61 | 8.37 x $10^{-6}$ |
| Iodine | I | 53 | 126.905 | 113.5 | ...... |
| Iridium | Ir | 77 | 192.22 | 2410 | 4.71 x $10^{-6}$ |
| Iron | Fe | 26 | 55.847 | 1535 | 9.71 x $10^{-6}$ |
| Krypton | Kr | 36 | 83.80 | - 156.6 | ...... |
| Lanthanum | La | 57 | 138.906 | 920 ± 5 | 57.0 x $10^{-6}$ |
| Lawrencium | Lw | 103 | (260)● | ...... | ...... |
| Lead | Pb | 82 | 207.2 | 327.5 | 20.65 x $10^{-6}$ |
| Lithium | Li | 3 | 6.941 | 180.54 | 8.55 x $10^{-6}$ |
| Lutetium | Lu | 71 | 174.967 | 1656 ± 5 | 79.0 x $10^{-6}$ |
| Magnesium | Mg | 12 | 24.305 | 648.8 | 4.45 x $1^{-6}$ |
| Manganese | Mn | 25 | 54.938 | 1244 ± 3 | 185.0 x $10^{-6}$ |
| Mendelevium | Md | 101 | (258)● | ...... | ...... |
| Mercury | Hg | 80 | 200.59 | - 38.87 | 95.78 x $10^{-6}$ |
| Molybdenum | Mo | 42 | 95.94 | 2617 | 5.2 x $10^{-6}$ |
| Neodymium | Nd | 60 | 144.24 | 1010 | 64.0 x $10^{-6}$ |
| Neon | Ne | 10 | 20.118 | - 248.67 | ...... |
| Neptunium | Np | 93 | 237.048 | 640 ± 1 | ...... |
| Nickel | Ni | 28 | 58.69 | 1453 | 6.84 x $10^{-6}$ |
| Niobium | Nb | 41 | 92.906 | 2468 ± 10 | 14.6 x $10^{-6}$ |
| Nitrogen | N | 7 | 14.007 | - 209.86 | ...... |
| Nobelium | No | 102 | (259)● | ...... | ...... |
| Osmium | Os | 76 | 190.2 | 3045 ± 30 | 9.5 x $10^{-6}$ |
| Oxygen | O | 8 | 15.999 | - 218.4 | ...... |
| Palladium | Pd | 46 | 106.42 | 1554 | 9.93 x $10^{-6}$ |
| Phosphorus | P | 15 | 30.974 | 44.1 | 1 x $10^{11}$ |
| Platinum | Pt | 78 | 195.08 | 1772 | 9.85 x $10^{-6}$ |
| Plutonium | Pu | 94 | (244)● | 641 | 146.45 x $10^{-6}$ |
| Polonium | Po | 84 | (209)● | 254 | 42 x $10^{-6}$ |
| Potassium | K | 19 | 39.098 | 63.25 | 6.15 x $10^{-6}$ |
| Praseodymium | Pr | 59 | 140.908 | 931 ± 4 | 68.0 x $10^{-6}$ |
| Promethium | Pm | 61 | (145)● | 1080 * | ...... |
| Protactinium | Pa | 91 | 231.036 | 1600 | ...... |
| Radium | Ra | 88 | 226.025 | 700 | ...... |
| Radon | Rn | 86 | (222)● | - 71.0 | ...... |
| Rhenium | Re | 75 | 186.207 | 3180 | 19.3 x $10^{-6}$ |

Note: ● values in parentheses are for the mass number of the most stable isotope known; ▪ at 28 Atm; ▲ sublimates; * approximate; ♦ at 26 Atm.

**TABLE 4.17** The Elements—Properties (*Continued*)

| Element Name | Symbol | Atomic Number | Atomic Weight | Melting Point Celcius | Electrical Resistivity Ohm-centimeters |
|---|---|---|---|---|---|
| Rhodium | Rh | 45 | 102.906 | 1965 ± 3 | $4.33 \times 10^{-6}$ |
| Rubidium | Rb | 37 | 85.468 | 38.89 | $12.5 \times 10^{-6}$ |
| Ruthenium | Ru | 44 | 101.07 | 2310 | $7.6 \times 10^{-6}$ |
| Samarium | Sm | 62 | 150.36 | 1072 ± 5 | $90.0 \times 10^{-6}$ |
| Scandium | Sc | 21 | 44.956 | 1539 | $61.0 \times 10^{-6}$ |
| Selenium | Se | 34 | 78.96 | 217 | $10.0 \times 10^{6}$ |
| Silicon | Si | 14 | 28.086 | 1410 | 10 |
| Silver | Ag | 47 | 107.868 | 961.9 | $1.59 \times 10^{-6}$ |
| Sodium | Na | 11 | 22.99 | 97.8 | $4.2 \times 10^{-6}$ |
| Strontium | Sr | 38 | 87.62 | 769 | $23.0 \times 10^{-6}$ |
| Sulfur | S | 16 | 32.06 | 112.8 | $2.0 \times 10^{18}$ |
| Tantalum | Ta | 73 | 180.948 | 2996 | $12.45 \times 10^{-6}$ |
| Technetium | Tc | 43 | (98)● | 2172 | ...... |
| Tellurium | Te | 52 | 127.60 | 449.5 | 0.43 |
| Terbium | Tb | 65 | 158.925 | 1360 ± 4 | $116 \times 10^{-6}$ |
| Thallium | Tl | 81 | 204.383 | 303.5 | $18.0 \times 10^{-6}$ |
| Thorium | Th | 90 | 232.038 | 1750 | $13.0 \times 10^{-6}$ |
| Thulium | Tm | 69 | 168.934 | 1545 ± 15 | $79.0 \times 10^{-6}$ |
| Tin | Sn | 50 | 118.71 | 231.97 | $11.5 \times 10^{-6}$ |
| Titanium | Ti | 22 | 47.88 | 1660 ± 10 | $42.0 \times 10_{-6}$ |
| Tungsten | W | 74 | 183.85 | 3410 ± 20 | $5.5 \times 10^{-6}$ |
| Uranium | U | 92 | 238.029 | 1132 | $30.0 \times 10^{-6}$ |
| Vanadium | V | 23 | 50.942 | 1890 ± 10 | $24.8 \times 10^{-6}$ |
| Xenon | Xe | 54 | 131.29 | - 111.9 | ...... |
| Ytterbium | Yb | 70 | 173.04 | 824 ± 5 | $28.0 \times 10^{-6}$ |
| Yttrium | Y | 39 | 88.906 | 1523 ± 8 | $57.0 \times 10^{-6}$ |
| Zinc | Zn | 30 | 65.39 | 419.6 | $5.92 \times 10^{-6}$ |
| Zirconium | Zr | 40 | 91.224 | 1852 ± 2 | |

Note: ● values in parentheses are for the mass number of the most stable isotope known; ▪ at 28 Atm; ▲ sublimates; * approximate; ♦ at 26 Atm.

DR. WM. B. D. PENNIMAN
1866-1938
DR. ARTHUR LEE BROWNE
1867-1933

EXECUTIVE STAFF

PHILIP M. AIDT
ALLEN W. THOMPSON
DANTE G. BERETTA
J. ADRIAN BUTT
DONALD W. SMITH

**PENNIMAN & BROWNE, INC.**
CHEMISTS-ENGINEERS-INSPECTORS
6252 FALLS ROAD
BALTIMORE, MARYLAND 21209

ESTABLISHED
1896

CABLE ADDRESS
"BALTEST"

TELEPHONE
825-4131
AREA CODE 301

ENGINEERING DIVISION

**REPORT OF TEST**

Attn: Mr. R.A. Walsh, R & D July 25, 1979

*No.* 791553

*Sample of* Aluminum Castings

*Client* Powercon Corp.

*Marks or Other Data* Physical & Chemical Analysis - two castings to verify Alloy 356 T 6 Condition.

Tensile Test

| Sample No. | 1 | 2 |
|---|---|---|
| Tensile Strength, psi. | 33,755 | 34,735 |
| Yield Strength, psi. | 21,960 | 24,025 |
| Elongation % in 1" | 7.5 | 7.5 |

Chemical Analysis

| Sample | 1 | 2 |
|---|---|---|
| Silicon | 6.55 | 6.68 |
| Iron | 0.098 | 0.10 |
| Copper | 0.15 | 0.17 |
| Manganese | 0.024 | 0.023 |
| Magnesium | 0.23 | 0.25 |
| Zinc | 0.047 | 0.066 |
| Titanium | 0.085 | 0.080 |

Both samples meet chemical & physical requirements for Aluminum Alloy 356 in the T6 Condition.

PENNIMAN & BROWNE, INC.

J.A. Butt

spl

FORM 30 L/B

**FIGURE 4.4** Laboratory analysis report for aluminum alloy specimens.

**TABLE 4.18*a*** Former Standard SAE Steels

| SAE no. | UNS no. | C | Mn | P, max* | S, max* | Si | Cr | Ni | Mo | V, min | AISI no. | Date |
|---|---|---|---|---|---|---|---|---|---|---|---|---|
| 1009 | — | 0.15 max | 0.60 max | 0.040 max | 0.050 max | — | — | — | — | — | 1009 | 1965 |
| 1013† | G10130 | 0.11–0.16 | 0.50–0.80 | 0.040 | 0.050 | — | — | — | — | | | |
| 1019† | G10190 | 0.15–0.20 | 0.70–1.00 | 0.040 | 0.050 | — | — | — | — | — | | |
| 1033 | — | 0.30–0.36 | 0.70–1.00 | 0.040 max | 0.050 max | — | — | — | — | — | 1033 | 1965 |
| 1034 | — | 0.32–0.38 | 0.50–0.80 | 0.040 max | 0.050 max | — | — | — | — | — | C1034 | 1968 |
| 1037† | G10370 | 0.32–0.38 | 0.70–1.00 | 0.040 | 0.050 | — | — | — | — | — | | |
| 1059† | — | 0.55–0.65 | 0.50–0.80 | 0.040 max | 0.050 max | — | — | — | — | — | — | 1968 |
| 1062 | — | 0.54–0.65 | 0.85–1.15 | 0.040 max | 0.050 max | — | — | — | — | — | C1062 | 1953 |
| 1064† | G10640 | 0.60–0.70 | 0.50–0.80 | 0.040 | 0.050 | — | — | — | — | — | — | |
| 1069† | G10690 | 0.65–0.75 | 0.40–0.70 | 0.040 | 0.050 | — | — | — | — | — | | |
| 1075† | G10750 | 0.70–0.80 | 0.40–0.70 | 0.040 | 0.050 | — | — | — | — | — | | |
| 1084† | G10840 | 0.80–0.93 | 0.60–0.90 | 0.040 | 0.050 | — | — | — | — | — | | |
| 1085 | G10850 | 0.80–0.93 | 0.70–1.00 | 0.040 | 0.050 | — | — | — | — | — | | |
| 1086† | G10860 | 0.80–0.94 | 0.30–0.50 | 0.040 max | 0.050 max | — | — | — | — | — | — | 1977 |
| 1108† | G11080 | 0.08–0.13 | 0.50–0.80 | 0.040 | 0.08–0.13 | — | — | — | — | — | | |
| 1109 | G11090 | 0.08–0.13 | 0.60–0.90 | 0.040 max | 0.08–0.13 | — | — | — | — | — | 1109 | 1977 |
| 1111 | — | 0.13 max | 0.60–0.90 | 0.07–0.12 | 0.10–0.15 | — | — | — | — | — | B1111 | 1969 |
| 1112 | — | 0.13 max | 0.70–1.00 | 0.07–0.12 | 0.16–0.23 | — | — | — | — | — | B1112 | 1969 |
| 1113 | — | 0.13 max | 0.70–1.00 | 0.07–0.12 | 0.24–0.33 | — | — | — | — | — | B1113 | 1969 |
| 1114 | — | 0.10–0.16 | 1.00–1.30 | 0.040 max | 0.08–0.13 | — | — | — | — | — | C1114 | 1952 |
| 1115 | — | 0.13–0.18 | 0.60–0.90 | 0.040 max | 0.08–0.13 | — | — | — | — | — | 1115 | 1965 |
| 1116 | — | 0.14–0.20 | 1.10–1.40 | 0.040 max | 0.16–0.23 | — | — | — | — | — | C1116 | 1952 |
| 1119 | G11190 | 0.14–0.20 | 1.00–1.30 | 0.040 max | 0.24–0.33 | — | — | — | — | — | 1119 | 1977 |
| 1120 | — | 0.18–0.23 | 0.70–1.00 | 0.040 max | 0.08–0.13 | — | — | — | — | — | 1120 | 1965 |
| 1126 | — | 0.23–0.29 | 0.70–1.00 | 0.040 max | 0.08–0.13 | — | — | — | — | — | 1126 | 1965 |
| 1132 | G11320 | 0.27–0.34 | 1.35–1.65 | 0.040 max | 0.08–0.13 | — | — | — | — | — | 1132 | 1977 |
| 1138 | — | 0.34–0.40 | 0.70–1.00 | 0.040 max | 0.08–0.13 | — | — | — | — | — | 1138 | 1965 |
| 1139† | G11390 | 0.35–0.43 | 1.35–1.65 | 0.040 | 0.13–0.20 | — | — | — | — | — | | |
| 1145 | G11450 | 0.42–0.49 | 0.70–1.00 | 0.040 max | 0.04–0.07 | — | — | — | — | — | 1145 | 1977 |
| 1151† | G11510 | 0.48–0.55 | 0.70–1.00 | 0.040 | 0.08–0.13 | — | | — | — | — | | |
| 1211† | G12110 | 0.13 max | 0.60–0.90 | 0.07–0.12 | 0.10–0.15 | — | — | — | — | — | | |
| 1320 | — | 0.18–0.23 | 1.60–1.90 | 0.040 | 0.040 | 0.20–0.35 | — | — | — | — | A1320 | 1956 |
| 1345 | G13450 | 0.43–0.48 | 1.60–1.90 | 0.035 | 0.040 | 0.15–0.35 | — | — | — | — | | |
| 1518 | G15180 | 0.15–0.21 | 1.10–1.40 | 0.040 max | 0.050 max | — | — | — | — | — | — | 1977 |
| 1525 | G15250 | 0.23–0.29 | 0.80–1.10 | 0.040 max | 0.050 max | — | — | — | — | — | — | 1977 |
| 1536 | G15360 | 0.30–0.37 | 1.20–1.50 | 0.040 | 0.050 | — | — | — | — | — | | |
| 1547 | G15470 | 0.43–0.51 | 1.35–1.65 | 0.040 max | 0.050 max | — | — | — | — | — | — | 1977 |
| 1551 | G15510 | 0.45–0.56 | 0.85–1.15 | 0.040 | 0.050 | — | — | — | — | — | | |
| 1561 | G15610 | 0.55–0.65 | 0.75–1.05 | 0.040 | 0.050 | — | — | — | — | — | | |
| 1572 | G15720 | 0.65–0.76 | 1.00–1.30 | 0.040 max | 0.050 max | — | — | — | — | — | — | 1977 |

| | | | | | | | | | | | | |
|---|---|---|---|---|---|---|---|---|---|---|---|---|
| 2317 | — | 0.15–0.20 | 0.40–0.60 | 0.040 | 0.040 | 0.20–0.35 | — | 3.25–3.75 | — | — | A2317 | 1956 |
| 2330 | — | 0.28–0.33 | 0.60–0.80 | 0.040 | 0.040 | 0.20–0.35 | — | 3.25–3.75 | — | — | A2330 | 1953 |
| 2340 | — | 0.38–0.43 | 0.70–0.90 | 0.040 | 0.040 | 0.20–0.35 | — | 3.25–3.75 | — | — | A2340 | 1953 |
| 2345 | — | 0.43–0.48 | 0.70–0.90 | 0.040 | 0.040 | 0.20–0.35 | — | 3.25–3.75 | — | — | A2345 | 1952 |
| 2512 | — | 0.09–0.14 | 0.45–0.60 | 0.025 | 0.025 | 0.20–0.35 | — | 4.75–5.25 | — | — | E2512 | 1953 |
| 2515 | — | 0.12–0.17 | 0.40–0.60 | 0.040 | 0.040 | 0.20–0.35 | — | 4.75–5.25 | — | — | A2515 | 1956 |
| 2517 | — | 0.15–0.20 | 0.45–0.60 | 0.025 | 0.025 | 0.20–0.35 | — | 4.75–5.25 | — | — | E2517 | 1959 |
| 3115 | — | 0.13–0.18 | 0.40–0.60 | 0.040 | 0.040 | 0.20–0.35 | 0.55–0.75 | 1.10–1.40 | — | — | A3115 | 1953 |
| 3120 | — | 0.17–0.22 | 0.60–0.80 | 0.040 | 0.040 | 0.20–0.35 | 0.55–0.75 | 1.10–1.40 | — | — | A3120 | 1956 |
| 3130 | — | 0.28–0.33 | 0.60–0.80 | 0.040 | 0.040 | 0.20–0.35 | 0.55–0.75 | 1.10–1.40 | — | — | A3130 | 1956 |
| 3135 | — | 0.33–0.38 | 0.60–0.80 | 0.040 | 0.040 | 0.20–0.35 | 0.55–0.75 | 1.10–1.40 | — | — | 3135 | 1960 |
| X3140 | — | 0.38–0.43 | 0.70–0.90 | 0.040 | 0.040 | 0.20–0.35 | 0.70–0.90 | 1.10–1.40 | — | — | A3141 | 1947 |
| 3140 | — | 0.38–0.43 | 0.70–0.90 | 0.040 | 0.040 | 0.20–0.35 | 0.55–0.75 | 1.10–1.40 | — | — | 3140 | 1964 |
| 3145 | — | 0.43–0.48 | 0.70–0.90 | 0.040 | 0.040 | 0.20–0.35 | 0.70–0.90 | 1.10–1.40 | — | — | A3145 | 1952 |
| 3150 | — | 0.48–0.53 | 0.70–0.90 | 0.040 | 0.040 | 0.20–0.35 | 0.70–0.90 | 1.10–1.40 | — | — | A3150 | 1952 |
| 3215 | — | 0.10–0.20 | 0.30–0.60 | 0.040 | 0.050 | 0.15–0.30 | 0.90–1.25 | 1.50–2.00 | — | — | — | 1941 |
| 3220 | — | 0.15–0.25 | 0.30–0.60 | 0.040 | 0.050 | 0.15–0.30 | 0.90–1.25 | 1.50–2.00 | — | — | — | 1941 |
| 3230 | — | 0.25–0.35 | 0.30–0.60 | 0.040 | 0.050 | 0.15–0.30 | 0.90–1.25 | 1.50–2.00 | — | — | — | 1941 |
| 3240 | — | 0.35–0.45 | 0.30–0.60 | 0.040 | 0.040 | 0.15–0.30 | 0.90–1.25 | 1.50–2.00 | — | — | A3240 | 1941 |
| 3245 | — | 0.40–0.50 | 0.30–0.60 | 0.040 | 0.040 | 0.15–0.30 | 0.90–1.25 | 1.50–2.00 | — | — | — | 1941 |
| 3250 | — | 0.45–0.55 | 0.30–0.60 | 0.040 | 0.040 | 0.15–0.30 | 0.90–1.25 | 1.50–2.00 | — | — | — | 1941 |
| 3310 | — | 0.08–0.13 | 0.45–0.60 | 0.025 | 0.025 | 0.20–0.35 | 1.40–1.75 | 3.25–3.75 | — | — | E3310 | 1964 |
| 3312 | — | 0.08–0.13 | 0.45–0.60 | 0.025 | 0.025 | 0.20–0.35 | 1.40–1.75 | 3.25–3.75 | — | — | — | 1948 |
| 3316 | — | 0.14–0.19 | 0.45–0.60 | 0.025 | 0.025 | 0.20–0.35 | 1.40–1.75 | 3.25–3.75 | — | — | E3316 | 1956 |
| 3325 | — | 20–30 | 0.30–0.60 | 0.040 | 0.050 | 0.15–0.30 | 1.25–1.75 | 3.25–3.75 | — | — | — | 1936 |
| 3335 | — | 30–40 | 0.30–0.60 | 0.040 | 0.050 | 0.15–0.30 | 1.25–1.75 | 3.25–3.75 | — | — | — | 1936 |
| 3340 | — | 35–45 | 0.30–0.60 | 0.040 | 0.050 | 0.15–0.30 | 1.25–1.75 | 3.25–3.75 | — | — | — | 1936 |
| 3415 | — | 0.10–0.20 | 0.30–0.60 | 0.040 | 0.050 | 0.15–0.30 | 0.60–0.95 | 2.75–3.25 | — | — | — | 1941 |
| 3435 | — | 0.30–0.40 | 0.30–0.60 | 0.040 | 0.050 | 0.15–0.30 | 0.60–0.95 | 2.75–3.25 | — | — | — | 1936 |
| 3450 | — | 0.45–0.55 | 0.30–0.60 | 0.040 | 0.050 | 0.15–0.30 | 0.60–0.95 | 2.75–3.25 | — | — | — | 1936 |
| 4012 | G40120 | 0.09–0.14 | 0.75–1.00 | 0.035 | 0.040 | 0.15–0.30 | — | — | 0.15–0.25 | — | 4012 | 1977 |
| 4024 | G40240 | 0.20–0.25 | 0.70–0.90 | 0.035 | 0.035–0.50 | 0.15–0.35 | — | — | 0.20–0.30 | — | | |
| 4032 | G40320 | 0.30–0.35 | 0.70–0.90 | 0.035 | 0.040 | 0.15–0.35 | — | — | 0.20–0.30 | — | | |
| 4042 | G40420 | 0.40–0.45 | 0.70–0.90 | 0.035 | 0.040 | 0.15–0.35 | — | — | 0.20–0.30 | — | | |
| 4053 | — | 0.50–0.56 | 0.75–1.00 | 0.040 | 0.040 | 0.20–0.35 | — | — | 0.20–0.30 | — | 4053 | 1956 |
| 4063 | G40630 | 0.60–0.67 | 0.75–1.00 | 0.040 | 0.040 | 0.20–0.35 | — | — | 0.20–0.30 | — | 4063 | 1964 |
| 4068 | — | 0.63–0.70 | 0.75–1.00 | 0.040 | 0.040 | 0.20–0.35 | — | — | 0.20–0.30 | — | A4068 | 1957 |
| 4119 | — | 0.17–0.22 | 0.70–0.90 | 0.040 | 0.040 | 0.20–0.35 | 0.40–0.60 | — | 0.20–0.30 | — | A4119 | 1956 |
| 4125 | — | 0.23–0.28 | 0.70–0.90 | 0.040 | 0.040 | 0.20–0.35 | 0.40–0.60 | — | 0.20–0.30 | — | A4125 | 1950 |
| 4135 | G41350 | 0.33–0.38 | 0.70–0.90 | 0.035 | 0.040 | 0.15–0.35 | 0.80–1.10 | — | 0.15–0.25 | — | | |
| 4161 | G41610 | 0.56–0.64 | 0.75–1.00 | 0.035 | 0.040 | 0.15–0.35 | 0.70–0.90 | — | 0.25–0.35 | — | | |

* Limits apply to semi-finished products for forgings, bars, wire rods, and seamless tubing.

† These grades remain standard for wire rods.

‡ Boron content 0.0005 to 0.003%.

***Source:*** Society of Automotive Engineers, 1986: *SAE Materials Handbook,* Vol. 1, *Materials.* Warrendale, Pa.

**TABLE 4.18*a*** Former Standard SAE Steels (*Continued*)

| SAE no. | UNS no. | C | Mn | P, max* | S, max* | Si | Cr | Ni | Mo | V, min | AISI no. | Date |
|---|---|---|---|---|---|---|---|---|---|---|---|---|
| 4317 | — | 0.15–0.20 | 0.45–0.65 | 0.040 | 0.040 | 0.20–0.35 | 0.40–0.60 | 1.65–2.00 | 0.20–0.30 | — | 4317 | 1953 |
| 4337 | G43370 | 0.35–0.40 | 0.60–0.80 | 0.040 | 0.040 | 0.20–0.35 | 0.70–0.90 | 1.65–2.00 | 0.20–0.30 | — | 4337 | 1964 |
| 4419 | — | 0.18–0.23 | 0.45–0.65 | 0.035 | 0.040 | 0.15–0.30 | — | — | 0.45–0.60 | — | 4520 | 1977 |
| 4419H | — | 0.17–0.23 | 0.35–0.75 | 0.035 | 0.040 | 0.15–0.30 | — | — | 0.45–0.60 | — | 4419H | 1977 |
| 4422 | G44220 | 0.20–0.25 | 0.70–0.90 | 0.035 | 0.040 | 0.15–0.35 | — | — | 0.35–0.45 | — | | |
| 4427 | G44270 | 0.24–0.29 | 0.70–0.90 | 0.035 | 0.040 | 0.15–0.35 | — | — | 0.35–0.45 | — | | |
| 4608 | — | 0.06–0.11 | 0.25–0.45 | 0.040 | 0.040 | 0.25 max | — | 1.40–1.75 | 0.15–0.25 | — | 4608 | 1956 |
| 46812[‡] | — | 0.10–0.15 | 0.45–0.65 | 0.040 | 0.040 | 0.20–0.35 | — | 1.65–2.00 | 0.20–0.30 | — | 46B12[a] | 1957 |
| 4615 | G46150 | 0.13–0.18 | 0.45–0.65 | 0.035 | 0.040 | 0.15–0.35 | — | 1.65–2.00 | 0.20–0.30 | — | | |
| 4617 | G46170 | 0.15–0.20 | 0.45–0.65 | 0.035 | 0.040 | 0.15–0.35 | — | 1.65–2.00 | 0.20–0.30 | — | | |
| X4620 | — | 0.18–0.23 | 0.50–0.70 | 0.040 | 0.040 | 0.20–0.35 | — | 1.65–2.00 | 0.20–0.30 | — | X4620 | 1956 |
| 4621 | G46210 | 0.18–0.23 | 0.70–0.90 | 0.035 | 0.040 | 0.15–0.30 | — | 1.65–2.00 | 0.20–0.30 | — | 4621 | 1977 |
| 4621H | — | 0.17–0.23 | 0.60–1.00 | 0.035 | 0.040 | 0.15–0.30 | — | 1.55–2.00 | 0.20–0.30 | — | 4621H | 1977 |
| 4626 | G46260 | 0.24–0.29 | 0.45–0.65 | 0.035 | 0.040 max | 0.15–0.35 | — | 0.70–1.00 | 0.15–0.25 | — | | |
| 4640 | — | 0.38–0.43 | 0.60–0.80 | 0.040 | 0.040 | 0.20–0.35 | — | 1.65–2.00 | 0.20–0.30 | — | A4640 | 1952 |
| 4718 | G47180 | 0.16–0.21 | 0.70–0.90 | — | — | — | 0.35–0.55 | 0.90–1.20 | 0.30–0.40 | — | | |
| 4812 | — | 0.10–0.15 | 0.40–0.60 | 0.040 | 0.040 | 0.20–0.35 | — | 3.25–3.75 | 0.20–0.30 | — | 4817 | 1956 |
| 4817 | G48170 | 0.15–0.20 | 0.40–0.60 | 0.035 | 0.040 | 0.15–0.35 | — | 3.25–3.75 | 0.20–0.30 | — | | |
| 5015 | G50150 | 0.12–0.17 | 0.30–0.50 | 0.035 | 0.040 | 0.15–0.30 | 0.30–0.50 | — | — | — | 5015 | 1977 |
| 50B40 | B50401 | 0.38–0.43 | 0.75–1.00 | 0.035 | 0.040 | 0.15–0.35 | 0.40–0.60 | — | — | — | | |
| 50B44 | G50441 | 0.43–0.48 | 0.75–1.00 | 0.035 | 0.040 | 0.15–0.35 | 0.40–0.60 | — | — | — | | |
| 5045 | — | 0.43–0.48 | 0.70–0.90 | 0.040 | 0.040 | 0.20–0.35 | 0.55–0.75 | — | — | — | 5045 | 1953 |
| 5046 | G50460 | 0.43–0.48 | 0.75–1.00 | 0.035 | 0.040 | 0.15–0.35 | 0.20–0.35 | — | — | — | | |
| 50B50 | G50501 | 0.48–0.53 | 0.75–1.00 | 0.035 | 0.040 | 0.15–0.35 | 0.40–0.60 | — | — | — | | |
| 5060 | G50600 | 0.56–0.64 | 0.75–1.00 | 0.035 | 0.040 | 0.15–0.35 | 0.40–0.60 | — | — | — | | |
| 50B60 | G50601 | 0.56–0.64 | 0.75–1.00 | 0.035 | 0.040 | 0.15–0.35 | 0.40–0.60 | — | — | — | | |
| 5115 | G51150 | 0.13–0.18 | 0.70–0.90 | 0.035 | 0.040 | 0.15–0.35 | 0.70–0.90 | — | — | — | | |
| 5117 | G51170 | 0.15–0.20 | 0.70–0.90 | 0.040 | 0.040 | 0.15–0.35 | 0.70–0.90 | — | — | — | | |
| 5135 | G51350 | 0.33–0.38 | 0.60–0.80 | 0.035 | 0.040 | 0.15–0.35 | 0.80–1.05 | — | — | — | | |
| 5145 | G51450 | 0.43–0.48 | 0.70–0.90 | 0.035 | 0.040 | 0.15–0.30 | 0.70–0.90 | — | — | — | 5145 | 1977 |
| 5145H | H51450 | 0.42–0.49 | 0.60–1.00 | 0.035 | 0.040 | 0.15–0.30 | 0.60–1.00 | — | — | — | 5145H | 1977 |
| 5147 | G51470 | 0.46–0.51 | 0.70–0.95 | 0.035 | 0.040 | 0.15–0.35 | 0.85–1.15 | — | — | — | | |
| 5152 | — | 0.48–0.55 | 0.70–0.90 | 0.040 | 0.040 | 0.20–0.35 | 0.90–1.20 | — | — | — | 5152 | 1956 |
| 5155 | G51550 | 0.51–0.59 | 0.70–0.90 | 0.035 | 0.040 | 0.15–0.35 | 0.70–0.90 | — | — | — | | |
| 50100 | G50986 | 0.98–1.10 | 0.25–0.45 | 0.025 | 0.025 | 0.15–0.35 | 0.40–0.60 | — | — | — | | |
| 6115 | — | 0.10–0.20 | 0.30–0.60 | 0.040 | 0.050 | 0.15–0.30 | 0.80–1.10 | — | — | 0.15 | — | 1936 |
| 6117 | — | 0.15–0.20 | 0.70–0.90 | 0.040 | 0.040 | 0.20–0.35 | 0.70–0.90 | — | — | 0.10 | 6117 | 1956 |
| 6118 | G61180 | 0.16–0.21 | 0.50–0.70 | 0.035 | 0.040 | 0.15–0.35 | 0.50–0.70 | — | V-0.10–0.15 | — | | |
| 6120 | — | 0.17–0.22 | 0.70–0.90 | 0.040 | 0.040 | 0.20–0.35 | 0.70–0.90 | — | — | 0.10 | 6120 | 1961 |

| | | | | | | | | | | | | |
|---|---|---|---|---|---|---|---|---|---|---|---|---|
| 6125 | — | 0.20–0.30 | 0.60–0.90 | 0.040 | 0.050 | 0.15–0.30 | 0.80–1.10 | — | — | 0.15 | — | 1936 |
| 6130 | — | 0.25–0.35 | 0.60–0.90 | 0.040 | 0.050 | 0.15–0.30 | 0.80–1.10 | — | — | 0.15 | — | 1936 |
| 6135 | — | 0.30–0.40 | 0.60–0.90 | 0.040 | 0.050 | 0.15–0.30 | 0.80–1.10 | — | — | 0.15 | — | 1941 |
| 6140 | — | 0.35–0.45 | 0.60–0.90 | 0.040 | 0.050 | 0.15–0.30 | 0.80–1.10 | — | — | 0.15 | — | 1936 |
| 6145 | — | 0.43–0.48 | 0.70–0.90 | 0.040 | 0.050 | 0.20–0.35 | 0.80–1.10 | — | — | 0.15 | 6145 | 1956 |
| 6195 | — | 0.90–1.05 | 0.20–0.45 | 0.030 | 0.035 | 0.15–0.30 | 0.80–1.10 | — | — | 0.15 | — | 1936 |
| 71360 | — | 0.50–0.70 | 0.30 max | 0.035 | 0.040 | 0.15–0.30 | 3.00–4.00 | 12.00–15.00W | — | — | — | 1936 |
| 71660 | — | 0.50–0.70 | 0.30 max | 0.035 | 0.040 | 0.15–0.30 | 3.00–4.00 | 15.00–18.00W | — | — | — | 1936 |
| 7260 | — | 0.50–0.70 | 0.30 max | 0.035 | 0.040 | 0.15–0.30 | 0.50–1.00 | 1.50–2.00W | — | — | — | 1936 |
| 8115 | G81150 | 0.13–0.18 | 0.70–0.90 | 0.035 | 0.040 | 0.15–0.35 | 0.30–0.50 | 0.20–0.40 | 0.08–0.15 | — | | |
| 81B45 | G81451 | 0.43–0.48 | 0.75–1.00 | 0.035 | 0.040 | 0.15–0.35 | 0.35–0.55 | 0.20–0.40 | 0.08–0.15 | — | | |
| 8625 | G86250 | 0.23–0.28 | 0.70–0.90 | 0.035 | 0.040 | 0.15–0.35 | 0.40–0.60 | 0.40–0.70 | 0.15–0.25 | — | | |
| 8627 | G86270 | 0.25–0.30 | 0.70–0.90 | 0.035 | 0.040 | 0.15–0.35 | 0.40–0.60 | 0.40–0.70 | 0.15–0.25 | — | | |
| 8632 | — | 0.30–0.35 | 0.70–0.90 | 0.040 | 0.040 | 0.20–0.35 | 0.40–0.60 | 0.40–0.70 | 0.15–0.25 | — | 8632 | 1951 |
| 8635 | — | 0.33–0.38 | 0.75–1.00 | 0.040 | 0.040 | 0.20–0.35 | 0.40–0.60 | 0.40–0.70 | 0.15–0.25 | — | 8635 | 1956 |
| 8641 | — | 0.38–0.43 | 0.75–1.00 | 0.040 | 0.040–0.60 | 0.20–0.35 | 0.40–0.60 | 0.40–0.70 | 0.15–0.25 | — | 8641 | 1956 |
| 8642 | G86420 | 0.40–0.45 | 0.75–1.00 | 0.035 | 0.040 | 0.15–0.35 | 0.40–0.60 | 0.40–0.70 | 0.15–0.25 | — | | |
| 86B45 | G86451 | 0.43–0.48 | 0.75–1.00 | 0.035 | 0.040 | 0.15–0.35 | 0.40–0.60 | 0.40–0.70 | 0.15–0.25 | — | | |
| 8650 | G86500 | 0.48–0.53 | 0.75–1.00 | 0.035 | 0.040 | 0.15–0.35 | 0.40–0.60 | 0.40–0.70 | 0.15–0.25 | — | | |
| 8653 | — | 0.50–0.56 | 0.75–1.00 | 0.040 | 0.040 | 0.20–0.35 | 0.50–0.80 | 0.40–0.70 | 0.15–0.25 | — | 8653 | 1956 |
| 8647 | — | 0.45–0.50 | 0.75–1.00 | 0.040 | 0.040 | 0.20–0.35 | 0.40–0.60 | 0.40–0.70 | 0.15–0.25 | — | 8647 | 1948 |
| 8655 | G86550 | 0.51–0.59 | 0.75–1.00 | 0.035 | 0.040 | 0.15–0.35 | 0.40–0.60 | 0.40–0.70 | 0.15–0.25 | — | | |
| 8660 | G86600 | 0.56–0.64 | 0.75–1.00 | 0.035 | 0.040 | 0.15–0.35 | 0.40–0.60 | 0.40–0.70 | 0.15–0.25 | — | | |
| 8715 | — | 0.13–0.18 | 0.70–0.90 | 0.040 | 0.040 | 0.20–0.35 | 0.40–0.60 | 0.40–0.70 | 0.20–0.30 | — | 8715 | 1956 |
| 8717 | — | 0.15–0.20 | 0.70–0.90 | 0.040 | 0.040 | 0.20–0.35 | 0.40–0.60 | 0.40–0.70 | 0.20–0.30 | — | 8717 | 1956 |
| 8719 | — | 0.18–0.23 | 0.60–0.80 | 0.040 | 0.040 | 0.20–0.35 | 0.40–0.60 | 0.40–0.70 | 0.20–0.30 | — | 8719 | 1952 |
| 8735 | G87350 | 0.33–0.38 | 0.75–1.00 | 0.040 | 0.040 | 0.20–0.35 | 0.40–0.60 | 0.40–0.70 | 0.20–0.30 | — | 8735 | 1952 |
| 8740 | G87400 | 0.38–0.43 | 0.75–1.00 | 0.035 | 0.040 | 0.15–0.35 | 0.40–0.60 | 0.40–0.70 | 0.20–0.30 | — | | |
| 8742 | G87420 | 0.40–0.45 | 0.75–1.00 | 0.040 | 0.040 | 0.20–0.35 | 0.40–0.60 | 0.40–0.70 | 0.20–0.30 | — | 8742 | 1964 |
| 8745 | — | 0.43–0.48 | 0.75–1.00 | 0.040 | 0.040 | 0.20–0.35 | 0.40–0.60 | 0.40–0.70 | 0.20–0.30 | — | 8745 | 1953 |
| 8750 | — | 0.48–0.53 | 0.75–1.00 | 0.040 | 0.040 | 0.20–0.35 | 0.40–0.60 | 0.40–0.70 | 0.20–0.30 | — | 8750 | 1956 |
| 9250 | — | 0.45–0.55 | 0.60–0.90 | 0.040 | 0.040 | 1.80–2.20 | — | — | — | — | 9250 | 1941 |
| 9254[†] | G92540 | 0.51–0.59 | 0.60–0.80 | 0.035 | 0.040 | 1.20–1.60 | 0.60–0.80 | — | — | — | | |
| 9255 | G92550 | 0.51–0.59 | 0.70–0.95 | 0.035 | 0.040 | 1.80–2.20 | — | — | — | — | 9255 | 1977 |
| 9261 | — | 0.55–0.65 | 0.75–1.00 | 0.040 | 0.040 | 1.80–2.20 | 0.10–0.25 | — | — | — | 9261 | 1956 |
| 9262 | G92620 | 0.55–0.65 | 0.75–1.00 | 0.040 | 0.040 | 1.80–2.20 | 0.25–0.40 | — | — | — | 9262 | 1961 |
| 9310 | G93100 | 0.08–0.13 | 0.45–0.65 | 0.025 | 0.025 | 0.15–0.35 | 1.00–1.40 | 3.00–3.50 | 0.08–0.15 | — | | |
| 9315 | — | 0.13–0.18 | 0.45–0.65 | 0.025 | 0.025 | 0.20–0.35 | 1.00–1.40 | 3.00–3.50 | 0.08–0.15 | — | E9315 | 1959 |
| 9317 | — | 0.15–0.20 | 0.45–0.65 | 0.025 | 0.025 | 0.20–0.35 | 1.00–1.40 | 3.00–3.50 | 0.08–0.15 | — | E9317 | 1959 |

* Limits apply to semi-finished products for forgings, bars, wire rods, and seamless tubing.

[†] These grades remain standard for wire rods.

[‡] Boron content 0.0005 to 0.003%.

***Source:*** Society of Automotive Engineers, 1986: *SAE Materials Handbook,* Vol. 1, *Materials.* Warrendale, Pa.

**TABLE 4.18*a*** Former Standard SAE Steels (*Continued*)

| SAE no. | UNS no. | C | Mn | P, max* | S, max* | Si | Cr | Ni | Mo | V, min | AISI no. | Date |
|---|---|---|---|---|---|---|---|---|---|---|---|---|
| 94B15 | G94151 | 0.13–0.18 | 0.75–1.00 | 0.035 | 0.040 | 0.15–0.35 | 0.30–0.50 | 0.30–0.60 | 0.08–0.15 | — | | |
| 94B17 | G94171 | 0.15–0.20 | 0.75–1.00 | 0.035 | 0.040 | 0.15–0.35 | 0.30–0.50 | 0.30–0.60 | 0.08–0.15 | — | | |
| 94B30 | G94301 | 0.28–0.33 | 0.75–1.00 | 0.035 | 0.040 | 0.15–0.35 | 0.30–0.50 | 0.30–0.60 | 0.08–0.15 | — | | |
| 9437 | — | 0.35–0.40 | 0.90–1.20 | 0.040 | 0.040 | 0.20–0.35 | 0.30–0.50 | 0.30–0.60 | 0.08–0.15 | — | 9437 | 1950 |
| 9440 | — | 0.38–0.43 | 0.90–1.20 | 0.040 | 0.040 | 0.20–0.35 | 0.30–0.50 | 0.30–0.60 | 0.08–0.15 | — | 9440 | 1950 |
| 94B40‡ | G94401 | 0.38–0.43 | 0.75–1.00 | 0.040 | 0.040 | 0.20–0.35 | 0.30–0.60 | 0.30–0.60 | 0.08–0.15 | — | 94B40 | 1964 |
| 9442 | — | 0.40–0.45 | 0.90–1.20 | 0.040 | 0.040 | 0.20–0.35 | 0.30–0.50 | 0.30–0.60 | 0.08–0.15 | — | 9442 | 1950 |
| 9445 | — | 0.43–0.48 | 0.90–1.20 | 0.040 | 0.040 | 0.20–0.35 | 0.30–0.50 | 0.30–0.60 | 0.08–0.15 | — | 9445 | 1950 |
| 9447 | — | 0.45–0.50 | 0.90–1.20 | 0.040 | 0.040 | 0.20–0.35 | 0.30–0.50 | 0.30–0.60 | 0.08–0.15 | — | 9447 | 1950 |
| 9747 | — | 0.45–0.50 | 0.50–0.80 | 0.040 | 0.040 | 0.20–0.35 | 0.10–0.25 | 0.40–0.70 | 0.15–0.25 | — | 9747 | 1950 |
| 9763 | — | 0.60–0.67 | 0.50–0.80 | 0.040 | 0.040 | 0.20–0.35 | 0.10–0.25 | 0.40–0.70 | 0.15–0.25 | — | 9763 | 1950 |
| 9840 | G98400 | 0.38–0.43 | 0.70–0.90 | 0.040 | 0.040 | 0.20–0.35 | 0.70–0.90 | 0.85–1.15 | 0.20–0.30 | — | 9840 | 1964 |
| 9845 | — | 0.43–0.48 | 0.70–0.90 | 0.040 | 0.040 | 0.20–0.35 | 0.70–0.90 | 0.85–1.15 | 0.20–0.30 | — | 9845 | 1950 |
| 9850 | G98500 | 0.48–0.53 | 0.70–0.90 | 0.040 | 0.040 | 0.20–0.35 | 0.70–0.90 | 0.85–1.15 | 0.20–0.30 | — | 9850 | 1961 |
| 438V12‡ | — | 0.08–0.13 | 0.75–1.00 | — | — | 0.20–0.35 | 0.40–0.60 | 1.65–2.00 | 0.20–0.30 | 0.03 | — | — |
| 438V14‡ | — | 0.10–0.15 | 0.45–0.65 | — | — | 0.20–0.35 | 0.40–0.60 | 1.65–2.00 | 0.08–0.15 | 0.03 | — | — |

* Limits apply to semi-finished products for forgings, bars, wire rods, and seamless tubing.
† These grades remain standard for wire rods.
‡ Boron content 0.0005 to 0.003%.
***Source:*** Society of Automotive Engineers, 1986: *SAE Materials Handbook,* Vol. 1, *Materials.* Warrendale, Pa.

**TABLE 4.18*b*** Former EX/MS Steels

| EX No. | Composition, % C | Mn | Cr | Mo | Other | Approximate SAE Grade | Deletion date |
|---|---|---|---|---|---|---|---|
| 1* | 0.15–0.21 | 0.35–0.60 | — | 0.20–0.30 | 4.80–5.30 Ni | 9310 | 1976 |
| 2* | 0.64–0.75 | 0.25–0.45 | 0.15–0.30 | 0.08–0.15 | 0.70–1.00 Ni | — | 1971 |
| 3 | 0.56–0.64 | 0.75–1.00 | 0.40–0.60 | — | — | 5060 | Made standard |
| 4 | 0.18–0.23 | 0.75–1.00 | 0.45–0.65 | 0.05–0.10 | — | 4118 | 1973 |
| 5 | 0.18–0.23 | 0.75–1.00 | 0.45–0.65 | 0.08–0.15 | 0.40–0.70 Ni | 8620 | 1971 |
| 6 | 0.20–0.25 | 0.75–1.00 | 0.45–0.65 | 0.08–0.15 | 0.40–0.70 Ni | 8622 | 1971 |
| 7 | 0.23–0.28 | 0.75–1.00 | 0.45–0.65 | 0.08–0.15 | 0.40–0.70 Ni | 8625 | 1971 |
| 8 | 0.25–0.30 | 0.75–1.00 | 0.45–0.65 | 0.08–0.15 | 0.40–0.70 Ni | 8627 | 1971 |
| 9* | 0.19–0.24 | 0.95–1.25 | 0.25–0.40 | 0.05–0.10 | 0.20–0.40 Ni | 8620 | 1976 |
| 11* | 0.38–0.43 | 0.75–1.00 | 0.25–0.40 | 0.05–0.10 | 0.20–0.40 Ni, 0.0005 B min | 8640 | 1976 |
| 12 | 0.38–0.43 | 0.75–1.00 | 0.25–0.40 | 0.05–0.10 | 0.20–0.40 Ni, 0.0005 B min | 8640 | 1976 |
| 13* | 0.66–0.75 | 0.80–1.05 | 0.25–0.40 | 0.05–0.10 | 0.20–0.40 Ni | — | 1976 |
| 14* | 0.66–0.75 | 0.80–1.05 | 0.25–0.40 | 0.05–0.10 | 0.20–0.40 Ni | — | 1976 |
| 15 | 0.18–0.23 | 0.90–1.20 | 0.40–0.60 | 0.13–0.20 | — | — | |
| 22 | 0.13–0.18 | 0.75–1.00 | 0.45–0.65 | 0.20–0.30 | — | 8615 | 1973 |
| 23 | 0.15–0.20 | 0.75–1.00 | 0.45–0.65 | 0.20–0.30 | — | 8617 | 1973 |
| 24 | 0.18–0.23 | 0.75–1.00 | 0.45–0.65 | 0.20–0.30 | — | — | |
| 25 | 0.20–0.25 | 0.75–1.00 | 0.45–0.65 | 0.20–0.30 | — | 8622 | 1973 |
| 26 | 0.23–0.28 | 0.75–1.00 | 0.45–0.65 | 0.20–0.30 | — | 8625 | 1973 |
| 27 | 0.25–0.30 | 0.75–1.00 | 0.45–0.65 | 0.20–0.30 | — | 8627 | 1976 |
| 28 | 0.16–0.21 | 0.75–1.00 | 0.45–0.65 | 0.30–0.40 | 0.40–0.70 Ni | 4718 | 1973 |
| 29 | 0.18–0.23 | 0.75–1.00 | 0.45–0.65 | 0.30–0.40 | 0.40–0.70 Ni | 4320 | 1976 |
| 30 | 0.13–0.18 | 0.70–0.90 | 0.45–0.65 | 0.45–0.60 | 0.70–1.00 Ni | — | |
| 35 | 0.35–0.40 | 0.90–1.20 | 0.45–0.65 | 0.13–0.20 | — | 8637 | 1976 |
| 37 | 0.40–0.45 | 0.90–1.20 | 0.45–0.65 | 0.13–0.20 | — | 8642 | 1976 |
| 41 | 0.56–0.64 | 0.90–1.20 | 0.45–0.65 | 0.13–0.20 | — | 8660 | 1976 |
| 42 | 0.13–0.18 | 0.95–1.25 | 0.25–0.40 | 0.05–0.10 | 0.20–0.40 Ni | 8615 | 1976 |
| 43 | 0.13–0.18 | 0.95–1.25 | 0.25–0.40 | 0.05–0.10 | 0.20–0.40 Ni, 0.0005 B min | — | 1976 |
| 44 | 0.15–0.20 | 0.95–1.25 | 0.25–0.40 | 0.05–0.10 | 0.20–0.40 Ni | 8617 | 1976 |
| 45 | 0.15–0.20 | 0.95–1.25 | 0.25–0.40 | 0.05–0.10 | 0.20–0.40 Ni, 0.0005 B min | — | 1976 |
| 46 | 0.20–0.25 | 0.95–1.25 | 0.25–0.40 | 0.05–0.10 | 0.20–0.40 Ni | 8622 | 1976 |
| 47 | 0.23–0.28 | 0.95–1.25 | 0.25–0.40 | 0.05–0.10 | 0.20–0.40 Ni | 8625 | 1976 |
| 48 | 0.25–0.30 | 0.95–1.25 | 0.25–0.40 | 0.05–0.10 | 0.20–0.40 Ni | 8627 | 1976 |
| 49 | 0.28–0.33 | 0.95–1.25 | 0.25–0.40 | 0.05–0.10 | 0.20–0.40 Ni | 8630 | 1976 |
| 50 | 0.33–0.38 | 0.95–1.25 | 0.25–0.40 | 0.05–0.10 | 0.20–0.40 Ni | 8635 | 1976 |
| 51 | 0.35–0.40 | 0.95–1.25 | 0.25–0.40 | 0.05–0.10 | 0.20–0.40 Ni | 8637 | 1976 |
| 52 | 0.38–0.43 | 0.95–1.25 | 0.25–0.40 | 0.05–0.10 | 0.20–0.40 Ni | 8640 | 1976 |
| 53 | 0.40–0.45 | 0.95–1.25 | 0.25–0.40 | 0.05–0.10 | 0.20–0.40 Ni | 8642 | 1976 |
| 60 | 0.20–0.25 | 1.00–1.30 | 0.70–0.90 | — | — | — | 1983 |
| 62 | 0.25–0.30 | 1.00–1.30 | 0.70–0.90 | — | — | — | 1983 |

* All steels contain (1) 0.035 P max except EX 1 (0.040 P max), and EX 2, EX 13, and EX 14 (0.025 P max); (2) all contain 0.040 S max except EX 2, EX 13, and EX 14 (0.025 S max); and (3) all contain 0.15 to 0.35 Si except EX 9, EX 11, and EX 13 (0.050 Si max).

***Source:*** Society of Automotive Engineers, 1986: *SAE Materials Handbook,* Vol. 1, *Materials.* Warrendale, Pa.

**TABLE 4.19** Identification of Plastics—Simple Test Procedures

| Plastic Material | No Flame | Burns but extinguishes on removal of flame | | | Continues to burn after removal of flame | | | | Remarks |
|---|---|---|---|---|---|---|---|---|---|
| | ■ Odor | Odor | Color of flame | Drips | Odor | Color of flame | Drips | Burn Speed | |
| ABS | ..... | Acrid | Yellow, blue edges | No | Acrid | Yellow, blue edges | Yes | Slow | Black smoke, soot in air |
| Acetals | ..... | ..... | ..... | ..... | Formaldehyde | Blue, no smoke | Yes | Slow | ..... |
| Acrylics | ..... | ..... | ..... | ..... | Fruity | Blue, yellow tip | No (cast) Yes (molded) | Slow | Flame may spurt if rubber molded. |
| Cellulosics (Acetate) | ..... | Vinegar | Yellow with sparks | No | Vinegar | Yellow | Yes | Slow | Flame may spark |
| Fluocarbons | | | | | | | | | |
| FEP | Burnt hair (Faint) | ..... | ..... | ..... | ..... | ..... | ..... | ..... | Deforms, no combustion but drips. |
| PTFE | " | ..... | ..... | ..... | ..... | ..... | ..... | ..... | Deforms but does not drip. |
| CTFE | Acetic acid | ..... | ..... | ..... | ..... | ..... | ..... | ..... | Deforms, no combustion but drips. |
| PVF | Acidic | ..... | ..... | ..... | ..... | ..... | ..... | ..... | Deforms. |
| Nylons Type 6 & 6/6 | ..... | Burnt wool | Blue, yellow tip | Yes | ..... | ..... | ..... | ..... | 6/6 more rigid |
| Polycarbonates | ..... | Faint sweet (aromatic) | Orange | Yes | ..... | ..... | ..... | ..... | Black smoke with soot. |
| Polyethylenes | ..... | ..... | ..... | ..... | Paraffin | Blue, yellow tip | Yes | Slow | Floats in water. |
| Polyimides | ◆ | ..... | ..... | ..... | ..... | ..... | ..... | ..... | Chars, material rigid. |
| Polypropylenes | ..... | Acrid | Yellow | No | Sweet | Blue, yellow tip | Yes | Slow | Floats in water. More difficult to scratch than polyethylene. |
| Polystyrenes | ..... | ..... | ..... | ..... | Illuminating gas | Yellow | Yes | Rapid | Black smoke with soot in air. |
| Polysulfones | ..... | ◆ | Orange | Yes | ..... | ..... | ..... | ..... | Black smoke. |
| Polyurethanes | ..... | ..... | ..... | ..... | ◆ | Yellow | No | Slow | Black smoke. |

| | | | | | | | | | |
|---|---|---|---|---|---|---|---|---|---|
| Vinyls: Flex - rigid | ..... | Hydrochloric Acid | Yellow with green spurts. | No | ..... | ..... | ..... | ..... | Chars, melts. |
| Polyblends: | | | | | | | | | |
| ABS/Polycarbonate | ..... | ..... | ..... | ..... | ♦ | Yellow, blue edges | No | ..... | Black smoke, soot in air. |
| ABS-PVC | ..... | Acrid | Yellow, blue edges | No | ..... | ..... | ..... | ..... | Black smoke, soot in air. |
| PVC/Acrylic | ..... | Fruity | Blue, yellow tip | No | ..... | ..... | ..... | ..... | ..... |
| Melamines | Formaldehyde and fish. | ..... | ..... | ..... | ..... | ..... | ..... | ..... | ..... |
| Phenolics | Formaldehyde and phenol | Phenol and wood/paper | Yellow | No | ..... | ..... | ..... | ..... | May crack. |
| Polyesters | ..... | Hydrochloric Acid | Yellow | No | ♦ | Yellow, blue edges | No | Slow | Cracks and breaks. |
| Silicones | ♦ | ..... | ..... | ..... | ..... | ..... | ..... | ..... | Deforms. |
| Ureas | Formaldehyde | ..... | ..... | ..... | ..... | ..... | ..... | ..... | ..... |

NOTES: ♦ = Nondescript. ■ = Freshly cut or filed to bring out odor.
Source: Laird Plastics, West Palm Beach, Florida.

## *FURTHER READING*


The Aluminum Association, 1979: *Aluminum Data and Standards,* 6th ed. Washington, D.C.

Aluminum Company of America, 1960: *Alcoa Structural Handbook.* Pittsburgh, Pa.

Copper Development Association, 1964: *Standards Handbook—Copper and Copper Alloys,* CDA Publication No. 101, 5th ed. New York.

Ryerson, Joseph T., 1975: *Steel and Aluminum Data Book.* Philadelphia, Pa.: Ryerson Company.

Weast, Robert C., ed., 1969: *Handbook of Chemistry and Physics,* 50th ed. Cleveland, Ohio: The Chemical Rubber Company.

Society of Automotive Engineers, 1986: *SAE Handbook,* Vol. 1, *Materials.* Warrendale, Pa.

Walsh, Ronald A., 1994, *McGraw-Hill Machining and Metalworking Handbook.* New York: McGraw-Hill.

# CHAPTER 5
# STRENGTH OF MATERIALS

*Strength of materials,* as it is used in this discussion, concerns the relationship between external forces applied to materials and the strains or deformations that occur as a result of these forces.

We will apply principles and equations outlined in this section along with those taken from the chapters on mechanics and machine elements to the design of structures, mechanisms, parts, and machines.

## 5.1 COMMON TERMS USED IN STRENGTH OF MATERIALS

**Cold flow** The gradual extrusion or deformation of a nonelastic material under load. Materials such as lead, other soft metals, and some plastics like Teflon exhibit cold flow under load. Some compensation must be made in design for this condition, such as spring loading the parts.

**Damping or elastic hysteresis** When tensile loads are applied to a body, complete elongation is not produced immediately, but is accompanied by a time lag that depends on the material type and the magnitude of the applied stresses. When the tensile load is removed, complete recovery of energy does not occur. This effect is known as *elastic hysteresis,* or, for vibrating stresses, *damping.* This occurs within the proportional limit of the material.

**Ductility** The ability of a material to be drawn into wire form.

**Elastic limit** The maximum stress that may be applied to a material that then returns to its original dimensions. The elastic limit and the proportional limit for steels may be taken as the same point for practical purposes.

**Elasticity** The ability of a material to return to its original state (dimensions) after the load has been removed.

**Factor of safety** Used to minimize the risk that a material's working stress will exceed the strength of the material,

$$S = \frac{S_1}{f_s}$$

where $S$ = working stress

$S_1$ = yield strength for ductile materials, ultimate strength for brittle materials, and fatigue strength for cyclically loaded members

$f_s$ = factor of safety

[*Note:* The factor of safety can sometimes lead to confusion and erroneous interpretation, so it is customary today for allowable design stresses to be determined by building codes or recognized authorities (ASTM, AISI, AISC, etc.).]

**Fatigue endurance limit** Defined as the breaking strength caused by repeated loadings (in psi). Loadings below the fatigue endurance limit can be applied indefinitely without part breakage.

**Hooke's law** A body acted on by external forces will deform in direct proportion to the stress incurred, as long as the unit stress does not exceed the proportional limit. This law is clearly seen in the function of springs.

**Izod impact test** A 10-mm square specimen with a 2-mm deep notch of 45° is struck by a weighted pendulum, and the impact, in ft-lb or N-m, required to break the specimen is recorded. The Charpy impact test is similar to the Izod test with minor variations.

**Malleability** The ability of a material to be rolled or hammered into a thin sheet.

**Modulus of elasticity** The ratio of unit stress to unit strain within the proportional limit of a material that is in tension or compression. Also known as *Young's modulus.*

**Modulus of elasticity in shear** The ratio of unit stress to unit strain within the proportional limit of a material that is in shear. In most metals, the modulus of elasticity in shear is approximately 0.4 times the modulus of elasticity in tension.

**Modulus of rigidity** The ratio of the shear stress to the shear strain, in radians, within the proportional limit. Also called the *modulus of elasticity in shear.*

**Offset method** Because many materials do not have an elastic range, yield strength is determined by the offset method. The stress value on the curve is taken as a definite amount of permanent set or strain, 0.2 percent of the original dimension (see Fig. 5.1).

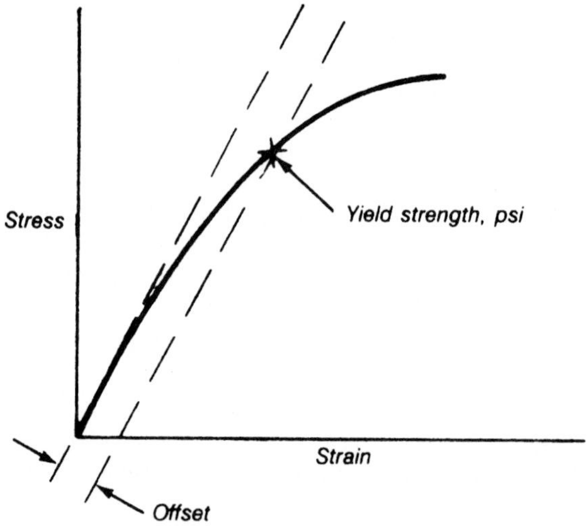


**FIGURE 5.1** Yield strength by offset method.

**Poisson's ratio** The ratio of lateral strain to longitudinal strain for a material subject to longitudinal stress within the proportional limit.

**Proportional limit** The point on a stress-strain curve where deviation occurs from the linear relation between stress and strain.

**Resilience** The recoverable strain energy of a deformed body when the load is removed.

**Strain** The amount of dimensional change produced in a body by a load, divided by the original dimension. A more precise term to use in place of strain is *deformation.*

**Stress rupture failure** This type of failure results in many aluminum alloys from the characteristic that the longer the time of loading, the lower the allowable ultimate strength. Carefully hardened aluminum alloys can be free from stress-rupture failure. Parts designed within the fatigue endurance limit are never subject to stress-rupture failure.

**Tensile strength** The maximum stress to which a material may be subjected prior to failure. Also known as *ultimate strength.*

**Tensile stress** The internal force that resists external forces that tend to cause elongation of a body.

**Unit deformation** The amount of deformation per unit length divided by the original length before an external load was applied. (*Example:* If a metal bar 6 in long is stretched to 6.25 in, the total deformation is 0.25/6 = 0.042 per in of the original length.)

**Yield point** A point on the stress-strain curve where there is a sudden increase in strain without a corresponding increase in stress. All materials do not have a yield point (e.g., glass, porcelain, plastics).

**Yield strength** The maximum stress that may be applied to a material without incurring a permanent strain or deformation.

## 5.2 IMPACT STRESSES

When a sudden load is applied to a bar or beam, the unit stress and unit strain (deformation) are two times those produced by an equal load applied gradually. The bar or beam will oscillate, dissipating energy by damping, and gradually resume the same condition of stress and strain equal to the static load.

### 5.2.1 Shock Loads and Developed Stresses

***Stresses in Structural Members Produced by Shocks.*** All elastic structures that are subjected to shock loads will deflect until the deflection distance or movement in the structure is proportional to the energy of the shock. For a given shock load, the average resisting stresses are inversely proportional to the deflection. Since there are no perfectly rigid bodies, there will always be a deflection for any given shock load. If the body were perfectly rigid, a shock load would, in effect, produce an infinite stress. The effect of a shock load on any structure is, to a great extent, dependent on the elastic properties of the structure subjected to the impact of the shock load.

The kinetic energy of a moving mass (weight) that produces a shock load on any elastic structure, when struck, is dissipated in the following fashions:

- Deformation of the falling or moving mass
- Deformation of the structure subjected to the shock load
- Partial deformation of both the moving mass and deformation of the stationary structure. (A large amount of the energy will be dissipated as heat.)
- A portion of the energy will be absorbed by the supporting members of the structure, if these supports are not too rigid or are inelastic.

Investigations into stresses produced by shock loads have shown that

- Suddenly applied loads (not shock loads) will produce the same deflection and stress as a static load that is twice as great (see Sec. 5.2).

- The unit stress for a load producing a shock varies directly as the square root of the modulus of elasticity $E$ and inversely as the square root of the length of the beam or structural member $L$.

The following simple equations will allow you to calculate the unit stress $s$ produced by a weight $W$ falling through a distance $h$.

*Case 1.* A structural beam supported at both ends; struck in the center span:

$$s = \frac{WyL}{I}\left(1 + \sqrt{1 + \frac{96hEI}{WL^3}}\right)$$

*Case 2.* A structural beam fixed at one end; struck on the free end:

$$s = \frac{WyL}{I}\left(1 + \sqrt{1 + \frac{6hEI}{WL^3}}\right)$$

*Case 3.* A structural beam fixed at both ends; struck in the center span:

$$s = \frac{WyL}{I}\left(1 + \sqrt{1 + \frac{384hEI}{WL^3}}\right)$$

where $s$ = unit stress, $lb/in^2$ (psi) or $N/mm^2$
$W$ = weight, lb or N (if the weight is given in kilograms, the kilograms must be multiplied by 9.81 to arrive at newtons; see Chap. 2)
$E$ = modulus of elasticity, $lb/in^2$ or $N/mm^2$
$I$ = moment of inertia, $in^4$ or $mm^4$
$y$ = distance of extreme fiber from neutral axis, in or mm
$L$ = length of beam or structure, in or mm
$h$ = distance the load drops, in or mm

The general equation for shock loads from which other specific equations for shock stresses in beams, springs, and other machine elements are derived is

$$s = S_1\left(1 + \sqrt{1 + \frac{2h}{f}}\right)$$

where $s$ = stress in $lb/in^2$ due to shock caused by impact of moving load
$S_1$ = stress in $lb/in^2$ when load is applied statically
$h$ = distance in inches that load falls before striking the beam or structural member
$f$ = deflection in inches resulting from application of static load $W$ (see beam equations referenced in Sec. 5.15)

By substituting the maximum stress formulas (in place of $S_1$) and deflection formulas (in place of $f$), which are shown in Figs. 5.23 and 5.24 (later in this chapter), you may customize the general equation for shock loads shown previously to meet many different design applications.

Before the advent and practical development of general shock equations, the mechanical designer often multiplied the static load times 10 to calculate the stress and deflection produced by a shock load. This often leads to overdesigning the structural member or making it stronger than is actually required. This practice is no longer considered to be effective in most applications.

## 5.3 FATIGUE

When parts or structures are subjected to repeated or varying loads, standard methods for calculating stresses under static loading conditions are not adequate. Parts such as rotating shafts, crank shafts, and piston rods fall into this category.

Fatigue failure of ductile materials such as iron or copper is similar to the static failure of a brittle material.

The highest unit stress to which a material may be repeatedly subjected without evidence of failure is known as the *fatigue endurance limit* of the material. This limit is usually determined by material tests. By way of example, the endurance limit of medium-carbon steel with a Brinell hardness of 150 to 200 ranges between 42,000 and 50,000 psi.

## 5.4 VIBRATORY STRESS IN BARS AND BEAMS

For a steady vibratory stress, the deflection of the bar or beam is increased by a dynamic magnification factor (DMF), according to the relationship

$$\Delta_f,\ \text{dynamic} = \Delta_f,\ \text{static} \times \text{DMF}$$

where DMF = dynamic magnification factor, and the two $\Delta_f$ terms represent deflections.

Further

$$\Delta_f,\ \text{dynamic} = \Delta_f,\ \text{static} \times \frac{1}{1 - (\phi/\phi_n)^2}$$

in which $\phi$ is the oscillating frequency of the load and $\phi_n$ is the natural frequency of the bar or beam defined as

$$\phi_n = \sqrt{\frac{3\ EIg}{l^3\ w}}$$

where $E$ = modulus of tensile elasticity
$I$ = moment of inertia of the section, in$^4$
$g$ = acceleration of gravity (32.2 ft/sec$^2$)
$l$ = length of the bar or beam, in
$w$ = weight of oscillating load, lb

SI units are, respectively, N/mm$^2$, mm$^4$, 9.8 m/sec$^2$, mm, and N.

## 5.5 BEAMS AND REACTIONS AT THE SUPPORTS

The reactions at support points are calculated using the following conditions of equilibrium.

- The algebraic sum of all vertical forces must equal zero; i.e., downward forces are equal to upward reactions.
- The algebraic sum of the moments of the vertical forces must equal zero.

*Example 1.* Referring to Fig. 5.2, we will by convention take clockwise moments as plus-signed and counterclockwise ones as minus-signed. By moments about $R_1$:

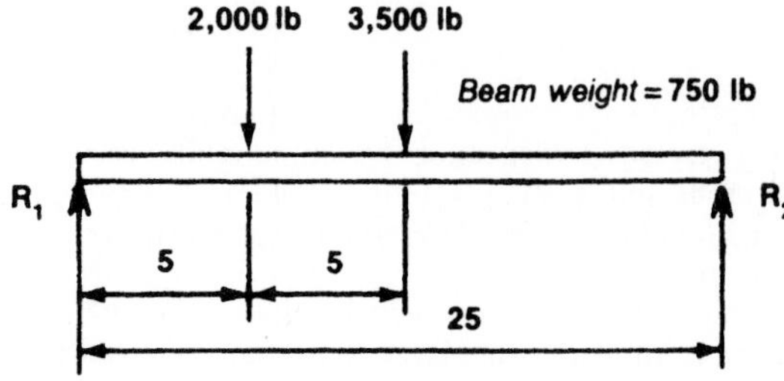


**FIGURE 5.2** Reactions at the supports.

$$(750 \times 12.5) + (2000 \times 5) + (3500 \times 10) - (R_2 \times 25) = 0$$

$$9375 + 10{,}000 + 35{,}000 - 25R_2 = 0$$

$$R_2\text{, the reaction at } R_2\text{,} = 2175$$

By moments about $R_2$

$$(R_1 \times 25) - (20{,}000 \times 20) - (3500 \times 15) - (750 \times 12.5) = 0$$

$$25R_1 - 40{,}000 - 52{,}500 - 9375 = 0$$

$$R_1\text{, the reaction at } R_1\text{,} = 4075$$

Also

$$R_1 + R_2 = 2000 + 3500 + 750 = 6250$$

The sum 4075 + 2175 = 6250, showing equilibrium.

Note that in the U.S. customary system, the units would be pounds and feet. In SI, units would be newtons and meters, with reactions in pounds-force and newtons, respectively.

*Example 2.* Referring to Fig. 5.3, about $R_1$

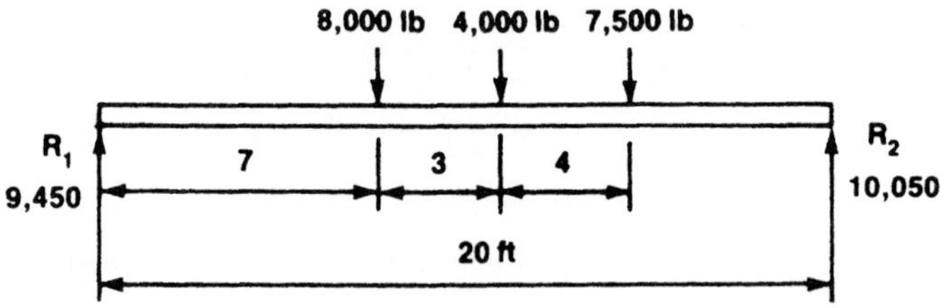


**FIGURE 5.3** Reactions at the supports.

$$(8000 \times 7) + (4000 \times 10) + (7500 \times 14) - (R_2 \times 20) = 0$$

$$-20R_2 + 56{,}000 + 40{,}000 + 105{,}000 = 0$$

$$R_2 = 10{,}050$$

About $R_2$

$$(R_1 \times 20) - (7500 \times 6) - (4000 \times 10) - (8000 \times 13) = 0$$

$$20R_1 = 189{,}000$$

$$R_1 = 9450$$

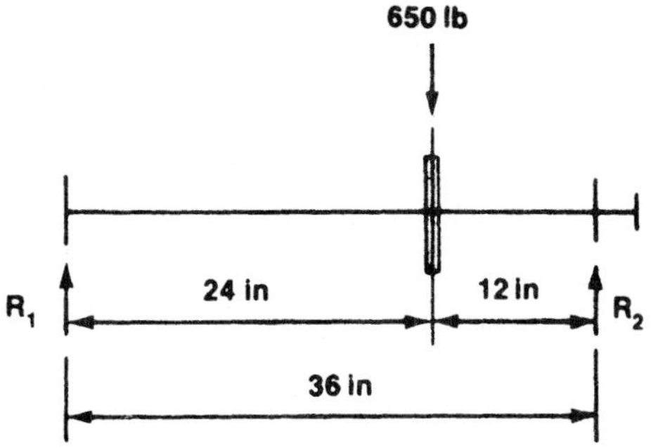


**FIGURE 5.4** Bearing reactions.

Check: 8000 + 4000 + 7500 = 10,050 + 9450 = 19,500.

*Example 3.* Referring to Fig. 5.4, find the reactions at the bearings $R_1$ and $R_2$. About $R_1$

$$(650 \times 24) - (R_2 \times 36) = 0$$

$$36R_2 - 15{,}600 = 0$$

$$R_2 = 433.333$$

About $R_2$

$$(R_1 \times 36) - (650 \times 12) = 0$$

$$36R_1 - 7800 = 0$$

$$R_1 = 216.666$$

To check: $R_1 + R_2 = 650$; 216.666 + 433.333 = 650 (neglecting roundoff difference).

## 5.6 SHEAR DIAGRAMS

A graphic representation of vertical shear at all cross sections of a bar or beam is given in Fig. 5.5.

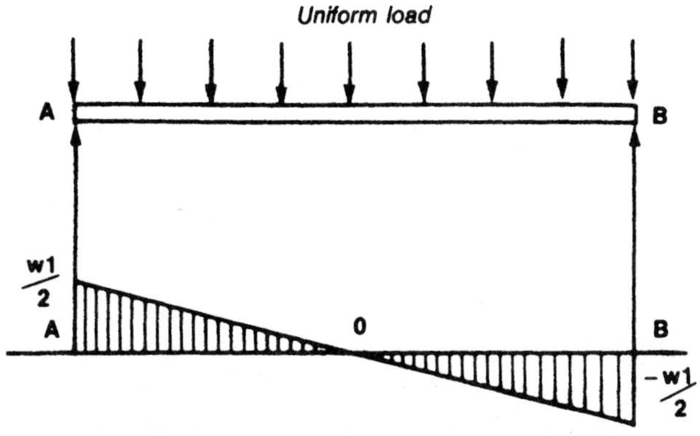


**FIGURE 5.5** Shear diagram for a uniform load.

At any cross section

$$V_{sr} = V_s = \Psi A \qquad \Psi = \frac{V_s}{A}$$

where $V_s$ = vertical shear, lb
$V_{sr}$ = resisting shear, lb
$\Psi$ = average unit shearing stress, lb/in$^2$
$A$ = area of section, in$^2$

At any point in any cross section, the vertical unit shearing stress is

$$\Psi = \frac{V_s A_1 c_1}{It}$$

where $V_s$ = total vertical shear of section
$A_1$ = area of cross section between the horizontal plane through the point where the shear is to be found and the extreme fiber (same side of neutral axis)
$c_1$ = distance from the neutral axis to the center of gravity of the area $A_1$
$I$ = moment of inertia of the section
$t$ = width of the section at the shear plane

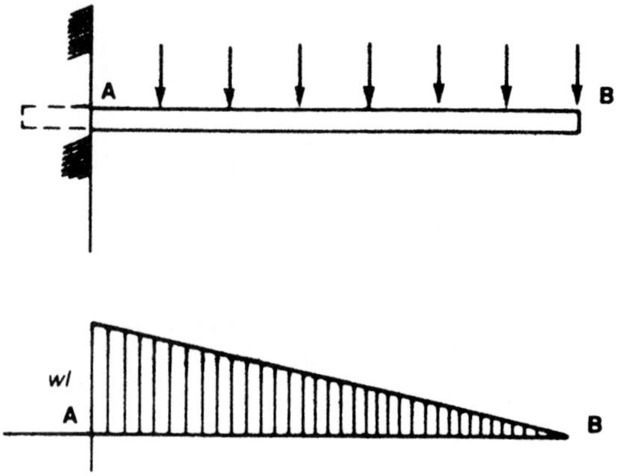


**FIGURE 5.6** Shear diagram for a cantilever load.

Refer to Figs. 5.6 and 5.7. For a rectangular bar or beam (solid)

$$\Psi = \frac{3\,V_s}{2\,A_s}$$

For a rod or beam (solid, circular)

$$\Psi = \frac{4\,V_s}{3\,A_s}$$

In SI distance is in meters, force is in newtons, moment of inertia is in meters to the fourth power, and area is in square meters.

## 5.7 MOMENT DIAGRAMS

Moment diagrams show the bending moments at all the cross sections of a bar or beam (Fig. 5.8). The cross section where the bending moment is greatest is known as the "dangerous sec-

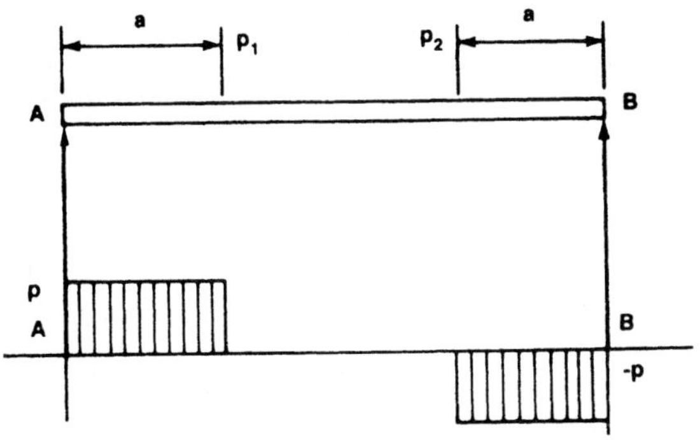


**FIGURE 5.7** Loads $P_1$ and $P_2$.

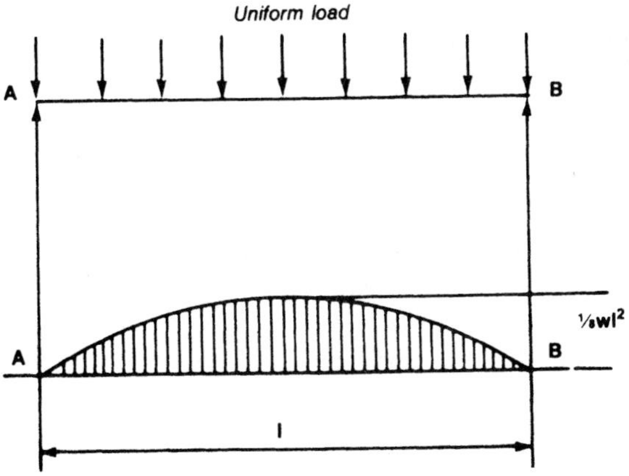


**FIGURE 5.8** Uniform load moment diagram.

tion." In a cantilever beam, it is at the section of the support (see Fig. 5.9). In a uniformly loaded, simple beam it is greatest at the center of the span. (See Figs. 5.8 and 5.10.)

## 5.8 BENDING

The extreme fiber stress in a beam is given by the flexure equation,

$$f = \frac{Mc}{I}$$

where $f$ = extreme fiber stress, klb (kip)/in$^2$, or 1000 lb $f$/in$^2$
$M$ = applied bending moment, kip
$c$ = distance from neutral axis to extreme fiber, in
$I$ = moment of inertia of cross section about its neutral axis, in$^4$

The *section modulus S* is equal to $I/c$ in$^3$.

The preceding equation for fiber stress can be used for the design of members that must resist bending moments and are sufficiently compact that no buckling problem exists.

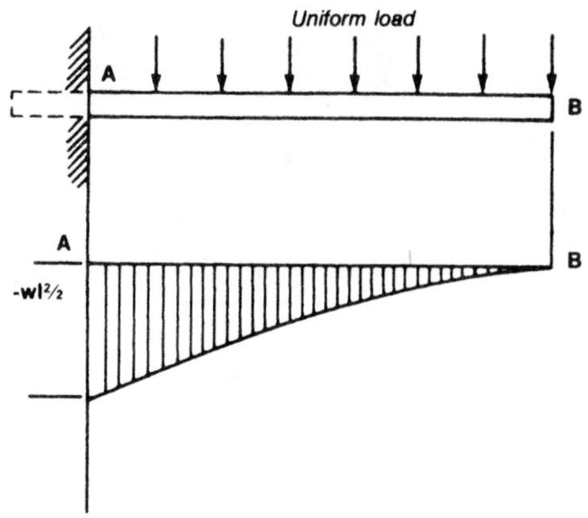


**FIGURE 5.9** Cantilever moment diagram.

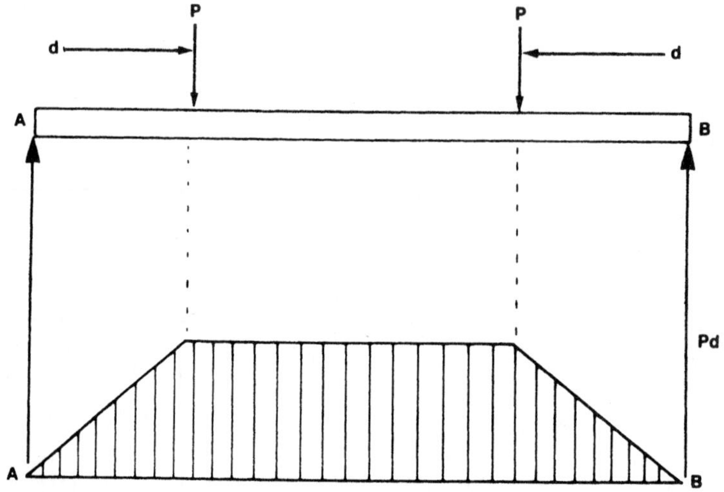


**FIGURE 5.10** Compound moment diagram.

## 5.9 ULTIMATE STRENGTH OF BEAMS

For the beam shapes illustrated here, use the adjacent equations

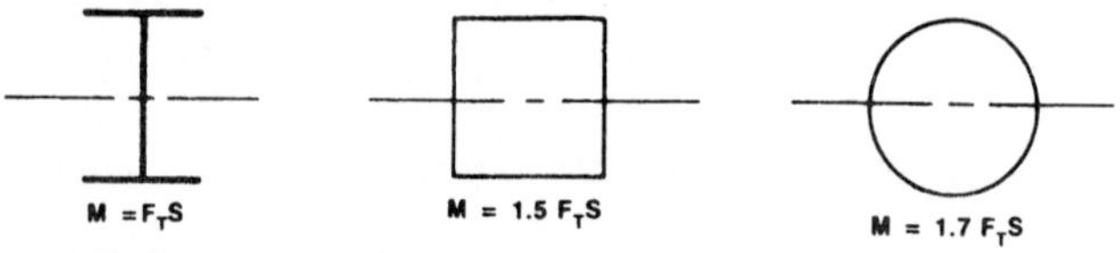


where $M$ = bending moment (approximate ultimate) in·kip
$F_T$ = ultimate tensile strength, kip/in$^2$
$S$ = section modulus ($I/c$), in$^3$
1 kip = 1000 lb

## 5.10 TORSION

For members that transmit torque, the following torsion equations may be used to find the approximate ultimate torque.

Referring to Fig. 5.11

$$S_f = \frac{2T}{\pi R^3} \quad \text{(at boundary)}$$

$$T_u = \frac{2\pi R^3 F_s}{3} \qquad T_c = \frac{\pi R^4}{2}$$

Referring to Fig. 5.12

$$S_f = \frac{2R_1 T}{\pi (R_1^4 - R_2^4)}$$

$$T_u = \frac{2\pi (R_1^3 - R_2^3) F_s}{3} \qquad T_c = \frac{\pi}{2} (R_1^4 - R_2^4)$$

where $S_f$ = shear stress (not exceeding the shear/yield strength), kip/in$^2$
$T$ = torque, in·kip
$T_u$ = approximate ultimate torque, in·kip
$F_s$ = ultimate shear strength, kip/in$^2$
$T_c$ = torsion constant, in$^4$

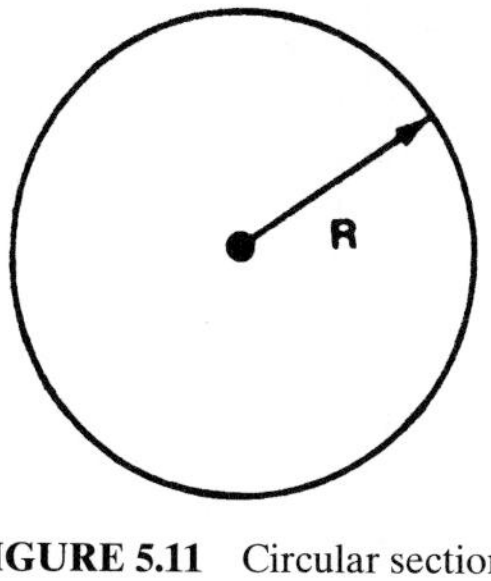


**FIGURE 5.11** Circular section.

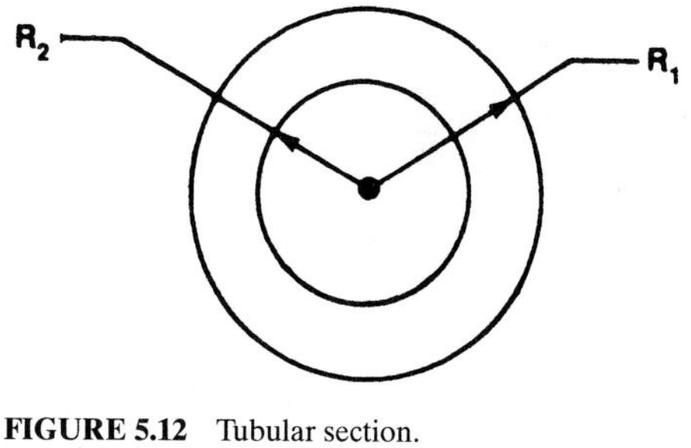


**FIGURE 5.12** Tubular section.

Check thin-walled sections for buckling (see Sec. 5.11).
Referring to Fig. 5.13

$$S_f = \frac{4.8T}{a^3} \quad \text{(at midpoints of sides)}$$

$$T_u = \frac{a^3 F_s}{3} \qquad T_c = \frac{a^4}{7}$$

Referring to Fig. 5.14

$$S_f = \frac{3T}{bt^2}\left(1 + \frac{t}{1.67b}\right) \quad \text{(at midpoints of longer sides)}$$

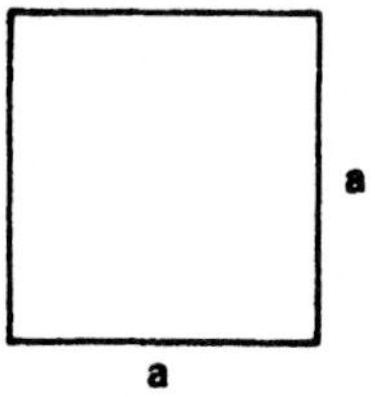


**FIGURE 5.13** Square section.

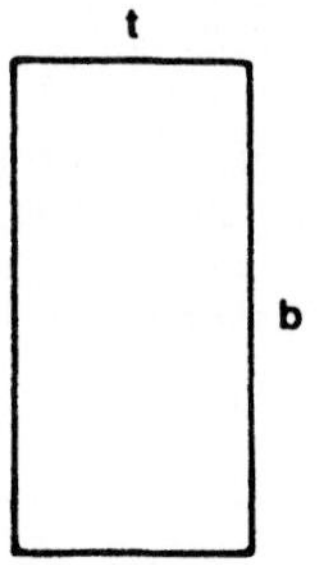


**FIGURE 5.14** Rectangular section.

$$T_u = \frac{bt^2F_s}{e}\left(\frac{1-t}{3b}\right)$$

$$T_c = \frac{bt^3}{3}\left[1 - \frac{t}{1.59b} + 0.05\left(\frac{t}{b}\right)^2\right]$$

Referring to Fig. 5.15

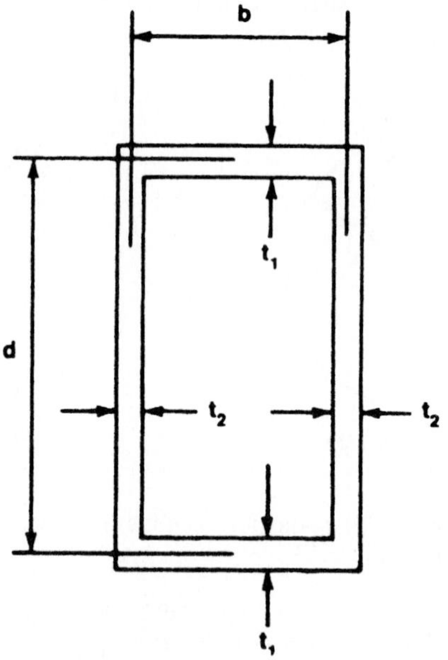


**FIGURE 5.15** Rectangular tubular section.

$$S_f = \frac{T}{2t_1bd} = \frac{T}{2t_2bd}$$

$$T_u = 2bd_{\min}F_s$$

($t_{min} = t_1$ or $t_2$, whichever is smaller). Finally

$$T_c = \frac{2b^2d^2}{(b/t_1) + (d/t_2)}$$

Again, check for buckling, which is the next topic.

## 5.11 BUCKLING

Column bending failure is the type referred to when the term *column strength* is used.

Column strength is a function of the properties of the material and the *effective slenderness ratio,* which equals the effective $KL/r$, where $L$ is the length of the column, $r$ is the radius of gyration, and $K$ is a factor determined by the end conditions of the column.

For aluminum only, typical $K$ values are

| | |
|---|---|
| Both ends fixed | 0.5 |
| One end fixed, other end pinned | 0.7 |
| Both ends pinned | 1.0 |
| One end fixed, the other end free | 2.0 |

The factor $K$ can be determined for *any* material by the equation:

$$K = \frac{S}{C\pi^2 E}$$

where $S$ = ultimate compressive strength of material, psi (in SI, N/mm$^2$)
$E$ = modulus of elasticity of material, psi (in SI, N/mm$^2$)
$C$ = 1 for round or pivoted ends, 4 for fixed ends, 2 for one end fixed, one end rounded, and ¼ for one end fixed, one end free

### 5.11.1 Rankine Equation

Use the equation for slenderness ratios between 20 and 100.

$$P = \frac{S}{1 + K(l/r)^2}$$

where $l/r$ = slenderness ratio
$P$ = ultimate unit load, psi or N/m$^2$
$S$ = ultimate compressive strength of material, psi or N/m$^2$
$l$ = length of column member, in or mm
$r$ = least radius of gyration, in or mm
$K$ = end factor (see $K$ values)

### 5.11.2 Euler Equation

The equation is used for very slender columns in which bending or buckling predominates. Compressive stresses are not considered.

$$P = \frac{C\pi^2 IE}{l^2}$$

where $P$ = total ultimate load, lbf or N
$C$ = values for types of ends (see preceding)
$I$ = least moment of inertia, $in^4$ or $mm^4$
$E$ = modulus of elasticity of material, $lb/in^2$ or $N/mm^2$
$l$ = length of column member, in or mm

## 5.12 MACHINE ELEMENTS USED AS COLUMNS

See Chap. 8 on machine elements.

## 5.13 MOMENTS OF INERTIA

The method that follows determines moments by areas and square of distance from the centroid.

*Example.* Find the moment of inertia of the wide-flange beam section shown in Fig. 5.16. Then find its section modulus and radius of gyration.

$$\left.\begin{aligned} [(2.221)(0.340)](1.111)^2 &= 0.9321 \\ [(2.221)(0.340)](3.332)^2 &= 8.3837 \\ [(0.558)(10)](4.721)^2 &= \underline{124.3662} \end{aligned}\right\} \text{Half the section}$$

$$133.6820 \times 2 = 267.364 \text{ in}^4$$

Now, add the four radius areas of Fig. 5.17.

$$\left.\begin{aligned} A &= (a \times b) - A_4 \\ A &= (a \times b) - \frac{\pi r^2}{4} \\ A &= 0.0625 - 0.0491 \\ A &= 0.013 \text{ in}^2 \end{aligned}\right\} \text{Moment of the radius areas} = (0.013)(4.317)^2 \times 4 = 0.969 \text{ in}^4$$

Moment of inertia of the complete section = 267.364 + 0.969 = 268.333 $in^4$. The book value of this section is 272.9 $in^4$. Therefore, the accuracy is

$$\frac{272.0 - 268.3}{268.3} = 0.017, \text{ or } 1.7\%$$

This method is accurate enough for all practical applications.

Moments of inertia of plane areas can be found accurately with this method, making irregular shapes and special shapes relatively easy to calculate. Examples illustrated here are

- Unsymmetrical shapes (see Fig. 5.18).
- Symmetrical shapes (see Fig. 5.19).
- Composite beams (see Fig. 5.20).
- "Beefed-up" members (see Fig. 5.21).

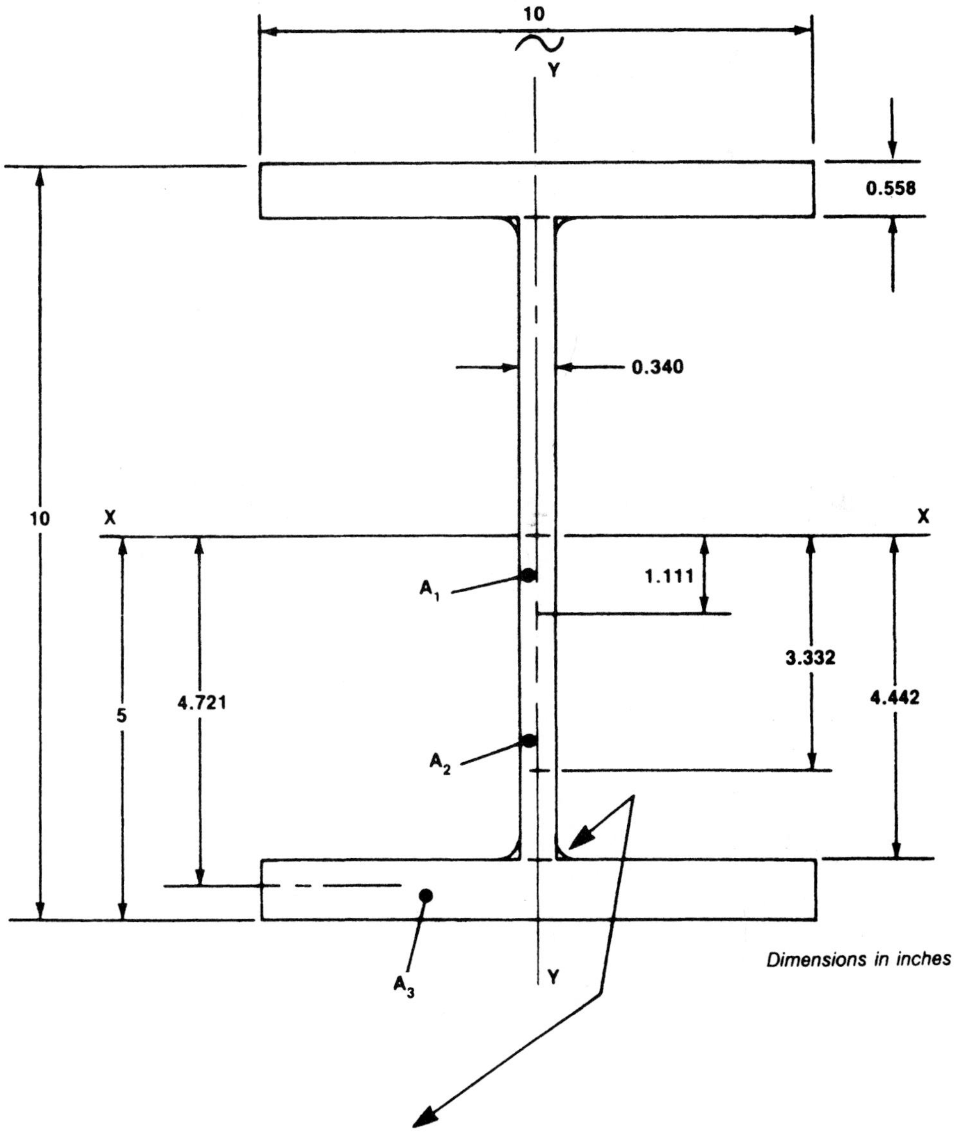


**FIGURE 5.16** Typical wide-flange beam section.

You can find the center of gravity (c.g.) by the scale-model–plumb-line method, or geometrically. The section modulus $Z$ may then be found by using

$$Z = \frac{I}{y}$$

where $y$ = distance from neutral axis to extreme fiber, in, and $I$ = moment of inertia of the section, $in^4$.

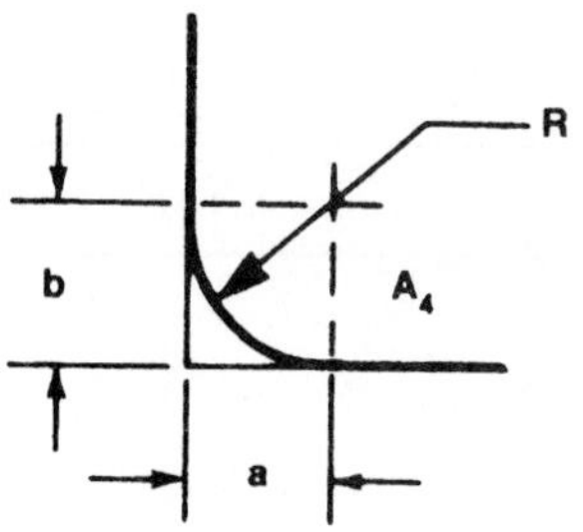


FIGURE 5.17 Radiused areas of beam fillets.

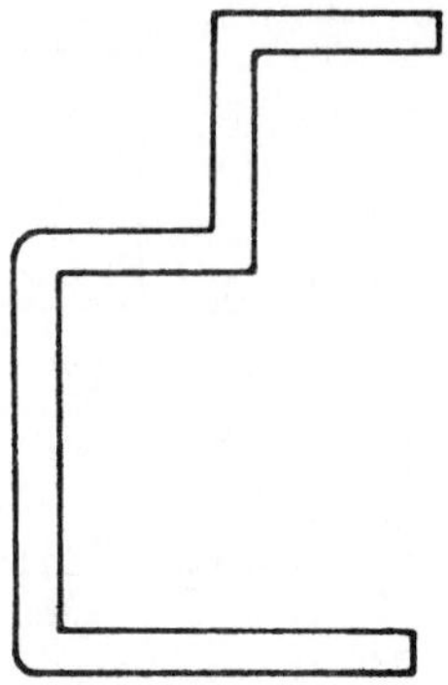

FIGURE 5.18 Unsymmetrical shape.

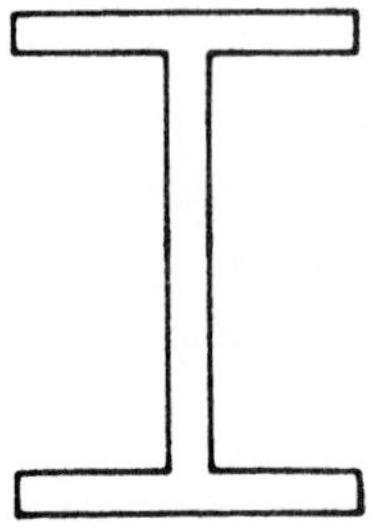

FIGURE 5.19 Symmetrical shape.

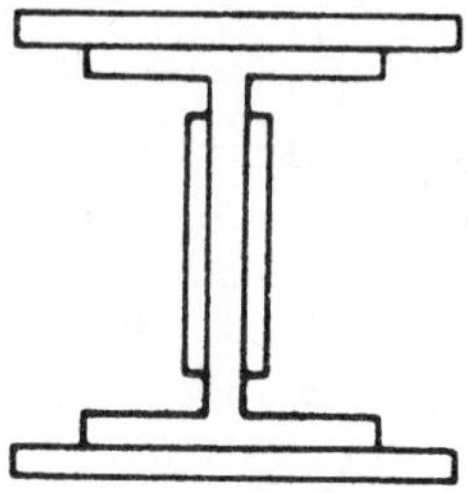

FIGURE 5.20 Composite beam.

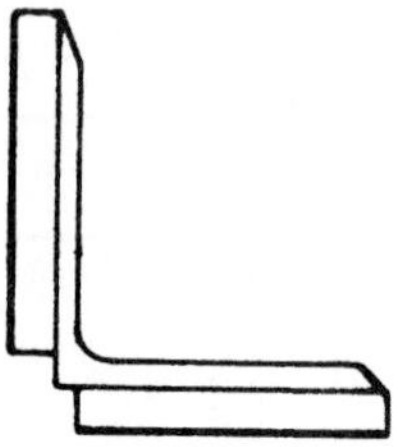

FIGURE 5.21 Reinforced structural member.

The radius of gyration $R$ may then be found by

$$R = \sqrt{\frac{I}{A}}$$

where $A$ = area of the section, $in^2$ or $mm^2$.

## 5.14 PROPERTIES OF SECTIONS

Included here are the most common and standard sections encountered in general design problems (see Sec. 5.13 for nonstandard shapes). Equations and geometries are displayed in Fig. 5.22, where $A$ = area, $c$ = axis of moments, $I$ = moment of inertia, $S$ = section modulus (elastic), $r$ = radius of gyration, and $Z$ = section modulus (plastic).

## PROPERTIES OF GEOMETRIC SECTIONS

### SQUARE
Axis of moments through center

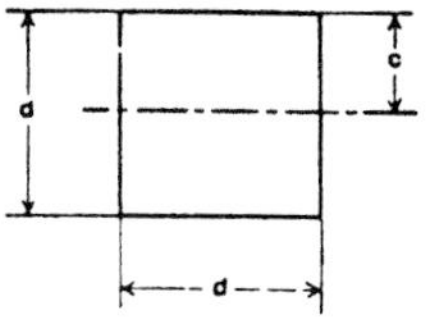


$$A = d^2$$
$$c = \frac{d}{2}$$
$$I = \frac{d^4}{12}$$
$$S = \frac{d^3}{6}$$
$$r = \frac{d}{\sqrt{12}} = .288675\ d$$
$$Z = \frac{d^3}{4}$$

### SQUARE
Axis of moments on base

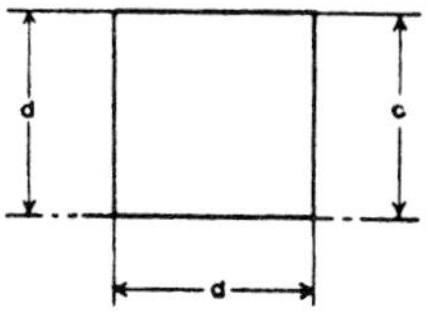


$$A = d^2$$
$$c = d$$
$$I = \frac{d^4}{3}$$
$$S = \frac{d^3}{3}$$
$$r = \frac{d}{\sqrt{3}} = .577350\ d$$

### SQUARE
Axis of moments on diagonal

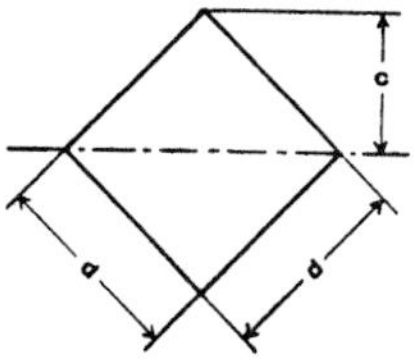


$$A = d^2$$
$$c = \frac{d}{\sqrt{2}} = .707107\ d$$
$$I = \frac{d^4}{12}$$
$$S = \frac{d^3}{6\sqrt{2}} = .117851\ d^3$$
$$r = \frac{d}{\sqrt{12}} = .288675\ d$$
$$Z = \frac{2c^3}{3} = \frac{d^3}{3\sqrt{2}} = .235702 d^3$$

### RECTANGLE
Axis of moments through center

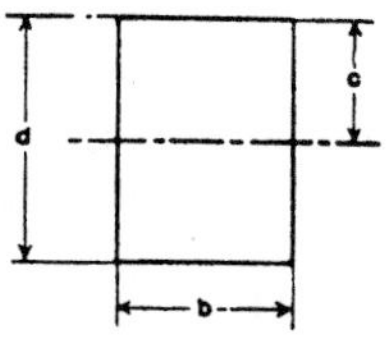


$$A = bd$$
$$c = \frac{d}{2}$$
$$I = \frac{bd^3}{12}$$
$$S = \frac{bd^2}{6}$$
$$r = \frac{d}{\sqrt{12}} = .288675\ d$$
$$Z = \frac{bd^2}{4}$$

**FIGURE 5.22** Properties of geometric sections.

## PROPERTIES OF GEOMETRIC SECTIONS

### RECTANGLE
Axis of moments on base

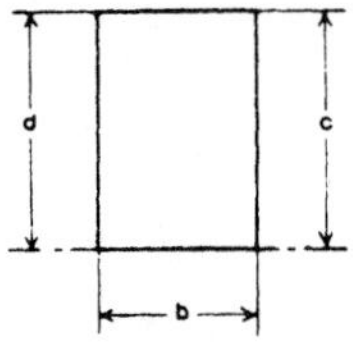


$$A = bd$$
$$c = d$$
$$I = \frac{bd^3}{3}$$
$$S = \frac{bd^2}{3}$$
$$r = \frac{d}{\sqrt{3}} = .577350\ d$$

### RECTANGLE
Axis of moments on diagonal

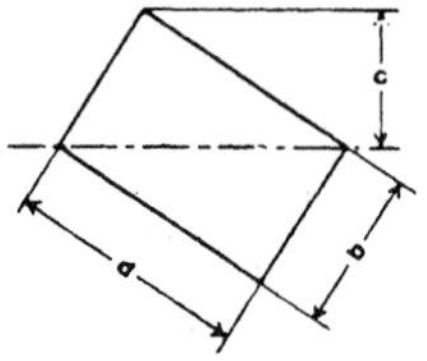


$$A = bd$$
$$c = \frac{bd}{\sqrt{b^2 + d^2}}$$
$$I = \frac{b^3d^3}{6\,(b^2 + d^2)}$$
$$S = \frac{b^2d^2}{6\sqrt{(b^2 + d^2}}$$
$$r = \frac{bd}{\sqrt{6\,(b^2 + d^2)}}$$

### RECTANGLE
Axis of moments any line through center of gravity

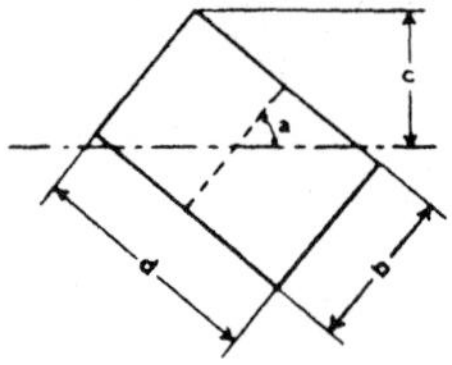


$$A = bd$$
$$c = \frac{b \sin a + d \cos a}{2}$$
$$I = \frac{bd\,(b^2 \sin^2 a + d^2 \cos^2 a)}{12}$$
$$S = \frac{bd\,(b^2 \sin^2 a + d^2 \cos^2 a)}{6\,(b \sin a + d \cos a)}$$
$$r = \sqrt{\frac{b^2 \sin^2 a + d^2 \cos^2 a}{12}}$$

### HOLLOW RECTANGLE
Axis of moments through center

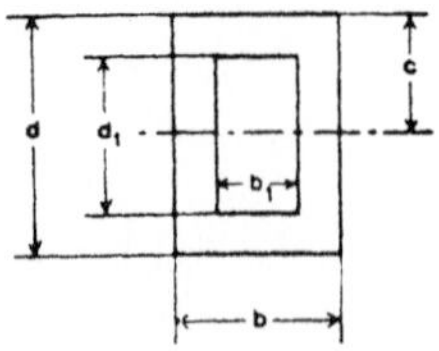


$$A = bd - b_1d_1$$
$$c = \frac{d}{2}$$
$$I = \frac{bd^3 - b_1d_1^3}{12}$$
$$S = \frac{bd^3 - b_1d_1^3}{6d}$$
$$r = \sqrt{\frac{bd^3 - b_1d_1^3}{12A}}$$
$$Z = \frac{bd^2}{4} - \frac{b_1d_1^2}{4}$$

**FIGURE 5.22** (*Continued*)

## PROPERTIES OF GEOMETRIC SECTIONS

### EQUAL RECTANGLES
Axis of moments through center of gravity

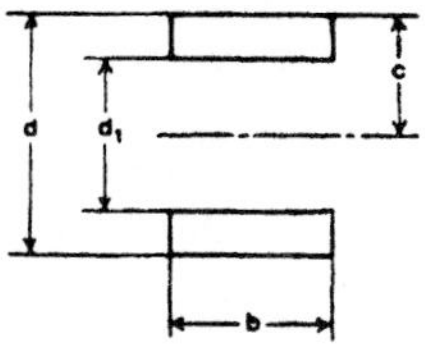


$$A = b(d - d_1)$$
$$c = \frac{d}{2}$$
$$I = \frac{b(d^3 - d_1^3)}{12}$$
$$S = \frac{b(d^3 - d_1^3)}{6d}$$
$$r = \sqrt{\frac{d^3 - d_1^3}{12(d - d_1)}}$$
$$Z = \frac{b}{4}(d^2 - d_1^2)$$

### UNEQUAL RECTANGLES
Axis of moments through center of gravity

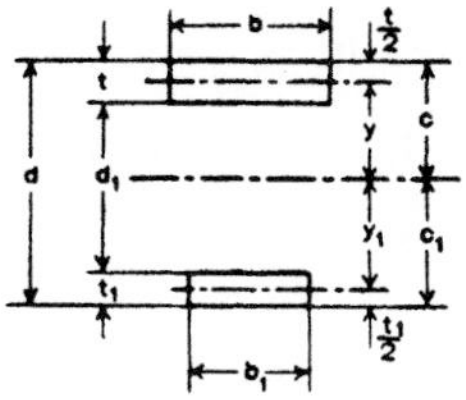


$$A = bt + b_1t_1$$
$$c = \frac{\frac{1}{2}bt^2 + b_1t_1(d - \frac{1}{2}t_1)}{A}$$
$$I = \frac{bt^3}{12} + bty^2 + \frac{b_1t_1^3}{12} + b_1t_1y_1^2$$
$$S = \frac{I}{c} \qquad S_1 = \frac{I}{c_1}$$
$$r = \sqrt{\frac{I}{A}}$$
$$Z = \frac{A}{2}\left[d - \left(\frac{t + t_1}{2}\right)\right]$$

### TRIANGLE
Axis of moments through center of gravity

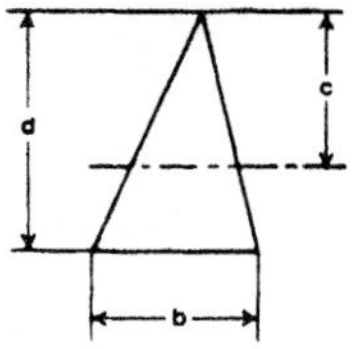


$$A = \frac{bd}{2}$$
$$c = \frac{2d}{3}$$
$$I = \frac{bd^3}{36}$$
$$S = \frac{bd^2}{24}$$
$$r = \frac{d}{\sqrt{18}} = .235702\ d$$

### TRIANGLE
Axis of moments on base

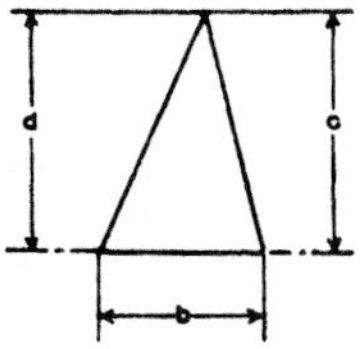


$$A = \frac{bd}{2}$$
$$c = d$$
$$I = \frac{bd^3}{12}$$
$$S = \frac{bd^2}{12}$$
$$r = \frac{d}{\sqrt{6}} = .408248\ d$$

**FIGURE 5.22** *(Continued)*

## PROPERTIES OF GEOMETRIC SECTIONS

### TRAPEZOID

Axis of moments through center of gravity

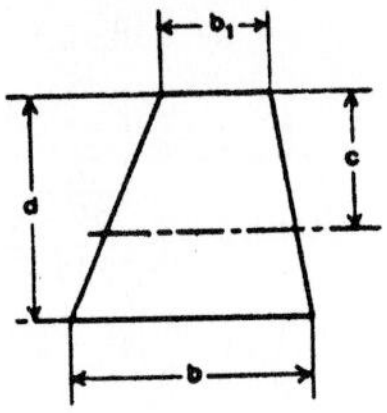


$$A = \frac{d(b + b_1)}{2}$$

$$c = \frac{d(2b + b_1)}{3(b + b_1)}$$

$$I = \frac{d^3 (b^2 + 4\,bb_1 + b_1^2)}{36\,(b + b_1)}$$

$$S = \frac{d^2 (b^2 + 4\,bb_1 + b_1^2)}{12\,(2b + b_1)}$$

$$r = \frac{d}{6(b + b_1)} \sqrt{2\,(b^2 + 4\,bb_1 + b_1^2)}$$

### CIRCLE

Axis of moments through center

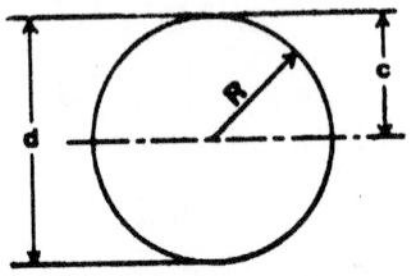


$$A = \frac{\pi d^2}{4} = \pi R^2\ .785398\ d^2 = 3.141593\ R^2$$

$$c = \frac{d}{2} = R$$

$$I = \frac{\pi d^4}{64} = \frac{\pi R^4}{4} = .049087\ d^4 = .785398\ R^4$$

$$S = \frac{\pi d^3}{32} = \frac{\pi R^3}{4} = .098175\ d^3 = .785398\ R^3$$

$$r = \frac{d}{4} = \frac{R}{2}$$

$$Z = \frac{d^3}{6}$$

### HOLLOW CIRCLE

Axis of moments through center

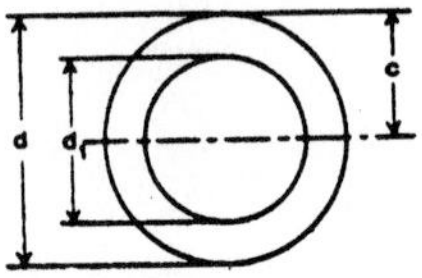


$$A = \frac{\pi(d^2 - d_1^2)}{4} = .785398\ (d^2 - d_1^2)$$

$$c = \frac{d}{2}$$

$$I = \frac{\pi(d^4 - d_1^4)}{64} = .049087\ (d^4 - d_1^4)$$

$$S = \frac{\pi(d^4 - d_1^4)}{32d} = .098175\ \frac{d^4 - d_1^4}{d}$$

$$r = \frac{\sqrt{d^2 + d_1^2}}{4}$$

$$Z = \frac{d^3}{6} - \frac{d_1^3}{6}$$

### HALF CIRCLE

Axis of moments through center of gravity

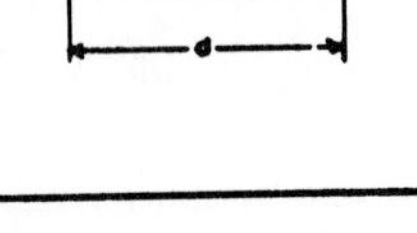

$$A = \frac{\pi R^2}{2} = 1.570796\ R^2$$

$$c = R\left(1 - \frac{4}{3\pi}\right) = .575587\ R$$

$$I = R^4\left(\frac{\pi}{8} - \frac{8}{9\pi}\right) = .109757\ R^4$$

$$S = \frac{R^3}{24}\,\frac{(9\pi^2 - 64)}{(3\pi - 4)} = .190687\ R^3$$

$$r = R\,\frac{\sqrt{9\pi^2 - 64}}{6\pi} = .264336\ R$$

**FIGURE 5.22** *(Continued)*

## PROPERTIES OF GEOMETRIC SECTIONS

### PARABOLA

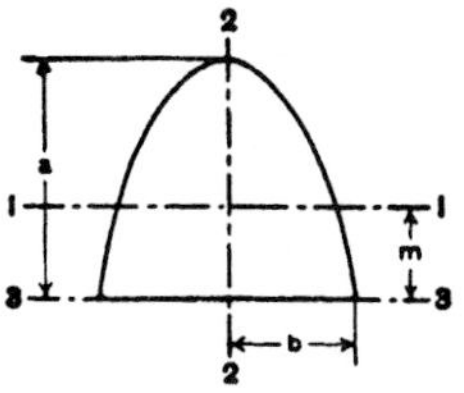

$$A = \frac{4}{3}ab$$

$$m = \frac{2}{5}a$$

$$I_1 = \frac{16}{175}a^3b$$

$$I_2 = \frac{4}{15}ab^3$$

$$I_3 = \frac{32}{105}a^3b$$

### HALF PARABOLA

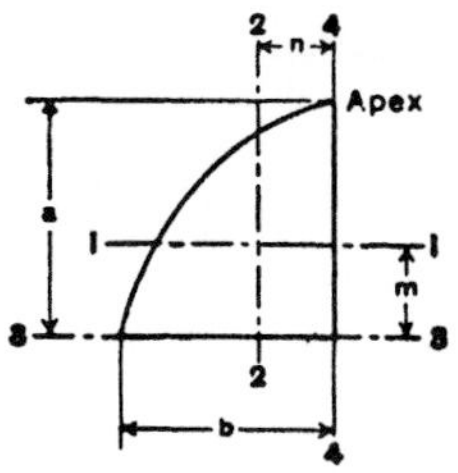


$$A = \frac{2}{3}ab$$

$$m = \frac{2}{5}a$$

$$n = \frac{3}{8}b$$

$$I_1 = \frac{8}{175}a^3b$$

$$I_2 = \frac{19}{480}ab^3$$

$$I_3 = \frac{16}{105}a^3b$$

$$I_4 = \frac{2}{15}ab^3$$

### COMPLEMENT OF HALF PARABOLA

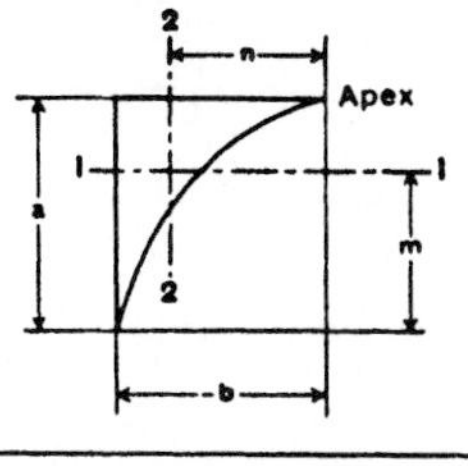


$$A = \frac{1}{3}ab$$

$$m = \frac{7}{10}a$$

$$n = \frac{3}{4}b$$

$$I_1 = \frac{37}{2100}a^3b$$

$$I_2 = \frac{1}{80}ab^3$$

### PARABOLIC FILLET IN RIGHT ANGLE

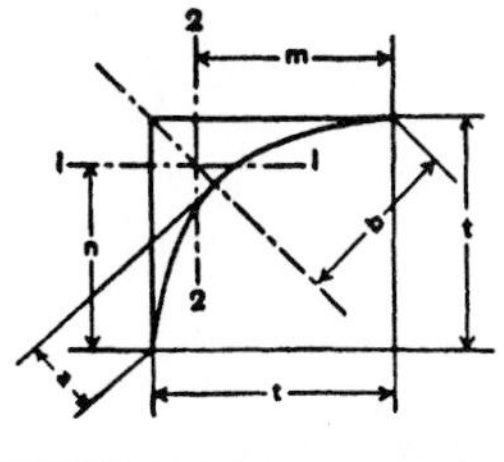

$$a = \frac{t}{2\sqrt{2}}$$

$$b = \frac{t}{\sqrt{2}}$$

$$A = \frac{1}{6}t^2$$

$$m = n = \frac{4}{5}t$$

$$I_1 = I_2 = \frac{11}{2100}t^4$$

**FIGURE 5.22** (*Continued*)

## PROPERTIES OF GEOMETRIC SECTIONS

*HALF ELLIPSE

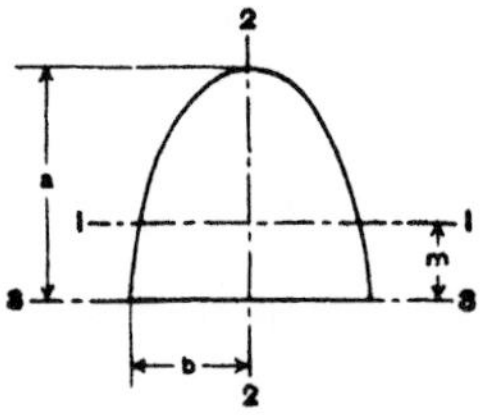


$$A = \frac{1}{2}\pi ab$$

$$m = \frac{4a}{3\pi}$$

$$I_1 = a^3b\left(\frac{\pi}{8} - \frac{8}{9\pi}\right)$$

$$I_2 = \frac{1}{8}\pi ab^3$$

$$I_3 = \frac{1}{8}\pi a^3b$$

*QUARTER ELLIPSE

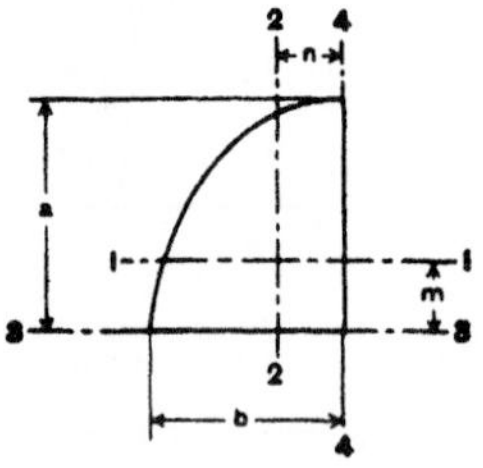


$$A = \frac{1}{4}\pi ab$$

$$m = \frac{4a}{3\pi}$$

$$n = \frac{4b}{3\pi}$$

$$I_1 = a^3b\left(\frac{\pi}{16} - \frac{4}{9\pi}\right)$$

$$I_2 = ab^3\left(\frac{\pi}{16} - \frac{4}{9\pi}\right)$$

$$I_3 = \frac{1}{16}\pi a^3b$$

$$I_4 = \frac{1}{16}\pi ab^3$$

*ELLIPTIC COMPLEMENT

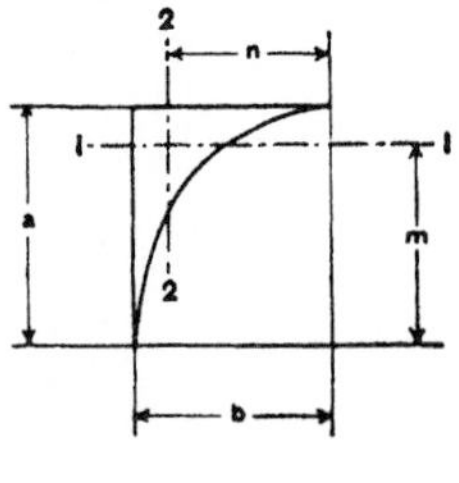


$$A = ab\left(1 - \frac{\pi}{4}\right)$$

$$m = \frac{a}{6\left(1 - \frac{\pi}{4}\right)}$$

$$n = \frac{b}{6\left(1 - \frac{\pi}{4}\right)}$$

$$I_1 = a^3b\left(\frac{1}{3} - \frac{\pi}{16} - \frac{1}{36\left(1 - \frac{\pi}{4}\right)}\right)$$

$$I_2 = ab^3\left(\frac{1}{3} - \frac{\pi}{16} - \frac{1}{36\left(1 - \frac{\pi}{4}\right)}\right)$$

*To obtain properties of half circle, quarter circle and circular complement substitute a = b = R.

**FIGURE 5.22** *(Continued)*

# PROPERTIES OF GEOMETRIC SECTIONS

## REGULAR POLYGON

Axis of moments through center

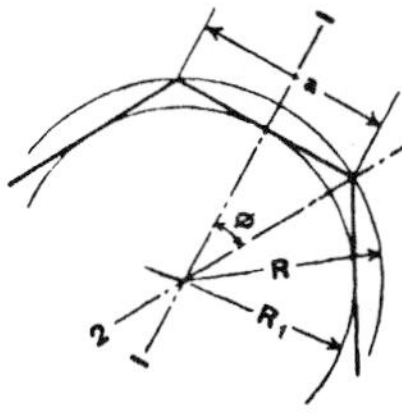


$$n = \text{Number of sides}$$

$$\phi = \frac{180°}{n}$$

$$a = 2\sqrt{R^2 - R_1^2}$$

$$R = \frac{a}{2 \sin \phi}$$

$$R_1 = \frac{a}{2 \tan \phi}$$

$$A = \frac{1}{4} na^2 \cot \phi = \frac{1}{2} nR^2 \sin 2\phi = nR_1^2 \tan \phi$$

$$I_1 = I_2 = \frac{A(6R^2 - a^2)}{24} = \frac{A(12R_1^2 + a^2)}{48}$$

$$r_1 = r_2 = \sqrt{\frac{6R^2 - a^2}{24}} = \sqrt{\frac{12R_1^2 + a^2}{48}}$$

## ANGLE

Axis of moments through center of gravity

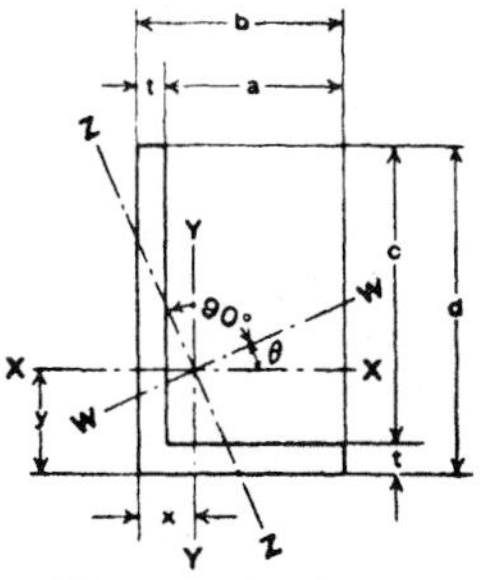


Z-Z is axis of minimum I

$$\tan 2\theta = \frac{2K}{I_Y - I_X}$$

$$A = t(b + c) \quad x = \frac{b^2 + ct}{2(b + c)} \quad y = \frac{d^2 + at}{2(b + c)}$$

K = Product of Inertia about X-X & Y-Y

$$= \mp \frac{abcdt}{4(b + c)}$$

$$I_X = \frac{1}{3}[t(d - y)^3 + by^3 - a(y - t)^3]$$

$$I_Y = \frac{1}{3}[t(b - x)^3 + dx^3 - c(x - t)^3]$$

$$I_Z = I_X \sin^2\theta + I_Y \cos^2\theta + K \sin 2\theta$$

$$I_W = I_X \cos^2\theta + I_Y \sin^2\theta - K \sin 2\theta$$

K is negative when heel of angle, with respect to c. g., is in 1st or 3rd quadrant, positive when in 2nd or 4th quadrant.

## BEAMS AND CHANNELS

Transverse force oblique through center of gravity

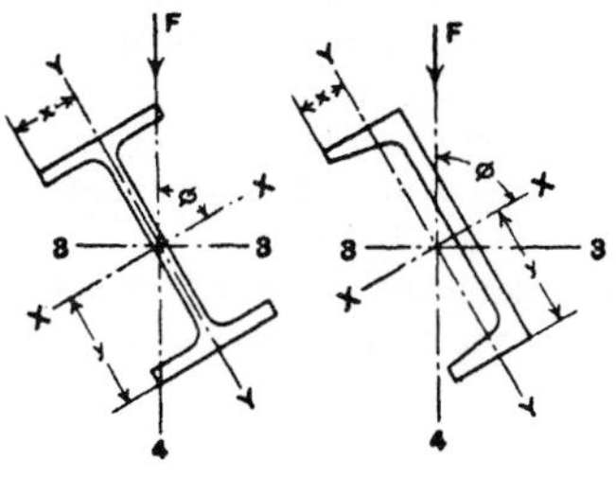


$$I_3 = I_X \sin^2\phi + I_Y \cos^2\phi$$

$$I_4 = I_X \cos^2\phi + I_y \sin^2\phi$$

$$f_b = M\left(\frac{y}{I_X} \sin\phi + \frac{x}{I_Y} \cos\phi\right)$$

where Mj is bending moment due to force F.

**FIGURE 5.22** *(Continued)*

## 5.15 BEAM EQUATIONS

Included here are the most common and standard conditions encountered in general design problems. These equations may be used for structures, machine elements, and various types of parts used in many design applications.

Standard beam equations (AISC) are displayed in Fig. 5.23.

Additional, simplified beam equations are displayed in Fig. 5.24.

### BEAM DIAGRAMS AND FORMULAS

**Nomenclature**

$E$ = Modulus of Elasticity of steel at 29,000 ksi. i.e. ($29 \times 10^6$ psi)
$I$ = Moment of Inertia of beam, $in.^4$.
$L$ = Total length of beam between reaction points ft.
$M_{max}$ = Maximum moment, kip in.
$M_1$ = Maximum moment in left section of beam, kip-in.
$M_2$ = Maximum moment in right section of beam, kip-in.
$M_3$ = Maximum positive moment in beam with combined end moment conditions, kip-in.
$M_x$ = Moment at distance x from end of beam, kip-in.
$P$ = Concentrated load, kips
$P_1$ = Concentrated load nearest left reaction, kips.
$P_2$ = Concentrated load nearest right reaction, and of different magnitude than $P_1$, kips.
$R$ = End beam reaction for any condition of symmetrical loading, kips.
$R_1$ = Left end beam reaction, kips.
$R_2$ = Right end or intermediate beam reaction, kips.
$R_3$ = Right end beam reaction, kips.
$V$ = Maximum vertical shear for any condition of symmetrical loading, kips.
$V_1$ = Maximum vertical shear in left section of beam, kips.
$V_2$ = Vertical shear at right reaction point, or to left of intermediate reaction point of beam, kips.
$V_3$ = Vertical shear at right reaction point, or to right of intermediate reaction point of beam, kips.
$V_x$ = Vertical shear at distance x from end of beam, kips.
$W$ = Total load on beam, kips.
$a$ = Measured distance along beam, in.
$b$ = Measured distance along beam which may be greater or less than a, in.
$l$ = Total length of beam between reaction points, in.
$w$ = Uniformly distributed load per unit of length, kips/in.
$w_1$ = Uniformly distributed load per unit of length nearest left reaction, kips/in.
$w_2$ = Uniformly distributed load per unit of length nearest right reaction and of different magnitude than $w_1$, kips/in.
$x$ = Any distance measured along beam from left reaction, in.
$x_1$ = Any distance measured along overhang section of beam from nearest reaction point, in.
$\Delta_{max}$ = Maximum deflection, in.
$\Delta_\alpha$ = Deflection at point of laod, in.
$\Delta_x$ = Deflection at any point x distance from left reaction, in.
$\Delta_{x1}$ = Deflection of overhang section of beam at any distance from nearest reaction point, in.

**FIGURE 5.23** Standard beam equations (AISC).

# BEAM DIAGRAMS AND FORMULAS
## For various static loading conditions

### 1. SIMPLE BEAM—UNIFORMLY DISTRIBUTED LOAD

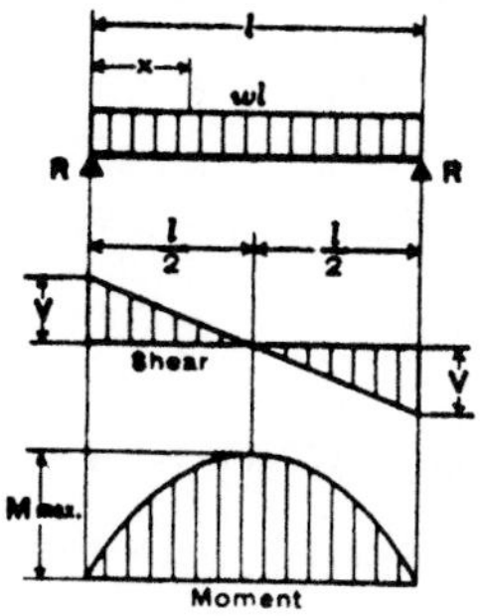


| | |
|---|---|
| Total Equiv. Uniform Load . . . . | $= wl$ |
| R = V . . . . . . . . . . . . | $= \frac{wl}{2}$ |
| $V_x$ . . . . . . . . . . . . . | $= w\left(\frac{l}{2} - x\right)$ |
| M max. (at center) . . . . . | $= \frac{wl^2}{8}$ |
| $M_x$ . . . . . . . . . . . . . | $= \frac{wx}{2}(l - x)$ |
| Δmax. (at center) . . . . . | $= \frac{5\,wl^4}{384\,EI}$ |
| $\Delta_x$ . . . . . . . . . . . . . | $= \frac{wx}{24EI}(l^3 - 2lx^2 + x^3)$ |

### 2. SIMPLE BEAM—LOAD INCREASING UNIFORMLY TO ONE END

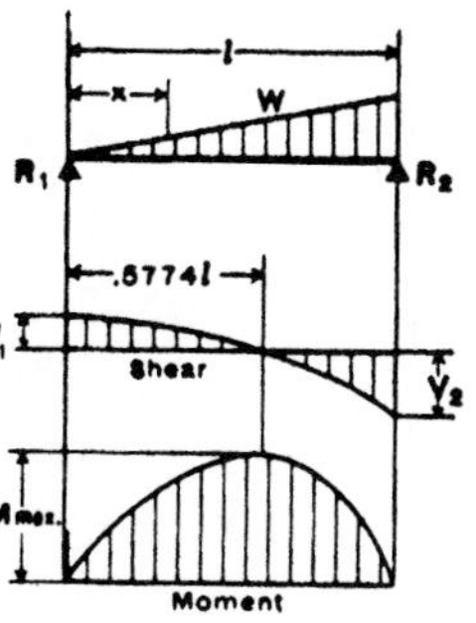


| | |
|---|---|
| Total Equiv. Uniform Load . . . . | $= \frac{16W}{9\sqrt{3}} = 1.0264W$ |
| $R_1 = V_1$ . . . . . . . . . . . | $= \frac{W}{3}$ |
| $R_2 = V_2$ max. . . . . . . . . . | $= \frac{2W}{3}$ |
| $V_x$ . . . . . . . . . . . . . | $= \frac{W}{3} - \frac{Wx^2}{l^2}$ |
| M max. $\left(\text{at } x = \frac{l}{\sqrt{3}} = .5774l\right)$ . . . | $= \frac{2Wl}{9\sqrt{3}} = .1283\,Wl$ |
| $M_x$ . . . . . . . . . . . . . | $= \frac{Wx}{3l^2}(l^2 - x^2)$ |
| Δmax. $\left(\text{at } x = l\sqrt{1 - \sqrt{\frac{8}{15}}} = .5193l\right)$ | $= .01304\,\frac{Wl^3}{EI}$ |
| $\Delta_x$ . . . . . . . . . . . . . | $= \frac{Wx}{180EI\,l^2}(3x^4 - 10l^2x^2 + 7l^4)$ |

### 3. SIMPLE BEAM—LOAD INCREASING UNIFORMLY TO CENTER

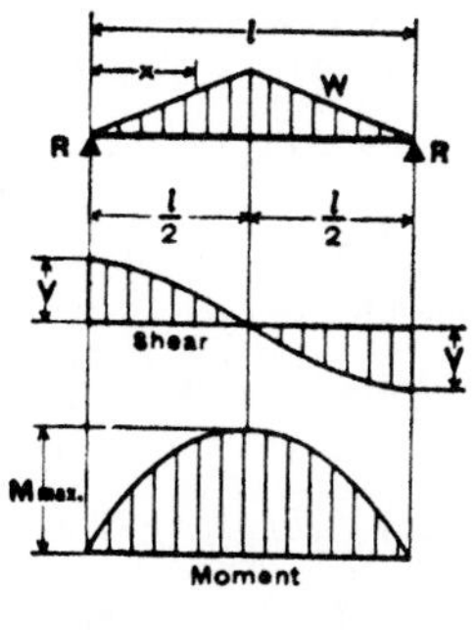


| | |
|---|---|
| Total Equiv. Uniform Load . . . . | $= \frac{4W}{3}$ |
| R = V . . . . . . . . . . . . | $= \frac{W}{2}$ |
| $V_x$ $\left(\text{when } x < \frac{l}{2}\right)$ . . . . . | $= \frac{W}{2l^2}(l^2 - 4x^2)$ |
| M max. (at center) . . . . . | $= \frac{Wl}{6}$ |
| $M_x$ $\left(\text{when } x < \frac{l}{2}\right)$ . . . . . | $= Wx\left(\frac{1}{2} - \frac{2x^2}{3l^2}\right)$ |
| Δmax. (at center) . . . . . | $= \frac{Wl^3}{60EI}$ |
| $\Delta_x$ $\left(\text{when } x < \frac{l}{2}\right)$ . . . . . | $= \frac{Wx}{480\,EI\,l^2}(5l^2 - 4x^2)^2$ |

**FIGURE 5.23** *(Continued)*

# BEAM DIAGRAMS AND FORMULAS

## For various static loading conditions

### 4. SIMPLE BEAM—UNIFORM LOAD PARTIALLY DISTRIBUTED

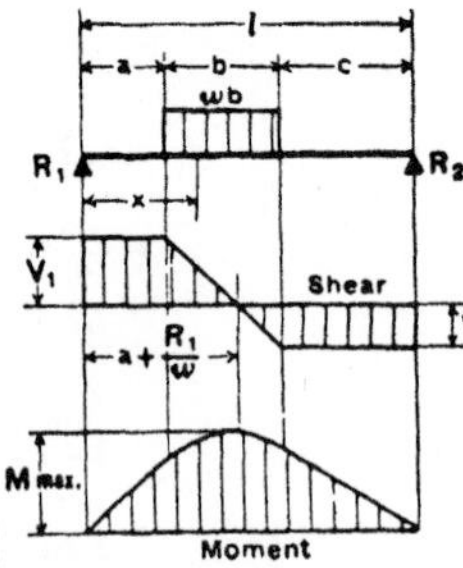


$$R_1 = V_1 \quad \left(\text{max. when } a < c\right) \quad \ldots = \frac{wb}{2l}(2c + b)$$

$$R_2 = V_2 \quad \left(\text{max. when } a > c\right) \quad \ldots = \frac{wb}{2l}(2a + b)$$

$$V_x \quad \left(\text{when } x > a \text{ and } < (a + b)\right) \quad . = R_1 - w(x - a)$$

$$M\ \text{max.}\left(\text{at } x = a + \frac{R_1}{w}\right) \quad \ldots = R_1\left(a + \frac{R_1}{2w}\right)$$

$$M_x \quad \left(\text{when } x < a\right) \quad \ldots = R_1 x$$

$$M_x \quad \left(\text{when } x > a \text{ and } < (a + b)\right) \quad . = R_1 x - \frac{w}{2}(x - a)^2$$

$$M_x \quad \left(\text{when } x > (a + b)\right) \quad \ldots = R_2(l - x)$$

### 5. SIMPLE BEAM—UNIFORM LOAD PARTIALLY DISTRIBUTED AT ONE END

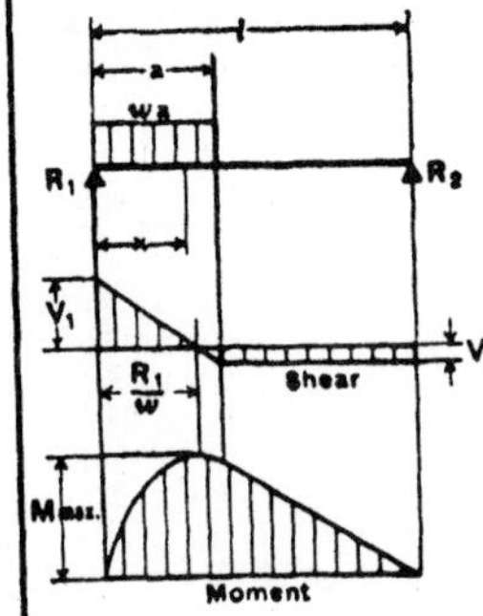

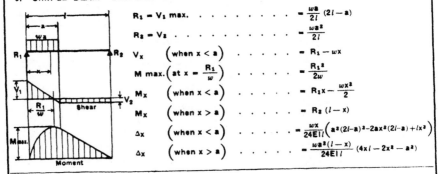

$$R_1 = V_1\ \text{max.} \quad \ldots = \frac{wa}{2l}(2l - a)$$

$$R_2 = V_2 \quad \ldots = \frac{wa^2}{2l}$$

$$V_x \quad \left(\text{when } x < a\right) \quad \ldots = R_1 - wx$$

$$M\ \text{max.}\left(\text{at } x = \frac{R_1}{w}\right) \quad \ldots = \frac{R_1^2}{2w}$$

$$M_x \quad \left(\text{when } x < a\right) \quad \ldots = R_1 x - \frac{wx^2}{2}$$

$$M_x \quad \left(\text{when } x > a\right) \quad \ldots = R_2(l - x)$$

$$\Delta x \quad \left(\text{when } x < a\right) \quad \ldots = \frac{wx}{24EIl}\left(a^2(2l-a)^2 - 2ax^2(2l-a) + lx^3\right)$$

$$\Delta x \quad \left(\text{when } x > a\right) \quad \ldots = \frac{wa^2(l - x)}{24EIl}(4xl - 2x^2 - a^2)$$

### 6. SIMPLE BEAM—UNIFORM LOAD PARTIALLY DISTRIBUTED AT EACH END

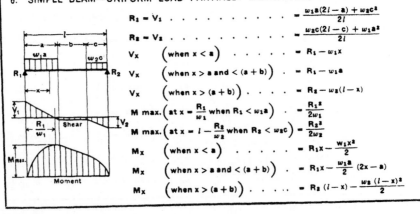

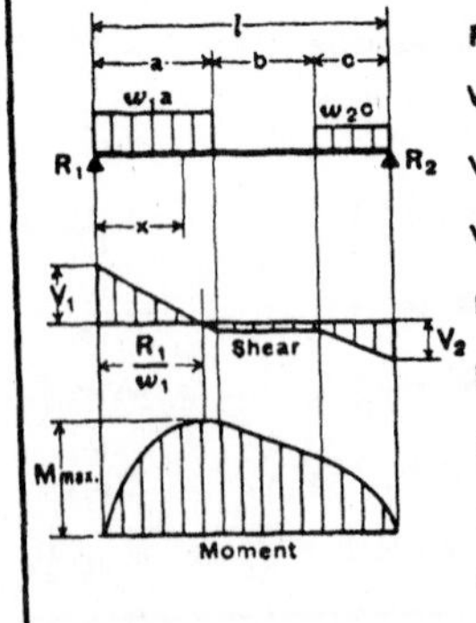


$$R_1 = V_1 \quad \ldots = \frac{w_1 a(2l - a) + w_2 c^2}{2l}$$

$$R_2 = V_2 \quad \ldots = \frac{w_2 c(2l - c) + w_1 a^2}{2l}$$

$$V_x \quad \left(\text{when } x < a\right) \quad \ldots = R_1 - w_1 x$$

$$V_x \quad \left(\text{when } x > a \text{ and } < (a + b)\right) \quad . = R_1 - w_1 a$$

$$V_x \quad \left(\text{when } x > (a + b)\right) \quad \ldots = R_2 - w_2(l - x)$$

$$M\ \text{max.}\left(\text{at } x = \frac{R_1}{w_1} \text{ when } R_1 < w_1 a\right) \quad . = \frac{R_1^2}{2w_1}$$

$$M\ \text{max.}\left(\text{at } x = l - \frac{R_2}{w_2} \text{ when } R_2 < w_2 c\right) = \frac{R_2^2}{2w_2}$$

$$M_x \quad \left(\text{when } x < a\right) \quad \ldots = R_1 x - \frac{w_1 x^2}{2}$$

$$M_x \quad \left(\text{when } x > a \text{ and } < (a + b)\right) \quad . = R_1 x - \frac{w_1 a}{2}(2x - a)$$

$$M_x \quad \left(\text{when } x > (a + b)\right) \quad \ldots = R_2(l - x) - \frac{w_2(l - x)^2}{2}$$

**FIGURE 5.23** *(Continued)*

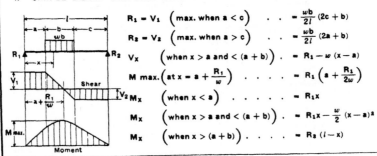

# BEAM DIAGRAMS AND FORMULAS
## For various static loading conditions

### 7. SIMPLE BEAM—CONCENTRATED LOAD AT CENTER

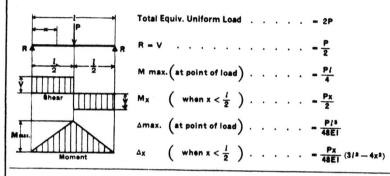

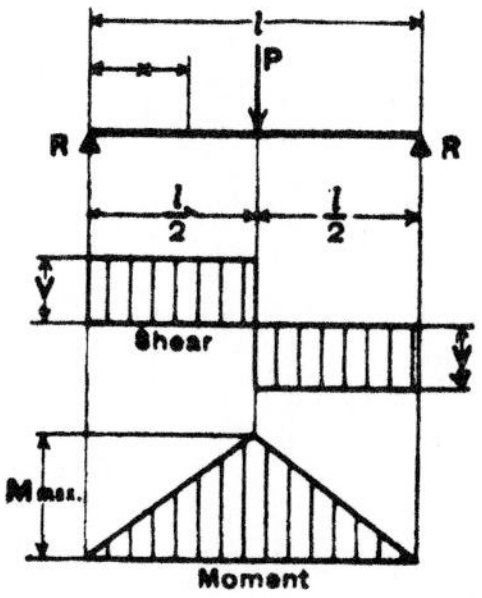


Total Equiv. Uniform Load . . . . . . $= 2P$

$R = V$ . . . . . . . . . . . . . $= \frac{P}{2}$

M max. (at point of load) . . . . . . $= \frac{Pl}{4}$

$M_x$ (when $x < \frac{l}{2}$) . . . . . . $= \frac{Px}{2}$

$\Delta$max. (at point of load) . . . . . . $= \frac{Pl^3}{48EI}$

$\Delta x$ (when $x < \frac{l}{2}$) . . . . . . $= \frac{Px}{48EI}(3l^2 - 4x^2)$

### 8. SIMPLE BEAM—CONCENTRATED LOAD AT ANY POINT

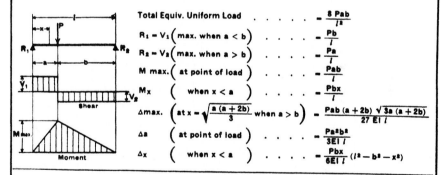

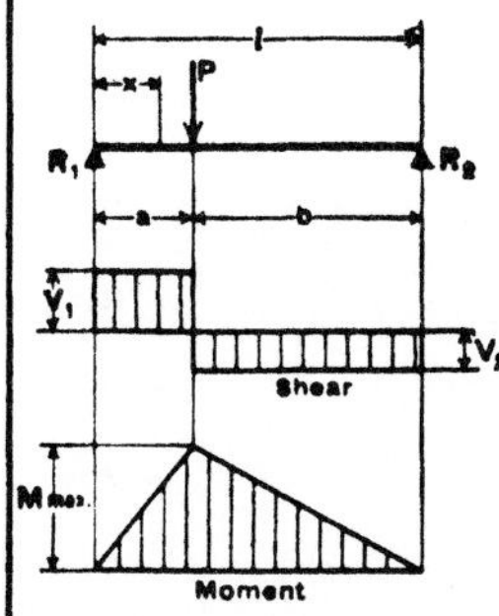


Total Equiv. Uniform Load . . . . . . $= \frac{8Pab}{l^2}$

$R_1 = V_1$ (max. when $a < b$) . . . . . $= \frac{Pb}{l}$

$R_2 = V_2$ (max. when $a > b$) . . . . . $= \frac{Pa}{l}$

M max. (at point of load) . . . . . $= \frac{Pab}{l}$

$M_x$ (when $x < a$) . . . . . $= \frac{Pbx}{l}$

$\Delta$max. (at $x = \sqrt{\frac{a(a+2b)}{3}}$ when $a > b$) $= \frac{Pab(a+2b)\sqrt{3a(a+2b)}}{27EIl}$

$\Delta a$ (at point of load) . . . . . $= \frac{Pa^2b^2}{3EIl}$

$\Delta x$ (when $x < a$) . . . . . $= \frac{Pbx}{6EIl}(l^2 - b^2 - x^2)$

### 9. SIMPLE BEAM—TWO EQUAL CONCENTRATED LOADS SYMMETRICALLY PLACED

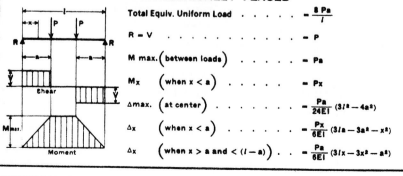

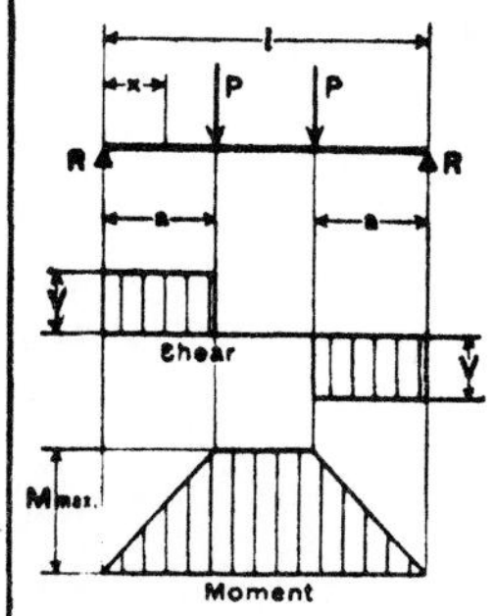


Total Equiv. Uniform Load . . . . . . $= \frac{8Pa}{l}$

$R = V$ . . . . . . . . . . . . . $= P$

M max. (between loads) . . . . . . $= Pa$

$M_x$ (when $x < a$) . . . . . . $= Px$

$\Delta$max. (at center) . . . . . . . . $= \frac{Pa}{24EI}(3l^2 - 4a^2)$

$\Delta x$ (when $x < a$) . . . . . . . $= \frac{Px}{6EI}(3la - 3a^2 - x^2)$

$\Delta x$ (when $x > a$ and $< (l - a)$) . . $= \frac{Pa}{6EI}(3lx - 3x^2 - a^2)$

**FIGURE 5.23** (*Continued*)

# BEAM DIAGRAMS AND FORMULAS

## For various static loading conditions

### 10. SIMPLE BEAM—TWO EQUAL CONCENTRATED LOADS UNSYMMETRICALLY PLACED

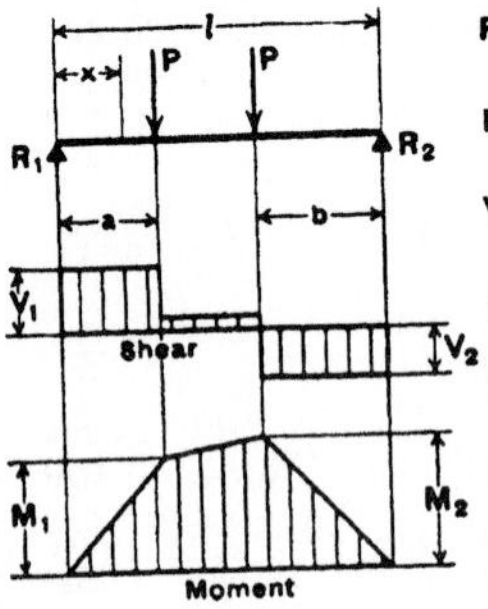

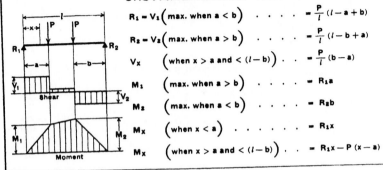

$R_1 = V_1 \left(\text{max. when } a < b\right) \quad \ldots\ldots \quad = \frac{P}{l}(l - a + b)$

$R_2 = V_2 \left(\text{max. when } a > b\right) \quad \ldots\ldots \quad = \frac{P}{l}(l - b + a)$

$V_x \quad \left(\text{when } x > a \text{ and } < (l - b)\right) \quad \ldots \quad = \frac{P}{l}(b - a)$

$M_1 \quad \left(\text{max. when } a > b\right) \quad \ldots\ldots \quad = R_1 a$

$M_2 \quad \left(\text{max. when } a < b\right) \quad \ldots\ldots \quad = R_2 b$

$M_x \quad \left(\text{when } x < a\right) \quad \ldots\ldots \quad = R_1 x$

$M_x \quad \left(\text{when } x > a \text{ and } < (l - b)\right) \quad \ldots \quad = R_1 x - P(x - a)$

### 11. SIMPLE BEAM—TWO UNEQUAL CONCENTRATED LOADS UNSYMMETRICALLY PLACED

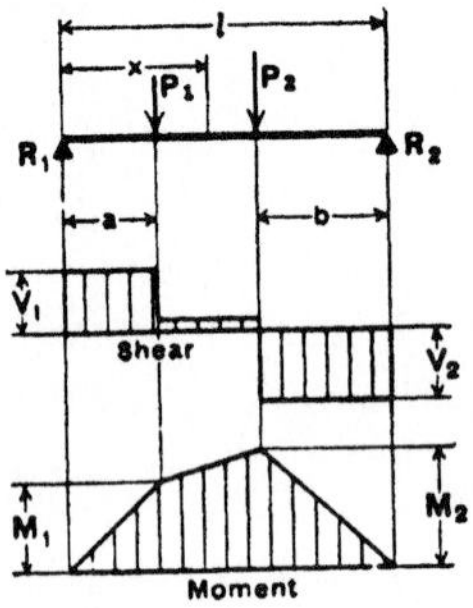


$R_1 = V_1 \quad \ldots\ldots \quad = \frac{P_1(l - a) + P_2 b}{l}$

$R_2 = V_2 \quad \ldots\ldots \quad = \frac{P_1 a + P_2(l - b)}{l}$

$V_x \quad \left(\text{when } x > a \text{ and } < (l - b)\right) \quad \ldots \quad = R_1 - P_1$

$M_1 \quad \left(\text{max. when } R_1 < P_1\right) \quad \ldots \quad = R_1 a$

$M_2 \quad \left(\text{max. when } R_2 < P_2\right) \quad \ldots \quad = R_2 b$

$M_x \quad \left(\text{when } x < a\right) \quad \ldots\ldots \quad = R_1 x$

$M_x \quad \left(\text{when } x > a \text{ and } < (l - b)\right) \quad \ldots \quad = R_1 x - P_1(x - a)$

### 12. BEAM FIXED AT ONE END, SUPPORTED AT OTHER—UNIFORMLY DISTRIBUTED LOAD

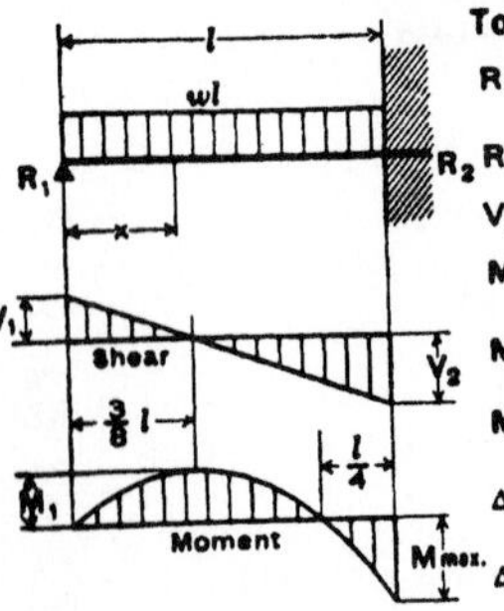

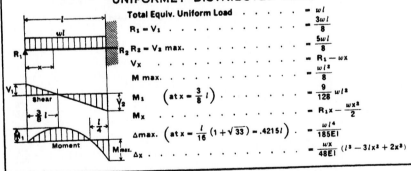

Total Equiv. Uniform Load $\ldots\ldots = wl$

$R_1 = V_1 \quad \ldots\ldots \quad = \frac{3wl}{8}$

$R_2 = V_2 \text{ max.} \quad \ldots\ldots \quad = \frac{5wl}{8}$

$V_x \quad \ldots\ldots \quad = R_1 - wx$

$M \text{ max.} \quad \ldots\ldots \quad = \frac{wl^2}{8}$

$M_1 \quad \left(\text{at } x = \frac{3}{8} l\right) \quad \ldots\ldots \quad = \frac{9}{128} wl^2$

$M_x \quad \ldots\ldots \quad = R_1 x - \frac{wx^2}{2}$

$\Delta\text{max.} \quad \left(\text{at } x = \frac{l}{16}(1 + \sqrt{33}) = .4215l\right) \quad \ldots \quad = \frac{wl^4}{185EI}$

$\Delta x \quad \ldots\ldots \quad = \frac{wx}{48EI}(l^3 - 3lx^2 + 2x^3)$

**FIGURE 5.23** *(Continued)*

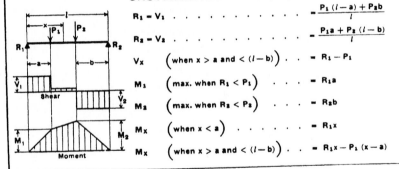

# BEAM DIAGRAMS AND FORMULAS

## For various static loading conditions

### 13. BEAM FIXED AT ONE END, SUPPORTED AT OTHER—CONCENTRATED LOAD AT CENTER

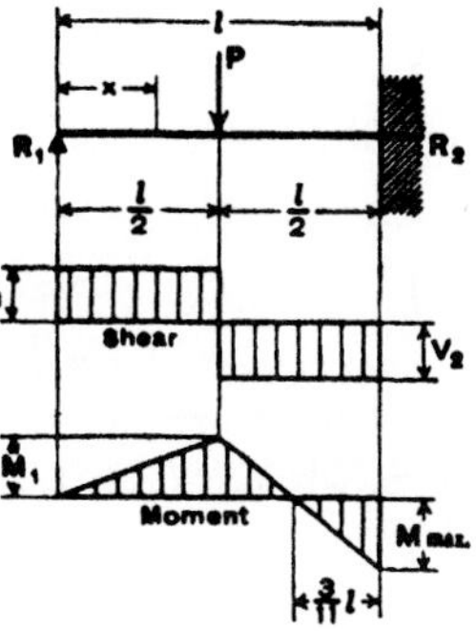


| | | |
|---|---|---|
| Total Equiv. Uniform Load | | $= \frac{3P}{2}$ |
| $R_1 = V_1$ | | $= \frac{5P}{16}$ |
| $R_2 = V_2$ max. | | $= \frac{11P}{16}$ |
| M max. | (at fixed end) | $= \frac{3Pl}{16}$ |
| $M_1$ | (at point of load) | $= \frac{5Pl}{32}$ |
| $M_x$ | $\left(\text{when } x < \frac{l}{2}\right)$ | $= \frac{5Px}{16}$ |
| $M_x$ | $\left(\text{when } x > \frac{l}{2}\right)$ | $= P\left(\frac{l}{2} - \frac{11x}{16}\right)$ |
| $\Delta$max. | $\left(\text{at } x = l\sqrt{\frac{1}{5}} = .4472l\right)$ | $= \frac{Pl^3}{48EI\sqrt{5}} = .009317\frac{Pl^3}{EI}$ |
| $\Delta x$ | (at point of load) | $= \frac{7Pl^3}{768EI}$ |
| $\Delta x$ | $\left(\text{when } x < \frac{l}{2}\right)$ | $= \frac{Px}{96EI}(3l^2 - 5x^2)$ |
| $\Delta x$ | $\left(\text{when } x > \frac{l}{2}\right)$ | $= \frac{P}{96EI}(x - l)^2(11x - 2l)$ |

### 14. BEAM FIXED AT ONE END, SUPPORTED AT OTHER—CONCENTRATED LOAD AT ANY POINT

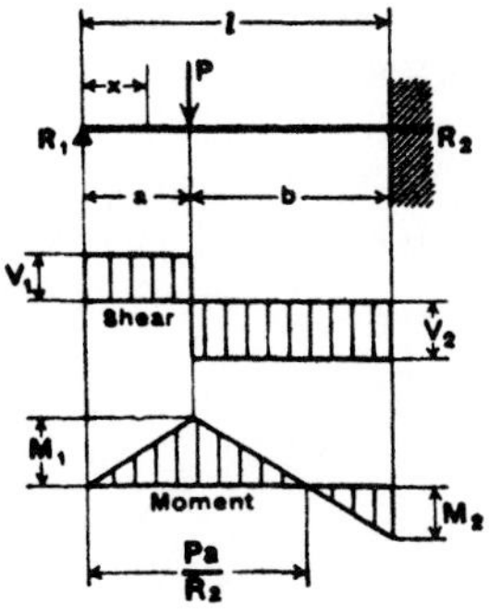


| | | |
|---|---|---|
| $R_1 = V_1$ | | $= \frac{Pb^2}{2l^3}(a + 2l)$ |
| $R_2 = V_2$ | | $= \frac{Pa}{2l^3}(3l^2 - a^2)$ |
| $M_1$ | (at point of load) | $= R_1a$ |
| $M_2$ | (at fixed end) | $= \frac{Pab}{2l^2}(a + l)$ |
| $M_x$ | (when $x < a$) | $= R_1x$ |
| $M_x$ | (when $x > a$) | $= R_1x - P(x - a)$ |
| $\Delta$max. | $\left(\text{when } a < .414l \text{ at } x = l\frac{l^2+a^2}{3l^2-a^2}\right)$ | $= \frac{Pa}{3EI}\frac{(l^2 - a^2)^3}{(3l^2 - a^2)^2}$ |
| $\Delta$max. | $\left(\text{when } a > .414l \text{ at } x = l\sqrt{\frac{a}{2l+a}}\right)$ | $= \frac{Pab^2}{6EI}\sqrt{\frac{a}{2l + a}}$ |
| $\Delta a$ | (at point of load) | $= \frac{Pa^2b^3}{12EIl^3}(3l + a)$ |
| $\Delta x$ | (when $x < a$) | $= \frac{Pb^2x}{12EIl^3}(3al^2 - 2lx^2 - ax^2)$ |
| $\Delta x$ | (when $x > a$) | $= \frac{Pa}{12EIl^3}(l - x)^2(3l^2x - a^2x - 2a^2l)$ |

**FIGURE 5.23** (*Continued*)

# BEAM DIAGRAMS AND FORMULAS

## For various static loading conditions

### 15. BEAM FIXED AT BOTH ENDS—UNIFORMLY DISTRIBUTED LOADS

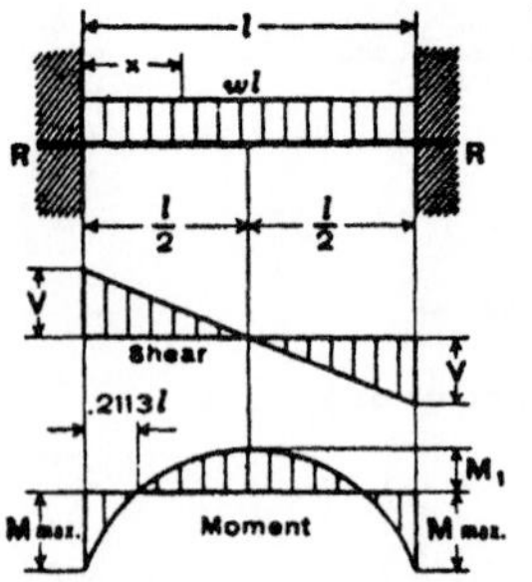


| | |
|---|---|
| Total Equiv. Uniform Load | $= \frac{2wl}{3}$ |
| R = V | $= \frac{wl}{2}$ |
| $V_x$ | $= w\left(\frac{l}{2} - x\right)$ |
| M max. (at ends) | $= \frac{wl^2}{12}$ |
| $M_1$ (at center) | $= \frac{wl^2}{24}$ |
| $M_x$ | $= \frac{w}{12}(6lx - l^2 - 6x^2)$ |
| $\Delta$max. (at center) | $= \frac{wl^4}{384EI}$ |
| $\Delta x$ | $= \frac{wx^2}{24EI}(l - x)^2$ |

### 16. BEAM FIXED AT BOTH ENDS—CONCENTRATED LOAD AT CENTER

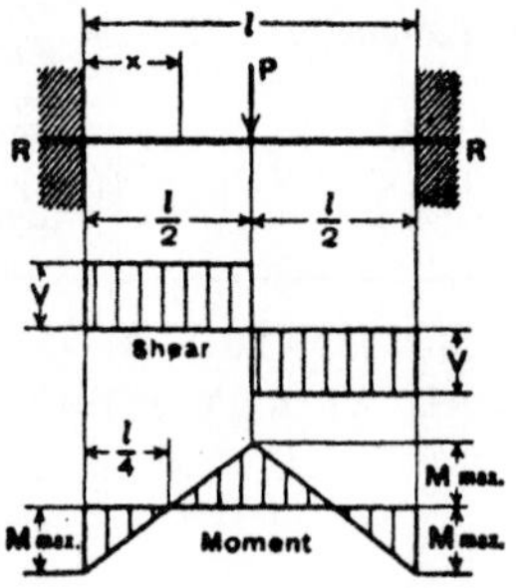


| | |
|---|---|
| Total Equiv. Uniform Load | = P |
| R = V | $= \frac{P}{2}$ |
| M max. (at center and ends) | $= \frac{Pl}{8}$ |
| $M_x$ (when $x < \frac{l}{2}$) | $= \frac{P}{8}(4x - l)$ |
| $\Delta$max. (at center) | $= \frac{Pl^3}{192EI}$ |
| $\Delta x$ (when $x < \frac{l}{2}$) | $= \frac{Px^2}{48EI}(3l - 4x)$ |

### 17. BEAM FIXED AT BOTH ENDS—CONCENTRATED LOAD AT ANY POINT

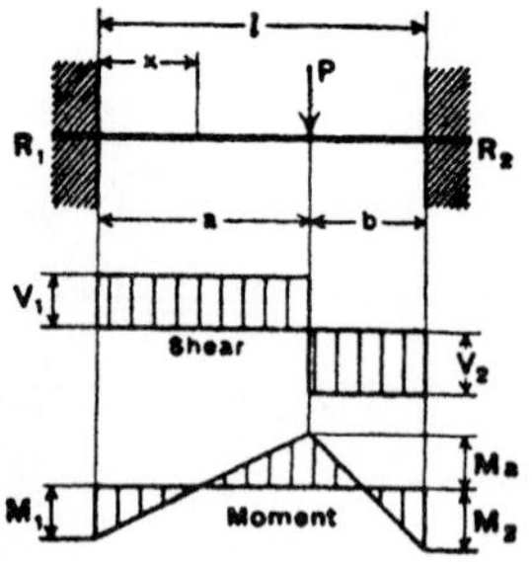


| | |
|---|---|
| $R_1 = V_1$ (max. when a < b) | $= \frac{Pb^2}{l^3}(3a + b)$ |
| $R_2 = V_2$ (max. when a > b) | $= \frac{Pa^2}{l^3}(a + 3b)$ |
| $M_1$ (max. when a < b) | $= \frac{Pab^2}{l^2}$ |
| $M_2$ (max. when a > b) | $= \frac{Pa^2b}{l^2}$ |
| $M_a$ (at point of load) | $= \frac{2Pa^2b^2}{l^3}$ |
| $M_x$ (when x < a) | $= R_1x - \frac{Pab^2}{l^2}$ |
| $\Delta$max. (when a > b at $x = \frac{2al}{3a + b}$) | $= \frac{2Pa^3b^2}{3EI(3a + b)^2}$ |
| $\Delta a$ (at point of load) | $= \frac{Pa^3b^3}{3EIl^3}$ |
| $\Delta x$ (when x < a) | $= \frac{Pb^2x^2}{6EIl^3}(3al - 3ax - bx)$ |

**FIGURE 5.23** (*Continued*)

# BEAM DIAGRAMS AND FORMULAS
## For various static loading conditions

### 18. CANTILEVER BEAM—LOAD INCREASING UNIFORMLY TO FIXED END

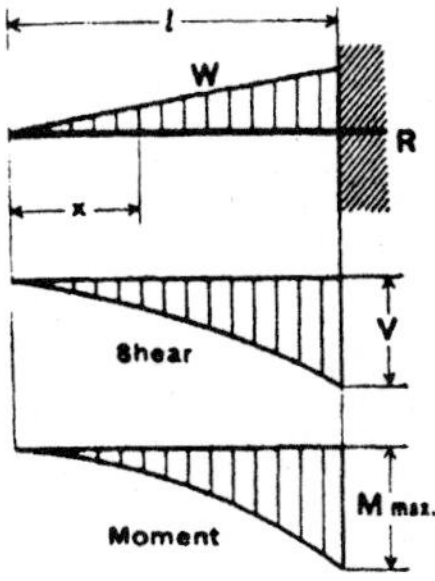


Total Equiv. Uniform Load . . . . $= \frac{8}{3}W$

R = V . . . . . . . . . . . $= W$

$V_x$ . . . . . . . . . . . . $= W\frac{x^2}{l^2}$

M max. (at fixed end) . . . . . . $= \frac{Wl}{3}$

$M_x$ . . . . . . . . . . . . $= \frac{Wx^3}{3l^2}$

$\Delta$max. (at free end) . . . . . . $= \frac{Wl^3}{15EI}$

$\Delta x$ . . . . . . . . . . . . $= \frac{W}{60EIl^2}(x^5 - 5l^4x + 4l^5)$

### 19. CANTILEVER BEAM—UNIFORMLY DISTRIBUTED LOAD

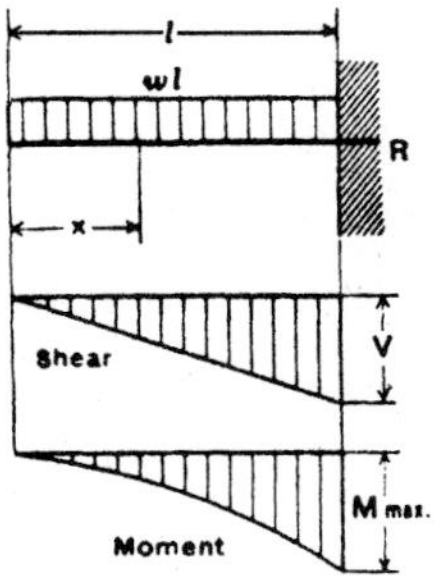


Total Equiv. Uniform Load . . . . $= 4wl$

R = V . . . . . . . . . . . $= wl$

$V_x$ . . . . . . . . . . . . $= wx$

M max. (at fixed end) . . . . . . $= \frac{wl^2}{2}$

$M_x$ . . . . . . . . . . . . $= \frac{wx^2}{2}$

$\Delta$max. (at free end) . . . . . . $= \frac{wl^4}{8EI}$

$\Delta x$ . . . . . . . . . . . . $= \frac{w}{24EI}(x^4 - 4l^3x + 3l^4)$

### 20. BEAM FIXED AT ONE END, FREE TO DEFLECT VERTICALLY BUT NOT ROTATE AT OTHER—UNIFORMLY DISTRIBUTED LOAD

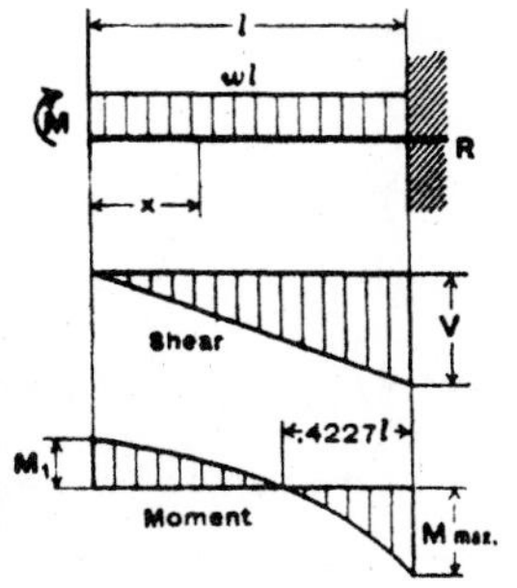


Total Equiv. Uniform Load . . . . $= \frac{8}{3}wl$

R = V . . . . . . . . . . . $= wl$

$V_x$ . . . . . . . . . . . . $= wx$

M max. (at fixed end) . . . . . . $= \frac{wl^2}{3}$

$M_1$ (at deflected end) . . . . . $= \frac{wl^2}{6}$

$M_x$ . . . . . . . . . . . . $= \frac{w}{6}(l^2 - 3x^2)$

$\Delta$max. (at deflected end) . . . . . $= \frac{wl^4}{24EI}$

$\Delta x$ . . . . . . . . . . . . $= \frac{w(l^2 - x^2)^2}{24EI}$

**FIGURE 5.23** (*Continued*)

# BEAM DIAGRAMS AND FORMULAS

## For various static loading conditions

### 21. CANTILEVER BEAM—CONCENTRATED LOAD AT ANY POINT

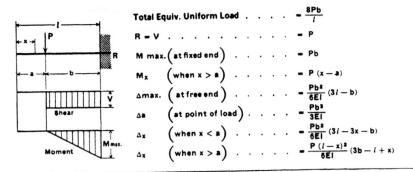

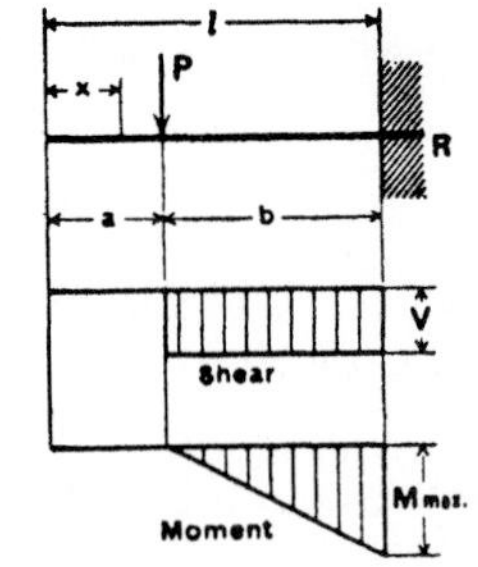


Total Equiv. Uniform Load . . . . . $= \frac{8Pb}{l}$

R = V . . . . . . . . . . . . . . $= P$

M max. (at fixed end) . . . . . . $= Pb$

$M_x$ (when $x > a$) . . . . . . $= P(x - a)$

$\Delta$max. (at free end) . . . . . . $= \frac{Pb^2}{6EI}(3l - b)$

$\Delta a$ (at point of load) . . . . . $= \frac{Pb^3}{3EI}$

$\Delta x$ (when $x < a$) . . . . . . $= \frac{Pb^2}{6EI}(3l - 3x - b)$

$\Delta x$ (when $x > a$) . . . . . . $= \frac{P(l - x)^2}{6EI}(3b - l + x)$

### 22. CANTILEVER BEAM—CONCENTRATED LOAD AT FREE END

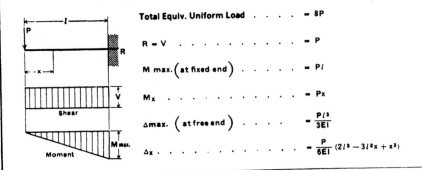

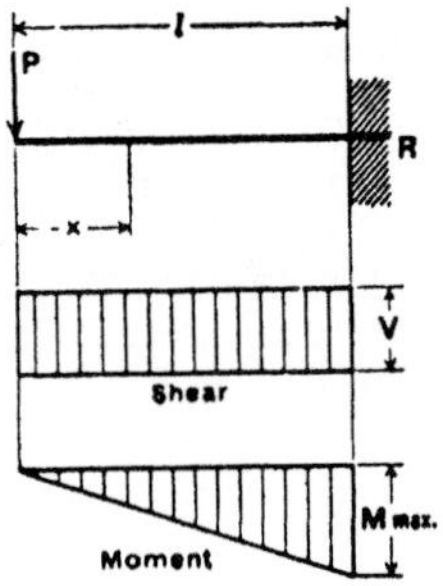


Total Equiv. Uniform Load . . . . . $= 8P$

R = V . . . . . . . . . . . . . . $= P$

M max. (at fixed end) . . . . . . $= Pl$

$M_x$ . . . . . . . . . . . . . . $= Px$

$\Delta$max. (at free end) . . . . . $= \frac{Pl^3}{3EI}$

$\Delta x$ . . . . . . . . . . . . . . $= \frac{P}{6EI}(2l^3 - 3l^2x + x^3)$

### 23. BEAM FIXED AT ONE END, FREE TO DEFLECT VERTICALLY BUT NOT ROTATE AT OTHER—CONCENTRATED LOAD AT DEFLECTED END

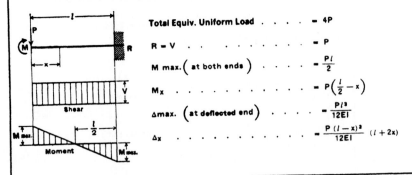

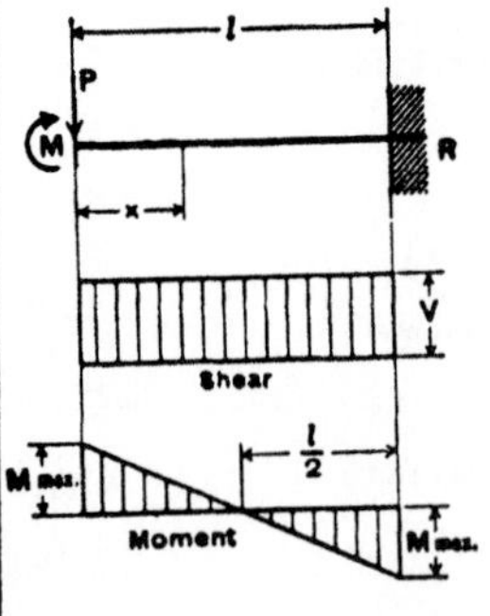


Total Equiv. Uniform Load . . . . . $= 4P$

R = V . . . . . . . . . . . . . . $= P$

M max. (at both ends) . . . . . $= \frac{Pl}{2}$

$M_x$ . . . . . . . . . . . . . . $= P\left(\frac{l}{2} - x\right)$

$\Delta$max. (at deflected end) . . . . $= \frac{Pl^3}{12EI}$

$\Delta x$ . . . . . . . . . . . . . . $= \frac{P(l - x)^2}{12EI}(l + 2x)$

FIGURE 5.23 *(Continued)*

# BEAM DIAGRAMS AND FORMULAS

## For various static loading conditions

### 24. BEAM OVERHANGING ONE SUPPORT—UNIFORMLY DISTRIBUTED LOAD

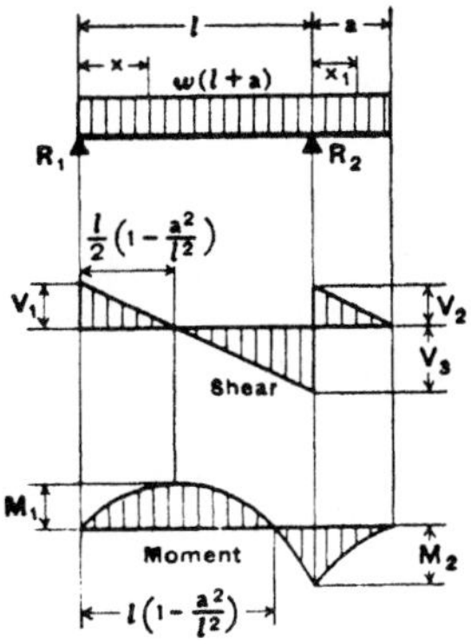


$R_1 = V_1$ . . . . . . . . . . $= \frac{w}{2l}(l^2 - a^2)$

$R_2 = V_2 + V_3$ . . . . . . . . $= \frac{w}{2l}(l + a)^2$

$V_2$ . . . . . . . . . . . . . $= wa$

$V_3$ . . . . . . . . . . . . . $= \frac{w}{2l}(l^2 + a^2)$

$V_x$ (between supports) . . $= R_1 - wx$

$V_{x_1}$ (for overhang) . . . . $= w(a - x_1)$

$M_1$ $\left(\text{at } x = \frac{l}{2}\left[1 - \frac{a^2}{l^2}\right]\right)$ . . $= \frac{w}{8l^2}(l + a)^2(l - a)^2$

$M_2$ (at $R_2$) . . . . . . . $= \frac{wa^2}{2}$

$M_x$ (between supports) . . $= \frac{wx}{2l}(l^2 - a^2 - xl)$

$M_{x_1}$ (for overhang) . . . . $= \frac{w}{2}(a - x_1)^2$

$\Delta x$ (between supports) . . $= \frac{wx}{24EIl}(l^4 - 2l^2x^2 + lx^3 - 2a^2l^2 + 2a^2x^2)$

$\Delta x_1$ (for overhang) . . . . $= \frac{wx_1}{24EI}(4a^2l - l^3 + 6a^2x_1 - 4ax_1^2 + x_1^3)$

### 25. BEAM OVERHANGING ONE SUPPORT—UNIFORMLY DISTRIBUTED LOAD ON OVERHANG

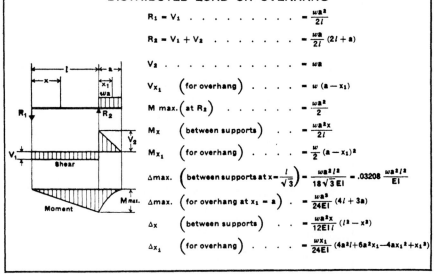

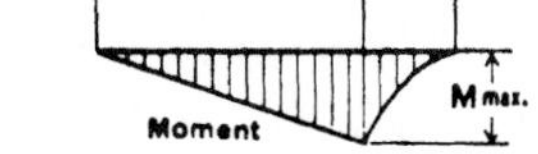


$R_1 = V_1$ . . . . . . . . . . $= \frac{wa^2}{2l}$

$R_2 = V_1 + V_2$ . . . . . . . . $= \frac{wa}{2l}(2l + a)$

$V_2$ . . . . . . . . . . . . . $= wa$

$V_{x_1}$ (for overhang) . . . . $= w(a - x_1)$

M max. (at $R_2$) . . . . . . . $= \frac{wa^2}{2}$

$M_x$ (between supports) . . $= \frac{wa^2x}{2l}$

$M_{x_1}$ (for overhang) . . . . $= \frac{w}{2}(a - x_1)^2$

$\Delta$max. $\left(\text{between supports at } x = \frac{l}{\sqrt{3}}\right) = \frac{wa^2l^2}{18\sqrt{3}EI} = .03208\,\frac{wa^2l^2}{EI}$

$\Delta$max. (for overhang at $x_1 = a$) . $= \frac{wa^3}{24EI}(4l + 3a)$

$\Delta x$ (between supports) . . $= \frac{wa^2x}{12EIl}(l^2 - x^2)$

$\Delta x_1$ (for overhang) . . . . $= \frac{wx_1}{24EI}(4a^2l + 6a^2x_1 - 4ax_1^2 + x_1^3)$

**FIGURE 5.23** (*Continued*)

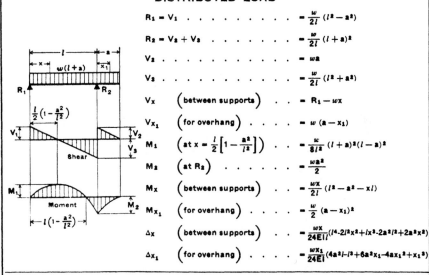

# BEAM DIAGRAMS AND FORMULAS

## For various static loading conditions

### 26. BEAM OVERHANGING ONE SUPPORT—CONCENTRATED LOAD AT END OF OVERHANG

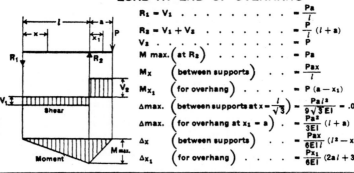

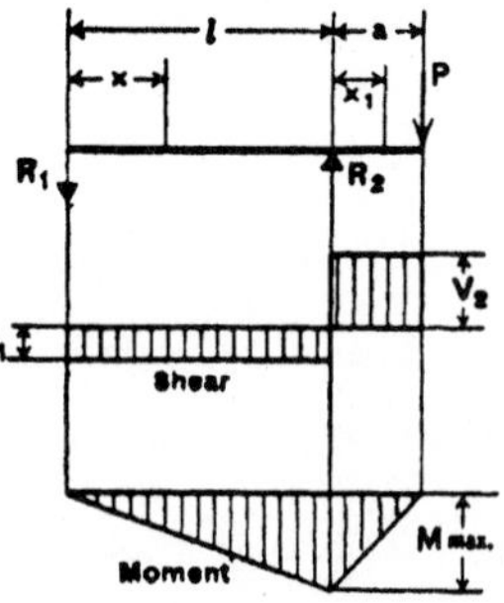


- $R_1 = V_1$ . . . . . . . . . . $= \frac{Pa}{l}$
- $R_2 = V_1 + V_2$ . . . . . . . . $= \frac{P}{l}(l + a)$
- $V_2$ . . . . . . . . . . . . $= P$
- M max. (at $R_2$) . . . . . . . $= Pa$
- $M_x$ (between supports) . . . $= \frac{Pax}{l}$
- $M_{x_1}$ (for overhang) . . . . . $= P(a - x_1)$
- $\Delta$max. (between supports at $x = \frac{l}{\sqrt{3}}$) $= \frac{Pal^2}{9\sqrt{3}EI} = .06415\,\frac{Pal^2}{EI}$
- $\Delta$max. (for overhang at $x_1 = a$) . $= \frac{Pa^2}{3EI}(l + a)$
- $\Delta x$ (between supports) . . . $= \frac{Pax}{6EIl}(l^2 - x^2)$
- $\Delta x_1$ (for overhang) . . . . . $= \frac{Px_1}{6EI}(2al + 3ax_1 - x_1^2)$

### 27. BEAM OVERHANGING ONE SUPPORT—UNIFORMLY DISTRIBUTED LOAD BETWEEN SUPPORTS

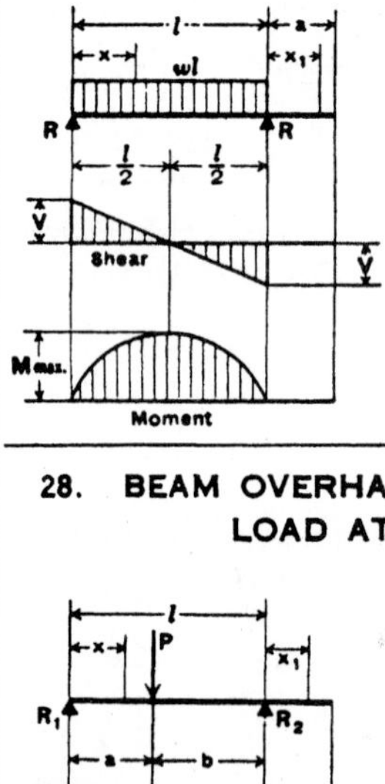

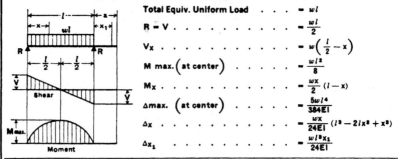

- Total Equiv. Uniform Load . . . $= wl$
- $R = V$ . . . . . . . . . . $= \frac{wl}{2}$
- $V_x$ . . . . . . . . . . . . $= w\left(\frac{l}{2} - x\right)$
- M max. (at center) . . . . . . $= \frac{wl^2}{8}$
- $M_x$ . . . . . . . . . . . . $= \frac{wx}{2}(l - x)$
- $\Delta$max. (at center) . . . . . . $= \frac{5wl^4}{384EI}$
- $\Delta x$ . . . . . . . . . . . . $= \frac{wx}{24EI}(l^3 - 2lx^2 + x^3)$
- $\Delta x_1$ . . . . . . . . . . . $= \frac{wl^3x_1}{24EI}$

### 28. BEAM OVERHANGING ONE SUPPORT—CONCENTRATED LOAD AT ANY POINT BETWEEN SUPPORTS

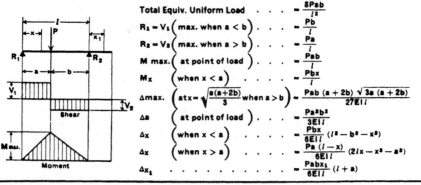

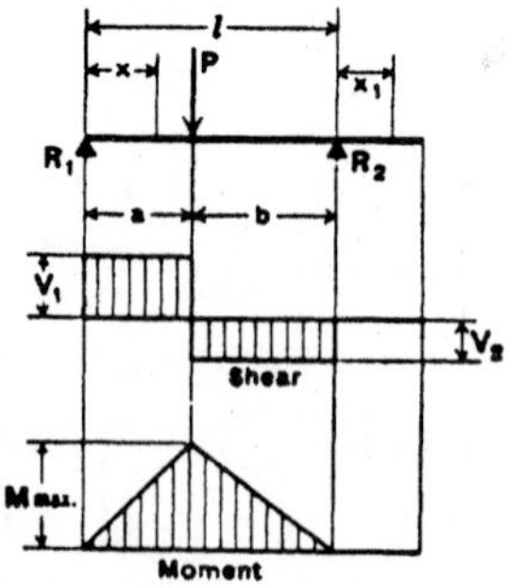


- Total Equiv. Uniform Load . . . $= \frac{8Pab}{l^2}$
- $R_1 = V_1$ (max. when $a < b$) . . . $= \frac{Pb}{l}$
- $R_2 = V_2$ (max. when $a > b$) . . . $= \frac{Pa}{l}$
- M max. (at point of load) . . . $= \frac{Pab}{l}$
- $M_x$ (when $x < a$) . . . . . $= \frac{Pbx}{l}$
- $\Delta$max. (at $x = \sqrt{\frac{a(a+2b)}{3}}$ when $a > b$) $= \frac{Pab\,(a + 2b)\,\sqrt{3a\,(a + 2b)}}{27EIl}$
- $\Delta a$ (at point of load) . . . $= \frac{Pa^2b^2}{3EIl}$
- $\Delta x$ (when $x < a$) . . . . $= \frac{Pbx}{6EIl}(l^2 - b^2 - x^2)$
- $\Delta x$ (when $x > a$) . . . . $= \frac{Pa\,(l - x)}{6EIl}(2lx - x^2 - a^2)$
- $\Delta x_1$ . . . . . . . . . . $= \frac{Pabx_1}{6EIl}(l + a)$

**FIGURE 5.23** *(Continued)*

# BEAM DIAGRAMS AND FORMULAS

## For various static loading conditions

### 29. CONTINUOUS BEAM—TWO EQUAL SPANS—UNIFORM LOAD ON ONE SPAN

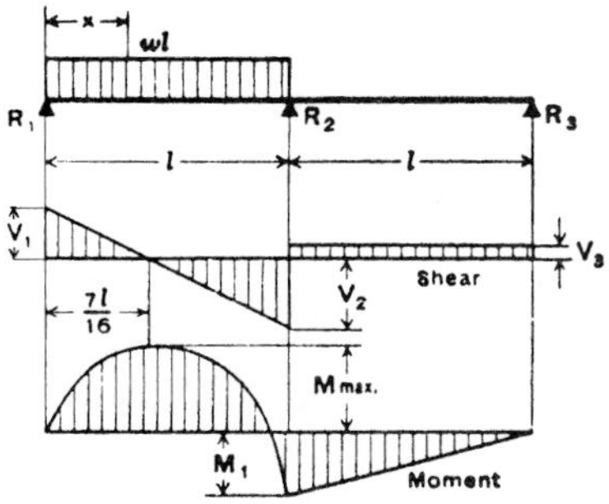


Total Equiv. Uniform Load = $\frac{49}{64} wl$

$R_1 = V_1$ . . . . . . . . = $\frac{7}{16} wl$

$R_2 = V_2 + V_3$ . . . . . = $\frac{5}{8} wl$

$R_3 = V_3$ . . . . . . . . = $-\frac{1}{16} wl$

$V_2$ . . . . . . . . . = $\frac{9}{16} wl$

M max. $\left(\text{at } x = \frac{7}{16} l\right)$ . . = $\frac{49}{512} wl^2$

$M_1$ (at support $R_2$) . = $\frac{1}{16} wl^2$

$M_x$ (when $x < l$) . . = $\frac{wx}{16}(7l - 8x)$

$\Delta$ Max. (0.472 $l$ from $R_1$) = $0.0092\ wl^4/EI$

### 30. CONTINUOUS BEAM—TWO EQUAL SPANS—CONCENTRATED LOAD AT CENTER OF ONE SPAN

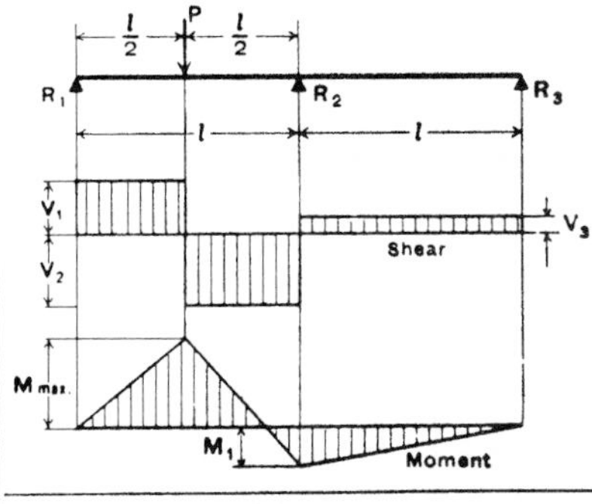


Total Equiv. Uniform Load . = $\frac{13}{8} P$

$R_1 = V_1$ . . . . . . . . = $\frac{13}{32} P$

$R_2 = V_2 + V_3$ . . . . . = $\frac{11}{16} P$

$R_3 = V_3$ . . . . . . . . = $-\frac{3}{32} P$

$V_2$ . . . . . . . . . = $\frac{19}{32} P$

M max. (at point of load) . = $\frac{13}{64} Pl$

$M_1$ (at support $R_2$) . = $\frac{3}{32} Pl$

$\Delta$ Max. (0.480 $l$ from $R_1$) = $0.015\ Pl^3/EI$

### 31. CONTINUOUS BEAM—TWO EQUAL SPANS—CONCENTRATED LOAD AT ANY POINT

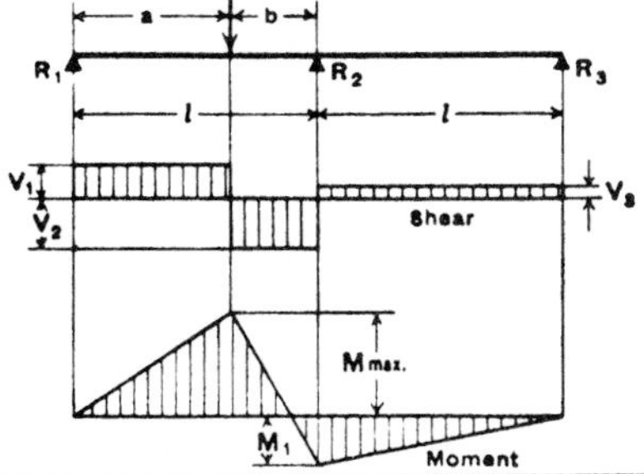


$R_1 = V_1$ . . . . . . . . = $\frac{Pb}{4l^3}\left(4l^2 - a(l+a)\right)$

$R_2 = V_2 + V_3$ . . . . . = $\frac{Pa}{2l^3}\left(2l^2 + b(l+a)\right)$

$R_3 = V_3$ . . . . . . . . = $-\frac{Pab}{4l^3}(l + a)$

$V_2$ . . . . . . . . . = $\frac{Pa}{4l^3}\left(4l^2 + b(l+a)\right)$

M max. (at point of load) . = $\frac{Pab}{4l^3}\left(4l^2 - a(l+a)\right)$

$M_1$ (at support $R_2$) . = $\frac{Pab}{4l^2}(l + a)$

**FIGURE 5.23** (*Continued*)

## 32. BEAM—UNIFORMLY DISTRIBUTED LOAD AND VARIABLE END MOMENTS

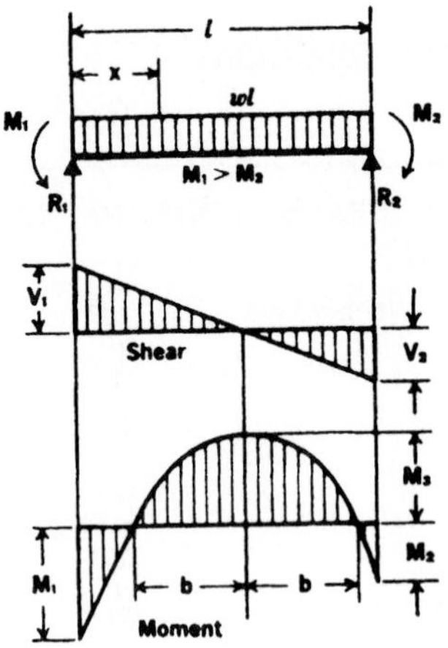

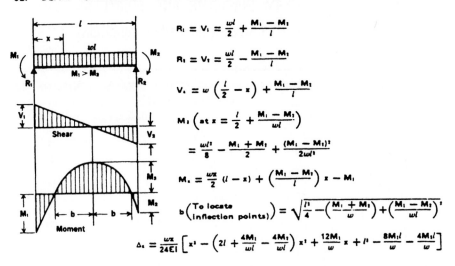

$$R_1 = V_1 = \frac{wl}{2} + \frac{M_1 - M_2}{l}$$

$$R_2 = V_2 = \frac{wl}{2} - \frac{M_1 - M_2}{l}$$

$$V_x = w\left(\frac{l}{2} - x\right) + \frac{M_1 - M_2}{l}$$

$$M_3\left(\text{at } x = \frac{l}{2} + \frac{M_1 - M_2}{wl}\right) = \frac{wl^2}{8} - \frac{M_1 + M_2}{2} + \frac{(M_1 - M_2)^2}{2wl^2}$$

$$M_x = \frac{wx}{2}(l - x) + \left(\frac{M_1 - M_2}{l}\right)x - M_1$$

$$b\left(\begin{array}{l}\text{To locate}\\ \text{inflection points}\end{array}\right) = \sqrt{\frac{l^2}{4} - \left(\frac{M_1 + M_2}{w}\right) + \left(\frac{M_1 - M_2}{wl}\right)^2}$$

$$\Delta_x = \frac{wx}{24EI}\left[x^3 - \left(2l + \frac{4M_1}{wl} - \frac{4M_2}{wl}\right)x^2 + \frac{12M_1}{w}x + l^3 - \frac{8M_1 l}{w} - \frac{4M_2 l}{w}\right]$$

## 33. BEAM—CONCENTRATED LOAD AT CENTER AND VARIABLE END MOMENTS

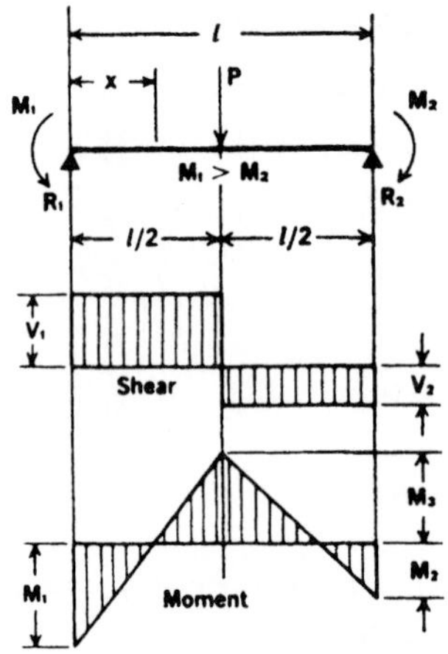

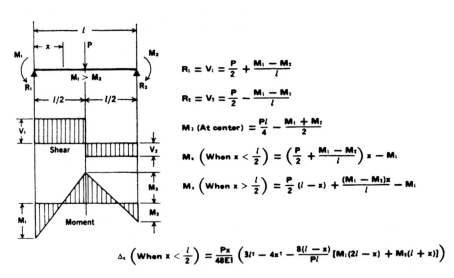

$$R_1 = V_1 = \frac{P}{2} + \frac{M_1 - M_2}{l}$$

$$R_2 = V_2 = \frac{P}{2} - \frac{M_1 - M_2}{l}$$

$$M_3 \text{ (At center)} = \frac{Pl}{4} - \frac{M_1 + M_2}{2}$$

$$M_x\left(\text{When } x < \frac{l}{2}\right) = \left(\frac{P}{2} + \frac{M_1 - M_2}{l}\right)x - M_1$$

$$M_x\left(\text{When } x > \frac{l}{2}\right) = \frac{P}{2}(l - x) + \frac{(M_1 - M_2)x}{l} - M_1$$

$$\Delta_x\left(\text{When } x < \frac{l}{2}\right) = \frac{Px}{48EI}\left(3l^2 - 4x^2 - \frac{8(l - x)}{Pl}[M_1(2l - x) + M_2(l + x)]\right)$$

**FIGURE 5.23** *(Continued)*

# Beam Formulas

## BENDING MOMENTS AND DEFLECTIONS OF BEAMS

### 1. CANTILEVER BEAM

*Concentrated Load at Free End*

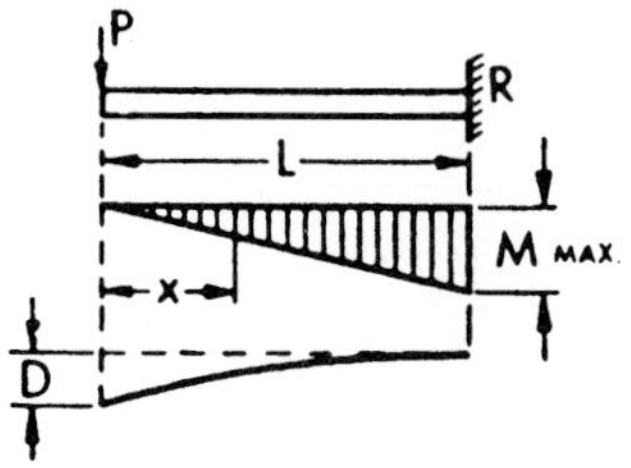


Reaction, $R = P$

Moment at any point: $M = Px$

Maximum moment, $M_{max} = PL$

Maximum deflection, $D = \dfrac{PL^3}{3\,EI}$

### 2. CANTILEVER BEAM

*Uniform Load, w per unit of length, total load W*

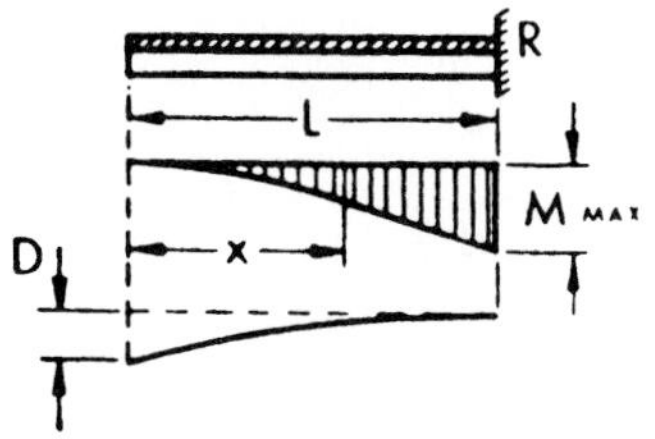


Reaction, $R = wL = W$

Moment at any point: $M = \dfrac{wx^2}{2} = \dfrac{Wx^2}{2L}$

Maximum moment, $M_{max} = \dfrac{wL^2}{2} = \dfrac{WL}{2}$

Maximum deflection, $D = \dfrac{wL^4}{8\,EI} = \dfrac{WL^3}{8\,EI}$

### 3. SIMPLE BEAM

*Concentrated Load at Center*

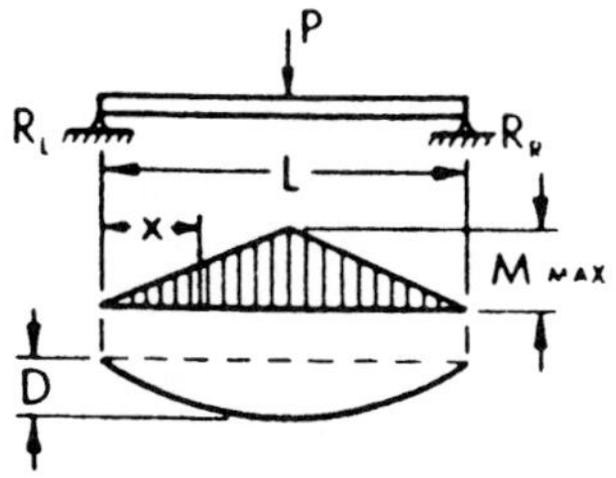


Reactions: $R_L = R_R = \dfrac{P}{2}$

Moment at any point:

$$x \leqq \frac{L}{2},\; M = \frac{Px}{2}$$

$$x \geqq \frac{L}{2},\; M = \frac{P(L-x)}{2}$$

Maximum moment, at center, $M_{max} = \dfrac{PL}{4}$

Maximum deflection, $D = \dfrac{PL^3}{48\,EI}$

**FIGURE 5.24** Simplified beam equations.

## 4. SIMPLE BEAM

*Concentrated Load at any point*

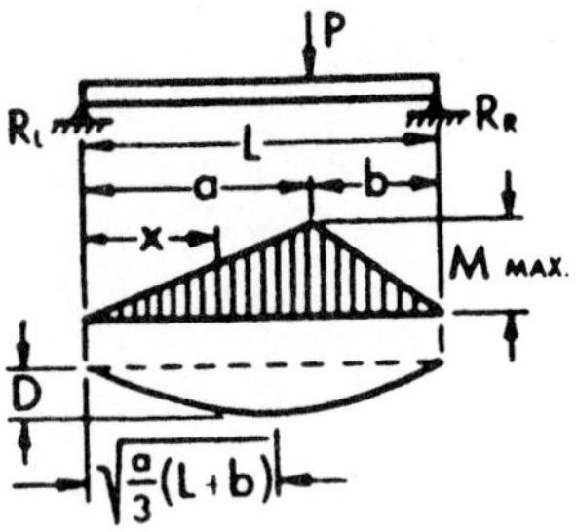


Reactions: $R_L = \frac{Pb}{L}$, $R_R = \frac{Pa}{L}$

Moment at any point:

$$x \leqq a, \; M = R_{LX} = \frac{Pbx}{L}$$

$$x \geqq a, \; M = R_R\,(L - x) = \frac{Pa(L - x)}{L}$$

Maximum moment, at $x = a$, $M_{max} = \frac{Pab}{L}$

Maximum deflection, $D = \frac{Pab(L + b)\sqrt{3a(L + b)}}{27\,EIL}$

## 5. SIMPLE BEAM

*Two equal, concentrated loads, symmetrically placed*

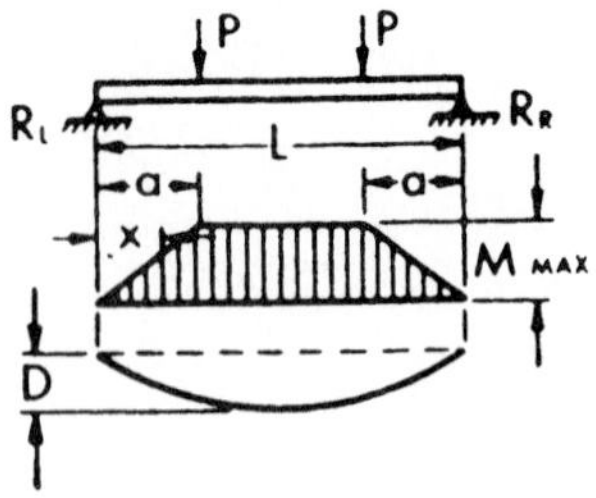

Reactions: $R_L = R_R = P$

Moment at any point:

$$x \leqq a, \; M = R_{LX} = Px$$

$$a \leqq x \leqq (L - a), \; M = Pa$$

$$x \geqq (L - a), \; M = P(L - x)$$

Maximum moment, $M_{max} = Pa$

Maximum deflection, $D = \frac{Pa}{24\,EI}(3L^2 - 4a^2)$

## 6. SIMPLE BEAM

*Uniform Load, w per unit of length, total load W*

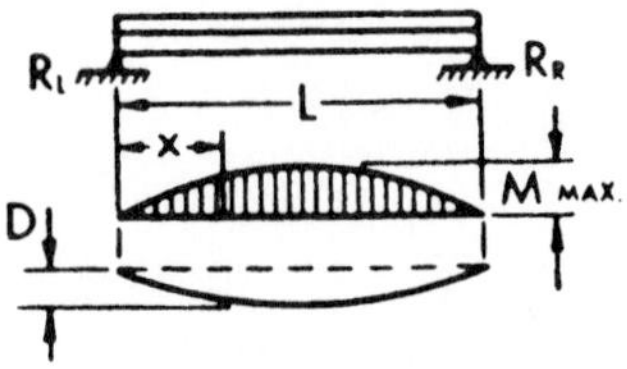

Reactions: $R_L = R_R = \frac{wL}{2} = \frac{W}{2}$

Moment at any point:

$$M = \frac{wx(L - x)}{2} = \frac{Wx(L - x)}{2L}$$

Maximum moment, at center, $M_{max} = \frac{wL^2}{8} = \frac{WL}{8}$

Maximum deflection, $D = \frac{5wL^4}{384\,EI} = \frac{5WL^3}{384\,EI}$

**FIGURE 5.24** (*Continued*)

**7. SIMPLE BEAM**

*Uniform Load, w per unit of length, on part of span*

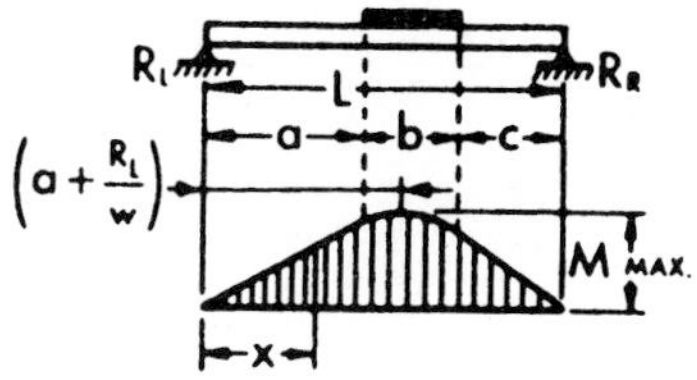


Reactions: $R_L = \frac{bw(2c+b)}{2L}$, $R_R = \frac{bw(2a+b)}{2L}$

Moment at any point:

$$x \leqq a,\ M = R_{LX} = \frac{bwx(2c+b)}{2L}$$

$$a \leqq x \leqq (a+b),\ M = R_{LX} - \frac{(x-a)^2 w}{2}$$

$$x \geqq (a+b),\ M = R_R\,(L-x)$$

Maximum moment, $M_{max} = R_L\left(a + \frac{R_L}{2w}\right)$

**8. BEAM FIXED AT ONE END, SIMPLE SUPPORT AT OTHER**

*Concentrated Load at any point*

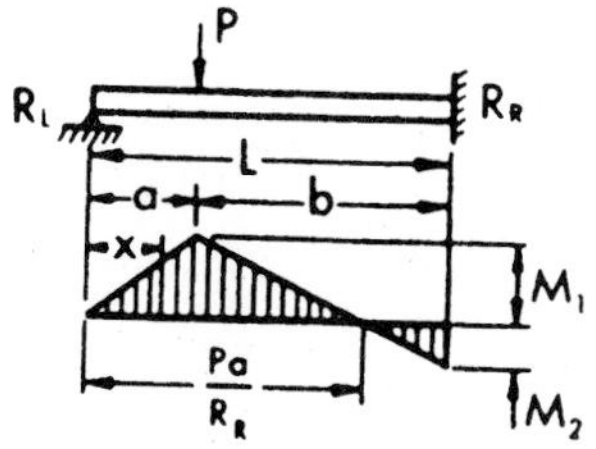


Reactions: $R_L = \frac{Pb^2}{2L^3}(2L+a)$, $R_R = P - R_L$

Moment at any point:

$$x \leqq a,\ M = R_{LX} = \frac{Pb^2x}{2L^3}(2L+a)$$

$$x \geqq a,\ M = R_{LX} - P(x-a)$$

Moment at $x = L$, $M_2 = \frac{-Pab}{2L^2}(L+a)$

Moment at $x = a$, $M_1 = \frac{Pab^2}{2L^3}(2L+a)$

**9. BEAM FIXED AT ONE END, SIMPLE SUPPORT AT OTHER**

*Uniform Load, w per unit of length, total load W*

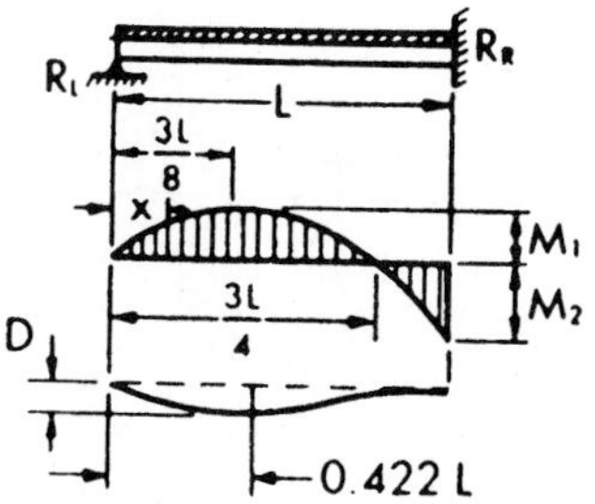


Reactions: $R_L = \frac{3wL}{8}$, $R_R = \frac{5wL}{8}$

Moment at any point:

$$x \leqq L,\ M = wx\left(\frac{3L}{8} - \frac{x}{2}\right)$$

Moment at $x = \frac{3L}{8}$, $M_1 = \frac{9wL^2}{128}$

Maximum moment, $x = L$, $M_2 = \frac{-wL^2}{8}$

Maximum deflection, $D = 0.00542\frac{wL^4}{EI} = 0.00542\frac{WL^3}{EI}$

**FIGURE 5.24** *(Continued)*

**10. BEAM FIXED AT BOTH ENDS**

*Concentrated Load at center*

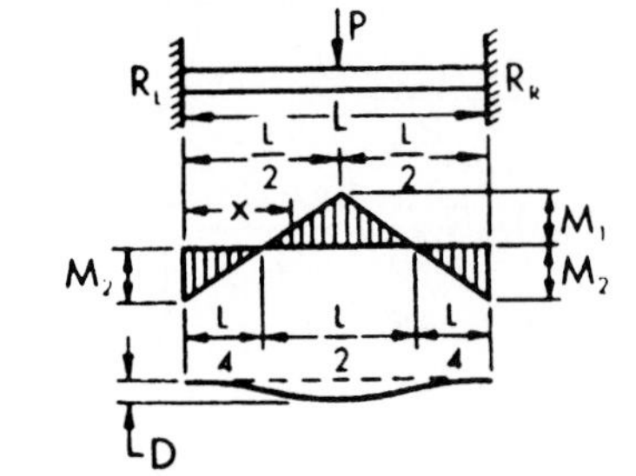


Reactions: $R_L = R_R = \dfrac{P}{2}$

Moment at any point:

$$x \leqq \frac{L}{2},\ M = \frac{-P}{2}\left(\frac{L}{4} - x\right)$$

$$x \geqq \frac{L}{2},\ M = \frac{P}{2}\left(\frac{3L}{4} - x\right)$$

Maximum moment:

$$x = 0 \text{ and } x = L,\ M_2 = \frac{-PL}{8}$$

$$x = \frac{L}{2},\ M_1 = \frac{PL}{8}$$

Maximum deflection, $D = \dfrac{PL^3}{192\ EI}$

**11. BEAM FIXED AT BOTH ENDS**

*Concentrated Load at any point*

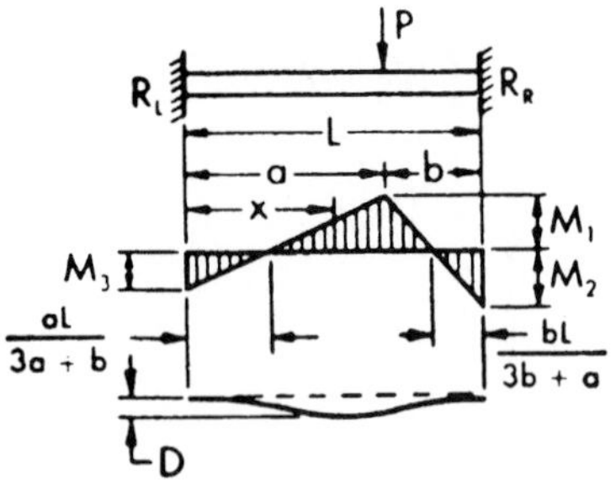


Reactions: $R_L = \dfrac{Pb^2}{L^3}(L + 2a)$, $R_R = \dfrac{Pa^2}{L^3}(L + 2b)$

Moment at any point:

$$x \leqq a,\ M = \frac{-Pab^2}{L^2} + R_{L}x$$

$$x \geqq a,\ M = \frac{-Pa^2b}{L^2} + R_R(L - x)$$

Moment at $x = 0$, $M_3 = \dfrac{-Pab^2}{L^2}$

Moment at $x = a$, $M_1 = \dfrac{2Pa^2b^2}{L^3}$

Maximum moment, $a \geqq b$, $M_2 = \dfrac{-Pa^2b}{L^2}$

Maximum deflection, $a \geqq b$, $D = \dfrac{2Pa^3b^2}{3\ EI\ (3a + b)^2}$

**12. BEAM FIXED AT BOTH ENDS**

*Uniform Load, w per unit of length, total load W*

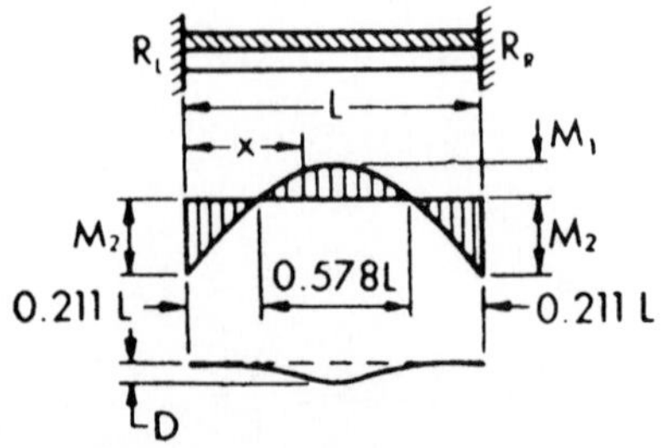


Reactions: $R_L = R_R = \dfrac{wL}{2} = \dfrac{W}{2}$

Moment at any point:

$$x \leqq L,\ M = \frac{-wL^2}{12} - \frac{wx^2}{2} + \frac{wLx}{2}$$

Maximum moment, $x = 0$ and $x = L$, $M_2 = \dfrac{-wL^2}{12} = \dfrac{-WL}{12}$

Moment at $x = \dfrac{L}{2}$, $M_1 = \dfrac{wL^2}{24} = \dfrac{WL}{24}$

Maximum deflection, $D = \dfrac{wL^4}{384\ EI} = \dfrac{WL^3}{384\ EI}$

**FIGURE 5.24** *(Continued)*

**13. BEAM**

*Constant Moment, M, applied at ends*

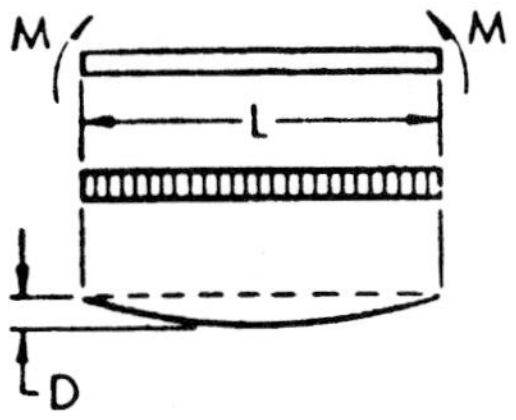


Moment at any point $= M$

Maximum deflection, $D = \dfrac{ML^2}{8\,EI}$

**14. SIMPLE BEAM**

*Moment applied at one end*

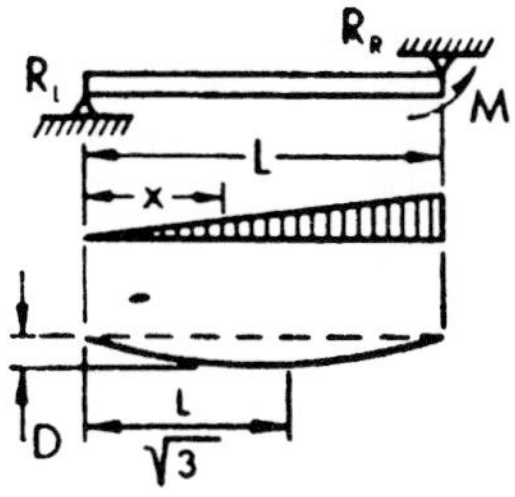


Reactions: $R_L = -R_R = \dfrac{M}{L}$

Moment at any point $= R_{LX}$

Maximum moment $= M$

Maximum deflection, $D = 0.0642\,\dfrac{ML^2}{8\,EI}$

**15. CONTINUOUS BEAM, TWO EQUAL SPANS**

*Concentrated Load, P, at any point*

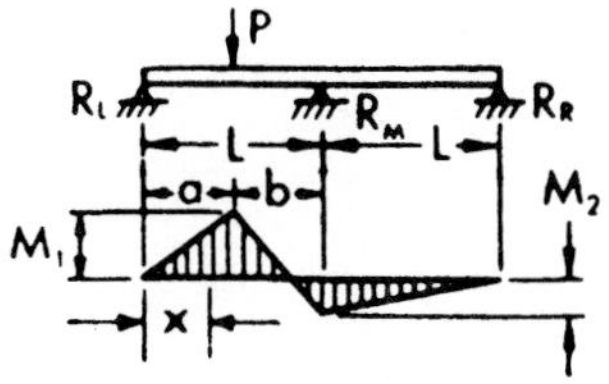


Reactions:

$$R_L = \frac{Pb}{4L^3}[4L^2 - a(L + a)]$$

$$R_M = \frac{Pa}{2L^3}(3L^2 - a^2)$$

$$R_R = \frac{Pa}{4L^3}(L^2 - a^2)$$

Moment at any point:

$x \leqq a,\ M = R_{LX}$
$a \leqq x \leqq L,\ M = R_{LX} - P(x - a)$
$x \geqq L,\ M = -R_R\,(2L - x)$

Moment at $x = L$, $M_2 = -R_RL$

Maximum moment, $M_1 = R_La$

**FIGURE 5.24** (*Continued*)

## 5.16 COMBINED STRESSES

See Chap. 8 on machine elements.

## 5.17 PRESSURE VESSELS

See Chap. 9 on basic pneumatics and hydraulics.

## 5.18 EYEBOLT SAFE LOADS

Referring to Fig. 5.25

$$P = \frac{20{,}000\, d^3}{DK} \qquad d = \sqrt[3]{\frac{PDK}{20{,}000}}$$

Values of $K$ for ratios of $a/D$ are as follows:

| $a/D$ | 0.2 | 0.3 | 0.4 | 0.5 | 0.6 | 0.7 | 0.8 |
|---|---|---|---|---|---|---|---|
| $K$ | 0.98 | 0.97 | 0.96 | 0.95 | 0.93 | 0.92 | 0.90 |

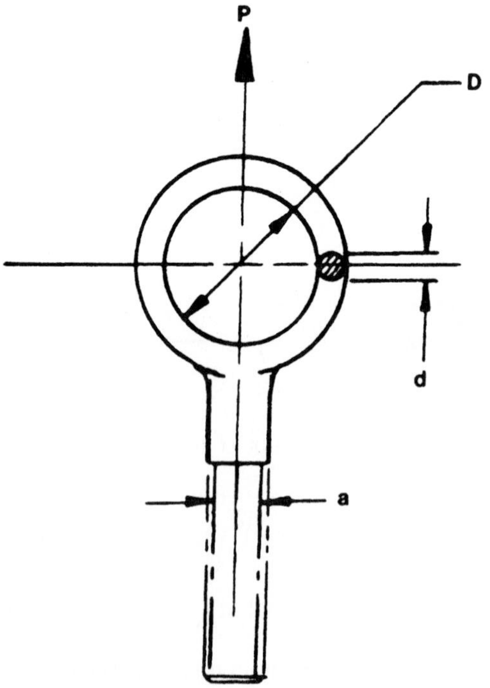


**FIGURE 5.25** Eyebolt.

where $P$ = safe load, lb
$D$ = inside diameter of eye, in
$d$ = ring section diameter, in
$a$ = root diameter of thread, in

This procedure can be used when the eyebolt does not have a listed safe load in the eyebolt catalog, or as a verification of the rating when the load is listed.

## 5.19 HARDNESS CONVERSION EQUIVALENTS

The most common hardness scales used in design engineering practice are the Rockwell and Brinell scales, for which equivalences are shown in Table 5.1.

## 5.20 SPECIAL JOINT APPLICATIONS

Shown here are three common types of bolted joints that occur frequently in structural and machine-design practice.

***Bracket-Type Joint.*** This special case is illustrated in Fig. 5.26. The $K$ Factor can be expressed as

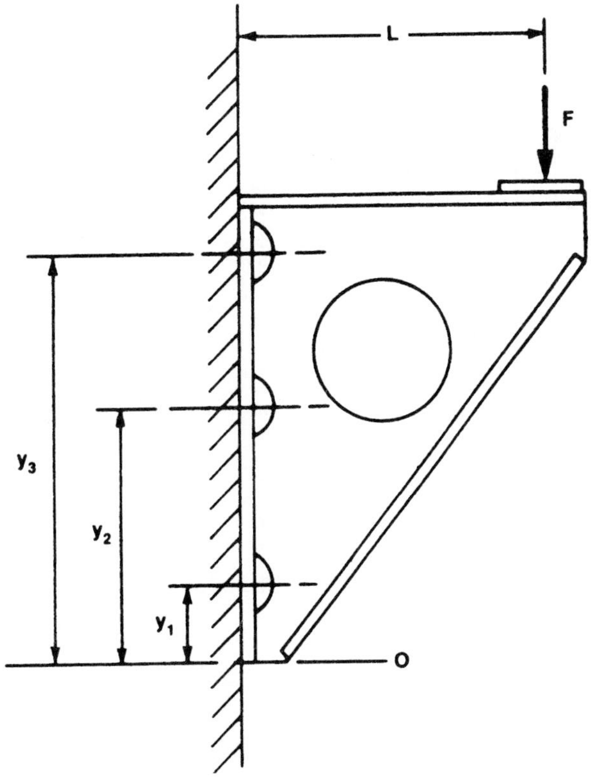


**FIGURE 5.26** Structural bracket.

**TABLE 5.1** Hardness Conversion Table (Rockwell and Brinell)

| Hardened Steel and Hard Alloys | | | | |
|---|---|---|---|---|
| SCALE | | | | |
| C | A | 15-N | 30-N | Brinnell |
| LOAD (Kg) | | | | |
| 150 | 60 | 15 | 30 | 3000 |
| 80 | 92.0 | 96.5 | 92.0 | – |
| 78 | 91.0 | 96.0 | 91.0 | – |
| 76 | 90.0 | 95.5 | 90.0 | – |
| 74 | 89.0 | 95.0 | 88.5 | – |
| 72 | 88.0 | 94.5 | 87.0 | – |
| 70 | 86.5 | 94.0 | 86.0 | – |
| 68 | 85.5 | – | 84.5 | – |
| 66 | 84.5 | 92.5 | 83.0 | – |
| 64 | 83.5 | – | 81.0 | – |
| 62 | 82.5 | 91.0 | 79.0 | – |
| 60 | 81.0 | 90.0 | 77.5 | 614 |
| 58 | 80.0 | – | 75.5 | 587 |
| 56 | 79.0 | 88.5 | 74.0 | 560 |
| 54 | 78.0 | 87.5 | 72.0 | 534 |
| 52 | 77.0 | 86.5 | 70.5 | 509 |
| 50 | 76.0 | 85.5 | 68.5 | 484 |
| 49 | 75.5 | 85.0 | 67.5 | 472 |
| 48 | 74.5 | 84.5 | 66.5 | 460 |
| 47 | 74.0 | 84.0 | 66.0 | 448 |
| 46 | 73.5 | 83.5 | 65.0 | 437 |
| 45 | 73.0 | 83.0 | 64.0 | 426 |
| 44 | 72.5 | 82.5 | 63.0 | 415 |
| 43 | 72.0 | 82.0 | 62.0 | 404 |
| 42 | 71.5 | 81.5 | 61.5 | 393 |
| 41 | 71.0 | 81.0 | 60.5 | 382 |
| 40 | 70.5 | 80.5 | 59.5 | 372 |
| 39 | 70.0 | 80.0 | 58.5 | 362 |
| 38 | 69.5 | 79.5 | 57.5 | 352 |
| 37 | 69.0 | 79.0 | 56.5 | 342 |
| 36 | 68.5 | 78.5 | 56.0 | 332 |
| 35 | 68.0 | 78.0 | 55.0 | 322 |
| 34 | 67.5 | 77.0 | 54.0 | 313 |
| 33 | 67.0 | 76.5 | 53.0 | 305 |
| 32 | 66.5 | 76.0 | 52.0 | 297 |
| 31 | 66.0 | 75.5 | 51.5 | 290 |
| 30 | 65.5 | 75.0 | 50.5 | 283 |
| 29 | 65.0 | 74.5 | 49.5 | 276 |
| 28 | 64.5 | 74.0 | 48.5 | 270 |
| 27 | 64.0 | 73.5 | 47.5 | 265 |
| 26 | 63.5 | 72.5 | 47.0 | 260 |
| 25 | 63.0 | 72.0 | 46.0 | 255 |
| 24 | 62.5 | 71.5 | 45.0 | 250 |
| 23 | 62.0 | 71.0 | 44.0 | 245 |
| 22 | 61.5 | 70.5 | 43.0 | 240 |
| 21 | 61.0 | 70.0 | 42.5 | 235 |
| 20 | 60.5 | 69.5 | 41.5 | 230 |

| Unhardened Steel, Soft Tempered Steel Gray and Malleable Cast Iron and most Non-ferrous metals | | |
|---|---|---|
| SCALE | | |
| B | Brinnell | |
| LOAD (Kg) | | |
| 100 | 500 | 3000 |
| 100 | 201 | 240 |
| 98 | 189 | 228 |
| 96 | 179 | 216 |
| 94 | 171 | 205 |
| 92 | 163 | 195 |
| 90 | 157 | 185 |
| 88 | 151 | 176 |
| 86 | 145 | 169 |
| 84 | 140 | 162 |
| 82 | 135 | 156 |
| 80 | 130 | 150 |
| 78 | 126 | 144 |
| 76 | 122 | 139 |
| 74 | 118 | 135 |
| 72 | 114 | 130 |
| 70 | 110 | 125 |
| 68 | 107 | 121 |
| 66 | 104 | 117 |
| 64 | 101 | 114 |
| 62 | 98 | 110 |
| 60 | 95 | 107 |
| 58 | 92 | 104 |
| 56 | 90 | 101 |
| 54 | 87 | – |
| 52 | 85 | – |
| 50 | 83 | – |
| 48 | 81 | – |
| 44 | 78 | – |
| 42 | 76 | – |
| 38 | 73 | – |
| 34 | 70 | – |
| 30 | 67 | – |
| 28 | 66 | – |
| 26 | 65 | – |
| 25 | 64 | – |
| 23 | 63 | – |
| 21 | 62 | – |
| 19 | 61 | – |
| 17 | 60 | – |
| 15 | 59 | – |
| 13 | 58 | – |
| 10 | 57 | – |
| 7 | 56 | – |
| 5 | 55 | – |
| 2 | 54 | – |
| 0 | 53 | – |

$$K = \frac{y_1^2 + y_2^2 + y_3^2}{y_3} \qquad KA = Z$$

where $A$ = stress area of one fastener, in$^2$ or mm$^2$

$Z$ = section modulus, in$^3$ or mm$^3$

$y_1$ to $y_3$ = distances from 0 reference point, in or mm

Then, the tensile stress in one bolt = $Fl/Z$, in psi or N/mm$^2$, where $F$ = force, lb or N; and $l$ = distance of force arm, in or mm.

***Shear Joints Eccentrically Loaded (Symmetrical).*** This type of joint, secured by threaded fasteners, rivets, or pins, should be carefully designed using the following procedures. Simplified or arbitrary selection of fastener locations in a heavily loaded joint of this nature can result in fastener failure.

The procedure for analyzing this type of joint consists of

**1.** Locating the centroid of the bolt pattern.
**2.** Finding the reaction moment force on a given fastener.
**3.** Finding the primary shear force on one fastener.
**4.** Combining the primary and secondary shear forces vectorially.
**5.** Computing the shear stress within a particular fastener from

$$\tau = \frac{F_R}{A_n}$$

where $\tau$ = shear stress, psi or MPa
$F_R$ = resultant of primary and secondary shear forces, lb or kN
$A_n$ = cross-sectional area of $n$th fastener, in$^2$ or mm$^2$

See Figs. 5.27 and 5.28*a* for location of the centroid.

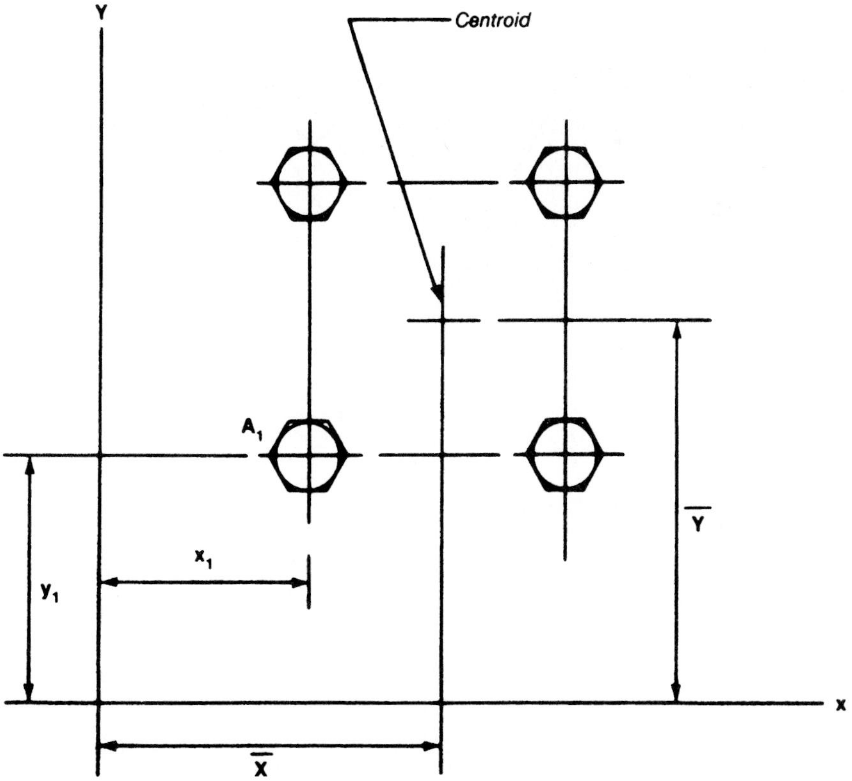


**FIGURE 5.27** Location of the centroid.

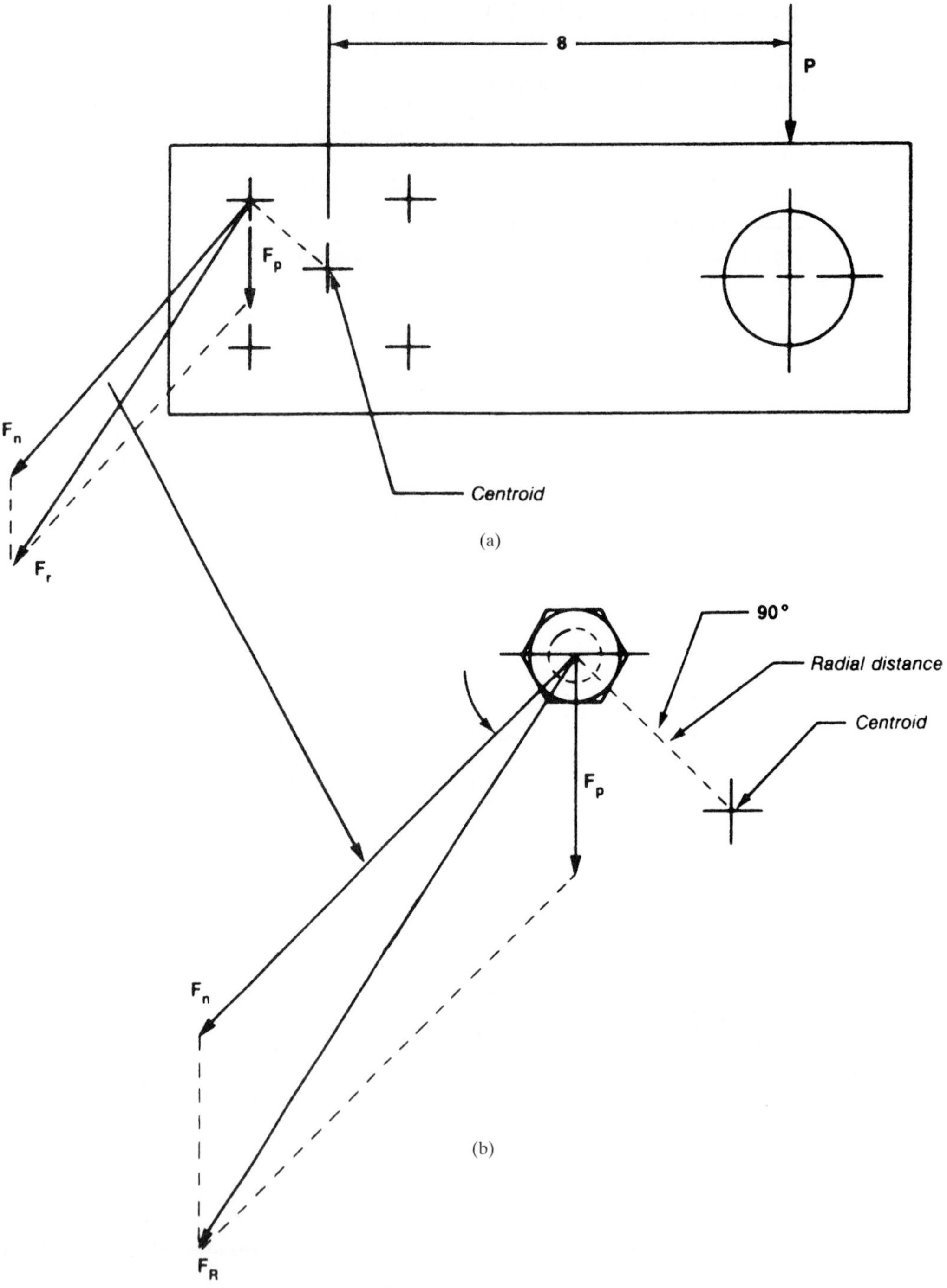


**FIGURE 5.28** (*a*) Location of the centroid; (*b*) solving the vector diagram.

Locate the reference axes arbitrarily. Then

$$\bar{x} = \frac{A_1x_1 + A_2x_2 + \cdots + A_nx_n}{A_1 + A_2 + \cdots + A_n}$$

$$\bar{y} = \frac{A_1y_1 + A_2y_2 + \cdots + A_ny_n}{A_1 + A_2 + \cdots + A_n}$$

where $x_1$ and $y_1$ are distances in inches or millimeters from the origin of the arbitrary axes.

Note that the centroid for simple, symmetrical patterns is the geometric center.

The reaction moment force is then (see Fig. 5.28*a* and 5.28*b*)

$$F_n = \frac{Mr_n}{r_1^2 + r_2^2 + r_3^2 + \cdots + r_n^2}$$

where $F_n$ = reaction moment force on a given bolt, lb or N

$r_n$ = radial distance to *n*th bolt, in or mm

$r_1, r_2, \ldots, r_n$ = radial distances from centroid to the individual fasteners, in or mm

$M$ = moment exerted on the shear joint (load × distance from application point of load to the centroid of the bolt or fastener pattern), lb·in or N·mm

***Shear Joint Eccentrically Loaded (Unsymmetrical).*** This case applies to the lifting bar shown in Fig. 5.29, which uses ½-in-diameter bolts.

Using the equations and symbols shown in the preceding case, we first determine the location of the centroid.

$$\bar{x} = \frac{(0.196)(1) + (0.196)(3.5) + (0.196)(6) + (0.196)(2.5) + (0.196)(4.75)}{5(0.196)} = \frac{3.43}{0.98} = 3.5 \text{ in}$$

$$\bar{y} = \frac{3(0.196)(1) + 2(0.196)(3)}{5(0.196)} = \frac{1.764}{0.98} = 1.8 \text{ in}$$

Reaction moment force for the worst-case bolt (Fig. 5.30) is

$$F_n = \frac{M(r_n)}{r_1^2 + r_2^2 + r_3^2 + r_4^2 + r_5^2}$$

$$M = 5000 \times 8$$

$$F_n = \frac{40{,}000(2.625)}{(2.625)^2 + (0.8)^2 + (2.265)^2 + (1.732)^2 + (1.732)^2} = \frac{105{,}000}{20.42} = 5142 \text{ lb}$$

Note that the primary, vertical shear force $F_v$ is the total load divided by the number of bolts; 5000/5 = 1000 lb.

Now find the resultant force $F_R$ by solving triangle $A_1$, $F_R$, $F_n$ shown in Fig. 5.31. Use the law of cosines; $F_R = 5687$ lb. Then find the shear stress in the worst-case bolt from

$$\tau = \frac{F_R}{A_n} = \frac{5687}{0.196} = 29{,}015 \text{ psi}$$

An ASTM A325 bolt of ½-in diameter will suffice in this application if no threads on the bolt are in the shear planes and the load is static. This joint configuration would not be suitable for a dynamically loaded condition, because the shear stress is too close to the allowable shear stress for dynamic loads. Specification of ⅝- to ¾-in-diameter bolts would be indicated in this case, and the joint would require another analysis for verification.

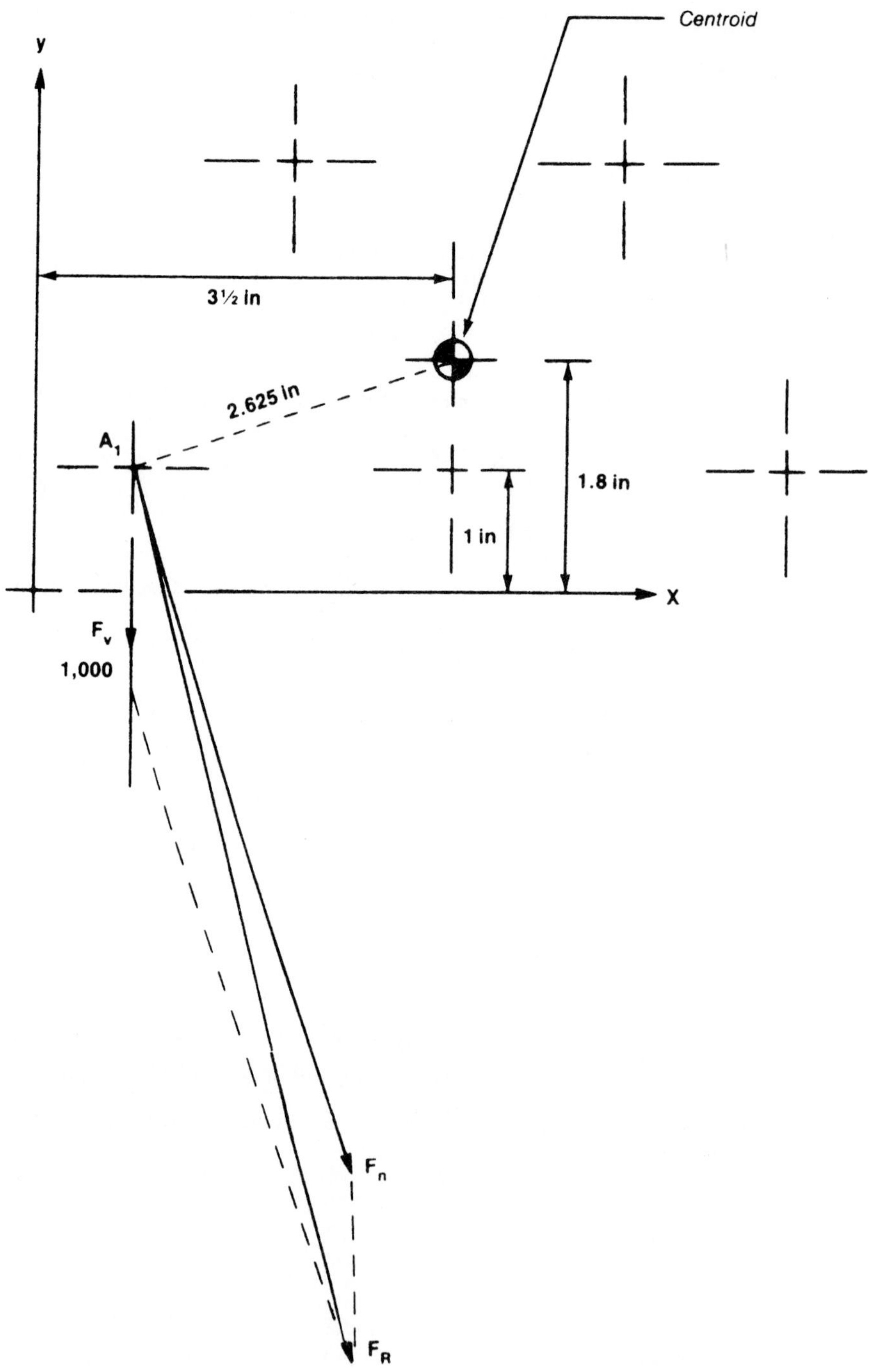


**FIGURE 5.29** Lifting-bar vector diagram.

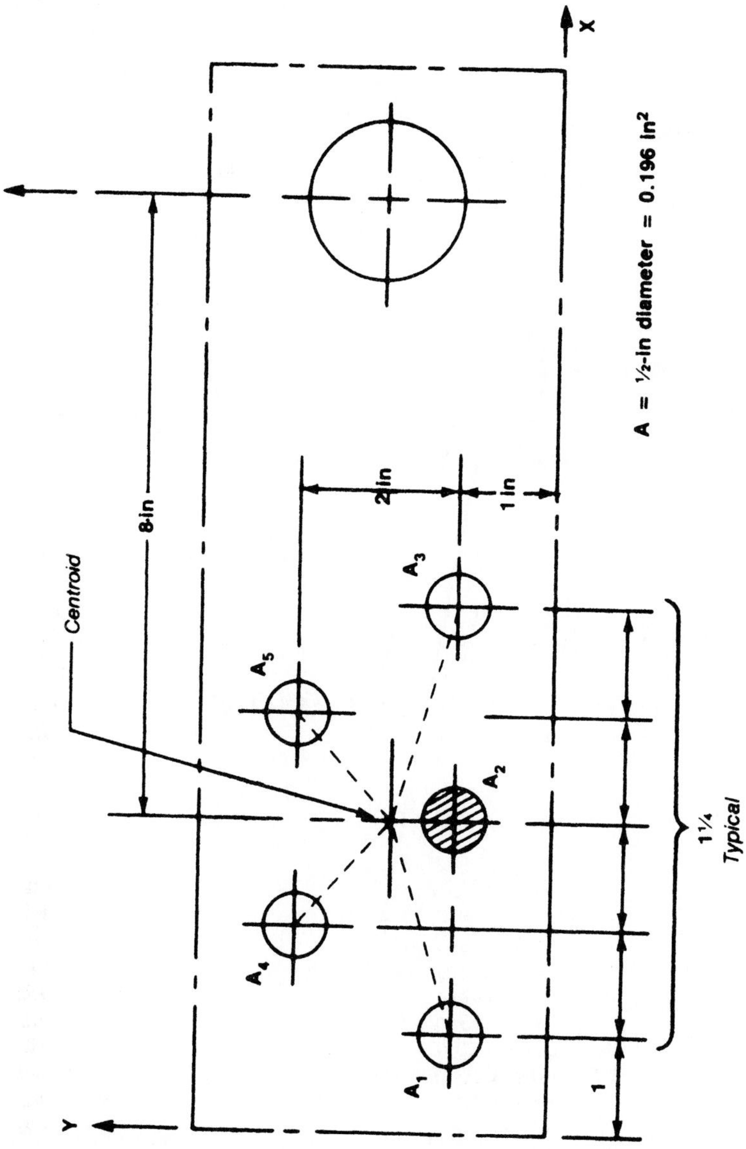


**FIGURE 5.30** Five-bolt lifting bar.

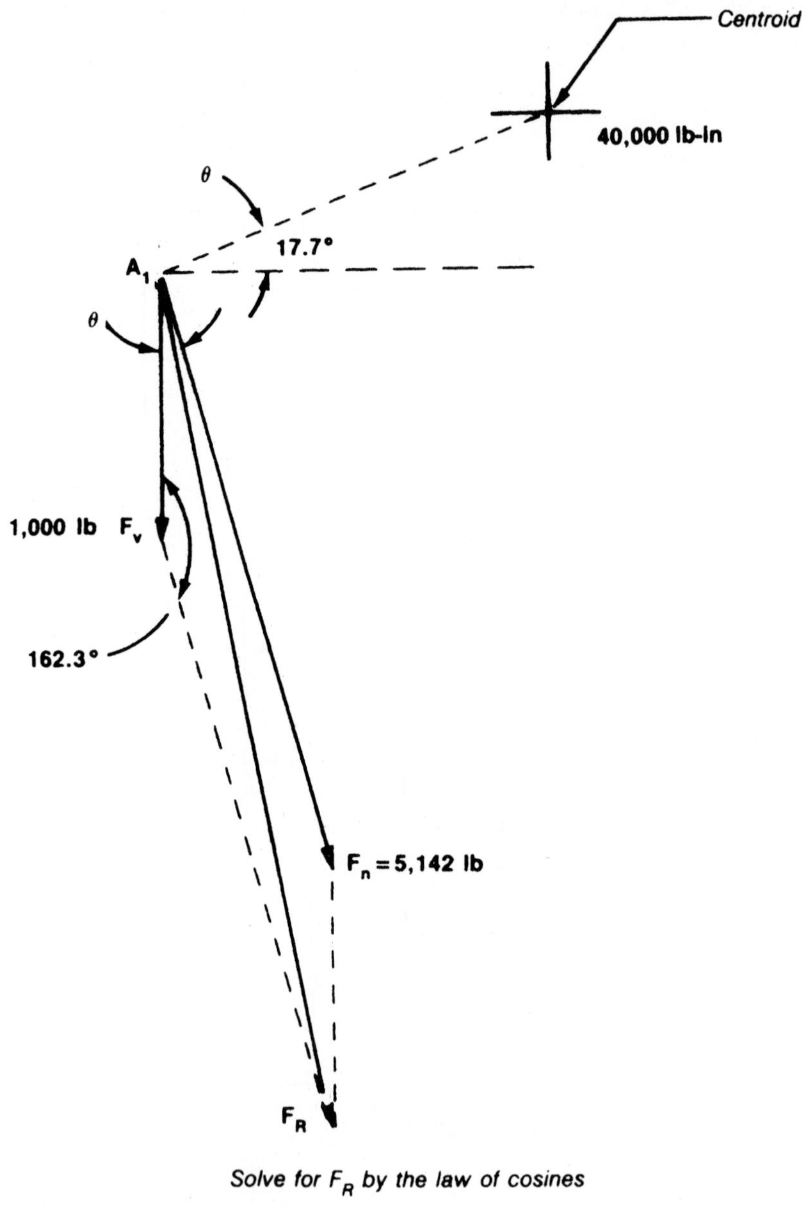


**FIGURE 5.31** Finding the resultant $F_R$.

Observe that reversal of the lifting bar shown in Fig. 5.30, so that the three bolts are above the two bolts, will make no difference in the final shear stress imposed on the worst-case bolt. This result holds even though the centroid shifts to a higher position relative to the $X$ axis. The radial distance $r_n$ will remain the same (2.625 in), as will all the other radial distances from the centroid to the bolt centerlines.

## 5.21 CONVERTING BEAM EQUATIONS FOR MAXIMUM STRESS AT ANY POINT

The bending moment equations for all the beam equations shown in Figs. 5.23 and 5.24 may be converted to find the maximum stress at any point, or the point shown, by including the *section modulus* in the divisor of the bending moment equation selected.

*Example.* In the case 1 simple beam of Fig. 5.23 in the beam equations, the maximum moment at the center is $wl^2/8$, so the stress at this point therefore is

$$s = \frac{wl^2}{8Z} \quad \text{or} \quad \frac{Wl}{8Z}$$

where $s$ = stress, lb/in$^2$
$w$ = load in lb/in of beam length
$W$ = total uniform load in lbs across the span of the beam

Also the moment at any point

$$M_x = \frac{wx}{2}(1 - x)$$

may be converted for the stress at any point by

$$s = \frac{wx}{2Z}(1 - x)$$

Note the section modulus $Z$ in the divisor. [*Note:* The beam equations (Figs. 5.23 and 5.24) and properties of geometric sections (Fig. 5.22) may be used to calculate structures, machine elements, sheet-metal shapes, and composite sections that are used in many design applications.]

## 5.22 PROPERTIES OF STRUCTURAL SECTIONS

Figure 5.32 shows the equations for calculating area, neutral axis, moment of inertia, section modulus, and radius of gyration for various structural and machine element shapes.

## FURTHER READING

*Structural Handbook,* 1960: Pittsburgh, Pa.: Aluminum Company of America.

*Manual of Steel Construction,* 8th ed. 1984: Chicago, Ill.: American Institute of Steel Construction (AISC).

Eshbach, O. W., and M. Souders, 1975: *Handbook of Engineering Fundamentals,* 3rd ed. New York: Wiley.

Najder, K. W., 1958: *Machine Designer's Guide,* 4th ed. Ann Arbor, Mich.: Edwards Bros.

Oberg, E., F. Jones, and H. Horton, 1980: *Machinery's Handbook,* 21st ed. New York: Industrial Press.

Walsh, Ronald A., 1994, *McGraw-Hill Machining and Metalworking Handbook.* New York: McGraw-Hill.

# Area, Neutral Axis, Moment of Inertia, Section Modulus & Radius of Gyration

| SECTION | Area | Neutral Axis | Moment of Inertia | Section Modulus | Radius of Gyration |
|---|---|---|---|---|---|
| 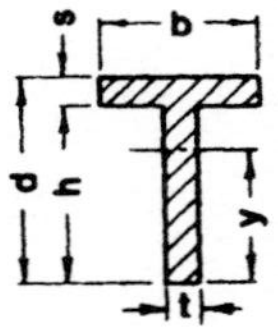 | $bs + ht$ | $d - \frac{d^2t + s^2(b - t)}{2(bs + ht)}$ | $\frac{1}{3}\left[ty^3 + b(d - y)^3 - (b - t)(d - y - s)^3\right]$ | $\frac{I}{y}$ | $\sqrt{\frac{I}{3(bs + ht)}}$ <br> $\left[ty^3 + b(d - y)^3 - (b - t)(d - y - s)^3\right]$ |
| 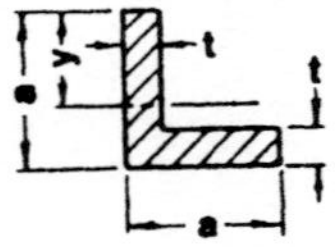 | $t(2a - t)$ | $a - \frac{a^2 + at - t^2}{2(2a - t)}$ | $\frac{1}{3}\left[ty^3 + a(a - y)^3 - (a - t)(a - y - t)^3\right]$ | $\frac{I}{y}$ | $\sqrt{\frac{I}{A}}$ |
| 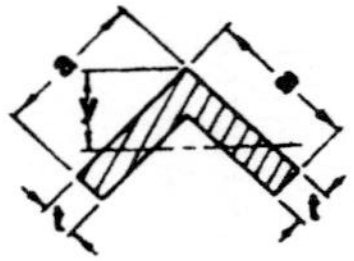 | $t(2a - t)$ | $\frac{a^2 + at - t^2}{2(2a - t)\cos 45}$ | $\frac{1}{3}\left[2x^4 - 2(x - t)^4 + t\left[a - (2x - 0.5t)^3\right]\right]$ <br> $x = \frac{a^2 + at - t^2}{2(2a - t)}$ | $\frac{I}{y}$ | $\sqrt{\frac{I}{A}}$ |
| 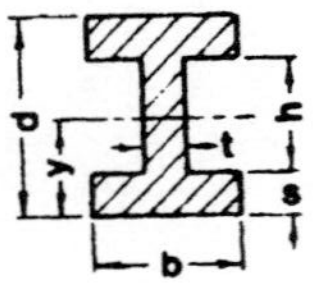 | $bd - h(b - t)$ | $\frac{d}{2}$ | $\frac{bd^3 - h^3(b - t)}{12}$ | $\frac{bd^3 - h^3(b - t)}{6d}$ | $\sqrt{\frac{bd^3 - h^3(b - t)}{12[bd - h(b - t)]}}$ |

**FIGURE 5.32** Properties of structural sections.

## Area, Neutral Axis, Moment of Inertia, Section Modulus & Radius of Gyration - (Continued)

| SECTION | Area | Neutral Axis | Moment of Inertia | Section Modulus | Radius of Gyration |
|---|---|---|---|---|---|
| 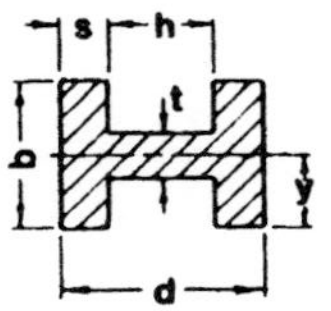 | $bd - h(b - t)$ | $\frac{b}{2}$ | $\frac{2sb^3 + ht^3}{12}$ | $\frac{2sb^3 + ht^3}{6b}$ | $\sqrt{\frac{2sb^3 - h^3(b - t)}{12[bd - h(b - t)]}}$ |
| 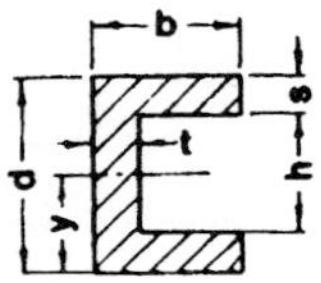 | $bd - h(b - t)$ | $\frac{d}{2}$ | $\frac{bd^3 - h^3(b - t)}{12}$ | $\frac{bd^3 - h^3(b - t)}{6d}$ | $\sqrt{\frac{bd^3 - h^3(b - t)}{12[bd - h(b - t)]}}$ |
| 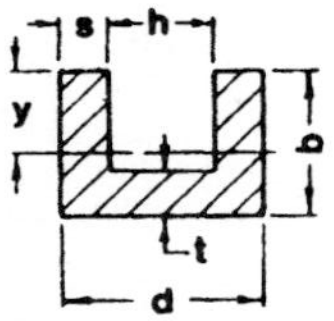 | $bd - h(b - t)$ | $b - \frac{2b^2s + ht^2}{2bd - 2h(b - t)}$ | $\frac{2sb^3 + ht^3}{3} - A(b - y)^2$ | $\frac{I}{y}$ | $\sqrt{\frac{I}{A}}$ |
| 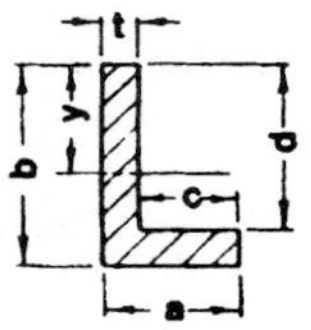 | $t(a + b - t)$ | $b - \frac{t(2d + a) + d^2}{2(d + a)}$ | $\frac{1}{3}[ty^3 + a(b - y)^3$<br>$- (a - t)(b - y - t)^3]$ | $\frac{I}{y}$ | $\sqrt{\frac{I}{3t(a + b - t)}}$<br>$[ty^3 + a(b - y)^3 -$<br>$(a - t)(b - y - t)^3]$ |

**FIGURE 5.32** *(Continued)*

## Area, Neutral Axis, Moment of Inertia, Section Modulus & Radius of Gyration - (Continued)

| SECTION | Area | Neutral Axis | Moment of Inertia | Section Modulus | Radius of Gyration |
|---|---|---|---|---|---|
| 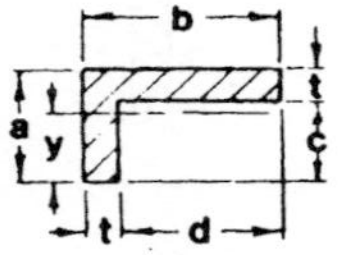 | $t(a + b - t)$ | $a - \frac{t(2c + b) + c^2}{2(d + a)}$ | $\frac{1}{3}\Big[ty^3 + b(a - y)^3$ $- (b - t)(a - y - t)^3\Big]$ | $\frac{I}{y}$ | $\sqrt{\frac{I}{3t(a + b - t)}}$ $\Big[ty^3 + b(a - y)^3 -$ $(b - t)(a - y - t)^3\Big]$ |
| 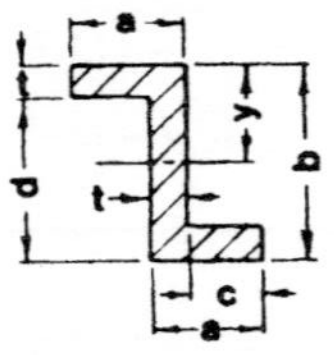 | $t\left[b + 2(a - t)\right]$ | $\frac{b}{2}$ | $\frac{ab^3 - c(b - 2t)^3}{12}$ | $\frac{ab^3 - c(b - 2t)^3}{6b}$ | $\sqrt{\frac{ab^3 - c(b - 2t)^3}{12t\left[b + 2(a - t)\right]}}$ |
| 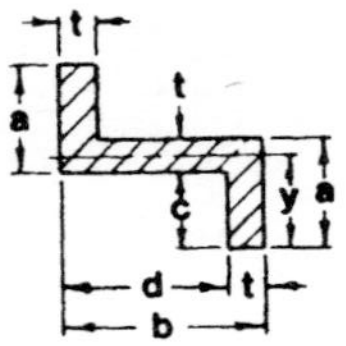 | $t\left[b + 2(a - t)\right]$ | $\frac{2a - t}{2}$ | $\frac{b(a + c)^3 - 2c^3d - 6a^2cd}{12}$ | $\frac{b(a + c)^3 - 2c^3d - 6a^2cd}{6(2a - t)}$ | $\sqrt{\frac{b(a + c)^3 - 2c^3d - 6a^2cd}{12t\left[b + 2(a - t)\right]}}$ |
| | | | | | |

**FIGURE 5.32** *(Continued)*

# CHAPTER 6

# ELECTRICAL AND ELECTRONIC ENGINEERING PRACTICES AND DESIGN DATA

This chapter of the handbook provides data and design procedures for the electrical and electronic aspects of electromechanical design practices. Industrial product designers and developers are required to perform all aspects of the design of a product in many small- and medium-sized companies today. A product designer who is knowledgeable in all the disciplines of electromechanical design practices is an asset to any company or organization.

Specialists in the various engineering disciplines are required to assist the electromechanical design engineers in formulating specific, specialized engineering functions and data for use by the electromechanical designer.

Electromechanical design practices include the integration of mechanical, electrical, and electronic engineering principles in the design of a consumer or industrial product. The term *mechatronics* has recently been used to indicate the same functions encountered in electromechanical design practices. Those individuals who use this term or try to imply that this is a new engineering subject should be aware of the fact that electromechanical design engineering practices have been used in the United States since the birth of electrical power usage more than 100 years ago. Since it is, in general, extremely difficult to separate electrical and electronic design practices for all applications, the term *mechatronics* should not be used to indicate the electromechanical design function. It must be understood that electrical components are always used in electronic systems, which also include purely electronic components. The integrated circuit itself, strictly speaking, is an electromechanical system component and is also considered an electronic component.

This section of the handbook will therefore be very useful to those individuals trained or currently being trained in the electromechanical disciplines or who use these principles on a daily or occasional basis.

## 6.1 TERMS, CONVERSIONS, AND CONSTANTS

### 6.1.1 Common Electrical Terms

**Ampere (A)** A unit of current or rate of flow of electricity equal to volts/ohms (Ohm's law).

**Coulomb (C)** Unit of electricity equal to the quantity that passes a cross section of a conductor in one second when the current is one ampere.

**Dielectric constant of an insulating material** Ratio of capacitance of a condenser having the material as the dielectric to the capacitance of the same condenser with vacuum as dielectric.

**Dielectric strength of an insulating material** Maximum voltage or potential gradient that the material can withstand without rupture.

**Farad (F)** Unit of electric capacitance = that capacity whose potential will be raised one volt by the addition of the charge of one coulomb. The practical unit is the microfarad (μF).

**Henry (H)** Unit of inductance = inductance of a closed circuit in which an electromotive force of one volt is produced when the electric current traversing the circuit varies uniformly at the rate of one ampere per second.

**Hertz (Hz)** The frequency of an alternating current in cycles per second.

**Kilowatt (kW)** 1000 watts (W).

**Kilowatthour (kWh)** 1000 watthours (Wh).

**Kilovolt-ampere (kVA)** 1000 volt-amperes (VA).

**Mho** Unit of electric conductance = 1/ohm (reciprocal ohm). Also called *siemen* (S).

**Ohm (Ω)** Unit of electric resistance.

**Power factor** The ratio of active power to apparent power.

**Specific inductive capacity (SIC)** Dielectric constant = permitivity.

**Volt (V)** Unit of electromotive force.

**Watt (W)** Volts × amperes (dc); volts × amperes × power factor (ac). Volt-amperes is the *apparent* power.

**Watthour (Wh)** Volts × amperes × hours (direct current); volts × amperes × power factor × hours (alternating current).

### 6.1.2 Physical Constants

| | |
|---|---|
| Avogadro constant | $6.02252 \times 10^{23}$ mol$^{-1}$ |
| Boltzmann constant | $1.38054 \times 10^{-23}$ J/K |
| Electric constant | $8.85419 \times 10^{-12}$ F/m |
| Elementary charge | $1.60210 \times 10^{-19}$ C |
| Faraday constant | $9.64870 \times 10^{4}$ C/mol |
| Gravitational constant | $6.67000 \times 10^{-11}$ N · m$^2$/kg$^2$ |
| Magnetic constant | $4\pi \times 10^{-7}$ H/m |
| Planck constant | $6.6256 \times 10^{-24}$ Js |
| Speed of electromagnetic waves in vacuum | $2.997925 \times 10^{8}$ m/s |
| Standard acceleration of free fall | 9.80665 m/s$^2$ |

### 6.1.3 Quantities of Electricity, Magnetism, and Light

1 gamma = $10^{-9}$ tesla (T)
1 gauss (G) = $10^{-4}$ T
1 gilbert (Gb) = 0.79577 A
1 maxwell (Mx) = $10^{-8}$ weber (Wb)
1 oersted (Oe) = 79.577 A/m
221 footcandles (fc) = 10.764 lux [lumens (lm)/m$^2$)
1 footlambert (ft-L) = 3.426 candelas/m$^2$
1 candela/ft$^2$ = 10.764 candelas/m$^2$

### 6.1.4 Electrical Conversion Factors

Refer to Table 6.1.

### 6.1.5 Resistivity and Conductivity Conversion

Refer to Table 6.2.

*Examples.* Copper, with a conductivity of 100 percent (International Annealed Copper Standard (IACS), will have a resistivity (mass basis) in microhm-inches (μΩ · in) of

$$\frac{1}{N} \times 603.45 \times \frac{1}{\alpha} = x \qquad \frac{1}{100} \times 603.45 \times \frac{1}{8.89} = x$$

Solving, $x = 0.6788\ \mu\Omega$, or copper conductivity (100 percent IACS) $= 0.679 \times 10^{-6}\ \Omega \cdot \text{in}$.

**TABLE 6.1** Electrical Conversion Factors

| To Convert | To | Multiply By |
|---|---|---|
| Amperes/cm$^2$ | Amperes/in$^2$ | 6.452 |
| Ampere-turns | Gilberts | 1.257 |
| Ampere-turns/cm | Ampere-turns/in | 2.540 |
| Circular mils | Square inches | $7.854 \times 10^{-7}$ |
| Circular mils | Square mils | 0.7854 |
| Gauss | Lines/in$^2$ | 6.452 |
| Gilberts | Ampere-turns | 0.7958 |
| Gilberts/cm | Ampere-turns/in | 2.0210 |
| Joules (Int.) | Btu | $9.480 \times 10^{-4}$ |
| | Ergs | $10^7$ |
| | Foot-pounds | 0.7378 |
| | Watt-hours | $2.778 \times 10^{-4}$ |
| Kilolines | Maxwells | $10^3$ |
| Kilowatts | Btu/min | 56.88 |
| | Ft-lb/min | $4.427 \times 10^{4}$ |
| | Horsepower | 1.341 |
| | Watts | $10^3$ |
| Kilowatt-hours | Btu | 3413 |
| | Ft-lb | $2.656 \times 10^{6}$ |
| | Hp-hours | 1.341 |
| | Joules | $3.6 \times 10^{6}$ |
| Lines/cm$^2$ | Gauss | 1 |
| Lines/in$^2$ | Gauss | 0.1550 |
| Lumens/ft$^2$ | Foot-candles | 1 |
| Maxwells | Kilolines | $10^{-3}$ |
| Megohms | Ohms | $10^6$ |
| Ohms | Megohms | $10^{-6}$ |
| | Microhms | $10^6$ |
| Watts | Btu/min | 0.05688 |
| | Ergs | $10^7$ |
| | Ft-lbs/min | 44.27 |
| | Hp | $1.341 \times 10^{-3}$ |
| | Kilowatts | $10^3$ |
| Watt-hours | Btu | 3.413 |
| | Foot-pounds | 2,656 |
| | Hp-hours | $1.341 \times 10^{-3}$ |
| | Kilowatt-hours | $10^{3}$ |

**TABLE 6.2** Resistivity and Conductivity Conversions

| Given (N) | Perform | To Obtain |
|---|---|---|
| Ohm-circular mil/ft | × 0.065450 | Microhm-inch |
| | × 0.16624 | Microhm-cm |
| | × 0.0016624 × α | Ohm-gram/m$^2$ |
| | 1/N × 1037.1 | % IACS (vol. basis) |
| | 1/N × 9220 × 1/α | % IACS (mass basis) |
| Ohm-mm$^2$/m | × 601.53 | Ohm-circular mil/ft |
| | × 39.37 | Microhm-inch |
| | × 100 | Microhm-cm |
| | N × α | Ohm-gram/m$^2$ |
| | 1/N × 1.7241 | % IACS (vol. basis) |
| | 1/N × 15.328 × 1/α | % IACS (mass basis) |
| Microhm-inch | × 15.279 | Ohm-circular mil/ft |
| | × 0.02540 | Ohm-mm$^2$/m |
| | × 2.540 | Microhm-cm |
| | × 0.02540 × α | Ohm-gram/m$^2$ |
| | 1/N × 67.879 | % IACS (vol. basis) |
| | 1/N × 603.45 × 1/α | % IACS (mass basis) |
| Microhm-cm | × 6.0153 | Ohm-circular mil/ft |
| | × 0.0100a | Ohm-mm$^2$/m |
| | × 0.3937 | Microhm-inch |
| | × 0.0100 × α | Ohm-gram/m$^2$ |
| | 1/N × 172.41 | % IACS (vol. basis) |
| | 1/N × 1532.8 × 1/α | % IACS (mass basis) |
| Ohm-gram/m$^2$ | N × 601.53 × 1/α | Ohm-circular mil/ft |
| | N × 1/α | Ohm-mm$^2$/m |
| | N × 39.37 × 1/α | Microhm-inch |
| | N × 100 × 1/α | Microhm-cm |
| | 1/N × 1.7241 × α | % IACS (vol. basis) |
| | 1/N × 15.328 | % IACS (mass basis) |
| % IACS (vol. basis) | 1/N × 1037.1 | Ohm-circular mil/ft |
| | 1/N × 1.7241 | Ohm-mm$^2$/m |
| | 1/N × 67.879 | Microhm-inch |
| | 1/N × 172.41 | Microhm-cm |
| | 1/N × 1.7241 × α | Ohm-gram/m$^2$ |
| | × 0.11249 × 1/α | % IACS (mass basis) |
| % IACS (mass basis) | 1/N × 9220 × 1/α | Ohm-circular mil/ft |
| | 1/N × 15.328 × 1/α | Ohm-mm$^2$/m |
| | 1/N × 603.45 × 1/α | Microhm-inch |
| | 1/N × 1532.8 × 1/α | Microhm-cm |
| | 1/N × 15.328/α | Ohm-gram/m$^2$ |
| | 1/N × 0.11249 × α | % IACS (vol. basis) |

*Note*: N = number to be converted, α = density of material, grams/cubic centimeter.

Other examples are summarized here.

| | Ohm-inches | Ohm-centimeters |
|---|---|---|
| Copper (100% IACS) 99.9% Pure ETP | $0.679 \times 10^{-6}$ | $1.724 \times 10^{-6}$ |
| Aluminum | | |
| 6101-T63 | $1.212 \times 10^{-6}$ | $3.078 \times 10^{-6}$ |
| 6063-T6 | $1.281 \times 10^{-6}$ | $3.254 \times 10^{-6}$ |
| EC (62% IACS) | $1.095 \times 10^{-6}$ | $2.781 \times 10^{-6}$ |

Converting copper from microhm-inches to microhm-centimeters, we obtain

$$\frac{1\ \mu\Omega \cdot \text{cm}}{0.679\ \mu\Omega \cdot \text{in}} = \frac{2.54}{x}$$

The result is $x = 1.724\ \mu\Omega \cdot \text{cm}$, or $= 1.724 \times 10^{-6}\ \Omega \cdot \text{cm}$.

### 6.1.5.1 Resistivity of Copper

The International Annealed Copper Standard for the resistance of annealed copper of 100 percent conductivity at 20°C is 0.15328 Ω for a uniform round wire 1 meter long and weighing 1 gram.

The annealed copper standard may be expressed in other units as follows:

*Mile, pound:* Wire 1 mile long weighing 1 pound: 875.20 Ω.

*Microhm, centimeter:* 1 square centimeter section, 1 centimeter long: 1.7241 μΩ.

*Microhm, inch:* 1 square inch section, 1 inch long: 0.67897 μΩ.

*Foot, inch*$^2$: 1 square inch section, 1 foot long: 8.1455 μΩ.

*Mil, foot:* Wire 1 circular mil section, 1 foot long: 10.371 Ω.

*Meter, square millimeter:* Wire 1 square millimeter section, 1 meter long: 0.017241 Ω.

The temperature coefficient of mass resistance of copper is proportional to its conductivity and at 20°C is given by $\alpha' = n\alpha$, where $n$ is the actual conductivity in terms of the annealed copper standard expressed as a decimal (for 99 percent conductivity, $n = 0.99$), $\alpha'$ is the coefficient at $n$ conductivity, and $\alpha$ is the coefficient of the annealed copper standard (0.00393 per °C at 20°C). At another temperature $T$ and conductivity $n$, the coefficient is given by

$$\alpha'_T = \frac{0.00393\, n}{1 + 0.00393\, n(T - 20)}$$

If resistance $R$ and the resistance temperature coefficient $\alpha'_T$ are known at any temperature $T$°C, the resistance $R_1$ at any other temperature $T_1$°C is

$$R_1 = R[1 + \alpha'_T(T_1 - T)]$$

where $R_1$ = resistance per square inch per foot.

### 6.1.5.2 Typical Physical Properties of Electrical Copper

Electrical conductivity: 98 to 99.9 percent IACS

Specific gravity: 8.89

Weight: 0.321 lb/in$^3$; 555 lb/ft$^3$

Tensile strength: 35,000 to 55,000 psi

Yield strength: 25,000 to 50,000 psi

Young's modulus of elasticity: 16,000,000 to 17,000,000 psi

Coefficient of thermal expansion (20 to 100°C): 0.0000166 per °C (°C$^{-1}$); 0.00000922 per °F (°F$^{-1}$)

Melting temperature: 1083°C; 1981°F

Thermal conductivity: 0.92 cal/cm$^2$/cm/sec/°C; 3.85 W/cm$^2$/cm/sec/°C; 9.78 W/in$^2$/in/sec/°C

Specific heat: 0.092 cal/g/°C

#### 6.1.5.3 Circular Mil Calculations

One circular mil = $(\pi/4) \times 10^{-6}$ in$^2$ = 0.0000007854 in$^2$ = $7.854 \times 10^{-7}$ in$^2$.
Therefore, a conductor of 0.75-in diameter is equivalent to

$$A = \pi r^2 = 3.14169\ (0.375)^2 = 0.4418 \text{ in}^2$$

So

$$\frac{0.4418}{7.854 \times 10^{-7}} = 562{,}516 \text{ circular mils, or 563 Mcm}$$

The M in Mcm represents a multiplier of a "thousand"; 1000 Mcm = 1,000,000 circular mils, and the diameter of 1000-Mcm cable or wire (solid), would be $1{,}000{,}000 \times 7.854 \times 10^{-7} = 0.7854$ in$^2$. (From this, the diameter in inches may be calculated using the equation for the area of a circle, $A = \pi r^2$.)

### 6.1.6 Prefixes for Powers of Ten

| Multiplier | Prefix | Symbol |
|---|---|---|
| $10^{18}$ | exa | E |
| $10^{15}$ | peta | P |
| $10^{12}$ | tera | T |
| $10^{9}$ | giga | G |
| $10^{6}$ | mega | M |
| $10^{3}$ | kilo | k |
| $10^{2}$ | hecto | h |
| 10 | deca | da |
| $10^{-1}$ | deci | d |
| $10^{-2}$ | centi | c |
| $10^{-3}$ | milli | m |
| $10^{-6}$ | micro | μ |
| $10^{-9}$ | nano | n |
| $10^{-12}$ | pico | p |
| $10^{-15}$ | femto | f |
| $10^{-18}$ | atto | a |

### 6.1.7 Metals' Weights, Melting Points, and Conductivities

Refer to Table 6.3.

### 6.1.8 Dielectric Constants

Refer to Table 6.4*a* and 6.4*b*.

## 6.2 CIRCUIT ELEMENTS

### 6.2.1 Basic Circuit Formulas

Resistance in *series:*

$$R_t = R_1 + R_2 + R_3 + \cdots + R_n$$

**TABLE 6.3** Properties of Metals

| Metal | Specific Gravity | lb/in³ | Melting Point, °C | Melting Point, °F | Electrical Conductivity* Copper = 100% |
|---|---|---|---|---|---|
| Aluminum | 2.708 | 0.097 | 658 | 1,217 | 61.0 |
| Antimony | 6.618 | 0.239 | 630 | 1,166 | 4.1 |
| Bismuth | 9.8 | 0.353 | 271 | 520 | 1.5 |
| Cadmium | 8.55 | 0.308 | 321 | 610 | 22.7 |
| Copper (annealed) | 8.89 | 0.321 | 1,083 | 1,981 | 100.0 |
| Gold | 19.33 | 0.696 | 1,063 | 1,945 | 70.7 |
| Iron | 7.86 | 0.284 | 1,530 | 2,786 | 17.2 |
| Lead | 11.37 | 0.411 | 327 | 621 | 8.5 |
| Lithium | .592 | 0.0214 | 186 | 366 | 20.1 |
| Magnesium | 1.74 | 0.0627 | 651 | 1,204 | 37.5 |
| Mercury | 14.38 | 0.518 | −38.87 | − 37.97 | 1.8 |
| Nickel | 8.90 | 0.320 | 1,452 | 2,646 | 22.1 |
| Platinum | 21.5 | 0.775 | 1,755 | 3,191 | 17.2 |
| Potassium | .863 | 0.0311 | 62 | 144 | 25.4 |
| Silver | 10.50 | 0.378 | 960 | 1,761 | 105.8 |
| Sodium | .978 | 0.0352 | 97 | 207 | 35.1 |
| Tantalum | 16.6 | 0.599 | 2,900 | 5,250 | 11.1 |
| Tin | 7.29 | 0.263 | 232 | 449 | 15.0 |
| Tungsten | 19.30 | 0.696 | 3,400 | 6,152 | 31.3 |
| Zinc | 7.10 | 0.256 | 419 | 787 | 27.8 |

*See Section 6.1.5 to convert conductivity into other terms/units.

**TABLE 6.4*a*** Dielectric Constants

| Dielectric Material | K value |
|---|---|
| Air (1 atmosphere) | 1.0 |
| Bakelite | 5.0 |
| Varnished cambric | 4.0 |
| Glass (common and flint) | 8.0 |
| Gutta percha | 4.0 |
| Mica | 6.0 |
| Paraffin | 2.5 |
| Porcelain (wet process) | 6.0 |
| Pyrex glass | 4.5 |
| Quartz | 5.0 |
| Rubber (natural) | 3.0 |
| Slate | 7.0 |
| Wood (common dry) | 5.0 |
| Nylon | 5.0 |
| Phenolic resin | 8.0 |
| Plexiglass | 3.2 |
| Polystyrene | 3.0 |
| Teflon | 2.0 |
| Water, distilled | 80.0 |

**TABLE 6.4b** Dielectric Constants (*Continued*)

| | k |
|---|---|
| Inorganic crystalline: | |
| NaCl, dry crystal | 5.5 |
| $CaCO_3$ (av) | 9.15 |
| $Al_2O_3$ | 10.0 |
| MgO | 8.2 |
| BN | 4.15 |
| $TiO_2$ (av) | 100 |
| $BaTiO_3$ crystal | 4,100 |
| Muscovite mica | 7.0–7.3 |
| Fluorophlogopite (synthetic mica) | 6.3 |
| Ceramics: | |
| Alumina | 8.1–9.5 |
| Steatite | 5.5–7.0 |
| Forsterite | 6.2–6.3 |
| Aluminum silicate | 4.8 |
| Typical high-tension porcelain | 6.0–8.0 |
| Titanates | 50–10,000 |
| Beryl | 4.5 |
| Zirconia | 8.0–10.5 |
| Magnesia | 8.2 |
| Glass-bonded mica | 6.4–9.2 |
| Glasses: | |
| Fused silica | 3.8 |
| Corning 7740 (common laboratory Pyrex) | 5.1 |
| Polymer resins: | |
| Nonpolar resins: | |
| Polyethylene | 2.3 |
| Polystyrene | 2.5–2.6 |
| Polypropylene | 2.2 |
| Polytetrafluoroethylene | 2.0 |
| Polar resins: | |
| Polyvinyl chloride (rigid) | 3.2–3.6 |
| Polyvinyl acetate | 3.2 |
| Polyvinyl fluoride | 8.5 |
| Nylon | 4.0–4.6 |
| Polyethylene terephthalate | 3.25 |
| Cellulose cotton fiber (dry) | 5.4 |
| Cellulose Kraft fiber (dry) | 5.9 |
| Cellulose cellophane (dry) | 6.6 |
| Cellulose triacetate | 4.7 |
| Tricyanoethyl cellulose | 15.2 |
| Epoxy resins unfilled | 3.0–4.5 |
| Methylmethacrylate | 3.6 |
| Polyvinyl acetate | 3.7–3.8 |
| Polycarbonate | 2.9–3.0 |
| Phenolics (cellulose-filled) | 4–15 |
| Phenolica (glass-filled) | 5–7 |
| Phenolics (mica-filled) | 4.7–7.5 |
| Silicones (glass-filled) | 3.1–4.5 |

Resistance in *parallel:*

$$\frac{1}{R_t} = \frac{1}{R_1} + \frac{1}{R_2} + \frac{1}{R_3} + \cdots + \frac{1}{R_n}$$

$$R_t = \frac{R_1 R_2}{R_1 + R_2} \quad \text{(2 in parallel)}$$

Capacitance in *series:*

$$\frac{1}{C_t} = \frac{1}{C_1} + \frac{1}{C_2} + \frac{1}{C_3} + \cdots + \frac{1}{C_n}$$

$$C_t = \frac{C_1 C_2}{C_1 + C_2} \quad \text{(2 in series)}$$

Capacitance in *parallel:*

$$C_t = C_1 + C_2 + C_3 + \cdots + C_n$$

The capacitance of a parallel-plate capacitor element is given as

$$C = 0.0885\,\frac{KA(n-1)}{d}$$

where $C$ = capacitance, picofarads (pF)
$K$ = dielectric constant of insulating medium
$A$ = area of one plate, $cm^2$
$n$ = number of plates
$d$ = thickness of the dielectric, cm

If $A$ and $d$ are given in inches, the constant (0.0885) becomes 0.224.

***Self-Inductance.*** In series:

$$L_t = L_1 + L_2 + L_3 + \cdots + L_n$$

In parallel:

$$\frac{1}{L_t} = \frac{1}{L_1} + \frac{1}{L_2} + \frac{1}{L_3} + \cdots + \frac{1}{L_n}$$

***Coupled Inductance.*** In series with fields *aiding:*

$$L_t = L_1 + L_2 + 2M$$

In series with fields *opposing:*

$$L_t = L_1 + L_2 - 2M$$

In parallel with fields *aiding:*

$$\frac{1}{L_t} = \frac{1}{L_1 + M} + \frac{1}{L_2 + M}$$

In parallel with fields *opposing:*

$$\frac{1}{L_t} = \frac{1}{L_1 - M} + \frac{1}{L_2 - M}$$

where $L_t$ = total inductance
$M$ = mutual inductance
$L_1, L_2$ = self-inductance of individual coils

***Mutual Inductance*** ($M$)

$$M = \frac{L_A - L_O}{4}$$

where $M$ = mutual inductance, same units as $L_A$ and $L_O$
$L_A$ = total inductance of coils $L_1$ and $L_2$, fields *aiding*
$L_O$ = total inductance of coils $L_1$ and $L_2$, fields *opposing*

***Resonance.*** The resonant frequency, or frequency at which inductive reactance ($X_L$) equals capacitive reactance ($X_C$) is given as

$$f_r = \frac{1}{2\pi\sqrt{LC}} \quad \textit{and} \quad L = \frac{1}{4\pi^2 f_r^2 C} \quad \text{and} \quad C = \frac{1}{4\pi^2 f_r^2 L}$$

where $f_r$ = resonant frequency, Hz
$L$ = inductance, H
$C$ = capacitance, F

***Reactance.*** Inductive:

$$X_L = 2\pi f L$$

Capacitive:

$$X_C = \frac{1}{2\pi f C}$$

where $X_L$ = inductive reactance, Ω (positive reactance)
$X_C$ = capacitive reactance, Ω (negative reactance)
$f$ = frequency, Hz (cycles per second)
$L$ = inductance, H
$C$ = capacitance, F

***Impedance.*** In an ac circuit where resistance and reactance values of the $R$, $L$, and $C$ components are given, the numerical magnitudes of *impedance* and *phase angle* are expressed by the following several equations. Series circuits:

$$Z_T = \sqrt{R_T^2 + X_T^2}$$

Parallel circuits:

$$Z_T = \frac{1}{\sqrt{G_T^2 + B_T^2}}$$

where $R_T$ = total resistance, Ω
$X_T$ = total reactance, Ω
$G_T$ = total conductance, mhos (or S)
$B_T$ = total susceptance, mhos (or S)

In series circuits

$$Z = \frac{R}{\cos\theta} \qquad Z = \frac{X}{\sin\theta}$$

$$R = Z\cos\theta \qquad X = Z\sin\theta$$

where $Z$ = impedance, Ω
$R$ = resistance, Ω
$X$ = reactance, Ω
$\theta$ = phase angle in degrees by which the current leads voltage in a capacitive circuit, or lags voltage in an inductive circuit

In a *resonant* circuit, $X_L = X_C$ and $\theta = 0$ degrees.

### 6.2.1.1 Numerical Magnitude of Impedance

For various circuit configurations illustrated in the figures, we can calculate case-by-case.

*Case 1.* Resistance alone:

$$Z = R \qquad \theta = 0°$$

*Case 2.* Resistance in series:

$$Z = R_1 + R_2 + R_3 + \cdots + R_n$$

$$\theta = 0°$$

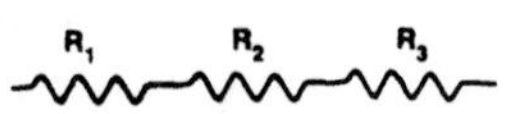

*Case 3.* Inductance alone:

$$Z = X_L \qquad \theta = 90°$$

*Case 4.* Inductance in series:

$$Z = X_{L1} + X_{L2} + \cdots + X_{Ln}$$
$$\theta = +90°$$

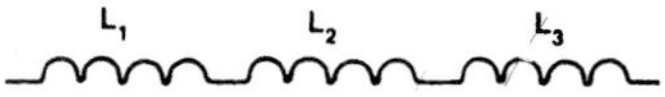


*Case 5.* Capacitance alone:

$$Z = X_C \qquad \theta = -90°$$

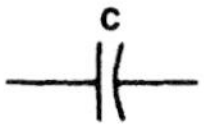


*Case 6.* Capacitances in series:

$$Z = X_{C_1} + X_{C_2} + \cdots + X_{C_n}$$
$$\theta = -90°$$

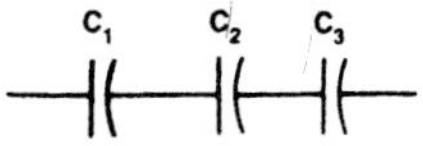


*Case 7.* Resistance and inductance in series:

$$Z = \sqrt{R^2 + X_L^2} \qquad \theta = \arctan \frac{X_L}{R}$$

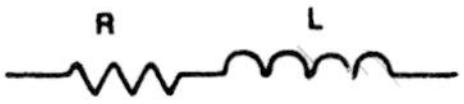


*Case 8.* Resistance and capacitance in series:

$$Z = \sqrt{R^2 + X_C^2} \qquad \theta = \arctan \frac{X_C}{R}$$

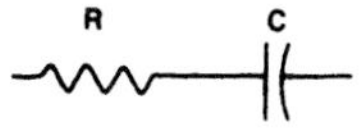


*Case 9.* Inductance and capacitance in series:

$$Z = X_L - X_C$$

$\theta = -90°$ when $X_L < X_C$

$\theta = 0°$ when $X_L = X_C$

$\theta = +90°$ when $X_L > X_C$

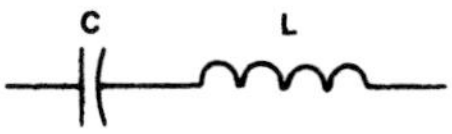


*Case 10.* Resistance, inductance, and capacitance in series:

$$Z = \sqrt{R^2 + (X_L - X_C)^2}$$
$$\theta = \arctan \frac{X_L - X_C}{R}$$

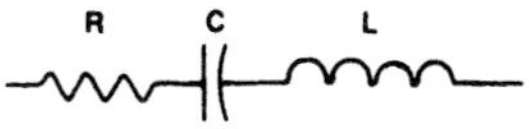


*Case 11.* Resistances in parallel:

$$Z = \frac{1}{\dfrac{1}{R_1} + \dfrac{1}{R_2} + \dfrac{1}{R_3} + \cdots + \dfrac{1}{R_n}}$$

$$\theta = 0°$$

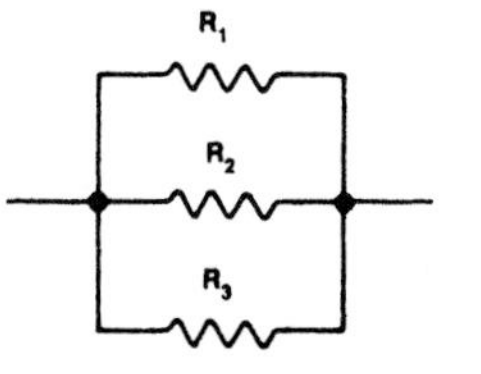


*Case 12.* Inductances in parallel:

$$Z = \frac{1}{\dfrac{1}{X_{L_1}} + \dfrac{1}{X_{L_2}} + \dfrac{1}{X_{L_3}} + \cdots + \dfrac{1}{X_{L_n}}}$$

$$\theta = +90°$$

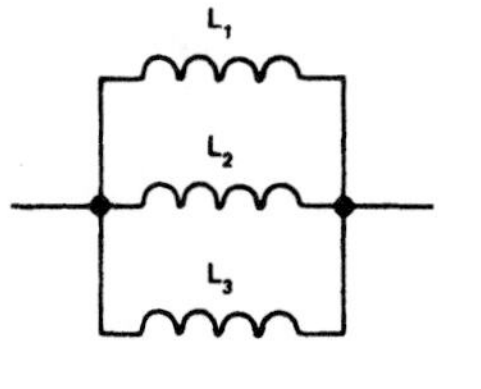

*Case 13.* Capacitances in parallel:

$$Z = \frac{1}{\dfrac{1}{X_{C_1}} + \dfrac{1}{X_{C_2}} + \dfrac{1}{X_{C_3}} + \cdots + \dfrac{1}{X_{C_n}}}$$

$$\theta = -90°$$

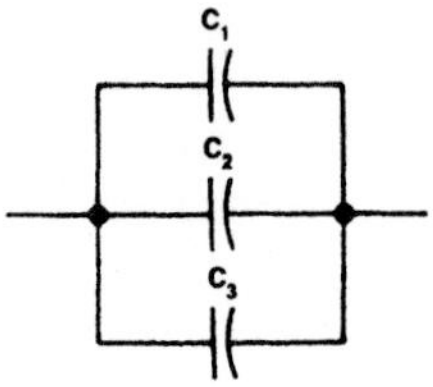

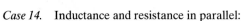

*Case 14.* Inductance and resistance in parallel:

$$Z = \frac{RX_L}{\sqrt{R^2 + X_L^2}}$$

$$\theta = \arctan \frac{X_L}{R}$$

$$\theta = -\arctan \frac{R}{X_L}$$

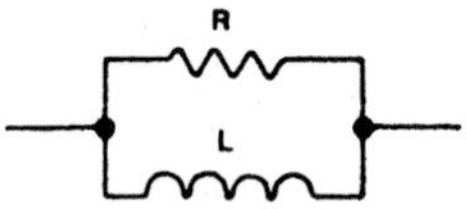


*Case 15.* Capacitance and resistance in parallel:

$$Z = \frac{RX_C}{\sqrt{R^2 + X_C^2}}$$

$$\theta = -\arctan \frac{R}{X_C}$$

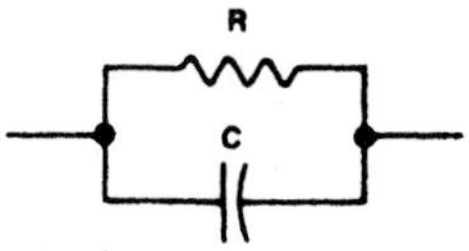


*Case 16.* Inductance and capacitance in parallel:

$$Z = \frac{X_L X_C}{X_L - X_C}$$

$$\theta = 0° \qquad \text{when} \qquad X_L = X_C$$

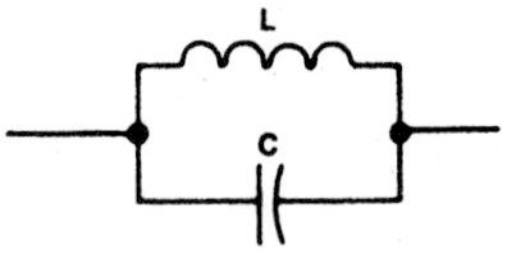


*Case 17.* Inductance, resistance, and capacitance in parallel:

$$Z = \frac{RX_L X_C}{\sqrt{X_L^2 X_C^2 + (RX_L - RX_C)^2}}$$

$$\theta = \arctan \frac{RX_C - RX_L}{X_L X_C}$$

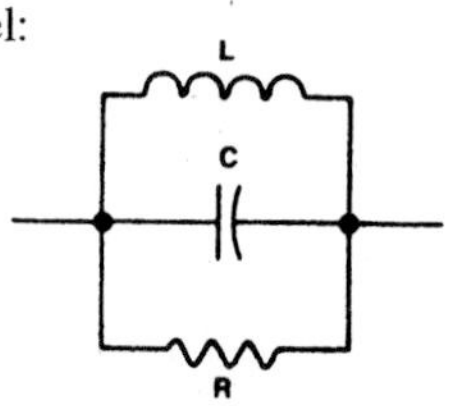


*Case 18.* Inductance and series resistance in parallel with capacitance:

$$Z = \sqrt{\frac{R^2 + X_L^2}{R^2 + (X_L - X_C)^2}}$$

$$\theta = \arctan \left( \frac{X_L X_C - X_L^2 - R^2}{RX_C} \right)$$

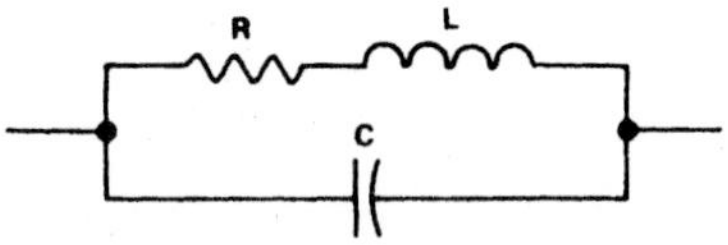


*Case 19.* Capacitance and series resistance in parallel with inductance and series resistance:

$$Z = \sqrt{\frac{(R_L^2 + X_L^2)(R_C^2 + X_C^2)}{(R_L + X_C)^2 + (X_L - X_C)^2}}$$

$$\theta = \arctan \frac{X_L(R_C^2 + X_C^2) - X_C(R_L^2 + X_L^2)}{R_L(R_C^2 + X_C^2) + R_C(R_L^2 + X_L^2)}$$

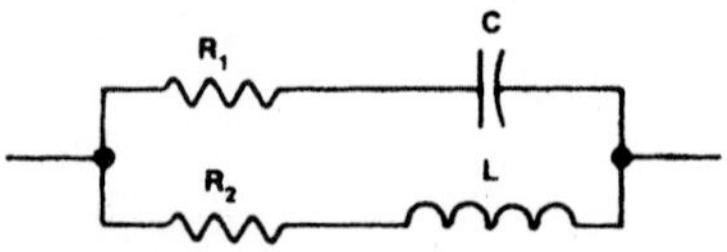

***Conductance.*** In dc circuits, conductance is given as $G = 1/R$, where $G$ = conductance, mhos (known as *reciprocal ohms* or S); and $R$ = resistance, Ω.

In a dc circuit with resistances $R_1$, $R_2$, etc., in parallel, conductance is given as

$$G_T = G_1 + G_2 + G_3 + \cdots + G_n$$

and total load current ($I_T$) as

$$I_T = EG_T$$

The current in a single, particular resistance is therefore

$$I_3 = \frac{I_T G_3}{G_1 + G_2 + G_3 + \cdots + G_n}$$

Ohm's law may be expressed in terms of conductance as follows.

$$R = \frac{1}{G} \qquad E = \frac{1}{G} \qquad I = EG$$

where $R$ = resistance, Ω
$E$ = potential, V
$I$ = current, A
$G$ = conductance, mhos (S)

***Susceptance.*** In an ac circuit, series susceptance is expressed as

$$B = \frac{X}{R^2 + X^2}$$

When resistance is 0, susceptance becomes the reciprocal of reactance:

$$B = \frac{1}{X}$$

where $R$ = resistance, Ω
$X$ = reactance, Ω
$B$ = susceptance, mhos (S)

***Admittance.*** In an ac circuit, series admittance is expressed as

$$Y = \frac{1}{\sqrt{R^2 + X^2}}$$

Admittance is expressed as the reciprocal of impedance.

$$Y = \frac{1}{Z}$$

where $R$ = resistance, Ω
$X$ = reactance, Ω
$Z$ = impedance, Ω
$Y$ = admittance, mhos (S)

Resistance and reactance in terms of conductance and susceptance are

$$R = \frac{G}{G^2 + B^2} \qquad X = \frac{B}{G^2 + B^2}$$

### 6.2.1.2 Steady-State Current Flow

In a capacitive circuit where resistance loss is considered negligible

$$I = \frac{E}{X_C} = \frac{E}{1/2\pi f C} = E\,(2\pi f C)$$

In an inductive circuit where resistance and capacitance are considered negligible

$$I = \frac{E}{X_L} + \frac{E}{2\pi f L}$$

where $I$ = current, A
$X_C$ = capacitive reactance, Ω
$X_L$ = inductive reactance, Ω
$E$ = applied potential, V
$f$ = constant frequency, Hz
$C$ = capacitance, F
$L$ = inductance, H

### 6.2.1.3 Alternating Current Values of *I* and *E*

Peak, root-mean-square (rms), and average values are listed here.

Peak = 1.414 × rms

Peak = 1.57 × average

Rms = 0.707 × peak

Rms = 1.11 × average

Average = 0.637 × peak

Average = 0.9 × rms

### 6.2.1.4 Time Constant in *LCR* Circuits

Discharging and charging in inductance-capacitance-resistance circuits is defined as the time in seconds for current or voltage to fall to $1/e$ (or 36.8 percent) of its initial value, or to rise to $1 - 1/e$ (or 63.2 percent) of its final value. The value $e = 2.71828\ldots$ (i.e., the base of the natural-logarithm system).

### 6.2.1.5 Alternating Current Circuit Formulas

The properties of circuits are treated for several categories. (Symbols employed here are defined at the end of Sec. 6.2.1.6.)

***Single-Phase Circuits***

$$\text{kW} = \frac{IE\cos\theta}{1000} \qquad \text{kVA} = \frac{IE}{1000}$$

$$\text{hp} = \frac{IE\cos\theta \times \text{Eff}}{746}$$

$$I = \frac{\text{hp} \times 746}{E\cos\theta \times \text{Eff}} \qquad I = \frac{\text{kW} \times 1000}{E\cos\theta} \qquad I = \frac{\text{kVA} \times 1000}{E}$$

***Two-Phase Three-Wire Circuits.*** In these circuits, the current in the common conductor is 1.41 times that in either phase conductor.

$$\text{kW} = \frac{2IE \cos\theta}{1000} \qquad \text{kVA} = \frac{2IE}{1000}$$

$$\text{hp} = \frac{2IE \cos\theta \times \text{Eff}}{746}$$

$$I = \frac{\text{hp} \times 746}{2E \cos\theta \times \text{Eff}} \qquad I = \frac{\text{kW} \times 1000}{2E \cos\theta} \qquad I = \frac{\text{kVA} \times 1000}{2E}$$

***Three-Phase Circuits***

$$\text{kW} = \frac{1.73IE \cos\theta}{1000} \qquad \text{kVA} = \frac{1.73IE}{1000}$$

$$\text{hp} = \frac{1.73IE \cos\theta \times \text{Eff}}{746}$$

$$I = \frac{\text{hp} \times 746}{1.73E \cos\theta \times \text{Eff}} \qquad I = \frac{\text{kW} \times 1000}{1.73E \cos\theta} \qquad I = \frac{\text{kVA} \times 1000}{1.73E}$$

***Volts at Receiving and Sending Ends in AC***

$$e_r = \sqrt{e_s^2 - I^2 (X \cos\theta \mp R \sin\theta)^2} - I(R \cos\theta \pm X \sin\theta)$$

$$e_s = \sqrt{(e_r \cos\theta + IR)^2 + (e_r \sin\theta + IX)^2}$$

*Note:* See Tables 6.5 and 6.6 for a convenient, condensed version of the preceding circuit equations.

### 6.2.1.6 Direct Current Circuit Formulas

$$\text{kW} = \frac{IE}{1000} \qquad \text{kVA} = \frac{IE}{1000}$$

$$\text{hp} = \frac{IE \times \text{Eff}}{746}$$

$$I = \frac{\text{hp} \times 746}{E \times \text{Eff}} \qquad I = \frac{\text{kW} \times 1000}{E} \qquad I = \frac{\text{kVA} \times 1000}{E}$$

$$e_r = e_s - IR \qquad e_s = e_r + IR$$

***Symbols.*** Variables appearing in equations pertaining to properties of circuits are defined as follows.

| | |
|---|---|
| $\cos\theta$ | Power factor of load |
| $E$ | Volts between conductors |
| $e$ | Volts to neutral |
| Eff | Efficiency of motor (i.e., 85 percent = 0.85) |
| $e_r$ | Volts at the receiving end |

**TABLE 6.5** Circuit Formulas

RELATIONS OF RESISTANCE, INDUCTANCE AND CAPACITANCE IN AC CIRCUITS

| Circuit Contains | Reactance | Impedance | "E" for a Current "I" | Power Factor |
|---|---|---|---|---|
| Resistance (R) | O | R | IR | 1 |
| Inductance (L) | $2\pi fL$ | $2\pi fL$ | $2\pi fLI$ | O |
| Capacitance (C) | $\frac{1}{2\pi fC}$ | $\frac{1}{2\pi fC}$ | $\frac{I}{2\pi fC}$ | O |
| Resistance & Inductance in series (R & L) | $2\pi fL$ | $\sqrt{R^2+(2\pi fL)^2}$ | $I\sqrt{R^2+(2\pi fL)^2}$ | $\frac{R}{\sqrt{R^2+(2\pi fL)^2}}$ |
| Resistance & Capacitance in series (R & C) | $\frac{1}{2\pi fC}$ | $\sqrt{R^2+\left(\frac{1}{2\pi fC}\right)^2}$ | $I\sqrt{R^2+\left(\frac{1}{2\pi fC}\right)^2}$ | $\frac{R}{\sqrt{R^2+\left(\frac{1}{2\pi fC}\right)^2}}$ |
| Resistance, Inductance & Capacitance in series (R & L & C) | $2\pi fL\frac{1}{2\pi fC}$ | $\sqrt{R^2+\left(2\pi fL-\frac{1}{2\pi fC}\right)^2}$ | $I\sqrt{R^2+\left(2\pi fL-\frac{1}{2\pi fC}\right)^2}$ | $\frac{R}{\sqrt{R^2+\left(2\pi fL-\frac{1}{2\pi fC}\right)^2}}$ |

$E$ = Pressure or Voltage in volts
$I$ = Current in amperes
$R$ = Resistance in ohms
$L$ = Inductance in henries
$C$ = Capacitance in farads
$f$ = Frequency in cycles per second
$\pi$ = 3.1416

**TABLE 6.6** Circuit Formulas

ELECTRIC PROPERTIES OF CIRCUITS

| Desired Data | Alternating Current | | | Direct Current |
|---|---|---|---|---|
| | Single-Phase | Two-Phase Four-Wire† | Three-Phase | |
| Kilowatts | $\frac{IE\cos\theta}{1000}$ | $\frac{2\,IE\cos\theta}{1000}$ | $\frac{1.73\,IE\cos\theta}{1000}$ | $\frac{IE}{1000}$ |
| Kilovolt-amperes | $\frac{IE}{1000}$ | $\frac{2\,IE}{1000}$ | $\frac{1.73\,IE}{1000}$ | $\frac{IE}{1000}$ |
| Horsepower Output | $\frac{IE\cos\theta\times Eff.}{746}$ | $\frac{2\,IE\cos\theta\times Eff.}{746}$ | $\frac{1.73\,IE\cos\theta\times Eff.}{746}$ | $\frac{IE\times Eff.}{746}$ |
| Amperes When Horsepower Is Known | $\frac{hp\times 746}{E\cos\theta\times Eff.}$ | $\frac{hp\times 746}{2\,E\cos\theta\times Eff.}$ | $\frac{hp\times 746}{1.73\,E\cos\theta\times Eff.}$ | $\frac{hp\times 746}{E\times Eff.}$ |
| Amperes When Kilowatts Are Known | $\frac{kw\times 1000}{E\cos\theta}$ | $\frac{kw\times 1000}{2\,E\cos\theta}$ | $\frac{kw\times 1000}{1.73\,E\cos\theta}$ | $\frac{kw\times 1000}{E}$ |
| Amperes When Kilovolt-amperes Are Known | $\frac{kva\times 1000}{E}$ | $\frac{kva\times 1000}{2\,E}$ | $\frac{kva\times 1000}{1.73\,E}$ | $\frac{kva\times 1000}{E}$ |
| $e_r$ When $e_s$, I, $\cos\theta$ Are Known | $\sqrt{e_s^2-I^2(X\cos\theta\mp R\sin\theta)^2}-I(R\cos\theta\pm X\sin\theta)$ | | | $e_s-IR$ |
| $e_s$ When $e_r$, $I$, $\cos\theta$ Are Known | $\sqrt{(e_r\cos\theta+IR)^2+(e_r\sin\theta+IX)^2}$ | | | $e_r+IR$ |

†In two-phase three-wire circuits, the current in the common conductor is 1.41 times that in either phase conductor.

NOTATION
$\cos\theta$ = Power factor of load
$E$ = Volts between conductors
$e$ = Volts to neutral
$Eff.$ = Efficiency of motor
$e_r$ = Volts at receiving end
$e_s$ = Volts at sending end
$I$ = Line current amperes
$R$ = Resistance in ohms to neutral
$\sin^2\theta = 1-\cos^2\theta$
$X$ = Reactance in ohms to neutral

Where double signs, such as ∓ or ± are shown, use upper one for lagging and lower one for leading power factor.

| | |
|---|---|
| $e_s$ | Volts at the sending end |
| $I$ | Line current, A |
| $R$ | Resistance in ohms to neutral |
| $\sin^2 \theta$ | $1 - \cos^2 \theta$ |
| $X$ | Reactance in ohms to neutral |

*Note:* When double signs ($\mp$ or $\pm$) are shown, use upper sign for lagging power factor and lower sign for leading power factor.

### 6.2.2 Ohm's Law for DC and AC Circuits

In a dc circuit

$$E = IR \qquad I = \frac{E}{R} \qquad R = \frac{E}{I}$$

$$P = I^2R \qquad P = EI \qquad P = \frac{E^2}{R}$$

$$R = \frac{P}{I^2} \qquad R = \frac{E^2}{P}$$

$$I = \sqrt{\frac{P}{R}} \qquad I = \frac{P}{E}$$

$$E = \frac{P}{I} \qquad E = \sqrt{PR}$$

In an ac circuit

$$E = IZ \qquad I = \frac{E}{Z} \qquad Z = \frac{E}{I}$$

$$P = I^2Z \cos \theta \qquad P = IE \cos \theta \qquad P = \frac{E^2 \cos \theta}{Z}$$

$$Z = \frac{P}{I^2 \cos \theta} \qquad Z = \frac{E^2 \cos \theta}{P}$$

$$I = \sqrt{\frac{P}{Z \cos \theta}} \qquad I = \frac{P}{E \cos \theta}$$

$$E = \frac{P}{I \cos \theta} \qquad E = \sqrt{\frac{PZ}{\cos \theta}}$$

***Phase Angle in AC Circuits.*** This angle is the difference in degrees by which the current leads voltage in a capacitive circuit, or lags voltage in an inductive circuit. In series circuits it is equal to the angle whose tangent is given by the ratio $X/R$; it is also expressed as

$$\theta = \arctan \frac{X}{R}$$

Therefore when $\theta = 0°$, $\cos \theta = 1$ and $P = EI$; when $\theta = 90°$, $\cos \theta = 0$ and $P = 0$.

***Power Factor.*** In an ac circuit, the power factor (PF) is equal to the true power in watts divided by the apparent power in volt-amperes, which is equal to the cosine of the phase angle.

$$PF = \frac{EI \cos \theta}{EI} = \cos \theta$$

$EI \cos \theta$ = true power in watts; $EI$ = apparent power in volt-amperes.

Therefore, in a strictly resistive circuit, $\theta = 0$ and PF = 1. In a reactive circuit, $\theta = 90°$ and PF = 0. In a *resonant* circuit, $\theta = 0$ and PF = 1.

### 6.2.2.1 Voltage Drop across Resistances in DC Circuits

***Resistances in Series.*** In a series circuit, the current is the same through each resistance, and the sum of the voltage drops across each resistance is equal to the supply voltage (Fig. 6.1).

$$E = R_T I \qquad I = \frac{E}{R_t}$$

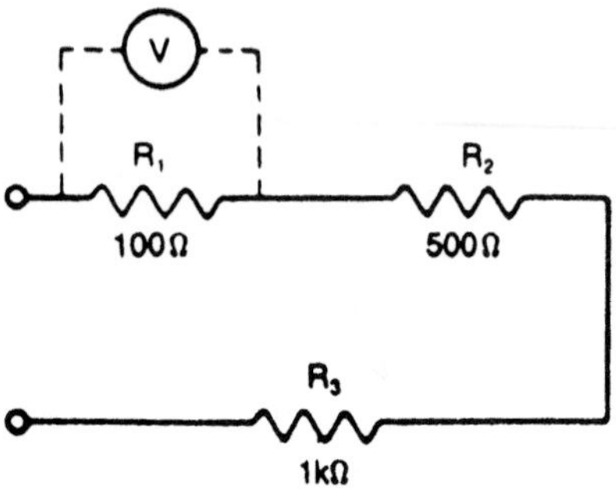


**FIGURE 6.1** Series circuit.

$$I = 125/1600 = 0.078 \text{ A} \quad \text{or} \quad 78 \text{ mA}$$

At $R_1$ $E = R_1 I = 100\,(0.078) = 7.8$ V across $R_1$

At $R_2$ $E = R_2 I = 500\,(0.078) = 39$ V across $R_2$

At $R_3$ $E = R_3 I = 1000\,(0.078) = 78$ V across $R_3$.

$E_T = 7.8 + 39 + 78 = 124.8$ V. Rounded to the nearest volt, $E_T$ = 125 V dc, which was the supply voltage.

***Resistances in Parallel.*** In a parallel circuit the voltage across each resistance is equal to the supply voltage, and the current through each resistance depends on its resistance value (Fig. 6.2).

The total load current is defined by several steps. First, find the total resistance in the circuit.

$$R_T = \frac{1}{\dfrac{1}{R_1} + \dfrac{1}{R_2} + \dfrac{1}{R_3}} = \frac{1}{\dfrac{1}{100} + \dfrac{1}{500} + \dfrac{1}{1000}} = \frac{1}{0.013} = 76.92\ \Omega$$

Then $\qquad E = RI \qquad 125 = 76.92I \qquad I = 1.625$ mm

This may be veriiied by finding the current in each resistance. At $R_1$

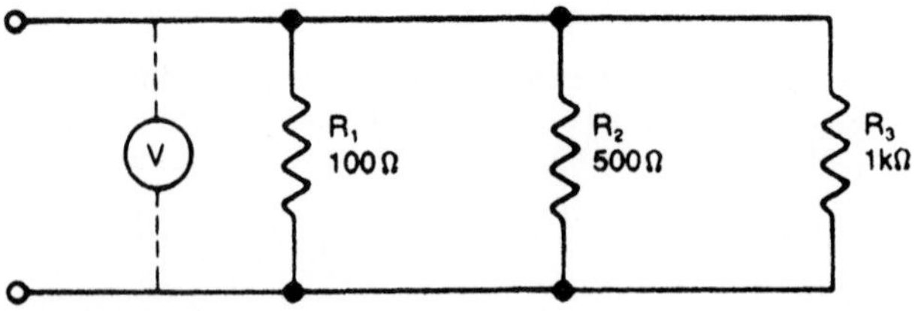


**FIGURE 6.2** Parallel circuit.

$$E = R_1 I \qquad 125 = 100\, I$$

$$I = \frac{125}{100} = 1.25 \text{ A}$$

At $R_2$ $\quad E = R_2 I \qquad 125 = 500I \qquad I = \frac{125}{500} = 0.25 \text{ A}$

At $R_3$ $\quad E = R_3 I \qquad 125 = 1000\, I$

$$I = \frac{125}{1000} = 0.125 \text{ A} \qquad I_t = 1.25 + 0.25 + 0.125 = 1.625 \text{ A}$$

## 6.3 DECIBELS

The number of decibels (dB) by which two power outputs $P_1$ and $P_2$ (in watts) may differ is given by

$$\text{dB, in terms of power} = 10 \log_{10} \frac{P_1}{P_2}$$

In terms of volts

$$\text{dB} = 20 \log_{10} \frac{E_1}{E_2}$$

In terms of current

$$\text{dB} = 20 \log_{10} \frac{I_1}{I_2}$$

Power ratios are independent of source and load impedance values. Voltage and current ratios in the preceding formulas hold true only when source and load impedances are equal. When these impedances differ, the ratios are then given by

$$\text{dB} = 20 \log_{10} \frac{E_1\sqrt{Z_1}}{E_2\sqrt{Z_2}} \qquad \text{and} \qquad 20 \log_{10} \frac{I_1\sqrt{Z_1}}{I_2\sqrt{Z_2}}$$

## 6.4 WAVELENGTH

Wavelength λ calculated in meters from frequency is

$$\lambda = \frac{3 \times 10^5}{f}$$

where $f$ = frequency, kHz. In centimeters

$$\lambda = \frac{3 \times 10^4}{f}$$

where $f$ = frequency, MHz.

Frequency $f$ in kilohertz from wavelength is

$$f = \frac{3 \times 10^5}{\lambda}$$

where $\lambda$ = wavelength, m. In megahertz, it is

$$f = \frac{3 \times 10^4}{\lambda}$$

where $\lambda$ = wavelength, cm.

## 6.5 PROPERTIES OF ELECTRICAL METALS

See Chap. 4, Sec. 6.1.5, and Table 6.3.

## 6.6 ELECTRICAL COMPONENTS

### 6.6.1 Resistors

*Resistors* are circuit elements that limit the flow of current in a circuit. The basic equation $W = I^2R$ relates the amount of heat dissipated by a resistor to the current flowing through the resistor. Thus, a resistor that passes 1.5 A of current and whose rating is 500 Ω dissipates $(1.5)^2 \times 500 = 1125$ W power.

There are various common forms of resistors in use:

- Carbon composition
- Metal film
- Wirewound (precision and power types)
- Filament or ribbon (power types)
- Cast-alloy element (power type)

Refer to Fig. 6.3 and the accompanying listing for an explanation of the color code for carbon composition resistors.

| Color | 1st digit (A) | 2d digit (B) | Multiplier (C) | Tolerance (D) |
|---|---|---|---|---|
| Black | 0 | 0 | 1 | |
| Brown | 1 | 1 | 10 | |
| Red | 2 | 2 | 100 | |
| Orange | 3 | 3 | 1000 | |
| Yellow | 4 | 4 | 10000 | |
| Green | 5 | 5 | 100000 | |
| Blue | 6 | 6 | 1000000 | |
| Violet | 7 | 7 | 10000000 | |
| Gray | 8 | 8 | 100000000 | |
| White | 9 | 9 | — | |
| Gold | — | — | 0.1 | ± 5% |
| Silver | — | — | 0.01 | ± 10% |
| No color | — | — | — | ± 20% |

Precision wirewound and carbon-film resistors have the resistance value and tolerance (usually 1 percent) marked on them.

***Maximum Continuous Working Voltage across a Resistor***

$$V_W = \sqrt{PR}$$

where $P$ = power, W ($I^2R$); $R$ = resistance, Ω.

*Example.* How many volts may be placed across a 100-Ω resistor with a power dissipation of 200 W?

$$V_W = \sqrt{200(100)} = \sqrt{20{,}000} = 141.4 \text{ V}$$

The maximum working voltage across the resistor is therefore 141.4 V

Carbon-composition, metal-film, and wirewound resistors are available in the sizes and wattage ratings listed in Tables 6.7*a* to 6.7*e*.

## 6.6.2 Capacitors

Standard disk and tubular capacitors have their values stamped on them. Tubular capacitors are generally marked in microfarads (0.01, 0.001 μF, etc.). A disk capacitor may be marked as just ".002," meaning a value in microfarads, and a small disk capacitor may be marked simply

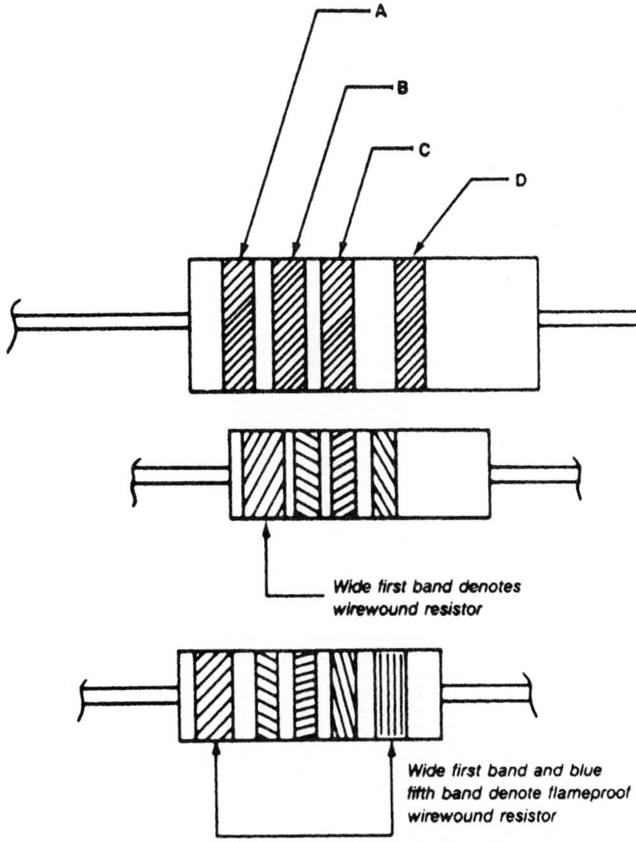


**FIGURE 6.3** Carbon composition resistors and color code.

**TABLE 6.7*a*** Standard Resistor Values

**Carbon Composition Resistor Values**
**(± 5% and ± 10%)**

| | |
|---|---|
| Industrial Grade<br>MIL-R-11 style RC | General purpose |
| MIL-R-39008 style RCR | High reliability |

| A -B Type | MIL-R-11 Type | MIL-R-39008 Type | Rating Watts | Size Lth x dia. | Volts |
|---|---|---|---|---|---|
| BB | RC05 | RCR05 | 1/8 | .145 x .062 | 150 |
| CB | RC07 | RCR07 | 1/4 | .250 x .090 | 250 |
| EB | RC20 | RCR20 | 1/2 | .375 x .140 | 350 |
| GB | RC32 | RCR32 | 1 | .562 x .225 | 500 |
| HB | RC42 | RCR42 | 2 | .688 x .312 | 750 |

Note: There is also a series of 2% carbon composition resistors of 1/4 and 1/2 watt ratings available from some of the resistor manufacturers.

**Resistor Values**

OHMS

| | | | | | | | |
|---|---|---|---|---|---|---|---|
| 1.0 | 5.1 | 27 | 130 | 600 | 3600 | 10000 | 91000 |
| 1.1 | 5.6 | 30 | 180 | 750 | 3800 | 20000 | 100000 |
| 1.2 | 6.2 | 33 | 180 | 820 | 4300 | 22000 | 110000 |
| 1.3 | 6.8 | 36 | 180 | 910 | 4700 | 24000 | 120000 |
| 1.5 | 7.5 | 38 | 200 | 1000 | 5100 | 27000 | 130000 |
| 1.6 | 8.2 | 43 | 220 | 1100 | 5800 | 30000 | 150000 |
| 1.8 | 9.1 | 47 | 240 | 1200 | 6200 | 33000 | 160000 |
| 2.0 | 10 | 51 | 270 | 1300 | 6800 | 36000 | 180000 |
| 2.2 | 11 | 56 | 300 | 1500 | 7500 | 38000 | 200000 |
| 2.4 | 12 | 62 | 330 | 1600 | 8200 | 43000 | 220000 |
| 2.7 | 13 | 66 | 360 | 1800 | 9100 | 47000 | |
| 3.0 | 15 | 75 | 380 | 2000 | 10000 | 51000 | |
| 3.3 | 16 | 82 | 430 | 2200 | 11000 | 56000 | |
| 3.6 | 18 | 91 | 470 | 2400 | 12000 | 62000 | |
| 3.9 | 20 | 100 | 510 | 2700 | 13000 | 66000 | |
| 4.3 | 22 | 110 | 560 | 3000 | 15000 | 75000 | |
| 4.7 | 24 | 120 | 620 | 3300 | 16000 | 82000 | |

MEGOHMS

| | | | | | | | |
|---|---|---|---|---|---|---|---|
| 0.24 | 0.43 | 0.75 | 1.3 | 2.4 | 4.3 | 7.5 | 13.0 |
| 0.27 | 0.47 | 0.82 | 1.5 | 2.7 | 4.7 | 8.2 | 15.0 |
| 0.30 | 0.51 | 0.91 | 1.6 | 3.0 | 5.1 | 9.1 | 16.0 |
| 0.33 | 0.58 | 1.0 | 1.8 | 3.3 | 5.6 | 10.0 | 18.0 |
| 0.36 | 0.62 | 1.1 | 2.0 | 3.6 | 6.2 | 11.0 | 20.0 |
| 0.38 | 0.68 | 1.2 | 2.2 | 3.9 | 6.8 | 12.0 | 22.0 |

**TABLE 6.7*b*** Standard Resistor Values (*Continued*)

**MIL TYPE RN OHMIC VALUES**

| | | | | | | | | | | | | | | |
|---|---|---|---|---|---|---|---|---|---|---|---|---|---|---|
| 10.0 | 11.5 | 13.3 | 15.4 | 17.8 | 20.5 | 23.7 | 27.4 | 31.6 | 36.5 | 42.2 | 48.7 | 56.2 | 64.9 | 75.0 | 86.6 |
| 10.2 | 11.8 | 13.7 | 15.8 | 18.2 | 21.0 | 24.3 | 28.0 | 32.4 | 37.4 | 43.2 | 49.9 | 57.6 | 66.5 | 76.8 | 88.7 |
| 10.5 | 12.1 | 14.0 | 16.2 | 18.7 | 21.5 | 24.9 | 28.7 | 33.2 | 38.3 | 44.2 | 51.1 | 59.0 | 68.1 | 78.7 | 90.9 |
| 10.7 | 12.4 | 14.3 | 16.5 | 19.1 | 22.1 | 25.5 | 29.4 | 34.0 | 39.2 | 45.3 | 52.3 | 60.4 | 69.8 | 80.6 | 93.1 |
| 11.0 | 12.7 | 14.7 | 16.9 | 19.6 | 22.6 | 26.1 | 30.1 | 34.8 | 40.2 | 46.4 | 53.6 | 61.9 | 71.5 | 82.5 | 95.3 |
| 11.3 | 13.0 | 15.0 | 17.4 | 20.0 | 23.2 | 26.7 | 30.9 | 35.7 | 41.2 | 47.5 | 54.9 | 63.4 | 73.2 | 84.5 | 97.6 |

Note: Standard resistance values are obtained from the decade table above, by multiplying by powers of 10. i.e. 13.3 can be 13.3, 133, 1.33K, 133K, etc.

**TABLE 6.7*c*** Standard Resistor Values (*Continued*)

MIL TYPE RN OHMIC VALUES

| | | | | | | | | |
|---|---|---|---|---|---|---|---|---|
| .008 | .5 | 6 | 22.5 | 56 | 150 | 400 | 910 | 3.5K |
| .01 | .75 | 7 | 25 | 60 | 160 | 430 | 1K | 4K |
| .02 | 1 | 7.5 | 27 | 62 | 180 | 450 | 1.2K | 5K |
| .03 | 1.5 | 8 | 30 | 70 | 200 | 470 | 1.3K | 10K |
| .05 | 2 | 10 | 33 | 75 | 220 | 500 | 1.5K | 15K |
| .1 | 2.5 | 12 | 35 | 80 | 250 | 560 | 1.8K | 20K |
| .15 | 3 | 15 | 40 | 82 | 270 | 600 | 2K | 25K |
| .2 | 3.3 | 16 | 45 | 100 | 300 | 680 | 2.2K | 40K |
| .25 | 4 | 20 | 47 | 110 | 330 | 700 | 2.5K | 50K |
| .3 | 5 | 22 | 50 | 120 | 390 | 750 | 3K | 100K |
| | | | | | | | | 150K |

MIL TYPE RN OHMIC VALUES

| | | | | | | | | |
|---|---|---|---|---|---|---|---|---|
| .1 | 1.2 | 10 | 50 | 220 | 500 | 825 | 2K | 10K |
| .13 | 1.5 | 15 | 60 | 250 | 510 | 900 | 2.2K | 12K |
| .15 | 2 | 20 | 75 | 270 | 560 | 909 | 2.5K | 15K |
| .2 | 3 | 25 | 100 | 300 | 600 | 1K | 3K | 20K |
| .25 | 4 | 30 | 110 | 330 | 619 | 1.1K | 3.3K | 25K |
| .3 | 5 | 33 | 120 | 332 | 620 | 1.2K | 3.5K | 30K |
| .33 | 6 | 35 | 125 | 350 | 681 | 1.3K | 4K | 40K |
| .5 | 7 | 40 | 150 | 400 | 700 | 1.5K | 5K | 50K |
| .75 | 7.5 | 45 | 180 | 450 | 750 | 1.78K | 7K | 100K |
| 1 | 8 | 47 | 200 | 470 | 800 | 1.8K | 8.2K | 150K |

**TABLE 6.7*d*** Standard Resistor Values (*Continued*)

## RG Series Precision MIL Metal Film Resistor Values (±5% tolerance)

| MIL Type | Type | MIL Power Rating (W) | Temp. Coeff. PPM/°C | Size (in.) BD | BL | LL | Range (Ohms) |
|---|---|---|---|---|---|---|---|
| RN-55D | TO-55 | ⅛ @ 70°C | ±100 | .09 | .25 | 1.50 | 10-301K |
| RN-55C | T2-55 | 1/10 @ 125°C | ±50 | .09 | .25 | 1.50 | 49.9-100K |
| RN-60D | TO-60 | ¼ @ 70°C | ±100 | .14 | .39 | 1.50 | 10-1M |
| RN-60C | T2-60 | ⅛ @ 125°C | ±50 | .14 | .39 | 1.50 | 49.9-499K |

RESISTANCE VALUE TABLE

| | | | | | | | | | | | |
|---|---|---|---|---|---|---|---|---|---|---|---|
| 1.00 | 1.21 | 1.47 | 1.78 | 2.15 | 2.61 | 3.16 | 3.83 | 4.64 | 5.62 | 6.81 | 8.25 |
| 1.02 | 1.24 | 1.50 | 1.82 | 2.21 | 2.67 | 3.24 | 3.92 | 4.75 | 5.76 | 6.98 | 8.45 |
| 1.05 | 1.27 | 1.54 | 1.87 | 2.26 | 2.74 | 3.32 | 4.02 | 4.87 | 5.90 | 7.15 | 8.66 |
| 1.07 | 1.30 | 1.58 | 1.91 | 2.32 | 2.80 | 3.40 | 4.12 | 4.99 | 6.04 | 7.32 | 8.87 |
| 1.10 | 1.33 | 1.62 | 1.96 | 2.37 | 2.87 | 3.48 | 4.22 | 5.11 | 6.19 | 7.50 | 9.09 |
| 1.13 | 1.37 | 1.65 | 2.00 | 2.43 | 2.94 | 3.57 | 4.32 | 5.23 | 6.34 | 7.68 | 9.31 |
| 1.15 | 1.40 | 1.69 | 2.05 | 2.49 | 3.01 | 3.65 | 4.42 | 5.36 | 6.49 | 7.87 | 9.53 |
| 1.18 | 1.43 | 1.74 | 2.10 | 2.55 | 3.09 | 3.74 | 4.53 | 5.49 | 6.65 | 8.06 | 9.76 |

**Note: All decade multiples of the values shown in the table above, are standard values for metal film resistors.**

as "25," meaning a value in picofarads. The former term *micro-micro* has been superseded by its equivalent, *picofarads.* The voltage rating is also stamped on the capacitor.

Electrolytic capacitors are *polarized,* and you must connect them in the circuit observing the correct polarity. Polarity is marked at the ends of the capacitor and indicated by + and – signs. Reversing the polarity in circuit can cause the capacitor to rupture. Some types of tubular capacitors are also polarized.

When using these components as filters in dc power supplies, you can apply a rule of thumb that allows 1000 μF per each ampere of output to the load. When power supply is critical, the value of the filters must be calculated and then verified by test.

**TABLE 6.7*e*** Standard Resistor Values (*Continued*)

**SPH Series Low Power Wirewound Resistor Values (Fail Safe Type)**
**(±5% tolerance)**
**MIL-R-11 style equivalent = RC32/RC42**

Power ratings: 2 watts @ 70°C
1 watt @ 115°C
1/2 watt @ 137°C

Working voltage = $(PR)^{1/2}$

RESISTANCE VALUE TABLE FOR WIREWOUND RESISTORS (OHMS)

| | | | | | | | | | | |
|---|---|---|---|---|---|---|---|---|---|---|
| 0.10* | 0.27* | 0.68* | 1.8* | 4.7* | 12* | 33* | 82* | 220* | 560* | 1.5K* |
| 0.11 | 0.30 | 0.75 | 2.0 | 5.1 | 13 | 36 | 91 | 240 | 620 | 1.6K |
| 0.12* | 0.33* | 0.82* | 2.2* | 5.6* | 15* | 39* | 100* | 270* | 680* | 1.8K* |
| 0.13 | 0.36 | 0.91 | 2.4 | 6.2 | 16 | 43 | 110 | 300 | 750 | 2.0K |
| 0.15* | 0.39* | 1.0* | 2.7* | 6.8* | 18* | 47* | 120* | 330* | 820* | |
| 0.16 | 0.43 | 1.1 | 3.0 | 7.5 | 20 | 51 | 130 | 360 | 910 | |
| 0.18* | 0.47* | 1.2* | 3.3* | 8.2* | 22* | 56* | 150* | 390* | 1K* | |
| 0.20 | 0.51 | 1.3 | 3.6 | 9.1 | 24 | 62 | 160 | 430 | 1.1K | |
| 0.22* | 0.56* | 1.5* | 3.9* | 10* | 27* | 68* | 180* | 470* | 1.2K* | |
| 0.24 | 0.62 | 1.6 | 4.3 | 11 | 30 | 75 | 200 | 510 | 1.3K | |

Connecting the capacitor to a voltage source in excess of its voltage rating will cause the capacitor to fail in service. The capacitor could "short" or "open" (short-circuit or open-circuit), and the failure action could be abrupt or prolonged, varying according to the magnitude of the overvoltage from the rating. (See Sec. 6.9 for filter calculations.)

***Basic Capacitor Formulas***

*Capacitance, F*

U.S. customary:

$$C = \frac{0.224\ \mathrm{kA}}{t_p}$$

Metric:

$$C = \frac{0.0884\ \mathrm{kA}}{t_p}$$

*Energy stored in capacitors, J, W · sec*

$$E = \frac{CV^2}{2}$$

*Linear charge of a capacitor, A*

$$I = C\,\frac{dV}{dt}$$

*Total impedance of a capacitor, Ω*

$$Z = \sqrt{R_s^2 + (X_C - X_L)^2}$$

where $R_s$ = equivalent series resistance (ESR).

*Capacitive reactance, Ω*

$$X_C = \frac{1}{2\pi fC}$$

*Inductive reactance, Ω*

$$X_L = 2\pi f L$$

*Phase angles*

*Ideal capacitors:* current leads voltage 90°
*Ideal inductors:* current lags voltage 90°
*Ideal resistors:* current in phase with voltage

*Dissipation factor*

$$\mathrm{DF} = \tan\delta\ (\text{loss angle}) = \frac{\mathrm{ESR}}{X_C} = (2\pi f C)\,(\mathrm{ESR})$$

*Power factor*

$$\mathrm{PF} = \sin\delta\ (\text{loss angle}) = \cos\phi\ (\text{phase angle})$$
$$= \mathrm{DF}\ (\text{when} <10\%)$$

*Equivalent series resistance, Ω*

$$\mathrm{ESR} = (\mathrm{DF})(X_c) = (\mathrm{DF})(\frac{1}{2}\ \pi f C)$$

*Power loss, W*

$$\text{Power loss} = (2\pi f C V^2)(\mathrm{DF})$$

*KVA, kW*

$$KVA = 2\pi f C V^2 \times 10^{-3}$$

Symbols for the preceding capacitor formulas are

$K$ = dielectric constant
$A$ = area of electrode overlap
$t_p$ = dielectric thickness
$V$ = voltage, V
$t$ = time, h
$R_s$ = series resistance, Ω
$f$ = frequency, Hz
$L$ = inductance, H
$\delta$ = loss angle
$\phi$ = phase angle

***Capacitors in AC Power Systems (Power Factor Improvement).*** Capacitors are used to improve the power factor of industrial loads. Table 6.8 lists the multipliers for calculating the kvar (kilovar) rating of capacitors used to accomplish this.

*Example.* The total system load is 200 kW at a 60 percent power factor. Find the kvar rating of the capacitor required to alter the power factor to 80 percent. Multiply the kW load times the multiplier listed in the table (60 to 80 percent).

$$200 \times 0.583 = 116.6 \text{ kvar}$$

Now select the nearest standard rating (kvar); for this system, 120 kvar would be recommended.

**TABLE 6.8** Capacitor Multipliers

| Original P.F., % | Desired Power Factor, % | | | | |
|---|---|---|---|---|---|
| | 100 | 95 | 90 | 85 | 80 |
| 60 | 1.333 | 1.004 | 0.849 | 0.713 | 0.583 |
| 62 | 1.266 | 0.937 | 0.782 | 0.646 | 0.516 |
| 64 | 1.201 | 0.872 | 0.717 | 0.581 | 0.451 |
| 66 | 1.138 | 0.809 | 0.654 | 0.518 | 0.388 |
| 68 | 1.078 | 0.749 | 0.594 | 0.458 | 0.328 |
| 70 | 1.020 | 0.691 | 0.536 | 0.400 | 0.270 |
| 72 | 0.964 | 0.635 | 0.480 | 0.344 | 0.214 |
| 74 | 0.909 | 0.580 | 0.425 | 0.289 | 0.159 |
| 76 | 0.855 | 0.526 | 0.371 | 0.235 | 0.105 |
| 77 | 0.829 | 0.500 | 0.345 | 0.209 | 0.079 |
| 78 | 0.802 | 0.473 | 0.318 | 0.182 | 0.052 |
| 79 | 0.776 | 0.447 | 0.292 | 0.156 | 0.026 |
| 80 | 0.750 | 0.421 | 0.266 | 0.130 | |
| 81 | 0.724 | 0.395 | 0.240 | 0.104 | |
| 82 | 0.698 | 0.369 | 0.214 | 0.078 | |
| 83 | 0.672 | 0.343 | 0.188 | 0.052 | |
| 84 | 0.646 | 0.317 | 0.162 | 0.206 | |
| 85 | 0.620 | 0.291 | 0.136 | | |
| 86 | 0.593 | 0.264 | 0.109 | | |
| 87 | 0.567 | 0.238 | 0.083 | | |
| 88 | 0.540 | 0.211 | 0.056 | | |
| 89 | 0.512 | 0.183 | 0.028 | | |
| 90 | 0.484 | 0.155 | | | |
| 91 | 0.456 | 0.127 | | | |
| 92 | 0.426 | 0.097 | | | |
| 93 | 0.395 | 0.066 | | | |
| 94 | 0.363 | 0.034 | | | |
| 95 | 0.329 | | | | |
| 96 | 0.292 | | | | |
| 97 | 0.251 | | | | |
| 99 | 0.143 | | | | |

***Ceramic Capacitors.*** Ceramic capacitors are generally used where small capacitances are required in an electrical circuit. Table 6.9 gives the ceramic capacitor codes, with capacitances given in picofarads (pF).

### 6.6.3 Inductors

*Inductors* are wirewound elements used in circuits as coils, chokes, reactors, and transformers. Their ratings are given in henrys, i.e., mH, μH, or H. (Refer to Sec. 6.6.8 and Sec. 6.9 for a description of coil-winding equations.)

A form of inductor, the *reactor,* is often used in power distribution equipment and also small electrical devices. A large air-core reactor is shown in Fig. 6.4*a* and 6.4*b*. In this reactor, the windings are aluminum wire and all parts of the core and structure are nonferrous material. This particular reactor core is part of a three-phase current-limiting reactor for power switchgear. The three-phase reactor, in this instance, is used to limit the short-circuit current in the power system into which it is installed. The core shown in Fig. 6.4*a* and 6.4*b* is in the process of being wound on a coil-winding machine. This particular reactor core weighs approximately 500 lb and was produced by the Powercon Corporation, Severn, Maryland.

### 6.6.4 Solenoids

*Solenoids* are electromechanical devices used to convert electrical energy into mechanical work. There are three basic types: pull-type, push-type, and rotary.

**TABLE 6.9** Ceramic Capacitor Codes

| | | | Tolerance | | | | | |
|---|---|---|---|---|---|---|---|---|
| | | | Class 1 | | | | | |
| Color | Digit | Multiplier | 10pF or less | Over 10pF | Class 2 | Temperature Coefficient ppm°C | Significant Figure | Multiplier |
| Black | 0 | 1 | ±2.0pF | ±20% | ±20% | 0 | 0.0 | −1 |
| Brown | 1 | 10 | ±0.1pF | ±1% | | −33 | | −10 |
| Red | 2 | 100 | | ±2% | | −75 | 1.0 | −100 |
| Orange | 3 | 1000 | | ±3% | | −150 | 1.5 | −1000 |
| Yellow | 4 | 10000 | | | +100% −0% | −220 | 2.0 | −10000 |
| Green | 5 | | ±0.5pF | ±5% | ±5% | −330 | 3.3 | +1 |
| Blue | 6 | | | | | −470 | 4.7 | +10 |
| Violet | 7 | | | | | −750 | 7.5 | +100 |
| Gray | 8 | 0.01 | ±0.25pF | | +80% −20% | +150 to −1500 | | +1000 |
| White | 9 | 0.1 | ±1.0pF | ±10% | ±10% | +100 to −750 | | +10000 |
| Silver | | | | | | | | |
| Gold | | | | | | | | |

The second and third types are variations of the first type. Fundamentally, all solenoids are actually pull-type devices, an iron core being pulled into a coil when electrical energy is passed through the coil windings. Solenoids are electromagnets.

Solenoids can be designed for ac or dc duty. The dc-type solenoid is usually equipped with a set of cutout contacts that open when the solenoid is energized, and just before seating. These contacts allow insertion of series resistors in the coil circuit that limit the hold-in current on the solenoid's coil. An arc-suppression capacitor may be placed in parallel across the cutout contacts to reduce arc energy across them.

Solenoids are rated by coil voltage, pull-in force, stroke, and power dissipation.

The *inrush* current of an open solenoid is several times greater than the *holding* current when the solenoid is closed, or seated.

A small solenoid can deliver a large pull or push when it is operated intermittently on overvoltage. That is, a large overvoltage can be used to operate the solenoid for a short time, provided the coil insulation rating is not exceeded. Thus if the coil is rated for 24 V, applying a voltage of 48 V will force more current to flow in the coil, making the solenoid more powerful mechanically. This method produces high pull-in forces from small solenoids in intermittent duty cycles. The off time must be considerably longer than the on time, in order to allow heat generated in the coil to dissipate. Heating of a solenoid coil follows the $W = I^2R$ rule for circuit power dissipation.

***Fuses in Solenoid Circuits.*** If a solenoid is blocked open, by malfunction, with a constant power source still connected to its coil, it will quickly overheat and burn out.

To protect solenoids against this hazard, a slow-blow fuse should be included in the coil circuit. The fuse rating should be 1.5 times the hold-in current rating. Check the solenoid specifications to determine the latter rating (solenoid seated). (See Fig. 6.5*a* to 6.5*d* for typical solenoid graphs.)

The dc solenoid is usually equipped with a set of auxiliary contacts, which "open" when the solenoid armature is pulled into the solenoid coil windings. The wiring diagram for this type of application is shown in Fig. 6.6. Contact 89ST opens when the solenoid is actuated, which in turn inserts the two 100-Ω power resistors in the solenoid coil circuit. These resistors limit the hold-in current of the solenoid and prevent overheating of the coil windings. (*Note:* The resistance and wattage values of the current-limiting power resistors must be determined by the size and current rating of the dc solenoid coil. Dc solenoid graphs usually indicate the

(a)

**FIGURE 6.4** Winding a 1200-A, 15-kV reactor core.

maximum allowable hold-in current for each particular size of device, thus making the required calculations simply by using Ohm's law and $I^2R$ heating effects.)

***Force of a Solenoid (Pull)***

*Case 1.* For a simple, cylinder-type solenoid, the maximum uniform pull is expressed by (refer to Fig. 6.7)

$$*F = \frac{CsNI}{l}$$

where $I$ = current, A
$N$ = number of turns in the coil
$s$ = cross-sectional area of core or plunger, $in^2$
$C$ = 0.01 $lb/in^2$ At · in (ampere-turn–inches)—when coil is 3 times the diameter of the plunger and length is 2 times coil diameter
$l$ = length of solenoid coil, in

(*Note:* $*F$ is valid when plunger is inserted from 0.4 to 0.8 times the coil length.)

*Case 2.* For an iron-clad solenoid (strong-box type), the maximum uniform pull is expressed by (see Fig. 6.8)

$$**F = sN\left[\mathrm{I}\left(\frac{NI}{l\hat{a}\,c^2} + \frac{C}{l}\right)\right]$$

where $la$ = air-gap length, in
$c$ = 2600 for soft iron cores
$C$ = 0.01 nominal factor

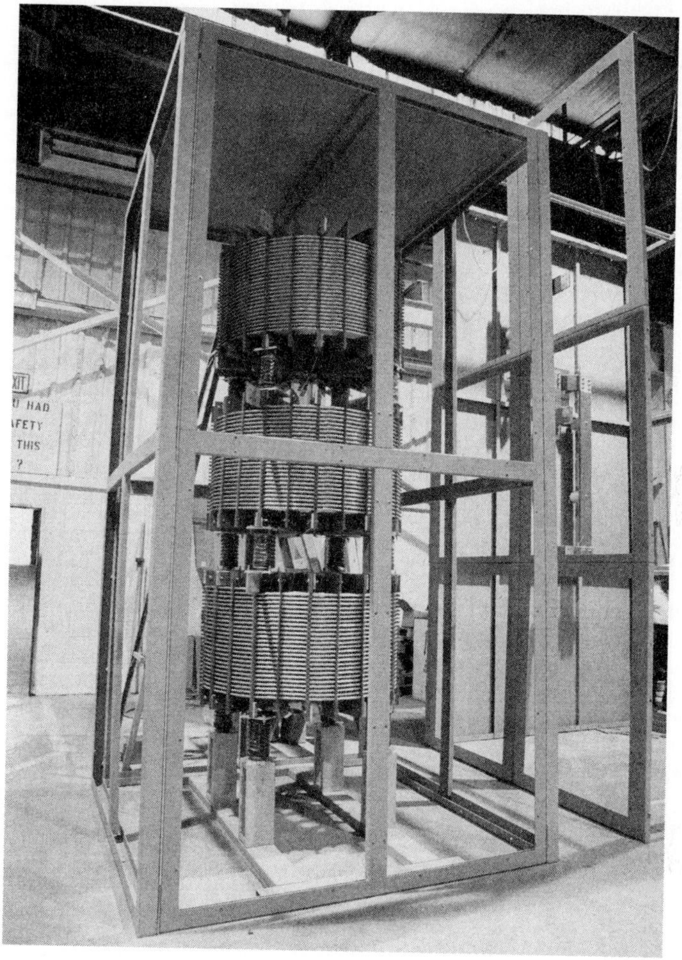

(b)

**FIGURE 6.4** (*Continued*)

(See symbol definitions following equation for other variables in case 1. *Note:* **$F$ is valid when plunger is inserted 0.4 or more times coil length.)

The two preceding simple equations may be used to set up initial design for simple solenoids of two types.

Three typical solenoid hold-in circuits are shown in Fig. 6.9*a* to 6.9*c*. For an approximation of the hold-in resistance value, use the graph shown in Fig. 6.10. Large, ceramic wirewound power resistors are commonly employed for this task. If you measure the solenoid coil resistance, you may use the graph to give an approximate resistance value for your particular application.

Approximately twice as much electrical energy must be supplied to a dc solenoid than an ac solenoid to accomplish the same work output. Conventional efficiency for a dc solenoid is therefore less than 50 percent.

The effective static mechanical force of a constant current ac electromagnet (solenoid) is 20 percent less than that of the same electromagnet when operated with dc for a given num-

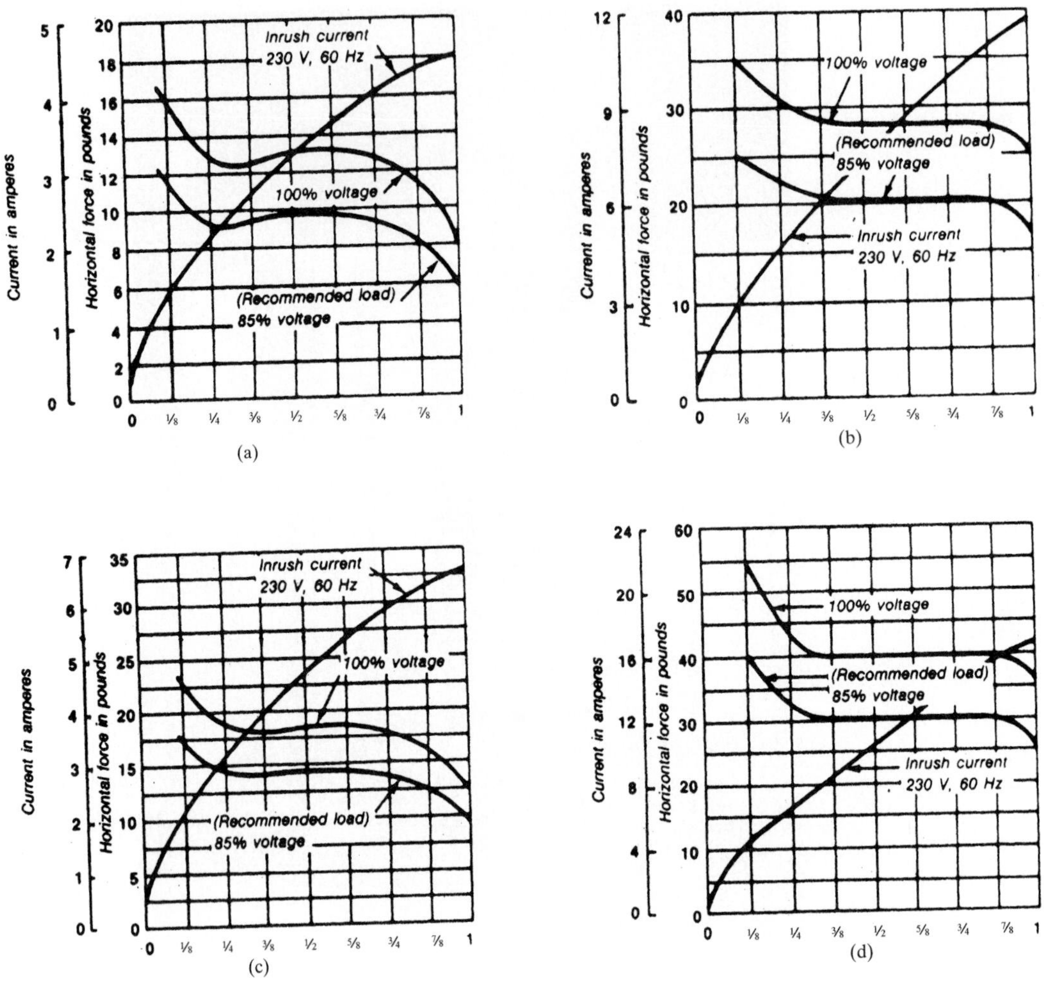


**FIGURE 6.5** Solenoid curves.

ber of ampere-turns. The efficiency of a well-designed ac electromagnet is approximately 85 percent. It is almost double that of a dc electromagnet. This is because energy stored in the magnetic circuit during the first half of an alternation (cycle) is returned to the circuit during the second half of the alternation.

***Heating in Electromagnets (Solenoids, etc.).*** If a working test shows that the coil of an electromagnet can dissipate 15 W at a given safe temperature, then a constant equivalent $I^2R$ power input may be applied to the coil without producing overheating. Less $I^2R$ input will result in a proportionally lower temperature rise.

***Mechanical Force of a Permanent Magnet.*** For a magnet with flat, parallel surfaces

$$F = \frac{KB^2S}{72.13}$$

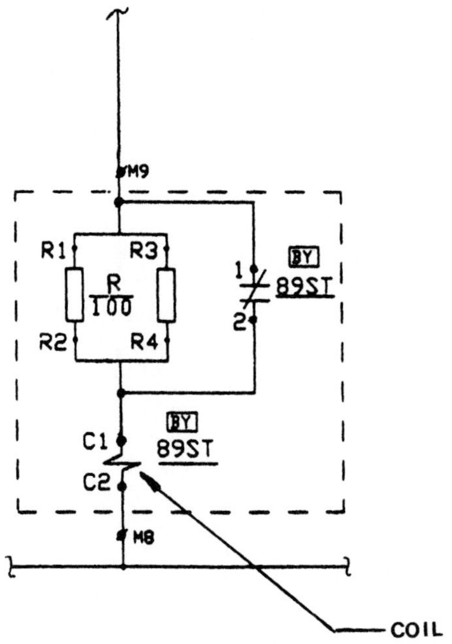


**FIGURE 6.6** Current-limiting resistor circuit for dc solenoid.

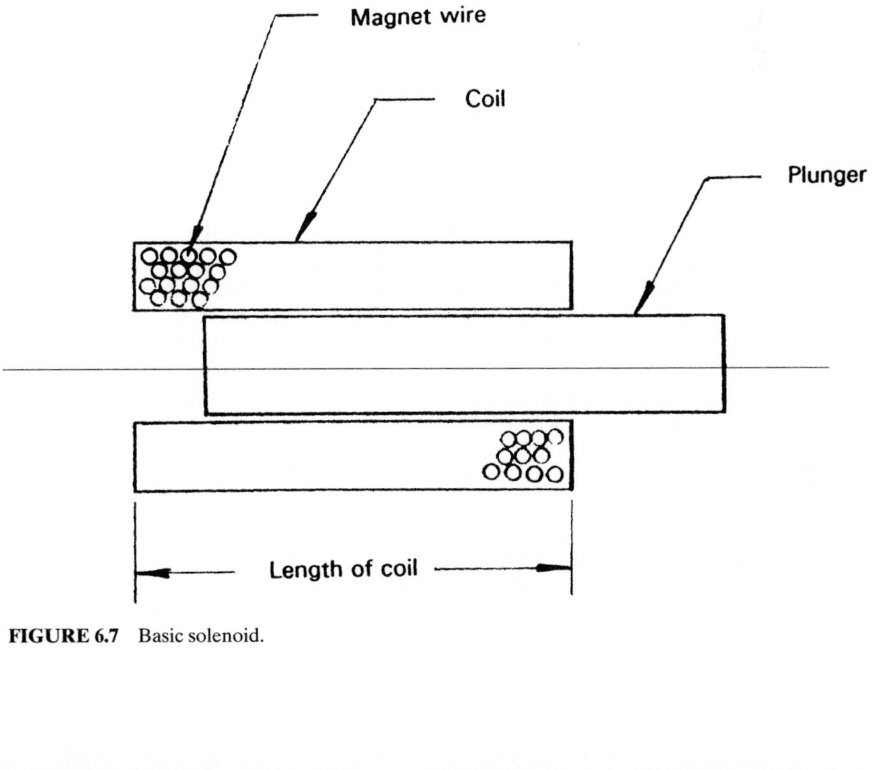


**FIGURE 6.7** Basic solenoid.

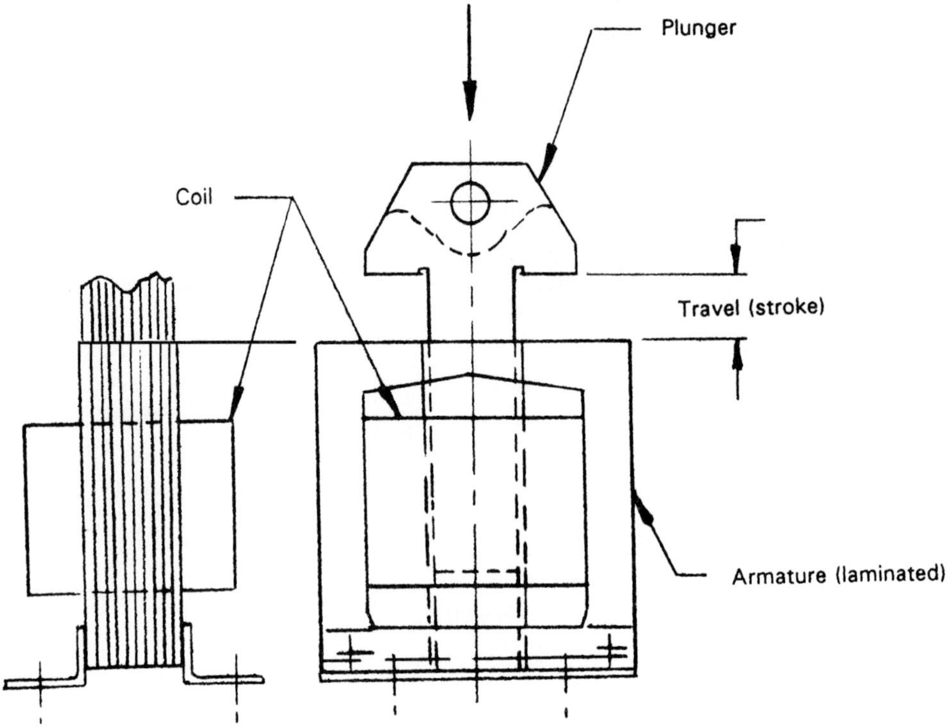


**FIGURE 6.8** Strong-box industrial solenoid.

where $F$ = force, lb
$B$ = flux density, kilolines (Klines)/in$^2$
$S$ = cross-sectional area of the air gap, in$^2$
$K$ = factor for short air gaps, 0.2 to 0.8

*Note*: Try a K factor of 0.5 as an initial estimate for approximate results.

***Coil-Suppression Components for Solenoid Applications.*** Coil suppressors are advised for all applications which use dc solenoids and relay coils. The suppressors protect associated circuitry and reduce electromagnetic interference (EMI). Switching the solenoid or relay coil on the ac side of the power supply will prevent voltage transients, but will increase the dropout time of the solenoid or relay. If dropout time is critical for the application, the solenoid or relay must be switched on the dc side, and a high-speed suppressor should be connected across (parallel or shunt wired) the coil. See Sec. 6.9.6 and Sec. 6.6.5 for arc-suppression circuits (dc and ac).

### 6.6.5 Relays

Relays are switching devices that are available with a multitude of contact arrangements, coil-voltage ratings, contact-voltage ratings, and contact-current ratings. "Make" and "break" ratings, both resistive and inductive, are also specified to indicate the recommended current range for safe operation.

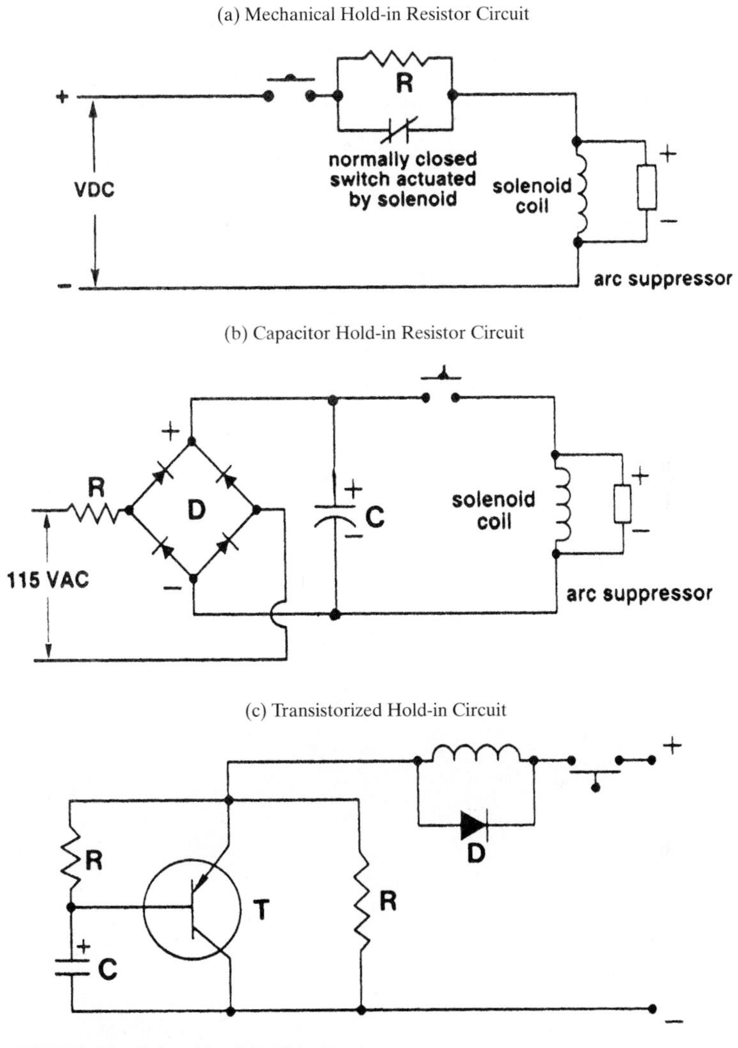


**FIGURE 6.9** Solenoid coil hold-in circuits.

Of prime importance to the modern designer is the class of miniature and subminiature relays now available from many sources. These small relays are produced for relatively high current ratings and supplied in dual-inline package (DIP) form for mounting in 16-pin sockets on printed circuit boards (Fig. 6.11). This allows the designer to produce small, high-power-handling control circuits for motor switching, lighting applications, control circuits, and so forth. The entire control circuit can be made on easily replaceable printed-circuit (PC) cards, affording economy, compact design and easier maintenance. (See Fig. 6.12*a* and 6.12*b*.) The relay shown in Fig. 6.11 is suitable for switching very small inductive loads only, on the order of $\frac{1}{20}$th horsepower or less.

Fig. 6.13 shows a miniature relay made by Aromat that is capable of switching relatively heavy inductive loads of 8 to 10 A. In the photograph, note the large contacts and large con-

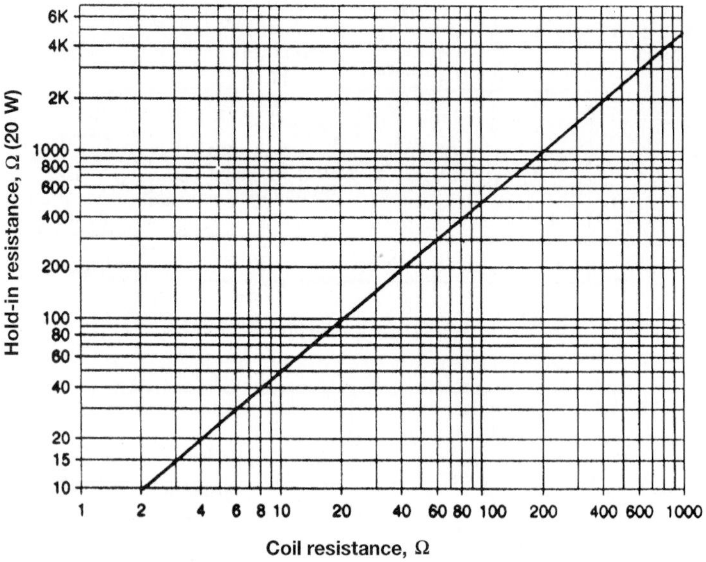


**FIGURE 6.10** Hold-in resistor selection graph for solenoids.

**FIGURE 6.11** Miniature DIP relay.

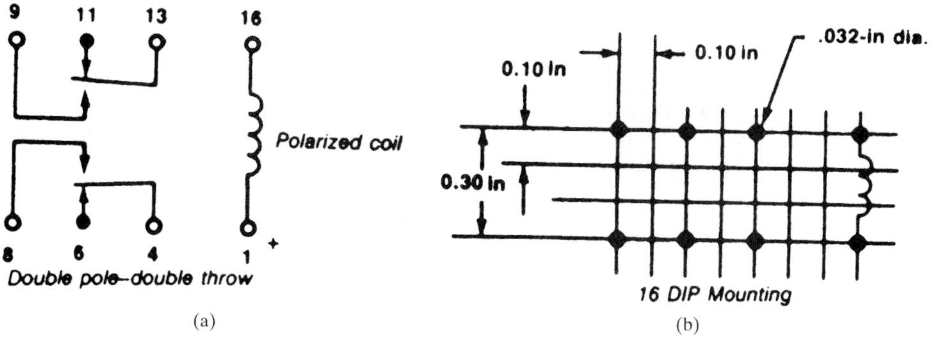


**FIGURE 6.12** (*a*) Miniature relay form; (*b*) 16-pin DIP mounting.

tact spacing. During inductive motor load switching, relay contacts will fail unless the points separate with enough air gap and speed. Permanent magnets placed around the contacts are found on some switch and relay manufacturer's devices. The circular permanent magnets help to collapse or dissipate the dc arc during the break switching cycle. The dc arcs developed at the separating switch or relay contacts are severe, even for relatively low horsepower motors, or other types of inductive loads. Scales are shown in Figs. 6.11 and 6.13 to show the relative size of these relays. Varistors may be used effectively to reduce or eliminate the inductive arcs across switch and relay contacts. Other methods of arc suppression across switching contacts are shown in the following section.

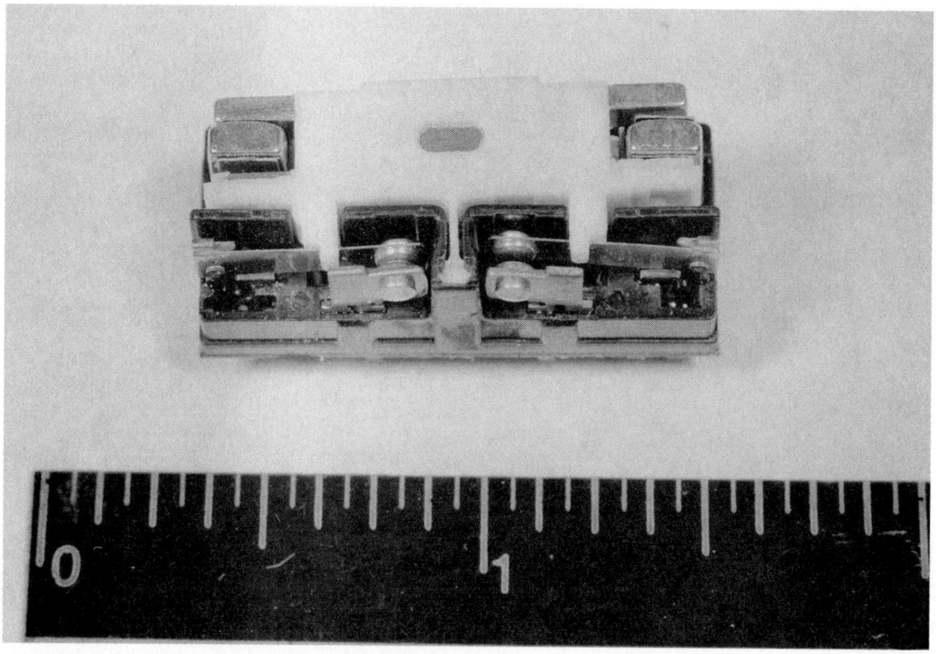


**FIGURE 6.13** Field-load switching relay rated 8 A, 24 V dc.

***Protecting Relay and Switch Contacts (Inductive Loads).*** To suppress or eliminate the destructive arc across relay or switch contacts during the switching of inductive loads, various methods are employed. The following methods will suppress or eliminate this destructive arcing, but at the same time will tend to increase the de-energizing time of the inductive load.

*Case 1.* Resistor across the inductive load. Peak transient voltage developed when the contacts open is determined by the resistance value. Note, however, that the resistor $R_1$ consumes $I^2R$ power while the load is energized (Fig. 6.14).

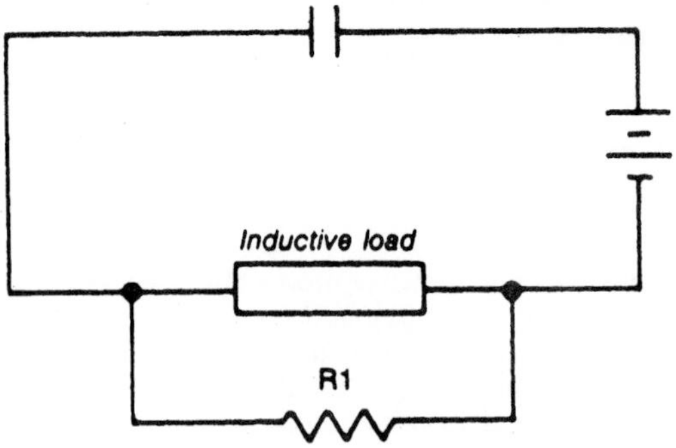


**FIGURE 6.14** Resistor across the inductive load.

*Case 2.* Resistor-capacitor across the contacts (Fig. 6.15). An initial value of $R$ and $C$ may be calculated from

$$R = \frac{E/10}{1 + (50/E)} \qquad \text{and} \qquad C = \frac{I^2}{10}$$

where $R$ = resistance, Ω
$C$ = capacitance, μF
$I$ = current before closing, mA
$E$ = voltage before closing, V

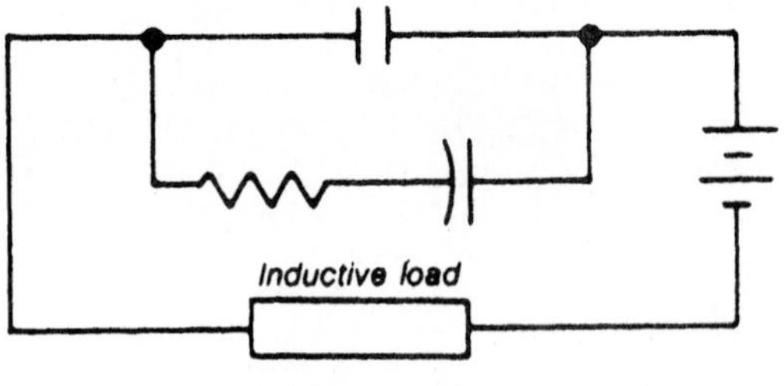


**FIGURE 6.15** *RC* across the contacts.

Peak values of voltage and current must be used to calculate values of $R$ and $C$ for arc suppression of ac loads. To ensure protection with adequate arc suppression, test the circuit and adjust $R$ and $C$ values as necessary to eliminate arcing at the contacts.

*Case 3.* Diode across the load (Fig. 6.16). The diode provides a low-resistance path for stored energy in the load when the contacts open.

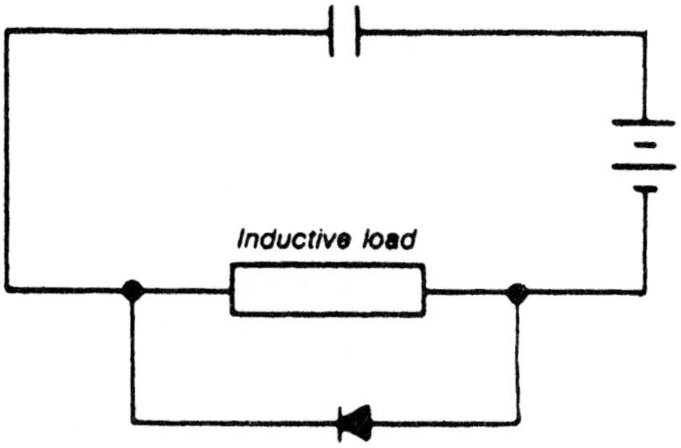


**FIGURE 6.16** Diode across the load.

*Case 4.* Diode–resistor (Fig. 6.17). Treat this case as similar to the diode–zener diode in case 5.

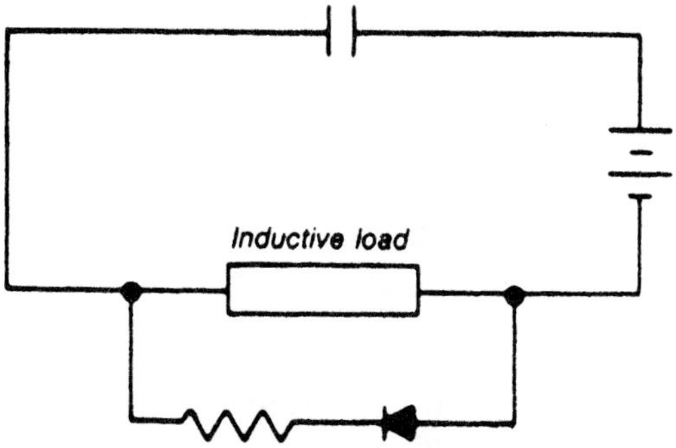


**FIGURE 6.17** Diode-resistor.

*Case 5.* Diode–zener diode (Fig. 6.18). This method speeds deenergization time of the inductive load.

*Case 6.* Varistor (Fig. 6.19). Allowing the varistor to carry approximately 10 percent of the load current will limit the transient voltage level to approximately two times the source voltage. This method is also used in ac circuit applications.

*Case 7.* Resistor-capacitor-diode for dc circuits. This method is for extremely inductive loads (Fig. 6.20). The voltage drop across the opening contacts is zero. $C$ is chosen so that the peak voltage to which it charges does not cause diode, contact gap, or capacitor breakdown.

*Case 8.* Resistor-capacitor-diode for ac circuits. Also for extremely inductive loads (Fig. 6.21), this arc-suppression circuit can be connected across the contact gap, or across the load. Typical component values for 115-V-ac applications are diode peak inverse voltage (PIV) = 400 V; $C_1$, $C_2$ voltage rating = 200 V ac; and $R_1$ = 100 kΩ, 1 W.

***Delay Operation of Relays.*** If a delay is required in relay release action, a diode or capacitor may be applied across the relay coil. A better method for delaying the opening or closing of contacts is to insert a 555 integrated-circuit (IC) timer (Fig. 6.22) delay circuit into the relay circuit. Circuits of this type can be found in the 555 timer applications book listed in the bibliography. A 555 IC timer delay circuit is shown in Fig. 6.23. This delay timer has many appli-

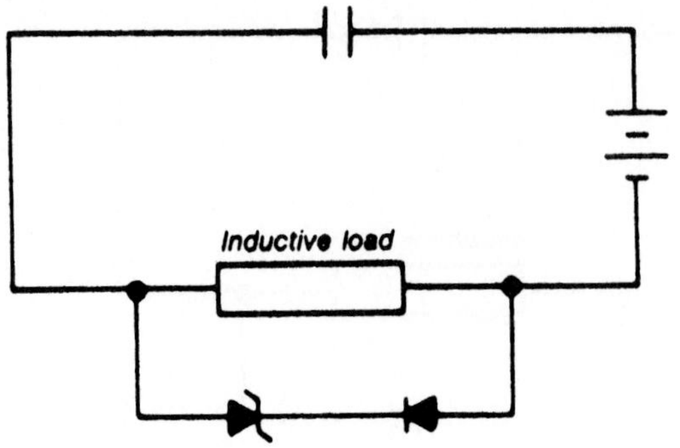


**FIGURE 6.18** Diode–zener diode.

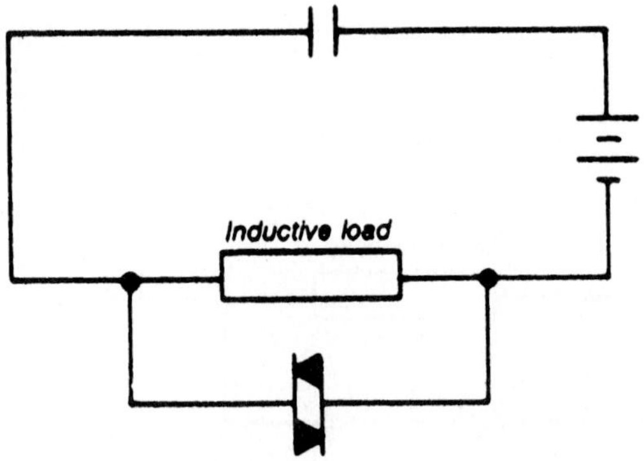


**FIGURE 6.19** Varistor across the inductive load.

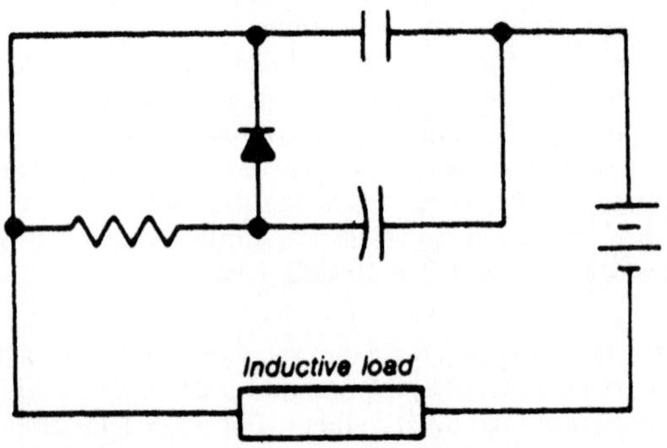


**FIGURE 6.20** Resistor-capacitor-diode, dc.

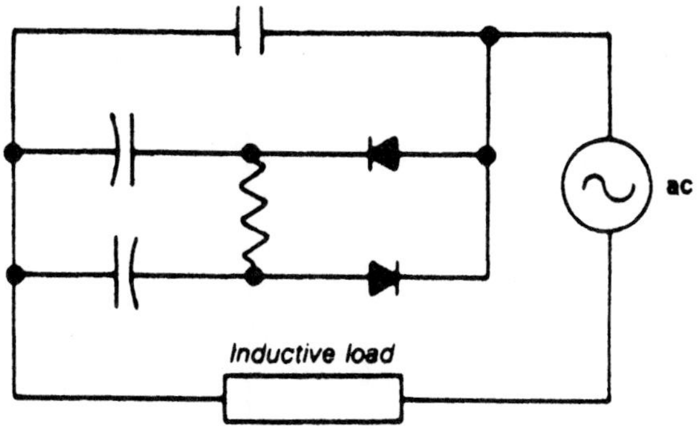


**FIGURE 6.21** Resistor-capacitor-diode, ac.

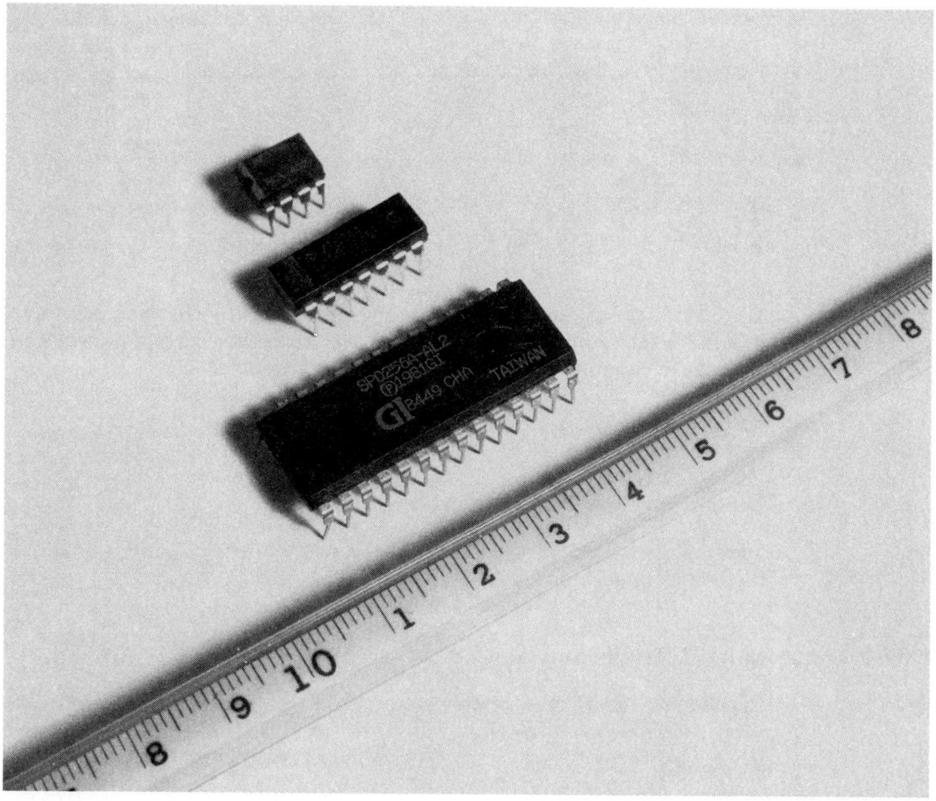


**FIGURE 6.22** Integrated circuits. Top: 555 timer; center: logic circuit; bottom: complex speech synthesizer.

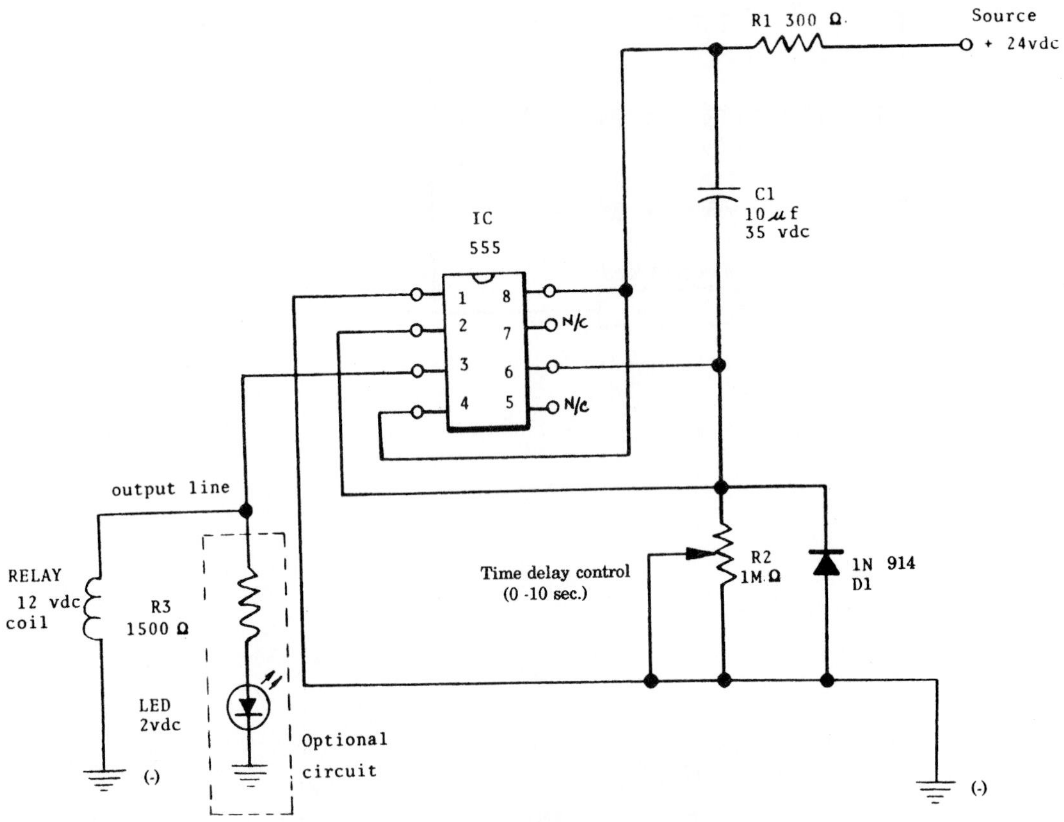


**FIGURE 6.23** 555 IC delay timer circuit (for relay control).

cations and can be powered by different dc voltages by inserting the appropriate dropping resistor after the source, at $R_1$. Refer to the electronics sections for more data on ICs and 555 timers. Potentiometer $R_2$ controls the delay time interval.

## 6.6.6 Electric Motors and Codes

### 6.6.6.1 Motor Dimensions for NEMA Frames

Refer to Figs. 6.24 to 6.26. These figures show the sizes, mountings, and interfaces for all the NEMA types and frame sizes.

### 6.6.6.2 Electric Motor Terminology

**Ambient (Amb)**: The temperature of the space around the motor. Most motors are designed to operate in an ambient not over 40°C (104°F). *Note:* A rating of 40°C ambient is not the same as a rating of 40°C rise; see **Temperature rise**.

**Airover**: Motors intended for fan and blower service. Must be located in the airstream to provide motor cooling.

**Efficiency**: The ratio of output power divided by input power; usually expressed as a percentage. A measure of how well the electrical energy input to a motor is converted into mechanical energy at the output shaft. The higher the efficiency, the better the conversion process and lower the operating costs.

**Enclosure (Encl)**: The motor's housing. Types:

**Dripproof (DP)**: Ventilation openings in end shields and shell placed so drops of liquid falling within an angle of 15° from vertical will not affect performance. Usually used indoors, in fairly clean locations.

**Totally enclosed (TE)**: No ventilation openings in motor housing (but not airtight). Used in locations which are dirty, damp, oily, etc.

*Totally enclosed, fan-cooled (TEFC):* Includes an external fan, in a protective shroud, to blow cooling air over the motor.

*Totally enclosed, nonventilated (TENV):* Not equipped with an external cooling fan. Depends on convection air for cooling.

*Totally enclosed, airover (TEAO):* Airflow from driven device provides cooling.

*Explosionproof (EX PRF):* A totally enclosed motor designed to withstand an internal explosion of specified gases or vapors, and not allow the internal flame or explosion to escape.

**Full-load amps (F/L amps)**: Line current (amperage) drawn by a motor when operating at rated load and voltage. Shown on motor nameplate. Important for proper wire size selection and motor starter heater selection.

**Frame**: Usually refers to the NEMA system of standardized motor mounting dimensions, which facilitates replacement.

**Bearings (Brgs)**: Basic types:

**Sleeve (Slv)**: Preferred where low noise level is important, as on fan and blower motors. Unless otherwise stated, sleeve bearing motors listed herein can be mounted in any position, including shaft-up or shaft-down (all-position mounting).

**Ball**: Used where higher load capacity is required or periodic lubrication is impractical. Two means used to keep out dirt:

*Shields:* Metal rings with close running clearance on one side (single-shielded) or both sides (double-shielded) of bearing.

*Seals:* Similar to shields, except have rubber lips that press against inner race, more effectively excluding dirt, etc.

**Unit**: Motors are constructed with a long, single sleeve bearing. For fan duty only. All-position mounting unless otherwise stated.

**Hertz (Hz)**: frequency in cycles per second, of ac power; usually 60 Hz in USA, 50 Hz overseas. (Abbreviated Cps or Cy in the past.)

**Insulation (Ins)**: In motors, usually classified by maximum allowable operating temperatures: Class A—105°C (221°F), Class B—130°C (266°F), Class F—155°C (311°F), Class H—180°C (356°F).

**Motor speeds**

**Synchronous**: The theoretical maximum speed at which an induction-type motor can operate. Synchronous speed is determined by the power line frequency and motor design (number of poles), and calculated by the formula

$$\text{Syn. rpm} = \frac{\text{Frequency in Hz} \times 120}{\text{no. of poles}}$$

## Motor Dimensions for NEMA Frames - Inches

| NEMA Frame | D ♦ | 2E | 2F | BA | H | N-W | U | V ■ | Wide | Key - Thick | Long |
|---|---|---|---|---|---|---|---|---|---|---|---|
| 42 | 2.625 | 3.500 | 1.688 | 2.063 | 0.281 slot | 1.125 | 0.375 | ..... | ..... | 0.328 flat | ..... |
| 48 | 3.000 | 4.250 | 2.750 | 2.500 | 0.344 slot | 1.500 | 0.500 | ..... | ..... | 0.453 flat | |
| 56 | 3.500 | 4.875 | 3.000 | 2.750 | 0.344 slot | 1.875* | 0.625* | ..... | 0.188* | 0.188* | 1.375* |
| 56H | 3.500 | 4.875 | 3 & 5 | 2.750 | 0.344 slot | 1.875* | 0.625* | ..... | 0.188* | 0.188* | 1.375* |
| 56HZ | 3.500 | 4.875 | 3 & 5 | 2.750 | 0.344 slot | 2.250 | 0.875 | 2.000 | 0.188 | 0.188 | 1.375 |
| 66 | 4.125 | 5.875 | 5.000 | 3.125 | 0.406 slot | 2.250 | 0.750 | ..... | 0.188 | 0.188 | 1.875 |
| 143T | 3.500 | 5.500 | 4.000 | 2.250 | 0.344 dia. | 2.250 | 0.875 | 2.000 | 0.188 | 0.188 | 1.375 |
| 145T | 3.500 | 5.500 | 5.000 | 2.250 | 0.344 dia. | 2.250 | 0.875 | 2.000 | 0.188 | 0.188 | 1.375 |
| 182 | 4.500 | 7.500 | 4.500 | 2.750 | 0.406 dia. | 2.250 | 0.875 | 2.000 | 0.188 | 0.188 | 1.375 |
| 184 | 4.500 | 7.500 | 5.500 | 2.750 | 0.406 dia. | 2.250 | 0.875 | 2.000 | 0.188 | 0.188 | 1.375 |
| 182T | 4.500 | 7.500 | 4.500 | 2.750 | 0.406 dia. | 2.750 | 1.125 | 2.500 | 0.250 | 0.250 | 1.750 |
| 184T | 4.500 | 7.500 | 5.500 | 2.750 | 0.406 dia. | 2.750 | 1.125 | 2.500 | 0.250 | 0.250 | 1.750 |
| 203# | 5.000 | 8.000 | 5.500 | 3.125 | 0.406 dia. | 2.250 | 0.750 | 2.000 | 0.188 | 0.188 | 1.375 |
| 204# | 5.000 | 8.000 | 6.500 | 3.125 | 0.406 dia. | 2.250 | 0.750 | 2.000 | 0.188 | 0.188 | 1.375 |
| 213 | 5.250 | 8.500 | 5.500 | 3.500 | 0.406 dia. | 3.000 | 1.125 | 2.750 | 0.250 | 0.250 | ..... |
| 215 | 5.250 | 8.500 | 7.000 | 3.500 | 0.406 dia. | 3.000 | 1.125 | 2.750 | 0.250 | 0.250 | ..... |
| 213T | 5.250 | 8.500 | 5.500 | 3.500 | 0.406 dia. | 3.375 | 1.375 | 3.125 | 0.312 | 0.312 | 2.375 |
| 215T | 5.250 | 8.500 | 7.000 | 3.500 | 0.406 dia. | 3.375 | 1.375 | 3.125 | 0.312 | 0.312 | 2.375 |
| 224# | 5.500 | 9.000 | 6.750 | 3.500 | 0.406 dia | 3.000 | 1.000 | 2.750 | 0.250 | 0.250 | 2.000 |
| 225# | 5.500 | 9.000 | 7.500 | 3.500 | 0.406 dia. | 3.000 | 1.000 | 2.750 | 0.250 | 0.250 | 2.000 |
| 254# | 6.250 | 10.000 | 8.250 | 4.250 | 0.656 dia. | 3.375 | 1.125 | 3.125 | 0.250 | 0.250 | 2.375 |
| 254U | 6.250 | 10.000 | 8.250 | 4.250 | 0.531 dia. | 3.750 | 1.375 | 3.500 | 0.312 | 0.312 | 2.750 |
| 256U | 6.250 | 10.000 | 10.000 | 4.250 | 0.531 dia. | 3.750 | 1.375 | 3.500 | 0.312 | 0.312 | 2.750 |
| 254T | 6.250 | 10.000 | 8.250 | 4.250 | 0.531 dia. | 4.000 | 1.625 | 3.750 | 0.375 | 0.375 | 2.875 |
| 256T | 6.250 | 10.000 | 10.000 | 4.250 | 0.531 dia. | 4.000 | 1.625 | 3.750 | 0.375 | 0.375 | 2.875 |
| 284# | 7.000 | 11.000 | 9.500 | 4.750 | 0.656 dia. | 3.750 | 1.250 | 3.500 | 0.250 | 0.250 | 2.750 |
| 284U | 7.000 | 11.000 | 9.500 | 4.750 | 0.531 dia. | 4.875 | 1.625 | 4.625 | 0.375 | 0.375 | 3.750 |
| 286U | 7.000 | 11.000 | 11.000 | 4.750 | 0.531 dia. | 4.875 | 1.625 | 4.625 | 0.375 | 0.375 | 3.750 |

| | | | | | | | | | | | |
|---|---|---|---|---|---|---|---|---|---|---|---|
| 284TS | 7.000 | 11.000 | 9.500 | 4.750 | 0.531 dia. | 3.250 • | 1.625 | 3.000 • | 0.375 | 0.375 | 1.875 • |
| 286T | 7.000 | 11.000 | 11.000 | 4.750 | 0.531 dia. | 4.625 | 1.875 | 4.375 | 0.500 | 0.500 | 3.250 |
| 324# | 8.000 | 12.500 | 10.500 | 5.250 | 0.656 dia. | 4.875 | 1.625 | 4.625 | 0.375 | 0.375 | 3.750 |
| 326# | 8.000 | 12.500 | 12.000 | 5.250 | 0.656 dia. | 4.875 | 1.625 | 4.625 | 0.375 | 0.375 | 3.750 |
| 324U | 8.000 | 12.500 | 10.500 | 5.250 | 0.656 dia. | 5.625 | 1.875 | 5.375 | 0.500 | 0.500 | 4.250 |
| 326U | 8.000 | 12.500 | 12.000 | 5.250 | 0.656 dia. | 5.625 | 1.875 | 5.375 | 0.500 | 0.500 | 4.250 |
| 324T | 8.000 | 12.500 | 10.500 | 5.250 | 0.656 dia. | 5.250 | 2.125 | 5.000 | 0.500 | 0.500 | 3.875 |
| 326T | 8.000 | 12.500 | 12.000 | 5.250 | 0.656 dia. | 5.250 | 2.125 | 5.000 | 0.500 | 0.500 | 3.875 |
| 326TS | 8.000 | 12.500 | 12.000 | 5.250 | 0.656 dia. | 3.750 • | 1.875 • | 3.500 • | 0.500 | 0.500 | 2.000 • |
| 364# | 9.000 | 14.000 | 11.250 | 5.875 | 0.656 dia. | 5.625 | 1.875 | 5.375 | 0.500 | 0.500 | 4.250 |
| 364S# | 9.000 | 14.000 | 11.250 | 5.875 | 0.656 dia. | 3.250 | 1.625 | 3.000 | 0.375 | 0.375 | 1.875 |
| 364T | 9.000 | 14.000 | 11.250 | 5.875 | 0.656 dia. | 5.875 | 2.375 | 5.625 | 0.625 | 0.625 | 4.250 |
| 365# | 9.000 | 14.000 | 12.250 | 5.875 | 0.656 dia. | 5.625 | 1.875 | 5.375 | 0.500 | 0.500 | 4.250 |
| 365T | 9.000 | 14.000 | 12.250 | 5.875 | 0.656 dia. | 5.875 | 2.375 | 5.375 | 0.625 | 0.625 | 4.250 |
| 364U | 9.000 | 14.000 | 11.250 | 5.875 | 0.656 dia. | 6.375 | 2.125 | 6.125 | 0.500 | 0.500 | 5.000 |
| 365U | 9.000 | 14.000 | 12.250 | 5.875 | 0.656 dia. | 6.375 | 2.125 | 6.125 | 0.500 | 0.500 | 5.000 |
| 404T | 10.000 | 16.000 | 12.250 | 6.625 | 0.812 dia. | 7.250 | 2.875 | 7.000 | 0.750 | 0.750 | 5.625 |
| 405T | 10.000 | 16.000 | 13.750 | 6.625 | 0.812 dia. | 7.250 | 2.875 | 7.000 | 0.750 | 0.750 | 5.625 |
| 444T | 11.000 | 18.000 | 14.500 | 7.500 | 0.812 dia | 8.500 | 3.375 | 8.250 | 0.875 | 0.875 | 6.938 |
| 445T | 11.000 | 18.000 | 16.500 | 7.500 | 0.812 dia. | 8.500 | 3.375 | 8.250 | 0.875 | 0.875 | 6.938 |

NOTES: ◆ Dimension D will never be greater than the above values on rigid mount motors, but it may be less, so that shims up to 0.063" thick may be required for coupled or geared machines. • Standard short shaft for direct-drive applications. # Discontinued NEMA frames. * Certain NEMA 56Z frame motors have 0.500" diameter × 1.500" long shaft with 0.045" flat. ■ Dimension V is shaft length available for coupling, pinion or pulley hub; this is a minimum value.

**FIGURE 6.24** (*Continued*)

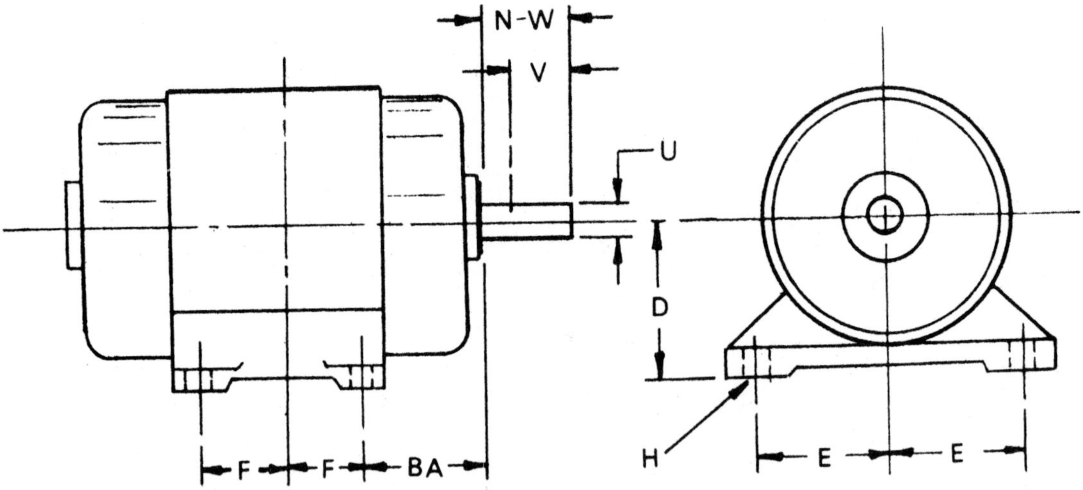


**FIGURE 6.25** NEMA standard motor dimensions.

**Full-load**: The nominal speed at which an induction motor operates under rated input and load conditions. This will always be less than the synchronous speed and will vary depending on the rating and characteristics of the particular motor. For example, four-pole 60-Hz fractional horsepower motors have a *synchronous* speed of 1800 rpm, a *nominal* full-load speed (as shown on the nameplate) of 1725 rpm, and an *actual* full-load speed ranging from 1715 to 1745 rpm.

**Motor types**: Classified by operating characteristics and/or type of power required:

**Induction motors for AC operation**: Most common type. Speed remains relatively constant as load changes. There are several kinds of induction motors:

**Single phase**: Available in these types:

*Shaded pole:* Low starting torque, low cost. Usually used in direct-drive fans and blowers, and in small gearmotors.

*Permanent split capacitor (PSC):* Performance and applications similar to shaded pole but more efficient, with lower line current and higher horsepower capabilities.

*Split-phase start—induction run (or simply split phase):* moderate starting torque, high breakdown torque. Used on easy-starting equipment, such as belt-driven fans and blowers, grinders, centrifugal pumps, gearmotors, etc.

*Capacitor-start, induction-run (or simply capacitor start or capacitor):* High starting and breakdown torque, medium starting current. Used on hard-starting applications; compressors, positive displ. pumps, farm equipment, etc.

*Capacitor-start, capacitor-run:* Similar to capacitor-start, induction-run, except have higher efficiency. Generally used in higher single phase hp ratings.

**Three phase**: Operate on three-phase power only. High starting and breakdown torque, high efficiency, medium starting current, simple, rugged design, long life. For industrial uses.

**Direct current (dc)**: Usable only if dc available and in adjustable speed applications.

**Ac/dc (ac series or universal)**: Operate on ac (60 or 50 Hz) or dc power. High speed, usually 5000 rpm or more. Brush type. Speed drops rapidly as load increases. Useful for drills, saws, etc., where high output and small size are desired and speed characteristic and limited life (primarily of brushes) is acceptable.

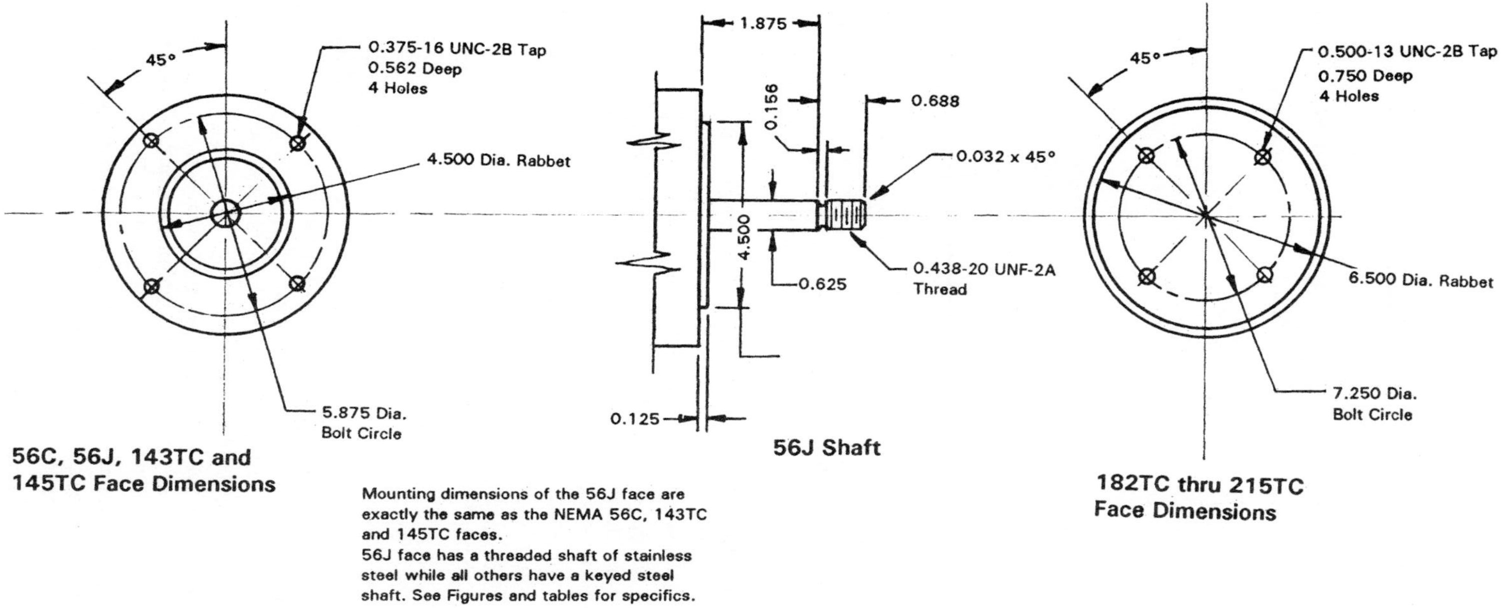
45°
0.375-16 UNC-2B Tap
0.562 Deep
4 Holes
4.500 Dia. Rabbet
5.875 Dia.
Bolt Circle
56C, 56J, 143TC and
145TC Face Dimensions
1.875
0.156
0.688
0.032 x 45°
4.500
0.625
0.438-20 UNF-2A
Thread
0.125
56J Shaft
Mounting dimensions of the 56J face are exactly the same as the NEMA 56C, 143TC and 145TC faces.
56J face has a threaded shaft of stainless steel while all others have a keyed steel shaft. See Figures and tables for specifics.
45°
0.500-13 UNC-2B Tap
0.750 Deep
4 Holes
6.500 Dia. Rabbet
7.250 Dia.
Bolt Circle
182TC thru 215TC
Face Dimensions

(a)

**FIGURE 6.26** NEMA C and J face mount dimensions.

## NEMA C and J-Face Mount Dimensions - Inches

| NEMA Face | Shaft Diameter (U) | Shaft Long (N-W) | Rabbet Diameter | Bolt Circle Diameter |
|---|---|---|---|---|
| 42C | 0.375 | 1.125 | 3.000 | 3.750 |
| 48C | 0.500 | 1.500 | 3.000 | 3.750 |
| 56C | 0.625 | 1.875 | 4.500 | 5.875 |
| 56J | 0.625 | 2.438 | 4.500 | 5.875 |
| 143TC & 145TC | 0.875 | 2.250 | 4.500 | 5.875 |
| 182TC & 184TC | 1.125 | 2.750 | 8.500 | 7.250 |
| 213TC & 215TC | 1.375 | 3.375 | 8.500 | 7.250 |
| 254TC & 256TC | 1.625 | 4.000 | 8.500 | 7.250 |
| 284TC & 286TC | 1.875 | 4.625 | 10.500 | 9.000 |

(b)

**FIGURE 6.26** (*Continued*)

**Mounting (Mtg)**: Basic types:

**Rigid**: Motor solidly fastened to equipment through metal base that is bolted or welded to motor shell.

**Resilient (Res)**: Sometimes called *rubber* or *rbr.*—motor shell isolated from base by vibration-absorbing material, such as rubber rings on the end shields, to reduce transmission of vibration to the driven equipment.

**Face or flange**: Shaft end has a flat mounting surface, machined to standard dimensions, with holes to allow easy, secure mounting to driven equipment. Commonly used on jet pumps, oil burners, and gear reducers.

**Studs**: Motor has bolts extending from front or rear, by which it is mounted. Often used on small, direct drive fans and blowers.

**Yoke**: Tabs or ears are welded to motor shell, to allow bolting to a fan column or bracket. Used on fan-duty motors.

**Power**: The energy used to do work. Also the rate at which work is done. Measured in watts, horsepower, etc.

**Power factor**: The ratio of real power (watts) divided by apparent power (volt-amperes). Do not confuse power factor with efficiency. A measure of the extent to which power transmission or distribution systems are fully utilized.

**Rotation (Rot)**: Direction in which shaft rotates: CW = clockwise; CCW = counterclockwise; Rev = reversible, rotation can be changed. Unless stated otherwise, rotation specified is as viewed facing shaft end of motor.

**Service factor (SF, Svc Fctr)**: A measure of the reserve margin built into a motor. Motors rated over 1.0 SF have more than normal margin, and are used where unusual conditions such as occasional high or low voltage, momentary overloads, etc., are likely to occur.

**Severe duty**: A totally enclosed motor with extra protection (shaft slinger, gasketed terminal box, etc.) to resist the entry of contaminants. Used in extra dirty, wet, or other contaminated environments.

**Temperature rise**: The amount by which a motor, operating under rated conditions, is hotter than its surroundings. On most motors, manufacturers have replaced the *rise* rating on the motor nameplate with a listing of the *ambient-temperature* rating, insulation class, and service factor.

**Thermal protector**: A temperature-sensing device built into the motor that disconnects the motor from its power source if the temperature becomes excessive for any reason. Basic types:

**Automatic-reset (Auto)**: After motor cools, protector automatically restores power. Should not be used where unexpected restarting would be hazardous.

**Manual-reset (Man)**: An external button must be pushed to restore power to motor. Preferred where unexpected restarting would be hazardous, as on saws, conveyors, compressors, etc.

**Impedance (Imp)**: Motor is designed so that it will not burn out in less than 15 days under locked rotor (stalled) conditions, in accordance with UL (Underwriter Laboratories) standard No. 73.

**Torque**: Twist, or turning ability, as applied to a shaft. Measured in foot-pounds (ft-lb), inch-pounds (in-lb), ounce-feet (oz-ft), or ounce-inches (oz-in). In a motor, two torque values are important:

**Locked rotor torque, or starting torque**: The torque produced at initial start.

**Breakdown torque**: The maximum torque a motor will produce while running, without an abrupt drop in speed and power.

**Voltage**: The pressure in an electrical system. The force pushing the electric current through the circuit, like pressure in a water system.

## 6.6.6.3 Electric Motor Engineering Applications

The proper application of an electric motor depends on the selection of a motor that meets the kinetic energy requirements of the *driven* machine without the motor overheating and without exceeding the torque rating of the motor.

***Motor Selection Governing Factors***

- Speed
- Horsepower
- Inertial requirements
- Torque

*Speed.* Variation in speed, from no load to full load, is greatest with motors having series field windings and is absent with synchronous motors.

*Horsepower.* The peak horsepower determines the maximum torque required by the driven machine, and the motor "maximum running torque" must be in excess of this value.

*Torque.* Starting torque requirements can vary from 5 percent of full load to over 250 percent of full load, depending on the driven machine requirements. The motor torque supplied to the machine must be well above that required by the driven machine at all points up to full speed. The greater the excess torque, the greater the motor acceleration.

***Electric Motor Calculations.*** The time required $t$ in seconds for acceleration from rest to full-load speed is given by

$$t = \frac{WK^2Nf}{308TA} \quad \text{sec (approximate)}$$

where $WK^2$ = inertia of rotating parts, lb · ft$^2$ and $W$ = weight, lb; $K$ = radius of gyration of rotating part

$Nf$ = full-load speed, rpm
$TA$ = torque, average lb · ft available for acceleration
308 = constant converting minutes to seconds, weight into mass, and radius into circumference

(*Note:* If acceleration time $t$ is more than 20 sec, special motors or motor starters may be required to prevent motor overheating.)

***Running Torque.*** Running torque is given by

$$TR = \frac{5250(\text{hp})}{Nr} \qquad \text{lb} \cdot \text{ft}$$

where hp = horsepower supplied to the driven machine
$Nr$ = running speed, rpm
5250 = constant, converting hp to ft · lb/min and work per revolution into torque, lb · ft

Transposing for horsepower:

$$\text{hp} = \frac{TR\,N}{r}$$

***Power.*** The term *motor load* refers to the horsepower required to drive a machine. When the load is cyclic, a horsepower–time curve for the driven machine is useful. Peak and rms horsepower can be determined from this curve, and rms horsepower indicates the required continuous motor rating. Peak load horsepower is not always an indication of the required motor rating except when peak load is maintained for a period of time, in which case the motor horsepower rating should not be less than the peak load horsepower.

***Intermittent Duty.*** The rms method is the usual technique for determining motor size (hp) for duty-cycle loads in which the motor is loaded for a time period and then idles for a time period. Therefore

$$\text{Rms load (hp)} = \text{peak load (hp)} \sqrt{tp}$$

where $tp$ = on time cycle period, percent of cycle period (see Fig. 6.27), i.e., 38 percent = 0.38, etc.

***Motor Curves.*** Different curves are available from the motor manufacturers, including locked-rotor–torque curves and motor speed–torque curves. Motor performance is determined from these curves. The design engineer needs to determine the following criteria:

**1.** Will the motor start under load?

**2.** Will the motor temperature rise to unacceptable levels during operation?

**3.** Can the motor attain running speed?

### 6.6.6.4 Types and Selection of Electric Motors

***Types***

*DC motors*

- Permanent magnet
- Series-wound
- Shunt-wound
- Compound-wound

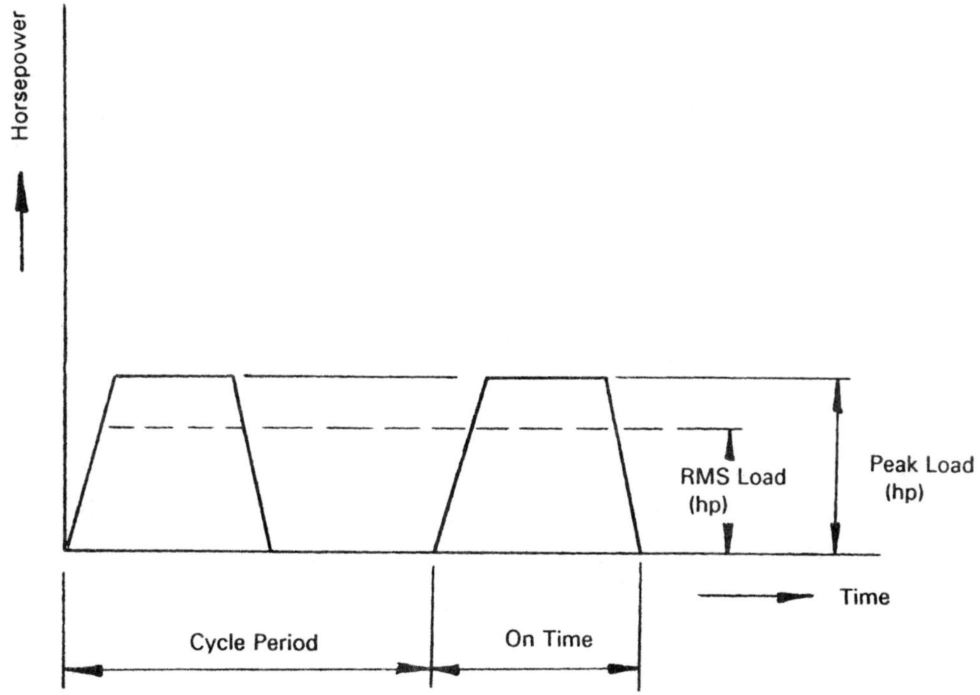


**FIGURE 6.27** Motor duty timing diagram.

*AC motors*

- **I.** Single-phase class
  - **A.** Induction
    - **1.** Squirrel cage
      - *a.* Split phase
      - *b.* Capacitor start
      - *c.* Permanent split capacitor
      - *d.* Shaded pole
      - *e.* Two-valve capacitor
    - **2.** Wound rotor
      - *a.* Repulsion
      - *b.* Repulsion start
      - *c.* Repulsion induction
  - **B.** Synchronous
    - **1.** Shaded pole
    - **2.** Hysteresis
    - **3.** Reluctance
    - **4.** Permanent magnet
- **II.** Polyphase class
  - **A.** Induction
    - **1.** Wound rotor
    - **2.** Squirrel cage
  - **B.** Synchronous
- **III.** Universal motors
  - **A.** Ac/dc wound rotor

***Selection of Electric Motors—General***

*Ac-type-motors.* Split-phase motors are used for light loads and where power is not applied until motor reaches top speed (belt-drive fans and blowers, table saws, and drill presses). Capacitor start motors and three-phase motors are used for heavy loads (conveyors, compressors, and other heavy starting loads). Universal motors (ac and dc) are used for high-energy, low-weight applications.

*Dc-type motors.* Dc series wound motors and traction motors are used when high starting torque is required.

### 6.6.6.5 Efficiency of Electric Motors

The relation between input, output, and efficiency for electric motors is expressed by the following equations, wherein hp = horsepower; $\eta$ = efficiency; $I$ = line current, A; $V$ = line voltage, V; PF = power factor. For dc motors:

$$\text{Hp} = \frac{\eta IV}{746}$$

For single-phase ac motors:

$$\text{Hp} = \frac{\eta IV\,(\text{PF})}{746}$$

For three-phase ac motors:

$$\text{Hp} = \frac{\sqrt{3}\,\eta IV\,(\text{PF})}{746} \qquad \eta = \frac{\text{Hp}\,(746)}{\sqrt{3}\,IV\,(\text{PF})} \qquad \text{(transposed for efficiency)}$$

To solve the preceding equations for efficiency ($\eta$), or any other variable, transpose the equations.

### 6.6.6.6 Measuring Motor Horsepower

Refer to Fig. 6.28, which shows the *prony brake* or dynamometer method of measuring motor horsepower. This method is valid for all types of motors including internal combustion engines, turbines, and all electric motors.

$$TQ = 24 \times 25 = 600 \text{ lb} \cdot \text{in or } 50 \text{ lb} \cdot \text{ft}$$

From this

$$TR = \frac{5250\,(\text{hp})}{Nr} \qquad \text{hp} = \frac{TR\,N}{r} = \frac{50\,(250)}{5250} = 2.38 \text{ hp}$$

where hp = horsepower (1 hp = 33,000 ft · lb/min = 550 ft · lb/sec).
$TR$ = running torque, lb · ft
$Nr$ = running speed, rpm

### 6.6.6.7 Bodies with Straight-Line Motion Connected to Rotating Drives

Bodies with straight-line motion connected to rotating drive units, such as electric motors, by rack and pinion, cable, or cam mechanisms, have the equivalent $WK^2$ value given by

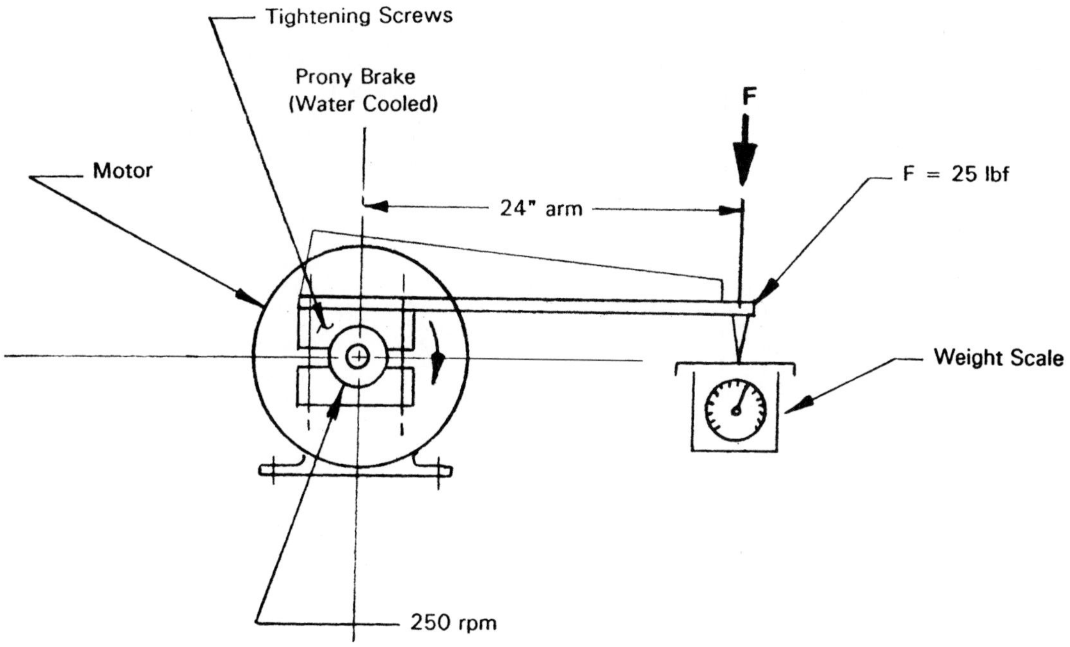


**FIGURE 6.28** Prony-brake dynamometer.

$$WK^2 = W_1 \left(\frac{S}{2\pi n}\right)^2$$

where $W_1$ = load weight, lb
$S$ = translation speed, ft/min
$n$ = rotational speed, rpm

[*Note:* The expression $WK^2$ = inertia of rotating parts, lb · ft$^2$ (see Sec. 6.6.6.3).]

For translating bodies, acceleration time $t_1$ is given by

$$t_1 = \frac{W(S_2 - S_1)}{1932\ Ft}$$

where $W$ = weight of load, lb
$S_2 - S_1$ = translation speed difference, ft/min
$Ft$ = translation force, lbf

### 6.6.6.8 Full-Load and Locked Rotor Currents of Electric Motors

Table 6.10 shows the full-load and locked-rotor currents for typical single-phase and dc electric motors. This table will assist in the selection of the appropriate fuses to use in typical electric motor circuits; however, the particular motor application determines to a large extent the type and rating of the appropriate fuses. Fusing applications are detailed in the various fuse manufacturers' catalogs.

**TABLE 6.10** Single-Phase and DC Motor Currents

| HP | Alternating Current | | | | Direct Current | | | |
|---|---|---|---|---|---|---|---|---|
| | 115 Volts | | 230 Volts | | 115 Volts | | 230 Volts | |
| | Full Load | Locked Rotor | Full Load | Locked Rotor | Full Load | Locked Rotor | Full Load | Locked Rotor |
| 2 | 24.0 | 144.0 | 12.0 | 72.0 | 17.0 | 170.0 | 8.5 | 85.0 |
| 1-1/2 | 20.0 | 120.0 | 10.0 | 60.0 | 13.2 | 132.0 | 6.6 | 66.0 |
| 1 | 16.0 | 96.0 | 8.0 | 48.0 | 9.6 | 96.0 | 4.8 | 48.0 |
| 3/4 | 13.8 | 82.8 | 6.9 | 41.4 | 7.4 | 74.0 | 3.7 | 37.0 |
| 1/2 | 9.8 | 58.8 | 4.9 | 29.4 | 5.4 | 54.0 | 2.7 | 27.0 |
| 1/3 | 7.2 | 43.2 | 3.6 | 21.6 | 3.8 | 38.0 | 1.9 | 19.0 |
| 1/4 | 5.8 | 34.8 | 2.9 | 17.4 | 3.0 | 30.0 | 1.5 | 15.0 |
| 1/6 | 4.4 | 26.4 | 2.2 | 13.2 | 2.4 | 24.0 | 1.2 | 12.0 |
| 1/8 | 3.8 | 22.8 | 1.9 | 11.4 | 2.2 | 22.0 | 1.1 | 11.0 |
| 1/10 | 3.0 | 18.0 | 1.5 | 9.0 | 2.0 | 20.0 | 1.0 | 10.0 |
| 1/20 | 1.5 | 9.0 | — | — | — | — | — | — |

The motor current table (Table 6.10) is general in nature and may not apply to all types of motors and specific applications; it is included in this section for general reference purposes only. Fusing requirements are usually only 1.5 to 2.5 times the full-load current. A slow-blow type of fuse may be employed to allow for motor inrush currents which occur during motor startup under load.

Any electric motor of reasonable size can be easily and quickly tested to determine motor characteristics such as full-load current, horsepower, efficiency, torque, and locked-rotor currents, provided the proper test and measuring equipment is available.

## 6.6.7 Switches

The term *switch,* in the electrical context, is defined as a circuit element for interrupting electric power from an electric power supply source to an electrical load circuit. Switches are also used for selecting power supply sources and for isolating load circuits.

Switches of many types are available for a vast array of electrical functions such as

- Limit switches
- Low-power switches (toggle), in various forms [single-pole single-throw, single-pole double-throw, double-pole double-throw (SPST, SPDT, DPDT), etc)]
- Rotary circuit-selective switches (for electronics and switchgear use)
- Line selector switches (low, medium, and high voltage)
- High-power interrupter switches (fused and nonfused)
- Isolator or non-load-break switches (low, medium, and high voltage)
- High-current switches (low, medium, and high voltage)
- Vacuum switches (medium and high voltage)
- Transformer tap-changer switches

Some of the more important types of switches used in electromechanical design practice are described in the subsections of the handbook that follow.

### 6.6.7.1 General-Purpose Switches and Limit Switches

***Limit Switches and Standard Switches.*** Electrical switches are available in a vast array of forms and ratings to perform many functions. Numerous contact forms and arrangements are produced to standards, and many can be ordered to meet the requirements of special applications.

Wire-connection points on switches are available in such forms as: turret lugs (solder connection), eye lugs (solder connection), screws (mechanical connection), crimp lugs (mechanical connection), spade lugs (mechanical connection), solder cup lugs (solder connection), and clamp lugs (mechanical connection).

A particular switch type that is important to the equipment designer is the *limit* switch. This type is generally used as a control element in many electromechanical systems for the control of electric motors, solenoids, lights, interlocks, and machine-control circuits. The limit switch is available in the following actuating forms:

- Lever (blade or wire)
- Roller lever (short and long)
- Plunger (spring)
- Roller plunger
- Button
- Magnetic proximity
- Rotary
- Wobble stick

Common contact forms are shown in Fig. 6.29 for SPST, DPDT, SPDT, and DPDT (eight-point).

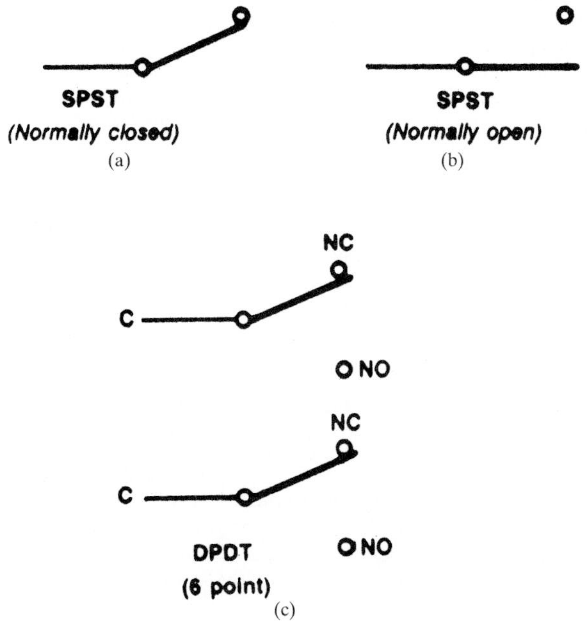


**FIGURE 6.29** Basic switch symbols and types.

**FIGURE 6.29** *(Continued)*

The ANSI symbols for relay switching contact forms are shown in Fig. 6.30. Relays are nothing more than solenoid-operated electromechanical switching devices.

Figure 6.31 shows the International Electrotechnical Commission (IEC), JIC, and NMTBA common switch symbols.

The limit switch is widely used in electromechanical design practice as a control element in electromechanical systems. Figure 6.32 shows the normally available actuator forms for limit switches. (*Source:* Microswitch Corp.)

Figure 6.33 shows the operating characteristics of the limit switch for inline plunger operation. The operation is similar for all actuator types. (*Source:* Microswitch Corp.)

A particularly important limit switch is the magnetic blowout type such as the "Microswitch" MT/MO Series. This style of limit switch is designed for switching large dc currents, on the order of 10 A at 125 V dc. The contact points of this type of switch are surrounded with a circular permanent magnet that helps to dissipate the dc arc and interrupt the dc circuit. Available forms for this limit switch are shown in Fig. 6.34.

This type of limit switch may be used for dc motor control in the appropriate arrangement, thus eliminating reversing contactors and other types of expensive relays. Although the horsepower rating of this class of switch is ¼ hp, larger motors may be controlled by using varistors with the limit switch (across the points). Also, on certain types of dc drive motor applications, the dc motor may not be under full load when the limit switch opens its contacts. In these types of applications, this small limit switch will effectively control a ¾-hp dc motor. A typical use for this class of limit switch is shown in a later section, where a high-power interrupter switch ("AutoSwitch," designed by the author), driven by a ¾-hp dc motor, is controlled by two of the MT/MO-type limit switches.

Designing your motor control circuits, wherein magnetic blowout limit switches break the motor loads, can save a considerable amount of money, in lieu of using reversing contactors and expensive high-power relays. Circuit-breaker and high-voltage switch electrical control circuits are a prime example for the use of the limit switch with high-current dc switching ability.

***Switching Diagrams.*** Figures 6.35*a* and 6.35*b* to 6.46 show various switching diagrams used for a large number of switching applications.

***Using AC Switches for DC Loads.*** Ac-rated switches may be used in some dc applications. The maximum dc voltage and current that may be effectively controlled is a function of the switch's contact-separation distance. Listed in Table 6.11 are dc loads (amperes) at the indicated dc voltages that can be effectively switched with a standard 250-V-ac 20-A switch.

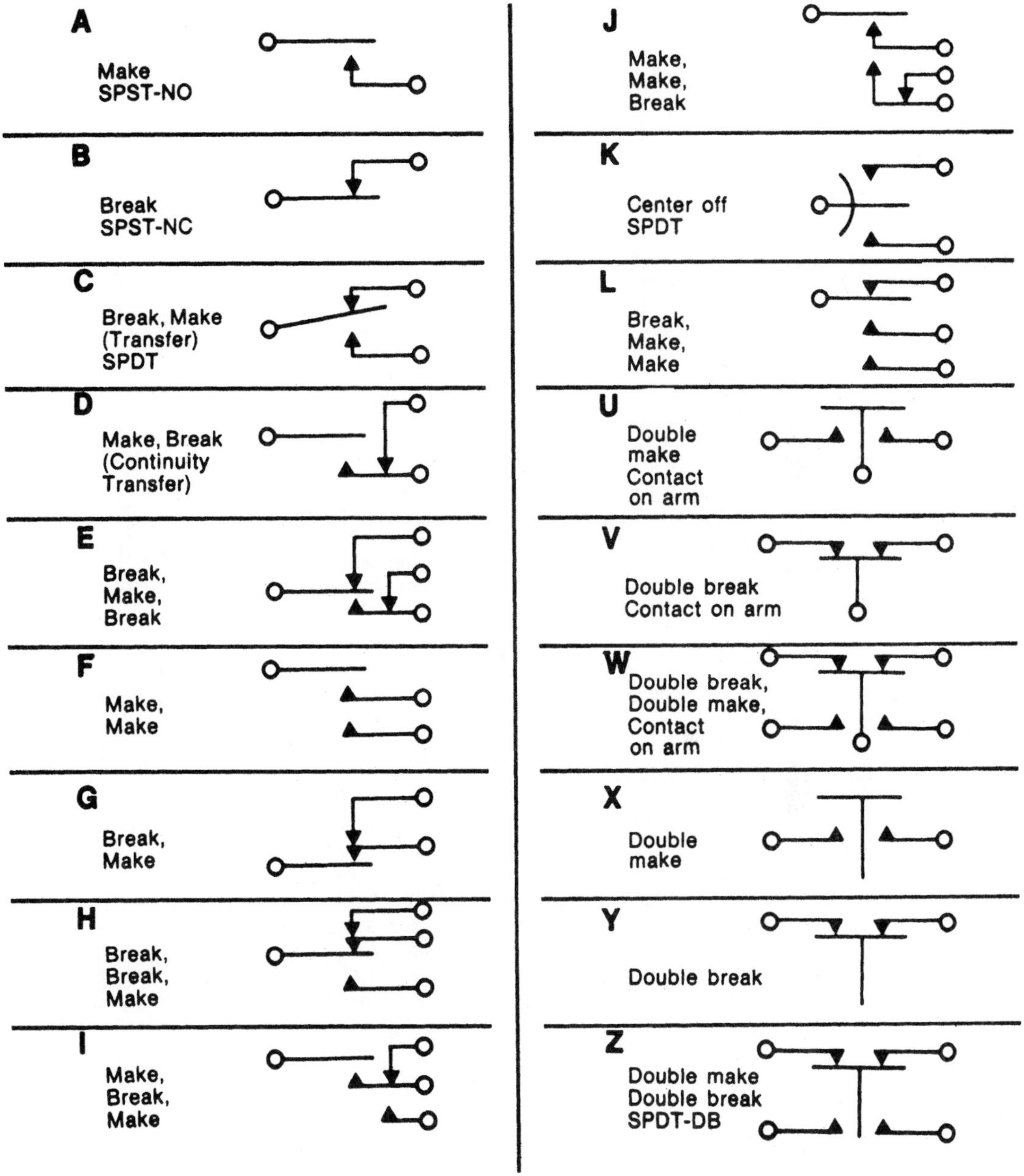


**FIGURE 6.30** ANSI relay contact forms.

| Symbol | Description | Form |
|---|---|---|
|  | Make or SPSTNO | A |
|  | Break or SPSTNC | B |
| 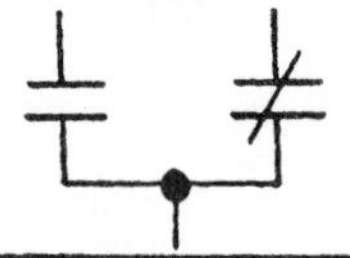 | Break, make or SPDT (B-M) or transfer | C |
| 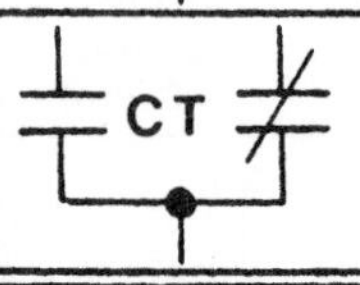  | Make, break or make-before break or SPDT (M-B) or continually transfer | D |

**FIGURE 6.31** IEC, JIC, and NMBTA symbols.

***DC Switches.*** Dc switches use different contact materials and gap-spacing dimensions to achieve their ratings. Some types of dc switches that are used to switch heavy motor loads are equipped with magnetic rings surrounding the contacts to allow the dc arc to be more effectively dissipated and interrupted.

When electric motors are too powerful to be controlled by small switches, contactors, motor starters, or circuit breakers are required to switch the motor loads. This is true for both ac and dc, single- and multiphase high-horsepower motors.

***Electrical Connectors and Receptacles (Wiring Devices).*** Heavy-duty caps, connectors, and receptacles are shown in Fig. 6.47. Receptacle wiring systems and diagrams are shown in Fig. 6.48. The architectural symbols for all common wiring devices are shown in Fig. 6.49.

Table 6.12 shows the chemical resistance of materials commonly used in wiring devices. Table 6.13 shows the mechanical and electrical properties of materials commonly used in wiring devices.

## 6.6.8 Transformers

In a transformer

$$\frac{Tp}{Ts} = \frac{Ep}{Es}$$

where $Tp$ = primary turns
$Ts$ = secondary turns
$Ep$ = primary voltage
$Es$ = secondary voltage

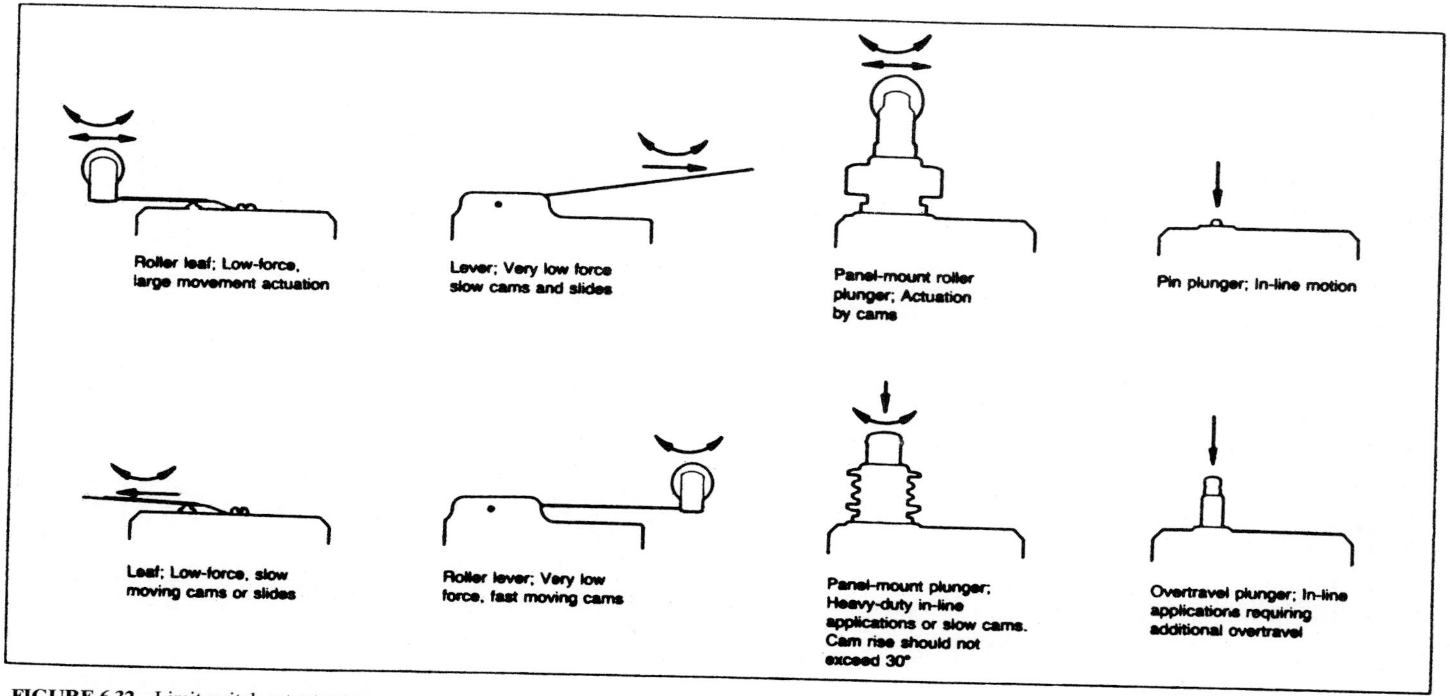
Roller leaf; Low-force, large movement actuation
Lever; Very low force slow cams and slides
Panel-mount roller plunger; Actuation by cams
Pin plunger; In-line motion
Leaf; Low-force, slow moving cams or slides
Roller lever; Very low force, fast moving cams
Panel-mount plunger; Heavy-duty in-line applications or slow cams. Cam rise should not exceed 30°
Overtravel plunger; In-line applications requiring additional overtravel

**FIGURE 6.32** Limit switch actuator types.

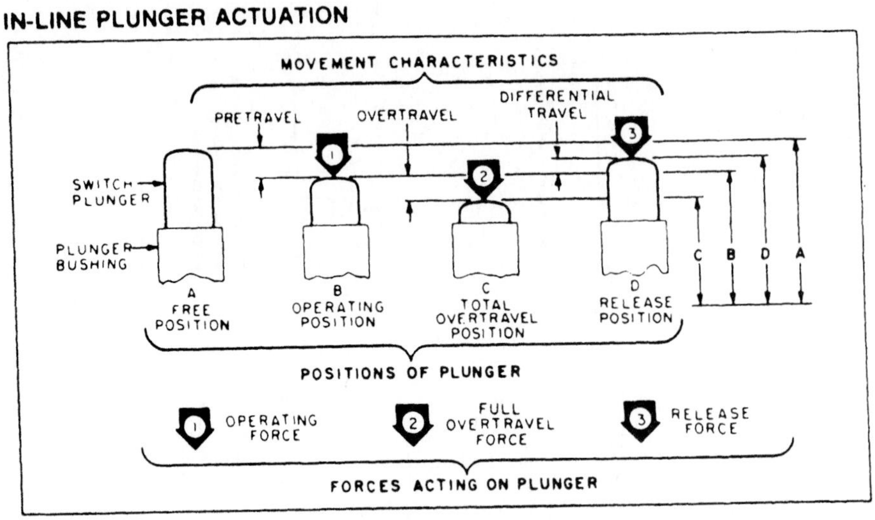


# Operating characteristics

**Differential Travel (D.T.)**—Plunger or actuator travel from point where contacts "snap-over" to point where they "snap-back."

**Free Position (F.P.)**—Position of switch plunger or actuator when no external force is applied (other than gravity).

**Full Overtravel Force**—Force required to attain full overtravel of actuator.

**Operating Position (O.P.)**—Position of switch plunger or actuator at which point contacts snap from normal to operated position. Note that in the case of flexible or adjustable actuators, the operating position is measured from the end of the lever or its maximum length. Location of operating position measurement shown on mounting dimension drawings.

**Operating Force (O.F.)**—Amount of force applied to switch plunger or actuator to cause contact "snap-over." Note in the case of adjustable actuators, the force is measured from the maximum length position of the lever.

**Overtravel (O.T.)**—Plunger or actuator travel safely available beyond operating position.

**Pretravel (P.T.)**—Distance or angle traveled in moving plunger or actuator from free position to operating position.

**Release Force (R.F.)**—Amount of force still applied to switch plunger or actuator at moment contacts snap from operated position to unoperated position.

**Total Travel (T.T.)**—Distance from actuator free position to overtravel limit position.

**FIGURE 6.33** Operating characteristics of limit switch plungers.

**MOUNTING DIMENSIONS**

PIN PLUNGER

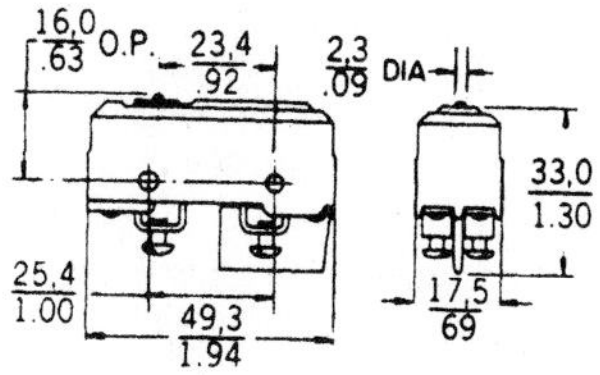


STRAIGHT LEVER

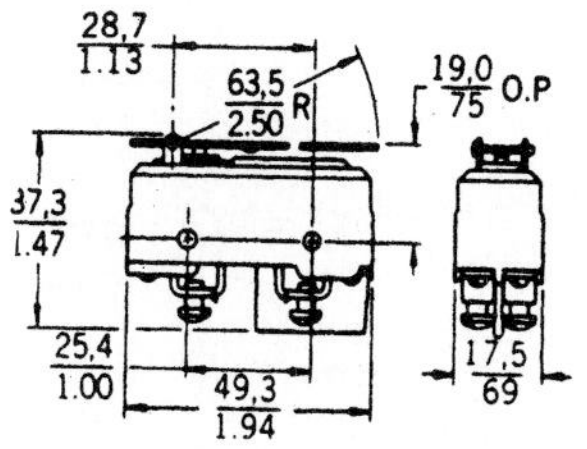


ROLLER LEVER

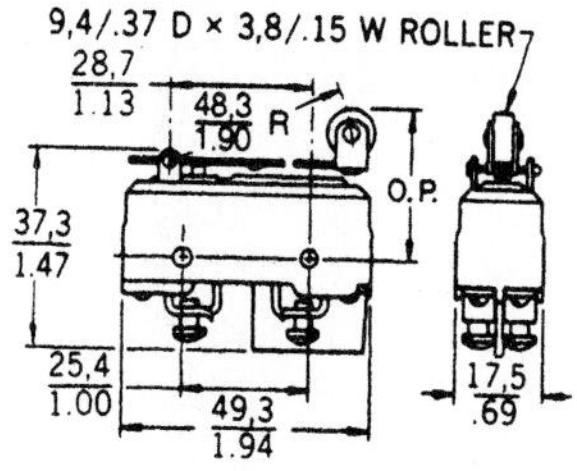


MT-4RV2-A28

FLEXIBLE LEAF

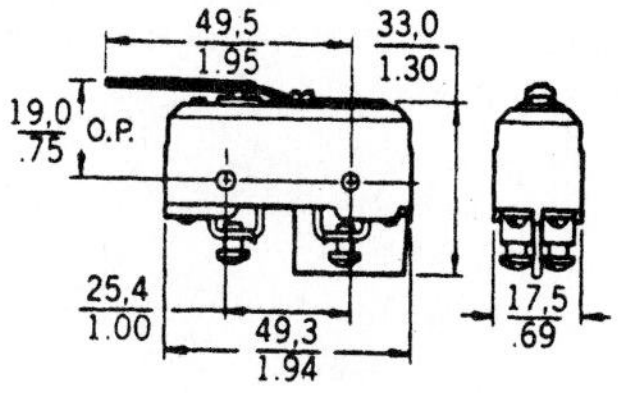


FLEXIBLE ROLLER LEAF

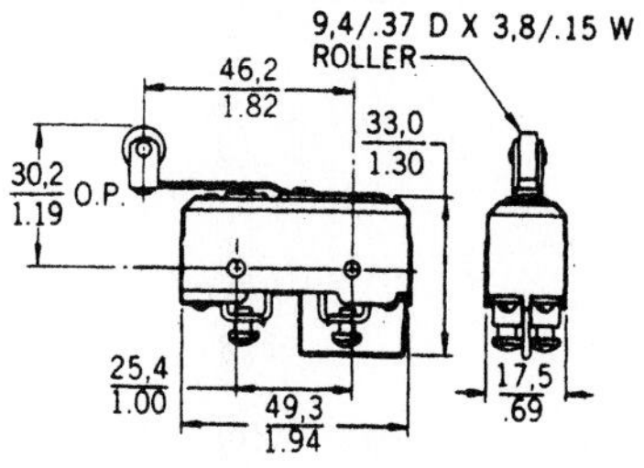


Mounting holes accept pins or screws of .139″ (3,53 mm) diameter.

**FIGURE 6.34** MT/MO magnetic blowout basic switches (Microswitch Corp.).

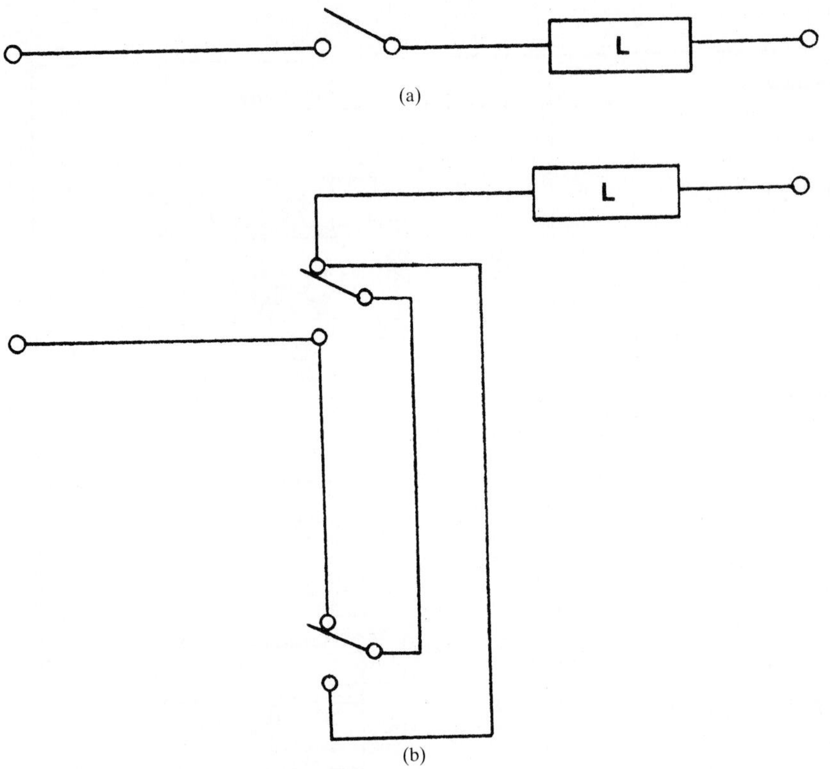


**FIGURE 6.35** (*a, b*) Conversion of an existing single point on-off control to two points of control as in (*b*).

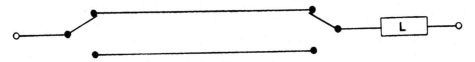


**FIGURE 6.36** Simple circuit giving on-off control at two points.

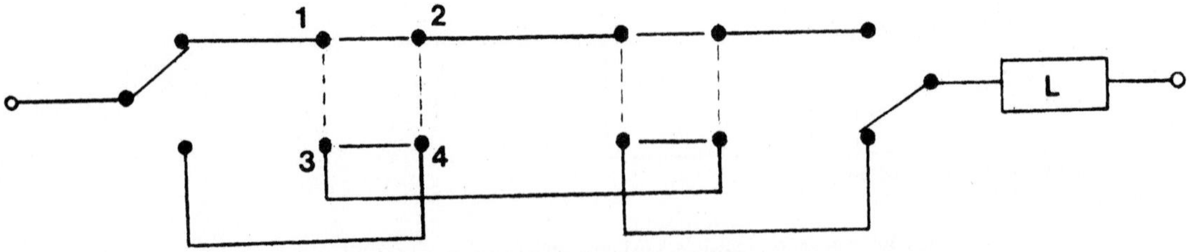


**FIGURE 6.37** Two-way intermediate circuit giving on-off control at any number of points.

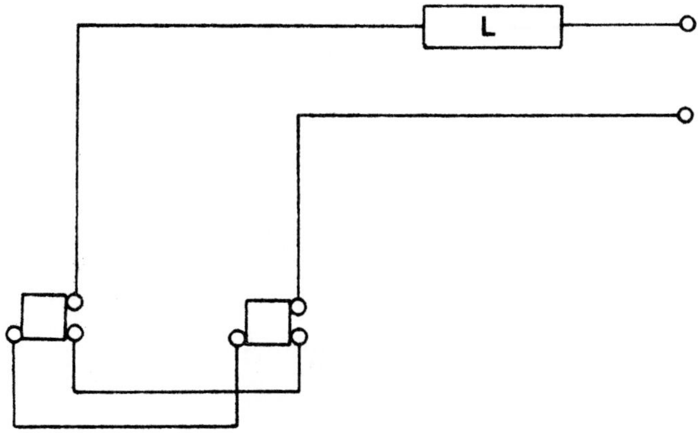


**FIGURE 6.38** Two 3-way switches wired with the load at the end of the circuit.

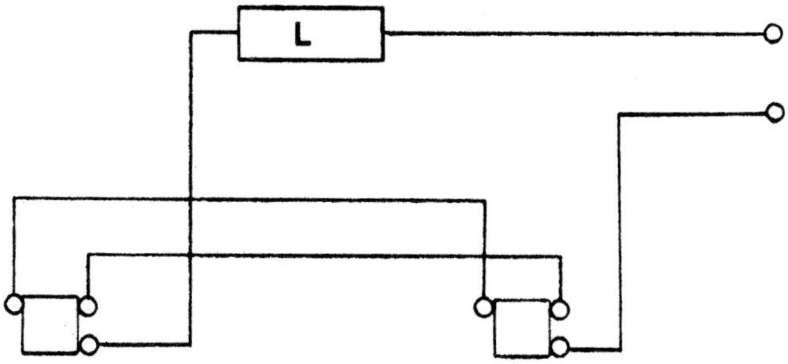


**FIGURE 6.39** Two 3-way switches wired with the load between the switches.

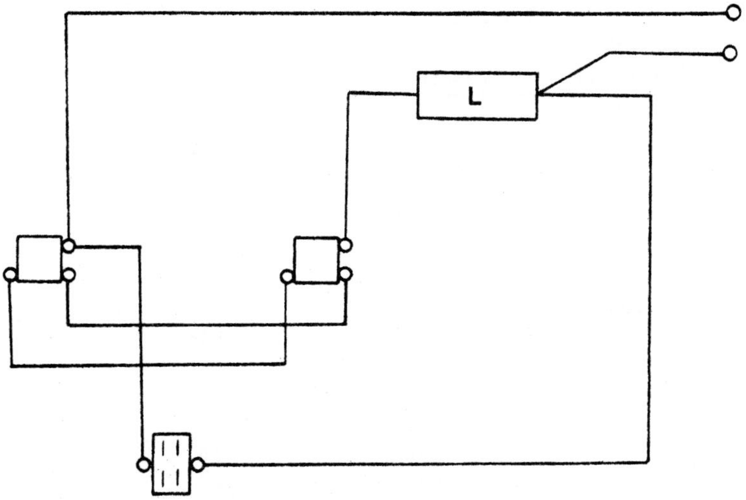


**FIGURE 6.40** Two 3-way switches with an intermediate convenience outlet which is always "hot."

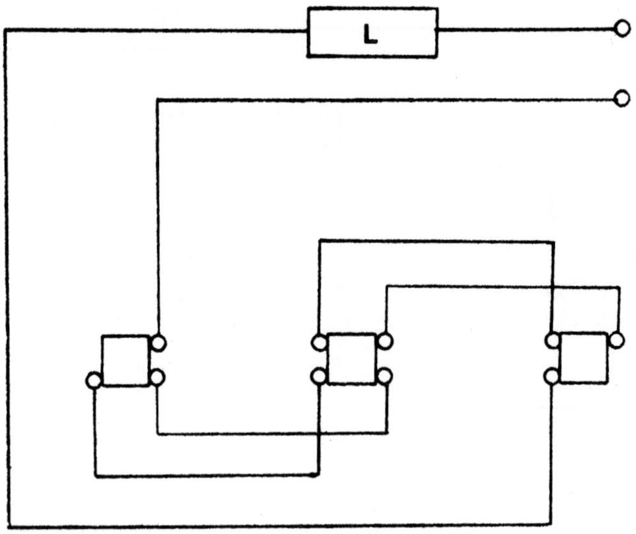


**FIGURE 6.41** Two 3-way and a 4-way switch wired for control of the load from three points.

**TABLE 6.11** DC Switching Loads for AC Switches

| Separation of Contacts, in | V dc (Voltage) | Resistive Load, A | Inductive Load, A* |
|---|---|---|---|
| 0.010 | 3–10 | 2.5 | 7.5 |
| | 12–20 | 2.5 | 4.5 |
| | 24–36 | 1.75 | 0.75 |
| | 110–125 | 0.3 | 0.02 |
| 0.020 | 3–10 | 2.75 | 18.0 |
| | 12–20 | 2.75 | 9.0 |
| | 24–36 | 2.50 | 4.0 |
| | 110–125 | 0.3 | 0.04 |
| 0.040 | 3–10 | 2.75 | 18.0 |
| | 12–20 | 2.75 | 18.0 |
| | 24–36 | 2.50 | 9.0 |
| | 110–125 | 0.5 | 0.09 |
| 0.070 | 3–10 | 2.75 | 18.0 |
| | 12–20 | 2.75 | 18.0 |
| | 24–36 | 2.5 | 9.0 |
| | 110–125 | 0.7 | 0.3 |

*The inductive loads shown are for sea level. The inductive load ratings shown should be downgraded by 25% for applications of altitude of 50,000 feet (15,240 meters).

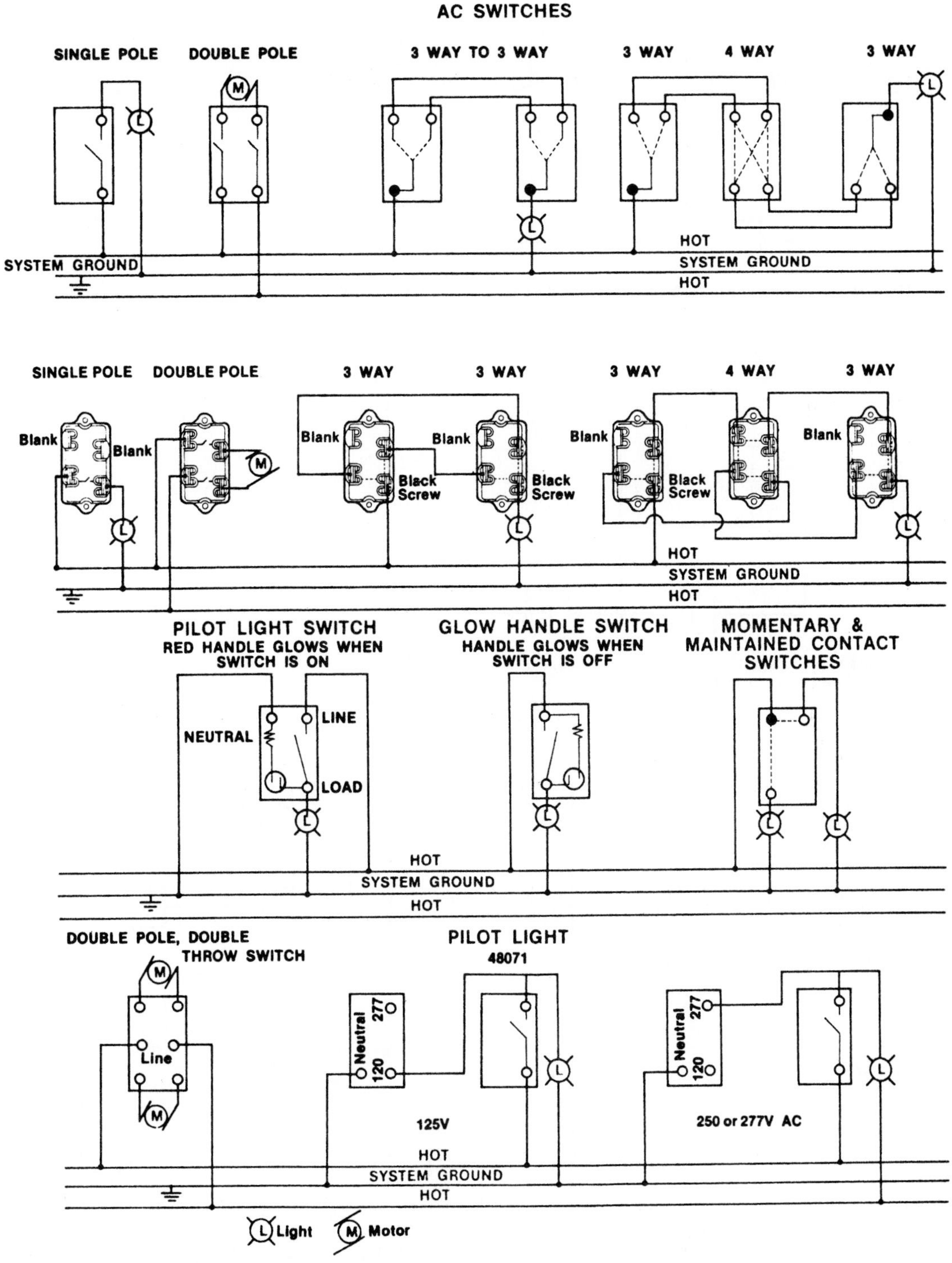


**FIGURE 6.42** Ac switch wiring diagrams.

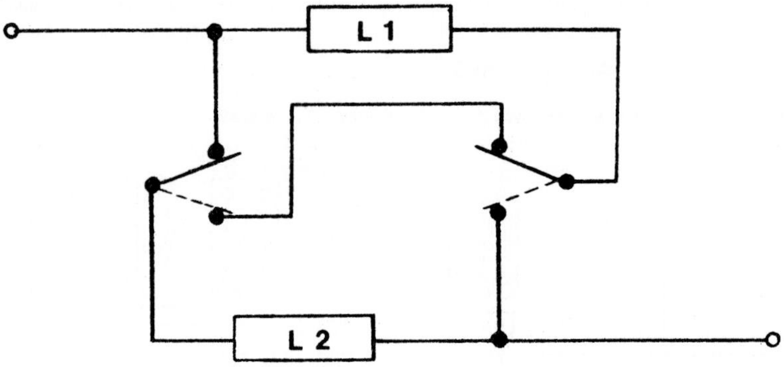


**FIGURE 6.43** Two independent 2-way switches at one control point enabling $L_1$ or $L_2$ to be switched on alone, in series or in parallel.

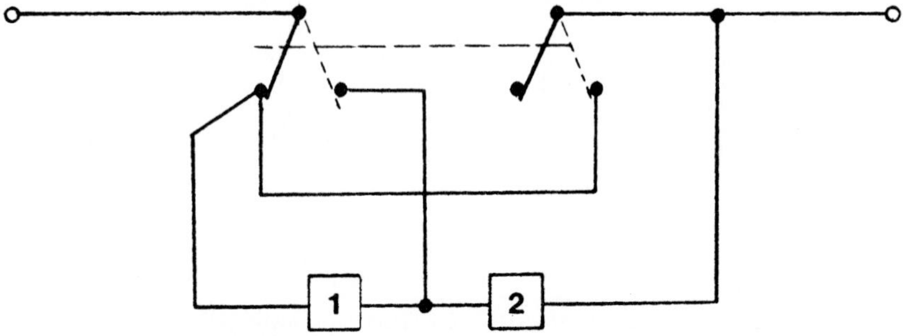


**FIGURE 6.44** Equivalent of a series-parallel switch; two coupled 2-way switches; loads 1 or 2 can be switched in series or parallel.

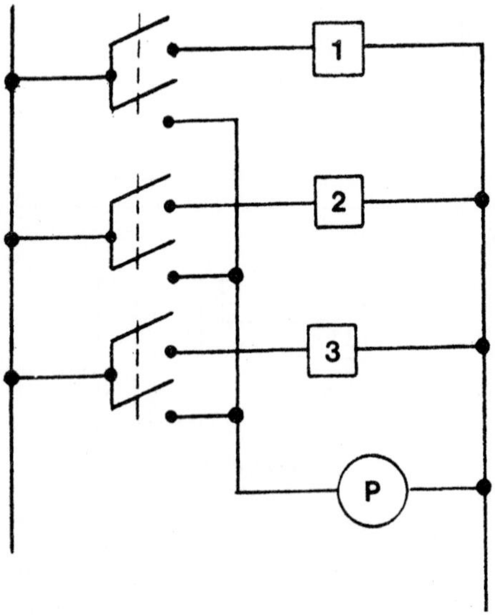


**FIGURE 6.45** Pilot circuit; pilot indication when any one of several circuits 1, 2, and 3 is closed.

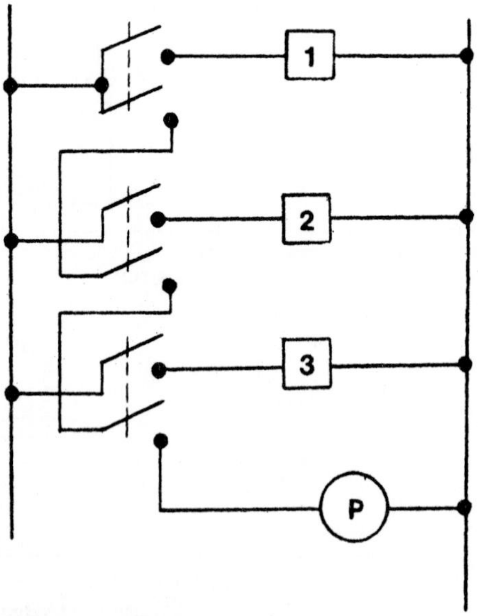


**FIGURE 6.46** Pilot circuit; pilot indication only when all circuits are closed.

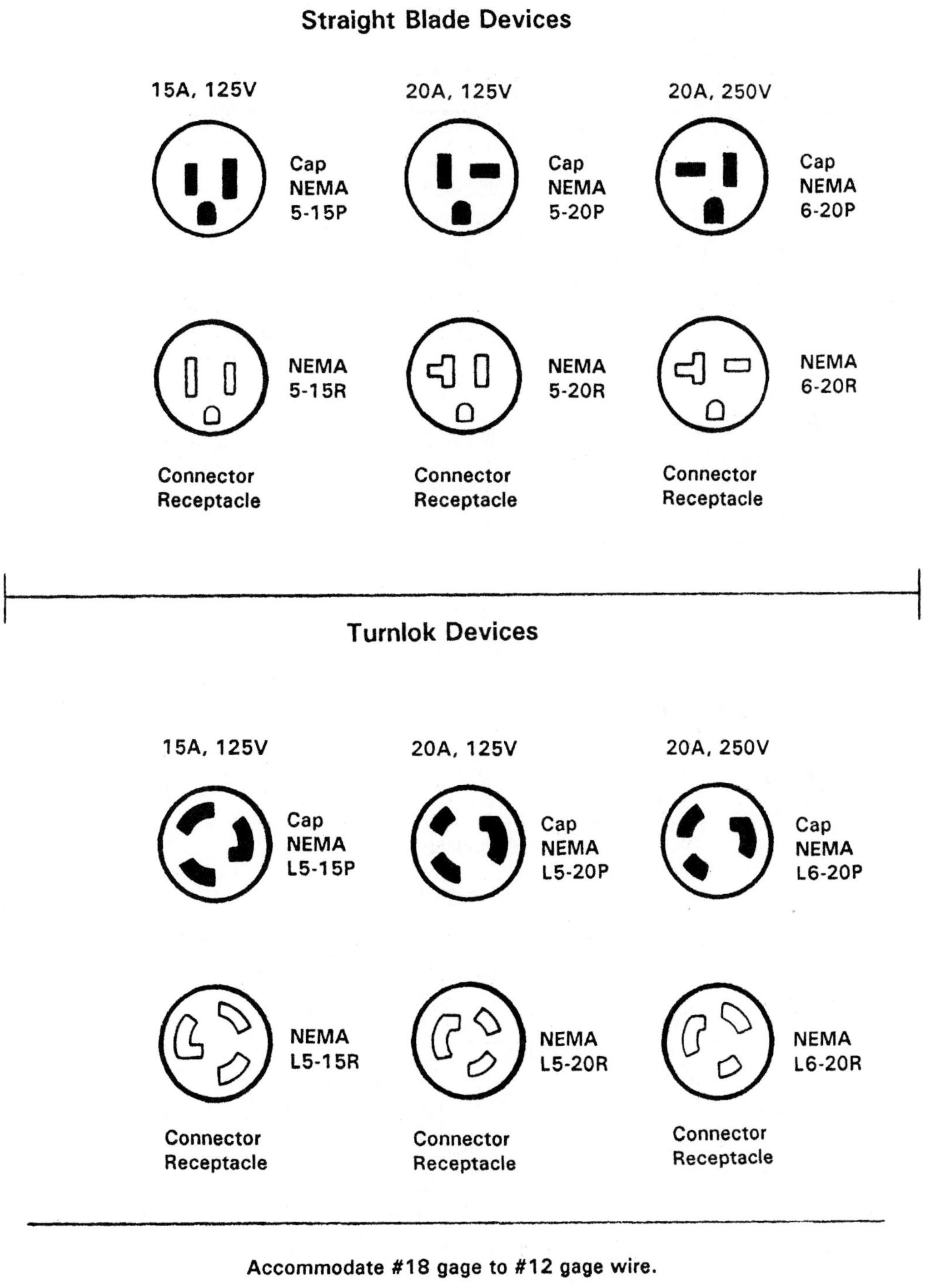


**FIGURE 6.47** Three-wire devices; caps, connectors, and receptacles.

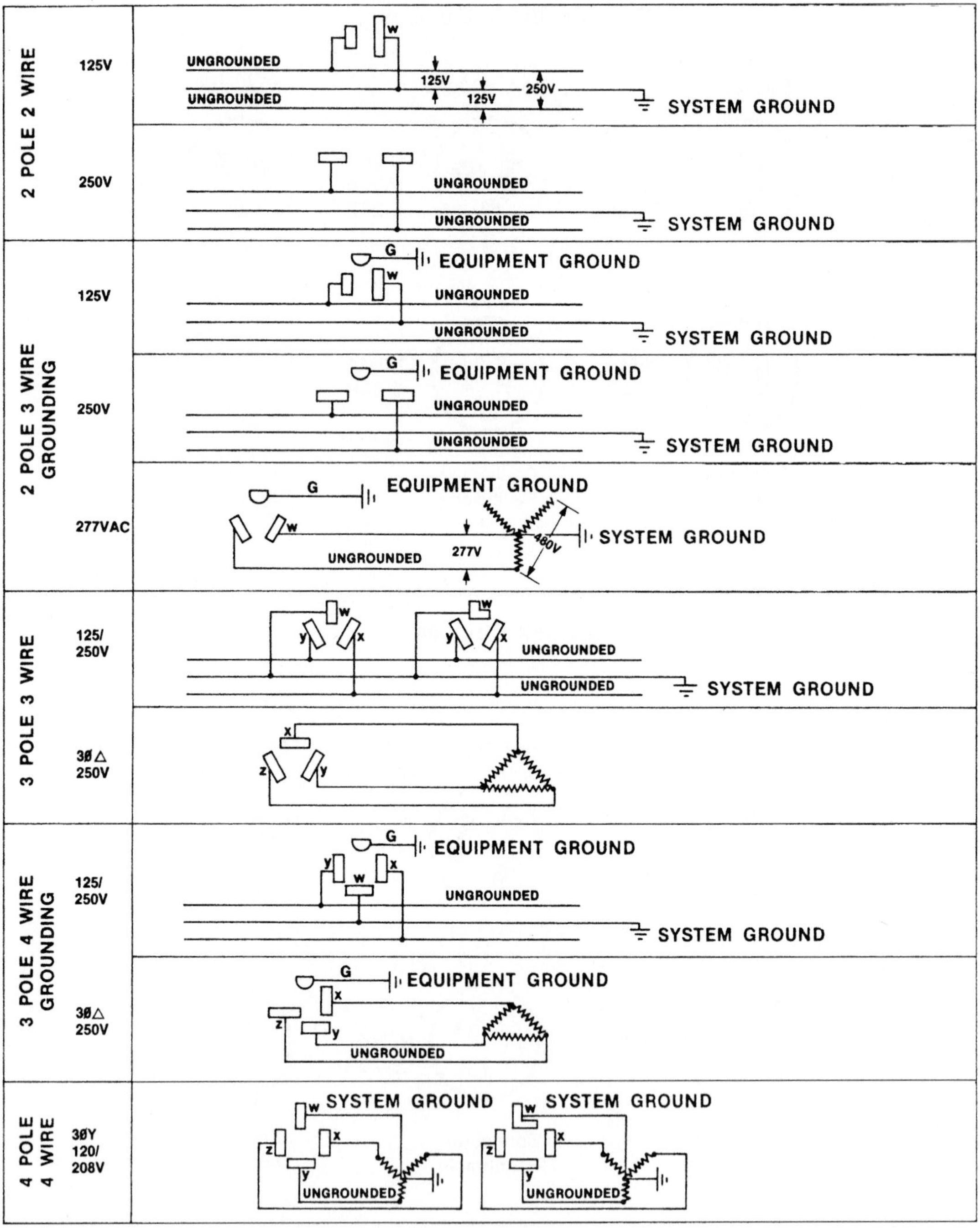


(Reproduced with permission of: Bryant Division of Westinghouse)

**FIGURE 6.48** Various wiring schemes.

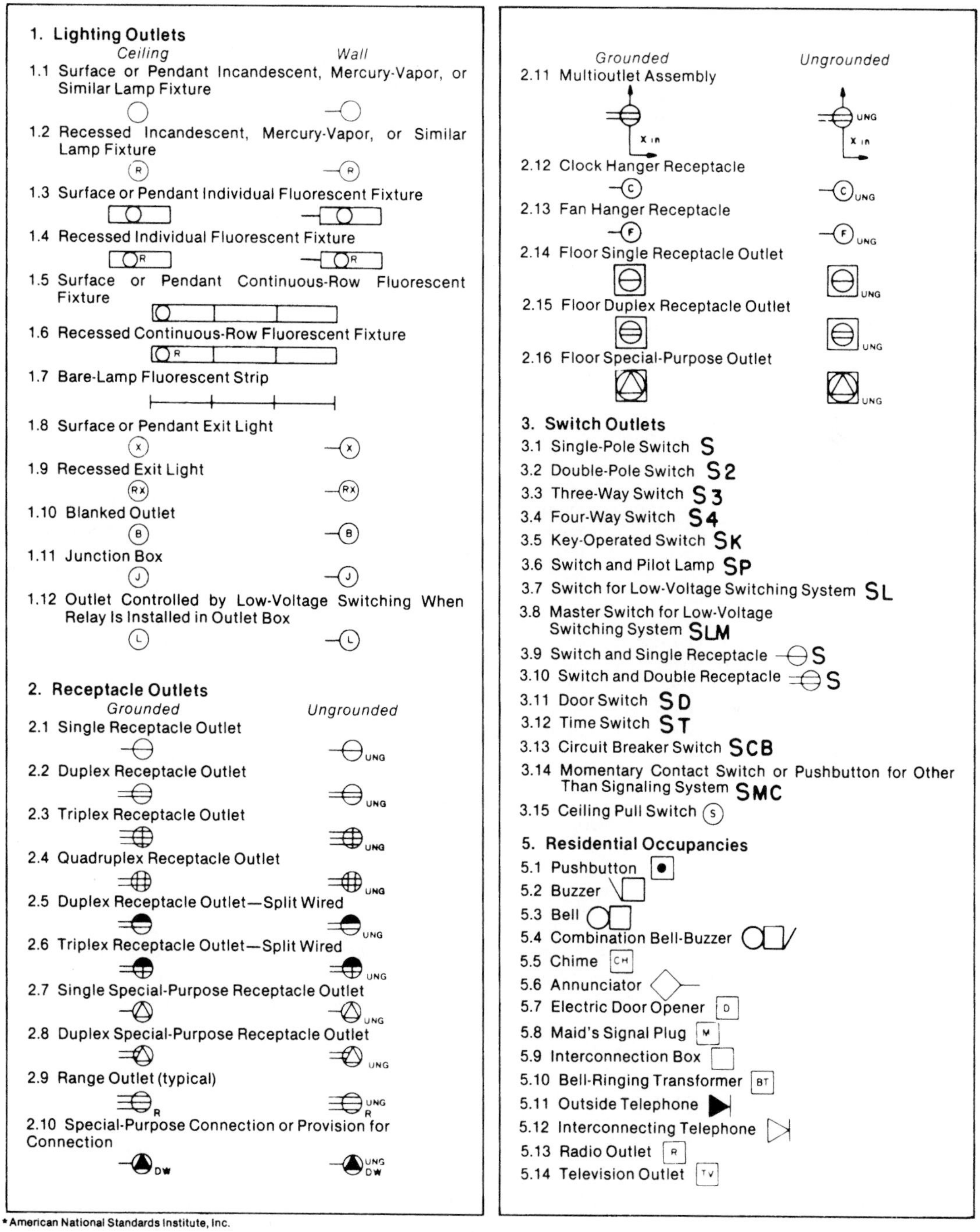
1. Lighting Outlets
Ceiling
Wall
1.1 Surface or Pendant Incandescent, Mercury-Vapor, or Similar Lamp Fixture
1.2 Recessed Incandescent, Mercury-Vapor, or Similar Lamp Fixture
1.3 Surface or Pendant Individual Fluorescent Fixture
1.4 Recessed Individual Fluorescent Fixture
1.5 Surface or Pendant Continuous-Row Fluorescent Fixture
1.6 Recessed Continuous-Row Fluorescent Fixture
1.7 Bare-Lamp Fluorescent Strip
1.8 Surface or Pendant Exit Light
1.9 Recessed Exit Light
1.10 Blanked Outlet
1.11 Junction Box
1.12 Outlet Controlled by Low-Voltage Switching When Relay Is Installed in Outlet Box
2. Receptacle Outlets
Grounded
Ungrounded
2.1 Single Receptacle Outlet
2.2 Duplex Receptacle Outlet
2.3 Triplex Receptacle Outlet
2.4 Quadruplex Receptacle Outlet
2.5 Duplex Receptacle Outlet—Split Wired
2.6 Triplex Receptacle Outlet—Split Wired
2.7 Single Special-Purpose Receptacle Outlet
2.8 Duplex Special-Purpose Receptacle Outlet
2.9 Range Outlet (typical)
2.10 Special-Purpose Connection or Provision for Connection
UNG
DW
2.11 Multioutlet Assembly
X in
2.12 Clock Hanger Receptacle
2.13 Fan Hanger Receptacle
2.14 Floor Single Receptacle Outlet
2.15 Floor Duplex Receptacle Outlet
2.16 Floor Special-Purpose Outlet
3. Switch Outlets
3.1 Single-Pole Switch S
3.2 Double-Pole Switch S2
3.3 Three-Way Switch S3
3.4 Four-Way Switch S4
3.5 Key-Operated Switch SK
3.6 Switch and Pilot Lamp SP
3.7 Switch for Low-Voltage Switching System SL
3.8 Master Switch for Low-Voltage Switching System SLM
3.9 Switch and Single Receptacle S
3.10 Switch and Double Receptacle S
3.11 Door Switch SD
3.12 Time Switch ST
3.13 Circuit Breaker Switch SCB
3.14 Momentary Contact Switch or Pushbutton for Other Than Signaling System SMC
3.15 Ceiling Pull Switch
5. Residential Occupancies
5.1 Pushbutton
5.2 Buzzer
5.3 Bell
5.4 Combination Bell-Buzzer
5.5 Chime
5.6 Annunciator
5.7 Electric Door Opener
5.8 Maid's Signal Plug
5.9 Interconnection Box
5.10 Bell-Ringing Transformer
5.11 Outside Telephone
5.12 Interconnecting Telephone
5.13 Radio Outlet
5.14 Television Outlet

*American National Standards Institute, Inc.

(Reproduced with permission of: Bryant Division of Westinghouse)

**FIGURE 6.49** Architectural wiring device symbols.

**TABLE 6.12** Chemical Resistances

| Chemical | Nylon | Melamine | Phenolic | Urea | Polyvinyl Chloride | Polycar-bonate | Rubber |
|---|---|---|---|---|---|---|---|
| Acids | C | B | B | B | A | A | B |
| Alcohol | A | A | A | A | A | A | B |
| Caustic Bases | A | B | B | B | A | C | C |
| Gasoline | A | A | A | C | A | A | B |
| Grease | A | A | A | A | A | A | B |
| Kerosene | A | A | A | A | A | A | A |
| Oil | A | A | A | A | A | A | A |
| Solvents | A | A | A | A | C | C | C |
| Water | A | A | A | A | A | A | B |

A—Completely resistant. Good to excellent, general use.
B—Resistant. Fair to good, limited service.
C—Slow attack. Not recommended for use.

(Reproduced with permission of: Bryant Division of Westinghouse)

**TABLE 6.13** Mechanical and Electrical Properties

| | Nylon | Melamine | Phenolic | Urea | Polyvinyl Chloride | Polycarbonate |
|---|---|---|---|---|---|---|
| Tensile Strength (psi) | 9000-12,000 | 7000-13,000 | 6500-10,000 | 5500-13,000 | 5000-9000 | 8000-9500 |
| Elongation (percent) | 60-300 | 0.6-0.9 | 0.4-0.8 | 0.5-1.0 | 2.0-4.0 | 60-100 |
| Tensile Modulus ($10^5$ psi) | 1.75-4.1 | 12-14 | 8-17 | 10-15 | 3.5-6 | 3.5 |
| Compressive Strength (psi) | 6700-12.500 | 25,000-45,000 | 22,000-36,000 | 25,000-45,000 | 8000-13,000 | 12,500 |
| Flexural Strength (psi) | No break | 10,000-16,000 | 8500-12,000 | 10,000-18,000 | 10,000-16,000 | 13,500 |
| Impact Strength (ft-lb/in.) | 1.0-2.0 | 0.24-0.35 | 0.24-0.60 | 0.25-0.40 | 0.4-20 | 12-16 |
| Rockwell Hardness | R109-R118 | M110-M125 | M96-M120 | M110-M120 | 70-90 shore | M70-R116 |
| Continuous Temperature Resistance (°F) | 180-200 | 210 | 350-360 | 170 | 150-175 | 250 |
| Heat Distortion, 66 psi (°F) | 360-365 | * | * | * | 179 | 285 |
| Dielectric Strength (volts/mil) | 385-470 | 300-400 | 200-400 | 300-400 | 425-1300 | 400 |
| Arc Resistance (sec) ASTM D495 | 100-105 | 110-180 | 0-7 | 110-150 | 60-80 | 10-120 |
| Burning Rate (in./min) | Self extinguishing | Self extinguishing | Very low | Self extinguishing | Self extinugishing | Self extinguishing |

*Not applicable because these are thermosetting materials.

(Reproduced with permission of: Bryant Division of Westinghouse)

A 3 to 5 percent loss in secondary voltage (va) occurs when a transformer is loaded. This is due to core losses, wire losses, and flux losses within the transformer.

Neglecting losses, the following holds true.

$$Ep \times Ip = Es \times Is$$

(primary va) (secondary va)

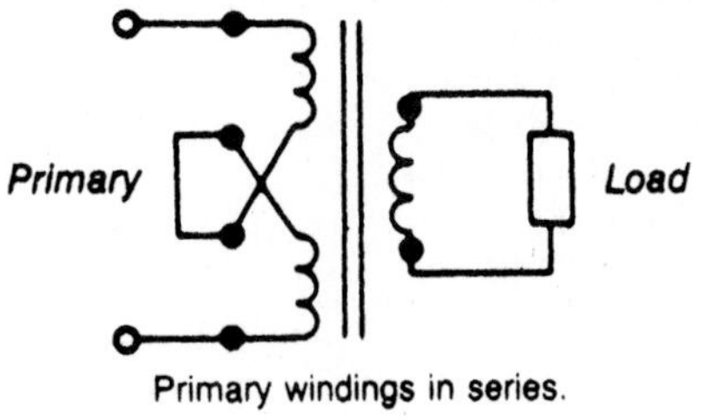


**FIGURE 6.50** Transformer primary windings in series.

The proportion of primary to secondary turns of a transformer is called the *turn ratio.*

*Example.* Refer to Fig. 6.50, showing primary winding in series. Each of the primary and secondary windings of a transformer has 300 turns. Therefore, this arrangement has a turn ratio of 2 to 1 and secondary voltage is half the supply voltage.

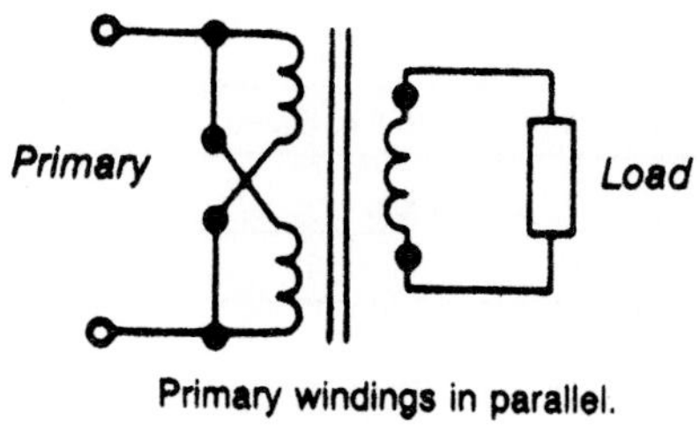


**FIGURE 6.51** Transformer primary windings in parallel.

*Example.* Refer to Fig. 6.51, showing primary windings in parallel. Each of the primary and secondary windings has 300 turns. Therefore, this arrangement has a turn ratio of 1 to 1 and secondary voltage is equal to primary voltage.

***Full-Load Currents in Transformers.*** Full-load current (in amperes) for single-phase circuits is

$$Ifl = \frac{\text{kVA} \times 1000}{\text{circuit voltage}}$$

Full-load current (in amperes) for three-phase circuits is

$$Ifl = \frac{\text{kVA} \times 1000}{\sqrt{3} \times \text{system voltage}}$$

*Example.* A 75-kVA, three-phase transformer will have how many amperes of full-load current at a system voltage of 4160 V?

$$Ifl = \frac{75 \times 1000}{\sqrt{3} \times 4160} = \frac{75{,}000}{7205.3} = 10.4 \text{ A}$$

### 6.6.9 Copper Wire Current-Carrying Capacity

See Table 6.14.

### 6.6.10 Fusing Time Current for Copper Connections

The following equation, developed by I. M. Onderdonk, may be used to calculate the fusing time current of copper conductors.

$$33\left(\frac{I}{A}\right)^2 S = \log_{10}\left(\frac{Tm - Ta}{234 + Ta} + 1\right) \qquad I = A\sqrt{\frac{\log[(Tm - Ta)/(234 + Ta) +}{1}}$$

where $I$ = current, A
$A$ = conductor area, circular mils
$S$ = time current applied, sec
$Tm$ = melting point of copper, °C
$Ta$ = ambient temperature, °C

See Table 6.15 for data for standard annealed bare-copper wire (B & S, American wire gauge).

## 6.7 ELECTRICAL POWER SYSTEM COMPONENTS

### 6.7.1 Current-Carrying Capacities, Copper Conductors

See Table 6.16 for dc capacities of copper bus bars.
See Tables 6.17*a*, 6.17*b*, and 6.17*c* for data on ac (60-Hz) current-carrying capacities for copper bus bars, copper round rods, and copper pipes.

**TABLE 6.14** Allowable Current-Carrying Capacities

(Not more than three conductors in raceway or direct burial, based on 30°C, 86°F ambient)

| Max. Oper. Temp. | 60°C | 75°C | 85°C | 110°C | 125°C | 200°C |
|---|---|---|---|---|---|---|
| | Types of Insulation | | | | | |
| AWG or MCM (size) | R, RW<br>RH-RW<br>T, TW<br>Flamenol | RH, RHW<br>RH-RW<br>THW<br>Flamenol | Paper<br>V-C (ABV)<br>MI<br>RHH Super | V-C (AVA)<br>V-C (AVL) | AI<br>AIA | A<br>AA |
| 14 | 15 | 15 | 25 | 30 | 30 | 30 |
| 12 | 20 | 20 | 30 | 35 | 40 | 40 |
| 10 | 30 | 30 | 40 | 45 | 50 | 55 |
| 8 | 40 | 45 | 50 | 60 | 65 | 70 |
| 6 | 55 | 65 | 70 | 80 | 85 | 95 |
| 4 | 70 | 85 | 90 | 105 | 115 | 120 |
| 3 | 80 | 100 | 105 | 120 | 130 | 145 |
| 2 | 95 | 115 | 120 | 135 | 145 | 165 |
| 1 | 110 | 130 | 140 | 160 | 170 | 190 |
| 0 | 125 | 150 | 155 | 190 | 200 | 225 |
| 00 | 145 | 175 | 185 | 215 | 230 | 250 |
| 000 | 165 | 200 | 210 | 245 | 265 | 285 |
| 0000 | 195 | 230 | 235 | 275 | 310 | 340 |
| 250 | 215 | 255 | 270 | 315 | 335 | ... |
| 300 | 240 | 285 | 300 | 345 | 380 | ... |
| 350 | 260 | 310 | 325 | 390 | 420 | ... |
| 400 | 280 | 335 | 360 | 420 | 450 | ... |
| 500 | 320 | 380 | 405 | 470 | 500 | ... |
| 600 | 355 | 420 | 455 | 525 | 545 | ... |
| 700 | 385 | 460 | 490 | 560 | 600 | ... |
| 750 | 400 | 475 | 500 | 580 | 620 | ... |
| 800 | 410 | 490 | 515 | 600 | 640 | ... |
| 900 | 435 | 520 | 555 | ... | ... | ... |
| 1000 | 455 | 545 | 585 | 680 | 730 | ... |

Correction Factors for Room Temperatures Over 30°C

| C | F | | | | | | |
|---|---|---|---|---|---|---|---|
| 40 | 104 | 0.82 | 0.88 | 0.90 | 0.94 | 0.95 | ... |
| 45 | 113 | 0.71 | 0.82 | 0.85 | 0.90 | 0.92 | ... |
| 50 | 122 | 0.58 | 0.75 | 0.80 | 0.87 | 0.89 | ... |

NOTE: Multiply the correction factor times the current-carrying capacity to arrive at the corrected current-carrying capacity at the elevated temperature.

### 6.7.2 Heating in Electrical Conductors

The heat caused by electrical losses in a conductor is dissipated by convection and radiation. Heat carried off by conduction through supports and connections is small and can be disregarded in most cases.

For steady-state conditions, the rate of heat dissipation (watts) is equal to the rate of heat supply, so

$$I^2R = Wc\,Ac + Wr\,Ar$$

where $I$ = current in conductor, A
$R$ = resistance of conductor, Ω
$Wc$ = wattage loss per square inch due to convection
$Wr$ = wattage loss per square inch due to radiation
$Ac$ = surface area of conductor, in$^2$ (convection)
$Ar$ = surface area of conductor, in$^2$ (radiation)

**TABLE 6.15** Properties of Standard Annealed Copper

| Bare Wire Gauge* AWG | Diameter (Nominal) in | Area, Cir. Mils | Ohms/1000 ft | Ft/ohm | Current Capacity (A), Rubber ins. |
|---|---|---|---|---|---|
| 0000 | 0.4600 | 211,600 | 0.04901 | 20,400 | 225 |
| 000 | 0.4096 | 167,800 | 0.06180 | 16,180 | 175 |
| 00 | 0.3648 | 133,100 | 0.07793 | 12,830 | 150 |
| 0 | 0.3249 | 105,500 | 0.09827 | 10,180 | 125 |
| 1 | 0.2893 | 83,690 | 0.1239 | 8,070 | 100 |
| 2 | 0.2576 | 66,370 | 0.1563 | 6,400 | 90 |
| 3 | 0.2294 | 52,640 | 0.1970 | 5,075 | 80 |
| 4 | 0.2043 | 41,740 | 0.2485 | 4,025 | 70 |
| 5 | 0.1819 | 33,100 | 0.3133 | 3,192 | 55 |
| 6 | 0.1620 | 26,250 | 0.3951 | 2,531 | 50 |
| 7 | 0.1443 | 20,820 | 0.4982 | 2,007 | — |
| 8 | 0.1285 | 16,510 | 0.6282 | 1,592 | 35 |
| 9 | 0.1144 | 13,090 | 0.7921 | 1,262 | — |
| 10 | 0.1019 | 10,380 | 0.9989 | 1,001 | 25 |
| 11 | 0.09074 | 8,234 | 1.260 | 794 | — |
| 12 | 0.08081 | 6,530 | 1.588 | 630 | 20 |
| 14 | 0.06408 | 4,107 | 2.525 | 396 | 15 |
| 16 | 0.05082 | 2,583 | 4.016 | 249 | 6 |
| 18 | 0.04030 | 1,624 | 6.385 | 157 | 3 |
| 20 | 0.03196 | 1,022 | 10.15 | 98.5 | — |
| 22 | 0.02535 | 642.4 | 16.14 | 61.95 | — |
| 24 | 0.02010 | 404.0 | 25.67 | 38.96 | — |
| 26 | 0.01594 | 254.1 | 40.81 | 24.50 | — |
| 28 | 0.01264 | 159.8 | 64.90 | 15.41 | — |
| 30 | 0.01003 | 100.5 | 103.2 | 9.691 | — |
| 32 | 0.00795 | 63.21 | 164.1 | 6.095 | — |
| 36 | 0.00500 | 25.00 | 414.8 | 2.411 | — |
| 40 | 0.003145 | 9.888 | 1,049 | 0.9534 | — |

*B&S (Brown & Sharpe), American Wire Gauge.

When calculating the temperature rise, you must select the proper value of *emissivity* for the conductor. Heat loss varies considerably with various conductor conditions. Values are summarized in Table 6.18.

This dissipation of heat by convection is brought about by natural air currents generated by rising, heated air surrounding the bus bar. The amount of heat dissipated depends on the size and shape of the conductor. Small conductors dissipate heat more efficiently than larger ones.

Heat loss by convection $Wc$ for the outside surfaces of a horizontal bus in still but unconfined air (indoors, natural convection) is given in wattage loss per square inch ($W/\text{in}^2$) by

$$Wc = \frac{0.0022\, P^{0.5}\theta^{1.25}}{h^{0.25}}$$

where $P$ = air pressure (1 atm at sea level and normal pressure)
$\theta$ = temperature rise, °C, of the metal above that of ambient air
$h$ = the vertical dimension of a horizontal bar (width), or the diameter of a round tube, in

**TABLE 6.16** DC Current-Carrying Capacities—Copper Bus

| Cross Section Size, in | DC Capacities 30°C Rise, 40°C Ambient | |
|---|---|---|
| | Current Carrying Capacity, A (amperes) | Weight, lb/ft |
| 1/8 × 1 | 245 | 0.4830 |
| 1/8 × 2 | 445 | 0.9660 |
| 1/8 × 3 | 695 | 1.449 |
| 1/8 × 4 | 900 | 1.932 |
| 1/4 × 1 | 400 | 0.9660 |
| 1/4 × 2 | 645 | 1.932 |
| 1/4 × 3 | 990 | 2.898 |
| 1/4 × 4 | 1,280 | 3.864 |
| 1/4 × 5 | 1,560 | 4.830 |
| 1/4 × 6 | 1,830 | 5.796 |
| 1/4 × 8 | 2,320 | 7.728 |
| 3/8 × 1 | 500 | 1.449 |
| 3/8 × 2 | 890 | 2.898 |
| 3/8 × 3 | 1,250 | 4.347 |
| 3/8 × 4 | 1,590 | 5.796 |
| 3/8 × 5 | 1,920 | 7.245 |
| 3/8 × 6 | 2,260 | 8.694 |
| 3/8 × 8 | 2,860 | 11.59 |
| 1/2 × 1 | 605 | 1.932 |
| 1/2 × 2 | 1,040 | 3.864 |
| 1/2 × 3 | 1,460 | 5.796 |
| 1/2 × 4 | 1,860 | 7.728 |
| 1/2 × 5 | 2,250 | 9.660 |
| 1/2 × 6 | 2,620 | 11.59 |
| 1/2 × 8 | 3,360 | 15.46 |

**TABLE 6.17*a*** AC Current-Carrying Capacities—Copper Bus

| Bar pattern** / Bar size | | 2.38" | 0.25" | | | 2.38" | | 2.38" | 2.38" |
|---|---|---|---|---|---|---|---|---|---|
| 1/8 × 1 | 245 | 460 | 390 | 496 | 566 | 670 | 466 | 790 | 1,075 |
| 1/8 × 2 | 445 | 815 | 706 | 896 | 1,025 | 1,215 | 845 | 1,430 | 1,945 |
| 1/8 × 3 | 696 | 1,296 | 1,100 | 1,390 | 1,600 | 1,896 | 1,320 | 2,225 | 3,025 |
| 1/8 × 4 | 900 | 1,636 | 1,420 | 1,800 | 2,070 | 2,445 | 1,710 | 2,880 | 3,915 |
| 1/4 × 1 | 385 | 685 | 575 | 730 | 840 | 995 | 695 | 1,170 | 1,590 |
| 1/4 × 2 | 645 | 1,175 | 1,020 | 1,295 | 1,485 | 1,760 | 1,225 | 2,070 | 2,815 |
| 1/4 × 3 | 970 | 1,770 | 1,540 | 1,945 | 2,236 | 2,646 | 1,845 | 3,115 | 4,230 |
| 1/4 × 4 | 1,220 | 2,220 | 1,925 | 2,440 | 2,800 | 3,315 | 2,315 | 3,906 | 5,305 |
| 1/4 × 5 | 1,460 | 2,666 | 2,300 | 2,920 | 3,356 | 3,970 | 2,775 | 4,670 | 6,350 |
| 1/4 × 6 | 1,660 | 3,020 | 2,620 | 3,320 | 3,815 | 4,515 | 3,155 | 5,310 | 7,220 |
| 1/4 × 8 | 2,020 | 3,675 | 3,190 | 4,040 | 4,645 | 5,495 | 3,835 | 6,465 | 8,785 |
| 3/8 × 1 | 500 | 915 | 790 | 1,005 | 1,155 | 1,365 | 955 | 1,605 | 2,185 |
| 3/8 × 2 | 865 | 1,575 | 1,365 | 1,730 | 1,990 | 2,355 | 1,645 | 2,765 | 3,765 |
| 3/8 × 3 | 1,180 | 2,145 | 1,860 | 2,360 | 2,715 | 3,210 | 2,240 | 3,775 | 5,135 |
| 3/8 × 4 | 1,440 | 2,620 | 2,280 | 2,880 | 3,310 | 3,915 | 2,735 | 4,605 | 6,285 |
| 3/8 × 5 | 1,685 | 3,065 | 2,660 | 3,370 | 3,875 | 4,585 | 3,200 | 5,390 | 7,330 |
| 3/8 × 6 | 1,960 | 3,575 | 3,100 | 3,920 | 4,505 | 5,330 | 3,725 | 6,270 | 8,525 |
| 3/8 × 8 | 2,420 | 4,405 | 3,820 | 4,840 | 5,565 | 6,580 | 4,595 | 7,745 | 10,525 |
| 1/2 × 1 | 605 | 1,095 | 955 | 1,205 | 1,385 | 1,640 | 1,145 | 1,930 | 2,625 |
| 1/2 × 2 | 990 | 1,800 | 1,560 | 1,980 | 2,275 | 2,695 | 1,880 | 3,165 | 4,305 |
| 1/2 × 3 | 1,325 | 2,410 | 2,090 | 2,650 | 3,045 | 3,605 | 2,515 | 4,240 | 5,765 |
| 1/2 × 4 | 1,630 | 2,965 | 2,570 | 3,260 | 3,750 | 4,435 | 3,095 | 5,215 | 7,090 |
| 1/2 × 5 | 1,935 | 3,520 | 3,060 | 3,870 | 4,450 | 5,265 | 3,675 | 6,190 | 8,415 |
| 1/2 × 6 | 2,220 | 4,040 | 3,500 | 4,440 | 5,105 | 6,035 | 4,215 | 7,105 | 9,645 |
| 1/2 × 8 | 2,760 | 5,025 | 4,350 | 5,220 | 6,345 | 7,505 | 5,245 | 8,830 | 12,005 |

*Capacity based on 40°C ambient - 30°C rise. For a 40°C ambient indoors with a 50°C rise, multiply current capacities for 30°C rise by a factor of 1.32 for approximate values. To convert indoor current carrying capacities in table to outdoor capacities, multiply by a factor of 1.27.

**Spacing between bars is 1/4in unless noted otherwise.

**TABLE 6.17*b*** AC Current-Carrying Capacities for Round Copper Rods

| Diameter, in. | Cross Section, $in^2$ | Weight, Lb/Ft | 60 Hz Current Rating (Amps) 30°C rise | 50°C rise |
|---|---|---|---|---|
| Electrolytic Tough Pitch Cu. (110) | | | | |
| 0.250 | 0.4909 | 0.1897 | 107 | 136 |
| 0.375 | 0.1104 | 0.4268 | 190 | 240 |
| 0.500 | 0.1964 | 0.7587 | 285 | 355 |
| 0.625 | 0.3068 | 1.185 | 385 | 480 |
| 0.750 | 0.4418 | 1.707 | 495 | 620 |
| 0.875 | 0.6013 | 2.324 | 610 | 760 |
| 1.000 | 0.7854 | 3.035 | 730 | 890 |
| 1.250 | 1.227 | 4.742 | 960 | 1150 |
| 1.500 | 1.767 | 6.828 | 1175 | 1400 |
| 1.750 | 2.405 | 9.294 | 1380 | 1650 |
| 2.000 | 3.142 | 12.14 | 1560 | 1850 |
| 2.500 | 4.909 | 18.97 | 1960 | 2300 |
| 3.000 | 7.069 | 27.31 | 2325 | 2750 |

NOTE: Current carrying capacities are based on 40°C ambient, bright surfaces, indoors and free from external magnetic influences. Rods of other copper alloys are lower in current carrying capacity, except silver bearing copper alloys of high purity. ETP 110 copper has a nominal capacity of 99.9% IACS.

**TABLE 6.17*c*** AC Current-Carrying Capacities for Copper Bus Tubes

| Pipe Size | Nominal Dimensions OD | ID | Wall | Wt/Ft | Area, $in^2$ | 60 Hz Current Rating (Amperes) 30°C rise | 50°C rise |
|---|---|---|---|---|---|---|---|
| 0.500 | 0.840 | 0.542 | 0.149 | 1.25 | 0.3235 | 420 | 590 |
| 0.750 | 1.050 | 0.736 | 0.157 | 1.71 | 0.4405 | 590 | 760 |
| 1.000 | 1.315 | 0.951 | 0.182 | 2.51 | 0.6478 | 750 | 1000 |
| 1.250 | 1.660 | 1.272 | 0.194 | 3.46 | 0.8935 | 975 | 1300 |
| 1.500 | 1.900 | 1.494 | 0.203 | 4.19 | 1.0820 | 1150 | 1500 |
| 2.000 | 2.375 | 1.933 | 0.221 | 5.80 | 1.4960 | 1500 | 1950 |
| 2.500 | 2.875 | 2.315 | 0.280 | 8.85 | 2.283 | 1975 | 2550 |
| 3.000 | 3.500 | 2.892 | 0.304 | 11.8 | 3.052 | 2425 | 3200 |
| 3.500 | 4.000 | 3.358 | 0.321 | 14.4 | 3.710 | 2875 | 3700 |
| 4.000 | 4.500 | 3.818 | 0.341 | 17.3 | 4.455 | 3100 | 4200 |
| 5.000 | 5.562 | 4.812 | 0.375 | 23.7 | 6.111 | 3850 | 5300 |
| 6.000 | 6.625 | 5.751 | 0.437 | 32.9 | 8.495 | 4500 | 6500 |
| 8.000 | 8.625 | 7.625 | 0.500 | 49.5 | 12.76 | 6500 | 8400 |
| 10.00 | 10.75 | 9.750 | 0.500 | 62.4 | 16.10 | 8000 | 10600 |

Note: Variations from these listed values are expected in practice. ASTM specification B 188 controls all dimensions, tolerances, tensile properties, conductivity, tests and inspection procedures. Tables for standard pipe sizes may be obtained from Anaconda Copper. Values in this table are used in switchgear practice because the volumetric efficiency is most favorable in the extra strong series of pipe sizes.

Heat loss ($Wc$) for the inner surfaces of "built-up" bus conductors (parts close together, natural ventilation indoors), is given in $W/in^2$ by

$$Wc = \frac{0.0017\, P^{0.5}\theta^{1.25}}{h^{0.25}}$$

For outdoor service, the heat loss by convection in $W/in^2$ is estimated from the general equation for low-velocity wind, applicable for diameters or widths from ½ to 5 in.

$$Wc = \frac{0.0128\, P^{0.5}V^{0.5}\theta}{Ta^{0.123}d^{0.5}}$$

**TABLE 6.18** Radiation Emissivity Coefficients

| Surfaces | Radiation Emissivity Coefficient |
|---|---|
| **Aluminum surfaces** | |
| New bus bar, extruded | 0.05 to 0.15 |
| New bus bar, cold rolled | 0.05 to 0.20 |
| Old bar, 2 years indoors | 0.25 to 0.45 |
| Old bar, 2 years outdoors | 0.50 to 0.90 |
| Flat painted (nonmetallic) | 0.90 to 0.95 |
| **Copper surfaces** | |
| Polished | 0.03 |
| Shiny | 0.07 |
| Slightly oxidized | 0.30 |
| Normally oxidized | 0.50 |
| Heavily oxidized | 0.70 |
| Flat painted (nonmetallic) | 0.95 |
| **The following are representative values:** | |
| Indoor current ratings, (partially oxidized) | 0.35 |
| Outdoor current ratings, (normally oxidized) | 0.50 |
| Dull painted surfaces | 0.90 |
| Openings between members of built-up bars | 0.90 |

where $V$ = air velocity crosswise to conductor, ft/sec
$Ta$ = average absolute temperature of conductor, K
$d$ = outside diameter of conductor, in

The heat dissipated in $W/\text{in}^2$ by radiation $Wr$ is given by the Stefan-Boltzmann law for heat loss by radiation from a body to the walls of a room or enclosure.

$$Wr = 36.9 \times 10^{-12} e\,(T_1^4 - T_2^4)$$

where $T_1$ = temperature of the warm body, K
$T_2$ = temperature of the walls or sides of the enclosure
$e$ = emissivity coefficient (see earlier discussion in this section)

### 6.7.2.1 Temperature Rise in Electrical Equipment

Electrical equipment is generally tested under load to determine the temperature rise during operation. Since the final temperature of the operating equipment must conform to various standards, such as those of ANSI and NEMA (National Electrical Manufacturers Association), it is important for the equipment to be tested for proper operation. Various methods are employed to test the equipment in what is commonly termed a "*heat run.*" The following equation is precise for tests in which the equipment is held at a constant temperature and in which temperatures are taken at two different times.

$$(Tf - T_0)\left[1 - \left(\frac{Tf - T_1}{Tf - T_0}\right)^{t_2/t_1}\right] - (T_2 - T_0) = 0$$

This equation cannot be solved explicitly for $Tf$ unless a computer is programmed employing iterative operations. However, final temperature can be determined with reasonable accuracy using a method presented by P. Bustamente of MACI Industries, Ontario, Canada (Bustamente, 1986) summarized as follows.

$$Tf = T_0 + \frac{T_2 - T_0}{1 - [(T_2 - T_1)/(T_1 - T_0)]^2\,[t_1/(t_2 - t_1)]^2}$$

When $t_2 = 2t_1$, this equation becomes

$$Tf = \frac{T_1^2 - T_0T_2}{2T_1 - T_2 - T_0}$$

Accuracy is compromised when values of $T_1$ and $T_2$ are taken too early in the test.

Nomenclature given for the foregoing temperature-rise equations is the following: $t_1$, $t_2$ = temperature-taking times during test, min; $T_1$, $T_2$ = temperatures at times $t_1$ and $t_2$, °C; $Tf$ = final temperature, °C; and $T_0$ = initial or ambient temperature, °C.

Note that the ambient temperature must be held constant during the test for meaningful results. Estimates from the second and third equations above give reasonable accuracy and can be used to save time, reduce power consumption, and prevent equipment damage by forecasting the final temperature early in a test. Temperature readings can be taken at 15- or 30-min intervals ($t_1$, $t_2$), although the mass of the equipment should be taken into consideration in selecting these intervals. Larger masses require longer time intervals.

#### 6.7.2.2 Electrical Equipment Enclosures

NEMA definitions have been extracted in Fig. 6.52.

### 6.7.3 Short-Circuit Calculations in High-Current Busses

Under short-circuit conditions, high-*fault* currents produce electromagnetic forces between conductors. These forces can reach magnitudes of thousands of pounds-force per linear foot, uniformly distributed along the conductors.

The instantaneous force in pounds per foot between two straight, parallel, round conductors is given by

$$F = 5.4\,\frac{i_1 i_2}{d}\,10^{-7}$$

where $i_1$ and $i_2$ = instantaneous currents, A, and $d$ = distance between conductor centerlines, in.

(*Note:* When the two currents flow in the same direction, the force is attractive, and when the currents flow in opposite directions, the force is repulsive.) (See Figs. 6.53 and 6.54.)

***Forces in Direct-Current Circuit Conductors.*** The electromagnetic force between positive and negative conductors tending to separate them, in pounds per foot, is given by

$$F = K\,5.4\,\frac{I^2}{d}\,10^{-7}$$

where $K$ = the shape correction factor (see Fig. 6.55). The $K$ factor is used to adjust the equation when the conductors are large compared to their center distances. For high-capacity busses, the general design layout is such that the $K$ factor may be considered as unity.

## ELECTRICAL ENCLOSURES

**NEMA DEFINITIONS**
*(Extracted from NEMA Standard ICS-110)*

| ENCLOSURES | DESCRIPTION |
|---|---|
| NEMA1 | **General Purpose—Indoor.** Intended for use indoors, primarily to prevent accidental contact of personnel with the enclosed equipment. In addition, they provide protection against falling dirt. |
| NEMA3 | **Dusttight, Raintight and Sleet (Ice) Resistant—Outdoor.** Intended for use outdoors to protect the enclosed equipment against windblow dust and water. They are not sleet (ice) proof. |
| NEMA 3R | **Rainproof and Sleet (Ice) Resistant—Outdoor** Intended for use outdoors to protect the enclosed equipment against rain. They are not dust, snow, or sleet (ice) proof. |
| NEMA 3S | **Dusttight, Raintight and Sleet (Ice) Proof—Outdoor** Intended for use outdoors to protect the enclosed equipment against windblow dust and water and to provide for its operation when the enclosure is covered by external ice or sleet. Do not protect the enclosed equipment against malfunction resulting from internal icing. |
| NEMA 4 | **Watertight and Dusttight—Indoor** Intended for use indoors to protect the enclosed equipment against splashing water, seepage of water, falling or hose-directed water, and severe external condensation. |
| NEMA 4X | **Watertight and Dusttight—Indoor** Same provisions as NEMA 4 enclosures and, in addition, are corrosion resistant. |
| NEMA 5 | **Superseded by NEMA 12.** |
| NEMA 6 | **Submersible, Watertight, Dusttight and Sleet (Ice) Resistant—Indoor and Outdoor** Intended for use indoors or outdoors where occasional submersion is encountered. |
| NEMA 12 | **Industrial Use—Dusttight and Driptight—Indoor** Intended for use indoors to protect the enclosed equipment against fibers, flyings, lint, dust and dirt, and light splashing, seepage, drippings and external condensation of noncorrosive liquids. |
| NEMA 13 | **Oiltight and Dusttight—Indoor** Intended for use indoors primarily to house pilot device such as limit switches, foot switches, pushbuttons, selector switches, pilot lights, etc., and to protect these devices against lint and dust, seepage, external condensation, and spraying of water, oil or coolant. |

| ENCLOSURES FOR HAZARDOUS LOCATIONS | DESCRIPTION |
|---|---|
| | The term "explosion-proof" has been so loosely applied that NEMA depreciates its use. As defined by the National Electrical Code, the term "explosion-proof" applies only to NEMA 7 and 10 enclosures which, when properly installed and maintained are designed to contain an internal explosion without causing external hazard. |
| NEMA 7, Class 1 Group A, B, C, or D | Intended for use indoors, in the atmospheres and locations defined as Class 1 and Group A, B, C or D in the National Electrical Code. The letters indicate the gas or vapor in the hazardous location. |
| NEMA 9, Class II Group E, F, or G | Intended for use indoors in the atmospheres defined as Class II and Group E, F or G in the National Electrical Code. The letters E, F or G indicate the dust in the hazardous location. |
| NEMA 10 | Designed to meet the requirements of the U.S. Bureau of Mines which relate to atmospheres containing mixtures of methane and air, with or without coal dust. |

**FIGURE 6.52** NEMA standard electrical equipment enclosures.

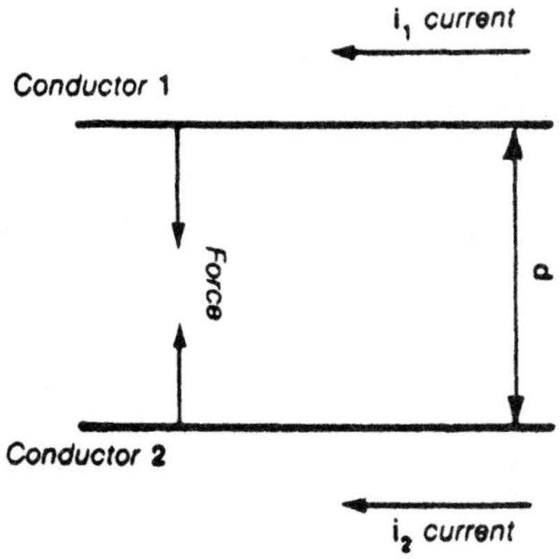


**FIGURE 6.53** Attractive electromagnetic forces on parallel conductors.

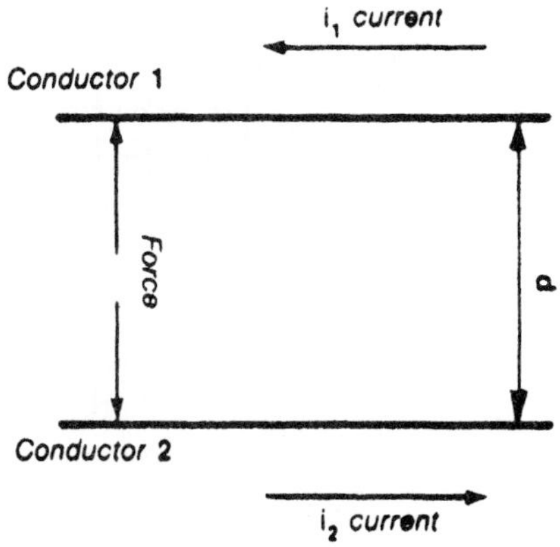


**FIGURE 6.54** Repulsive electromagnetic forces on parallel conductors.

***Forces in Alternating-Current Circuit Conductors.*** For buses or conductors carrying ac current, the maximum electromagnetic force depends on the point in the voltage wave at which the fault takes place and on the arrangement of the conductors.

Equations for calculating the maximum forces on conductors carrying ac currents are shown in Table 6.19. The value for current $I$ in the formulas in the table is the initial rms value, in amperes, of the alternating, symmetric, portion of the current. However, in asymmetric faults, a dc component is also present and its effect is included in the equations.

(*Note:* The *subtransient* reactance of rotating machinery is used in computing short-circuit currents in networks.)

### 6.7.4 Temperature Rise during Short Circuits

The following equations assume no heat loss from the conductor. Increase in resistance with rising temperature was taken into consideration.

$$I = 0.144 \times 10^6 A \sqrt{\frac{1}{t} \log \frac{\theta_2 + 228}{\theta_1 + 228}} \quad \text{for 61\% IACS EC aluminum}$$

$$I = 0.220 \times 10^6 A \sqrt{\frac{1}{t} \log \frac{\theta_2 + 234}{\theta_1 + 234}} \quad \text{for 99\% IACS ETP copper}$$

where $A$ = area of cross section of conductor, in$^2$
$I$ = rms steady-state current, A
$t$ = time, sec
$\theta_1$ = initial temperature of the conductor, °C
$\theta_2$ = final temperature of the conductor, °C

Standard practice allows a temperature of 300°C as the final ($\theta_2$) allowable for determining *short-time* rating of copper bus conductors, and 260°C for aluminum bus conductors.

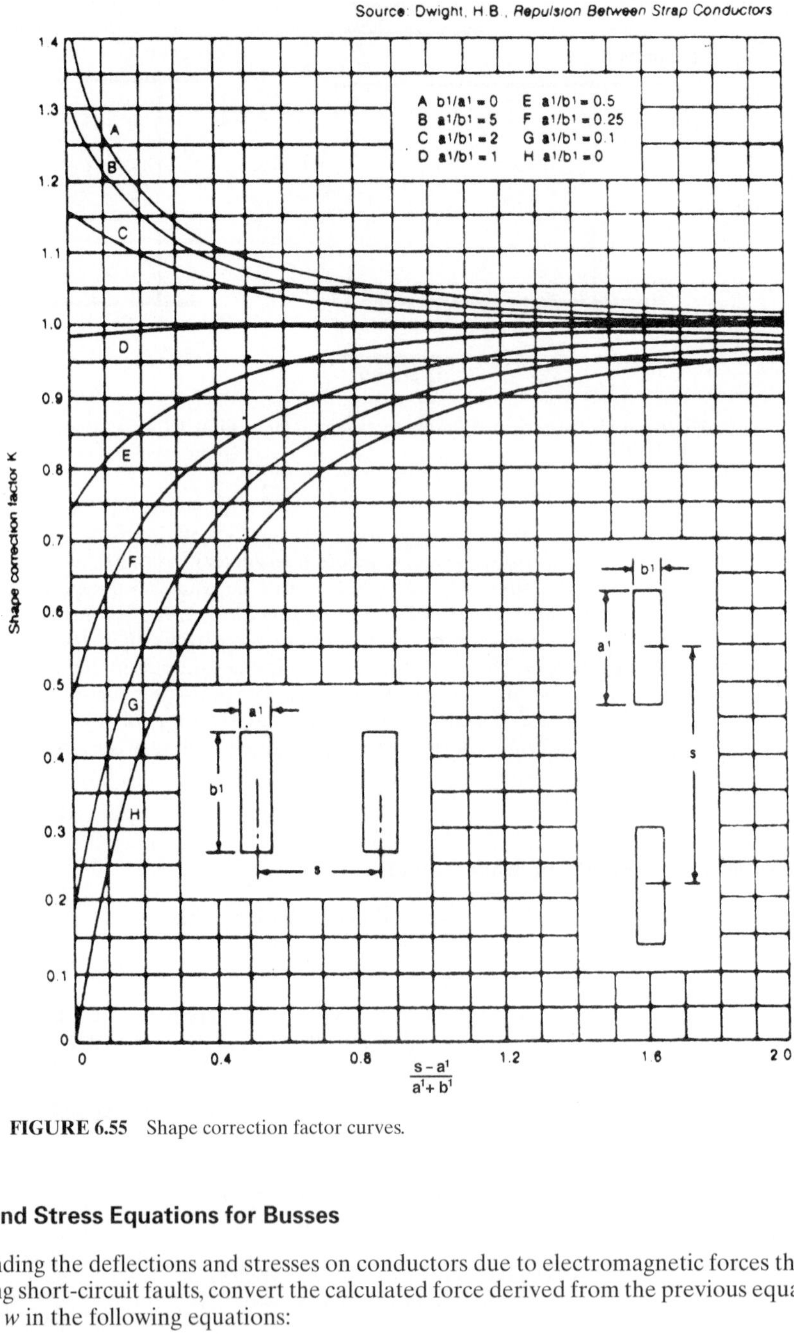


**FIGURE 6.55** Shape correction factor curves.

### 6.7.5 Deflection and Stress Equations for Busses

In finding the deflections and stresses on conductors due to electromagnetic forces that occur during short-circuit faults, convert the calculated force derived from the previous equations to *W* or *w* in the following equations:

***Simple Beam.*** Referring to Fig. 6.56, maximum deflection is

$$D = \frac{5wl^4}{384\ EI}$$

**TABLE 6.19** Indoor Switch and Bus Insulators

## *Classes A and B——2", 3" and 5" Bolt Circle Units—USASI Standard*

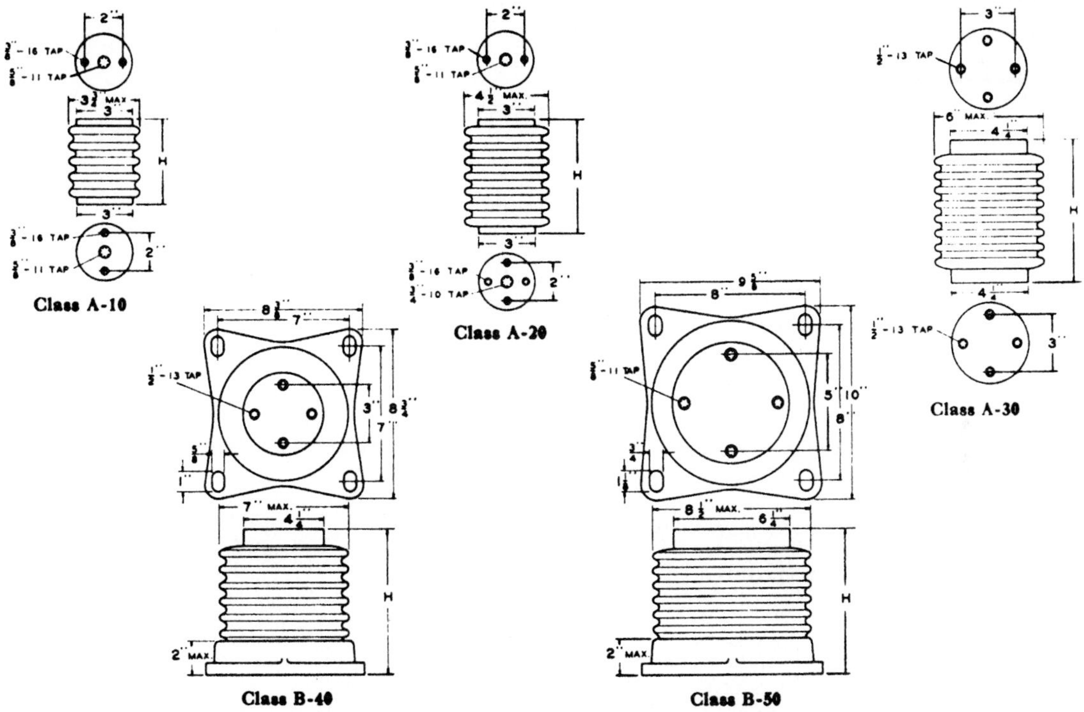


| Catalog* Number | Class | Rating kv | Height Inches H | Withstand Ratings Impulse 1½x40 kv | Withstand Ratings 60~ 1 min. Dry kv | Strength Ratings Canti-lever Lbs. 2½" above top | Strength Ratings Torsion Inch-Lbs. | Strength Ratings Tension Lbs. | Strength Ratings Com-pression Lbs. | Pkd. Wt. Lbs. Each |
|---|---|---|---|---|---|---|---|---|---|---|
| | | | | *Class A USASI Standard* | | | | | | |
| 53164 | A-10 | 2.5 | 2½ | 45 | 15 | 750 | 1,500 | 1,500 | 10,000 | 3.5 |
| 48933 | A-10 | 5 | 3½ | 60 | 19 | 750 | 1,500 | 1,500 | 10,000 | 3.75 |
| 48934 | A-10 | 7.5 | 4½ | 75 | 26 | 750 | 1,500 | 1,500 | 10,000 | 5 |
| 48935 | A-20 | 5 | 3½ | 60 | 19 | 1,000 | 2,500 | 2,000 | 20,000 | 4.5 |
| 48936 | A-20 | 7.5 | 4½ | 75 | 26 | 1,500 | 3,500 | 3,000 | 20,000 | 5.75 |
| 48937 | A-20 | 15L | 6 | 95 | 36 | 1,250 | 3,500 | 3,000 | 20,000 | 8 |
| 48938 | A-20 | 15H | 7½ | 110 | 50 | 1,000 | 3,500 | 3,000 | 20,000 | 10.5 |
| 48939 | A-30 | 5 | 3½ | 60 | 19 | 2,000 | 4,500 | 3,500 | 30,000 | 9.5 |
| 48940 | A-30 | 7.5 | 4½ | 75 | 26 | 3,000 | 6,000 | 5,000 | 30,000 | 10 |
| 48941 | A-30 | 15L | 6 | 95 | 36 | 2,500 | 6,000 | 5,000 | 30,000 | 16 |
| 48942 | A-30 | 15H | 7½ | 110 | 50 | 2,000 | 6,000 | 5,000 | 30,000 | 20 |
| 48943 | A-30 | 23 | 10½ | 150 | 60 | 1,500 | 6,000 | 5,000 | 30,000 | 26.5 |
| 48944 | A-30 | 34.5 | 15 | 200 | 80 | 1,250 | 6,000 | 5,000 | 30,000 | 35 |
| | | | | *Class B USASI Standard* | | | | | | |
| 48945 | B-40 | 7.5 | 6 | 75 | 26 | 6,000 | 10,000 | 8,000 | 50,000 | 26.5 |
| 48946 | B-40 | 15L | 7½ | 95 | 36 | 5,000 | 10,000 | 8,000 | 50,000 | 30 |
| 48947 | B-40 | 15H | 9 | 110 | 50 | 4,000 | 10,000 | 8,000 | 50,000 | 34 |
| 48948 | B-40 | 23 | 12 | 150 | 60 | 3,000 | 10,000 | 8,000 | 50,000 | 46 |
| 48949 | B-40 | 34.5 | 16½ | 200 | 80 | 2,500 | 10,000 | 8,000 | 50,000 | 58 |
| 48950 | B-50 | 7.5 | 6 | 75 | 26 | 12,000 | 15,000 | 12,000 | 80,000 | 44 |
| 48951 | B-50 | 15L | 7½ | 95 | 36 | 10,000 | 15,000 | 12,000 | 80,000 | 49 |
| 48952 | B-50 | 15H | 9 | 110 | 50 | 8,000 | 15,000 | 12,000 | 80,000 | 54 |
| 48953 | B-50 | 23 | 12 | 150 | 60 | 6,000 | 15,000 | 12,000 | 80,000 | 67.5 |
| 48954 | B-50 | 34.5 | 16½ | 200 | 80 | 5,000 | 15,000 | 12,000 | 80,000 | 75 |

*Hardware ferrous, black enameled or hot-dip galvanized as specified. Chocolate brown glaze is standard.

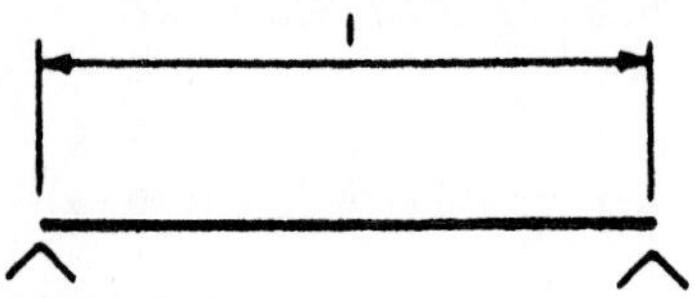


FIGURE 6.56 Simple beam.

Maximum moment is

$$M = \frac{wl^2}{8}$$

Fiber stress is

$$f' = \frac{wl^2}{8S}$$

Maximum load is

$$W = \frac{8fS}{l}$$

Maximum span is

$$l = \sqrt{\frac{8fS}{w}}$$

***Beam Fixed at Both Ends.*** Referring to Fig. 6.57, maximum deflection is

$$D = \frac{wl^4}{384\ EI}$$

Maximum moment is

$$M = \frac{wl^2}{12}$$

Fiber stress is

$$f' = \frac{wl^2}{12S}$$

Maximum load is

$$W = \frac{12fS}{l}$$

Maximum span is

$$l = \sqrt{\frac{2fS}{w}}$$

***Continuous Beam, Two Spans.*** Referring to Fig. 6.58, maximum deflection is

$$D = \frac{wl^4}{185\ EI}$$

Maximum moment is

$$M = \frac{wl^2}{8}$$

Fiber stress is

$$f' = \frac{wl^2}{8S}$$

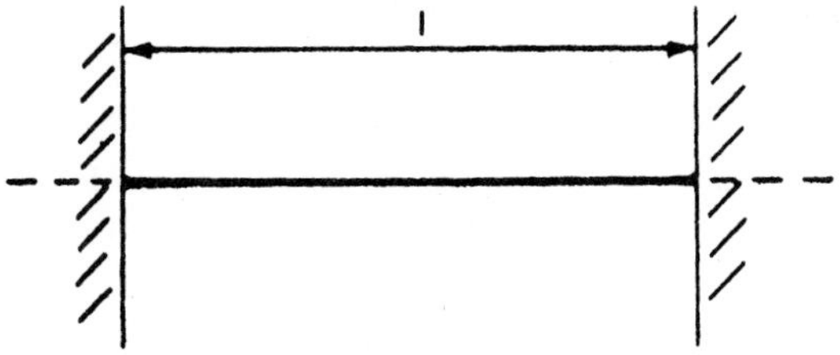


**FIGURE 6.57** Beam fixed at both ends.

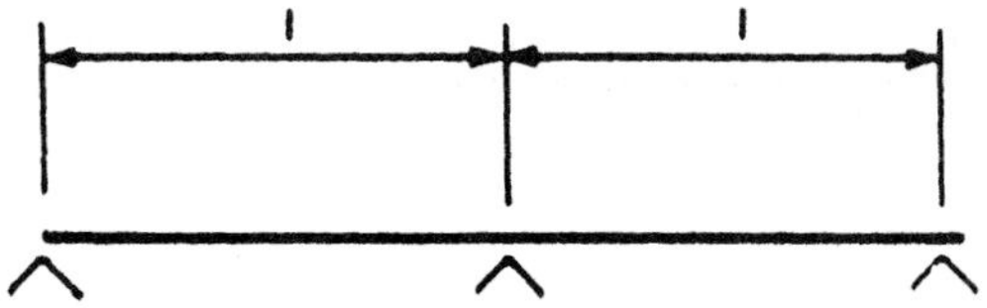


**FIGURE 6.58** Continuous beam, two spans.

Maximum load is

$$W = \frac{8fS}{l}$$

Maximum span is

$$l = \sqrt{\frac{8fS}{w}}$$

***Continuous Beam, More Than Two Spans.*** Maximum deflection occurs in the end spans and is slightly less than for two spans.

Maximum moment is

$$M = 0.107\ wl^2 \quad \text{or} \quad \frac{wl^2}{9.35}$$

Fiber stress is

$$f' = \frac{0.107\ wl^2}{S} \quad \text{or} \quad \frac{wl^2}{9.35\ S}$$

Maximum load is

$$W = \frac{fS}{0.107\ l} \quad \text{or} \quad \frac{9.35\ fS}{l}$$

Maximum span is

$$l = \sqrt{\frac{fS}{0.107\ w}} \quad \text{or} \quad \sqrt{\frac{9.35\ fS}{w}}$$

In the preceding deflection and stress equations, $D$ = deflection, in; $w$ = load, pounds per linear inch of length = lb/ft/12; $W$ = total uniform load ($w \times l$), lb; $l$ = span, in; $E$ = modulus of elasticity of conductor material, psi; $I$ = moment of inertia of conductor section, in$^4$; $M$ = bending moment, lb · in; $S$ = section modulus of conductor, in$^3$; $f'$ = fiber stress, lb/in$^2$; $f$ = maximum allowable fiber stress of the conductor material. The minimum yield point is commonly used; some examples for materials are listed here:

| | | |
|---|---|---|
| ETP (99.9% IACS) copper | | = 12,000–45,000 psi |
| Minimum yield | soft bar | = 12,000 psi |
| | hard bar | = 45,000 psi |

$$\text{Hard-copper bus bar tensile strength} = 48{,}000\text{–}50{,}000 \text{ psi}$$

$$\text{EC (61\% IACS) aluminum} = 14{,}000\text{–}24{,}000 \text{ psi}$$

$$\text{Minimum yield} \begin{cases} \text{H14} & = 14{,}000 \text{ psi} \\ \text{H19} & = 24{,}000 \text{ psi} \end{cases}$$

### 6.7.5.1 Stresses in Switchgear Components Due to Electromagnetic Forces

Switchgear components such as bus ducts, interconnecting buses, and switching devices are all subject to mechanical forces induced by electromagnetic fields. Fusing time currents, current-carrying capacities, heating in electrical conductors, temperature rise in electrical equipment, and short-circuit forces were all detailed in previous subsections, with the appropriate equations for calculating these phenomena.

The following disclosures will present some of the less familiar aspects of switchgear component design.

***Simplified Bus Duct Force Nomographs.*** Figures 6.59 and 6.60 are nomographs for determining forces, brace distances, and section moduluses for single-phase symmetrical ac fault currents. Figures 6.61 and 6.62 are nomographs for determining forces, brace distances, and section moduluses for three-phase symmetrical ac fault currents.

Figure 6.63 shows a typical 15-kV metal enclosed switchgear and its associated bus work at the interface end of the gear. This piece of equipment consists of a load-interrupter switch and a transition compartment.

Figure 6.64 and 6.65 are typical nonsegregated phase bus duct sections showing the bus bracing and enclosures. Figure 6.64 is a 5-kV, 2000-A bus duct with polyester fiberglass bus braces that terminate in a horizontal elbow section. Figure 6.65 is another section of the same bus duct system at a vertical elbow section. Notice the clear lexan (polycarbonate) bus "boots" at the 90° joints.

Figure 6.66 is a 15-kV, 2000-A high-current disconnect switch (isolator), showing the vertical bus terminating at the switch contact terminals. This switch is of the "Torque-Lok" type, designed by the author, and is produced at the Powercon Corporation in Severn, Maryland. This series of switching devices is available in voltage ranges of 2.4 to 34.5 kV, current ratings from 2000 to 4000 A, and impulse levels to 150-kV BIL. The "X" marks on the bolted connections are the quality control inspector's marks, indicating that the bolts have been torqued to their proper range.

***Break-Jaw Forces on AC Switching Devices.*** When an ac switching device's main current-carrying blades close on a load circuit, a minor force is generated by the load current as it is established through the prestriking arc, preceding contact. However, when a switching device is closed against a short circuit, currents on the order of many thousands of amperes are produced by the fault condition. The electromagnetic forces on such occasions can be very powerful. The results of such actions may be that the switching device will not close or will be blown off the contacts, with great damage to the switching device and associated equipment. If the switching device does not close rapidly enough, the prestriking arc will do great damage to the switching device during the arc-time interval.

Closing blade angular velocities of 25 to 30 radians per second (rad/sec) are required to prevent excessive damage to the blades and contacts during a fault-closing operation of a switching device.

The break-jaw forces which the switching device must overcome in order to close on a fault may be calculated using H. B. Dwight's break-jaw equation.

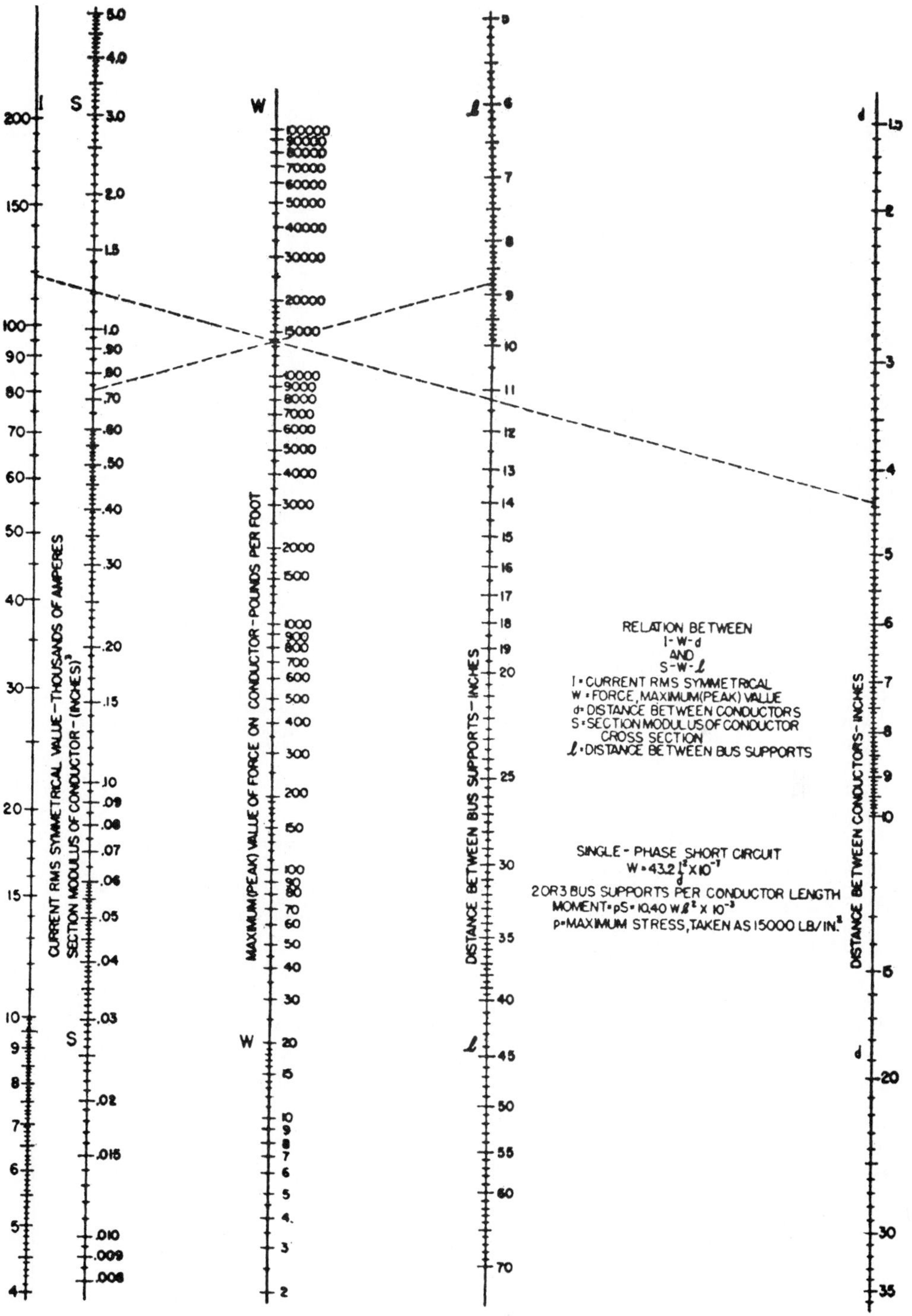


**FIGURE 6.59** Nomograph for single-phase bus short-circuit conditions.

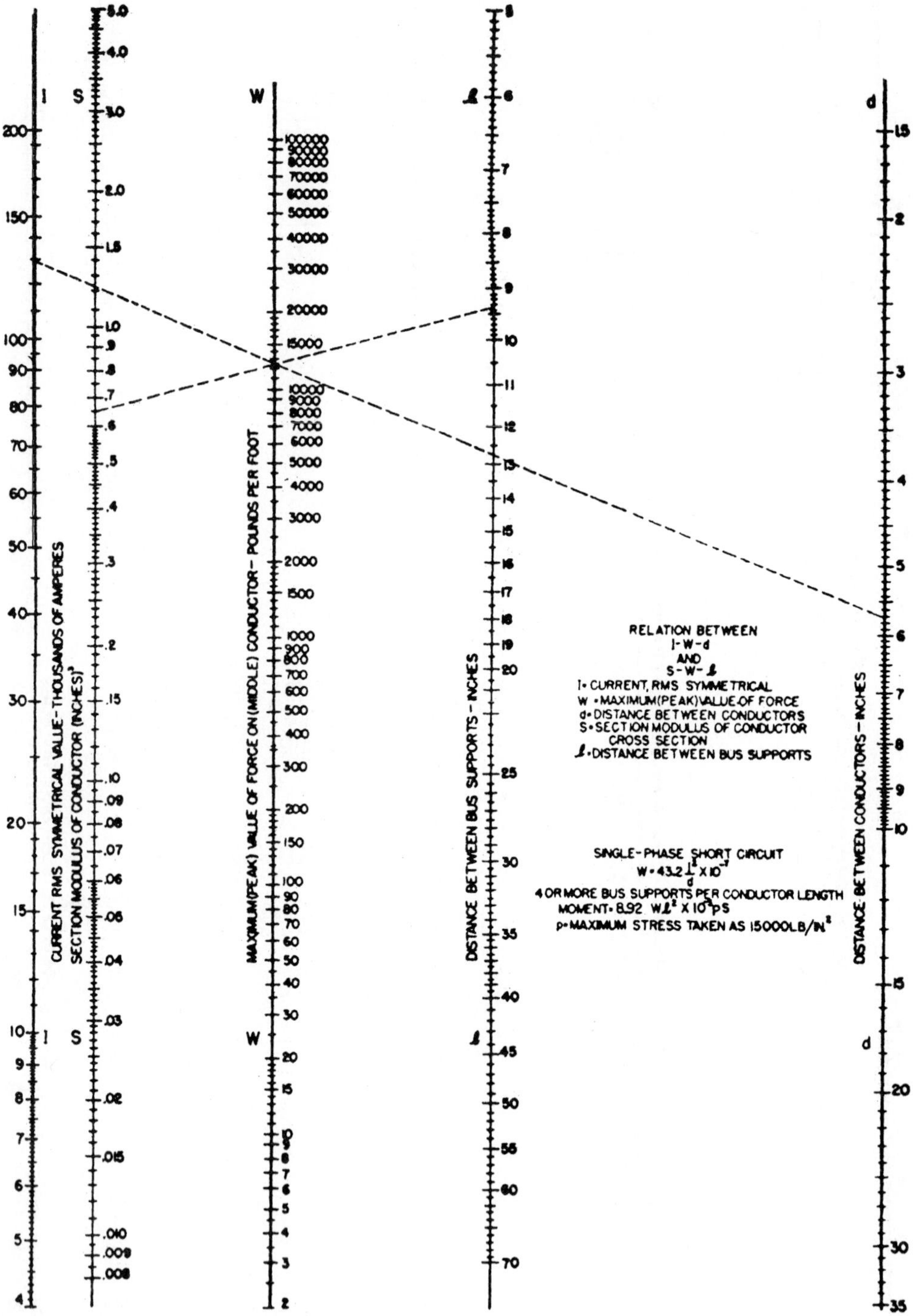


**FIGURE 6.60** Nomograph for single-phase bus short-circuit conditions.

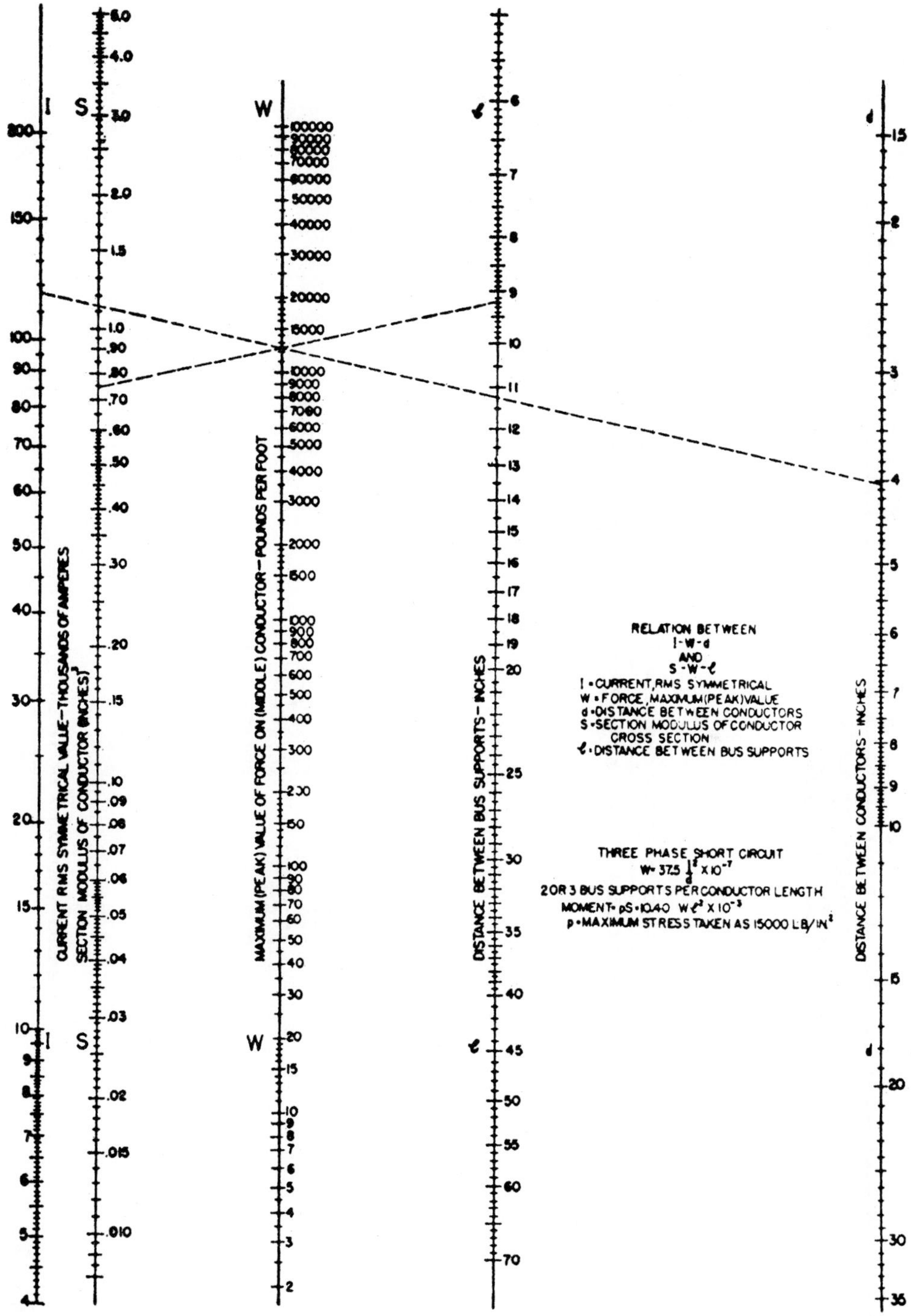


**FIGURE 6.61** Nomograph for three-phase bus short-circuit conditions.

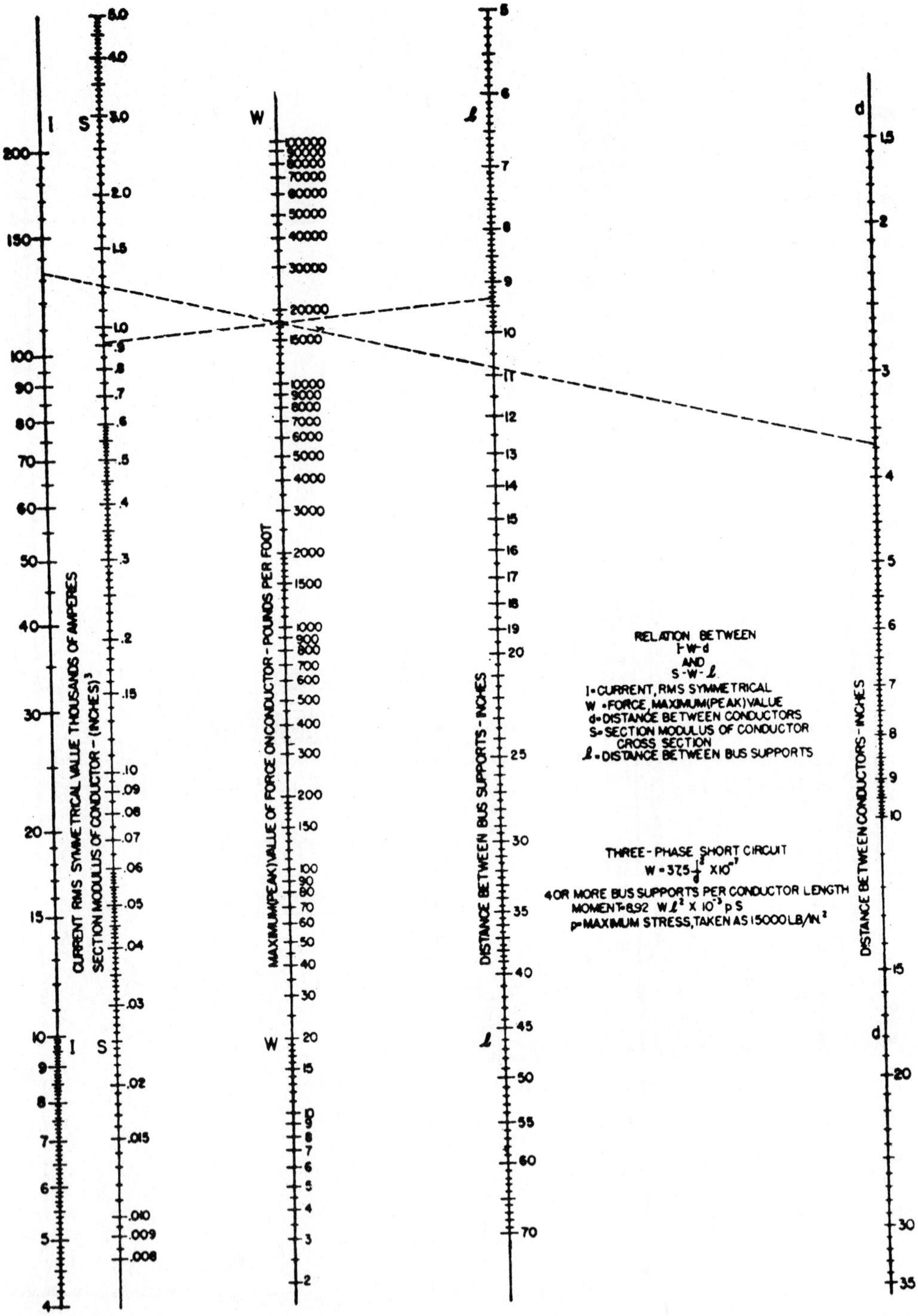


**FIGURE 6.62** Nomograph for three-phase bus short-circuit conditions.

**FIGURE 6.63** Switchgear section showing main 3000-A bus transition.

**FIGURE 6.64** Three-phase, 3000-A, 15-kV bus duct sections with horizontal elbow.

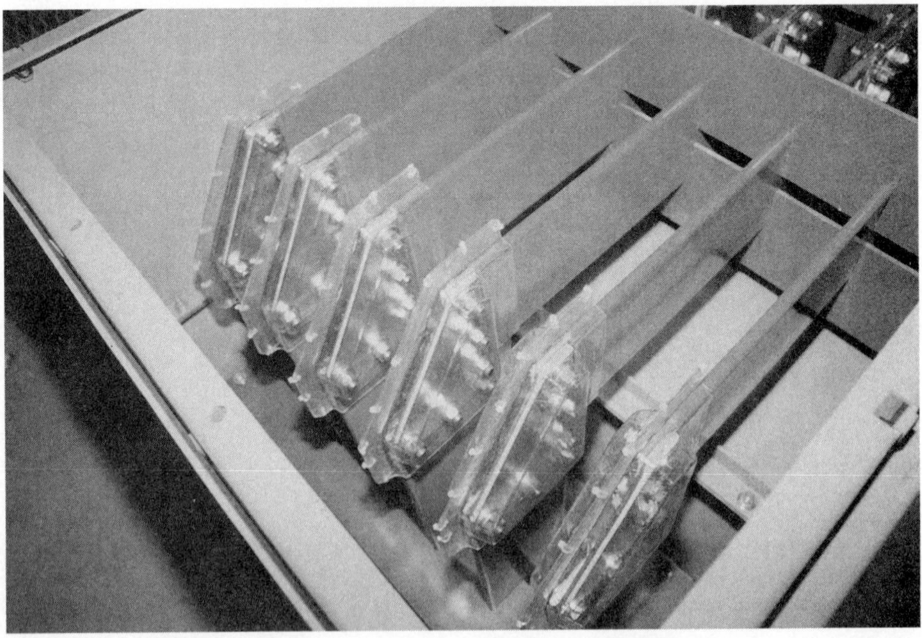

**FIGURE 6.65** Three-phase, 3000-A, 15-kV bus duct vertical elbow showing insulated bus joints.

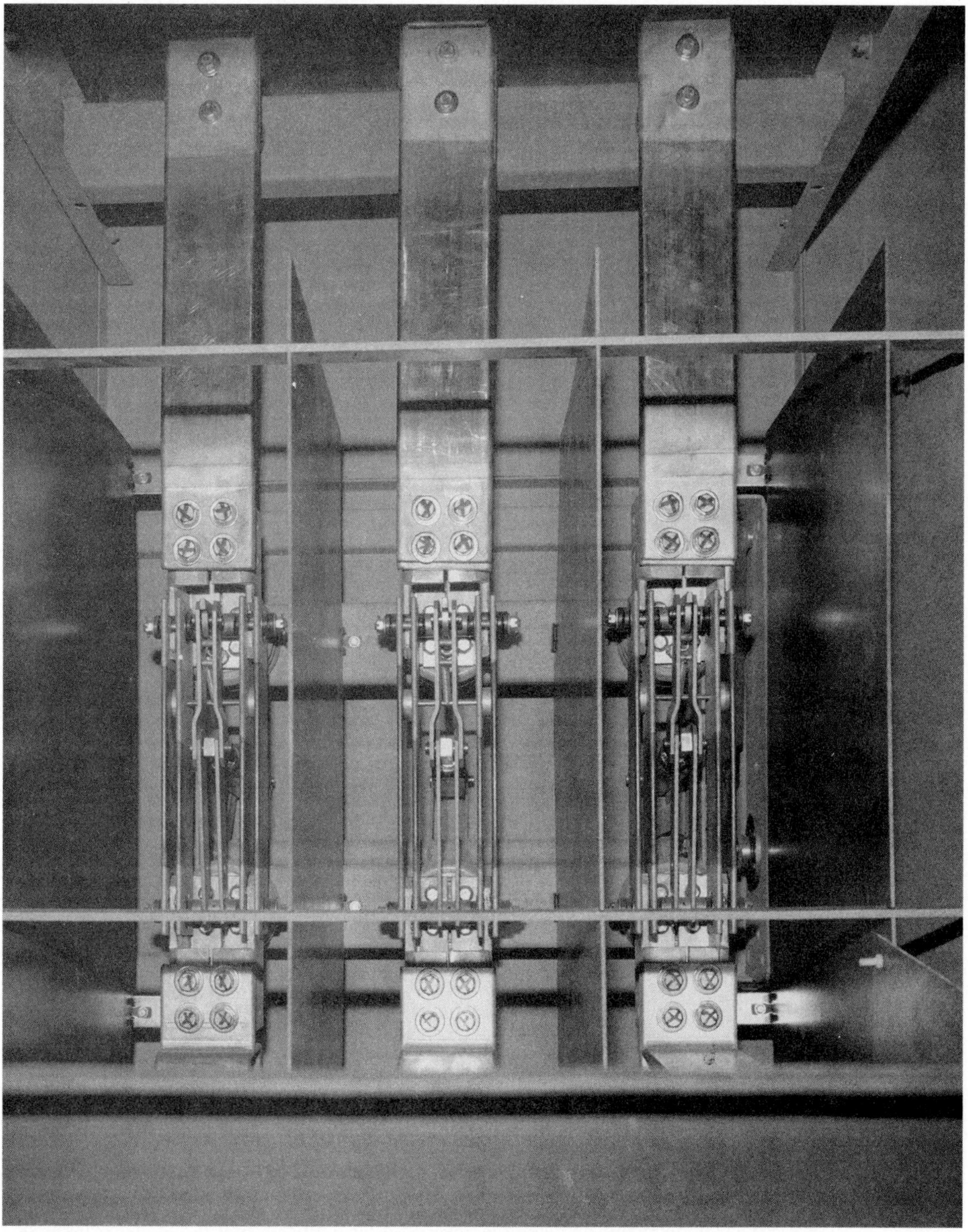

**FIGURE 6.66** Isolator switch showing 3000-A bus; 3000 A, 15 kV.

Typical system short-circuit current ratings (ANSI standard) are

| | |
|---|---|
| 5-kV, 600-A system | Fault current: 40,000 A asymmetrical |
| 5-kV, 1200-A system | Fault current: 61,000 A asymmetrical |
| 15-kV, 600-A system | Fault current: 40,000 A asymmetrical |
| 15-kV, 1200-A system | Fault current: 61,000 A asymmetrical |

(*Note:* The asymmetrical current is generally taken as approximately 1.6 times the symmetrical current rating. The asymmetrical current takes into account the dc component in the circuit, which causes the current wave to offset, resulting in higher peak current values.)

***Break-Jaw Force Calculations.*** Figure 6.67 shows a modern-design medium-voltage interrupter switch. This type of switching device is made for voltages ranging from 2.4 to 34.5 kV, with maximum currents of 1200 A for switches below 34.5 kV and 600 A at 34.5 kV. This particular switch belongs to a series of interrupter switches, designed by the author, and produced by the Powercon Corporation in Severn, Maryland.

Figure 6.68 is a detail view of one pole of the switch and is marked with the dimensional letters used in the following equation for calculating the break-jaw forces.

The following equation is known as "Dwight's equation for break-jaw forces in switching devices." This equation is used by the design engineer to ascertain the maximum force that is generated on the jaw end of a switch when the switch closes on a short circuit (fault). On switching systems that generate more than 10,000 A of short-circuit current, this force must be taken into consideration when designing the switching device. The physical dimensions of the switch also determine, to a degree, the extent of the force. The blades of the switch must overcome these forces when closing into a fault current or short circuit and remain in the closed position, preferably with a locking device or latch. The "momentary" forces must also be calculated when the switch is in the closed position, to make certain that the supporting insulators do not break and the blades themselves do not deform under the induced electromagnetic forces generated during a system short circuit or fault. Previous subsections detailed the equations and procedures involved in determining the forces between poles or conductors during fault currents.

***Dwight's Break-Jaw Equation.*** The break-jaw force $K$, measured in pounds-force (lbf), may be determined from the following equation (see Fig. 6.68 for dimensional parameters):

$$K = \frac{I^2}{4.45 \times 10^7}\left[2.30 \log_{10}\left(\frac{2A}{r}\right) - \frac{2}{3} - \frac{A}{2B} - \frac{C^2}{6A^2} + \frac{3r^2}{20A^2} + \frac{A^3}{24B^3} + \frac{AC^2}{24B^3} + \frac{B}{S}\right]$$

where $K$ = maximum momentary force, lbf
$I$ = peak value of the fault current, A
$S$ = distance of return bus behind switch, in or mm
$r$ = ½ $D$, in or mm
$A$ = offset of center of blades to conductor, in or mm
$B$ = contact centers (centerline CL of hinge to CL of jaw) in or mm

The equation holds true when $A$ is less than $B$. Also, when the bus does not return behind the switch, $B/S = 0$. Also, if $I$ represents the peak value of the current, the equation gives the maximum momentary value of the force $K$, lbf. The dimensions of the switch may be given in inches or millimeters, since the equation uses only ratios of the dimensions.

Since the last three terms in the equation are usually quite small, they may be omitted for quicker calculation. The dimensions $A$, $B$, $2C$, $S$, $D$, and $r$ are determined by the design of the switch pole and its members. Note the dimensions $2C$ and $D = 2r$. If the blade is 2 in wide, $C$ will be 1 in, and if $D = 2$ in, $r$ will be 1 in; i.e., $D = 2r$, so $r = D/2$. The dimensions may be manipulated to minimize the break-jaw force, if this is of advantage in the initial design of the

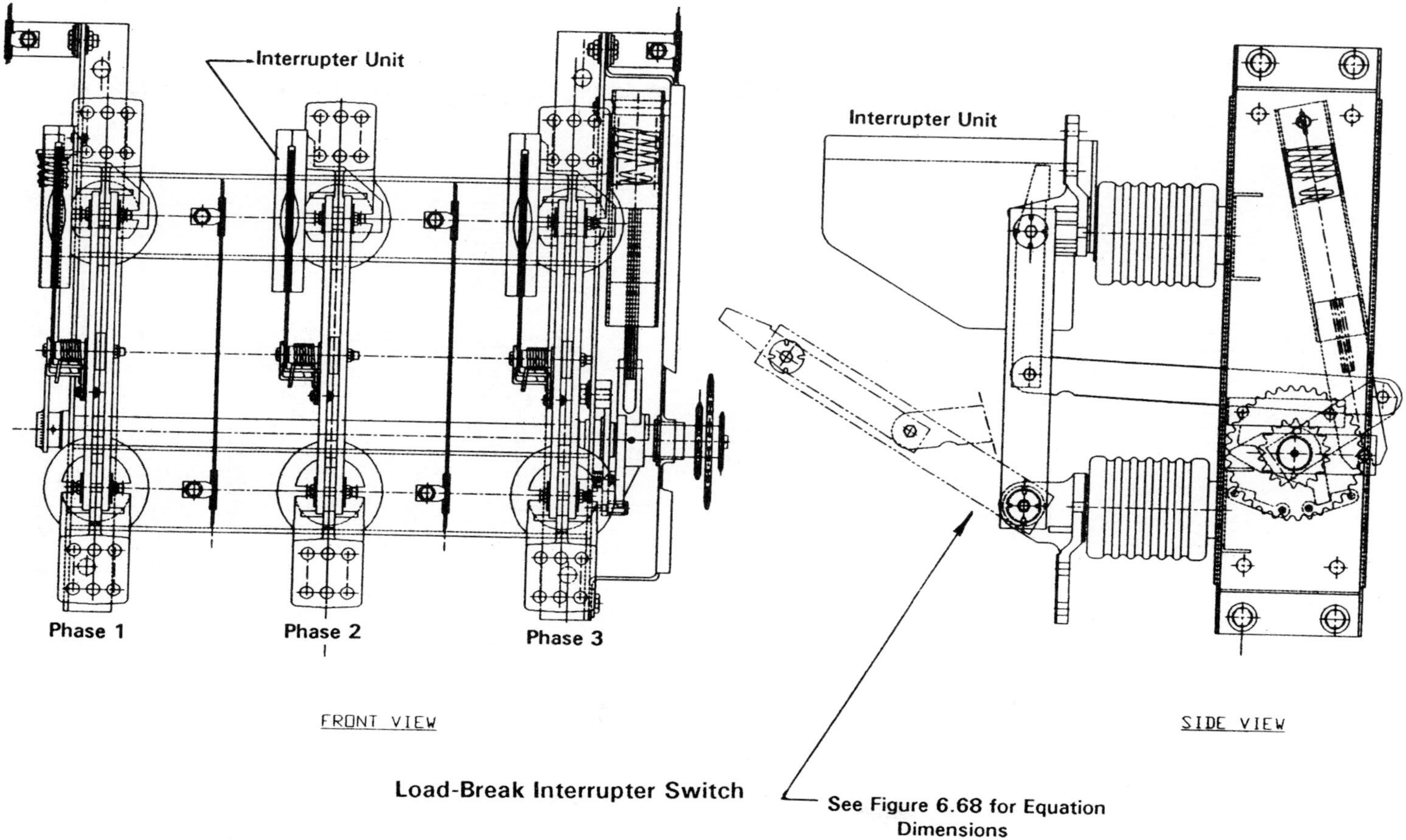


**FIGURE 6.67** Typical frame-mounted load-interrupter switch, three-phase, 15 kV, 95-kV BIL.

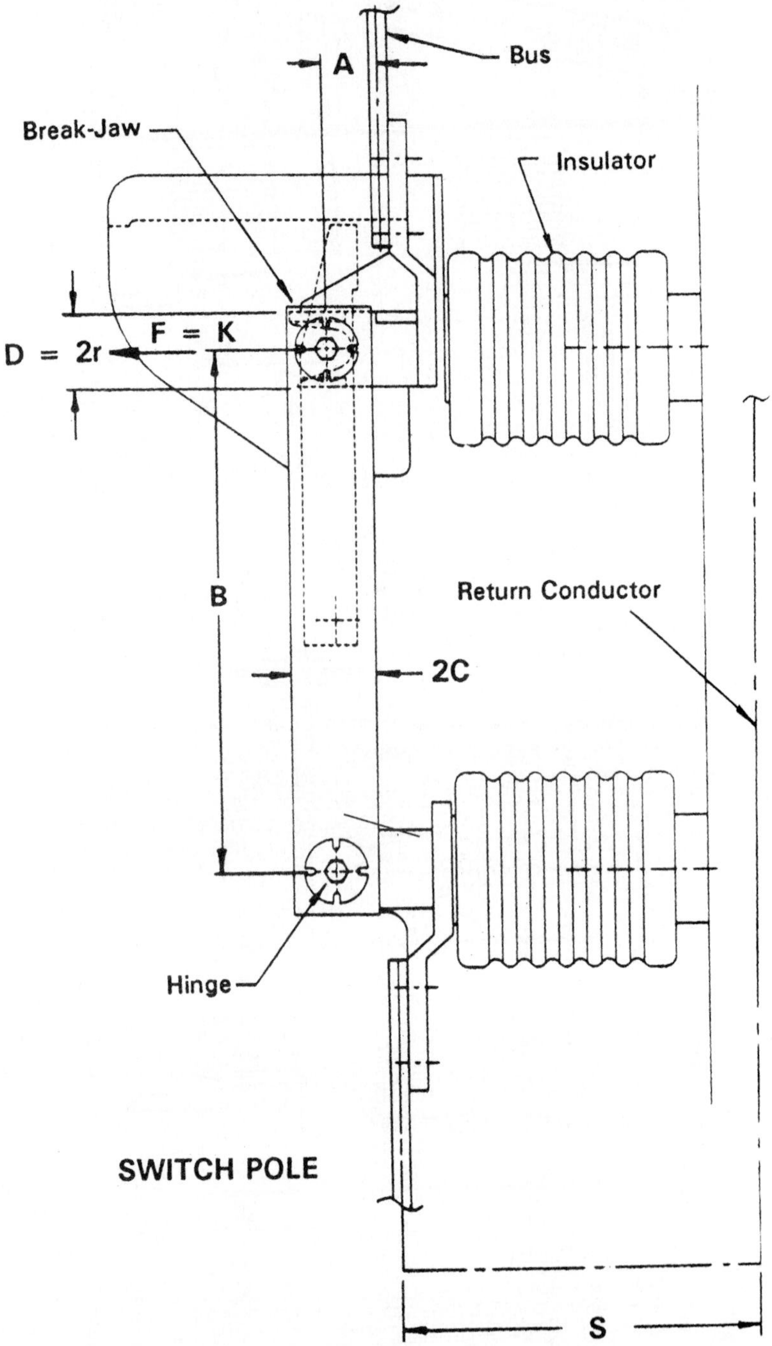


**FIGURE 6.68** Switch pole geometry for break-jaw force calculations.

switch. Often, the dimensions are determined by the voltage and current ratings and the *basic impulse level* (BIL) rating.

Many switch design parameters are stipulated in the ANSI standards for American switchgear and switching devices. (Refer to ANSI C37 series.)

### 6.7.5.2 Switching Device Contact Pressures

On high-pressure, small-area electrical contacts such as disconnect fingers, the small-area pressure should be on the order of 3000 to 9000 psi. This will produce current densities as high as 30,000 A/in$^2$ of contact area. As an example, the AKR-type low-voltage contact fingers, when tightened to a clamp load of 50 lbf, will result in contact-area pressures of approximately 7000 psi. This amount of pressure will ensure that the contacts will wipe clean when being slid into the connected position and maintain the high pressure. Silver-to-silver contact surfaces are recommended. The electrodeposited silver should be in the range of 0.2 to 3 mil (0.0002 to 0.003 in), with the heaver thickness preferred on high-pressure sliding contacts such as switches and circuit-breaker primary disconnect contacts.

On low-pressure, large-area contacts, pressures of 0.5 lb/A are adequate. These types of contact surfaces are found on some interrupter and high-pressure, high-current disconnect switches.

*Example.* On a Torque-Lok-type switch rated 3000 A, a total clamp pressure of 0.5 × 1500 = 750 lbf would be adequate on each web of the split contact, although pressures of 75 percent of this amount will produce good results because of the formation of a higher-pressure area in the vicinity of the switches' belleville spring washers at the jaw and hinge cross-bolts. [*Note:* In all sliding or moving electrical contacts, the copper conductors and fingers should be heavily silver plated to 0.2 to 3 mils (0.0002 to 0.003 in). Bare copper contact surfaces can corrode or oxidize, causing hot spots and uncontrollable temperature rises at the contact points or areas, resulting in electrical failure of the equipment. Silver-to-silver contacts will always run at a cooler temperature than copper-to-copper contacts and provide the additional safety feature of nondetrimental oxidation and corrosion resistance at the high-pressure or low-pressure contact points. Exception is taken to the preceding comments on silver plating when the electrical equipment is commissioned at a paper mill or petrochemical processing plant. In these environments, hydrogen sulfide gas is present in the atmosphere and will corrode silver-plated current-carrying parts, rendering them nonconductive (silver sulfide). The use of heavy tin plating is required for service at most corrosive atmosphere locations. The plating thickness of the tin is usually specified as 3 mils (0.003 in), although thicknesses of 1 mil (0.001 in) are usually sufficient.]

A typical low-voltage, high-current spring-loaded breaker contact is shown in Fig. 6.69. The contact shown is half of a set of two, which are rated for 3000-A continuous current-carrying capacity.

### 6.7.5.3 Modern Switching Devices (Alternating Current)

Power switching devices are classified as either switches or circuit breakers. Contactors, reclosers, and other specialized switching devices are included in the two main categories. The switching devices commonly used on modern ac electric power systems and switchgear include switch and breaker types. Switch types are

- Interrupter or load-break (fused or nonfused)
- Isolator or non-load-break
- Selector, two- and three-position
- Tap changer (transformer applications)

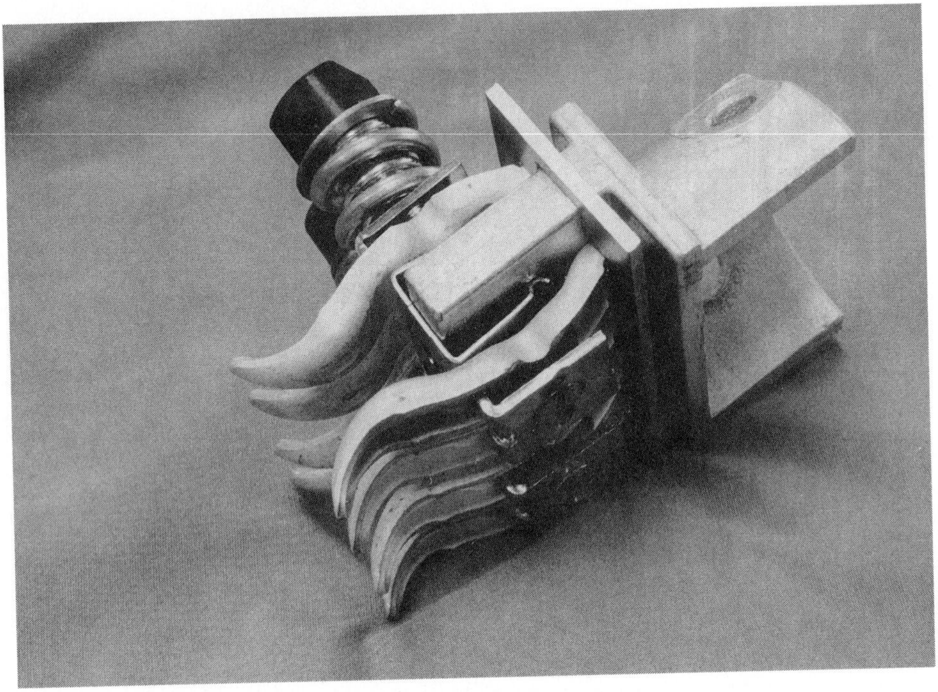

**FIGURE 6.69** Typical low-voltage, high-pressure 3000-A contact half.

[*Note:* The medium used for interruption of the ac current may be "hard gas" (an arc chute type with ablative properties), vacuum, air-blast, or gas.]

Breaker types are

- Air-magnetic (Magneblast)
- Vacuum
- Oil
- Gas [sulfur hexafluoride ($SF_6$), etc.]
- Air blast

***Interruption of AC Current.*** Alternating current, as supplied by power companies in the United States, is a sinusoidal wave current of 60 Hz frequency at 120 or 140 V or higher for industrial applications. Most European countries are supplied with 50-Hz-frequency ac current. Transportation systems often use 25-Hz ac current for tractive power applications such as electric trains and trolleys.

The theory of ac current interruption is relatively complex but can be summarized here for simplification. The pure ac sine wave is familiar to most design engineers and can be seen to pass from positive to negative, while going through two zero points per cycle (termed *current zeros*). All ac interrupting devices attempt to extinguish the interruption arc as the current wave passes through a current-zero point. In many switches, an insulative gas such as formaldehyde is ablated from the plastic material that forms the interrupting device containment. The evolved gas prevents the arc current from restriking as the system voltage recovers, after the current passes the current-zero point, thus interrupting the arc which sustains the circuit. The circuit is therefore interrupted or "opened."

In high-power electrical service applications, failure of a switching device to interrupt the load or fault current produces great damage to the device and associated equipment enclosures. The damage occurs very rapidly (on the order of 6 to 10 cycles' duration), unless a downstream device interrupts the circuit. One cycle of 60 Hz frequency is 0.0167 sec or 16.7 msec.

Switches normally interrupt load currents on the order of 600 to 1200 A but are not capable of interrupting fault currents, which may be on the order of thousands of amperes. Fuses are employed with switches to interrupt fault currents, as well as system overloads.

Circuit breakers are capable of interrupting load currents as well as fault currents, are automatically reset, and are ready for immediate continued service.

***Ablative Materials for Interruption Devices.*** Most common medium-voltage switches use an arc chute or interruption chamber to extinguish the interruption arc during load-break operations. The following materials have all proved to be excellent for this application and all are *ablative* (they evolve an insulative gas such as formaldehyde in the presence of the high-temperature interruption arc).

- *Acetal:* thermoplastic (Delrin, etc.)
- *Acrylics:* thermoplastic (Plexiglas, Lucite, etc.)
- *Urea formaldehyde:* thermoset plastic

***Tightening Electrical Connections—High-Power Buses and Terminals.*** Electrical connections such as bus joints and switch terminal connections should be made in the following manner.

*Bus Joints.* Copper-to-copper bus-joint connections are made using plated-steel, stainless-steel, or high-strength bronze bolts and nuts, with flat and lock washers. The bolt is torqued to its listed torque rating for each particular grade (such as grades 2, 5, or 8). For high-pressure joints, grade 5 or 8 fasteners are indicated, although grade 2 fasteners are often used for common joints where the joint thickness does not exceed 2 bars. The tightening torques required for each particular fastener grade are found in Chap. 10 of this handbook.

Aluminum-to-aluminum bus-joint connections are made using plated-steel, stainless-steel, or high-strength aluminum alloy bolts and nuts, with belleville washers. The bolt is torqued until the belleville washers are completely flattened, while not applying additional torque, and are left in this condition. This procedure allows the bus bars to remain under the full compressive load of the flattened belleville washers when the bars are thermally cycled during the current-carrying and non-current-carrying phases. To prevent the different rates of expansion encountered when steel bolts are used with aluminum bus, the use of high-strength aluminum alloy fasteners is recommended.

Plain carbon steel fasteners should not be used on copper-to-copper bus systems or electrical terminal connections that carry more than 2000 A. The use of austenitic stainless steel or high-strength bronze (Everdur) bolts is required to prevent hot spots created by the induction heating due to the large circulating currents. Aluminum-to-aluminum connections, in this case, should be made using austenitic stainless-steel or high-strength aluminum alloy fasteners.

Aluminum-to-aluminum bus or terminal connections must be specially plated and a thin coat of "No Oxid" applied between the joints to prevent corrosion, which will cause overheating and possible destruction of the connection.

*Modern High-Power Switching Devices.* Figure 6.70 shows a typical interrupter switch for 15-kV, 600-A power system applications. The interrupter units are located to the left side of each pole or phase and are made of urea formaldehyde. This is a typical hard-gas interrupter which is common on many industrial three-phase electric power systems.

Figure 6.71*a* and 6.71*b* shows a special type of interrupter switch that is used for specialized applications requiring a double-tripping action. Its mechanical operation is similar to the action of a circuit breaker, except that it cannot interrupt fault currents. Figure 6.72 shows a

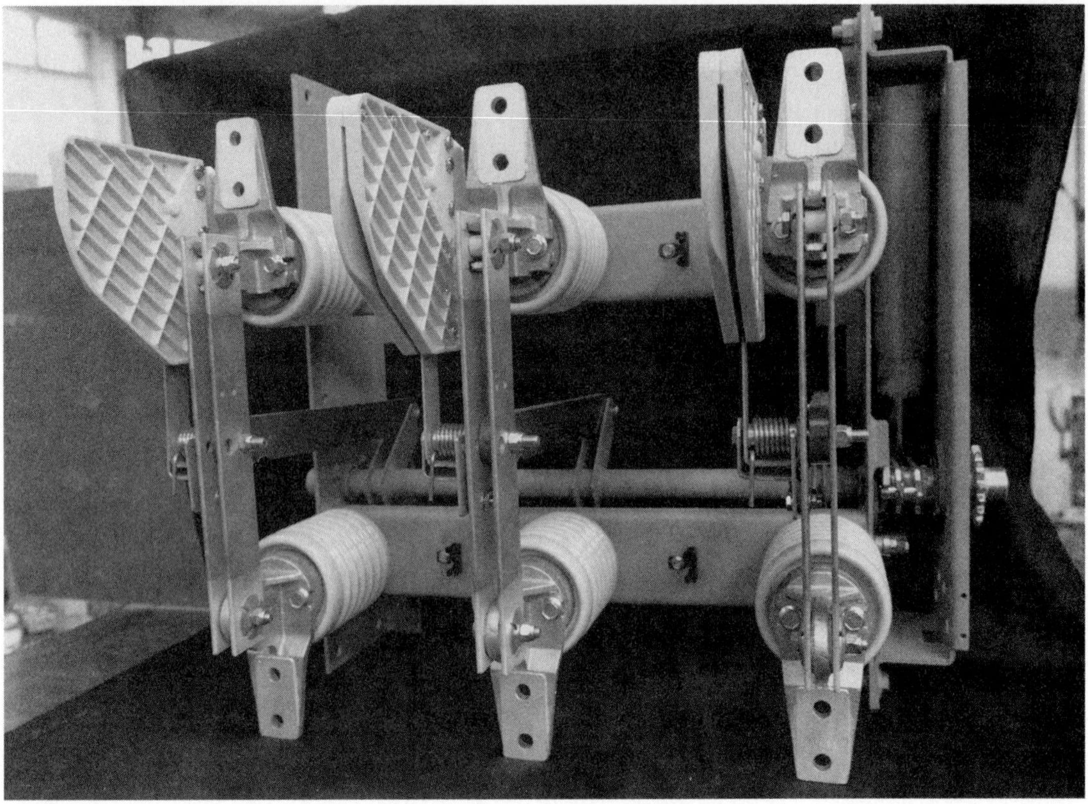

**FIGURE 6.70** Typical frame-mounted load interrupter switch, 600 A, 15 kV, 95-kV BIL.

prototype model of the special switch shown in Fig. 6.71*a* and 6.71*b*. In Fig. 6.72 the control system and auxiliary switches are shown on the control side of the switch, together with the solenoids used to trip the device open and closed.

Figure 6.73*a* and 6.73*b* shows a high-current, medium-voltage isolator switch in the closed and open positions. This type of switch is often used to isolate a power circuit and is used in conjunction with a circuit breaker, both units being interlocked so that the circuit breaker must be locked open before the isolator switch can be opened. Figure 6.74 is an exploded-view drawing showing the detail parts of one pole of the isolator switch of Fig. 6.73*a* and 6.73*b*.

Figure 6.75 shows a small vacuum interrupter unit that may be used on an interrupter switch or low-power circuit breaker, 15-kV, 600-A, continuous current, and 2000- to 4000-A interrupting capability.

Figure 6.76 shows a General Electric Magneblast circuit breaker installed and elevated to the connected position in its cell within a metal-clad switchgear unit. The electrical controls and instruments are shown on the backside of the doors. This breaker is rated 15 kV, 3000 A, 1000 MVA, 95-kV BIL and is of the air-magnetic interruption type. This type of breaker has been in industrial service for more than 40 years, having replaced most of the oil-type breakers previously in service.

Figure 6.77 shows the connection end of a 34.5-kV, 1200-A vacuum circuit breaker. This circuit breaker was manufactured by Siemens in Germany. The three vacuum interrupter units can be seen between the poles of this three-phase interrupting device. The secondary

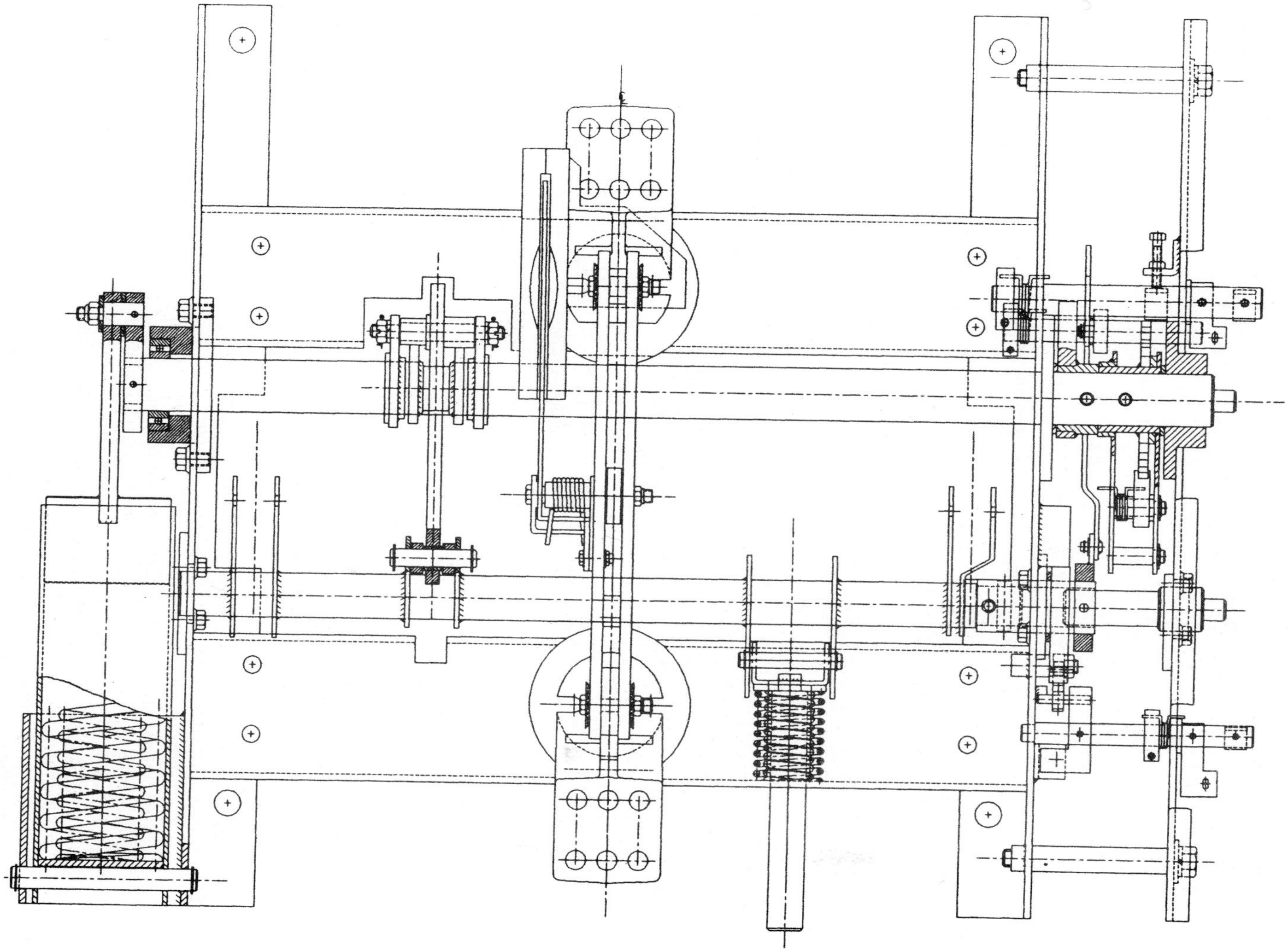

(a)

**FIGURE 6.71 (*a*)** Automatic load-interrupter switch with breaker type mechanical action.

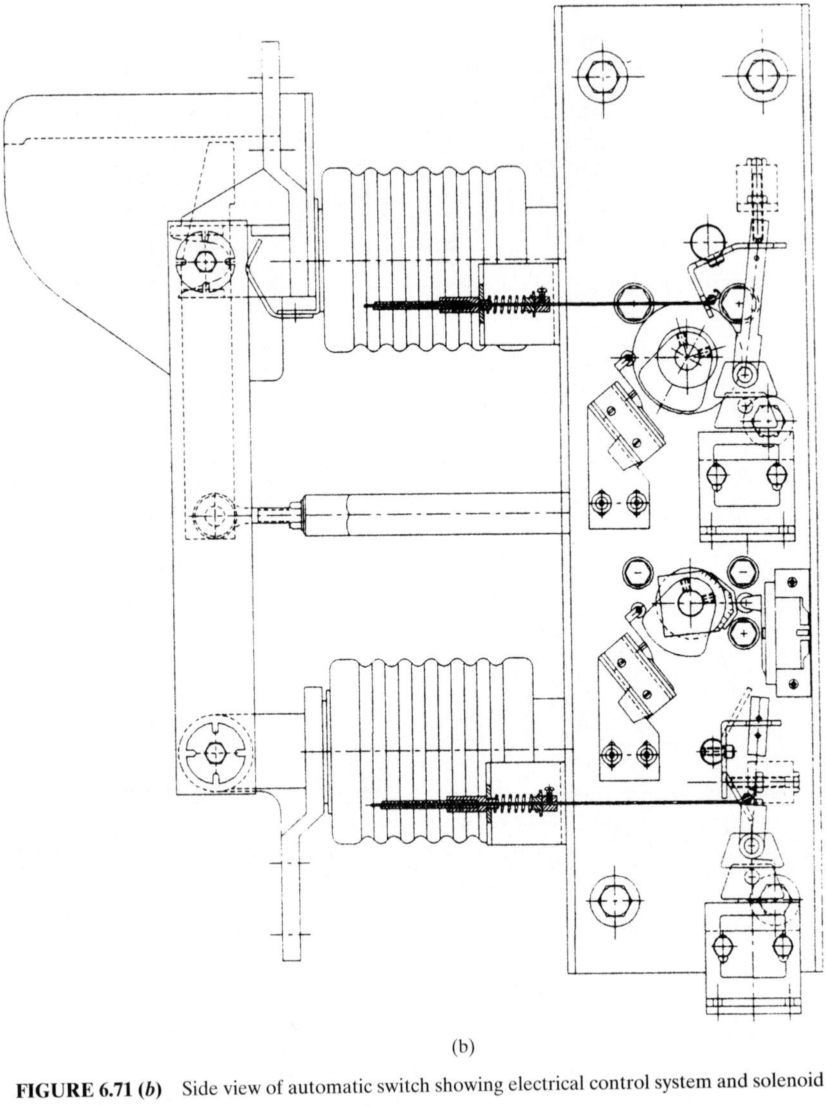

(b)

**FIGURE 6.71 (*b*)** Side view of automatic switch showing electrical control system and solenoid trips.

disconnect can be seen at the lower right side of the unit. The secondary contacts are used to supply the breaker with control power and other electrical control functions.

Figure 6.78 is a time-lapse, front-view photograph of an interrupter switch at the instant of current interruption. The arc is shown emanating from the λ phase 3 arc chute. This was a low-power interruption, used as a preliminary prior to the final test sequence.

Figure 6.79 is a time-lapse photograph of an interrupter switch attempting to interrupt magnetizing current. Magnetizing current is greatly out of phase with the system voltage, making interruption difficult. A switch of this type will normally interrupt 1200 A at 15,000 V with a power factor of 70 percent (i.e., voltage and current out of phase by 45°). With a phase angle of 45° (cosine 45° = 0.707), the power factor is 70.7 percent. The current of the interruption arc in Fig. 6.79 is only 30 A. The photograph shows a failure to interrupt this amount of magnetizing current, with an unfavorable phase angle.

**FIGURE 6.72** Prototype of automatic switch shown in Fig. 6.71*a* and 6.71*b*.

Figure 6.80 shows an interrupter switch after a failure to meet the momentary requirements of the power system being tested. The phase 1 arc interrupting device is completely destroyed as well as the jaw contact and the top of the main blades. The distortion of the bus bars in the unit show the effects of the electromagnetic forces imposed during the time the fault current flows in the circuit. Damage of this magnitude takes about six to eight cycles or 0.10 to 0.13 sec during a fault current of 40,000 A asymmetrical.

### 6.7.5.4 Modern Switchgear (Electric Power Distribution Equipment)

Figure 6.81 shows the front doors and control equipment and instruments on a typical metal-clad switchgear section. The Magneblast air-magnetic circuit breaker is contained in the right-hand unit, while the left-hand unit contains drawout potential transformers and control power transformers. The unit rating is 15 kV, 3000 A, 1000 MVA, 95-kV BIL.

Figure 6.82 shows a dual-compartment metal-clad switchgear section for General Electric Power-Vac circuit breakers. The advent of the vacuum circuit breaker has reduced the size of the equipment, making the switchgear substation buildings smaller and more economical. This particular piece of equipment is rated 15 kV, 1200 A, 500 MVA, 95-kV BIL and is of outdoor aisleless construction.

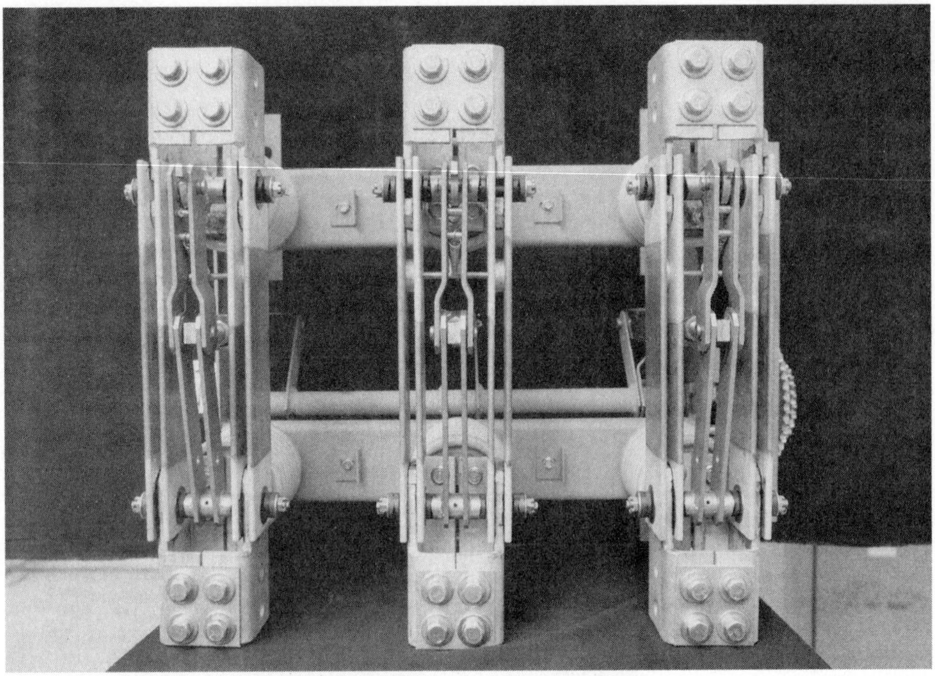

(a)

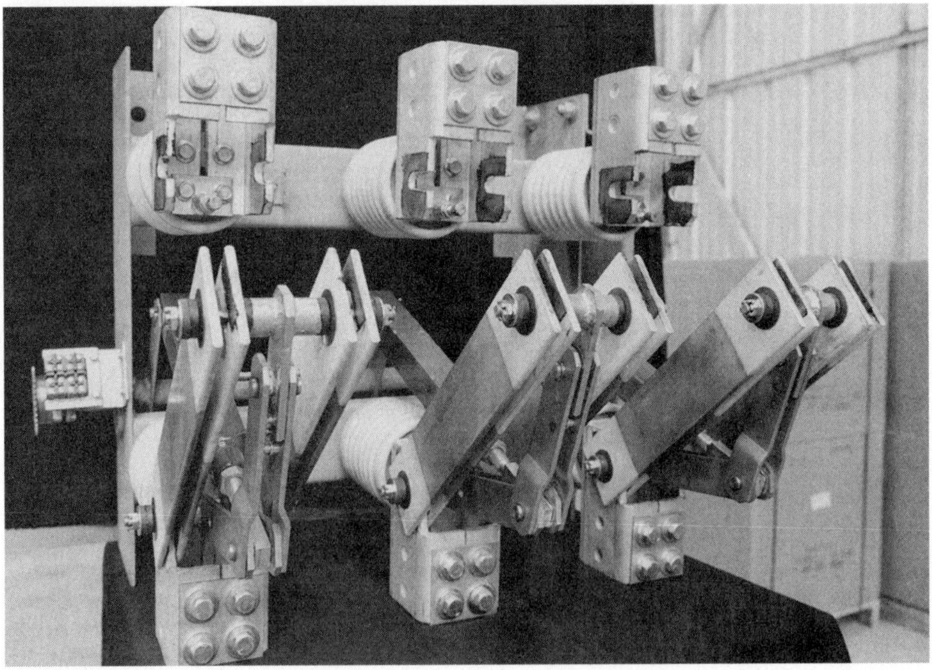

(b)

**FIGURE 6.73** (*a*) High-pressure, high-current isolator switch, 3000 A, 15 kV, 95-kV BIL; (*b*) isolator switch of Fig. 6.73*a*, with main blades in open position.

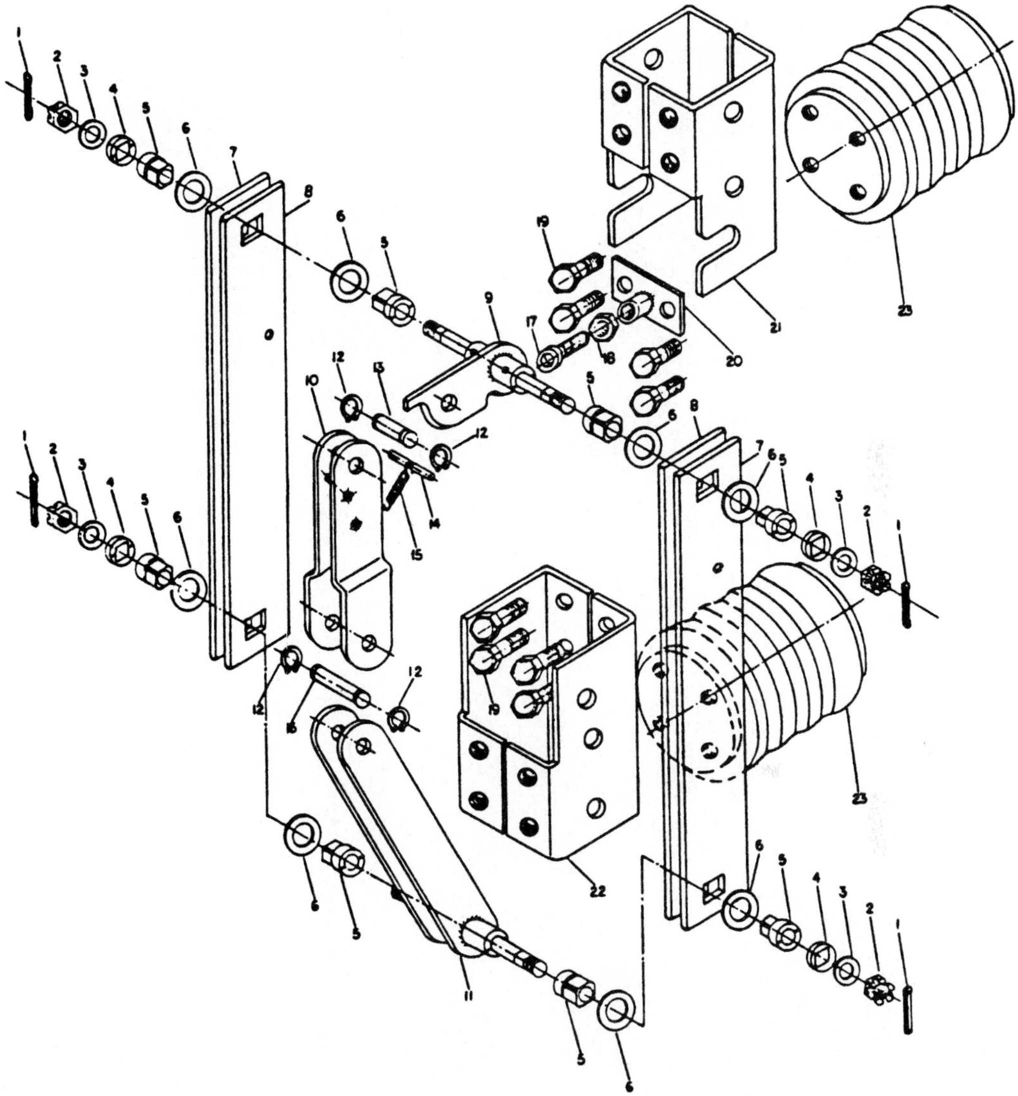

**FIGURE 6.74** Exploded view of main pole parts of isolator switch shown in Fig. 6.73*a* and 6.73*b*.

Figure 6.83 shows a complete metal-clad, metal-enclosed substation consisting of a dual incoming service with associated interrupter switch feerers. This equipment is rated 15 kV, 1200 A, 500 MVA, 95-kV BIL and is of indoor construction.

Figure 6.84 is a section of metal-clad vacuum breaker switchgear with overhead disconnect switches that are operated with hook sticks. The circuit breakers are of the vacuum type, with two breakers being installed in each separate unit. The markings on the front of the switchgear doors show the single line diagram of the power system and is referred to as a "mimic bus."

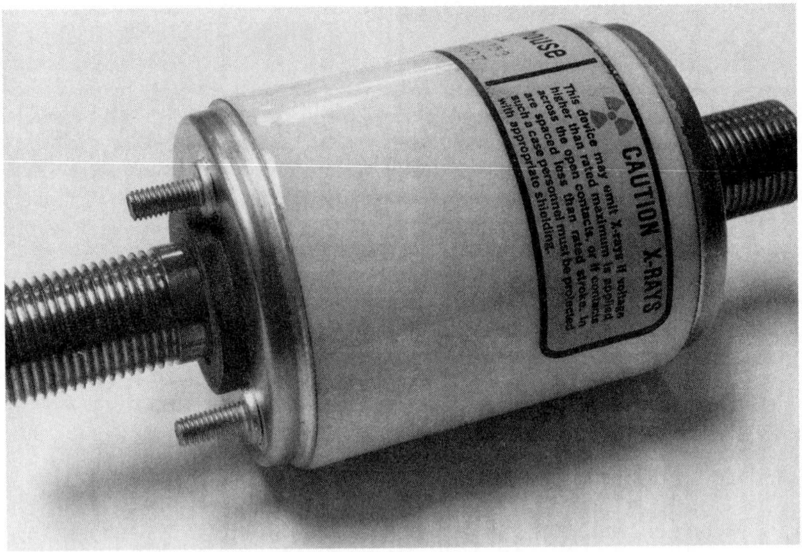


**FIGURE 6.75** Vacuum interrupter unit, 15-kV, 600-A, 2000- to 4000-A fault interrupter.

**FIGURE 6.76** Vertical-lift air-magnetic circuit breaker rated 15 kV, 3000 A, 1000 MVA, 95-kV BIL, showing control panels and instrumentation.

**FIGURE 6.77** Drawout-type vacuum breaker; connection end. Rated 34.5 kV, 1200 A, 150-kV BIL (Siemens 3AF type).

**FIGURE 6.78** View looking into the phase C interrupter unit of an interrupter switch during low current interruption. The arc is emanating from the arc chute. Time-lapse photograph.

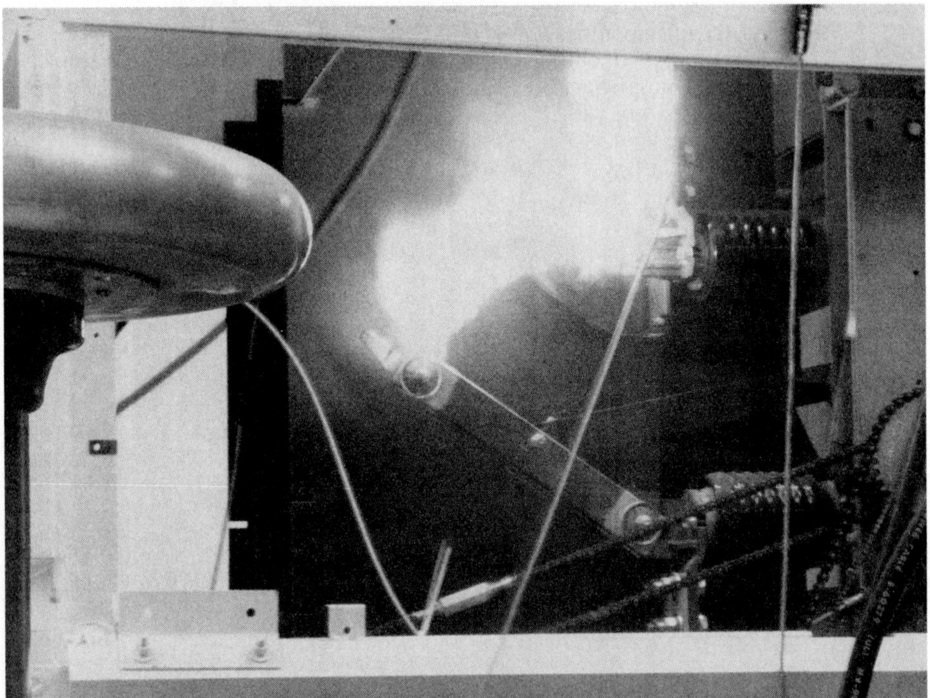

**FIGURE 6.79** Test laboratory photograph showing failure to interrupt high magnetizing current.

**FIGURE 6.80** View of an interrupter switch that failed the "momentary" test procedure. Notice bus that was bent as a result of electromagnetic forces during high current. The phase A switch pole was destroyed.

Figure 6.85 shows an outdoor, protected aisle switchgear house with Fig. 6.86 showing some of the metal-clad, vacuum breaker units inside the protective housing. Switchgear housings of this type are often air-conditioned and filtered. Panic safety handles are provided on the inside housing entrance doors to prevent personnel from accidentally being locked inside and to allow an operator to escape quickly if an accident occurs.

Figure 6.87 shows a stationary circuit breaker switchgear lineup with the associated capacitor banks and three phase reactors. The unit on the far right contains a nonfused interrupter switch.

Figure 6.88 shows a complete metal-enclosed switchgear lineup consisting of seven interrupter switches and the transition at the left side of the lineup for connection to a power transformer. Switch 1 is the line switch, while the remaining six switches are used as feeders.

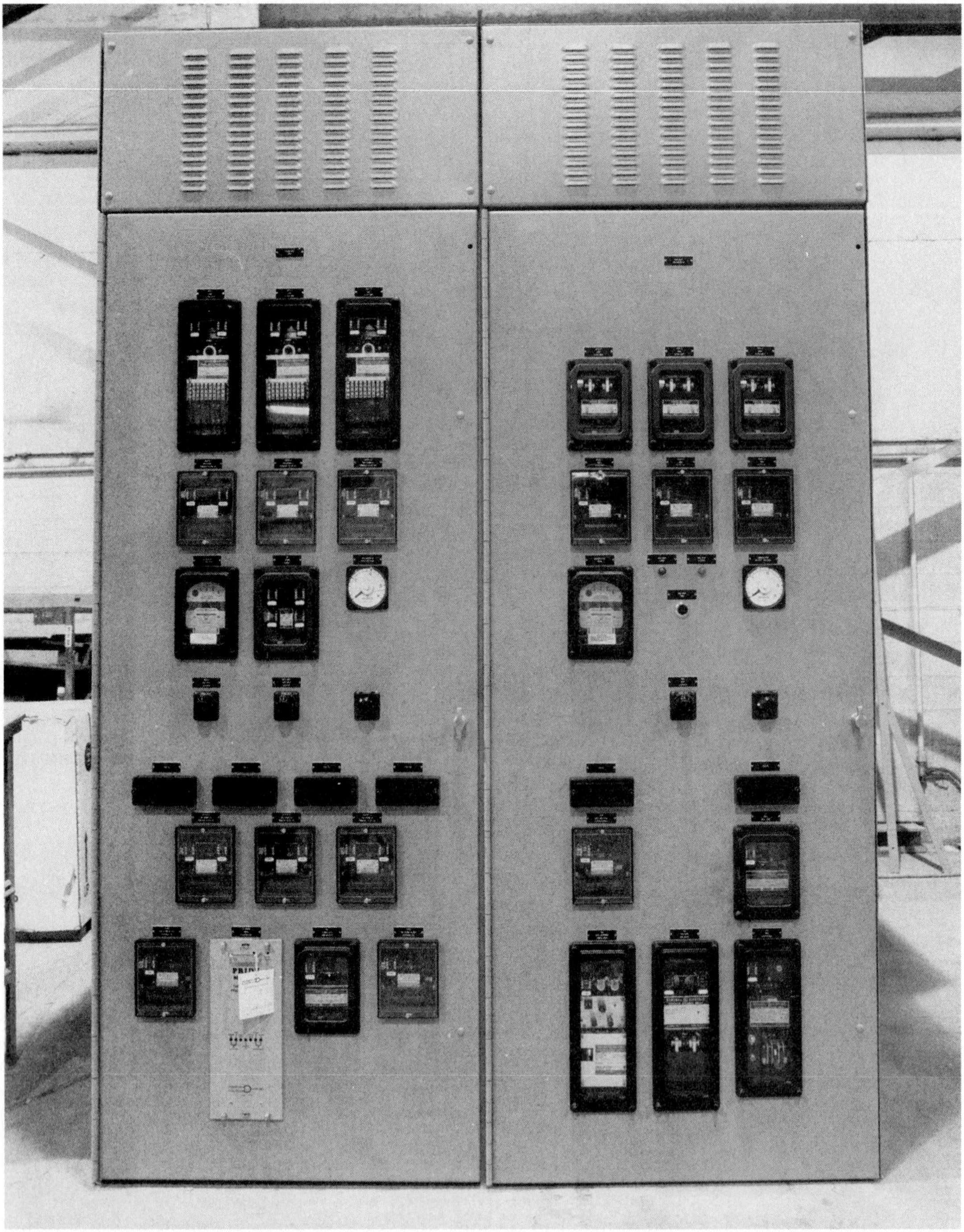

**FIGURE 6.81** View of front control panels of metal-clad breaker unit showing protective relaying devices.

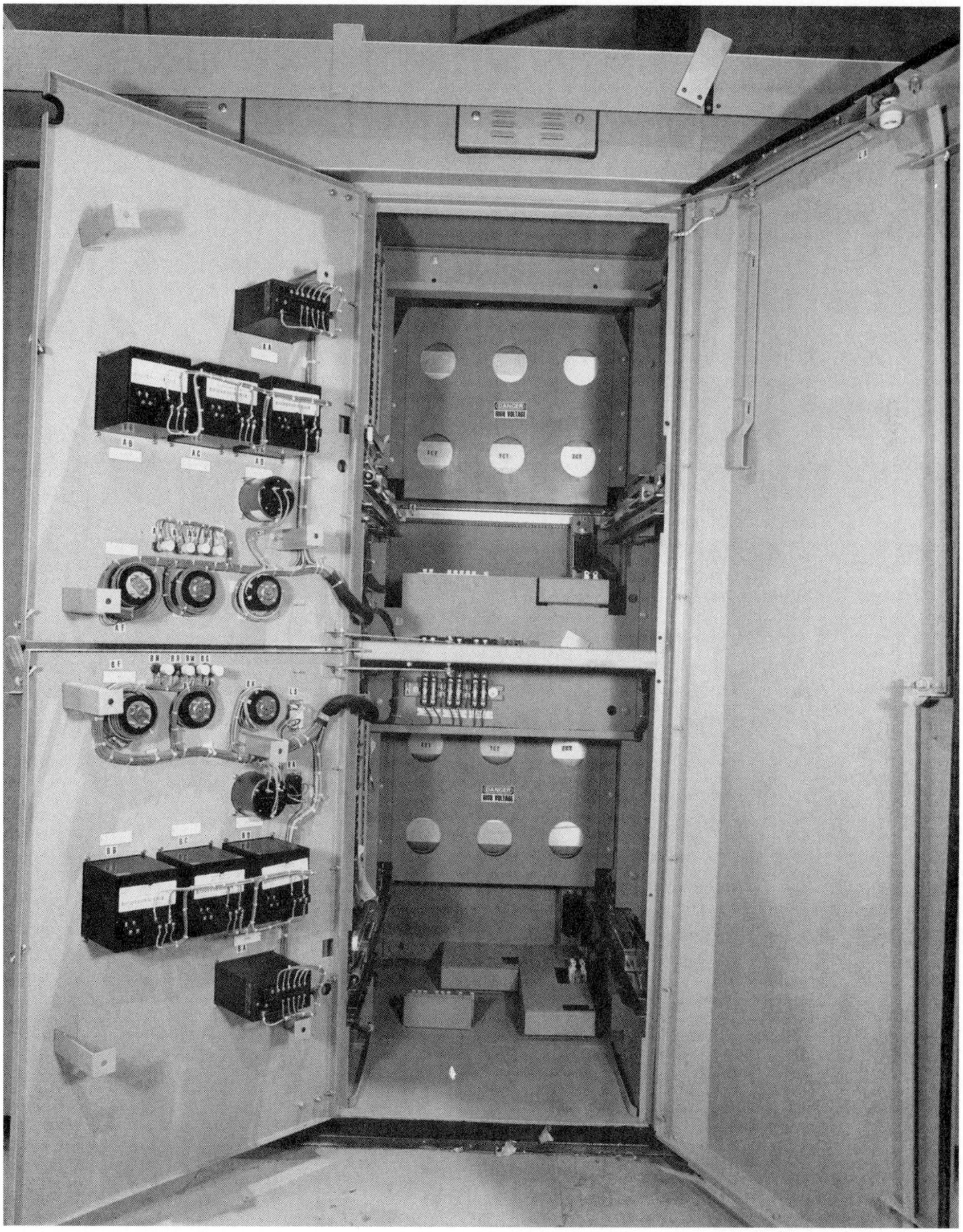

**FIGURE 6.82** View of modern-type vacuum metal-clad breaker units. Two breakers per unit and control panel wiring.

**FIGURE 6.83** Typical dual incoming power switchgear showing breaker units and load-interrupter switch units.

**FIGURE 6.84** Vacuum switchgear with isolator switch compartments mounted above. Notice the "mimic" bus on the front of this switchgear.

**FIGURE 6.85** An outdoor, walk-in-type metal-clad switchgear weather-protected house.

### 6.7.5.5 Locating a Fault in a Feeder or Line Cable

The location of a fault in interconnecting power cables may be done using either one of two generally accepted methods: the *Murray loop test* or the *Varley loop test.* The two methods are outlined in the following disclosures.

***Murray Loop Test.*** The Murray loop test is a simple bridge method for locating cable failures or faults between conductors and ground. It is applicable for any cable system where there is a second conductor of the same size as the one with the fault. This is an excellent method for locating cable faults in low-resistance power circuits. In the Murray loop test (see Fig. 6.89)

$$X = 2L\,\frac{a}{a+b}$$

***Varley Loop Test.*** In the Varley loop test (see Fig. 6.90)

$$X = \frac{L}{a+b}\left(2a - \frac{br}{Rc}\right)$$

where $a, b, r$ = resistances of bridge arms, Ω
$L$ = length of cable, ft
$X$ = distance to fault, ft
$Rc$ = resistance of good conductor, Ω

**FIGURE 6.86** Vacuum breaker units inside the switch house shown in Fig. 6.85.

### 6.7.6 Insulator Systems

Various types of insulating materials are used in contact with or to support live-current-carrying parts such as bus bars, electrical contacts, and electric cables. Whereas materials like melamine, diallyl phthalate, mica, Benelex, and phenolic laminates are adequate for voltages up to 600 V, other materials are required for medium- to high-voltage systems. Voltages of these systems range from 2400 V to 125 kV and higher and exhibit or produce *corona,* which can break down inadequate materials and lead to tracking, with eventual insulation failure.

The following materials have proved to be efficient in supporting live, high-voltage parts.

- Wet-process porcelain
- Glass

**FIGURE 6.87** Stationary breaker switchgear showing reactors and capacitor banks.

- Polyester-fiberglass, NEMA grade GPO-3
- Epoxy-glass laminates, polyimide-glass laminates
- Teflon [polytetrafluoroethylene (PTFE)]
- Cycloaliphatic resin, an epoxy compound

An insulation system must be strong mechanically and *nontracking* electrically, with high dielectric strength. A material that allows tracking will build up carbonaceous material between the live part and ground in the presence of corona or high potential stress and will *flash* to ground when an impulse or transient voltage "spike" occurs. This happens if the distance between the live part and ground or other phase of a three-phase system is inadequate.

Therefore, good high-voltage insulating material must withstand the BIL (basic impulse level) rating of the system on which it is used. Thus, a 95-kV BIL insulator material must withstand an impulse of 95,000 V tested as specified in the applicable ANSI electrical standards.

Properties of insulating materials such as arc resistance, flame retardance, dielectric strength, specific gravity, and water absorption can be obtained from the manufacturer of the material in the form of a specification sheet. (See Fig. 4.1*a* for a sample specification sheet from a material manufacturer.) When an application is critical, do not rely on general material handbook data. Materials specs are subject to changes, and only the manufacturer of a material is qualified to present them accurately for that material.

With proper fillers, some organic resins can be made nontracking to a large extent. Some resins such as polymethyl methacrylate and polymethylene oxide burst into flame under elec-

**FIGURE 6.88** Line-up of metal enclosed switches; incoming line and feeders.

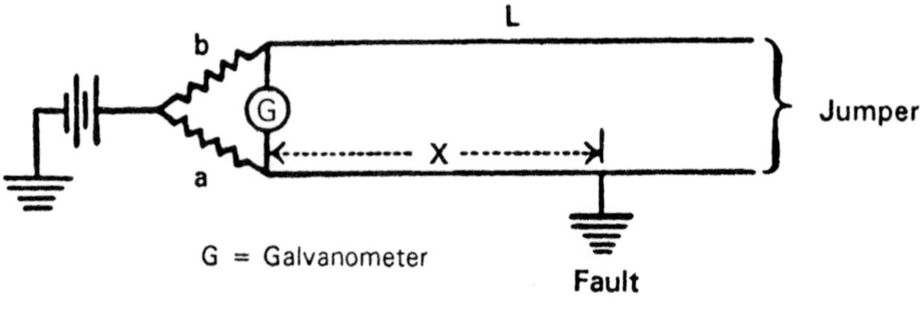


**FIGURE 6.89** Fault location circuit; Murray loop test.

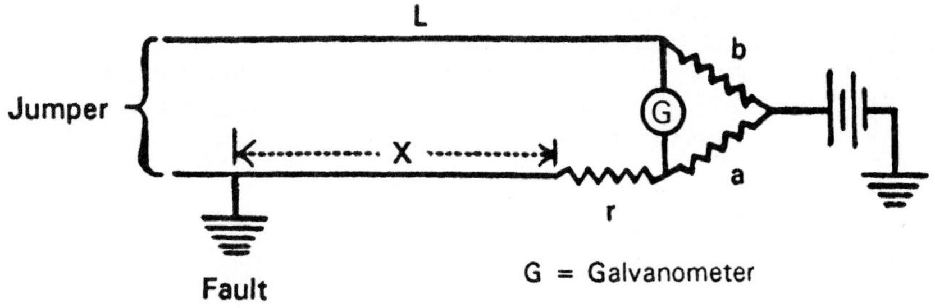


**FIGURE 6.90** Fault location circuit; Varley loop test.

tric arcing conditions. Organic resinous insulating materials are subject in varying degrees to deteriation due to thermal aging and may become poor insulating materials with time.

Care must be exercised in selecting the proper insulating material for your particular application. All the materials manufacturers produce data sheets for their materials showing the physical, electrical, and chemical properties for each particular material they produce. Consultation with the materials manufacturer can help in the initial design stages for any application for which the material is suitable.

One common type of insulator used on electrical apparatus is the wet-process porcelain standoff insulator, which is described in Table 6.19. This type of insulator is used to mount switch and circuit breaker live parts and poles and also for bus supports in switchgear and bus ducts. The different classes for these insulators represent the different strength categories as shown under strength columns for cantilever, torsion, tension, and compression.

### 6.7.7 Electrical Clearances of Conductors through Air

Electrical clearances (Table 6.20) for medium- to high-voltage conductors are defined in the ANSI electrical standards. These clearances are not always maintainable because of size restrictions on certain types of electrical equipment and switchgear. The following electrical clearances are absolute minimums as determined by appropriate tests on various types of electrical equipment. The ANSI standards stipulate that the clearances listed in the standards may be reduced when verified by the appropriate dielectric tests.

Table 6.21 shows bus clearances and test voltages for switchgear. These may differ from some of the standards but have been verified through the proper test procedures.

## *6.8 COMMON ELECTRONIC AND ELECTRICAL TERMS AND DEFINITIONS*

**Admittance** The ease with which current flows in an ac circuit; the reciprocal of impedance, measured in mhos (siemens).

**Ambient temperature** The normal or average temperature associated with a particular location.

**Amperehour** The unit measuring the charging or discharging of a battery; 1 ampere of current flow for 1 hour.

**Ampere-turns** Equal to amperes of current flow multiplied by the number of turns of wire on a coil or winding (transformer).

**TABLE 6.20** Electrical Clearances

**Bare conductors without barriers**

| Nominal Voltage, kV | Bil, kV | Minimum Clearances, Line to Ground | in Line to Line |
|---|---|---|---|
| 4.16 | 60 | 3.5 | 4.5 |
| 7.2 | 75 | 4.5 | 6.0 |
| 13.8 | 95 | 6.0 | 7.0 |
| 14.4 | 110 | 7.0 | 7.5 |
| 23.0 | 125 | 8.0 | 9.0 |
| 34.5 | 150 | 10.5 | 12.0 |
| 34.5 | 200 | 15.0 | 15.0 |

**Bare conductors with barriers**

| Nominal Voltage, kV | Bil, kV | Minimum Clearances Live Part to Barrier | in Barrier to Ground* (Near Live Part) |
|---|---|---|---|
| 4.16 | 60 | 0.75 | 0.75 |
| 13.8 | 95 | 1.5 | 1.5 |
| 23.0 | 125 | 2.5 | 2.5 |
| 34.5 | 150 | 3.5 | 3.5 |
| 34.5 | 200 | 5.0 | 5.0 |

*Barriers are of insulating material; i.e., polyester fiberglass type GPO-3.

**Fully insulated conductors**

| Nominal Voltage, kV | Bil, kV | Minimum Clearance, Line to Ground | in Line to Line |
|---|---|---|---|
| 4.16 | 60 | 2.0 | 2.0 |
| 13.8 | 95 | 3.0 | 3.0 |
| 23.0 | 125 | 5.0 | 5.0 |
| 34.5 | 150 | 6.0 | 6.0 |

**Attenuation** A reduction of signal strength, expressed in decibels.

**Autotransformer** A transformer that has only one coil or winding; the single winding serves as both primary and secondary.

**Breadboard** A temporarily constructed circuit that is used for tests.

**Cascade** An alternate name for the series connection.

**Choke** An induction used to pass large amounts of dc current; used in power supplies to improve regulation and reduce *ripple,* and as blocking devices to prevent radio-frequency interference from feeding into a circuit.

**Conductance** The ease of conducting current; the reciprocal of resistance in a dc circuit, and the real part of an impedance in ac circuits; expressed in mhos.

**Coulomb** The unit of electrical charge consisting of $6.25 \times 10^{18}$ electrons.

**Ground** The common return path for electric current in electric/electronic circuits; a reference point assumed to be at zero potential with respect to the Earth.

**Induction** The process by which a voltage is induced into a conductor by a changing magnetic field, causing current to flow in the conductor in a closed circuit.

**Kirchhoff's current law** The algebraic sum of all the currents flowing to a given point in a circuit is equal to the algebraic sum of all the currents flowing away from that point.

**TABLE 6.21** Bus Clearances and Test Voltages—Switchgear

**MINIMUM BUS CLEARANCES* IN INCHES BETWEEN RIGID LIVE PARTS OR BARE CONDUCTORS IN AIR**

| Voltage Class kV (BIL) | Indoors | | |
|---|---|---|---|
| | Creepage Distance** | Phase-to-Phase | Phase-to-Ground |
| 0.240 | $1\frac{3}{4}$ | $\frac{3}{4}$ | $\frac{1}{2}$ |
| 0.600 | 2 | 1 | 1 |
| 1.2 (45)[1] | 4 | $2\frac{1}{2}$ | 2 |
| 1.2 (30)[2] | $2\frac{1}{2}$ | $1\frac{1}{2}$ | $1\frac{1}{2}$ |
| 1.2 (10)[3] | $1\frac{1}{2}$ | 1 | 1 |
| 2.5 (60)[1] | $4\frac{1}{2}$ | 3 | $2\frac{1}{2}$ |
| 2.5 (45)[2] | 4 | $2\frac{1}{2}$ | 2 |
| 2.5 (20)[3] | $1\frac{1}{2}$ | 1 | 1 |
| 5 (75)[1] | 7 | $4\frac{1}{2}$ | $3\frac{1}{2}$ |
| 5 (60)[2] | $4\frac{1}{2}$ | 3 | $2\frac{1}{2}$ |
| 5 (25)[3] | $1\frac{1}{2}$ | 1 | 1 |
| 8.66 (95)[1] | 10 | $6\frac{1}{2}$ | $5\frac{1}{2}$ |
| 8.66 (75)[2] | 7 | $4\frac{1}{2}$ | $3\frac{1}{2}$ |
| 8.66 (35)[3] | $2\frac{1}{2}$ | $1\frac{1}{2}$ | $1\frac{1}{2}$ |
| 15 (110)[1] | 11 | $7\frac{1}{2}$ | 6 |
| 15 (95)[2] | 10 | $6\frac{1}{2}$ | $5\frac{1}{2}$ |
| 15 (50)[3] | 4 | $2\frac{1}{2}$ | 2 |
| 23 (150) | 17 | $11\frac{1}{2}$ | 9 |
| 34.5 (200) | 24 | 16 | 13 |
| 46 (250) | 31 | $20\frac{1}{2}$ | 17 |
| 69 (350) | 45 | $30\frac{1}{2}$ | 25 |

* Note that dimensions are phase-to-phase (or ground) clearances not center-to-center dimensions

** Measurement is vertical plus $\frac{1}{2}$ of horizontal

(1) BIL of switchgear and power class transformer (liquid filled)

(2) BIL of associated distribution class transformer (liquid filled)

(3) BIL of associated dry-type transformer

**SWITCHGEAR DIRECT-CURRENT TEST VOLTAGES (INITIAL AND MAINTENANCE)**

| Rated kV | BIL kV | Switchgear Assemblies* Including Bus and Internal Curernt Transformers and Potential Transformers |
|---|---|---|
| | | Test, kV - DC |
| 0.6 | | 2.3 |
| 1.2 | 30 | 10 |
| 2.4 | 45 | 15 |
| 4.2 | 60 | 20 |
| 7.2 | 75 | 27 |
| 7.2 | 95 | 38 |
| 13.8 | 95 | 38 |
| 14.4 | 110 | 53 |
| 23 | 150 | 64 |
| 34.5 | 200 | 85 |
| 46 | 250 | 105 |
| 69 | 350 | 150 |

* Disconnect potential transformers before dc high potential test is started

**TRANSFORMER DIRECT-CURRENT TEST VOLTAGE (INITIAL AND MAINTENANCE)**

| Rated kV | BIL kV | Test, kV - DC | |
|---|---|---|---|
| | | Liquid or Gas Immersed | Air Immersed |
| 1.2 | 10 | 4 | 4 |
| 1.2 | 30 | 10 | - |
| 1.2 | 45 | 10 | - |
| 2.5 | 20 | 10 | 10 |
| 2.5 | 45 | 15 | - |
| 2.5 | 60 | 15 | - |
| 5.0 | 25 | 12 | 12 |
| 5.0 | 60 | 19 | - |
| 5.0 | 75 | 19 | - |
| 8.66 | 35 | 19 | 19 |
| 8.66 | 65 | 19 | 19 |
| 8.66 | 75 | 26 | 26 |
| 8.66 | 95 | 26 | - |
| 15 | 50 | 31 | 31 |
| 15 | 65 | 31 | 31 |
| 15 | 95 | 34 | 34 |
| 15 | 110 | 34 | - |
| 23 | 150 | 50 | - |
| 34.5 | 200 | 70 | - |
| 46 | 250 | 90 | - |
| 69 | 350 | 138 | - |

**Kirchhoff's voltage law** The algebraic sum of all the voltage rises and voltage drops around any closed loop is equal to zero.

**Magnetic flux density** The number of lines of magnetic flux per unit area, measured in a plane perpendicular to the direction of the flux.

**Mil** One one-thousandth of an inch.

**Permeability** The measure of the ease with which magnetic lines of force can move through a material, compared to air.

**Phase** The time displacement expressed as an angular displacement between two ac quantities such as the voltage and current in an ac circuit.

**Phase shift** The time difference between two signals measured between reference points on each signal and expressed in angular-displacement degrees.

**Reactance** The impedance that a pure inductance or capacitance provides to current flowing in an ac circuit.

**Resistance** A characteristic of a material that opposes the flow of electrons through it; the result is a loss of energy in a circuit, dissipated as heat ($I^2R$).

**Shunt** A parallel circuit branch, or to wire or connect a circuit in parallel.

**Susceptance** The reciprocal of the reactance, in mhos.

**Time constant** The time in seconds required for a capacitor to attain 63.2 percent of its final charge; the time in seconds required for current flowing in an inductance to attain 63.2 percent of its final value.

**Voltage drop** The difference in potential between two points caused by a current flow through an impedance or resistance.

### 6.8.1 Common Electrical and Electronic Abbreviations and Acronyms

**ac** alternating current
**ABS** acrylonitrile butadiene-styrene
**A/D** analog to digital
**ADC** analog-to-digital converter
**ADP** automatic data processing
**AFC** audio-frequency control
**AFT** automatic filter test
**ALGOL** algorithmic language
**AM** a logical operator
**AQL** acceptable quality level
**ASCII** American national standard code for information interchange
**ASIC** application-specific IC
**ATE** automatic test equipment
**ATG** automatic test generation
**AWG** American wire gauge

**BALUN** balanced to unbalanced
**BASIC** beginners' all-purpose symbolic instruction
**BCD** binary-coded decimal
**BNC** baby N connector
**BUS** basic utility system

**CAD** computer-aided design
**CAE** computer-aided engineering
**CAM** computer-aided manufacture
**CAMAC** computer-automated measurement and control
**CAPP** computer-aided process planning
**CAT** computer-aided test
**CATV** cable television or community antenna television
**CCC** (leaded) ceramic chip carriers
**CCD** computer-controlled display
**CERMET** ceramic metal element
**CIE** computer-integrated engineering
**CIM** computer-integrated manufacturing
**CLCC** ceramic leadless chip carriers
**CMOS** complementary metal-oxide semiconductor
**Comm.** communications
**CPS** characters per second
**CPU** central processing unit

**D/A** digital to analog
**DAC** digital-to-analog converter
**DBMS** database management system

**dc** direct current
**DFT** design for testability
**DIAC** bidirectional trigger diode
**DIL** dual in-line
**DIN** Deutsche Industrie Normenausschuss
**DIP** dual in-line package
**DPM** digital panel meter
**DRAM** dynamic RAM
**DTL** diode transistor logic

**EDA** electronic design automation
**EDP** electronic data processing
**EBCDIC** extended binary-coded decimal interchange code
**EEPROM** electrically erasable PROM
**EIA** Electronic Industries Association
**EMC** electromagnetic compatibility
**EMI** electromagnetic interference
**EMP** electromagnetic pulse
**EPROM** erasable programmable ROM
**ESR** equivalent series resistance

**FEP** fluorinated ethylene propylene Copolymar, 200° Teflon
**FET** field-effect transistor
**FM** frequency modulation
**FORTRAN** formula translator

**GaAsFET** gallium arsenide FET
**CMIP** common management information protocol
**GPIB** general-purpose interface bus
**GPS** global positioning system

**HF** high frequency (3 to 30 MHz)
**HIPOT** test, measurement, and diagnostic equipment
**HLL** high-level language
**HV** high voltage
**Hz** hertz

**IAB** Internet Activities Board
**IC** integrated circuit
**IDC** insulation displacement contact
**IF** intermediate frequency
**IO** input/output (devices)
**IR** insulation resistance (or) infrared
**ISDN** integrated services digital network
**ISO** International Standards Organization

**JEDEC** Joint Electronic Device Engineering Council
**JFET** junction FET
**JMOS** junction MOS

**kHz** kilohertz ($10^3$ hertz)
**KSR** keyboard send/receive

**LAN** local area network
**LC** (filters) inductance/capacitance
**LCC** leaded chip carrier
**LCD** liquid-crystal display
**LED** light-emitting diode
**LF** low frequency (30 to 300 kHz)
**LIF** low insertion force
**LNA** low-noise amplifier
**LNB** low-noise band
**LSI** large-scale integration
**LVDT** linear velocity displacement transformer

**MASER** microwave amplification by stimulated emission of radiation
**MATV** master antenna television
**μf** microfarad
**MHz** megahertz
**MIC** microwave IC
**MIPS** million instructions per second
**MIB** microcomputer interface board
**MOS** metal oxide semiconductor
**MOSFET** metal oxide semiconductor FET
**MOSIGT** metal oxide semiconductor insulated-gate transistor
**MOV** metal oxide varistor
**MPU** microprocessor unit
**MSI** medium-scale integration

**NBS** National Bureau of Standards
**NEMP** nuclear electromagnetic pulse
**NIM** national instrumentation module
**NOR** not OR
**NTC** negative temperature coefficient

**OCR** optical character reader
**OEM** original equipment manufacturer
**OR** a logical operator
**OSI** open systems interconnection

**PAI** programmable array logic
**PC** printed circuit
**PCB** printed circuit board
**PCM** pulse code modulation
**PCS** plastic clad silica
**pF** picofarad
**PGA** pin grid array
**PIN** positive-intrinsic-negative (transistor)
**PLA** programmable logic array
**PLCC** plastic leaded chip carrier
**PLL** phase-locked loop
**PPI** plan position indicator
**PROM** programmable ROM
**PTC** positive temperature coefficient
**PUT** programmable unijunction transistor

**QAM** quadrature amplitude modulation

**Q (meter)** quality-factor meter

**RAM** random access memory
**RC** resistance-capacitance
**RFI** radio-frequency interference
**rms** root mean square
**ROM** read-only memory
**RF** radio frequency
**RFC** radio-frequency choke
**RFI** radio-frequency interference
**RTD** resistance-temperature detector
**RTL** resistor transistor logic
**RTV** room temperature vulcanizing
**RW** read/write

**SAS** silicon asymmetrical switch
**SBC** single board (micro) computer
**SBS** silicon bilateral switch
**SCR** silicon controlled rectifier
**SCS** silicon controlled switch
**SDA** source and detector assemblies
**SGMP** simple gateway monitoring protocol
**SIDAC** bidirectional voltage-triggered switch
**SIMM** single inline memory module
**SIP** single inline package
**SMD** surface-mount device
**SMT** surface-mount technology
**SNMP** simple network management protocol
**SO** small outline
**SOIC** small-outline IC
**SOT** small-outline transistor
**SSI** small-scale integration
**SSR** solid-state relay
**STL** Schottky transistor logic
**SUS** silicon unilateral switch

**TC** temperature coefficient
**TCE** thermal coefficient of expansion
**TFE** tetrafluoride ethylene propylene Copolymer, 250° Teflon
**TRIAC** bidirectional AC switch
**TTL** transistor-transistor logic
**TTY** teletypewriter
**TXCO** temperature compensated crystal oscillator

**UHF** ultrahigh frequency (300 MHz to 3 GHz)
**UJT** silicon unijunction transistor
**UL** Underwriters Laboratory
**UPS** uninterruptible power supply

**V ac** volts (of) alternating current
**VAN** value-added network
**VAR** volt-ampere reactive
**VCR** videocassette recorder
**V dc** volts (of) direct current
**VHF** very high frequency (30 kHz to 300 MHz)
**VHSIC** very high-speed integrated circuit
**VLF** very low frequency (3 to 30 kHz)
**VLSI** very large-scale integration
**VME** virtual machine environment
**VRM** voice recognition module
**VTVM** vacuum tube volt meter
**VU** volume unit
**VSWR** voltage standing wave ratio

**WPS** word-processing software

**YIG** yttrium iron garnet

**ZIF** zero insertion force

## 6.9 ELECTRONICS AND ELECTRONIC COMPONENTS

This section will detail the principles of electronic components, electronic design procedures, and practical electronic circuits and their construction, as applicable to consumer and industrial products.

Some of the electronic components such as resistors, capacitors, and inductors are described in the preceding sections of this chapter. The following sections will detail transistors, integrated circuits, and other specialized electronic devices and formulas (e.g., for coils as shown in Figs. 6.91 to 6.93) commonly used in electromechanical design practice.

Figure 6.94 shows an array of modern electronic components including resistors, capacitors, diodes, trim pots, miniature switches, sockets, miniature circuit fuses, neon indicators, light-emitting diodes (LEDs), and integrated circuit packages.

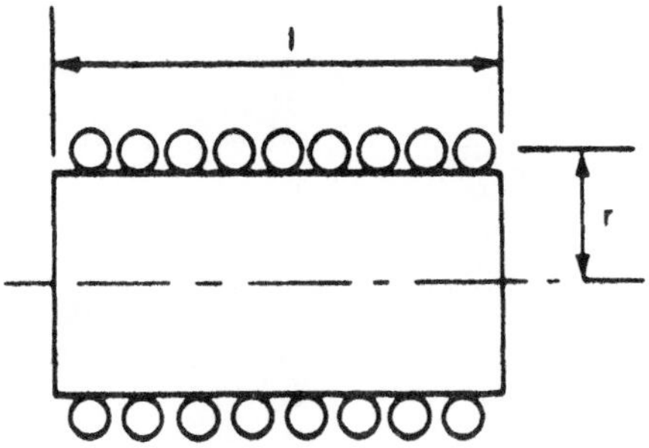


FIGURE 6.91 Single-layer coil.

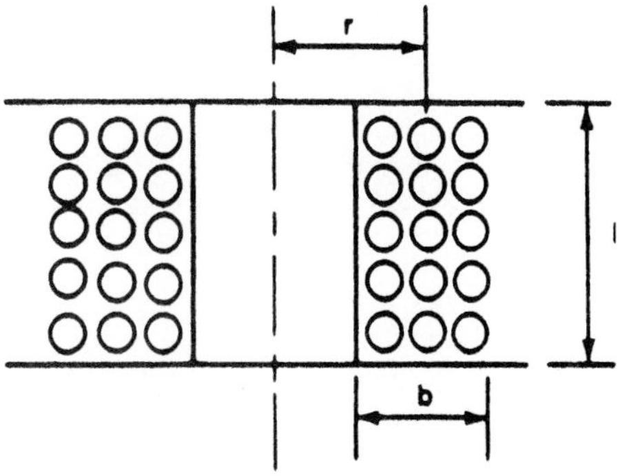


FIGURE 6.92 Multilayer coil.

Figure 6.95 shows some of the available linear and digital integrated circuits and a typical triac. Note in the figure that some of the devices are mounted on a conductive sponge material. These are complementary metal oxide substrate (CMOS) integrated circuits and should be handled with care to prevent a static electric charge from entering the pins of the devices. A static charge will quickly destroy a CMOS device because of its extremely high input impedance. Some of the ICs shown in Fig. 6.95 are used in the IC circuits sections to follow. The 28-pin IC shown on the left in Fig. 6.95 is a digital speech synthesizer.

### 6.9.1 Resistors

See Sec. 6.6.1.

***Sizing for Wattage.*** When sizing a resistor for use in an electrical or electronic circuit, the calculated heat dissipation in watts must be carefully considered. This is particularly true if the circuit or printed-circuit board on which the resistor is mounted is contained in a relatively airtight package, such as a steel or aluminum container. Tests may show that when uncased, the circuit will perform adequately, but when contained and operative for many hours, the resistor marginally selected for wattage will cause burning on the surface of the printed-circuit board, eventually failing.

This problem can be solved by increasing the wattage size of the resistor or by ventilating the enclosure if possible. An operational test can be performed on the packaged circuit, or temperature-rise calculations can be performed using the equations and methods shown in Sec. 6.7.2.1. See also Sec. 9.4, on heat transfer. Heat calculations on complex circuit packages

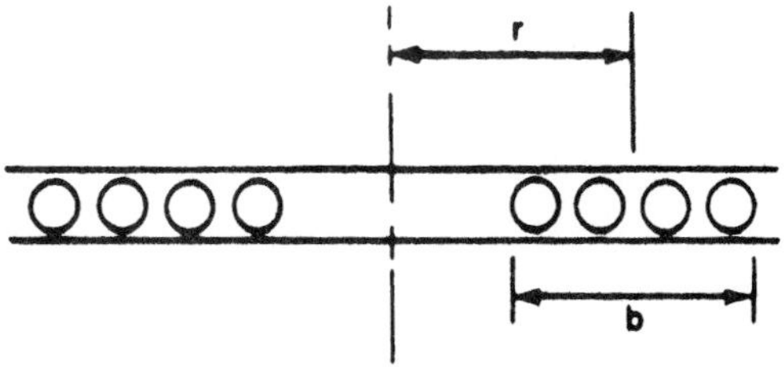


FIGURE 6.93 Spiral coil.

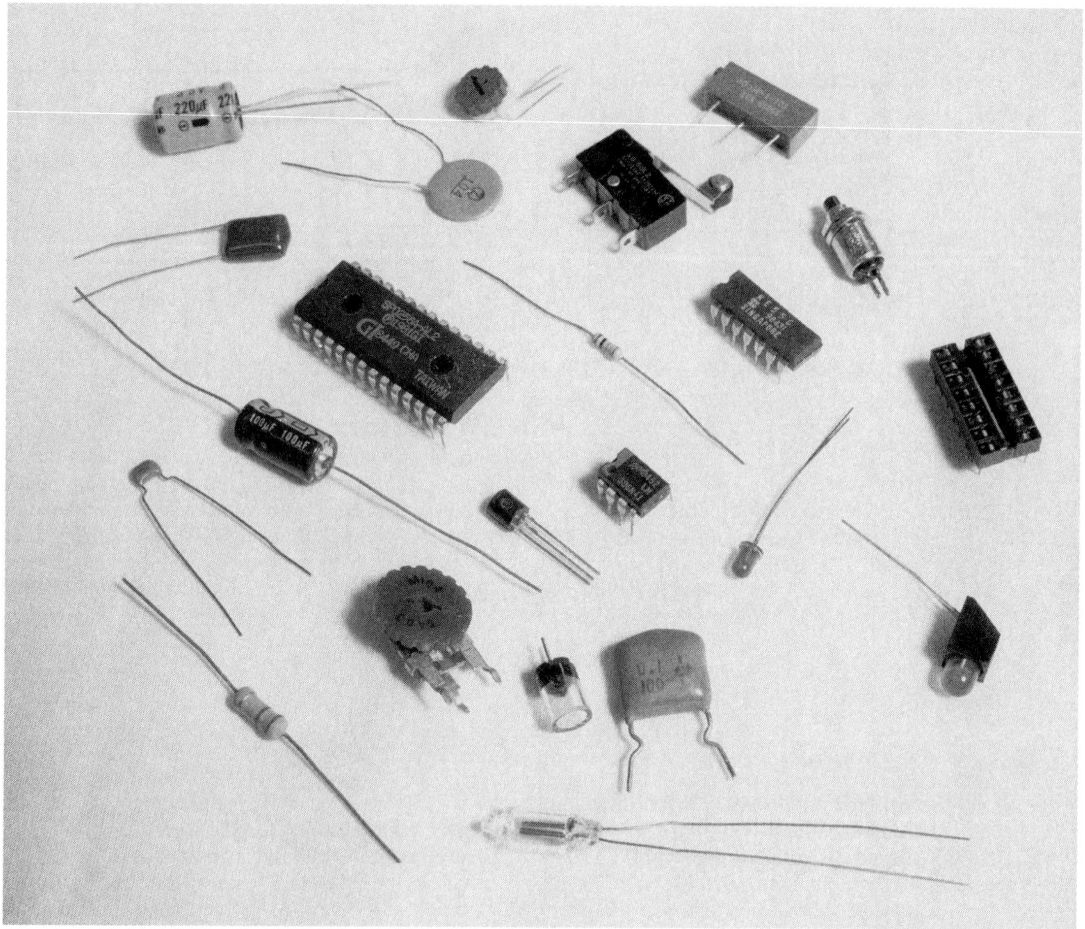

FIGURE 6.94 Electronic components.

as are found in aerospace equipment and systems are best solved through definitive testing. Attempts to "calculate" the temperature requirements of a complex electronics package system are often futile.

### 6.9.2 Capacitors

See Sec. 6.6.2.

### 6.9.3 Inductors

See also Sec. 6.6.3.

**FIGURE 6.95** Integrated circuit components.

### 6.9.3.1 Coil-Winding Formulas

For most small air-core coils, the following formulas are accurate to within 1 percent. Single-layer coils (Fig. 6.91):

$$L = \frac{(rN)^2}{9r + 10l} \qquad N = \sqrt{\frac{L\,(9r + 10l)}{r}}$$

Multilayer coils (Fig. 6.92):

$$L = \frac{0.8\,(rN)^2}{6r + 9l + 10b} \quad \text{(transpose for } N\text{)}$$

Single-layer spiral coils (Fig. 6.93):

$$L = \frac{(rN)^2}{8r + 11b} \quad \text{(transpose for } N\text{)}$$

In the preceding formulas, $L$ = self-inductance, μH; $N$ = total number of turns; $r$ = mean radius, in; $l$ = length of coil, in; and $b$ = depth of coil, in.

## 6.9.4 Transistors

*Transistors* are semiconductor devices used in circuits as current amplifiers, switching devices, logic components, etc. The two basic types of bipolar transistor are negative-positive-negative and positive-negative-positive (npn and pnp).

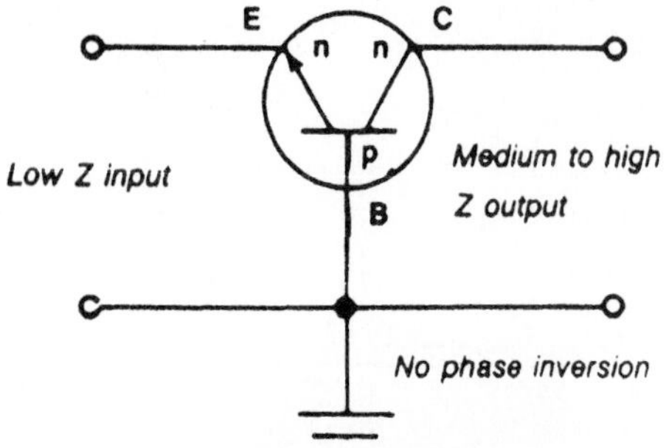


**FIGURE 6.96** Common base.

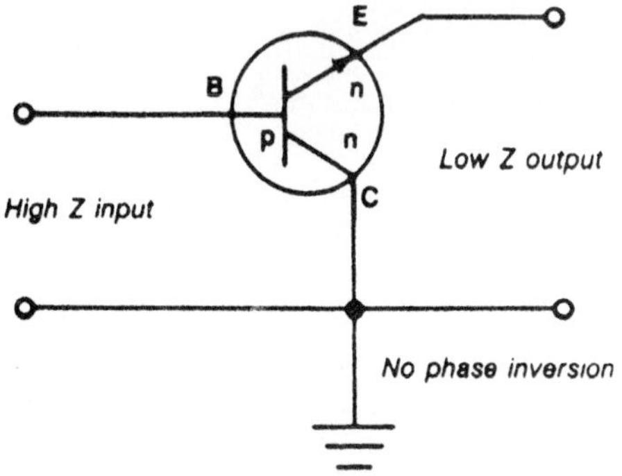


**FIGURE 6.97** Common collector.

### 6.9.4.1 Basic Transistor Configurations (Bipolar)

The common-base npn transistor is shown in Fig. 6.96. The common-collector npn transistor is shown in Fig. 6.97. The common-emitter npn transistor is shown in Fig. 6.98. The most important parameter describing transistors is current gain β, which is the ratio of collector to base current $Ic/Ib$. The ratio of collector to emitter current is called alpha (α) and is the ratio $Ic/Ie$. The beta for an amplifying transistor may range from 20 to 200, while alpha is less than 1. The sum of the base and collector currents must equal the emitter current.

***Common-Base Configuration.*** The common-base circuit provides no current gain, but couples a very low input impedance to a high output impedance. But there is considerable power gain because of this impedance difference between input and output. Since *Ie* is almost equal to *Ic* and since $P = I^2R$, it follows that $Ic^2R_{out} > Ie^2R_{in}$.

***Common-Collector Configuration.*** The common-collector circuit, also called the *emitter-follower*, provides high input impedance and low output impedance. There is little or no voltage gain, but there is considerable power gain. Since $E_{in} \cong E_{out}$ and $P = E^2/R$, then $E^2_{out}/R_{out} > E^2_{in}/R_{in}$ (due to the low output impedance).

***Common-Emitter Configuration.*** The common-emitter circuit provides current gain, power gain, and moderately low input impedance to moderately high output impedance. The common-emitter circuit is the only configuration that *inverts* the signal from input to output.

***Single Power Source for Transistor Circuits.*** The bias and power supply for transistor circuits may be provided from a single source as shown in Fig. 6.99*a* to 6.99*c*. Note that $R_3$ in each case shown provides a stabilizing effect by negative feedback.

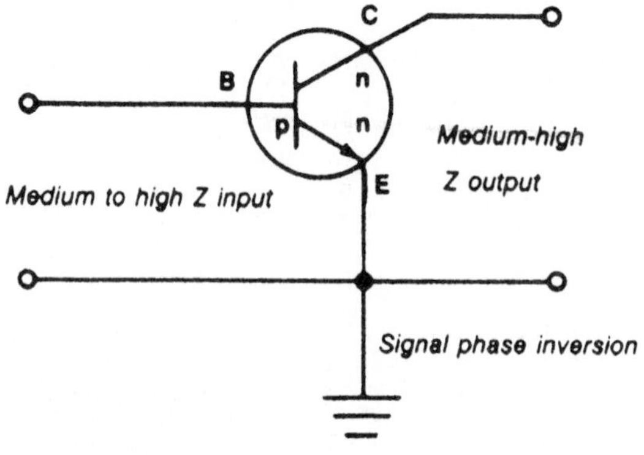


**FIGURE 6.98** Common emitter.

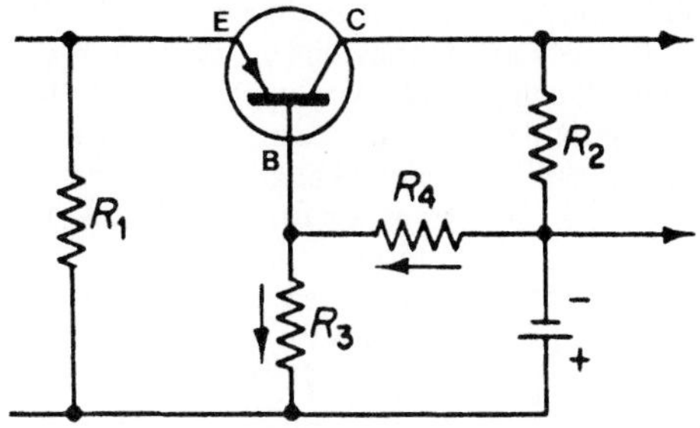


(a) Base

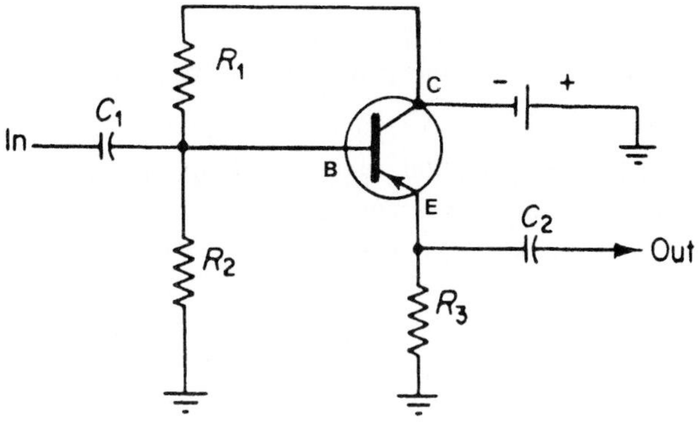


(b) Collector

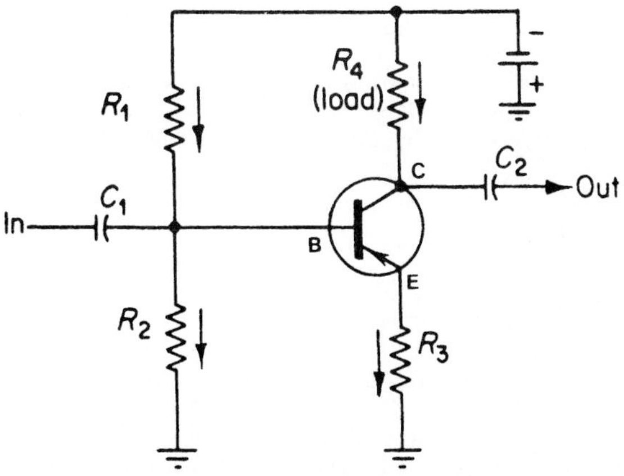


(c) Emitter

**FIGURE 6.99** (*a*) Single-source bias for base configuration; (*b*) single-source bias for collector configuration; (*c*) single-source bias for emitter configuration.

***High-Frequency Performance of Transistors.*** The high-frequency performance of a transistor circuit depends on the type of transistor and its circuit configuration. The common-emitter circuit can provide bandwidths from 15 to 2000 kHz.

The common-base circuit can provide better high-frequency performance from 700 kHz to 100 MHz. But this is achieved at a loss of current gain.

The common emitter, when driven from a low-impedance source, can improve bandwidth by 25 times. It is also helpful to place a resistance in the emitter.

Temperature variations affect the collector current to a marked degree; the collector cutoff current $I_{co}$ will increase by two times for every 9°C rise in temperature. This rising-temperature effect can clip and distort the signal passing through a common-emitter amplifier and may eventually lead to thermal runaway with destruction of the transistor. Thermal resistance is therefore an important characteristic of transistors and is listed on the transistor data sheet (degrees of temperature rise per milliwatt of power fed into the transistor). Maximum allowable operating temperatures are also listed on the device data sheet.

Transistors are used in some of the circuits shown in Sec. 6.9.9.

***General Rules for Analyzing Transistor Circuits.*** The following rules will be helpful for analyzing transistor circuit applications, especially class A amplifiers:

- The dc electron current flow is always opposite the direction of the arrow on the emitter.
- When electron flow is out from the emitter, electron flow will be into the collector.
- When electron flow is into the emitter, electron flow will be out from the collector.
- The collector-base junction is always reverse-biased.
- The emitter-base junction is always forward-biased.
- A base-input voltage that opposes or decreases the forward bias also decreases the emitter and collector currents.
- A base-input voltage that aids or increases the forward bias also increases the emitter and collector currents.
- The middle letter in npn or pnp always applies to the base.
- The first two letters in npn or pnp (np or pn) refer to the relative bias polarities of the emitter with respect to either the base or collector. For instance, the letters pn in pnp indicate that the emitter is positive with respect to both the base and collector. The letters np in npn indicate that the emitter is negative with respect to both the base and collector.

Note that the common-emitter circuit is the most widely used configuration. The arrow on the transistor symbol always points to the n-type material in its construction. Designations in the figures are $E$ = emitter, $B$ = base, $C$ = collector, $Z$ = impedance, $n$ = negative, and $p$ = positive.

### 6.9.5 Diodes

Diodes are semiconductor devices used variously as rectifiers of ac current, voltage doublers, and logic-circuit components. With a positive anode or negative cathode, diodes conduct in the *forward* direction. Silicon diodes are the most common in use and have a high *front-to-back ratio,* meaning that they conduct much more readily in one direction than the other. Figure 6.100 shows a basic half-wave rectifier circuit. (*Note:* The pn diode is forward-biased when the supply dc voltage is as shown in Fig. 6.101. The forward current moves opposite the direction of the arrow symbol of the diode. On a reverse-biased diode, most of the source voltage appears across the diode, in most cases.)

Observe that when a diode is used as an ac rectifier, the plus (+) voltage is located at the cathode junction. When a diode is used in a dc circuit, the electron flow is against the arrow

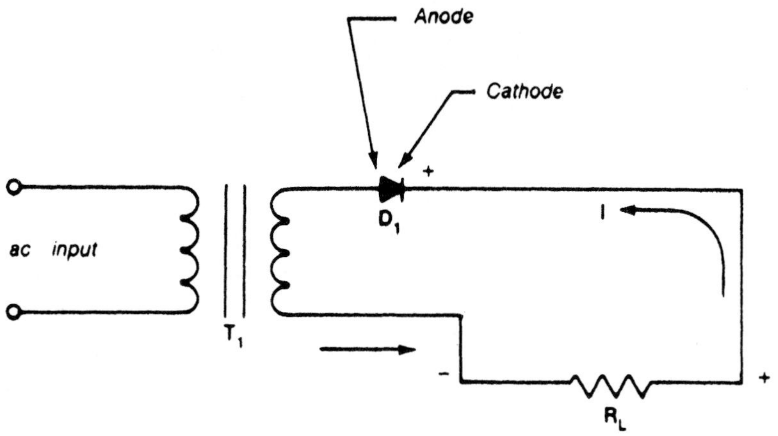


**FIGURE 6.100** Basic half-wave rectifier circuit.

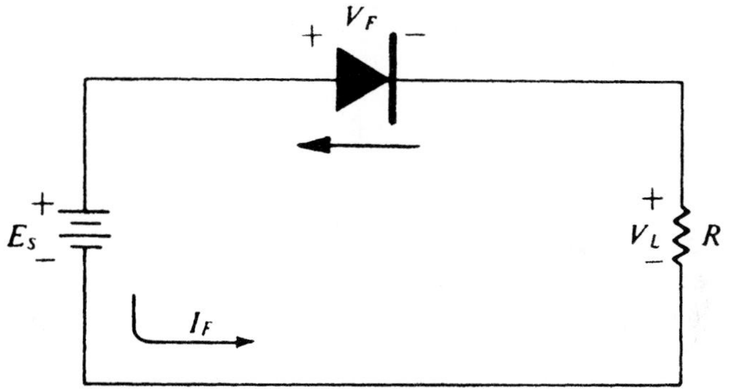


**FIGURE 6.101** Biasing a diode for forward current.

symbol and travels through the diode as shown in Fig. 6.101. When the voltage is reversed from that shown in Fig. 6.101 [cathode (+)], the diode blocks the current and does not conduct. On the symbol for a diode, the arrow is the anode and the bar is the cathode.

The silicon diode proves most useful in circuits used to rectify or convert ac to dc. The following three circuit configurations are the most widely used for single-phase rectification:

- Half-wave (see Fig. 6.100)
- Full-wave center tap (see Fig. 6.102)
- Full-wave bridge (see Figs. 6.103 and 6.104)

The following two circuits are used for three-phase rectification of ac:

- Half-wave three-phase rectifier (see Fig. 6.105)
- Full-wave three-phase rectifier (see Fig. 6.106)

See Sec. 6.9.5.1 for the average output and ripple frequency.

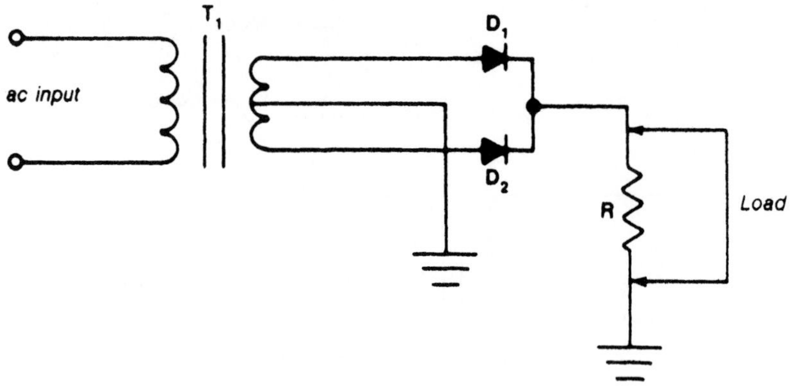


**FIGURE 6.102** Full-wave center-tapped rectifier circuit.

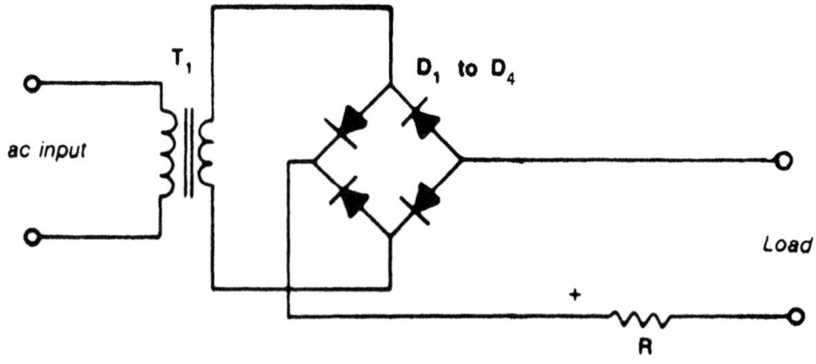


**FIGURE 6.103** Full-wave bridge rectifier.

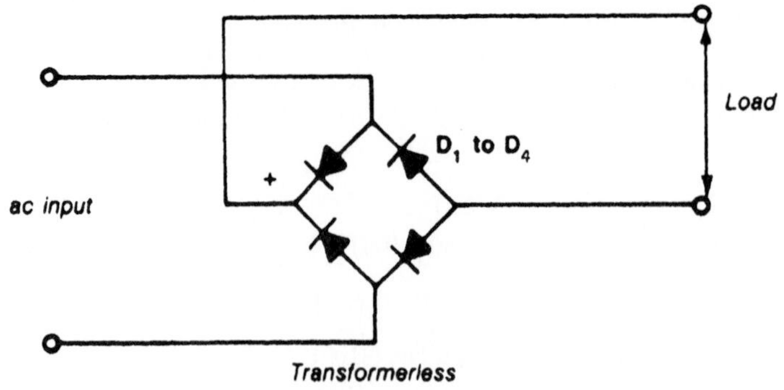


**FIGURE 6.104** Transformerless full-wave bridge rectifier.

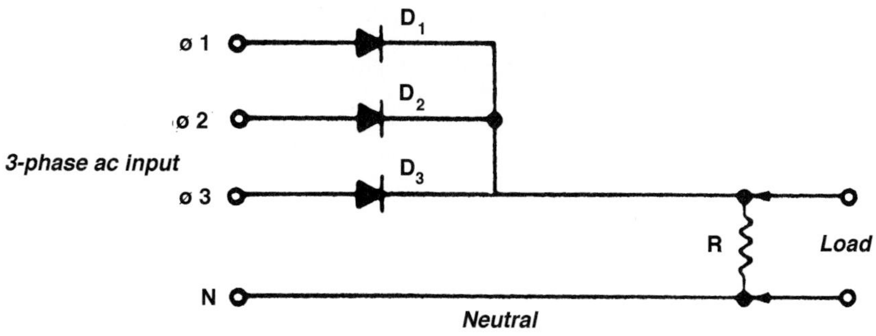


**FIGURE 6.105** Three-phase, half-wave rectifier.

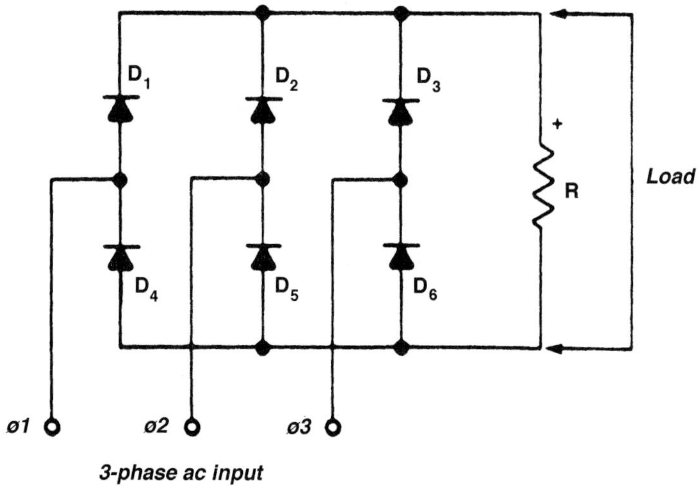


**FIGURE 6.106** Three-phase, full-wave rectifier.

### 6.9.5.1 Rectifier Circuit Data

In the following rectifier circuits, the important diode specifications are (1) peak surge current, (2) forward dc current, and (3) reverse breakdown voltage.

| Circuit type | Average output | Ripple frequency |
|---|---|---|
| Half-wave | $0.318Vp$ | $f_{\text{input}}$ |
| Full-wave center-tap | $0.636Vp$ | $2f_{\text{input}}$ |
| Full-wave bridge | $0.636Vp$ | $2f_{\text{input}}$ |
| Half-wave, 3-phase | $0.826Vp$ | $3f_{\text{input}}$ |
| Full-wave, 3-phase bridge | $0.955Vp$ | $6f_{\text{input}}$ |

### 6.9.5.2 Filter Capacitors for Rectifiers (Unregulated)

For simple, unregulated power supplies, the value of the filter capacitor may be calculated from

$$C = \frac{IL}{0.417\ V_{\text{rms}}}$$

where $C$ = capacitance, μF
$IL$ = load current, mA
$V_{\text{rms}}$ = rms ripple voltage, V

*Example.* What value of filter capacitor is required to keep the ripple below 0.25 $V_{\text{rms}}$ for a 6-V power supply and a load resistance of 1.5 kΩ? Load current is first found from Ohm's law.

$$IL = \frac{V_{\text{dc}}}{RL} = \frac{6\text{ V}}{1500\ \Omega} = 0.004\text{ A} = 4\text{ mA}$$

Therefore the capacitor value, using the previous equation, is

$$C = \frac{4}{0.417\,(0.25)} = \frac{4}{0.10425} = 38.4\ \mu\text{F}$$

Use a 40-μF capacitor in this filter circuit, rated 6 or 12 V.

For frequencies other than 60 Hz, use the following equation:

$$V_{\text{rms}} = \frac{IL}{0.00694\,fC} \qquad \text{or} \qquad C = \frac{IL}{0.00694\,f\,V_{\text{rms}}}$$

where $f$ = frequency, Hz.

Three common filter circuits for rectifier use on unregulated power supplies are

- Simple capacitor filter (see Fig. 6.107)
- *RC* filter (see Fig. 6.108)
- Pi (π) filter (see Fig. 6.109)

### 6.9.5.3 Special Diodes—Characteristics and Applications

Diodes are made in various types for different applications. The special-purpose diodes are shown in Table 6.22 with their applications. Some of the special-purpose diodes are also described in the following subsections.

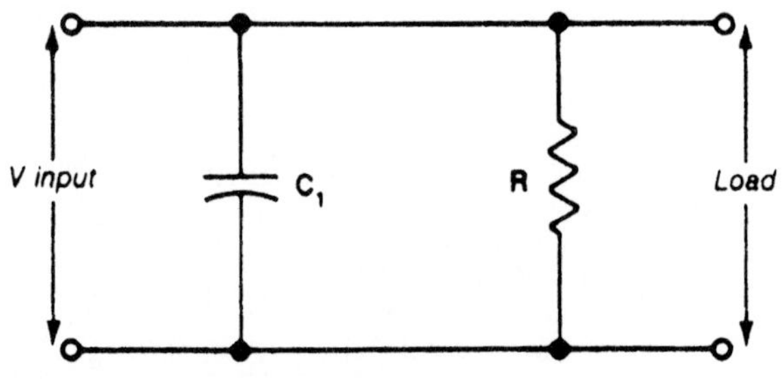


**FIGURE 6.107** Simple capacitor filter.

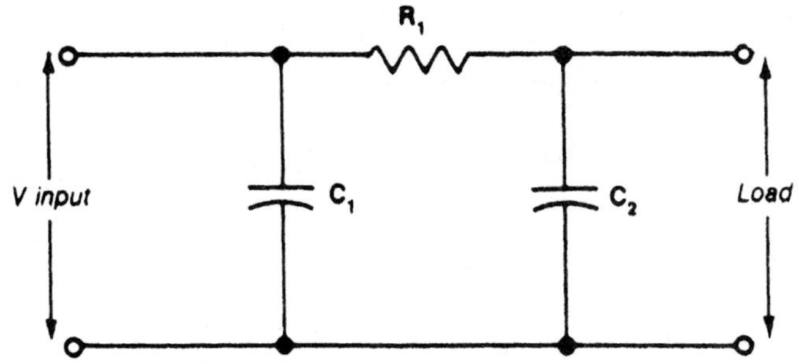


**FIGURE 6.108** *RC* filter.

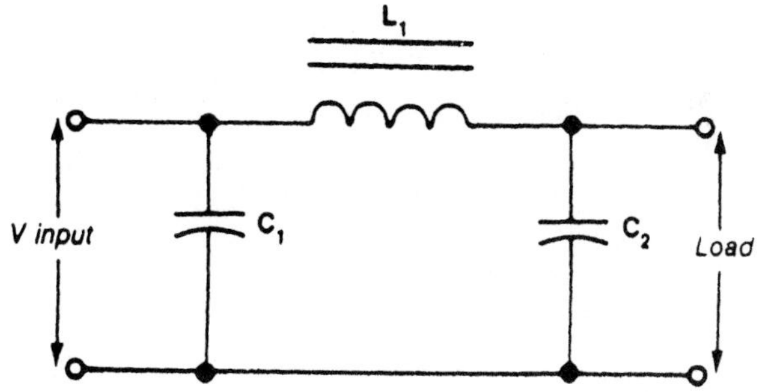


**FIGURE 6.109** Pi filter with choke regulation.

### 6.9.5.4 Zener Diodes

The zener diode is similar to the basic junction diode but is designed to function at breakdown voltage when reverse-biased. The breakdown voltage is called the zener or avalanche voltage. Zener diodes are operated in the breakdown region, and maximum zener current is limited only by the power dissipation allowed for the diode, in accordance with its specifications. Zener diodes are used as voltage regulation (see Fig. 6.110) and voltage-reference devices in many circuits.

Notice the polarity of the zener diode's cathode in Fig. 6.110. When the output voltage in the circuit shown in Fig. 6.110 goes above the zener breakdown voltage, the diode conducts in the reverse direction and clamps or regulates the output voltage of the circuit. Regulation of the output voltage can be held to closer tolerances or limits using an IC regulator such as the 723. See Sec. 6.9.9 for electronic circuits using IC regulators.

### 6.9.5.5 Light-Emitting Diodes

Light-emitting diodes emit visible light when current is passed through them. They are manufactured in clear form and in various colors.

The typical values of the forward operating voltage ranges between 1.7 to 5 V dc with currents of 2 to 20 mA. Currents of 10 to 20 mA are more common and have higher light output with voltages from 2.2 to 5 V dc. The power dissipation range is 4 to 65mW, with 40 to 50mW being the more common range. On an LED, the operating voltage polarity is the same as that of a forward conducting diode; that is, the anode is connected to the positive side of the dc voltage source (see Fig. 6.101).

**TABLE 6.22** Special Diodes

| Diode Type | Property/Characteristic | Application |
|---|---|---|
| Tunnel | Negative resistance | UHF-Microwave oscillators and amplifiers |
| Zener | Automatic variable resistance | Voltage reference, regulation, limiting |
| Varactor | Voltage variable capacitance | Resonant circuits and oscillator control |
| Pin | Nonrectification of high frequency | UHF switching |
| Step recovery | Delayed current flow | Frequency multiplication |
| Schottky | High speed recovery | UHF switching, detection |
| Backward | Negative avalanche knee at zero bias | UHF, low level, low impedance detector |
| Photo | Conductive or generative modulation | Light sensors and counters |

The wire leads of an LED are generally supplied in different lengths on each LED, with the shorter lead being the cathode.

Light intensity values are typically between 1.5 and 10 microcandela (μcd). The LED is used as an indicating light in various control, logic, and other circuits. LEDs can be mounted on PC boards or inserted into special sockets for easy replaceability.

Figure 6.111*a* and 6.111*b* shows the various forms of the LEDs in common use. Figure 6.111a shows the typical LED which may be mounted in sockets or soldered directly into the circuit. Figure 6.111*b* shows the type of LED assemblies used on PC (printed circuit) boards, where the LED is positioned so that it protrudes from the edge of the PC board and may be observed directly when the PC board is mounted on a chassis and the LED shows from the front of the electronic panel. Note the letter "c" on the LED leads, indicating the cathode. The cathode location for panel-mounted LEDs is indicated on the manufacturer's device data sheet.

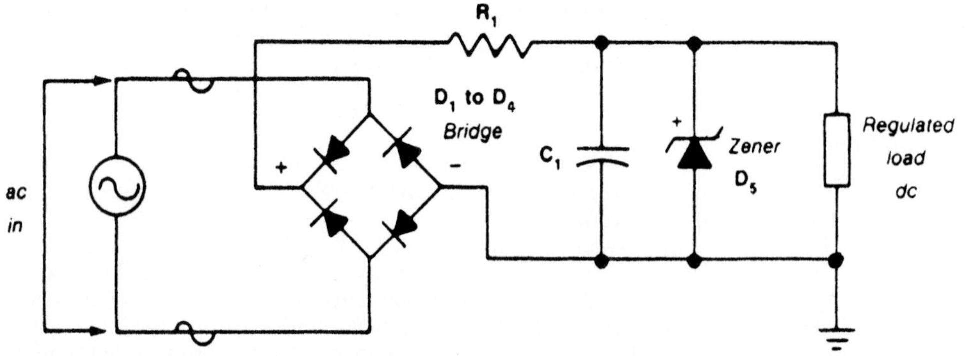


**FIGURE 6.110** Full-wave bridge rectifier circuit with Zener voltage regulation.

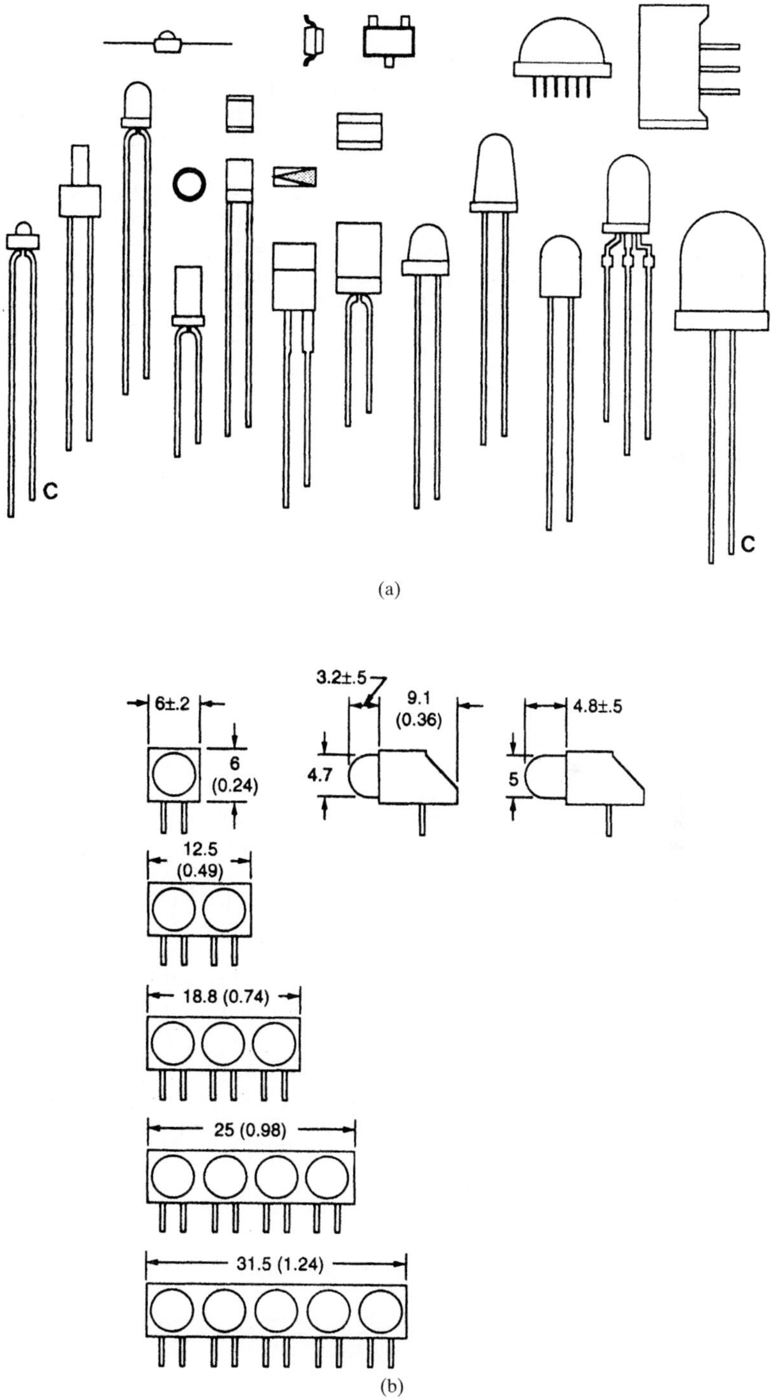


**FIGURE 6.111** (*a*) Standard LED configurations; (*b*) PC-board mount-type LEDs.

### 6.9.6 Varistors

*Varistors* are semiconductor devices whose resistance decreases rapidly as the voltage across them increases. Varistors are used to suppress arcs across relay and switch contacts and to remove or absorb voltage transients from power sources feeding a circuit or device.

Metal-oxide varistors have symmetrical, sharp breakdown characteristics, which are capable of suppressing transients from the sudden release of stored electrical energy. The devices may be used for power supply protection, line transients absorption, and protection of ICs and computers that are vulnerable to power line surges. Some metal-oxide varistors have response times of less than 35 ns (nanoseconds). They are rated for steady-state applied voltage ($V_{rms}$ and $V_{dc}$), average power dissipation (watts), peak current, and joules. See Section 6.6.5 and Fig. 6.19 for varistor application.

#### 6.9.6.1 Varistor Applications

Varistors are often used in electrical circuits as transient-voltage-suppression devices. Voltage transients in circuits result from the sudden release of previously stored electrical energy, either within the circuit or outside the circuit when the circuit steady-state condition is altered, as either a repeatable or random action.

A sudden change in the electrical conditions of any electrical circuit will cause a transient voltage to be generated from energy stored in circuit inductance and capacitance. The rate of change of current $di/dt$ in an inductor $L$ will generate a voltage equal to $-L(di/dt)$ and be of a polarity that causes current to continue to flow in the same direction. This effect accounts for most switching-induced transient overvoltages.

In a transformer, when the primary windings are energized at the peak of the supply voltage, the coupling of this voltage to the stray inductance and capacitance of the secondary winding can generate an oscillating transient voltage with a peak amplitude up to two times the normal peak secondary voltage. Thus, it is possible for the secondary circuit to see a large fraction of the peak applied primary voltage.

Deenergizing or opening the primary circuit of a transformer also generates high-voltage transients, especially if the transformer drives a high-impedance load. Unless a low-impedance discharge path is provided, this surge of transient energy appears across the load and can destroy components that may have a limited operating voltage rating.

Other sources of transient voltages include

- Fuse blowing during power faults (devices parallel to the load "see" the transient voltage).
- Switch or breaker arcing during opening and contact "bounce" (voltage also escalates during "restrikes").

Random transients are caused by switching parallel loads on the same branch of a power distribution system and also by lightning strikes. To perform the voltage-limiting function, three types of suppression devices have been used successfully, and include:

- Selenium cells
- Zener diodes
- Varistors [silicon-carbide and metal-oxide (zinc oxide, etc.)]

Here, we are concerned with only metal-oxide varistors. Applications of metal-oxide varistors extend from the low-power electronic circuits to the largest power distribution, utility-type surge arrestors.

In selecting a suitable varistor for your particular application, a five-step process is normally implemented:

1. Determination of working voltage (steady state)
2. Transient energy absorbed by the varistor (joules or watt-seconds)
3. Calculation of peak transient current through the varistor (amperes)
4. Power dissipation requirements
5. Selection of a device that will provide the required voltage-clamping characteristics

Selection of the correct varistor voltage is straightforward, but selection of the proper energy (joules) rating is more difficult and presents a degree of uncertainty. Choosing a varistor with a high energy rating is expedient but seldom cost-effective. ANSI/IEEE standard C62.41-1980 covers many considerations in practical varistor applications.

***Simple Circuit Applications of the Varistor***

*For single-phase (complete varistor protection) circuits:* A varistor is connected between the line and neutral; another is connected between line and ground. Another varistor is sometimes connected between neutral and ground together with the others.

*For single-phase, two-wire, 120- and 240-V circuits:* A single varistor is connected between line and ground.

*For single-phase, three-wire, 120- and 240-V circuits:* A varistor is connected between line 1 and ground or neutral, and another varistor is connected between line 2 and ground or neutral, while the third varistor is connected between line 1 and line 2.

*For dc circuits:* The single varistor is connected between the positive and negative lines or between the positive supply voltage and ground or between the negative supply voltage and ground.

***Varistor Testing.*** For inspection and verification purposes, the varistor voltage should be measured (at 1 mA). This can be done using the test circuit shown in Fig. 6.112. With the switch in position I, adjust the power supply so that the digital voltmeter (DVM) reads 100 V (this corresponds to a current of 1 mA). Then place the switch in the V position and the varistor voltage will be indicated on the voltmeter (DVM).

### 6.9.7 Thermistors

*Thermistors* are semiconductor devices that change resistance with changing temperature. They are used as temperature sensing devices. There are two types of standard thermistors:

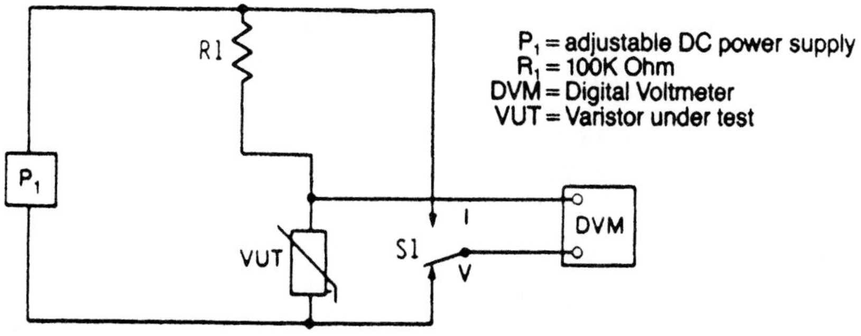


**FIGURE 6.112** Varistor test circuit.

negative-temperature-coefficient (NTC), and positive-temperature-coefficient (PTC) thermistors. The NTC type decreases in resistance as temperature increases (maximum operating temperature is 125°C). The PTC type is characterized by an extremely large resistance change in a small temperature span or range. The temperature at which resistance begins to increase rapidly is called the *transition temperature* (°C). See PTC and NTC curves in Fig. 6.113.

PTC thermistors are also used as resettable fuse elements. The minimum limit current is the minimum amount of current required by the PTC to guarantee switching at the minimum ambient temperature. The maximum limit current is the maximum amount of current the PTC must be able to pass without switching at the maximum ambient temperature.

***Thermistor Applications.*** NTC thermistors have many industrial applications, including:

- Industrial process control
- Photographic processing
- Hot-mold equipment
- Oven temperature control
- Thermostats
- Fire detection
- Myocardial probes
- Differential temperature controls
- Chemical analysis
- Calorimetry
- Spectrophotometers
- Transistor temperature compensation

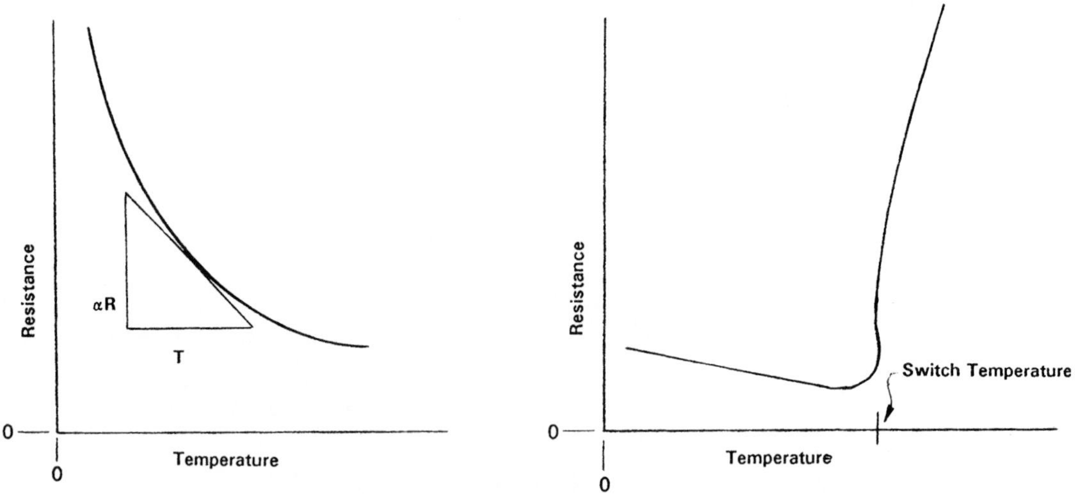


$R = Ae^{B/T}$

$\alpha = -B/T^2 = -2\%/°C$ to $-5\%/°C$

(a) (b)

**FIGURE 6.113** (*a*) NTC thermistor curve; (*b*) PTC thermistor curve.

In the NTC thermistor, the resistance of the thermistor decreases as the temperature value increases. A typical commercial example is $R = 28.53\ \Omega$ at 0°C (32°F) and $R = 1.58\ \Omega$ at 100°C (212°F). Many resistance-versus-temperature ranges are available.

The applications listed above are all based on the resistance-temperature characteristics of thermistors, demonstrating their versatility, usefulness, and importance as electronic components. The primary function of the thermistor is to show a resistance change as a function of temperature.

The simple voltage divider circuit is used in many thermistor applications, (see Fig. 6.114). In this circuit, the output voltage is taken across the fixed resistor. This circuit provides an increasing output voltage for increasing temperatures, which allows the loading effect of any external measurement circuitry to be included into the calculations for the resistor $R$. Thus, the loading will not affect the output voltage as the temperature varies.

The output voltage as a function of temperature is expressed as follows:

$$eo\ (T) = es\ \frac{R}{(R + RT)} \qquad \text{and} \qquad \frac{eo\ (T)}{es} = \frac{R}{R + RT} = \frac{1}{1 + RT/R}$$

where $RT$ = zero power resistance of the thermistor at the given temperature, Ω
$RT_o$ = zero power resistance of the thermistor at standard reference temperature (25°C), Ω
$R$ = fixed resistance, Ω

From the plot of the output voltage it can be seen that a range of temperatures exists where the circuit is reasonably linear with good sensitivity. Therefore, the objective will be to solve

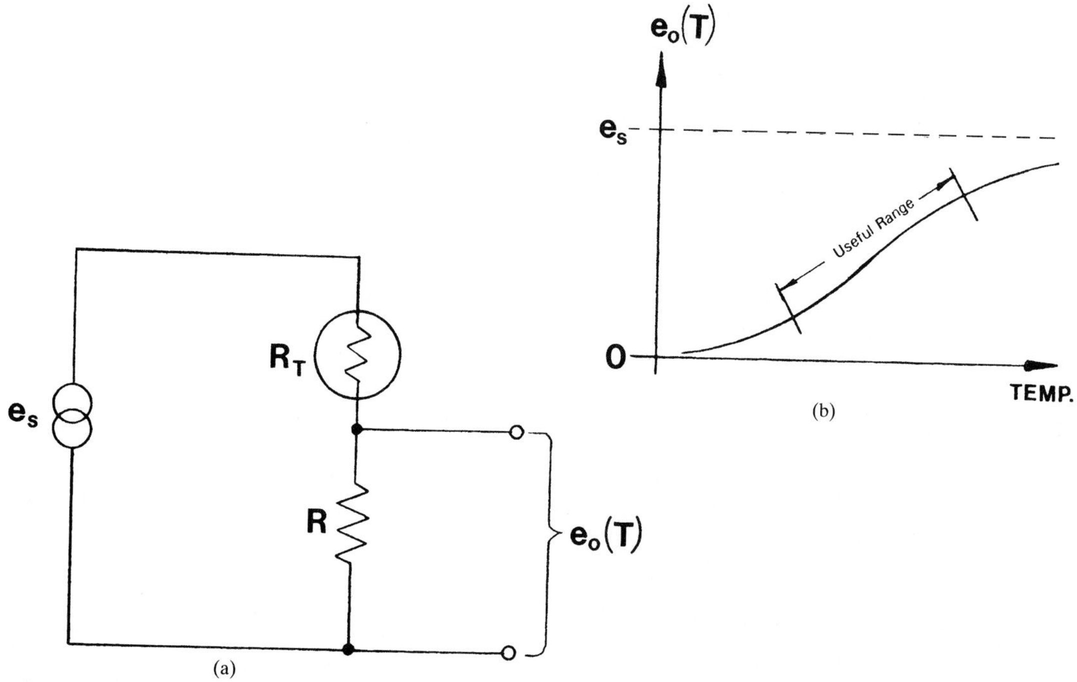


**FIGURE 6.114** (*a*) Voltage-divider thermistor circuit; (*b*) typical thermistor response curve.

for a fixed resistor value $R$ that provides optimum linearity for a given resistance-temperature characteristic and a given temperature range. Refer to Fig. 6.114*a* and 6.114*b* for symbolism and clarity of the explanation for the equations.

The usual practice is to furnish resistance ratio versus temperature information for each type of thermistor and material system listed in the thermistor catalogs. The ratio of the zero power resistance of the thermistor at the desired reference temperature to the fixed-value resistor is a constant we can call $s$. Thus

$$s = \frac{RT_o}{R}$$

The standard function is then

$$F(T) = \frac{eo(T)}{es} = \frac{1}{1 + srT}$$

The standard function $F(T)$ is dependent on the circuit constant $s$ and the resistance ratio temperature characteristic, $rT$. If we allow the circuit constant to assume a series of constant values and solve for the standard function, we will generate a family of S-shaped curves similar to those in Fig. 6.115. This figure of curves for "A"-type material is described in MIL-T-23648A, *Military Specification, Thermistor.* Design curves such as shown in Fig. 6.115 can be used to provide a graphical solution or a first approximation for many thermistor applications. Also, from the design curves, a value for the circuit constant $s$ exists such that optimum linearity can be achieved for the divider network (circuit) of Fig. 6.114*a* over a specified temperature range.

More accurate solutions for a specific design problem requiring great accuracy may be approached analytically. The two analytical methods often used are the *inflection-point method* and the *equal-slope method.**

Figure 6.116 shows two thermistors with a scale for size comparison. The thermistor marked "A" is a high-quality, extremely accurate and linear thermistor, covered with a Teflon sheath for corrosion resistance. This thermistor is used in the IC-thermistor temperature controller shown in the IC circuit section. It allows accurate temperature control of photographic bath solutions held to ±0.5°F. The thermistor marked "B" is a low-cost unit that can be effectively used where great accuracy is not required, such as oven temperature control. The maximum operating temperature of most thermistors is 125°C (257°F).

### 6.9.8 Varactors

*Varactors* are semiconductor devices in which the reverse junction capacitance varies with reverse voltage. They are used as voltage-controlled tuning devices to control oscillator frequency.

### 6.9.9 Integrated Circuits

*Integrated circuits* are complex semiconductor devices or structures that contain complete analog or digital circuits interconnected together on a single silicon chip. Integrated circuits are produced in the following forms:

- Small-scale integration (SSI)
- Medium-scale integration (MSI)

* These analytical methods may be obtained from the *Thermistor Sensor Handbook,* by Thermometrics, Inc., Edison, New Jersey.

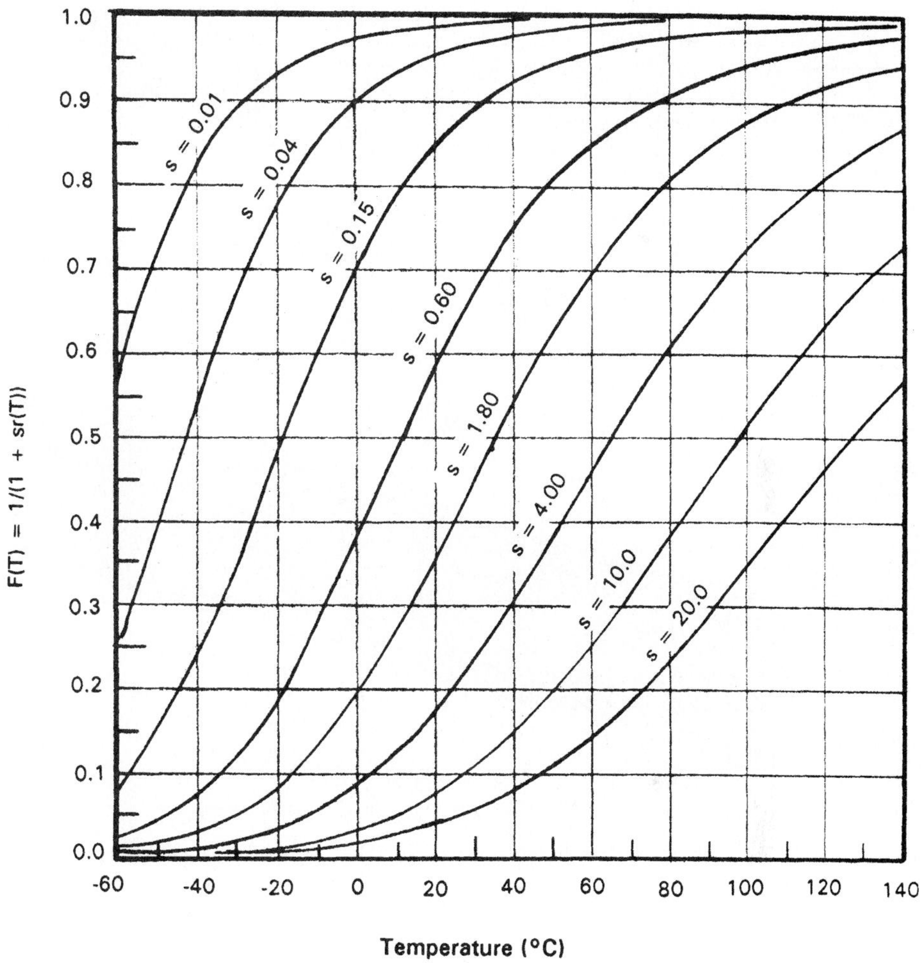


**FIGURE 6.115** Thermistor curves for type A material.

- Large-scale integration (LSI)
- Very large scale integration (VLSI)

Some common types and uses for ICs are shown in Tables 6.23 to 6.25. This listing of linear and digital ICs covers a broad range of devices which may be used in countless applications. Integrated circuits are produced as linear as well as digital electronic devices. A photograph of IC packages is shown in Fig. 6.95, and some of the common case forms are shown in Fig. 6.117.

As an introduction to the practical application of integrated circuits, Fig. 6.118 shows the internal components contained in a typical linear IC. The CA3094 IC is classified as an *operational transconductance amplifier* (OTA) and is commonly used as a programmable power switch-amplifier. The schematic diagram in Fig. 6.118 shows that many transistors, diodes, and resistors are incorporated in the circuitry of this particular device. This is all contained in a small 8-lead package in either a TO-5 case, an 8-lead mini-DIP, or as a chip. The CA3094 is

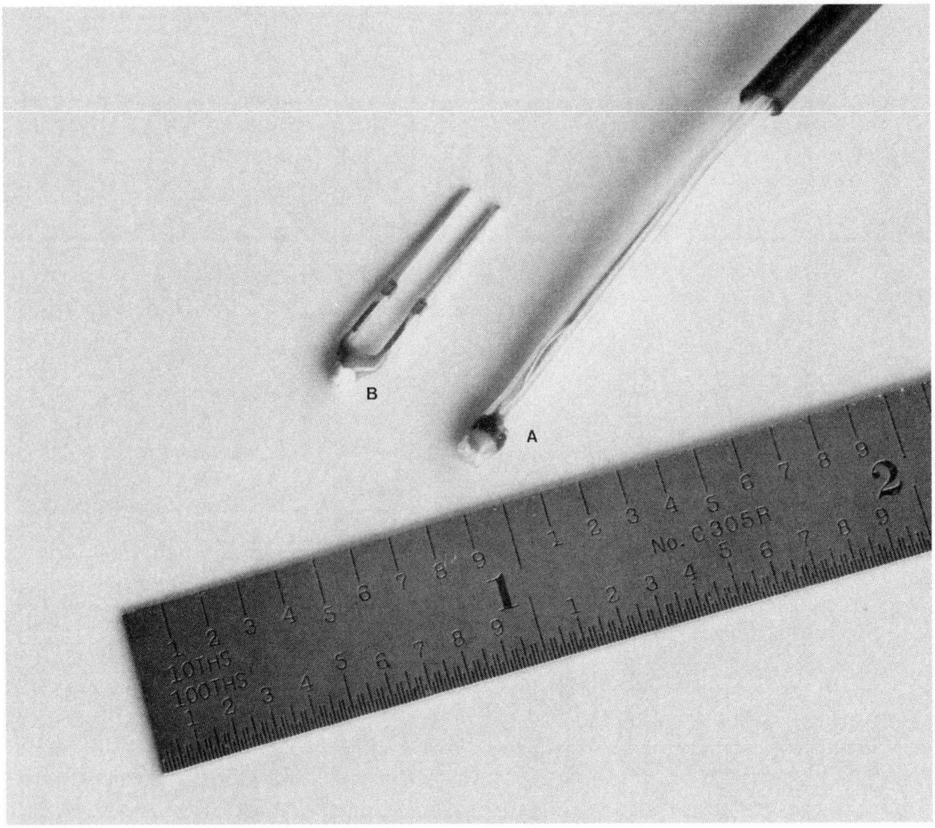


**FIGURE 6.116** Typical thermistors. A—high-quality thermistor; B—low-cost standard thermistor.

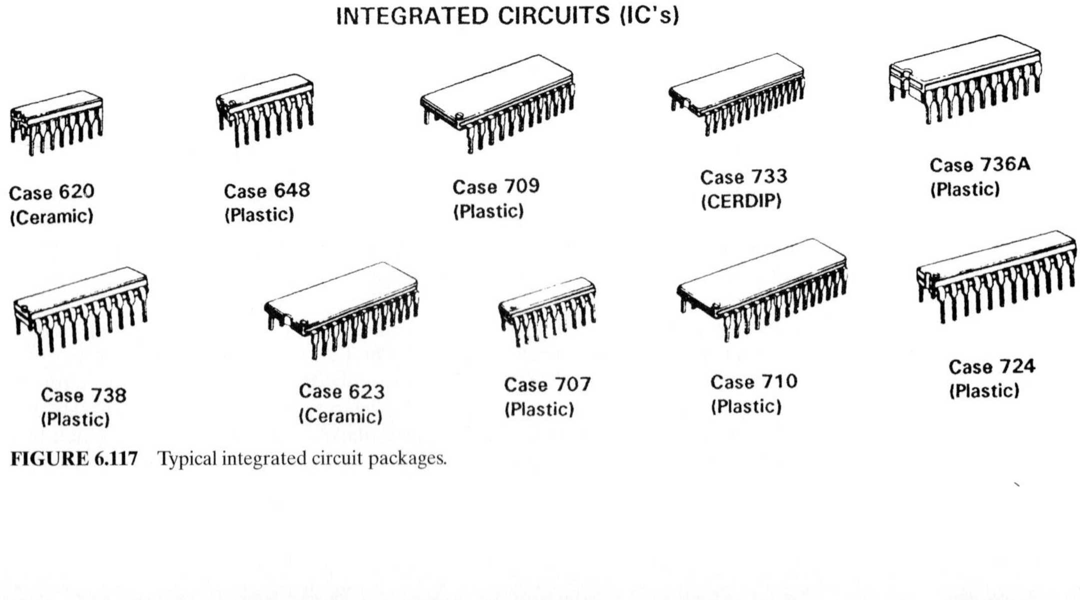


**FIGURE 6.117** Typical integrated circuit packages.

**TABLE 6.23** Common Usage–Integrated Circuits

| Integrated Circuit Designation (Generic) | Used in or for: |
|---|---|
| 555 | Monostable devices, astable devices, power supply circuits, measurement and control circuits. (8-pin mini DIP) Timers. |
| 556 | This is the dual 555 in a 14-pin DIP. |
| MOC 3010 | Optoelectronic (driver) for isolated triac triggering, isolated ac switching and high electrical isolation. |
| 7400 | Quad two-input nand gate (digital systems) |
| 7404 | Hex inverter (digital TTL) |
| 7408 | Quad two-input and gate (digital TTL) |
| 7490 | Divide by two or five, BCD counter (TTL) |
| 4001 | Quad two-input or gate (digital CMOS) |
| 4011 | Quad two-input nand gate (digital CMOS) |
| 4013 | Dual type D flip-flop (digital CMOS) |
| 4017 | Decade counter/divider (digital CMOS) |
| 4023 | Triple 3-input nand gate (digital CMOS) |
| 4049 | Inverting hex buffer (digital CMOS) |
| 4066 | Quad bilateral switch (digital CMOS) |
| 4116 | 16K dynamic RAM (digital memory) |
| TL 317 | Positive voltage regulator, linear |
| 723 | Adjustable voltage regulator, linear |
| 7812 | 12-volt regulator, linear |
| 7905 | 3-terminal negative regulator, linear |
| 324 | Quad op amp, linear |
| 741 | Operational amplifier, linear |
| 1458 | Dual operational amplifier, linear |
| 3900 | Quad op Norton amplifier, linear |
| uA2240C | Programmable timer/counter, linear |
| SN94281 | Complex sound generator, (analog and digital) |
| 380 | Audio power amplifier, linear |
| 383 | 8-watt audio power amplifier, linear |
| MC1350 | IF amplifier, linear |
| ICL7660 | Voltage converter, linear |
| 339 | Quad comparator, linear |
| 565 | Phase-locked loop, linear |
| 3909 | LED flasher/oscillator, linear |
| ICL8069 | Low-voltage reference. |

Note: MOS devices have extremely high input resistance and are susceptible to damage when exposed to static electrical charges. Therefore, caution should be used when handling these devices. Do not allow a static discharge into the pins or leads of these devices. Keep your body and soldering devices grounded while working with these ICs.

used in the accurate dual voltage tracking regulated dc power supply shown in the section for IC practical circuits and products to follow.

To give the reader an idea of the power and versatility of linear IC technology, the following list shows the application capabilities of the CA3094 IC.

- Temperature control with thermistor sensor
- Speed control for shunt-wound dc motor
- Error signal detector
- Overcurrent, overvoltage, and overtemperature protectors
- Dual-tracking power supplies

**TABLE 6.24** Linear Integrated Circuits

| Type | Description | Pins |
|---|---|---|
| CA723 | Power control circuits - voltage regulator | 14 |
| CA3085 | Positive voltage regulation - 1.7 to 46v | 8 |
| CA3020 | Wide-band power amplifier | 12 |
| CA3059 | Zero-voltage switch - temp. control circuits | 14 |
| CA3079 | Zero-voltage switch, thyristor control | 14 |
| CA3088 | AM receiver subsystem/amplifier array | 16 |
| CA3089 | FM IF system (intermediate frequency) | 16 |
| CA3002 | IF amp for communications equip. | 10 |
| CA555 | Analog timer, for delay and oscillator use | 8 |
| CA3046 | Transistor array | 14 |
| CA3082 | 7-transistor common collector | 16 |
| CA3039 | Diode array | 12 |
| CA3026 | Dual independent differential amplifiers | 12 |
| CA3010 | Operational amplifier (OPAMP) | 12 |
| CA3130 | Operational amplifier | 8/BIMOS |
| CA3094 | Operational amplifier - high current | 8 |
| CA3060 | Operational transconductance amp. (OTA) | 16 |
| CA3098 | Voltage comparator - Schmitt trigger | 8/E |
| CA3000 | Differential amplifier | 10 |

NOTE: The above listed linear IC devices have wide use in many consumer products. Some of the listed devices are shown in the practical electronic circuit sections along with photographs of actual finished products and PC boards.

**TABLE 6.25** Digital Integrated Circuits

| Type | Function | Pins | Package |
|---|---|---|---|
| ICL7106CPL | Analog to digital convertor<br>LCD direct drive - 3.5 digit | 40 | Plastic DI |
| ICL7107CPL | Analog to digital convertor<br>LED common anode - 3.5 digit | 40 | Plastic DI |
| ICL7129CPL | Analog to digital convertor<br>LCD triplexed - 4.5 digit | 40 | Plastic DI |
| AD7520JN | Digital to analog convertor<br>10 bit | 16 | Plastic DI |
| AD7521JN | Digital to analog convertor<br>12 bit | 16 | Plastic DI |
| AD590JH | Current output temp. transducer | 3 leads | TO-52 |
| ICL8038ACJD | Waveform generator - Oscillator<br>120ppm/C typ. | 14 | CERDIP |
| ICL8038CCJD | Waveform generator - Oscillator<br>250ppm/C typ. | 14 | CERDIP |

NOTE: Some of the listed devices are shown in actual circuits and products in the practical IC circuits section of the handbook.

# CA3094
# CA3094A
# CA3094B

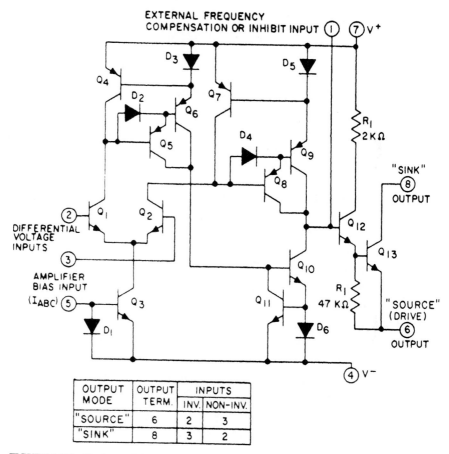

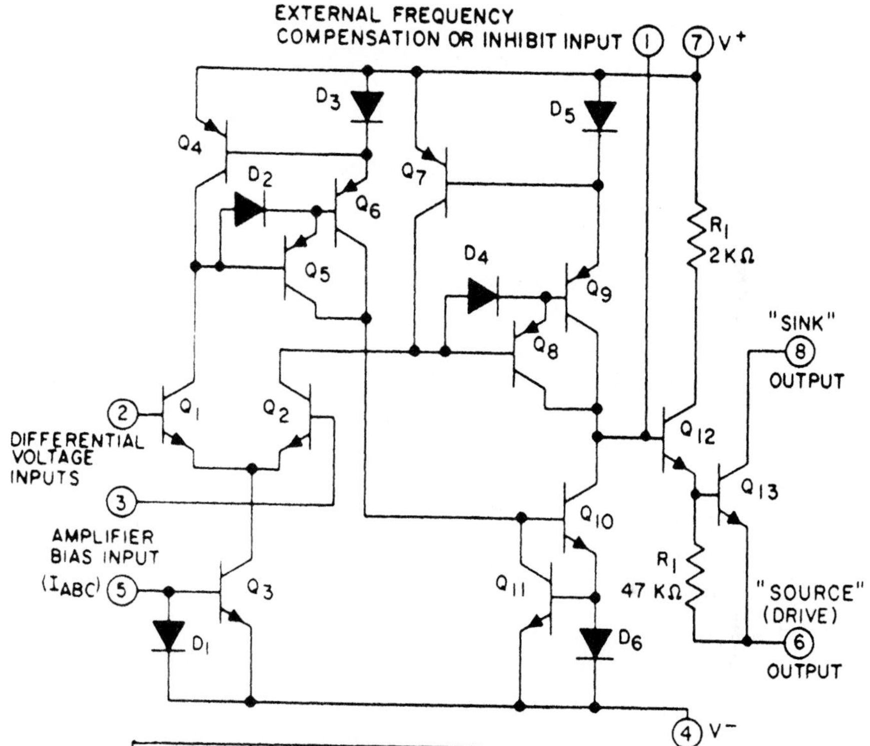


| OUTPUT MODE | OUTPUT TERM. | INPUTS | |
|---|---|---|---|
| | | INV. | NON-INV. |
| "SOURCE" | 6 | 2 | 3 |
| "SINK" | 8 | 3 | 2 |

**FIGURE 6.118** The internal circuitry of CA3094-type IC.

- Wide-frequency-range oscillators
- Analog timer
- Level detector
- Alarm systems
- High-power comparator
- Ground-fault interrupter circuits (GFIs)

The CA3094 is available in three voltage ranges:

- CA3094 (≤24 V)
- CA3094A (≤36 V)
- CA3094B (≤44 V)

The development of ICs has made possible the implementation of many electromechanical and electronic devices and systems which were previously too complex and uneconomical to develop and produce using conventional electronic components.

The classes of ICs commonly referred to as *linear* and *digital* are available in the following configurations:

- Operational amplifiers (op amps)
- Operational transconductance amplifiers (OTAs)
- Comparators
- Voltage regulators
- Power amplifiers
- Zero-voltage switches
- Schmitt triggers
- A/D convertors (analog to digital)
- D/A convertors (digital to analog)
- Speed controllers
- Decoders
- Volume-tone controls
- Preamplifiers
- Timers and counters

Using these forms of integrated circuits, many types of complete electronic and electromechanical systems may be more economically and reliably developed for the following uses:

- All types of amplifier systems
- Power supplies with excellent regulation and control
- Temperature sensors, indicators, and controllers
- Timers
- Flashers
- Medical instrumentation
- Alarm systems
- Liquid level sensors and controls
- Telecommunications applications
- Motor controls
- Waveform generators
- Controllers
- Measuring instruments
- Remote controls and robotic systems
- Machine tool controls
- Switchgear controls and devices

- Battery chargers and eliminators
- Ground-fault detection
- Annunciators
- Pressure and strain sensors

and a host of other applications.

The electromechanical designer may implement many types of electronic systems using the ICs in conjunction with external electronic support components. A vast array of complete circuits are available in the component manufacturers application handbooks which are available from sources such as Harris, Motorola, Texas Instruments, Intersil, Phillips, NTE, Sprague, and PMI.

This leaves the product designer with the work of efficiently packaging these modern IC systems for countless commercial applications.

The electromechanical work usually consists of the following stages:

1. Circuit selection
2. Breadboarding for functionality and testing
3. Development of the appropriate PC board
4. Packaging of the electronic system in a suitable enclosure
5. Development and coordination of auxiliary devices required per system
6. Prototype construction
7. Testing and quality-control procedures
8. Required tooling effort, when required
9. Completion of the engineering documentation/drawings for production
10. Manufacturing the system

Included in the following sections are some typical linear and digital IC systems and applications for your reference and possible use. The final sections of this chapter contain simplified methods used by modern designers to aid in the task of electrical and electronic circuit analysis. These methods, theorems, and laws are shown for reference and to give the electromechanical designer a background for further work. The bibliography section for this chapter lists many reference texts used for circuit analysis and other aspects of electrical and electronic engineering.

The IC is in effect a "black box" type of device that is supported by external electronic and electrical components such as resistors, capacitors, inductors, thermistors, varactors, varistors, diodes, transistors, and crystal oscillators. It is not necessary for the electromechanical designer to understand the principles of the network within the integrated circuits in order to utilize them in a complete electromechanical or electronic system. The IC design engineer and circuit analyst have thus given the product designer a very powerful and useful array of tools with which to work. Many modern products would be extremely difficult and too costly to manufacture for commercial use without the versatile and valuable modern IC devices.

### 6.9.9.1 Practical IC Applications and Products

This section shows some of the many applications for the linear and digital IC. One of the earliest and most often used IC is the 555 timer. This device is manufactured by all the leading IC component manufacturers and has a very broad usage base throughout industry. The 555 IC timer pinout and designation list appears in Fig. 6.119.

Figures 6.120 and 6.121 show the 555 timer used in two versions of the monostable mode to start logic circuits in their proper mode when the supply voltage is turned on, or if it is inter-

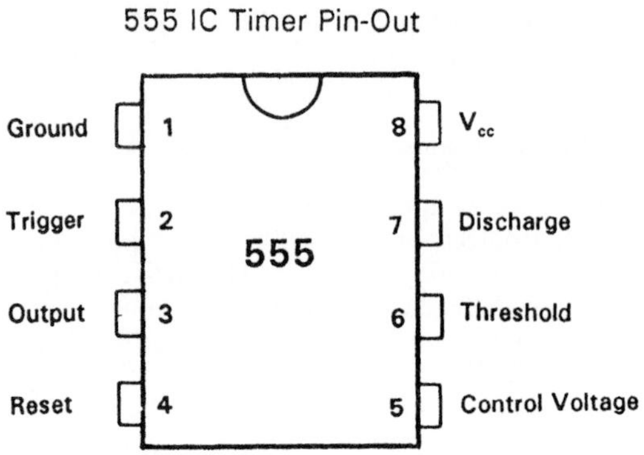


**DESIGNATIONS FOR THE 555 IC TIMER**

**CA555/CA555C - Harris**
**SN52555/SN72555 - Texas Instruments**
**SE555/NE555 - Intersil**
**NE555 - Fairchild**
**MC1455/MC1555 - Motorola**
**LM555/LM555C - National**
**RM555/RC555 - Raytheon**

**FIGURE 6.119** Pin-out and designations for the 555 IC timer.

rupted. These circuits are used to automatically reset transistor-transistor-logic (TTL) counters such as the 7490, 7492, and 7493. In these two circuits, the location of the timing capacitor determines whether a positive or negative output pulse is generated. In Fig. 6.120, the capacitor is connected to ground and will generate a positive output pulse. In Fig. 6.121, the circuit operates similarly, except that the capacitor is returned to *Vcc*, the supply voltage. If immediate triggering is not required, the diode may be omitted.

Figure 6.122 shows the basic circuit for a thermostat controller that will maintain temperature within given upper and lower limits. This is used with a thermistor-resistor divider network that will produce a voltage that is directly proportional to temperature. A divider network for thermistor applications is shown in Fig. 6.114*a* and explained in that section of the handbook. To prevent noise signals from triggering the 555 timer prematurely, bypass pins 2 and 6 with 0.01-μF disk capacitors.

For a relay control application, refer to Fig. 6.23 in the section covering relays. This type of application is a good example where the 555 timer can be used to eliminate expensive time-delay relays. In Fig. 6.123 this same timer delay network is incorporated into a miniature, sophisticated motor-control circuit that was designed by the author for switchgear applications. In Fig. 6.123, the 555 timer can be seen in the lower left side of the circuit, where it drives a relay that controls the reclose cycle of the motor-control circuit. This task was formerly appointed to expensive relay-delay timers that were settable for specific time delay intervals. The 555 timer circuit performs exactly the same function by incorporating a 1-MΩ potentiometer that may be adjusted for a time delay of 0 to 10 sec prior to the relay being

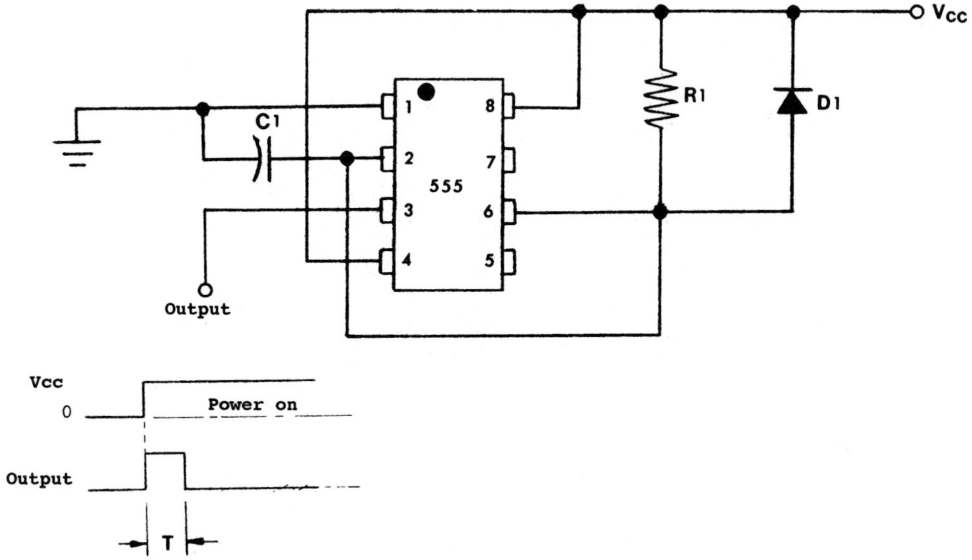


**FIGURE 6.120** Monostable mode circuit for the 555 timer IC.

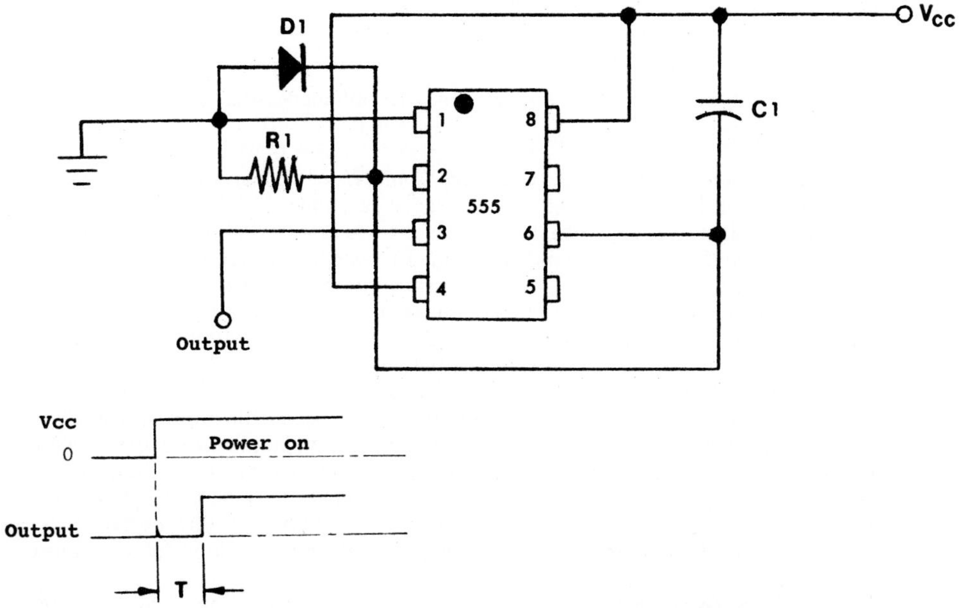


**FIGURE 6.121** Monostable mode circuit for the 555 timer; capacitor returned to *Vcc*.

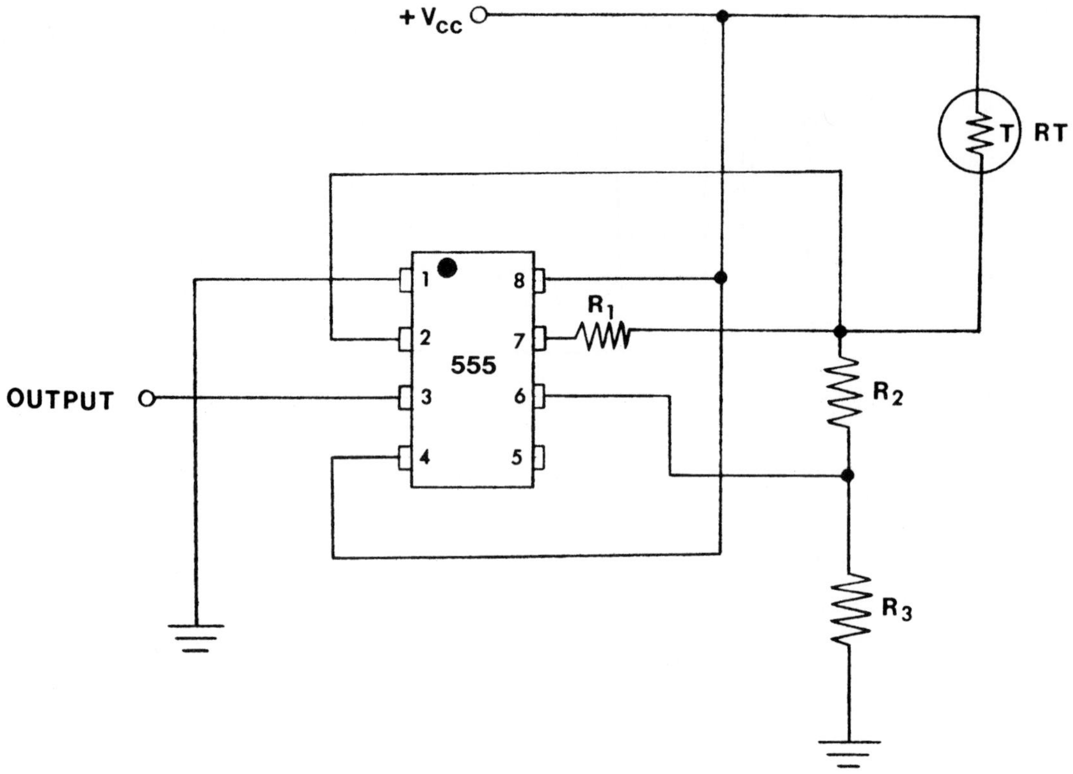


**FIGURE 6.122** 555 timer IC circuit for thermostatic control of temperature.

energized. On-time or off-time functions may be obtained using the simple and inexpensive 555 timer IC.

***Precision IC and Thermistor Temperature Controller.*** A high-precision temperature controller system is illustrated in Figs. 6.124, 6.125, 6.126, and 6.127. These figures show the complete project from circuit schematic to finished product. A PC foil pattern is included for those who would like to produce this system. This temperature controller will maintain a circulating water bath for photographic processing of film to within ±0.5°F. The water bath may contain from 3 to 25 gal of water that is circulated with a submersible pump, when the correct size heating element is selected. Size of the heating element is determined by the triac used in the circuit. This circuit will control a water bath up to 5 gal.

The temperature is maintained when the thermistor probe (Fig. 6.116, thermistor A) senses the temperature and changes resistance. The thermistor probe is also shown in Fig. 6.127 of the control cabinet. The change in resistance unbalances the divider network show in Fig. 6.124 (right side of schematic) and causes the CA3098 to trigger the gate of the triac, which in turn supplies power to the external 600-W heating element. A heating element that will draw as much as 45 amperes at 120 V ac can be used by selecting a triac that is rated for 45 A. The water-bath temperature can be maintained within the limits of 62 to 130°F.

***Dual-Voltage Tracking Regulated DC Power Supply.*** An accurate and useful dc tracking regulated power supply may be made from Figs. 6.128 to 6.130. Figure 6.128 is the schematic

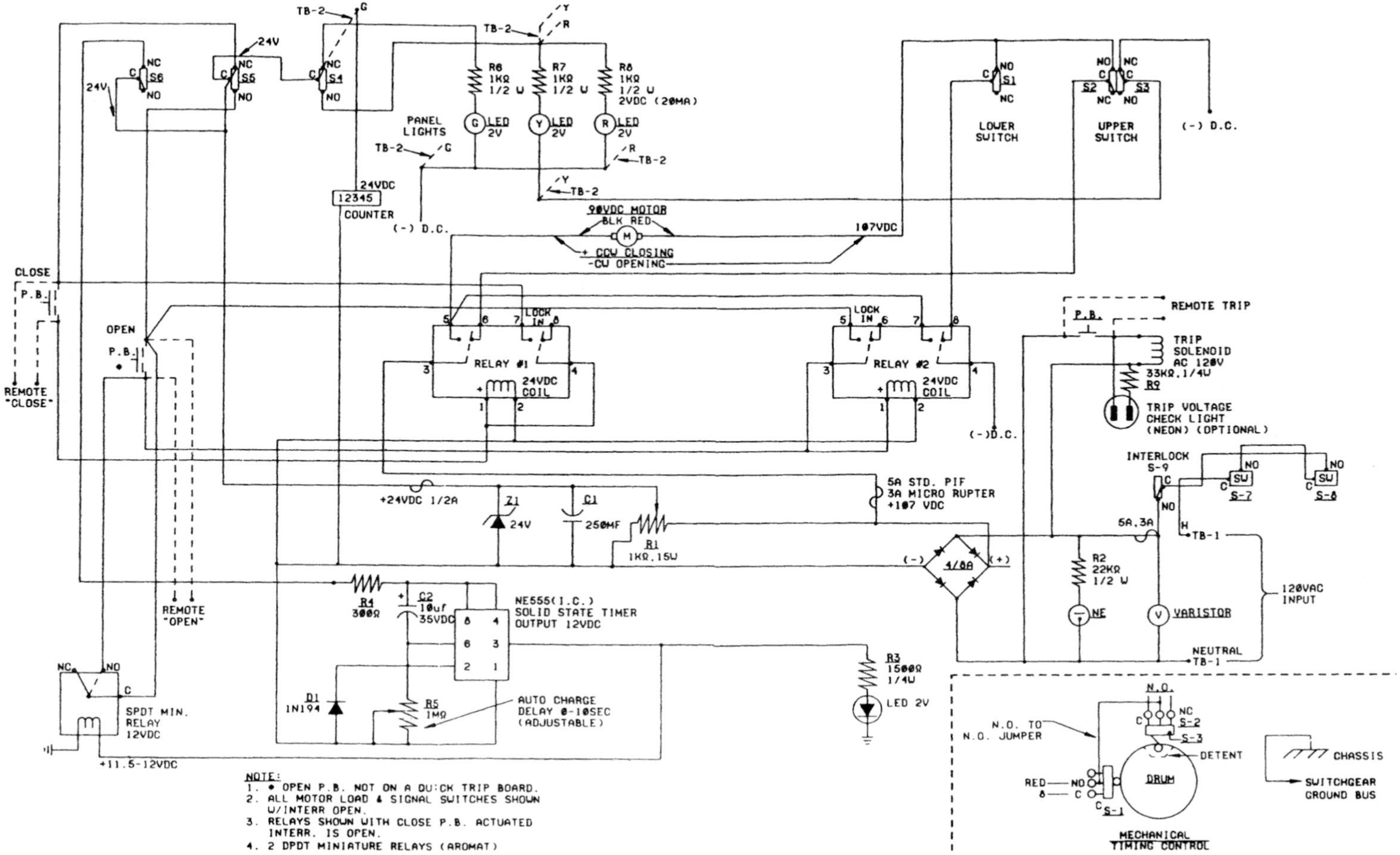


**FIGURE 6.123** Complex motor control circuit with 555 IC timer used for time delay of relay.

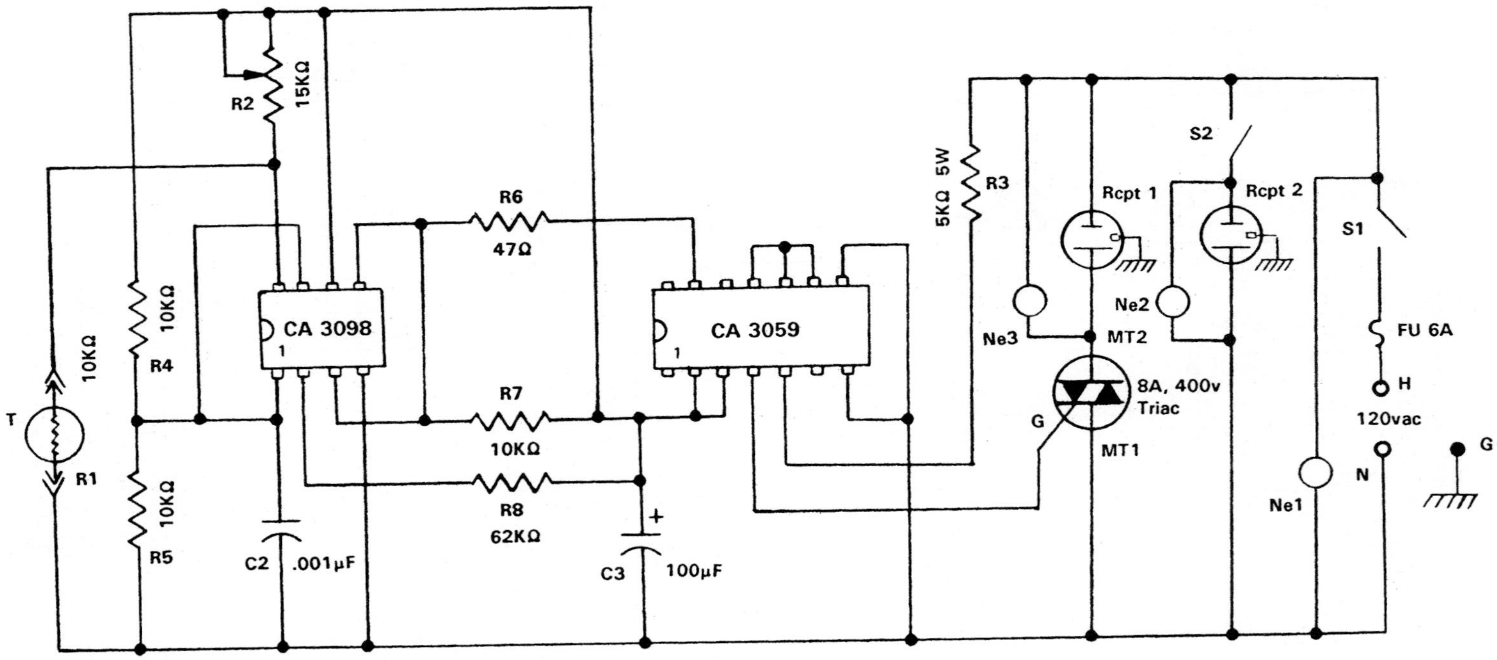


**FIGURE 6.124** IC precision temperature controller circuit with 10-kΩ precision thermistor.

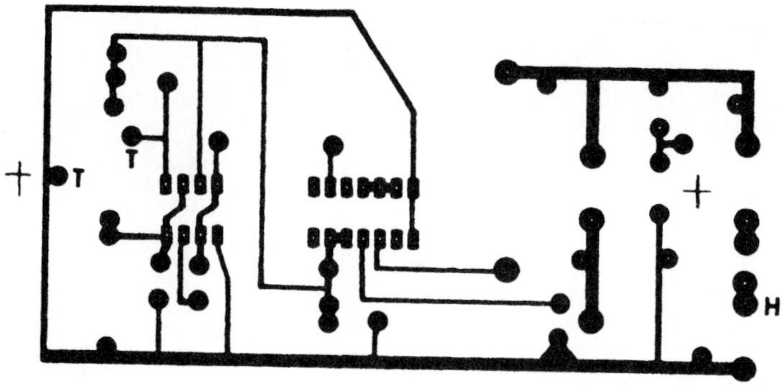


**FIGURE 6.125** Printed-circuit pattern for circuit shown in Fig. 6.124 (temperature controller).

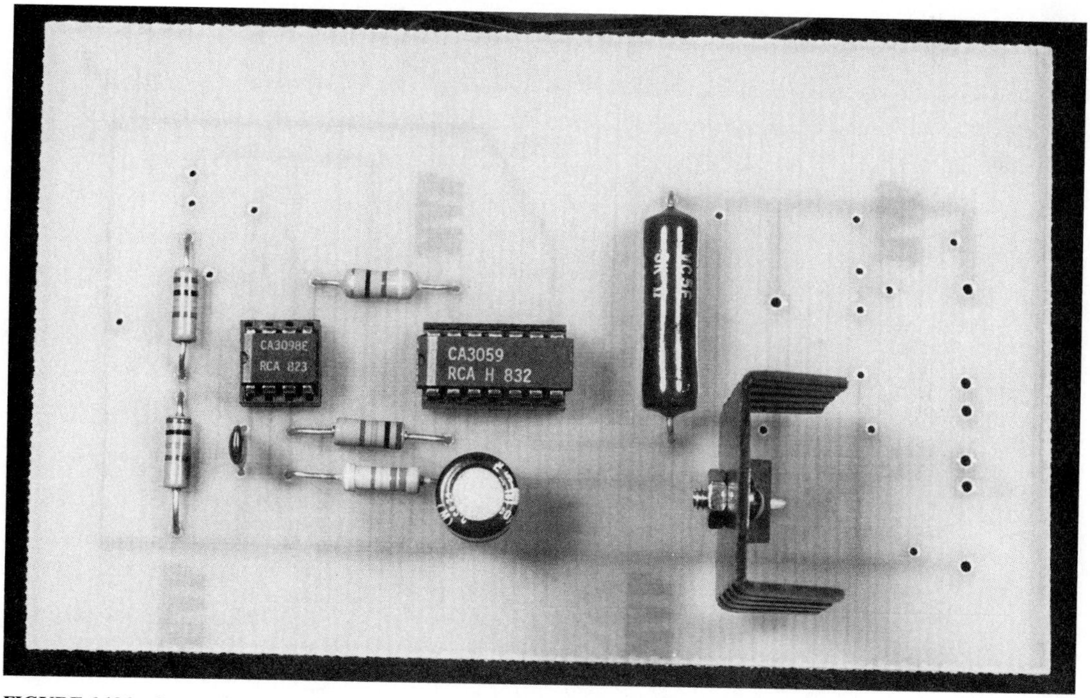


**FIGURE 6.126** Components mounted on PC board for temperature controller circuit of Fig. 6.124.

diagram that shows the CA3085A and CA3094A linear ICs as used in this practical circuit. Figure 6.129 is the foil pattern of the circuit that can be used to produce the PC board. Figure 6.130 is a photograph of the complete unit, ready for mounting and application. The rating of this power supply is ±15 V dc at 100 mA. Both positive and negative voltages are available for using the supply to power operational amplifiers, where a positive and negative voltage source is required.

Figure 6.131 is a closeup view showing a CA3094E IC in its mounting socket. The CA3094E is the 8-pin mini-DIP version of the CA3094A, which is good for voltages of up to 36 V.

**FIGURE 6.127** Control console for precision temperature controller (see Figs. 6.124 and 6.125).

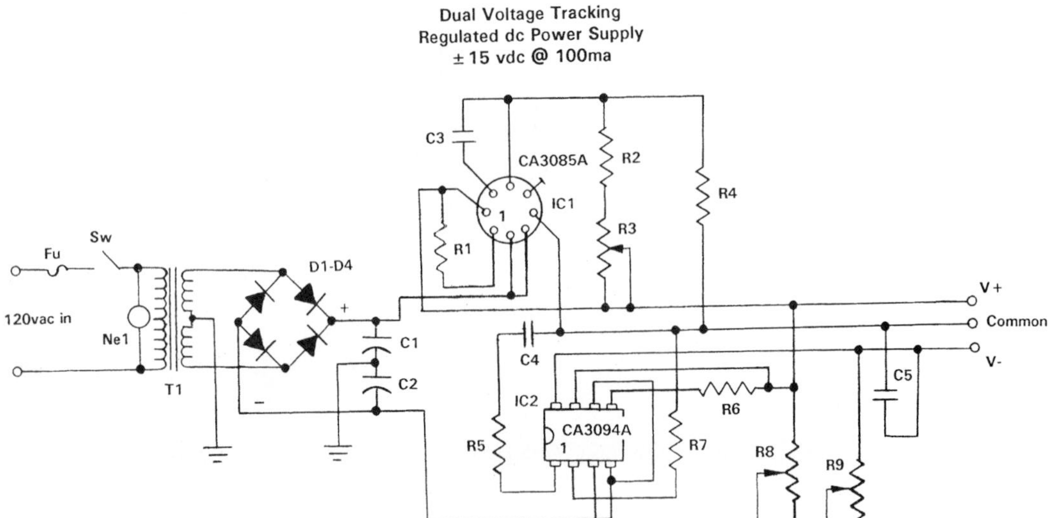


Fu 0.1A
Ne1 120vac neon
T1 = Stancor Lb656 120/56, c.t.
D1-D4 = Bridge diode, 200v, 1A
IC1 = CA3085A
IC2 = CA3094A
C1 = 100μF, 50v
C2 = 100μF, 50v
C3 = 0.01μF
C4 = 0.0056μF
C5 = 0.033μF

R1 = 5.6Ω
R2 = 10KΩ
R3 = 5KΩ
R4 = 1.5KΩ
R5 = 100Ω
R6 = 200KΩ
R7 = 5.1KΩ
R8 = 10KΩ (Trim-pot) 1%
R9 = 10KΩ (Trim-pot) 1%

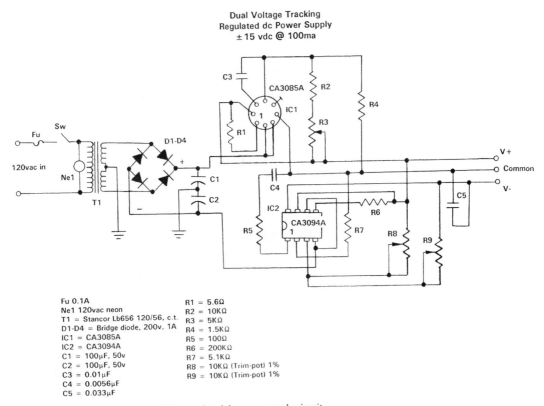

**FIGURE 6.128** Dual-voltage tracking regulated dc power supply circuit.

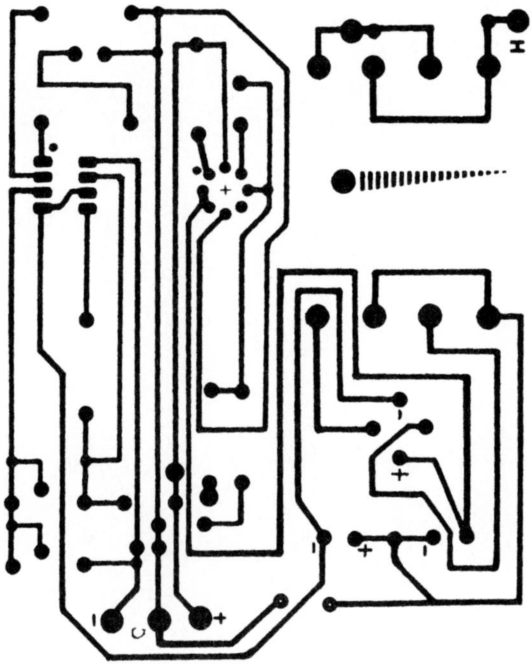

**FIGURE 6.129** Printed-circuit pattern for power supply shown in Fig. 6.128.

**FIGURE 6.130** Components mounted on PC board for power supply shown in Fig. 6.128.

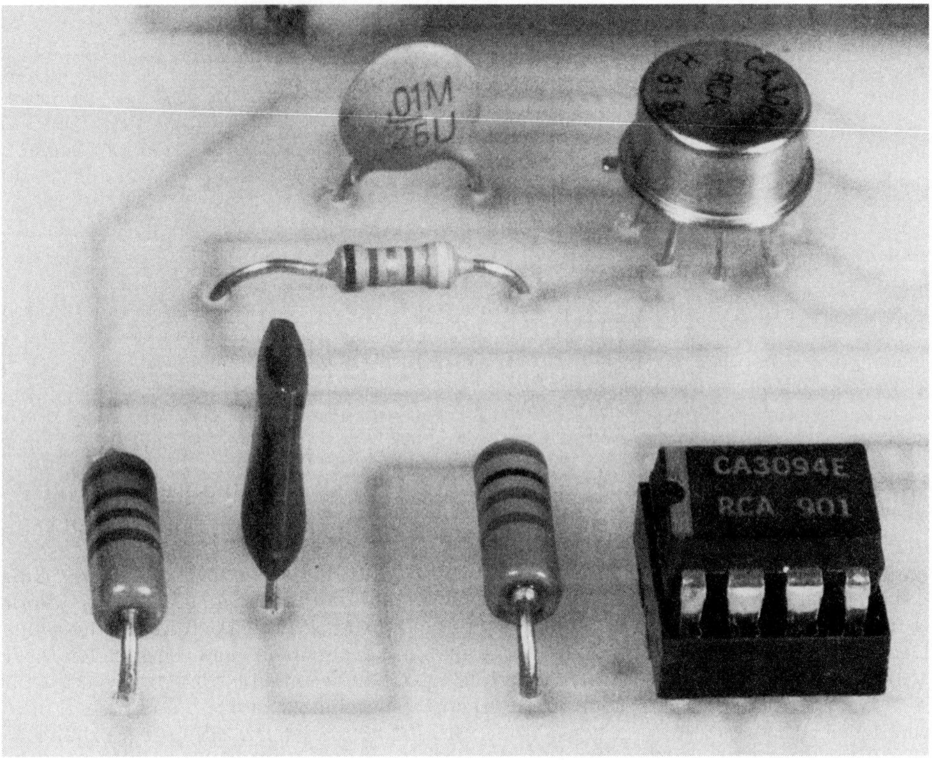


**FIGURE 6.131** Closeup view of components of circuit shown in Fig. 6.128.

***Digital IC Portable Thermometer.*** A digital IC thermometer circuit is shown in Figs. 6.132 and 6.133. This device uses the ICL7106CPL analog-to-digital converter to drive the liquid crystal display (LCD). The AD590JH transducer produces a variable current output in relation to its temperature, which is picked up and converted to a digital signal by the ICL7106CPL. The digital signal is then sent from the ICL7106CPL to the LCD unit, where the temperature may be directly read.

Figure 6.134 shows the AD590JH transducer converted to a probe for use in this device. Figure 6.135 is a foil pattern of the complete circuit. Figure 6.136 shows a photograph of the finished unit. Note the jumper wires that are used to complete the circuit so that a two-sided PC board would not need to be made for use in this device. The potentiometers used to calibrate the device to the proper temperature reading tolerance are shown in Fig. 6.136. The device is calibrated by placing the probe in ice to calibrate the lower temperature limit and in boiling water to calibrate the upper temperature limit. This device has an upper temperature limit of 199.9°F due to the limit of the 3½-digit LCD display. A larger digit display will allow the unit to indicate higher temperatures. The author uses this device to monitor the temperatures of photographic processing tanks used for film processing, where a maximum temperature of 130°F is sufficient.

***IC Schematics.*** The following schematic diagrams are for useful electronic device systems that have a broad range of use in industrial and consumer products. The values of the circuit components are shown so that the circuits may be easily constructed into a finished product.

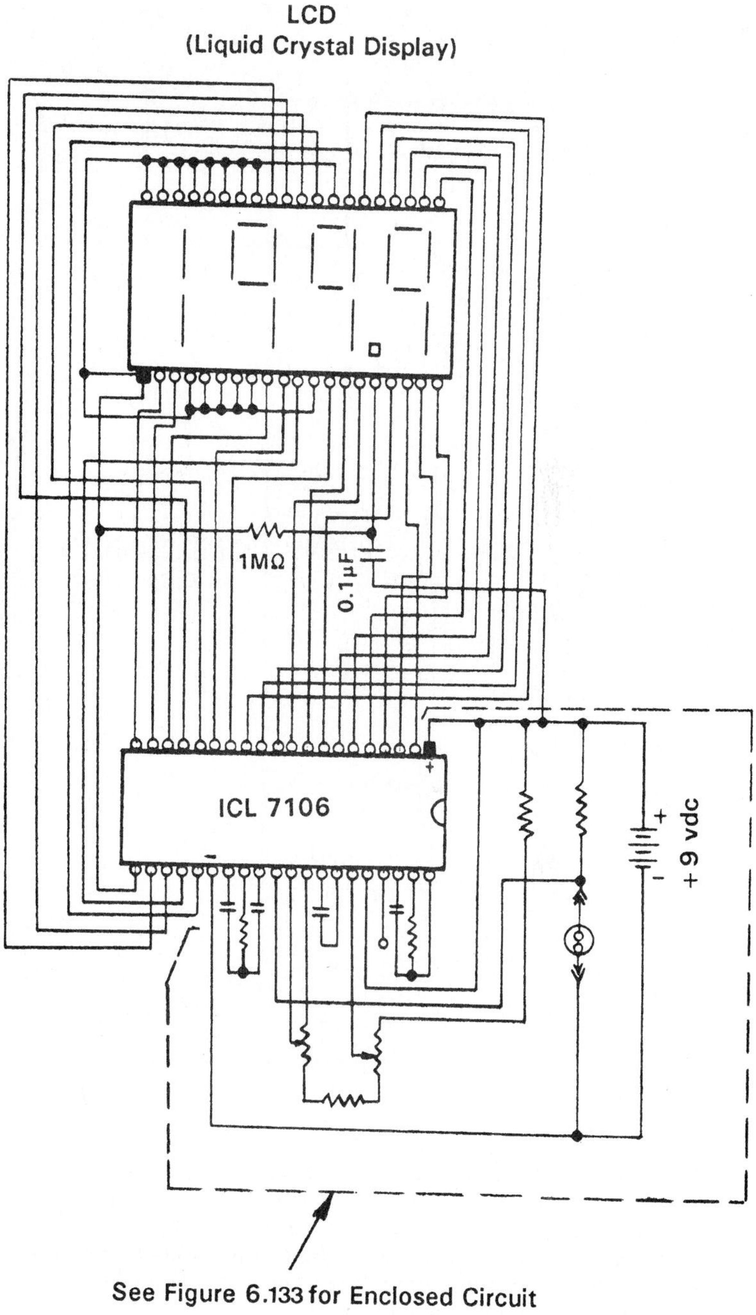


**FIGURE 6.132** Digital temperature indicating circuit using ICL7106CPL analog-to-digital converter.

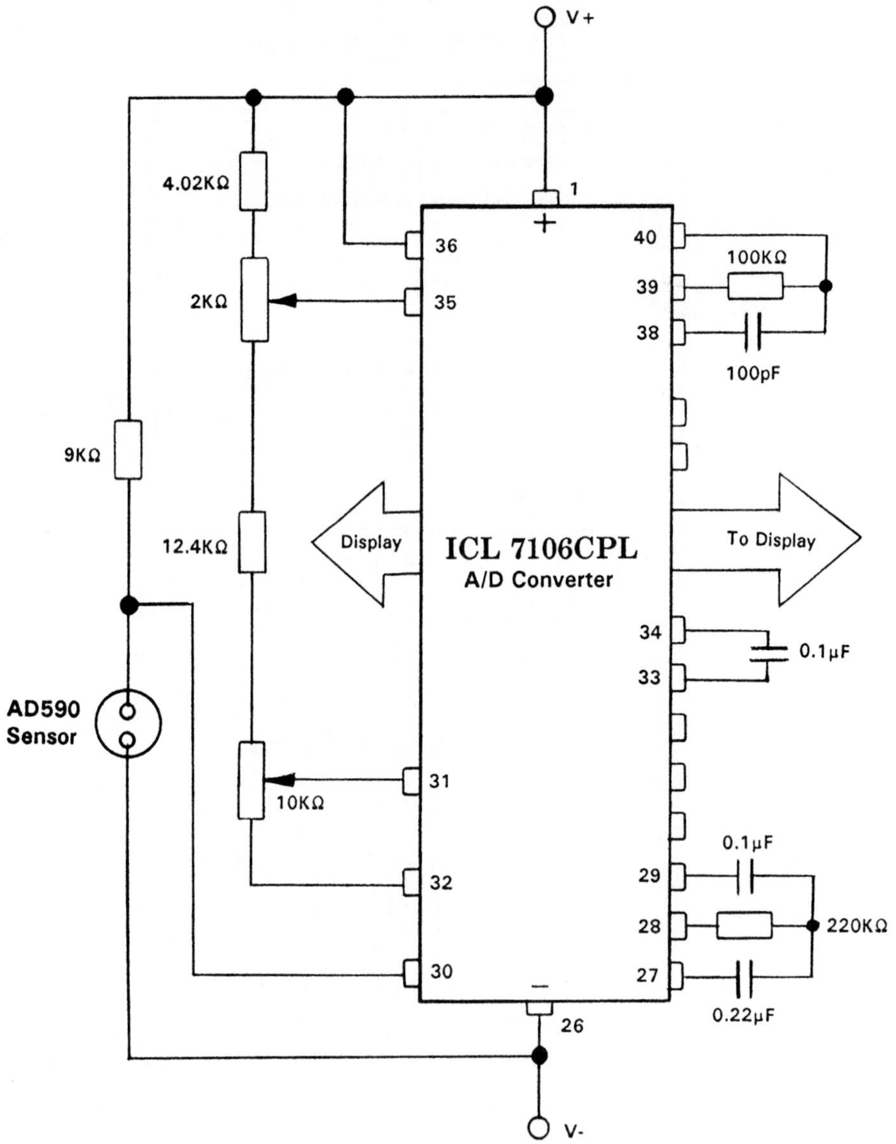


**FIGURE 6.133** Detailed circuit for ICL7106CPL shown in Fig. 6.132.

Figure 6.137 is the schematic for a popular IC regulated power supply rated 1.8 to 30 V dc at 5 A output. The output current rating may be increased by using another type of series-pass power transistor in the output stage.

Figure 6.138 is the schematic for a low-voltage regulator using the 723 IC.

Figure 6.139 is the schematic for an IC 386 amplifier with a gain of 200.

Figure 6.140 is the schematic of a typical operational amplifier.

Figures 6.141 and 6.142 are schematics for inverting and noninverting operational amplifiers, where the resistances may be calculated from the formulas shown on the diagrams.

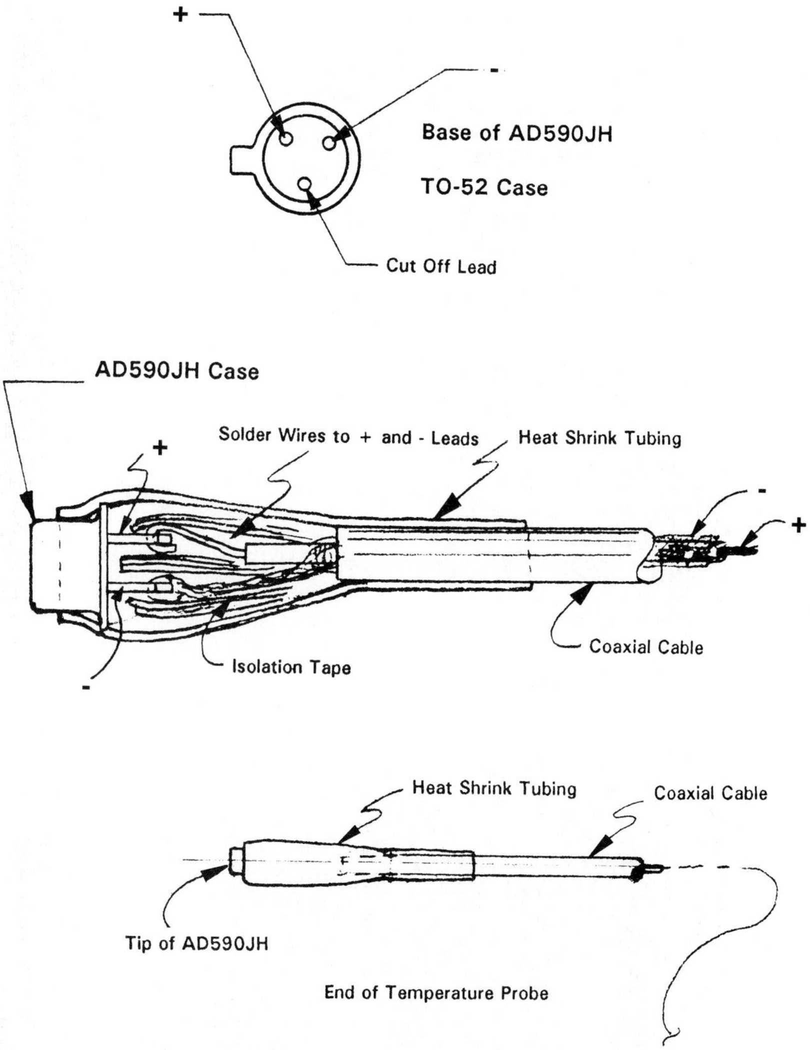


**FIGURE 6.134** Temperature transducer probe construction for Fig. 6.132.

Figure 6.143 is the schematic for a CA3098 timer used for the OFF mode.
Figure 6.144 is the schematic for a CA3098 timer used for the ON mode.
(*Note:* The CA3098 IC may have certain advantages over the 555 IC timer, which may be ascertained according to the particular application.)

***Application-Specific Integrated Circuits (ASICs).*** The application-specific IC is a custom-engineered device, designed for a specific application or function. ASICs are available from the large IC component manufacturers to meet a specific electronic design function in an economical manner. The component manufacturers will develop and produce these ICs when the purchased quantity justifies the manufacturing and special engineering costs involved in producing these devices.

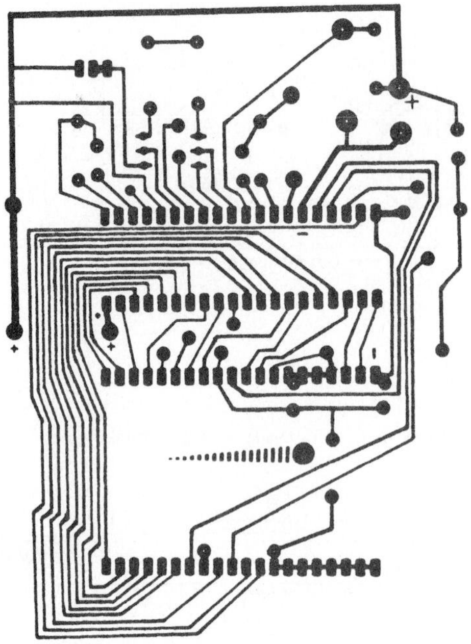

**FIGURE 6.135** Printed-circuit pattern for temperature indicating circuit shown in Fig. 6.132.

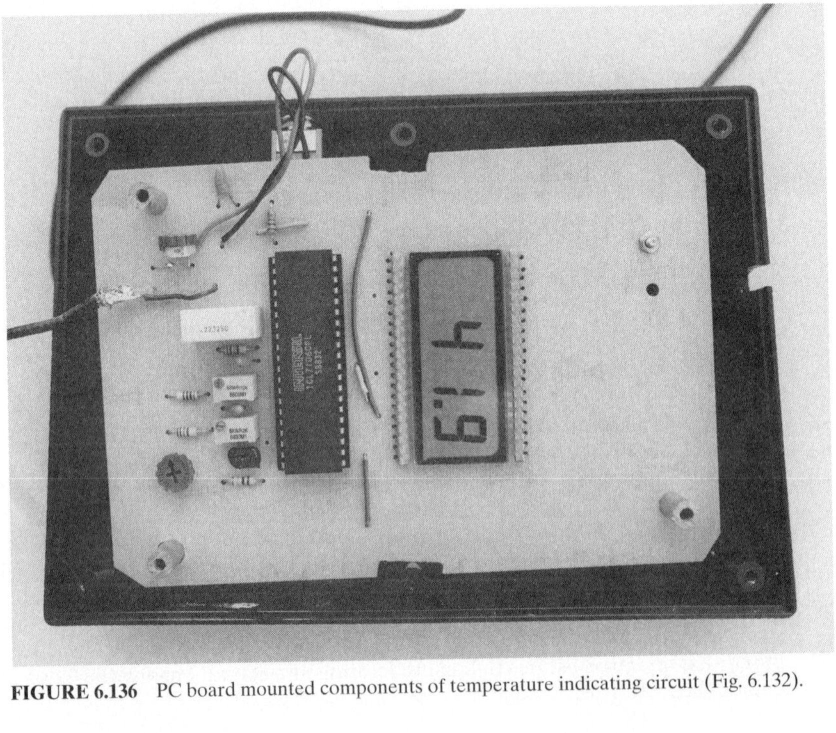

**FIGURE 6.136** PC board mounted components of temperature indicating circuit (Fig. 6.132).

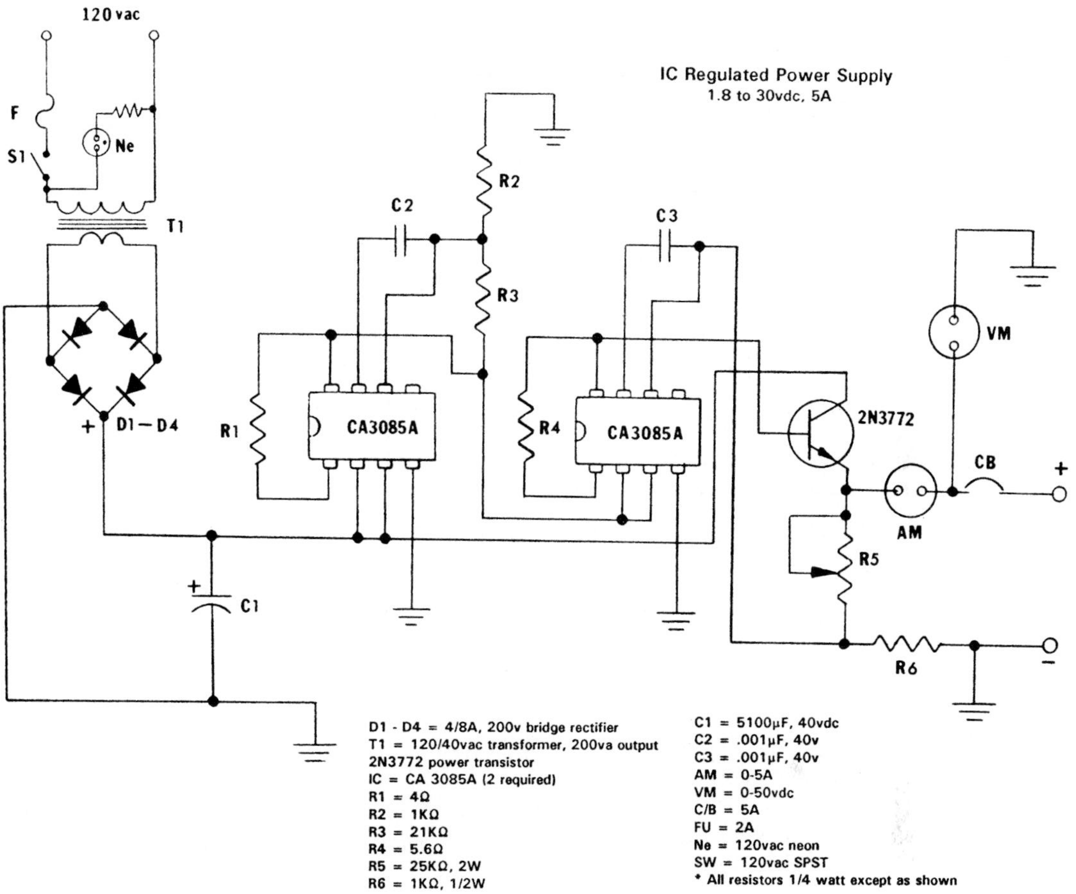


**FIGURE 6.137** IC regulated power supply circuit, 1.8 to 30 V dc, 5 A.

The ASIC is a custom-engineered unit made to meet special requirements of electronic system producers. ASICs are used in many devices such as electronic calculators, spelling and thesaurus devices, and a host of other special products where justification for the use of these component systems is indicated. Figure 6.145 is a typical example of a product PC board on which ASIC devices are installed.

***Breadboarding Electronic Circuits.*** The *breadboard* is a device used to lay out the electronic components and ICs with their associated wiring and interconnections so that the complete circuit may be tested and evaluated prior to manufacturing of the system. Complete systems or subsystem circuits are constructed and wired on the breadboard, where they are tested and changes are made to correct the circuit for proper operation. A typical large breadboard is shown in Fig. 6.146, where a relatively large and complex circuit may be layed out and tested for evaluation. Jumper-wire sets are available for interconnecting the electronic components that make up the circuit. Different voltage power supplies may be fed onto the breadboard shown in the figure. This breadboard has isolated sections that may be interwired or kept isolated from one another.

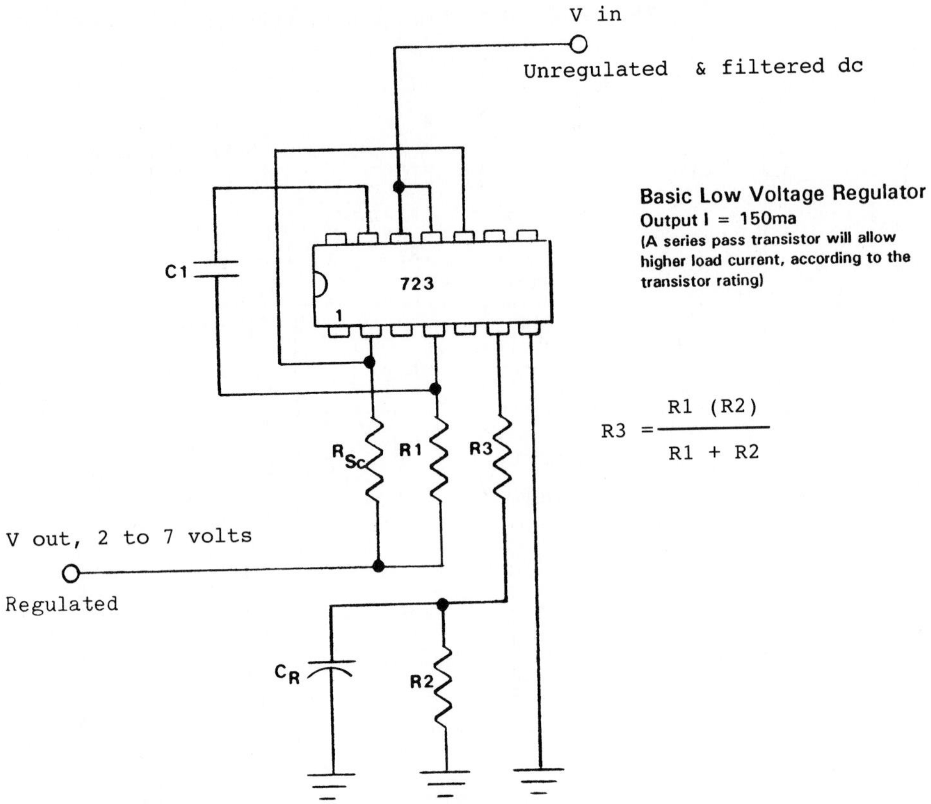


**FIGURE 6.138** Basic low-voltage regulator using 723 IC.

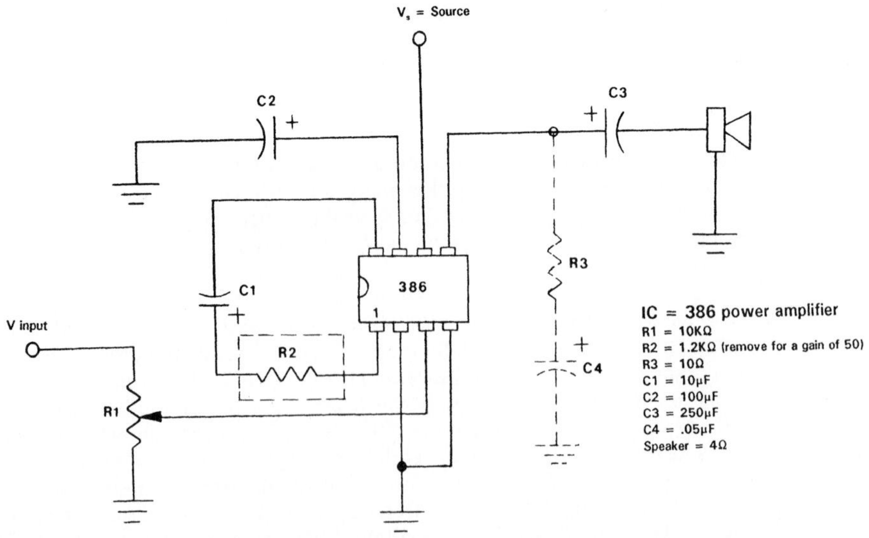


**FIGURE 6.139** Amplifier circuit using 386 IC.

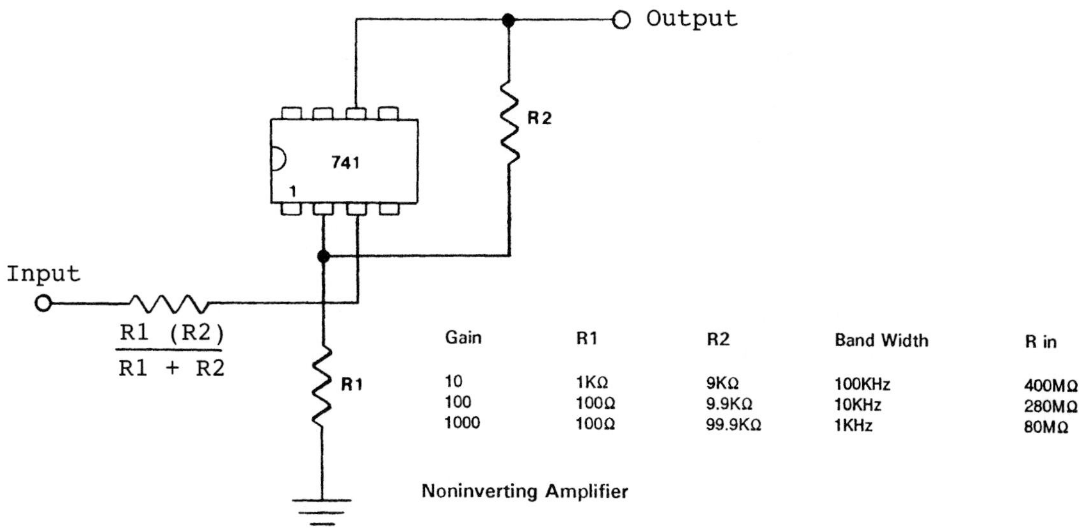


| Gain | R1 | R2 | Band Width | R in |
|---|---|---|---|---|
| 10 | 1KΩ | 9KΩ | 100KHz | 400MΩ |
| 100 | 100Ω | 9.9KΩ | 10KHz | 280MΩ |
| 1000 | 100Ω | 99.9KΩ | 1KHz | 80MΩ |

**FIGURE 6.140** Basic operational-amplifier circuit.

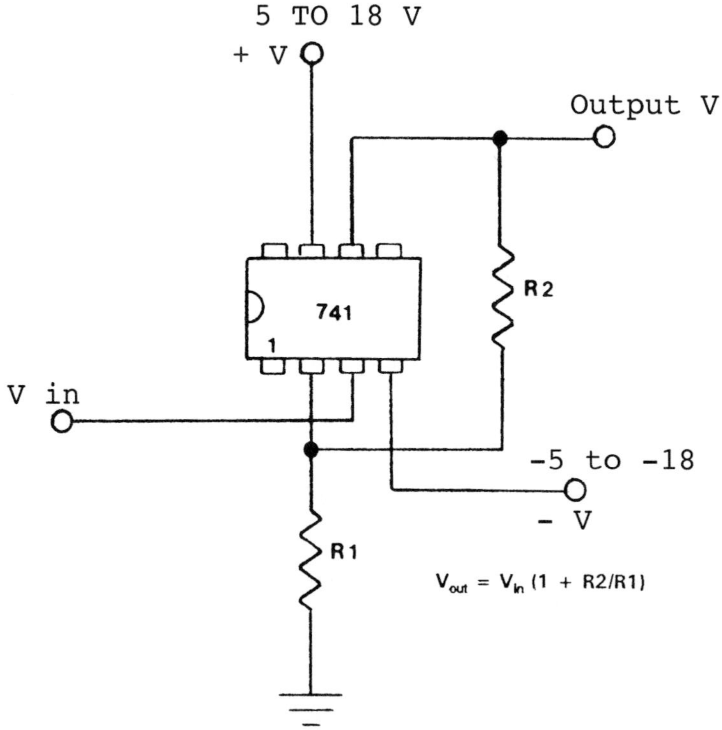


**FIGURE 6.141** Noninverting op-amp circuit.

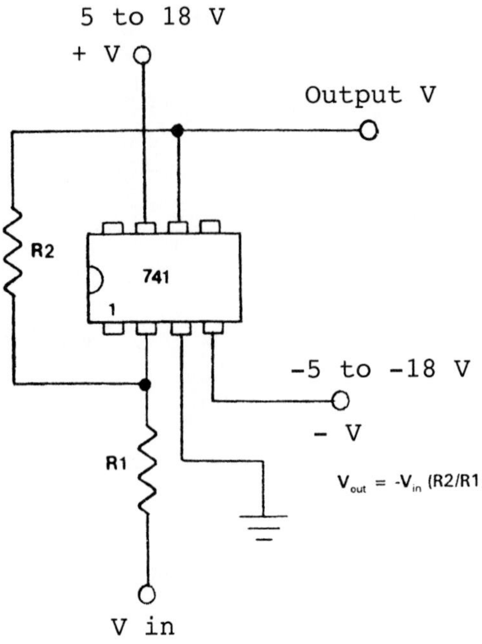


**FIGURE 6.142** Inverting op-amp circuit.

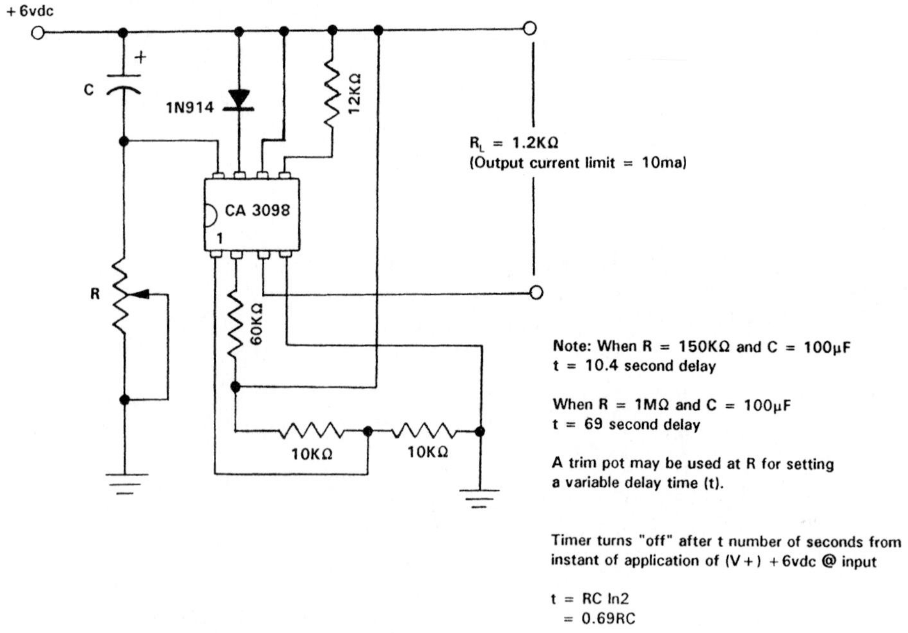


**FIGURE 6.143** CA9038 IC timer used for OFF mode.

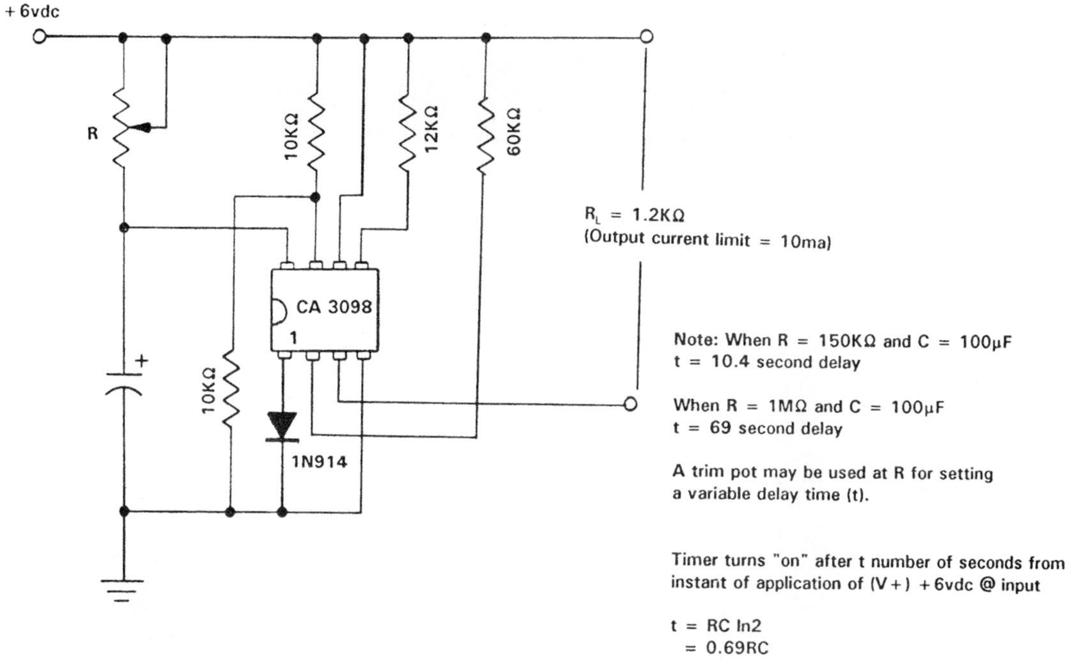


**FIGURE 6.144** CA9038 IC timer used for ON mode.

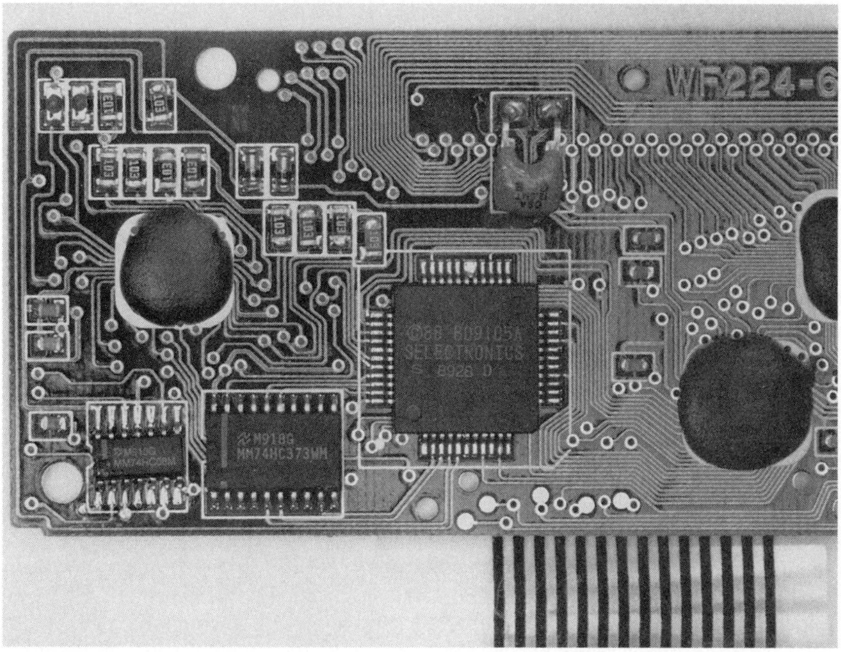

**FIGURE 6.145** Closeup view of ASICs and surface-mount components.

**FIGURE 6.146** A standard breadboard for test wiring of circuits.

Note that the components shown in Fig. 6.145 are of the surface-mount type. Surface-mount components do not have lead wires, which are typical of standard electronic components. The surface-mount component is soldered directly onto the printed circuit board after being bonded in place.

***Surface-Mount Components.*** The surface-mount electronic component was developed some years ago to allow smaller PC boards to be used in a final product. The surface-mount component is also well adapted for robotic or automatic placement and soldering of the components for high-speed production.

Figure 6.147*a* and 6.147*b* show two types of surface-mount components: the resistor (*a*) and capacitor (*b*). Other components such as inductors (coils, chokes, transformers, etc.) are also produced in surface-mount configurations.

An excellent example of the use of surface-mount components can be seen in videocassette recorders (VCRs), camcorders, and photographic cameras, where their small size and economic production allow such complex devices to be economically produced in small sizes, with high reliability.

Figure 6.148 shows the surface-mount component styles, while Fig. 6.149*a* and 6.149*b* show the conventional IC with its plug-in socket (*a*) and a surface-mount IC component (*b*).

Specialized books on surface-mount technology are available today for use by the electronic and electrical and electromechanical design engineer.

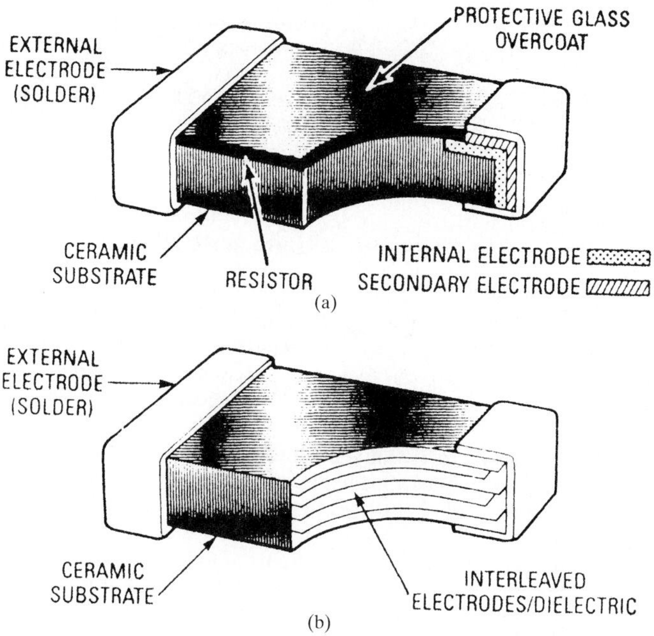


**FIGURE 6.147** (*a*) Surface-mount resistor; (*b*) surface-mount capacitor.

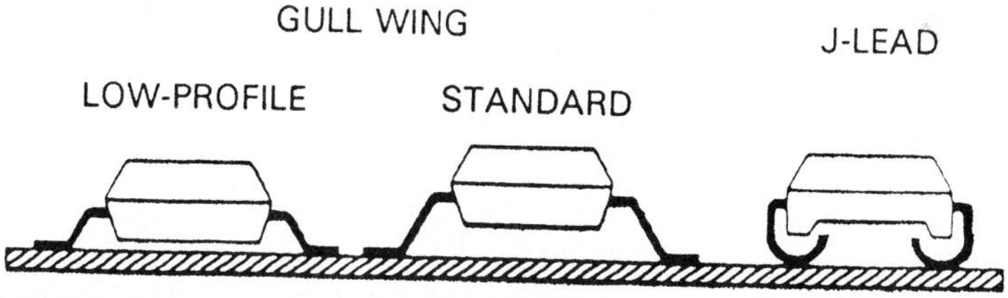


**FIGURE 6.148** Surface-mount component styles.

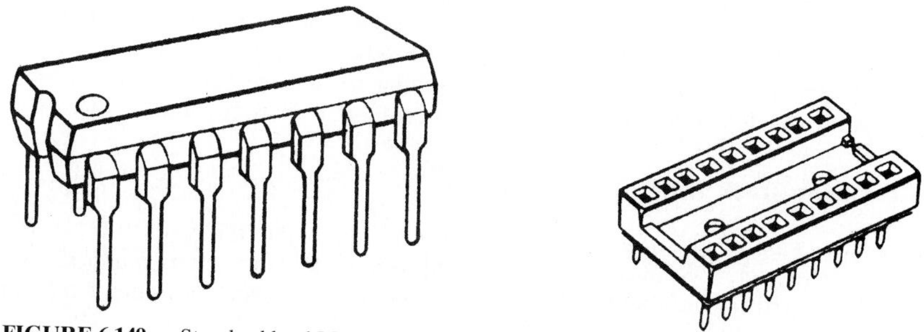

**FIGURE 6.149*a*** Standard lead IC with PC mount socket.

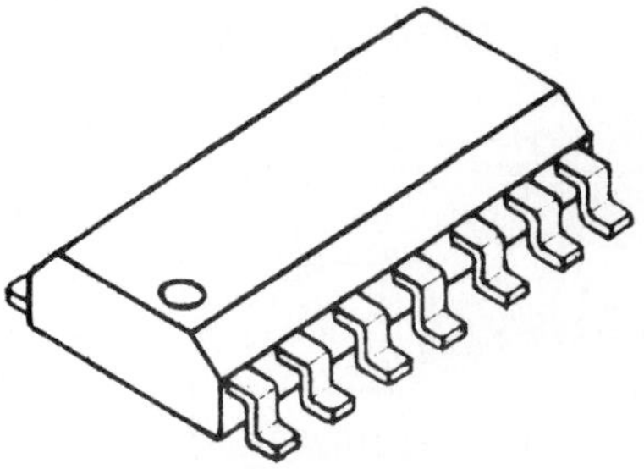

**FIGURE 6.149*b*** Surface-mount IC.

### 6.9.10 Solid-State Semiconductor and IC Package Identification

See Fig. 6.150*a* through 6.150*d* for semiconductor and IC package styles and identifications. Also refer to Figure 6.117.

### 6.9.11 Electronic Component Data and IC Circuit Manuals

A vast amount of circuit design data is available from the large producers of semiconductors, electronic components, and IC components. Circuit design data and circuit analysis techniques are contained in most of the books listed in the bibliography at the end of this chapter. Books of electronic circuits and applications are also listed.

Some of the large manufacturers of semiconductor devices and ICs, who produce reference manuals and bulletins/publication sheets for circuit applications, include General Electric Co., Harris, Motorola, Texas Instruments, Sylvania, National Semiconductor, Intersil, Fairchild, and Raytheon.

It is a wise decision to obtain the circuit manuals of the large IC and component producers in lieu of using other published works on practical circuits. The author has found that in many cases the circuits shown in some of the non-industry-published works simply do not function. The schematic diagrams are sometimes incorrect, and many do not show the values of the electronic components which make up the circuit. A circuit which does not show component values is of little practical use or may need extensive development before it can be implemented.

For an IC cross-reference and other data, the *IC Master* index is also available (see Further Reading).

## 6.10 PRINTED-CIRCUIT BOARDS

### 6.10.1 Materials

Standard PC boards are generally fabricated from epoxy-fiberglass, phenolic fiber, or polyimide. The copper surface conductor is bonded to the surface of the board and is available in single- or double-sided form.

The copper conductor surface is available in the following standard weights and thicknesses: ½ oz, 0.00067 in; 1 oz, 0.00135 in; 2 oz, 0.0027 in.; and 3 oz, 0.004 in.

Table 6.26 shows the current-carrying capacity for different track widths in copper weights of ½, 1, and 2 oz. Note that the relationship between track width and current capacity is not strictly linear, but is close enough to being so that we can interpolate between the values shown to determine track widths for intermediate current values.

### 6.10.2 Determining Track Width on Copper PCs

This example illustrates the use of the current-capacity table.

*Example.* What track width should be used to carry a maximum of 8.5 A on a 1-oz-clad PC board if you allow a temperature rise of 30°C over an ambient temperature of 25°C? Charted values are

7 A with 30°C rise = 0.100-in track width

10 A with 30°C rise = 0.150-in track width

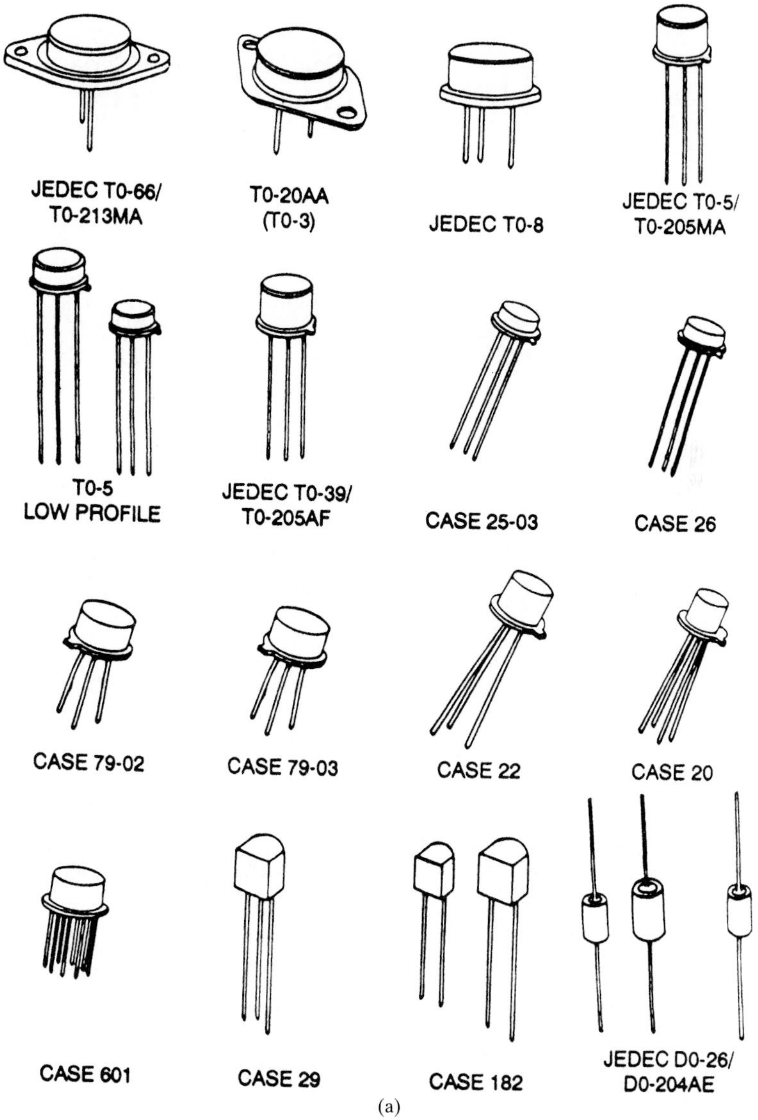


(a)

**FIGURE 6.150** (*a–d*) Electronic component package styles—ICs, diodes, and transistors.

Interpolating

$$\begin{array}{llcl} & 7 \ldots\ldots 0.100 & & \\ 1.5 & 8.5 \ldots\ldots \quad x & & x \\ 3.0 & 10 \ldots\ldots 0.150 & & 0.050 \end{array}$$

By proportion

$$\frac{1.5}{3} = \frac{x}{0.050} \qquad 3x = 0.075 \qquad x = 0.025 = \text{increment to be added to } 0.100$$

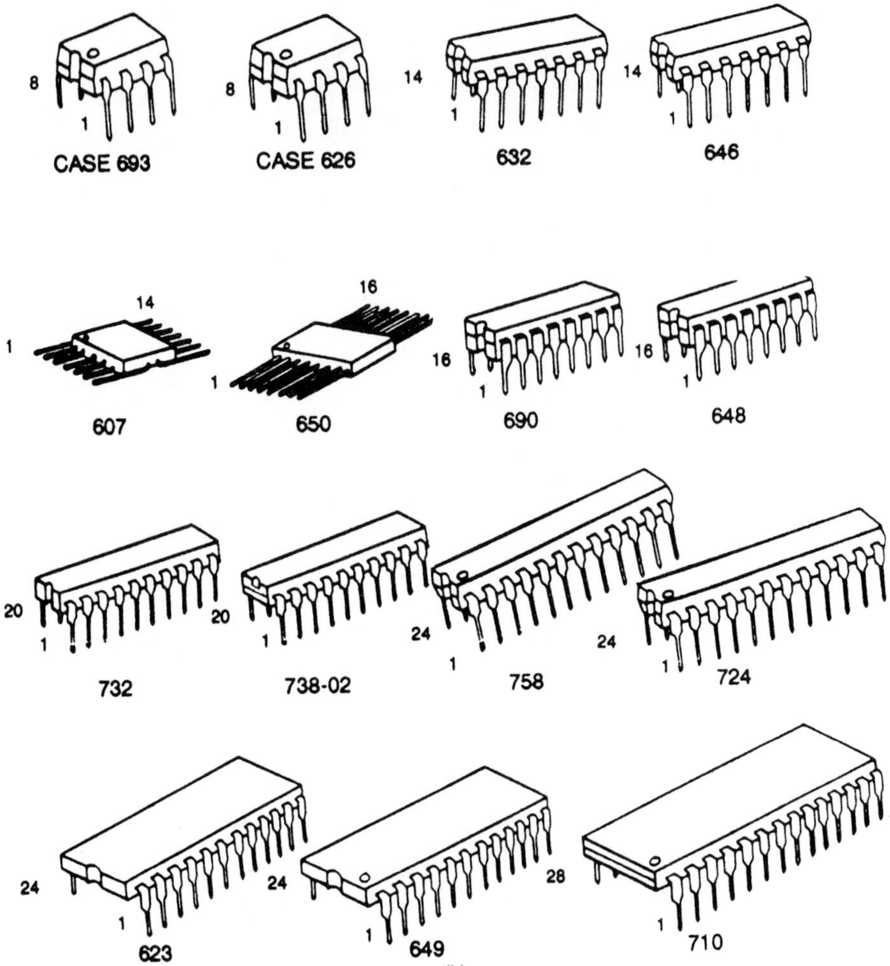

(b)

**FIGURE 6.150** (*Continued*)

So 0.100 + 0.025 = 0.125. This is the desired track width required to carry 8.5 A with a 30°C rise.

(*Design note:* A track width of 0.022 in on a real circuit board would be rather difficult to maintain consistently. In practice, when doing the graphic art work for the foil conductor pattern, select a standard width pen or press-on line that in all cases should be at least 10 percent wider that the estimate to allow for etchant undercut when the PC board is etched during fabrication. As a last check prior to production of the boards, a current-heat test should be made on the board to verify its performance.)

### 6.10.3 Prototype Fabrication of PC Boards

To prepare a PC board for prototype use or for small production runs, the following outline may be used. Instructions are given in this and the following section for the production of "quality" PC boards.

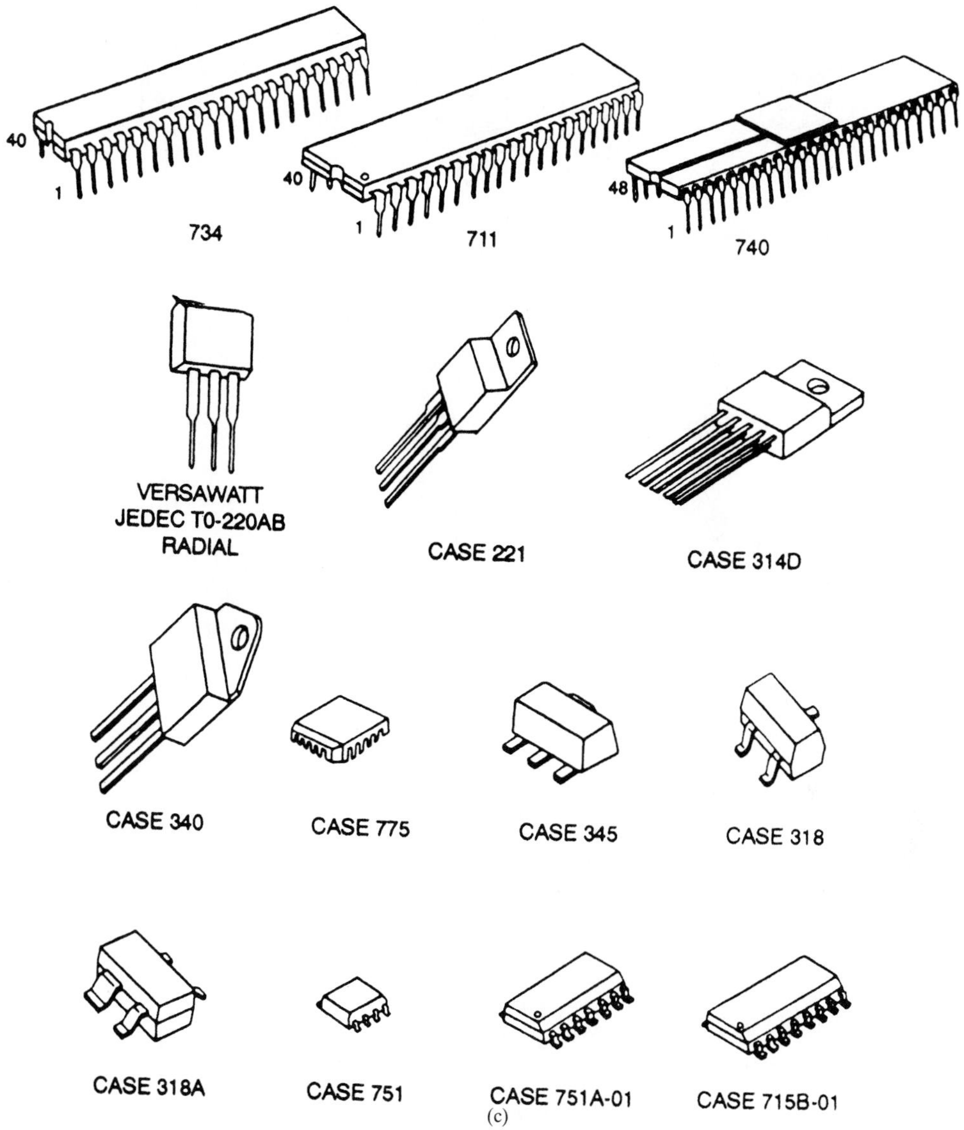

(c)

**FIGURE 6.150** (*Continued*)

The principal steps in PC board fabrication are

1. Select the proper type of material and weight of copper cladding to handle the current-carrying requirements of your circuit (see Sec. 6.10.2).
2. Have your graphic art work of the PC-board conductor pattern complete, checked, and ready for photographing.
3. Photograph the conductor pattern for negative production.
4. Coat the blank PC board with photo-resist and prebake at 175°F for 6 min. Cool and store away from light.

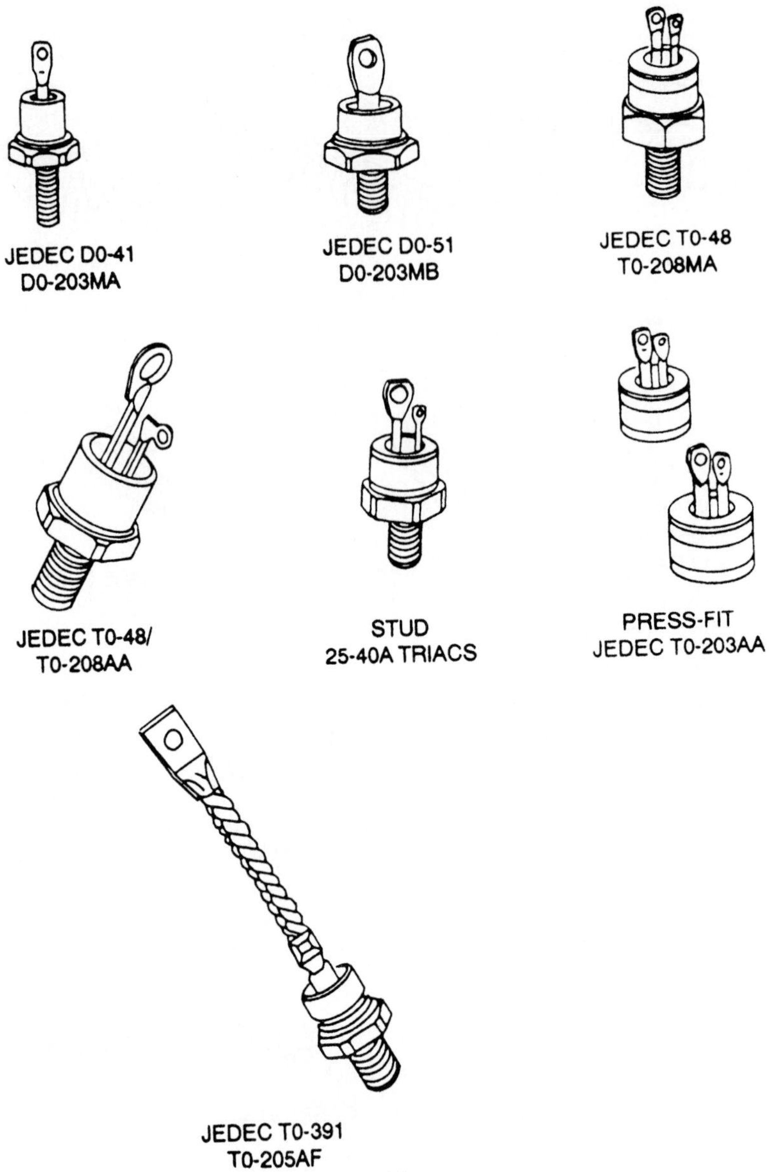


(d)

**FIGURE 6.150** *(Continued)*

5. Expose the sensitized PC board under ultraviolet light from the negative (use a clamping frame).
6. Develop the PC board in resist developer.
7. Apply resist dye to the surface and inspect the foil pattern for flaws. (Repair it with shellac "stopout" as required.)
8. Etch the developed board in a ferric chloride solution; wash it with clean water and dry.

**TABLE 6.26** Track Widths for PC Boards

| Current | Track Width, in | | | | | | | | |
|---|---|---|---|---|---|---|---|---|---|
| | ½ oz (0.00067 in) | | | 1 oz (0.00135 in) | | | 2 oz (0.0027 in) | | |
| Amps | 10°C | 30°C | 60°C | 10°C | 30°C | 60°C | 10°C | 30°C | 60°C |
| .125 | .003 | .003 | .003 | .002 | .002 | .002 | .001 | .001 | .001 |
| .200 | .005 | .005 | .005 | .004 | .004 | .004 | .003 | .003 | .003 |
| .500 | .015 | .010 | .007 | .008 | .006 | .004 | .005 | .004 | .004 |
| .750 | .030 | .020 | .016 | .015 | .010 | .007 | .008 | .006 | .006 |
| 1.00 | .040 | .030 | .025 | .025 | .015 | .010 | .012 | .008 | .008 |
| 1.50 | .065 | .040 | .034 | .030 | .020 | .013 | .015 | .010 | .011 |
| 2.00 | .090 | .050 | .042 | .045 | .025 | .016 | .022 | .013 | .013 |
| 3.00 | .125 | .060 | .050 | .060 | .030 | .020 | .030 | .017 | .015 |
| 5.00 | .250 | .125 | .095 | .130 | .060 | .040 | .070 | .034 | .025 |
| 7.00 | — | .220 | .150 | .220 | .100 | .070 | .120 | .060 | .035 |
| 10.0 | — | .300 | .200 | .340 | .150 | .090 | .175 | .080 | .050 |
| 15.0 | — | — | .375 | — | .260 | .175 | .300 | .140 | .090 |
| 20.0 | — | — | — | — | .375 | .250 | — | .200 | .140 |

9. Remove the photo-resist left on the PC board with paint remover.
10. Wash the board in warm water and air-dry it.
11. Trim the board to size if required and drill all component mounting holes. Deburr the board of sharp edges.
12. Plate the finished board with the desired metal (tin, etc.).

### 6.10.3.1 Details of PC-Board Production

Table 6.27 lists materials required for PC-board production. A step-by-step description of the process is given here.

1. Assuming the PC-board artwork is complete, begin by either photographing the pattern or direct negative production from the pattern by contact exposing. The artwork is assumed to be positive. With Kodalith film, the pattern may be photographed directly using two 500-W photoflood lamps in reflectors, with the lights at a 45° angle and approximately 42 to 46 in from the copy (see Fig. 6.151). The aperture on the lens should be set to *f* 8 and the exposure time should be 6 to 8 sec. The lens is assumed to be 210-mm focal length. For contact printing, an exposure of 2 to 3 sec with a 100-W lamp at 3 to 5 ft distance should be tried.
2. Develop the film in Kodalith developer (either fine-line or standard Kodalith, catalog no. 146 5152) by agitating the film according to the instructions given in the film pack. Use the Kodak gray-scale exposure guide for optimum results.
3. Stop bath and fix the film per instructions for Kodalith film.
4. Wash the negative in cold, running water for 10 min. Then dip the negative in "Foto-Flo" solution for 10 sec and hang to dry. If the negative is cloudy or overexposed, it may be cleared using Kodak Ceric Sulfate Dot Etching Solution R-20. When dry, touch up the negative with opaque as required.
5. Clean the PC board in a 10% aqueous solution of glacial acetic acid. Use powdered pumice or other mild abrasive. Redip the board in the acetic acid solution and then rinse in cold running water. Dry with paper towel and remove lint from the copper surface.

**TABLE 6.27** Materials for PC-Board Fabrication

| Material | Description/Quantity |
|---|---|
| Photoresist | Kodak "KPR" solution, cat. no. 189 2074 |
| "KPR" developer | Photo resist developer, cat. 176 3572 |
| "KPR" dye | Photoresist dye (1 gallon) |
| Kodalith ortho film | Graphic arts film, type 3 #2556 × (size) |
| Kodalith "fine-line" | Film developer, part A and part B |
| Ferric chloride | (Anhydrous), 5 pounds $FeCl_3$ |
| Kodak rapid fixer | Plate film fixer (1 gallon) |
| Kodak stop bath | Indicator stop bath, cat. no. 140 8731 |
| Kodak "foto-flo" | Film-wetting solution |
| Kodak "gray scale" | Exposure guide, cat. no. 131 4319 |
| Negative "opaque" | Negative touchup compound |
| Red safe light | Type 1A for orthographic film |
| Yellow safe light | 40 to 60-watt bug-repellent light |
| Electric oven | Min. 300°F with thermostat control |
| Oven thermometer | Commercial, ± 10%°F |
| Lab thermometer | Glass type, ± ½°F |
| Paint remover | Methylene chloride type, commercial |
| Distilled water | 3 gallons |
| Glacial acetic acid | 1 quart, commercial grade |
| Concen. hydrochloric acid | 1 quart, commercial grade |
| Ultra-violet lamp | Commercial sun lamp, 250 watts |
| Exposure frame | Clamping frame for PC board and negative |
| Trays, glass | 2 for "KPR" developer |
| Trays, plastic | 4 for film development and washing |
| Etching tank, glass | For ferric chloride etching |
| Miscellaneous | Tongs, hangers, storage bottles, etc. |
| Paper towels | Soft, absorbent type. |

**6.** Dip the cleaned PC board into an 8% aqueous solution of hydrochloric acid for 10 to 15 sec. Water-rinse and air-dry the board.

**7.** Equip the darkroom with a 40- to 60-W yellow bug-repellent light and work in this light during the Kodak Photo Resist (KPR) stages to follow.

**8.** With the PC board held horizontal, pour a small amount of KPR resist on the copper surface and rock back and forth to cover completely the surface with KPR resist. Use the KPR resist full strength. Drain along the edge to a point, with the excess going back into the photo-resist bottle.

**9.** Set the PC board vertically and allow the photo-resist to drain onto a paper towel.

**10.** Leave the PC board to air-dry for 20 to 30 min (make sure that lint and dust particles do not get on the drying KPR surface).

**11.** Place the PC board in the electric oven (preheated to 175 to 180°F) with the KPR surface up, and bake for 6 min. Remove and air-cool the board, then store it in a lightproof box. Boards sensitized in this way will keep for a period of one year or longer before use.

**12.** When ready to produce a PC board, place the PC-pattern negative against the sensitized surface and clamp it in a plate-glass frame. Expose the board for 7 min to an ultraviolet sunlamp (250 W), at 33 to 36 in from the PC-board surface.

**13.** Develop the board in a glass or stainless-steel tray with the KPR developer for 30 sec with a rocking-motion agitation. Transfer the board to another tray with fresh KPR developer for another 30 sec with a rocking motion. Remove the board from the developer and drain the board vertically on edge until it dries (10 to 20 min). You may dye the surface of the resist when it is removed from the developer, before it hardens. To do this,

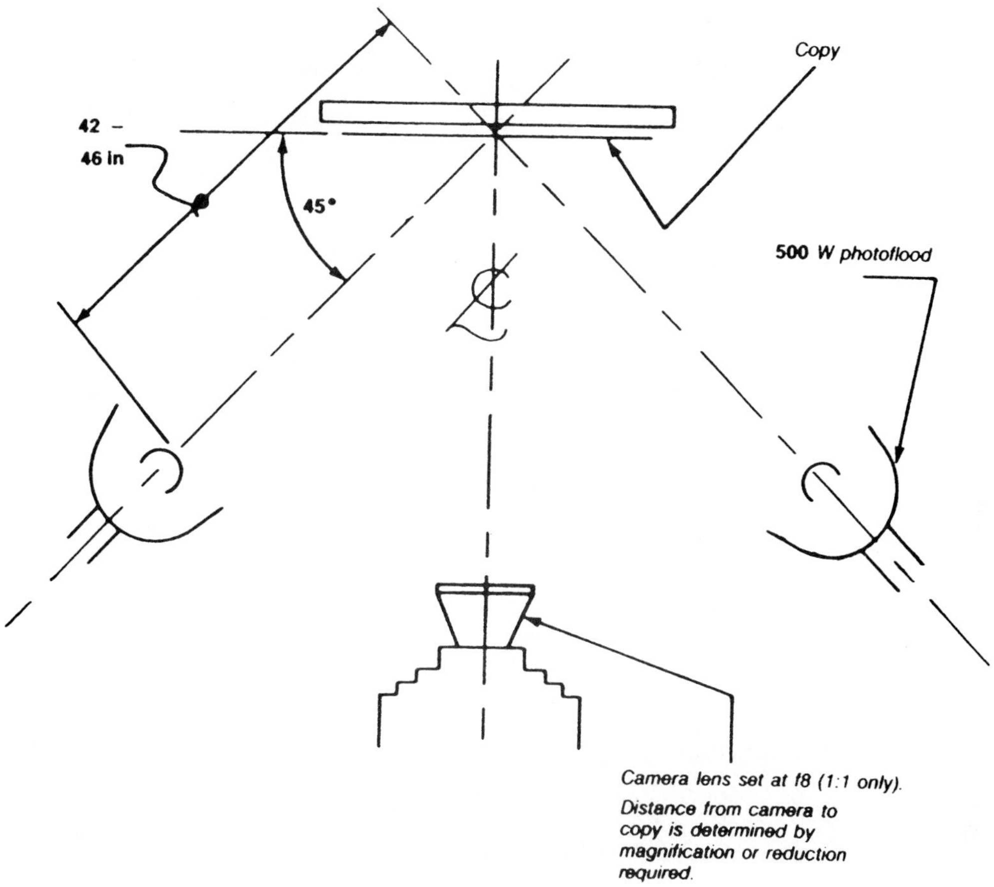


**FIGURE 6.151** Camera setup for PC-board negative production.

pour the KPR dye onto the surface and let it set for 15 sec; follow with a gentle water rinse. In this manner, you may inspect the resist coating for flaws more easily.

14. Place the developed, air-dried board in the electric oven (preheated to 240°F) for 6 min. Remove the board and cool to room temperature by laying it horizontally on a paper surface. Do not force-cool the board.
15. The PC board is now ready for etching (stripping off the unwanted copper).
16. Make a 30 to 35°Bé aqueous solution of ferric chloride. Use rubber gloves.
17. Submerge the developed, postbaked PC board in the ferric chloride solution and gently agitate for approximately 20 min for ½-oz boards, or more until all traces of unwanted copper are etched away. *Do not* overetch the board, or else undercutting and ragged edges will result on the copper foil conductor lines. Wash in warm running water and set aside to dry.
18. Heavily coat the etched, dry PC board with methylene chloride paint remover and allow to set for 10 min. Wash the board under warm running water and air-dry or wipe with a paper towel.

After this stage is complete, the board may then be trimmed and drilled for components. After the drilling, deburr the board of sharp edges and around the drilled holes. The finished PC board may then be tin-plated using the plating kits available from electronic-components distributors nationwide.

When a through-board connection is required, use eyelets or circuit-board plugs through the holes. In this way, through-hole plating may be eliminated for prototype work and also for small production runs (see Fig. 6.152).

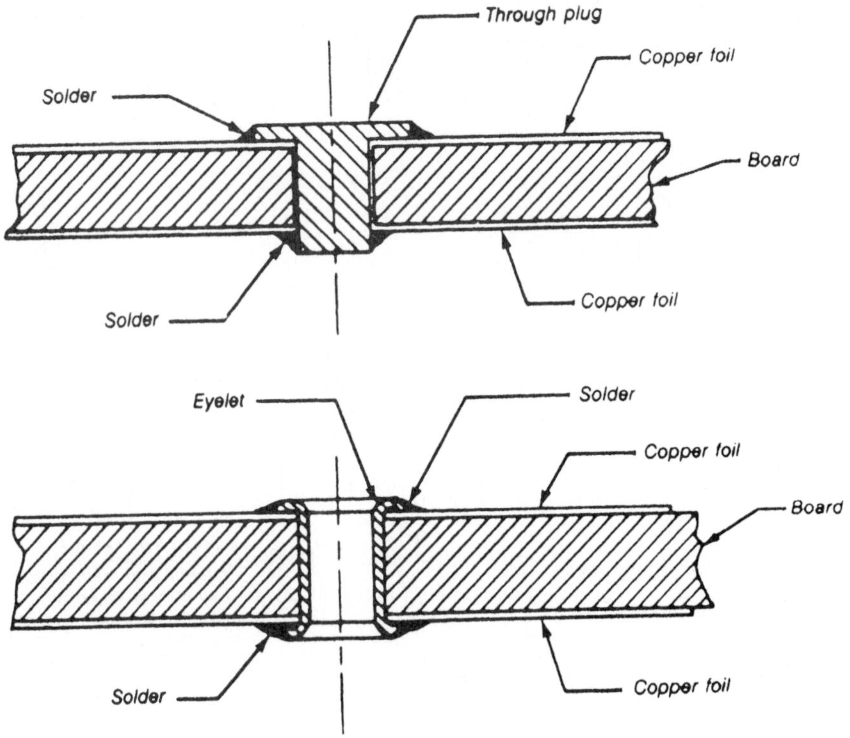


**FIGURE 6.152** Types of PC-board through connections.

(*Caution:* The chemicals involved in the PC-board production process shown here are corrosive and toxic. Use caution and follow the manufacturers' directions and warnings supplied with each material. The author and publisher assume no responsibility for injuries or damages incurred using the preceding PC-board processing instructions. The process described herein has been used many times by the author with excellent results.)

Figure 6.153 is the foil pattern for the double-sided PC board used for the special motor-control circuit shown in Fig. 6.123. That particular motor-control circuit was developed by the author to drive a ⅛-hp dc motor for electrically operating a high-power load interrupter switch for switchgear equipment.

Figure 6.154 shows the three epoxy-glass PC boards that were used for the IC circuits projects shown in the IC circuit section. The foil patterns show the bare copper foil, prior to being tin-plated.

Figure 6.155 is a high-quality PC board that was produced by Intersil for one of their ICL7106PCL analog-to-digital converter kits used to drive an LCD display. Note the sharp-

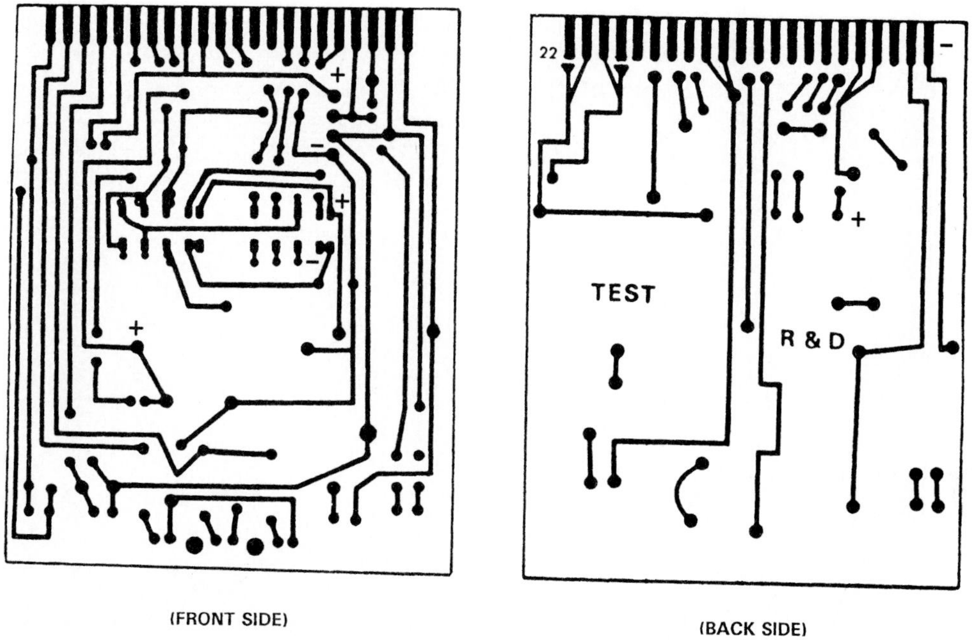


**FIGURE 6.153** Printed-circuit pattern for circuit shown in Fig. 6.123 (motor control).

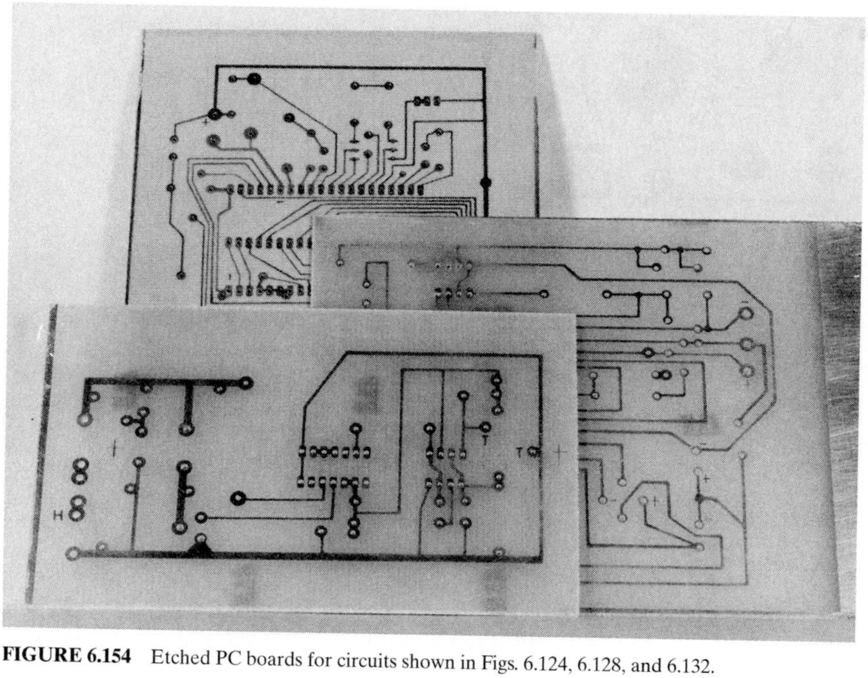

**FIGURE 6.154** Etched PC boards for circuits shown in Figs. 6.124, 6.128, and 6.132.

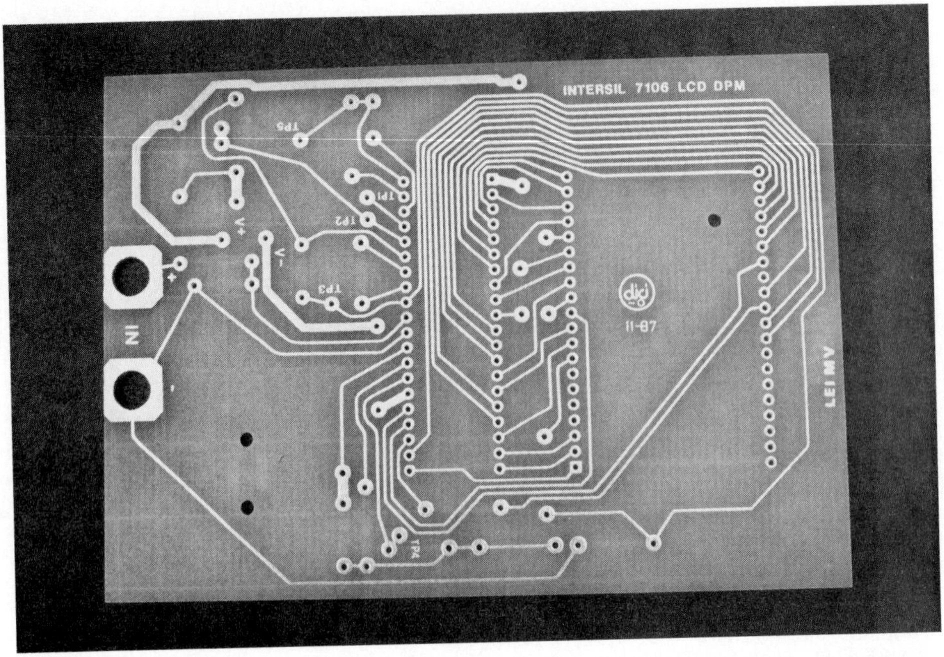


**FIGURE 6.155** A high-quality PC board, etched and tin-plated.

ness and consistency of the foil lines, indicating a quality PC board. The PC-board material is epoxy-fiberglass, type G-10, which is a high-quality PC board material.

Figure 6.156 shows two different PC boards connected with a flexible circuit. The board on the left is a ceramic (porcelain) PC board on which are attached four trim-pots and printed resistors. This PC-board assembly was taken from a K-2 camera produced by one of the leading Japanese camera manufacturers. The camera was produced more than 15 years ago and shows the advanced techniques and high-quality electronic products that were produced at that early time period prior to the advent of the integrated circuit, which was invented in the United States and quickly exploited by every major country in the world.

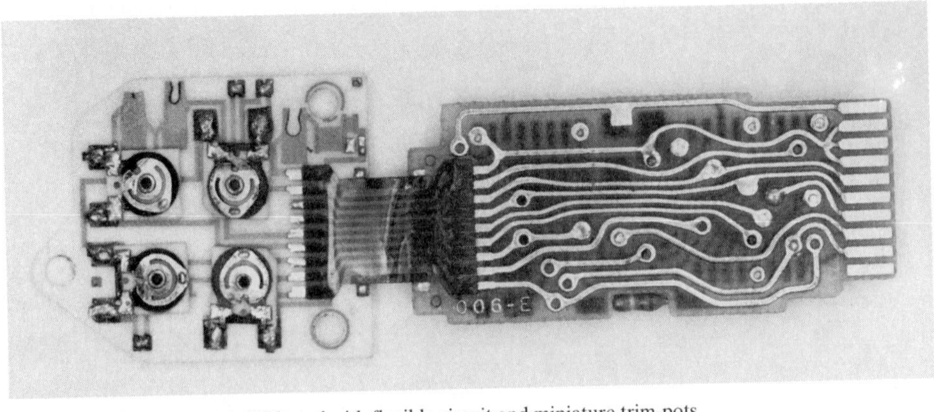

**FIGURE 6.156** A ceramic PC board with flexible circuit and miniature trim-pots.

## 6.11 ELECTROCHEMICAL BATTERY SYSTEMS

There are many types of electrochemical batteries used today. The most common forms are

- Carbon-zinc
- Alkaline
- Zinc chloride
- Lithium
- Nickel-cadmium, sealed and wet
- Lead-acid, sealed and wet
- Mercury
- Silver-oxide, silver-cadmium
- Lead-calcium
- Nickel-iron, Edison cell

### 6.11.1 Battery Design Applications

Select the proper electrochemical battery by this procedure.

1. Determine the voltage requirement and select a battery series that may be suitable for your application.
2. Determine the initial drain in amperes or milliamperes required for your application.
3. Estimate the operating-time schedule (i.e., continuous service or daily service periods of 4 h, 8 h, etc.).
4. Determine the closed-circuit endpoint voltage below which the circuit will not operate.
5. Select a battery from the voltage group that appears to meet the requirements (size, weight, etc.).
6. Refer to the service-life graphs for your selected battery. These graphs are available from the battery manufacturers (see Fig. 6.157). From these graphs, determine the number of hours of continuous service provided at a specific drain in milliamperes.

*Example.* For a 1.5-V application, the service-life curve for an end voltage of 1.1 V per cell can be used. Here, a continuous 20-mA drain will provide a service life of approximately 200 h. Curves are also available for service of 4 h per day. In the case of this cell, 275 h of service life can be expected at the 4-hr/day drain rate. Longer service life is provided when cells are arranged in parallel.

Using two cells from the graph shown in Fig. 6.157 will provide $2 \times 200 = 400$ h of continuous service, or $2 \times 275 = 550$ h at a 4-hr/day drain of 20 mA. Two cells in series would provide 3.0 V with the service life of one cell, or 200 h of continuous service.

***Rechargeable Nickel-Cadmium Batteries.*** These types of batteries are hermetically sealed, requiring no maintenance except recharging when their stored energy is depleted. Used in many consumer products and in industrial equipment, these batteries have numerous qualities to recommend them, including:

- Rechargeability
- Near-constant discharge
- Excellent charge retention

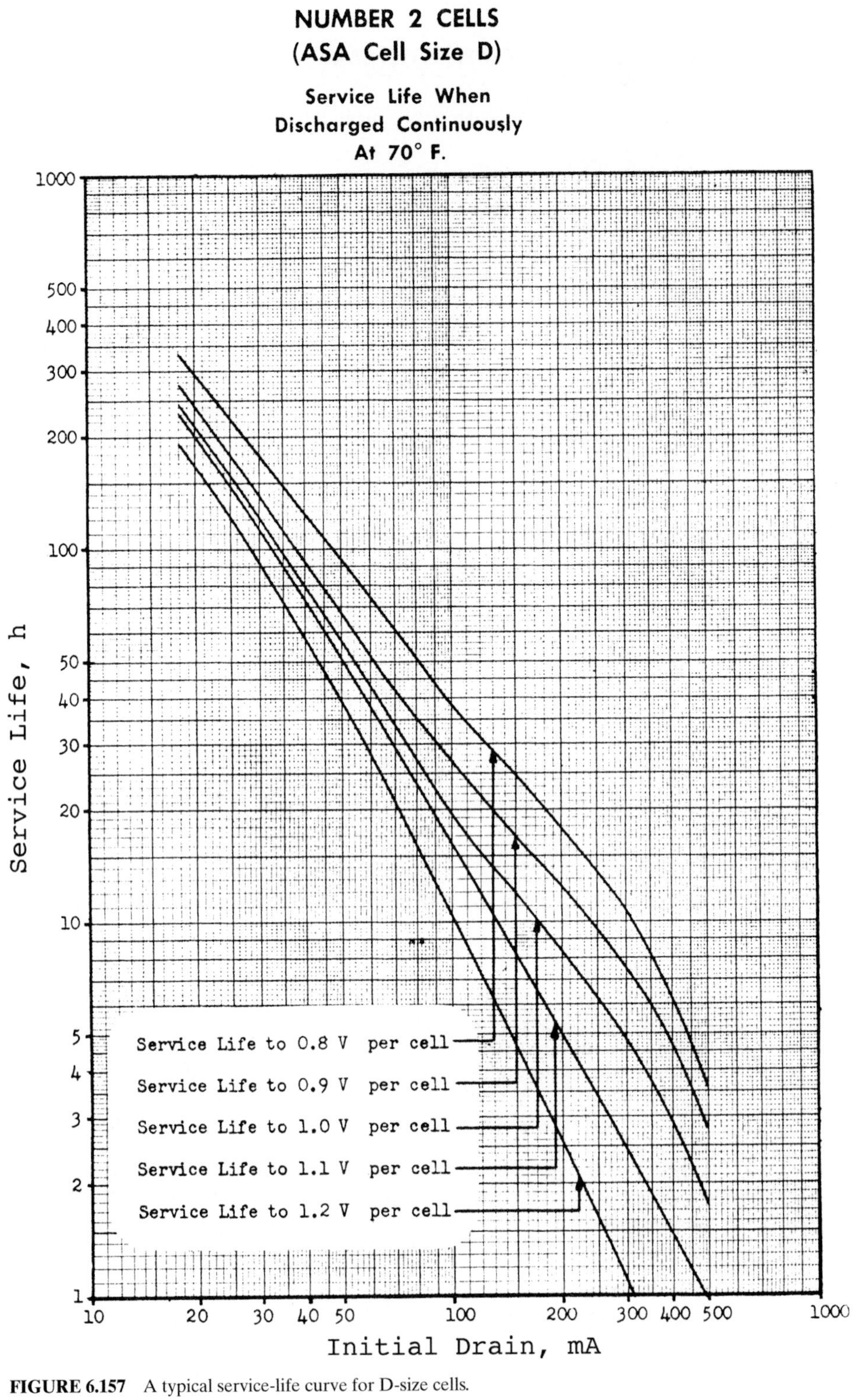


**FIGURE 6.157** A typical service-life curve for D-size cells.

- Low cost per use-hour
- No corrosive fumes
- Good operation at low temperatures
- Rugged construction

The size of sealed nickel-cadmium cells is expressed in terms of a nominal mAh (milliampere hour) service life, such as 100- or 450-mAh size.

***Expressing Rate of Discharge.*** The mAh capacity depends on the rate of discharge. This is usually taken as the capacity obtained when the battery is discharged to its endpoint voltage (≈1.1 V) in 10 h.

This nominal capacity is represented by *C;* the current required for the discharge is then *C*/10, known as the *standard rate of discharge.* Other rates of discharge are expressed as follows:

*C*/5 = 2 times the standard rate (current)
*C*/2.5 = 4 times the standard rate (current)
*C*/2 = 5 times the standard rate (current)
*C*/1 = 10 times the standard rate (current)
*C*/0.5 = 20 times the standard rate (current)

***Temperature Characteristics.*** Discharging of nickel-cadmium cells in relation to temperature is approximated here.

| Discharge temperature, °F | % of 70°F discharge |
|---|---|
| 113* | 93 |
| 70 | 100 |
| 40 | 93 |
| 28 | 88 |
| –4* | 60 |

The asterisks adjacent to these temperature limits indicate that they should not be exceeded.

***Estimating Service Life.*** Figure 6.158 shows service-life curves for nickel-cadmium cells.

*Example 1.* How long will a cell rated at 450 mAh discharge at 25 mA to 1.1 V at an ambient temperature of 32°F?

Capacity at 32°F = 0.88 × 450 = 396 mAh; the number 0.88 is from the discharge temperature chart.

$$\text{Standard rate (current)} = \frac{396}{10} = 39.6\text{ mA}$$

$$\frac{\text{Required rate, mA}}{\text{Standard rate, mA}} = \frac{25}{39.6} = 0.63 \times \text{standard rate}$$

Note that *C*/20 = 0.5 × standard rate. Referring to Fig. 6.158, we find that ≈*C*/20 is not on the service-life curve. So, to 1.1 V it will be ≥100 percent of the normal mAh capacity and 1.00 × 396 = 396 mAh. The result is 396/25 = 15.8 h of service, minimum, before recharging is required.

*Example 2.* What cell will be required to furnish 10 h of service at 75 mA to an endpoint voltage of 1.2 V at 30°F?

Milliampere output of the cell = 10 h × 75 mA = 750 mAh.

$$\frac{1}{0.88} \times 750 = 852\text{ mAh capacity required}$$

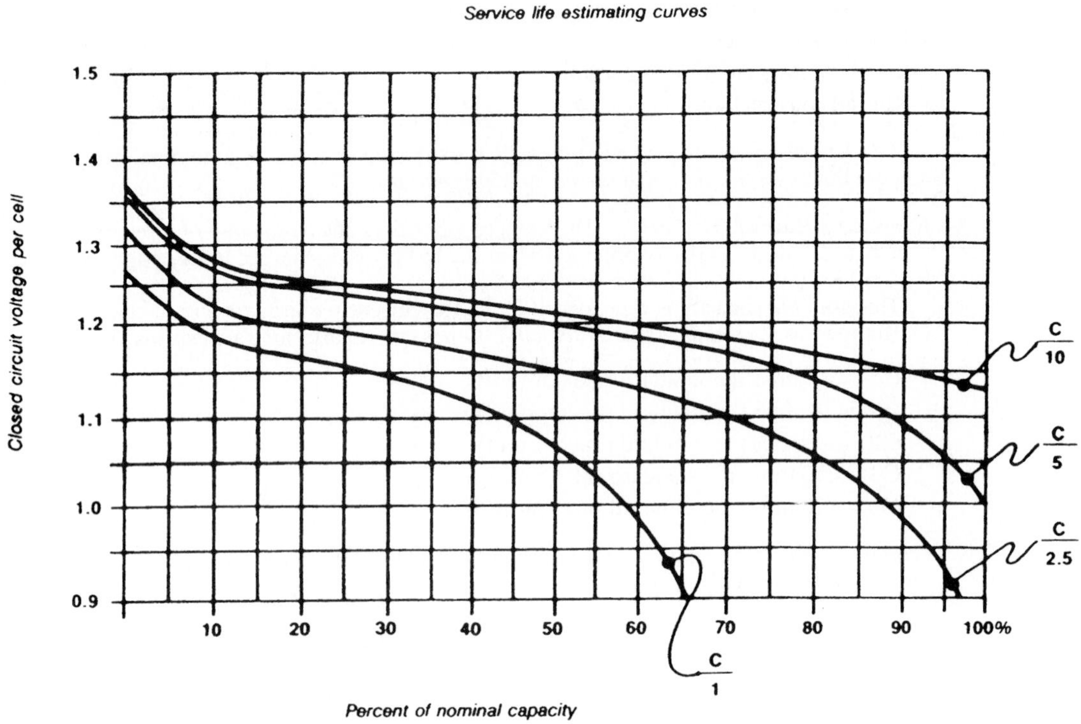


**FIGURE 6.158** Typical service-life estimation curve; nickel-cadmium batteries.

A standard 900-mAh cell may meet the requirement.

Checking

$$\text{Standard rate (current)} = \frac{900 \times 0.88}{10} = 79.2 \text{ mA}$$

$$\frac{75 \text{ mA}}{79.2 \text{ mA}} = 0.95 \times \text{standard rate}$$

Using the *C*/10 curve, *C*/10 to 1.2 V is 65 percent nominal capacity. So $0.65 \times 900 \times 0.88 =$ 515 mAh available from the 900-mAh cell to 1.2 V at 30°F.

The required capacity was 852 mAh, but the 900-mAh cell will only deliver 515 mAh under the conditions stated in the problem. So, a larger cell must be selected. Try a 1500-mAh-capacity cell.

## 6.11.2 Characteristics of Battery Systems

The characteristics of the modern electrochemical battery systems are shown in Table 6.28. Advances in electrochemical battery technology are being made on a steady basis. The capacities of electrochemical cells and batteries are being increased with advances in the technology. A prime example is in the nickel-cadmium cell system, where capacity has been increased by at least 40 percent by using a new process for making the plates of the cell. The Matsushita Electric Company in Japan use a new process that they developed, called the "foamed-plate

**TABLE 6.28** Characteristics of Battery Systems

| System | Anode | Cathode | Typical Voltage | Capacity Wh/Kg | Wh/$dm^3$ |
|---|---|---|---|---|---|
| Primary: | | | | | |
| Leclanche | Zn | $MnO_2$ | 1.2 | 65 | 175 |
| Magnesium | Mg | $MnO_2$ | 1.5 | 100 | 195 |
| Alkaline | Zn | $MnO_2$ | 1.15 | 65 | 200 |
| Mercury | Zn | HgO | 1.2 | 80 | 370 |
| Mercad | Cd | HgO | 0.85 | 45 | 175 |
| Silver oxide | Zn | AgO | 1.5 | 130 | 310 |
| Zinc-Air | Zn | $O_2$ (air) | 1.1 | 200 | 190 |
| Secondary: | | | | | |
| Lead-acid | Pb | $PbO_2$ | 2.0 | 37 | 70 |
| Edison | Fe | Ni oxides | 1.2 | 29 | 65 |
| Nickel-cadmium • | Cd | Ni oxides | 1.2 | 44• | 60 +• |
| Silver-zinc | Zn | AgO | 1.5 | 100 | 170 |
| Silver-cadmium | Cd | AgO | 1.05 | 55 | 120 |
| Zinc-nickel oxide | Zn | Ni oxides | 1.6 | 55 | 110 |
| Zinc-$O_2$ | Zn | $O_2$ | 1.1 | 130 | 120 |
| $H_2$-$O_2$ | $H_2$ | $O_2$ | 0.8 | 45 | 65 |
| Reserve: | | | | | |
| Silver-chloride | Mg | AgCl | 1.5 | 60 | 95 |
| Zinc-silver oxide | Zn | AgO | 1.5 | 30 | 75 |
| Cuprous chloride | Mg | CuCl | 1.4 | 45 | 65 |

NOTE: • Nickel-cadmium battery system capacities have increased substantially due to developments by the Matsushita Electric Company, Japan In these advanced nickel-cadmium systems, a process known as "Foamed-plate", is employed to increase storage capacities by 40%.

process," which allows the capacity of the nickel-cadmium system to be increased 40 percent or more. Advances such as this will allow the development of electricity-powered automotive systems in the very near future, or today.

A typical electric-battery-powered automotive system is shown in a later section in this chapter. Calculation procedures are also shown for estimating horsepower and battery capacity requirements for electricity-powered vehicles. The disclosures of this nickel-cadmium battery-powered system are enlightening and should prove interesting to many electromechanical design engineers who are interested in battery-powered automotive systems.

### 6.11.3 Battery Cross-Referencing

Table 6.29 identifies commonly interchanged batteries.

### 6.11.4 Typical Curves and Procedures for Standby Battery Systems

Standby service batteries such as those used for switchgear equipment and other critical systems requiring standby battery power need to be analyzed more precisely than other types of battery systems.

**TABLE 6.29** Battery Cross-reference

| Eveready | Neda | Duracell | Rayovac |
|---|---|---|---|
| 1015 | 15F | M15F | 7AA |
| 1209 | 908D | — | 944 |
| 1215 | 15D | M15HD | 5AA |
| 1231 | 918D | — | 928 |
| 1235-2 | 14D | M14HD | 4C |
| 1250-2 | 13D | M13HD | 6D |
| 186BP | 1167A | — | RW84 |
| 189BP | 1168A | — | RW89 |
| 206 | 1611 | — | — |
| 216 | 1604 | M1604 | 1604 |
| 226 | 1600 | — | — |
| 228 | 1610 | — | — |
| 266 | 1605 | — | 1605 |
| 276 | 1603 | — | — |
| 1222 | 1604D | M1604HD | D1604 |
| 246 | 1602 | — | — |
| 301BP | 1132SO | D301 | RW34 |
| 303BP | 1130SO | D303 | — |
| 343BP | 1154M | D343 | RW56 |
| 357BP | 1131SO | D357 | RW42 |
| 362BP | 1158SO | D362 | RW310 |
| 364BP | 1175SO | D364 | RW320 |
| 381BP | 1170SO | — | RW30 |
| 386BP | 1133SO | D386 | RW44 |
| 391BP | 1160SO | D391 | RW40 |
| 392BP | 1135SO | D392 | RW47 |
| 393BP | 1137SO | D393 | RW48 |
| 396BP | 1163SO | D396 | RW411 |
| 410/735 | 900 | M900 | 900 |
| 413 | 210 | — | — |
| 4156/763 | 710 | — | — |
| 416 | 217 | — | — |
| 420 | 225 | — | — |

The Exide Corporation, which produces battery systems, uses a system of analysis where the *S* curve is employed. A typical *S* curve for use in these types of systems is shown in Fig. 6.159. The use of the *S* curves is explained in section 50.50 of the Exide standby service battery manual. Very precise information may be derived using the *S* curves that are provided for each of the Exide electrochemical battery systems.

## 6.11.5 Battery Charger Systems

***Nickel-Cadmium Charger.*** A typical nickel-cadmium battery charger circuit is shown in Fig. 6.160*a*. The usual technique is to charge the depleted batteries at the *C*/10 rate. This type of charger generally requires between 12 and 14 h to recharge a fully depleted battery. The current limiter shown in the schematic diagram can be a resistive element or an electronic circuit. An LED is usually provided to indicate that charging is taking place.

As will be seen in a later section on nickel-cadmium battery-powered automotive systems, this C/10 charging rate can be accelerated to a great degree, with charging times taking only 10 min.

**TABLE 6.29** Battery Cross-reference (*Continued*)

| Eveready | Neda | Duracell | Rayovac |
|---|---|---|---|
| 455 | 201 | — | — |
| 457 | 203 | — | — |
| 482 | 202 | — | — |
| 484 | 207 | — | — |
| 489 | 728 | PF489 | N150 |
| 491 | 729 | — | — |
| 497 | 741 | PF497 | 1012 |
| 715 | 903 | — | 903 |
| 504 | 220 | M504 | 220 |
| 521 | 918A | MN918 | — |
| 522 | 1604A | MN1604 | A1604 |
| 523 | 1306AP | PX21 | RPX21 |
| 529 | 908A | MN908 | — |
| 531 | 1307AP | PX19 | RPX19 |
| 532 | 1308AP | PX24 | RPX24 |
| 544BP | 1406SOP | PX28 | RPX28 |
| 706 | 902 | — | — |
| 711 | 700 | — | — |
| 716 | 904 | — | 903 |
| 717 | 9 | — | — |
| 731 | 918 | M918 | 918 |
| 732 | 926 | — | — |
| 904 | 910F | M910 | 716 |
| 912 | 24F | M24F | 400 |
| 935 | 14F | M14F | 1C |
| 950 | 13FM13F | 2D | — |
| ACC75 | — | BC-1 | CH666 |
| ACC50 | — | — | CH222 |
| CH1.2 | 10022 | — | — |
| CH15 | 10015 | NC15 | 615 |
| CH35 | 10014 | NC14 | 614 |
| CH50 | 10013 | NC13 | 613 |
| ECR2016 | 5000L | DL2016 | — |
| ECR2025 | 5003L | DL2025 | CR2025 |
| ECR2032 | 5004L | DL2032 | CR2032 |
| E1 | 100M | RM-1 | — |
| E126 | 1611M | TR-126 | — |
| E132 | 1200M | TR132H | R132 |
| E132N | 1203M | TR132R | — |
| E133 | 1206M | TR133 | — |

***Lead-Acid Charger.*** A typical lead-acid battery charger circuit is shown in Fig. 6.160*b*. Voltage-limited charging techniques are required for lead-acid batteries to extend their life. At 72°F lead-acid cells require 2.3 V per cell to remain at 100 percent charge. The voltage tolerance of the charging circuit should be kept to ±1 percent by using voltage trimming by way of a linear voltage regulator circuit within the charging circuit. Charge times for depleted batteries require approximately 3 to 5 h.

### 6.11.6 Battery Systems for Automotive Power

As a review of the general meaning of rate, as applicable to cells and batteries, the following applies. The "rate" is the current $I$ that would completely discharge the cell in a specified

**TABLE 6.29** Battery Cross-reference (*Continued*)

| Eveready | Neda | Duracell | Rayovac |
|---|---|---|---|
| E133N | 1314M | TR133R | — |
| E134N | 1409M | TR134R | — |
| E135N | 1505M | TR135R | — |
| E146X | 1604M | TR146X | — |
| E164 | 1404M | TR164 | — |
| E165 | 1500M | TR165 | — |
| E177/TR177 | 1606M | TR177 | — |
| E1N | 1109M | RM1R | — |
| E4 | 1112M | RM4R | — |
| E42N | 1115M | RM42R | — |
| E431BP | 1801M | TR431 | T431 |
| E9 | 15M | ZM9 | — |
| E90 | 910A | MN9100 | 810 |
| E91 | 15A | MN1500 | 815 |
| E92 | 24A | MN2400 | 824 |
| E93 | 14A | MN1400 | 814 |
| E95 | 13A | MN1300 | 813 |
| EP175 | 1501M | TR175 | T175 |
| EP401E-2 | 1118M | MP401H | RP401 |
| EP675E-6 | 1127MD | MP675H | RH675 |
| EPX13BP | 1114MP | PX13 | RPX13 |
| EPX625BP | 1124MP | PX625 | RPX625 |
| EPX76BP | 1107SO | MS76 | RS76 |
| 389BP | 1138SO | D389 | RW49 |
| EXP640BP | 1126MP | PX640 | — |
| EXP675BP | 1128MP | PX675 | — |
| 509 | 908 | M908 | 941 |
| 744 | 6 | — | — |
| HS91 | 15AC | 1D1500 | AL-AA |
| HS93 | 14AC | 1D1400 | AL-C |
| HS95 | 13AC | 1D1300 | AL-D |
| HS22 | 1604AC | 1D1604 | AL-9V |
| 417 | 224 | — | — |
| 411 | 208 | — | — |
| 412 | 215 | M215 | 215 |
| 415 | 213 | — | — |
| 490 | 204 | — | — |
| 505 | 221 | M505 | 221 |

SOURCE: Allied Electronics

number of hours. The current then becomes the ampere hour (Ah) capacity divided by the number of hours. Conversely, if the current is known, then the rate in hours is calculated by dividing the capacity in ampere hours by the current.

*Example.* A 20-h rate capacity of 140 Ah = 7A/h for 20 h. A 5-h rate capacity of 4 Ah = 0.8A/h for 5 h. (Batteries for these capacities would or should be marked as follows: 7 Ah at 20-h rate and 4 Ah at 5-h rate).

Few batteries are marked with the proper data to show their actual energy capacity or the amount of power they will deliver before becoming discharged. The policy of some companies not to mark their batteries with the energy capacity is to their advantage and not the consumer's.

***Nickel-Cadmium Battery Systems (Cells).*** The normal nickel-cadmium cell voltage drops during the discharge period, when under load, from a starting voltage of approximately 1.28

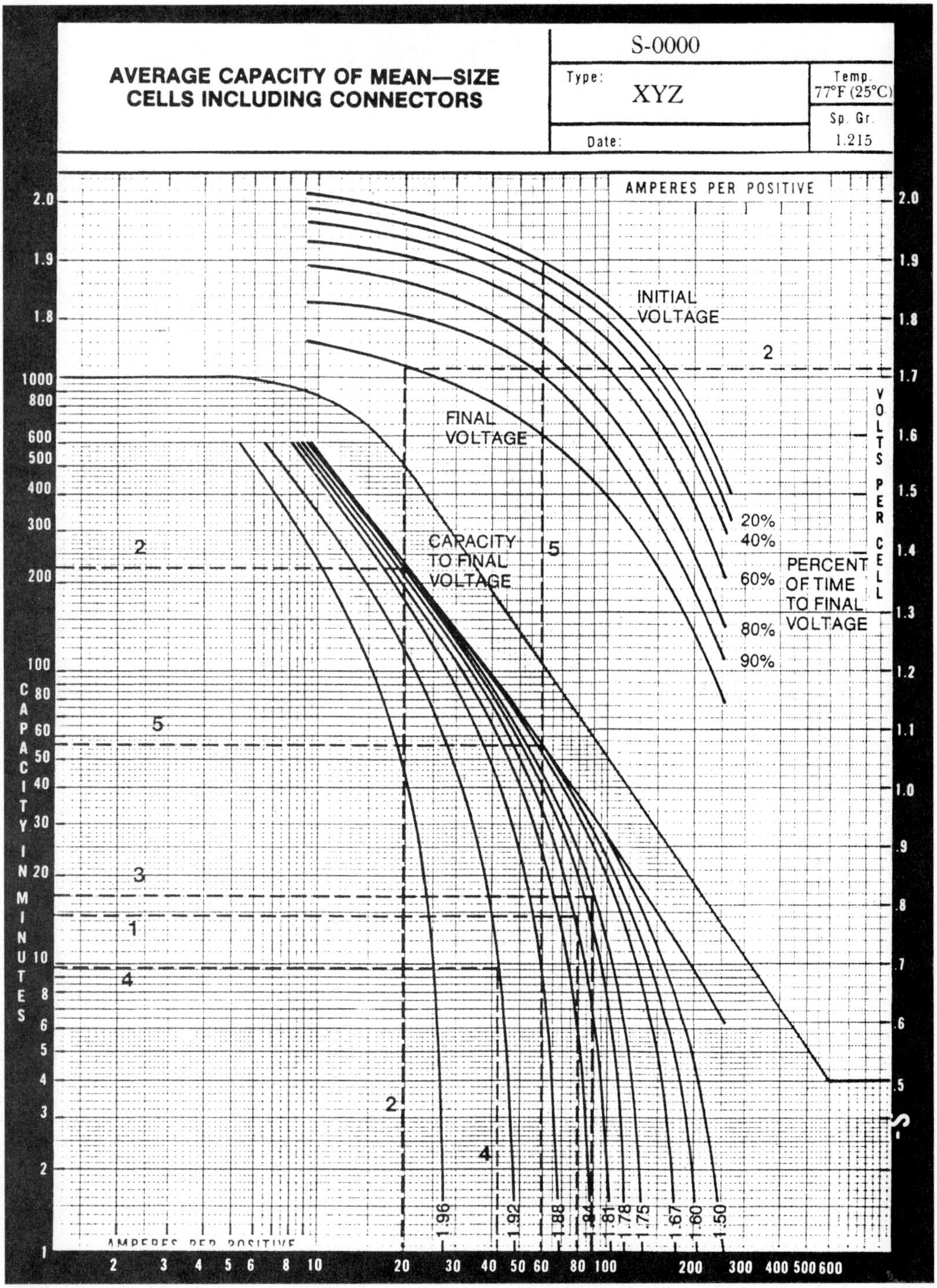


**FIGURE 6.159** An *S*-curve data sheet; Exide battery systems. (Source: Exide Corp.)

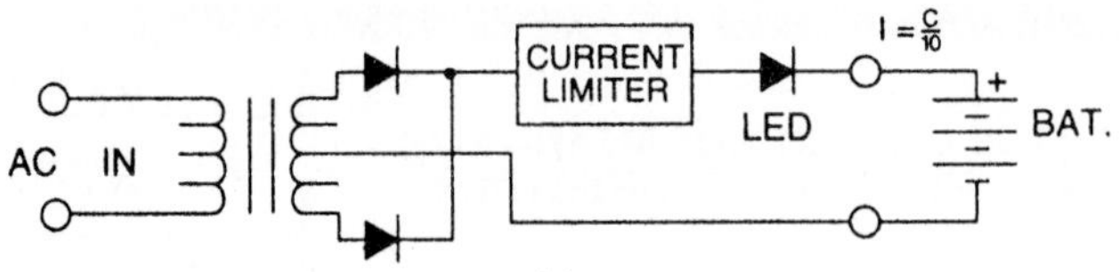


(a)

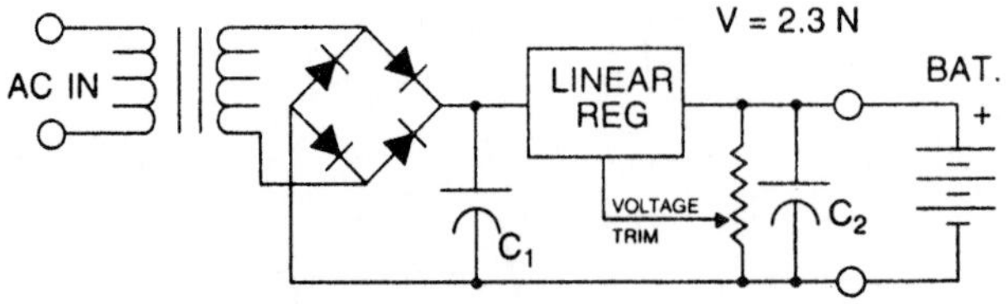


(b)

**FIGURE 6.160** (*a*) Charging circuit for nickel-cadmium batteries; (*b*) charging circuit for lead-acid batteries.

to 1.29 V to a final voltage level of approximately 1.1 V at the end of its useful output capacity. A typical discharge curve for a commercial, inexpensive sealed nickel-cadmium cell is shown in Fig. 6.161. This C-size, sealed nickel-cadmium cell was discharged through a 1-Ω resistor, taking 50 min to discharge the cell to 1.145 V, at which point the cell would no longer deliver power to the load. The actual energy delivered by this small cell was 1.25 Wh. (*Note:* Watt is a power measurement and watt-hour is an energy measurement.)

Photographs of this cell (C-size cylindrical) are shown in Figs. 6.162 and 6.163, where the construction can be easily seen. Note that the collector (D) in Fig. 6.162 is spot-welded to the edges of the plates at both ends of the cell. One collector terminates at the positive and the other, at the negative end of the cell. The collectors are spot-welded to the edges of the plates so that the cell can be charged quicker, by allowing current to enter the plates at many distribution points, thus effectively reducing the internal resistance of the cell and allowing more uniform current distribution throughout the electrode material. Current entering the edges of the cell positive and negative plates, electrodes, at many points allows the cell to be charged much faster than normally would be the case for a standard-construction cell.

A cell of this type of construction may be safely recharged in 10 min without overheating the cell. This quick recharge rate may be repeated hundreds of times without adverse effects on the cell structure or chemistry.

Cells exactly the same as that shown in the photographs were charged and rapidly discharged 500 times before the cells deteriorated. It should be noted that the test cells were inexpensive cells made in Taiwan, ROC (Republic of China). A quality cell using the foamed-plate process of construction would perform much better.

In Fig. 6.162, A is the positive electrode, B is the negative electrode, C is the electrolyte separator pad, D is the collector, and E shows one of the multiple spot welds used to join the collector to the edges of the plates or electrodes. The collector joining the negative plates is joined to the cell's nickel-plated steel casing. The positive collector is isolated from the case and terminates at the tip end of the cell (+).

Figure 6.163 shows the electrodes unrolled together with the electrolytic separator or pad. A is the positive electrode, B is the negative electrode, and C is the electrolyte separator pad that holds the electrolyte solution.

***Chemistry of the Nickel-Cadmium Sealed Cell.*** The active material of the positive electrode (plate) is nickel hydrate with graphite or metallic flakes added to aid conductivity. The

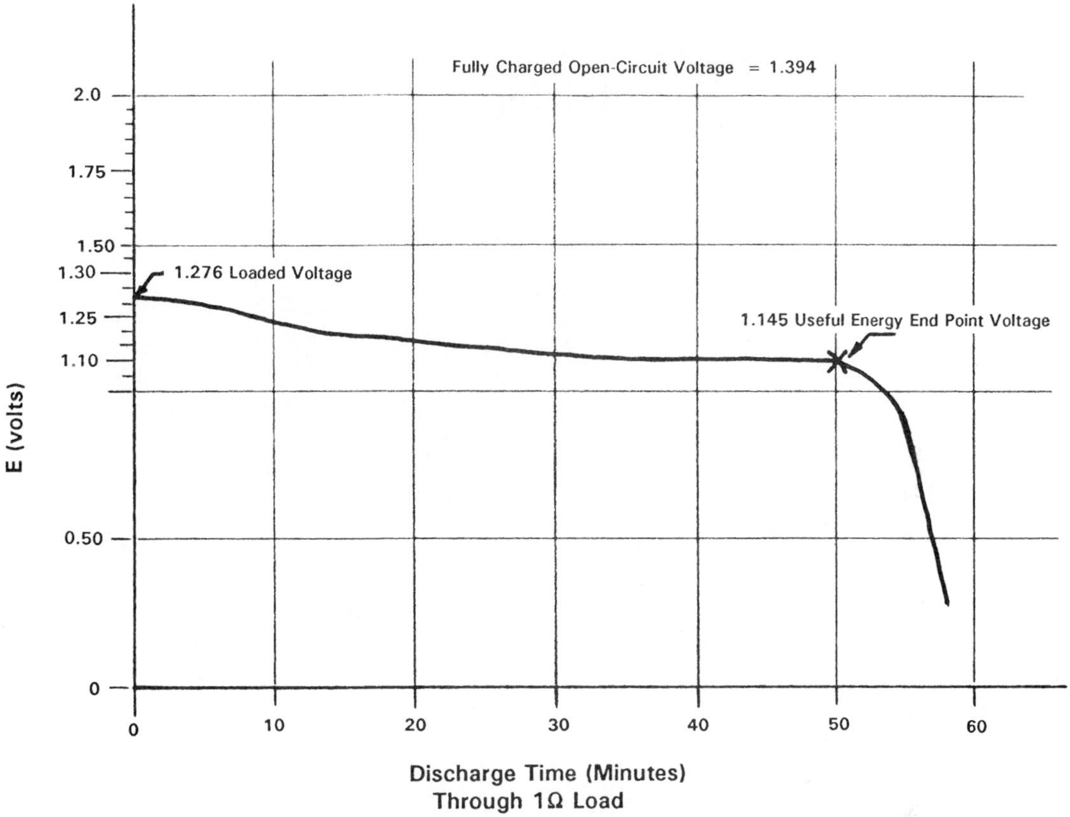


**FIGURE 6.161** Typical discharge curve for sealed nickel-cadmium battery systems.

metallic flakes may be seen on the positive plate in Fig. 6.162 (A) as small, bright spots. The active material of the negative electrode is cadmium metal sponge with additives to aid conductivity. The electrolyte is potassium hydroxide (KOH) with a specific gravity of 1.160 to 1.190 at 77°F. A small amount of lithium hydroxide is added to the electrolyte to improve capacity.

The chemical reaction within the cell is represented by

$$\underset{\text{charged}}{2\,Ni(OH)_3 + Cd} \rightleftharpoons \underset{\text{discharged}}{2\,Ni(OH)_2 + Cd(OH)_2}$$

Fully charged, the nickel hydrate of the positive electrode is oxidized, while the negative electrode is cadmium metal sponge. When discharged, the positive electrode is reduced to a lower oxide, while the negative electrode (cadmium metal sponge) is oxidized. In Fig. 6.163, the negative electrode (B) shows the white formation of the metallic oxide $(OH)_2$ beginning at the pocket areas.

The core of the electrodes is made of nickel-plated steel with holes used as pockets for holding the active materials. The negative electrode is made 10 to 20 percent longer than the positive electrode to absorb the hydrogen gas evolved during prolonged recharging, after the cell is fully charged. The quick-charge cells generally have the negative electrode 20 percent longer than the positive electrode. These types of sealed cells also incorporate a safety vent

**FIGURE 6.162** Rolled electrodes from a C-size sealed nickel-cadmium cell.

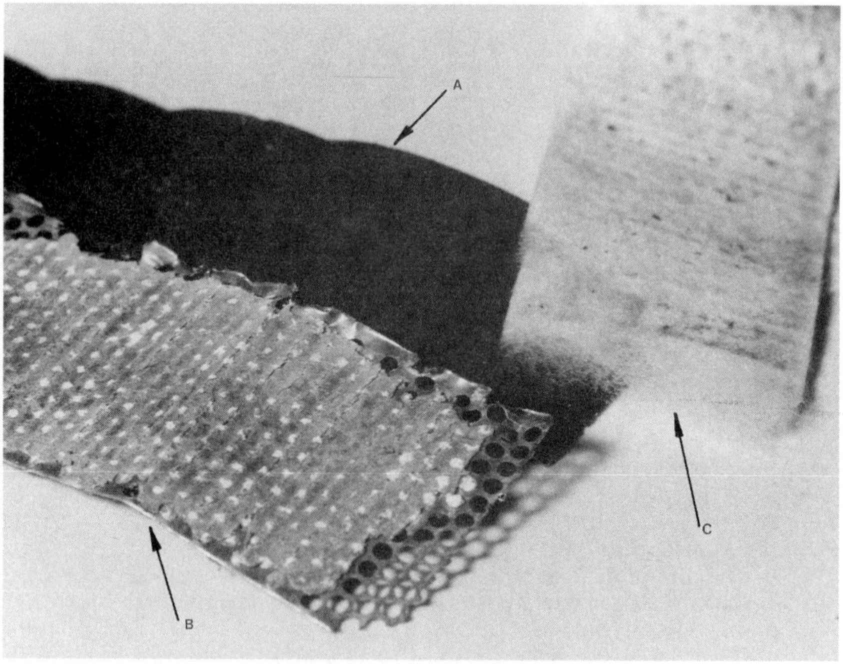


**FIGURE 6.163** Flattened electrodes from nickel-cadmium cell showing construction.

under the positive cap to prevent explosion of the cell in case of short circuit or extreme overcharging current.

***Nickel-Cadmium Cell and Battery Design for Automotive Electric Power Systems.*** The following disclosures show some of the methods that may be employed in developing high-power, quick-charge battery systems for automotive vehicle power. Bear in mind that the systems shown rely on the increased capacity of the latest nickel-cadmium cell systems that use foamed plates. A standard nickel-cadmium system cannot economically or practicably contain enough stored electrochemical energy to produce a realistic automotive electrochemical battery power system.

***Details of an Electrochemical Battery System for Automotive Power.*** Figures 6.164 through 6.173 show the details of a typical nickel-cadmium, rechargeable battery system for automotive power. The captions to the illustrations are self-explanatory, and the dimensions shown are nominal, practical dimensions of a typical system, dimensioned to fit a small- to medium-size automobile.

The author has been studying electrochemical battery systems for automotive power for over 20 years. Economical, quick-recharge electric automobiles are a reality today, using the presently available technology. Cell systems, electronic controls, and efficient dc electric tractive-power motors are all available for development into a sensible, economical system for electricity-powered vehicles that may be recharged in 10 min or less.

The single cells shown in the system of Figs. 6.164 through 6.173 will have a capacity of 43 to 45 Ah at 1.3 V at the 2-h discharge rate. That is equivalent to 58 Wh per cell. If the system shown carries 20 cells per battery and contains 14 batteries, the available system energy is 15.3 kWh or 20.5 horsepower hours (hPh). The energy density of these cells is approximately 62 Wh/kg. This system may be fully recharged at a current rate of 310 to 325 A at 175 to 185 V dc or 54 to 60 kW for 10 min. Even distribution of the charging currents in this system makes this high charging rate possible.

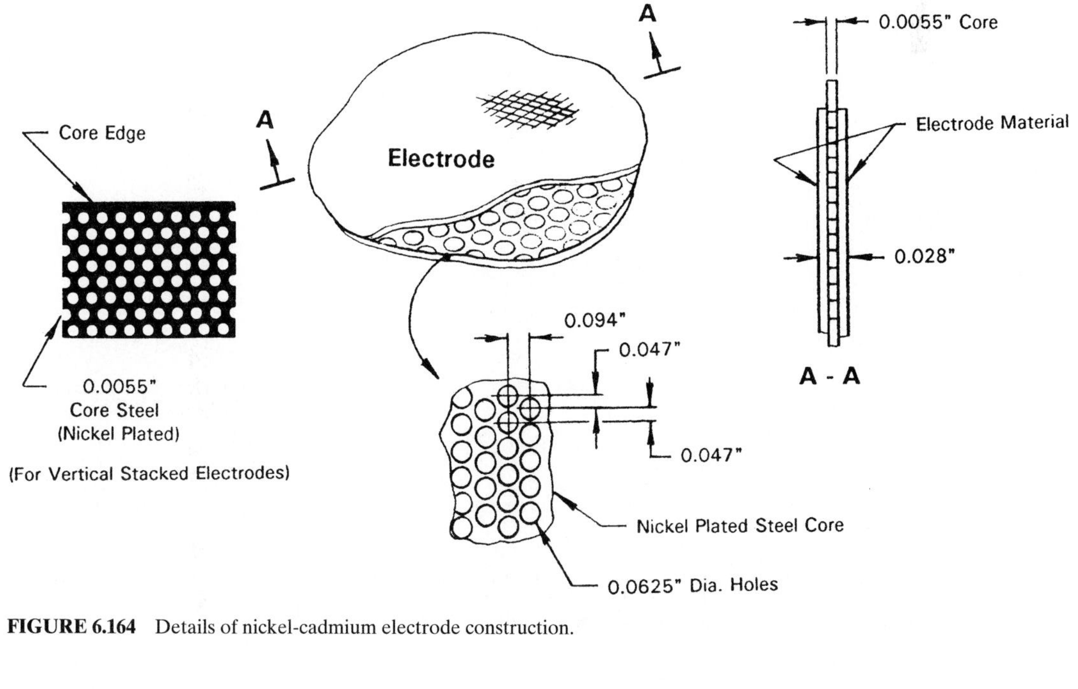


**FIGURE 6.164** Details of nickel-cadmium electrode construction.

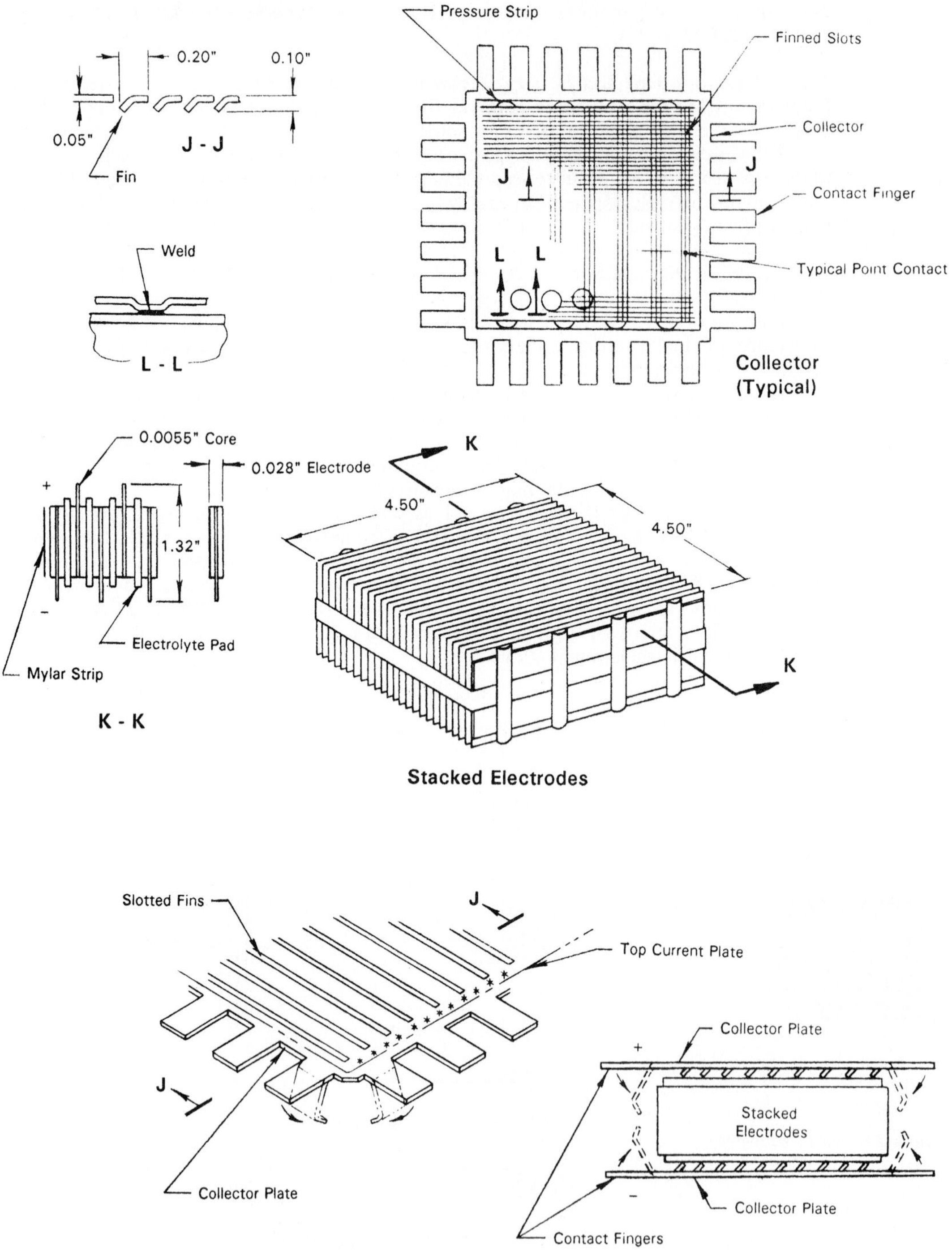


**FIGURE 6.165** High-density, quick-charge cell construction details.

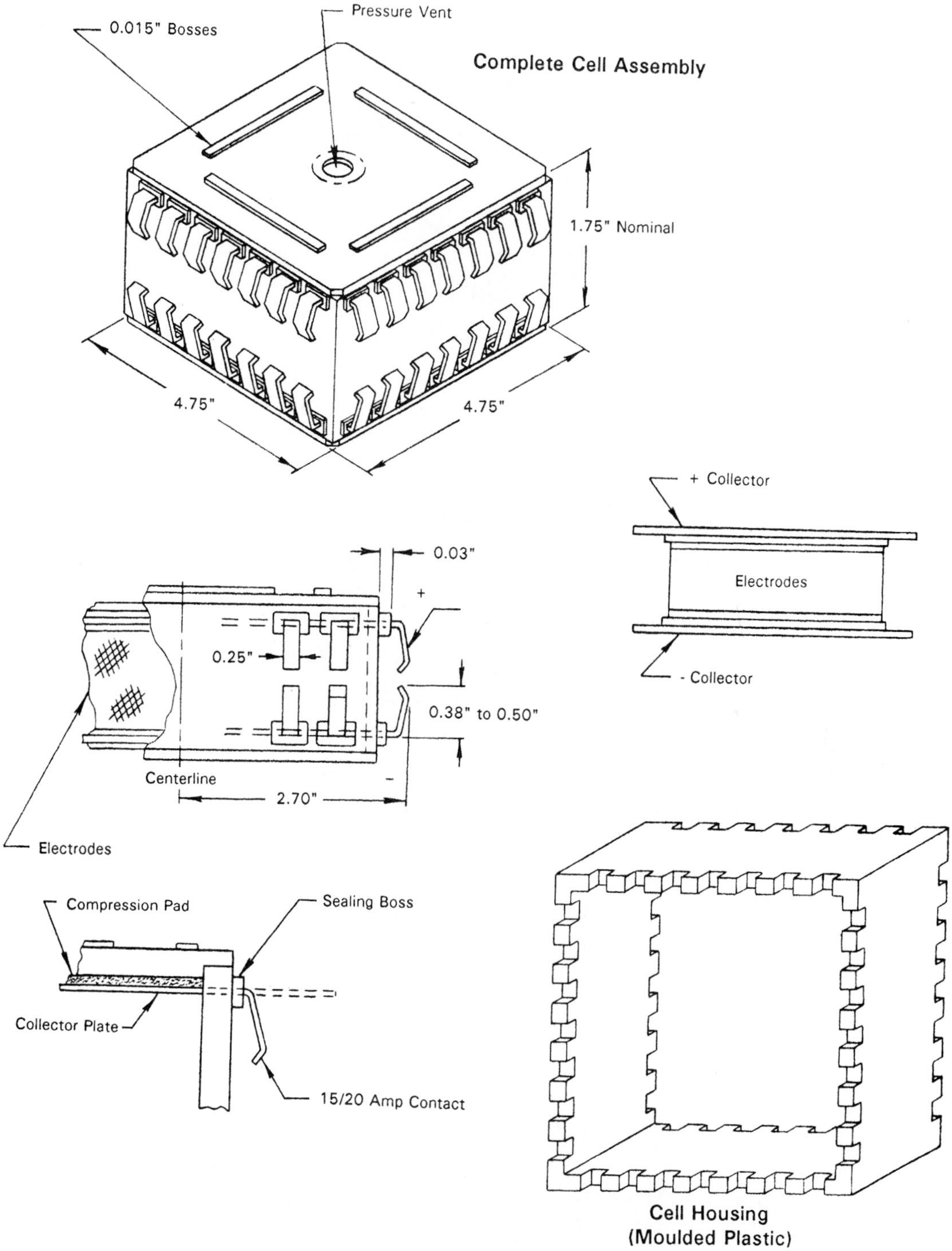


**FIGURE 6.166** Typical modern design for quick-charge nickel-cadmium cell system.

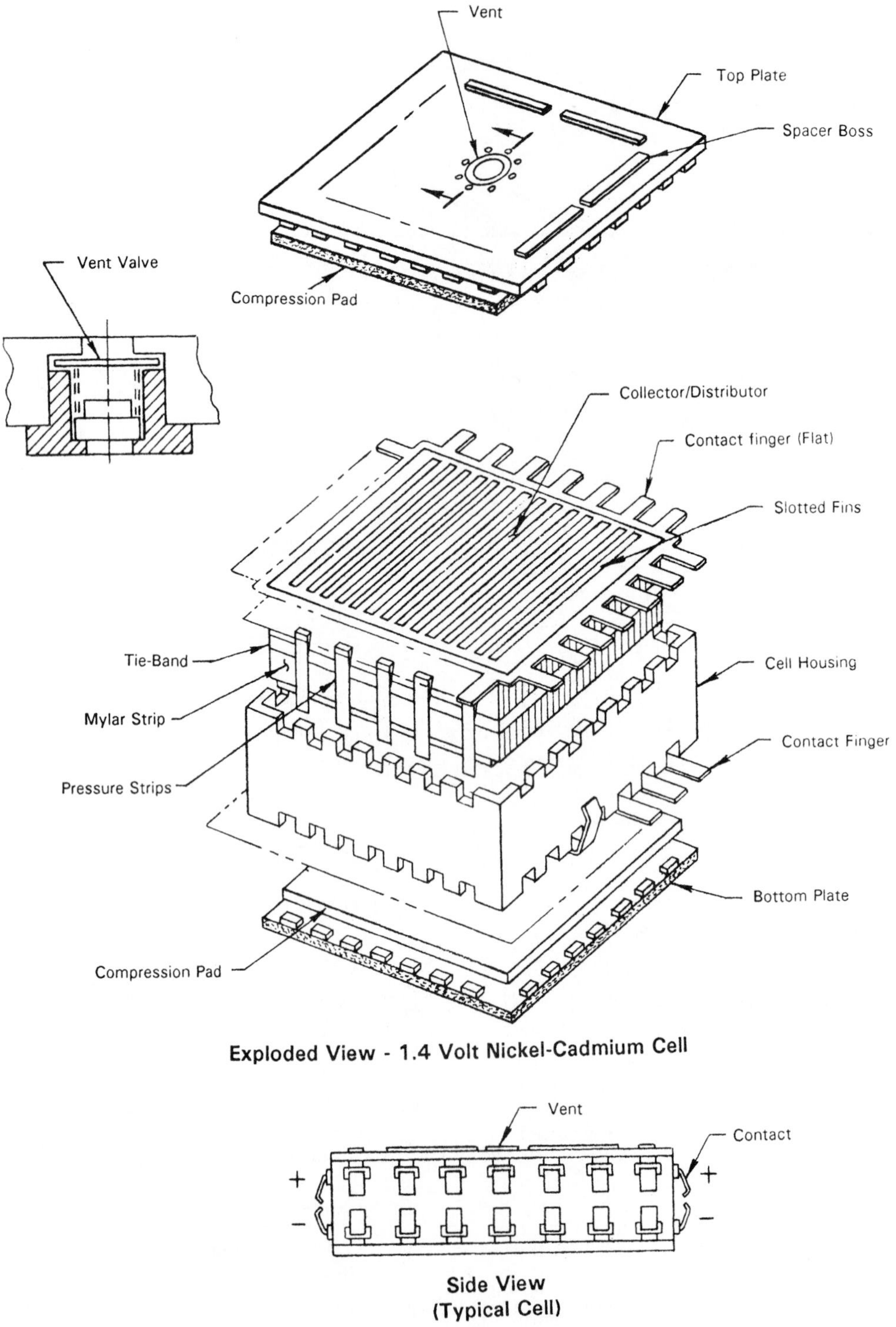


**FIGURE 6.167** Exploded detail view of sealed nickel-cadmium quick-charge, high-density cell.

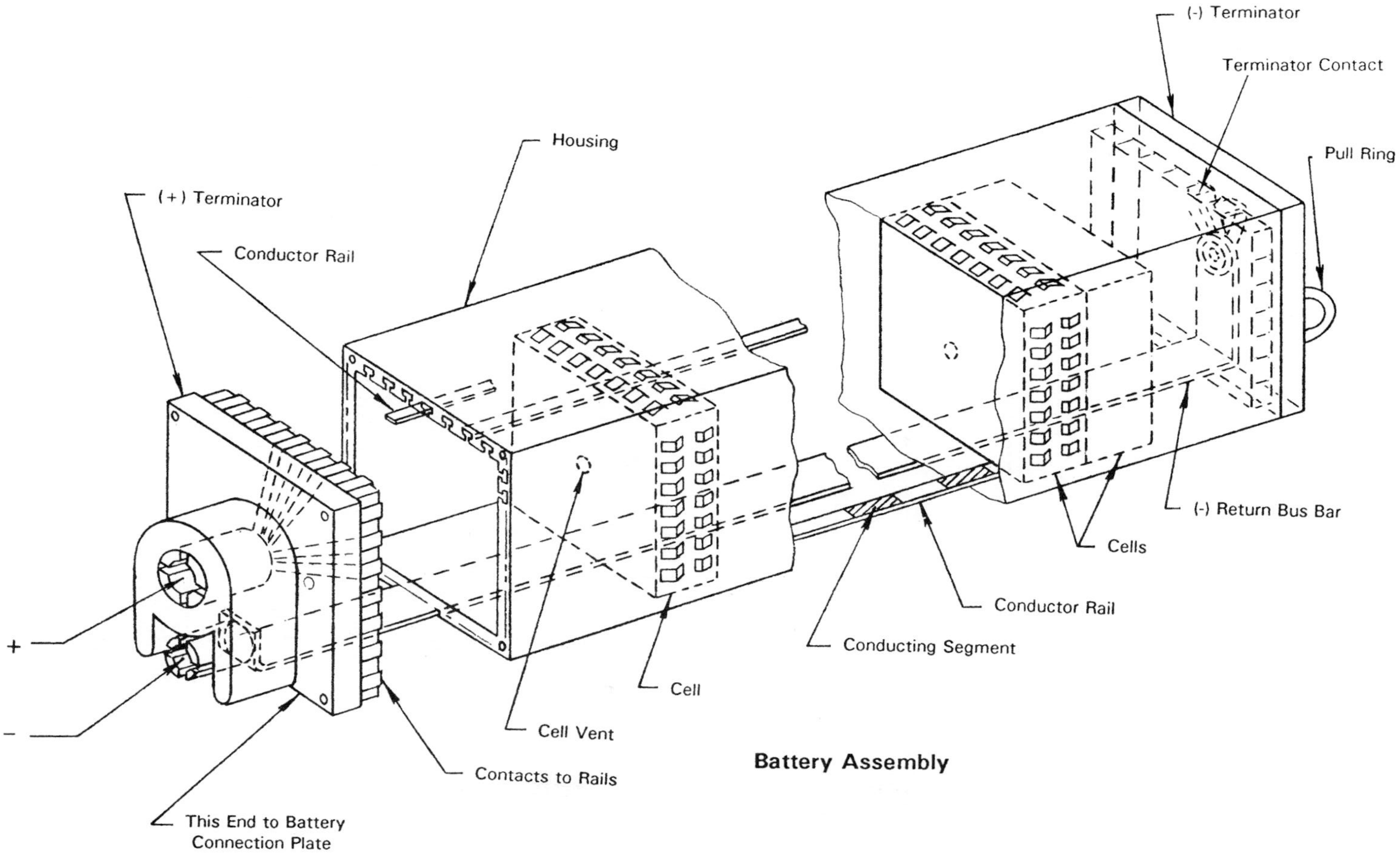


**FIGURE 6.168** Cells forming a quick-charge battery system for automotive power use.

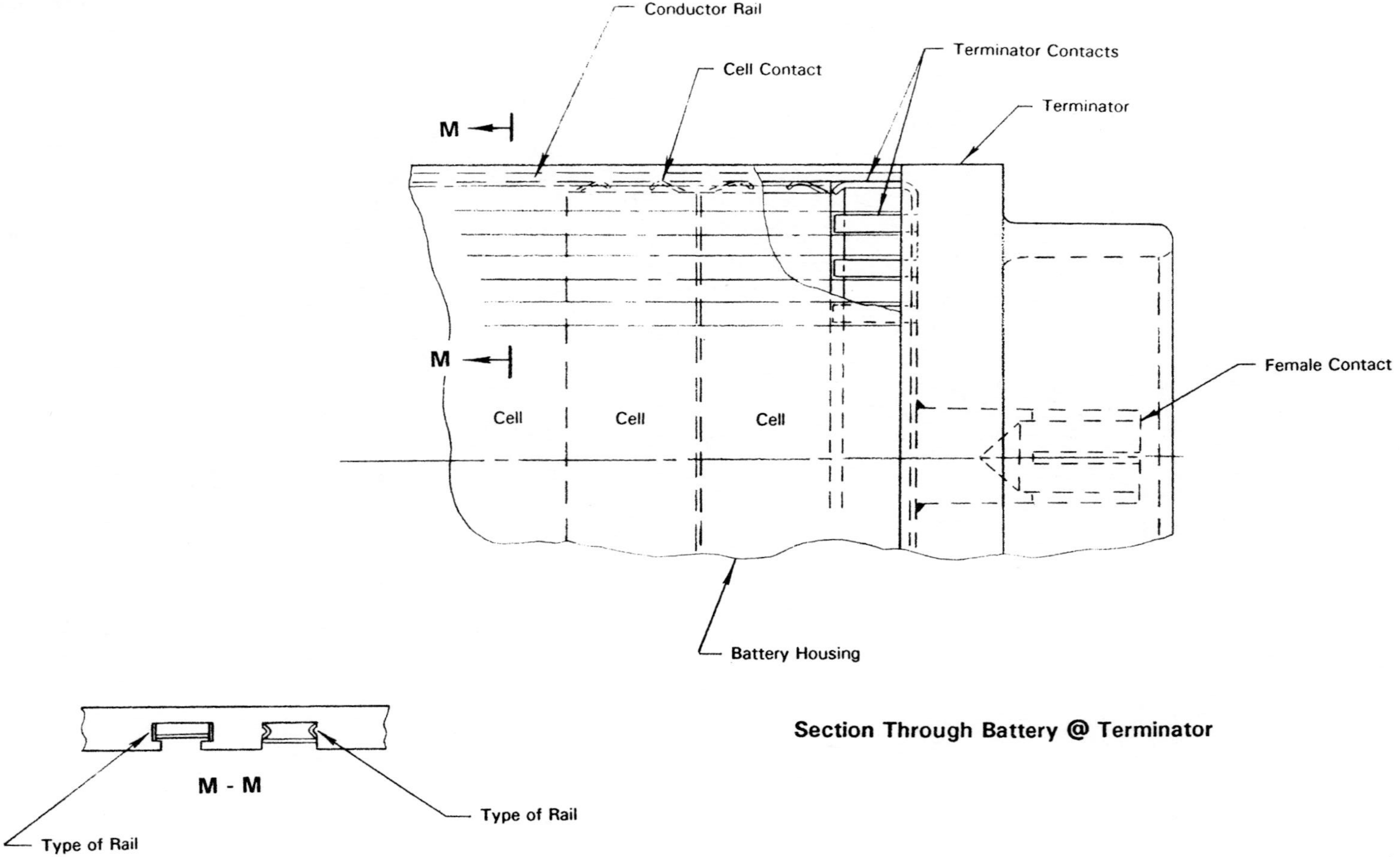


**FIGURE 6.169** Section through the battery terminator (see Fig. 6.168).

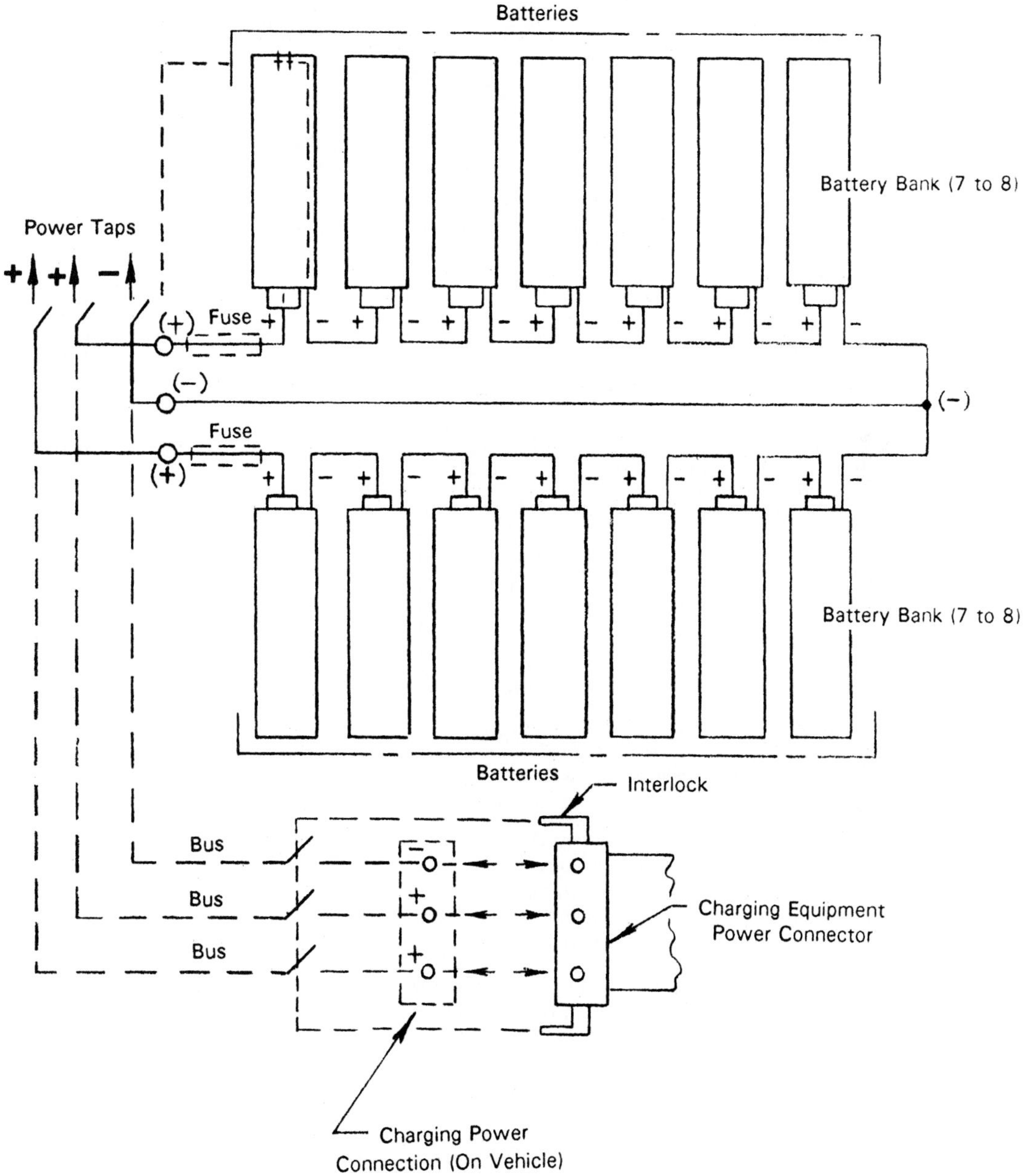


**Recharging Schematic Diagram**

**FIGURE 6.170** Example of automotive battery charging system.

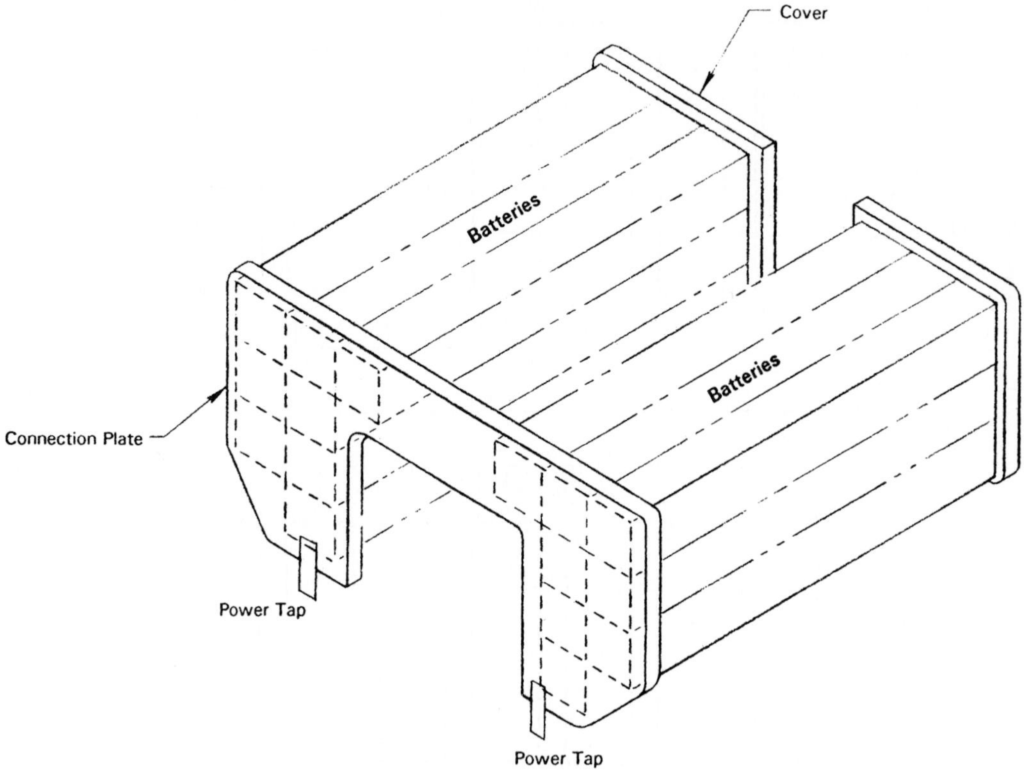


**FIGURE 6.171** An automotive battery module consisting of 16 batteries connected in series and parallel.

***Calculations for Electricity-Powered Vehicles.*** With the following simplified equations and explanations, the horsepower requirements, speed requirements, and battery energy capacity requirements for a given travel range may be calculated for practical electricity-powered vehicles. The results of the calculations are nominal, without variable tolerances, and may be considered as good first-order estimates.

***Energy, Horsepower, and Range Calculations for Electricity-Powered Vehicles.*** The following simplified calculations show the approximate power requirements and probable approximate range of an electricity-powered surface vehicle traveling on a smooth, level surface or road. In order to perform these simple calculations, we must make some assumptions based on experience, practicality, and other engineering works.

*Assumptions.* Vehicle parameters:

Gross vehicle weight = 2450 lb (two passengers)

Coefficient of drag = 0.25

Battery array weight = 900 to 1000 lb

Battery system energy rating = 15 kWh

Rolling resistance = 0.02 lb/lb of vehicle weight

Air resistance* = $CA_m v^2$

* The air-resistance equation was derived from Nadjer, K. W., *Machine Design,* John Emanuelson, ed., Copenhagen Polytechnic Institute, Denmark, 1958.

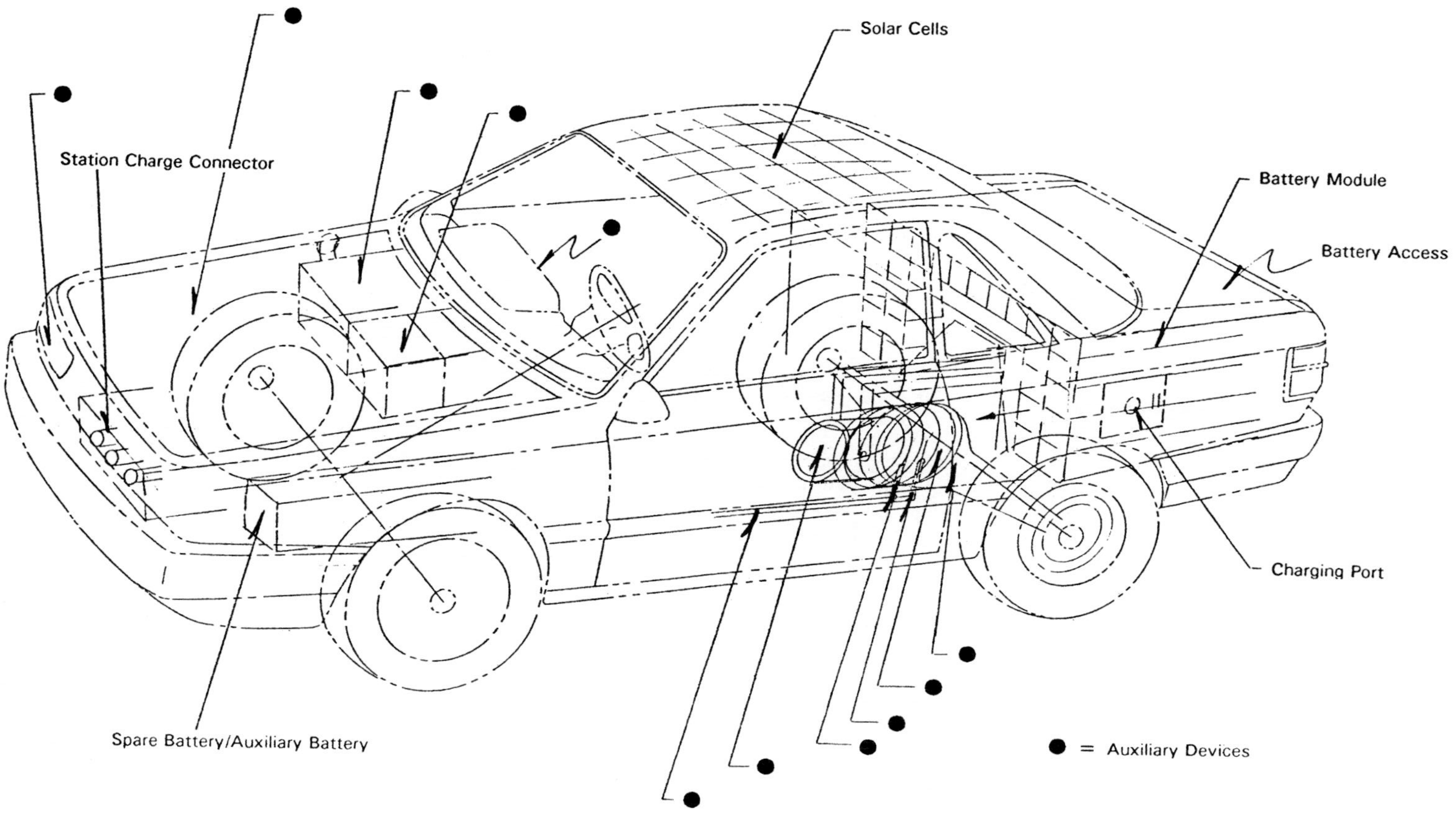


**FIGURE 6.172** Layout of sample battery-powered automotive system.

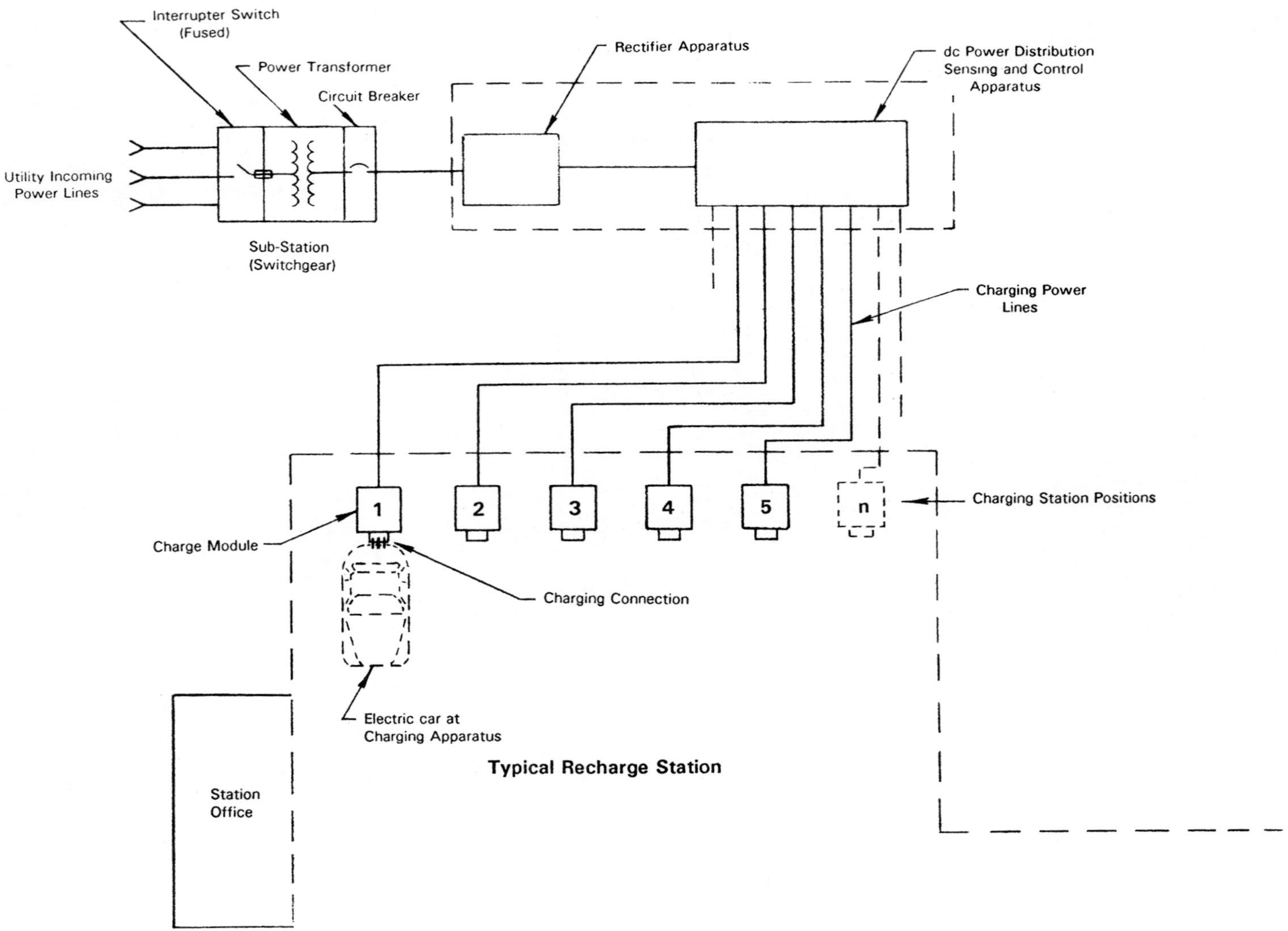


**FIGURE 6.173** Typical battery recharge station layout for electric-powered automotive vehicles.

In this equation, $C$ is a mass constant derived from

$$C = \frac{\text{weight of 1 ft}^3\text{ of air}}{32.16} = \frac{0.077}{32.16} = 0.0024 \text{ slugs}$$

where $A_m$ is the frontal area of a typical automobile in $ft^2 \times$ the drag coefficient 0.25 ($21 \times 0.25 = 5.25$ $ft^2$) corrected effective area and $v$ is the velocity of travel in miles per hour. Take this as 55 mph, constant velocity.

So rolling resistance = 2450 × 0.02 = 49 lb, and air resistance = $0.0024 \times 5.25 \times (55)^2 = 38$ lb. Thus the total retardation is 87 $lb_F = F$. Then, from the basic horsepower equation where $F$ = total resistive force, $v$ = velocity, and $T$ = time, we have

$$\text{Hp} = \frac{Fv}{550\text{T}} = \frac{87 \times 55 \times 5280}{550 \times 3600} = 12.76 \text{ hp required at 55 mph}$$

By contrast, some of the electric van developers in the United States and Europe are taking a standard steel-bodied van and loading it with *1800 lb* of lead-acid or nickel-iron batteries connected with cables and are arriving at low traveling ranges, using 60- to 80-hp electric motors. These vehicles weigh as much as 5200 lb. This type of battery-powered system is not only clumsy and grossly heavy but still has the basic problem of "What do you do with the batteries when they are discharged?" The modular battery system as disclosed by the author solves this problem. Horsepower requirements for this type of vehicle using the same parameters and the same simplified equations as the previous problem are on the order of 38 to 42 hp, and that is a great deal of raw tractive power requiring a battery energy supply of 28 to 31 kWh, minimum. Those who doubt the validity of these simplified equations and the answers to the preceding problems should be aware of the fact that current test vans with total weights and other parameters that the author used to calculate the 38- to 42-hp figure noted above, currently being tested in the United States, are equipped with at least 60-hp electric motors.

In the sample problem shown above, using the parameters listed, the approximate range for the 2450-lb vehicle, containing 1000 lb of modular nickel-cadmium batteries, would be as follows, assuming 15 kWh available from batteries. Since 1 kWh = 1.341 hph, we have available 1.341 × 15 = 20.1 hph to propel the vehicle, thus giving it a theoretical range of 20.1/12.76 = 1.58 h running time at 55 mph and 1.58 × 55 = ~87-mi range. When the vehicle is driven at an average speed of 30 mph, the probable range increases to approximately 125 mi. The figures shown do not take into account steep grades on the road. Since 90 percent of the average motorists in the United States, according to the latest statistical survey, travel only 60 mi per day or less, we have a viable surface vehicle system described herein.

(*Note:* In contrast to the American electric vehicles, BMW of West Germany has developed an electric vehicle which is lightweight, cruises at 55 mph with a top speed of 63 mph, has a 93-mi range in city driving, and is equipped with *only* a 22.7-hp electric motor. It is powered by sodium-sulfur cells, but it also has the basic problem of "What do you do when the batteries are discharged?" It takes too long to recharge the batteries.)

***The Solar Constant.*** In electric-powered automotive systems, it would be advantageous to obtain free energy from the sun during the summer months and days when the sun is intense at other seasons. On electric vehicles, this is possible using solar conversion cells located on the roof of the vehicle.

The solar constant is equal to approximately 1.92 g · cal/min per square centimeter of incident sunlight. This converts to 1784 g · cal/($ft^2$) (min) or 125 W/$ft^2$. Since most modern solar conversion cells are only 12 percent efficient, the approximate energy gain would be 125 × .12 = 15 W/ft$^2$. For every hour the sun's energy is converted, you will receive 15 Wh of energy per square foot of solar cell area.

Silicon solar cells operate on the solar photon flux, which in the vicinity of the Earth is equal to approximately 1400 W/m$^2$. The above figures are derived from this constant.

## 6.12 ELECTRICAL MEASURING INSTRUMENTS

Modern electrical instruments that measure temperature and forces are important to the designer and include:

- Temperature indicating devices, such as thermocouples, thermistors, resistance temperature detectors (RTDs), and IC sensors
- Strain gauges
- Pressure transducers
- Load cells (incorporating strain gauges)
- Accelerometers

These instruments allow the design engineer to monitor temperature, stress, strain, load, and forces and pressures in all phases of component, structure, and assembly design, especially in the prototype stage. Complex structures and elements such as wing spars in aircraft and other types of complex structures can be more effectively analyzed using the above-indicated instruments. Large companies that produce complex products such as aircraft, missiles, space vehicles, and automotive equipment usually incorporate an instrumentation engineering group as part of the design engineering structure of the company. The instrumentation group is responsible for designing, installing, and conducting measurement systems and tests, as well as for product evaluation and in-depth engineering studies.

### 6.12.1 Thermocouples and Their Characteristics

Table 6.30 summarizes the designations and ranges of many thermocouples. Table 6.31 lists sensitivities.

### 6.12.2 Thermocouple Principles

Thomas Seebeck discovered in 1821 that when two wires of dissimilar metals are joined at both ends and heat is applied to one end, an electric current flows in the circuit. When this circuit is broken at the center, the open-circuit voltage is a function of junction temperature and the composition of the two metals constituting the thermoelectric wires. All dissimilar metals exhibit this characteristic, but the most common metals suitable for thermocouple construction are listed in Table 6.30.

The open-circuit voltage of a thermocouple cannot be accurately measured with a voltmeter because the voltmeter leads create a new thermoelectric circuit. The open-circuit voltage is known as the *Seebeck voltage.*

When a current flows through a thermocouple junction, heat is either absorbed or flows out of the junction, depending on the current flow direction. This phenomenon is called the *Peltier effect* and is independent of the $I^2R$ joule heating. A simple thermocouple circuit is shown in Fig. 6.174.

Refrigeration units can be manufactured based on the Peltier effect. Low-power thermoelectric generators operate using the reverse effect: heating a thermoelectric junction will

**TABLE 6.30** Thermocouple Characteristics

| ANSI Designation | | Material and Trade Names | Range |
|---|---|---|---|
| B | BP + | Platinum 30% rhodium | 32 to 3,095°F |
| | BN− | Platinum 6% rhodium | 0 to 1,700°C |
| C* | CP + | Tungsten 5% rhenium | 32 to 4,210°F |
| | CN− | Tungsten 26% rhenium | 0 to 2,320°C |
| D* | DP + | Tungsten 3% rhenium | 32 to 4,210°F |
| | DN− | Tungsten 25% rhenium | 0 to 2,320°C |
| E | EP + | Chromel, Tophel | − 328 to 1,655°F |
| | EN− | Constantan, Advance | − 200 to 900°C |
| G* | GP+ | Tungsten | 32 to 4,210°F |
| | GN− | Tungsten 26% rhenium | 0 to 2,320°C |
| J | JP + | Iron | 32 to 1,385°F |
| | JN− | Constantan, Advance | 0 to 750°C |
| K | KP + | Chromel, Tophel | − 328 to 2,285°F |
| | KN− | Alumel, Nial | − 200 to 1,250°C |
| R | RP+ | Platinum 30% rhodium | 32 to 2,645°F |
| | RN− | Pure platinum | 0 to 1,450°C |
| S | SP + | Platinum 10% rhodium | 32 to 2,645°F |
| | SN− | Pure platinum | 0 to 1,450°C |
| T | TP + | Copper | −328 to 662°F |
| | TN− | Constantan, Advance | − 200 to 350°C |

*Not ANSI standard

**TABLE 6.31** Sensitivities of Thermocouple Types

| | |
|---|---|
| B = 7.6 μv/°C | J = 52.6 μv/°C |
| C = 16.6 μv/°C | K = 38.8 μv/°C |
| D = 17.0 μv/°C | R = 12.0 μv/°C |
| E = 67.9 μv/°C | S = 10.6 μv/°C |
| G = 16.0 μv/°C | T = 40.5 μv/°C |

SOURCE: Omega Engineering, Inc.

produce electric power. The efficiency of most thermoelectric generators is too low to be of practical use, although atomic-powered thermoelectric generators have been built and are in use for various applications. The atomic heat source for this type of generator is usually plutonium.

A group of thermocouples connected in series, termed a *thermopile,* multiplies the thermoelectromotive force and exhibits great sensitivity.

Figure 6.174*a* shows the use of multiple thermocouples with a selector switch shown on the right side of the photograph. This application of the thermocouple was used to measure the temperatures at different parts on the high-current isolator switch shown inside a switchgear unit. A test of this type is called a "heat run" and is used to check the final temperature rise of electrical apparatus per ANSI standard test conditions for continuous current-carrying ability of switching devices. The final temperature for this class of electrical equipment cannot exceed 90°C (194°F), with compensation for ambient temperatures.

### 6.12.3 Strain Gauges

The most common measuring device for electrical measurement of mechanical quantities is the strain gauge. The strain gauge operates on the principle of proportional variance of elec-

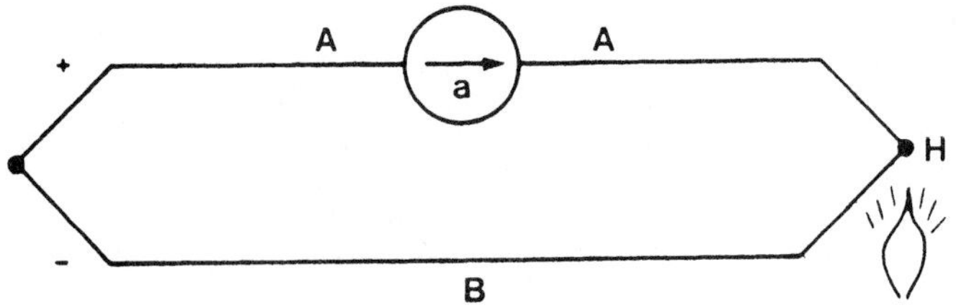


**FIGURE 6.174*a*** A thermocouple circuit.

**FIGURE 6.174*b*** Multiple thermocouples being used to monitor temperature during a test of electrical apparatus.

trical resistance to strain (movement). The types of strain gauges are the piezoresistive or semiconductor gauge, the bonded-metallic-wire gauge, the carbon-resistive gauge, and the foil resistance gauge.

Strain gauges are bonded to the surface of the structure in which strain is to be measured. Movement or strain in the structure is transmitted to the grid in the gauge through the adhesive that bonds the gauge to the structure. The Wheatstone bridge circuit is most frequently used for static strain measurements, with the strain gauge forming one of the resistive elements of the bridge. In this application, the Wheatstone bridge exhibits outstanding sensitivity, allowing accurate measurements to be made. See Fig. 6.175*a* and 6.175*b* for diagrams of single- and multiple-element gauges. Figure 6.175*a* shows two measuring grid strain gauges with ribbon leads parallel to the grids. Figure 6.175*b* shows two 10-element strain gauges with a compensation grid. The nominal resistance of these types of gauges is 120 Ω.

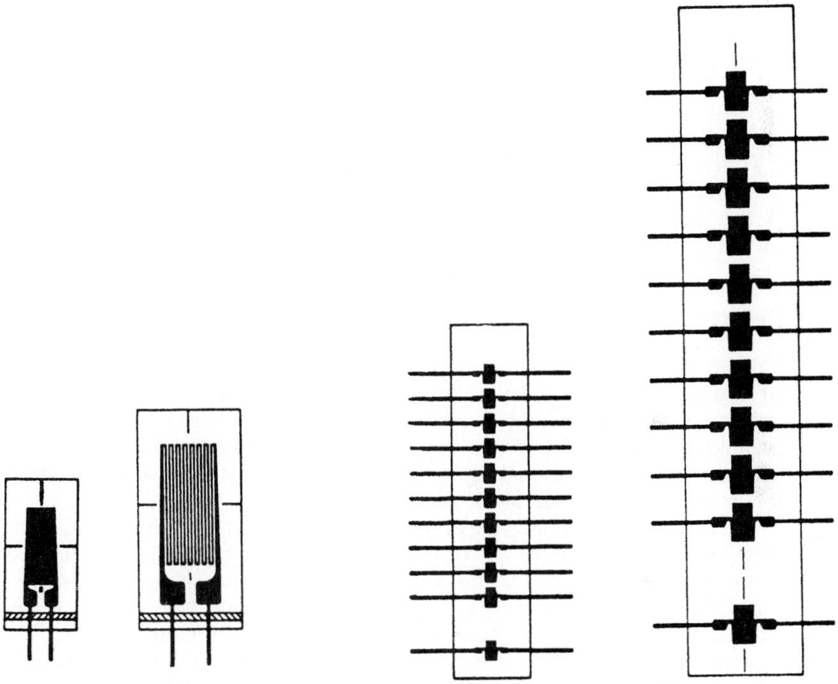

(a) (b)

**FIGURE 6.175** (*a*) Grid-type strain gauges; (*b*) multiple grid-type strain gauges.

Because of its outstanding sensitivity, the Wheatstone bridge circuit shown in Fig. 6.176*a* is the most frequently used circuit for static strain measurements. The use of a computer with the measurement instrumentation can simplify use of the bridge circuit, increase accuracy, and compile large quantities of data from multichannel systems. The computer eliminates the requirement of balancing the bridge and compensates for nonlinearities in output.

In Fig. 6.176*a*, $V_{in}$ is the bridge input voltage; $R_g$ is the resistance of the strain gauge; $R_1$, $R_2$, and $R_3$ are the resistances of the bridge arm resistors; and $V_{out}$ is the bridge output voltage. A ¼-bridge configuration exists when one arm of the bridge is an active gauge and the other arms are fixed resistors. We have an ideal condition when the strain gauge $R_g$ is the only resistor in the bridge circuit that varies. $V_{out}$ is a function of $V_{in}$, $R_1$, $R_2$, $R_3$, and $R_g$. This relationship is defined by

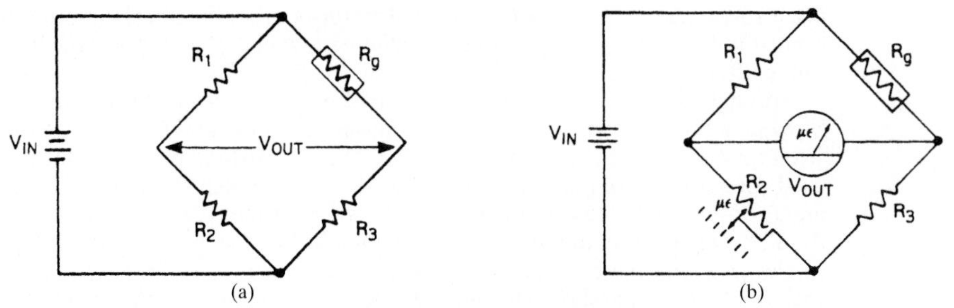


**FIGURE 6.176** (*a*) Wheatstone bridge circuit used with strain gauges; (*b*) Wheatstone bridge with adjustable resistance for strain gauges.

$$V_{\text{out}} = V_{\text{in}} \frac{R_3}{R_3 + R_g} - \frac{R_2}{R_1 + R_2}$$

When $(R_1/R_2) = (R_g/R_3)$, $V_{\text{out}}$ becomes zero and the bridge is balanced. If we could adjust one of the resistor values, such as $R_2$, we could then balance the bridge for various values of the other resistors in the bridge arms. Figure 6.176*b* shows the schematic diagram of this type of bridge circuit configuration. The equation written for $V_{\text{out}}$ previously shown can be rewritten in the form of the ratio of $V_{\text{out}}$ to $V_{\text{in}}$ for unbalanced bridge strain-gauge measurements.

### 6.12.4 Load Cells

A load cell converts a load acting on it into an electrical signal, thereby classifying itself as a transducer. The conversion of the load into an electrical signal is accomplished by the deformation of strain gauges, which are bonded to a load cell beam that deflects under a load. The strain gauges are wired into the sensitive Wheatstone bridge circuit, where the gauge resistivities are converted into load measurements. There are two basic load cell configurations, the *shear beam* load cell and the *bending beam* load cell.

A simple substitute for a load cell can be made using a hydraulic pressure gauge and piston arrangement as shown in Fig. 6.177. The accuracy of this device is limited by the pressure gauge's accuracy and the frictional drag created by the sealing O rings. The diameter of the piston is made so that the area at the end of the piston is exactly 1 in$^2$. Different load ranges can be measured by using pressure gauges with various pressure ranges.

The details of this device are shown in Fig. 6.178. Any good grade of cold-drawn steel may be used in its construction, such as C-1018, but stainless steel (type 304) is preferred because of its corrosion resistance. These devices are useful where great accuracy is not required and when a load cell for measurements is not available to the designer. Attachments can be made on the tapped end of the piston for convenience in certain applications.

## 6.13 ELECTRICAL AND ELECTRONIC TEST INSTRUMENTS

To aid the circuit designer, technician, test departments, test laboratories, and quality control in the design and fabrication phases of modern electrical and electronic systems, a variety of test instruments are employed, including

- Digital multimeters (DMMs)
- Digital frequency counters

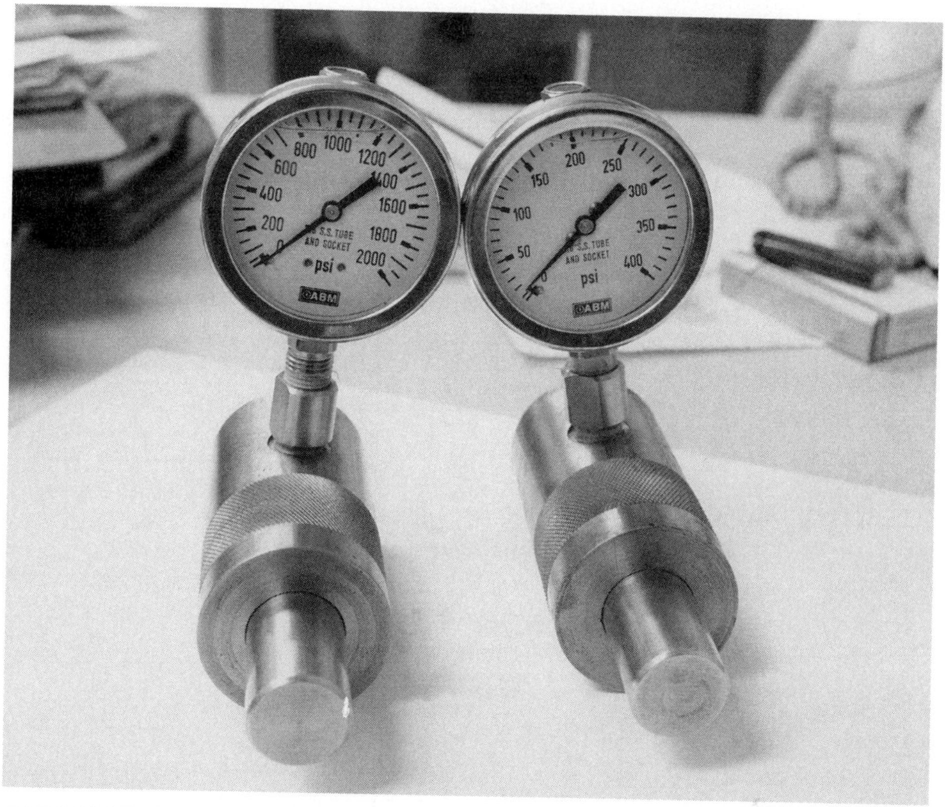


**FIGURE 6.177** Load cell substitutes; hydraulic pressure gauges.

- Audio- and radio-frequency generators
- Sweep frequency generators
- Oscilloscopes
- Regulated power supplies
- Capacitor testors
- Programmable IC testers
- *LCR* meters
- Network analyzers
- Wattmeters
- Transistor testers
- Pulse generators
- Temperature test meters and digital thermometers
- Logic probes and monitors
- Vectorscopes and waveform analyzers and other specialized equipment

Perhaps the two most basic, important, and widely used test instruments for basic circuit design and testing are the multitrace digital oscilloscope and the DMM.

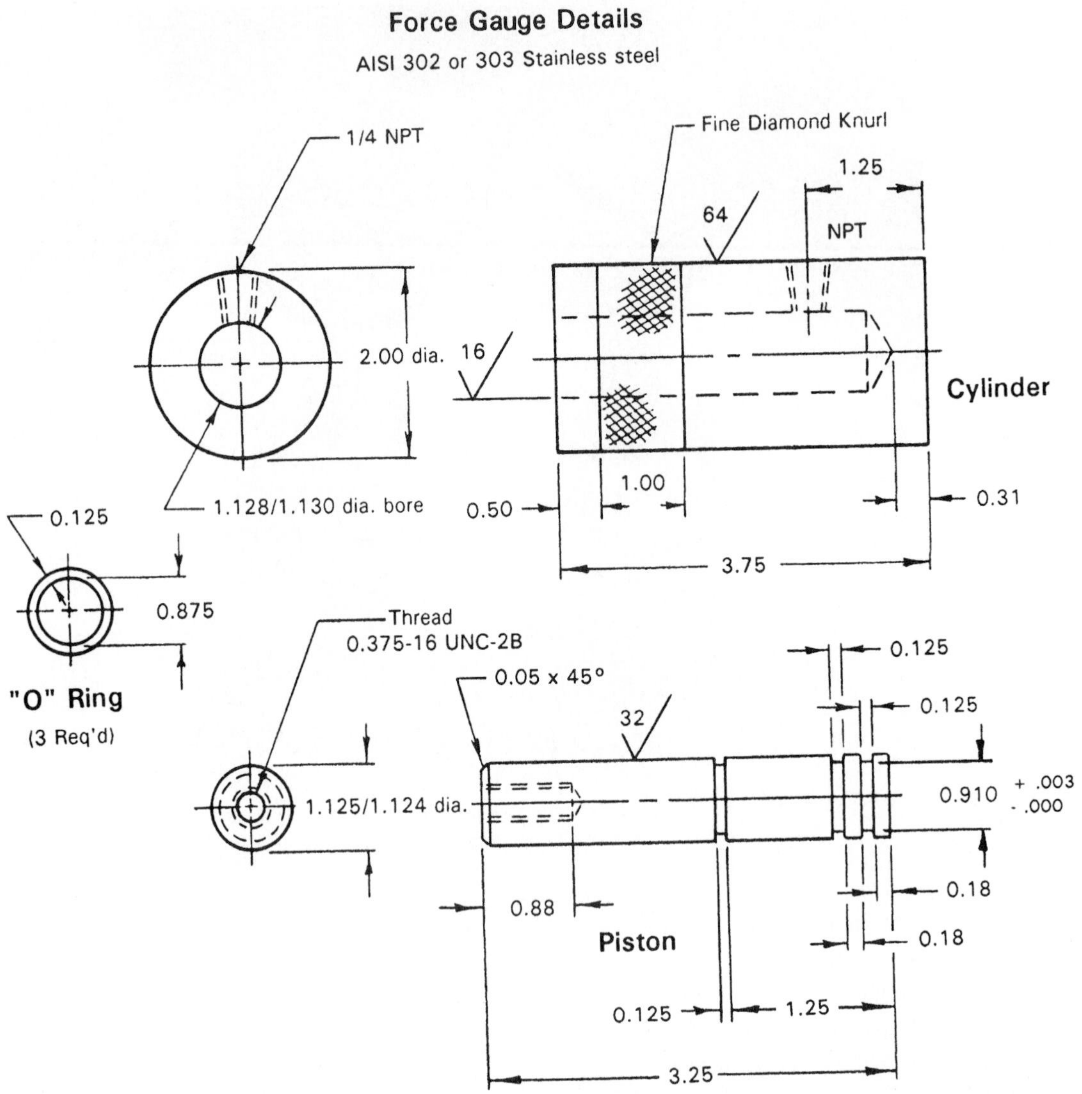


**FIGURE 6.178** Dimensioned drawing for hydraulic pressure gauges shown in Fig. 6.177.

### 6.13.1 The Versatile Oscilloscope

A well-designed multitrace oscilloscope is capable of measuring ac and dc voltages, frequency of a waveform, rise times, phase relationships between waves, and digital pulse timing, and can be used as a null indicator for balancing bridge measuring circuits. No other single electrical or electronic test instrument is capable of all these tasks. With the addition of a camera, one-time, externally triggered events may be recorded, or an oscillograph may be used.

A modern dual-trace, two-channel, 35-MHz oscilloscope is shown in Fig. 6.179. The advent and large-scale use of ICs has made possible the economic production of instruments such as this. Although not suitable for measuring some of the events in sophisticated, high-speed digital circuits, this type of "scope" has broad use for measurements in circuits employing linear ICs and other electrical and electronic circuits operating within the frequency range and sensitivity of this instrument.

### 6.13.2 Voltage and Frequency Measurements Using the Oscilloscope

In Fig. 6.179, we need refer only to the controls labeled 1, 2, and 3, which are, respectively:

1. Variable sweep time per division
2. Channel A volts per division
3. Channel B volts per division

Since we are including here only a single-waveform measurement procedure, only controls 1 and 3 will be used for the following example. With a waveform as depicted in Fig. 6.180 appearing on the scope screen through channel B, the voltage and frequency measurement is obtained as follows:

1. With the scope controls set to view a signal on channel B, rotate the variable volts per division until the waveform is approximately 3 to 4 cm in height. Check the voltage value on

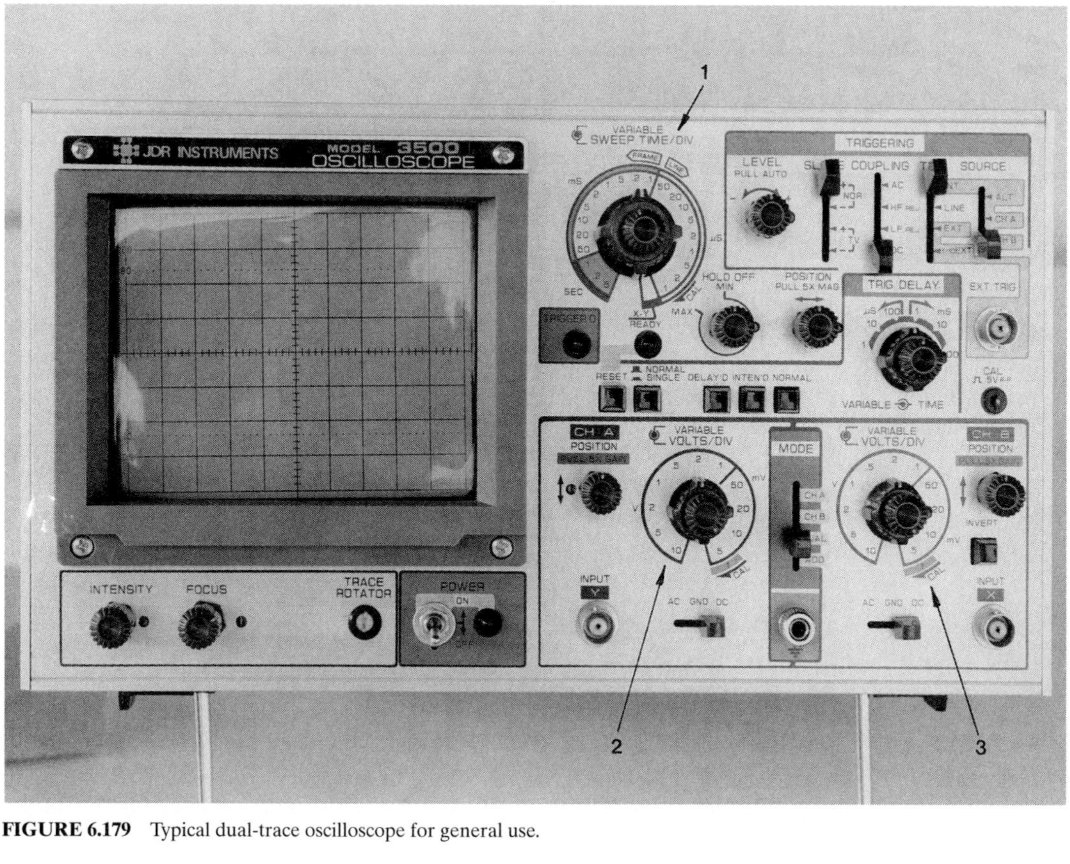


**FIGURE 6.179** Typical dual-trace oscilloscope for general use.

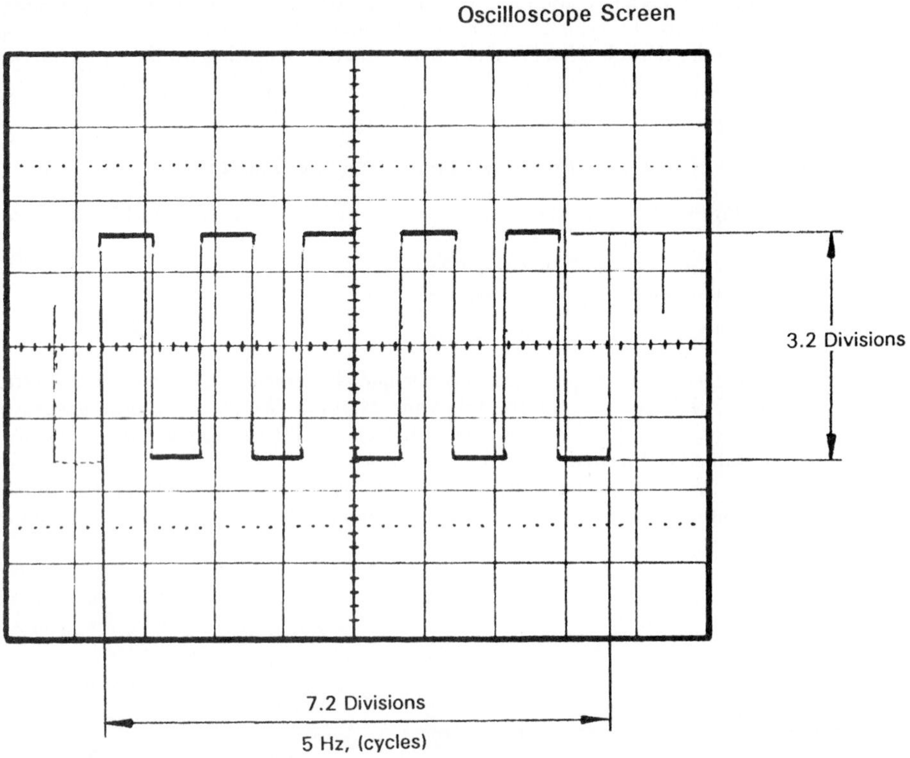


**FIGURE 6.180** A scope trace for measuring frequency with oscilloscope.

the variable volts per division scale. For example, let us say it is 2 V per division. Now, accurately measure the vertical height of the waveform, which is 3.2 divisions per Fig. 6.180. The peak-to-peak (P-P) ac voltage of the waveform is then 2 V per division times 3.2 divisions = 6.4 V P-P.

2. Having previously rotated the variable sweep time per division to allow 5 Hz (cycles) of the wave to be seen on the scope screen, read the sweep time per division. Let this be 0.5 msec ($0.5 \times 10^{-3}$ sec). Now, find the divisions for 1 Hz (cycle) from 7.2 divisions per 5 cycles = 1.44 divisions per cycle.

The frequency $f$ of the observed waveform is then

$$f = \frac{1}{\text{divisions per cycle} \times \text{sec}} = \frac{1}{1.44 \times 0.0005} = 1388.9 \text{ Hz}$$

or 1.389 kHz.

### 6.13.3 The Oscilloscope as a Bridge Null Indicator

Refer to Sect. 6.12.3 on bridge circuits for strain-gauge applications. The oscilloscope, in this case, is used so that no voltage level appears on the scope screen, indicating that the bridge circuit is balanced (zero volts) output, $V_{out} = 0$.

***Other Basic Test Instruments.*** A high-quality 100-MHz digital storage oscilloscope is shown in Fig. 6.181. A sample trace and reading is shown on the scope screen. This instrument is a test laboratory-grade device finding widespread industrial and scientific use worldwide.

Figure 6.182 shows a popular digital multimeter used for accurate voltage and current measurements, both ac and dc. Resistance may also be measured on this instrument as well as continuity testing and diode checking. Figure 6.183 shows a handheld DVM for voltage, resistance, and continuity testing. Small instruments such as this have widespread use for field service and benchwork, where a light, easy-to-use DVM is applicable.

## 6.14 ELECTRONIC PACKAGING TECHNIQUES

The discipline of electronic packaging has advanced considerably in the past 20 years. Circuit components have changed from discrete, to miniature, and to ICs and surface-mount miniature, discrete styles. ASICs are becoming popular and economical, while miniature surface-mount components have made electronic products smaller and more reliable.

An early form of electronic packaging is shown in Fig. 6.184. This package contained an electronic computer system, which today could be packaged in a volume vastly smaller than used in this design. This was a rugged, possibly depressurized package for aerospace applications.

Electronic or electrical packages used externally on aerospace vehicles such as rockets and missiles are usually hermetically sealed to prevent contamination. If sealed, the container must withstand the internal pressure generated when the space vehicle leaves the atmosphere, unless the package was vacuumed. The air or nitrogen purge-gas in the container will generate a minimum of 14.7 psi pressure on the inside of the container when the space vehicle leaves the atmosphere. Unless the container is extremely rigid and strong, this amount of

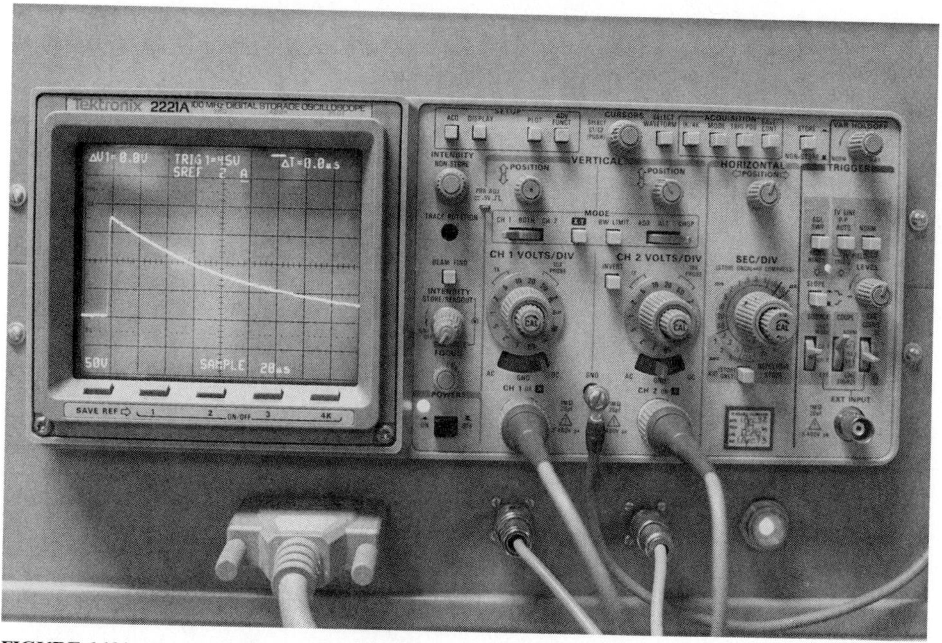


**FIGURE 6.181** A high-quality, 100-MHz laboratory-grade digital storage oscilloscope.

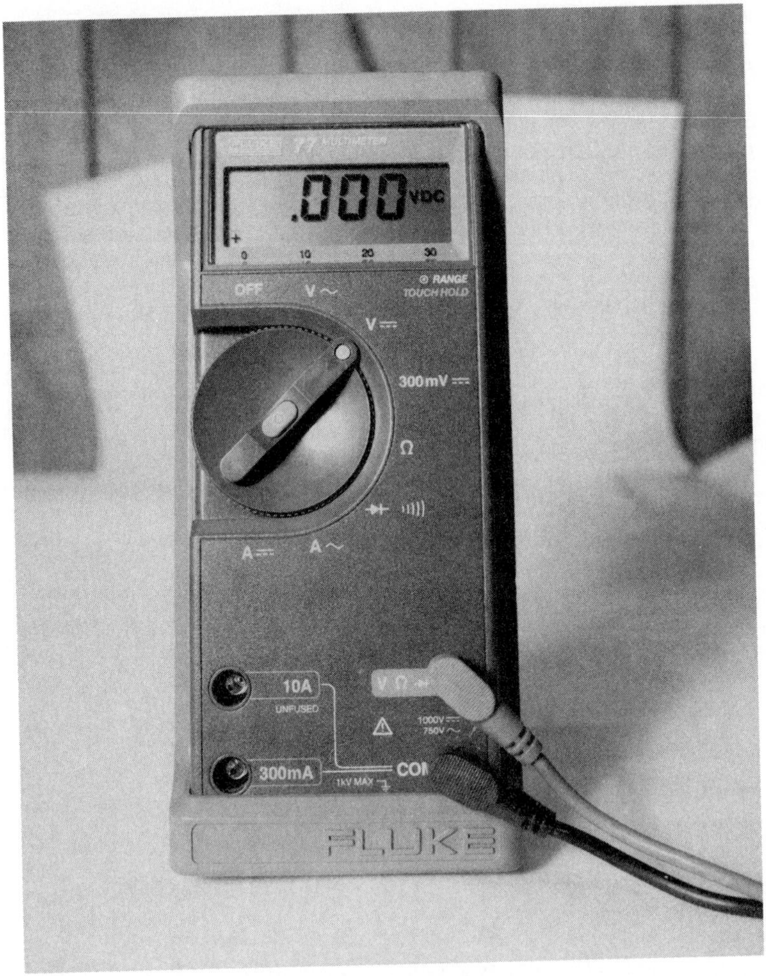


FIGURE 6.182 A high-quality digital multimeter, DVM.

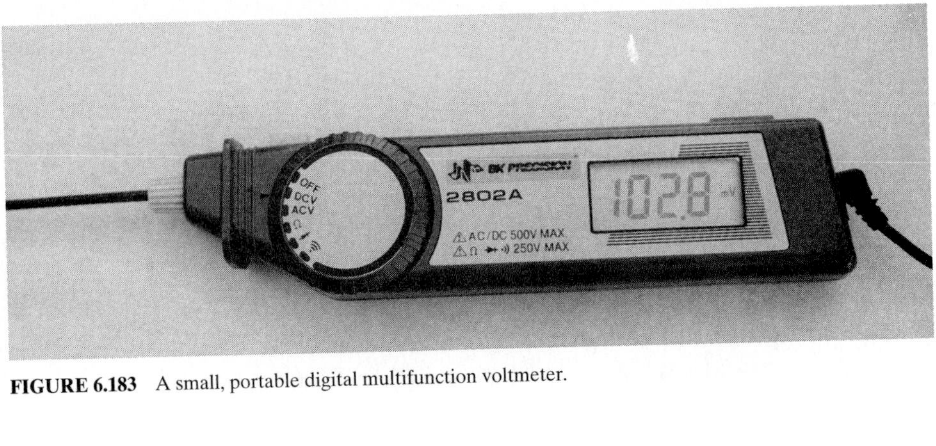


FIGURE 6.183 A small, portable digital multifunction voltmeter.

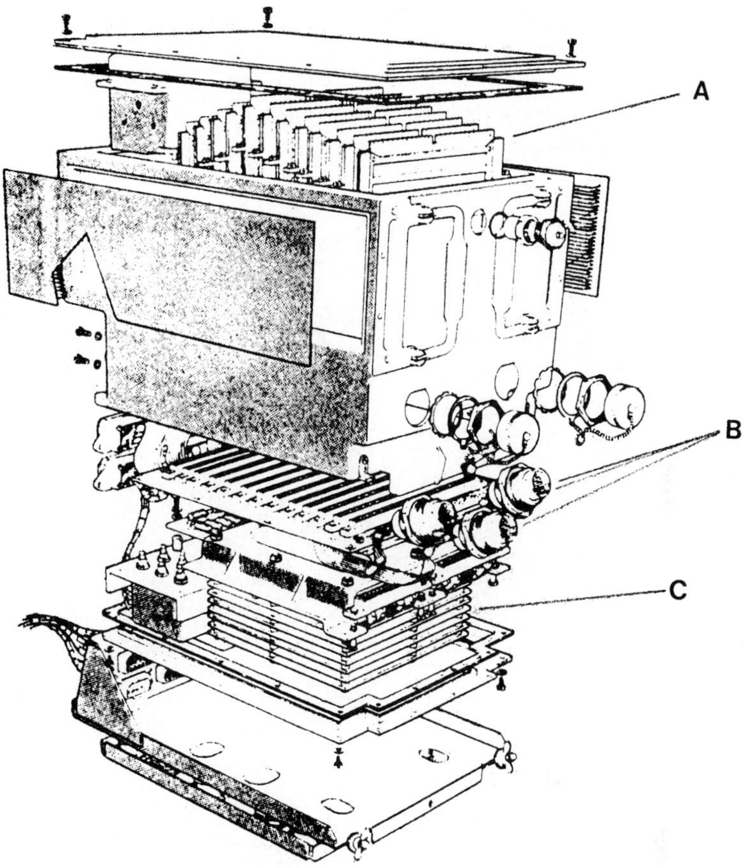


**FIGURE 6.184** An example of high-density electronic packaging.

internal pressure will deform or rupture a plane-surfaced container such as a cube or parallelopiped (rectangular solid). Since aerospace equipment is designed to be as lightweight as possible, the only logical containment would be cylindrical or spherical in shape, wherein high pressures could be safely contained. Spherical shapes are not practical, so the only logical choice was a cylindrical shape. Many of the aerospace electronic packaging designs were purged with nitrogen gas prior to or during the final assembly stages before deployment.

Figure 6.185*a* and 6.185*b* shows electronic packaging designs that were patented by the author more than 20 years ago, for use on aerospace vehicles and military aircraft. This type of packaging is high-density, with shock mount provisions, hermetic sealing, and high-strength container features. This type of electronic packaging was anticipated by the inventor while employed in the aerospace industry.

Figure 6.186*a* and 6.186*b* shows derivatives of the same type of packaging design principles as shown in Fig. 6.185*a* and 6.185*b*. In this style of high-density electronic packaging, the use of ICs was anticipated together with miniature printed circuit boards and the multicontact principles shown in U.S. Patent 3,596,140, wherein thousands of electrical connections could automatically be made at assembly, without recourse to soldering.

Figure 6.187*a* and 6.187*b* shows a form of discrete electrical packaging that was to be used for producing power distribution junction boxes (J boxes), used on aerospace vehicles and

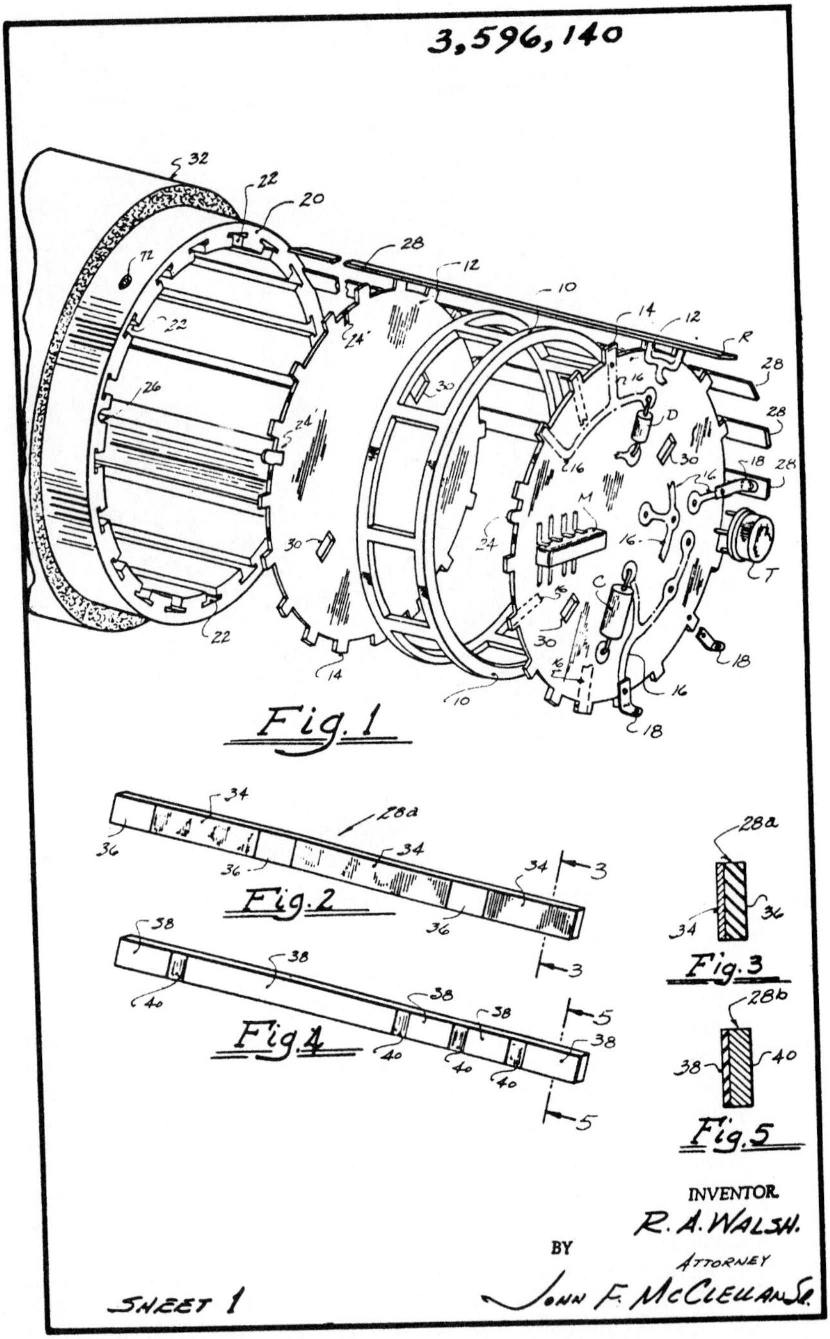


FIGURE 6.185*a* Sheet 1 of high-density electronic packaging letters patent.

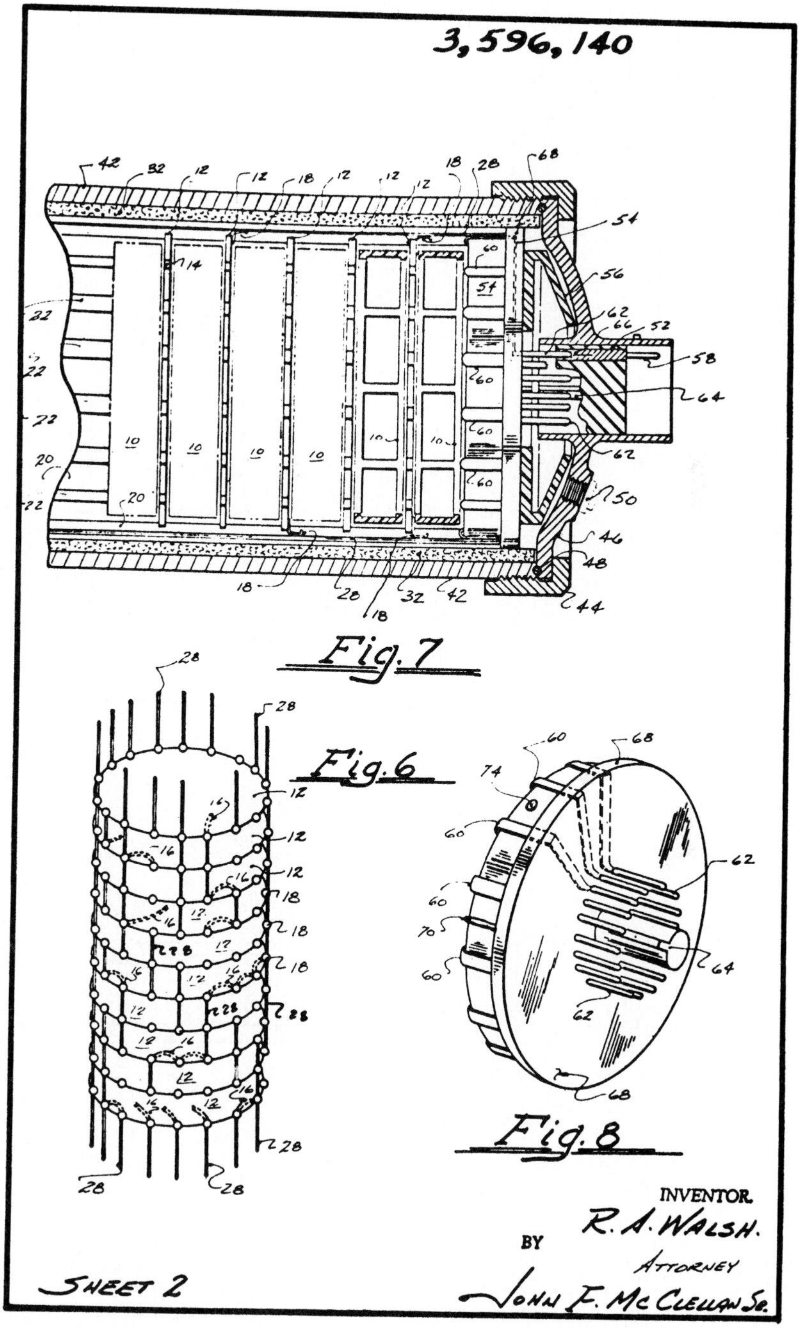


**FIGURE 6.185*b*** Sheet 2 of high-density electronic packaging letters patent.

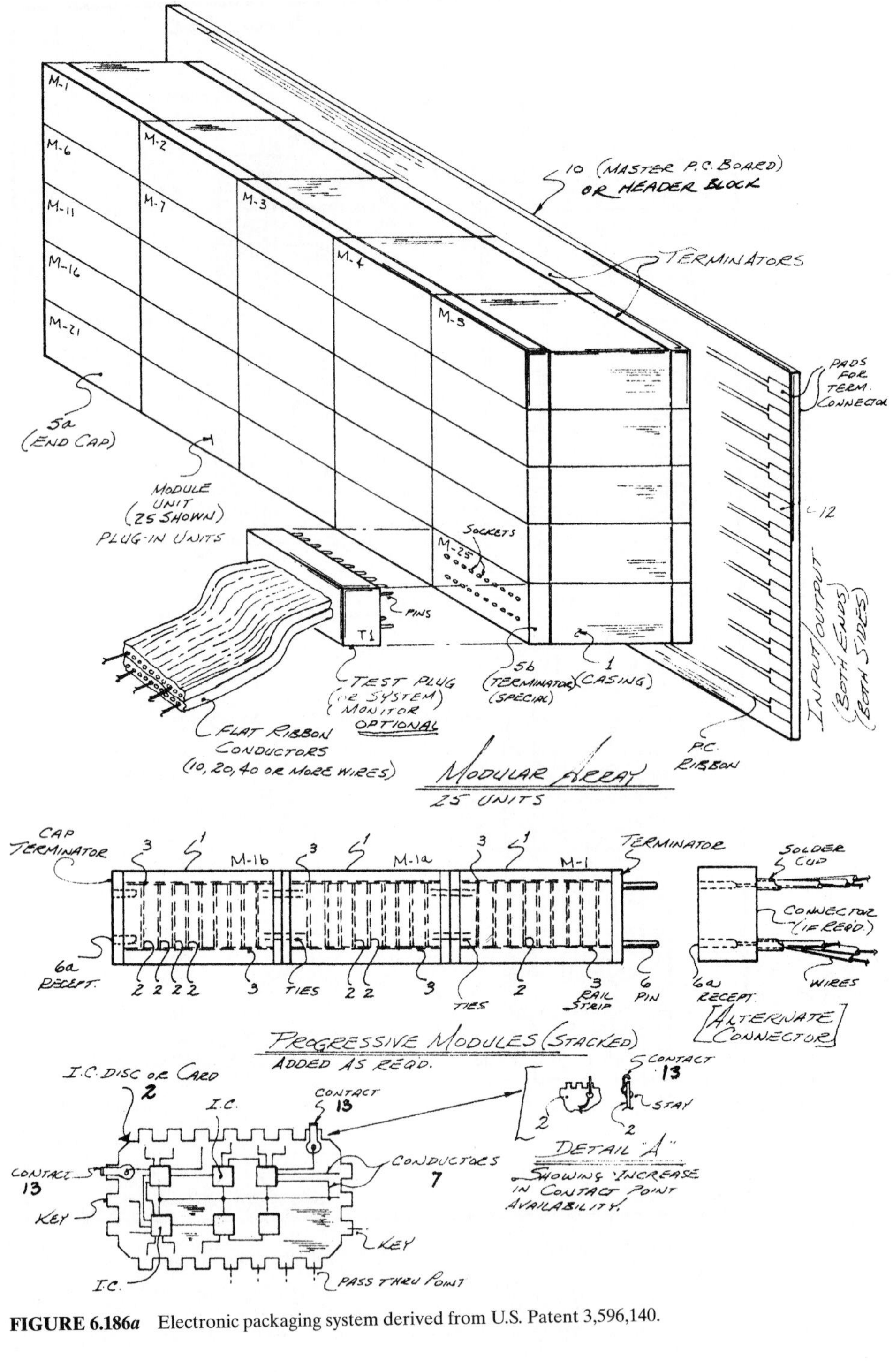


**FIGURE 6.186*a*** Electronic packaging system derived from U.S. Patent 3,596,140.

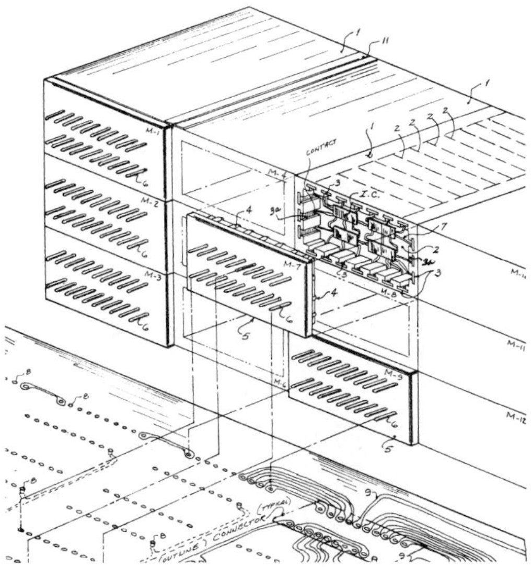

**FIGURE 6.186*b*** Detail view of (*a*).

military aircraft. This is a hard-wired unit type of package with hermetically sealed electrical connectors used to bring the internal wiring to the distribution wire harnesses outside the container. The author secured the rights to these packaging design principles through U.S. Patent 3,596,139, issued in 1971. Both types of packaging principles had been used and are perhaps still used today in various industries.

New electronic packaging designs are constantly evolving, in order to suit the particular applications encountered in the industrial and military environments.

The packaging designer must overcome many obstacles in the initial design phases of electronic packaging, such as function, circuit layout and components, heat, cold, radiation, pressure, shock, size limitations, weight limitations, assembly considerations, economics, producibility, interchangeability, maintainability, hermetic sealing, safety, and risk considerations. Effective design for electronic and electrical packaging units for critical applications is a difficult phase of electromechanical design practice requiring experience and insight.

The electronic packaging design principles shown in U.S. Patents 3,596,139 and 3,596,140 can be improved on for many future applications in electrical and electronic systems. Some of these principles can be seen in Sec. 6.11.6, concerning electrochemical cells and battery systems.

## 6.15 BASIC ELECTRIC CIRCUIT ANALYSIS AND PROCEDURES—DC AND AC

In this section, we will review some of the basic characteristics of *RCL* (resistance, capacitance, and inductance) components in electric circuits and the methods commonly used to analyze both dc and ac circuits.

Assigning the positive current direction is simple for simple dc circuits. The direction is taken as moving from the negative point of the power source, through the load, and then toward the positive source of the dc power supply. The current in a circuit can be designated as either negative or positive, depending on which direction it is flowing. To do this, we must assign one particular current direction as the *positive direction.* The current direction is obvious in a simple dc circuit as explained previously: moving from the negative point in a direction toward the positive point. In complex circuits it is not always possible to know the current directions until a detailed circuit analysis is performed. In these types of circuits, we can assign one arbitrary direction as the positive direction. If after making the calculations or measuring

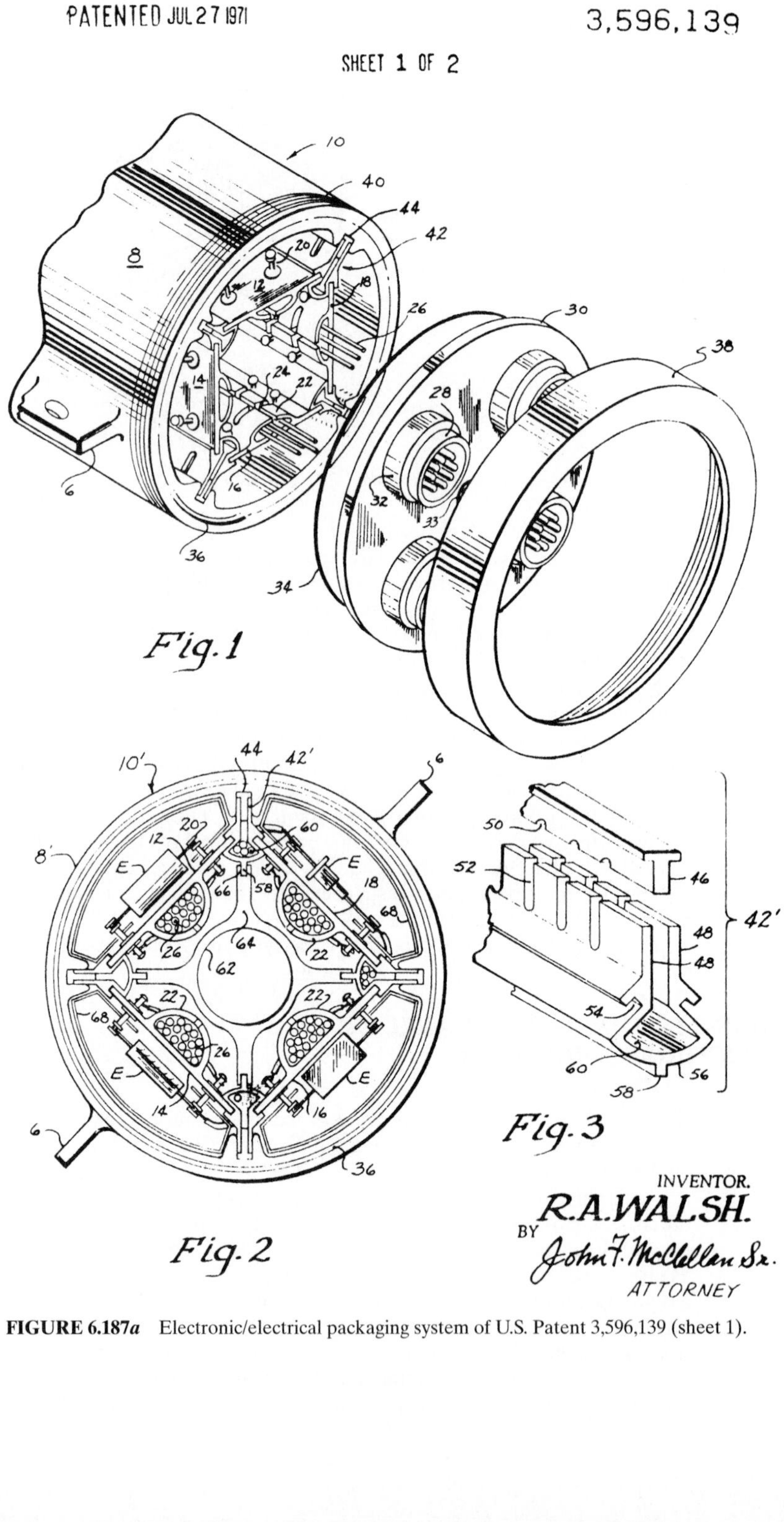


**FIGURE 6.187*a*** Electronic/electrical packaging system of U.S. Patent 3,596,139 (sheet 1).

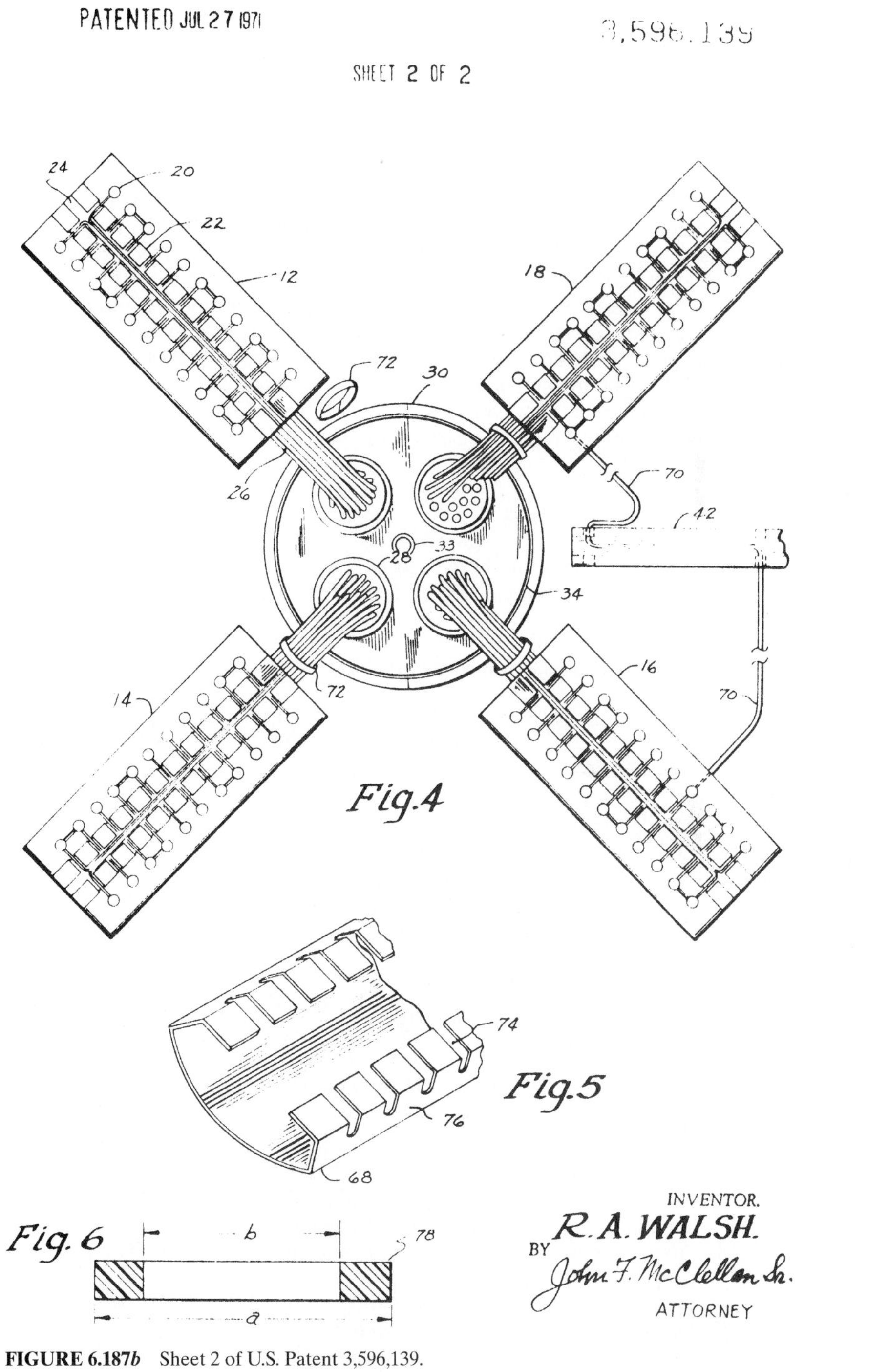


**FIGURE 6.187*b*** Sheet 2 of U.S. Patent 3,596,139.

the current direction we find that the direction actually flows opposite that which was chosen as the positive direction, we simply reverse the current directions on the circuit diagram. Or, we can leave the current direction arrows as they were assigned, and indicate that the current direction is negative. In an ac circuit, the current direction periodically alternates from one direction to the opposite direction. We then assign one of the directions as the positive direction and the other direction automatically becomes the negative direction. Actual electron flow in a dc circuit is from the negative point toward the positive point.

Only a relatively small number of circuits operate with voltage and current values that are constant. In general, the voltages and currents in most electric circuits are *time-varying*. Time-varying voltages and currents are represented by a graph showing how the voltage or current varies with time. This time-varying graph is called a *waveform* and is what an oscilloscope trace represents on the oscilloscope screen. The most common waveform is sinusoidal in shape and is representative of the ac waveform of the voltage and current as supplied by the electric power companies. This waveform may be easily seen on any oscilloscope screen simply by probing the voltage of an ac power source and adjusting the sweep time and sensitivity controls, until the waveform appears on the screen. Waveforms assume an infinite number of shapes, according to the generating source and its associated circuit design.

### 6.15.1 Basic *RCL* Characteristics

Table 6.32 shows the three basic circuit properties of resistance, capacitance, and inductance. Table 6.33 Shows the ac circuit characteristics of resistance, inductance, and capacitance.

### 6.15.2 Circuit Analysis Laws, Theorems, and Methods

The circuit analysis laws, theorems, and methods that are commonly used by the circuit analyst to simplify the processes involved in the detailed analysis of dc and ac circuits are represented by the following:

- Ohm's law
- Kirchhoff's voltage and current laws
- Maxwell's loop (or mesh) equations
- The Thévenin theorem
- The Norton theorem
- The superposition theorem
- Maximum power-transfer theorem
- Reciprocity theorem
- Delta-Y and Y-delta conversions and transforms
- The branch current method
- The node voltage method (nodal analysis)

Circuit analysis utilizes the preceding laws, theorems, and methods to help simplify circuits and make their analysis easier and quicker. Shown here are some of these methods in a simplified form to familiarize the electromechanical designer with these methods. In doing so, we hope this will encourage further study by those electromechanical designers who wish to increase their knowledge of circuit analysis. Only the basics are shown, but even this allows the electromechanical designer to better understand the more basic types of electric circuits comprised of support components surrounding the ICs, which were emphasized earlier, as an important tool in many electromechanical devices and systems.

Earlier sections of this chapter have shown many of the equations and calculation procedures involved in industrial electrical and electronic design work, as well as the basic laws and definitions.

***Ohm's Law.*** In an electrical circuit, the current $I$ is directly proportional to the applied emf (electromotive force), or voltage, and is inversely proportional to the resistance $R$ or impedance $Z$. Thus

$$E = \frac{I}{R} \qquad I = \frac{E}{R} \qquad R = \frac{E}{I} \qquad \text{(dc circuits)}$$

See Sec. 6.2.2 for Ohm's law applications in dc and ac circuits.

**TABLE 6.32** Resistance, Capacitance, and Inductance Properties

| Resistance, R | Capacitance, C | Inductance, L |
|---|---|---|
| $i = v/R$ | $Q = Cv$<br>$i = C\,(dv/dt)$ | $v = L\,(di/dt)$ |
| Opposes the flow of current | Opposes any sudden change in voltage with $\tau = RC$ | Opposes any sudden change in current with $\tau = L/R$ |
| Dissipates energy as heat<br>$(p = i^2R = v^2/R)$ | Stores energy $= 1/2Cv^2$ | Stores energy $= 1/2Li^2$ |
| $R_T = R_1 + R_2 \ldots + R_n$ | $1/C_T = 1/C_1 + 1/C_2 \ldots + 1/C_n$<br>Series Combinations | $L_T = L_1 + L_2 \ldots + L_n$ |
| $1/R_T = 1/R_1 + 1/R_2 \ldots + 1/R_n$ | $C_T = C_1 + C_2 \ldots + C_n$<br>Parallel Combinations | $1/L_T = 1/L_1 + 1/L_2 \ldots + 1/L_n$ |
| Ideal R has fixed resistance for any input voltage. Practical R will have its resistance change at higher voltages. | Ideal C dissipates no energy. Practical C has leakage resistance $R_l$ which does dissipate some power. | Ideal L has no resistance and dissipates no energy. Practical L dissipates energy due to its winding resistance and also due to hysteresis losses in the core. |
| R acts the same during transient interval and in steady-state. | C opposes changes in $v$ during transient interval and acts as an open circuit in dc steady-state. | L opposes change in $i$ during transient interval and acts as a short-circuit in dc steady-state. |

**TABLE 6.33** Ac Characteristics of Resistance, Capacitance, and Inductance

| Element | R (Resistance) | L (Inductance) | C (Capacitance) |
|---|---|---|---|
| Basic Units. | ohms - Ω | henries - H | farads - F |
| Opposition to dc: | resistance R<br>(I = E/R) | zero (short-circuit) | infinite (open circuit) |
| Opposition to ac: | resistance R<br>(I = E/R) | reactance $X_L$<br>$I = E/X_L$ | reactance $X_C$<br>$I = E/X_C$ |
| Effects of Frequency: | none: resistance = R at all $f$ | $X_L$ increases with $f$:<br>$X_L = 2\pi fL$ | $X_C$ decreases with $f$:<br>$X_C = 1/2\pi fC$ |
| Series Combination: | $R_T = R_1 + R_2 \ldots\ldots + R_n$ | $X_{LT} = X_{L1} + X_{L2} \ldots + X_{Ln}$ | $X_{CT} = X_{C1} + X_{C2} \ldots\ldots + X_{Cn}$ |
| Parallel Combination: | $1/R_T = 1/R_1 + 1/R_2 \ldots + 1/R_n$ | $1/X_{LT} = 1/X_{L1} + 1/X_{L2} \ldots + 1/X_{Ln}$ | $1/X_{CT} = 1/X_{C1} + 1/X_{C2} \ldots + 1/X_{Cn}$ |
| Phase Angle: | $i$ in phase with $e$ | $i$ lags $e$ by 90° | $i$ leads $e$ by 90° |
| Voltage and Current: | $e = E_M \sin 2\pi ft$<br>$i = (E_M/R) \sin 2\pi ft$ | $i = (E_M/X_L) \sin(2\pi ft - \pi/2)$ | $i = (E_M/X_C) \sin(2\pi ft + \pi/2)$ |
| Phasor Diagram: | 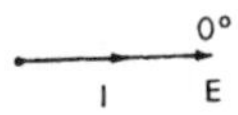  | 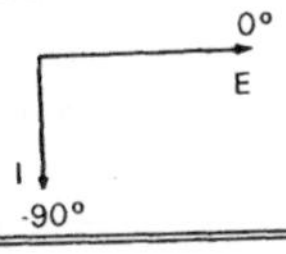  | 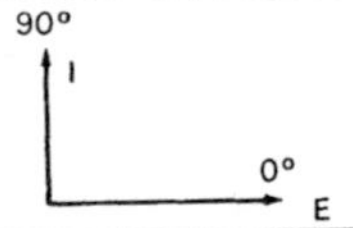  |

***Kirchhoff's Laws.*** The three basic Kirchhoff laws shown herein apply to voltages in *series* circuits and to currents in *parallel* circuits.

1. Kirchhoff's voltage law: "The sum of the voltage drops across components in a series circuit is equal to the source or supply voltage." See Fig. 6.188, where $V_s = V_1 + V_2 + V_3$. By transposition, the voltage drop across any resistive component may be determined when the source voltage and other voltage drops are known:

$$V_1 = V_s - V_2 - V_3 \qquad V_2 = V_s - V_1 - V_3 \qquad V_3 = V_s - V_1 - V_2$$

2. Kirchhoff's current law 1: "The sum of the currents in the branches of a parallel circuit is equal to the total current drawn from the source or supply current." See Fig. 6.189, where $I_T = I_1 + I_2 + I_3 + I_4$. By transposition, $I_1 = I_T - I_2 - I_3 - I_4$, etc.
3. Kirchhoff's current law 2: "At any junction in an electrical circuit, the total current flowing toward the junction is equal to the total current flowing away from the junction" (see Fig. 6.190).

The three preceding laws may be used to solve many simple circuit problems. As the circuits become more complex, these laws still apply, but easier solutions to circuit analysis are obtained using Thévenin's theorem, Norton's theorem, the superposition theorem, and others, which we will review in the following sections.

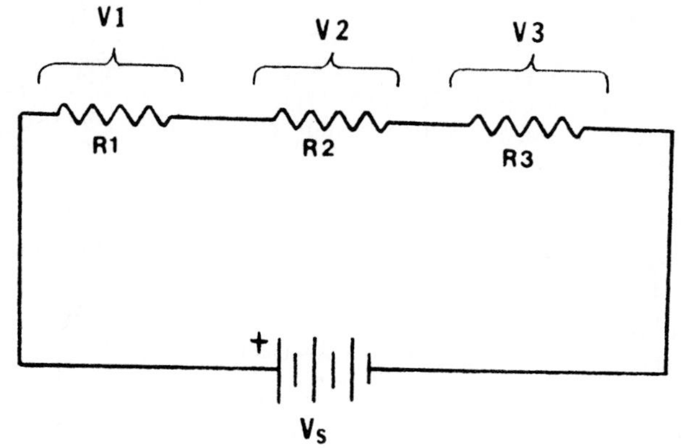


**FIGURE 6.188** Series circuit.

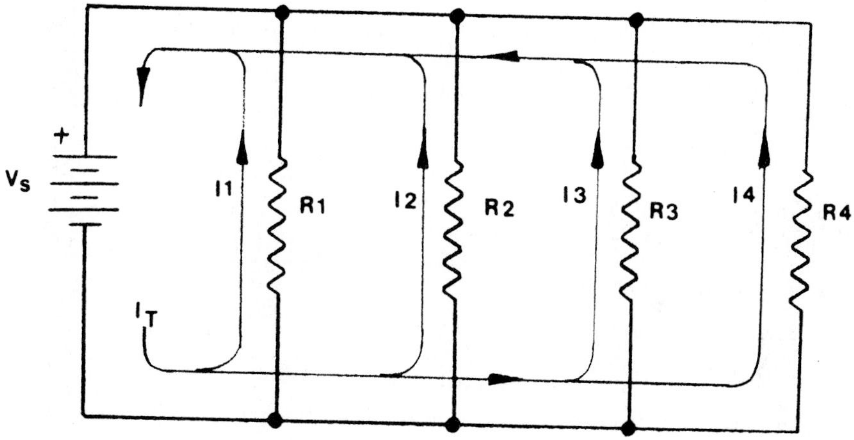


**FIGURE 6.189** Parallel circuit.

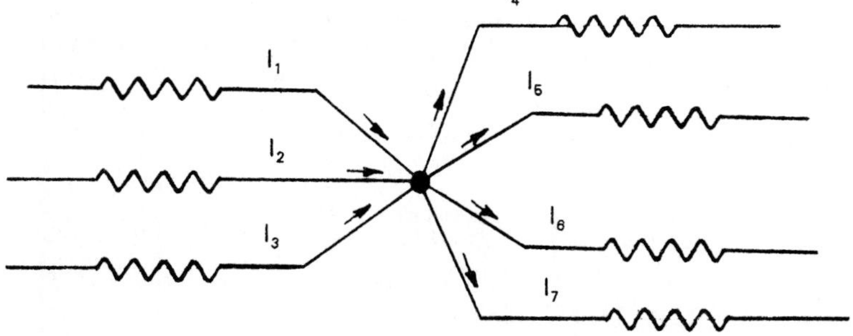


$$I_1 + I_2 + I_3 = I_4 + I_5 + I_6 + I_7$$

$$I_1 + I_2 + I_3 - I_4 - I_5 - I_6 - I_7 = 0$$

**FIGURE 6.190** Kirchhoff's current law.

***The Thévenin Theorem.*** "The current that will flow through an impedance $Z_1$ when connected to any two terminals of a linear network between which an open-circuit voltage $E$ and impedance $Z$ previously existed is equal to the voltage $E$ divided by the sum of $Z$ and $Z_1$."

*Method.* To determine the current flow of a branch circuit containing a resistance $R$ of an active network containing a voltage source $E$, thévenize the circuit by using the following steps (see Figs. 6.191 to 6.195).

1. Remove the resistance $R$ from that branch.
2. Determine the open-circuit voltage $E$ across the break at $A$ and $B$, using Ohm's law and Kirchhoff's law. This is the Thévenin equivalent voltage, $V_{TH}$.
3. Remove each source of voltage and replace it by a resistance equal to the sources' internal resistance $r$ "looking in" at the break. The total resistance is the Thévenin equivalent resistance $R_{TH}$.

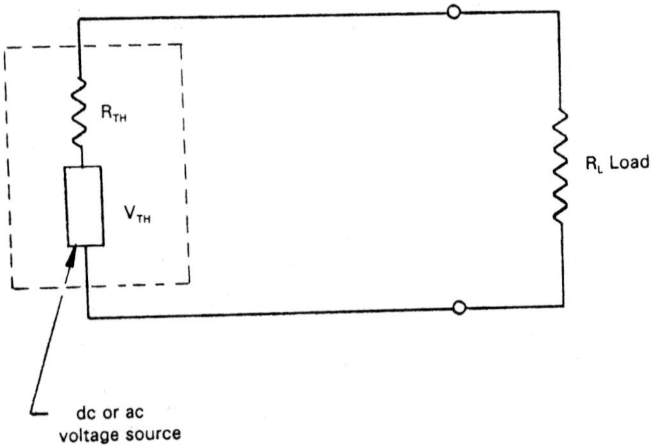


**FIGURE 6.191** Thévenin equivalent circuit.

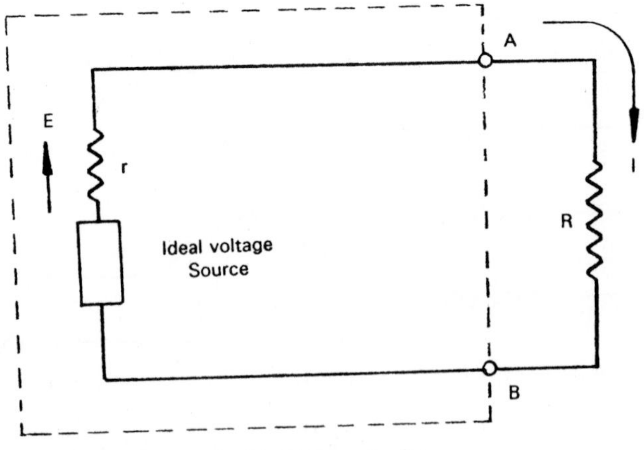


**FIGURE 6.192** Step in thévenizing a circuit.

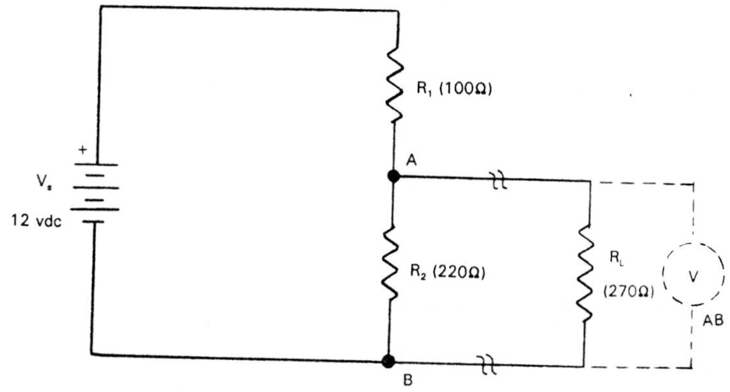


**FIGURE 6.193** Step in thévenizing a circuit.

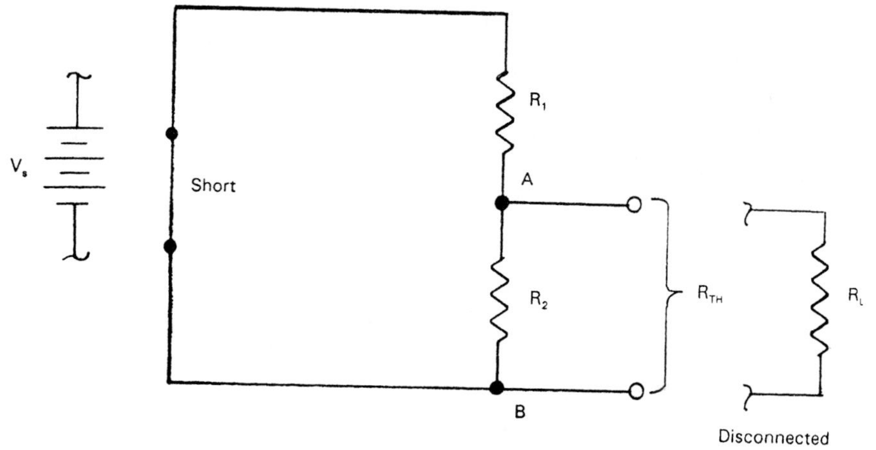


**FIGURE 6.194** Step in thévenizing a circuit.

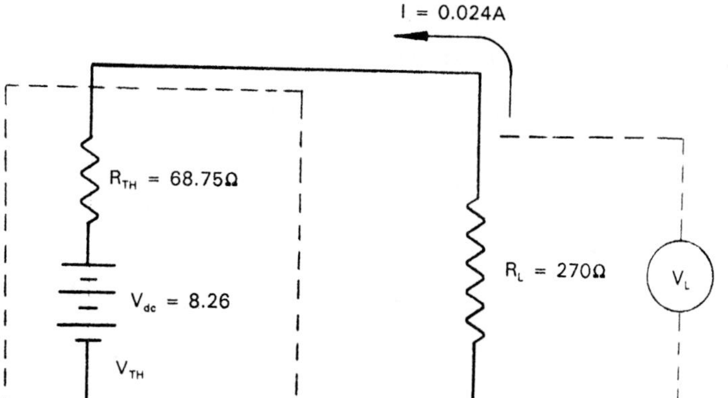


**FIGURE 6.195** Step in thévenizing a circuit.

4. Determine current $I$ flowing through resistance $R$ from the Thévenin equivalent circuit shown in Fig. 6.192.

Then
$$I = \frac{E}{R + r}$$

Then the load voltage is
$$V_L = IR_L$$

where
$$R_L = R_T - R_{\mathrm{TH}} \quad \text{and} \quad R_T = \frac{Rr}{R + r}$$

*Example.* To thévenize the circuit shown in Fig. 6.193

1. Remove the load resistance.
2. Calculate the Thévenin equivalent voltage $V_{\mathrm{TH}} = V_{AB}$. The voltage across points $A$ and $B$ follows the voltage-divider rule:

$$V_{AB} = V_s \frac{R_2}{R_1 + R_2}$$
$$= V_{\mathrm{TH}} = V_s\left(\frac{220}{100 + 220}\right)$$
$$V_{\mathrm{TH}} = 12\,(0.688) = 8.26 \text{ V dc across points } A \text{ and } B$$

3. Compute the Thévenin equivalent resistance $R_{\mathrm{TH}}$ by leaving the load disconnected and short-circuiting the voltage source. The circuit now assumes the form shown in Fig. 6.194. Since the resistors are in parallel

$$\frac{1}{R_{\mathrm{TH}}} = \frac{1}{R_1} + \frac{1}{R_2} \quad \text{or} \quad R_{\mathrm{TH}} = \frac{R_1 R_2}{R_1 + R_2}$$
$$= \frac{100(220)}{100 + 220} = 68.75\ \Omega$$

The total Thévenin equivalent resistance is then 68.75 Ω.

4. The Thévenin equivalent circuit now appears as shown in Fig. 6.195. Now, calculate the total circuit resistance $R_{\mathrm{TC}}$:

$$R_{\mathrm{TC}} = R_{\mathrm{TH}} + R_L = 68.75 + 270 = 338.75\ \Omega$$

Then, calculate the current from

$$I = \frac{V_{\mathrm{TH}}}{R_{\mathrm{TC}}} \quad \text{(from step 2)}$$
$$= \frac{8.26}{338.75}$$
$$= 0.024 \text{ A} \quad \text{or} \quad 24 \text{ mA}$$

The load voltage is then

$$V_L = I\,(R_L)$$
$$= 0.024\,(270) = 6.48 \text{ V dc} \quad \text{(verified by experiment)}$$

The Thévenin theorem is widely used to reduce a circuit to an equivalent voltage source as explained in the preceding example.

***The Superposition Theorem.*** When a number of voltages, distributed in any manner throughout the network, are applied to the network simultaneously, the current that flows is the algebraic sum of the component currents that would flow if the same voltages had acted individually. Also, the potential difference that exists between any two points is the component potential difference that would exist there under the same conditions.

*Method*

1. Disable all but one voltage source in the circuit. Short-circuit all other voltage sources.
2. Calculate the total circuit resistance. Calculate the currents and voltages acting on each component using Kirchhoff's and Ohm's laws.
3. Steps 1 and 2 are repeated for each voltage source in the circuit.
4. Combine the currents or voltages per step 2 for the selected component by algebraic addition.

In the practical application of the superposition and Thévenin theorems, the power sources in the circuit must be voltage sources. A perfect voltage source has zero output impedance or zero internal resistance. No voltage source is perfect as there will always be internal resistance. An electrochemical battery is one of the best voltage sources. A well-regulated, low-output-impedance power supply is also an excellent voltage source. Internal resistance of a voltage source must be taken into consideration while performing Thévenin and superposition analyses.

***The Voltage-Divider Equation.*** The voltage divider is one of the most useful of all electric circuits. A typical voltage divider circuit is shown in Fig. 6.196, and the voltage division may be calculated as follows:

$$V_0 = V_1 \frac{R_2}{R_1 + R_2}$$

***The Norton Theorem.*** The current in any impedance $Z_R$, connected to two terminals of a network, is the same as though $Z_R$ were connected to a constant-current generator, the gen-

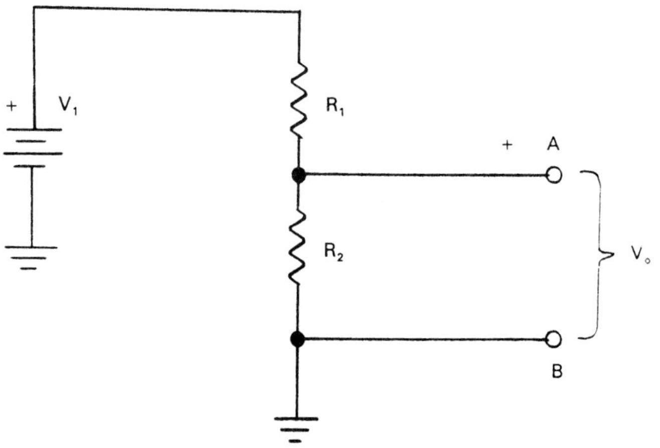


**FIGURE 6.196** Voltage-divider circuit.

erated current of which is equal to the current that flows through the two terminals when these terminals are short-circuited; the constant-current generator is in parallel with an impedance equal to the impedance of the network, looking back from the terminals.

*Method*

1. Apply a short circuit across terminals $A$ and $B$.
2. Determine the short-circuit current $I_s$.
3. Remove each source of voltage and replace them by resistances equal to their internal resistances, or, if a current source exists, replace with an open circuit. Then determine the resistance $R$ looking in at a break made between points $A$ and $B$.
4. Determine the current $I$ flowing in resistance $R_L$ from the Norton equivalent circuit shown in Fig. 6.197.

Then

$$I = I_s \frac{R}{R + R_L}$$

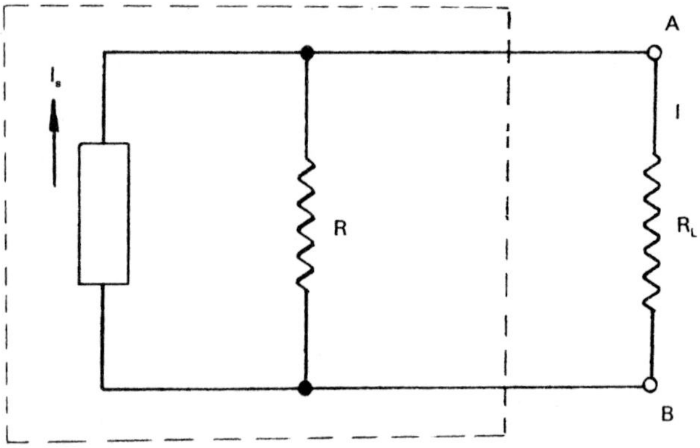


**FIGURE 6.197** Norton equivalent circuit.

(*Note:* A Thévenin equivalent circuit having a voltage $E$ and internal resistance $r$ can be replaced by a Norton equivalent circuit containing a current source $I_s$ and internal resistance $R$, where $R = r$, $E = I_sR$, and $I_s = E/r$. See Fig. 6.198 for the Thévenin and Norton equivalent circuits.)

***The Maximum Power-Transfer Theorem.*** The power transfered from an electric power source to a load is at a maximum when the load resistance is equal to the supply sources' internal resistance, or output impedance.

***The Reciprocity Theorem.*** In any system composed of linear bilateral impedances, if an electromotive force $E$ is applied between any two terminals and the current $I$ is measured in any branch, their ratio, called *transfer impedance,* will be equal to the ratio obtained if the positions of $E$ and $I$ are interchanged.

(*Note:* A circuit *node* is a junction point, a terminal of any branch of a network, or a terminal common to two or more branches of a circuit.)

***Simple Circuit Analysis Examples.*** Using the superposition theorem, analyze the circuit shown in Fig. 6.199 to find the output voltage of a double-voltage source circuit. The circuit shown in Fig. 6.199 is a voltage-divider network with two voltage sources. Voltage source $V_1$

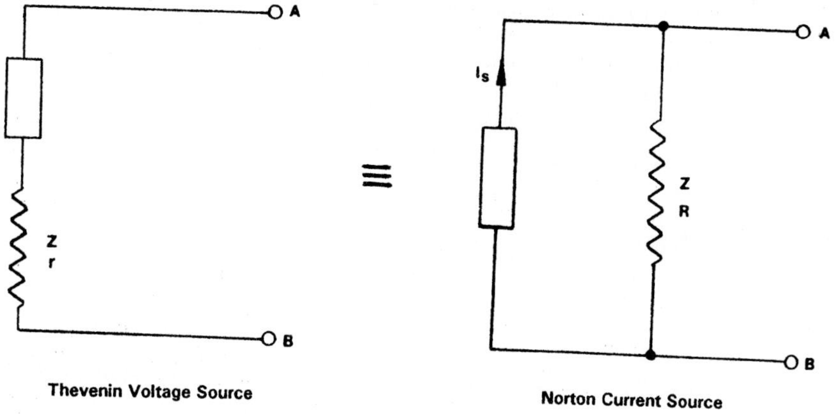


**FIGURE 6.198** Thévenin and Norton equivalent circuits.

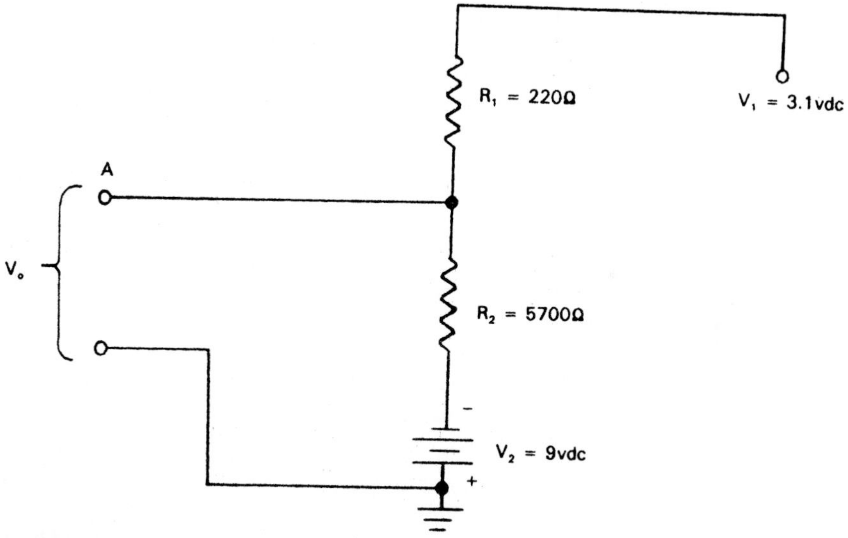


**FIGURE 6.199** Circuit for superposition theorem.

is +3.1 V dc, and voltage source $V_2$ is −9.0 V dc. What is the voltage from terminal $A$ to ground $V_0$?

*Procedure*

**1.** Replace the −9.0-V-dc source $V_1$ with a short circuit. Then

$$V_{A1} = \frac{V_1 R_2}{(R_1 + R_2)} \quad \text{(voltage-divider equation)}$$

$$= \frac{3.1(5700)}{(220 + 5700)} = \frac{17{,}670}{5920} = 2.98 \text{ V dc}$$

**2.** Reconnect $V_2$ and short-circuit $V_1$. Then

$$V_{A2} = \frac{V_2 R_1}{R_1 + R_2}$$

$$= \frac{-9(220)}{(220 + 5700)} = \frac{-1980}{5920} = -0.33 \text{ V dc}$$

Therefore

$$V_0 = V_{A1} + V_{A2}$$

$$= (2.98) + (-0.33)$$

$$= +2.65 \text{ V dc from point } A \text{ to ground}$$

***Measuring Internal Resistance or Output Impedance of a Voltage Source or Power Supply***

*Procedure.* Refer to Fig. 6.200.

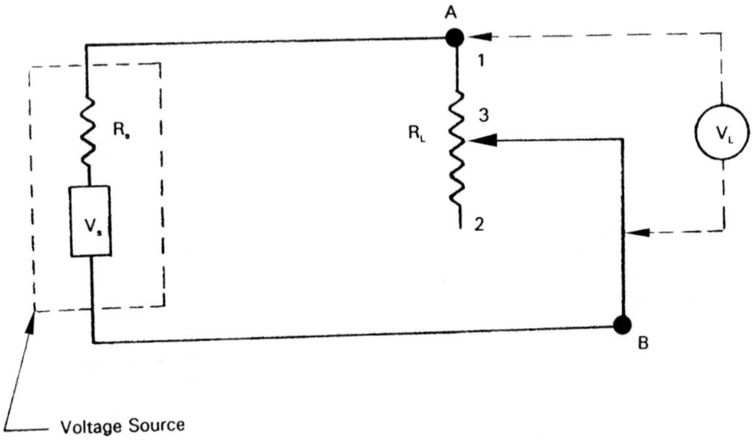


**FIGURE 6.200** Measuring internal resistance or output impedance.

1. Select a linear-taper potentiometer whose resistance value is 10 kΩ, that is, $R_L = 10$ kΩ.
2. Measure the open-circuit voltage between points $A$ and $B$ as in Fig. 6.200.
3. Connect the potentiometer between points 1 and 3.
4. Connect a voltmeter or oscilloscope across the potentiometer, points 1 to 3.
5. Adjust the potentiometer until the load voltage $V_L$ is exactly half the open-circuit voltage of the source, as in step 2. As an example, you would adjust $R_L$ so that $V_L = 5$ V.
6. Remove the potentiometer $R_L$ from the circuit and measure the resistance in the resistance leg, points 1 to 3. As an example, if the ohmeter reads 420 Ω, this is the voltage source output impedance or internal resistance.

(*Note:* Ohm's law for series circuits makes this technique possible. For higher accuracy, repeat the procedure using a 500-Ω potentiometer.)

***Series-Parallel Resistances.*** The combination series-parallel resistive circuit shown in Fig. 6.201 may be solved for total resistance between points $A$ and $B$ as follows. To find the equivalent resistance between points $A$ and $B$:

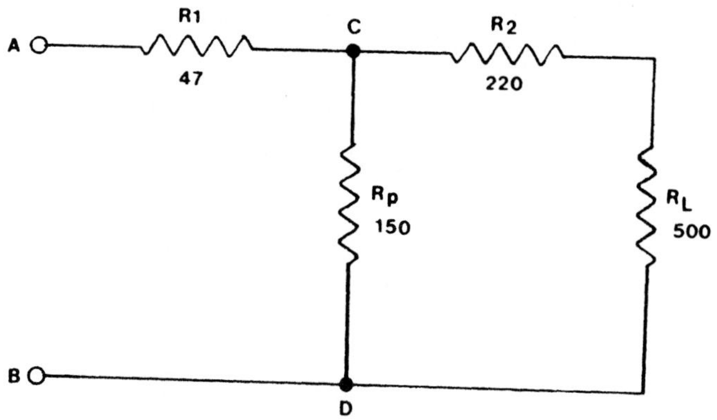


**FIGURE 6.201** Series-parallel resistances.

1. Combine the resistances in series.
2. Convert the series resistance into conductance.
3. Add these with other conductances in parallel.

*Procedure*

1. Begin by combining the resistances $R_2$ and $R_L$ into one resistance and determine the corresponding conductance:

$$\text{Conductance} = \frac{1}{R_2 + R_L} = \frac{1}{650} = 0.00154 \text{ S}$$

2. Add this conductance to $1/R_p$: $0.00154 + 1/150 = 0.00154 + 0.00667 = 0.00821$.
3. This is equal to the conductance between points $C$ and $D$ (0.00821 S).
4. The reciprocal of this conductance gives the equivalent resistance between points $C$ and $D$ ($1/0.00821 = 121.8\ \Omega$).
5. Add the resistance $R_1$ to the resistance determined in step 4; this will be the equivalent resistance between points $A$ and $B$ ($121.8 + 47 = 168.8\ \Omega$).

This procedure has many uses in circuit analysis.

***The Resistive Current-Divider Rule.*** In the circuit shown in Fig. 6.202, the current through $R_1$ may be found from

$$I_1 = \left(\frac{R_2}{R_1 + R_2}\right) I_T$$

where $I_1$ = current through $R_1$
$R_1$ = resistance 1
$R_2$ = resistance 2
$I_T$ = main supply current (known value)

***The Resistive Voltage Divider for Producing a Desired Output Voltage.*** From the voltage-divider equation, with $R_2$ unknown, we solve for $R_2$. Let $R_2 = x$; then (see Fig. 6.203)

$$V_0 = V_s \left(\frac{R_2}{R_1 + R_2}\right)$$

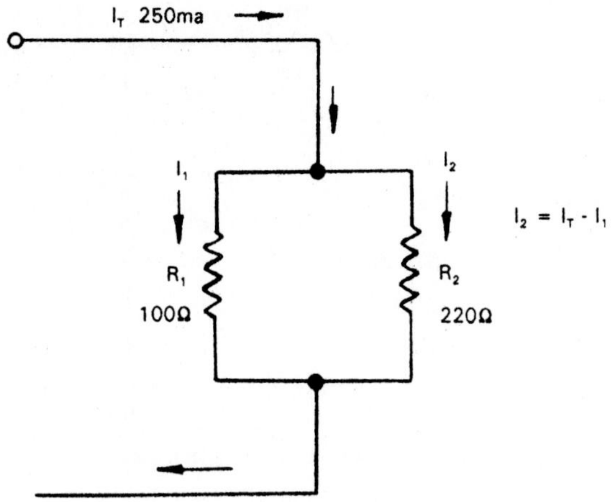


**FIGURE 6.202** Resistor current divider.

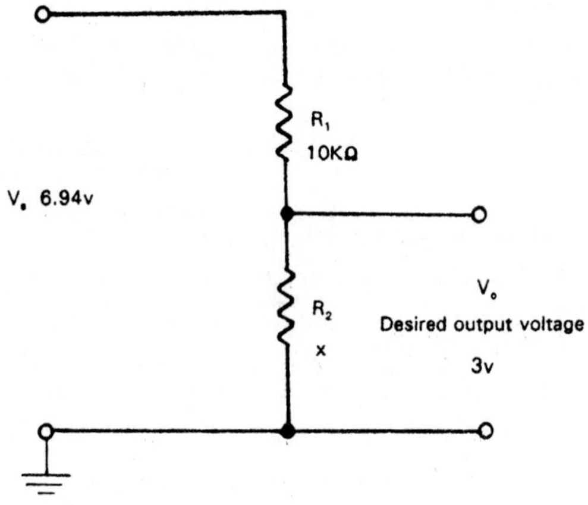


**FIGURE 6.203** Resistor voltage divider.

$$3 = 6.94\left(\frac{x}{10{,}000 + x}\right)$$

$$3 = \frac{6.94x}{10{,}000 + x}$$

$$3x + 30{,}000 = 6.94x$$

$$-3.94x = -30{,}000$$

$$x = 7614\ \Omega = R_2$$

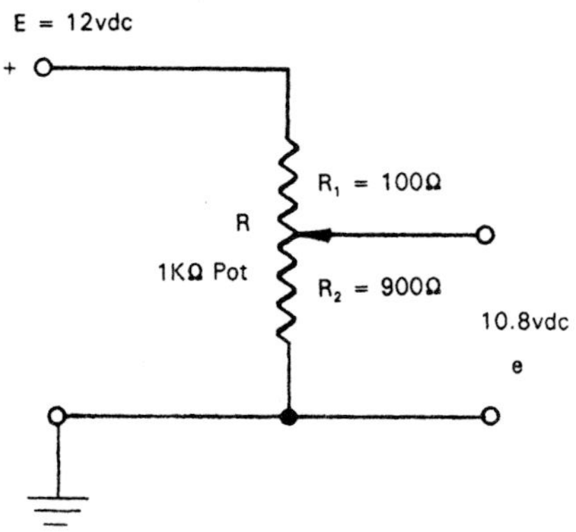


**FIGURE 6.204** Potentiometer circuit.

***The Potentiometer Circuit (for Obtaining Partial Voltage).*** In the potentiometer circuit shown in Fig. 6.204, we may change the output voltage by varying the resistance legs of the potentiometer $R$. In this circuit, we used a 1-kΩ potentiometer and trimmed it so that the two resistance legs were 100Ω and 900 Ω, respectively. The output voltage may be calculated from

$$e = E\left(\frac{R_2}{R}\right) = 10.8 \text{ V dc}$$

***The Loaded Voltage Divider Circuit—Resistive Type.*** When the voltage divider circuit has a load connected across the output resistor, the load resistance or impedance of the device or circuit that the voltage divider powers should be approximately 10 times greater than the value of the output resistor. If this rule is not adhered to, the voltage output of the voltage divider, when under load, will not be as calculated using the voltage-divider equation. In cases where this 10:1 ratio is not applicable, the voltage divider must be recalculated using a different value for the output resistor, and a circuit analysis performed, as shown later. The input voltage of the voltage divider may need to be increased as a solution to the problem.

However, an analysis of the load circuit and output resistor should be made to determine the values required. Note that the output resistor and the load circuit or network form a parallel resistance circuit in series with the other resistor in the voltage-divider basic circuit. Also, the current available from the voltage-divider circuit divides when passing through the output resistor and the load resistance or impedance, and the sum of these currents is *limited* by the resistors in the voltage-divider circuit, which are in series with the main input voltage source. See the section on current-divider circuits for methods to calculate the branch currents.

(*Note:* The maximum load current must be considered because this corresponds to the minimum load resistance, and therefore the greatest effect on the output voltage $V_{out}$.)

Breadboarding the voltage-divider circuit and the load resistance or impedance and using potentiometers (as the resistors in the voltage divider) and a variable dc or ac power supply make the solution to the problem quite simple. Balance the circuit while monitoring the voltages with a DVM to arrive at precise values for the voltages and then measure the potentiometer's resistances with the DVM for the values of the resistive components $R_1$ and $R_2$. Also, monitor the load current.

When the load resistance or impedance is variable, a regulated voltage source must be used at the input of the voltage divider. This is easily accomplished using IC regulator circuits, some of which are shown in the section on practical IC circuits.

The standard general equation for the voltage-divider circuit is

$$V_{out} = \frac{R_{out}}{R_{total}} V_{in} \quad \text{(no-load condition)}$$

Changing the values of the resistors obviously alters the output voltage, as does the value of the load resistance or impedance.

When the voltage-divider circuit has a load across the output resistor $R_2$, as explained earlier, the output resistor should have a value equal to 10 times the load resistance $R_L$. If we know the load resistance or impedance, we will know the value of the load resistor $R_2$. We may now calculate the value of the other resistor in the voltage divider $R_1$ from the following equation:

$$R_1 = \frac{V_{in}\,(R_{eff}) - V_{out}\,(R_{eff})}{V_{out}}$$

where $R_L$ = load resistance across $R_2$, Ω
$V_{in}$ = input voltage, V
$V_{out}$ = voltage divider output voltage, V
$R_2 = 10R_L$, Ω

Find $R_1$, $I$, and $I_L$ and verify output voltage across the load $V_{out}$.

$$R_{eff} = \frac{R_L^2}{R_L + 10R_L}$$

where $I$ = total current through circuit, A
$I_L$ = current through the load, A
$R_1$ = other resistor in the voltage divider, Ω
$V_{out}$ = voltage-divider output voltage across the load $R_L$, V

*Solution.* Refer to Fig. 6.205*a* and 6.205*b*. The load resistance $R_L$ and resistor $R_2$ are in parallel and also in series with resistor $R_1$. The combined parallel resistance of $R_L$ and $R_2$ is the effective resistance $R_{eff}$ and is calculated per the preceding equation. The effective resistance $R_{eff}$ is then considered as one of the resistances in the voltage-divider circuit, while $R_1$ is the other resistance. We next find the value of resistor $R_1$ using the preceding equation. The true voltage output of the divider may now be calculated. Then, the current $I$ through the main branch of the network can be calculated. Last, the load current $I_L$ is calculated.

As the load resistance decreases, the wattage values of the divider resistors $R_1$ and $R_2$ increase. As an example, with a supply voltage of 125 V dc and a resistive load of 50 Ω at 24 V dc, the wattage values of $R_1$ and $R_2$ become quite large and impractical, while energy in the form of heat is wasted. This must be taken into account when a resistive voltage divider is used to drop a large voltage and feed a low-resistance or low-impedance load.

The loss of a few hundred watts in the form of heat may or may not be tolerable for your application. In any event, the resistive voltage-divider circuit is one of the most useful circuits available to the electromechanical designer from the standpoint of simplicity and ease of calculation or analysis.

Figure 6.206 shows a practical circuit that may be used in place of an expensive relay delay timer (delay-on). This circuit uses the common resistive voltage divider, an integrated circuit

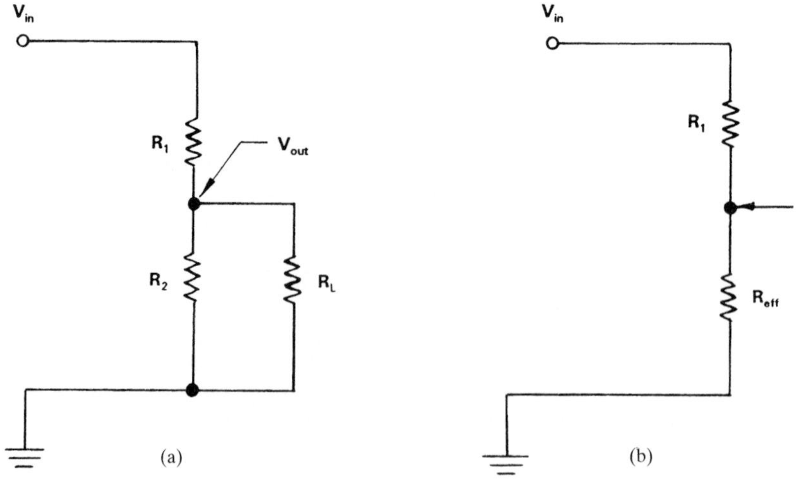


**FIGURE 6.205** (*a*) The loaded divider; (*b*) the effective resistance in series with $R_1$.

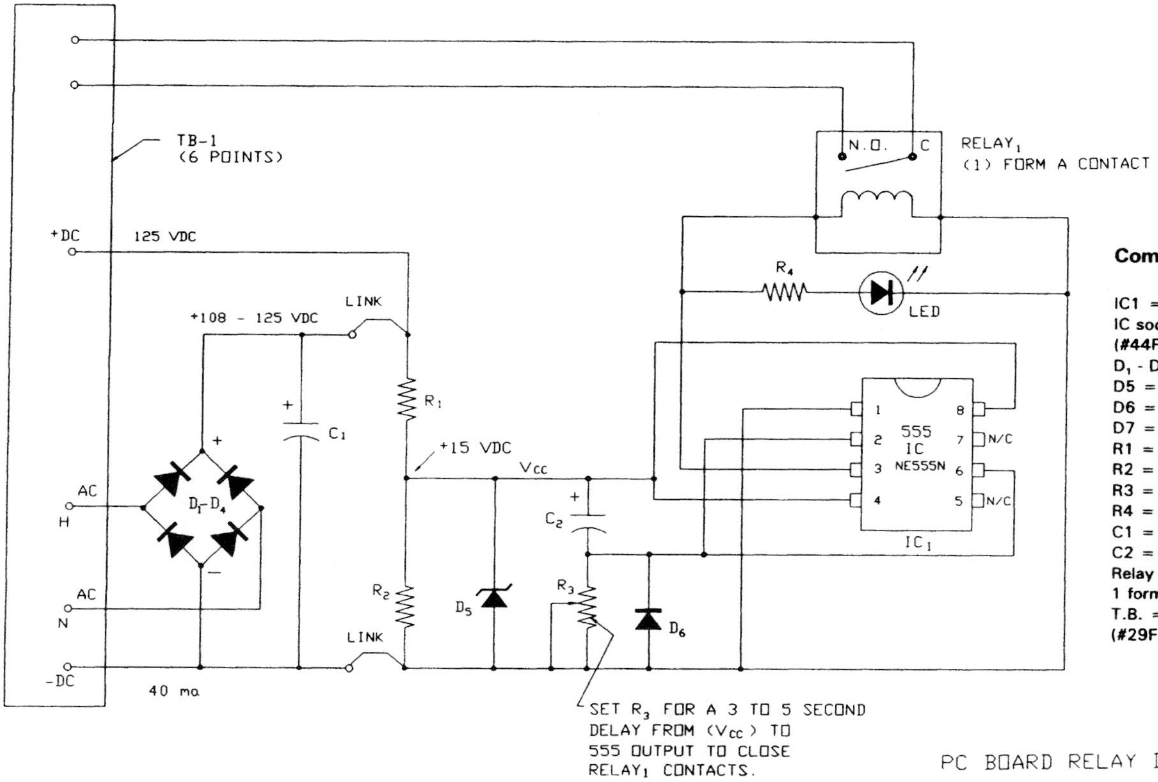

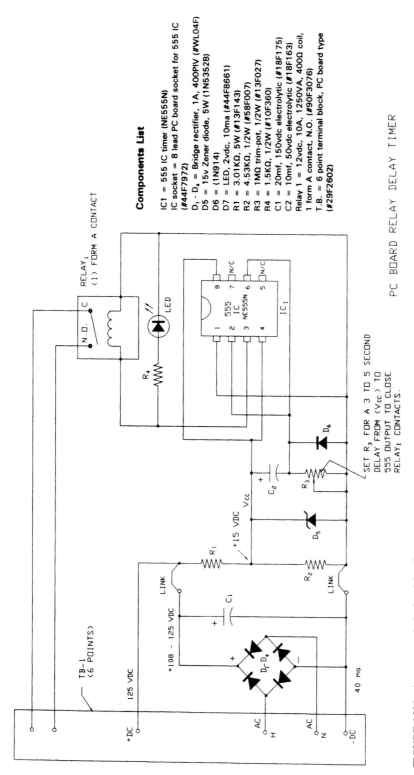

**Components List**

IC1 = 555 IC timer (NE555N)
IC socket = 8 lead PC board socket for 555 IC (#44F7972)
$D_1$ - $D_4$ = Bridge rectifier, 1A, 400PIV (#WL04F)
D5 = 15v Zener diode, 5W (1N5352B)
D6 = (1N914)
D7 = LED, 2vdc, 10ma (#44F8661)
R1 = 3.01KΩ, 5W (#13F143)
R2 = 4.53KΩ, 1/2W (#58F007)
R3 = 1MΩ trim-pot, 1/2W (#13F027)
R4 = 1.5KΩ, 1/2W (#10F360)
C1 = 20mf, 150vdc electrolytic (#18F175)
C2 = 10mf, 50vdc electrolytic (#18F163)
Relay 1 = 12vdc, 10A, 1250VA, 400Ω coil, 1 form A contact, N.O. (#90F3076)
T.B. = 6 point terminal block, PC board type (#29F2602)

PC BOARD RELAY DELAY TIMER

**FIGURE 6.206** A more practical relay delay timer.

(555 IC timer), a zener diode to regulate the output voltage supplying the 555 timer's input, and a low-cost high-power relay rated for 1250 VA, 12 V dc. The zener diode must be used in this circuit because when the 555 IC is not conducting, the unloaded divider voltage to the 555 input would be high enough to damage it. The relay in this circuit may be any one of a large assortment to suit the application.

***The Resistor Voltage Divider with Current Drain at the Taps.*** A resistive voltage divider with current drain at the taps is shown schematically in Fig. 6.207.

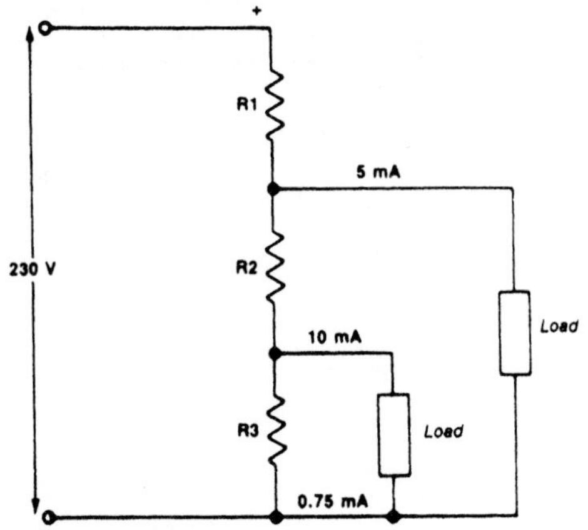


**FIGURE 6.207** Resistor voltage divider.

In the circuit shown, a 230-V supply is available, and we wish to have a 10-mA tap at 150 V and a 5-mA tap at 225 V. The bleeder current $I_B$ is 0.75 mA.

Only the bleeder current $I_B$ passes through $R_3$. The 10-mA current plus the bleeder current will pass through $R_2$. The three currents all pass through $R_1$.

Solution by Ohm's law:

$$R_3 = \frac{150}{0.75 \times 10^{-3}} = 200\ \text{k}\Omega \qquad R_2 = \frac{E_2}{I_2}$$

$$E_2 = 225 - 150 = 75\ \text{V} \qquad I_2 = 0.75 + 10 = 10.75\ \text{mA}$$

yields

$$R_2 = \frac{75}{10.75 \times 10^{-3}} = 6976.7\ \Omega \qquad R_1 = \frac{225}{15.75 \times 10^{-3}} = 14{,}285.7\ \Omega$$

This procedure is summarized as follows:

1. Arrange the current diagram of the circuit to determine the current through each resistor.
2. Calculate the potential difference across each resistor.
3. Use Ohm's law to calculate the resistor values.
4. Determine the wattage value of each resistor by $W = EI = I^2R$.

***The Capacitor Voltage Divider.*** A capacitive voltage divider is shown schematically in Fig. 6.208. In many communications circuits a series connection of two capacitors may be used for voltage division. In the figure there are two capacitors connected in series with 20 V across them. $C_1 = 150$ pF and $C_2 = 25$ pF (pF = picofarads, formerly μμF). Then

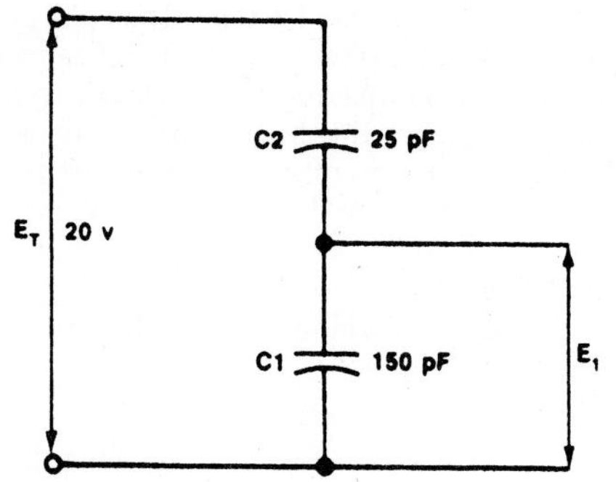


**FIGURE 6.208** Capacitor voltage divider.

$$\frac{E_1}{E_T} = \frac{X_1}{X_T}$$

Since $X_T = X_1 + X_2$, we can write

$$\frac{X_1}{X_T} = \frac{X_1}{X_1 + X_2}$$

Using the capacitive reactance equation

$$\frac{X_1}{X_1 + X_2} = \frac{1}{2\pi f C_1} \div \left[\frac{1}{2\pi f C_1} + \frac{1}{2\pi f C_2}\right]$$

This reduces to

$$\frac{1}{C_1} \div \left[\frac{1}{C_1} + \frac{1}{C_2}\right] = \frac{1}{C_1} \div \frac{C_1 + C_2}{C_1 C_2}$$

Reciprocating $$\frac{1}{C_1}\left[\frac{C_1 C_2}{C_1 + C_2}\right]$$

Cancel $C_1$; then $$\frac{C_2}{(C_1 + C_2)}$$

Now $$\frac{E_1}{E_T} = \frac{C_2}{C_1 + C_2}$$

The voltage is directly proportional to reactance with the smaller voltage across the larger capacitor. The solution then is for $E_1$:

$$\frac{E_1}{20} = \frac{25 \times 10^{-12}}{150 \times 10^{-12}} = \frac{25}{150}$$

$$E_1 = 3.33 \text{ V}$$

You may select the capacitors to arrive at the voltage you desire at $E_1$, within the limits of the circuit supply voltage.

***Limiting Resistor Selection for LEDs.*** The light-emitting diode is normally fed from a supply voltage source that is higher than the LED can sustain without burnout. A limiting resistor, in this case, must be inserted in the circuit to drop the voltage across the diode to an acceptable limit. The forward voltage drop across an LED is generally between 1.4 to 5 V, and the limiting resistor must be placed in series with the LED to produce the correct voltage drop and current through the diode. If we used an LED with a forward voltage of 2.2 V at 10 mA and the supply voltage were 12 V dc, what value limit resistor would be required to efficiently operate the LED?

*Solution.* The LED is connected into the circuit as shown in Fig. 6.209 and the value of the limit resistor would be

$$R_d = \frac{V_s - V_f}{I_f} \qquad \left(\text{from Ohm's law, } R = \frac{E}{I}\right)$$

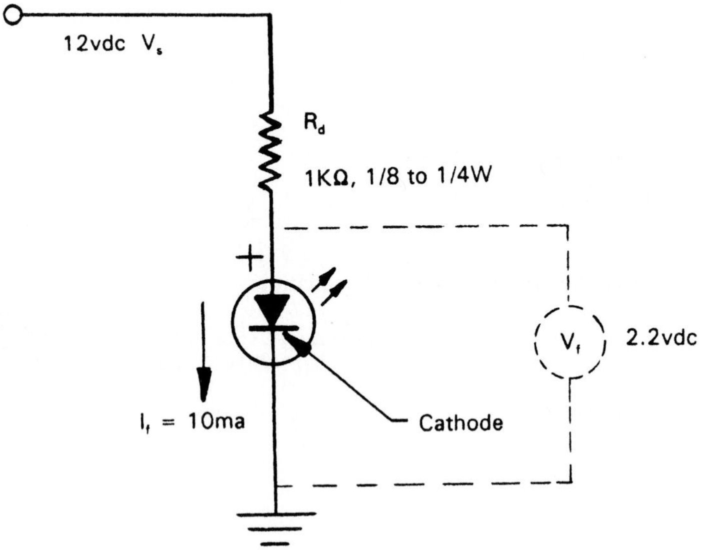


**FIGURE 6.209** LED Circuit.

where $V_s$ = supply voltage, 12 V dc
$V_f$ = LED forward voltage drop, 2.2 V dc
$I_f$ = LED forward current, 10 mA (0.01 A)

With the 12-V-dc supply voltage as shown in Fig. 6.209, the limiting resistor $R_d$ would have a value of

$$R_d = \frac{12 - 2.2}{0.01} = 980\ \Omega \text{ or } 1\text{-k}\Omega \text{ standard resistor}$$

For maximum life of the LED, the limit resistor should be carefully selected per the preceding calculations to limit the forward current flowing through the LED and to limit the voltage across the LED terminals. Note the position of the LED cathode in the circuit. If the positive voltage is connected to the opposite end of the LED, it will not conduct and will produce a light output.

***Resistive Heating Elements.*** A resistive heating element is rated at 2500 W at 240 V ac and 2000 W at 208 V ac. What will be the power rating in watts if the operating voltage is changed to 120 V ac?

From Ohm's law, $W = EI$ or $I^2R$. Then

$$W = EI \qquad 2500 = 240I$$

$$I = 10.42 \text{ A}$$

and

$$E = IR \qquad 240 = 10.42R$$

$$R = 23\ \Omega \text{ resistance of the element}$$

The current at 120 V ac would then be

$$E = IR \qquad 120 = 23I$$

$$I = 5.22 \text{ A}$$

Then the wattage rating at 120 V ac would be

$$W = EI = 120 \times 5.22 = 626.4 \text{ W}$$

Also

$$W = I^2R = (5.22)^2 \times 23 = 626.7 \text{ W}$$

which checks with the above wattage rating.

***Basic Circuits for Voltage Increases.*** The circuit shown in Fig. 6.210 is a voltage doubler. The circuit shown in Fig. 6.211 is a voltage quadrupler. These circuits are comprised of electrolytic capacitors and diodes of the appropriate ratings. Simple circuit analyses shown in other sections will allow the selection of the ratings of the components.

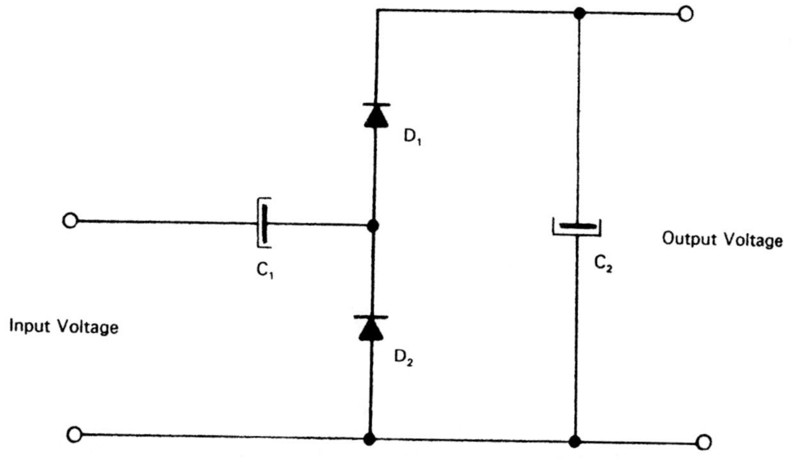


**FIGURE 6.210** Voltage doubler circuit.

***Developing a Negative-Voltage Power Supply.*** Figure 6.212 shows a solid-state power supply with a negative-voltage branch. This is possible by using a center-tapped transformer with a center-tap terminal provision. Figure 6.213 shows another configuration of negative-voltage power supply. Figure 6.214 shows a vacuum-tube configuration. Electrolytic capacitors and a

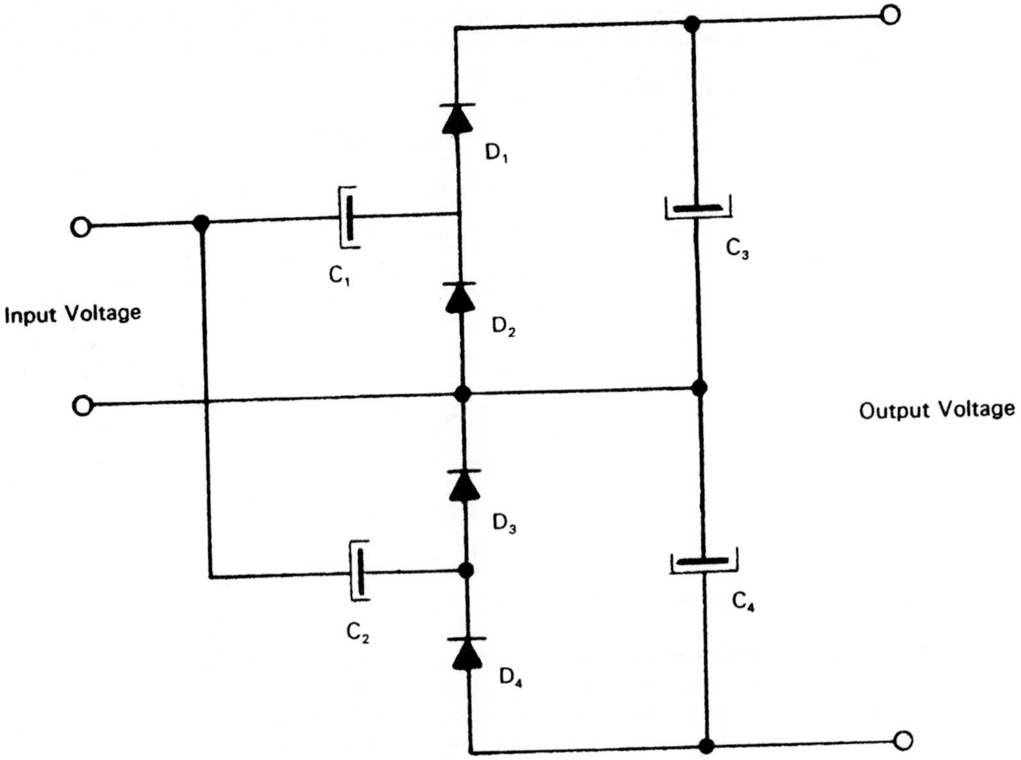


**FIGURE 6.211** Voltage quadrupler circuit.

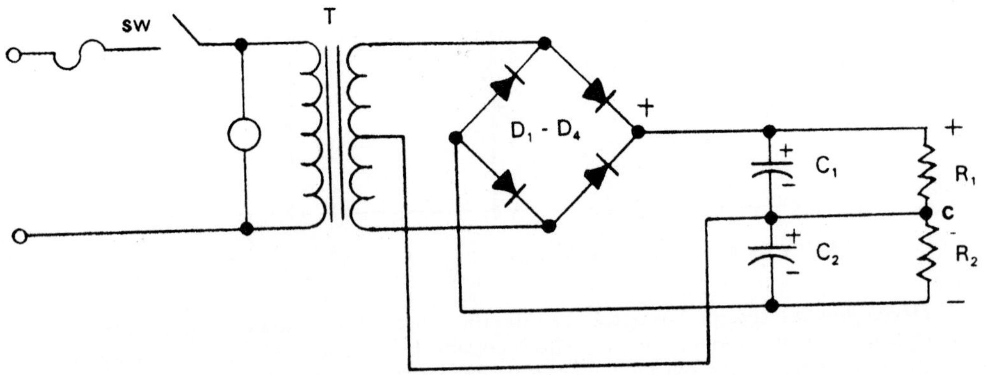


**FIGURE 6.212** Negative-voltage power supply.

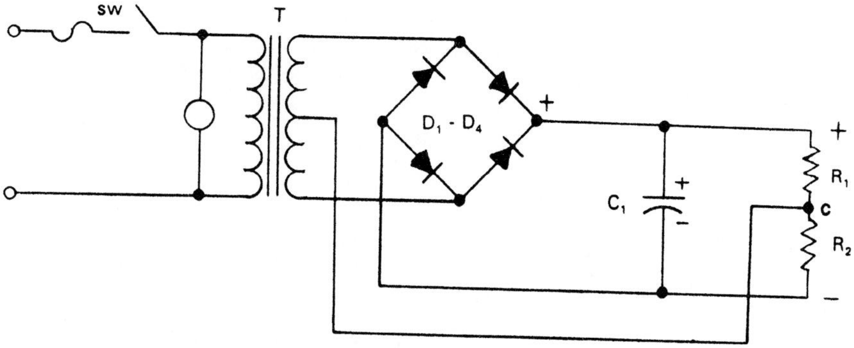


**FIGURE 6.213** Negative-voltage power supply.

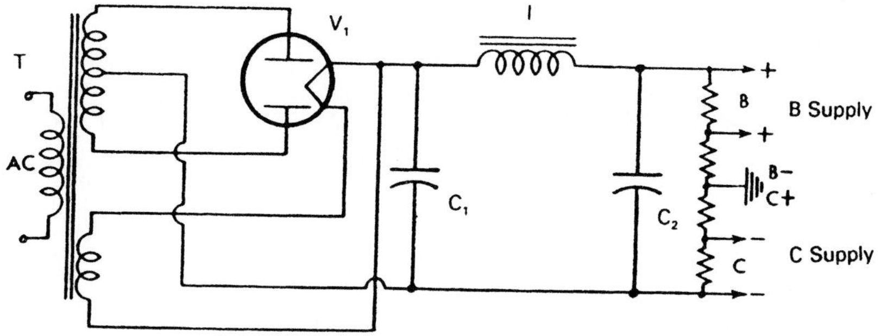


**FIGURE 6.214** Vacuum-tube negative-voltage power supply.

full-wave bridge rectifier are shown in each of these negative-voltage power supplies. The negative-voltage power supply is required for operating IC operational amplifiers and other types of ICs. Regulation devices may be added into the circuitry for full voltage regulation in critical applications.

### 6.15.3 Basic Three-Phase AC Electrical Power Circuits

The star or Y system is depicted in Fig. 6.215.

$$I_L = I_P$$

where $I_L$ = line current and $I_P$ = phase current. For balanced systems:

$$I_A = I_B = I_C$$

$$V_A = V_B = V_C$$

$$V_{AB} = V_{BC} = V_{AC}$$

$$Z_A = Z_B = Z_C \qquad \text{and} \qquad I_N = 0$$

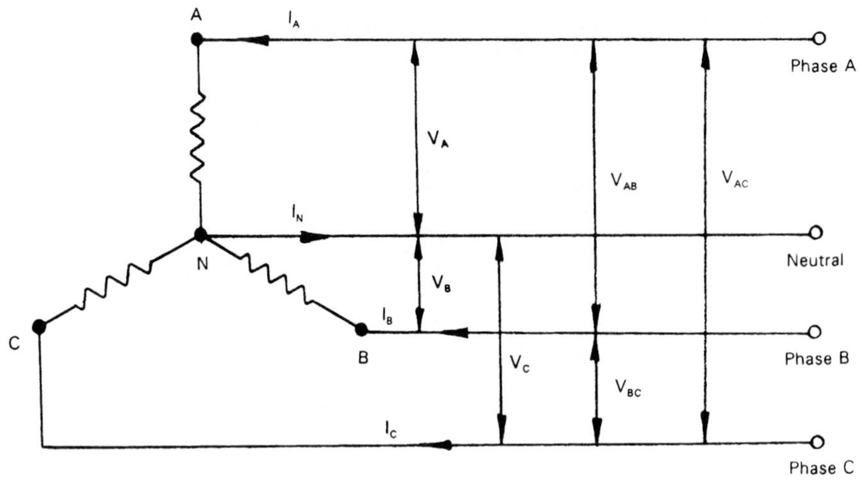


**FIGURE 6.215** Y system.

For balanced Y connections:

$$V_L = \sqrt{3} V_P \qquad V_P = \frac{V_L}{\sqrt{3}}$$

where $V_L$ = line voltage and $V_P$ = phase voltage.

The delta or mesh system is depicted in Fig. 6.216.

$$V_L = V_P$$

where $V_L$ = line voltage and $V_P$ = phase voltage.

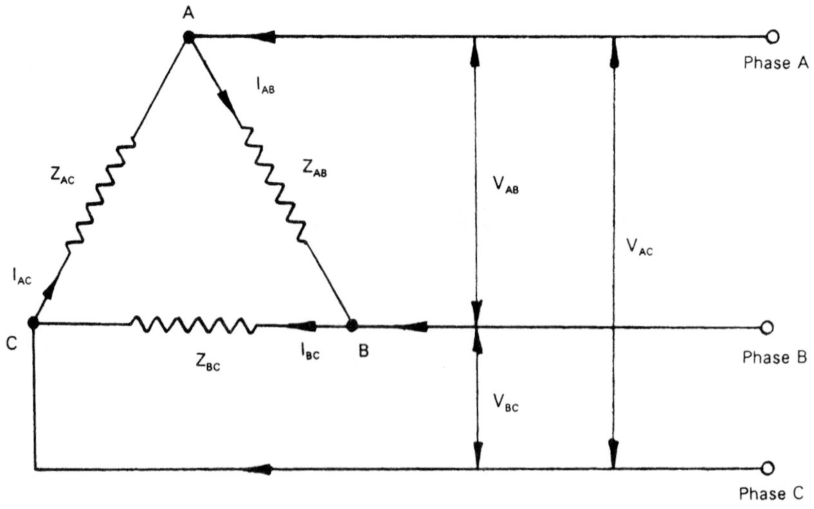


**FIGURE 6.216** Delta system.

In a balanced delta connection:

$$I_L = \sqrt{3} I_P \qquad I_P = \frac{I_L}{\sqrt{3}}$$

where $I_L$ = line current and $I_P$ = phase current.

Power dissipated in a three-phase load for either Y or delta connections is

$$P = \sqrt{3} V_L I_L \cos \phi, \quad \text{W}$$

The power factor is

$$\tan \phi = \sqrt{3}\, \frac{P_1 - P_2}{P_1 + P_2}$$

where $P_1$ and $P_2$ are wattmeter readings using the "two-wattmeter" method.

***Y-to-Delta Conversions.*** It is often convenient to convert a Y network (or T) to a delta network (or π), or the reverse. Figure 6.217 shows the Y network in solid lines and the delta equivalent network as dotted lines. The conversion equations are as shown below.

$$R_W = \frac{R_A R_B + R_B R_C + R_A R_C}{R_C} \qquad R_X = \frac{R_A R_B + R_B R_C + R_A R_C}{R_B} \qquad R_Z = \frac{R_A R_B + R_B R_C + R_A R_C}{R_A}$$

***Delta-to-Y Conversions.*** Here, we are converting from the delta network to the Y network. Figure 6.218 shows the delta network in solid lines and the equivalent Y network in dotted lines. The delta conversion to a Y network is accomplished using the equations shown below.

$$R_A = \frac{R_W R_X}{R_W + R_X + R_Z} \qquad R_B = \frac{R_W R_Z}{R_W + R_X + R_Z} \qquad R_C = \frac{R_X R_Z}{R_W + R_X + R_Z}$$

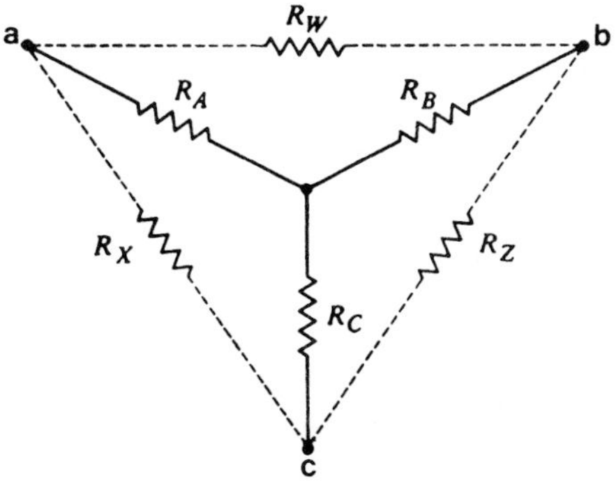


**FIGURE 6.217** Y-to-delta conversion.

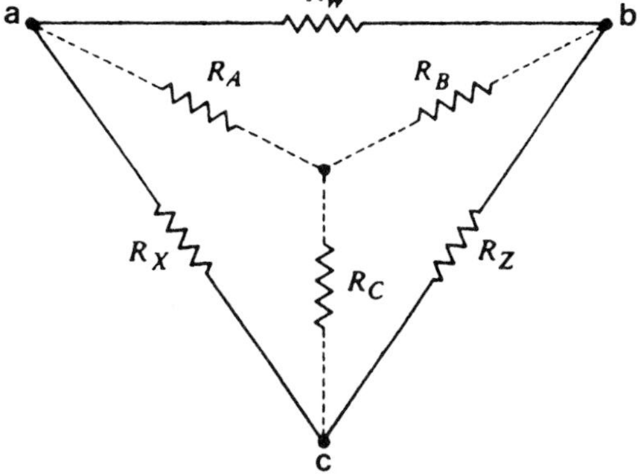


**FIGURE 6.218** Delta-to-Y conversion.

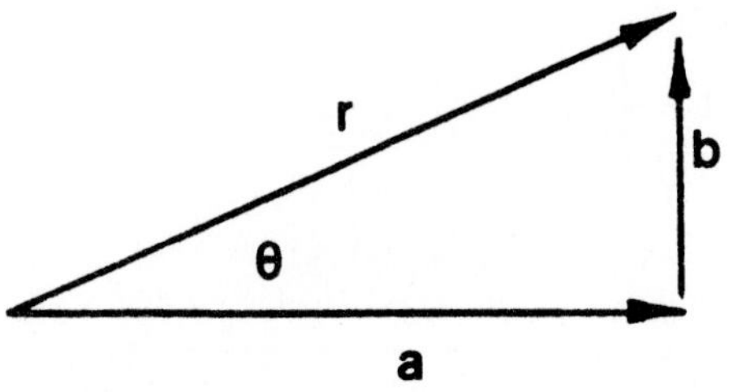


**FIGURE 6.219** Vector diagram.

### 6.15.4 The Fundamental Vector Analysis Relationships and Coordinate Transformations

Refer to Fig. 6.219.

***Vector Relationships***

$$\text{Vector r} = \mathbf{r} = r\angle\theta \text{ polar form}$$

$$\mathbf{r} = (a + jb) \text{ complex form}$$

$$\mathbf{r} = (a, b) \text{ rectangular form}$$

$$\text{Length of } \mathbf{r} = \sqrt{a^2 + b^2}$$

$$a = r\cos\theta \qquad b = r\sin\theta$$

$$\tan\theta = \frac{b}{a}$$

***Coordinate Transformation***

$$\text{Polar} \leftrightarrow \text{rectangular: } r\angle\theta = \sqrt{a^2 + b^2}\ \angle\tan^{-1}\left(\frac{b}{a}\right)$$

$$\text{Polar} \leftrightarrow \text{complex: } r\angle\theta = a + jb = r\cos\theta + jr\sin\theta$$

*Example*

$$15\angle 220° = 15\cos 220° + 15j\sin 220°$$

$$\cos 220° = -\cos 40° \qquad \sin 220° = -\sin 40°$$

$$15\angle 220° = -15\cos 40° - 15j\sin 40°$$

$$\cos 40° = 0.6428 \qquad \sin 40° = 0.7660$$

$$a = -11.5 \qquad b = -9.63$$

$$15\angle 220° = -11.5 - j\,9.63$$

*Answer*

$$15\angle 220° = (-11.5, -9.63) = -11.5 - j\,9.63$$

*Example*

$$-3 + j^2 = \sqrt{-3^2 + 2^2}\ \angle\tan^{-1}\left(\frac{2}{-3}\right)$$

$$-3 + j^2 = 3.6\angle\tan^{-1}(-0.6667)$$

$$\tan^{-1}(-0.6667) = 33.7° \qquad 180° - 33.7° = 146.3°$$

$$-3 + j^2 \text{ is in second quadrant}$$

*Answer*

$$-3 + j^2 = 3.6\angle 146.3°$$

### 6.15.5 Complex Quantities

***Imaginary Roots and Numbers***

$$\sqrt{-1} = I \text{ or } j \qquad i^2 = -1 \qquad j^2 = -1$$

where $j$ = active component in electrical circuits; $i$ = for use in nonelectrical work.

$$(a + jb) + (c + jd) = (a + c) + j(b + d) = (ac - bd) + j(bc + ad)$$

$$a + jb = \sqrt{a^2 + b^2} \cdot \varepsilon^{j\theta} \qquad \text{where } \sqrt{a^2 + b^2} > 0$$

$$a + jb = \sqrt{a^2 + b^2}\,[\cos\theta + j\sin\theta] \qquad \sin\theta = \frac{b}{\sqrt{a^2 + b_2}}$$

$$\sin\theta = \frac{\varepsilon^{j\theta} - \varepsilon^{-j\theta}}{2j} = I_m[\varepsilon^{j\theta}] \qquad \cos\theta = \frac{a}{\sqrt{a^2 + b^2}}$$

$$\cos\theta = \frac{\varepsilon^{j\theta} + \varepsilon^{-j\theta}}{2} = \text{real}[\varepsilon^{j\theta}]$$

$$(a + jb)\,(c + jd) = ac + j^2bd + j(ad + bc)$$

$$= ac - bd + j(ad + bc)$$

### 6.15.6 Basic Vector Algebra

For addition and subtraction, vectors must be in complex form.

Addition:

$$(a + jb) + (c + jd) = (a + c) + j(b + d)$$

Subtraction:

$$(a + jb) - (c + jd) = (a - c) + j(b - d)$$

Multiplication:

$$(a + jb)\,(c + jd) = ac + j(bc + ad) - \text{bd} \qquad (j^2 = -1)$$

$$(r_1 \angle\theta_1)\,(\text{r}_2 \angle\theta_2) = r_1r_2 \underline{/(\theta_1 + \theta_2)}$$

Division:

$$\frac{(a + jb)}{(c + jd)} = \frac{(a + jb)}{(c + jd)} \cdot \frac{(c - jd)}{(c - jd)} = \frac{ac - j(bc + ad) + bd}{c^2 + d^2}$$

$$\frac{r_1\angle\theta_1}{r_2\angle\theta_2} = \frac{r_1}{r_2} \underline{/(\theta_1 - \theta_2)}$$

## 6.16 BRIDGE CIRCUITS FOR ELECTRICAL ELEMENT MEASUREMENTS: R, C, AND L

Bridge circuits for measuring various unknown electrical properties of components such as resistors, inductors, and capacitors have been developed into various circuit arrangements.

The more commonly known and used bridge circuit configurations and their functions include the following.

*Wheatstone bridge:* Ac- and dc-type bridges for accurately measuring unknown resistances and capacitances. The dc Wheatstone bridge measures resistance and the ac Wheatstone bridge measures resistance and capacitance on two-terminal components.

*Kelvin bridge:* A dc bridge for the accurate measurement of very low resistance values, typically less than 0.1 Ω. For four-terminal components.

*Maxwell bridge:* For measuring inductance and resistance.

*Wein bridge:* For resistance, capacitance, and frequency.

*Wein-DeSauty bridge:* For capacitance.

*Schering bridge:* Loss angles of high-voltage power cables or capacitors for low-voltage applications.

*Hay bridge:* Similar to the Maxwell bridge.

*Wein-Maxwell bridge:* For inductance and capacitance.

*Owen's bridge:* For inductance and capacitance.

*Anderson's bridge:* For inductance.

*Campbell's bridge:* For mutual inductance.

*Inductance bridge:* For inductance.

*Resonance bridge:* For inductance, impedance, resistance, and frequency.

*Transformer bridge:* For accurate determination of capacitance.

*Slide-wire or meter bridge:* For resistance; moderate accuracy.

The operation and balance equations for the dc Wheatstone bridge, ac Wheatstone bridge, Wein bridge, Maxwell bridge, Wein-Maxwell bridge, and Wein-DeSauty bridge follow.

*Note:* The letters "G" and "D" at the center of the bridges denote a galvanometer or detector, respectively. An oscilloscope is frequently used as a bridge detector.

***The Dc Wheatstone Bridge.*** Refer to Fig. 6.220.

**1.** Adjust $R_1$, $R_2$, and $R_3$ until the galvanometer indicates a null (no current).

**2.** $R_1$, $R_2$, and $R_3$ are then the known resistors. The unknown resistor is then found from

$$R = R_2 \frac{R_3}{R_1}$$

[*Notes:* (1) if $R$ is also inductive, such as a motor field winding, the dc source switch $S_1$ should be closed before the galvanometer switch $S_2$ to protect the galvanometer from the initial transient or inrush current; (2) $R_{c1}$ is a current-limiting resistor which may be optionally added to the dc supply leg to prevent possible damage to the resistors in the bridge so that their $I^2R$ dissipation capacities are not exceeded; (3) the dc voltage is normally 3 to 9 V.]

***The Ac Wheatstone Bridge.*** Refer to Fig. 6.221. With $R$ and $C$ the unknown values, the bridge is balanced through the detector $D$ at the center of the bridge. Then

$$C = \frac{C_3 R_2}{R_1} \quad \text{and} \quad R = \frac{R_3 R_1}{R_2}$$

(*Note:* $R_1$, $R_2$, $R_3$, and $C_3$ are normally standard decade resistances and capacitance, adjustable in increments; ac source voltage is determined by the bridge leg resistance and the detector sensitivity. A "variac" may be used as the source and its output adjusted accordingly, taking into account the $I^2R$ component dissipation capacities.)

***The Wein Bridge—Ac.*** Refer to Fig. 6.222. With this bridge

$$\frac{C}{C_3} = \frac{R_2}{R_1} - \frac{R_3}{R}$$

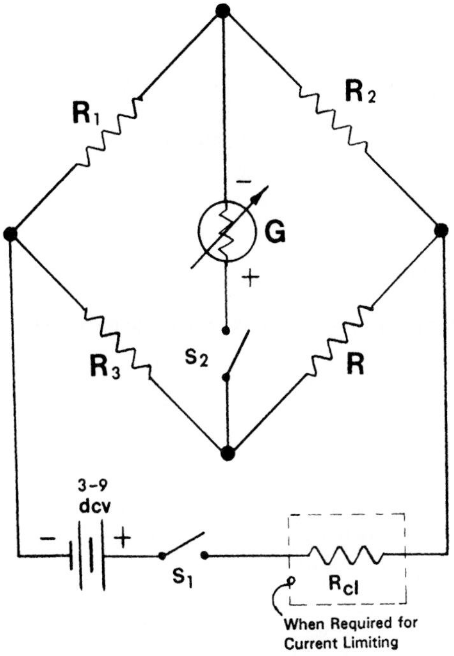


**FIGURE 6.220** Dc Wheatstone bridge circuit.

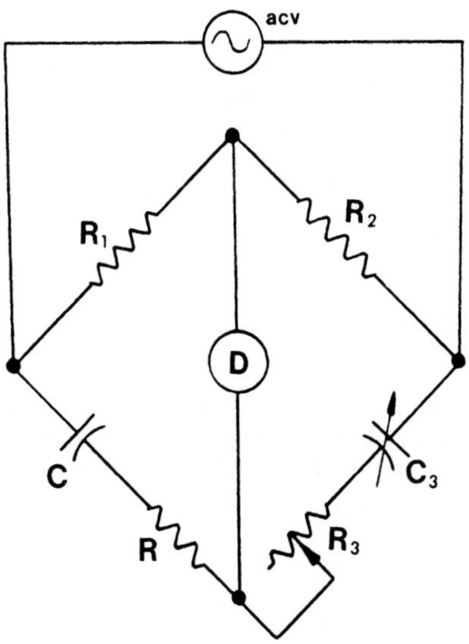


**FIGURE 6.221** Ac Wheatstone bridge circuit.

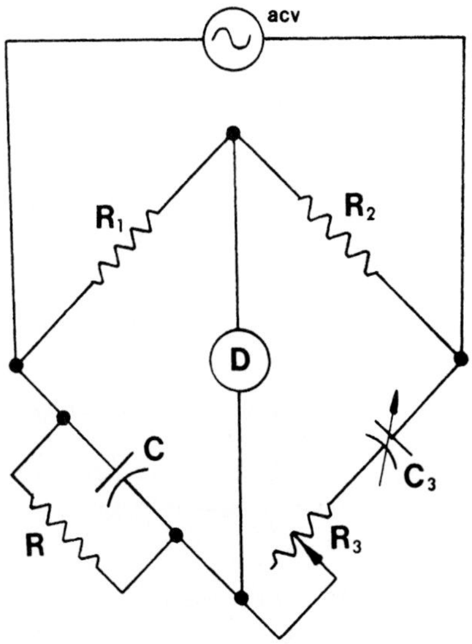


**FIGURE 6.222** Wein bridge circuit.

Then, by transposition

$$R = \frac{C_3 (RR_2 - R_1R_3)}{CR_1} \qquad \text{and} \qquad C = \frac{C_3 (RR_2 - R_1R_3)}{R_1R}$$

when used to determine frequency $f$; if $C = C_3$, $R = R_3$, and $R_2 = 2R_1$, then

$$f = \frac{1}{2\pi R_3C_3}$$

where $f$ = frequency in Hz (cycles per second).

***The Maxwell Bridge—Ac.*** Refer to Fig. 6.223. In this bridge, the balance equation is given by

$$L = R_1R_3C_2 \qquad \text{and} \qquad R = \frac{R_1R_3}{R_2}$$

where $L$ = inductance, H, and $R$ = resistance, Ω.

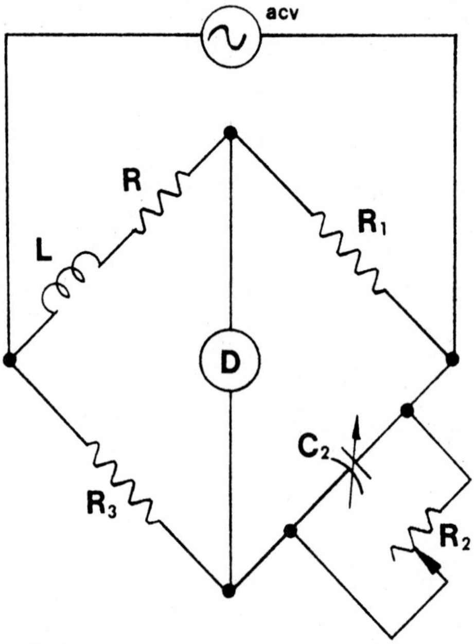


**FIGURE 6.223** Maxwell bridge circuit.

***The Wein-Maxwell Bridge—Ac.*** Refer to Fig. 6.224. In this bridge, the accurate measurement of inductance may be made and the bridge is most easily balanced by adjustment of $C_2$ and $R_2$, as these elements are in quadrature, making their adjustments noninteractive. The balance equations are

$$L = R_1R_3C_2 \qquad \text{and} \qquad R = \frac{R_1R_3}{R_2}$$

where $L$ = inductance, H, and $R$ = resistance, Ω.

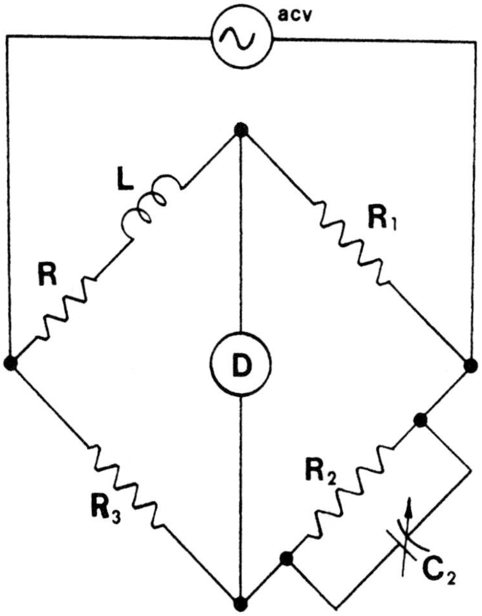


**FIGURE 6.224** Wein-Maxwell bridge circuit.

***The Wein-DeSauty Bridge—Ac.*** Refer to Fig. 6.225. In this bridge, capacitances may be compared in terms of a resistance ratio. Note that the loss angles of the two capacitance arms must be equal and that a series resistor $R_s$ is added in the branch with the smaller loss angle. The resistance $R_3$ is in series with the reference capacitor $C_s$. The balance equation is

$$C = C_s \frac{R_2}{R_1}$$

***Bridge Circuit Detectors.*** Detectors used in bridge measurement circuits are selected in regard to impedance and frequency response. Some of the common devices used as bridge detectors are:

*Vibration galvanometers:* Low-impedance devices.

*Wave analyzers:* Crystal controlled with tuned preamplifiers. Used for audible frequency, 20 to 20,000 Hz.

*Oscilloscopes used as null detectors:* By feeding a phase-adjustable voltage from the bridge supply to the horizontal plates of the scope ($x$-axis input) and the unbalanced signal in the detector branch of the bridge to the vertical plates of the scope ($y$-axis input), an elliptical "Lissajous" figure is generated on the scope screen. The Lissajous figure will close down to a straight line when the bridge is balanced. Bridge input must be sinusoidal for best results. The Lissajous figure informs the scope operator of the individual magnitudes of in-phase and quadrature unbalance.

*Audio headphones:* When the bridge is balanced, no tone is heard in the headphones. The headphone method is not always accurate because their response is very broad (20 to 20,000 Hz), and the balance point may be masked by harmonics generated within the circuit.

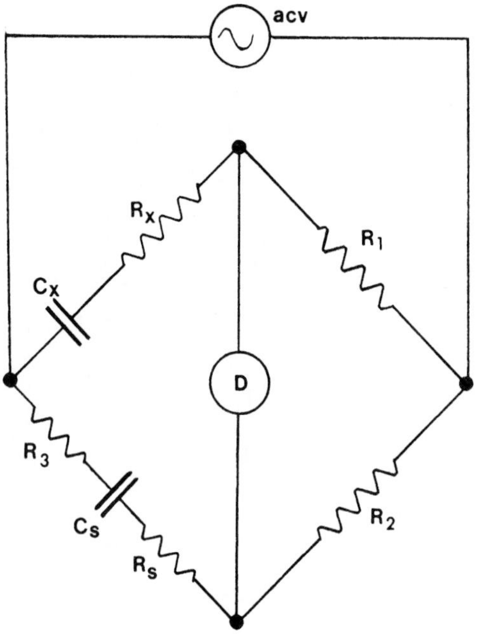


**FIGURE 6.225** Wein-DeSauty bridge circuit.

In the preceding bridge circuit balance equations:

$R$ = resistance, Ω
$L$ = inductance, H
$C$ = capacitance, F
$f$ = frequency, Hz

## 6.17 ELECTRICAL AND ELECTRONIC SYMBOLS

Figure 6.226 is a compilation of most of the common electrical and electronic symbols used in the United States. Also consult the ANSI standards for other electrical symbols not shown here.

## 6.18 INTERNATIONAL ELECTRICAL AND ELECTRONIC COMPONENT TRADEMARKS

Figure 6.227 shows the important international symbols used on electrical and electronic components and equipment. The country of origin may be readily identified by using this listing.

| | |
|---|---|
| 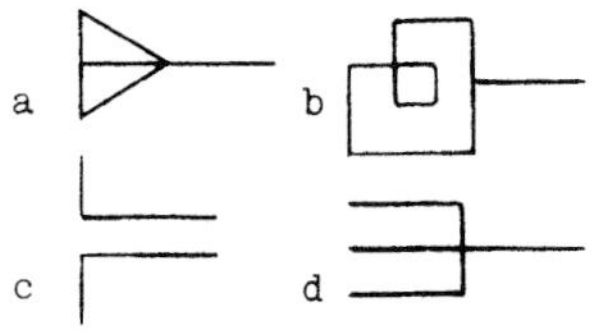 | **ANTENNAS**<br>a) General; b) Loop; c) Dipole; d) Counterpoise |
| 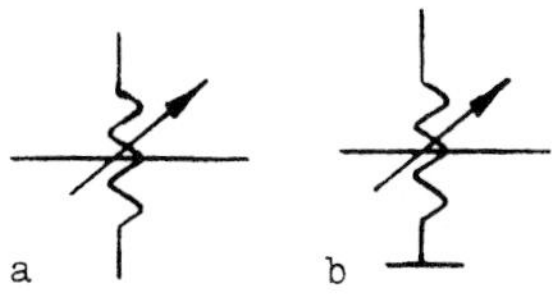 | **ATTENUATORS**<br>a) Balanced; b) Unbalanced |
| 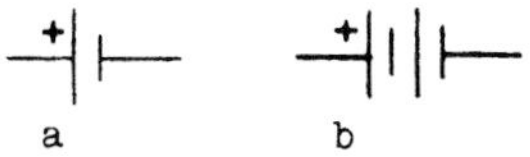 | **BATTERIES**<br>a) DC source, 1 cell; b) Multiple cells |
| 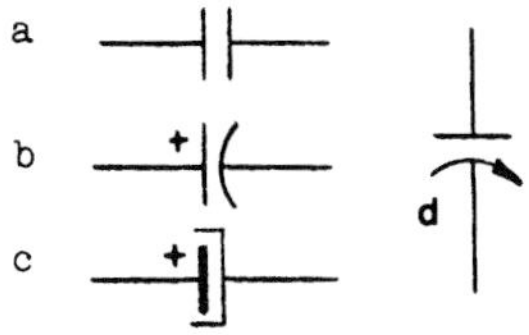 | **CAPACITORS**<br>a) Fixed; b) Polarized; c) Electrolytic; d) Trimmer |
| 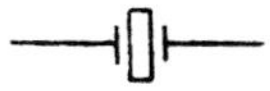 | **PIEZOELECTRIC CRYSTAL**<br>Frequency determining |
| 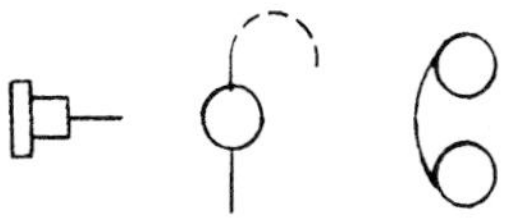 | **EARPHONES & HEARING AIDS** |
| 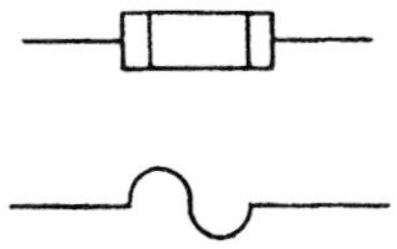 | **FUSES** |

**FIGURE 6.226** Electrical and electronic symbols.

| | |
|---|---|
| 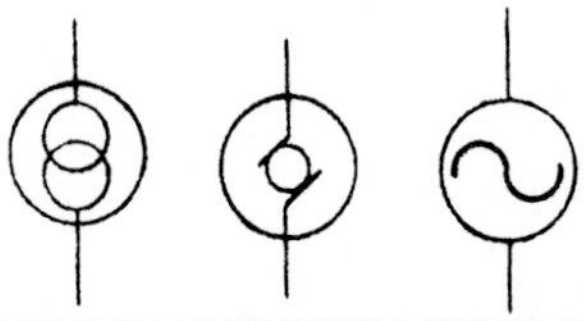 | **ALTERNATING CURRENT SOURCES** |
| 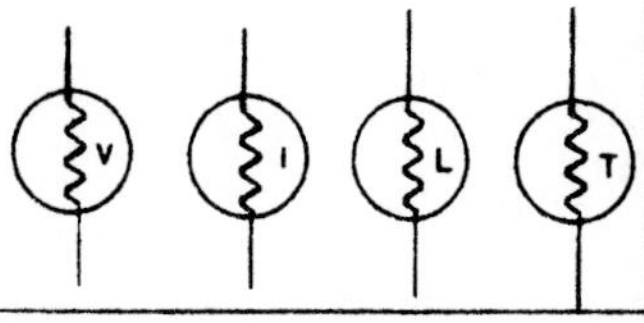 | **RESISTORS**<br>Voltage dependent; Current dependent; Light dependent; Tempertaure dependent |
| 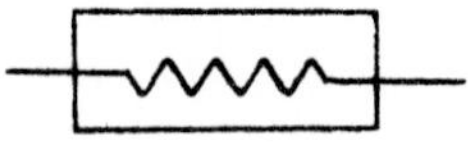 | **RESISTOR**<br>Fusible |
| 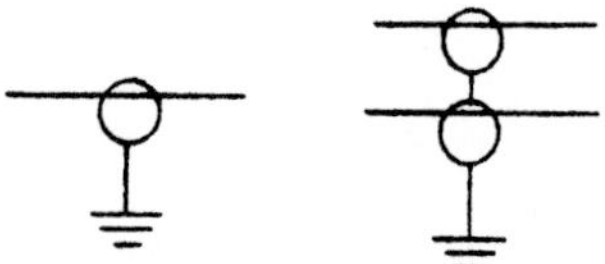 | **SHIELDS**<br>Shielded wire; Common ground |
| 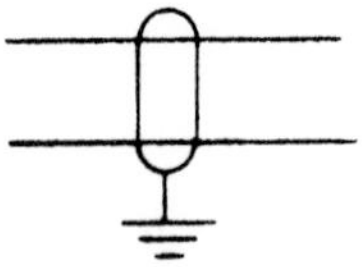 | **SHIELD**<br>Shielded pair |
| 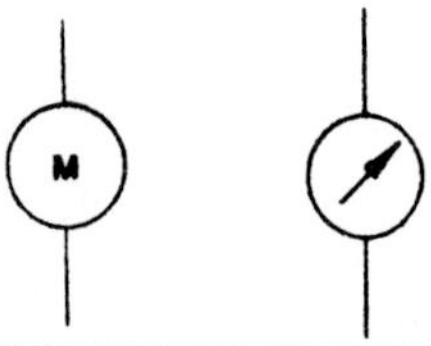 | **METERS** |
| 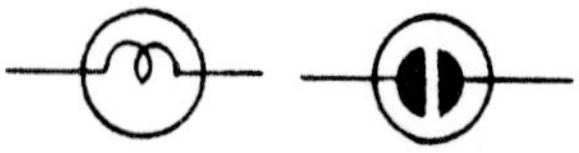 | **LAMPS**<br>Filament; Neon |

**FIGURE 6.226** (*Continued*)

| | |
|---|---|
| 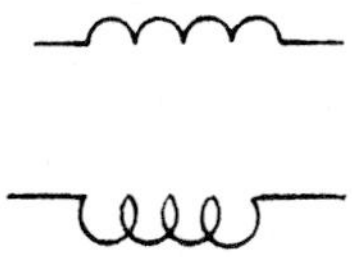 | **INDUCTOR WINDINGS**<br>General |
| 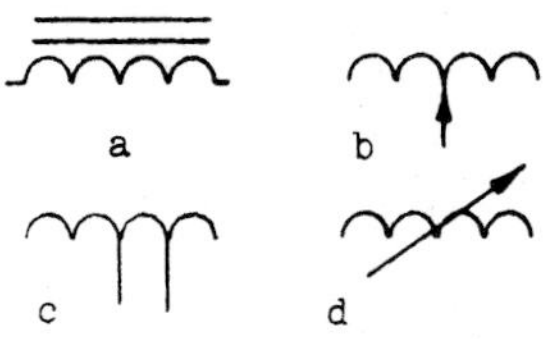 | **INDUCTOR WINDINGS**<br>a) Magnetic core; b) Adjustable; c) Tapped<br>d) Continuously adjustable |
| 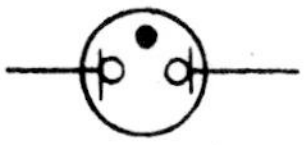 | **LAMP**<br>AC glow; cold cathode or neon |
| 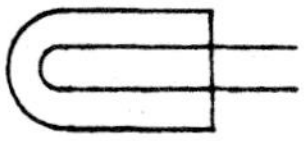 | **LAMP**<br>Incandescent |
| 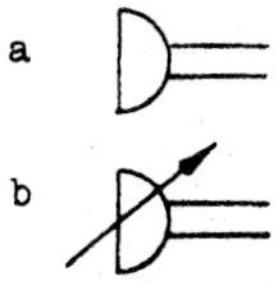 | **MICROPHONES**<br>a) General; b) Directional |
| 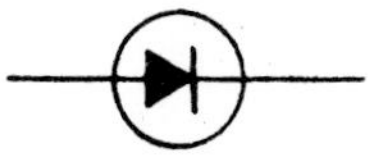 | **RECTIFIER, METALLIC, GENERAL** |
|  | **CIRCUIT BREAKER** |

**FIGURE 6.226** (*Continued*)

| | |
|---|---|
| a, b 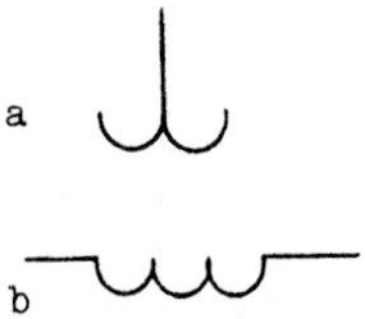 | **TRANSFORMERS**<br>a) Potential (PT); b) Current (CT) |
|  | **DISCONNECT JOINT** |
| a, b 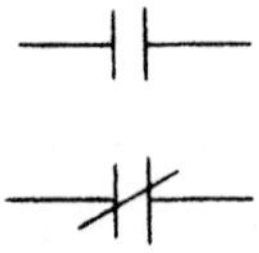 | **RELAY CONTACTS**<br>a) Normally open; b) Normally closed |
| a, b  | **GROUNDS**<br>a) Earth or common; b) Chassis |
| a, b 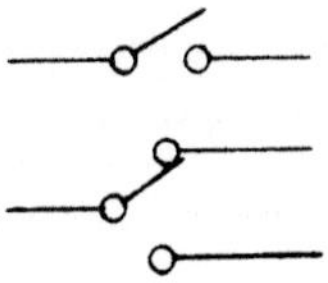 | **SWITCHES**<br>a) SPST; b) Two position DPDT |
| a, b  | **SWITCHES**<br>a) Pushbutton, Make; b) Pushbutton, Break<br>Circuit closing; Circuit opening |
| 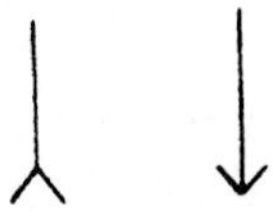 a, b | **POWER LEADS**<br>a) Incoming; b) Outgoing |

**FIGURE 6.226** (*Continued*)

| | |
|---|---|
|  | **CIRCUIT TERMINATION POINTS** |
| 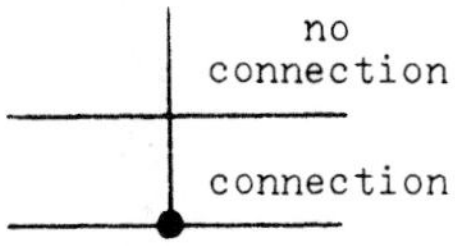  | **ELECTRICAL CONNECTIONS** |
| 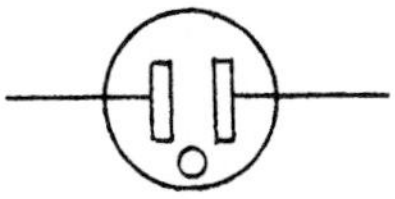 | **NEON INDICATOR** |
|   | **KEY INTERLOCK** |
| 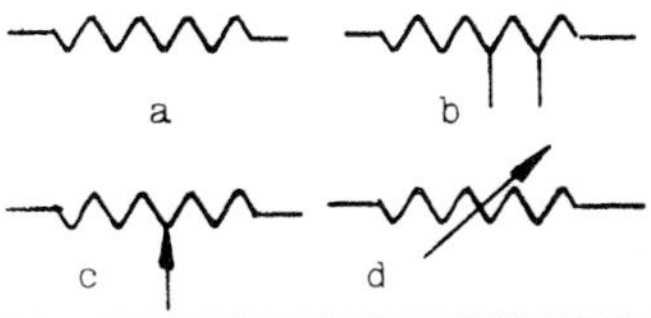  | **RESISTORS**<br>**a) Fixed; b) Tapped; c) Adjustable contact; d) Variable** |
|  | **SHIELDING**<br>**Electric or Magnetic** |
|   | **RELAY** |

**FIGURE 6.226** (*Continued*)

| | |
|---|---|
| 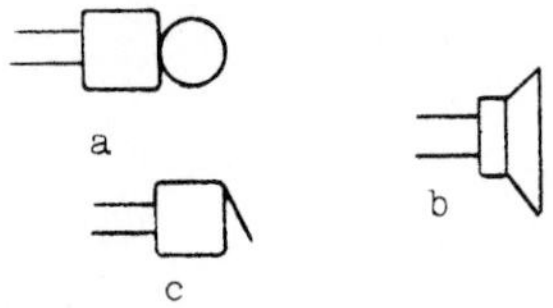  | SIGNALLING DEVICES<br>a) Bell; b) Buzzer; c) Horn, loudspeaker or siren |
| 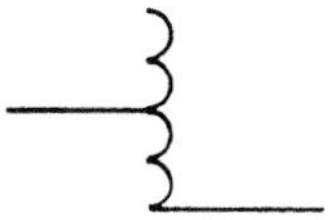 | AUTOTRANSFORMER |
| 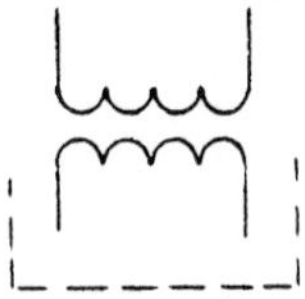 | SHIELDED TRANSFORMER |
| 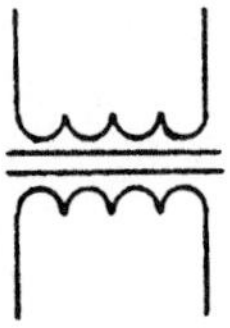 | MAGNETIC CORE TRANSFORMER |
| 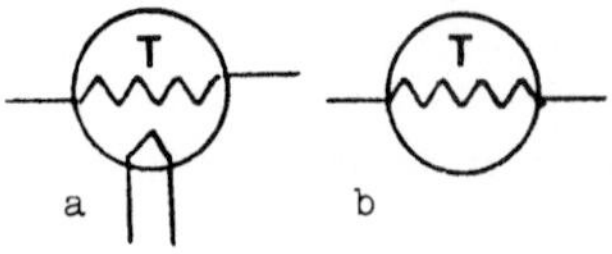  | THERMISTORS<br>a) General with independent heater; b) General (complete) |
| 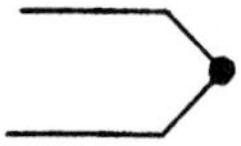 | THERMOCOUPLE |
| 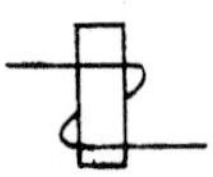 | RELAY COIL |

**FIGURE 6.226** (*Continued*)

| | |
|---|---|
| 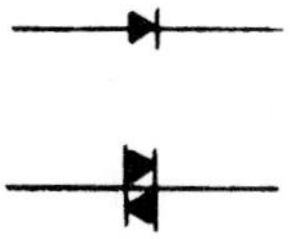 | VARISTORS<br>a) asymmetrical; b) Symmetrical |
| 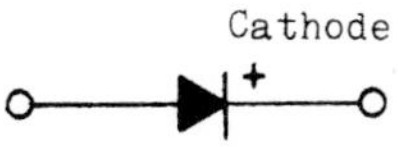 | DIODE, METALLIC |
| 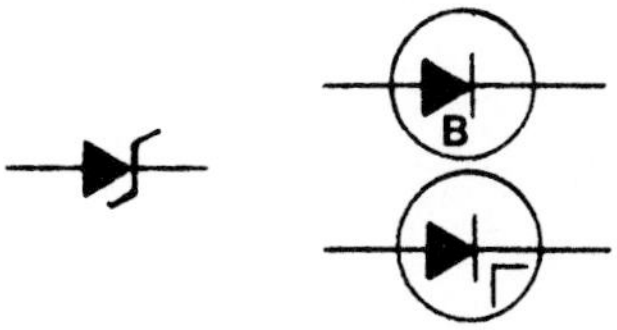 | ZENER (AVALANCHE) DIODES |
| 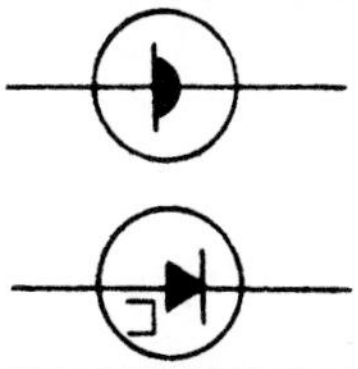 | TUNNEL DIODES |
| 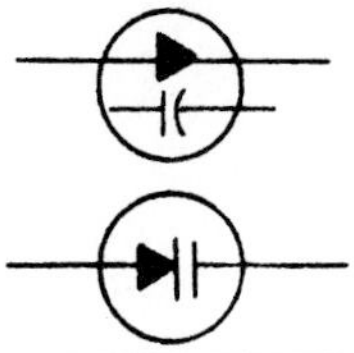 | VARACTOR DIODES |
| 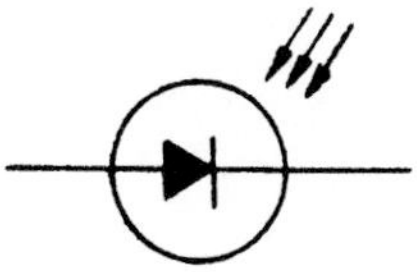 | LIGHT SENSOR DIODE |
| 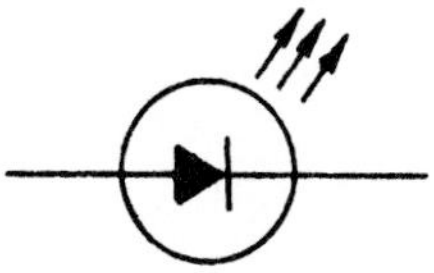 | LIGHT EMISSIVE DIODE<br>(Injection Lasers) |

**FIGURE 6.226** (*Continued*)

| | |
|---|---|
| 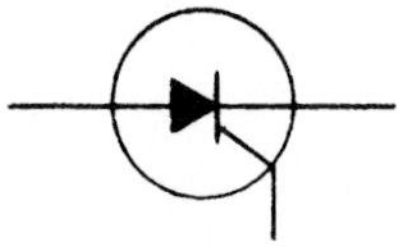 | SILICON CONTROLLED RECTIFIER (SCR) |
| 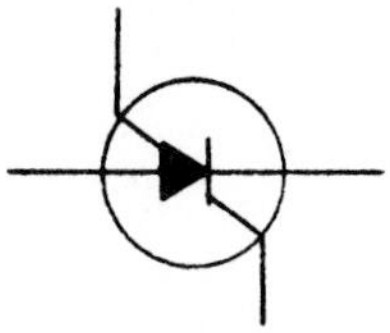 | SILICON CONTROLLED SWITCH (SCS) |
| 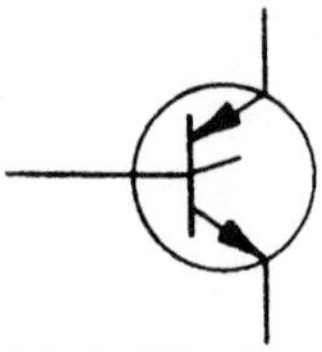 | GATE TURN-OFF SWITCH (GTO) |
| 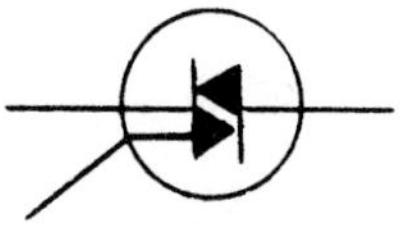 | TRIAC<br>Bidirectional triode |
| 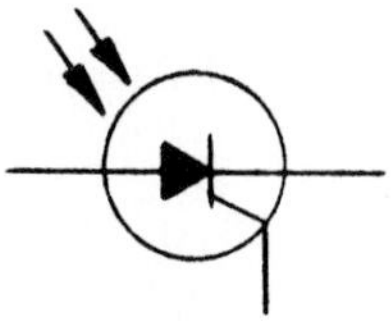 | LASCR<br>Light activated SCR |
| 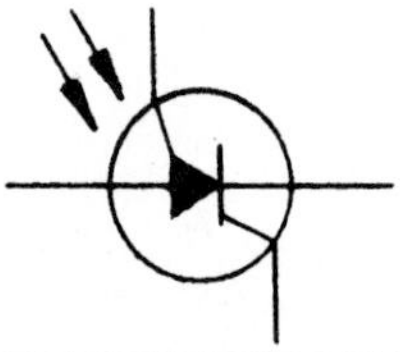 | LASCS<br>Light activated switch |
| 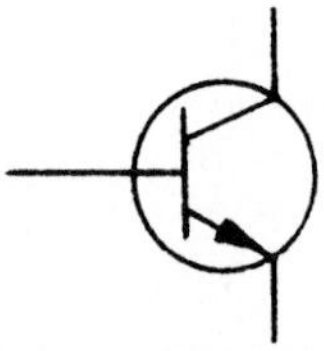 | NPN TRANSISTOR |

**FIGURE 6.226** (*Continued*)

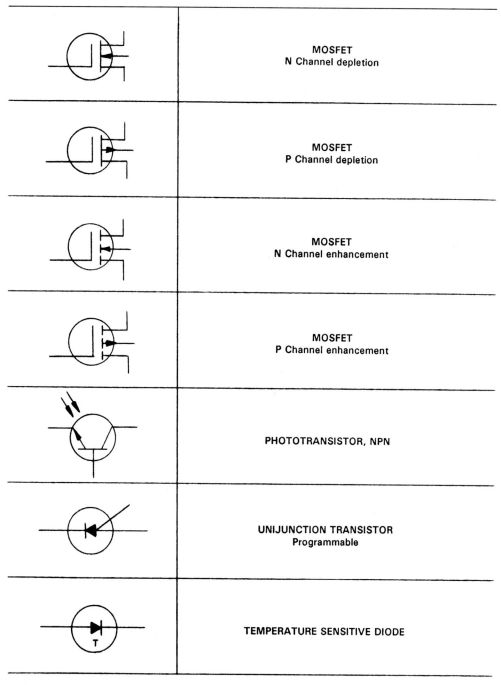

| | |
|---|---|
| 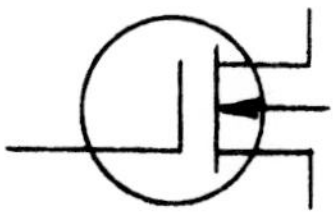 | MOSFET<br>N Channel depletion |
| 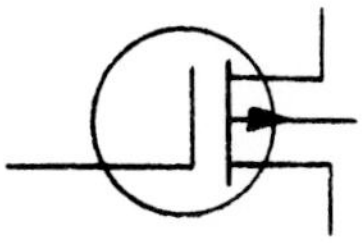 | MOSFET<br>P Channel depletion |
| 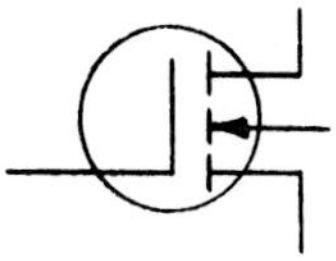 | MOSFET<br>N Channel enhancement |
| 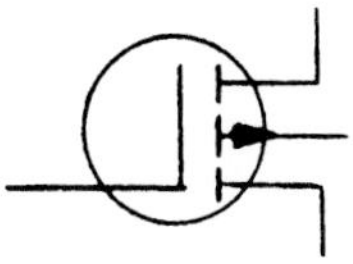 | MOSFET<br>P Channel enhancement |
| 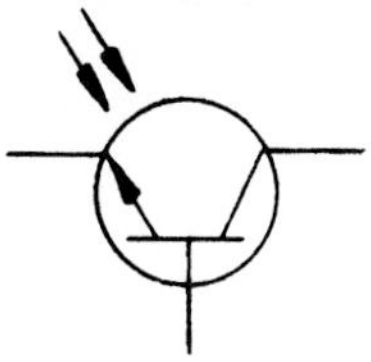 | PHOTOTRANSISTOR, NPN |
| 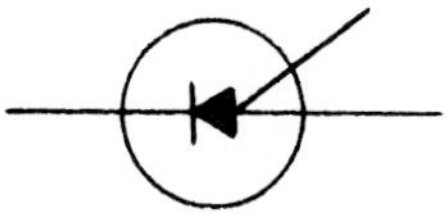 | UNIJUNCTION TRANSISTOR<br>Programmable |
| 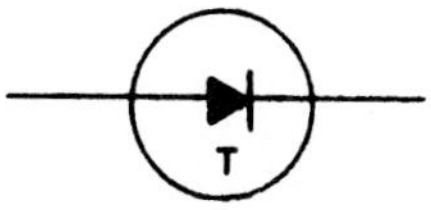  | TEMPERATURE SENSITIVE DIODE |

**FIGURE 6.226** (*Continued*)

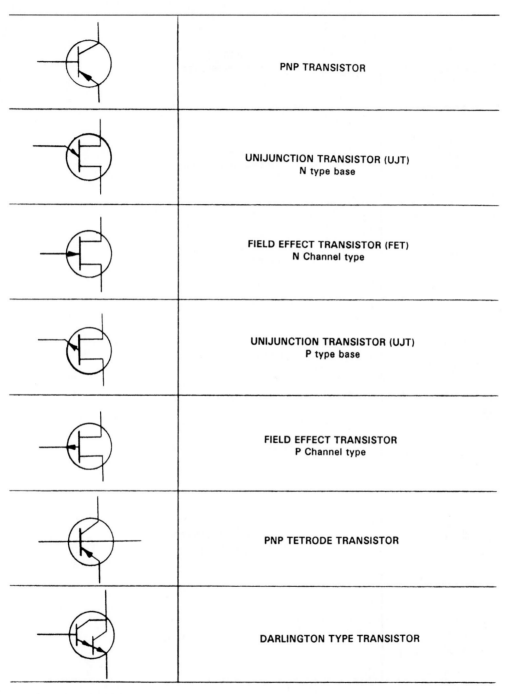

| | |
|---|---|
| 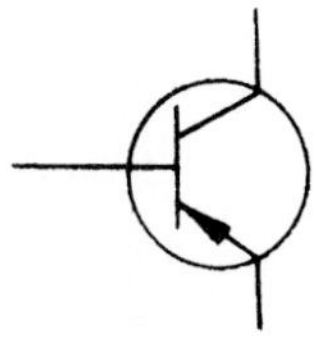 | PNP TRANSISTOR |
| 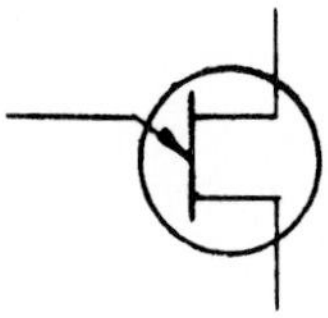 | UNIJUNCTION TRANSISTOR (UJT)<br>N type base |
| 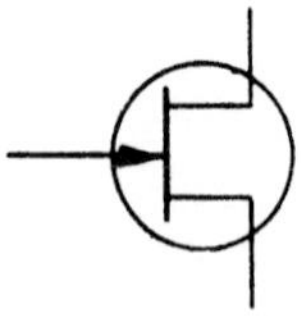 | FIELD EFFECT TRANSISTOR (FET)<br>N Channel type |
| 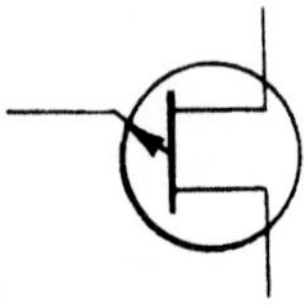 | UNIJUNCTION TRANSISTOR (UJT)<br>P type base |
| 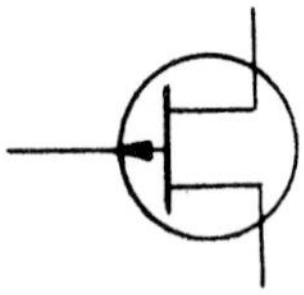 | FIELD EFFECT TRANSISTOR<br>P Channel type |
| 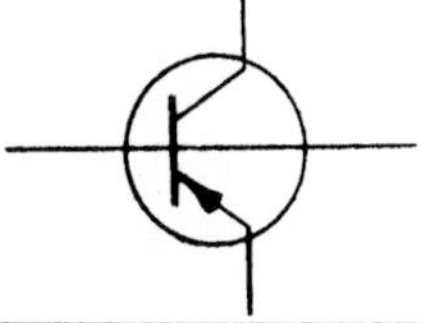 | PNP TETRODE TRANSISTOR |
| 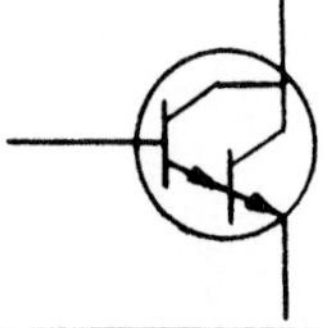 | DARLINGTON TYPE TRANSISTOR |

**FIGURE 6.226** (*Continued*)

| | |
|---|---|
| 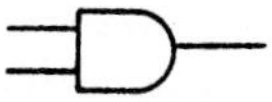 | AND GATE |
| 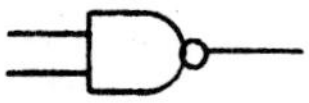 | NAND GATE |
| 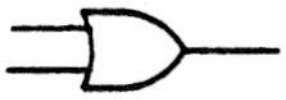 | OR GATE |
|  | NOR GATE |
|  | EXCLUSIVE OR GATE |
| 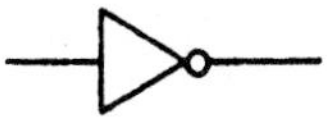 | INVERTER |
| 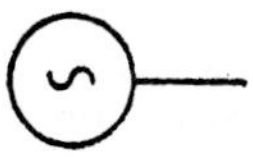 | OSCILLATOR |

**FIGURE 6.226** (*Continued*)

**Argentina (IRAM)** Instituto Argentino de Racionalization de Materiales

**Australia (SAA)** Standards Association of Australia

**Austria (ÖVE)** Öesterreichischer Verband für Elektrotechnik

**Belgium (CEB)** Comité Electrotechnique Belge

**Brazil (ABNT)** Associação Brasileira de Normas Teonicas

**Canada** (CSA) Canadian Standards Association

**Czechoslovakia** Urad pro Normalizaci a Mereni (Office for Standards and Measurements)

**Denmark (DEK)** Dansk Elektroteknisk Komite (Danish Electrotechnical Committee)

**Egypt (EOS)** Egyptian Organization for Standardization and Quality Control

**Finland (SETI)** Electrical Inspectorate

**France (UTE)** Union Technique de l'Électricité

**Germany, Democratic Republic of (East) (ASMW)** Amt für Standardi-sierung Messwesen und Warenprufung (Office for Standardization, Metrology and Quality Control)

**Germany, Federal Republic of (West) (DKE)** Deutsche Electrotechnische Kommission im DIN und VDE (German Electrotechnical Commission of DIN and VDE)

**Greece (ELOT)** The Hellenic Organization of Standardization

**Hong Kong** Hong Kong Standards & Testing Centre

**Hungary (MSZH)** Magyar Szabvanyugi Hivatal (Hungarian Office for Standardization)

**India (ISI)** Indian Standards Institute "Manak Bhavan"

**Indonesia** Badan Kerjasama Standardisasi LIPI-YDNI (LIPI-YDNI Joint Standardization Committee)

**Iran** Institute of Standards and Industrial Research of Iran

**Ireland (IIRS)** Institute for Industrial Research and Standards

**Israel (SII)** Standards Institution of Israel

**Italy (CEI)** Comitato Elettrotecnico Italiano (Italian Electrotechnical Committee)

**Japan (JISC)** Japanese Industrial Standards Committee Agency of Industrial Science and Technology

**Korea, Republic of (South)** Industrial Advance Administration

**Mexico (DGN)** Dirección General de Normas Secretaria de Patrimonio y Fomento Industrial Puente de Tecamachalco

**Netherlands (NEC)** Nederlands Elektrotechnisch Comite

**New Zealand (SANZ)** Standards Association of New Zealand

**Norway (NEK)** Norsk Elektroteknisk Komite (Norwegian Electrotechnical Committee)

**Pakistan** Pakistan Standards Institution

**Poland** Polski Komitet Normalizacji Miar i Jakosci (Polish Committee for Standardization, Measures and Quality Control)

**Portugal (CEP)** Comissão Electrotécnica Portuguesa

**Romania** Institutul Roman de Standardizare (Romanian Standards Institute)

**Singapore** Singapore Institute of Standards and Industrial Research

**South Africa (SABS)** South African Bureau of Standards

**Spain (IRANOR)** Instituto Nacional de Racionalización y Normalización

**Sweden (SEK)** Svenska Elektriska Kommissionen (Swedish Electrotechnical Commission)

**Switzerland (SEV)** Schweizerischer Elektrotechnischer Verein Postfach

**Turkey (TSE)** Turk Standardlari Enstitusu (Turkish Standards Institution)

**United Kingdom (BSI)** British Standards Institution

**United States** Underwriters Laboratories Inc. Recognized Component Mark

**United States** Underwriters Laboratories Inc.

**Venezuela** CODELECTRA en Conjunto con COVENIN Avda. Ppal.

**Yugoslavia** Federal Institute of Standardization

**FIGURE 6.227** International electrical and electronic component and equipment trademarks. (Source: EEM)

**TABLE 6.34** Fuse Wire Sizes and Materials

| Fusing Current (Amperes) | Copper | | Tin | | Lead | |
|---|---|---|---|---|---|---|
| | Gauge* | Dia. | Gauge* | Dia. | Gauge* | Dia. |
| 1 | 47 | 0.0021 | 37 | 0.0072 | 35 | 0.0081 |
| 2 | 43 | 0.0034 | 31 | 0.0113 | 30 | 0.0128 |
| 3 | 41 | 0.0044 | 28 | 0.0149 | 27 | 0.0168 |
| 4 | 39 | 0.0053 | 26 | 0.0181 | 25 | 0.0203 |
| 5 | 38 | 0.0062 | 25 | 0.0210 | 23 | 0.0236 |
| 10 | 33 | 0.0098 | 21 | 0.0334 | 20 | 0.0375 |
| 15 | 30 | 0.0129 | 19 | 0.0437 | 18 | 0.0491 |
| 20 | 28 | 0.0156 | 17 | 0.0529 | 17 | 0.0595 |
| 25 | 26 | 0.0181 | 16 | 0.0614 | 15 | 0.0690 |
| 30 | 25 | 0.0205 | 15 | 0.0694 | 14 | 0.0779 |
| 40 | 23 | 0.0248 | 14 | 0.0840 | 13 | 0.0944 |
| 50 | 22 | 0.0288 | 13 | 0.0975 | 12 | 0.1095 |
| 70 | 20 | 0.0360 | 10 | 0.1220 | 9 | 0.1371 |
| 100 | 18 | 0.0457 | 8 | 0.1548 | 7 | 0.1739 |

Note: * Wire gauges are approximate and are for standard wire gauge sizes.

## 6.19 FUSING WIRE SIZES AND MATERIALS

Table 6.34 shows the wire size, diameter, and material for fusing wires. This table is useful when a particular fuse element is required for emergency use on electrical and electronic equipment and apparatus. These fusing wires are not intended for permanent insertion in an electrical circuit. On voltages up to and including 600 V, the length of the fusing wire should not be less than 1 in. Fusing wires are not recommended for voltages higher than 600 V.

## 6.20 SHORT CIRCUITS IN ELECTRIC POWER DISTRIBUTION EQUIPMENT (SWITCHGEAR)

The analysis of the ac short-circuit capabilities of an electric power system is a complex process, requiring many calculations. The amount of current available in a short-circuit fault is determined by the capacity of the system voltage sources and the impedances in the system, including the fault.

The voltage sources consist of the power supply (utility or on-site generators) plus all rotating machines connected in the system during the time of the fault. A fault may be either an arcing or bolted fault. In an arcing fault, part of the circuit voltage is consumed across the fault and the total fault current is less than for a bolted fault, so the latter condition is the value sought in the fault calculations.

Basically, the short-circuit current is determined by Ohm's law, except that the impedance is not constant since some reactance is included in the system. The effect of reactance in an ac system is to cause the initial current to be high and then to decay toward steady-state (the Ohm's law) value. The fault current then consists of an exponentially decreasing dc component superimposed on a decaying alternating current. The rate of decay of both the dc and ac components depends on the ratio of reactance to resistance $X/R$ of the circuit. The greater this ratio, the longer the current remains higher than the steady-state value which it would eventually reach.

The total fault current is not symmetrical with respect to the time axis because of the dc components; therefore it is called *asymmetrical current.* The dc component depends on the point on the voltage wave at which the fault is initiated.

The ac component is not constant if rotating machines are connected to the system because the impedance of this apparatus is not constant. The rapid variation of motor and generator impedance is due to the following factors:

- Subtransient reactance
- Transient reactance
- Synchronous reactance
- Transformer impedance

The maximum current a transformer can deliver to a fault condition is the quantity of 100 divided by percent impedance times the transformer rated secondary current. Synchronous reactance has no effect as far as short-circuit calculations are concerned but is useful in the determination of protective relay settings.

When the fault capability of a circuit is given in megavolt-amperes (MVA), the rms symmetrical short-circuit current may be calculated from the following simple equation:

$$I_S = \frac{\text{MVA}}{V_S\sqrt{3}}$$

where MVA = system available fault power, i.e., 250 MVA = $250 \times 10^6$ VA
$V_S$ = system voltage (2.4, 4.16, 13.8 kV, etc.)
$I_S$ = short-circuit current, rms amperes

In the equation 250 MVA is written 250,000,000; 2.4 kV is written 2400; and so on; then, the current will be in whole number amperes (rms). Powers-of-ten notation is convenient to work with in this type of calculation. Then, 250,000,000 would be expressed as $250 \times 10^6$ and 2.4 kV as $2.4 \times 10^3$.

As a rule-of-thumb method, the symmetrical current may be multiplied by 1.6 to arrive at a general, average asymmetrical value for average circuit conditions. So, a short-circuit current of 20,000 A symmetrical can be considered to have an average asymmetrical value of $1.6 \times 20{,}000 = 32{,}000$ A in most power system circuits or networks.

## FURTHER READING


Berlin, H. M., 1985: *The 555 Timer Applications Sourcebook,* 1st ed., 7th printing. Indianapolis, Ind.: Howard W. Sams.

Berlin, H. M., 1988: *Design of Op-Amp Circuits.* Indianapolis, Ind.: Howard W. Sams.

Buchsbaum, W. H., 1984: *Handbook of Practical Electronic Reference Data,* 2nd ed. New York: Prentice-Hall.

*Burgess Engineering Manual.* Freeport, Ill.: Clevite Corp.

Bustamente, P., 1986: *Machine Design,* March 26, 1987, p. 127.

Cassell, W. L., 1964: *Linear Electric Circuits.* New York: Wiley.

*Contemporary Electronics Series,* McGraw-Hill Continuing Education Center. Washington, D.C.: McGraw-Hill.

Evans, A. J., J. D. Mullen, and D. H. Smith, 1985: *Basic Electronics Technology.* Dallas, Tex.: Texas Instruments Information Publishing Center.

Fink, D. G., and J. M. Carroll, 1968: *Standard Handbook for Electrical Engineers,* 10th ed. New York: McGraw-Hill.

Graf, R. F., 1983: *Electronic Data Book,* 3d ed. Blue Ridge Summit, Pa.: Tab Books.

*IC Master* (Index), 3 vols., 1988: Garden City, N.Y.: Hearst Business Communications.

Kaufman, M., 1982: *Electronics Technology, Theory and Problems* (Schaum's Outline Series). New York: McGraw-Hill.

Mantell, C. L., 1970: *Batteries and Energy Systems.* New York: McGraw-Hill.

Markus, J., 1982: *Popular Circuits—Ready Reference.* New York: McGraw-Hill.

Mendelson, R. M., 1984: *178 I.C. Designs and Applications.* Hasbrouck, N.J.: Hayden.

Omega Engineering, Inc., 1985: *Temperature Handbook.* Stamford, Conn.: Omega Engineering.

Reed, M. B., 1955: *Electric Network Synthesis.* Englewood Cliffs, N.J.: Prentice-Hall.

Tocci, R. J., 1983: *Introduction to Electric Circuit Analysis,* 2d ed. Columbus, Ohio: Charles E. Merrill.

Van Valkenberg, L., 1960: *Introduction to Modern Network Synthesis.* New York: Wiley.

Weinberg, L. 1962: *Network Analysis and Synthesis.* New York: McGraw-Hill.

# CHAPTER 7
# COMPREHENSIVE SPRING DESIGN

## 7.1 INTRODUCTION TO SPRING DESIGN

Springs are among the most important and most often used mechanical components. Many mechanisms and assemblies would be virtually impossible to design and manufacture without the use of springs in one form or another. There are many different types of springs and spring materials. We will cover all the most important and common types and forms which are designed and manufactured today. See Fig. 7.1 for typical springs.

Spring types or forms are summarized as follows.

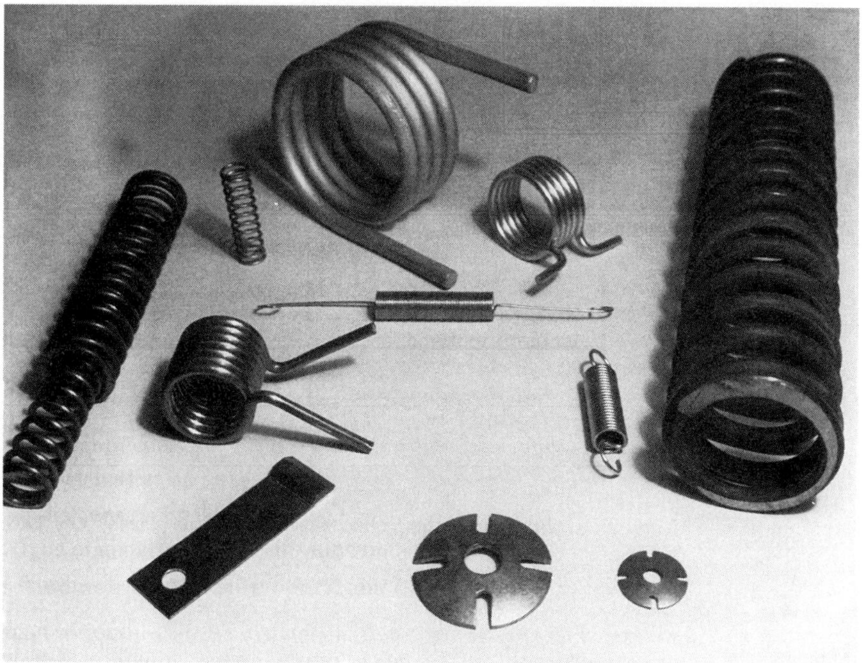

**FIGURE 7.1** Samples of typical springs.

1. Compression springs, straight and conical
   *a.* Round wire
   *b.* Square wire
   *c.* Rectangular wire
2. Extension springs, straight
   *a.* Round wire
3. Torsion springs
   *a.* Round wire
   *b.* Square wire
4. Coil springs, spiral
5. Leaf springs
   *a.* Cantilever
   *b.* Both ends supported (beam)
6. Spring washers
   *a.* Curved
   *b.* Wave
7. Hair springs, very light load
8. Torsion bars
9. Belleville washers (disk springs)

***Considerations Prior to the Spring Design Process.*** It is important when designing springs to adhere to proper procedures and design considerations. Some of the important design considerations in spring work are outlined here.

1. *Selection of material for spring construction*
   *a. Space limitations:* Do you have adequate space in the mechanism to use economical materials such as oil-tempered ASTM A229 spring wire? If your space is limited by design and you need maximum energy and mass, you should consider using materials such as music wire, ASTM A228 chrome vanadium or chrome silicon steel wire.
   *b. Economy:* Will economical materials such as ASTM A229 wire suffice for the intended application?
   *c. Corrosion resistance:* If the spring is used in a corrosive environment, you may select materials such as 17-7 PH stainless steel or the other stainless steels (301, 302, 303, 304, etc.).
   *d. Electrical conductivity:* If you require the spring to carry an electric current, materials such as beryllium copper and phosphorous bronze are available.
   *e. Temperature range:* Whereas low temperatures induced by weather are seldom a consideration, high-temperature applications will call for materials such as 301 and 302 stainless steel, nickel chrome A286, 17-7 PH, Inconel 600, and Inconel X750. Design stresses should be as low as possible for springs designed for use at high operating temperatures.
   *f. Shock loads, high endurance limit, and high strength:* Materials such as music wire, chrome vanadium, chrome silicon, 17-7 stainless steel, and beryllium copper are indicated for these applications.
2. General spring design recommendations
   *a.* Try to keep the ends of the spring, where possible, within such standard forms as closed loops, full loops to center, closed and ground, and open loops.
   *b. Pitch.* Keep the coil pitch constant unless you have a special requirement for a variable-pitch spring.
   *c.* Keep the spring index [mean coil diameter, in/wire diameter, in $(D/d)$] between 6.5 and 10 wherever possible. Stress problems occur when the index is too low, and entanglement and waste of material occur when the index is too high.
   *d.* Do not electroplate the spring unless it is required by the design application. The spring will be subject to hydrogen embrittlement unless it is processed correctly after electroplating. Hydrogen embrittlement causes abrupt and unexpected spring failures. Plated springs must be baked at a specified temperature for a definite time interval immediately

after electroplating to prevent hydrogen embrittlement. For cosmetic and minimal corrosion protection, zinc electroplating is generally used, although other plating such as chromium, cadmium, and tin are also used per the application requirements. Die springs usually come from the die spring manufacturers with colored enamel paint finishes for identification purposes. Black oxide and blueing are also used for spring finishes.

3. *Special processing either during or after manufacture:*
   *a.* Shot peening improves surface qualities from the standpoint of reducing stress concentration points on the spring wire material. This process can also improve the endurance limit and maximum allowable stress on the spring.
   *b.* Subjecting the spring to a certain amount of permanent *set* during manufacture eliminates the set problem of high energy versus mass on springs that have been designed with stresses in excess of the recommended values. This practice is *not* recommended for springs that are used in critical applications.
4. *Stress considerations:* Design the spring to stay within the allowable stress limit when the spring is fully compressed or "bottomed." This can be done when there is sufficient space available in the mechanism, and economy is not a consideration. When space is not available, design the spring so that its maximum working stress at its maximum working deflection does not exceed 40 to 45 percent of its minimum yield strength for compression and extension springs and 75 percent for torsion springs. Remember that the minimum tensile strength allowable is different for differing wire diameters; higher tensile strengths are indicated for smaller wire diameters. See Table 7.3 indicating the minimum tensile strengths for different wire sizes and different materials. See also Chap. 4.

***Direction of Winding on Helical Springs.*** Confusion is sometimes caused by different interpretations of what constitutes a right-hand- or left-hand-wound spring. Standard practice recognizes that the winding hand of helical springs is the same as standard right-hand screw thread and left-hand screw thread. A right-hand-wound spring has its coils going in the same direction as a right-hand screw thread and the opposite for a left-hand spring. On a right-hand helical spring, the coil helix progresses away from your line of sight in a clockwise direction, when viewed on end. This seems like a small problem, but it can be quite serious when designing torsion springs, where the direction of wind is critical to proper spring function. In a torsion spring, the coils must "close down" or tighten when the spring is deflected during normal operation, returning to its initial position when the load is removed. If a torsion spring is operated in the wrong direction, or "opened" as the load is applied, the working stresses become much higher and the spring could fail. The torsion spring coils also increase in diameter when operated in the wrong direction and likewise decrease in diameter when operated in the correct direction. See equations under torsion springs for calculations which show the final diameter of torsion springs when they are deflected during operation.

Also note that when two helical compression springs are placed one inside the other for a higher combined rate, the coil helixes must be wound on hands opposite from one another. This prevents the coils from jambing or tangling during operation. Compression springs employed in this manner are said to be in *parallel,* with the final rate equal to the combined rate of the two springs added together. Springs that are employed one atop the other or in a straight line are said to be in *series,* with their final rate equal to 1 divided by the sum of the reciprocals of the separate spring rates.

*Example.* Springs in parallel:

$$R_f = R_1 + R_2 + R_3 + \cdots + R_n$$

Springs in series:

$$\frac{1}{R_f} = \frac{1}{R_1} + \frac{1}{R_2} + \frac{1}{R_3} + \cdots + \frac{1}{R_n}$$

where $R_f$ = final combined rate; $R_{1,2,3}$ = rate of each individual spring.

In the following subsections, you will find all the design equations, tables, and charts required to do the majority of spring design work today. Special springs such as irregular-shaped flat springs and other nonstandard forms are calculated using the standard beam equations and column equations found in other sections of the handbook, or they must be analyzed using involved stress calculations or prototypes made and tested for proper function.

### 7.1.1 Spring Design Procedures

1. Determine what spring rate and deflection or spring travel is required for your particular application. Also, determine the energy requirements.
2. Determine the space limitations in which the spring is required to work and try to design the spring accordingly, using a parallel arrangement if required, or allow space in the mechanism for the spring per its calculated design dimensions.
3. Make a preliminary selection of the spring material that is dictated by the application or economics.
4. Make preliminary calculations to determine wire size or other stock size, mean diameter, number of coils, length, and so forth.
5. Perform the working stress calculations with the Wahl stress correction factor applied, to see if the working stress is *below* the *allowable* stress.
   *a.* *The working stress* is calculated using the appropriate equation with the working load applied to the spring. The load on the spring is found by multiplying the spring rate times the deflection length of the spring. Thus, if the spring rate was calculated to be 25 lbf/in and the spring is deflected 0.5 in, the load on the spring is $25 \times 0.5 = 12.5$ lbf.
   *b.* The *maximum allowable stress* is found by multiplying the minimum tensile strength allowable for the particular wire diameter or size used in your spring times the appropriate multiplier. (See Table 7.3 in this section for minimum tensile strength allowables for different wire sizes and materials and the appropriate multipliers.) For example, you are designing a compression spring using 0.130-in-diameter music wire, ASTM A228. The allowable maximum stress for this wire size is

   $$0.45 \times 258{,}000 = 116{,}100 \text{ psi} \qquad \text{(see Table 7.3)}$$

   (*Note:* A more conservatively designed spring would use a multiplier of 40 percent (0.40), while a spring that is not cycled frequently can use a multiplier of 50 percent (0.50), with the spring possibly taking a slight *set* during repeated operations or cycles. The multiplier for torsion springs is 75 percent (0.75), in all cases and is conservative.
   *c.* If the working stress in the spring is below the maximum allowable stress, the spring is properly designed relative to its stress level during operation.
   *d.* Remember that the modulus of elasticity of spring materials diminishes as the working temperature rises. This factor causes a decline in the spring rate. Also, working stresses should be decreased as the operating temperature rises. Table 7.2 in this section shows the maximum working temperature limits for different spring and spring wire materials. Only appropriate tests will determine to what extent these recommended limits may be altered.

## 7.2 COMPRESSION AND EXTENSION SPRINGS

This section contains formulas for various kinds of compression and extension springs. Symbols contained in the formulas are explained in Sec. 7.2.5.1.

***Round Wire.*** Rate:

$$R\text{, lb/in} = \left.\frac{Gd^4}{8ND^3}\right\} \quad \text{transpose for } d, N, \text{ or } D$$

Torsional stress:

$$S\text{, total corrected stress} = \left.\frac{8K_aDP}{\pi d^3}\right\} \quad \text{transpose for } D, P, \text{ or } d$$

Wahl curvature-stress correction factor:

$$K_a = \frac{4C-1}{4C-4} + \frac{0.615}{C} \qquad \text{where} \qquad C = \frac{D}{d}$$

***Square Wire.*** Rate:

$$R = \left.\frac{Gt^4}{5.6ND^3}\right\} \quad \text{transpose for } t, N, \text{ or } D$$

Torsional stress:

$$S\text{, total corrected stress } \left.\frac{2.4K_{a1}DP}{t^3}\right\} \quad \text{transpose for } D, P, \text{ or } t$$

Wahl curvature-stress correction factor:

$$K_{a1} = 1 + \frac{1.2}{C} + \frac{0.56}{C^2} + \frac{0.5}{C^3} \qquad \text{where} \qquad C = \frac{D}{t}$$

***Rectangular Wire.*** Rate (see Fig. 7.2 for factors $K_1$ and $K_2$):

$$R = \left.\frac{Gbt^3}{ND^3} K_2\right\} \quad \text{transpose for } b, t, N, \text{ or } D$$

Torsional stress, corrected:

$$S = \left.\frac{PD}{bt\sqrt{bt}}\beta\right\} \quad \text{transpose for } P \text{ or } D$$

$\beta$ is obtained from Fig. 7.2

### 7.2.1 Solid Height of Compression Springs

***Round Wire.*** Refer to Sec. 7.2.3.

***Square and Rectangular Wire.*** Because of distortion of the cross section of square or rectangular wire when the spring is formed, the compressed solid height can be determined from

$$t' = 0.48\,t\left(\frac{\text{o.d.}}{D} + 1\right)$$

where $t'$ = new thickness of inner edge of section in the axial direction, after coiling
$t$ = thickness of section before coiling
$D$ = mean diameter of the spring

**FACTORS FOR SQUARE AND RECTANGULAR SECTIONS**

| b/t | 1 | 1.2 | 1.5 | 2 | 2.5 | 3 | 5 | 10 | ∞ |
|---|---|---|---|---|---|---|---|---|---|
| Factor $K_1$ | 0.416 | 0.438 | 0.462 | 0.492 | 0.516 | 0.534 | 0.582 | 0.624 | 0.666 |
| Factor $K_2$ | 0.180 | 0.212 | 0.250 | 0.292 | 0.317 | 0.335 | 0.371 | 0.398 | 0.424 |

**STRESS FACTOR β FOR RECTANGULAR WIRE (b and t as shown)**

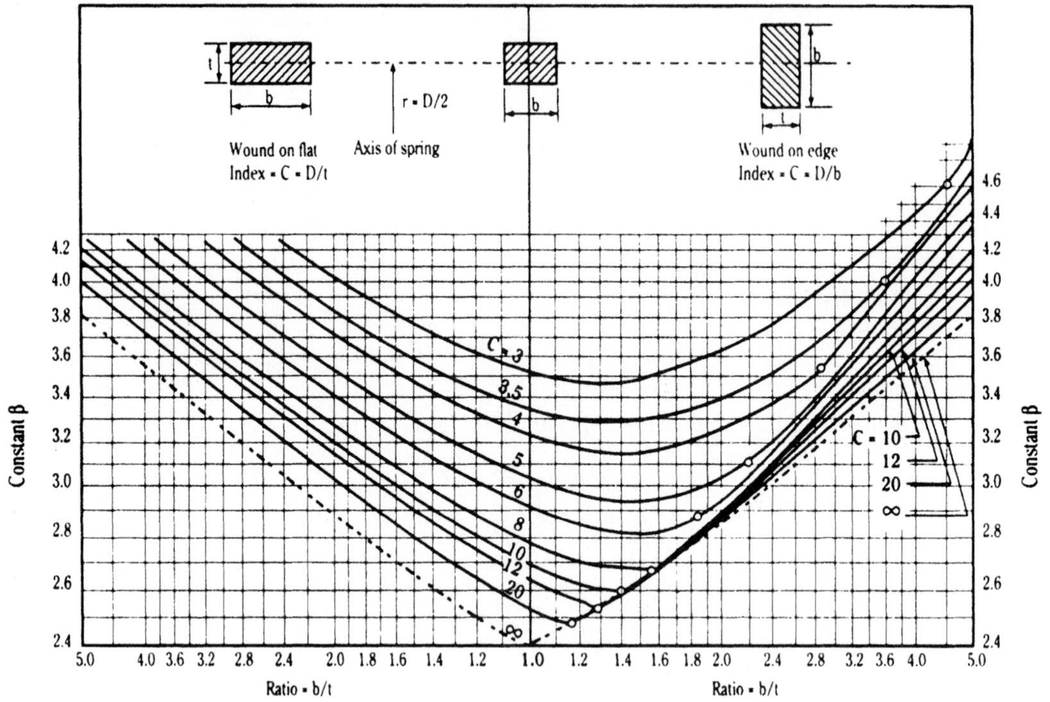


**FIGURE 7.2** Design factors for $K_1$, $K_2$, and β.

### 7.2.2 Initial Tension in Close-Wound Extension Springs

First, calculate torsional stress $S_i$ due to initial tension $P_1$ in

$$S_i = \frac{8DP_1}{\pi d^3}$$

where $P_1$ = initial tension, lbs.

Second, for the value of $S_i$ calculated and the known spring index $D/d$, determine on the graph in Fig. 7.3 whether $S_i$ appears in the preferred (shaded) area.

If $S_i$ falls in the shaded area, the spring can be produced readily. If $S_i$ is above the shaded area, reduce it by increasing the wire size. If $S_i$ is below the shaded area, select a smaller wire size. In either case, recalculate stress and alter the number of coils, axial space, and initial tension as necessary.

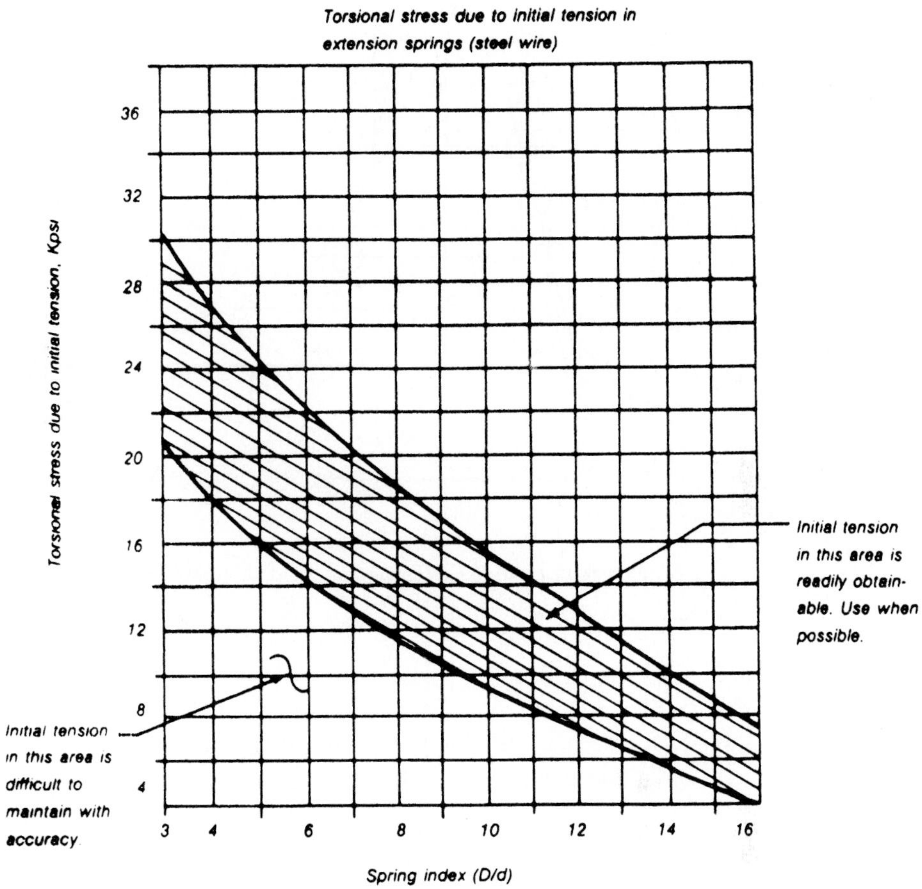


**FIGURE 7.3** Torsional stress due to initial tension in extension springs.

## 7.2.3 Active Coils (Compression Springs)

Style of ends may be selected as follows:

- Open ends, not ground; all coils are active.
- Open ends, ground; one coil inactive; $N - 1$ coils (where $N$ = the total number of coils in the spring).
- Closed ends, not ground; two coils inactive; $N - 2$ coils.
- Closed ends, ground; two coils inactive; $N - 2$ coils.

## 7.2.4 Conical Compression Springs

Conical compression springs are calculated as follows (Fig. 7.4):

1. Assuming that the spring is to have equal pitch (distance between coils), find the average geometric mean diameter from

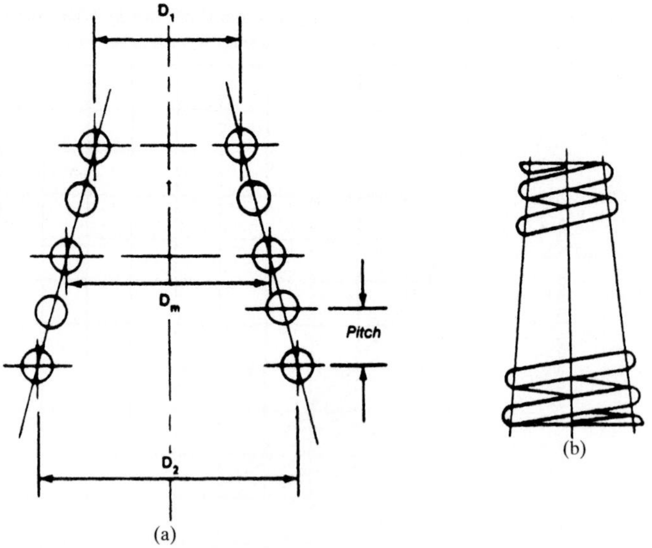


**FIGURE 7.4** (*a*) Conical spring dimensions; (*b*) typical conical spring.

$$D_m = \frac{D_1 + D_2}{2}$$

where $D_1$ = mean diameter of the top coil and $D_2$ = mean diameter of the bottom coil.

2. The spring rate may now be found from

$$R = \frac{Gd^4}{8ND_m^{\ 3}} \Bigg\} \quad \text{transpose for } d, N, \text{ or } D_m$$

and the stress from

$$S = \frac{8K_a D_m P}{\pi d^3} \Bigg\} \quad \text{transpose for } d, P, \text{ or } D_m$$

where $D_m$ = geometric mean diameter. Note that when the spring is deflected until the working bottom coil is bottomed, the equations are no longer valid because the rate will change to a higher value.

### 7.2.5 Spring Energy Content

The potential energy which may be stored in a deflected compression or extension spring is given as

$$P_e, \text{in} \cdot \text{lb} = \frac{Rs^2}{2} \text{ by integration}$$

where $R$ = rate of the spring in lb/in, lb/ft, N/m
$s$ = distance spring is compressed or extended, in, ft, m
$P_e$ = ft · lb, J

*Example.* A compression spring with a rate of 50 lb/in is compressed 4 in. What is the potential energy contained in the loaded spring?

$$P_e = \frac{50\,(4)^2}{2} = 400 \text{ in} \cdot \text{lb or } 33.3 \text{ ft} \cdot \text{lb}$$

Thus, the spring will perform 33.3 ft · lb of work energy when released from its loaded position. Internal losses are negligible. This procedure is useful to designers who need to know the work a spring will produce in a mechanism, and the input energy requirement to load the spring.

### 7.2.5.1 Symbols for Compression and Extension Springs

$R$ = rate, pounds of load per inch of deflection
$P$ = load, lb
$F$ = deflection, in
$D$ = mean coil diameter, in
$d$ = wire diameter, in
$t$ = side of square wire or thickness of rectangular wire, in
$b$ = width of rectangular wire, in
$G$ = torsional modulus of elasticity, psi
$N$ = number of active coils, determined by the type of ends on a compression spring; equal to all the coils on an extension spring (see Sec. 7.2.3)
$S$ = torsional stress, psi
o.d. = outside diameter of coils, in
i.d. = inside diameter of coils, in
$C$ = spring index, $D/d$
$L$ = spring length, in
$H$ = solid height of spring, in

### 7.2.5.2 Derivation of Spring Energy Equations

The equation for expressing the potential energy that may be stored in a compressed or extended helical coil spring was previously given as

$$P_e = \frac{Rs^2}{2} \text{ or } \frac{1}{2\,Rs^2}$$

From Fig. 7.5, we see that the force $F$, required to deflect a spring is $F = Rs$, where $R$ = spring rate, lb/ft, N/m, and $s$ = distance deflected, ft, m. So

$$\int_0^s F\,ds = \int_0^s Rs\,ds \qquad \text{or} \qquad P_e = \frac{1}{2}\,Rs^2$$

taking the antiderivative of $Rs$, and the work done in deflecting the spring from position $s_1$ to $s_2$ is

$$\Delta P_e = \int_{s1}^{s2} Rs\,ds = \frac{1}{2}\,R\,(s_2^2 - s_1^2)$$

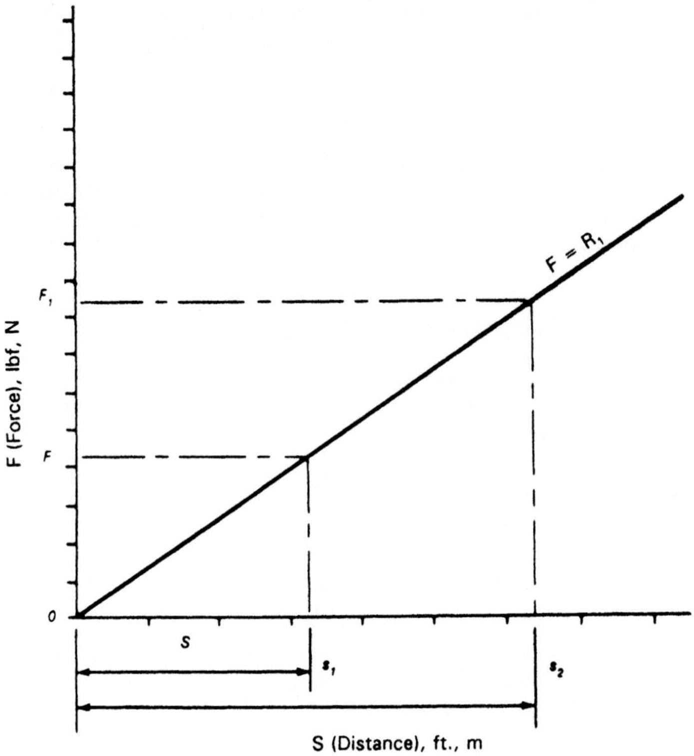


**FIGURE 7.5** Representation of work deflecting a spring.

In the case of a torsion or spiral spring

$$M = R\theta$$

where $M$ = resisting torque, lb · ft, N · m
$R$ = spring rate, lb/rad, N/rad
$\theta$ = angle of deflection, rad

Then

$$P_e = \int_0^\theta R\theta \, d\theta \qquad \text{or} \qquad P_e = \frac{1}{2} R\theta^2 \quad (\text{antiderivative } R\theta)$$

Units of elastic potential energy are the same as those for work and are expressed in foot-pounds in the U.S. customary system and in joules (J) in the SI system.

In the preceding derivations of spring potential-energy equations, the spring rate is of course assumed to be linear. In practice, spring rates are slightly nonlinear for standard spring materials, although constant-rate springs are manufactured from special materials. Calculations involving spring energy content using the above equations are satisfactory for most standard engineering problems. A close, average spring rate may be established by checking the spring force at various deflections under test.

**TABLE 7.1** Features of Compression Springs

| Feature | Type of End: Open or Plain (not ground) | Open or Plain (with ends ground) | Squared or Closed (not ground) | Closed and Ground |
|---|---|---|---|---|
| | Formula | | | |
| Pitch ($p$) | $\frac{FL-d}{N}$ | $\frac{FL}{TC}$ | $\frac{FL-3d}{N}$ | $\frac{FL-2d}{N}$ |
| Solid Height ($SH$) | $(TC+1)d$ | $TC \times d$ | $(TC+1)d$ | $TC \times d$ |
| Number of Active Coils ($N$) | $N = TC$ or $\frac{FL-d}{p}$ | $N = TC - 1$ or $\frac{FL}{p} - 1$ | $N = TC - 2$ or $\frac{FL-3d}{p}$ | $N = TC - 2$ or $\frac{FL-2d}{p}$ |
| Total Coils ($TC$) | $\frac{FL-d}{p}$ | $\frac{FL}{p}$ | $\frac{FL-3d}{p}+2$ | $\frac{FL-2d}{p}+2$ |
| Free Length ($FL$) | $(p \times TC)+d$ | $p \times TC$ | $(p \times N)+3d$ | $(p \times N)+2d$ |

$d$ = wire dia.

### 7.2.6 Expansion of Compression Springs When Deflected

A compression spring o.d. will expand when the spring is compressed. This may pose a problem if the spring must work within a tube or cylinder and its o.d. is close to the inside diameter of the containment. The following equation may be used to calculate the amount of expansion that takes place when the spring is compressed to solid height. For intermediate values, use the percent of compression multiplied by the total expansion.

$$\text{Total expansion} = \text{o.d., solid} - \text{o.d.}$$

Expanded diameter is

$$\text{o.d., solid} = \sqrt{D^2 + \frac{p^2 - d^2}{\pi^2}} + d$$

where $p$ = pitch (distance between adjacent coil center lines), in
$d$ = wire diameter, in
$D$ = mean diameter of the spring, in
o.d., solid = expanded diameter when compressed solid, in

### 7.2.7 Compression Spring Features

Refer to Table 7.1, in which $d$ = wire diameter.

## *7.3 TORSION SPRINGS*

Refer to Fig. 7.6.

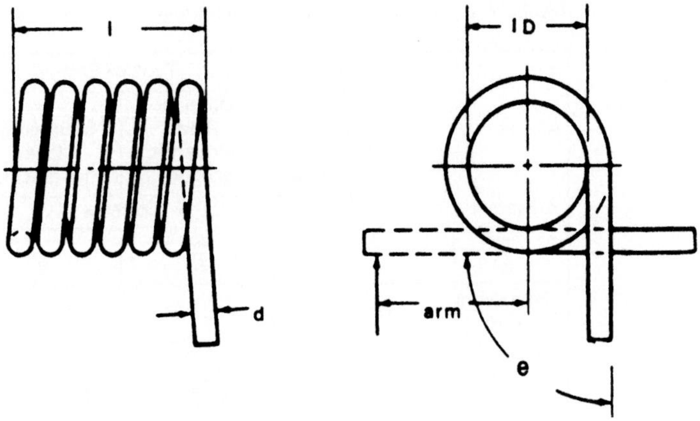


**FIGURE 7.6** Torsion spring.

***Round Wire.*** Moment (torque):

$$M = \frac{Ed^4T}{10.8\,ND} \quad \text{transpose for } d, T, N, \text{ or } D$$

Tensile stress

$$S = \frac{32\,M}{\pi d^3} K \quad \text{transpose for } M \text{ or } d$$

***Square Wire.*** Moment (torque):

$$M = \frac{Et^4T}{6.6ND} \quad \text{transpose for } t, T, N, \text{ or } D$$

Tensile stress:

$$S = \frac{6M}{t^3} K_1 \quad \text{transpose for } M \text{ or } t$$

The stress correction factor for torsion springs with round and square wire is applied according to spring index.

$$\left.\begin{aligned} \text{When the spring index} &= 6, \quad K = 1.15 \\ &= 8, \quad K = 1.11 \\ &= 10, \; K = 1.08 \end{aligned}\right\} \quad \text{round wire}$$

$$\left.\begin{aligned} \text{When the spring index} &= 6, \quad K_1 = 1.13 \\ &= 8, \quad K_1 = 1.09 \\ &= 10, \; K_1 = 1.07 \end{aligned}\right\} \quad \text{square wire}$$

For spring indexes that fall between the values shown, interpolate the new correction factor values.

***Rectangular Wire.*** Moment (torque):

$$M = \frac{Ebt^3T}{6.6\,ND} \quad \text{transpose for } b, t, T, N, \text{ or } D$$

Tensile stress:

$$S = \frac{6M}{bt^2} \quad \text{transpose for } M \text{ or } t$$

### 7.3.1 Symbols for Torsion Springs

$D$ = mean coil diameter, in
$d$ = diameter of round wire, in
$N$ = number of coils
$E$ = tension modulus of elasticity, psi
$T$ = revolutions through which the spring works (e.g., 90° arc = 90/360 = .25 revolutions)
$S$ = bending stress, psi
$M$ = moment or torque, lb · in
$b$ = width of rectangular wire, in (Fig. 7.8), in
$t$ = thickness of rectangular wire (Fig. 7.8), in
$K, K_1$ = stress correction factor for round and square wire

### 7.3.2 Torsion Spring Reduction of Diameter

$$ID_r = \frac{360N\,(ID_f)}{360N + R^\circ}$$

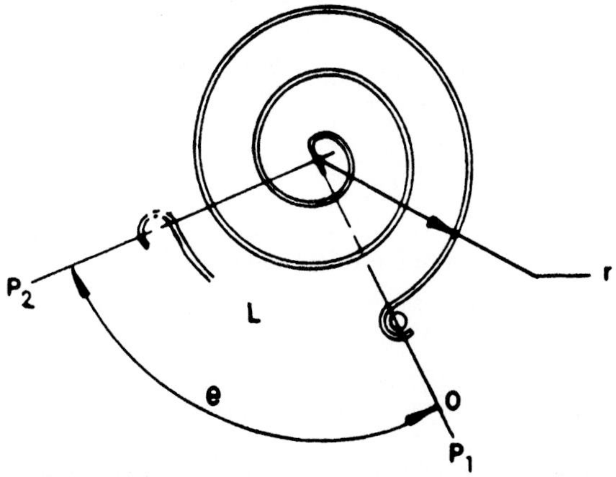


**FIGURE 7.7** Spiral coil spring.

where $ID_r$ = i.d. after deflection (closing)
$ID_f$ = i.d. before deflection (free)
$N$ = number of coils
$R°$ = number of degrees rotated in the closing direction

## 7.4 SPIRAL TORSION SPRINGS

For these coil springs, moment (torque) is

$$M,\ \text{lb}\cdot\text{in} = \frac{\pi E b t^3 \theta}{6L} \quad \Big\} \text{ transpose for } b,\ t, \text{ or } L$$

Bending stress:

$$S,\ \text{psi} = \frac{6M}{bt^2} \quad \Big\} \text{ transpose for } M,\ b, \text{ or } t$$

Space occupied by the spring:

$$OD_f = \frac{2L}{\pi[(\sqrt{A^2 + 1.27\ Lt} - A)/2t] - \theta}$$

This equation is based on concentric circles with a uniform space between coils and gives a close approximation of the minimum $OD_f$ (see Fig. 7.7).

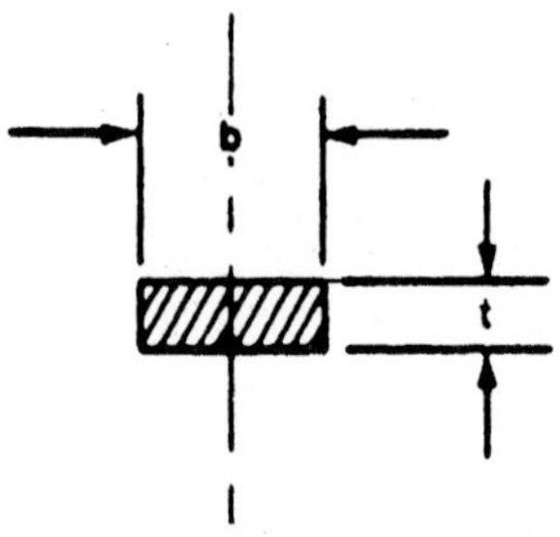


**FIGURE 7.8** Cross section, rectangular wire.

### 7.4.1 Spring Energy (Torsion and Coil Springs)

Referring to Fig. 7.7, $P_1$ = initial torque, $P_2$ = final torque, $r$ = radius of action, $L$ = action arc length, and $\theta$ = degrees of deflection.

To calculate the potential energy that may be stored in a torsion or coil spring, proceed as follows, assuming that the spring rate is linear.

$P_1$ is 0 when the spring is relaxed. Measure or calculate $P_2$ in lb/in of torque when deflected $\theta$ degrees. Translate the torque at $P_2$ to force, in pounds.

$$\frac{P_2}{r} = P_{f1},\ \text{lbf}$$

Now, the arc distance $L$ is calculated from

$$L = \frac{\pi r \theta}{180°}$$

Then, the apparent rate of the spring would be

$$R_a = \frac{P_f}{L},\ \text{lb per inch of arc}$$

The potential energy can be calculated from

$$P_e = \frac{R_a L^2}{2}$$

*Example.* A spiral spring develops 85 lb · in of torque when deflected through a 120° arc. What is the potential energy stored in the spring if its coil radius is 2 in?

$$P_f = \frac{P_2}{r} = \frac{85}{2} = 42.5 \text{ lb}$$

and

$$L = \frac{\pi r \theta}{180} = \frac{3.1416\,(2)\,(120)}{180} = 4.19 \text{ in}$$

and

$$R_a = \frac{P_f}{L} = \frac{42.5}{4.19} = 10.14 \text{ lb/in}$$

So

$$P_e = \frac{R_a L^2}{2} = 10.14\,\frac{(4.19)^2}{2} = 89 \text{ in} \cdot \text{lb or } 7.4 \text{ ft} \cdot \text{lb}$$

The value calculated for the energy content is within ±10 percent actual.

In a similar manner, the potential-energy content of leaf and beam springs can be derived by finding the apparent rate and the distance through which the spring moves. The potential energy contained in belleville spring washers is covered in Sec. 7.7.

### 7.4.2 Symbols for Spiral Torsion Springs

$E$ = bending modulus of elasticity, psi
$\theta$ = angular deflection, revolutions
$L$ = length of active spring material
$M$ = moment or torque, lb · in
$b$ = material width, in
$t$ = material thickness, in
$A$ = arbor diameter, in
$OD_f$ = outside diameter in the "free" condition

## 7.5 FLAT SPRINGS

Refer to Figs. 7.9 and 7.10.

***Cantilever Spring.*** Load (refer to Fig. 7.9):

$$P, \text{lb} = \left.\frac{EFbt^3}{4L^3}\right\} \quad \text{transpose for } F, b, t, \text{ or } L$$

Stress:

$$S, \text{psi} = \left.\frac{3EFt}{2L^2} = \frac{6PL}{bt^2}\right\} \quad \text{transpose for } F, t, L, b, \text{ or } P$$

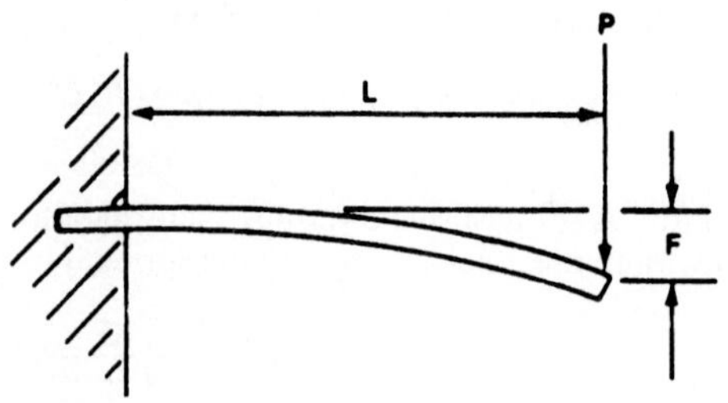


**FIGURE 7.9** Flat spring in cantilever.

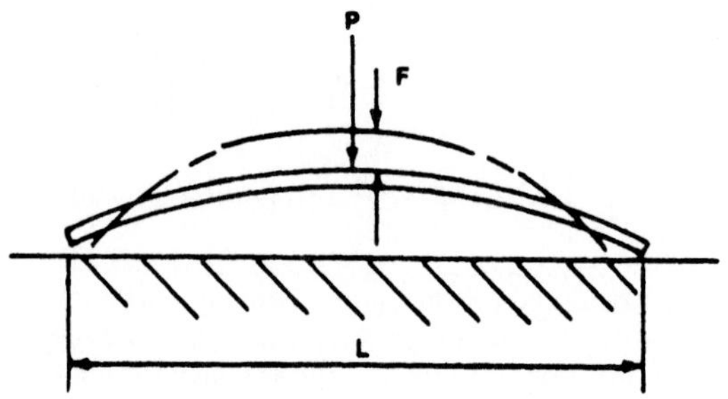


**FIGURE 7.10** Flat spring as a beam.

***Simple Beam Spring.*** Load (refer to Fig. 7.10):

$$P, \text{lb} = \left.\frac{4EFbt^3}{L^3}\right\} \quad \text{transpose for } F, b, t, \text{ or } L$$

Stress:

$$S, \text{psi} = \left.\frac{6EFt}{L^2} = \frac{3PL}{2bt^2}\right\} \quad \text{transpose for } F, b, t, \text{ or } L$$

Symbols used in the preceding discussion of cantilever and beam springs are summarized here.

$P$ = load, lb
$E$ = tension modulus of elasticity, psi
$F$ = deflection (see figure), in
$t$ = thickness of material, in
$b$ = width of material, in
$S$ = design bending stress, psi
$L$ = active spring length, in

## 7.6 SPRING WASHERS

Refer to Fig. 7.11*a* and 7.11*b*.

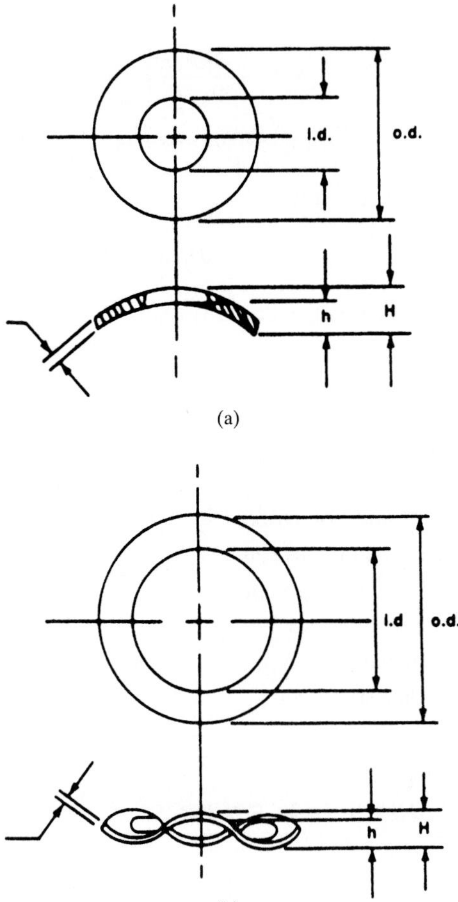


**FIGURE 7.11** (*a*) Curved spring washer; (*b*) wave washer.

***Curved Washers.*** Load (refer to Fig. 7.11*a*):

$$P\text{, lb} = \left.\frac{4EFt^3\,(\text{o.d.} - \text{i.d.})}{(\text{o.d.})^3}\right\} \quad \text{transpose for } F \text{ or } t$$

Stress.

$$S\text{, psi} = \left.\frac{1.5P\,(\text{o.d.})}{t^2\,(\text{o.d.} - \text{i.d.})}\right\} \quad \text{transpose for } P \text{ or } t$$

The preceding two equations yield approximate results.

***Wave Washers.*** Load (refer to Fig. 7.11*b*):

$$P\text{, lb} = \left.\frac{EFbt^3N^4}{2.4\,D^3}\left(\frac{\text{o.d.}}{\text{i.d.}}\right)\right\} \quad \text{transpose for } F, b, t, N, \text{ or } D$$

Stress:

$$S, \text{psi} = \left.\frac{3\pi PD}{4bt^2N^2}\right\} \quad \text{transpose for } P, D, b, t, \text{ or } N$$

These equations yield approximate results.

For deflections between 0.25 $h$ and 0.75 $h$, the o.d. increases and the new mean diameter $D_1$ can be calculated from

$$D_1, \text{in} = \sqrt{D^2 + 0.458\, h^2N^2}$$

Symbols for curved and wave washers:

$P$ = load, lb
$E$ = tensile modulus of elasticity, psi
$f$ = deflection, in
$t$ = material thickness, in
$b$ = radial width of material, in
$h$ = free height minus $t$, in
$H$ = free overall height, in
$N$ = number of waves
$D$ = mean diameter, inches = (o.d. + i.d.)/2
$S$ = bending stress, psi

## 7.7 BELLEVILLE WASHERS (DISK SPRINGS)

Equations for load $P$, stress at the convex inner edge $S$, and constants $M$, $C_1$, and $C_2$ are given as

$$P, \text{lb} = \frac{4Ef}{M(1-\mu^2)(\text{o.d.})^2}\left[\left(h - \frac{f}{2}\right)(h-f)t + t^3\right]$$

$$S, \text{psi} = \frac{4Ef}{M(1-\mu^2)(\text{o.d.})^2}\left[C_1\left(h - \frac{f}{2}\right) + C_2 t\right]$$

$$M = \frac{6}{\pi \ln a}\left[\frac{(a-1)^2}{a^2}\right] \quad \text{(constant } M\text{)}$$

$$C_1 = \frac{6}{\pi \ln a}\left[\frac{(a-1)}{\ln a} - 1\right] \quad \text{(constant } C_1\text{)}$$

$$C_2 = \frac{6}{\pi \ln a}\left[\frac{a-1}{2}\right] \quad \text{(constant } C_2\text{)}$$

Refer to Fig. 7.12, which diagrams a belleville spring; Fig. 7.13, which displays a load deflection chart; and Fig. 7.14, which gives stress constants.

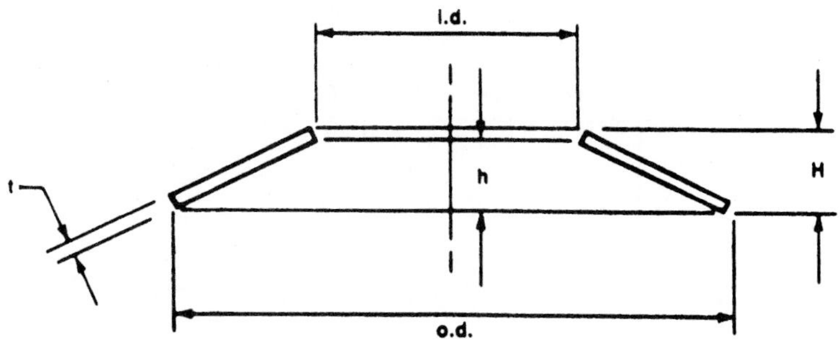


**FIGURE 7.12** Belleville spring geometry.

Symbols used in these equations are as follows:

$a$ = o.d./i.d.
$\mu$ = Poisson's ratio (see Poisson ratio chart, Sec. 7.12.3)
$f$ = deflection, in
$h$ = inner height of washer, in
$E$ = tensile modulus of elasticity, psi (Young's modulus)
$P$ = load, lb
$S$ = stress at the convex inner edge, psi
$t$ = thickness of washer, in
$M$, $C_1$, $C_2$ = constants (see equations)
ln = natural or hyperbolic logarithm (base $e$)

The preceding equations are presented with permission of the Spring Manufacturers Institute.

### 7.7.1 Simple Belleville Applications

***Parallel Stacking.*** Refer to Fig. 7.15. Placing belleville washers in this configuration will double the load for a given deflection.

***Series Stacking.*** Refer to Fig. 7.16. Placing belleville washers in this configuration will equal the load for a given deflection of one washer.

Belleville washers may be stacked in various series and parallel arrangements to produce varying results.

A form of belleville washer that is useful is the slotted spring washer made with a radiused section. Note that the equations for belleville washers apply only to those washers that have a conical section, as in Fig. 7.12. A washer with a radiused section will exhibit a different rate value on deflection. On critical applications for these types of spring washers, a load cell may be used to find the load value for a particular deflection. This also holds true for critical applications of standard section belleville washers. In any event, the equations will predict load results to approximately ±20 percent for standard section belleville washers.

A loading of belleville washers is illustrated in Fig. 7.17.

*Example.* If it requires 200 lb to flatten one belleville washer, flattening both as in the figure will produce a 200-lb clamping load or tension in the bolt. The reactions at $R_1$ and $R_2$ = 200 lb, and the clamp load is thus 200 lb.

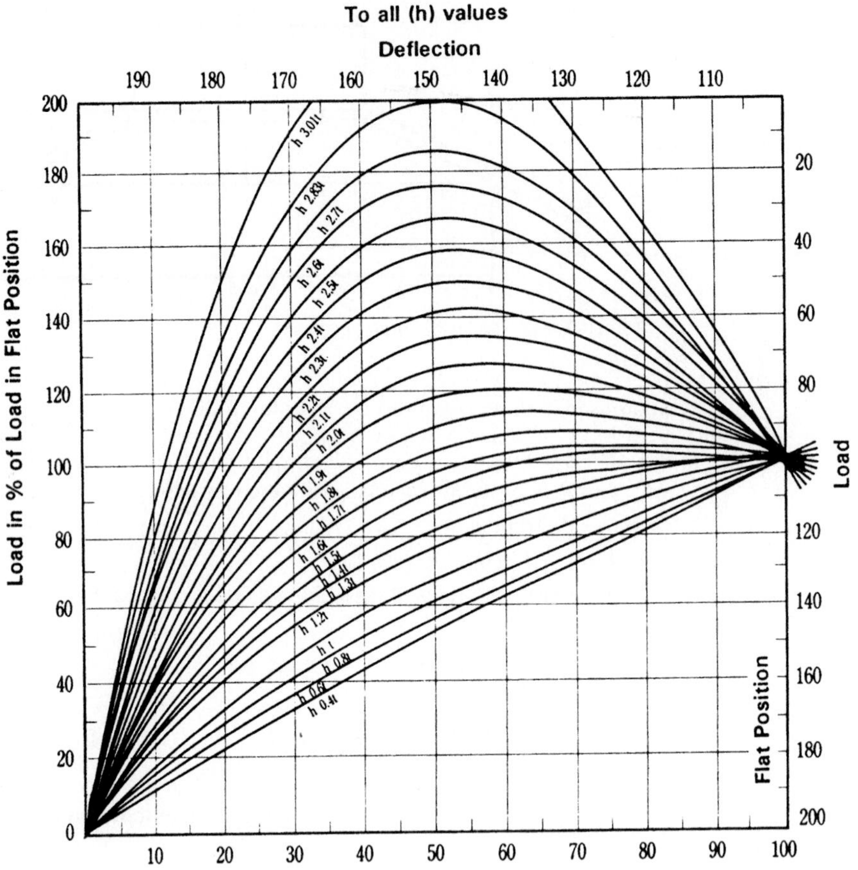


**FIGURE 7.13** Belleville load deflection chart.

## 7.8 HAIR SPRINGS

These springs find application in clocks, meters, gauges, and instruments. Moment (torque) is

$$M = \frac{\pi b t^3 E \theta}{6L} \quad \text{transpose for } b, t, L, \text{ or } \theta$$

Stress:

$$S = \frac{6M}{bt^2} \quad \text{transpose for } b, t, \text{ or } M$$

These equations are for light loads and low stresses.

## 7.9 TORSION BARS

See the chart in Fig. 7.2 for values of $K_1$ and $K_2$ used here.

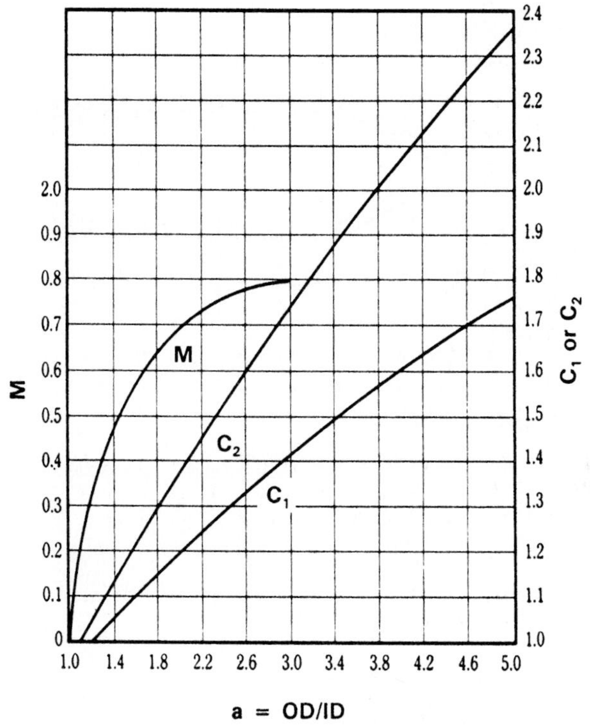


**FIGURE 7.14** Stress constants for belleville spring washers.

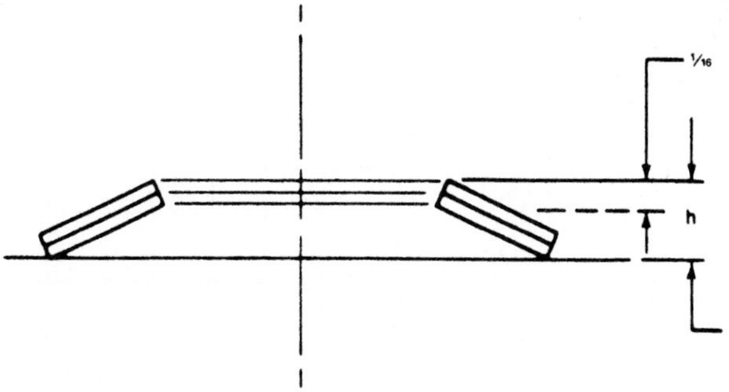


**FIGURE 7.15** Belleville washers in parallel stack.

***Round.*** Moment is

$$M = \frac{\pi^2 d^4 G\,\theta}{16L} \Bigg\} \quad \text{transpose for } d,\ L, \text{ or } \theta$$

Stress:

$$S = \frac{16M}{\pi d^3} = \frac{\pi d G\,\theta}{L} \Bigg\} \quad \text{transpose for } d,\ M,\ L, \text{ or } \theta$$

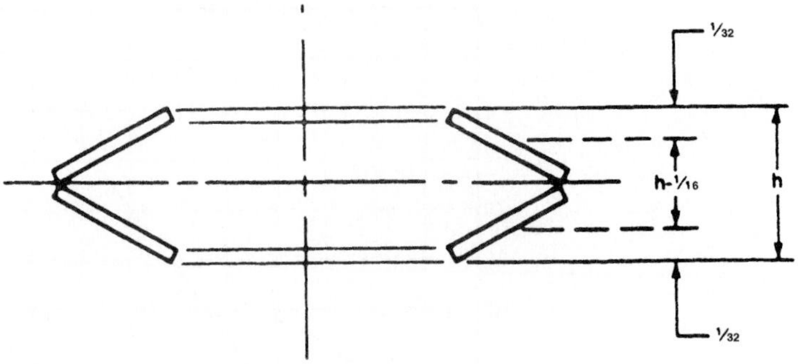


**FIGURE 7.16** Belleville washers in series stack.

***Rectangular.*** Moment is

$$M = \frac{\pi^2 Gbt^3\theta}{2L}(K_2)\Bigg\} \quad \text{transpose for } b, t, L, \text{ or } \theta$$

Stress:

$$S = \frac{2M}{bt^3K_2}\Bigg\} \quad \text{transpose for } b, t, \text{ or } M^*$$

---

* Note. Use $K_1$ for square wire.

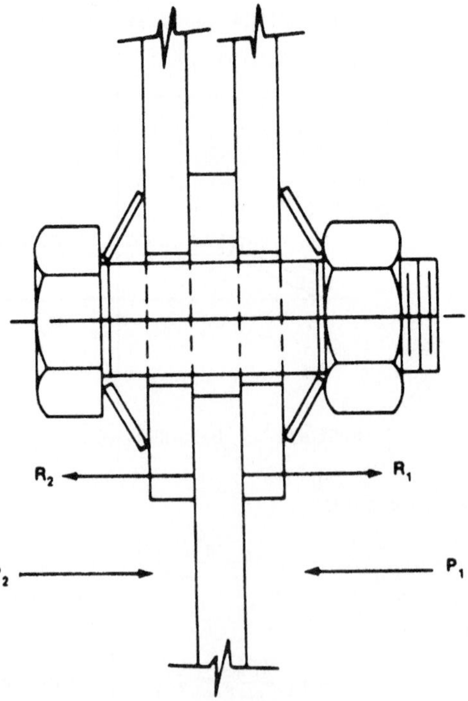


**FIGURE 7.17** Belleville washers under load.

## 7.10 ALLOWABLE WORKING STRESSES IN SPRINGS

Maximum design-stress allowables taken as a percentage of the minimum tensile strength are set.

1. Extension and compression springs.
   *a.* Hard-drawn steel (ASTM A227)
   *b.* Stainless steel (ASTM A313)
   Allowable stress (torsional) = 40 percent of the minimum tensile strength for each particular wire size (see stress-allowable tables)

   *c.* Oil-tempered (ASTM A229)
   *d.* Music wire (ASTM A228)
   *e.* Chrome vanadium (AISI 6150)
   *f.* Chrome silicon (AISI 9254)
   *g.* 17-7 PH stainless (AMS 5673B)
   *h.* Beryllium copper (ASTM B197)

   Allowable stress (torsional) = 45 percent of the minimum tensile strength for each particular wire size (see stress-allowable tables)
2. Torsion and flat springs—all materials

   Allowable stress (bending) = 75 percent of the minimum tensile strength for each particular wire size or gauge (see stress-allowable tables)

The listed values of the stress allowables are for average design applications. For cyclic loading and high repetition of loading and unloading, lower values (in percent) should initially be used. Life testing in critical applications is indicated and should be performed.

It should be noted that these values may also be increased in cases where permanent *set* is performed during spring manufacture, and where only occasional deflection of the spring is encountered. Also, higher values of permissible stress are sometimes used on statically loaded springs.

It should be understood by the designer that allowable stress values for spring materials cannot be "generalized." The minimum tensile or torsional strength of *each* wire size must be known accurately in order to calculate and design springs accurately. The minimum tensile strengths for spring materials are presented in Sec. 7.12.2 and should be used for accurate spring design calculations. Note that smaller wire sizes have higher tensile strength *rates* than larger wire sizes. If necessary, accurate torsional and bending stress values may be obtained directly from the spring wire manufacturers.

In highly stressed spring designs, spring manufacturers may also be consulted and their recommendations followed.

Whenever possible in mechanism design, space for a moderately stressed spring should be allowed. This will avoid the problem of marginally designed springs—that is, springs that tend to be stressed close to or beyond the recommended maximum allowable stress. This, of course, is not always possible and adequate space for moderately stressed springs is not always available. Music wire is commonly used when high stress is a factor in design.

## 7.11 SPRING END TYPES

### 7.11.1 Preferred Ends

Refer to Fig. 7.18.

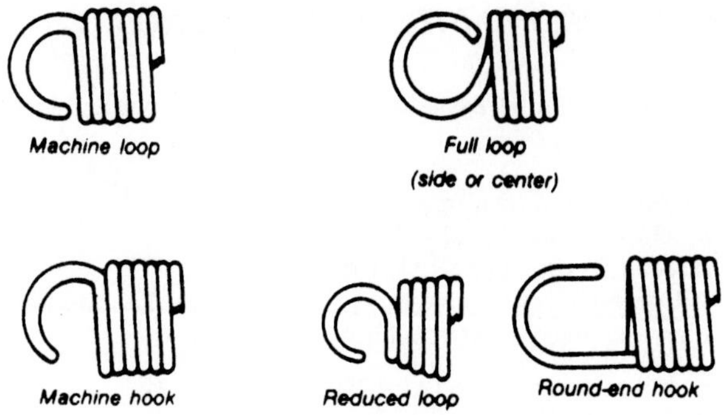


**FIGURE 7.18** Preferred ends, extension springs.

### 7.11.2 Special Ends

Refer to Fig. 7.19.

## 7.12 SPRING MATERIALS DATA

### 7.12.1 Materials and Properties

The chart in Table 7.2 shows physical properties of spring wire and strip that are used for spring design calculations.

### 7.12.2 Minimum Yield Strength of Spring Wire

Refer to Table 7.3.

### 7.12.3 Poisson's Ratios for Spring Materials

Refer to Table 7.4.

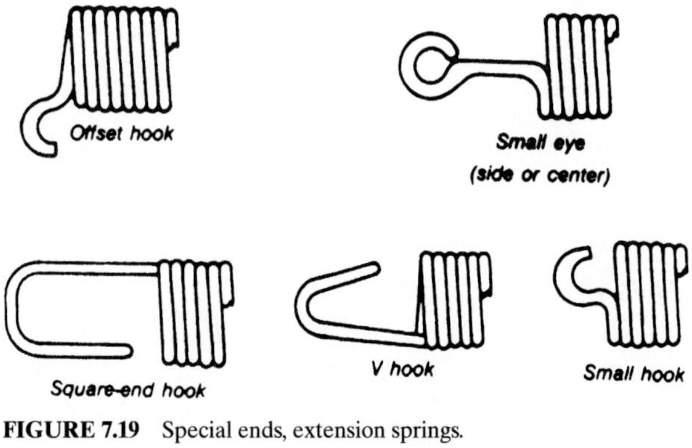


**FIGURE 7.19** Special ends, extension springs.

**TABLE 7.2** Properties of Spring Materials

| Material and specification | E $10^6$ psi | G $10^6$ psi | Design stress % min. yield | Cond. % IACS | Density lb/in$^3$ | Max. oper. temp. °F | F.A.* | S.A.* |
|---|---|---|---|---|---|---|---|---|
| **High-carbon wire** | | | | | | | | |
| Music | | | | | | | | |
| ASTM A228 | 30 | 11.5 | 45 | 7 | 0.284 | 250 | E | H |
| Hard-drawn | | | | | | | | |
| ASTM A227 | 30 | 11.5 | 40 | 7 | 0.284 | 250 | P | M |
| ASTM A679 | 30 | 11.5 | 45 | 7 | 0.284 | 250 | P | M |
| Oil-tempered | | | | | | | | |
| ASTM A229 | 30 | 11.5 | 45 | 7 | 0.284 | 300 | P | M |
| Carbon valve | | | | | | | | |
| ASTM A230 | 30 | 11.5 | 45 | 7 | 0.284 | 300 | E | H |
| **Alloy steel wire** | | | | | | | | |
| Chrome-vanadium | | | | | | | | |
| ASTM A231 | 30 | 11.5 | 45 | 7 | 0.284 | 425 | E | H |
| Chrome-silicon | | | | | | | | |
| ASTM A401 | 30 | 11.5 | 45 | 5 | 0.284 | 475 | F | H |
| Silicon-manganese | | | | | | | | |
| AISI 9260 | 30 | 11.5 | 45 | 4.5 | 0.284 | 450 | F | H |
| **Stainless wire** | | | | | | | | |
| AISI 302/304 | | | | | | | | |
| ASTM A313 | 28 | 10 | 35 | 2 | 0.286 | 550 | G | M |
| AISI 316 | | | | | | | | |
| ASTM A313 | 28 | 10 | 40 | 2 | 0.286 | 550 | G | M |
| 17-7 PH | | | | | | | | |
| ASTM A313 (631) | 29.5 | 11 | 45 | 2 | 0.286 | 650 | G | H |
| **Non-ferrous alloy wire** | | | | | | | | |
| Phosphor-bronze | | | | | | | | |
| ASTM B159 | 15 | 6.25 | 40 | 18 | 0.320 | 200 | G | M |
| Beryllium-copper | | | | | | | | |
| ASTM B197 | 18.5 | 7 | 45 | 21 | 0.297 | 400 | E | H |
| Monel 400 | | | | | | | | |
| AMS 7233 | 26 | 9.5 | 40 | — | — | 450 | F | M |
| Monel K 500 | | | | | | | | |
| QQ-N-286 | 26 | 9.5 | 40 | — | — | 550 | F | M |
| **High temperature alloy wire** | | | | | | | | |
| Nickel-chrome | | | | | | | | |
| ASTM A286 | 29 | 10.4 | 35 | 2 | 0.290 | 510 | — | L |
| Inconel 600 | | | | | | | | |
| QQ-W-390 | 31 | 11 | 40 | 1.5 | 0.307 | 700 | F | L |
| Inconel X750 | | | | | | | | |
| AMS 5698, 5699 | 31 | 12 | 40 | 1 | 0.298 | 1100 | F | L |
| **High carbon steel strip** | | | | | | | | |
| AISI 1065 | 30 | 11.5 | 75 | 7 | 0.284 | 200 | F | M |
| AISI 1075 | 30 | 11.5 | 75 | 7 | 0.284 | 250 | G | H |
| AISI 1095 | 30 | 11.5 | 75 | 7 | 0.284 | 250 | E | H |
| **Stainless steel strip** | | | | | | | | |
| AISI 301 | 28 | 10.5 | 75 | 2 | 0.286 | 300 | G | M |
| AISI 302 | 28 | 10.5 | 75 | 2 | 0.286 | 550 | G | M |
| AISI 316 | 28 | 10.5 | 75 | 2 | 0.286 | 550 | G | M |
| 17-7 PH | | | | | | | | |
| ASTM A693 | 29 | 11 | 75 | 2 | 0.286 | 650 | G | H |
| **Non-ferrous alloy strip** | | | | | | | | |
| Phosphor-bronze | | | | | | | | |
| ASTM B103 | 15 | 6.3 | 75 | 18 | 0.320 | 200 | G | M |
| Beryllium-copper | | | | | | | | |
| ASTM B194 | 18.5 | 7 | 75 | 21 | 0.297 | 400 | E | H |

**TABLE 7.2** Properties of Spring Materials (*Continued*)

| | | | | | | | | |
|---|---|---|---|---|---|---|---|---|
| Monel 400 | | | | | | | | |
| AMS 4544 | 26 | — | 75 | — | — | 450 | — | — |
| Monel K 500 | | | | | | | | |
| QQ-N-286 | 26 | — | 75 | — | — | 550 | — | — |
| **High temperature alloy strip** | | | | | | | | |
| Nickel-chrome | | | | | | | | |
| ASTM A286 | 29 | 10.4 | 75 | 2 | 0.290 | 510 | — | L |
| Inconel 600 | | | | | | | | |
| ASTM B168 | 31 | 11 | 40 | 1.5 | 0.307 | 700 | F | L |
| Inconel X750 | | | | | | | | |
| AMS 5542 | 31 | 12 | 40 | 1 | 0.298 | 1100 | F | L |

*Letter designations of the last two columns indicate: F.A. = fatigue applications, S.A. = strength applications, E = excellent, G = good, F = fair, L = low, H = high, M = medium, P = poor.

**TABLE 7.3** Minimum Tensile Strength of Spring Wire Materials

**Stainless steels**

| Wire Size, In. | Type 302 | Type* 17-7 PH | Wire Size, In. | Type 302 | Type* 17-7 PH | Wire Size, In. | Type 302 | Type* 17-7 PH |
|---|---|---|---|---|---|---|---|---|
| .008 | 325 | 345 | .033 | 276 | | .060 | 256 | |
| .009 | 325 | | .034 | 275 | | .061 | 255 | 305 |
| .010 | 320 | 345 | .035 | 274 | | .062 | 255 | 297 |
| .011 | 318 | 340 | .036 | 273 | | .063 | 254 | |
| .012 | 316 | | .037 | 272 | | .065 | 254 | |
| .013 | 314 | | .038 | 271 | | .066 | 250 | |
| .014 | 312 | | .039 | 270 | | .071 | 250 | 297 |
| .015 | 310 | 340 | .040 | 270 | | .072 | 250 | 292 |
| .016 | 308 | 335 | .041 | 269 | 320 | .075 | 250 | |
| .017 | 306 | | .042 | 268 | 310 | .076 | 245 | |
| .018 | 304 | | .043 | 267 | | .080 | 245 | 292 |
| .019 | 302 | | .044 | 266 | | .092 | 240 | 279 |
| .020 | 300 | 335 | .045 | 264 | | .105 | 232 | 274 |
| .021 | 298 | 330 | .046 | 263 | | .120 | 225 | 272 |
| .022 | 296 | | .047 | 262 | | .125 | | 272 |
| .023 | 294 | | .048 | 262 | | .131 | | 260 |
| .024 | 292 | | .049 | 261 | | .148 | 210 | 256 |
| .025 | 290 | 330 | .051 | 261 | 310 | .162 | 205 | 256 |
| .026 | 289 | 325 | .052 | 260 | 305 | .177 | 195 | |
| .027 | 287 | | .055 | 260 | | .192 | | |
| .028 | 286 | | .056 | 259 | | .207 | 185 | |
| .029 | 284 | | .057 | 258 | | .225 | 180 | |
| .030 | 282 | 325 | .058 | 258 | | .250 | 175 | |
| .031 | 280 | 320 | .059 | 257 | | .375 | 140 | |
| .032 | 277 | | | | | | | |

*After aging

**Chrome silicon/chrome vanadium**

| Wire Size, In. | Chrome Silicon | Chrome Vanadium |
|---|---|---|
| .020 | | 300 |
| .032 | 300 | 290 |
| .041 | 298 | 280 |
| .054 | 292 | 270 |
| .062 | 290 | 265 |
| .080 | 285 | 255 |
| .092 | 280 | |
| .105 | | 245 |
| .120 | 275 | |
| .135 | 270 | 235 |
| .162 | 265 | 225 |
| .177 | 260 | |
| .192 | 260 | 220 |
| .218 | 255 | |
| .250 | 250 | 210 |
| .312 | 245 | 203 |
| .375 | 240 | 200 |
| .437 | | 195 |
| .500 | | 190 |

**Copper-base alloys**

| Phosphor Bronze (Grade A) | |
|---|---|
| Wire Size Range—in. | |
| .007–.025 | 145 |
| .026–.062 | 135 |
| .063 and over | 130 |
| Beryllium Copper (Alloy 25 pretemp) | |
| .005–.040 | 180 |
| .041 and over | 170 |
| Spring Brass all sizes | 120 |

**Nickel-base alloys**

| Inconel (Spring Temper) | |
|---|---|
| Wire Size Range—in. | |
| up to .057 | 185 |
| .057–.114 | 175 |
| .114–.318 | 170 |
| Inconel X Spring Temper | After Aging |
| 190 | 220 |

**TABLE 7.3** Minimum Tensile Strength of Spring Wire Materials (*Continued*)

**Ferrous**

| Wire Size, in. | Music Wire | Hard Drawn | Oil Temp. | Wire Size, in. | Music Wire | Hard Drawn | Oil Temp. | Wire Size, in. | Music Wire | Hard Drawn | Oil Temp |
|---|---|---|---|---|---|---|---|---|---|---|---|
| .008 | 399 | 307 | 315 | .046 | 309 | 249 | | .094 | 274 | | |
| .009 | 393 | 305 | 313 | .047 | 309 | 248 | 259 | .095 | 274 | 219 | |
| .010 | 387 | 303 | 311 | .048 | 306 | 247 | | .099 | 274 | | |
| .011 | 382 | 301 | 309 | .049 | 306 | 246 | | .100 | 271 | | |
| .012 | 377 | 299 | 307 | .050 | 306 | 245 | | .101 | 271 | | |
| .013 | 373 | 297 | 305 | .051 | 303 | 244 | | .102 | 270 | | |
| .014 | 369 | 295 | 303 | .052 | 303 | 244 | | .105 | 270 | 216 | 225 |
| .015 | 365 | 293 | 301 | .053 | 303 | 243 | | .106 | 268 | | |
| .016 | 362 | 291 | 300 | .054 | 303 | 243 | 253 | .109 | 268 | | |
| .017 | 362 | 289 | 298 | .055 | 300 | 242 | | .110 | 267 | | |
| .018 | 356 | 287 | 297 | .056 | 300 | 241 | | .111 | 267 | | |
| .019 | 356 | 285 | 295 | .057 | 300 | 240 | | .112 | 266 | | |
| .020 | 350 | 283 | 293 | .058 | 300 | 240 | | .119 | 266 | | |
| .021 | 350 | 281 | | .059 | 296 | 239 | | .120 | 263 | 210 | 220 |
| .022 | 345 | 280 | | .060 | 296 | 238 | | .123 | 263 | | |
| .023 | 345 | 278 | 289 | .061 | 296 | 237 | | .124 | 261 | | |
| .024 | 341 | 277 | | .062 | 296 | 237 | 247 | .129 | 261 | | |
| .025 | 341 | 275 | 286 | .063 | 293 | 236 | | .130 | 258 | | |
| .026 | 337 | 274 | | .064 | 293 | 235 | | .135 | 258 | 206 | 215 |
| .027 | 337 | 272 | | .065 | 293 | 235 | | .139 | 258 | | |
| .028 | 333 | 271 | 283 | .066 | 290 | | | .140 | 256 | | |
| .029 | 333 | 267 | | .067 | 290 | 234 | | .144 | 256 | | |
| .030 | 330 | 266 | | .069 | 290 | 233 | | .145 | 254 | | |
| .031 | 330 | 266 | 280 | .070 | 289 | | | .148 | 254 | 203 | 210 |
| .032 | 327 | 265 | | .071 | 288 | | | .149 | 253 | | |
| .033 | 327 | 264 | | .072 | 287 | 232 | 241 | .150 | 253 | | |
| .034 | 324 | 262 | | .074 | 287 | 231 | | .151 | 251 | | |
| .035 | 324 | 261 | 274 | .075 | 287 | | | .160 | 251 | | |
| .036 | 321 | 260 | | .076 | 284 | 230 | | .161 | 249 | | |
| .037 | 321 | 258 | | .078 | 284 | 229 | | .162 | 249 | 200 | 205 |
| .038 | 318 | 257 | | .079 | 284 | | | .177 | 245 | 195 | 200 |
| .039 | 318 | 256 | | .080 | 282 | 227 | 235 | .192 | 241 | 192 | 195 |
| .040 | 315 | 255 | | .083 | 282 | | | .207 | 238 | 190 | 190 |
| .041 | 315 | 255 | 266 | .084 | 279 | | | .225 | 235 | 186 | 188 |
| .042 | 313 | 254 | | .085 | 279 | 225 | | .250 | 230 | 182 | 185 |
| .043 | 313 | 252 | | .089 | 279 | | | .3125 | | 174 | 183 |
| .044 | 313 | 251 | | .090 | 276 | 222 | | .375 | | 167 | 180 |
| .045 | 309 | 250 | | .091 | 276 | | 230 | .4375 | | 165 | 175 |
| | | | | .092 | 276 | 220 | | .500 | | 156 | 170 |
| | | | | .093 | 276 | | | | | | |

## 7.13 SPRING CALCULATIONS AND SAMPLE DESIGNS

### 7.13.1 Calculation Examples

A compression spring is needed to power a toggle mechanism. The spring must work inside of a 2-in-diameter bore. Free length cannot exceed 8 in, and a 4-in compression limit is desired. The spring is to have maximum energy for the limited space, while the stress level is not to exceed 50 percent of the minimum yield strength of the wire. The spring operates periodically with long intervals of rest.

*Procedure*

1. Look at the space available when the spring is compressed. The space = 4 in if a free length of 8 in is assumed and the spring is compressed 4 in.

**TABLE 7.4** Poisson's Ratios for Spring Materials

| Material | Poisson's Ratio, $\mu$ |
|---|---|
| Music wire ASTM A228 | 0.30 |
| Hard drawn ASTM A227 | 0.30 |
| Oil tempered ASTM A229 | 0.30 |
| AISI 1065 carbon steel | 0.30 |
| AISI 1075 carbon steel | 0.30 |
| AISI 1095 carbon steel | 0.30 |
| AISI 6150 vanadium steel | 0.30 |
| AISI 5160 chromium steel | 0.30 |
| Inconel 600 | 0.28 |
| Inconel 718 | 0.28 |
| Inconel X750 | 0.29 |
| AISI 301/302 stainless steel | 0.31 |
| 17-7 PH stainless steel | 0.34 |
| Carpenter 455 stainless steel | 0.30 |
| Phosphor-bronze ASTM B103 and B159 | 0.20 |
| Beryllium-copper ASTM B194 and B197 | 0.33 |

2. Look for maximum wire diameter in music wire and see how many coils the spring can contain. So, 4 in/0.250 = 16 coils. To keep from jamming the spring, select 15 coils and if the ends are to be closed and ground, this will become 13 working coils.
3. The bore is 2 in diameter, so the spring o.d. should be about 0.06 in less in diameter to keep the spring from jamming in the bore when it is compressed. That will give an o.d. of 1.94 in.
4. If we try 0.250-in-diameter wire, the trial mean diameter will be equal to 1.94 – 0.250 = 1.69 in.
5. This will produce a spring index of 1.69/0.25 = 6.76, which is close to the ideal index.
6. Calculate a trial rate from

$$R = \frac{Gd^4}{8ND^3}$$

where $G = 11.5 \times 10^6$, $d = 0.250$ in, $N = 13$, and $D = 1.69$ in.

$$R = \frac{11.5 \times 10^6(0.25)^4}{8(13)(1.69)^3} = 89 \text{ lb/in}$$

7. Check the stress when the spring is deflected 4 in.

$$S = \frac{8K_a DP}{\pi d^3} = \frac{8(1.221)(356)1.69}{3.1416(0.25)^3} = 119{,}695 \text{ psi}$$

8. The stress allowable for 0.25 diameter music wire (see Sec. 7.12.2, Table 7.3) is

$$0.50 \times 230{,}000 = 115{,}000 \text{ psi}$$

This is a little lower than the corrected stress, but since the spring is operated periodically, it should suffice for this application. Note that the Wahl stress correction factor $K_a$ of 1.221 in the above example was derived from the Wahl equation for compression springs with round wire.

$$K_a = \frac{4C-1}{4C-4} + \frac{0.615}{C}$$

Also, the load $P$ was derived from 4 in × 89 lb/in = 356 lb.

**9.** Since the stress on this spring is slightly overextended, the spring manufacturer may need to set it during manufacture or have it shot-peened. You may reduce the stress by compressing the spring 3¾ in instead of 4 in.

**10.** From the rate, we can calculate the available potential energy $P_e$ that can be stored in the spring.

$$P_e = \frac{Rs^2}{2} = \frac{89\,(4)^2}{2} = 712 \text{ in} \cdot \text{lb or } 59.3 \text{ ft} \cdot \text{lb}$$

**11.** If the energy of the application is more than the spring can store, you may place another spring inside of the existing spring. The two will then be in parallel, with their rates added to produce the total rate. If another spring is added inside the existing spring, it must be wound opposite-hand to the existing spring, so that the coils do not tangle or jamb. The added spring is calculated in a similar manner to produce the maximum rate.

The preceding design description was taken from an actual production spring used in approximately 50,000 mechanisms. It is to be noted that no failures occurred in these springs, which have been in service for 10 years at temperatures ranging from –20 to 150°F. Maximum use for this spring was 3000 to 4000 loading and unloading cycles. If hard-drawn ASTM A227 or oil-tempered ASTM A229 steel wire were to be used in this application, instead of music wire, the springs would fail by taking a permanent set of ¼ to ½ in. This illustrates the importance of music wire, chrome vanadium, or chrome silicon in high-energy/stress applications.

If the spring were continuously cycled, a stress level of 40 percent of minimum yield strength or less would be required, and the spring would require a redesign using more coils and smaller wire diameter, which in turn would reduce the rate $R$, load $P$, and stored energy $P_e$.

Designing springs is thus a tradeoff of one variable with another. Since there is a definite limit of energy versus mass that a spring can contain, designing a maximum energy spring that will not fail in service is a rigorous design problem.

It should be noted by designers who are not thoroughly familiar with spring design that much trial and error is involved in the actual design of springs, due to the high number of variables that are encountered. This fact leads us to the next section, which deals with spring design using a programmable calculator.

### 7.13.2 Spring Design by Programmable Calculator

Due to the high number of calculations that are frequently encountered in initial spring design, it is advisable for the designer who deals with springs on a regular basis to use a programmable calculator. There are a number of excellent handheld calculators available, such as Texas Instruments TI-66, various Hewlett-Packard models, Casio Fx-7000G, and Sharp and Tandy machines.

You will need to program the calculator in such a way that you will be able to assign the variables of the spring equations to the different memory positions in the calculator, so that different numerical values assigned to the memories can be changed easily as you proceed from each set of trial values.

The procedure is quite simple when using a calculator such as the TI-66. First, the basic spring equation is programmed so that unknown variables, such as *d, D, R, S, E, G, t, a,* and *b,* are assigned to different memory positions and are recalled back into the basic equation programmed for solution. In this manner, you may change the wire diameter, number of coils, mean diameter, and so forth, while keeping the program running until a satisfactory result is obtained. This procedure saves a vast amount of time as compared to running the program using only a basic calculator, where you must run the procedure through all the key strokes each time you change the variables.

The TI-66 also runs an optional printer so that you can keep a record of the calculations. On calculators such as these you can program many different equations and run each equation by striking one key, such as A, B, C, and A′. The TI-66 will allow you to program 10 equations for instant rerun. In this manner, you may assign program A to a solution for compression spring rate, program B to a solution for stress of the compression spring, program C to a solution for torsion spring rate, and so on.

The other programmable calculators function in a similar manner. With the Casio Fx-7000G, you can run programs in the fashion described or graph an equation and find the roots of cubic, quartic, and logarithmic equations. These calculators make the designer's work not only easier, but much more accurate. You will note that there are no tables of logarithms or trigonometric functions in this handbook. The reason is that these once tedious methods are now automatically carried out on modern handheld scientific calculators.

## 7.14 SPRING RATE CURVES

If in your design work you deal with many springs that are similar, you can easily construct a set of spring rate curves for standard models of springs. The procedure is to select a particular mean diameter and set number of coils, then plot a curve on graph paper with wire-size graduations running horizontally and spring-rate graduations running vertically. Using this type of curve or graph, you can get a quick idea of the rate versus the wire size for a particular mean diameter and number of coils.

This procedure is easier to implement and use than a composite spring chart, which can be confusing and difficult to read. To develop each curve, you will need to solve between 10 and 15 rate equations to describe the curve accurately. When you have a number of these curves covering mean diameters from ½ in to 2 in, you will have a good reference point to start from when an initial spring design problem is begun.

Sample curves are shown in Figs. 7.20 and 7.21.

## 7.15 SPRING DRAWINGS OR FORMS

Refer to Fig. 7.22*a* to 7.22*c*.

### 7.15.1 Simplified Spring Drawings

When a spring drawing is submitted to a spring manufacturer, it should contain all pertinent information required to manufacture it, including a drawing of the actual spring.

The following data should be indicated for each spring type as shown. Compression springs:

1. Wire diameter
2. Wire material and specification (e.g., music wire, ASTM A228).

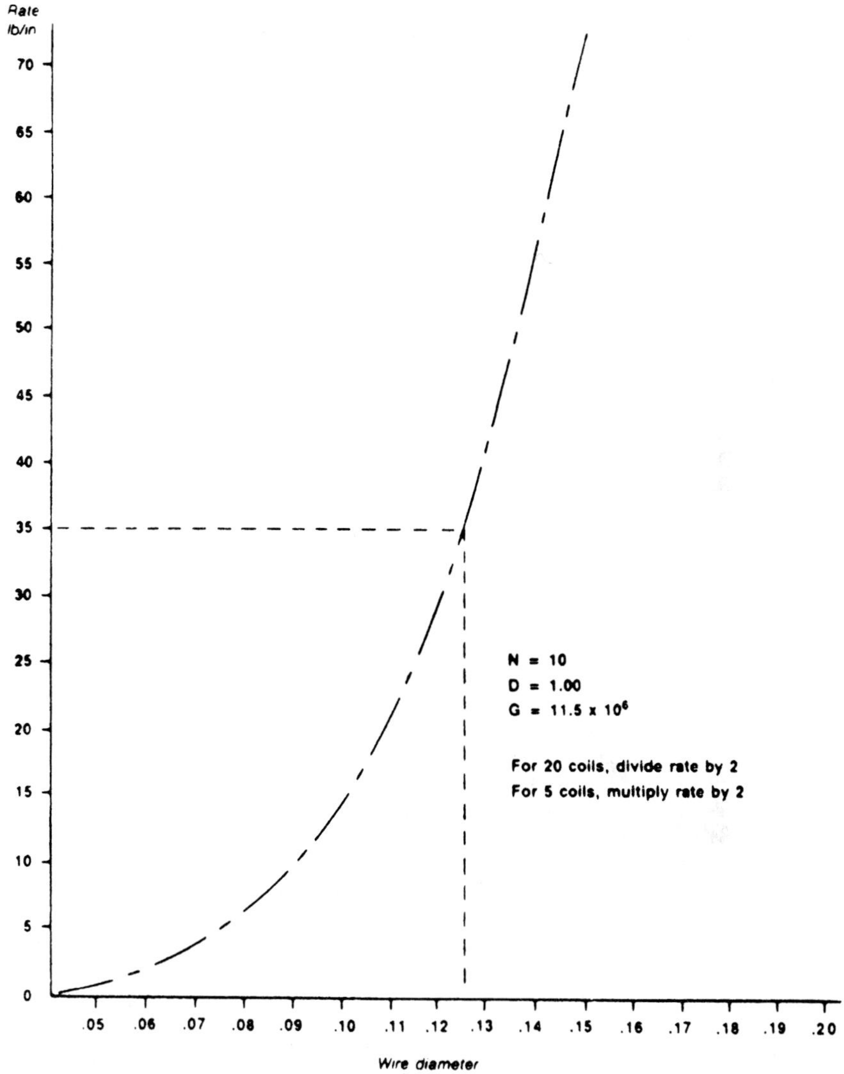


**FIGURE 7.20** Typical spring rate curve.

3. Inside or outside diameter
4. Free length
5. Spring rate or load value at a particular deflection
6. Direction of coil wind, right-hand, left-hand, or optional
7. Type of ends
8. Finish, if any
9. Number of active coils and total coils
10. Mean coil diameter

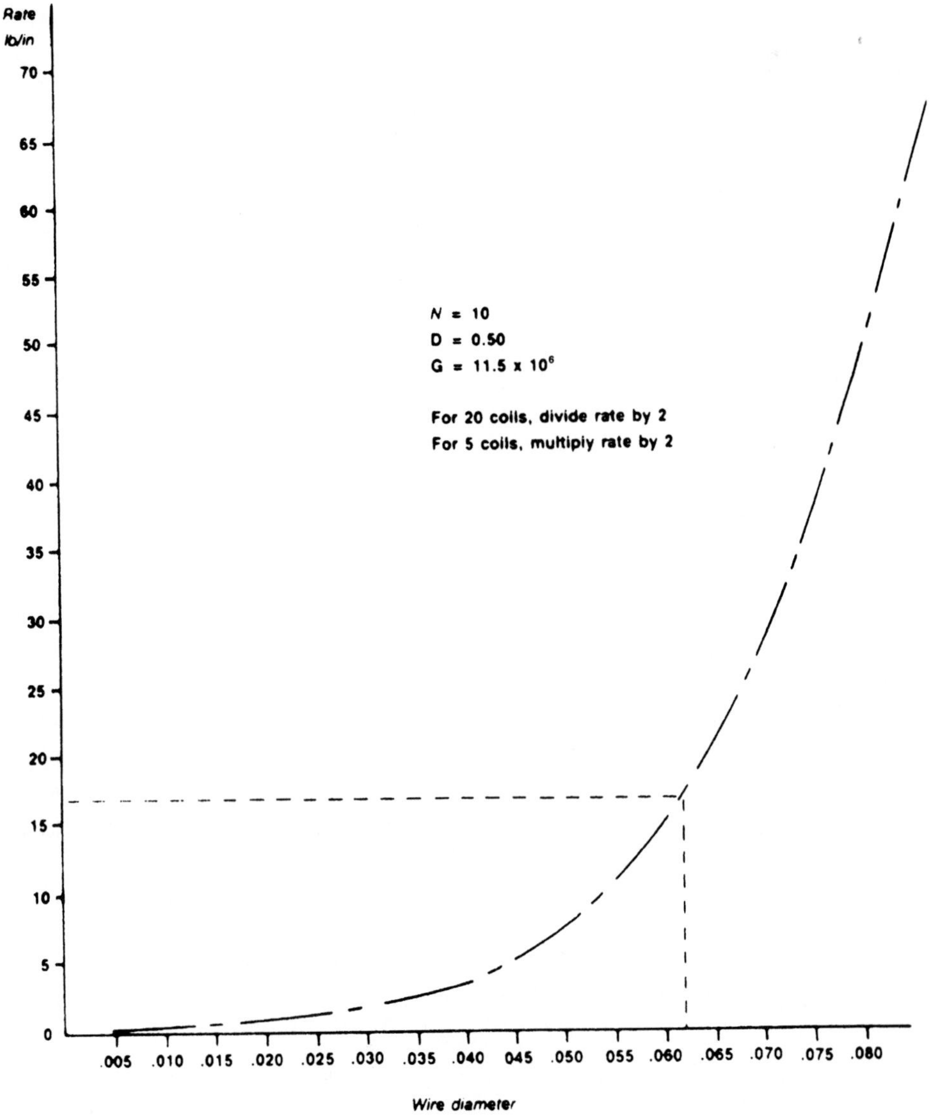


**FIGURE 7.21** Typical spring rate curve.

Extension springs:

1. Wire diameter
2. Length inside of the ends
3. Inside or outside diameter
4. Rate or loads at different deflections
5. Maximum extended length without permanent set
6. Loop positions, angular

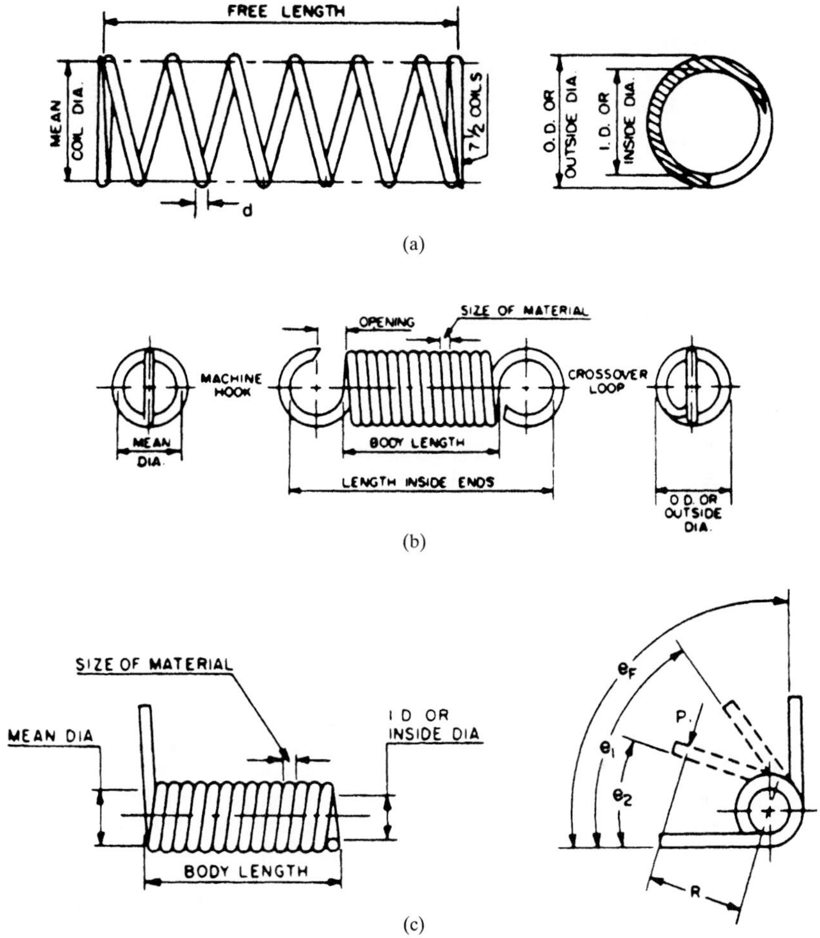


**FIGURE 7.22** (*a*) Compression spring drawing form; (*b*) extension spring drawing form; (*c*) torsion spring drawing form.

7. Direction of coil wind, right-hand, left-hand, or optional
8. Type of ends
9. Mean coil diameter
10. Number of coils
11. Body length
12. Initial tension, preload
13. Type of material and specification number
14. Finish

Torsion springs:

1. To work over ____ in diameter shaft
2. Inside diameter, minimum

3. Torque at a specified angular deflection and maximum deflection
4. Body length
5. Length of moment arm
6. Wire diameter
7. Mean coil diameter
8. Number of coils
9. Direction of coil wind, right-hand, left-hand, or optional
10. Type of material and specification number
11. Finish
12. Type of ends

### 7.15.2 Spring Tolerances

Charts showing tolerances on all spring wire and strip available commercially are shown in the various handbooks published by the Spring Manufacturers Institute (see Chap. 15, Sec. 15.1). Tolerances on coil diameters, loads, free lengths, squareness, and solid height are also published by the institute. Tolerances on spring materials are shown in Chap. 5 of the *McGraw-Hill Machining and Metalworking Handbook* (1994).

Any specified value, including rate and load, for a particular spring can also be discussed with the spring manufacturer. Certain dimensions such as a critical outside diameter, free length, and load at a specified deflection may be held to close limits by the spring manufacturer, but such requirements will incur higher cost for the particular spring involved. Unless you have a critical application, the tolerance variations allowed on wire and strip will not affect your spring to a significant extent. Also, tolerances on many other spring features seldom have a significant effect on a common usage spring.

If you have a special condition or requirement, however, the allowed tolerances can affect your design and should be referred to the spring manufacturer. Tolerances that must be maintained must be shown on the formal spring drawing that is presented to the spring manufacturer. No one is in a better position than the spring manufacturer to know exactly what tolerances and tolerance limits can be maintained when manufacturing a spring. Assigning arbitrary tolerances to a spring usually incurs additional costs that are often unwarranted for the application.

### 7.15.3 Spring Material Analysis

Designers and manufacturers should be aware that *all* spring materials can be precisely analyzed for proper identification of the material used in the spring. There is a possibility of materials being accidentally substituted during manufacture. If this occurs with a highly stressed spring, the spring may fail in service. This can happen if hard-drawn or oil-tempered wire is accidentally used in place of music wire or other high-strength material. If you design a spring with a high stress level that has been functional but suddenly fails in service, have the spring analyzed at a materials and engineering laboratory before you attempt to redesign it.

Materials such as the stainless steels, beryllium copper, phosphor bronze, and chrome vanadium can be quickly identified by chemical analysis. Materials such as music wire (ASTM A228), hard-drawn steel (ASTM A227), and oil-tempered spring steel (ASTM A229) *cannot* be identified chemically because their chemistries overlap because of compositional tolerances of their constituent elements. But these materials *can* be differentiated by microscopic analysis. Oil-tempered spring steel will show a definite martensitic structure under 100 to 400× magnification, while music wire will show a definite cold-work structure due to cold drawing during its manufacture. In this manner, it may be ascertained which type of steel was used to make the spring being analyzed. The microstructures for these listed steels differ and can be identified at the test laboratory using the ASTM book of microstructures of steels. Figure 7.23 shows a photomicrograph of a section of a spring made of oil-tempered spring steel and shows in detail the martensitic structure of this type of steel.

The laboratory report shown in Fig. 7.24 was made as a consequence to the photomicrograph shown in Fig. 7.23. The analyzed spring sample was supposed to have been manufactured from ASTM A228 music wire, but was in fact manufactured accidentally from oil-tempered ASTM A229 steel wire. The consequence of this mistake in material selection during spring manufacture caused the rejection of 5000 large compression springs, which were returned to the spring manufacturer as they were not capable of withstanding the working stresses without taking a permanent *set.* The springs, in effect, were unsuitable for the intended application and did not satisfy the spring specifications shown on the spring drawing.

## 7.16 THE HEAT TREATMENT AND POSTBAKING OF ELECTROPLATED SPRINGS

### 7.16.1 Heat Treatment of Springs

When a helical compression, extension, or torsion spring is formed on the spring-coiling machine during manufacture, residual stresses induced during the coiling operation must be relieved soon after the spring is completed. The normal heat-treatment procedures are as follows:

Music wire (ASTM A228), oil-tempered (ASTM A229), and hard-drawn (ASTM A227 and A679): oven-heated at 500°F for 30 to 40 min

Stainless steels—17-7 PH, types 301, 302, 304 (ASTM A313): oven-heated at 600 to 650°F for 30 to 40 min

Inconel 600 and X750: oven-heated at 700 to 1200°F for as long as 4 h (Inconel X750)

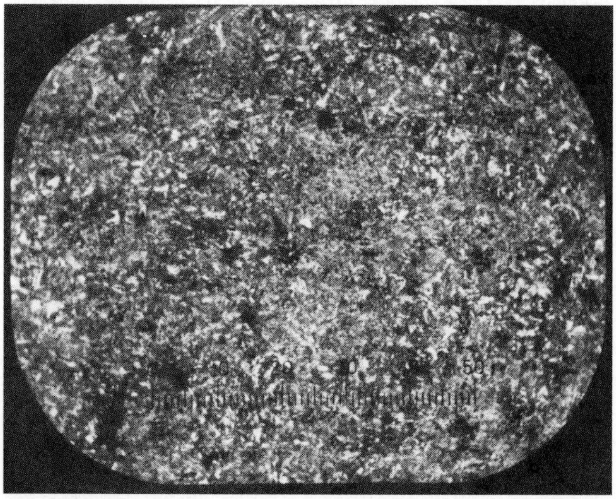

**FIGURE 7.23** Photomicrograph of oil-tempered martensitic structure at 200×.

DR. WM. B. D. PENNIMAN
1866-1938
DR. ARTHUR LEE BROWNE
1867-1933

EXECUTIVE STAFF
PHILIP M. AIDT
ALLEN W. THOMPSON
DANTE G. BERETTA
J. ADRIAN BUTT
DONALD W. SMITH

PENNIMAN & BROWNE, INC.
CHEMISTS-ENGINEERS-INSPECTORS
6252 FALLS ROAD
BALTIMORE, MARYLAND 21209

ESTABLISHED
1896

CABLE ADDRESS
"BALTEST"

TELEPHONE
825-4131
AREA CODE 301

ENGINEERING DIVISION

**REPORT OF TEST**

Attn: R. A. Walsh May 27, 1981

*No.* 811265

*Sample of* Compression Spring

*Client*

*Marks or Other Data* Verify material to be Music Wire (ASTM A 228)
Service failures suggest possibility that material might be oil tempered (ASTM A 229)

The above indicated ASTM specifications were checked and found to have overlapping chemistries preventing the use of chemical analysis for identification.

The ASTM Handbook of Microstructures showed the difference between the hard drawn structure (ASTM A 228) and the oil tempered microstructure.

Samples were cut from the spring, mounted for examination in transverse and longitudinal directions, ground, polished, etched and examined at 100X to 400X magnification. The attached micrograph at 200X magnification clearly shows a martensitic, oil-tempered structure indicating an A229 material.

No hardness requirements are listed for these materials.

PENNIMAN & BROWNE, INC.

J. A. Butt

FORM 30 L/B

**FIGURE 7.24** Laboratory test report of spring material shown in Fig. 7.23.

### 7.16.2 Electroplating Springs

Springs may be electroplated with a number of different metals such as cadmium, chromium, zinc, copper, tin, and nickel. The most common and widely used metal for plating springs is bright zinc.

The zinc plating Federal Specification is QQ-P-416 and includes

| | |
|---|---|
| Type 1 | bright zinc |
| Type 2 | bright zinc with chromate |

| | |
|---|---|
| Class 1 plating thickness | 0.2 mils (0.0002 in) |
| Class 2 plating thickness | 0.3 mils (0.0003 in) |
| Class 3 plating thickness | 0.5 mils (0.0005 in) |

### 7.16.3 Postbaking Electroplated Springs

Immediately after the electroplating process is completed, the plated spring must be postbaked at a specified temperature for a specific time interval to prevent *hydrogen embrittlement,* which will invariably occur if the postbake operation is not performed. The postbake operation usually consists of oven heating the spring at 500°F for 3 h.

## *7.17 DYNAMICS OF HELICAL COMPRESSION AND EXTENSION SPRINGS*

When a helical compression spring is rapidly loaded or unloaded, a surge wave is generated within the spring. This surge wave limits the rate at which the spring can release or absorb energy by limiting the impact velocity.

The impact velocity is defined by

$$V \cong 10.1S\sqrt{\frac{g}{2\rho\, G}} \quad \text{(approximate m/sec)}$$

$$\cong S\sqrt{\frac{g}{2\rho\, G}} \quad \text{(approximate in/sec)}$$

where $S$ = maximum stress in the spring, psi, MPa
$g$ = gravity constant, 9.8 m/sec
$\rho$ = density of spring material, g/cm$^3$, lb/in$^3$
$G$ = shear modulus of elasticity, MPa, psi
$G$ = 11,500,000 psi for steel

When a compressed helical compression spring is released instantaneously and the stress is known, the maximum spring velocity is a function of the maximum stress $S$ and the spring material. When the impact velocity is known, the maximum stress induced in the spring may be calculated approximately. High-spring loading velocities limit spring performance and often cause *resonance;* for instance, valve springs in high-performance internal-combustion engines may "float" or bounce due to this effect at high engine speeds.

***Dynamic Loading—Resonance (Compression and Extension Springs).*** A spring will exhibit resonance when the cyclic loading/unloading rate is near the springs' natural frequency or multiple of the natural frequency. Resonance can cause spring bounce or floating, resulting in lower loads than those calculated. The natural frequency should be a minimum of 13 times the operating frequency. If the operating frequency is 100 cycles per second, the design natural frequency should be 13 × 100 = 1300 cycles (Hz).

For helical compression springs with both ends fixed, the natural frequency is given by

$$n = \frac{1120d}{ND^2}\sqrt{\frac{Gg}{\rho}} \quad \text{(for SI units)}$$

$$= \frac{0.111d}{ND^2}\sqrt{\frac{Gg}{\rho}} \quad \text{(for U.S. customary units)}$$

***Helical Extension Springs (Resonance).*** With one end fixed, the natural frequency (Hz) of a helical extension spring is given by

$$n \; \frac{560d}{ND^2}\sqrt{\frac{Gg}{\rho}} \quad \text{(for SI units)}$$

$$= \frac{0.056d}{ND^2}\sqrt{\frac{Gg}{\rho}} \quad \text{(for U.S. customary units)}$$

In the preceding equations

$n$ = natural frequency, Hz (cycles per second)
$d$ = wire diameter, mm, in
$D$ = mean diameter, mm, in
$G$ = shear modulus, psi, MPa (11,500,000 psi for steels)
$g$ = acceleration of gravity, 386 in/sec, 9.81 m/sec
$\rho$ = density of spring material, lb/in$^3$, g/cm$^3$
$N$ = number of active coils in the spring (all coils active in an extension spring; $N$ in compression springs determined by type of ends)

(*Note:* To prevent resonance, energy-damping devices are sometimes used on springs when the natural frequency cannot be made more than 13 times the operating frequency.)

***Bending and Torsional Stresses in Ends of Extension Springs.*** Bending and torsional stresses are developed at the bends in the ends of extension springs when the spring is stretched under load. These stresses should be checked by the spring designer after the spring has been designed and dimensioned. Alterations to the ends and radii may be required to bring the stresses into their allowable range (see Fig. 7.25).

The bending stress may be calculated from

$$\text{Bending stress at point } A = S_b = \frac{16PD}{\pi d^3}\left(\frac{r_1}{r_2}\right)$$

The torsional stress may be calculated from

$$\text{Torsional stress at point } B = S_t = \frac{8PD}{\pi d^3}\left(\frac{r_3}{r_4}\right)$$

Check the allowable stresses for each particular wire size of the spring being calculated from the wire tables in this section. The calculated bending and torsional stresses cannot exceed the allowable stresses for each particular wire size. As a safety precaution, take 75 percent of the allowable stress shown in the tables as the minimum allowable when using the preceding equations.

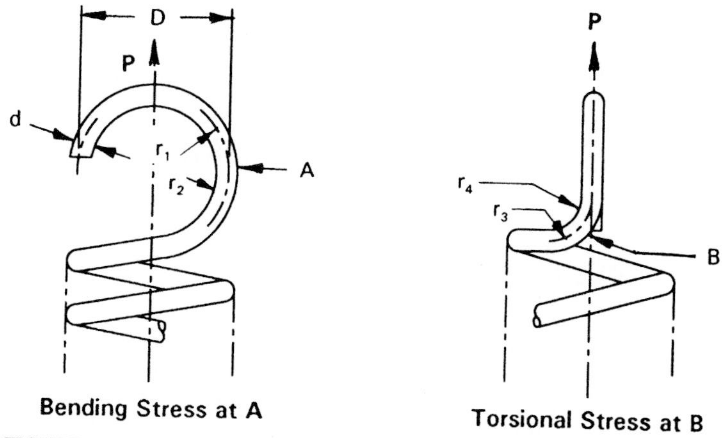


**FIGURE 7.25** Bending and torsional stresses at ends on extension springs.

***Buckling of Unsupported Helical Compression Springs.*** Unsupported or unguided helical compression springs become unstable in relation to their slenderness ratio and deflection percentage of their free length. Figure 7.26 may be used to determine the unstable condition of any particular helical compression spring under a particular deflection load or percent of free length.

## 7.18 SPRING RATES FOR SPRING COMBINATIONS

The combined spring rates for springs in series and parallel are calculated the same as capacitances in series and parallel; that is, the final rate in parallel

$$R_t = R_1 + R_2 + R_3 + \cdots + R_n$$

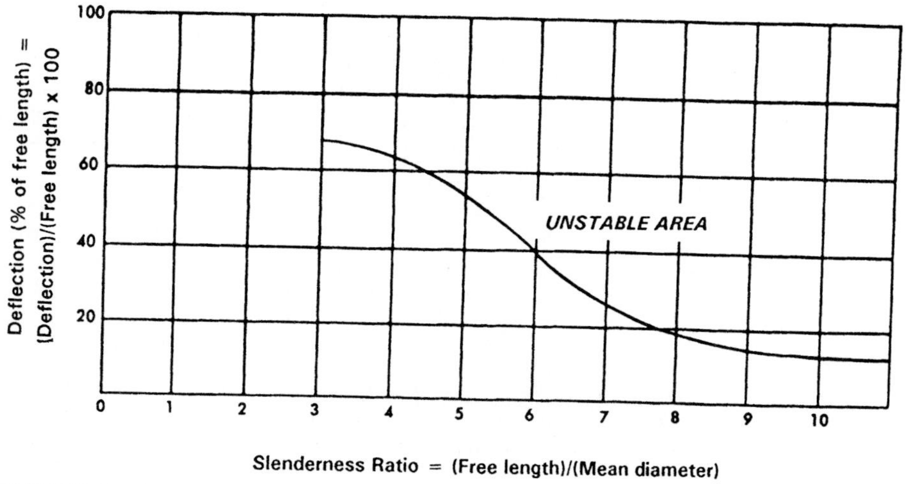


**FIGURE 7.26** Buckling of helical compression springs.

To find the final rate in series

$$\frac{1}{R_t} = \frac{1}{R_1} + \frac{1}{R_2} + \cdots + \frac{1}{R_n}$$

where $R_1$, $R_2$, $R_n$ = rates of the individual springs and $R_t$ = final combined rate.

## 7.19 FINAL NOTES ON SPRING DESIGN

***Spring Drawings.*** Engineering drawings for springs should contain all the pertinent engineering information needed by the spring manufacturer in order to produce the spring. Figure 7.27 is a typical AutoCad drawing of a helical compression spring. The information contained on the drawing follows the guidelines shown in Sec. 7.15.1 concerning spring drawings.

***Manufacturing Defects on Springs.*** Springs that are manufactured containing abnormal processing marks may fail in service. This is due to the abnormally high stress concentration that generates from the point of defect. Figure 7.28 shows a torsion spring with the defect mark located at the arrow. This mark caused this particular spring to fail in service after 1000 cycles of operation at its design load. The mark was caused by the bending mandrel used by the spring manufacturer to make the bend in the leg of the spring. This was not a manufacturing error, but an engineering error caused by specifying a bend radius that was too small for the wire diameter.

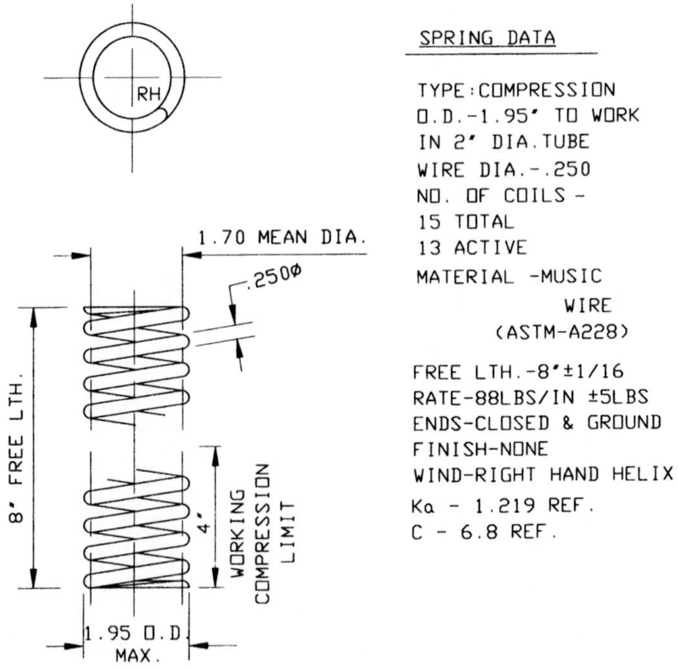


**FIGURE 7.27** Typical AutoCad spring drawing.

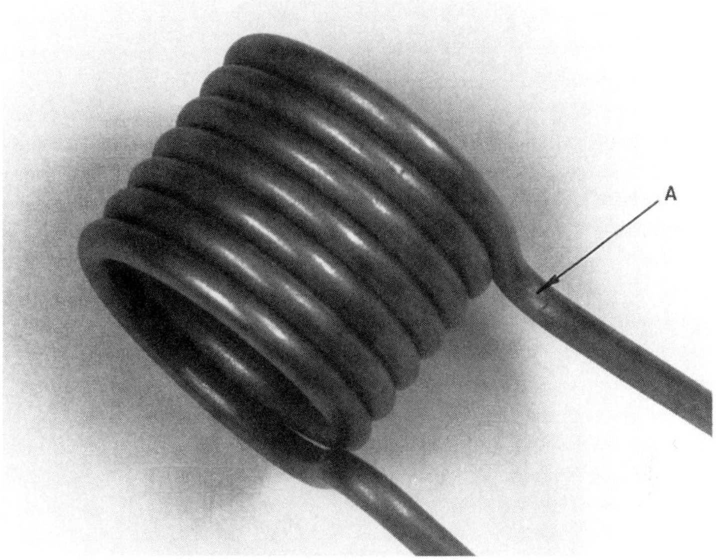


**FIGURE 7.28** Photograph of manufacturing defect mark on torsion spring.

To correct this problem, the engineering drawing was changed to increase the bend radius as shown in Fig. 7.29. Figure 7.27 (1) is the spring with the small radius; (2) shows the radius as increased in order to eliminate the stress marks caused by the mandrel being too small for the wire diameter. After the radius was enlarged, the spring performed satisfactorily, with no sign of fatigue after thousands of cycles of operation.

Nicks, cuts, indents, and other abrasions and defects on the surface of a spring will eventually cause problems if the spring wire is highly stressed in service. This is especially true if bend radii are not selected with care. Check with the spring manufacturer if in doubt about any particular radius on the spring wire at a bend junction on the spring. This problem is most prevalent on extension and torsion springs.

Improper heat treatment can also cause a spring to fail in service. Another cause of spring failure is improper postbaking of the spring immediately after the spring has been electroplated. Improper or *no* postbaking after electroplating will lead to hydrogen embrittlement with subsequent spring failure. Hydrogen embrittlement will occur on springs even though they are not under load. Springs have been known to break and fall apart in their shipping boxes as a result of hydrogen embrittlement caused by improper processing.

***Spring Design Programs for the Personal Computer.*** Spring design programs are available for use on the personal computer (PC), either as specialized programs or general mathematics programs such as MathCAD. Figure 7.30 is a laser printout of a spring problem as solved by MathCAD 2.5, which the author uses for extended spring problems involving many calculations. Using such a program, the spring calculations may be printed as they appear on the computer screen and kept for record purposes. A record of spring calculations is almost mandatory today because of the liability problems associated with equipment failures. Extensive testing of a critical spring component on dangerous equipment is absolutely necessary and must be performed to prove the adequacy of the design. Failure of a spring such as shown in Fig. 7.28 could cause a dangerous high-power electrical fault to occur on the device in which this spring is used.

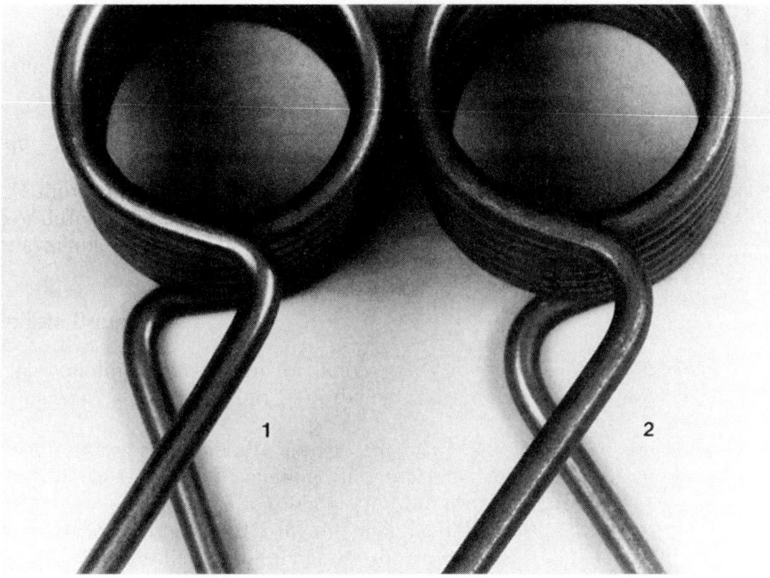


**FIGURE 7.29** Small radius at (1); increased radius at (2).

## 7.20 SPRING TERMINOLOGY FOR DESIGN AND MANUFACTURING

The following list of spring terms and meanings will be useful to spring designers and those support personnel who handle and order springs for the engineering department. It is important to know these specialized meanings for spring terms when communicating with the spring manufacturers. This terminology is standard throughout the industry, and the appropriate term should be applied so that there is no misunderstanding of terms between the engineering departments and spring manufacturers.

Be specific and accurate about spring materials, terms, dimensions, loads, rates, and any other parameters shown in this section of the handbook concerning the different types of springs. Springs are such critical mechanical components in many mechanisms that their design and manufacture must be carefully controlled and executed.

**Active coils** $n_a$ Those coils which are free to deflect under load.

**Angular relationship of ends** The relative position of the plane of the hooks or loops of extension springs to each other.

**Baking** Heating of electroplated springs to relieve hydrogen embrittlement.

**Buckling** Bowing or lateral deflection of compression springs when compressed, related to the slenderness ratio *L*/*D*.

**Closed ends** Ends of compression springs where pitch of the end coils is reduced so that the end coils touch.

**Closed and ground ends** As with closed ends, except that the end is ground to provide a flat plane.

**Closed length** See *Solid height.*

**Close-wound** Coiled with adjacent coils touching.

MATHCAD 2.5

Analysis of main power spring of interrupter switch

G := 11500000 d := 0.250 D := 1.700 N := 13 $\frac{D}{d} = 6.8$ INDEX

C := 6.8 P := 250,260 ..400 K := 1.220

$$\frac{4\cdot C - 1}{4\cdot C - 4} + \frac{0.615}{C} = 1.22$$

RATE = R

$$\frac{G\cdot d^{4}}{8\cdot N\cdot D^{3}} = 87.918$$

| $\frac{8\cdot K\cdot D\cdot P}{3.1416\cdot d^{3}}$ = STRESS | LOAD lb-f |
|---|---|
| $8.45\cdot 10^{4}$ | 250 |
| $8.788\cdot 10^{4}$ | 260 |
| $9.126\cdot 10^{4}$ | 270 |
| $9.464\cdot 10^{4}$ | 280 |
| $9.802\cdot 10^{4}$ | 290 |
| $1.014\cdot 10^{5}$ | 300 |
| $1.048\cdot 10^{5}$ | 310 |
| $1.082\cdot 10^{5}$ | 320 |
| $1.115\cdot 10^{5}$ | 330 |
| $1.149\cdot 10^{5}$ | 340 |
| $1.183\cdot 10^{5}$ | 350 ** |
| $1.217\cdot 10^{5}$ | 360 |
| $1.251\cdot 10^{5}$ | 370 |
| $1.284\cdot 10^{5}$ | 380 |
| $1.318\cdot 10^{5}$ | 390 |
| $1.352\cdot 10^{5}$ | 400 |

Minimum tensile strength of 0.250" dia. music wire, ASTM A-228 = 230,000 psi.

Allowable stress = .50 x 230,000 = 115,000 psi (.50 used due to intermittent spring duty cycle)

Spring is compressed 4", so: 4 x 88 = 352 lb-f

At 350 pounds load, the stress on this spring is: 118,300 psi.

The spring is stressed slightly above the allowable of 50% of tensile strength.

This spring has performed satisfactorily in over 50,000 mechanisms with only a very slight tendency to take "permanent set." When the spring is properly heat treated by the spring manufacturer, no problems occurred. The operating temperature range of this spring is from -40 degrees F to 150 degrees F.

R. A. Walsh, Mgr.
R & D Department
POWERCON Corporation
Severn, Maryland 21061
March 25, 1991

**FIGURE 7.30** Laser printout of MathCAD solution for compression spring design.

**Coils per inch** See *Pitch.*

**Deflection *F*** Motion of spring ends or arms under the application or removal of an external load *P.*

**Elastic limit** Maximum stress to which a material may be subjected without permanent set.

**Endurance limit** Maximum stress at which any given material will operate indefinitely without failure for a given minimum stress.

**Free angle** Angle between the arms of a torsion spring when the spring is not loaded.

**Free length *L*** The overall length of a spring in the unloaded position.

**Frequency (natural)** The lowest inherent rate of free vibration of a spring itself (usually in cycles per second) with ends restrained.

**Gradient** See *Rate R.*

**Heat setting** Fixturing a spring at elevated temperature to minimize loss of load at operating temperature.

**Helix** The spiral form (open or closed) of compression, extension, and torsion springs.

**Hooke's law** Load is proportional to displacement.

**Hooks** Open loops or ends of extension springs.

**Hot pressing** See *Heat setting.*

**Hydrogen embrittlement** Hydrogen absorbed in electroplating or pickling of carbon steels, tending to make the spring material brittle and susceptible to cracking and failure, particularly under sustained loads.

**Hysteresis** The mechanical energy loss that always occurs under cyclic loading and unloading of a spring, proportional to the area between the loading and unloading load-deflection curves within the elastic range of a spring.

**Initial tension $P_i$** The force that tends to keep the coils of an extension spring closed and which must be overcome before the coils start to open.

**Load *P*** The force applied to a spring that causes a deflection *F.*

**Loops** Coil-like wire shapes at the ends of extension springs that provide for attachment and force application.

**Mean coil diameter *D*** Outside spring diameter (o.d.) minus one wire diameter *d.*

**Modulus in shear or torsion *G*** Coefficient of stiffness for extension and compression springs.

**Modulus in tension or bending *E*** Coefficient of stiffness used for torsion and flat springs (Young's modulus).

**Moment *M*** See *Torque.*

**Open ends, not ground** End of a compression spring with a constant pitch for each coil.

**Open ends ground** Open ends, not ground followed by an end-grinding operation.

**Passivating** Acid treatment of stainless steel to remove contaminants and improve corrosion resistance.

**Permanent set** A material that is deflected so far that its elastic properties have been exceeded and it does not return to its original condition on release of load is said to have taken a "permanent set."

**Pitch *p*** The distance from center to center of the wire in adjacent active coils (recommended practice is to specify number of active coils rather than pitch).

**Poisson's ratio** The ratio of the strain in the transverse direction to the strain in the longitudinal direction.

**Preset** See *Remove set.*

**Rate *R*** Change in load per unit deflection, generally given in pounds per inch (N/mm).

**Remove set** The process of closing to solid height a compression spring which has been coiled longer than the desired finished length, so as to increase the apparent elastic limit.

**Residual stress** Stresses induced by set removal, shot peening, cold working, forming, or other means. These stresses may or may not be beneficial, depending on the application.

**Set** Permanent distortion which occurs when a spring is stressed beyond the elastic limit of the material.

**Shot peening** A cold-working process in which the material surface is peened to induce compressive stresses and thereby improve fatigue life.

**Slenderness ratio** Ratio of spring length $L$ to mean coil diameter $D$.

**Solid height $H$** Length of a compression spring when under sufficient load to bring all coils into contact with adjacent coils.

**Spring index** Ratio of mean coil diameter $D$ to wire diameter $d$.

**Squared and ground ends** See *Closed and ground ends.*

**Squared ends** See *Closed ends.*

**Squareness of ends** Angular deviation between the axis of a compression spring and a normal to the plane of the ends.

**Squareness under load** As in *Squareness of ends,* except with the spring under load.

**Stress range** The difference in operating stresses at minimum and maximum loads.

**Stress-relieve** To subject springs to low-temperature heat treatment so as to relieve residual stresses.

**Torque $M$** A twisting action in torsion springs which tends to produce rotation, equal to the load multiplied by the distance (or moment arm) from the load to the axis of the spring body. Usually expressed in oz · in, lb · in, lb · ft, or in N · mm.

**Total number of coils $N_t$** Number of active coils $n_a$ plus the coils forming the ends.

**Wahl factor** A factor to correct stress in helical springs—the effects of curvature and direct shear.

## FURTHER READING

American Society for Metals, 1972: *Metals Handbook,* 8th ed. Atlas of Microstructures, Vol. 7.

Oberg, E., F. Jones, and H. Horton, 1990: *Machinery's Handbook,* 23d ed. New York: Industrial Press.

Shigley, J., and C. R. Mischke, 1986: *Standard Handbook of Machine Design.* New York: McGraw-Hill.

Spring Manufacturers Institute, 1991: *Handbook of Spring Design,* Wheeling, Ill.

# CHAPTER 8
# MACHINE ELEMENT DESIGN AND MECHANISMS

The basic machines from which all complex machines and mechanisms may be constructed are the lever, the inclined plane (wedge), and the wheel and axle. The screw is a variation of the inclined plane (in its helix), and the pulley is a variation of the wheel and axle. Add to these the modern simple machines which include the gear-wheel and hydraulic press, and this list represents all the simple machines presently used. (The Rolomite device was declared a simple machine by the U.S. Patent Office years ago, but its use is presently limited in respect to common machine elements.)

In this chapter we will deal with the basic elements used in designing complex machines, mechanisms, and machine elements, including power transmission equipment and associated devices. Readers using the machine elements as columns or beams are referred to the strength of materials sections. Readers needing other machine design calculations may be referred to the chapters on basic engineering mechanics and other applicable sections. The theory of machine elements is not presented in this chapter. Instead, readers will find practical design data and design procedures for all the basic machine elements commonly encountered in industrial design practice. Most of the design data and procedures shown in this section were developed by the major machine element manufacturers throughout the United States.

## 8.1 POWER TRANSMISSION COMPONENTS AND DESIGN PROCEDURES

Power transmission components detailed in this section include the following:

- V-belts, standard and narrow, single and multiple
- Flat belts
- Ribbed and timing belts
- Chains and sprockets
- Gears (see Sec. 8.8)
- Shafts and couplings
- Clutches
- Power screws and ratchet systems

### 8.1.1 Belts and Sheaves

The most common belts used today include

- Flat
- Classic V
- Narrow V
- Cogged V
- Variable-speed
- Synchronous
- V-ribbed

Although flat belts are still used today, the classic V-belt is usually the first choice of selection when a belt drive is designed, either single or multiple.

***Sheaves and Pulleys.*** The *sheave* is the grooved wheel in which the V-belt runs. For a flat belt, the wheel on which the belt runs is commonly called the *pulley.* (See Fig. 8.1 for typical belt types.)

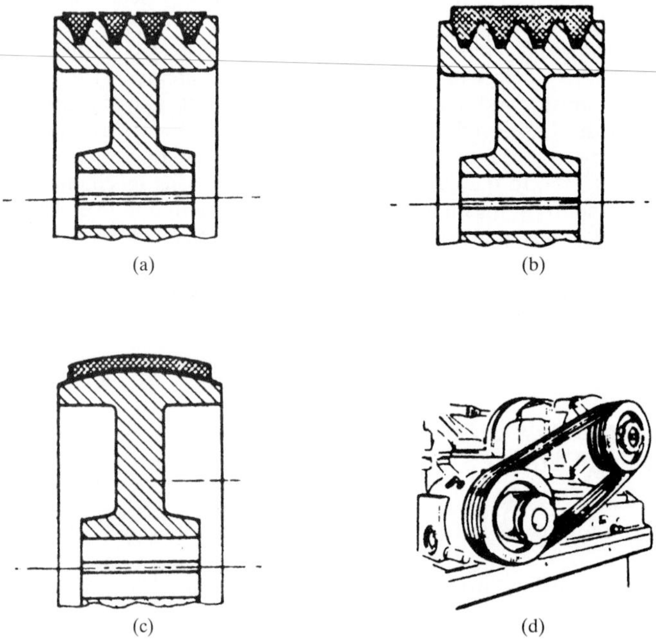


**FIGURE 8.1** Belts: (*a*) flat; (*b*) V-ribbed; (*c*) multiple-V; (*d*) multiple-V belt motor drive.

### 8.1.2 Standard V-Belts

The classic or conventional V-belt is available in standard sizes: A, B, C, D, and E. V-belts are commonly used individually in sizes A and B, but multiple belt drives in sizes A and B are also more economical than multiple drives in sizes C, D, and E.

The *narrow* V-belt transfers the applied loads in the belt cords more directly to the sides of the sheave, producing better force distribution than the classic V-belts. For a given width, the narrow V-belts have higher power ratings than do the classic types of V-belts. Narrow V-belts are standardized with the designations 3v, 5v, and 8v. Some of these are *cogged* to allow use over smaller diameter sheaves.

V-ribbed belts are a combination of flat and V-belt designs offering high power capacity with tensioning requirements only 20 percent greater than V-belts. These belts are used in high-power automotive applications as well as in mass-produced consumer products. V-ribbed belts are available in sizes H, J, K, L, and M.

Synchronous belts are available in sizes XL, L, and H as well as dual-drive sections termed DXL, DL, and DH.

### 8.1.3 V-Belt Drive Calculations

When power is transmitted by a V-belt, the belt is tight on one side and slack on the other side. The difference between these two tensions is the effective "pull" applied to the rim of the sheave. This pull produces work, and we may use the equation

$$\text{hp} = \frac{(T_T - T_S)d}{33{,}000t}$$

where $T_T - T_S$ = effective pull, lb
$T_T$ = tight-side tension
$T_S$ = slack-side tension
hp = horsepower
$d$ = distance moved, ft
$t$ = time, min

Belt speed $V$ is then calculated from

$$V = \frac{D\ (\text{rpm})}{3.82}$$

where $V$ = belt speed, ft/min; $D$ = sheave pitch diameter, in. Then

$$\text{hp} = \frac{(T_T - T_S)\ V}{33{,}000}$$

The tension ratio is

$$R = \left.\frac{T_T}{T_S}\right\} \quad \text{(as the ratio increases, the \% of belt slip increases)}$$

Using the *arc of contact* correction factor $G$ from Fig. 8.2, you can calculate tight-side and slack-side tensions using the following simplified equations:

$$T_T = \frac{41{,}250\ \text{hp}}{GV} \qquad T_S = \frac{33{,}000\ (1.25 - G)\ \text{hp}}{GV}$$

where $G$ = arc of contact correction factor; $V$ = belt speed (see previous equation for $V$), ft/min.

The arc of contact $A_c$ on the smaller sheave can be calculated from

$$A_c = \frac{(D - d)\ 60°}{C}$$

**Arc of contact correction factors for V-groove Belts.**

| (D−d)/C | Arc of contact, deg | Correction factor (G) |
|---|---|---|
| 0.0 | 180 | 1.00 |
| 0.1 | 174 | 0.99 |
| 0.2 | 169 | 0.97 |
| 0.3 | 163 | 0.96 |
| 0.4 | 157 | 0.94 |
| 0.5 | 151 | 0.93 |
| 0.6 | 145 | 0.91 |
| 0.7 | 139 | 0.89 |
| 0.8 | 133 | 0.87 |
| 0.9 | 127 | 0.85 |
| 1.0 | 120 | 0.82 |
| 1.1 | 113 | 0.80 |
| 1.2 | 106 | 0.77 |
| 1.3 | 99 | 0.73 |
| 1.4 | 91 | 0.70 |
| 1.5 | 83 | 0.65 |

**FIGURE 8.2** Correction factor table for arc of contact.

where $D$ = pitch diameter of larger sheave, in
$d$ = pitch diameter of smaller sheave, in
$C$ = center distance between sheaves, in

***Speed Ratio.*** Most belt drives provide speed reduction, and

$$\text{Speed ratio} = \frac{\text{faster sheave rpm}}{\text{slower sheave rpm}}$$

Belt drive ratio is always considered to be equal to or greater than 1, whether the drive is for speed-increasing or speed-reducing applications. As a general rule, for belt speeds over 5000 ft/min, the sheaves should be statically and dynamically balanced.

***Belt Length.*** For two-sheave drives,

$$L = 2C + 1.57\,(D + d) + \frac{(D - d)^2}{4C}$$

where $D$ = pitch diameter of large sheave, in
$d$ = pitch diameter of small sheave, in
$L$ = length of belt, in
$C$ = center distance of sheaves, in

***Application of Belt Drives.*** The designer must select the belt size, number of belts, and sheave sizes in order to satisfy these stipulations:

- Correct shaft speed
- Proper center distance
- Acceptable belt life

The *horsepower rating* method presumes an acceptable value for belt life in industrial applications. The *life-in-hours* of a belt drive can be considered to be from 200 to 2000 hours, depending on the application. What the designer should strive for in the belt drive system is the maximum life probability, weighed against the cost of the drive. Accurate life-in-hours for a belt drive system is determined by testing with simulated operating conditions.

### 8.1.4 Belt Drive Design Procedures

For a V-belt drive, you must know the following factors:

- The horsepower requirement of the drive
- Speed of the *driver* machine
- Speed of the *driven* machine
- Center distance for the sheaves

Proceed as follows:

**1.** Find the design horsepower from

$$\text{Design horsepower} = \text{service factor} \times \text{hp required}$$

Then approximate the service factor (ranging from 1.0 to 1.8) for heavier duty as follows:
*a.* Intermittent service (3 to 5 h/day): service factor 1.0 to 1.3
*b.* Normal service (8 to 10 h/day): service factor 1.3 to 1.5
*c.* Continuous duty (18 to 24 h/day): service factor 1.3 to 1.8

**2.** Select the proper V-belt (see Fig. 8.3*a* and 8.3*b*).
**3.** Find the speed ratio.
**4.** Select the sheave diameters:
*a.* Use one standard diameter for the larger sheave.
*b.* Find the sheave rim speed from

$$R_s = \frac{1}{3.82} \times \text{pitch diameter of sheave} \times \text{rpm of sheave}$$

Keep rim speeds below 6500 ft/min for narrow V-belts and 6000 ft/min for classic V-belts (types A through E)

**5.** Using a selected center distance, calculate the tentative belt length from

$$L_T = 1.57\,(D + d) + (\text{c.d.} \times 2)$$

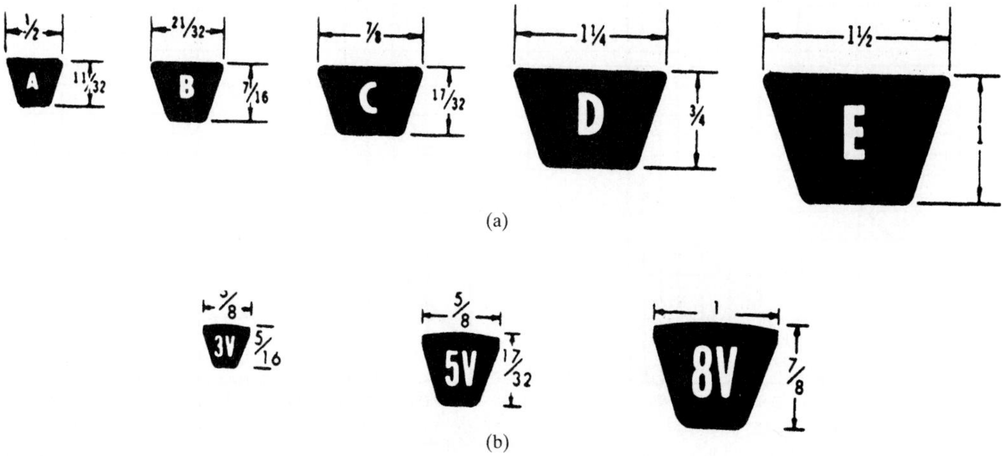


**FIGURE 8.3** (*a*) Standard V-belts; (*b*) narrow V-belts.

where $D$ = pitch diameter of larger sheave, in
$d$ = pitch diameter of smaller sheave, in
c.d. = center distance of the sheaves.

When you have selected the belt by choosing the closest standard belt length from the belt catalog, you may calculate the real center distance using the following equation:

$$C = 0.0625\left[b + \sqrt{b^2 - 32(D-d)^2}\right]$$

where $b = 4L - 6.28(D + d)$
$L$ = V-belt pitch length, in
$C$ = center distance of the sheaves, in
$D$ = pitch diameter of larger sheave, in
$d$ = pitch diameter of smaller sheave, in

Note that V-belt pitch lengths for drives with more than two sheaves are determined geometrically or by scaled layout.

**6.** Find the number of belts required.
   ***a.*** From manufacturers' tables, find the rated hp per belt.
   ***b.*** Divide the design horsepower by the hp per belt to obtain the number of belts. This figure is usually a fraction, so advance to the next highest whole number for the number of belts required.

Selection of the belt type and size can also be determined with the use of Figs. 8.4 and 8.5 when the design horsepower and the speed of the faster sheave are known.

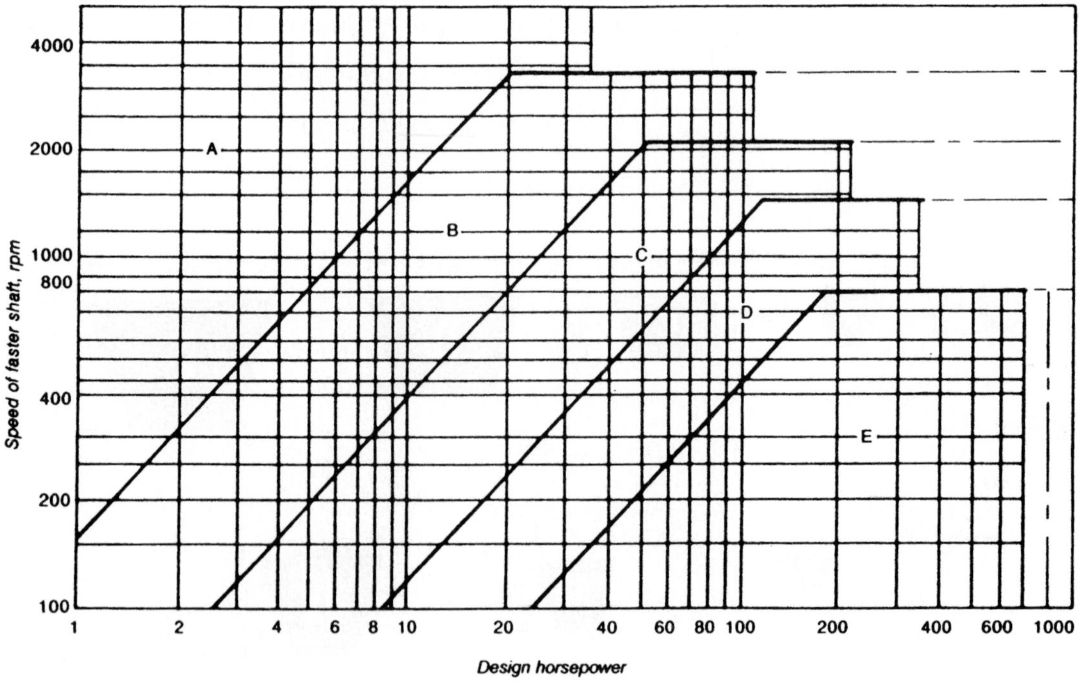


**FIGURE 8.4** Standard V-belt selection chart.

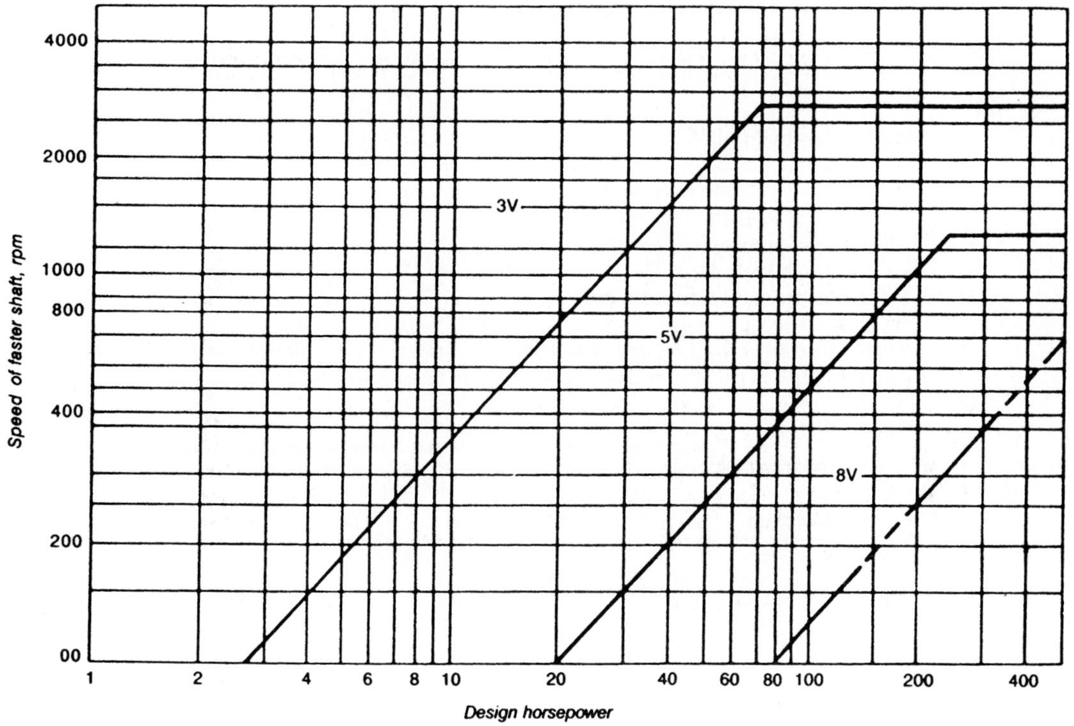


**FIGURE 8.5** Narrow V-belt selection chart.

## 8.1.5 Timing Belts

Timing belt sheaves must be in accurate alignment with both sheave shafts parallel. Avoid *preload* on timing belts. The belt should fit snug and not tight. Provision for sheave-center distance adjustment should be made in the drive design.

***Center Distance for Timing Belts.*** The center distance for timing belts is given approximately by

$$C = \frac{P}{4}\left[\text{n.b.} - \frac{N_1 + N_2}{2} + \sqrt{\left(\text{n.b.} - \frac{N_1 + N_2}{2}\right) - 2\left(\frac{N_1 + N_2}{\pi}\right)^2}\,\right]$$

where $C$ = center distance, in
$P$ = belt pitch, in
n.b. = number of teeth in the belt
$N_1$ = number of grooves in larger sheave
$N_2$ = number of grooves in smaller sheave

***V-Belt Drive Shaft Loads.*** The shaft loading used to calculate bearing loads on V-belt drives with 180° arc of contact is given as

$$P_B = 1.5\,(T_1 - T_2) = \frac{1.5 \times 63{,}025 \times \text{hp}}{\text{rpm} \times r}$$

For belts with other than 180° arc of contact, use

$$P_B = \left(\frac{2.5 - A_c}{A_c}\right)(T_1 - T_2) = \left(\frac{2.5 - A_c}{A_c}\right)\left(\frac{33{,}000 \times \text{hp}}{\text{ft/min}}\right)$$

where $P_B$ = belt pull, lb
$A_c$ = arc of contact factor (Fig. 8.2)
$T_1$ = tight side tension, lb
$T_2$ = slack side tension, lb
$r$ = sheave pitch radius, in
hp = horsepower
rpm = revolutions per minute

***Sheave Diameters for Electric Motor V-Belt Drives.*** The minimum recommended sheave diameters for electric motor drive service, in inches, are shown in Fig. 8.6.

**Minimum recommended sheave diameters for electric motors, in.**

| Motor hp | Motor rpm | | | | | |
|---|---|---|---|---|---|---|
| | 575 | 695 | 870 | 1160 | 1750 | 3450 |
| ½ | — | — | 2¼ | — | — | — |
| ¾ | — | — | 2½ | 2¼ | — | — |
| 1 | 3 | 2½ | 2½ | 2½ | 2¼ | — |
| 1½ | 3 | 2½ | 2½ | 2½ | 2¼ | 2¼ |
| 2 | 3¾ | 3 | 3 | 2½ | 2½ | 2½ |
| 3 | 4½ | 3¾ | 3 | 3 | 2½ | 2½ |
| 5 | 4½ | 4½ | 3¾ | 3 | 3 | 2¼ |
| 7½ | 5¼ | 4½ | 4½ | 3¾ | 3 | 3 |
| 10 | 6 | 5¼ | 4½ | 4½ | 3¾ | 3 |
| 15 | 6¾ | 6 | 5¼ | 4½ | 4½ | 3¾ |
| 20 | 8¼ | 6¾ | 6 | 5¼ | 4½ | 4½ |
| 25 | 9 | 8¼ | 6¾ | 6 | 4½ | 4½ |
| 30 | 10 | 9 | 6¾ | 6¾ | 5¼ | — |
| 40 | 10 | 10 | 8¼ | 6¾ | 6 | — |
| 50 | 11 | 10 | 9 | 8¼ | 6¾ | — |
| 60 | 12 | 11 | 10 | 9 | 7½ | — |
| 75 | 14 | 13 | 10 | 10 | 9 | — |
| 100 | 18 | 15 | 13 | 13 | 10 | — |

**FIGURE 8.6** Sheave diameters for electric motor drives.

***Conclusion.*** After designing the belt drive, you should additionally provide an adjustable means in the belt drive to take up belt slack. One adjustable sheave should be provided, but this is not always possible. An idler sheave or pulley is sometimes used for belt tensioning when other means of adjusting sheave centers are difficult or not possible. The use of an idler sheave or pulley is generally not recommended by the belt manufacturers. Idlers reduce the horsepower capacity of a belt drive system and also cause premature belt wear problems. Use an idler *only* when unavoidable. Correct placement and size of an idler are important points in design. Refer to the belt manufacturers' catalogs for application data on idler placement.

Conventional V-belt drive tensioning procedures are as follows:

**Step 1.** With all belts in their proper grooves, adjust the centers to take up all slack and until the belts are fairly taut.

**Step 2.** Start the drive and continue to adjust the tension until the belts have a slight bow on the slack side of the drive while operating under load.

**Step 3.** After a few days of operation the belts will "seat" themselves in the sheave grooves. It may then become necessary to readjust the belt tension until there is again a slight bow on the slack side under load.

Occasional readjustment may be necessary at periodic intervals as the belts and sheaves wear in service.

## 8.2 CHAINS AND SPROCKETS

Roller chain normally is supplied in both riveted and detachable types. The detachable types are assembled with cotter pins and are generally supplied in sizes larger than 1-in pitch.

See Fig. 8.7 for ANSI standard roller chain and Fig. 8.8 for dimensions of the various chain sizes. Figure 8.9 gives the ultimate strength (in tension) of ANSI standard roller chain in carbon and stainless steels.

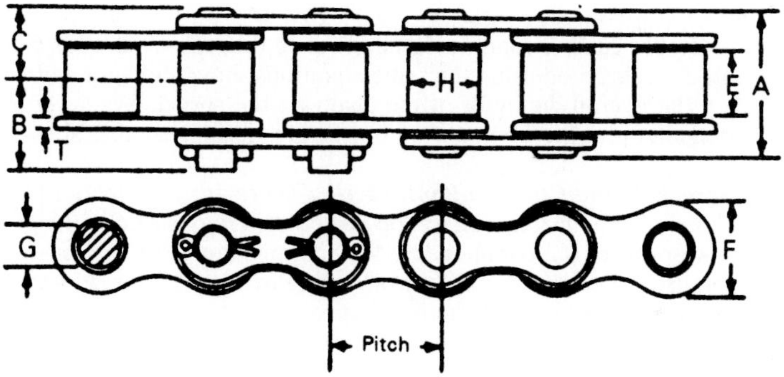


**FIGURE 8.7** ANSI standard roller chain.

| Chain number | Pitch | E | H | A | B | C | T | F | G | Weight/ ft, lb |
|---|---|---|---|---|---|---|---|---|---|---|
| 25* | ¼ | 0.125 | 0.130 | 0.31 | 0.19 | 0.15 | 0.030 | 0.23 | 0.0905 | 0.104 |
| 35* | ⅜ | 0.187 | 0.200 | 0.47 | 0.34 | 0.23 | 0.050 | 0.36 | 0.141 | 0.21 |
| 40 | ½ | 0.312 | 0.312 | 0.65 | 0.42 | 0.32 | 0.060 | 0.46 | 0.156 | 0.41 |
| S41 | ½ | 0.250 | 0.306 | 0.51 | 0.37 | 0.26 | 0.050 | 0.39 | 0.141 | 0.28 |
| S43 | ½ | 0.125 | 0.306 | 0.39 | 0.31 | 0.20 | 0.050 | 0.39 | 0.141 | 0.22 |
| 50 | ⅝ | 0.375 | 0.400 | 0.79 | 0.56 | 0.40 | 0.080 | 0.59 | 0.200 | 0.69 |
| 60 | ¾ | 0.500 | 0.468 | 0.98 | 0.64 | 0.49 | 0.094 | 0.70 | 0.234 | 0.96 |
| 80 | 1 | 0.625 | 0.625 | 0.128 | 0.74 | 0.64 | 0.125 | 0.93 | 0.312 | 1.60 |
| 100 | 1¼ | 0.750 | 0.750 | 1.54 | 0.91 | 0.77 | 0.156 | 1.16 | 0.375 | 2.56 |
| 120 | 1½ | 1.00 | 0.875 | 1.94 | 1.14 | 0.97 | 0.187 | 1.38 | 0.437 | 3.60 |
| 140 | 1¾ | 1.00 | 1.00 | 2.08 | 1.22 | 1.04 | 0.218 | 1.63 | 0.500 | 4.90 |
| 160 | 2 | 1.25 | 1.12 | 2.48 | 1.46 | 1.24 | 0.250 | 1.88 | 0.562 | 6.40 |
| 180 | 2¼ | 1.41 | 1.41 | 2.81 | 1.74 | 1.40 | 0.281 | 2.13 | 0.687 | 8.70 |
| 200 | 2½ | 1.50 | 1.56 | 3.02 | 1.86 | 1.51 | 0.312 | 2.32 | 0.781 | 10.30 |
| 240 | 3 | 1.88 | 1.88 | 3.76 | 2.27 | 1.88 | 0.375 | 2.80 | 0.937 | 16.90 |

**FIGURE 8.8** ANSI standard chain dimensions (see Fig. 8.7).

| Chain number | Carbon steel (pounds) | Stainless steel (pounds) |
|---|---|---|
| 25* | 925 | 700 |
| 35* | 2,100 | 1,700 |
| 40 | 3,700 | 3,000 |
| S41 | 2,000 | 1,700 |
| S43 | 1,700 | — |
| 50 | 6,100 | 4,700 |
| 60 | 8,500 | 6,750 |
| 80 | 14,500 | 12,000 |
| 100 | 24,000 | 18,750 |
| 120 | 34,000 | 27,500 |
| 140 | 46,000 | — |
| 160 | 58,000 | — |
| 180 | 80,000 | — |
| 200 | 95,000 | — |
| 240 | 130,000 | — |

*Rollerless

**FIGURE 8.9** Ultimate strength of roller chain in tension.

Roller chain is manufactured in single through six strands. Speed ratios for chain sprockets are calculated using the same equations shown for spur gears in Sec. 8.8.6.

The overall diameter of the chain on the sprocket is determined from the sum of pitch diameter plus the *F* dimension for the given chain number shown in Fig. 8.8.

***Sprockets and Classes of Sprockets.*** There are two standard classes of sprockets for roller chain, commercial and precision. When very high speed and high loads are required, a precision sprocket is recommended. Other applications for a precision sprocket include those that require fixed centers and critical timing or registration. A standard commercial sprocket is applicable for all other chain and sprocket uses.

Note that all chain made in the same chain number designation is *not* the same. All U.S.-manufactured chain which meets ANSI requirements and ASTM specifications is a high-quality product which will meet the strength specifications shown in Fig. 8.9. Imported roller chain must be carefully inspected and tested *before* applying it to your product. Various imported roller chain has performed below American standard specifications in a similar manner to the "counterfeit" hardware such as bolts and screws, which have been plaguing the industry since their introduction into this country some years ago. See Chap. 10 for more information on counterfeit hardware.

***Sprocket-to-Shaft Assembly.*** Sprockets should be assembled to shafts with keys and setscrews. The setscrew should be secured over the flat of the key and sized according to those shown below (see the shaft section for key sizes).

| Shaft diameter sprocket bore (range) | Setscrew size (diameter, in) |
|---|---|
| 0.5–0.875 | 0.25 |
| 0.937–1.75 | 0.375 |
| 1.812–2.25 | 0.50 |
| 2.312–3.25 | 0.625 |

Permanent assemblies may be made by welding the sprocket to the shaft if the design permits. Note that some of the smaller size sprockets are made using *powder-metal technology* and that this type of sprocket is *not* recommended for welding. Welding a powder-metal sprocket usually results in the sprocket breaking at the weld area. Brazing such sprockets to a shaft may be possible if the loads are not excessive.

Figure 8.10 shows the two standard types of sprockets in common use.

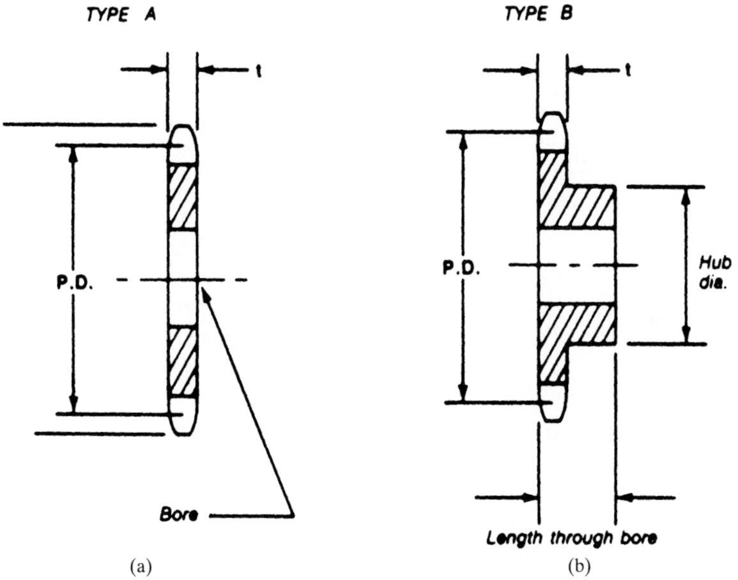


**FIGURE 8.10** Types of roller chain sprockets: (*a*) plate sprocket, (*b*) hubbed sprocket.

***Types of Chains for Chain Drives.*** Commercial standard chain types available include

- Roller, single- and multiple-strand
- Rollerless
- Double-pitch
- Detachable link
- Silent, inverted-tooth
- Offset sidebar
- Bead

***Allowable Speed Ratios.*** The speed ratio for roller chain or silent chain should not exceed 10:1, and for other types should not exceed 6:1. Use a double reduction drive to overcome these limitations. Double reduction ratios are calculated the same as those for spur gears (see Sec. 8.8.6).

### 8.2.1 Horsepower Capacity of Chain Drives

In the sample comparison chart shown in Fig. 8.11, horsepower ratings versus rotating speeds are shown for 15-tooth sprockets as a reference only. Actual capacities and ratings for your different applications are calculated per the preceding detailed procedures.

| rpm sprocket is | Roller chain | |
|---|---|---|
| | Single strand | Four strand |
| 100 | 100 | 330 |
| 500 | 180 | 600 |
| 700 | 100 | 335 |
| 1,000 | 60 | 195 |
| 1,400 | 30 | 98 |
| 1,800 | 18 | 55 |
| 2,200 | 11 | 37 |
| 2,600 | 7 | 25 |
| 3,000 | 6 | 19 |
| 3,500 | 3.5 | 12.5 |
| 4,500 | 2 | 7 |
| 5,500 | 1.75 | 6.1 |
| 6,500 | 1.3 | 4.7 |
| 7,500 | 1.2 | 3.8 |
| 8,500 | 1 | 3.2 |
| 10,000 | .75 | 2.4 |

**FIGURE 8.11** Maximum horsepower capacity for standard roller chains.

### 8.2.2 Conveyor Applications for Roller Chain

For conveyor applications, the chain pull $W$ can be calculated from

$$W = \frac{2T}{D} = \frac{33{,}000P}{V} = \frac{126{,}050\,P}{nD}$$

where $W$ = chain pull, lb
$T$ = torque, lb · in
$D$ = sprocket pitch diameter, in
$P$ = horsepower
$V$ = chain velocity, ft/min
$n$ = sprocket speed, rpm

When the chain pull is determined by one of these equations, select the chain size required from the table in Fig. 8.12.

When the center distance of the sprockets is known, the number of pitches required in the chain can be calculated per the following:

$$\text{Chain length, pitches} = N_p = \left(\frac{2\text{ c.d.}}{P}\right) + \left(\frac{T_{DR} + T_{DN}}{2}\right)$$

where c.d. = center distance of sprockets, in
$P$ = pitch of chain, in
$T_{DR}$ = number of teeth in driver sprocket
$T_{DN}$ = number of teeth in driven sprocket
$N_p$ = number of pitches in chain length

The center distance between sprockets should be greater than half the sum of the outside diameters of the sprockets.

***Slow-Speed Partial Drive.*** Chains and sprockets are used in many applications where a high torque is transmitted at a partial revolution or small number of turns of the driving sprocket. In such cases, the minimum tensile strength of the chain, along with a factor of

| | CHAIN NUMBERS | | | | | | | | |
|---|---|---|---|---|---|---|---|---|---|
| Single Pitch | 35* | 40 | 50 | 60 | 80 | 100 | 120 | 140 | 160 |
| Double Pitch | | C2040 | C2050 | C2060 | C2080 | C2100 | C2120 | | C2160 |
| Velocity of Chain (FPM) | MAXIMUM WORKING LOAD OR CHAIN PULL (Lbs.) | | | | | | | | |
| 25 | 250 | 443 | 690 | 995 | 1770 | 2760 | 3990 | 5430 | 7100 |
| 50 | 243 | 432 | 675 | 970 | 1730 | 2690 | 3880 | 5290 | 6900 |
| 75 | 233 | 414 | 645 | 930 | 1660 | 2580 | 3720 | 5060 | 6630 |
| 100 | 220 | 391 | 610 | 880 | 1570 | 2440 | 3520 | 4800 | 6250 |
| 125 | 206 | 366 | 570 | 820 | 1460 | 2280 | 3290 | 4470 | 5850 |
| 150 | 190 | 338 | 528 | 760 | 1350 | 2110 | 3040 | 4140 | 5400 |
| 175 | 175 | 311 | 485 | 700 | 1240 | 1940 | 2800 | 3810 | 4970 |
| 200 | 160 | 284 | 444 | 640 | 1140 | 1770 | 2560 | 3480 | 4550 |
| 225 | 146 | 259 | 405 | 584 | 1040 | 1620 | 2340 | 3180 | 4150 |
| 250 | 133 | 236 | 368 | 530 | 940 | 1470 | 2120 | 2890 | 3770 |
| 275 | 120 | 214 | 333 | 480 | 855 | 1330 | 1920 | 2610 | 3310 |
| 300 | 110 | 195 | 305 | 440 | 780 | 1220 | 1760 | 2390 | 3120 |
| | STANDARD PITCH BOSTON SPROCKETS TO OPERATE WITH ABOVE CHAIN | | | | | | | | |
| Pitch | 3/8" | 1/2" | 5/8" | 3/4" | 1" | 1-1/4" | 1-1/2" | 1-3/4" | 2" |

*No. 35 Chain is a Rollerless Chain.

**FIGURE 8.12** Chain load-rating chart.

safety or multiplier, is used to determine the chain size. So, you may calculate the maximum tensile load on the chain and multiply this figure by 2 to 3 and determine the chain size from this value.

*Example.* For the sample application (Fig. 8.13), let the reduction ratio be 1.5. The sprocket $S_1$ must rotate 1.5 times to revolve $S_2$ once. If the torque required at the driven sprocket $S_2$ is 2000 lb · in, the torque at sprocket $S_1$ will be

$$\frac{2000}{1.5} = 1333 \text{ lb} \cdot \text{in}$$

If you limit the tension in the chain to approximately 700 lb, you will need to calculate the pitch radius (p.r.) of the sprocket $S_2$, so

$$T_2 = \text{p.r.} \times 700 \qquad 2000 = \text{p.r.} \times 700$$

$$\text{p.r.} = \frac{2000}{700} = 2.86 \text{ in}$$

Therefore the pitch diameter of sprocket $S_2 = 2 \times 2.86 = 5.72$ in. Now, select a sprocket for the appropriate chain size from Fig. 8.9. Multiplying 700 × 2.5 safety factor = 1750 lb, you will find that S43 chain will suffice for this application. Select an S43 sprocket whose pitch diameter is close to 5.72 in and then select the driver sprocket with the correct number of teeth to give the reduction ratio of 1.5:1.

$$\frac{1.5}{1} = \frac{\text{no. of teeth in driven } S_2}{\text{no. of teeth in driver } S_1}$$

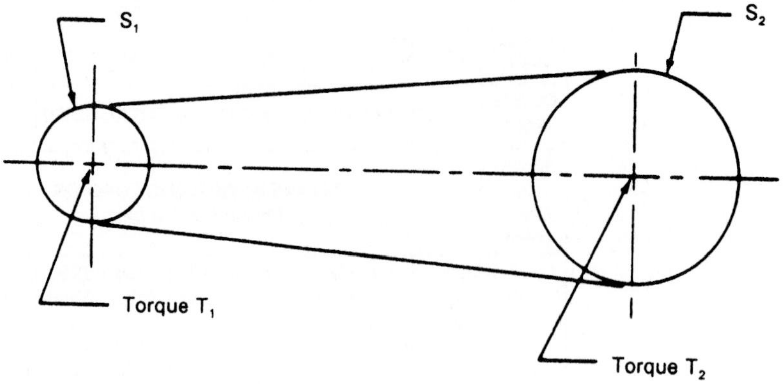


**FIGURE 8.13** Sprocket and chain drive diagram.

An S43 sprocket whose pitch diameter is close to 5.72 in is the 36-tooth sprocket, whose pitch diameter is 5.737 in. Now, we may determine the number of teeth in the driver sprocket $S_1$ from this simple relation:

$$\frac{1.5}{1} = \frac{36}{x} \qquad 1.5x = 36 \qquad x = 24 \text{ teeth in } S_1$$

This then is the method of properly selecting a chain drive system for partial-drive applications. A chain drive system selected using this conservative method will provide many hours of trouble-free service. Selection of the bearings for this type of system is detailed in other sections of the handbook pertaining to antifriction bearings.

## 8.3 SHAFTS AND SHAFTING MATERIALS

In operation, shafts are subject to different types of loads and stresses such as angular deviation (torque displacement α), bending (linear deflection), bending stresses, and torsional stresses. Shafts are also subject to the condition known as *critical speed.* A shaft carrying a number of loads or a distributed load can have an infinite number of critical speeds, but the *first critical speed* is of importance in the design of shafts and will be detailed in this section. The selection and processing of shafting and shafting materials will also be detailed.

### 8.3.1 Torsion in Shafts

The angle of torsional deflection α of a shaft subject to a torsional moment $M_t$ is given as follows. For *any* cross section:

$$\alpha = \frac{180\, M_t l}{\pi I_p G}$$

For a round cross-section shaft:

$$\alpha = \frac{583.6\, M_t l}{D^4 G} \qquad I_p \text{ for a round shaft} = \frac{\pi D^4}{32}$$

where $D$ = diameter, in
$l$ = length being twisted, in
$I_p$ = polar moment of inertia of the section through the shaft, $in^3$
$\alpha$ = angle of torsional deflection, degrees
$M_t$ = torsional moment, lb · in
$G$ = torsional modulus of elasticity, psi (for mild-carbon steel $G = 11.5 \times 10^6$ psi)

See the materials section for other materials. The torsional modulus of elasticity is approximately 40 percent of the modulus of elasticity in tension (Young's modulus) for many metals.

The degree of torsional deflection permissible for line shafting is determined by the application. Shafts carrying gears should be particularly stiff, with a recommended angular deflection not to exceed 0.08° per foot of length.

The angle of torsional deflection ($\alpha$) for different shaft forms includes the following. For square shaft:

$$\alpha = \frac{343.7\, M_t\, l}{h^4\, G}$$

where $h$ = length of side of square. For hollow round shafts or tubes:

$$\alpha = \frac{583.6\, M_t l}{(D^4 - d^4)G}$$

where $D$ = outside diameter; $d$ = inside diameter, in.

### 8.3.2 Bending in Shafts (Linear Deflection)

The shaft shown in Fig. 8.14 is taken as a beam of uniform cross section loaded transversely with a concentrated load $P$. Here

$$\Delta_x = \frac{P\, a^2 b^2}{3\, EI\, l}$$

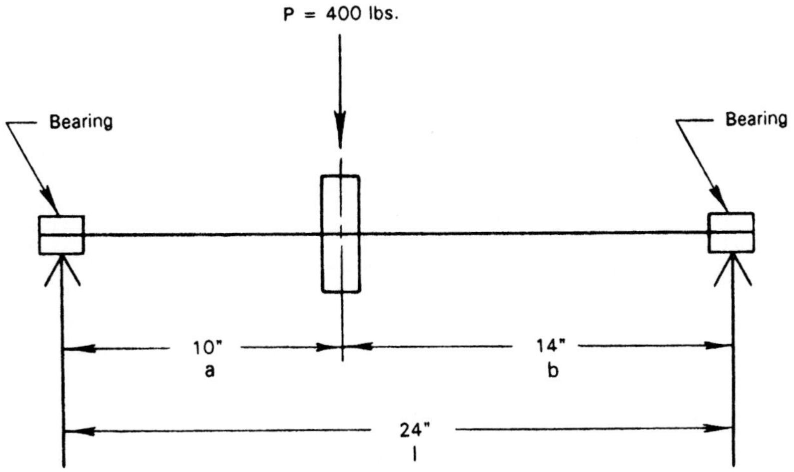


**FIGURE 8.14** Loaded shaft.

(See Beam equations, Index.). If the load $P$ on the gear is 400 lb, the shaft diameter $d = 1.25$ in; $a = 10$ in; $b = 14$ in, $E = 30 \times 10^6$ psi; $I$ = moment of inertia, in$^4$; and $\Delta_x$ = deflection at point $P$, in, we may proceed as follows. Find the moment of inertia of the shaft cross section $I$ (see Properties of geometric sections, Index) from

$$I = \frac{\pi d^4}{64} = \frac{3.14\,(1.25)^4}{64} = \frac{7.666}{64} = 0.12 \text{ in}^4 \quad \text{(moment of inertia)}$$

From the preceding equation for bending of a uniform beam with a concentrated load:

$$\Delta_x = \frac{P\,a^2b^2}{3\,EI\,l} = \frac{400\,(10)^2(14)^2}{3\,(30 \times 10^6)(0.12)(24)} = \frac{7{,}840{,}000}{2.592 \times 10^8} = 0.0302 \text{ in}$$

From this example, it can be seen that the deflection of the 1.25-in-diameter shaft will be 0.0302 in. For a gearing application, this deflection is unacceptable and the shaft size would need to be increased in diameter or the support points moved closer together. Moving the support points or increasing the shaft size would require recalculation.

### 8.3.3 Bending Stresses in Shafts

Analyzing the shaft system shown in Fig. 8.15, we begin by taking moments about $R_1$:

$$3000 \times 6 = R_2 \times 24 \qquad R_2 = \frac{3000 \times 6}{24} = 750\text{-lb reaction at } R_2$$

Taking moments about $R_2$:

$$R_1 \times 24 = 3000 \times 18 \qquad R_1 = \frac{3000 \times 18}{24} = 2250\text{-lb reaction at } R_1$$

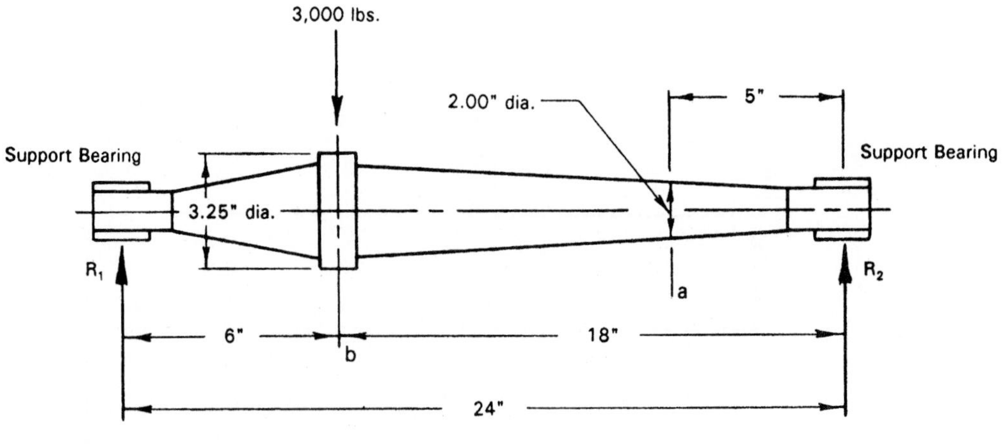


**FIGURE 8.15** Compound shaft under load.

The bending moment at $a$ is

$$M_a = 750 \times 5 = 3750 \text{ lb} \cdot \text{in}$$

The section modulus at $a$ is

$$Z = 0.098\, d^3 = 0.098\,(2)^3 = 0.784 \text{ in}^3$$

The stress at point $a$ is

$$S_a = \frac{M_a}{Z} = \frac{3750}{0.784} = 4783 \text{ psi}$$

Bending moment at $b$ is

$$M_b = 2250 \times 6 = 13{,}500 \text{ lb} \cdot \text{in}$$

Section modulus at $b$ is

$$Z = 0.098\, d^3 = 0.098\,(3.25)^3 = 3.364 \text{ in}^3$$

The stress at point $b$ is

$$S_b = \frac{M_b}{Z} = \frac{13{,}500}{3.364} = 4013 \text{ psi}$$

Note that in the preceding example

$$S = \frac{M}{Z}$$

or $\quad$ Stress on section at a given point $= \dfrac{\text{bending moment at the point}}{\text{section modulus of cross section at the point}}$

### 8.3.4 Torsional Stresses in Shafts

Shaft torsional stress $S_{ts}$ for solid round shaft is given as

$$S_{ts} = \text{hp} \times \frac{321{,}000}{\text{rpm} \times d^3}$$

Shaft diameter $d_s$ for solid shaft is given as

$$d_s = \sqrt[3]{\frac{\text{hp} \times 321{,}000}{\text{rpm} \times \text{allowable stress}}}$$

Shaft torsional stress $S_{ts}$ for tubular shaft is

$$S_{ts} = \frac{\text{hp} \times 321{,}000}{\text{rpm}\,(D^4 - d^4)}$$

where $d$ = inside diameter. Outside diameter $D$ when inside diameter is known is

$$D = \sqrt[3]{\frac{321{,}000 \times \text{hp}}{\text{rpm} \times \text{S}_\text{a}} + d^4}$$

where $S_a$ = allowable stress.

***Torque, Force, and Horsepower Relationships***

$$F = \frac{M_t}{r} \qquad M_t = \frac{63{,}000 \times \text{hp}}{n} \qquad \text{hp} = \frac{M_t n}{63{,}000} \qquad n = \frac{63{,}000 \times \text{hp}}{M_t}$$

where hp = horsepower
$n$ = rpm (revolutions per minute)
$M_t$ = torque, lb · in
$F$ = force, lb
$r$ = radius

## 8.3.5 Critical Speeds of Shafts

The following equations for the critical speeds of rotating shafts are according to S. H. Weaver and apply to shafts with single concentrated loads and shafts carrying uniformly distributed loads, with the shaft mounted horizontally or vertically. These equations apply to materials with a modulus of elasticity of $29 \times 10^6$ psi, which is normal for most steel shafting. A shaft carrying a number of loads or a distributed load can have an infinite number of critical speeds, but the *first critical speed* is of importance in designing shafts and is the speed obtained by using the equations given for the following shaft loading conditions.

For single concentrated loads, use Figs. 8.16 through 8.21. For uniformly distributed loads, use Figs. 8.22 through 8.24.

$$N = 387{,}000 \frac{d^2}{ab} \sqrt{\frac{l}{W}} \qquad \text{(Fig. 8.16)}$$

$$N = 1{,}550{,}500 \frac{d^2}{l\sqrt{Wl}} \qquad \text{(Fig. 8.17)}$$

$$N = 387{,}000 \frac{d^2}{ab} \sqrt{\frac{l}{Wab}} \qquad \text{(Fig. 8.18)}$$

$$N = 3{,}100{,}850 \frac{d^2}{l\sqrt{Wl}} \qquad \text{(Fig. 8.19)}$$

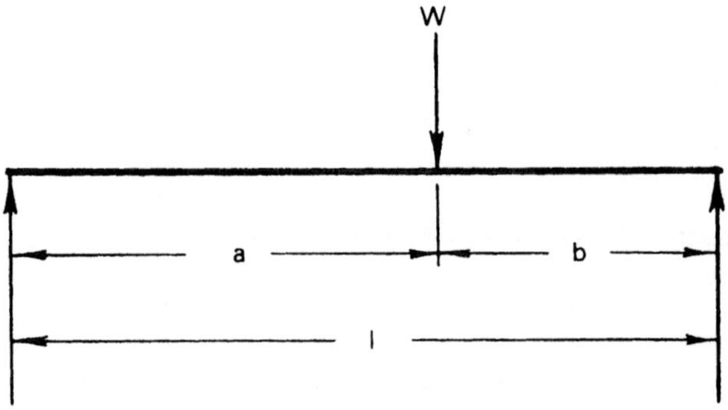


**FIGURE 8.16** Single concentrated load.

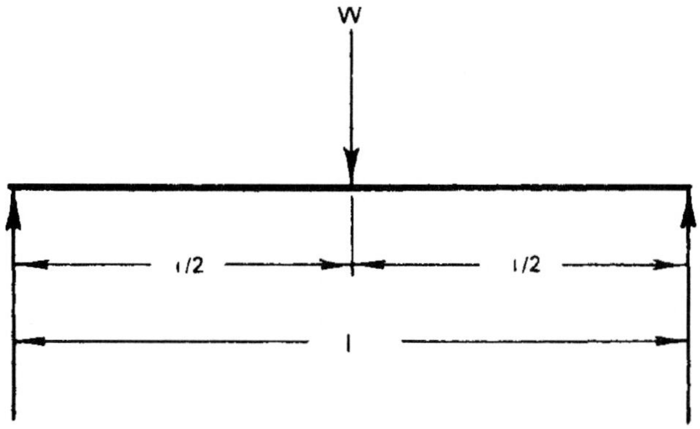


**FIGURE 8.17** Single concentrated load at center.

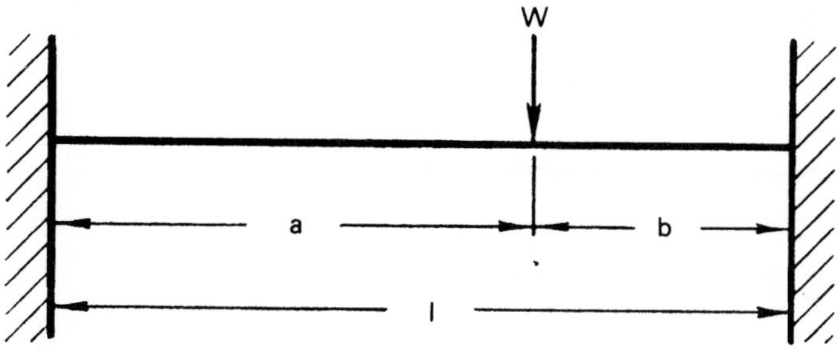


**FIGURE 8.18** Single concentrated load, ends fixed.

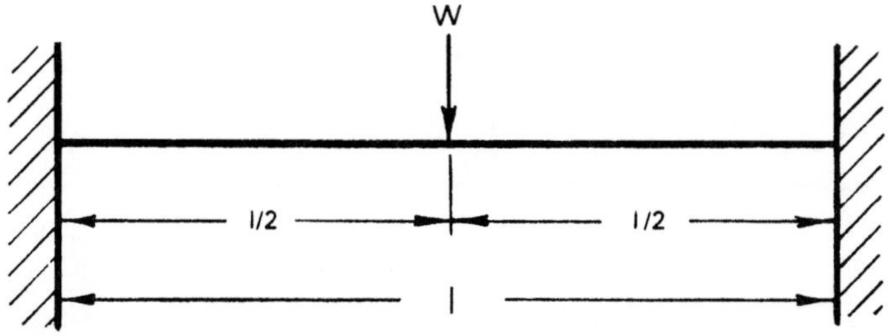


**FIGURE 8.19** Single concentrated load at center, ends fixed.

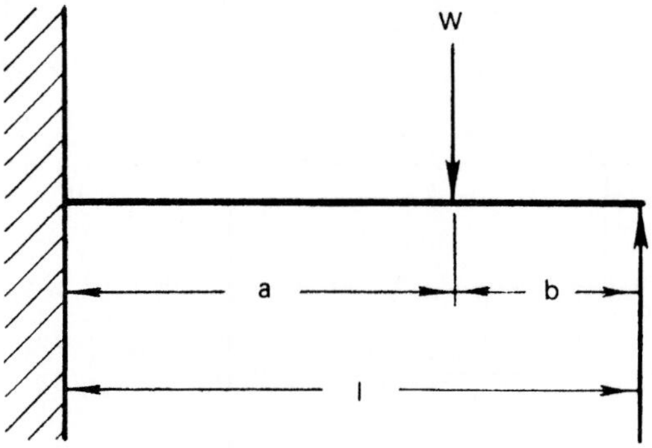


**FIGURE 8.20** Cantilever load.

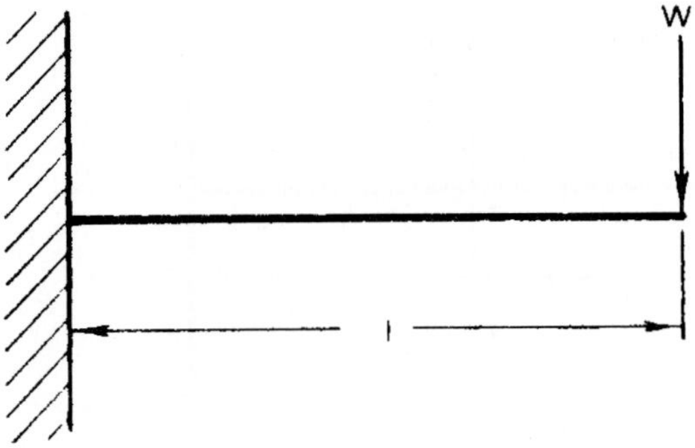


**FIGURE 8.21** Cantilever load at end.

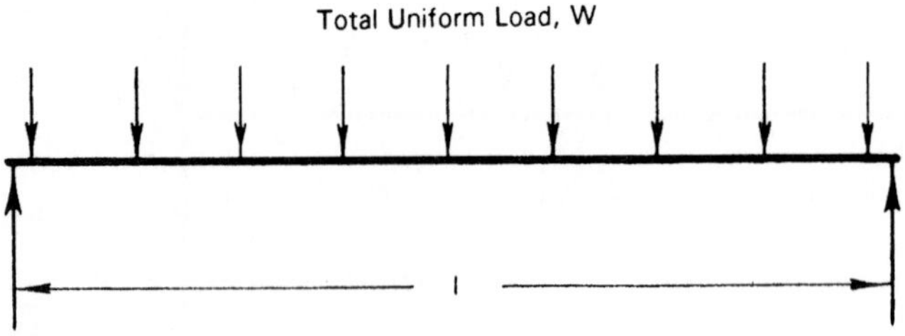


**FIGURE 8.22** Uniformly distributed load.

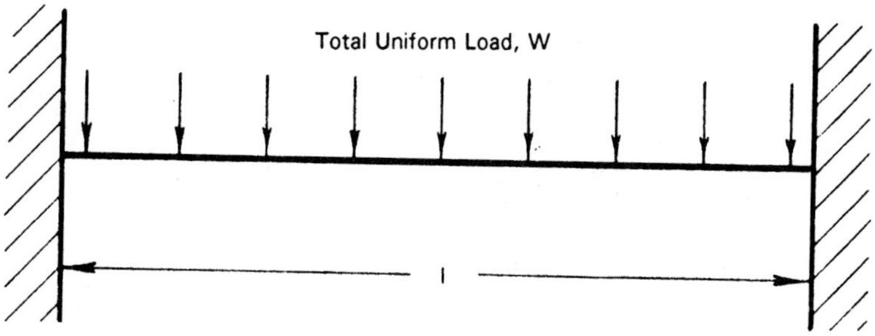


**FIGURE 8.23** Uniformly distributed load, ends fixed.

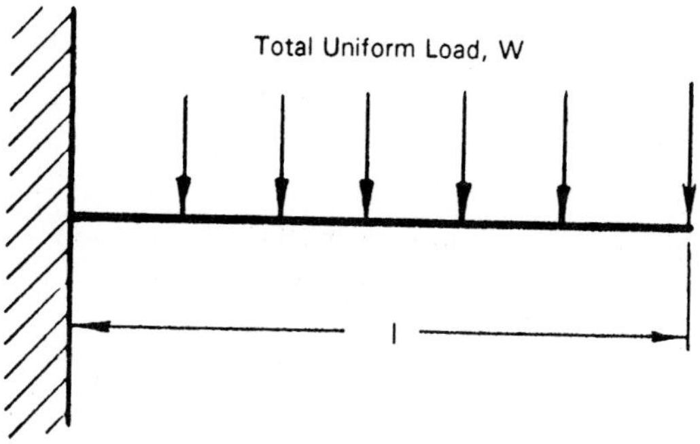


**FIGURE 8.24** Uniformly distributed cantilever load.

$$N = 775{,}200\,\frac{d^2 l}{ab}\sqrt{\frac{l}{Wa\,(3l+b)}} \qquad \text{(Fig. 8.20)}$$

$$N = 387{,}000\,\frac{d^2}{l\sqrt{Wl}} \qquad \text{(Fig. 8.21)}$$

$$N = 2{,}232{,}500\,\frac{d^2}{l\sqrt{Wl}} \qquad \text{(Fig. 8.22)}$$

$$N_1 = 4{,}760{,}000\,\frac{d}{l^2}$$

$$N = 4{,}979{,}250\,\frac{d^2}{l\sqrt{Wl}} \qquad \text{(Fig. 8.23)}$$

$$N_1 = 10{,}616{,}740\,\frac{d}{l^2}$$

$$N = 795{,}200\,\frac{d^2}{l\sqrt{Wl}} \qquad \text{(Fig. 8.24)}$$

$$N_1 = 1{,}695{,}500\,\frac{d}{l^2}$$

where $d$ = diameter of shaft, in
$W$ = load applied to shaft, lb
$l$ = distance between centers of bearings, in
$a,b$ = distances from bearings to load, in
$N$ = critical speed, rpm
$N_1$ = critical speed of shaft alone, rpm

### 8.3.6 Shaft Polar Moment and Polar Section Modulus

Polar moment of inertia $I_p$ and polar section modulus $Z_p$ are given for the shaft cross sections illustrated in Fig. 8.25.

### 8.3.7 Standard Keyways and Setscrews for Shafts

Refer to Figs. 8.26 and 8.27. In Fig. 8.27, $x$ and $x'$ for the key slots are as follows:

$$x = \sqrt{\left(\frac{D}{2}\right)^2 - \left(\frac{W}{2}\right)^2} + d + \frac{D}{2} \qquad x' = 2x - D$$

***Keyway Stress for Nonmetallic Gears.*** For phenolic laminated gears or pinions, the keyway stress $S$ should not exceed 3000 psi, and

$$S = \frac{33{,}000 \times \text{hp}}{VA}$$

where $S$ = unit stress, psi
hp = transmitted horsepower
$V$ = surface speed of the shaft, ft/min
$A$ = area of keyway (length × height), in$^2$

### 8.3.8 Shaft Overhung Loads

Refer to Fig. 8.28.

$$\text{Overhung load, lb} = \frac{(1.26 \times 10^5)\text{hp} \times F_c L_f}{\text{p.d.} \times \text{rpm}}$$

$$\text{Overhung load, N} = \frac{(1.91 \times 10^7)\text{hp} \times F_c L_f}{\text{p.d.} \times \text{rpm}}$$

where $L_f$ = load location factor (for $d = D$, $L_f = 1$)
$F_c$ = load connection factor (see table below)
hp = horsepower
kW = kilowatts
rpm = revolutions per minute
p.d. = pitch diameter, in or mm

***Values of $F_c$***

| | |
|---|---|
| Sprocket or timing belt | 1.00 |
| Machined pinion and gear | 1.25 |
| V-belt | 1.50 |
| V-ribbed belt | 2.00 |
| Flat belt | 2.50 |

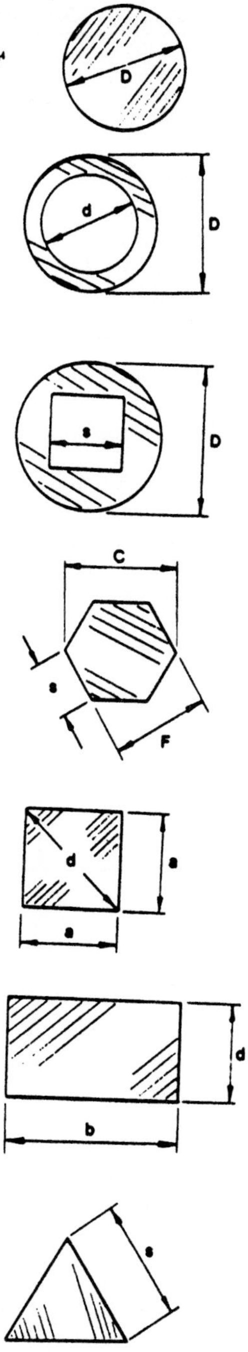


| | Polar Moment of Inertia, $I_p$ | Polar Section Modulus, $Z_p$ |
|---|---|---|
| 1. | $\frac{\pi D^4}{32} = 0.098\, D^4$ | $\frac{\pi D^3}{16} = 0.196\, D^3$ |
| 2. | $\frac{\pi}{32}(D^4 - d^4)$ | $\frac{\pi}{16}\left(\frac{D^4 - d^4}{D}\right)$ |
| 3. | $\frac{\pi D^4}{32} - \frac{s^4}{6}$ | $\frac{\pi D^3}{16} - \frac{s^4}{3D}$ |
| 4. | $\frac{5\sqrt{3}}{8} s^4$ | $0.20\, F^3$ |
| 5. | $\frac{a^4}{6}$ | $0.208\, a^3$ |
| 6. | $\frac{bd\,(b^2 + d^2)}{12}$ | $\frac{bd^2}{3 + 1.8\frac{d}{b}}$ |
| 7. | $\frac{\sqrt{3}\, s^4}{48}$ | $\frac{s^3}{20}$ |

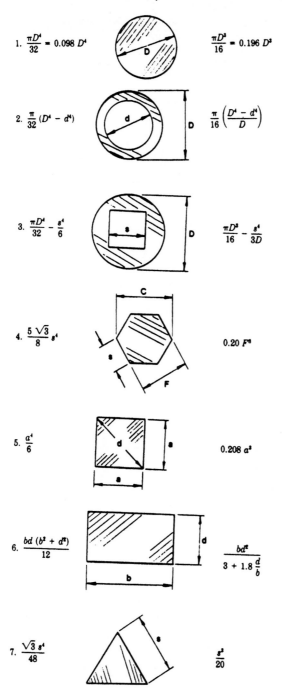

**FIGURE 8.25** Shaft polar moment and polar section modulus.

| Diameter of hole, in | Standard Keyway W, in | Standard Keyway d, in | Setscrew size |
|---|---|---|---|
| 5/16 to 7/16 | 3/32 | 3/64 | #10–32 |
| 1/2 to 9/16 | 1/8 | 1/16 | 1/4–20 |
| 5/8 to 7/8 | 3/16 | 3/32 | 5/16–18 |
| 15/16 to 1 1/4 | 1/4 | 1/8 | 3/8–16 |
| 1 5/16 to 1 3/8 | 5/16 | 5/32 | 7/16–14 |
| 1 7/16 to 1 3/4 | 3/8 | 3/16 | 1/2–13 |
| 1 13/16 to 2 1/4 | 1/2 | 1/4 | 9/16–12 |
| 2 5/16 to 2 3/4 | 5/8 | 5/16 | 5/8–11 |
| 2 13/16 to 3 1/4 | 3/4 | 3/8 | 3/4–10 |
| 3 5/16 to 3 3/4 | 7/8 | 7/16 | 7/8–9 |
| 3 13/16 to 4 1/2 | 1 | 1/2 | 1–8 |
| 4 9/16 to 5 1/2 | 1 1/4 | 7/16 | 1 1/8–7 |
| 5 9/16 to 6 1/2 | 1 1/2 | 1/2 | 1 1/4–6 |

**FIGURE 8.26** Standard keyways and setscrews for shafts.

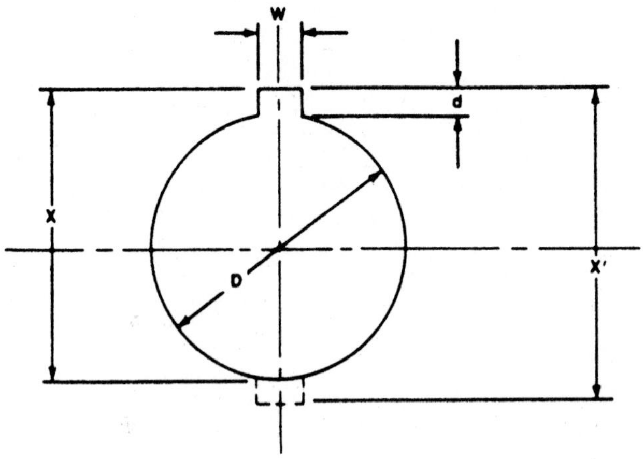


**FIGURE 8.27** Keyways, single and double.

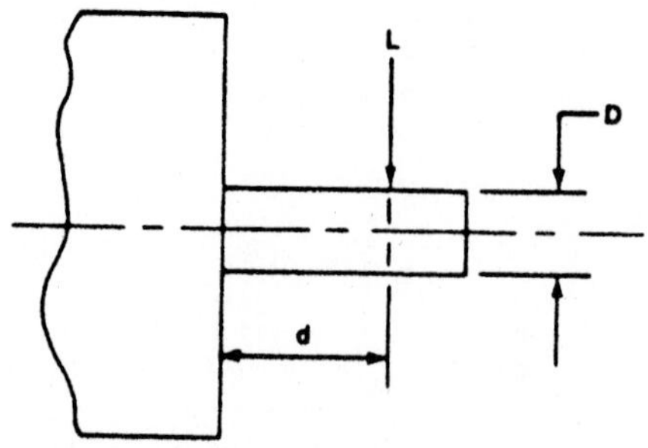


**FIGURE 8.28** Shaft overhung load.

### 8.3.9 Shafting Applications, Materials, and Heat Treatment

Shafts fabricated from carbon and alloy steels may require case hardening or through hardening, according to the application. Shafts that must fit standard-size bearings are normally purchased as "ground and polished shafting." This type of shafting always has a minimum tolerance on the basic diameter. Standard bearings are normally manufactured with a plus tolerance on their bore diameters. The condition of a minus tolerance on the shafting and a plus tolerance on the bearing bore assures a clearance fit at assembly of the shaft into the bearing. [*Note:* If the shaft is plated, the plating thickness must be taken into consideration. Plating thicknesses on common shafting applications should not exceed 0.2 mil (0.0002 in). Also, the type of plating must be considered so that the bearing will not destroy the plating.]

Some of the commonly used shafting stock materials are

| | |
|---|---|
| Potomac and 1045 (UNS G10450) | Medium-carbon steel (direct-hardening) flame and induction; yield strength = 59,000 psi. |
| Cumsco and 1140 (UNS G11410) | Medium-carbon steel (direct-hardening) flame and induction; yield strength = 61,000 psi (good "as-rolled" strength and toughness). |
| "Stressproof" and 1144 SRA-100 | Medium-carbon steel (high-manganese); yield = 100,000 psi (does not require heat treatment). |
| 1040 and 1042 | Medium-carbon (direct-hardening) flame and induction. |
| 1144 | Similar to 1141 except higher carbon and sulfur for improved machining and response to heat treatment. Used for induction hardening to $R_c$ 55 (Rockwell). |
| "Fatigueproof" | High manganese, induction hardenable; yield strength = 125,000 psi. |

(*Note:* 1040, 1042, 1144, and stressproof steels may need tolerance specifications for shafting service when ordering.)

Specialty steel shafting E4130, E4140, and E4340 are aircraft alloy steels of high strength and are used for critical applications. See Chap. 5 for compositions and heat-treatment procedures.

### 8.3.10 Hardness Ranges—Shafting

The general Rockwell hardness range for the previously listed carbon and alloy steel shafting is $R_c$40 to 58, depending on the specific steel, the hardness requirement of the application, and the ability of the heat treater to attain the required hardness with a minimum of distortion on the shaft. The shaft designer should consult the heat treater to arrive at a satisfactory solution. Small parts are generally no problem, but large parts and long shafts pose heat-treating problems. Flame and induction hardening of small areas or sections of long shafts running on far spread bearings is a common practice. This method eliminates distortion caused by hardening the entire length of the shaft. The heat treater can eliminate distortion in shafts by "hanging" the shaft during the heat treating process, but this practice is limited by the physical length of the shaft and the size of the heat treater's equipment.

Shafting such as stressproof and fatigueproof and 1144 SRA100 do not require heat treatment as they are severely cold-worked. Readily machinable, with minimum distortion, these high-manganese steels attain yield strengths to 125,000 psi with a Brinell hardness range between 270 to 285 ($R_c$30). Additional hardening may be attained through heat treatment (induction hardening). The type of antifriction bearing used in the shafting application will determine the extent of the hardening required. As stated previously, the Rockwell hardness range for shafting may be from $R_c40$ to 58. See Table 5.1 in Chap. 5.

## 8.4 COUPLINGS

The common couplings used to connect driving machines such as electric motors and internal-combustion engines or turbines to their drive shafts consist of the basic types listed here:

- Hooke's coupling (Cardan universal joint)
- Simple flanged couplings
- Sleeve couplings
- Flexible couplings

### 8.4.1 Hooke's Coupling (Cardan Universal Joint)

The common universal joint is used to connect two shafts whose axes are not in line, but which intersect at a point. The typical universal joint is generally used at an angle $\alpha$ between 5 and 30° (see Fig. 8.29).

The angular velocity of the driven shaft through a universal joint will not be the same as that of the *driving* shaft. That is, if the driving shaft has uniform motion, the *driven* shaft will *not.* The speed relationship is as follows:

$$\text{Minimum speed of driven shaft} = \text{driver shaft speed} \times \cos \alpha$$

$$\text{Maximum speed of driven shaft} = \text{driver shaft speed} \times \sec \alpha$$

*Example.* If the driver shaft rotates at a constant speed of 250 rpm and the shaft angle $\alpha$ is 23°, then the speed difference is 41.5 rpm.

This mechanical characteristic of the universal joint would be detrimental to many mechanisms and machine drives. The solution to this problem is solved with the use of two univer-

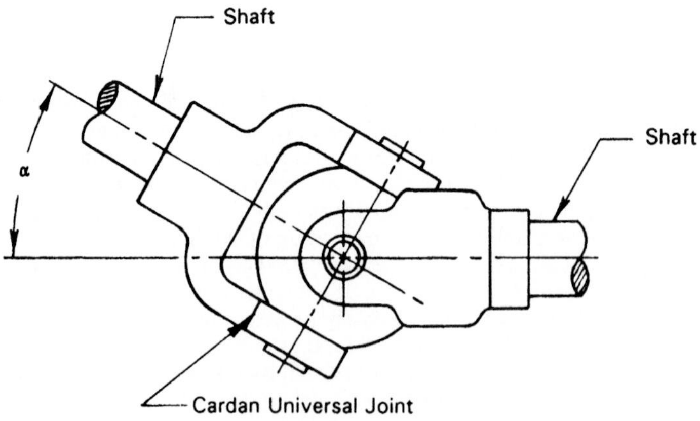


**FIGURE 8.29** Universal (Cardan) joint.

sal joints and an intermediate shaft. In this arrangement, two conditions must be met to achieve a constant speed ratio between driver and driven shafts:

- Driver and driven shafts must be parallel
- The forks on the intermediate shaft must be arranged in the same plane as shown in Fig. 8.30. Figure 8.30*a* shows the correct connection of two typical universal joint systems, and Fig. 8.30*b* shows the *incorrect* connection.

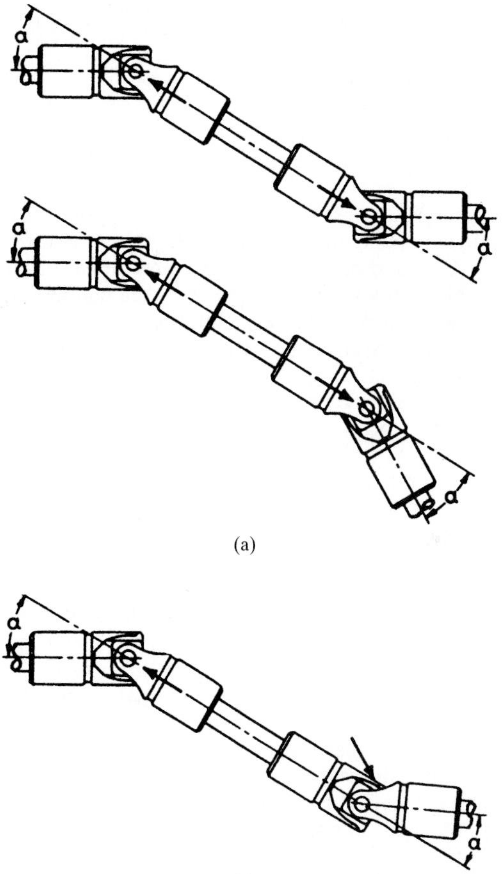

(a)

(b)

**FIGURE 8.30** Universal joint connections: (*a*) correct; (*b*) incorrect.

***Selection of a Cardan Universal Joint.*** The universal shaft torque can be calculated from

$$\text{Torque load, lb} \cdot \text{in} = \frac{63{,}025 \times \text{hp}}{\text{rpm}}$$

The equivalent *static load* = torque load × speed angle factor. The universal joint is rated for static torque, lb · in.

After calculating the torque load of your application, check the universal-joint factor from the manufacturers' catalog. Calculate the equivalent static load and select a universal joint rated for this calculated load.

### 8.4.2 Sleeve Couplings

Figure 8.31 shows a typical clamping-sleeve coupling which may be used to connect shafts of the same or different sizes, according to the table shown in the figure. Larger-size sleeve couplings are available from different manufacturers.

### 8.4.3 Flanged and Flexible Couplings

Flange-type couplings are chosen for heavy machinery and high-torque loads. American shaft couplings are standardized to 10-in-diameter drive shafts, with flange diameters reaching 18 in.

Flexible couplings are generally used to connect electrical machinery. The flexible insert coupling is popular for light- to medium-duty machinery and allows for shaft misalignment. The flexible insert material may be rubber or various tough plastics such as urethane or acetal.

Flexible couplings are applied using the manufacturers' installation instructions.

Types of flexible and flange couplings are shown in Fig. 8.32*a*–8.32*g*, while Fig. 8.33 shows the Falk-type flexible ribbon coupling. Falk-type coupling is used in moderate to heavy load applications.

The basic factors a designer must evaluate when specifying a coupling are allowable shaft misalignment, torque rating, lubrication and maintenance, conditions of service, service life, and cost.

## 8.5 CLUTCHES

The two types of clutch commonly used throughout industry are the toothed or positive clutch and the friction clutch. Positive clutches are connected by interlocking teeth of various forms, while friction clutches transmit power through frictional surfaces in the clutch. The positive tooth clutch is normally spring-loaded, wherein the teeth disengage after a predetermined amount of torque is encountered. The friction clutch styles include disk, conical, and centrifugal shoe. Another form of friction clutch includes the electromagnetic plate clutch, which has vast applications throughout industry.

Figure 8.34*a*–8.34*d* shows some of the basic clutch types such as one-way locking clutch, positive toothed clutch, ball-detent clutch, and centrifugal shoe clutch.

Figure 8.35*a*–8.35*c* shows a heavy-duty positive tooth clutch, a cam clutch, and a typical friction-cone clutch. Figure 8.36 indicates the geometry of the friction-cone clutch which is described in the text.

Figure 8.37 illustrates a heavy-duty friction-disk clutch used to drive a pulley for flat belt power transmission applications. All the aforementioned clutches are mechanical types. The application of friction-disk clutches is detailed in the following text.

### 8.5.1 Power Transmitted by Disk Clutches

Referring to Fig. 8.37, we approximate

$$\text{hp} = \frac{\mu n R_m F S}{63{,}000}$$

| DIMENSIONS | | | |
|---|---|---|---|
| BORE 'A' +002 -000 | BORE 'B' +002 -000 | O.D. | LENGTH |
| .250<br>.250 | .1875<br>.250 | 5/8" | 15/16" |
| .375<br>.375■ | .250<br>.375■ | 7/8 | 1-3/8 |
| .500<br>.500■ | .375■<br>.500■ | 1-1/8 | 1-3/4 |
| .625<br>.625<br>.625■ | .375■<br>.500■<br>.625■ | 1-5/16 | 2 |
| .750<br>.750<br>.750■ | .500■<br>.625■<br>.750■ | 1-1/2 | 2-1/4 |
| .875<br>.875<br>.875■ | .625■<br>.750■<br>.875■ | 1-5/8 | 2-1/2 |
| 1.000<br>1.000<br>1.000<br>1.000■ | .500■<br>.750■<br>.875■<br>1.000■ | 1-3/4 | 3 |
| 1.125<br>1.125<br>1.125■ | .875■<br>1.000■<br>1.125■ | 1-7/8 | 3-1/8 |
| 1.250<br>1.250<br>1.250<br>1.250■ | .750■<br>1.000■<br>1.125■<br>1.250■ | 2-1/16 | 3-1/4 |

**STANDARD KEY WAYS**

| BORE | K'WAY |
|---|---|
| 3/8" | 3/32" |
| 1/2 | 1/8 |
| 5/8 | 3/16 |
| 3/4 | 3/16 |
| 7/8 | 3/16 |
| 1 | 1/4 |
| 1-1/8 | 1/4 |
| 1-1/4 | 1/4 |

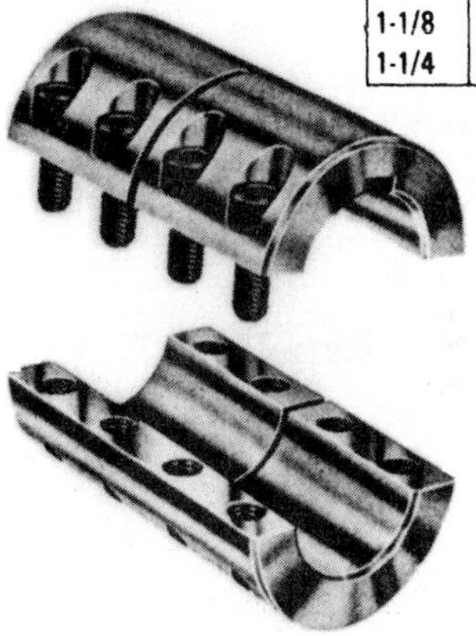

**■ THESE BORES HAVE KEY WAYS**

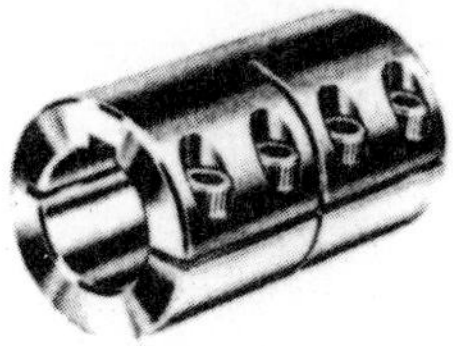

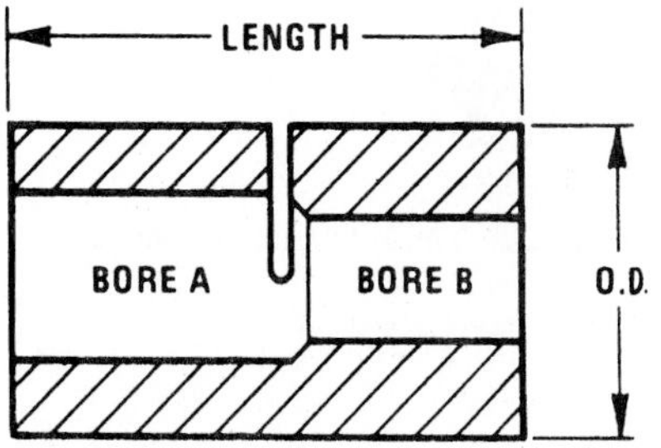


**FIGURE 8.31** Typical clamping-sleeve coupling. (*Source: Ruland Company.*)

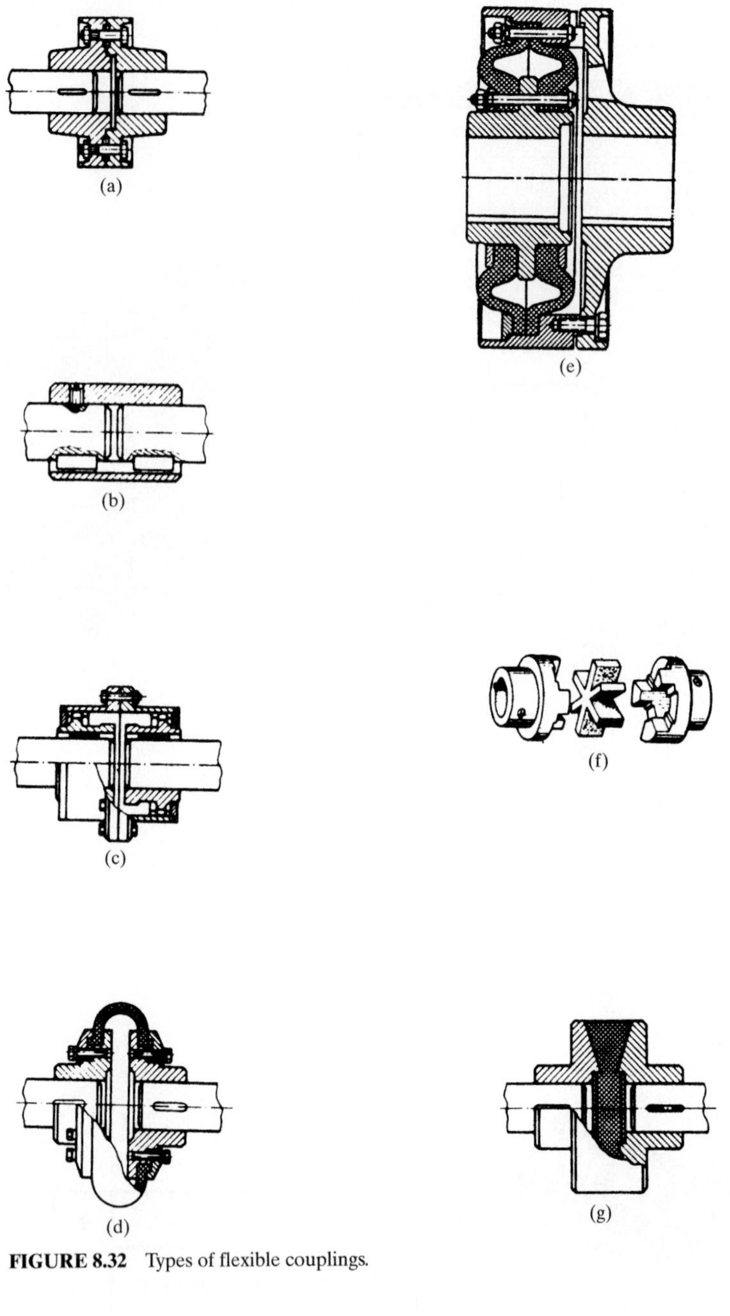


**FIGURE 8.32** Types of flexible couplings.

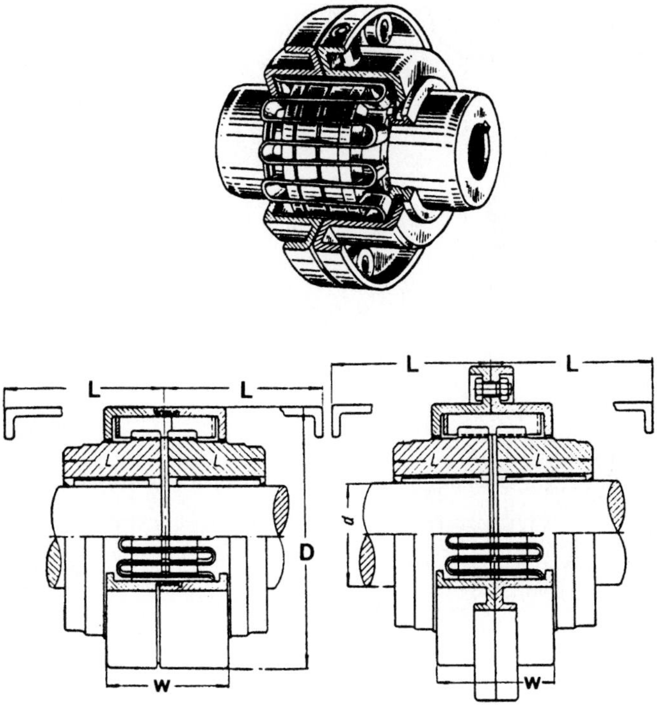


**FIGURE 8.33** Falk-type flexible coupling.

where $\mu$ = coefficient of friction (static)
$R_m$ = mean radius of engaged frictional surfaces, in
$F$ = axial force holding the disks in contact, lb
$n$ = number of frictional surfaces
$S$ = speed of drive shaft, rpm

Values for the coefficient of static friction ($\mu$) are as follows:

| | |
|---|---|
| Cork on metal, oiled | 0.32 |
| Cork on metal, dry | 0.35 |
| Metal on metal, dry | 0.15 |
| Disks, metal, oiled | 0.10 |
| Asbestos to steel, dry | 0.45 |
| Phenolic fiber to cast iron, dry | 0.25 |
| Cast iron on brass, dry | 0.21 |

***Slipping Torque.*** Single-plate clutch-slipping torque is given as

$$T\text{, lb} \cdot \text{in} = PR\mu n$$

$$T\text{, N} \cdot \text{m} = \frac{R\mu n}{1000}$$

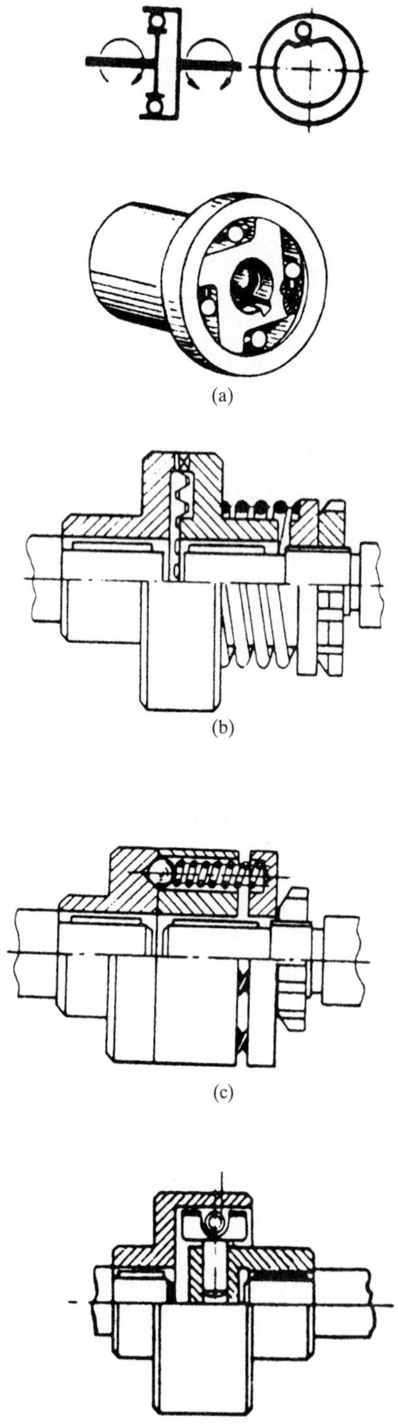


**FIGURE 8.34** Types of clutches: (*a*) one-way locking clutch; (*b*) spring-detent clutch; (*c*) ball-detent clutch; (*d*) centrifugal friction clutch.

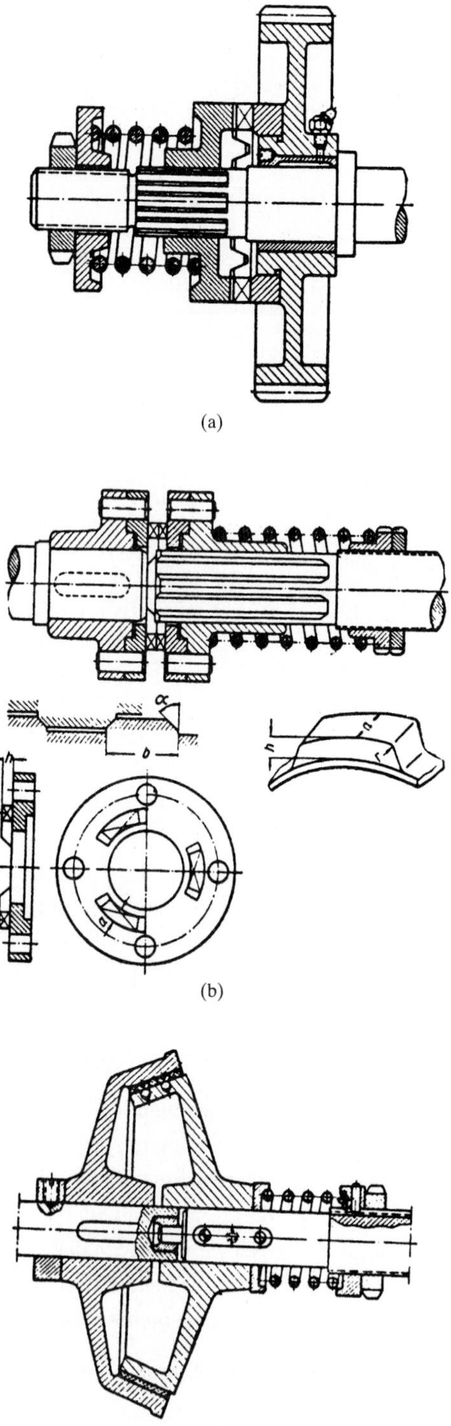


**FIGURE 8.35** Heavy-duty clutches: (*a*) positive tooth clutch; (*b*) cam-detent clutch; (*c*) friction-cone clutch.

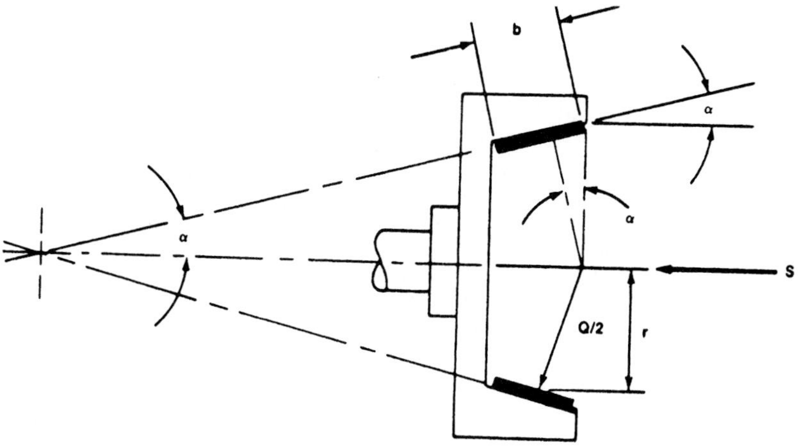


**FIGURE 8.36** Geometry of the friction-cone clutch.

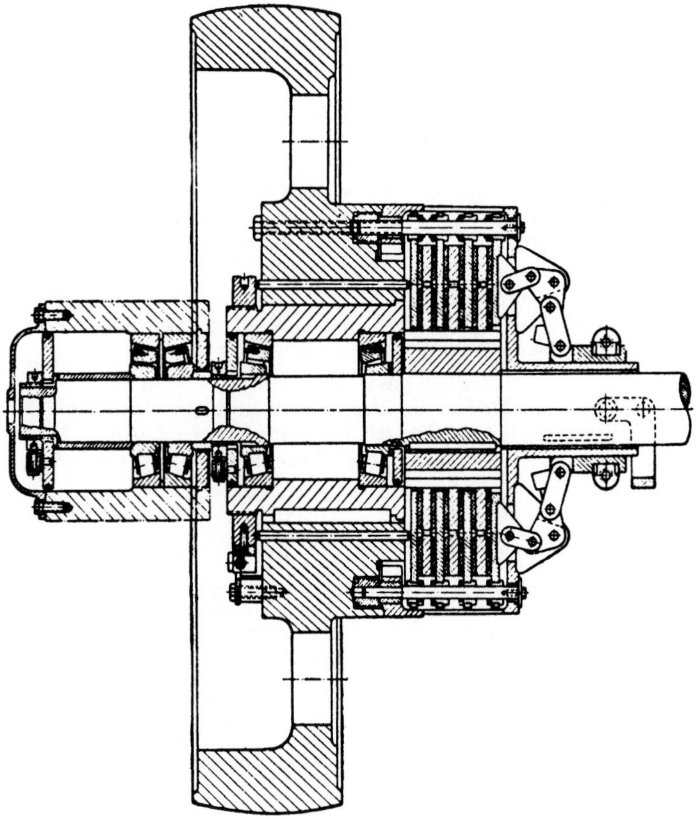

**FIGURE 8.37** Heavy-duty friction-disk clutch.

and spring pressure or actuating force as

$$P, \text{lb} = \frac{T}{R\mu n}$$

$$P, \text{N} = \frac{1000T}{R\mu n}$$

where $T$ = torque capacity, lb · in or N · m
$\mu$ = coefficient of static friction
$P$ = spring pressure or actuating force, lb or N
$R$ = effective friction radius, in or mm
$n$ = number of frictional surfaces

***Proportions of Disk Linings.*** The average radius or effective frictional radius is

$$R = \frac{d_o + d_i}{4}$$

where $R$ = effective frictional radius, in or mm
$d_o$ = outside diameter of disk lining, in or mm
$d_i$ = inside diameter of disk lining, in or mm

(*Note:* For any selected outside diameter, the torque capacity is maximized when ratio $d_o/d_i = 1.73$.)

***Slipping-Torque Capacities (Plate Clutches).*** The slipping-torque capacity of a friction plate clutch is the amount of torque the clutch is capable of transmitting when the clutch is just in the process of slipping.

In order to prevent slipping in service, a safety factor or multiplier must be applied to the slipping-torque capacity of the clutch. This factor or multiplier should be kept from 1.40 to 1.60 above the maximum torque of the driving element or motor supplying torque through the clutch.

*Example.* If you wish to transmit 2500 lb · in of torque through the clutch, the clutch should be designed so that its slipping-torque capacity is ≈1.50 × 2500 = 3750 lb · in. Use the preceding equations shown to do the simple calculations.

***Spring Pressure or Actuating Force.*** The clutch spring pressure or actuating force should be applied so that the pressure on the clutch disk material is kept at 20 to 35 psi when the clutch is engaged (automotive-type clutches). For allowable pressures of other clutch materials, see table in Fig. 8.38.

| Clutch materials | Allowable Pressure | |
|---|---|---|
| | psi | MPa |
| Steel to steel | 99 | 0.68 |
| Steel to cork | *7.25 to 14.5 | *0.05 to 0.10 |
| Phenolic to steel or cast iron | 99 | 0.68 |
| Carbon-graphite to steel | 297 | 2.05 |

*Low figure, wet; high figure, dry.

**FIGURE 8.38** Allowable pressures on clutch materials.

### 8.5.2 Cone Clutches

Refer to Figs. 8.35*c* and 8.36.

When the horsepower to be transmitted and the rpm of the shaft are given,

$$M_t = \frac{63{,}025\ \text{hp}}{n} \qquad P = \frac{M_t}{r} \qquad S = \frac{\sin\alpha + \mu\cos\alpha}{\mu}$$

$$\rho = \frac{Q}{b2\pi r} \quad \text{and} \quad b = \frac{Q}{\rho 2\pi r}$$

where $M_t$ = torque of shaft, lb · in
hp = horsepower
$n$ = rpm of shaft
$S$ = spring pressure or axial force, lb
$r$ = mean radius of cone, in
$\mu$ = coefficient of static friction
$Q$ = total force normal to conical surface, lb
$b$ = width of cone section, in
$P$ = tangential force at rim of cone, lb
$\alpha$ = half cone angle, degrees
$\rho$ = pressure normal to cone surface, psi

Values of $\rho$ are taken to be 7 to 10 psi for leather-equivalent to steel, 3 to 5 psi for cork to steel, and 50 to 60 psi for steel to steel. For other types of clutches, see Figs. 8.34 and 8.35.

(*Note:* 1 Pa = 0.000145034 lb/in$^2$ and 0.68 MPa = $(0.68 \times 10^6) \times 0.000145034$ = 98.6 psi.)

### 8.5.3 Electromagnetic Clutches

The electromagnetic clutch has a vast number of applications in industry. These types of clutches are manufactured in many different sizes and torque ratings. Figure 8.39 shows an exploded view of a small-size electromagnetic clutch. These clutches also are available in different operating voltage ranges. Although the small clutch shown in Fig. 8.39 is rated for 280 lb · in of torque at 90 V dc, it may be used to disconnect a large load by placing the clutch between the drive motor and the output gearbox, when the gearbox has a high reduction ratio. If the gearbox has a reduction ratio of 8:1, the clutch will drive a load whose torque value is $8 \times 280 = 2240$ lb · in or 186.7 lb · ft.

The electromagnetic clutch catalogs show the sizes, ratings, and dimensions of the various clutches available. These catalogs also contain magnetic brakes which are similar to electromagnetic clutches, except that the design calculations required to select the proper size for a particular application are different. In the design of magnetic brakes, heat energy and dissipation factors are taken into consideration.

## *8.6 POWER SCREWS*

Power screws with square threads, acme threads, and V threads have many uses in a large number of mechanical applications. Power screws used to lift a load vertically are termed *jackscrews.* Figure 8.40*a* and 8.40*b* shows the basic types of thread forms used in power screws. In this Figure, $l$ = lead of the thread; $C$ = half angle between V-thread faces, degrees; and $r$ = mean radius of the thread.

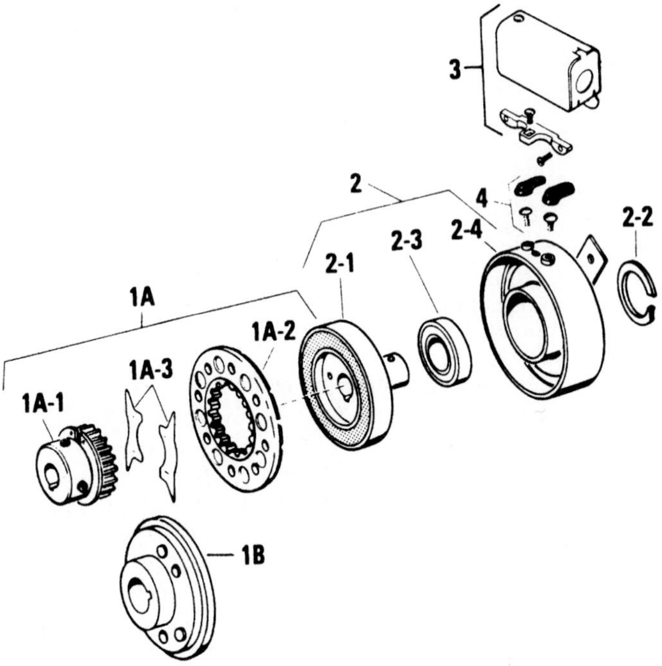


ELECTROMAGNETIC CLUTCH PARTS

| | |
|---|---|
| 1A | Armature and hub |
| 1A-1 | Armature hub |
| 1A-2 | Armature |
| 1A-3 | Release spring |
| 1B | Antibacklash armature |
| 2 | Field and rotor assembly |
| 2-1 | Rotor |
| 2-2 | Retainer ring |
| 2-3 | Ball bearing |
| 2-4 | Field |
| 3 | Conduit box |
| 4 | Terminal screws and covers |

**FIGURE 8.39** Electromagnetic clutch assembly—exploded view.

Turning force (torque) and resulting loads on power or jackscrews may be calculated from the following equations. For motion in the direction of the load *L*, assisted by the load (i.e., lowering a raised jack or unclamping a load)

$$F = L\,\frac{2\pi\mu r - l}{2\pi r + \mu l}\left(\frac{r}{R}\right)$$

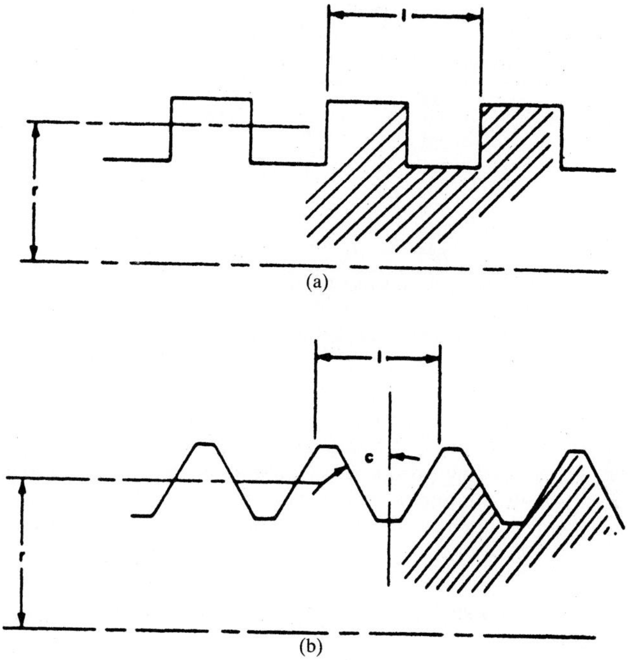


**FIGURE 8.40** Power-screw thread forms: (*a*) square; (*b*) acme or V thread.

For motion opposite the load $L$ (i.e., raising a jack or clamping a load)

$$F = L\,\frac{2\pi\mu r + l}{2\pi r - \mu l}\left(\frac{r}{R}\right)$$

where $F$ = force at end of handle or lever arm
$R$ = lever arm of $F$, in
$r$ = pitch radius of screw (see screw thread sections of the handbook), in
$l$ = lead of thread = $1/n$ when $n$ is the number of threads per inch, in
$L$ = load, lb
$\mu$ = coefficient of friction (dynamic)

Figure 8.41 lists coefficients of dynamic friction at various pressures for various material combinations.

Note that the values for the coefficients of dynamic or static friction are general in nature due to the following:

- Surface condition (rms value)
- Temperature
- Pressure between faying surfaces
- Type of lubricant
- Surface hardness
- Metallic plating (if present)

| Pressure Lb/In$^2$ | Cast iron / Wrought iron | Steel / Cast iron | Brass / Cast iron |
|---|---|---|---|
| 125 | 0.17 | 0.17 | 0.16 |
| 225 | 0.29 | 0.33 | 0.22 |
| 300 | 0.33 | 0.34 | 0.21 |
| 400 | 0.36 | 0.35 | 0.21 |
| 500 | 0.37 | 0.36 | 0.22 |
| 700 | 0.43 | Seized | 0.23 |
| 785 | Seized | — | 0.23 |
| 825 | — | — | 0.27 |

Surfaces lightly lubricated, Rennie.

**FIGURE 8.41** Coefficients of dynamic friction.

Various tests may be conducted by the designer to arrive at more meaningful values for these coefficients, when required on critical applications.

The preceding power screw equations are greatly simplified and will give approximate results with a minimum amount of calculation. The following equations are more precise, as they were written for the specific type of thread form indicated, preceding the equations.

### 8.6.1 Power Screw Loads and Efficiencies for Square, V, and Acme Threads

For square threads

$$P = L \tan(b+a) = L\,\frac{l + 2\pi r f}{2\pi r - l f} \quad \text{(for motion opposed to load)}$$

$$\text{Efficiency } e = \frac{\tan b}{\tan(b+a)} \quad \text{(for motion opposed to load)}$$

$$P = L \tan(b-a) = L\,\frac{l - 2\pi r f}{2\pi r + l f} \quad \text{(for motion assisted by load)}$$

$$\text{Efficiency } e = \frac{\tan(b-a)}{\tan b} \quad \text{(for motion assisted by load)}$$

For V and acme threads

$$P = L\,\frac{l + 2\pi r f \sec C}{2\pi r - l f \sec C} \quad \text{(for motion opposed to load)}$$

$$\text{Efficiency } e = \frac{\tan b\,(1 - f \tan b \sec C)}{(\tan b + f \sec C)} \quad \text{(for motion opposed to load)}$$

$$P = L\,\frac{l - 2\pi r f \sec C}{2\pi r + l f \sec C} \quad \text{(for motion assisted by load)}$$

$$\text{Efficiency } e = \frac{(\tan b - f \sec C)}{\tan b\,(1 + f \tan b \sec C)} \quad \text{(for motion assisted by load)}$$

where $a$ = friction angle ($\tan a = f$), degrees
$f$ = coefficient of friction, dynamic

$l$ = lead or thread advance in one revolution, in
$b$ = lead angle ($\tan b = \frac{1}{2}\pi r$), degrees
$r$ = mean radius of thread = ½ (root radius + outside radius), in
$P$ = equivalent driving force at radius $r$ from screw axis, lb
$L$ = axial load, lb
$e$ = efficiency
$C$ = half angle between V and acme thread faces, degrees

The coefficients of dynamic friction $f$ have different values at different surface pressures (see Fig. 8.41)

(*Note:* The screw equations shown in the preceding section may be used to calculate the tightening torque required on bolts when the clamp-load objective is specified or determined by the designer.)

Figure 8.42 shows a typical application for a power screw. This power screw assembly is used on certain types of milling machines to control backlash when the table feed screw rotates.

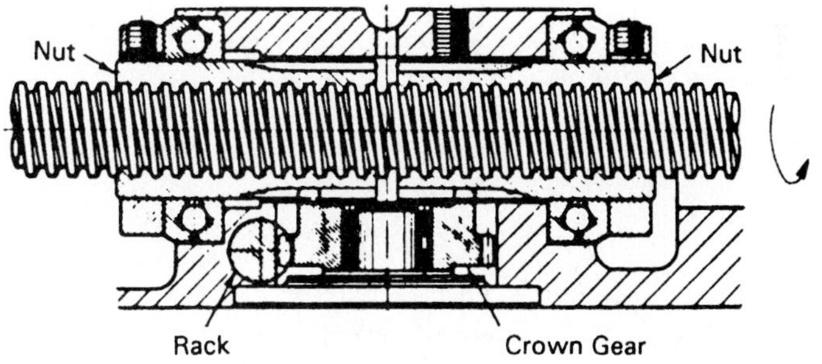


**FIGURE 8.42** Application of a power screw.

### 8.6.2 Load Capabilities of Power Screws

After the initial stages of power screw design shown in the preceding equations, the power screw system must be checked to see if it will carry the design load for the application. The following simple equations are used, together with the preceding equations, to finalize the design. Or, you may wish to use the following equations to help establish the basic thread size and diameter before using the previous equations. In any case, repeated calculations may be required before the final power screw system is completed.

Bending stress $S_b$ at the thread mean root radius is given as

$$S_b \approx \frac{3Wh}{2\pi n r_m b^2}, \text{ psi} \qquad \text{and} \qquad * \ (S_{bf} = S_b \times \text{f.s.})$$

Mean transverse shear stress is given as

$$S_s = \frac{W}{2\pi n r_m b}, \text{ psi} \qquad \text{and} \qquad * \ (S_{sf} = S_s \times \text{f.s.})$$

where $n$ = number of thread turns subject to load (thread engagement)
$b$ = width of thread section at root, in
$r_m$ = mean thread radius, in
$W$ = load parallel to screw axis, lb
$h$ = height of thread, in
f.s. = factor of safety (1.25 to 1.35)

(*Note:* In the preceding equations, multiply $S_b$ and $S_s$ by the f.s. range of 1.25 to 1.35 to arrive at final stress levels, $S_{bf}$ and $S_{sf}$.)

***Surface Bearing Pressure.*** The pressure $P$ between surfaces of the screw and nut threads is a critical design factor in power screws and is given approximately by

$$P = \frac{W}{2\pi n r_m h}, \text{ psi} \qquad \text{(see Fig. 8.43)}$$

Since this simple equation gives an estimate of a pressure which is lower than the actual real value, a factor of safety must be applied such as

$$P_f = P \times \text{f.s.}$$

where $P_f$ = final pressure between surfaces, psi
$P$ = approximate pressure from the equation shown
f.s. = 1.25 to 1.35 (recommended)

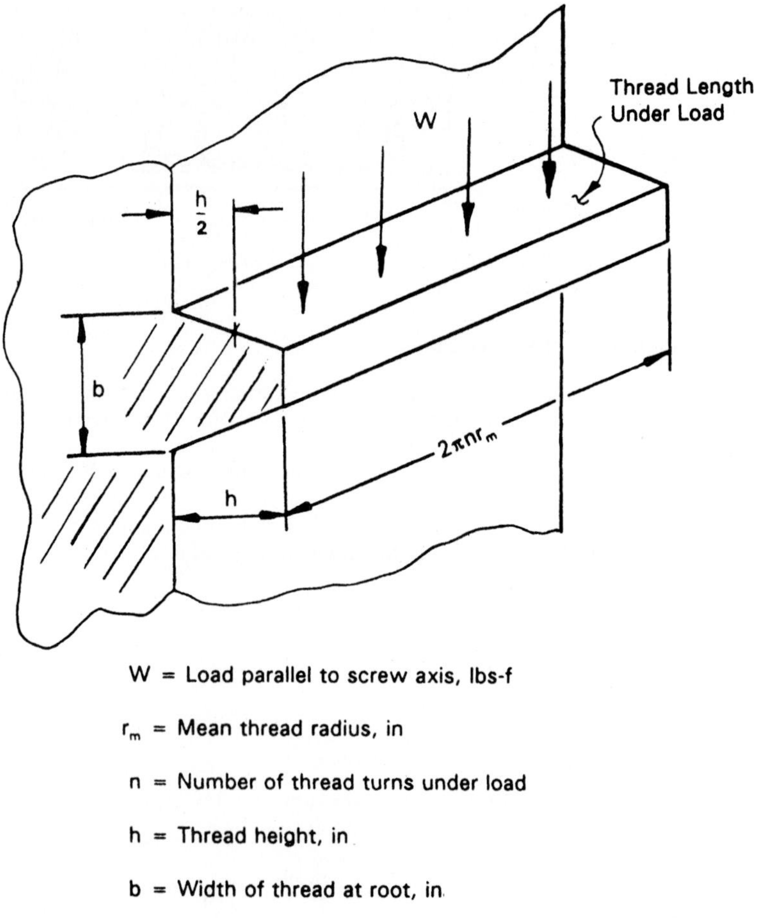


**FIGURE 8.43** Power screw thread geometry.

The allowable stresses (tensile and shear) for various materials are shown in Chap. 4. As a rule-of-thumb practice, the shear stress of many metals and alloys is taken as 45 to 55 percent of the value of the maximum allowable tensile strengths. This rule is appropriate for most ferrous metals and alloys (irons and steels).

## 8.7 RATCHETS AND RATCHET GEARING

A ratchet is a form of gear in which the teeth are cut for one-way operation or to transmit intermittent motion. The ratchet wheel is widely used in machinery and many mechanisms. Ratchet-wheel teeth can be either on the perimeter of a disk or on the inner edge of a ring.

The *pawl,* which engages the ratchet teeth, is a beam member pivoted at one end; the other end is shaped to fit the ratchet-tooth flank.

### 8.7.1 Ratchet Gear Design

In the design of ratchet gearing, the shape of the teeth must be designed so that the pawl will remain engaged under ratchet-wheel loading. In ratchet gear systems, the pawl will either push the ratchet wheel or the ratchet wheel will push on the pawl and/or the pawl will pull the ratchet wheel or the ratchet wheel will pull on the pawl.

See Fig. 8.44*a* (case 1) and 8.44*b* (case 2) for the four variations of ratchet and pawl action.

In Fig. 8.44*a* and 8.44*b*, *F* indicates the origin and direction of the force and *R* indicates the reaction direction. See Fig. 8.45 for tooth geometry for case 1 in Fig. 8.44*a*.

A line perpendicular to the face of the ratchet-wheel tooth must pass between the center of the ratchet wheel and the center of the pawl pivot point. See Fig. 8.46 for tooth geometry for case 2 in Fig. 8.44*b*.

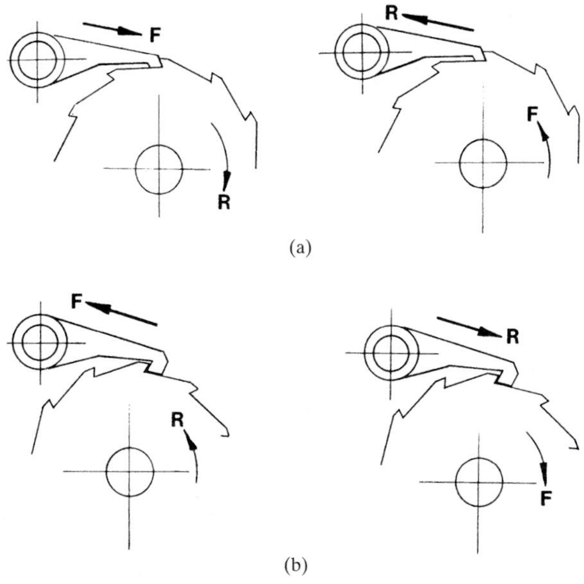


**FIGURE 8.44** Variation of ratchet and pawl action: (*a*) pushing; (*b*) pulling ($F$ = force; $R$ = reaction).

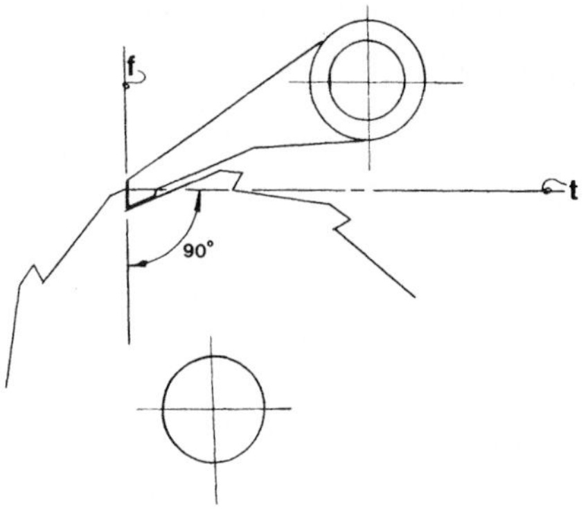


FIGURE 8.45 Tooth geometry for case 1 pawl action.

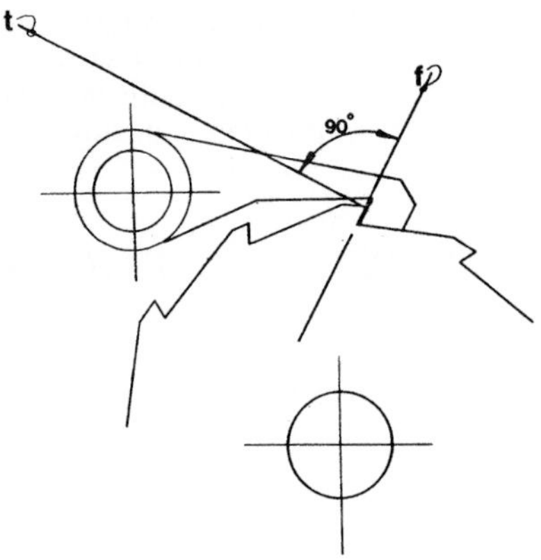


FIGURE 8.46 Tooth geometry for case 2 pawl action.

A line perpendicular from the face of the ratchet wheel tooth must fall outside the pivot center of the pawl and the ratchet wheel.

Spring loading the pawl is usually employed to maintain constant contact between the ratchet wheel and pawl. (Gravity or weight on the pawl is also sometimes used.) The pawl should be automatically pulled in and kept engaged with the ratchet wheel, independent of the spring or weight loading imposed on the pawl.

The swinging arc described by the point of the pawl must be designed to avoid excessive clearance between the point of the pawl and the root of the ratchet tooth when the pawl swings inward into the teeth of the ratchet for engagement. If this clearance is excessive, motion will be lost in driving the ratchet due to the backlash which will be present because of this excessive clearance between the point of the pawl and the root of the ratchet tooth. This is an important design consideration on reciprocating pawl actions, where the pawl movement may be limited because of the available reciprocating linear motion or stroke of the drive. The linear stroke or arc travel of the pawl point should be enough so that 1.25 to 1.50 teeth of the ratchet are indexed during each reciprocating stroke of the drive. This will ensure positive ratchet operation. Also, the holding pawl for the ratchet drive must be coordinated with the driving pawl stroke; in other words, its physical location in respect to the driving pawl must be correct.

## 8.7.2 Methods for Laying out Ratchet Gear Systems

***Type A: External Teeth Ratchet Wheels.*** Refer to Fig. 8.47.

1. Determine the pitch, tooth size, and radius $R$ to meet the strength and mechanical requirements of the ratchet gear system (see Sec. 8.7.3).
2. Select the position points $O$, $O_1$ and $A$ so that they all fall on a circle $C$ with angle $O\text{-}A\text{-}O_1$ equal to 90°.
3. Determine angle $\phi$ through the relationship $\tan \phi = r/c$ = a value greater than the coefficient of static friction of the ratchet wheel and pawl material; 0.25 is sufficient for standard low– to medium-carbon steel. Or, $r/R = 0.25$, since the sine and tangent of angle $\phi$ are close for angles from 0 to 30°. *Note:* The value $c$ is determined by the required ratchet wheel geometry, therefore, you must solve for $r$, so

$$r = c \tan \phi = c\ (0.25) \qquad \text{or} \qquad r = R \tan \phi = R\ (0.25)$$

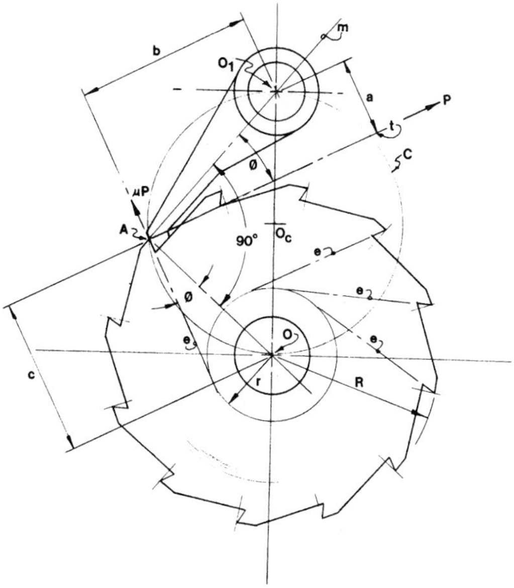


**FIGURE 8.47** Ratchet-wheel geometry, external teeth.

4. Angle ϕ is also equal to arctan ($a/b$), and to keep the pawl as small as practical, the center pivot point of the pawl $O_1$ may be moved along line $t$ toward point $A$ to satisfy space requirements.
5. The pawl is then self-engaging. This follows the principle stated earlier that a line perpendicular to the tooth face must fall *between* the centers of the ratchet wheel and pawl pivot points.

***Type B—Internal Teeth Ratchet Wheels.*** Refer to Fig. 8.48

1. Determine the pitch, tooth size, and radii $R$ and $R_1$ to meet the strength and mechanical requirements of the ratchet gear system. For simplicity, let points $O$ and $O_1$ be on the same centerline.
2. Select $r$, so that $f/g \geq 0.20$.
3. A convenient angle for β is 30° and $\tan \beta = f/g = 0.557$ which is greater than the coefficient of static friction for steel (0.15). This makes angle $\alpha = 60°$ because $\measuredangle\alpha + \measuredangle\beta = 90°$.

(*Note:* Locations of tooth faces are generated by element lines *e*. For self-engagement of the pawl, note that a line *t* perpendicular to the tooth face must fall outside the pawl pivot point $O_1$.)

### 8.7.3 Calculating the Pitch and Face Width of Ratchet-Wheel Teeth

The following equation may be used in calculating the pitch or the length of the tooth face (thickness of ratchet wheel) and is applicable to most general ratchet-wheel designs. Note that the selection of the values for $S_s$ (safe stress, psi) may be made more or less conservative, according to the requirements of the application. Low values for $S_s$ are selected for applications involving safety conditions. Note also that the shock stress allowable levels (psi) are 10 times less than for normal loading applications, where the safety factor is not a consideration.

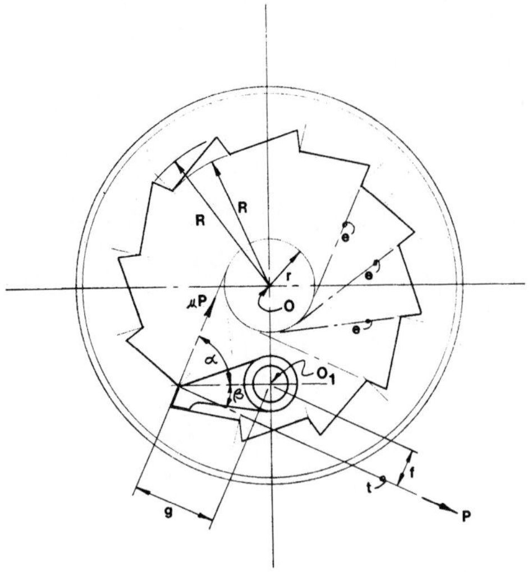


**FIGURE 8.48** Ratchet-wheel geometry, internal teeth.

The general pitch design equation and transpositions are given as

$$P = \sqrt{\frac{\alpha m}{lS_sN}} \qquad P^2 = \frac{\alpha m}{lS_sN} \qquad ; N = \frac{\alpha m}{lS_sP^2}; l = \frac{\alpha m}{NS_sP^2}$$

where $P$ = circular pitch, in, measured at the outside circumference.
$m$ = turning moment (torque) at ratchet-wheel shaft, lb·in
$l$ = length of tooth face; thickness of ratchet wheel, in
$S_s$ = safe stress (steel C1018; 4000 psi shock and 25,000 psi static)
$N$ = number of teeth in ratchet wheel
$\alpha$ = coefficient (50 for ≤12 teeth; 35 for 13 to 20 teeth; 20 for >20 teeth)

For other materials such as brass, bronze, stainless steel, and zinc castings, the $S_s$ rating may be proportioned to the values given for C1018 steel.

## 8.8 GEARING AND DESIGN PROCEDURES

Gears are used in most types of machinery and countless mechanisms throughout industry, including the automotive, machine tool, mechanical, and aerospace industries. Gears are a basic machine element and one of the common basic machines such as the lever, wheel and axle, inclined plane, or wedge. Gears have been in use by man for over 3000 years in one form or another. Figure 8.49*a* is a typical example of a common spur gear section with the terminology of the various features listed. More detailed illustrations, equations, and explanations of the different gear types will be given in the following subsections.

The methods used to manufacture gears depend on available machinery, design specifications or requirements, cost of production, and type of material from which the gear is to be made. Some of the methods used for producing different types of gears are

- *Low-load, low-speed gears:* die-stamped gears
- *Small, low-cost gears:* die-cast of zinc, aluminum, or brass

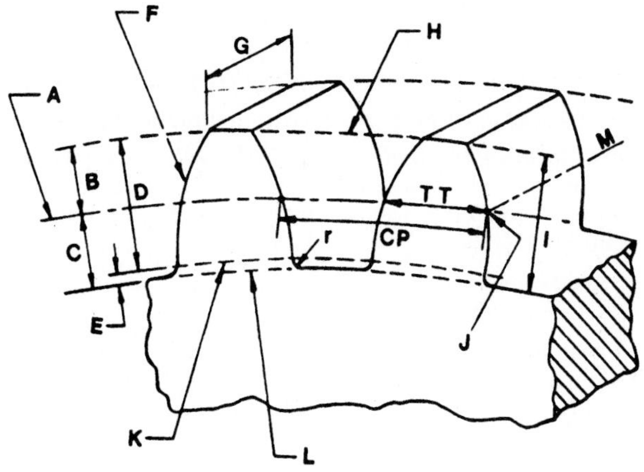


A Pitch Circle
B Addendum
C Dedendum
D Working Depth
E Clearance
F Tooth Profile (involute)
G Face Width
H Addendum Circle (O.D. gear)
I Whole Depth
J Pitch Point
K Working Depth Circle
L Root or Dedendum Circle
CP Circular Pitch
M Line of Contact
TT Tooth Thickness

(a)

| Parallel Axes | Intersecting Axes | Nonintersecting Nonparallel Axes |
|---|---|---|
| Spur, external<br>Spur, internal<br>Helical, external<br>Helical, internal | Straight bevel<br>Zerol bevel<br>Spiral bevel<br>Face gear | Crossed-helical<br>Single-enveloping worm<br>Double-enveloping worm<br>Hypoid<br>Spiroid |

(b)

**FIGURE 8.49** (*a*) Spur gear terminology; (*b*) common gear types.

- *Small, low-load, precision gears:* molded plastics such as acetal (Delrin), nylon (Zytel), and polycarbonate (Lexan)
- *Medium to small, moderate-load, quiet-operation gears:* phenolic fiber materials, machined
- *Small, accurate metallic gears:* extruded-metal shapes which are cut off to size
- *Small to medium, moderate- to high-load gears:* powder-metal technology for producing iron or mild steel and stainless-steel gears which are self-lubricating because of the porous nature of sintered powder-metal components, wherein the lubricant is infused into the gear material
- *Accurate and high-load-capacity gears:* for machine tool uses, shaving, and grinding the gear teeth are usually employed
- *Control gears:* usually made from medium-alloy, medium-carbon steels; the shaving and/or grinding methods are usually used to produce these accurate gears
- *Automotive gears:* usually cut from low-alloy steel forgings by hobbing, shaping, shaving, or grinding
- *Aerospace power gears:* usually made of high-alloy steel and fully hardened by case carburizing or nitriding
- *General industrial gears:* modern trends call for fully hardened steel or alloy steel gears made by hobbing, shaving, planing, milling, or grinding, according to the application.

Many methods are used to manufacture gears, including hobbing, shaping, grinding, shaving, lapping, honing, broaching, die-casting, cold forming, electric-discharge-machine (EDM) cutting, laser cutting, powder metallurgy, molding, die stamping, water-jet cutting, milling, forging, rolling, fly cutting, and extruding.

## 8.8.1 Gears in Common Use and Manufacturing Methods

The types of gears which are in common use are shown in Fig. 8.49*b*, which shows the shaft or axes configurations and the types of gears employed.

The method of manufacturing gears varies according to many parameters, including size of the gear, material, heat treatment, quantity to be produced, accuracy, tooth surface texture or finish, sound-producing characteristics, load-carrying capacity, speed of operation, and configuration. These parameters must be considered in selecting the most economical method or methods, consistent with the desired end results. Some of the methods used to manufacture gears are illustrated in Fig. 8.50*a*–8.50*h*.

A detailed description of each of the major gear production methods is described below.

***Hobbing.*** *Hobbing* is the generation of a toothed gear by the advancing of a rack-profiled tool in relation to the rotation of the gear blank. The curve of the gear tooth profile is developed by successive advancing positions of the rack teeth of the hobbing tool. The material of the gear is removed by a series of progressive rack-shaped cutting faces which are positioned around the tool as elements of a cylinder. The rack sections around the tool advance in a helical path so that the rotation of the tool about its axis provides the linear motion of the rack section with respect to the gear being cut (see Fig. 8.50*d*). The generation of the gear face width is accomplished by the continuous rotation of the tool while it is advanced across the face of the gear. Hobbing can be used as either a roughing or finishing operation and is suited for the manufacture of a large segment of the different types of gears produced such as spur, helical, worm, and wormgears.

***Shaping.*** *Shaping* of gear teeth is accomplished by using a circular form cutter with properly relieved teeth. The cutter rotates in timed relationship with the workpiece, while each stroke of the cutter produces a small portion of the tooth form or profile. Shaping can be used to pro-

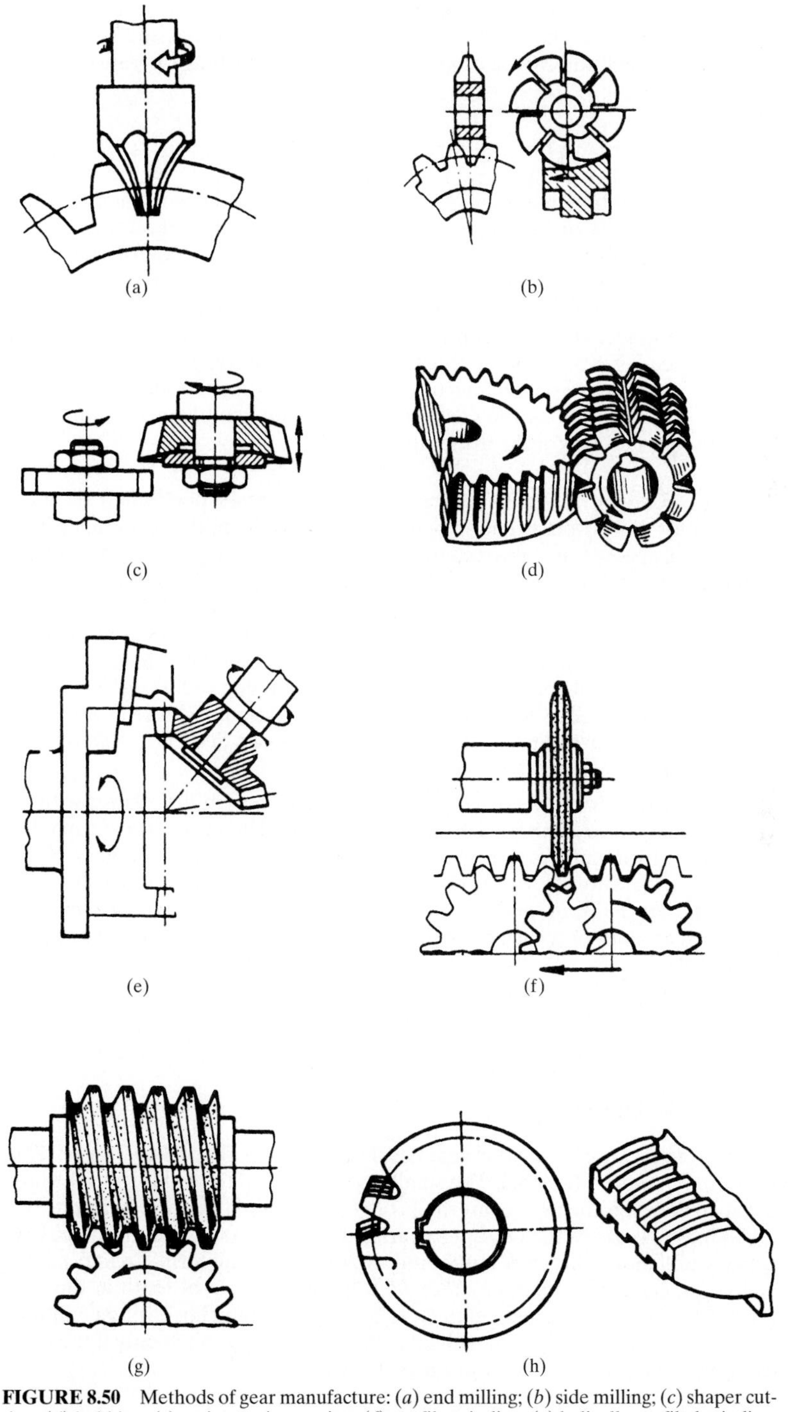


**FIGURE 8.50** Methods of gear manufacture: (*a*) end milling; (*b*) side milling; (*c*) shaper cutting; (*d*) hobbing; (*e*) reciprocating cutting; (*f*) profile grinding; (*g*) helically profiled grinding; (*h*) shaving.

duce many different types of gearing, including internal and external gears of either the spur or helical types, racks, noncircular, face gears, double helicals, and gear teeth which end close to a shoulder. Shaping can be used as a roughing or finishing operation. See Fig. 8.50*C* for the shaving motions.

***Grinding.*** There are three different methods of *grinding* gears. In the first method, a continuously formed grinding wheel resembling a large-diameter worm is fed into the workpiece similar to the hobbing method. As the wheel is fed into the workpiece, it generates both sides of the tooth profile or form (see Fig. 8.50*g*). The second method uses a grinding wheel with the proper tooth space form dressed on the perimeter of the grinding wheel. A gear tooth space is formed by the action of the grinding wheel turning on its own axis while traversing along the face of the gear in a reciprocating motion. The grinding wheel is progressively fed into the workpiece until full depth is obtained. In this process, the gear is indexed one tooth at a time (see Fig. 8.50*f*). In the third process, two disk-type grinding wheels are positioned so that their axes are perpendicular to the profiles of a rack tooth and the working faces of the grinding wheels serve as the rack profiles. Involute tooth profiles are generated by three simultaneous motions of the grinding wheels.

***Shaving.*** *Shaving* is a free-cutting machining process performed on a machine specially designed for this purpose. The shaving cutter is usually in the form of a precision-quality gear with several grooves or slots from tip to root in the normal plane of the teeth. Each tooth of the cutter thus has multiple cutting edges. The helix angle of the cutter teeth is usually different from the helix angle of the workpiece (gear blank). During the shaving process, the workpiece and cutter are rotated in tight mesh; the cutter is the driving member. Variations of this basic process are used by various designs of gear shaving machines. Shaving is most commonly used for finishing external spur and helical gears, but can be used for internal gears, when the configuration and size allow for proper shaving cutter design. Gears to be finished by shaving may be produced by any of the usual gear cutting processes. See Fig. 8.50*h* for shaving.

***Milling.*** *Milling* of gears is accomplished by a form of milling cutter with a profile matching that of a single space on a gear that is to be cut. The cutter travels axially across the gear blank, thus form-cutting the adjacent tooth flanks or sides. The gear blank is indexed one tooth at a time with a device such as a universal dividing head until all teeth in the gear have been cut. External spur, helical, and bevel gears may be produced using this method (see Fig. 8.50*b*). Gear teeth may also be formed by special end mills with the proper profile as shown in Fig. 8.50*a*.

Bevel gears may also be produced by the action shown in Fig. 8.50*e*, where a bevel gear is being cut by the action of reciprocating cutting tools.

### 8.8.2 Gear Action and Definitions

As an example, take two spur gears as a set, with a ratio of 4 to 3 (4:3). This means that the gear:pinion tooth ratio is 4:3. The smaller gear is called the *pinion* and the larger, the *gear*. The distance between the gear centerlines or shafts is called the *center distance*.

To mesh properly, the two spur gears must have the same tooth spacing. The *tooth spacing* is measured along the *pitch circle*, and the tooth spacing term *circular pitch* is defined as the circumference of the pitch circle divided by the number of teeth in the pinion or gear. The pitch circles of two meshed gears must be *tangent* (touching at one common point called the *pitch point*). The two-gear set will have the same circular pitch only if the ratio of the pitch circle radii is the same as the tooth ratio. As such

$$\frac{\text{Radius of gear pitch circle}}{\text{Radius of pinion pitch circle}} = \frac{\text{number of teeth in gear}}{\text{number of teeth in pinion}}$$

Dividing the pitch circle circumference of a gear by the number of teeth in the gear determines the circular pitch.

To keep the angular velocity of the driven gear to the angular velocity of the driving gear constant, a tooth form curve called the *involute* is normally used in gearing practice. The involute curve is well suited for designing and manufacturing gears because of the relative ease of generating and producing this curve on many machine tools, using various methods outlined in the previous section. Other tooth form curves will also produce uniform angular velocity between meshed gears, but the involute curve system is in common use for the reasons given above. When the gear tooth profiles are correct to produce uniform velocity between driver and driven gears, the tooth profiles are said to be *conjugate.* See Sec. 1.12, which details involute geometry and construction.

Any line perpendicular to the involute tooth profile is called a *line of pressure* because it is in the direction of the force when the mating drive gear tooth is in contact at that point. All the lines of pressure for each involute gear tooth are *tangent* to the base circle. When the involute gear and pinion rotate together in proper mesh, their two lines of pressure through the point of contact always lie along the same line. This line, called the *line of action,* is tangent to both base circles and contains the pitch point.

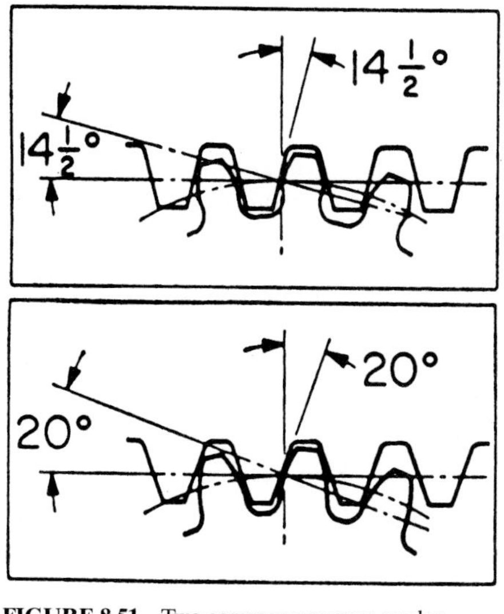


**FIGURE 8.51** Two common pressure angles.

The *pressure angle* is defined as the angle at a pitch point between the line perpendicular to the involute tooth surface and the line tangent to the pitch circle. The pressure angle of an involute gear tooth is determined by the size ratio between the base circle and the pitch circle (see Fig. 8.51). Two common pressure angles used by the gear industry are 14.5 and 20°. The pressure angle is equal to the profile angle between the tangent to the involute profile and the radial line through the pitch point. The cosine of the pressure angle is equal to the ratio of the base circle radius to the pitch circle radius. When the base circle is chosen to be approximately 94 percent of the size of the pitch circle, the pressure angle becomes approximately 20° (cos 20° = 0.9397). Many modern gears often use the 20° pressure angle.

Dynamic loads imposed on gear teeth in actual operation must be accounted for in gear design procedures. Usually, an allowance is made for the effective dynamic load and then specifying close controls on gear tooth accuracy. Inaccurate and poorly finished teeth and backlash account for peak dynamic loads on gear teeth during operation. Also, the gears should be operated above or below the resonance point, especially in spur gear systems. See Secs. 8.1 and 8.3 for calculation of the critical speeds of various shafts.

### 8.8.3 Pressure Angles and the Diametral Pitch System

*Pressure angles:* The two most common pressure angles used by many gear manufacturers are the 14.5 and 20° pressure angle. The 20° pressure angle gears are generally recognized as having a higher load-carrying capacity, but the 14.5° pressure angle gears are less sensitive to backlash because of center distance variations and are smoother and quieter in operation, provided the teeth are not undercut.

*Diametral pitch system:* The *diametral pitch* of a gear is the number of teeth in the gear for each inch of pitch diameter. Therefore, the diametral pitch determines the size of the gear teeth on each particular gear. Figure 8.52 shows the various tooth dimensions for different diametral pitches of spur gears. Figure 8.53 is a full-scale representation of the various involute teeth of the 14.5 and 20° diametral pitch systems. This figure may be used to check the diametral pitch of a gear you wish to compare or measure.

For convenience, tooth proportions of various standard diametral pitches of spur gears are given below.

| Diametral pitch | Circular pitch, in | Thickness of tooth on pitch line, in | Depth to be cut in gear, in (hobbed gears) | Addendum, in |
|---|---|---|---|---|
| 3 | 1.0472 | 0.5236 | 0.7190 | 0.3333 |
| 4 | 0.7854 | 0.3927 | 0.5393 | 0.2500 |
| 5 | 0.6283 | 0.3142 | 0.4314 | 0.2000 |
| 6 | 0.5236 | 0.2618 | 0.3565 | 0.1667 |
| 8 | 0.3927 | 0.1963 | 0.2696 | 0.1250 |
| 10 | 0.3142 | 0.1571 | 0.2157 | 0.1000 |
| 12 | 0.2618 | 0.1309 | 0.1798 | 0.0833 |
| 16 | 0.1963 | 0.0982 | 0.1348 | 0.0625 |
| 20 | 0.1571 | 0.0785 | 0.1120 | 0.0500 |
| 24 | 0.1309 | 0.0654 | 0.0937 | 0.0417 |
| 32 | 0.0982 | 0.0491 | 0.0708 | 0.0312 |
| 48 | 0.0654 | 0.0327 | 0.0478 | 0.0208 |
| 64 | 0.0491 | 0.0245 | 0.0364 | 0.0156 |

**FIGURE 8.52** Gear cutting table.

(*Note:* Spur and helical gears are normally made with involute teeth and are normally interchangeable, size for size. Involute-profile gears are generally interchangeable, while noninvolute gears are not.)

***Ratios for the Various Gear Systems or Arrangements.*** The ratios that are possible using the various gear systems or forms are shown in Fig. 8.54, but the power-transmitting capacity of different gear arrangements are variable to a marked extent. It is difficult to determine the upper power limit of gear systems because of many variables such as type of material, form of gear, size of equipment available to produce the proposed gear, life requirements, and lubrication limits.

### 8.8.4 Module Gear System (Metric Standard)

Since the SI or metric system is used throughout the world, American gear manufacturers have been cutting metric gears using the *module* system of measuring and producing gears for some years. *Module* is defined as the pitch diameter, in millimeters, divided by the number of teeth in the gear. The module is equal to the circular pitch in inches converted to millimeters multiplied by 0.3183. To find the outside diameter of a metric gear, add 2 to the number of teeth and multiply the sum by the module number (the diameter will be given in millimeters). Figure 8.55 shows the equivalents of diametral pitch, circular pitch, and module. Both the *diametral pitch* and the *module* are actually tooth size dimensions which cannot be directly measured on a gear. Diametral pitch and module are reference values which are used to calculate other dimensions on a gear, the values of which can be measured with measuring tools and instruments.

The following relationships are valid for all spur gears:

- Circular pitch = $\pi \times$ module (metric)
- Circular pitch = $\dfrac{\pi}{\text{diametral pitch}}$ (U.S. customary)

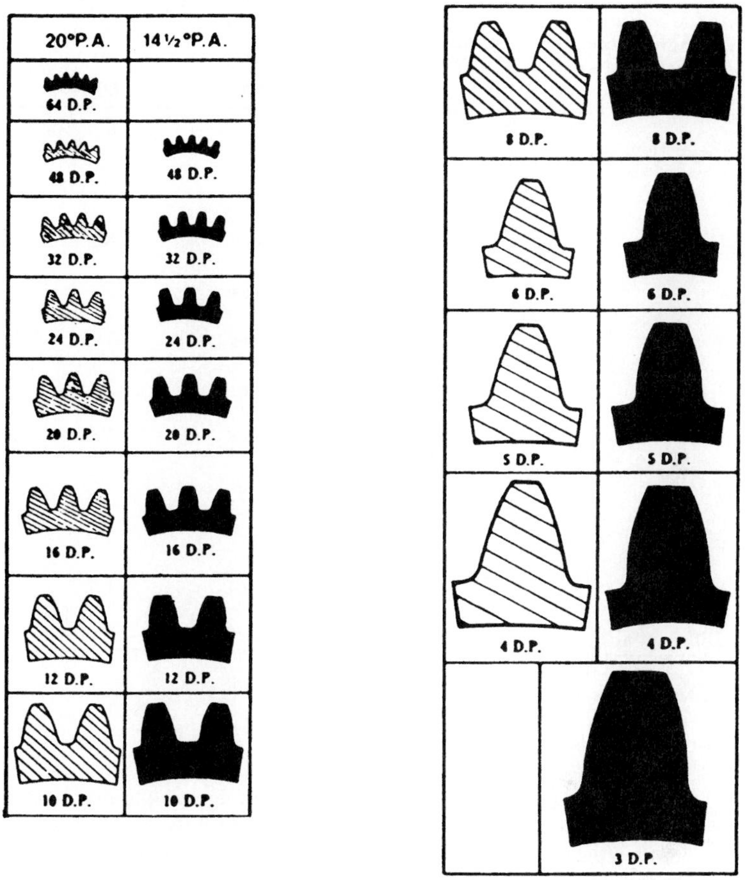


**FIGURE 8.53** Full-scale gear teeth, comparison gauge.

- Pitch diameter = number of teeth × module (metric)
- Pitch diameter = $\dfrac{\text{number of teeth}}{\text{diametral pitch}}$ (U.S. customary)

Mathematically:

$$\text{Module} = \frac{25.400}{\text{diametral pitch}}$$

$$\text{Diametral pitch} = \frac{25.400}{\text{module}}$$

## 8.8.5 Gear Types, Geometries, Measurements, and Calculations

The following subsections will detail all the commonly used gear forms or types, with their individual geometries, measurements, and basic calculations which may be used to select the gear type and system required for your particular application. These subsections on the vari-

| Arrangement Type | Min. No. of Toothed Parts | Ratio Range | | |
|---|---|---|---|---|
| | | 5:1 | 50:1 | 100:1 |
| Single Reduction: | | | | |
| Spur | 2 | Yes | No | No |
| Helical | 2 | Yes | No | No |
| Bevel | 2 | Yes | No | No |
| Hypoid | 2 | Yes | Yes | Yes |
| Face | 2 | Yes | No | No |
| Worm | 2 | Yes | Yes | Yes |
| Spiroid | 2 | No | Yes | Yes |
| Planoid | 2 | Yes | No | No |
| Simple Planetary | 3 | Yes | No | No |
| Fixed Differential | 5 | No | Yes | Yes |

**FIGURE 8.54** Gear ratio chart.

ous types of gears are intended for use by designers and machine shop personnel or machinists who wish to know the basic facts about the different gear types or systems. These subsections are not intended for use to produce or manufacture all the different types of gears representing the many types of specialized machinery, complex procedures, and tooling that is required in the production of complex gears and gear systems.

The design and production of accurate and complex gears is a complex science as well as an art. Different gear manufacturers use different machinery, tooling, and specialized procedures to produce their gears.

We will therefore present only those aspects of gear design, processing, and calculations which are of interest and general need to the designer, machinist, and other personnel within the metalworking industries who are not gear design engineers and specialists in gearing systems. Gear design engineering texts are listed in the bibliography for reference.

## 8.8.6 Spur Gears

Spur gears are the most common type of gears found in industry. They are manufactured in both external and internal forms; the external form is the most widely used. Simple and complex forms of planetary gearing use the internal and external forms of the spur gear in different combinations for a wide variety of planetary (epicyclic) gear systems. Helical gears are also used in epicyclic gear systems.

Refer to Fig. 8.56 for the complete terminology for external spur gears. The pitch diameter for the tooth cutting operation on a spur gear is given in the preceding section.

**Equivalent DP, CP, and module.**

| Diametral pitch | Circular pitch | Module |
|---|---|---|
| ¾ | 4.1888 | 33.8661 |
| 0.7854 | 4 | 32.3397 |
| 0.8467 | 3.7106 | 30 |
| 1 | 3.1415 | 25.3995 |
| 1.0160 | 3.0922 | 25 |
| 1.0472 | 3 | 24.2548 |
| 1¼ | 2.5133 | 20.3196 |
| 1.2700 | 2.4737 | 20 |
| 1.4111 | 2.2264 | 18 |
| 1½ | 2.0944 | 16.9330 |
| 1.5708 | 2 | 16.1698 |
| 1.5875 | 1.9790 | 16 |
| 1.6933 | 1.8553 | 15 |
| 1¾ | 1.7952 | 14.5140 |
| 1.8143 | 1.7316 | 14 |
| 1.9538 | 1.6079 | 13 |
| 2 | 1.5708 | 12.6998 |
| 2.0944 | 1½ | 12.1274 |
| 2.1166 | 1.4842 | 12 |
| 2¼ | 1.3963 | 11.2887 |
| 2.3090 | 1.3606 | 11 |
| 2½ | 1.2560 | 10.1598 |
| 2.5400 | 1.2369 | 10 |
| 2.8222 | 1.1132 | 9 |
| 3 | 1.0472 | 8.4665 |
| 3.1416 | 1 | 8.0849 |
| 3.1749 | 0.9895 | 8 |
| 3½ | 0.8976 | 7.2570 |
| 3.6285 | 0.8658 | 7 |
| 4 | 0.7854 | 6.3499 |
| 4.1888 | ¾ | 6.0637 |
| 4.2333 | 0.7421 | 6 |
| 5 | 0.6283 | 5.0799 |
| 5.0799 | 0.6184 | 5 |
| 6 | 0.5236 | 4.2333 |
| 6.2832 | ½ | 4.0425 |
| 6.3499 | 0.4947 | 4 |
| 8 | 0.3927 | 3.1749 |
| 8.4665 | 0.3711 | 3 |
| 10 | 0.3142 | 2.5400 |

**FIGURE 8.55** Gear measurement systems.

In a spur gear system, the operating pitch diameter is

$$\text{Pitch diameter (operating) of pinion} = \frac{2 \times \text{operating center distance}}{\text{ratio} + 1}$$

$$\text{Pitch diameter of gear} = \text{ratio} \times \text{pitch diameter (operating) of pinion}$$

The ratio is

$$\text{Ratio} = \frac{\text{number of teeth in gear}}{\text{number of teeth in pinion}}$$

***Spur Gear Formulas for Full-Depth Involute Teeth.*** Refer to Fig. 8.57.

All the basic dimensions for standard spur gears may be calculated and then specified by using the equations shown in Fig. 8.57.

Of interest to the designer who specifies the gear dimensions are the basic equations used to determine the tooth loads on the gears, both static and dynamic. Gear failure can result from tooth breakage or surface failure in operation. An important equation for calculating the tooth loads on spur gears is the Barth revision to the Lewis equation, which considers beam strength but not wear. The minimum load for wear on gear teeth is shown in a later subsection, where the *K* factor for determining the limiting load may be calculated when the physical properties of the gear material are accurately known.

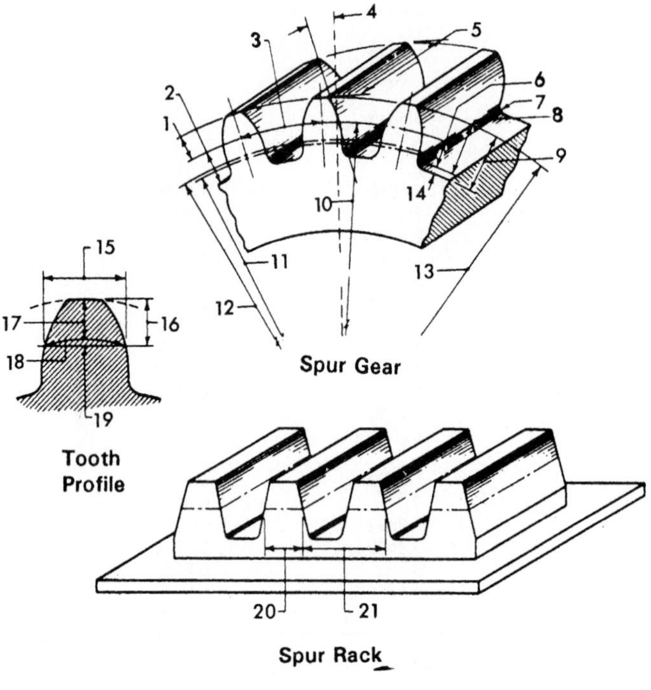


**Spur Gear Terminology**

1 Addendum
2 Dedendum
3 Circular pitch
4 Pressure angle (PA)
5 Face width
6 Addendum of mating gear
7 Tooth fillet
8 Working depth
9 Whole depth
10 Pitch diameter
11 Base circle diameter
12 Form diameter
13 Outside diameter
14 Clearance
15 Chordal thickness
16 Chordal addendum
17 Addendum
18 Arc thickness
19 Rise of arc
20 Tooth thickness
21 Circular pitch

**FIGURE 8.56** Spur gear terminology, external.

## SPUR GEAR FORMULAS

FOR FULL DEPTH INVOLUTE TEETH

| To Obtain | Having | Formula |
|---|---|---|
| Diametral Pitch (P) | Circular Pitch (p) | $P = \frac{3.1416}{p}$ |
| | Number of Teeth (N) & Pitch Diameter (D) | $P = \frac{N}{D}$ |
| | Number of Teeth (N) & Outside Diameter ($D_o$) | $P = \frac{N + 2}{D_o}$ (Approximate) |
| Circular Pitch (p) | Diametral Pitch (P) | $p = \frac{3.1416}{P}$ |
| Pitch Diameter (D) | Number of Teeth (N) & Diametral Pitch (P) | $D = \frac{N}{P}$ |
| | Outside Diameter (D) & Diametral Pitch (P) | $D = D_o - \frac{2}{P}$ |
| Base Diameter ($D_b$) | Pitch Diameter And Pressure Angle | $D_b = D\cos\phi$ |
| Number of Teeth (N) | Diametral Pitch (P) & Pitch Diameter (D) | $N = P \times D$ |
| Tooth Thickness (t) @Pitch Diameter (D) | Diametral Pitch (P) | $t = \frac{1.5708}{P}$ |
| Addendum (a) | Diametral Pitch (P) | $a = \frac{1}{P}$ |
| Outside Diameter ($D_o$) | Pitch Diameter (D) & Addendum (a) | $D_o = D + 2a$ |
| Whole Depth ($h_t$) (20P & Finer) | Diametral Pitch (P) | $h_t = \frac{2.2}{P} + .002$ |
| Whole Depth ($h_t$) (Coarser than 20P) | Diametral Pitch (P) | $h_t = \frac{2.157}{P}$ |
| Working Depth ($h_K$) | Addendum | $h_K = 2(a)$ |
| Clearance (c) | Whole Depth ($h_t$) Addendum (a) | $c = h_t - 2a$ |
| Dedendum (b) | Whole Depth ($h_t$) & Addendum (a) | $b = h_t - a$ |
| Contact Ratio ($M_c$) | Outside Radii, Base Radii, Center Distance and Pressure Angle | $M_c = \frac{\sqrt{R_o^2 - R_b^2} + \sqrt{r_o^2 - r_b^2} - C\sin\phi}{P_c\cos\phi}$ * |
| Root Diameter ($D_r$) | Pitch Diameter and Dedendum | $D_r = D - 2b$ |
| Center Distance (C) | Pitch Diameter or No. of Teeth and Pitch | $C = \frac{D_1 + D_2}{2}$ or $\frac{N_1 + N_2}{2P}$ |

*$R_o$ = Outside Radius, Gear
$r_o$ = Outside Radius, Pinion
$R_b$ = Base Circle Radius, Gear
$r_b$ = Base Circle Radius, Pinion

**FIGURE 8.57** Spur gear equations.

### *Lewis Equation (Barth Revision) for Spur Gears*

$$W = \frac{SFY}{P}\left(\frac{600}{600 + V}\right) \quad \text{(for metallic spur gears)}$$

where $W$ = tooth load, lb (along the pitch line)
$S$ = safe material stress allowable (static), psi (see Fig. 8.58)
$F$ = face width, in
$Y$ = tooth form factor, $Y$ (see Fig. 8.59)
$P$ = diametral pitch
$V$ = pitch line velocity, ft/min = $0.262 \times PD \times$ rpm (the equation for $W$ is valid for pitch line velocities to 1500 ft/min)
$D$ = pitch diameter, in

| Material | | S, psi |
|---|---|---|
| Plastic | | 5000 |
| Bronze | | 10000 |
| Cast Iron | | 12000 |
| Steel | .20 Carbon (Untreated) | 20000 |
| | .20 Carbon (Case-hardened) | 25000 |
| | .40 Carbon (Untreated) | 25000 |
| | .40 Carbon (Heat-treated) | 30000 |
| | .40 C. Alloy (Heat-treated) | 40000 |

**FIGURE 8.58** Safe static stress for gear materials.

| Number of teeth | 14½° full depth involute | 20° full depth involute |
|---|---|---|
| 10 | 0.176 | 0.201 |
| 11 | 0.192 | 0.226 |
| 12 | 0.210 | 0.245 |
| 13 | 0.223 | 0.264 |
| 14 | 0.236 | 0.276 |
| 15 | 0.245 | 0.289 |
| 16 | 0.255 | 0.295 |
| 17 | 0.264 | 0.302 |
| 18 | 0.270 | 0.308 |
| 19 | 0.277 | 0.314 |
| 20 | 0.283 | 0.320 |
| 22 | 0.292 | 0.330 |
| 24 | 0.302 | 0.337 |
| 26 | 0.308 | 0.344 |
| 28 | 0.314 | 0.352 |
| 30 | 0.318 | 0.358 |
| 32 | 0.322 | 0.364 |
| 34 | 0.325 | 0.370 |
| 36 | 0.329 | 0.377 |
| 38 | 0.332 | 0.383 |
| 40 | 0.336 | 0.389 |
| 45 | 0.340 | 0.399 |
| 50 | 0.346 | 0.408 |
| 55 | 0.352 | 0.415 |
| 60 | 0.355 | 0.421 |
| 65 | 0.358 | 0.425 |
| 70 | 0.360 | 0.429 |
| 75 | 0.361 | 0.433 |
| 80 | 0.363 | 0.436 |
| 90 | 0.366 | 0.442 |
| 100 | 0.368 | 0.446 |
| 150 | 0.375 | 0.458 |
| 200 | 0.378 | 0.463 |
| 300 | 0.382 | 0.471 |
| Rack | 0.390 | 0.484 |

**FIGURE 8.59** *Y* factors, spur gears.

For nonmetallic spur gears, the modified Lewis equation is

$$W = \frac{SFY}{P}\left(\frac{150}{200 + V} + 0.25\right) \quad \text{(for nonmetallic spur gears)}$$

*Note:* $S$ values of >6000 psi may be used for phenolic gears and 5000 psi, for some of the thermoset types of plastics commonly used for molding gears. $S$ is the safe static stress allowable, the values of which have a safety factor incorporated (see Fig. 8.58 and Sec. 8.9).

Most stock spur gears are cut to operate at standard center distances, which were defined previously. For mounting the spur gears at the calculated center distance, Fig. 8.60*a* lists the average *backlash.*

Alterations of the calculated center distance will occur during manufacturing processes, which in turn alters the backlash. The approximate relationship between center distance and backlash change for 14.5 and 20° pressure angle gears becomes:

- For 14.5°: change in center distance = 1.933 × change in backlash, i.e., change in backlash = change in center distance/1.933
- For 20°: change in center distance = 1.374 × change in backlash, i.e., change in backlash = change in center distance/1.374

Thus, it is apparent that 14.5°-pressure-angle (PA) gears will have a smaller change in backlash than 20°-PA gears for a given change in the center distance.

| Diametral pitch | Backlash, in | Diametral pitch | Backlash, in |
|---|---|---|---|
| 3 | .013 | 8-9 | .005 |
| 4 | .010 | 10-13 | .004 |
| 5 | .008 | 14-32 | .003 |
| 6 | .007 | 33-64 | .0025 |
| 7 | .006 | | |

(a)

(b)

**FIGURE 8.60** (*a*) Spur gear backlash; (*b*) spur gear milling cutter.

***Spur Gear Milling Cutters.*** Spur gears may be cut on a milling machine or a CNC machining center using the appropriate milling cutter, which has an involute profile. These cutters are available in 14.5 and 20° pressure angles and are available in different diametral pitch sizes and a cutter series (from 1 to 8) for the number of teeth required in the particular gear. The diametral pitch (DP) sizes are available from 1 through 48 DP, with 8 cutters available in each diametral pitch for cutting spur gears with different numbers of teeth. Therefore, there are more than 400 cutters available to cover all standard sizes of spur gears from the largest to the smallest. The machine tool catalogs list the cutters with their DP and series number so that you may select the proper cutter for your application. Figure 8.60*b* shows a typical spur gear milling cutter.

The individual cutters for each diametral pitch have the following teeth cutting ranges:

| | |
|---|---|
| Cutter 1 | 135 teeth to a rack |
| Cutter 2 | 55 to 134 teeth |
| Cutter 3 | 35 to 54 teeth |
| Cutter 4 | 26 to 34 teeth |
| Cutter 5 | 21 to 25 teeth |
| Cutter 6 | 17 to 20 teeth |
| Cutter 7 | 14 to 16 teeth |
| Cutter 8 | 12 to 13 teeth |

Figure 8.60*b* shows a spur gear milling cutter in operation, where the cutter axis is at right angles to the axis of the spur gear being cut.

***Undercut.*** When the number of teeth in a gear is small, the tip of the mating gear tooth may interfere with the lower portion of the tooth profile. To prevent this, the generating process removes material at this point. This results in the loss of a portion of the involute adjacent to the tooth base, reducing tooth contact and tooth strength.

On 14.5°-PA gears undercutting occurs during tooth generation where the number of teeth is less than 32 and on 20°-PA gears where the number of teeth is less than 18. This condition becomes worse as the number of teeth decrease, so the recommended minimum number of teeth is 16 for 14.5°-PA and 13 for 20°-PA spur gears. See Fig. 8.61 for a typical spur rack or gear and Fig. 8.49*a* for spur gear terminology.

In a similar manner, *internal* spur gear teeth may interfere when the pinion gear is too near the size of the mating internal gear. Therefore, for 14.5° PA, the difference in teeth numbers between the gear and pinion should not be less than 15. For 20° PA, the difference in teeth numbers should not be less than 12. See Fig. 8.63 (in Sec. 8.8) for internal spur gear terminology.

### 8.8.7 Stem Pinions

When a spur stem pinion is required to have a small number of teeth (5 to 10), undercutting of the teeth is minimized by using special enlarged pitch diameters. A spur stem pinion with a small number of teeth allows for high ratios, which are seldom available in standard spur gear sets. These special spur stem pinions are not intended for operation with internal spur gears or 11-tooth pinions, but will operate satisfactorily with all other standard spur gears of the same pressure angle. See Fig. 8.62 for a typical spur stem pinion with special enlarged pitch diameter.

### 8.8.8 Internal Gears

An external spur or helical gear may be meshed with an internal spur or helical gear, respectively. The external gear must not be larger than 66 to 67 percent of the pitch diameter of the

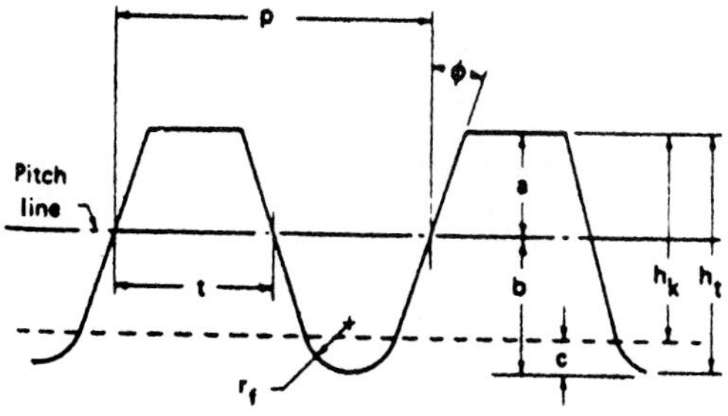


**Spur Rack or Gear Teeth**

- a Addendum
- b Dedendum
- c Clearance
- $h_k$ Working Depth
- $h_t$ Whole Depth
- p Circular Pitch
- $r_f$ Fillet Radius
- t Circular Tooth Thickness
- $\phi$ Pressure Angle

**FIGURE 8.61** Spur rack terminology.

internal gear. If the internal gear is to have 88 teeth, the external gear should have $0.66 \times 88 = 58$ teeth maximum. See the previous subsection on undercut. This generalization is not valid on epicyclic gear systems (planetary gears).

An internal gear is necessary for epicyclic gear systems and the short center distances afforded are beneficial to compact gear systems, where space is limited (see Sec. 8.12). Internal gears cannot be normally hobbed, but they can be shaped, milled, cast, or broached (size permitting). Figure 8.63 shows the basic terminology used for internal spur gears.

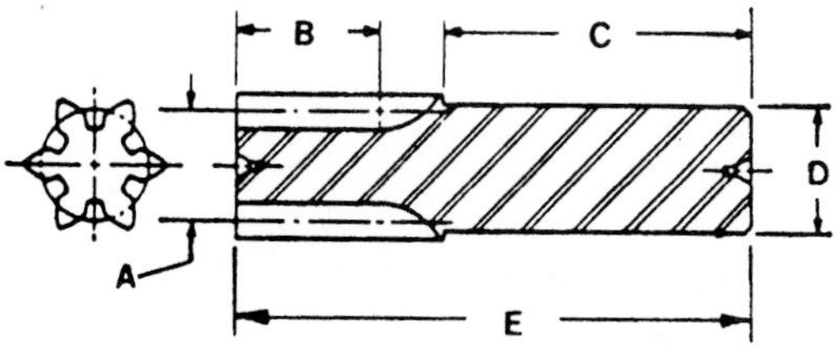


- A Pitch Diameter (special)
- B Face
- C Stem Length
- D Stem Diameter
- E Overall Length

**FIGURE 8.62** Spur stem pinion.

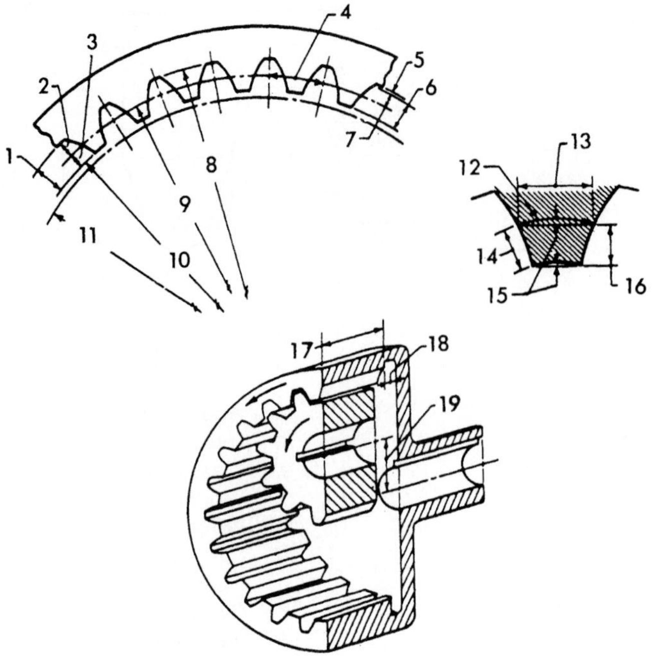


| | |
|---|---|
| 1 Whole depth | 11 Base circle diameter |
| 2 Dedendum | 12 Arc thickness |
| 3 Addendum | 13 Chordal thickness |
| 4 Circular pitch | 14 Addendum |
| 5 Clearance | 15 Rise of arc |
| 6 Working depth | 16 Chordal addendum |
| 7 Addendum of mating gear | 17 Width of face |
| 8 Root diameter | 18 Cutter clearance |
| 9 Pitch diameter | 19 Center distance |
| 10 Inside diameter | |

**FIGURE 8.63** Internal spur gear terminology.

All the simple equations used for external spur gears apply to internal spur gears except those which apply to center distance. The equations for internal spur gear center distance are

$$\text{Center distance} = \frac{\text{pitch diameter of gear} - \text{pitch diameter of pinion}}{2}$$

$$\text{Pitch diameter (operating) of pinion} = \frac{2 \times \text{operating center distance}}{\text{ratio} - 1}$$

$$\text{Pitch diameter (operating) of gear} = \frac{2 \times \text{operating center distance} \times \text{ratio}}{\text{ratio} - 1}$$

### 8.8.9 Helical Gears

Helical gears are used to transmit power or motion and force between shafts that are parallel. When helical gears are used in the parallel shaft arrangement, they must be opposite hand one from the other; that is to say, if a left-hand pinion is used, a right-hand gear must be selected, and if a right-hand pinion is used, a left-hand gear must be selected.

Single helical gear sets impose both thrust and radial loads on their support bearings. Helical gear teeth are normally made with an involute profile in the transverse plane. (The *transverse plane* is a cross-sectional plane which is perpendicular to the gear axis.) Small changes in center distance do not affect the action of helical gear sets. Helical gears may be made by hobbing, milling, shaping, or casting, with powder-metal technology occasionally applied successfully. Finishing is accomplished by grinding, shaving, lapping, rolling, or burnishing.

The size of helical gear teeth may be specified by module for the metric system or by diametral pitch (DP) for the U.S. customary system. Helical gears can be hobbed with standard spur gear hobs. Helical gears are produced mainly with a 20° pressure angle, although pressure angles of 22.5 and 25° are also used for higher load-transmitting capabilities.

Figure 8.64 shows the terminology of a helical gear and rack. In the transverse plane, the elements of a helical gear are the same as those as the spur gear. The equations for helical gear calculations are shown in Fig. 8.65.

Additional helical gear relational equations are

Normal circular pitch = circular pitch × cosine helix angle

Normal module = transverse module × cosine helix angle

Normal diametral pitch = transverse diametral pitch/cosine helix angle

Axial pitch = circular pitch/tangent helix angle

Axial pitch = normal circular pitch/sine helix angle

Figure 8.66 shows the relationship between transverse diametral pitch and normal diametral pitch for 45°-helix-angle helical gears.

Figure 8.67 illustrates the helix angle, normal and axial planes, axial circular pitch $p$, and normal circular pitch $p_n$.

An important equation for calculating the tooth loads on helical gears operating on parallel shafts is the Lewis equation modified to compensate for the difference between spur and helical gears with modified tooth form factors $Y$ as shown in Fig. 8.68.

Lewis equation for helical gears (modified):

$$W = \frac{SFY}{P_N}\left(\frac{600}{600 + V}\right)$$

where $W$ = tooth load, lb (along pitch line)
$S$ = safe material stress (static), psi (see Fig. 8.69)
$F$ = face width, in
$Y$ = tooth form factor (see Fig. 8.68)
$P_N$ = normal diametral pitch (refer to Fig. 8.66)
$D$ = pitch diameter, in
$V$ = pitch line velocity, ft/min = $0.262\ D \times$ rpm

The data contained in the spur gear section also pertain to helical gears which are cut to the diametral pitch system. The helix angle $\psi$ is the angle between any helix and an element of its cylinder. In helical gears, it is at the pitch diameter unless otherwise specified. The helical tooth form is normally involute in the plane of rotation and can be developed in a similar manner to that of the spur gear. The lead $L$ is the axial advance of a helix for one complete turn of 360°. The normal diametral pitch $P_N$ is the diametral pitch as calculated in the normal

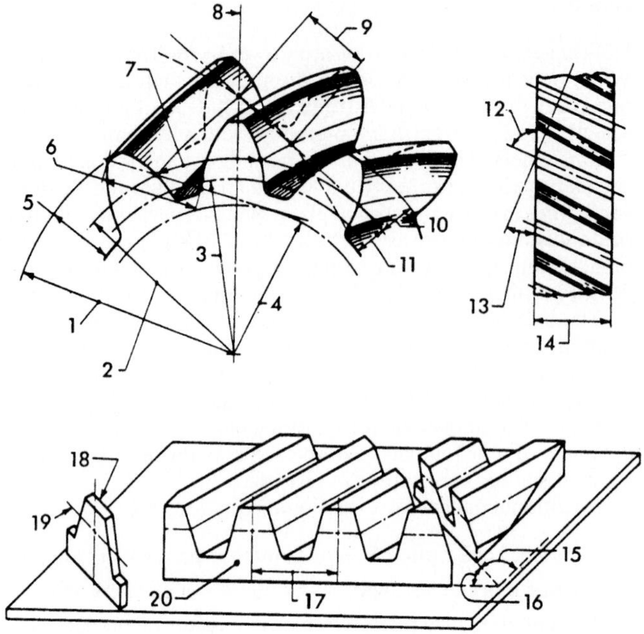

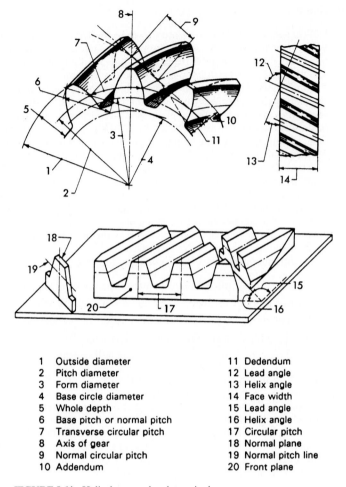

| | | | |
|---|---|---|---|
| 1 | Outside diameter | 11 | Dedendum |
| 2 | Pitch diameter | 12 | Lead angle |
| 3 | Form diameter | 13 | Helix angle |
| 4 | Base circle diameter | 14 | Face width |
| 5 | Whole depth | 15 | Lead angle |
| 6 | Base pitch or normal pitch | 16 | Helix angle |
| 7 | Transverse circular pitch | 17 | Circular pitch |
| 8 | Axis of gear | 18 | Normal plane |
| 9 | Normal circular pitch | 19 | Normal pitch line |
| 10 | Addendum | 20 | Front plane |

**FIGURE 8.64** Helical gear and rack terminology.

plane. (*Note:* Helical gears of the same hand operate with their axes at right angles, while helical gears of opposite hand operate with their axes parallel.)

When helical gears are operated with their axes at right angles, the tooth load is concentrated at a point, with the result that small loads produce very high pressures. The tooth load which may be applied to these types of drives is limited to a large degree.

As a result of the design of the helical gear tooth, an axial or thrust load is developed, which must be absorbed by the use of the proper bearings at the appropriate ends of the gears (see Fig. 8.70*a* for location of thrust bearings for the various helical gear arrangements). The magnitude of the thrust load is based on the calculated horsepower to be transmitted and may be calculated from

$$\text{Axial thrust load, lb} = \frac{126{,}050 \times \text{hp}}{\text{rpm} \times D}$$

where $D$ = pitch diameter.

| To obtain | Having | Formula |
|---|---|---|
| Transverse Diametral Pitch (P) | Number of Teeth (N) & Pitch Diameter (D) | $P = \frac{N}{D}$ |
| | Normal Diametral Pitch ($P_n$) Helix Angle ($\psi$) | $P = P_N \cos\psi$ |
| Pitch Diameter (D) | Number of Teeth (N) & Transverse Diametral Pitch ( p) | $D = \frac{N}{P}$ |
| Normal Diametral Pitch ($P_N$) | Transverse Diametral Pitch (P) & Helix Angle ($\psi$) | $P_N = \frac{P}{\cos\psi}$ |
| Normal Circular Tooth Thickness ($\tau$) | Normal Diametral Pitch ($P_N$) | $\tau = \frac{1.5708}{P_N}$ |
| Transverse Circular Pitch ($p_t$) | Diametral Pitch (P) (Transverse) | $p_t = \frac{\pi}{P}$ |
| Normal Circular Pitch ($p_n$) | Transverse Circular Pitch (p ) | $p_n = p_t \cos\psi$ |
| Lead (L) | Pitch Diameter and Pitch Helix Angle | $L = \frac{\pi D}{\tan\psi}$ |

**FIGURE 8.65** Helical gear equations.

Helical gears with a helix angle of 45°, which is common, produce a tangential force equal in magnitude to the axial thrust load. A separating force is also developed in the gear set, which is also based on the calculated horsepower, as follows:

$$\text{Separating force, lb} = \text{axial thrust load, (lb)} \times 0.26$$

(*Note:* The two preceding equations are based on a helix angle of 45° and a normal pressure angle of 14.5°, which is common in most helical gears.)

| Transverse diametral pitch $P$ | Normal diametral pitch $P_N$ |
|---|---|
| 24 | 33.94 |
| 20 | 28.28 |
| 16 | 22.63 |
| 12 | 16.97 |
| 10 | 14.14 |
| 8 | 11.31 |
| 6 | 8.48 |

*Example:* Normal $P_N = P/\cos 45° = 24/0.707 = 33.94$.

**FIGURE 8.66** Transverse and normal diametral pitch relations, helical gears.

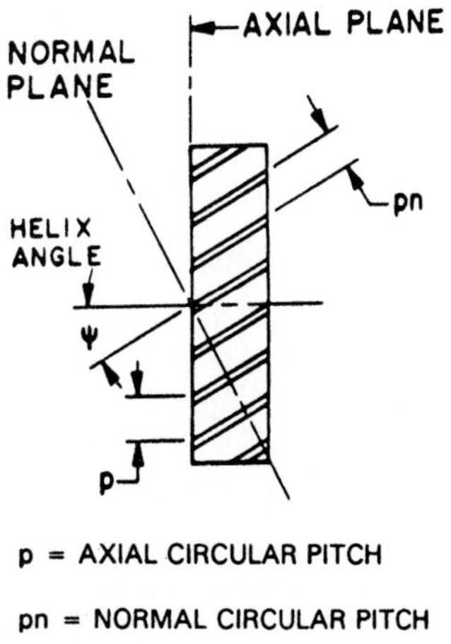


**FIGURE 8.67** Axial and normal circular pitch, helical gears.

| For $14\frac{1}{2}°$-PA, 45°-helix-angle gear | | | |
|---|---|---|---|
| No. of teeth | Factor $Y$ | No. of teeth | Factor $Y$ |
| 8 | .295 | 25 | .361 |
| 9 | .305 | 30 | .364 |
| 10 | .314 | 32 | .365 |
| 12 | .327 | 36 | .367 |
| 15 | .339 | 40 | .370 |
| 16 | .342 | 48 | .372 |
| 18 | .345 | 50 | .373 |
| 20 | .352 | 60 | .374 |
| 24 | .358 | 72 | .377 |

**FIGURE 8.68** *Y* tooth form factors for helical gears.

| Material | | $S$, lb/in$^2$ |
|---|---|---|
| Cast iron | | 10,000 |
| Steel | 0.20 carbon (untreated) | 12,000 |
| | 0.20 carbon (case-hardened) | 25,000 |
| | 0.40 carbon (untreated) | 25,000 |
| | 0.40 carbon (heat-treated) | 30,000 |
| | 0.40 carbon alloy (heat-treated) | 40,000 |

**FIGURE 8.69** Values of safe static stress, gear materials.

Helical gear systems have advantages over, or offer the following benefits relative to, spur gear systems:

- Improved tooth strength
- Increased contact ratio due to axial tooth overlap
- Greater load-carrying capacity, size for size
- Smoother and quieter operating characteristics

A typical helical gear set is shown in Fig. 8.70*b*.

## 8.8.10 Straight Miter and Bevel Gears

Bevel gear teeth are tapered in both tooth thickness and tooth height. At one end the tooth is large, while at the other end it is small. The actual tooth dimensions are usually specified for the large end of the teeth, which is at the far end of the gear set intersection, away from the pitch apex (see Fig. 8.71*a*). Most straight tooth bevel and miter gears are cut with a generated tooth form having a localized lengthwise tooth bearing, known as the *Coniflex* tooth form. The localization of tooth contact permits minor adjustment of the gears at assembly and allows for some displacement due to deflection under operating loads, without concentration of the load at the end of the tooth. The Coniflex system results in an increase in gear life and quieter operation.

Most miter and bevel gears are mounted with their shaft axes at 90°, although other angles of intersection are occasionally used. Figure 8.71*a* shows the terminology of the miter and bevel gear set.

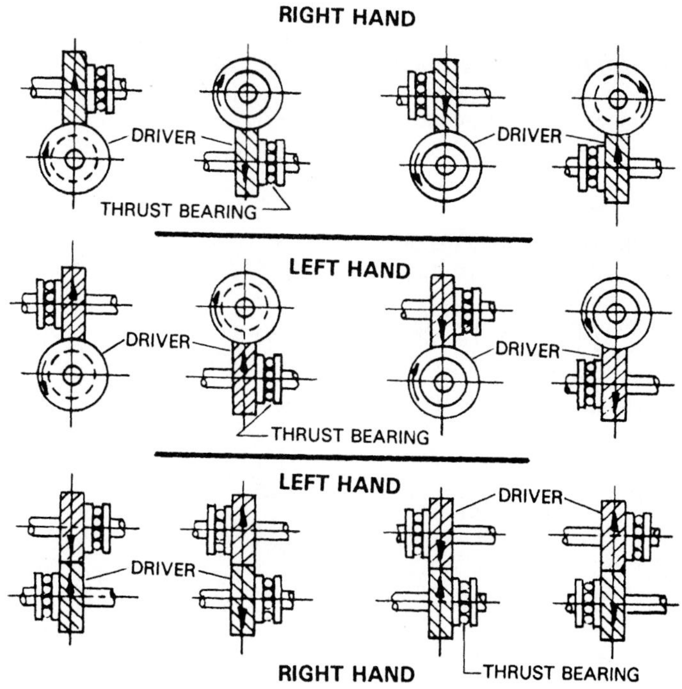


(a)

(b)

**FIGURE 8.70** (*a*) Locations for thrust bearings for helical gear sets; (*b*) typical helical gear set.

A typical straight bevel gear set is shown in Fig. 8.71*b*.

***Straight Miter and Bevel Gear Formulas.*** Refer to Fig. 8.72.

The tooth load equations for straight miter and bevel gears are as follows. Lewis equation (modified) for miter and bevel gears:

$$W = \frac{SFY}{P}\left(\frac{600}{600+V}\right)0.75$$

where $W$ = tooth load, lb (along the pitch line)
$S$ = safe material stress (static), psi (see Fig. 8.73)
$F$ = face width, in
$Y$ = tooth form factor (see Fig. 8.74)
$P$ = diametral pitch
$D$ = pitch diameter, in
$V$ = pitch line velocity, ft/min = $0.26D \times$ rpm

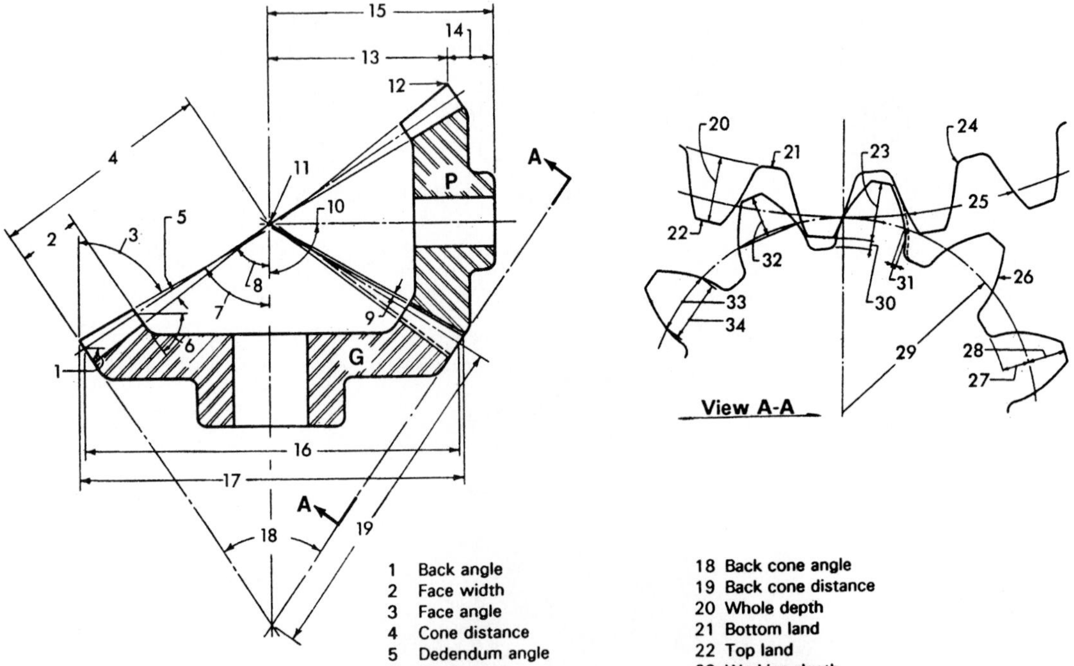


1 Back angle
2 Face width
3 Face angle
4 Cone distance
5 Dedendum angle
6 Front angle
7 Pitch angle
8 Root angle
9 Uniform clearance
10 Shaft angle
11 Pitch apex
12 Crown
13 Pitch apex to crown
14 Crown to back
15 Pitch apex to back
16 Pitch diameter
17 Outside diameter

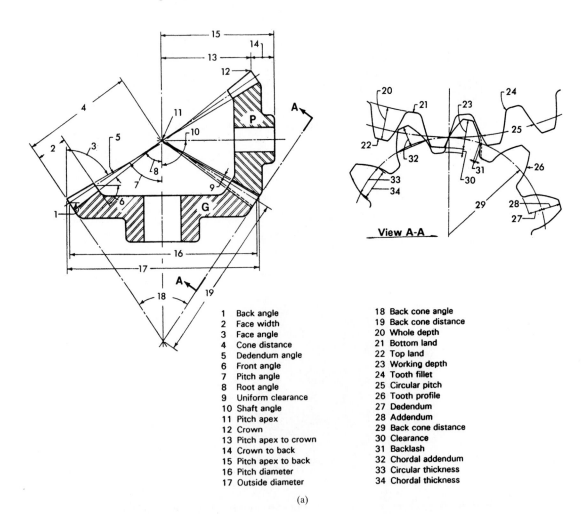

18 Back cone angle
19 Back cone distance
20 Whole depth
21 Bottom land
22 Top land
23 Working depth
24 Tooth fillet
25 Circular pitch
26 Tooth profile
27 Dedendum
28 Addendum
29 Back cone distance
30 Clearance
31 Backlash
32 Chordal addendum
33 Circular thickness
34 Chordal thickness

(a)

**FIGURE 8.71** (*a*) Bevel gear terminology.

(b)

**FIGURE 8.71** (*b*) A typical straight bevel gear set.

When the miter or bevel gear set is mounted at the exact mounting distance, the average backlash is as given in Fig. 8.75.

***Thrust Loads.*** The axial thrust loads developed by straight tooth miter and bevel gears always tend to separate the gears according to the following equations. Thrust loads for straight miter and bevel gears:

| To obtain | Having | Formula | |
|---|---|---|---|
| | | Pinion | Gear |
| Pitch diameter $D$, $d$ | Number of teeth and diametral pitch $P$ | $d = \frac{n}{P}$ | $D = \frac{n}{P}$ |
| Whole depth $h_1$ | Diametral pitch $P$ | $h_1 = \frac{2.188}{P} + 0.002$ | $h_1 = \frac{2.188}{P} + 0.002$ |
| Addendum $a$ | Diametral pitch $P$ | $a = \frac{1}{P}$ | $a = \frac{1}{P}$ |
| Dedendum $b$ | Whole depth $h_1$ and addendum $a$ | $b = h_1 - a$ | $b = h_1 - a$ |
| Clearance | Whole depth $a_1$ and addendum $a$ | $c = h_1 - 2a$ | $c = h_1 - 2a$ |
| Circular tooth thickness $\tau$ | Diametral pitch $P$ | $\tau = \frac{1.5708}{P}$ | $\tau = \frac{1.5708}{P}$ |
| Pitch angle | Number of teeth in pinion $N_p$ and gear $N$ | $L_p = \tan^{-1}\left(\frac{N_p}{N_g}\right)$ | $L_g = 90 - L_p$ |
| Outside diameter $D_o$, $d_o$ | Pinion and gear pitch diameter $(D_p + D_g)$ addendum $a$ and pitch angle $(L_p + L_g)$ | $d_o = D_p + 2a(\cos L_p)$ | $D_o = D_g + 2a(\cos L_g)$ |

**FIGURE 8.72** Miter and bevel gear equations.

| Material | | s, lb/in$^2$ |
|---|---|---|
| Bronze | | 10,000 |
| Cast iron | | 12,000 |
| Steel | 0.20 carbon (untreated) | 20,000 |
| | 0.20 carbon (case-hardened) | 25,000 |
| | 0.40 carbon (untreated) | 25,000 |
| | 0.40 carbon (heat-treated) | 30,000 |
| | 0.40 carbon alloy (heat-treated) | 40,000 |

**FIGURE 8.73** Values of safe static stress, gear materials.

| No. teeth in pinion | Ratio 1 | | 1.5 | | 2 | | 3 | | 4 | | 6 | |
|---|---|---|---|---|---|---|---|---|---|---|---|---|
| | Pinion | Gear | Pinion | Gear | Pinion | Gear | Pinion | Gear | Pinion | Gear | Pinion | Gear |
| 12 | — | | — | — | .345 | .283 | .355 | .302 | .358 | .305 | .361 | .324 |
| 14 | — | | .349 | .292 | .367 | .301 | .377 | .317 | .380 | .323 | .405 | .352 |
| 16 | .333 | | .367 | .311 | .386 | .320 | .396 | .333 | .402 | .339 | .443 | .377 |
| 18 | .342 | | .383 | .328 | .402 | .336 | .415 | .346 | .427 | .364 | .474 | .399 |
| 20 | .352 | | .402 | .339 | .418 | .349 | .427 | .355 | .456 | .386 | .500 | .421 |
| 24 | .371 | | .424 | .364 | .443 | .368 | .471 | .377 | .506 | .405 | — | — |
| 28 | .386 | | .446 | .383 | .462 | .386 | .509 | .396 | .543 | .421 | — | — |
| 32 | .399 | | .462 | .396 | .487 | .402 | .540 | .412 | — | — | — | — |
| 36 | .408 | | .477 | .408 | .518 | .415 | .569 | .424 | — | — | — | — |
| 40 | .418 | | — | — | .543 | .424 | .594 | .434 | — | — | — | — |

20° PA—long-addendum pinions; short-addendum pinions.

**FIGURE 8.74** *Y* tooth form factors, bevel gears.

| Diametral pitch | Backlash, in |
|---|---|
| 4 | 0.008 |
| 5 | 0.007 |
| 6 | 0.006 |
| 8 | 0.005 |
| 10 | 0.004 |
| 12–20 | 0.003 |
| 24–48 | 0.002 |

**FIGURE 8.75** Backlash in bevel gears.

$$\text{Gear thrust, lbf} = \frac{126{,}050 \times \text{hp}}{\text{rpm} \times D} \times \tan \alpha \cos \beta$$

$$\text{Pinion thrust, lbf} = \frac{126{,}050 \times \text{hp}}{\text{rpm} \times D} \times \tan \alpha \sin \beta$$

where $\alpha$ = tooth pressure angle; $\beta$ = ½ pitch angle (see Fig. 8.71*a*). Also, pitch angle (PA) is determined from

$$\text{PA}_{\text{pinion}} = \arctan\left(\frac{N_P}{N_G}\right) \quad \text{(for the pinion)}$$

$$\text{PA}_{\text{gear}} = 90° - \text{PA}_{\text{pinion}}$$

The 20°-PA straight miter and bevel gear is the most common and popular in present use. The correctly rated thrust bearings must be selected from the calculated loads, using the preceding equations for thrust loads.

The circular pitch and the pitch diameters of miter and bevel gears are calculated the same as those for spur gears. The pitch cone angles may be calculated by the use of one of the following equations:

$$\tan \text{PA}_{\text{pinion}} = \frac{\text{number of teeth in pinion}}{\text{number of teeth in gear}}$$

$$\tan \text{PA}_{\text{gear}} = \frac{\text{number of teeth in gear}}{\text{number of teeth in pinion}}$$

When the shaft angle is less than 90°:

$$\tan \text{PA}_{\text{pinion}} = \frac{\sin(\text{shaft angle})}{\text{ratio} + \cos(\text{shaft angle})}$$

$$\tan \text{PA}_{\text{gear}} = \frac{\sin(\text{shaft angle})}{1/\text{ratio} + \cos(\text{shaft angle})}$$

When the shaft angle is more than 90°:

$$\tan \text{PA}_{\text{pinion}} = \frac{\sin(180° - \text{shaft angle})}{\text{ratio} - \cos(180° - \text{shaft angle})}$$

$$\tan \text{PA}_{\text{gear}} = \frac{\sin(180° - \text{shaft angle})}{1/\text{ratio} - \cos(180° - \text{shaft angle})}$$

In all these cases for tan PA:

$$\text{PA}_{\text{pinion}} + \text{PA}_{\text{gear}} = \text{shaft angle}$$

### 8.8.11 Spiral Miter and Bevel Gears

Spiral miter and bevel gears have a lengthwise curvature similar to that of zerol gears. The difference is that spiral bevel gears have an appreciable angle with the axis of the gear. Spiral bevel gear teeth do not have a true helical spiral and resemble helical bevel gears. Spiral bevel gears are generated by the same machines that cut zerol gears, except that the cutting tool is set at an angle to the axis of the spiral bevel gear. Spiral bevel gears are made to 16, 17.5, 20, and 22.5° pressure angles, but the 20° pressure is considered an industry standard.

For spiral miter and bevel gears, the direction of axial thrust loads developed in operation are determined by the hand and the direction of rotation. Figure 8.76 shows the arrangements possible for spiral miter and bevel gears and the appropriate locations for the thrust bearings. The thrust bearings must be rated to withstand the imposed loads in operation (see Sec. 8.26).

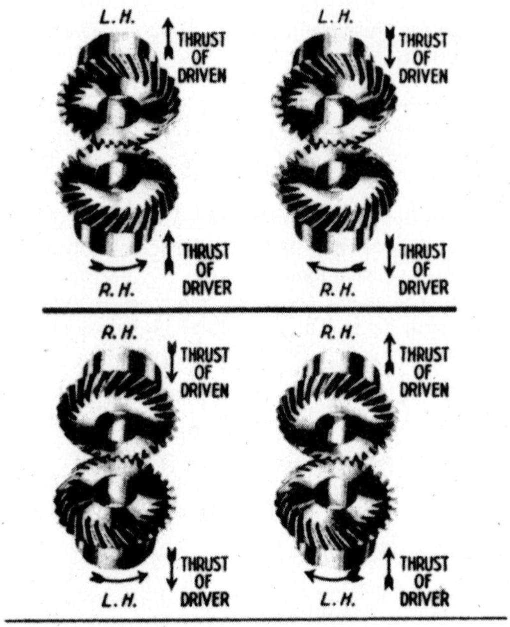


**FIGURE 8.76** Direction of thrust loads on spiral bevel gear sets.

Thrust loads for spiral miter and bevel gear sets are calculated using the equations shown in Fig. 8.77 (see also Fig. 8.71*a* for bevel gear terminology).

The spiral angle $\gamma$ for spiral miter and bevel gears is normally 35°, as produced by many gear manufacturers.

## 8.8.12 Worm Gears

Worms and worm gears are used to transmit power or motion between nonintersecting shafts at right angles (90°) to one another. Worm gear drives are the smoothest and quietest form of gearing when properly applied and maintained. Worm gear drives should be considered for the following requirements:

- High ratio speed reduction
- Limited space available
- Right angle, nonintersecting shafts
- Good resistance to "back-driving"
- Transmission of high torques

The terminology of standard worm gearing is illustrated in Fig. 8.78.

A typical worm gear set is illustrated in Fig. 8.79.

**Spiral Bevel and Miters**

Thrust values for Pinions and Gears are given for four possible combinations

| | |
|---|---|
| R.H. SPIRAL / CLOCKWISE | $T_p = \frac{126{,}050 \times HP}{RPM \times D}\left(\frac{\tan\alpha \sin\beta}{\cos\gamma} - \tan\gamma \cos\beta\right)$ |
| L.H. SPIRAL / C. CLOCKWISE | $T_G = \frac{126{,}050 \times HP}{RPM \times D}\left(\frac{\tan\alpha \cos\beta}{\cos\gamma} + \tan\gamma \sin\beta\right)$ |
| L.H. SPIRAL / CLOCKWISE | $T_p = \frac{126{,}050 \times HP}{RPM \times D}\left(\frac{\tan\alpha \sin\beta}{\cos\gamma} + \tan\gamma \cos\beta\right)$ |
| R.H. SPIRAL / C. CLOCKWISE | $T_G = \frac{126{,}050 \times HP}{RPM \times D}\left(\frac{\tan\alpha \cos\beta}{\cos\gamma} + \tan\gamma \sin\beta\right)$ |

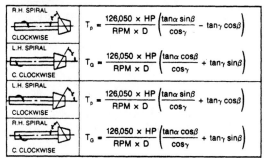

$\alpha$ = Tooth Pressure Angle
$\beta$ = 1/2 Pitch Angle
Pitch Angle = $\tan^{-1}\left(\frac{N_P}{N_G}\right)$
$\gamma$ = Spiral Angle = 35°

**FIGURE 8.77** Thrust load equations, spiral bevel gears.

The formulas for worm gearing are shown in Fig. 8.80.

When operating, worm gearing produces thrust loadings. Figure 8.81 indicates the direction of thrust of worms and worm gears when they are rotated as shown (in Fig. 8.81). To absorb the thrust loads, thrust bearings should be located as indicated in the figure.

***Efficiency of Worm Gearing.*** A commonly used efficiency equation used for worm gears is

$$\text{Efficiency} = e = \frac{\tan\gamma\,(1 - f\tan\gamma)}{f + \tan\gamma}$$

where $\gamma$ = worm lead angle, degrees; $f$ = coefficient of sliding or dynamic friction. For a bronze worm gear and hardened steel worm, $f$ = 0.03 to 0.05 for initial estimates.

Another commonly used efficiency equation for worm gearing is

$$\text{Efficiency} = e = \frac{1 - f\tan\gamma}{1 + (f/\tan\gamma)}$$

where $\gamma$ = lead angle, degrees; $f$ = coefficient of dynamic friction. *Note:* $\tan\gamma = \text{lead}/(\pi D_w)$, where lead = number of threads in worm × circular pitch and $D_w$ = pitch diameter of worm, in. The number of leads in a worm is normally 1 to 4 and determines the gear ratio of the worm gear set as follows:

$$R = \frac{N_g}{L}$$

where $N_g$ = number of teeth in the worm wheel; $L$ = number of leads on the worm, usually 1 to 4. Therefore, if a worm has two leads and the worm wheel has 40 teeth, the ratio is 40/2 = 20 and the ratio is 20:1.

The output torque of the worm gear shaft will be

$$\text{Output torque} = e\ (\text{efficiency}) \times \text{worm torque} \times \text{ratio } R$$

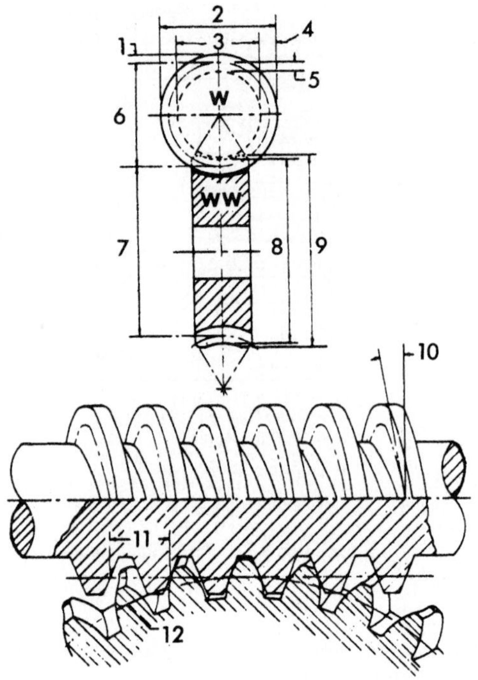


| | |
|---|---|
| 1 Addendum | 7 Pitch diameter (worm wheel) |
| 2 Outside diameter | 8 Throat diameter |
| 3 Root diameter | 9 Maximum diameter |
| 4 Whole depth | 10 Lead angle of worm |
| 5 Dedendum | 11 Linear pitch |
| 6 Pitch diameter (worm) | 12 Circular pitch |
| W Worm | WW Worm wheel |

**FIGURE 8.78** Worm gear terminology.

*Example.* If 100 lb · in of torque is applied to the worm shaft and the efficiency is 85 percent, the output torque of the worm wheel shaft will be

$$T_o = (0.85) \times 100 \times 20 = 1700 \text{ lb} \cdot \text{in or } 142 \text{ lb} \cdot \text{ft}$$

The output rpm ($\text{rpm}_o$) is then

$$\text{rpm}_o = \frac{\text{worm rpm}}{\text{ratio } R}$$

If the worm were rotating at 600 rpm, the worm wheel rpm ($\text{rpm}_o$) would be

$$\text{rpm}_o = \frac{600}{20} = 30 \text{ rpm}$$

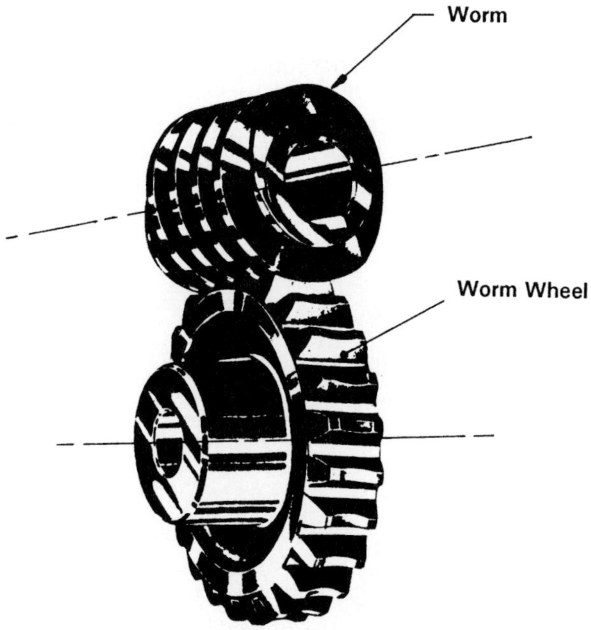


**FIGURE 8.79** Typical worm gear set.

| TO OBTAIN | HAVING | FORMULA |
|---|---|---|
| Circular Pitch (p) | Diametral Pitch (P) | $P = \frac{3.1416}{P}$ |
| Diametral Pitch (P) | Circular Pitch (p) | $P = \frac{3.1416}{P}$ |
| Lead (of Worm) ($\ell$) | Number of Threads in Worm & Circular Pitch (p) | L = P (Number of Threads) |
| Addendum (a) | Diametral Pitch (P) | $a = \frac{1}{P}$ |
| Pitch Diameter of Worm ($D_W$) | Outside Diameter ($d_o$) & Addendum (a) | $D_W = d_o - 2a$ |
| Pitch Diameter of Worm Gear ($D_G$) | Circular Pitch (p) & Number of Teeth (N) | $D_G = \frac{N_P}{3.1416}$ |
| Center Distance Between Worm and Worm Gear (CD) | Pitch Diameter of Worm ($D_W$) & Worm Gear ($D_G$) | $CD = \frac{P_W + D_G}{2}$ |
| Whole Depth of Teeth ($h_T$) | Circular Pitch (p) | $h_T = .6866\ p$ |
| | Diametral Pitch (P) | $h_T = \frac{2.157}{P}$ |
| Bottom Diameter of Worm ($d_r$) | Whole Depth (hT) & Outside Diameter ($d_w$) | $d_r = d_o - 2\ h_T$ |
| Throat Diameter of Worm Gear ($D_T$) | Pitch Diameter of Worm Gear (D) & Addendum (a) | $D_T = D + 2a$ |
| Lead Angle of Worm ( ) | Pitch Diameter of Worm (D) & The Lead (L) | $= \tan^{-1}\left(\frac{L}{3.1416d}\right)$ |
| Ratio | No. of Teeth on Gear ($N_G$) and Number of Threads on Worm | $\text{Ratio} = \frac{N_G}{\text{No. of Threads}}$ |

**FIGURE 8.80** Worm and worm gear equations.

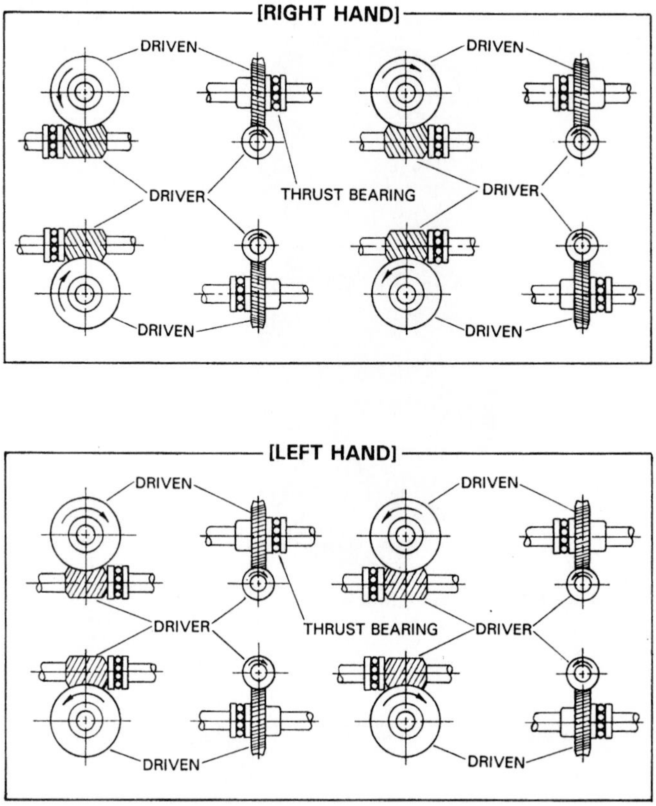


**FIGURE 8.81** Locations of thrust bearings on worm gear sets.

The torques (forces) and speeds can be altered at the input or output; however, there is no gain of power or energy (work), but only a loss which is determined by the efficiency of the worm gear drive. Efficiencies are calculated using one or the other of the preceding efficiency equations.

***Strength of Worm Gears.*** The Lewis equation for the strength of worm gears is

$$F = sbY\,P_c = \frac{sbY\pi}{P_d}$$

where $F$ = permissible tangential load, lbf (see equations for gear loads on bearings in Sec. 8.10.1)
$P_c$ = normal circular pitch
$P_d$ = normal diametral pitch
$b$ = face width of gear, in
$Y$ = tooth form factor (see Fig. 8.82)
$s$ = allowable bending stress, psi

Allowable bending stress $s$ is calculated from

$$s = s_1 \frac{1200}{1200 + V_g}$$

| Number of teeth | 14½° full-depth involute or composite | 20° full-depth involute | 20° stub involute |
|---|---|---|---|
| 12 | 0.067 | 0.078 | 0.099 |
| 13 | 0.071 | 0.083 | 0.103 |
| 14 | 0.075 | 0.088 | 0.108 |
| 15 | 0.078 | 0.092 | 0.111 |
| 16 | 0.081 | 0.094 | 0.115 |
| 17 | 0.084 | 0.096 | 0.117 |
| 18 | 0.086 | 0.098 | 0.120 |
| 19 | 0.088 | 0.100 | 0.123 |
| 20 | 0.090 | 0.102 | 0.125 |
| 21 | 0.092 | 0.104 | 0.127 |
| 23 | 0.094 | 0.106 | 0.130 |
| 25 | 0.097 | 0.108 | 0.133 |
| 27 | 0.099 | 0.111 | 0.136 |
| 30 | 0.101 | 0.114 | 0.139 |
| 34 | 0.104 | 0.118 | 0.142 |
| 38 | 0.106 | 0.122 | 0.145 |
| 43 | 0.108 | 0.126 | 0.147 |
| 50 | 0.110 | 0.130 | 0.151 |
| 60 | 0.113 | 0.134 | 0.154 |
| 75 | 0.115 | 0.138 | 0.158 |
| 100 | 0.117 | 0.142 | 0.161 |
| 150 | 0.119 | 0.146 | 0.165 |
| 300 | 0.122 | 0.150 | 0.170 |
| Rack | 0.124 | 0.154 | 0.175 |

**FIGURE 8.82** Form factor $Y$ for use in Lewis strength equation, worm gears.

where $s_1 = 0.30 \times$ ultimate strength of worm gear material, psi; $V_g$ = pitch line velocity of the gear, ft/min.

***AGMA Horsepower Rating for Worm Gears.*** Permissible horsepower inputs for worm gear units, which produce normal wear rates, are given by

$$\text{hp} = \frac{n}{R} KJV_f$$

where hp = horsepower of input
$n$ = worm speed, rpm
$R$ = transmission ratio = rpm worm/rpm gear
$K$ = pressure constant (see Fig. 8.83)
$J = R/(R + 2.5)$
$V_f$ = velocity factor

The velocity factor may be determined from

$$V_f = \frac{450}{450 + V_p + (3V_p/R)}$$

In the velocity factor relationship, $V_p$ = pitch line velocity of worm, ft/min.

The AGMA recommendation for input horsepower limit of plain worm gear units for worm gear speeds of 2000 rpm is given by

$$\text{hp} = \frac{9.5\,C^{1.7}}{R + 5}$$

| Center distance $C$, in | $K$ |
|---|---|
| 1 | 0.0125 |
| 2 | 0.025 |
| 3 | 0.04 |
| 4 | 0.09 |
| 5 | 0.17 |
| 6 | 0.29 |
| 7 | 0.45 |
| 8 | 0.66 |
| 9 | 0.99 |
| 10 | 1.20 |
| 15 | 4.0 |
| 20 | 8.0 |

*Note:* Interpolate for intermediate values.

**FIGURE 8.83** Pressure constants $K$ for worm gears.

where hp = permissible input power
$C$ = center distance, in
$R$ = transmission ratio = rpm worm/rpm gear

***Dynamic Load $F_d$.*** This may be approximated from the following equation:

$$F_d = \frac{1200 + V_g}{1200} F$$

where $F$ = actual tangential load, lbf (see equations for forces in gear systems in Sec. 8.10.1); $V_g$ = pitch line velocity of the gear, ft/min.

*Note:* The dynamic load should not exceed the endurance load $F_1$ as set out in the following Lewis equation:

$$F_1 = \frac{s_1 bY\pi}{P_d}$$

where $P_d$ = diametral pitch and $s_1 = 0.33 \times$ ultimate strength of worm gear material, psi. See previous matter for other symbols.

(*Note:* Quality worm gear systems for power transmission should have ground and polished steel worms and tough phosphor bronze worm gears. The less efficient, more economical worm gear set consists of a steel worm and cast iron worm gear.)

### 8.8.13 Other Gear Systems

The preceding data and information were presented for the gear systems most commonly used today. Other gear systems are used in industry, although not as frequently as the previously detailed systems. The other systems of gearing in use include the following:

- Zerol bevel gears
- Hypoid gears
- Face gears

- Crossed-helical gears
- Double-enveloping worm gears
- Spiroid gears

The following subsections give descriptions of these gear systems.

### 8.8.14 Zerol Bevel Gears

Zerol bevel gears are similar to straight bevel gears except that they have a curved tooth. Zerol bevel gears have 0° spiral angle. The machinery used to make zerol gears is different from that used to make straight bevel gears. Zerol gear teeth may be finished by grinding or lapping and are favored over straight bevel gears where high accuracy and full hardness are important. These gears are preferred in high-speed applications.

Zerol gears are usually made with a 20° pressure angle; 22.5 and 25° PAs are used when the numbers of teeth are small.

The calculations for pitch diameter and pitch cone angle are the same as those for straight bevel gears.

### 8.8.15 Hypoid Gears

Hypoid gears resemble bevel gears and spiral bevel gears and are used on crossed-axis shafts. The distance between a hypoid pinion axis and the axis of a hypoid gear is called the *offset.* Hypoid pinions may have as few as five teeth in a high ratio set. Ratios can be obtained with hypoid gears that are not available with bevel gears. High ratios are easy to obtain with the hypoid gear system.

Hypoid gears are matched to run together, just as zerol or spiral bevel gear sets are matched. The geometry of hypoid teeth is defined by the various dimensions used to set up the machines to cut the teeth.

### 8.8.16 Face Gears

Face gears have teeth cut on the face end of a gear. They are seldom regarded as bevel gears, but functionally they are related most closely to bevel gears.

A spur pinion and a face gear are mounted with intersecting shafts, usually at 90°. The pinion bearings carry radial loads, while the face gear bearings have both radial and thrust loads.

The formulas for determining the dimensions of a pinion to run with a face gear are the same as those of a pinion running with a mating gear on parallel axes. The pressure angles and pitches used are similar to spur gear or helical gear practices.

### 8.8.17 Crossed-Helical Gears

Crossed-helical gears are essentially nonenveloping worm gears. Both gears are cylindrical shaped. Crossed-helical gears are mounted on axes that do not intersect and are usually at 90°. The bearings for crossed-helical gears have both radial and thrust loads (see Fig. 8.70*a*).

A point contact is made between two spiral gears at 90° to one another. As the gears revolve, this point travels across the tooth in a sloping line. After the gears have "worn in" for a period of time, their load-carrying capacity increases considerably as the original point contact spreads to a line contact along the length of the gear faces. When the crossed-helical gearset is new, the load carrying capacity is very limited until this wear-in process has taken effect.

Some of the basic formulas for crossed-helical gears are

$$\text{Shaft angle} = \text{helix angle of driver} \pm \text{helix angle of driven}$$

$$\text{Normal module} = \frac{\text{normal circular pitch}}{\pi}$$

$$\text{Normal diametral pitch} = \frac{\pi}{\text{normal circular pitch}}$$

$$\text{Pitch diameter} = \frac{\text{number of teeth} \times \text{normal module}}{\cos(\text{helix angle})}$$

$$\text{Center distance} = \frac{\text{pitch diameter driver} + \text{pitch diameter driven}}{2}$$

$$\cos(\text{helix angle}) = \frac{\text{number of teeth} \times \text{normal circular pitch}}{\pi \times \text{pitch diameter}}$$

### 8.8.18 Double-Enveloping Worm Gears

The double-enveloping worm gear is like the single-enveloping worm gear (see Sec. 8.8.12), except that the worm envelops the worm gear. This worm gear system has more tooth surface in contact (area contact), and allows it to carry higher loads than single-enveloping worm gears.

The only double-enveloping worm gear drive system in current use is the Cone-Drive design (Ex-Cell-O Corp.).

In both single- and double-enveloping worm gears, it is generally recommended that the worm or pinion diameter be made a function of the center distance. Therefore

$$\text{Pitch diameter of worm} = \frac{(\text{center distance})^{0.875}}{2.2}$$

This equation merely recommends a good proportion of worm to gear diameter to obtain the best power capacity. Instrument gear designers often use different proportions, according to their intended application.

The helix angle of a single- or double-enveloping worm gear may be obtained from the following general formula:

$$\tan(\text{center helix angle of gear}) = \frac{\text{pitch diameter of gear}}{\text{pitch diameter of worm} \times \text{ratio}}$$

### 8.8.19 Spiroid Gears

The spiroid gear systems operate on nonintersecting, nonparallel shafts or axes. The spiroid pinion is tapered and resembles a worm, while the gear member is a face gear with teeth curved in a lengthwise direction; the inclination to the tooth is similar to that of a helix angle, but is not a true helical spiral. Figure 8.84 illustrates a spiroid gear set and its terminology.

The spiroid gear family has helicon, planoid, and spiroid gears. The *helicon* is a spiroid with no taper in the pinion. The *planoid* is used for lower ratios than the spiroid, and its offset is less.

The *spiroid* pinion may be made by hobbing, milling, rolling, or thread chasing. The spiroid gear is usually made by hobbing, using specially adapted machines and hobbing tools.

Spiroid gears are used in a wide variety of applications from aerospace actuators to automotive systems and appliances. High ratio in compact arrangements, low cost when mass-pro-

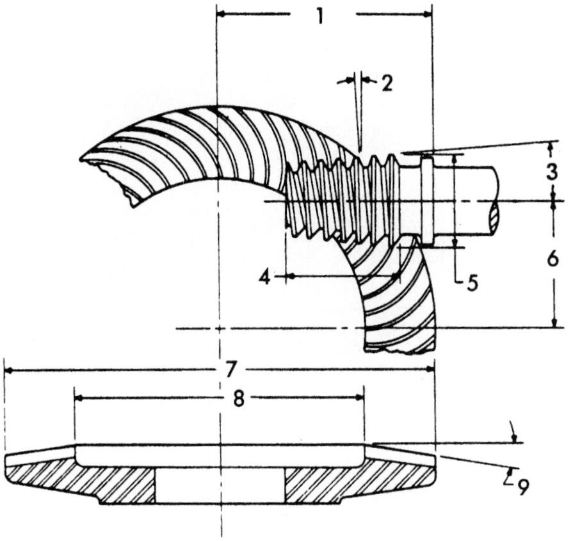


1 Pinion mounting distance
2 Thread angle
3 Pinion taper angle
4 Pinion length
5 Pinion O.D.
6 Center distance
7 Gear O.D.
8 Gear I.D.
9 Gear taper angle

**FIGURE 8.84** Spiroid gear terminology.

duced, and good load-carrying characteristics make the spiroid gear system attractive in many applications. The fact that this type of gearing system can be made with lower-cost machine tools and manufacturing processes is an important design and manufacturing consideration.

## 8.8.20 Gear Teeth Gauges

When a replacement gear is required for a machine or mechanism, you must determine the diametral pitch or relative size of the tooth and its pressure angle. In common use throughout industry, the 14.5°- and 20°-PA involute tooth form will be found. If you do not know the manufacturer or the cross-reference to the catalog number usually found on the gear hub, you may use a gear tooth gauge to find the diametral pitch and pressure angle. A typical gear tooth gauge is shown in Fig. 8.85 for both 14.5°- and 20°-PA involute teeth.

To measure a bevel or miter gear, the gauge must be used at the large end of the tooth, on the perimeter of the gear. To measure a helical gear, the gear tooth gauge must be held perpendicular to the gear axis or shaft. When the gauge can be rolled along the perimeter of the gear being measured without a mismatch, you have found the correct DP (diametral pitch) or relative tooth size. Counting the number of teeth in the gear or pinion and measuring the approximate pitch diameter will allow you to find a matching gear to replace the one in question, or the entire set.

Figure 8.86 shows some of the common gear types found in wide use for a variety of applications. The following gear types are represented: *a*—miter gear, 45°; *b*—pinion of a straight

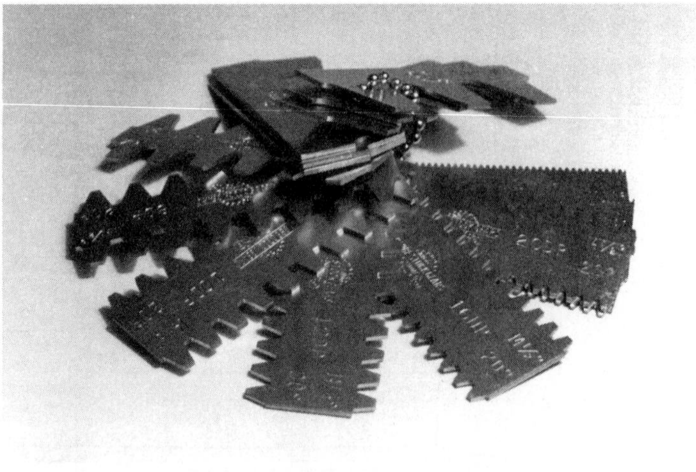

**FIGURE 8.85** Typical gear tooth gauge.

bevel gear set; *c*—typical spur gear; *d*—spur stem pinion, eight-tooth, 12-DP special modified involute tooth; *e*—helical pinion, left-hand 45° helix angle; *f*—worm from a worm gear set, right hand.

## 8.9 GEAR MATERIALS AND HARDNESS RANGES

### 8.9.1 Plastic Gears and Materials

Plastic gears find widespread use in many applications where low loads and speeds are specified. Thermoplastics such as polycarbonates, polyamides, and acetals are popular for low-strength applications where lubrication may be minimal or where the gears come into contact with the ambient surroundings such as in food- or drug-handling machinery. Thermoset plastics such as the laminated phenolic fabric materials may be used in relatively high-strength applications such as automotive timing gears, air compressors, household appliances, bottling

**FIGURE 8.86** Typical common gears.

machinery, and calculating machinery. Bakelite, another thermoset plastic, is also used to compression-mold gears for low-strength, high-volume applications. The thermoplastics are usually injection-molded, while the thermoset plastics are usually compression-molded. Thermoplastics may be remelted for remolding, while thermoset plastics have a one-way chemistry which does not allow their remelting for remolding. In general, thermoset plastics are harder than thermosetting plastics and usually more rigid and heat-resistant.

Figure 8.87 lists the physical properties for some of the commonly used thermoplastic and thermoset plastic materials for gearing applications.

## 8.9.2 Ferrous and Nonferrous Gear Materials and Their Characteristics and Processes

Typical gear materials—wrought steel (see Table 8.1)

Commonly used quenchants for ferrous gear materials (see Table 8.2)

Typical Brinell hardness ranges and strengths for annealed, normalized, and tempered steel gearing (see Table 8.3)

Typical Brinell hardness ranges and strengths for quenched and tempered alloy steel gearing (see Table 8.4)

Typical effective case depth specifications for carburized gearing (see Table 8.5)

Minimum hardness and tensile strength requirements for gray cast-iron gearing (see Table 8.6)

Mechanical properties of ductile iron gearing (see Table 8.7)

Machinability of common gear materials (see Table 8.8)

Mechanical properties of cast bronze alloys for gearing (see Table 8.9)

Refer to the materials section of the *McGraw-Hill Machining and Metalworking Handbook*, 1994, for extensive data relative to heat treatment of steels, alloy steels, and nonferrous alloys.

## 8.9.3 Gear Quality Numbers (AGMA)

The AGMA *gear quality number* refers to the accuracy and allowable tolerances which are permissible in manufacturing each particular gear intended for its specialized use. The AGMA quality numbers for racks and gears are shown in Table 8.10 for each particular application. The permissible tolerances for the different quality numbers may be obtained from

| Properties | Polycarbonate | Polyamide | Acetal | Phenolic Fabric - LE Crosswise | Phenolic Fabric - LE Lengthwise |
|---|---|---|---|---|---|
| Tensile strength, psi x $10^3$ | 9 - 11 | 8.5 - 11 | 10 | 9.5 | 13.5 |
| Flex. strength, psi x $10^3$ | 11 - 13 | 14.5 | 13.5 | 15 | |
| Elongation % | 60 - 100 | 60 - 300 | 15 - 75 | - | - |
| Impact strength, ft-lbs/in | 12 - 16 | 0.9 - 2 | 1.4 - 2.3 | 1 | 1.25 |
| Water absorbtion %/24 Hrs | 0.3 | 1.5 | 0.4 | See note 2 | |
| Heat resistance °F (continuous) | 250 - 275 | 250 | 175 | 250 | |
| Trade names | Lexan | Nylon, Zytel | Delrin | Phenolite, Ryertex, Textolite | |

NOTES: 1 Tabulated data are average values (see material specification sheet for exact data)
2 0.125", 1.3%; 0.25", 0.95%; 0.50", 0.70%; 1" and over, 0.55%
3 Source: General Electric Company, Plastics Division.

**FIGURE 8.87** Physical properties of plastics used for gears.

**TABLE 8.1** Typical Gear Materials

| Common alloy steel grades | Common heat-treatment practice* | General remarks and application |
|---|---|---|
| 1045 | T-H, I-H, F-H | Low hardenability |
| 4130 | T-H | Marginal hardenability |
| 4140 | T-H, T-H&N, I-H, F-H | Fair hardenability |
| 4145 | T-H, T-H&N, I-H, F-H | Medium hardenability |
| 8640 | T-H, T-H&N, I-H, F-H | Medium hardenability |
| 4340 | T-H, T-H&N, I-H, F-H | Good hardenability in heavy sections |
| Nitralloy 135 mod. | T-H&N | Special heat treatment |
| Nitralloy G | T-H&N | Special heat treatment |
| 4150 | I-H, F-H, T-H, TH&N | Quench-crack-sensitive Good hardenability |
| 4142 | I-H, F-H, T-H&N | Used when 4140 exhibits marginal hardenability |
| 4350† | T-H, I-H, F-H | Quench-crack-sensitive, excellent hardenability in heavy sections |
| 1020 | C-H | Very low hardenability |
| 4118 | C-H | Fair core hardenability |
| 4620 | C-H | Good case hardenability |
| 8620 | C-H | Fair core hardenability |
| 4320 | C-H | Good core hardenability |
| 8822 | C-H | Good core hardenability in heavy sections |
| 3310† | C-H | Excellent hardenability |
| 4820 | C-H | (in heavy sections) |
| 9310 | C-H | for all three grades |

* C-H = carburize-harden; F-H = flame-harden; I-H = induction-harden; T-H = through-harden; T-H&N = through-harden, then nitride
† Recognized, but not current standard grade.
SOURCE: Extracted from AGMA standard: ANSI/AGMA 2004-B89 (revision of AGMA 240.01) with the permission of the publisher, American Gear Manufacturers Association, 1500 King Street, Suite 201, Alexandria, Virginia 22314.

the AGMA standards, which show the type of gear and the permissible tolerances and inspection dimensions.

### 8.9.4 AGMA Gear Specification Sample Sheet

For indicating the basic gear data and manufacturing and inspection data when specifying a particular gear, the gear drawing should contain this specification data in a form as shown or similar to that shown in Table 8.11. See Table 8.11 for a sample gear specification sheet, AGMA standard 370.01.

## 8.10 FORCES AND WEAR LOADS IN GEARING SYSTEMS

When gears are in operation, different types of mechanical loads are produced by the different types of gears. Loads encountered in the various gear systems include radial loads, axial loads (tangential), and thrust loads. Heavy wearing loads on the tooth surfaces are also produced. The wearing loads and calculation procedures for the wearing loads are shown in the next section. All the basic types of gear loads will be presented in the next section, which will enable the designer to calculate the forces produced on the gear system shaft support bearings in order to select the properly rated bearings. Calculation procedures in the following section will allow for the proper selection of gear material and hardness, which will satisfy the maximum wear loads permissible for the various gear systems and materials.

**TABLE 8.2** Quenchants for Ferrous Gear Materials

| Material grade | Quenchant | Remarks |
|---|---|---|
| 1020 | Water or brine | Carburized and quenched with good quench agitation. |
| 4118<br>4620<br>8620<br>8822<br>4320 | Oil | Carburized and quenched in well-agitated conventional oil at 80–160°F (27–71°C) is normally required. For finer pitched gearing, hot oil at 275–375°F (135–190°C) may be used to minimize distortion. Some loss in core hardness will also result from hot-oil quench. |
| 3310<br>9310 | Oil | Carburized and quenched in hot oil at 275–375°F (135–190°C). This is the preferred quench. In larger sections, conventional oil can be used. |
| 1045<br>4130<br>8630 | Water, oil, or polymer | Type of quenchant depends on chemistry and section size. Large sections normally require water or low-concentration polymer. Smaller sections can be processed in well-agitated oil. |
| 1141<br>1541 | Oil or polymer | Good response in well-agitated conventional oil or polymer. Induction- or flame-hardened parts normally quenched in polymer. |
| 4140<br>4142<br>4145 | Oil or polymer | Same as above; however, thin sections or sharp corners can represent a crack hazard. Hot oil should be considered in these cases. With proper equipment, air quench can be used for flame-hardened parts.<br>These are high-hardenability steels which can be crack-sensitive in moderate to thin sections. Hot oil is often used. High-concentration polymer should be used with caution. |
| 4150<br>4340<br>4345<br>4350 | Oil or polymer | If conventional oil is used, parts are often removed warm and tempered promptly after quench.<br>Crack sensitivity applies also to flame- or induction-hardened parts, with high-concentration polymer being the usual quenchant. Oil is sometimes used, and air quench can be applied for flame hardening with proper equipment. |
| Gray or ductile iron | Oil, polymer, or air | Quench medium depends on alloy content. High-alloy irons can be air-quenched to moderate hardness levels.<br>Unalloyed or low-alloy irons require oil or polymer.<br>In this section parts and flame- or induction-hardened surfaces can be crack-sensitive. |

SOURCE: Extracted from AGMA standard: ANSI/AGMA 2004-B89 (revision of AGMA 240.01) with the permission of the publisher, American Gear Manufacturers Association, 1500 King Street, Suite 201, Alexandria, Virginia 22314.

### 8.10.1 Forces in Gear Systems (Bearing Loads)

In determining the forces developed by machine elements commonly encountered in bearing applications, the following equations are used.

***Spur Gearing.*** See Fig. 8.88. Tangential force:

$$F_{3g}, \text{N} = \frac{(1.91 \times 10^7)P}{P_{dg}\,\text{rpm}_g}$$

$$F_{3g}, \text{lbF} = \frac{(1.26 \times 10^5)P}{P_{dg}\,\text{rpm}_g}$$

**TABLE 8.3** Brinell Hardness Ranges and Strengths for Steel Gearing

| | Annealed heat treatment† | | | Normalized and tempered‡ | | |
|---|---|---|---|---|---|---|
| Typical alloy steels specified* | Brinell hardness range, HB | Tensile strength minimum, ksi (MPa) | Yield strength minimum, ksi (MPa) | Brinell hardness range, HB | Tensile strength minimum, ksi (MPa) | Yield strength minimum, ksi (MPa) |
| 1045 | 159–201 (550) | 80 (345) | 50 | 159–201 (550) | 80 (345) | 50 |
| 4130<br>8630 | 156–197 | 80 (550) | 50 (345) | 167–212 | 90 (620) | 60 (415) |
| 4140<br>4142<br>8640 | 187–229 | 95 (655) | 60 (415) | 262–302 | 130 (895) | 85 (585) |
| 4145<br>4150 | 197–241 | 100 (690) | 60 (415) | 285–331 | 140 (965) | 90 (620) |
| 4340<br>4350 type | 212–255 | 110 (760) | 65 (450) | 302–341 | 150 (1035) | 95 (655) |

* Steels shown in order of increased hardenability.
† Hardening by quench and tempering results in a combination of properties generally superior to that achieved by anneal or normalizing and tempering (impact, ductility, etc.).
‡ Hardness and strengths able to be obtained by normalizing and tempering are also a function of controlling section size and tempering temperature considerations.
SOURCE: Extracted from AGMA standard: ANSI/AGMA 2004-B89 (revision of AGMA 240.01) with the permission of the publisher, American Gear Manufacturers Association, 1500 King Street, Suite 201, Alexandria, Virginia 22314.

**TABLE 8.4** Brinell Hardness for Alloy Steel Gearing

| Alloy steel grade* | Heat treatment | Hardness range, HB† | Tensile strength minimum, ksi (MPa) | Yield strength minimum, ksi (MPa) |
|---|---|---|---|---|
| 4130<br>8630 | Water-quench and temper | 212–248 up to 302–341 | 100 (690)<br>145 (1000) | 75 (515)<br>125 (860) |
| 4140<br>8640 | Oil-quench and temper | 241–285‡ up to 341–388 | 120 (830) | 95 (655) |
| 4142<br>4145<br>4150 | | 341–388 | 170 (1170) | 150 (1035) |
| 4340<br>4350 | Oil-quench and temper | 277–321 up to 363–415§ | 135 (930)<br>180 (1240) | 110 (760)<br>145 (1000) |

* Steels shown in order of increased hardenability, 4350 being the highest. These steels can be ordered to H band hardenability ranges.
† Hardness range is dependent on controlling section size and quench severity.
‡ It is difficult to cut teeth in 4100 series steels above 341 HB and 4300 series steels above 375 HB (4340 and 4350 provide advantage because of higher tempering temperatures and microstructure considerations).
§ High specified hardness is used for special gearing, but costs should be evaluated due to reduced machinability.
SOURCE: Extracted from AGMA standard: ANSI/AGMA 2004-B89 (revision of AGMA 240.01) with the permission of the publisher, American Gear Manufacturers Association, 1500 King Street, Suite 201, Alexandria, Virginia 22314.

**TABLE 8.5** Effective Case Depths for Carburized Gearing

| Normal diametral pitch* | Normal tooth thickness† | Range of normal diametral pitch | Range of normal circular pitch | Effective case depth (inches) to RC 50‡: Spur, helical bevel, and miter§ | Effective case depth (inches) to RC 50‡: Worms with ground threads¶ |
|---|---|---|---|---|---|
| 16 | 0.098 | 17.5–13.7 | 0.180–0.230 | 0.010–0.020 | 0.020–0.030 |
| 14 | 0.112 | 17.5–13.7 | 0.180–2.300 | 0.010–0.020 | 0.020–0.030 |
| 12 | 0.131 | 13.7–10.5 | 0.230–0.300 | 0.015–0.025 | 0.025–0.040 |
| 10 | 0.157 | 10.5–8.5 | 0.300–0.370 | 0.020–0.030 | 0.035–0.050 |
| 8 | 0.198 | 8.5–7.5 | 0.370–0.480 | 0.025–0.040 | 0.040–0.055 |
| 7 | 0.224 | 7.5–6.5 | 0.370–0.480 | 0.025–0.040 | 0.040–0.055 |
| 6 | 0.251 | 6.5–5.2 | 0.480–0.600 | 0.030–0.050 | 0.045–0.060 |
| 5 | 0.314 | 5.2–4.3 | 0.600–0.728 | 0.040–0.060 | 0.045–0.060 |
| 4 | 0.393 | 4.3–3.7 | 0.728–0.860 | 0.050–0.070 | 0.045–0.060 |
| 3.5 | 0.449 | 3.7–3.1 | 0.860–1.028 | 0.060–0.080 | 0.060–0.075 |
| 3.0 | 0.523 | 3.1–2.8 | 1.026–1.200 | 0.070–0.090 | 0.075–0.090 |
| 2.75 | 0.571 | 2.8–2.6 | 1.026–1.200 | 0.070–0.090 | 0.075–0.090 |
| 2.5 | 0.628 | 2.6–2.3 | 1.200–1.400 | 0.080–0.105 | 0.075–0.090 |
| 2.25 | 0.698 | 2.3–2.2 | 1.200–1.400 | 0.080–0.105 | 0.075–0.090 |
| 2.0 | 0.785 | 2.2–1.9 | 1.428–1.676 | 0.090–0.125 | 0.075–0.090 |
| 1.75 | 0.897 | 1.9–1.6 | 1.676–1.976 | 0.105–0.140 | 0.075–0.090 |
| 1.5 | 1.047 | 1.6–1.3 | 1.976–2.400 | 0.120–0.155 | 0.075–0.090 |
| 1.25 | 1.256 | 1.3–1.1 | 2.400–2.828 | 0.145–0.180 | 0.075–0.090 |
| 1.0 | 1.570 | ≤1.1 | ≥2.828 | 0.170–0.205 | 0.075–0.090 |
| 0.75 | 2.094 | ≤1.1 | ≥2.325 | 0.170–0.205 | 0.075–0.090 |

* All case depths are based on normal diametral pitch. All other pitch measurements should be converted before specifying a case depth.
† Gears with thin top lands may be subject to excessive case depth at the tips. Land width should be calculated before a case is specified.
‡ Case at root is typically 50 to 70 percent of case at midtooth. The case depth for bevel and miter gears is calculated from the thickness of the tooth's small end. For gearing requiring maximum performance, detailed studies must be made of the application, loading, and manufacturing procedures to determine the required effective case depth. For further details, refer to AGMA 2001-B88.
§ To convert above data to metric, multiply values given by 25.4 to determine millimeter equivalent.
¶ Worm and ground-thread case depths allow for grinding. Unground worm gear cases may be decreased accordingly. For very heavily loaded coarse pitch ground thread worms, heavier case depth than shown in table may be required.
SOURCE: Extracted from AGMA standard: ANSI/AGMA 2004-B89 (revision of AGMA 240.01) with the permission of the publisher, American Gear Manufacturers Association, 1500 King Street, Suite 201, Alexandria, Virginia 22314.

**TABLE 8.6** Specification for Gray Cast-Iron Gearing

| ASTM class number* | Brinell hardness | Tensile strength, ksi (MPa) |
|---|---|---|
| 20 | 155 | 20 (140) |
| 30 | 180 | 30 (205) |
| 35 | 205 | 35 (240) |
| 40 | 220 | 40 (275) |
| 50 | 250 | 50 (345) |
| 60 | 285 | 60 (415) |

* See ASTM A48 for additional information.
SOURCE: Extracted from AGMA standard: ANSI/AGMA 2004-B89 (revision of AGMA 240.01) with the permission of the publisher, American Gear Manufacturers Association, 1500 King Street, Suite 201, Alexandria, Virginia 22314.

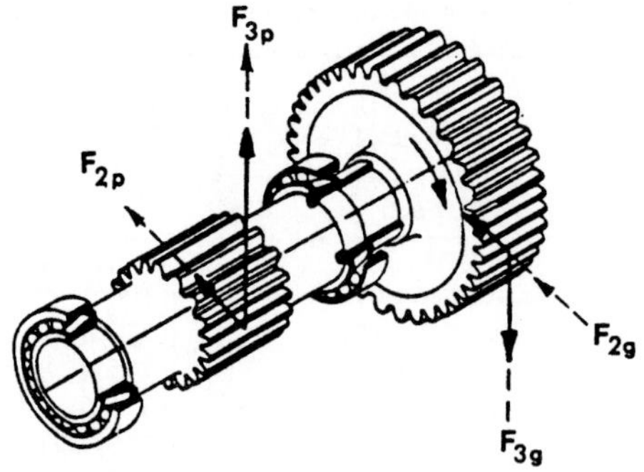


**FIGURE 8.88** Forces in spur gears.

Separating force:

$$F_{2g} = F_{3g} \tan \phi_p \cos\gamma_f$$

***Single Helical Gearing.*** See Fig. 8.89. Tangential force:

$$F_{3g}, \text{N} = \frac{(1.91 \times 10^7)P}{P_{dg} \text{ rpm}_g}$$

$$F_{3g}, \text{lbF} = \frac{(1.26 \times 10^5)P}{P_{dg} \text{ rpm}_g}$$

Separating force:

$$F_{2g} = \frac{F_{3g} \tan \phi_g}{\cos \psi_g}$$

Thrust force:

$$F_{1g} = F_{3g} \tan \psi_g$$

Note that for double helical (herringbone) gearing, $F_{1g} = 0$.

**TABLE 8.7** Mechanical Properties for Ductile Iron Gearing

| ASTM grade designation* | Former AGMA class | Recommended heat treatment | Brinell hardness range | Minimum tensile strength, ksi (MPa) | Minimum yield strength, ksi (MPa) | Elongation in 2 in (50 mm), percent min. |
|---|---|---|---|---|---|---|
| 60-40-18 | A-7-a | Annealed ferritic | 170 max. | 60 (415) | 40 (275) | 18.0 |
| 65-45-12 | A-7-b | As-cast or annealed ferritic-pearlitic | 156–217 | 65 (450) | 45 (310) | 12.0 |
| 80-55-06 | A-7-c | Normalized ferritic-pearlitic | 187–255 | 80 (550) | 55 (380) | 6.0 |
| 100-70-03 | A-7-d | Quench and tempered pearlitic | 241–302 | 100 (690) | 70 (485) | 3.0 |
| 120-90-02 | A-7-e | Quench and tempered martensitic | Range specified | 120 (830) | 90 (620) | 2.0 |

* See ASTM A536 or SAE J434 for further information.

*Note:* Other tensile properties and hardnesses should be used only by agreement between gear manufacturer and casting producer.

SOURCE: Extracted from AGMA standard: ANSI/AGMA 2004-B89 (revision of AGMA 240.01) with the permission of the publisher, American Gear Manufacturers Association, 1500 King Street, Suite 201, Alexandria, Virginia 22314.

**TABLE 8.8** Machinability of Common Gear Materials

| Material grades | Remarks |
|---|---|
| | Low-Carbon Carburizing Steel Grades |
| 1020 | Good machinability, as rolled, as forged, or normalized. |
| 4118<br>4620<br>8620<br>8822 | Good machinability, as rolled or as forged. However, normalized is preferred. Inadequate cooling during normalizing can result in gummy material, reduced tool life, and poor surface finish. Quench and temper as a prior treatment can aid machinability. The economics of the pretreatments must be considered. |
| 3310<br>4320<br>4820<br>9310 | Fair to good machinability if normalized and tempered, annealed or quenched and tempered. Normalizing without tempering results in reduced machinability. |
| | Medium-Carbon Through-Hardened Steel Grades |
| 1045<br>1141<br>1541 | Good machinability if normalized. |
| 4130<br>4140<br>4142 | Good machinability if annealed, or normalized and tempered to approximately 255 HB or quenched and tempered to approximately 321 HB. Over 321 HB, machinability is fair. Above 363 HB, machinability is poor. Inadequate (slack) quench with subsequent low tempering temperature may produce a part which meets the specified hardness but produces a mixed microstructure which results in poor machinability. |
| 4145<br>4150<br>4340<br>4345<br>4350 | Remarks for medium-carbon alloy steel (above) apply. However, the higher carbon results in lower machinability. Sulfur additions aid the machinability of these grades. 4340 machinability is good up to 363 HB. The higher carbon level in 4145, 4150, 4345, and 4350 makes them more difficult to machine and should be specified only for heavy sections. Inadequate (slack) quench can seriously affect machinability in these steels. |
| | Other Gear Material |
| Gray irons | Gray cast irons have good machinability. Higher-strength gray cast irons [above 50 ksi (345 MPa) tensile strength] have reduced machinability. |
| Ductile irons | Annealed or normalized ductile cast iron has good machinability. The "as cast" (not heat-treated) ductile iron has fair machinability. Quenched and tempered ductile iron has good machinability up to 285 HB and fair machinability up to 352 HB. Above 352 HB, machinability is poor. |
| Gear bronzes and brasses | All gear bronzes and brass have good machinability. The very high strength heat-treated bronzes [above 110 ksi (760 MPa) tensile strength] have fair machinability. |
| Austenitic stainless steel | All austenitic stainless steel grades only have fair machinability. Because of work-hardening tendencies, feeds and speeds must be selected to minimize work hardening. |

*Note:* Coarse-grain steels are more machinable than fine-grain. However, gear steels are generally used in the fine-grain condition since mechanical properties are improved and distortion during heat treatment is reduced. Increasingly cleaner steels are now also being specified for gearing. However, if sulfur content is low, less than 0.015%, machinability may decrease appreciably.

SOURCE: Extracted from AGMA standard: ANSI/AGMA 2004-B89 (revision of AGMA 240.01) with the permission of the publisher, American Gear Manufacturers Association, 1500 King Street, Suite 201, Alexandria, Virginia 22314.

**TABLE 8.9** Mechanical Properties for Cast Bronze Alloys

| Copper alloy UNS no.† | Former AGMA type | Casting method and condition‡ | Minimum tensile strength,§ ksi (MPa) | Minimum yield strength,§ ksi (MPa) | Minimum percent elongation in 2 in (50 mm) | Typical hardness,¶ kgf | |
|---|---|---|---|---|---|---|---|
| | | | | | | HB 500 | HB 3000 |
| C86200 | MNBR 3 | Sand, centrifugal, continuous | 90 (620) | 45 (310) | 18 | — | 180 |
| C86300 | MNBR 4 | Sand, centrifugal, | 110 (760) | 60 (415) | 12 | — | 225 |
| | | continuous | 110 (760) | 62 (425) | 14 | — | 225 |
| C86500 | MNBR 2 | Sand, centrifugal | 65 (450) | 25 (170) | 20 | 112 | — |
| C86500 | MNBR 2 | Continuous | 70 (485) | 25 (170) | 25 | 112 | — |
| C90700 | BRONZE 2 | Sand | 35 (240) | 18 (125) | 10 | 70 | — |
| C90700 | BRONZE 2 | Continuous | 40 (275) | 25 (170) | 10 | 80 | — |
| C90700 | BRONZE 2 | Centrifugal | 50 (345) | 28 (195) | 12 | 100 | — |
| C92500 | BRONZE 5 | Sand | 35 (240) | 18 (125) | 10 | 70 | — |
| C92500 | BRONZE 5 | Continuous | 40 (275) | 24 (165) | 10 | 80 | — |
| C92700 | BRONZE 3 | Sand | 35 (240) | 18 (125) | 10 | 70 | — |
| C92700 | BRONZE 3 | Continuous | 38 (260) | 20 (140) | 8 | 80 | — |
| C92900 | — | Sand, continuous | 45 (310) | 25 (170) | 8 | 90 | — |
| C95200 | ALBR 1 | Sand, centrifugal | 65 (450) | 25 (170) | 20 | — | 125 |
| C95200 | ALBR 1 | Continuous | 68 (470) | 26 (180) | 20 | — | 125 |
| C95300 | ALBR 2 | Sand, centrifugal | 65 (450) | 25 (170) | 20 | — | 140 |
| C95300 | ALBR 2 | Continuous | 70 (485) | 26 (180) | 25 | — | 140 |
| C95300 | ALBR 2 | Sand, centrifugal | 80 (550) | 40 (275) | 12 | — | 160 |
| C95300 | ALBR 2 | Continuous (HT) | 80 (550) | 40 (275) | 12 | — | 160 |
| C95400 | ALBR 3 | Sand, centrifugal (HT) | 75 (515) | 30 (205) | 12 | — | 160 |
| C95400 | ALBR 3 | Continuous | 85 (585) | 32 (220) | 12 | — | 160 |
| C95400 | ALBR 3 | Sand, centrifugal (HT) | 90 (620) | 45 (310) | 6 | — | 190 |
| C95400 | ALBR 3 | Continuous (HT) | 95 (655) | 45 (310) | 10 | — | 190 |
| C95500 | ALBR 4 | Sand, centrifugal | 90 (620) | 40 (275) | 6 | — | 190 |
| C95500 | ALBR 4 | Continuous | 95 (655) | 45 (290) | 10 | — | 190 |
| C95500 | ALBR 4 | Sand, centrifugal (HT) | 110 (760) | 60 (415) | 5 | — | 200 |
| C95500 | ALBR 4 | Continuous (HT) | 110 (760) | 62 (425) | 8 | — | 200 |

* For rating of worm gears in accordance with AGMA 6034-A87, the materials factor $k_a$ will depend on the particular casting method employed.

† Unified numbering system. For cross-reference to SAE, former SAE & ASTM, see SAE Information Report SAE J461. For added copper alloy information, also see SAE J462.

‡ Refer to ASTM B427 for sand and centrifugal cast C90700 alloy and sand cast C92900.

§ Minimum tensile strength and yield strength shall be reduced 10 percent for continuous cast bars having a cross section of 4 in (102 mm) or more (see ASTM B505, Table 3 footnote).

¶ BHN at other load levels (1000 or 1500 kgf) may be used if approved by purchaser.

SOURCE: Extracted from AGMA standard: ANSI/AGMA 2004-B89 (revision of AGMA 240.01) with the permission of the publisher, American Gear Manufacturers Association, 1500 King Street, Suite 201, Alexandria, Virginia 22314.

**TABLE 8.10** AGMA Applications and Quality Numbers for Racks and Gears

| Gearing application | Quality no.* |
|---|---|
| Aerospace | |
| — Actuators | 7–11 |
| — Control gearing | 7–11 |
| — Engine accessories | 10–13 |
| — Engine power | 10–13 |
| — Engine starting | 10–13 |
| — Loading hoist | 7–11 |
| — Propeller feathering | 10–13 |
| — Small engines | 12–13 |
| Agriculture | |
| — Baler | 3–7 |
| — Beet harvester | 5–7 |
| — Combine | 5–7 |
| — Corn picker | 5–7 |
| — Cotton picker | 5–7 |
| — Farm elevator | 3–7 |
| — Field harvester | 5–7 |
| — Peanut harvester | 3–7 |
| — Potato digger | 5–7 |
| Air compressor | 10–11 |
| Automotive industry | 10–11 |
| Baling machine | 5–7 |
| Bottling industry | |
| — Capping | 6–7 |
| — Filling | 6–7 |
| — Labeling | 6–7 |
| — Washer, sterilizer | 6–7 |
| Brewing industry | |
| — Agitator | 6–8 |
| — Barrel washer | 6–8 |
| — Cookers | 6–8 |
| — Filling machine | 6–8 |
| — Mash tubs | 6–8 |
| — Pasteurizer | 6–8 |
| — Racking machine | 6–8 |
| Brick-making machinery | 5–7 |
| Bridge machinery | 5–7 |
| Briquette machines | 5–7 |
| Cement industry | |
| — Quarry operation | |
| — — Conveyor | 5–6 |
| — — Crusher | 5–6 |
| — — Diesel electric locomotive | |
| — — Electric dragline | 8–9 |
| — — — Cast gear | 3 |
| — — — Cut gear | 6–8 |
| — — Electric locomotive | 6–8 |
| — — Electric shovel | |
| — — — Cast gear | 3 |
| — — — Cut gear | 6–8 |
| — — Elevator | 5–6 |
| — — Locomotive crane | |
| — — — Cast gear | 3 |
| — — — Cut gear | 5–6 |
| Cement industry (continued) | |
| — (Plant operation) | |
| — — Air separator | 5–6 |
| — — Ball mill | 5–7 |
| — — Compeb mill | 5–6 |
| — — Conveyor | 5–6 |
| — — Cooler | 5–6 |
| — — Elevator | 5–6 |
| — — Feeder | 5–6 |
| — — Filter | 5–6 |
| — — Kiln | 5–6 |
| — — Kiln slurry agitator | 5–6 |
| — — Overhead crane | 5–6 |
| — — Pug, rod, and tube mills | 5–6 |
| — — Pulverizer | 5–6 |
| — — Raw and finish mill | 5–6 |
| — — Rotary dryer | 5–6 |
| — — Slurry agitator | 5–6 |
| Chewing gum industry | |
| — Chicle grinder | 6–8 |
| — Coater | 6–8 |
| — Mixer-kneader | 6–8 |
| — Molder-roller | 6–8 |
| — Wrapper | 6–8 |
| Chocolate industry | |
| — Glazer, finisher | 6–8 |
| — Mixer, mill | 6–8 |
| — Molder | 6–8 |
| — Presser, refiner | 6–8 |
| — Tampering | 6–8 |
| — Wrapper | 6–8 |
| Clay-working machinery | 5–7 |
| Construction equipment | |
| — Backhoe | 6–8 |
| — Cranes | |
| — — Open gearing | 3–6 |
| — — Enclosed gearing | 6–8 |
| — Ditch digger | 3–8 |
| — — Transmission | 6–8 |
| — Drag line | 5–8 |
| — Dumpster | 6–8 |
| — Paver loader | 3 |
| — — Transmission | 8 |
| — — Mixer | 3–5 |
| — — Swing gear | 3–5 |
| — — Mixing bucket | 3 |
| — Shaker | 8 |
| — Shovels | |
| — — Open gearing | 3–6 |
| — — Enclosed gearing | 6–8 |
| — Stationary mixer | |
| — — Transmission | 8 |
| — — Drum gears | 3–5 |

* Quality numbers are inclusive from the highest to the lowest number shown.

**TABLE 8.10** AGMA Applications and Quality Numbers for Racks and Gears (*Continued*)

| Gearing application | Quality no.* | Gearing application | Quality no.* |
|---|---|---|---|
| Stone crusher | | Electronic instrument control and guidance systems (continued) | |
| Transmission | 8 | Altimeter-stabilizer | 9–11 |
| Conveyor | 6 | Analog computer | 10–12 |
| Truck mixer | | Antenna assembly | 7–9 |
| Transfer case | 9 | Antiaircraft detector | 12 |
| Drum gears | 3–5 | Automatic pilot | 9–11 |
| Commercial meters | | Digital computer | 10–12 |
| Gas | 7–9 | Gun data computer | 12–14 |
| Liquid, water, milk | 7–9 | Gyro caging mechanism | 10–12 |
| Parking | 7–9 | Gyroscope-computer | 12–14 |
| Computing and accounting machines | | Pressure transducer | 12–14 |
| Accounting-billing | 9–10 | Radar, sonar, tuner | 10–12 |
| Adding machine-calculator | 7–9 | Recorder, telemeter | 10–12 |
| Addressograph | 7 | Servo system component | 9–11 |
| Bookkeeping | 9–10 | Sound detector | 9 |
| Cash register | 7 | Transmitter, receiver | 10–12 |
| Comptometer | 6–8 | Engines | |
| Computing | 10–11 | Combustion | |
| Data processing | 7–9 | Engine accessories | 10–12 |
| Dictating machine | 9 | Supercharger | 10–12 |
| Typewriter | 8 | Timing gearings | 10–12 |
| Cranes | | Transmission | 8–10 |
| Boom hoist | 5–6 | Farm equipment | |
| Gantry | 5–6 | Milking machine | 6–8 |
| Load hoist | 5–7 | Separator | 8–10 |
| Overhead | 5–6 | Sweeper | 4–6 |
| Ship | 5–7 | Flour mill industry | |
| Crushers | | Bleacher | 7–8 |
| Ice, feed | 6–8 | Grain cleaner | 7–8 |
| Portable and stationary | 6–8 | Grinder | 7–8 |
| Rock, ore, coal | 6–8 | Hulling | 7–8 |
| Dairy industry | | Milling, scouring | 7–8 |
| Bottle washer | 6–7 | Polisher | 7–8 |
| Homogenizer | 7–9 | Separator | 7–8 |
| Separator | 7–9 | Foundry industry | |
| Dish washer | | Conveyor | 5–6 |
| Commercial | 5–7 | Elevator | 5–6 |
| Distillery industry | | Ladle | 5–6 |
| Agitator | 5–7 | Molding machine | 5–6 |
| Bottle filler | 5–7 | Overhead cranes | 5–6 |
| Conveyor, elevator | 6–7 | Sand mixer | 5–6 |
| Grain pulverizer | 6–8 | Sand slinger | 5–6 |
| Mash tub | 5–7 | Tumbling mill | 5–6 |
| Mixer | 5–7 | Home appliances | |
| Yeast tub | 5–7 | Blender | 6–8 |
| Electric furnace | | Mixer | 7–9 |
| Tilting gears | 5–7 | Timer | 8–10 |
| Electronic instrument control and guidance systems | | Washing machine | 8–10 |
| Accelerometer | 10–12 | Machine-tool industry | |
| Airborne temperature recorder | 12–14 | Hand motion (but not indexing and positioning) | 6–9 |
| Aircraft instrument | 12 | | |

* Quality numbers are inclusive from the highest to the lowest number shown.

**TABLE 8.10** AGMA Applications and Quality Numbers for Racks and Gears (*Continued*)

| Gearing application | Quality no.* |
|---|---|
| Machine-tool industry (continued) | |
| Power drives | |
| 0–800 ft/min | 6–8 |
| 800–2000 ft/min | 8–10 |
| 2000–4000 ft/min | 10–12 |
| Over 4000 ft/min | 12 & up |
| Indexing and positioning | |
| Approximate positioning | 6–10 |
| Accurate indexing and positioning | 12 & up |
| Marine industry | |
| Anchor hoist | 6–8 |
| Cargo hoist | 7–8 |
| Conveyor | 5–7 |
| Davit gearing | 5–7 |
| Elevator | 6–7 |
| Small propulsion | 10–12 |
| Steering gear | 8 |
| Winch | 5–8 |
| Metalworking | |
| Bending roll | 5–7 |
| Draw bench | 6–8 |
| Forge press | 5–7 |
| Punch press | 5–7 |
| Roll lathe | 5–7 |
| Mining and preparation | |
| Breaker | 5–6 |
| Car dump | 5–6 |
| Concentrator | 5–6 |
| Continuous miner | 6–7 |
| Conveyor | 5–7 |
| Cutting machine | 6–10 |
| Drag line | |
| Open gearing | 3–6 |
| Enclosed gearing | 6–8 |
| Drills | 5–6 |
| Drier | 5–6 |
| Electric locomotive | 6–8 |
| Elevator | 5–6 |
| Feeder | 6–8 |
| Flotation | 5–6 |
| Grizzly | 5–7 |
| Hoists, skips | 7–8 |
| Loader (underground) | 5–8 |
| Rock drill | 5–6 |
| Rotary car dump | 6–8 |
| Screen (rotary) | 7–8 |
| Screen (shaking) | 7–8 |
| Separator | 5–6 |
| Sedimentation | 5–6 |
| Shaker | 6–8 |
| Shovel | 3–8 |
| Tipple gearing | 5–7 |
| Washer | 6–8 |
| Paper and pulp | |
| Bag machines | 6–8 |
| Box machines | 6–8 |
| Building paper | 6–8 |
| Calender | 6–8 |
| Chipper | 6–8 |
| Coating | 6–8 |
| Envelope machines | 6–8 |
| Food container | 6–8 |
| Glazing | 6–8 |
| Log conveyor-elevator | 5–7 |
| Mixer, agitator | 6–8 |
| Paper machine | |
| Auxiliary | 8–9 |
| Main drive | 10–12 |
| Press, couch, drier rolls | 6–8 |
| Slitting | 10–12 |
| Steam drum | 6–8 |
| Varnishing | 6–8 |
| Wall paper machines | 6–8 |
| Paving industry | |
| Aggregate drier | 5–7 |
| Aggregate spreader | 5–7 |
| Asphalt mixer | 5–7 |
| Asphalt spreader | 5–7 |
| Concrete batch mixer | 5–7 |
| Photographic equipment | |
| Aerial | 10–12 |
| Commercial | 8–10 |
| Printing industry | |
| Press | |
| Book | 9–11 |
| Flat | 9–11 |
| Magazine | 9–11 |
| Newspaper | 9–11 |
| Roll reels | 6–7 |
| Rotary | 9–11 |
| Pump industry | |
| Liquid | 10–12 |
| Rotary | 6–8 |
| Slush-duplex-triplex | 6–8 |
| Vacuum | 6–8 |
| Quarry industry | |
| Conveyor-elevator | 6–7 |
| Crusher | 5–7 |
| Rotary screen | 7–8 |
| Radar and missile | |
| Antenna elevating | 8–10 |
| Data gear | 10–12 |
| Launch pad azimuth | 8 |
| Ring gear | 9–12 |
| Rotating drive | 10–12 |

* Quality numbers are inclusive from the highest to the lowest number shown.

**TABLE 8.10** AGMA Applications and Quality Numbers for Racks and Gears (*Continued*)

| Gearing application | Quality no.* |
|---|---|
| Railroads | |
| Construction hoist | 5–7 |
| Wrecking crane | 6–8 |
| Rubber and plastics | |
| Boot and shoe machines | 6–8 |
| Drier, press | 6–8 |
| Extruder, strainer | 6–8 |
| Mixer, tuber | 6–8 |
| Refiner, calender | 5–7 |
| Rubber mill, scrap cutter | 5–7 |
| Tire building | 6–8 |
| Tire chopper | 5–7 |
| Washer, banbury mixer | 5–7 |
| Small power tools | |
| Bench grinder | 6–8 |
| Drills-saws | 7–9 |
| Hair clipper | 7–9 |
| Hedge clipper | 7–9 |
| Sander, polisher | 8–10 |
| Sprayer | 6–8 |
| Space navigation | |
| Sextant and star tracker | 14 & up |
| Steel industry | |
| Miscellaneous drives | |
| Bessemer tilt-car dump | 5–6 |
| Coke pusher, distributor | 5–6 |
| Conveyor, door lift | 5–6 |
| Electric furnace tilt | 5–6 |
| Hot metal car tilt | 5–6 |
| Hot metal charger | 5–6 |
| Jib hoist, dolomite machine | 5–6 |
| Larry car, mud gun | 5–6 |
| Mixing bin, mixer tilt | 5–6 |
| Ore crusher, pig machine | 5–6 |
| Pulverizer, quench car | 5–6 |
| Shaker, sinter conveyor | 5–6 |
| Sinter machine skip hoist | 5–6 |
| Slag crusher, slag shovel | 5–6 |
| Primary and secondary rolling mill drives | |
| Blooming and plate mill | 5–6 |
| Heavy-duty hot mill drives | 5–6 |
| Slabbing and strip mill | 5–6 |
| Hot mill drives | |
| Sendzimer-stekel | 7–8 |
| Tandem-temper-skin | 6–7 |
| Cold mill drives | |
| Bar, merchant, rail, rod | 5–6 |
| Structural, tube | 5–6 |
| Auxiliary and miscellaneous drives | |
| Annealing furnace car | 5–6 |
| Bending roll | 5–6 |
| Blooming mill manipulator | 5–6 |
| Steel industry (continued) | |
| Blooming mill rack and pinion | 5–6 |
| Blooming mill side guard | 5–6 |
| Car haul | 5–6 |
| Coil conveyor | 5–6 |
| Coil dump | 5–6 |
| Crop conveyor | 5–6 |
| Edger drives | 5–6 |
| Electrolytic line | 6–7 |
| Flange machine ingot buggy | 5–6 |
| Leveler | 6–7 |
| Magazine pusher | 6–7 |
| Mill shear drives | 6–7 |
| Mill table drives (under 800 ft/min) | 5–6 |
| Mill table drives (over 800–1800 ft/min) | 6–7 |
| Mill table drives (over 1800 ft/min) | 8 |
| Nail and spike machine | 5–6 |
| Piler | 5–6 |
| Plate mill rack and pinion | 5–6 |
| Plate mill side guards | 5–6 |
| Plate turnover | 5–6 |
| Preheat furnace pusher | 5–6 |
| Processor | 6–7 |
| Pusher rack and pinion | 5–6 |
| Rotary furnace | 5–6 |
| Shear depress table | 5–6 |
| Slab squeezer | 5–6 |
| Slab squeezer rack and pinion | 5–6 |
| Slitter, side trimmer | 6–7 |
| Tension reel | 6–7 |
| Tilt table, upcoiler | 5–6 |
| Transfer car | 5–6 |
| Wire drawing machine | 6–7 |
| Precision gear drives | |
| Diesel electric gearing | 8–9 |
| Flying shear | 9–10 |
| Shear timing gears | 9–10 |
| High speed reels | 8–9 |
| Locomotive timing gears | 9–10 |
| Pump gears | 8–9 |
| Tube reduction gearing | 8–9 |
| Turbine | 9–10 |
| Overhead cranes | |
| Billet charger, cold mill | 5–6 |
| Bucket handling | 5–6 |
| Car repair shop | 5–6 |
| Cast house, coil storage | 5–6 |
| Charging machine | 5–6 |
| Cinder yard, hot top | 5–6 |
| Coal and ore bridges | 5–6 |

* Quality numbers are inclusive from the highest to the lowest number shown.

**TABLE 8.10** AGMA Applications and Quality Numbers for Racks and Gears (*Continued*)

| Gearing application | Quality no.* | Gearing application | Quality no.* |
|---|---|---|---|
| Steel industry (continued) | | Steel industry (continued) | |
| Electric furnace charger | 5–6 | Slab handling | 5–6 |
| Hot metal, ladle | 5–6 | Miscellaneous | |
| Hot mill, ladle house | 5–6 | Clocks | 6 |
| Jib crane, motor room | 5–6 | Counters | 7–9 |
| Mold yard, rod mill | 5–6 | Fishing reel | 6 |
| Ore unloader, stripper | 5–6 | Gauges | 8–10 |
| Overhead hoist | 5–6 | IBM card puncher, sorter | 8 |
| Pickler building | 5–6 | Metering pumps | 7–8 |
| Pig machine, sand house | 5–6 | Motion-picture equipment | 8 |
| Portable hoist | 5–6 | Popcorn machine, comm. | 6–7 |
| Scale pit, shipping | 5–6 | Pumps | 5–7 |
| Scrap balers and shears | 5–6 | Sewing machine | 8 |
| Scrap preparation | 5–6 | Slicer | 7–8 |
| Service shops | 5–6 | Vending machines | 6–7 |
| Skull cracker | 5–6 | | |

* Quality numbers are inclusive from the highest to the lowest number shown.
SOURCE: Extracted from AGMA standard 390.03 with the permission of the publisher, American Gear Manufacturers Association, 1500 King Street, Suite 201, Alexandria, Virginia 22314.

***Straight Bevel and Zerol Gearing (0° Spiral).*** In straight bevel and zerol gearing, the gear forces tend to push the pinion and gear out of mesh so that the direction of the thrust and separating forces is always the same regardless of the direction of rotation.

In calculating the tangential force $F_{3p}$ or $F_{3g}$ for bevel gearing, use the pinion or gear *mean* diameter $M_{dp}$ or $M_{dg}$ instead of the pitch diameter $P_{dp}$ or $P_{dg}$. The mean diameter is calculated from

$$M_{dg} = P_{dg} - b \sin \gamma_g \qquad M_{dp} = P_{dp} - b \sin \gamma_p$$

In straight bevel and zerol gearing, $F_{3p} = F_{3g}$. (See Fig. 8.90.)

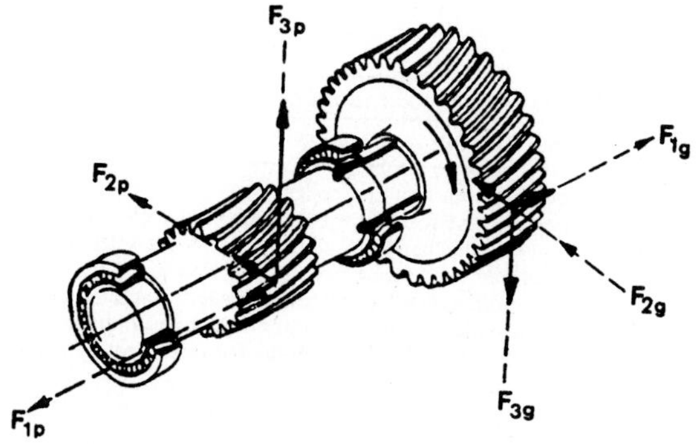


**FIGURE 8.89** Forces in helical gears.

**TABLE 8.11** AGMA Gear Specification Sheet

AGMA Design Manual for Fine-Pitch Gearing: Recommended Minimum Fine-Pitch Spur and Helical Gear Specifications for General Applications (See Note 1)

Arranged for printing on a standard form drawing or for application to drawing by rubber stamp; for proper use of capital or lowercase lettering, see note 2.

| | Spur Gear Data | | |
|---|---|---|---|
| Basic specification data | Number of teeth $N$ | | |
| | Diametral pitch $P$ | | |
| | Pressure angle $\phi$ | | |
| | Standard pitch diameter $N/P$ | (Ref.) | |
| | Tooth form (per AGMA 207.05) | | |
| | Max. calc. cir. thickness on standard pitch circle | | |
| Manufacturing and inspection data | Gear testing radius | | |
| | AGMA quality number | | |
| | Total composite tolerance | | |
| | Tooth-to-tooth composite tolerance | | |
| | Outside diameter | | |
| | Master gear basic circular tooth thickness at standard pitch circle | (Ref.) | |
| | Master gear number of teeth | (Ref.) | |
| | **Helical Gear Data*** | | |
| Basic specification data | Number of teeth $N$ | | |
| | Normal diametral pitch $P_n$ | | |
| | Normal pressure angle $\phi_n$ | | |
| | Helix angle—hand $\psi$ | | |
| | Standard pitch diameter $N/P_n \cos \psi$ | (Ref.) | |
| | Tooth form (per AGMA 207.05) | | |
| | Maximum calculated normal circular thickness on standard pitch circle | | |
| Manufacturing and inspection data | Gear testing radius | | |
| | AGMA quality number | | |
| | Total composite tolerance | | |
| | Tooth-to-tooth composite tolerance | | |
| | Lead | | |
| | Outside diameter | | |
| | Master gear basic normal circular tooth thickness at standard pitch circle | (Ref.) | |
| | Master gear number of teeth | (Ref.) | |

* If desired, a combination format covering both spur and helical gears can be used by specifying the helix angle equal to zero degrees. This permits standardization on the helical drawing format for both spur and helical gears.

*Note 1:* For data on the determination of spur and helical tooth proportions, see Section 4 or Standard AGMA 207.05. For data on inspection, see Section 10. For data on quality number, see Section 6 or AGMA 390.03.

*Note 2:* The use of all uppercase letters or both uppercase and lowercase letters is optional. The spur gear format illustrates the proper use of both uppercase and lowercase letters. The helical format illustrates the use of all uppercase (capital) letters.

SOURCE: Extracted from AGMA standard 370.01 (R-1978) with the permission of the publisher, American Gear Manufacturers Association, 1500 King Street, Suite 201, Alexandria, Virginia 22314.

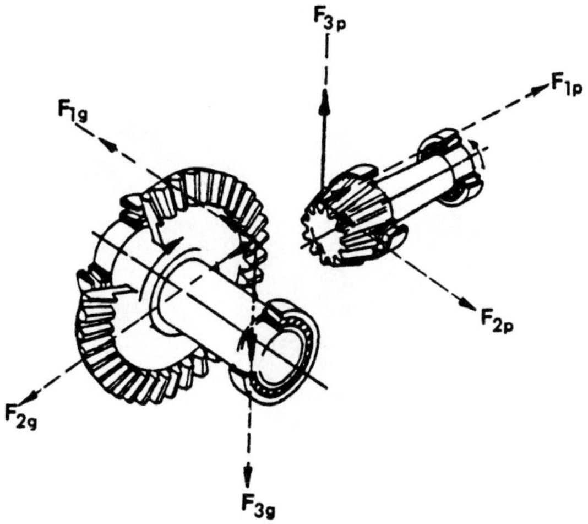


**FIGURE 8.90** Forces in straight bevel and zerol gears.

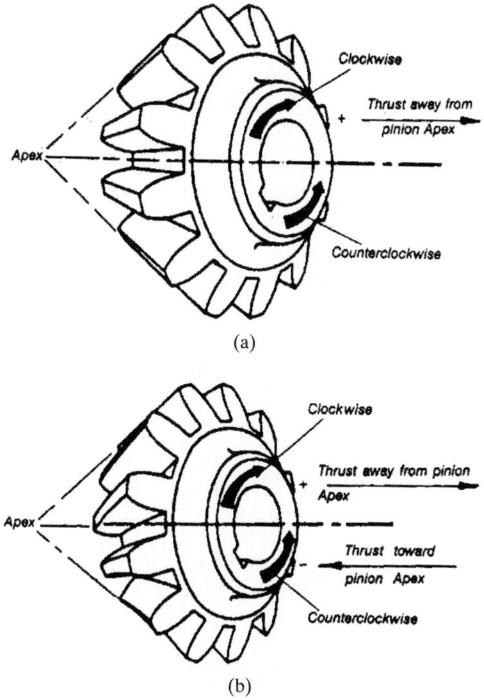


**FIGURE 8.91** (*a*) Direction of thrust, straight bevel gears; (*b*) direction of thrust, spiral bevel gears.

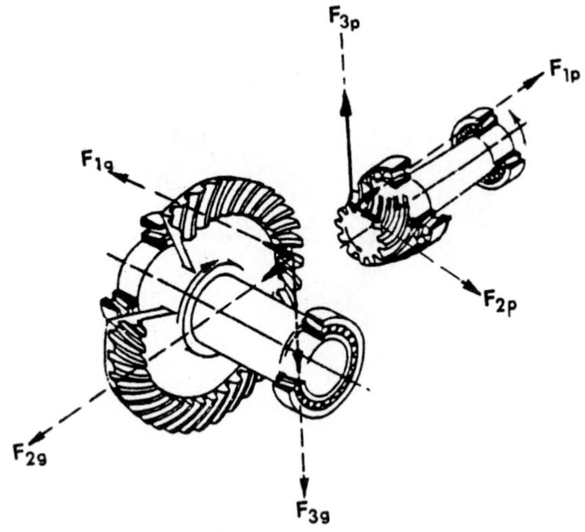


**FIGURE 8.92** Forces in spiral bevel gears.

At the *pinion,* tangential force is

$$F_{3p}, \mathrm{N} = \frac{(1.91 \times 10^7)P}{M_{dp}\ \mathrm{rpm}_p}$$

$$F_{3p}, \mathrm{lbF} = \frac{(1.26 \times 10^5)P}{M_{dp}\ \mathrm{rpm}_p}$$

Thrust force:

$$F_{1p} = F_{3p} \tan \phi_p \cos \gamma_p$$

| Driving-member rotation | Thrust force | Separating force |
|---|---|---|
| Right-hand spiral—clockwise | $F_{1p} = \frac{F_{3p}}{\cos \psi_p}(\tan \phi_p \sin \gamma_p - \sin \psi_p \cos \gamma_p)$, driving gear | $F_{2p} = \frac{F_{3p}}{\cos \psi_p}(\tan \phi_p \cos \gamma_p + \sin \psi_p \sin \gamma_p)$, driving gear |
| Left-hand spiral—counterclockwise | $F_{1g} = \frac{F_{3g}}{\cos \psi_g}(\tan \phi_g \sin \gamma_g + \sin \psi_g \cos \gamma_g)$, driven gear | $F_{2g} = \frac{F_{3g}}{\cos \psi_g}(\tan \phi_g \cos \gamma_g - \sin \psi_g \sin \gamma_g)$, driven gear |
| Right-hand spiral—counterclockwise | $F_{1p} = \frac{F_{3p}}{\cos \psi_p}(\tan \phi_p \sin \gamma_p + \sin \psi_p \cos \gamma_p)$, driving gear | $F_{2p} = \frac{F_{3p}}{\cos \psi_p}(\tan \phi_p \cos \gamma_p - \sin \psi_p \sin \gamma_p)$, driving gear |
| Left-hand spiral—clockwise | $F_{1g} = \frac{F_{3g}}{\cos \psi_g}(\tan \phi_g \sin \gamma_g - \sin \psi_g \cos \gamma_g)$, driven gear | $F_{2g} = \frac{F_{3g}}{\cos \psi_g}(\tan \phi_g \cos \gamma_g + \sin \psi_g \sin \gamma_g)$, driven gear |

**FIGURE 8.93** Thrust and separating force equations, spiral bevel and hypoid gearing.

Separating force:

$$F_{2p} = F_{3p} \tan \phi_p \cos \gamma_p$$

At the *gear*, tangential force is

$$F_{3g}, \text{N} = \frac{(1.91 \times 10^7)P}{M_{dg} \text{ rpm}_g}$$

$$F_{3g}, \text{lbF} = \frac{(1.26 \times 10^5)}{M_{dg} \text{ rpm}_g}$$

Thrust force:

$$F_{1g} = F_{3g} \tan \phi_g \sin \gamma_g$$

Separating force:

$$F_{2g} = F_{3g} \tan \phi_g \cos \gamma_g$$

***Spiral Bevel and Hypoid Gearing.*** In spiral bevel and hypoid gearing, the direction of the thrust and separating forces depends on spiral angle, hand of spiral, direction of rotation, and whether the gear is *driving* or *driven.*

The hand of the spiral is determined by noting whether the tooth curvature on the near face of the gear (Fig. 8.91*a* and 8.91*b*), inclines to the left or right from the shaft axis. Direction of rotation is determined by viewing toward the gear or pinion apex.

In spiral bevel gearing (Fig. 8.92)

$$F_{3p} = F_{3g}$$

In hypoid gearing

$$F_{3p} = \frac{F_{3g} \cos \psi_p}{\cos \psi_g}$$

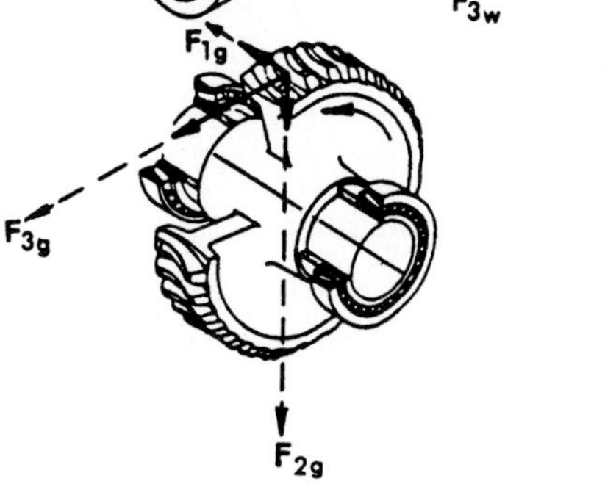

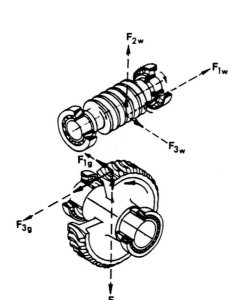


**FIGURE 8.94** Forces in worm gears.

The hypoid-pinion effective working diameter is

$$M_{dp} = M_{dg} \left(\frac{n_p}{n_g}\right)\left(\frac{\cos \psi_g}{\cos \psi_p}\right)$$

The hypoid-gear effective working diameter is

$$M_{dg} = P_{dg} - b \sin \psi_g$$

Thrust and separating-force equations are summarized in Fig. 8.93. The cases listed include driving member rotating either clockwise or counterclockwise.

***Straight Worm Gearing.*** See Fig. 8.94. For the worm, tangential force is

$$F_{3w}, \text{N} = \frac{(1.91 \times 10^7)P}{P_{dw} \text{ rpm}_w}$$

$$F_{3w}, \text{lbF} = \frac{(1.26 \times 10^5)P}{P_{dw} \text{ rpm}_w}$$

Thrust force:

$$F_{1w}, \text{N} = \frac{(1.91 \times 10^7)P\eta}{P_{dg} \text{ rpm}_g}$$

$$F_{1w}, \text{lbF} = \frac{(1.26 \times 10^5)P\eta}{P_{dg} \text{ rpm}_g}$$

or

$$F_{1w} = \frac{F_{3w}\eta}{\tan \lambda}$$

Separating force is

$$F_{2w} = \frac{F_{3w} \sin \phi}{\cos \phi \sin \lambda + \mu \cos \lambda}$$

For the worm gear, tangential force is

$$F_{3g}, \text{N} = \frac{(1.91 \times 10^7)\text{P}\eta}{P_{dg} \text{ rpm}_g}$$

$$F_{3g}, \text{lbF} = \frac{(1.26 \times 10^5)P\eta}{P_{dg} \text{ rpm}_g}$$

or

$$F_{3g} = \frac{F_{3w}\eta}{\tan \lambda}$$

Thrust force:

$$F_{1g}, \text{N} = \frac{(1.91 \times 10^7)P}{P_{dw} \text{ rpm}_w}$$

$$F_{1g}, \text{lbF} = \frac{(1.26 \times 10^5)P}{P_{dw} \text{ rpm}_w}$$

Separating force:

$$F_{2g} = \frac{F_{3w} \sin \phi}{\cos \phi \sin \lambda + \mu \cos \lambda}$$

where

$$\lambda = \tan^{-1}\left(\frac{P_{dg}}{g_r P_{dw}}\right) \text{ or } \lambda = \tan^{-1}\left(\frac{L}{\pi P_{dw}}\right)$$

and

$$\eta = \frac{\cos \phi - \mu \tan \lambda}{\cos \phi + \mu \cot \lambda}$$

In the SI metric system, the coefficient of friction $\mu$ is approximately

$$\mu = (5.34 \times 10^{-7})V_s^3 + \frac{0.146}{V_s^{0.09}} - 0.103$$

and

$$V_s, \text{m/sec} = \frac{P_{du} \text{rpm}_w}{(1.91 \times 10^4) \cos \lambda}$$

In the U.S. customary system (inches)

$$\mu = (7 \times 10^{-14})\, V_s^3 + \frac{0.235}{V_s^{0.09}} - 0.103$$

and

$$V_s,\ \text{ft/min} = \frac{P_{dw}\,\text{rpm}_w}{3.82 \cos \lambda}$$

The approximations are for the coefficient of friction as given in AGMA standard 440.04, October 1971, Table 4, for the rubbing-velocity range 0.015 to 15 m/sec (3 to 3000 ft/min).

## 8.10.2 Nomenclature (Bearing Forces)

$C$ = center distance of gears, in or mm
$F_b$ = belt or chain pull, lbf or N
$F_1$ = thrust force, lbf or N
$F_{1g}, F_{1p}, F_{1w}$ = thrust force on gear, pinion, or worm, lbf or N
$F_2$ = separating force, lbf or N
$F_{2g}, F_{2p}, F_{2w}$ = separating force on gear, pinion, or worm, lbf or N
$F_3$ = tangential force, lbf or N
$F_{3g}, F_{3p}, F_{3w}$ = tangential force on gear, pinion, or worm, lbf or N
$L$ = lead or axial advance of a helix for one complete revolution, in or mm
$M_d$ = mean diameter or effective working diameter, of sprocket, pulley, or sheave, in or mm
$M_{dg}, M_{dp}, M_{dw}$ = mean diameter or effective working diameter of gear, pinion, or worm, respectively, in or mm
$P$ = power, hp or kW
$P_{dg}, P_{dp}, P_{dw}$ = pitch diameter of gear, pinion or worm, in or mm
$V_s$ = surface or rubbing velocity, ft/min or m/sec
$b$ = tooth length, in or mm
$f_{cb}$ = belt or chain pull factor
$g$ = subscript for gear
$g_r$ = gearing ratio
$n_g, n_p, n_s$ = number of teeth in gear, pinion, or sprocket
$p$ = subscript for pinion
$p'$ = pitch or distance between equally spaced tooth surfaces along the pitch line, in or mm
$w$ = subscript for worm
$\gamma$ = for bevel gearing—pitch angle of gear, degrees; for hypoid gearing—face angle of pinion and root angle of gear, degrees
$\eta$ = efficiency
$\lambda$ = worm gearing lead angle, degrees
$\mu$ = coefficient of dynamic friction
$\phi$ = normal tooth pressure angle for gear or pinion, degrees

| | |
|---|---|
| $\psi$ | = for helical gearing, helix angle for gear or pinion, degrees spiral angle for gear or pinion, degrees |
| rpm | = rotation, revolutions per minute |

### 8.10.3 Vertical, Horizontal, Radial Reactions (Shaft on Two Supports)

The following equations will be useful for determining the resultant reactions at the bearing points in gear, V-belt, or chain-sprocket systems. The bearings may be selected using values obtained from these equations.

Referring to Fig. 8.95, the vertical-reaction component force at bearing B is

$$R_{LBv} = \frac{1}{x}\left[x_1\,(F_{2g}\cos\theta_1 + F_{3g}\sin\theta_1) + \frac{1}{2}\,(P_{dg} - b\sin\gamma_g)\,F_{1g}\cos\theta_1 + x_2\,F\cos\theta_2 + M\cos\theta_3\right]$$

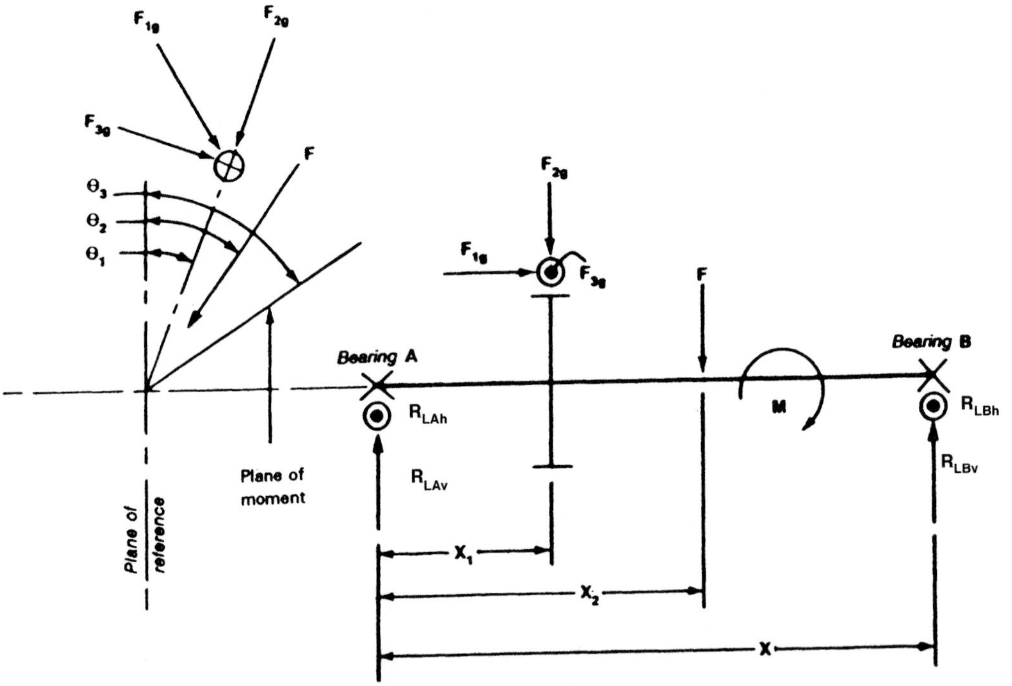


**Note:** Shown in this section are equations for the case of a shaft on two supports with gear forces $F_1$ (tangential), $F_2$ (separating), and $F_3$ (thrust), an external radial load F, and an external moment M. The loads are applied at arbitrary angles ($\theta_1$, $\theta_2$, and $\theta_3$) relative to the reference plane indicated in the above figure. Using the principles of superposition, the equations for radial and horizontal reactions ($R_{LA}$ and $R_{LB}$) can be expanded to include any number of gears, external forces, or moments. Use signs as determined from the gear force equations.

**FIGURE 8.95** Application of load diagram, gearing systems.

Horizontal-reaction component force at bearing B is

$$R_{LBh} = \frac{1}{x}\left[x_1 (F_{2g} \sin \theta_1 + F_{3g} \cos \theta_1) + \frac{1}{2} (P_{dg} - b \sin \gamma_g)\, F_{1g} \sin \theta_1 + x_2\, F \sin \theta_2 + M \sin \theta_3\right]$$

Vertical-reaction component force at bearing A is

$$R_{LAv} = F_{2g} \cos \theta_1 + F_{3g} \sin \theta_1 + F \cos \theta_2 - R_{LBv}$$

Horizontal-reaction component force at bearing A is

$$R_{LAh} = F_{2g} \sin \theta_1 - F_{3g} \cos \theta_1 + F \sin \theta_2 - R_{LBh}$$

The resultant radial reactions are

$$TR_{LA} = \sqrt{(R_{LAv})^2 + (R_{LAh})^2}$$

and

$$TR_{LB} = \sqrt{(R_{LBv})^2 + (R_{LBh})^2}$$

Nomenclature for preceding reaction equations is as follows:

$x$ = bearing spread, in or mm

$A$, $B$ = bearing-position subscripts

$x_1$, $x_2$ = linear distance (positive or negative), in or mm

$F_{1g}$, $F_{2g}$, $F_{3g}$ = Thrust, separating and tangential forces on the gear, sheave, or sprocket; i.e., $F_1$ = thrust, $F_2$ = separating force, $F_3$ = tangential force; $g$ = gear, sheave, or sprocket; lbf or N

$R_L$ = reaction component of force, lbf or N

$TR_L$ = resultant radial reaction, lbf or N

$h$ = subscript for horizontal

$M$ = moment, N·mm or lbf·in

$v$ = subscript for vertical

$\gamma$ = bevel gearing pitch angle of gear or pinion

$\theta_1$, $\theta_2$, $\theta_3$ = gear mesh angle relative to the plane of reference as shown in Fig. 8.95

## 8.11 GEAR LOADS AND DESIGN PROCEDURES

The following sections will aid the designer in selecting the correct gear sizes and gear materials for most general applications. The design procedures consist of a series of equations which are used to size the gears and then check the dynamic and wear loads to make certain that these loads are within the allowable stress limits for each particular design and material. The gear systems for which the equations have been developed include spur gears, helical gears, bevel gears, spiral bevel gears, and worm gears.

### 8.11.1 Spur Gear Loads and Design Procedures

A typical set of spur gears is shown in Fig. 8.96, from which the gear terminology and geometric features can be seen.

The proportions for standard gear teeth are shown in Fig. 8.97.

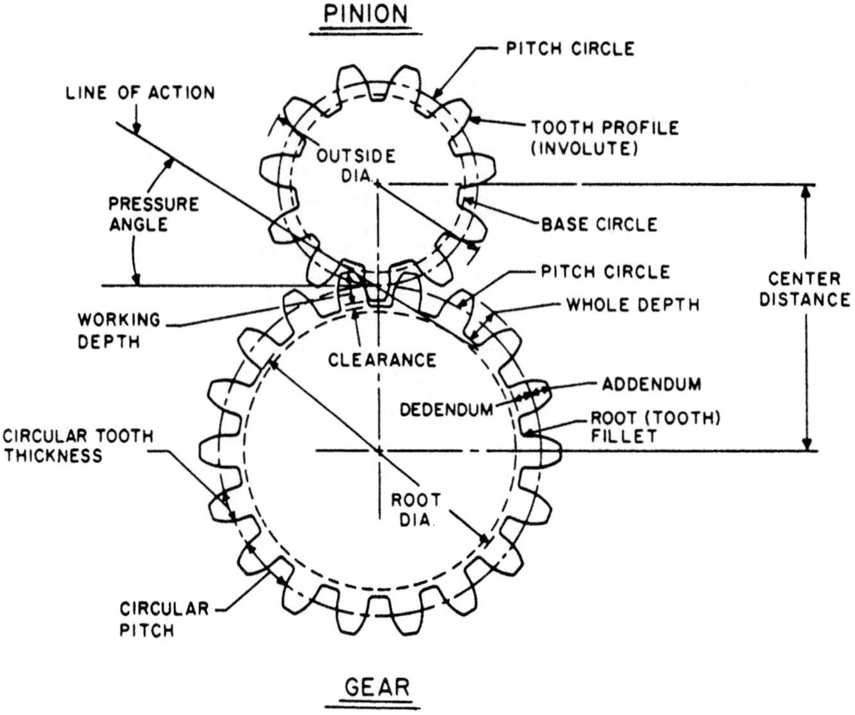


**FIGURE 8.96** Typical set of spur gears and terminology.

| Tooth Type | 14.5° Composite | 14.5° Full Depth Involute | 20° Full Depth Involute | 20° Stub Involute |
|---|---|---|---|---|
| Addendum | $1/P_d$ | $1/P_d$ | $1/P_d$ | $0.8/P_d$ |
| Minimum dedendum | $1.157/P_d$ | $1.157/P_d$ | $1.157/P_d$ | $1/P_d$ |
| Whole depth | $2.157/P_d$ | $2.157/P_d$ | $2.157/P_d$ | $1.8/P_d$ |
| Clearance | $0.157/P_d$ | $0.157/P_d$ | $0.157/P_d$ | $0.2/P_d$ |

Note: In the composite tooth form, the middle third of the tooth profile has an involute shape, while the remainder is cycloidal.

**FIGURE 8.97** Proportions of standard gear teeth.

*Note:* $P_c = \dfrac{\pi D}{N}$ $\quad P_d = \dfrac{N}{D}$ $\quad P_c P_d = \pi$

where $P_c$ = circular pitch, in
$P_d$ = diametral pitch
$N$ = number of teeth
$D$ = pitch diameter, in

***Interference of Spur Gear Teeth.*** To avoid involute profile *overlap* of teeth in spur gears, the maximum addendum radius for each gear must be equal to or less than

$$\sqrt{(\text{radius of base circle})^2 + (\text{center distance})^2 (\sin \phi)^2}$$

where $\phi$ = pressure angle, degrees.

***Allowable Tooth Stresses.*** The allowable tooth stress depends on the gear material and the pitch line velocity $V$ as follows. Allowable stress $s_a$:

$$s_a = \begin{cases} s_0 \dfrac{600}{600 + V} & \text{(for } V < 2000 \text{ ft/min)} \\[2ex] s_0 \dfrac{1200}{1200 + V} & \text{(for } V \text{ 2000–4000 ft/min)} \\[2ex] s_0 \dfrac{78}{78\sqrt{V}} & \text{(for } V > 4000 \text{ ft/min)} \end{cases}$$

where $s_a$ = allowable stress, psi; $s_0 = 0.33 \times$ ultimate (endurance) strength of material. Precise $s_0$ values may be obtained from the materials tables shown in Tables 8.3, 8.4, 8.6, and 8.7; also from the materials section in Chap. 4.

*Note:* $s_0$ values are approximate for the following materials:

Cast iron = 8000 psi

Bronze = 12,000 psi

Steels = 10,000 to 50,000 psi

***Actual Operating Tooth Stress.*** If the diametral pitch is *known*

$$\frac{P_d^2}{Y} = \frac{sk\pi^2}{F} \qquad \text{and} \qquad s = \frac{F(P_d)^2}{Yk\pi^2}$$

where $s$ = actual stress (must be $\leqq s_a$), psi (see equation above)
$k$ = 4 (as an upper limit)
$F$ = transmitted force, lbf
$= 2M_t/D$, where $M_t$ = torque of weaker gear, lb·in and $D$ = pitch diameter, in
$P_d$ = diametral pitch
$Y$ = tooth form factor (see Fig. 8.98)

If the pitch diameter is *unknown*

$$s = \frac{2M_t P_d^3}{k\pi^2 YN}$$

| Number of Teeth | 14.5° Full-Depth Involute or Composite | 20° Full-Depth Involute | 20° Stub Involute |
|---|---|---|---|
| 12 | 0.067 | 0.078 | 0.099 |
| 13 | 0.071 | 0.083 | 0.103 |
| 14 | 0.075 | 0.088 | 0.108 |
| 15 | 0.078 | 0.092 | 0.111 |
| 16 | 0.081 | 0.094 | 0.115 |
| 17 | 0.084 | 0.096 | 0.117 |
| 18 | 0.086 | 0.098 | 0.120 |
| 19 | 0.088 | 0.100 | 0.123 |
| 20 | 0.090 | 0.102 | 0.125 |
| 21 | 0.092 | 0.104 | 0.127 |
| 23 | 0.094 | 0.106 | 0.130 |
| 25 | 0.097 | 0.108 | 0.133 |
| 27 | 0.099 | 0.111 | 0.136 |
| 30 | 0.101 | 0.114 | 0.139 |
| 34 | 0.104 | 0.118 | 0.142 |
| 38 | 0.106 | 0.122 | 0.145 |
| 43 | 0.108 | 0.126 | 0.147 |
| 50 | 0.110 | 0.130 | 0.151 |
| 60 | 0.113 | 0.134 | 0.154 |
| 75 | 0.115 | 0.138 | 0.158 |
| 100 | 0.117 | 0.142 | 0.161 |
| 150 | 0.119 | 0.146 | 0.165 |
| 300 | 0.122 | 0.150 | 0.170 |
| Rack | 0.124 | 0.154 | 0.175 |

**FIGURE 8.98** Gear tooth form factor *Y*.

where $s$ = actual stress, psi (must be $\leqq s_a$)
$M_t$ = torque of weaker gear, lb·in
$k$ = 4 (as an upper limit)
$N$ = number of teeth on the weaker gear (~15 minimum)
$P_d$ = diametral pitch
$Y$ = tooth form factor (see Fig. 8.98)

(*Note:* The largest possible $P_d$ will provide the most economical design. In general, when diameters are known, design for the largest number of teeth possible; when diameters are unknown, design for the smallest pitch diameters possible.)

***Dynamic Tooth Loads—Buckingham Equation.*** The dynamic load $F_d$ equation approximating a detailed dynamic analysis is given as

$$F_d = \frac{0.05V\,(bC + F)}{0.05V + \sqrt{bC + F}} + F$$

where $F_d$ = dynamic load, lbf
$V$ = pitch line velocity, ft/min
$F$ = transmitted force, lbf = $2M_t/D$ (also = gear torque/pitch radius of gear)
$b$ = face width of gear, in
$C$ = constant (deformation factor) (see Fig. 8.99)

| Materials: Pinion | Materials: Gear | Involute tooth form | Tooth Error - inches: 0.0005 | 0.001 | 0.002 | 0.003 |
|---|---|---|---|---|---|---|
| Cast iron | Cast iron | 14.5° | 400 | 800 | 1600 | 2400 |
| Steel | Cast iron | 14.5° | 550 | 1100 | 2200 | 3300 |
| Steel | Steel | 14.5° | 800 | 1600 | 3200 | 4800 |
| Cast iron | Cast iron | 20° Full Depth | 415 | 830 | 1660 | 2490 |
| Steel | Cast iron | 20° Full Depth | 570 | 1140 | 2280 | 3420 |
| Steel | Steel | 20° Full Depth | 830 | 1660 | 3320 | 4980 |
| Cast iron | Cast iron | 20° Stub | 430 | 860 | 1720 | 2580 |
| Steel | Cast iron | 20° Stub | 590 | 1180 | 2360 | 3540 |
| Steel | Steel | 20° Stub | 860 | 1720 | 3440 | 5160 |

**FIGURE 8.99** Values of the deformation factor *C*.

***Wear Tooth Loads.*** To ensure durability in a gear *pair*, the profiles of the teeth must not have excessive contact stress as determined by the wear load $F_w$:

$$F_w = D_p b K_1 Q$$

where $D_p$ = pitch diameter of smaller gear (pinion), in
$b$ = face width of gear, in
$K_1$ = stress factor for fatigue (see Fig. 8.100 and next equation)

**Values for ($s_{es}$) and ($K_1$) for Various Materials - for use in wear load equations**

| Average Brinell Hardness Number of Steel Pinion and Steel Gear | | Surface Endurance Limit ($s_{es}$) | Stress Fatigue Factor $K_1$: 14.5° | 20° |
|---|---|---|---|---|
| 150 | | 50,000 | 30 | 41 |
| 200 | | 70,000 | 58 | 79 |
| 250 | | 90,000 | 96 | 131 |
| 300 | | 110,000 | 144 | 196 |
| 350 | | 130,000 | 206 | 281 |
| 400 | | 150,000 | 268 | 366 |
| **Brinell Hardness Number (BHN) Steel Pinion** | **Gear** | | | |
| 150 | Cast Iron | 50,000 | 44 | 60 |
| 200 | Cast Iron | 70,000 | 87 | 119 |
| 250 | Cast Iron | 90,000 | 144 | 196 |
| 150 | Phos. Bronze | 50,000 | 46 | 62 |
| 200 | Phos. Bronze | 65,000 | 73 | 100 |
| Cast Iron Pinion | Cast Iron Gear | 80,000 | 152 | 208 |
| Cast Iron Pinion | Cast Iron Gear | 90,000 | 193 | 284 |

**FIGURE 8.100** Constants for various gear materials.

$Q = 2N_g/(N_p + N_g)$
$N_g$ = number of teeth on gear
$N_p$ = number of teeth on pinion

$$K_1 = \frac{(s_{es})^2 \sin \phi}{1.4} \left[ \frac{1}{E_p} + \frac{1}{E_g} \right]$$

where $s_{es}$ = surface endurance limit of a gear pair, psi (see Fig. 8.100)
$E_p$ = modulus of elasticity of pinion material, psi (see Chap. 4)
$E_g$ = modulus of elasticity of gear material, psi (see Chap. 4)
$\phi$ = pressure angle, degrees

*Note:* $s_{es}$ may be estimated from

$$s_{es} = 400 \times (\text{Bhn}) - 10{,}000$$

and Bhn is the average Brinell hardness number of the pinion and gear for values up to 350 Bhn.

Then the wear load $F_w$ must be equal to or greater than the dynamic load $F_d$. Also, ($K_1$ factors for various materials and tooth forms are shown in Fig. 8.100 as recommended by Buckingham.

The preceding calculation procedures will establish the tentative or preliminary gear design. Final design decisions should be determined by the appropriate tests on the gear system in actual operation when weighed against the tentative design and expected or specified performance criteria.

## 8.11.2 Helical Gear Loads and Design Procedures

The following procedures are for helical gears operating on parallel shafts only, which is the normal function for helical gears. Remember that in helical gears operating on parallel shafts the pinion and the gear must be of opposite hands (i.e., LH pinion—RH gear; RH pinion—LH gear).

***Virtual Number of Teeth.*** The *virtual* number of teeth $N_f$ on a helical gear is the number of teeth that *could* be generated on the ellipse formed by taking a cross section through the gear in the *normal* plane (see Figs. 8.64 and 8.67).

$$N_f = \frac{N}{\cos^3 \psi}$$

where $N$ = actual number of teeth; $\psi$ = helix angle, degrees (see Figs. 8.64 and 8.67).

***Strength Design.*** These procedures are similar to those for spur gears (see Sec. 8.11.1). Analyzing the tooth normal to the helix, the normal load $F_n$ using the Lewis equation is given as

$$F_n = s\left(\frac{b}{\cos \psi}\right)\frac{\pi Y}{P_d/\cos \psi}$$

Then $F$, the tangential forces on the teeth, are

$$F = \frac{sbY\pi}{(P_d / \cos \psi)} = \frac{sk\pi^2 Y}{(P_d / \cos \psi)^2 \cos \psi} \quad \text{(when standard pitch is in the normal plane)}$$

or $$F = \frac{sbY\pi \cos \psi}{P_d} = \frac{sk\pi^2 Y \cos \psi}{P_d^2} \quad \text{(when standard pitch is in the diametral plane)}$$

where $F$ = tangential force, lbf
$K = b/P_c$ (limited to a maximum of ~6)
$P_d$ = diametral pitch in plane of rotation
$Y$ = tooth form factor based on the virtual number of teeth ($N_f$) above*
$s$ = allowable stress $s$, psi

$$s = s_0 \left( \frac{78}{78 + \sqrt{V}} \right)$$

where $s_0 = 0.33 \times$ ultimate strength of material, psi; $V$ = pitch line velocity, ft/min.

In checking the gear design strength, the pitch diameter is either known or unknown. If the pitch diameter $P_d$ is *known,*

$$\frac{P_d^2}{Y} = \frac{s_0 \, k\pi^2 \cos \psi}{F} \left( \frac{78}{78 + \sqrt{V}} \right)$$

where $k = b/P_c$
$F$ = tangential force = torque/pitch radius, lbf
$V$ = pitch line velocity, ft/min
$s_0$ = as in the following equation

$$s_0 = \frac{s}{78/(78 + \sqrt{V})}$$

where $s$ = allowable stress, psi.

If the pitch diameter $P_d$ is *unknown,*

$$s_1 = \frac{2TP_d^3}{kY\pi^2 N \cos \psi}$$

where $s_1$ = actual induced stress, psi
$T$ = resisting torque of weaker gear, lb·in
$N$ = actual number of teeth on the weaker gear

The preceding helical gear equations are a first approximation for determining possible pitch $P_d$ and face width $b$, which are then checked for dynamic and wear loads per the following equations.

***Limiting Endurance Beam Strength.*** The beam strength $F_0$ is derived from the Lewis equation, with no velocity factor, and is given as

$$F_0 = \frac{s_0 \, bY\pi \cos \psi}{P_d} \quad \text{(an allowed value)}$$

where symbols are as defined in the preceding equations.

Then the limiting endurance strength $F_0$ must be equal to or greater than the dynamic load $F_d$, which is given as

$$F_d = F + \frac{0.05V\,(Cb \cos^2\psi + F) \cos \psi}{0.05V + \sqrt{Cb \cos^2 \psi + F}} \quad \text{(use preceding symbols)}$$

and values of $C$ may be obtained from the spur gear section (Fig. 8.99).

* When the pressure angle ($\phi$) is standard in the normal plane, the $Y$ factors from Fig. 8.98 for spur gears may be used. When the pressure angle $\phi$ is standard in the diametral plane, the $Y$ factors of Fig. 8.98 are considered approximate. A more accurate value of $Y$ is seldom necessary when using these procedures, as corrections can be made as you proceed through the calculations.

***Limiting Wear Load.*** The wear load $F_w$ for helical gears is calculated from the Buckingham equation:

$$F_w = \frac{D_p bQK_1}{\cos^2 \psi} \quad \text{(an allowed value)}$$

where $D_p$ = pitch diameter of the pinion, in
$Q$ = as in the following equation

$$Q = \frac{2D_g}{D_p + D_g} = \frac{2N_g}{N_p + N_g} \quad (N_p \text{ and } N_g = \text{actual number of teeth})$$

$K_1$ = as in the following equation

$$K_1 = \frac{s_{es}^2 \sin \phi_n}{1.4}\left(\frac{1}{E_p} + \frac{1}{E_g}\right)$$

$s_{es}$ = surface endurance limit (see Sec. 8.11.1)

Then

- The limiting wear load $F_w$ must be equal to or greater than the dynamic load $F_d$.
- The endurance load $F_0$ must be equal to or greater than the dynamic load $F_d$.

(*Note:* $F_0$ and $F_w$ are allowable values which cannot be exceeded.)

## 8.11.3 Bevel Gear Loads and Design Procedures

Strength design of straight-tooth bevel gears may be based on a modified Lewis equation which takes into account the tapered teeth of the bevel gear. The permissible force $F$ that may be transmitted is then given as

$$F = \frac{sbY\pi}{P_d}\left(\frac{L-b}{L}\right)$$

where $F$ = force transmitted, lbf
$s$ = allowable bending stress, psi (see the following equation)
$Y$ = tooth form factor for bevel gears (use 75 percent of the values shown for spur gears in Fig. 8.98 as an approximation)
$L$ = cone distance, in (see Fig. 8.71*a* and the following equation)

$$L = \sqrt{(R_p)^2 + (R_g)^2}, \text{ in} \quad (\text{shaft angle} = 90°)$$

where $R_p$ = pitch radius of pinion, in
$R_g$ = pitch radius of gear, in
$b$ = face width of gear, in
$P_d$ = diametral pitch based on the largest part of the tooth cross section (at the perimeter of the bevel gear; the point where a gear tooth gauge is used to measure the diametral pitch of a straight-tooth bevel gear)

For improved producibility and operation of straight-tooth bevel gears, the face width $b$ should generally not be greater than $L/3$, where $L$ is the cone distance.

When designing for strength, the diameter of the gear may be either known or unknown. When the diameter is *known,* use the modified Lewis equation (the terms on the right side of this equation can be determined after the gear material has been specified) as shown:

$$\frac{P_d}{Y} = \frac{sb\pi}{F}\left(\frac{L-b}{L}\right) \quad \text{(an allowed value)}$$

where $F$ = transmitted force, lbf = torque of weaker gear/weaker gear pitch radius. The allowed stress $s$ is calculated as explained in the following equations, while the preceding equation then yields an allowable value for $P_d/Y$, which must be satisfied (must equal right side of equation numerically) by selecting an appropriate value for $P_d$.

When the diameter is *unknown,* use the following form of the Lewis equation for calculating the actual stress $s$:

$$s = \frac{2P_d^2 T}{\pi bYN}\left(\frac{L}{L-b}\right) \quad \text{(actual stress, psi; must be less than allowed stress)}$$

In this equation, you may substitute values for $b$ and $L/(L - b)$ as follows. Let

$$b = \frac{N_p}{6P_d}\sqrt{1+R^2} \qquad \text{and} \qquad \frac{L}{L-B} = \frac{3}{2}$$

where $N$ = actual number of teeth in weaker gear

$N_p$ = number of teeth on the pinion

$R$ = ratio of the angular velocity ($A_{vp}$) of the pinion to the angular velocity of the gear ($A_{vg}$), i.e., $R = A_{vp}/A_{vg}$

The design procedures shown for strength are a first approximation which must be checked for wear loads and dynamic loads as shown in the equations that follow.

The *allowable stress s* for average conditions may be calculated as follows:

$$s = s_0\left(\frac{1200}{1200+V}\right), \text{psi} \quad \text{(for cut teeth gears)}$$

or

$$s = s_0\left(\frac{78}{78+\sqrt{V}}\right), \text{psi} \quad \text{(for generated teeth)}$$

where $s_0 = 0.33 \times$ ultimate strength of material, psi; $V$ = pitch line velocity, ft/min ($0.262 \times$ pitch diameter $\times$ rpm).

The *virtual number of teeth* $N_f$ in a straight-tooth bevel gear is

$$N_f = \frac{N}{\cos\alpha}$$

where $N$ = actual number of teeth; $\alpha$ = pitch angle or half-cone angle, degrees.

***Limiting Wear Load.*** The limiting wear load $F_w$ may be approximated from

$$F_w = \frac{0.75 D_p b K_1 Q}{\cos\alpha} \quad \text{(an allowed value)}$$

where $D_p$, $b$, $K_1$, and $Q$ are the same as for spur gears, except that $Q$ is based on the virtual number of teeth $N_f$ and $\alpha$ is the pitch angle of the pinion in degrees.

***Limiting Endurance Load.*** The limiting endurance load $F_0$ may be calculated from

$$F_0 = \frac{s_0 bY\pi}{P_d}\left(\frac{L-b}{L}\right) \quad \text{(an allowed value)}$$

***Dynamic Load.*** The dynamic load $F_d$ may be approximated from

$$F_d = \frac{0.05V\,(bC+F)}{0.05V+\sqrt{bC+F}}$$

Symbols are the same as those for spur gears, and $Y$ is taken as 75 percent of the values shown in Fig. 8.98.

Finally, $F_d$ must be equal to or less than both $F_w$ and $F_0$. (*Note:* $F_w$ and $F_0$ are allowed values which must not be exceeded by the dynamic load $F_d$.)

AGMA standard recommended horsepower rating and wear equations for bevel gears are as follows. For peak load:

$$\text{hp} = \frac{snD_p\,bY\pi\,(L-0.5b)}{126{,}000\,P_d L}\left(\frac{78}{78+\sqrt{V}}\right)$$

where $s$ = 250 times Bhn of weaker gear (for gears hardened and also not hardened after cutting)
$s$ = 300 times Bhn of weaker gear if it is case-hardened
$n$ = speed of the pinion, rpm

Other symbols were defined in the preceding sections.

For wear (durability):

$$\text{hp} = 0.8\,C_m C_B\,b \quad \text{(for straight bevel gears)}$$

$$= C_m C_B\,b \quad \text{(for spiral bevel gears)}$$

where $C_m$ = material factor from Fig. 8.101
$b$ = face width of gear, in
$C_B$ = as in the following equation

$$C_B = \frac{D_p^{1.5}\,n}{233}\left(\frac{78}{78+\sqrt{V}}\right) \qquad (n = \text{rpm of pinion})$$

### 8.11.4 Worm Gear Loads and Design Procedures

The worm drive consists of a threaded worm in mesh with a gear called the *worm wheel.* The worm normally contains one to four leads or threads. The axial pitch of the worm $P_a$ is equal

| Gear | | Pinion | | |
|---|---|---|---|---|
| **Material** | **Brinell (BHN)** | **Material** | **Brinell (BHN)** | $C_m$ |
| Annealed Steel | 160-200 | Heat-Treated Steel | 210-245 | 0.30 |
| Heat-Treated Steel | 245-280 | Heat-Treated Steel | 285-325 | 0.40 |
| Heat-Treated Steel | 285-325 | Heat-Treated Steel | 335-360 | 0.50 |
| Heat-Treated Steel | 210-245 | Oil or Surface Hdn'd St'l | 500 | 0.40 |
| Heat-Treated Steel | 285-325 | Case-Hardened Steel | 550 | 0.60 |
| Oil or Surface Hdn'd St'l | 500 | Case-Hardened Steel | 550 | 0.90 |
| Case-Hardened Steel | 550 | Case-Hardened Steel | 550 | 1.00 |

Note: If cast iron teeth are strong enough, they will not fail by wear. If steel teeth satisfy the wear requirements, they will have enough strength.

**FIGURE 8.101** Material factors for various gear material combinations.

to the circular pitch $P_c$ of the gear or worm wheel. The lead is the axial distance a point on the worm helix advances per revolution.

The following relationships exist in worm gearing:

$$\tan \alpha = \frac{\text{lead}}{\pi D_w} = \frac{P_c N_w}{\pi D_w} \quad \text{and} \quad \frac{(\text{rpm})\, w}{(\text{rpm})\, g} = \frac{N_g}{N_w} = \frac{D_g}{D_w \tan \alpha}$$

where $\alpha$ = lead angle of worm, degrees
$D_w$ = worm pitch diameter, in
$D_g$ = gear pitch diameter, in
$N_w$ = number of threads on the worm
$N_g$ = number of teeth in the gear (worm wheel)

***Worm Gear Strength Design.*** The following strength equation is based on the Lewis equation:

$$F = sbYP_{nc} = \frac{sbY\pi}{P_{nd}}$$

where $P_{nc}$ = normal circular pitch, in
$P_{nd}$ = normal diametral pitch in plane normal to tooth
$F$ = permissible tangential load, lbf
$s$ = allowable stress, psi

$$s = s_0 \frac{1200}{1200 + V_g}$$

where $s_0 = 0.33 \times$ ultimate strength of material, psi; $V_g$ = pitch line velocity of the gear, ft/min.

***Dynamic Load.*** The dynamic load $F_d$ for the worm gear is estimated from

$$F_d = \frac{1200 + V_g}{1200} F$$

where $F$ = actual transmitted tangential load, lbf.

***Endurance Load.*** The endurance load $F_0$ for the worm gear is

$$F_0 = \frac{s_0\, bY\pi}{P_{nd}}$$

***Wear Load.*** The wear load $F_w$ for the worm gear is

$$F_w = D_g bB$$

where $D_g$ = pitch diameter of gear, in
$b$ = gear face width, in
$B$ = constant from Fig. 8.102

The values of constant $B$ listed in Fig. 8.102 are for worm lead angles up to 10°. For lead angles between 10 and 25°, increase $B$ by 25 percent (multiply the given values by 1.25). For lead angles over 25°, increase $B$ by 50 percent (multiply given values by 1.50). (*Note:* $F_0$ and $F_w$ are allowable values which must not be exceeded by the dynamic load $F_d$.)

AGMA horsepower rating equations for worm gear wear are

$$\text{hp} = \frac{n}{R} KQm \quad \text{(wear check)}$$

| Worm | Gear | B |
|---|---|---|
| Hardened Steel | Cast Iron | 50 |
| Steel, 250 BHN | Phosphor Bronze | 60 |
| Hardened Steel | Phosphor Bronze | 80 |
| Hardened Steel | Chilled Phos. Bronze | 120 |
| Hardened Steel | Antimony Bronze | 120 |
| Cast Iron | Phosphor Bronze | 150 |

Note: The tabulated values for (B) are for lead angles up to 10°.
For lead angles between 10° and 25°, multiply (B) value by 1.25.
For lead angles greater than 25°, multiply (B) values by 1.50.

**FIGURE 8.102** Material combination constants, worm gears.

where hp = input horsepower
$n$ = rpm of worm
$R$ = transmission ratio = rpm worm/rpm gear
$K$ = pressure constant per center distance (see Fig. 8.103)
$Q = R/(R + 2.5)$
$m$ = velocity factor estimated from

$$m = \frac{450}{450 + V_w + (3V_w/R)}$$

where $V_w$ = pitch line velocity of the worm, ft/min.

| Center Distance C (inches) | K |
|---|---|
| 1 | 0.0125 |
| 2 | 0.025 |
| 3 | 0.04 |
| 4 | 0.09 |
| 5 | 0.17 |
| 6 | 0.29 |
| 7 | 0.45 |
| 8 | 0.66 |
| 9 | 0.99 |
| 10 | 1.20 |
| 15 | 4.0 |
| 20 | 8.0 |
| 30 | 29.0 |
| 40 | 66.0 |
| 50 | 120.0 |

Note: To find values of (K) for intermediate values of center distance (C), use interpolation.

**FIGURE 8.103** Pressure constants *K*.

AGMA design equations are

$$D_w \approx \frac{C}{2.2} \approx 3P_c \qquad b \approx 0.73D_w \qquad L \approx P_c\left(4.5 + \frac{N_g}{50}\right)$$

where $D_w$ = pitch diameter of worm, in
$C$ = center distance between worm and gear, in
$b$ = face width of gear, in
$P_c$ = circular pitch of gear, in
$L$ = axial length of worm, in

The preceding equations are useful for approximating the preliminary proportions of the worm gear unit.

The AGMA horsepower rating (for gear speeds ≤2000 rpm) equation for plain worm gear heat dissipation is

$$\text{hp} \approx \frac{9.5C^{1.5}}{R+5} \qquad \text{(heat check)}$$

where hp = permissible input horsepower
$C$ = center distance, in
$R$ = transmission ratio (see previous symbols)

(*Note:* For efficiency equations of worm gear sets, see Sec. 8.8.12.)

## 8.12 EPICYCLIC GEARING

Epicyclic gearing, also called *planetary gearing,* is found in many variations and arrangements to meet a broad range of speed-ratio requirements. Some of the basic, most often used epicyclic systems are described in this section, together with some basic calculations used to specify the gears employed in these systems.

In general, the epicyclic gear train consists of a central "sun" gear, several "planets" meshing with the sun gear and spaced uniformly around the sun gear, and an "annulus" or ring gear meshing with the planet gears. The sun gear and planet gears are externally toothed, while the ring gear is internally toothed. The term *epicyclic* is derived from the fact that points on the planet gears describe epicycloidal curves in space during rotation.

Epicyclic gear systems consist of either external or internal spur gears or external or internal helical gears. When helical gears are used in epicyclic systems, the hand of the internal teeth is the same as that of the external teeth; internal left-hand gears mesh with external left-hand gears, and internal right-hand gears mesh with external right-hand gears.

A simple epicyclic gear system is shown in Fig. 8.104, together with the six arrangements possible for this system. The table in Fig. 8.104 lists the six variations, with their corresponding speed-ratio equations. The diagram to the right, called the *gear-train schematic diagram,* is the simplified system used to describe the action of the gear system and to show its parts with the gear number used in the ratio equations. In the table $N_1$ represents gear number 1, and the number of teeth in the gear is substituted in the ratio equation. You may select the ratio first and then the number of teeth in the first gear to solve for the number of teeth in the second gear. Or, you may select the number of teeth in both gears $N_1$ and $N_2$ and solve for the ratio. Epicyclic gear systems will not assemble unless the number of teeth in each gear is properly selected. If the planet gears are equally spaced around the sun gear, the following equation must be satisfied for assembly to be possible:

$$N_R = N_S + 2N_P \qquad \frac{N_R + N_S}{N_P} = \text{an integer}$$

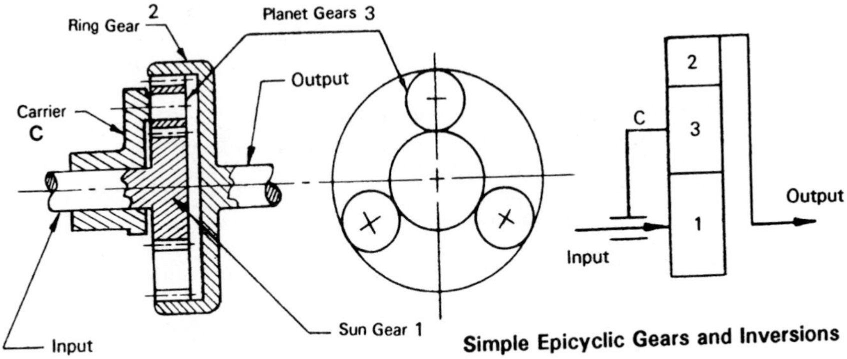


| Input Member | Fixed Member | Output Member | Speed-ratio Equation |
|---|---|---|---|
| 1 | C | 2 | $R = -N_2/N_1$ |
| 2 | C | 1 | $R = -N_1/N_2$ |
| 1 | 2 | C | $R = 1 + (N_2/N_1)$ |
| 2 | 1 | C | $R = 1 + (N_1/N_2)$ |
| C | 2 | 1 | $R = 1/(1 + (N_2/N_1))$ |
| C | 1 | 2 | $R = 1/(1 + (N_1/N_2))$ |

Note: The minus sign indicates opposite rotation from input.

**FIGURE 8.104** Planetary (epicyclic) gear systems.

where $N_R$ = number of ring or annulus teeth
$N_S$ = number of sun gear teeth
$N_P$ = number of planet gear teeth

The sum of the ring gear and sun gear teeth must be equally divisible by the number of planet gears in the system, or the gears will not assemble. The tooth load at each planet gear is balanced between the sun-planet mesh and the planet-ring mesh. This tangential driving load is calculated from

$$F_t = \frac{126{,}050\, P_h}{n_s d_s \text{ (no. of planets)}}$$

where $F_t$ = tangential driving force, lbf
$n_s$ = rpm of the sun gear
$d_s$ = pitch diameter of the sun gear, in
$P_h$ = horsepower transmitted

This equation shows the theoretical value of the tangential force, but in actual practice, this load may increase greatly because of dimensional tooth errors. With modern machinery, cutting tool materials, and CNC controls, this may be less of a problem than in the past.

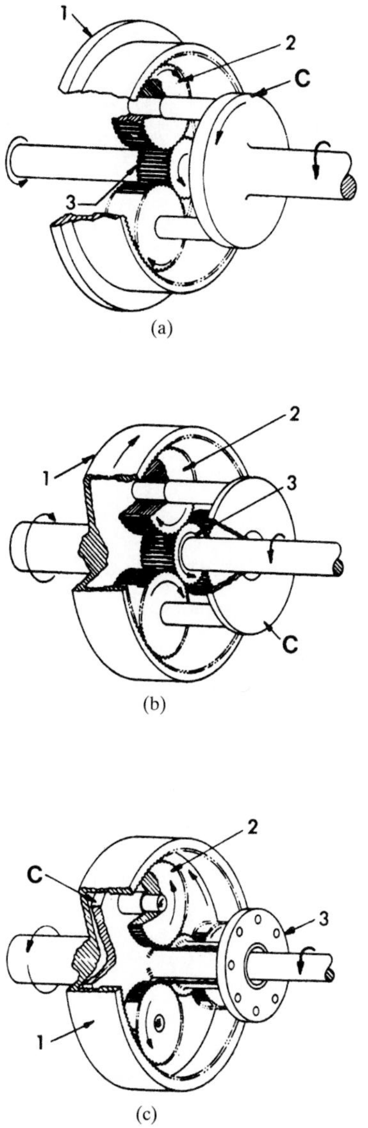


**FIGURE 8.105** Three common epicyclic gear systems.

Figure 8.105 shows the three most common forms of epicyclic gearing: the *planetary system* (*a*), the *star* system (*b*), and the *solar system* (*c*). Numeral 1 indicates the ring gear; numeral 2, the planet gear; numeral 3, the sun gear; and letter *C* (appearing in panels *a–c*) indicates the carrier of the planet gears. If you refer back to Fig. 8.104, you can see that by fixing or holding a specified gear motionless, the system will react or perform per the table and the speed-ratio equations.

Another type of epicyclic gearing system is shown in Fig. 8.106. In this system, gears 2 and 4 are joined on a common shaft with the carrier *C*, with gears 1 and 3 being the input or output. The carrier is also an output in two of the configurations shown in the table, where the speed ratios of the different configurations of this system of gearing are shown.

The derivations of the speed ratios of the different types of epicyclic gear systems are rather complicated and will not be explained in this section. Rather, samples of different types of epicyclic systems are shown together with their speed-ratio equations for your applications or reference. Some of the texts listed in the bibliography section cover epicyclic gear systems in greater detail.

An interesting epicyclic system known as *Humpage's bevel gears* is illustrated in Fig. 8.107 together with the speed-ratio equation of this system. In the figure, the input is at gear 1 and the output is at gear 4, whose shaft passes through a hole in gear 5 (which is fixed). Gears 2 and 3 are the planets, and only one set is shown, although the system must have at least two or three sets for dynamic balance, if the system is to be used for high-speed applications.

### 8.12.1 Speed-Ratio Definition and Number of Teeth in Epicyclic Systems

The speed ratios, by convention, are usually taken as greater than unity. Referring back to Fig. 8.104, in the first tabular line (row), input member is gear number 1, fixed member is the carrier, output is gear number 2, and there are three planet gears. If the ring gear has 48 teeth and the sun gear has 15 teeth, this will satisfy the previous equation stating that in a simple epicyclic system, the sum of the teeth in the ring gear and the sun gear must be equally divisible by the number of planet gears. This tells us if the system can be assembled properly:

$$\frac{\text{Number of teeth in ring gear} + \text{number of teeth in sun gear}}{\text{3 sun gears}} = 21 \text{ (an integer)}$$

As a better example of ratios and numbers of teeth in epicyclic systems, refer to Fig. 8.108. This figure represents an actual production sample of an epicyclic system used in a portable electric screwdriver. This system is the same as that shown in Fig. 8.104, tabular line 3, where the sun gear is the input, the ring gear is fixed (integral with the housing), and the carrier *C* is

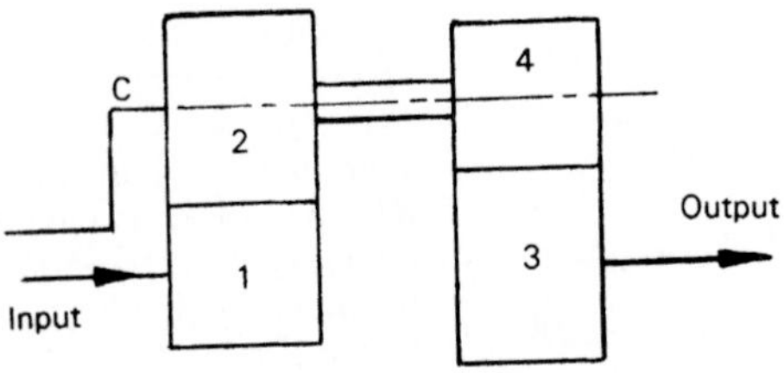


| Input Member | Fixed Member | Output Member | Speed-ratio Equation |
|---|---|---|---|
| 1 | C | 3 | $R = N_2N_3/N_1N_4$ |
| 1 | 3 | C | $R = 1 - [(N_2N_3)/(N_1N_4)]$ |
| 3 | 1 | C | $R = 1 - [(N_1N_4)/(N_2N_3)]$ |
| 3 | C | 1 | $R = N_4N_1/N_3N_2$ |
| C | 1 | 3 | $R = 1/\{1 - [(N_1N_4)/(N_2N_3)]\}$ |
| C | 3 | 1 | $R = 1/\{1 - [(N_2N_3)/(N_1N_4)]\}$ |

**FIGURE 8.106** Commonly used epicyclic gear system.

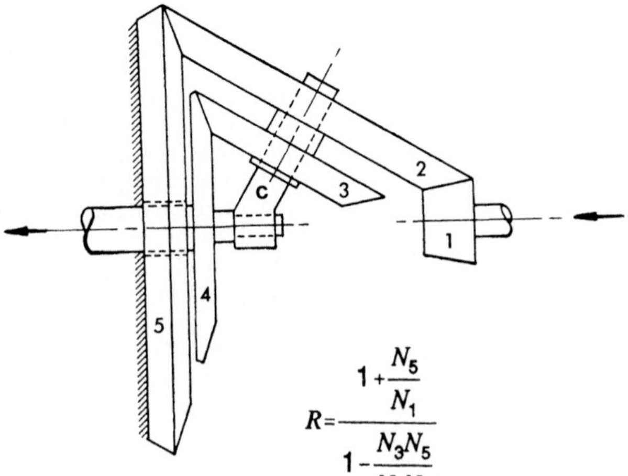


$$R = \frac{1 + \dfrac{N_5}{N_1}}{1 - \dfrac{N_3N_5}{N_4N_2}}$$

**FIGURE 8.107** Humpage's epicyclic gear system.

**FIGURE 8.108** Internal view of an actual epicyclic gear system.

the output. The sun gear in this system has 6 teeth and the internal ring gear has 42 teeth and satisfies the previous equation. The speed ratio of this system is given from Fig. 8.104 as

$$R = 1 + \frac{N_2}{N_1} = 1 + \frac{42}{6} = 1 + 7 = 8$$

The speed ratio is then 8:1 and the configuration of the system tells us that for each eight revolutions of the sun gear (the input), the carrier (output) will rotate one revolution. This system is thus a speed-reduction type. The equation also tells us that the output will rotate in the same direction as the input (since its sign is positive). Figure 8.108 shows only one stage of the entire system. If you look at Fig. 8.109, you will see the first stage, which consists of three planet gears attached to the carrier plate. The six-tooth sun gear, which is the input for the second stage, is attached at the center of the carrier plate of the first stage and is barely visible in Fig. 8.109. Figure 8.110 is a cross section through the housing of this two-stage epicyclic system, showing the ring gear (integral to the housing), the carrier plates with planets attached, and the sun gears. Since this is a two-stage system the total speed ratio is $8 \times 8 = 64$:1. The total speed ratio of a multistage system is equal to the product of the ratios of the stages. If this were a three-stage system, the final ratio would be $8 \times 8 \times 8 = 512$:1. It can then be realized that the epicyclic gear systems or trains offer large speed ratios in limited spaces (either speed increasing or speed reduction, including changes in rotational direction from input to output).

Following is a selection of epicyclic gear systems with different configurations and speed ratios. Figures 8.111 and 8.112 are examples of *coupled planetary drives.* Figures 8.113 and 8.114 are examples of *fixed-differential drives.* The gear train schematic diagram and speed-ratio equation is shown in each of the figures, together with a typical cross section taken through the epicyclic drive housing.

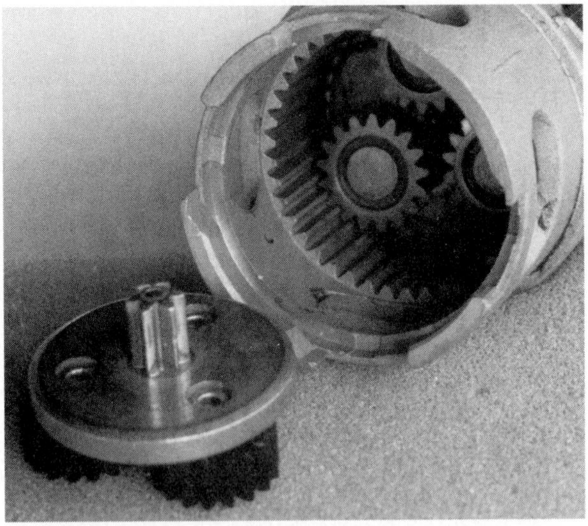

**FIGURE 8.109** Epicyclic gear system showing two stages.

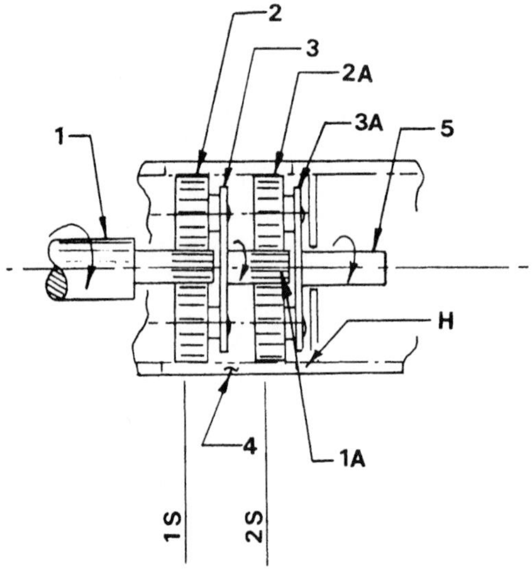


1 Input and sun gear
1A 2nd stage sun gear
2 1st stage planet gears
2A 2nd stage planet gears
3 1st stage carrier plate
3A 2nd stage carrier plate
4 Internal ring gear
5 Output
H Housing
1S 1st stage
2S 2nd stage

**FIGURE 8.110** Cross section through a two-stage epicyclic gear system.

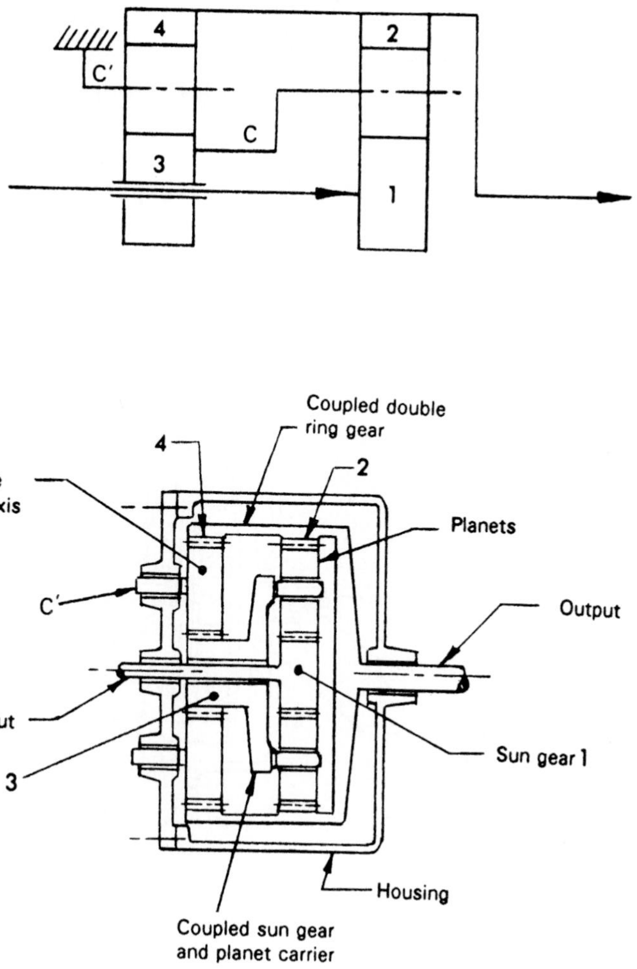


$$R = \left(1 + \frac{N_2}{N_1}\right)\left(-\frac{N_4}{N_3}\right) - \frac{N_2}{N_1}$$

**FIGURE 8.111** Sectional view of epicyclic system, coupled planetary drive.

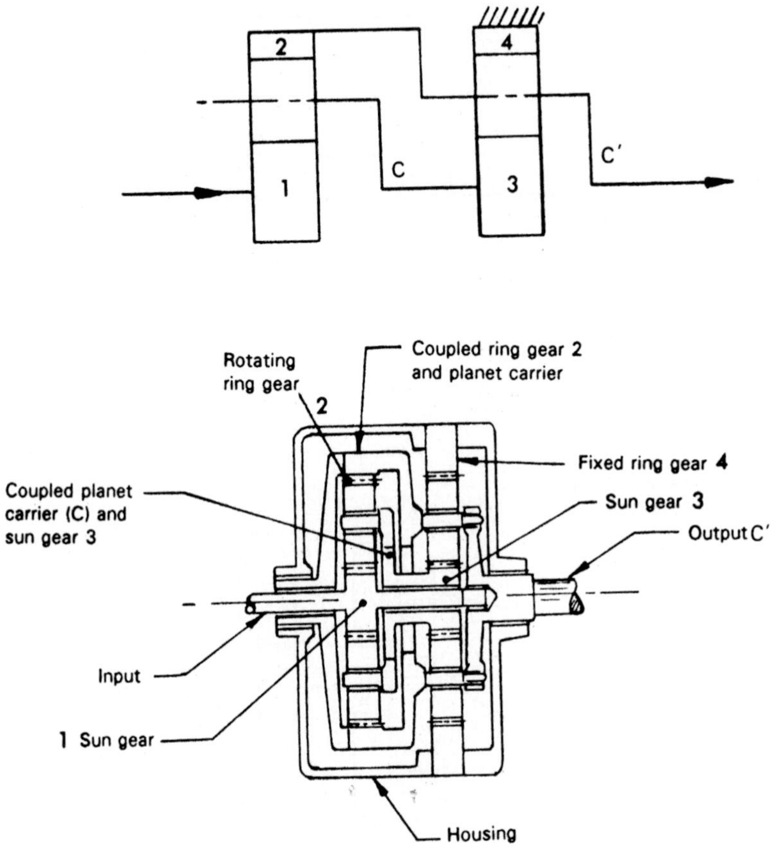


$$R = 1 + \frac{N_4}{N_3}\left(1 + \frac{N_2}{N_1}\right)$$

**FIGURE 8.112** Sectional view of epicyclic system, coupled planetary drive.

### 8.12.2 Epicyclic Drive Train Calculations for Simple Planetary Systems

Figure 8.115 is a representation of the epicyclic system shown in Fig. 8.104, tabular line 3, where the sun gear is the input, the ring gear is held stationary, and the output is taken from the rotating carrier. This is also shown in Fig. 8.110 (one stage of the system) and in the photographs of Figs. 8.108 and 8.109. The ratio equation was given as $R = 1 + (N_2/N_1)$. We will substitute the symbols of the ratio equation as follows:

$$N_1 = A \qquad N_2 = C$$

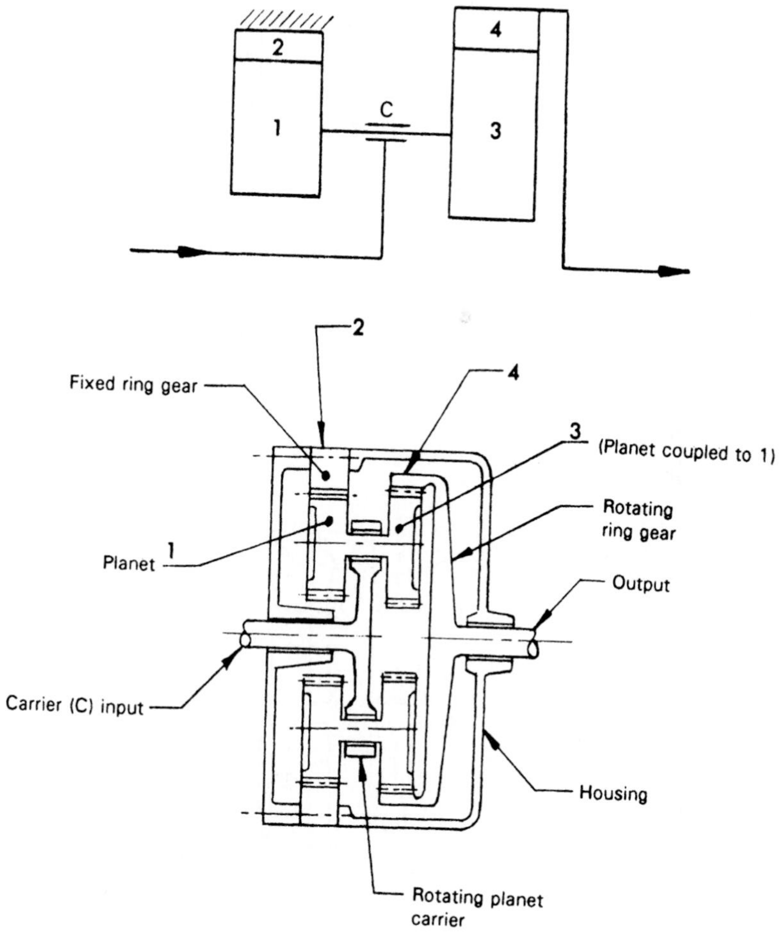


$$R = \frac{1}{1 - \frac{N_3 N_2}{N_4 N_1}}$$

**FIGURE 8.113** Sectional view of epicyclic system, fixed-differential drive.

where $A$ = size of driver (sun), pitch diameter, in, or number of teeth
$C$ = size of fixed ring gear, pitch diameter, in, or number of teeth
$R$ = rotation of driver (sun gear) for each revolution of the carrier (output)

*Example.* We require a planetary spur gear drive with a 6:1 ratio and teeth of 20 diametral pitch. We would like to limit the number of teeth in the driver (sun gear) to 12. The 6:1

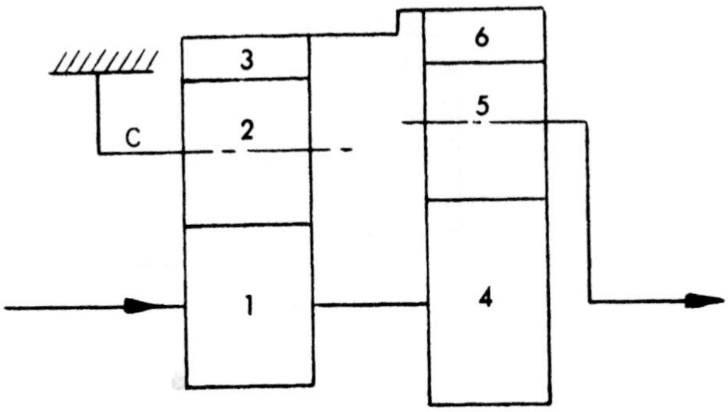


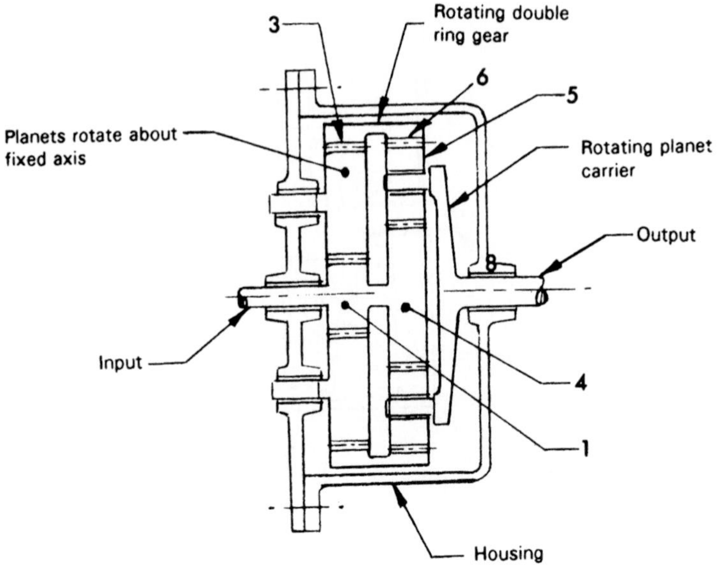


$$R=\frac{1+\frac{N_4}{N_6}}{\frac{N_4}{N_6}-\frac{N_1}{N_3}}$$

**FIGURE 8.114** Sectional view of epicyclic system, fixed-differential drive.

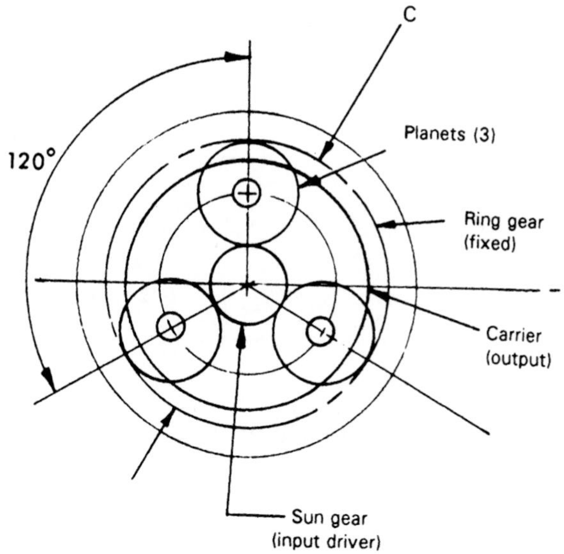


**FIGURE 8.115** Epicyclic gear system.

ratio means that the driver or sun gear must make six revolutions for each revolution of the output (carrier). Given

Ratio = 6:1

Diametral pitch required = 20

Desired number of teeth in the driver (sun gear) = 12

The pitch diameter of the driver is (use spur gear relations)

$$D_p = \frac{N}{P_d} = \frac{12}{20} = 0.60 \text{ in}$$

where $N$ = number of teeth in sun gear
$D_p$ = pitch diameter, in
$P_d$ = diametral pitch

Therefore, the pitch diameter of the 12-tooth sun gear is 0.60 in.

Then calculate the number of teeth required in the ring gear (fixed) for the 6:1 ratio from

$$R = 1 + \frac{C}{A} \qquad 6 = 1 + \frac{C}{12}$$

$$\frac{C}{12} = 5 \qquad C = 12 \times 5 = 60 \text{ teeth required in the ring gear}$$

The pitch diameter of the ring gear with 60 teeth and 20 diametral pitch is

$$D_p = \frac{N}{P_d} = \frac{60}{20} = 3.00 \text{ in pitch diameter of ring gear}$$

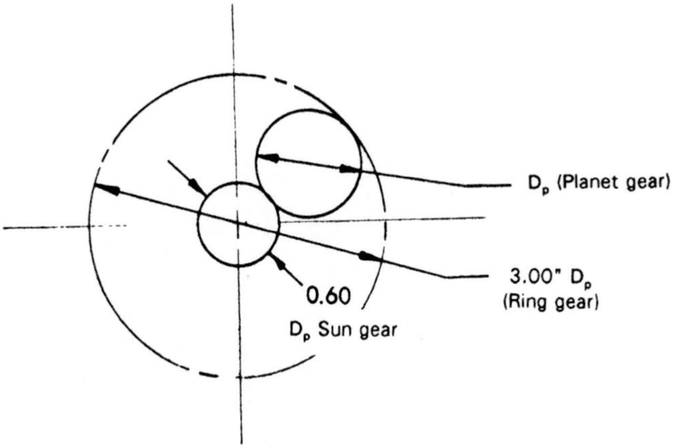


**FIGURE 8.116** Finding the pitch diameters.

The pitch diameter of the planet gears must now be found from (see Fig. 8.116)

$$\frac{3.00 - 0.6}{2} = D_p \qquad \frac{2.4}{2} = d \qquad D_p = 1.2 \text{ in pitch diameter of planet gears}$$

The number of teeth in the planet gears may now be found from

$$D_p = \frac{N}{P_d} \qquad 1.2 = \frac{N}{20} \qquad N = 24 \text{ teeth}$$

As a check, refer to Fig. 8.117 and the following:

$$R = 1 + \frac{C}{A} = 1 + \frac{60}{12} = 1 + 5 = 6 \quad \text{(so that the ratio is 6:1 as was required)}$$

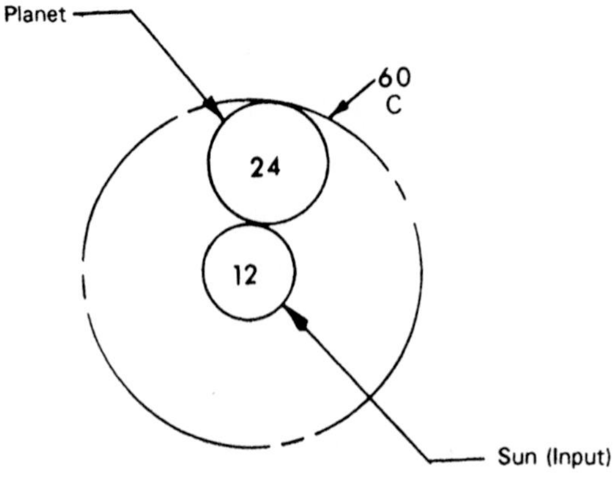


**FIGURE 8.117** Checking the pitch diameters.

As a check for assembly of this system

$$\frac{\text{Number of teeth in ring gear + number of teeth in sun gear)}}{\text{Number of planets}}$$

must be equal to an integer:

$$\frac{60 + 12}{3} = 24 \quad \text{(so that the system will assemble)}$$

Another example of the epicyclic system calculation procedure is to design the system for a limited-space application. Knowing the maximum desired pitch diameter of the ring gear and the required ratio, we may proceed as follows.

*Problem.* We wish to employ an epicyclic system as depicted in Fig. 8.104, tabular line 3 and in Figs. 8.108 and 8.109. The reduction or ratio is to be 20:1, driver to follower. The maximum pitch diameter of the ring gear (fixed) is not to exceed 2.00 in because of space limitations and a low power handling ability. The recommended minimum number of teeth in the driver (sun gear) is to be 13.

First, with a 20:1 ratio and minimum number of sun gear teeth of 13, we proceed to find the number of teeth required in the fixed gear from

$$R = 1 + \frac{C}{A} \qquad 20 = 1 + \frac{C}{13} \qquad 19 = \frac{C}{13} \qquad C = 247 \text{ teeth}$$

(See preceding symbol definitions.) If we allow this number of teeth in the fixed gear whose pitch diameter is not to excede 2.00 in, then the diametral pitch of the teeth must be

$$D_p = \frac{N}{P_d} \qquad 2.00 = \frac{247}{P_d} \qquad P_d = \frac{247}{2.00} = 123.5 \quad \text{(which is too small)}$$

Next, we will try a two-stage reduction. If the two-stage final reduction is 20:1, then each stage must have a reduction of

$$\sqrt{20} = 4.47 \text{ (for 2 stages)} \qquad \text{and} \qquad \sqrt[3]{20} = 2.71 \quad \text{(for 3 stages)}$$

Then

$$R = 1 + \frac{C}{A} \quad \text{(for 2-stage reduction)}$$

$$4.47 = \frac{C}{13} \qquad C = 45.11 \text{ (round off to 45 teeth)}$$

Now, the diametral pitch with 45 teeth on a 2.00 in pitch diameter is

$$D_p = \frac{N}{P_d} \qquad 2.00 = \frac{45}{P_d}$$

$$P_d = 22.5 \quad \text{(use 24 as a diametral pitch)}$$

Then the actual pitch diameter of the ring gear (fixed) will be

$$D_p = \frac{N}{P_d} = \frac{45}{24} = 1.875 \text{ in} \quad \text{(the actual pitch diameter of our fixed ring gear)}$$

Now, we know the number of teeth in the sun gear (driver), the ring gear, the pitch diameter of the ring gear, and the number of stages we need.

Next, we proceed to find the pitch diameter $D_p$ of the sun gear (driver), the pitch diameter of the planet gears, and the number of teeth in the planet gears. (All gears have a diametral pitch of 24.) The pitch diameter of the sun gear is

$$D_p = \frac{N}{P_d} = \frac{13}{24} = 0.542 \text{ in pitch diameter}$$

Then we find the pitch diameter of the planet gears from

$$D_p = \frac{1.875 - 0.542}{2} = 0.667 \text{ in (see Fig. 8.118)}$$

The number of teeth in each planet gear is therefore

$$D_p = \frac{N}{P_d} \qquad 0.667 = \frac{N}{24}$$

$N = 0.667 \times 24 = 16.008$ (call this 16 teeth) (see Fig. 8.119)

Check the ratio equation against the number of teeth in the sun gear (driver) and the ring gear (fixed) from

$$R = 1 + \frac{C}{A} = 1 + \frac{45}{13} = 1 + 3.462 = 4.462 \text{ ratio of each stage}$$

The actual final ratio will then be

$$R \times R = \text{final reduction} \qquad 4.462 \times 4.462 = 19.91$$

The actual fit of the pitch diameters is then (see Fig. 8.120)

$$0.667 + 0.667 + 0.542 = 1.876 \quad (1.875 \text{ was calculated})$$

So, the fit of the pitch diameters is very close to the theoretical value.

As a last check, we will see if the epicyclic system described previously can be assembled. The assembly check was previously given as

$$\frac{\text{Number of teeth in ring gear + number of teeth in sun gear}}{\text{Number of planets}}$$

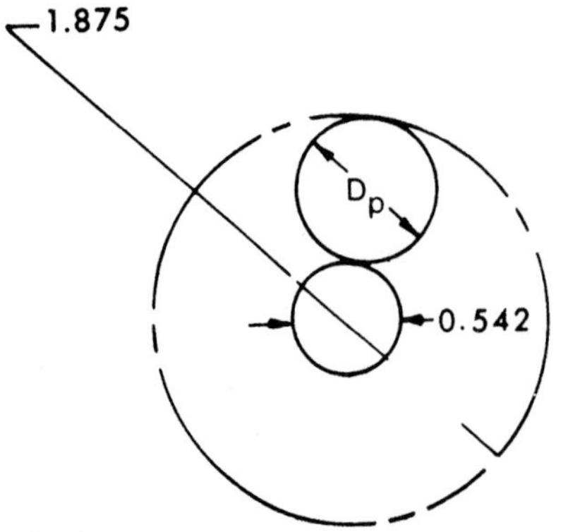


**FIGURE 8.118** Finding the pitch diameters of the planet gears.

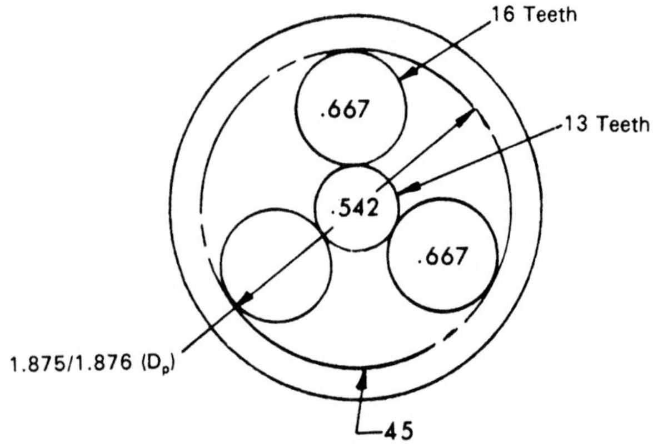


**FIGURE 8.119** Finding the number of teeth in the gears.

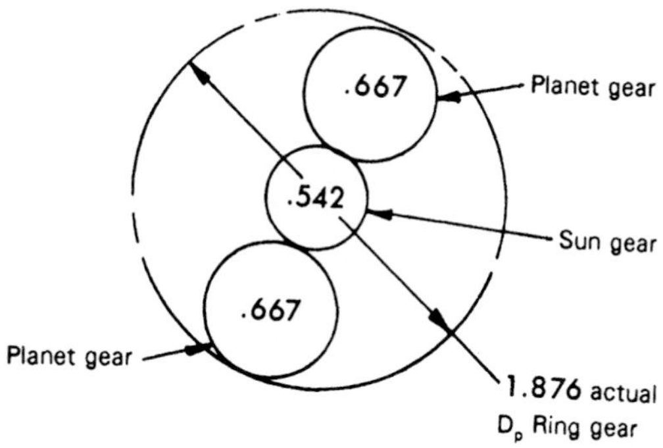


**FIGURE 8.120** Fit of the pitch diameters.

must equal an integer. So

$$\frac{45 + 13}{3} = \frac{58}{3} = 19.333$$

Therefore, we cannot use three planet gears in this system. We may use two planet gears only because (45 + 13)/2 = 29 (an integer). To make this system work with three planet gears, the number of teeth in the ring gear and sun gear, and possibly the ratio, must be adjusted and the system reevaluated to keep the pitch diameter of the ring gear a maximum of 2.00 in and at the same time maintaining a standard diametral pitch, so that the gears may be cut using standard hobs or milling cutters.

To keep a standard diametral pitch and maintain the other requirements of this system, helical gears may be used. Since the helical gears' transverse diametral pitch is a function of the cosine of the helix angle, altering the helix angle directly affects the pitch diameter. (See Fig. 8.65 for helical gear equations.)

***Conclusion.*** Because of the double-stage reduction of the preceding epicyclic system, the actual load on the teeth of the stage 1 planet gears may be low enough to use a material such as acetal (Delrin), nylon (Zytel), or polycarbonate (Lexan) for the planet gears. In the case of the electric screwdriver epicyclic system shown in Figs. 8.108 and 8.109, plastic planet gears were used. Note that the forces on the teeth are higher in the second-stage gears and are thus made of metal, as is the entire internal tooth ring gear, which is part of the gear system housing. As the reduction stages progress away from the input sun gear, the forces on the gear teeth of the succeeding stages increase.

## 8.13 GEAR-TRAIN CALCULATIONS

The speed ratios of gears are inversely proportional to their pitch diameters (see Fig. 8.121).

$$\text{Speed ratio} = \frac{\text{product of driven}}{\text{product of drivers}} = \frac{D_1 D_2 D_3}{d_1 d_2 d_3}$$

The speed ratio of a gear train can also be found if the number of teeth in each gear is known (see Fig. 8.122).

$$\frac{n_4}{n_1} = \frac{N_1}{N_2} \cdot \frac{N_3}{N_4}$$

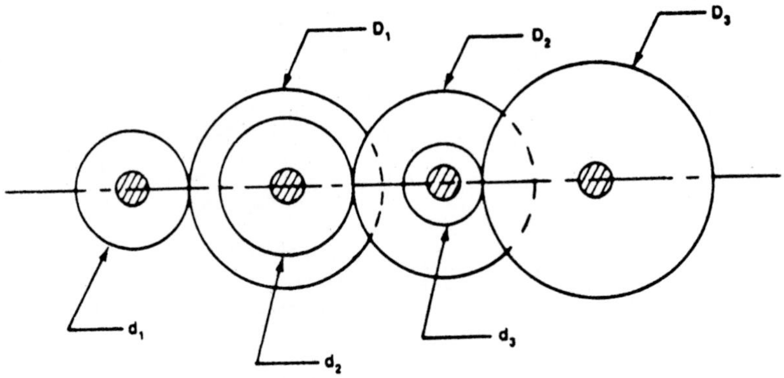


**FIGURE 8.121** Gear-train speed ratio.

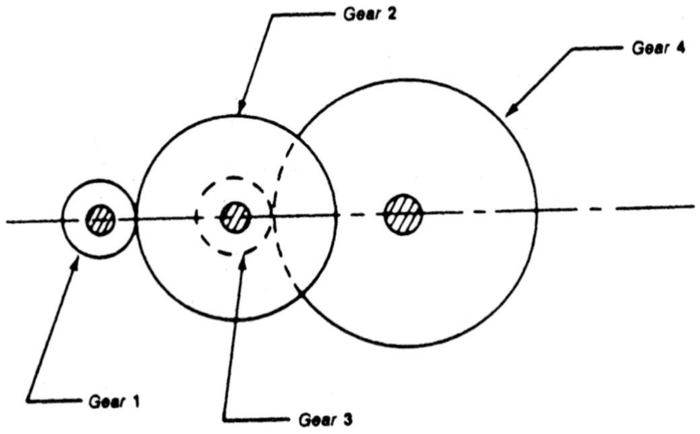


**FIGURE 8.122** Gear-train speed ratio.

where $n_1$ = speed of gear 1, rpm
$n_4$ = speed of gear 4, rpm
$N_1, N_2, N_3, N_4$ = number of teeth in gears 1 through 4

Additional ratios can be inserted into the tooth ratio side of the equation for longer gear trains, comprising more gears.

### 8.13.1 Gear-Train Design Procedures

The *reverse* problem of finding the number of teeth each gear must have if the ratio of the gear train is to equal an arbitrarily chosen value is more difficult. All given ratios cannot be exactly produced by a gear train because the individual gears must contain an integral number of teeth. We can approximate the desired ratio with a degree of accuracy that is suitable for almost all applications. The method used to accomplish this was shown by Spotts (1953) and is shown in detail by the following.

The given arbitrary ratio between the first and last shaft is represented by $G$. Let $u/w$ be a common fraction whose value is close to $G$. Then the ratio $G$ may be exactly represented by

$$G = \frac{uh - j}{wh}$$

where $j$ is an integer and $h$ has the value obtained by solving the equation

$$h = \frac{j}{u - wG}$$

In the equation for $G$, numerator and denominator must be integers that may be factored into terms suitable as the numbers of teeth in the gears of the train. Note, however, that the equation for $h$ seldom yields an integer. The numerator and denominator in the equation for $G$ will also not be integers generally. However, if an integer $h'$ whose value is close to $h$ is used, an approximate value of $G'$ of the ratio is obtained that may be very close to the exact value. Then

$$G' = \frac{uh' - j}{wh'}$$

*Example.* The three preceding equations are illustrated as follows. If the shaft on the left in Fig. 8.122 is to revolve 3.62 revolutions for each revolution of the output shaft on the right, what is the proper number of teeth for each gear in the train $N_1$, $N_2$, $N_3$, and $N_4$?

$$G = \frac{1}{3.62} = 0.2762431$$

Suitable values for $u/w$ can be found on a calculator by placing the given value for $G$ on the display and multiplying successive integers by $G$ until the product is very close to an integer. So $wG \approx u$.

*Example.* $40 \times G$ ($40G$) is close to 11, so a suitable fraction value for $u/w$ is 11/40. Let $j = 1$. Then

$$h = \frac{1}{11 - 40\,(0.2762431)} = -20.111$$

Now, if we suppose that $h' = -20$, then

$$G' = \frac{uh' - j}{wh'}$$

$$G' = \frac{(11)(-20) - 1}{40\,(-20)} = \frac{-220 - 1}{-800} = \frac{221}{800} = \frac{(13 \times 17)}{(20 \times 40)} = 0.27265$$

(See Sec. 1.16 tables for factors and primes.)

In the preceding equation for $G$ the numbers 221 and 800 are to be factored as $13 \times 17$ and $20 \times 40$, respectively. Use the factors and primes tables to find the factors of the numerator and denominator. In the preceding equation for $G$ which was solved and factored as $13 \times 17$ and $20 \times 40$, this represents the number of teeth in the four gears in the train shown in Fig. 8.122. Gears:

$$\frac{13 \text{ teeth}}{17 \text{ teeth}} \quad (N_1 \text{ and } N_3) \qquad \frac{20 \text{ teeth}}{40 \text{ teeth}} \quad (N_2 \text{ and } N_4)$$

This method for finding the number of teeth required in a preselected ratio gear train has many practical uses when working with gear systems.

We will now substitute the calculated values for the number of teeth into the basic ratio equation to see what the actual ratio is, as compared to what we originally desired (we wanted a input:output ratio of 3.62)

$$\frac{n_4}{n_1} = \frac{N_1}{N_2} \cdot \frac{N_3}{N_4} \qquad \frac{1}{3.62} = \frac{13 \times 17}{20 \times 40} \qquad 0.2762 = 0.2762$$

As you can see, the solution for the preselected ratio of 1/3.62 is satisfied.

(*Note:* More accuracy could be obtained by using a more appropriate value of $u/w = 11/40$, if one exists, or by going to a six-gear system and finding three factors in the numerator and denominator of $G'$.)

### 8.13.2 Force Ratios for Gear Trains

Refer to Fig. 8.123, which is a gear train with a weight suspended from one end and an applied force at the other end which is required to balance the weight load.

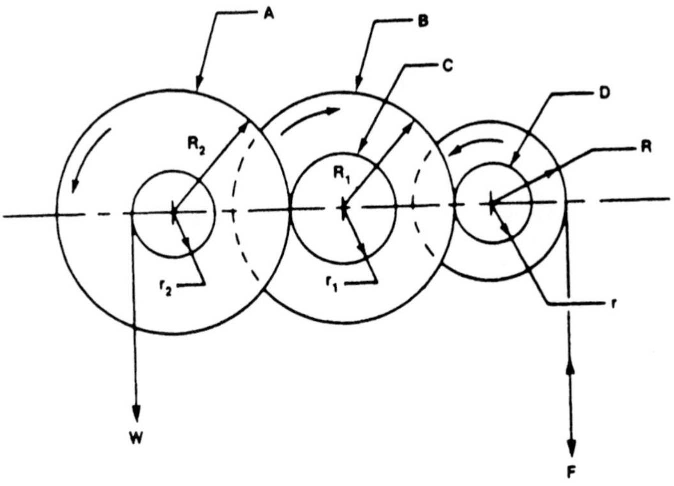


**FIGURE 8.123** Gear-train force ratio.

Here, *A, B, C,* and *D* are the pitch circles of the gears in the train. Then

$$F = \frac{Wrr_1r_2}{Rr_1R_2} \qquad \text{and} \qquad W = \frac{FRR_1R_2}{rr_1r_2}$$

where
$F$ = force, lbF or N
$W$ = weight, lb or (kg × g) N
$r, r_1, r_2$ = pitch radius, in or mm
$R, R_1, R_2$ = pitch radius, in or mm

[*Note:* If the weight at $W$ is 50 kg, the SI force is $50 \times 9.81 = 490.5$ N, which is equivalent to a weight of 50 kg suspended at point $W$; that is, the weight of a mass of $W$ kilograms is equal to a force of $W \times g$ newtons, where $g = 9.81$ ms/sec$^2$ (the acceleration due to gravity).]

## 8.14 DIFFERENTIAL GEARING

Differential gear systems are in a general sense an arrangement where the normal ratio of the unit can be changed by driving into the unit with a second drive. There may be two inputs and one output or one input and two outputs. One of the most common usages of the differential gear drive is in the rear end drives on automotive vehicles. When an automobile goes around a corner, one wheel must make more turns than the other (the wheel outside the curve makes more revolutions than the inside wheel). Fixed-differential drives were shown in a previous section (see Figs. 8.113 and 8.114).

A typical automotive differential is shown in Figs. 8.124 and 8.125.

In Fig. 8.125, the gears in the differential are of the spiral bevel type, which allows a higher load capability over conventional straight bevel gears and quieter operation.

## 8.15 SPROCKETS: GEOMETRY AND DIMENSIONING

Figure 8.126 shows the geometry of ANSI standard roller chain sprockets and derivation of the dimensions for design engineering or tool engineering use. With the following relational data and equations, dimensions may be derived for input to CNC machining centers or EDM machines for either manufacturing the different-size sprockets or producing the dies to stamp and shave the sprockets.

The equations for calculating sprockets are as follows:

$P$ = pitch ($ae$)

$N$ = number of teeth

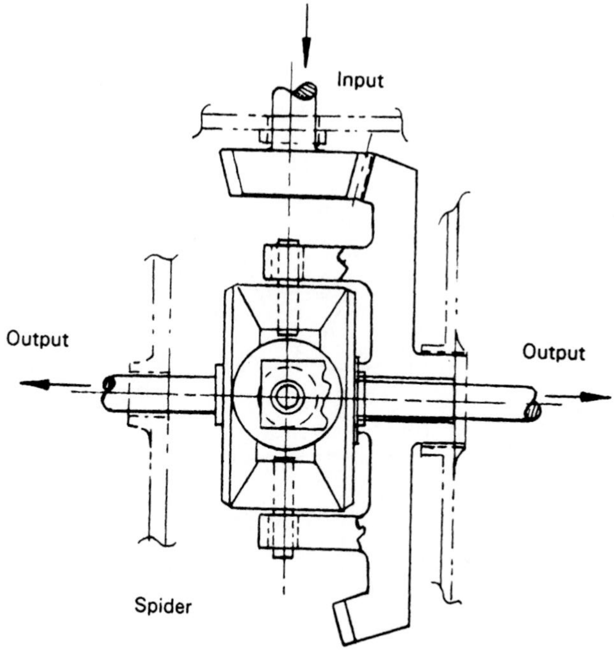


**FIGURE 8.124** Typical differential gear system.

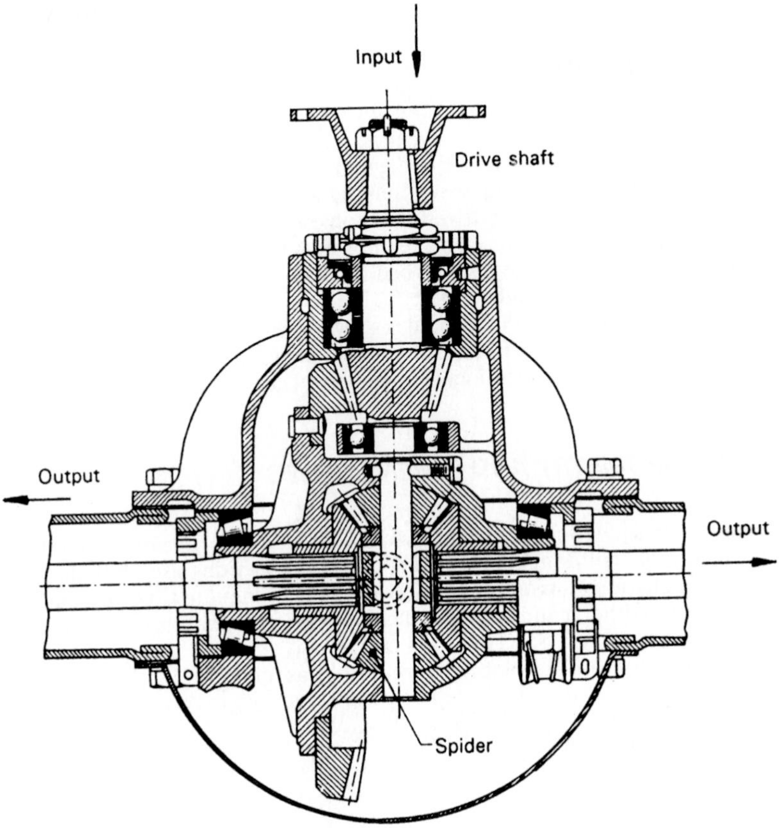


**FIGURE 8.125** Typical automotive differential gear train.

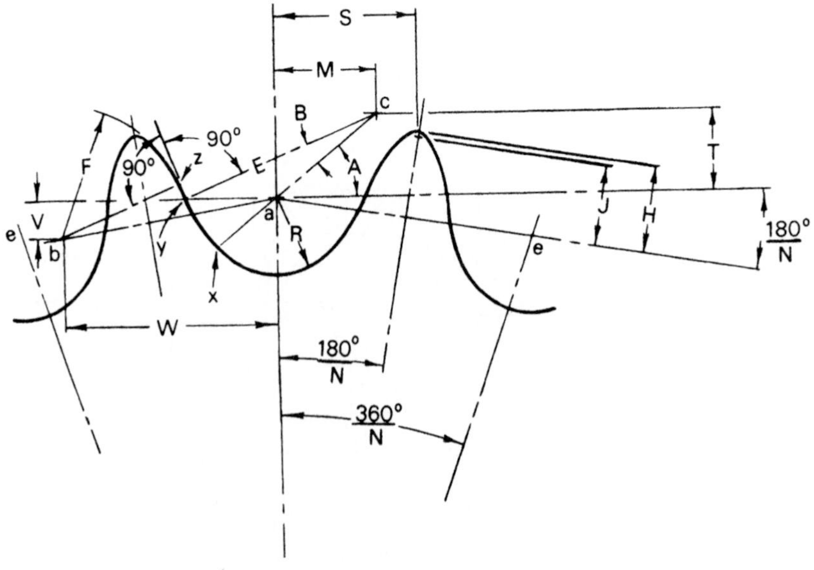


**FIGURE 8.126** Geometry of ANSI roller chain sprockets.

$D_r$ = nominal roller diameter
$D_s$ = seating curve diameter = $1.005D_r + 0.003$, in
$R = 1/2D_s$
$A = 35° + (60°/N)$
$B = 18° - (56°/N)$
$ac = 0.8D_r$
$M = 0.8D_r \cos[(35° + (60°/N)]$
$T = 0.8D_r \sin(35° + (60°/N))$
$E = 1.3025D_r + 0.0015$, in
Chord $xy = (2.605\ D_r + 0.003) \sin(9° - (28°/N)$, in
$yz = D_r\ \{1.4 \sin[17° - (64°/N) - 0.8 \sin(18° - (56°/N)]\}$
Length of line between $a$ and $b = 1.4D_r$
$W = 1.4D_r \cos(180°/N)$
$V = 1.4D_r \sin(180°/N)$
$F = D_r\{0.8 \cos[18° - (56°/N)] + 1.4 \cos[17° - (64°/N)] - 1.3025\} - 0.0015$ in
$H = \sqrt{F^2 - (1.4\ D_r - 0.5\ P)^2}$
$S = 0.5P \cos(180°/N) + H \sin(180°/N)$
Approximate o.d. of sprocket when $J$ is $0.3P = P[0.6 + \cot(180°/N)]$
Outer diameter of sprocket with tooth pointed $= p \cot(180°/N) + \cos(180°N)(D_s - D_r) + 2H$
Pressure angle for new chain $= xab = 35° - (120°/N)$
Minimum pressure angle $= xab - B = 17° - (64°/N)$
Average pressure angle $= 26° - (92°/N)$

The seating curve data for the preceding equations are shown in Fig. 8.127.

| P | $D_r$ | R Min | $D_s$ Min | $D_s$ Tolerance * |
|---|---|---|---|---|
| 1/4 | 0.130 | 0.0670 | 0.134 | 0.0055 |
| 3/8 | 0.200 | 0.1020 | 0.204 | 0.0055 |
| 1/2 | 0.306 | 0.1585 | 0.317 | 0.0060 |
| 1/2 | 0.312 | 0.1585 | 0.317 | 0.0060 |
| 5/8 | 0.400 | 0.2025 | 0.405 | 0.0060 |
| 3/4 | 0.469 | 0.2370 | 0.474 | 0.0065 |
| 1 | 0.625 | 0.3155 | 0.631 | 0.0070 |
| 1-1/4 | 0.750 | 0.3785 | 0.757 | 0.0070 |
| 1-1/2 | 0.875 | 0.4410 | 0.882 | 0.0075 |
| 1-3/4 | 1.000 | 0.5040 | 1.008 | 0.0080 |
| 2 | 1.125 | 0.5670 | 1.134 | 0.0085 |
| 2-1/4 | 1.406 | 0.7080 | 1.416 | 0.0090 |
| 2-1/2 | 1.562 | 0.7870 | 1.573 | 0.0095 |
| 3 | 1.875 | 0.9435 | 1.887 | 0.0105 |

Note: * Denotes plus tolerance only.

**FIGURE 8.127** Seating curve data for ANSI roller chain (inches).

## 8.16 RATCHETS

Ratchets are treated as gears and are described in Sec. 8.7.

## 8.17 GEAR DESIGN PROGRAMS FOR PCs AND CAD STATIONS

The Gleason Machine Division offers a service for gear system calculations when you require a computer analysis of the gear tooth design.

Also available from the Gleason Machine Division are computer programs which will assist you with a gear tooth design analysis:

- *Dimension sheet:* Calculation of the basic tooth geometry, contact ratios, stress analysis and data, bearing thrust loads, etc.
- *Summary:* Calculation of cutting and grinding machine setup data to produce the selected tooth geometry.
- *Tooth contact analysis:* A special analysis program that determines the tooth contact pattern and transmission motion errors based on specified cutting tools and gear tooth geometry.
- *Undercut check program.*
- *Loaded tooth contact analysis.*
- *Finite element analysis* (FEA).

Programs are also available from various sources for use on a personal computer (PC) to perform various gear design functions and to solve various other gear problems.

## 8.18 KEYWAYS AND SETSCREWS FOR GEAR SHAFTS

Standard keyways and setscrews for various shaft diameters are shown in Fig. 8.128.

For more design data on shafts and shafting, see Sec. 8.3.

## 8.19 CALCULATIONS FOR POWER, TORQUE, FORCE, VELOCITY, AND RPM

A convenient chart for calculating the various power, force, torque, rpm, and velocities for gearing is shown in Fig. 8.129.

## 8.20 ADDITIONAL GEAR DATA AND REFERENCES

When working with gears and gear systems, keep the following points in mind:

- *Module* is an index of metric gear tooth sizes and is measured in millimeters. Module is equal to the pitch diameter, in millimeters, divided by the number of teeth in the gear and is thus given in millimeters.
- *Diametral pitch* is an index of U.S. customary gear tooth sizes, but is a *ratio.* It is the ratio of the number of teeth in the gear per inch of pitch diameter. The smaller the number of the diametral pitch, the larger are the teeth.

| Diameter of hole, in | Standard Keyway W, in | Standard Keyway d, in | Setscrew size |
|---|---|---|---|
| 5/16 to 7/16 | 3/32 | 3/64 | #10–32 |
| 1/2 to 9/16 | 1/8 | 1/16 | 1/4–20 |
| 5/8 to 7/8 | 3/16 | 3/32 | 5/16–18 |
| 15/16 to 1 1/4 | 1/4 | 1/8 | 3/8–16 |
| 1 5/16 to 1 3/8 | 5/16 | 5/32 | 7/16–14 |
| 1 7/16 to 1 3/4 | 3/8 | 3/16 | 1/2–13 |
| 1 13/16 to 2 1/4 | 1/2 | 1/4 | 9/16–12 |
| 2 5/16 to 2 3/4 | 5/8 | 5/16 | 5/8–11 |
| 2 13/16 to 3 1/4 | 3/4 | 3/8 | 3/4–10 |
| 3 5/16 to 3 3/4 | 7/8 | 7/16 | 7/8–9 |
| 3 13/16 to 4 1/2 | 1 | 1/2 | 1–8 |
| 4 9/16 to 5 1/2 | 1 1/4 | 7/16 | 1 1/8–7 |
| 5 9/16 to 6 1/2 | 1 1/2 | 1/2 | 1 1/4–6 |

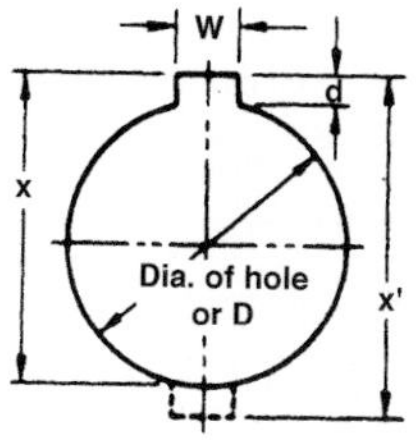


FORMULA:

$$X = \sqrt{(D/2)^2 - (W/2)^2} + d + D/2$$
$$X' = 2X - D$$

**FIGURE 8.128** Standard keyways and setscrew dimensions.

***Special Stem Pinions.*** Figure 8.130 is a photograph of a standard spur stem pinion, which has been specially designed for a double reduction, compact gearbox. This stem pinion has eight teeth, which are specially cut for proper mesh with standard involute spur gears of 14.5° pressure angle. The diameters of this special-purpose stem pinion, marked *A* and *B*, are the needle roller bearing support surfaces. This gear was machined, case-hardened, and then abrasively finished by sand blasting with special abrasive material. The far end of this stem-pinion gear is machined square, for the drive member to engage the gear in operation. The sandblasted, satin finish on the bearing sections of this gear are of a sufficiently fine surface texture that the needles in the bearings will quickly wear in the bearing surfaces.

## 8.21 GEAR WEAR AND FAILURE

Figure 8.131 is a photograph of a stem-pinion–spur gear set which failed in service because of improper heat treatment. The gear material was substantially softer than the needle rollers in the support bearings and galled and eroded after a short time in operation, causing the needles in the bearings to fracture. As soon as the needles fractured, the shaft was very quickly destroyed. This particular gear set was loaded very heavily in operation, but was rated for intermittent duty cycles. With the correct heat treatment, no problems were apparent after prolonged service.

| TO OBTAIN | HAVING | FORMULA |
|---|---|---|
| Velocity (V) Feet Per Minute | Pitch Diameter (D) of Gear or Sprocket — Inches & Rev. Per Min. (RPM) | $V = .2618 \times D \times RPM$ |
| Rev. Per Min. (RPM) | Velocity (V) Ft. Per Min. & Pitch Diameter (D) of Gear or Sprocket — Inches | $RPM = \frac{V}{.2618 \times D}$ |
| Pitch Diameter (D) of Gear or Sprocket — Inches | Velocity (V) Ft. Per Min. & Rev. Per Min. (RPM) | $D = \frac{V}{.2618 \times RPM}$ |
| Torque (T) In. Lbs. | Force (W) Lbs. & Radius (R) Inches | $T = W \times R$ |
| Horsepower (HP) | Force (W) Lbs. & Velocity (V) Ft. Per Min. | $HP = \frac{W \times V}{33000}$ |
| Horsepower (HP) | Torque (T) In. Lbs. & Rev. Per Min. (RPM) | $HP = \frac{T \times RPM}{63025}$ |
| Torque (T) In. Lbs. | Horsepower (HP) & Rev. Per Min. (RPM) | $T = \frac{63025 \times HP}{RPM}$ |
| Force (W) Lbs. | Horsepower (HP) & Velocity (V) Ft. Per Min. | $W = \frac{33000 \times HP}{V}$ |
| Rev. Per Min. (RPM) | Horsepower (HP) & Torque (T) In. Lbs. | $RPM = \frac{63025 \times HP}{T}$ |

**FIGURE 8.129** Mechanics equations for gears and sprockets.

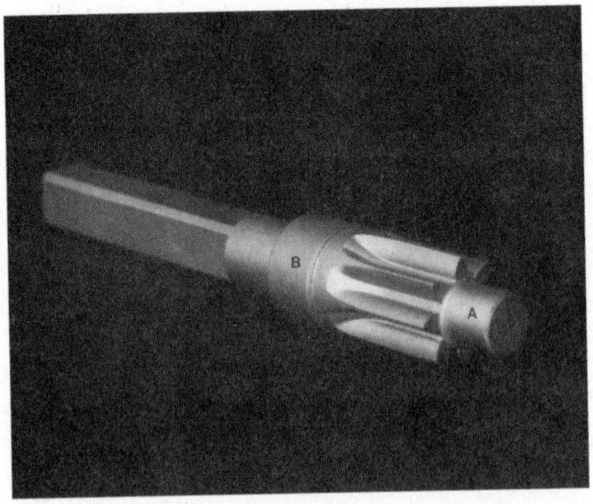


**FIGURE 8.130** A specially machined spur stem pinion.

**FIGURE 8.131** Example of gear failure due to improper heat treatment.

## 8.22 GEAR MANUFACTURING PROCESSES

Figure 8.132 shows in detail the methods used to manufacture different types of gears. Some of these methods were shown in previous sections, but not as clearly as the illustrations of this figure depict.

*Top left:* milling a spur gear
*Top right:* end-milling a spur gear
*Center left:* hobbing a spur gear
*Center right:* shaping an internal spur gear
*Bottom left:* shaping a bevel gear
*Bottom right:* milling a bevel gear on a special machine

Note the directions of feed and cut indicated by the arrows in the figures.

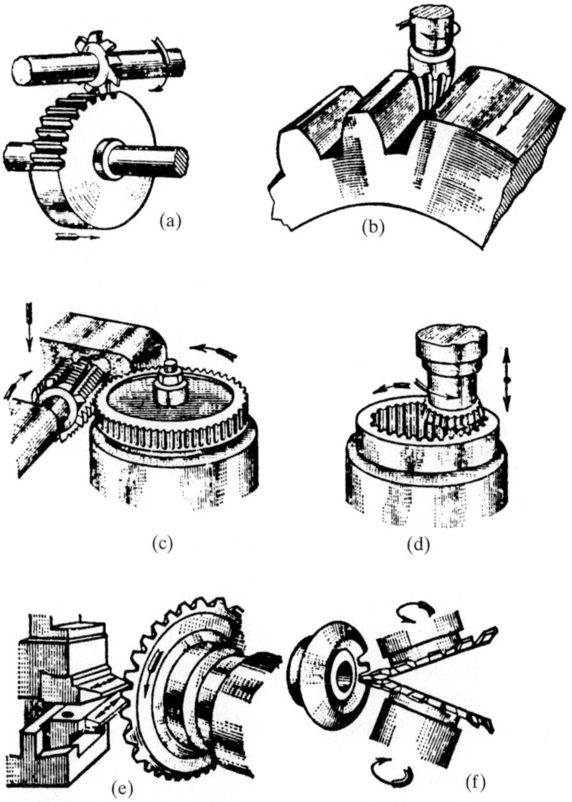


**FIGURE 8.132** Details of the methods of gear production: (*a*) milling a spur gear; (*b*) end-milling a spur gear; (*c*) hobbing a spur gear; (*d*) shaping an internal spur gear; (*e*) shaping a bevel gear; (*f*) milling a bevel gear on a special machine. Arrows indicate the direction of feed and cut.

## 8.23 GEAR LUBRICATION

All the major oil companies and lubrication specialty companies provide lubricants for gearing and other applications to meet a very broad range of operating conditions. General gear lubrication consists of a high-quality machine oil when there are no temperature extremes or other adverse ambient conditions. Many of the automotive greases and oils are suitable for a broad range of gearing applications.

For adverse temperatures, environmental extremes, and high-pressure applications, consult the lubrication specialty companies or the major oil companies to meet your particular requirements or specifications.

The science and technology of lubrication and its applications to various machinery and conditions are complex and should be referred to the lubrication specialists.

## 8.24 GEAR SUMMARY SHEETS

Table 8.12 shows a typical gear summary sheet which is used for manufacturing the gears. This summary sheet is for a set of spiral bevel gears from the Dudley Engineering Company. This sheet was extracted from Darle W. Dudley, 1984. *Handbook of Practical Gear Design,* McGraw-Hill, New York.

## 8.25 CAMS: DEVELOPMENT, LAYOUT, AND DESIGN

*Cams* are mechanical components which convert rotary motion into a selective or controlled translating or oscillating motion or action by way of a cam follower which bears against the working surface of the cam profile or perimeter. As the cam rotates, the cam follower rises and falls according to the motions described by the displacement curve.

Cams can be used to translate power and motion, such as the cams on the camshaft of an internal combustion engine, or for selective motions as in timing devices or generating functions. The operating and timing cycles of many machines are controlled by the action of cams.

There are basically two classes of cams; uniform-motion cams and accelerated-motion cams.

### 8.25.1 Cam Motions

The most important cam motions and displacement curves in common use are

- *Uniform-velocity motion:* for low speeds
- *Uniform acceleration:* for moderate speeds
- *Parabolic motion used in conjunction with uniform motion or uniform acceleration:* low to moderate speeds
- *Cycloidal:* for high speeds

The design of a typical cam is initiated with a *displacement curve* as shown in Fig. 8.133. Here, the $Y$ dimension corresponds to the cam rise or fall, and the $X$ dimension corresponds either to degrees, radians, or time displacement. The slope lines of the rise and fall intervals should be terminated with a parabolic curve to prevent shock loads on the follower. The total length of the displacement ($X$ dimension) on the displacement diagram represents one complete revolution of the cam. Standard graphical layout methods may be used to develop the dis-

**TABLE 8.12** Typical Gear Specification Sheet for Spiral Bevel Gears

| | Pinion | Gear |
|---|---|---|
| Number of teeth | 16 | 49 |
| Part number | | |
| Module | | 5.000 |
| Face width | 38.00 | 38.00 |
| Pressure angle | 20D OM | |
| Shaft angle | 90D OM | |
| Transverse contact ratio | | 1.192 |
| Face contact ratio | | 2.006 |
| Modified contact ratio | | 2.333 |
| Outer cone distance | | 128.87 |
| Mean cone distance | | 109.87 |
| Pitch diameter | 80.00 | 245.00 |
| Circular pitch | 15.71 | |
| Working depth | 8.28 | |
| Whole depth | 9.22 | 9.22 |
| Clearance | 0.94 | 0.94 |
| Addendum | 5.91 | 2.38 |
| Dedendum | 3.32 | 6.85 |
| Outside diameter | 91.23 | 246.48 |
| Face angle junction diameter | | |
| Theoretical cutter radius | 3.751" | |
| Cutter radius | 3.750" | |
| Calc. gear finish. pt. width | | 0.100" |
| Gear finishing point width | | 0.100" |
| Roughing point width | 0.045" | 0.080" |
| Outer slot width | 0.071" | 0.100" |
| Mean slot width | 0.083" | 0.100" |
| Inner slot width | 0.073" | 0.100" |
| Finishing cutter blade point | 0.045" | 0.065" |
| Stock allowance | 0.026" | 0.020" |
| Max. radius—cutter blades | 0.043" | 0.074" |
| Max. radius—mutilation | 0.061" | 0.076" |
| Max. radius—interference | 0.045" | 0.096" |
| Cutter edge radius | 0.025" | 0.025" |
| Calc. cutter number | 3 | 9 |
| Max. no. blades in cutter | | 11.295 |
| Cutter blades required | Std depth | Std depth |
| Gear angular face—concave | | 26D 47M |
| Gear angular face—convex | | 29D 30M |
| Gear angular face—total | | 31D 52M |

NOTE: All dimensions are in metric unless denoted otherwise. Angles are in degrees (D) and minutes (M).

placement curves and simple cam profiles. The placement of the parabolic curves at the terminations of the rise/fall intervals on uniform-motion and uniform-acceleration cams is depicted in the detail view of Fig. 8.133. The graphical construction of the parabolic curves which begin and end the rise/fall intervals may be accomplished using the principles of geometric construction shown in drafting manuals or in the *McGraw-Hill Machining and Metalworking Handbook* (1994).

The layout of the cam shown in Fig. 8.134 is a development of the displacement diagram shown in Fig. 8.133. In this cam, we have a dwell interval followed by a uniform-motion and uniform-velocity rise, a short dwell period, a uniform fall, and then the remainder of the dwell to complete the cycle of one revolution.

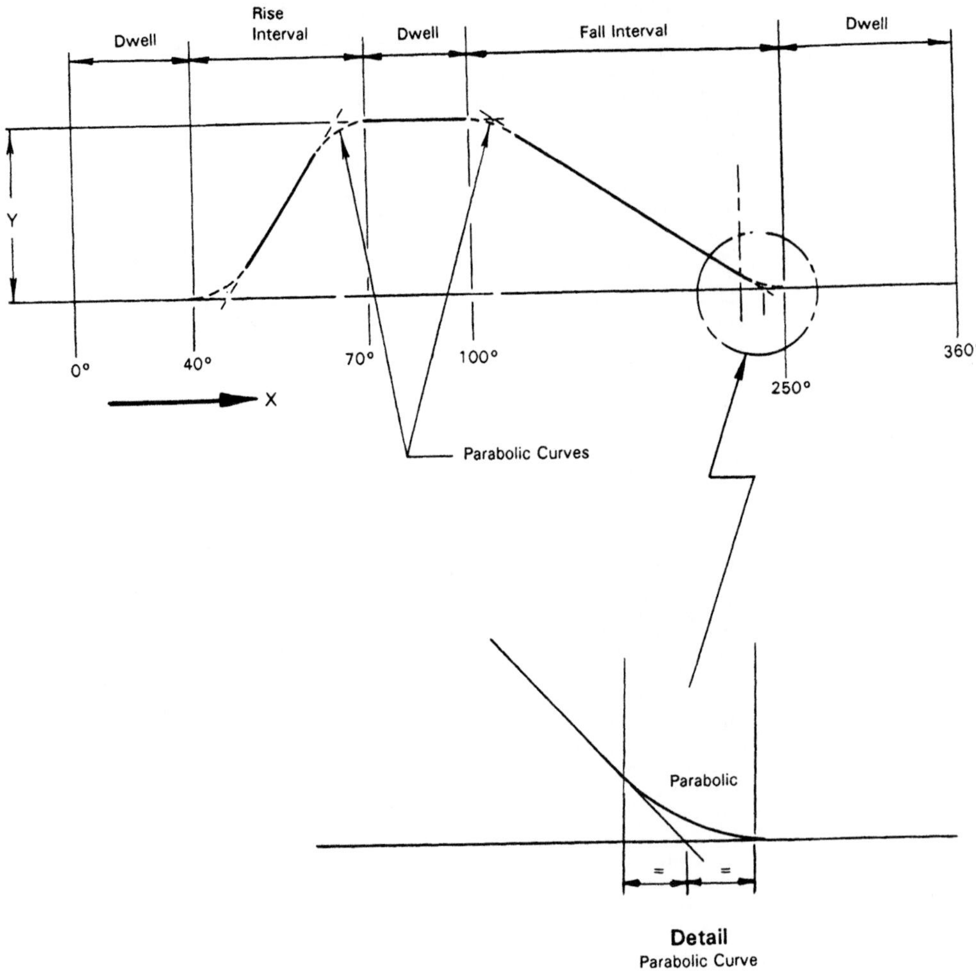


**FIGURE 8.133** Cam displacement diagram (the developed cam is as shown in Fig. 8.134).

The layout of a cam such as shown in Fig. 8.134 is relatively simple. The rise/fall periods are developed by dividing the rise or fall into the same number of parts as the angular period of the rise and fall. The points of intersection of the rise/fall divisions with the angular divisions are then connected by a smooth curve, terminating in a small parabolic curve interval at the beginning and end of the rise/fall periods. Cams of this type have many uses in industry and are economical to manufacture because of their simple geometries.

## 8.25.2 Uniform-Motion Cam Layout

The cam shown in Fig. 8.135 is a uniform or harmonic-motion cam, often called a "heart" cam because of its shape. The layout of this type of cam is simple as the curve is a development of

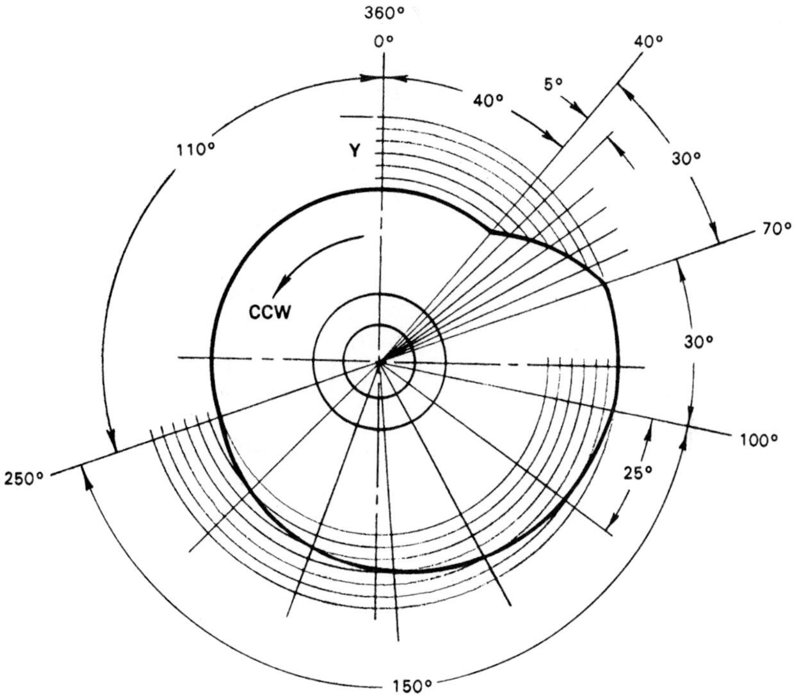


**FIGURE 8.134** Development of a cam whose displacement diagram is as shown in Fig. 8.133.

the intersection of the rise intervals with the angular displacement intervals. The points of intersection are then connected by a smooth curve.

### 8.25.3 Accelerated-Motion Cam Layout

The cam shown in Fig. 8.136 is a uniform-acceleration cam. The layout of this type of cam is also simple. The rise interval is divided into increments of 1-3-5-5-3-1 as shown in the figure. The angular rise interval is then divided into six equal angular sections as shown. The intersection of the projected rise intervals with the radial lines of the six equal angular intervals are then connected by a smooth curve, completing the section of the cam described. The displacement diagram that is generated for the cam follower motion by the designer will determine the final configuration of the complete cam.

### 8.25.4 Cylindrical Cam Layout

A cylindrical cam is shown in Fig. 8.137 and is layed out in a similar manner described for the cams of Figs. 8.134 and 8.135. A displacement diagram is made first, followed by the cam stretchout view shown in Fig. 8.137. The points describing the curve that the follower rides in may be calculated mathematically for a precise motion of the follower. Four- and five-axis machining centers are used to cut the finished cams from a computer program generated in the engineering department and fed into the controller of the machining center.

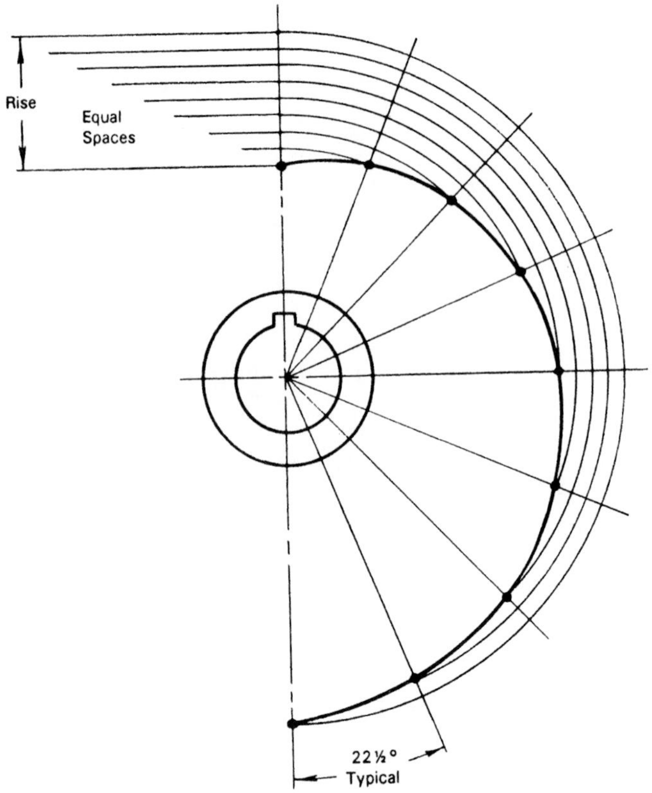


**FIGURE 8.135** Uniform-motion cam layout (harmonic motion).

Tracer cutting and incremental cutting are also used to manufacture cams, but are seldom used when the manufacturing facility is equipped with four- and five-axis machining centers, which do the work faster and more accurately than previously possible.

The design of cycloidal motion cams is not discussed in this section of the handbook because of their mathematical complexity and many special requirements. Cycloidal cams are also expensive to manufacture because of the requirements of the design and programming functions required in the engineering department. The bibliography section at the end of the chapter lists handbooks covering cycloidal cam design.

### 8.25.5 Eccentric Cams

A cam which is required to actuate a roller limit switch in a simple application or to provide a simple rise function may be made from an eccentric shape as shown in Fig. 8.138. The rise, diameter, and offset are calculated as shown in the figure. This type of cam is the most simple to design and economical to manufacture and has many practical applications. Materials used for this type of cam design can be steel, alloys, or plastics and compositions. Simple functions and light loads at low to moderate speeds are limiting factors for these types of cams.

In Fig. 8.138*a* and 8.138*b*, the simple relationships of the cam variables are as follows:

$$R = (x + r) - a \qquad a = r - x \qquad \text{rise} = D - d$$

The eccentric cam may be designed using the relationships shown above.

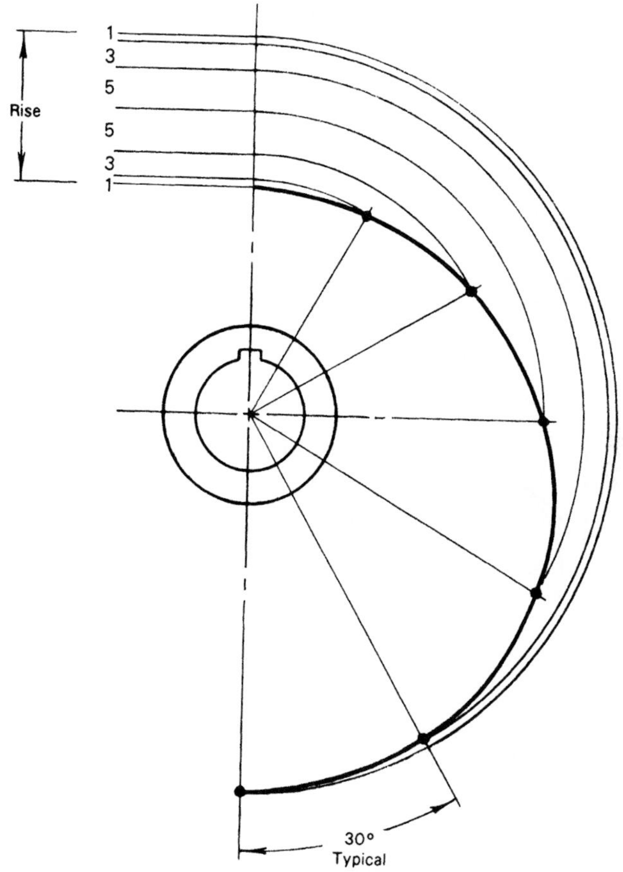


**FIGURE 8.136** Uniform-acceleration cam layout.

Figure 8.139 shows the assembly drawing for a typical application of eccentric cams being used to actuate limit switches for electric timing controls.

### 8.25.6 The Cam Follower

The most common types of cam follower systems are the radial translating, offset translating, and swinging roller as depicted in Figs. 8.140*a*–8.140*c*.

The cams in Figs. 8.140*a*–8.140*c* are *open-track cams,* in which the follower must be held against the cam surface at all times, usually by a spring. A *closed-track cam* is one in which a roller follower travels in a slot or groove cut in the face of the cam. The cylindrical cam shown in Fig. 8.137 is a typical example of a closed-track cam. The closed-track cam follower system is termed positive because the follower translates in the track without recourse to a spring holding the follower against the cam surface. The positive, closed-track cam has wide use on machines in which the breakage of a spring on the follower could otherwise cause damage to the machine.

Note that in Fig. 8.140*b,* where the cam follower is offset from the axis of the cam, the offset must be in a direction opposite that of the cam's rotation.

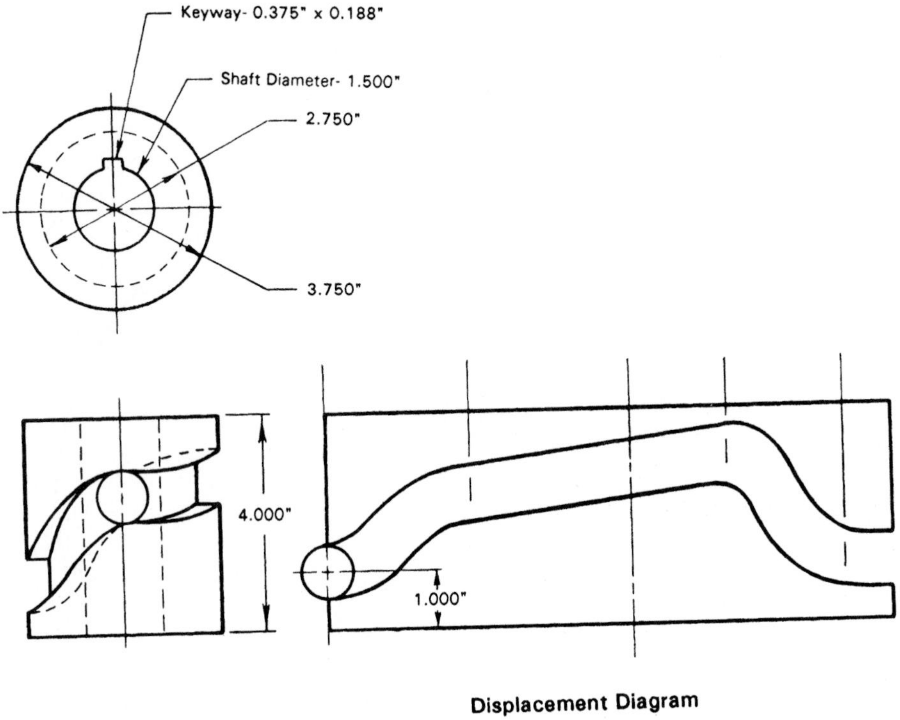


**FIGURE 8.137** Development of a cylindrical cam.

On cam follower systems which use a spring to hold the cam follower against the working curve or surface of the cam, the spring must be designed properly to prevent "floating" of the spring during high-speed operation of the cam. The cyclic rate of the spring must be kept below the natural frequency of the spring in order to prevent floating. Chapter 7 of the handbook shows procedures for the design of high-pressure, high-cyclic-rate springs in order to prevent this phenomenon from occurring. When you know the cyclic rate of the spring used on the cam follower and its working stress and material, you can design the spring to have a natural frequency which is below the cyclic rate of operation. The placement of springs in *parallel* is often required to achieve the proper results. The valve springs on high-speed automotive engines is a good example of this practice, wherein we wish to control natural frequency and at the same time have a spring with a high spring rate to keep the engine valves tightly closed. The spring rate must also be high enough to prevent separation of the follower from the cam surface during acceleration, deceleration, and shock loads in operation. The cam follower spring is often preloaded to accomplish this.

## 8.25.7 Pressure Angle of the Cam Follower

The *pressure angle* $\phi$ (see Fig. 8.141) is generally made 30° or less for a reciprocating cam follower and 45° or less for an oscillating cam follower. These typical pressure angles also depend on the cam mechanism design and may be more or less than indicated above.

The pressure angle $\phi$ is the angle between a common *normal* to both the roller and the cam profile and the direction of the follower motion, with one leg of the angle passing

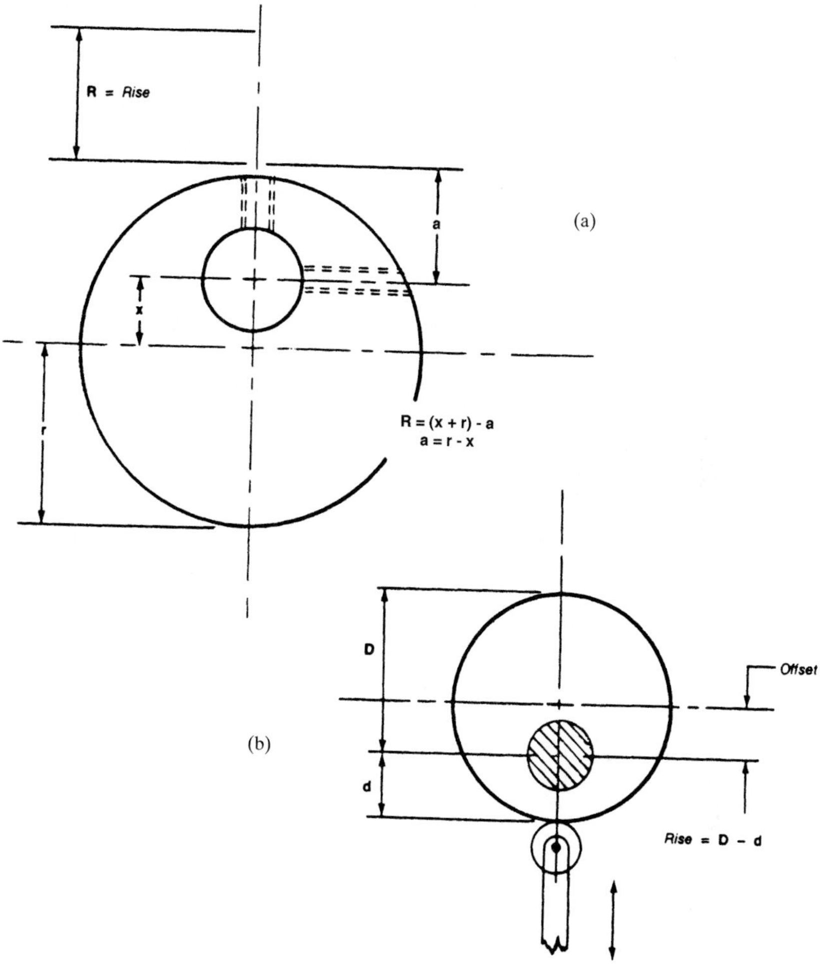


**FIGURE 8.138** (*a, b*) Eccentric cam geometry.

through the axis of the follower roller axis. This pressure angle is easily found using graphical layout methods.

To avoid undercutting cams with a roller follower, the radius $r$ of the roller must be less than $C_r$, which is the minimum radius of curvature along the cam profile.

***Pressure Angle Calculations.*** The pressure angle is an important factor in the design of cams. Variations in the pressure angle affect the transverse forces acting on the follower.

The simple equations which define the maximum pressure angle $\alpha$ and the cam angle $\theta$ at $\alpha$ are as follows (see Fig. 8.142*a*):

For simple harmonic motion:

$$\alpha = \arctan \frac{\pi}{2\beta} \left( \frac{S/R}{\sqrt{1 + (S/R)}} \right) \qquad \theta = \frac{\beta}{\pi} \arccos \left( \frac{S/R}{2 + (S/R)} \right)$$

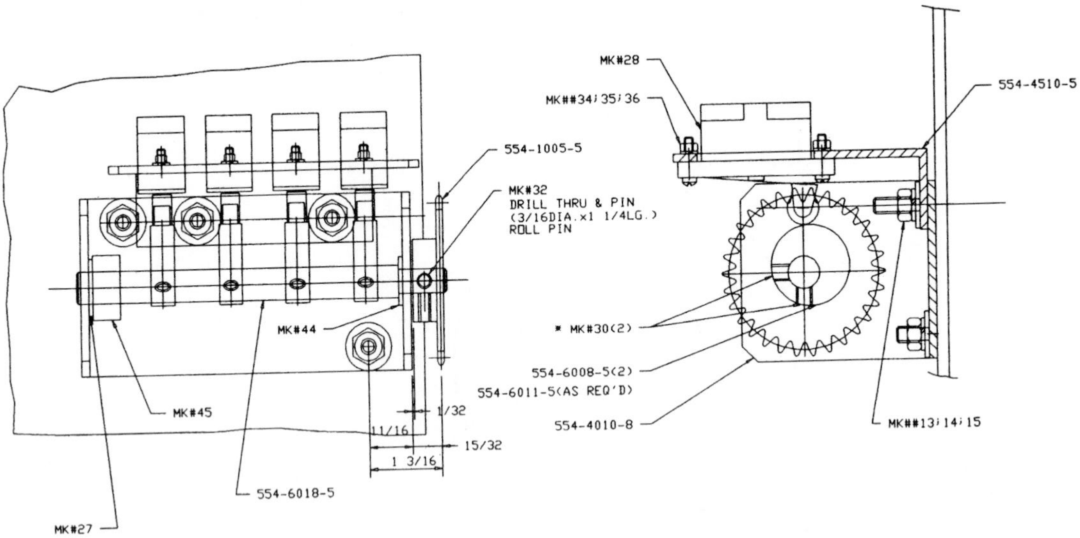


**FIGURE 8.139** Application of the eccentric cam (control timing switch assembly).

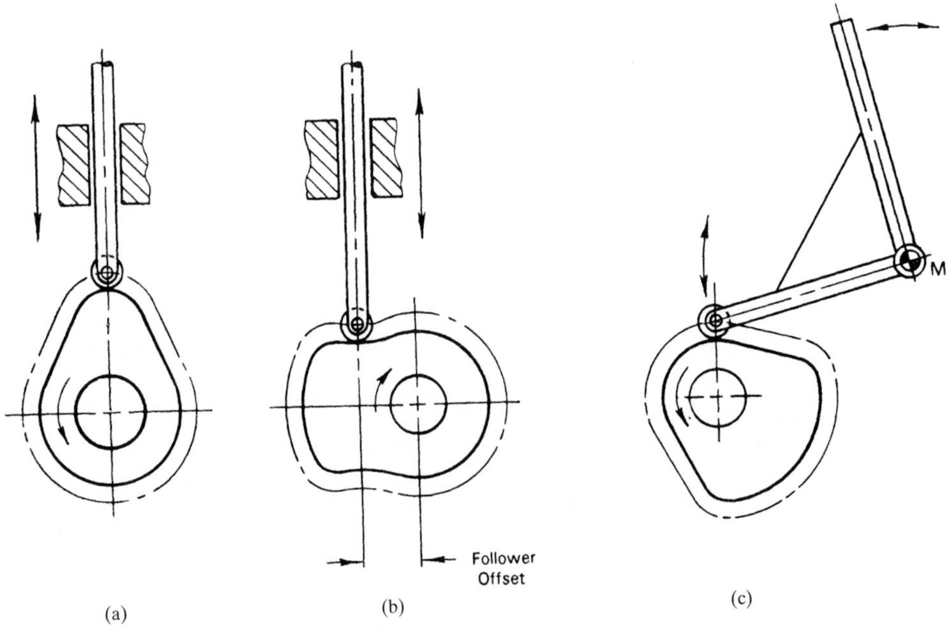


**FIGURE 8.140** (*a*) In-line follower; (*b*) offset follower; (*c*) swinging-arm follower.

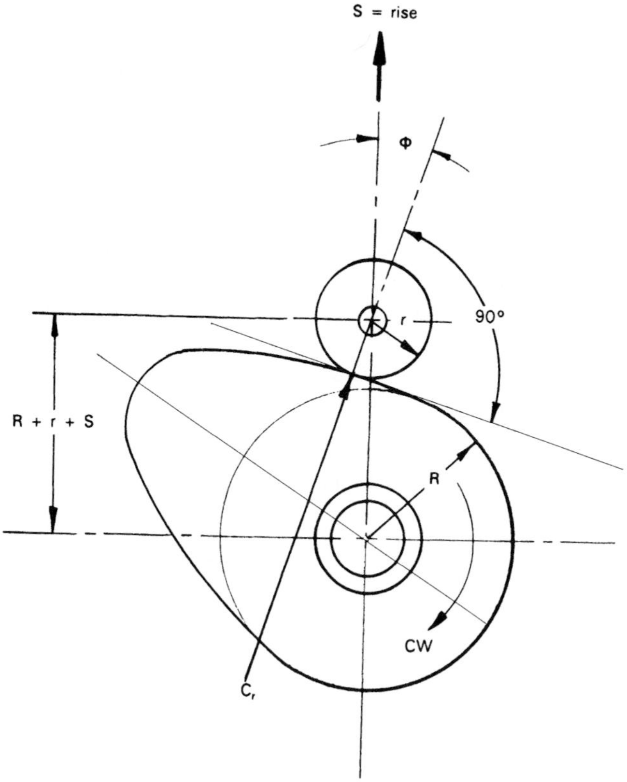


**FIGURE 8.141** The pressure angle of the cam follower.

For constant-velocity motion:

$$\alpha = \arctan \frac{1}{\beta}\left(\frac{S}{R}\right) \qquad \theta = 0$$

For constant-acceleration motion:

$$\alpha = \arctan \frac{2}{\beta}\left(\frac{S/R}{1+(S/R)}\right) \qquad \theta = \beta$$

For cycloidal motion:

$$\alpha = \arctan \frac{1}{2\beta}\left(\frac{S}{R}\right) \qquad \theta = 0$$

where $\alpha$ = maximum pressure angle of the cam, degrees
$S$ = total lift for a given cam motion during cam rotation, in
$R$ = initial base radius of cam; center of cam to center of roller, in
$\beta$ = cam rotation angle during which the total lift $S$ occurs for a given cam motion, rad
$\theta$ = cam angle at pressure angle $\alpha$

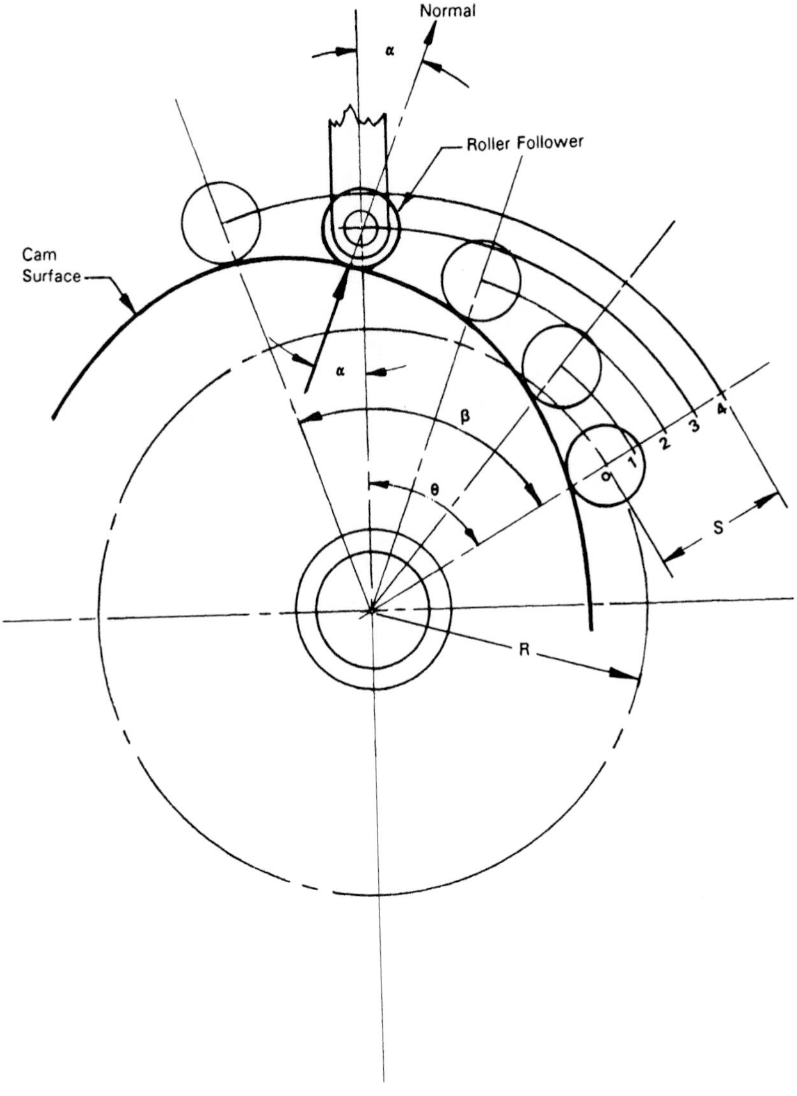


(a)

**FIGURE 8.142** (*a*) Diagram for pressure angle calculations.

### 8.25.8 Contact Stresses between Follower and Cam

To calculate the approximate stress $S_s$ developed between the roller and the cam surface, we can use the simple equation

$$S_s = C\sqrt{\frac{f_n}{w}\left(\frac{1}{r_f}+\frac{1}{R_c}\right)}$$

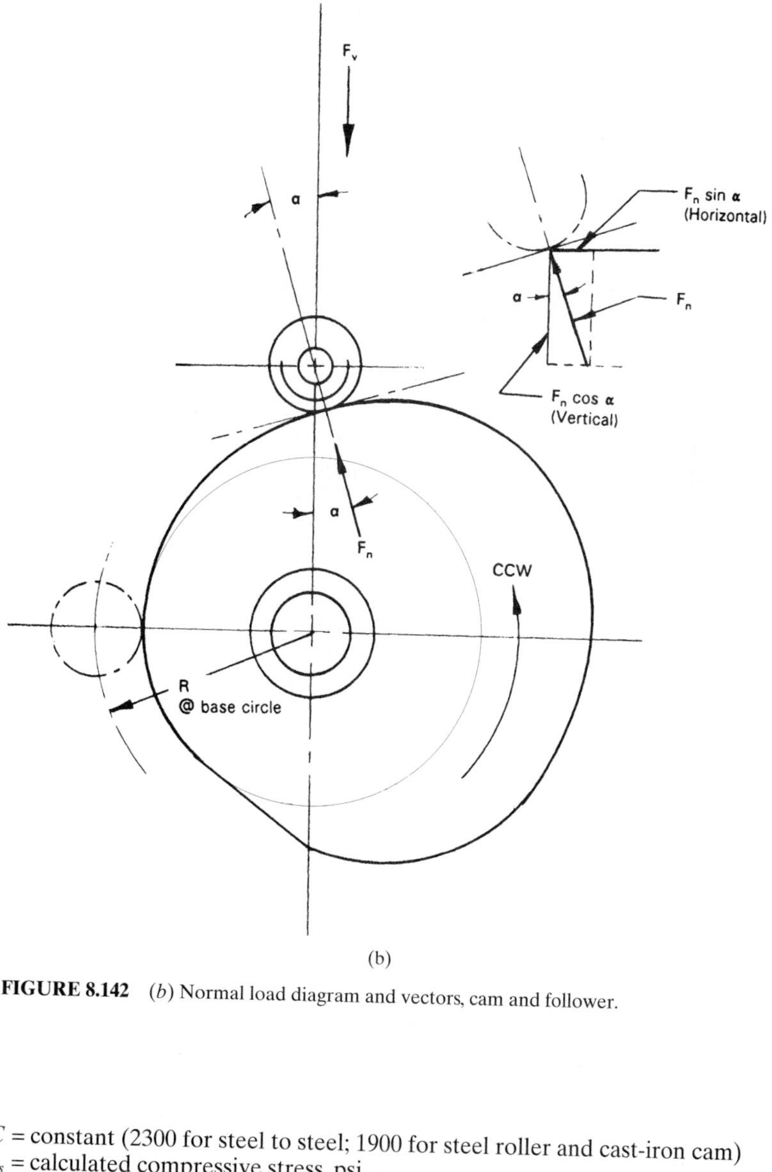


(b)

**FIGURE 8.142** (*b*) Normal load diagram and vectors, cam and follower.

where $C$ = constant (2300 for steel to steel; 1900 for steel roller and cast-iron cam)
$S_s$ = calculated compressive stress, psi
$f_n$ = normal load between follower and cam surface, lbF
$w$ = width of cam and roller common contact surface, in
$R_c$ = minimum radius of curvature of cam profile, in
$r_f$ = radius of roller follower, in

The highest stress is developed at the minimum radius of curvature of the cam profile. The calculated stress $S_s$ should be less than the maximum allowable stress of the weaker material of the cam or roller follower. The roller follower would normally be the harder material.

Cam or follower failure is usually due to fatigue when the surface endurance limit (permissible compressive stress) is exceeded.

Some typical maximum allowable compressive stresses for various materials used for cams, when the roller follower is hardened steel (Rockwell C45 to C55) include

| | |
|---|---|
| Gray iron—cast (200 Bhn) | 55,000 psi |
| ASTM A48-48 | |
| SAE 1020 steel (150 Bhn) | 80,000 psi |
| SAE 4150 steel HT (300 Bhn) | 180,000 psi |
| SAE 4340 steel HT ($R_c$ 50) | 220,000 psi |

(*Note:* Bhn designates Brinnel hardness number; $R_c$ is Rockwell C scale.)

### 8.25.9 Cam Torque

As the follower bears against the cam, resisting torque develops during rise *S*, and assisting torque develops during fall or return. The maximum torque developed during cam rise operation determines the cam drive requirements.

The instantaneous torque values $T_i$ may be calculated using the equation

$$T_i = \frac{9.55 v F_n \cos \alpha}{N}$$

where $T_i$ = instantaneous torque, lb · in
$v$ = velocity of follower, in/sec
$F_n$ = normal load, lb
$\alpha$ = maximum pressure angle, degrees
$N$ = cam speed, rpm

The normal load $F_n$ may be found graphically or calculated from the vector diagram shown in Fig. 8.142*b*. Here, the horizontal or lateral pressure on the follower = $F_n \sin \alpha$ and the vertical component or axial load on the follower = $F_n \cos \alpha$.

When we know the vertical load (axial load) on the follower, we solve for $F_n$ (the normal load) on the follower from

$$F_n \cos \alpha = F_v$$

given $\alpha$ = pressure angle, degrees
$F_v$ = axial load on follower (from preceding equation), lbf
$F_n$ = normal load at the cam profile and follower, lbf

*Example.* Spring load on the follower is 80 lb and the pressure angle $\alpha$ is 17.5°. Then

$$F_v = F_n \cos \alpha \qquad F_n = \frac{F_v}{\cos \alpha} = \frac{80}{\cos 17.5} = \frac{80}{0.954} = 84 \text{ lb}$$

Knowing the normal force $F_n$, we can calculate the pressure (stress) in pounds per square inch between the cam profile and roller on the follower (see Sec. 8.25.8).

### 8.25.10 Cam Manufacture

The cam type and motion curve should be kept as simple as the application will allow, in order to keep manufacturing costs low. Use the most economical material for each specific application. Plastic cams are indicated in low-speed, light-load systems and may be molded if the required quantity is large enough to justify the costs for tooling and molds. Cams are sometimes made of copper alloys such as high-strength aluminum bronze. Refer to Chap. 4 for other materials which may be used for cam applications.

The developed cam profile may be cut on the electric-discharge machine (EDM) using a computer program with high accuracy and the cam thus produced used as the master cam for tracer or CNC-programmed by milling large numbers of cams. This may be the most accurate and cost-effective method of producing radial face cams. The EDM machine is capable of accurately cutting hardened tool steels as thick as 2 in.

### 8.25.11 Dynamic Analysis of Cams during Operation

With the advent of high-speed motion-picture photography, it is now possible to analyze and measure the actions of high-speed cams in operation by way of a prototype of the cam system under study. Cameras with speeds of 1000 to 12,000 frames per second are commercially available for the study of high-speed mechanisms, allowing design corrections to be made which would be very difficult to accomplish mathematically. The solutions to many difficult engineering mechanics problems are thus available to the modern designer of mechanical systems with minimal recourse to advanced mathematical manipulations (which are usually approximations at best). The high-speed analyses are also accomplished with the additional use of oscillographs, accelerometers, and other transducers including strain gauges.

Figure 8.143 shows some typical, low-speed, low-force, high-pressure, and economical cams which are used for the applications indicated below:

(*a*) For roller limit switch actuation in an electrical timing device

(*b*) For general use as a limit switch actuator cam (see Fig. 8.139)

(*c*) A pair of profile cams used in a clamping mechanism; made from 7075-T651 aluminum alloy which has been hardcoat-anodized

### 8.25.12 Various Cam Types or Designs

The study of high-speed, high-load cams is mathematically complex and will not be undertaken in the handbook, although the cam systems shown in the beginning of this section are perfectly suitable for many low- and intermediate-speed applications. Graphical layout pro-

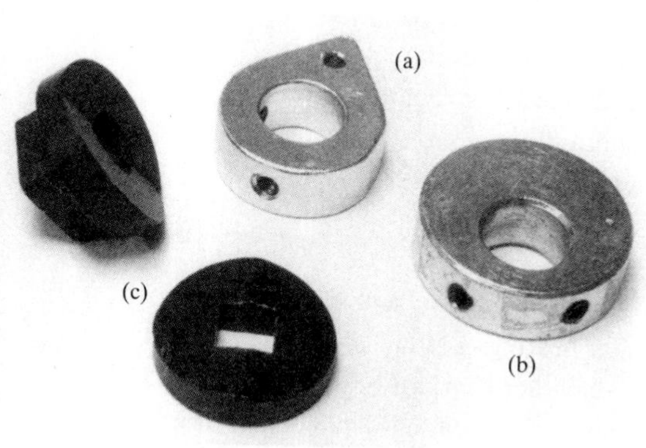


**FIGURE 8.143** Typical cams: (*a*) quick-rise cam; (*b*) eccentric cam; (*c*) profile cam set.

cedures and blending of the parabolic curves to the rise/fall intervals of simple cams is a preferred method of producing the simpler types of cam and follower systems shown in this section.

In advanced cam design, two classes of cams are generally recognized: the trigonometric family and the polynomial family. Included in these systems are cams such as the polydyne cams used in high-speed automotive systems. The *desmodromic* system, used many years ago, employed a positive-action cam design wherein springs were not used to hold the follower against the cam profile, making it suitable for ultra-high-speed applications such as the cams in early racing engines, such as the Mercedes-Benz around the 1950s.

In conclusion, the more common types of cams include radial plate cams, positive-action cams, face cams, and cylindrical cams. Cam followers take many forms such as roller followers, flat-plate followers, roller bearing followers, offset disk followers, and chisel or knife-point followers.

The design of complex, high-speed cam systems also may be undertaken using the high-speed computer designer stations available in some of the larger companies, wherein the mechanics, motions, and stresses may be analyzed quickly and accurately by way of computer-aided design (CAD) and finite element analysis (FEA).

## 8.26 ANTIFRICTION BEARINGS

Antifriction bearings are among the most important mechanical elements. Most mechanisms, apparatus, and machinery contain antifriction bearings of one type or another. By providing reduced frictional drag on moving parts, antifriction bearings allow machinery and mechanisms to operate at their most efficient levels.

Many factors are taken into account in the selection of any particular type of bearing for any particular application. One type of bearing may be more suitable for a given application than any other type. The factors which influence the selection of a bearing for a particular application include

- Available size
- Load-bearing characteristics
- Life of the bearing
- Permissible loads
- Efficiency
- Type of application
- Operating temperature
- Speed—linear or rotational
- Mounting
- Cost

The typical, common bearing types include

- Ball bearings—radial and thrust
- Roller bearings—radial and thrust
- Needle roller bearings—radial and thrust
- Tapered roller bearings—radial and thrust
- Cylindrical roller bearings
- Plain bearings—sleeve and flanged; radial and thrust

Figure 8.144 shows sections through the main typical bearing types, which include ball, roller, tapered roller, needle, and cylindrical roller.

Figure 8.145 depicts a typical caged-roller ball bearing of the type used in many mechanisms and machines.

Figure 8.146 shows the detailed construction of the widely used tapered roller bearing shown in Fig. 8.144*c*. This bearing is made with a pressed-steel cage for the rollers or the pin-type cage.

Figure 8.147 shows the detailed construction of the main types of bearings used throughout industry. Both radial and thrust types are shown.

Industry rolls on antifriction bearings and are manufactured in sizes ranging from near microscopic to over 100 in in diameter in order to carry loads of minute value to many tons. The simplest type of antifriction bearing is the plain sleeve or flanged type, with which we will begin our bearing section.

### 8.26.1 Plain Bearings: Sleeve and Flanged (Journal)

The plain bearing is available in either sleeve or flanged type. These types of bearings are most commonly made of sintered bronze with impregnated lubrication. Minute pores are pro-

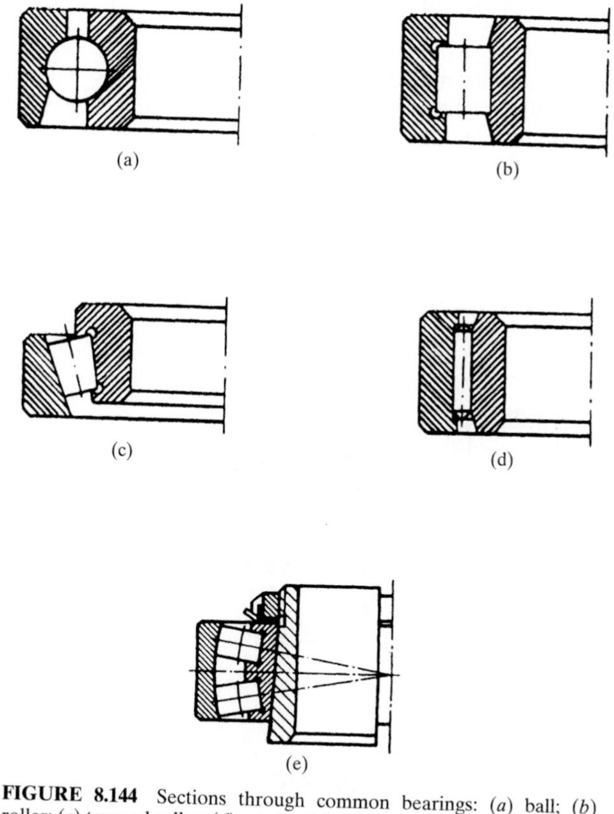


**FIGURE 8.144** Sections through common bearings: (*a*) ball; (*b*) roller; (*c*) tapered roller; (*d*) needle; (*e*) spherical roller.

**FIGURE 8.145** Typical caged roller bearing.

duced in the bearing material during the sintering process, and these are impregnated with a high-quality, long-lasting lubricant, making the bearings self-lubricating. Plain bearings are also made of aluminum alloys, copper-lead alloys, babbitt alloys, and various types of thermoplastics and thermoset plastics such as acetal, polycarbonates, Teflons, nylons, and phenolic impregnated cloth thermosets.

Most of the plain sleeve and flanged bearings are designed for a force fit into their receiving housing bore. The outside and inside diameters of these types of bearings are manufactured to close tolerances so that a force fit is easily made using the properly bored bearing mounting hole. It should be noted that if a bearing mounting hole is "overbored" and the bearing is a loose fit, the assembly can be saved by using a compound called Quick-Metal by Locktite Corporation. The Quick-Metal is applied over the outside diameter of the bearing and then inserted into the loose fitting hole. The Quick-Metal will permanently bond the bearing in the hole after a setting time of approximately 20 min to 1 h. An otherwise ruined bearing assembly may thus be saved. This application may be limited to noncritical assemblies and in certain tooling practices (tool repairs) and maintenance procedures. (See Fig. 8.148 for plain bearing press fits.)

Plain bearing stock is available for manufacturing your own bearings to a particular set of dimensions. The following bearing rules should be kept in mind when designing a plain cylindrical bearing. The length of a cylindrical plain bearing should be one to two times the shaft diameter, and the outer diameter should be approximately 25 percent larger than the shaft diameter. The bearing stock material must have adequate compressive strength for the application.

**TYPE TS SINGLE-ROW BEARINGS**

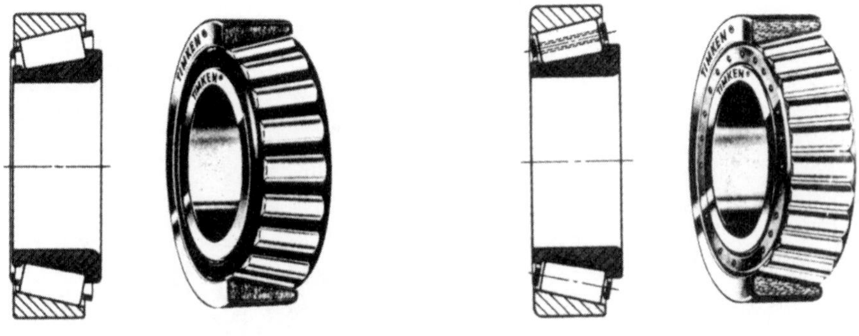

TS (PRESSED STEEL CAGE)

TS (PIN TYPE CAGE)

**FIGURE 8.146** Detailed construction, tapered roller bearings. (*Source: The Timken Company.*)

**FIGURE 8.147** Detailed construction of common bearings: (*a*) spherical roller; (*b*) ball; (*c*) tapered roller; (*d*) ball thrust; (*e*) caged needle; (*f*) roller thrust; (*g*) caged roller; (*h*) drawn-cup needle, open and closed end; (*i*) needle thrust. (*Source: The Torrington Company.*)

| Housing diameter in | Press fit (interference), in |
|---|---|
| 0.500 | 0.0005 |
| 1.000 | 0.0008 |
| 1.500 | 0.0010 |
| 2.000 | 0.0015 |
| 2.500 | 0.0020 |
| 3.000 | 0.0025 |

**FIGURE 8.148** Press-fit allowances for plain bearings (metallic).

### 8.26.2 Selection of Plain Bearings by *PV* Calculations

Load and velocity value limits are established for plain bearings by the use of *PV* calculations. The term *PV* represents a pressure and velocity factor and is a means of measuring the performance capabilities of plain bearings; *P* is expressed as a pressure or pounds per square inch on the projected area of a bearing, and *V* is the velocity in feet per minute of the wear surface (surface feet per minute). *PV* is expressed by the following relational equations:

$$PV = \left(\frac{W}{Ld} \times \frac{\pi dn}{12}\right) = \frac{\pi Wn}{12\text{L}} = \frac{0.262\ Wn}{L} \qquad P = \frac{W}{A\ (\text{bearing i.d.} \times L)}$$

where $V$ = surface velocity of the shaft, ft/min (0.262 × rpm × shaft diameter)
$W$ = bearing load, lb
$L$ = bearing length, in
$d$ = inner diameter (i.d.) of bearing, in
$n$ = shaft speed, rpm

| Material | PV value |
|---|---|
| Lead-base babbitt | 35,000 |
| Tin-base babbitt | 40,000 |
| Cadmium base | 90,000 |
| Copper-lead | 90,000 |

**FIGURE 8.149** *PV* values of plain metallic bearings.

Each bearing material has a specific maximum *PV* rating, as shown in Figs. 8.149 and 8.150. In addition, the bearing material also has a maximum pressure *P* and velocity *V* limitation. At no time can all maximum values be utilized. The selection of a plain bearing by *PV* calculation is a balance of the *P*, *V*, and *PV* values. The following example will illustrate the selection procedures.

When selecting a bearing, you must know the *PV*, *P*, and *V* values of the bearing material (see Figs. 8.149 and 8.150) (the bearing manufacturers' catalogs also list these values), noting that

$$PV_m = P_m V_m \qquad P_m = \frac{PV_m}{V_m} \qquad V_m = \frac{PV_m}{P_m}$$

where $PV_m$ = maximum *PV* value
$P_m$ = maximum pressure, psi
$V_m$ = maximum velocity, surface feet per minute

Do not exceed the maximum values.

*Example.* Select a plain bearing to satisfy the following conditions, using any of the previous equations and the values given in Figs. 8.149 and 8.150. The following is known: 0.625-in-diameter shaft, $n$ = 500 rpm $W_1$ = load on bearing 1 = 60 lb, $W_2$ = load on bearing 2 = 100 lb, and $L$ = length of bearing, in.

For bearing 1:

$$PV = \frac{0.262 W_1 n}{L} = \frac{0.262 \times 60 \times 500}{1} = 7860$$

For bearing 2:

$$PV = \frac{0.262 W_2 n}{L} = \frac{0.262 \times 100 \times 500}{1} = 13{,}100$$

| Material | PV limit unlub. | P Load limit psi | Max. speed, fpm V | Max. temp, °F |
|---|---|---|---|---|
| Acetal | 3,000 | 2,000 | 600 | 200 |
| Carbon-graphite | 15,000 | 600 | 2,500 | 600 |
| Nylon | 3,000 | 2,000 | 600 | 200 |
| Phenolic | 15,000 | 6,000 | 2,500 | 200 |
| Polycarb. | 3,000 | 1,000 | 1,000 | 225 |
| PTFE | 1,000 | 500 | 50 | 500 |
| Filled PTFE | 10,000 | 2,500 | 1,000 | 500 |
| PTFE fabric | 25,000 | 60,000 | 150 | 500 |

(SOURCE: *Machine Design*, June, 1984.)

**FIGURE 8.150** *PV* values of nonmetallic plain bearings.

With a calculated *PV* for bearing 1 of 7860 and for bearing 2 of 13,100, it can be seen from Fig. 8.150 that carbon graphite, phenolic, or PTFE fabric could possibly be used for these bearings. You must now check the maximum *P* and maximum *V* values using the relational equations shown previously to see which bearing material will be adequate for the job. If the plastic bearing materials cannot handle the required loads and speeds, the selection of a metallic bearing will be necessary. *PV* values for some of the metallic bearing materials are shown in Fig. 8.149.

The press-fit allowances for the plain metallic bearings are shown in Fig. 8.148.

### 8.26.3 Plain Bearing Wear Life Calculations

Wear life cannot be applied to sintered-bronze bearings because under ideal conditions, the shaft rides on a film of lubricant and will give almost infinite life, provided the lubricating film is not disturbed, causing metal-to-metal contact. A shaft which runs in a metallic or nonmetallic plain bearing should have a surface finish between 4 and 16 rms. Any finish rougher than this will tear the lubricating film and cause metal-to-metal contact, with subsequent bearing wear of an undeterminable nature.

*Wear rate* is usually defined as the volumetric loss of bearing material over a definite unit of time. Once a wear-rate factor *K* has been established, it can be used by the designer, engineer, or tooling engineer to calculate wear rates of plain bearings accurately. As a relative measure of the performance of one material versus another at the same operating conditions, the *K* factors have proved to be very reliable. The following relational equations will prove useful for determining plain bearing life.

$$t = K\,(PVT)$$

where $t$ = wear, in.

$$P = \frac{W \text{ (total load)}}{A \text{ (bearing i.d.} \times L)}$$

where $V$ = velocity, ft/min (0.262 × rpm × shaft diameter)

$$T = \frac{t}{KPV}$$

where $T$ = running time, h; $K$ = wear factor.

The $K$ factors for common nonmetallic bearing materials are shown below.

| Material | $K$ factor |
|---|---|
| Acetal (Delrin or Celcon) | $50 \times 10^{-10}$ |
| Nylatron GS | $35 \times 10^{-10}$ |
| Teflon-filled acetal | $17 \times 10^{-10}$ |
| Teflon-filled nylon | $13 \times 10^{-10}$ |
| Glass-filled Teflon | $12 \times 10^{-10}$ |
| Nylon | $12 \times 10^{-10}$ |

You may proceed to calculate the wear life of the nonmetallic bearings by using the preceding equations by calculating the $PV$ value first and then proceding as shown below:

After determining the $PV$ value, transpose the simple equation:

$$t = K\,(PVT) \qquad \therefore \qquad T = \frac{t}{K(PV)}$$

Remember that $PV$ is a single-valued expression. [*Note:* A low-viscosity lubricant applied initially or periodically during operation will extend the bearing life by several times the calculated value. For glass-filled Teflon bearings, the shaft material should be hardened to approximately $R_c$ 45 to 55 (Rockwell hardness C scale). Brass and aluminum shafts will exhibit a high rate of wear using glass-filled bearings. Hardened steel, stainless steels, and chromium-plated steels will exhibit the lower coefficients of dynamic friction, with subsequent less wear.]

### 8.26.4 Heat Dissipation in Plain Bearings

It is often required to know the rate of heat dissipated by the bearing in operation. The heat dissipation in plain bearings is expressed by the simple equation

$$R = pvf$$

where $p$ = bearing pressure on projected area, psi
$v$ = rubbing velocity, ft/min
$f$ = coefficient of dynamic friction
$R$ = rate of radiation given in $\text{ft} \cdot \text{lb/min}^{-1}\ \text{in}^2$ of projected bearing area

Note that the work equivalent foot-pounds may be converted to calories or British thermal units using the conversions shown in Table 1.4 of Chap. 1.

***Values of Dynamic Coefficients of Friction for Various Materials.*** See Figs. 8.151 and 8.152 for various coefficients of dynamic friction for various materials.

### 8.26.5 Standard Specifications for Babbitt Metal-Bearing Materials

(ASTM B23-83; reapproved 1988) reproduced with the permission of the American Society for Testing and Materials.

Table 8.13 lists the composition and physical properties of white-metal-bearing alloys (babbitt metal).

Table 8.14 lists the chemical compositions of white-metal-bearing alloys shown in Table 8.13.

| Pressure, Lb/in² | Cast iron / Wrought iron | Steel / Cast iron | Brass / Cast iron |
|---|---|---|---|
| 125 | 0.17 | 0.17 | 0.16 |
| 225 | 0.29 | 0.33 | 0.22 |
| 300 | 0.33 | 0.34 | 0.21 |
| 400 | 0.36 | 0.35 | 0.21 |
| 500 | 0.37 | 0.36 | 0.22 |
| 700 | 0.43 | Seized | 0.23 |
| 785 | Seized | — | 0.23 |
| 825 | — | — | 0.27 |

Surfaces lightly lubricated, Rennie.

**FIGURE 8.151** Coefficients of dynamic friction.

| Shaft material | |
|---|---|
| Hardened steel | 0.15 |
| Stainless steel | 0.15 |
| Chromium-plated steel | 0.16 |
| Cast iron | 0.19 |
| Hard-anodized aluminum | 0.20 |
| Monel | 0.23 |
| Cold-rolled steel | 0.25 |
| Brass* | 0.33 |
| Aluminum* | 0.35 |

* High rate of shaft wear.

**FIGURE 8.152** Coefficients of dynamic friction—glass-filled Teflon.

## *8.27 ROLLING-ELEMENT ANTIFRICTION BEARINGS*

Rolling-element antifriction bearings consist of the ball, roller, needle, tapered roller, and cylindrical roller types. The standard specifying authority for antifriction bearings is the Anti-Friction Bearing Manufacturers Association (AFBMA). They have consolidated their standards with the American National Standards Institute (ANSI). Some of the important AFBMA standards include

- *ANSI/AFBMA (Standard) 1—1990:* Termonology for antifriction ball and roller bearings and parts.
- *ANSI/AFBMA 4—1984:* Tolerance definitions and gauging practices for ball and roller bearings.
- *ANSI/AFBMA 7—1988:* Shaft and housing fits for radial ball and roller bearings (except tapered roller bearings) conforming to basic boundary plans.
- *ANSI/AFBMA 9—1990:* Load ratings and fatigue life for ball bearings.
- *ANSI/AFBMA 11—1990:* Load ratings and fatigue life for roller bearings.
- *ANSI/AFBMA 12.1:* Instrument ball bearings; metric design.
- *ANSI/AFBMA 12.2:* Instrument ball bearings, inch design.

**TABLE 8.13** Physical Properties for White Metal Bearing Alloys

| Alloy number[b] | Specified nominal composition of alloys, %: Tin | Antimony | Lead | Copper | Arsenic | Specific gravity[c] | Composition of alloys tested, %: Tin | Antimony | Lead | Copper | Yield point, psi[d] (MPa): 68°F (20°C) | 212°F (100°C) |
|---|---|---|---|---|---|---|---|---|---|---|---|---|
| 1 | 91.0 | 4.5 | | 4.5 | | 7.34 | 90.9 | 4.52 | None | 4.56 | 4400 (30.3) | 2650 (18.3) |
| 2 | 89.0 | 7.5 | | 3.5 | | 7.39 | 89.2 | 7.4 | 0.03 | 3.1 | 6100 (42.0) | 3000 (20.6) |
| 3 | 84.0 | 8.0 | | 8.0 | | 7.46 | 83.4 | 8.2 | 0.03 | 8.3 | 6600 (45.5) | 3150 (21.7) |
| 7 | 10.0 | 15.0 | Remainder | | 0.45 | 9.73 | 10.0 | 14.5 | 75.0 | 0.11 | 3550 (24.5) | 1600 (11.0) |
| 8 | 5.0 | 15.0 | Remainder | | 0.45 | 10.04 | 5.2 | 14.9 | 79.4 | 0.14 | 3400 (23.4) | 1750 (12.1) |
| 15 | 1.0 | 16.0 | Remainder | | 1.0 | 10.05 | | | | | | |

| Alloy number[b] | Johnson's apparent elastic limit psi (MPa)[e]: 68°F (20°C) | 212°F (100°C) | Ultimate strength in compression[f]: 68°F (20°C) | 212°F (100°C) | Brinell hardness[g]: 68°F (20°C) | 212°F (100°C) | Melting point, °F (°C) | Temperature of complete liquefaction, °F (°C) | Proper pouring temperature, °F (°C) |
|---|---|---|---|---|---|---|---|---|---|
| 1 | 2450 (16.9) | 1050 (7.2) | 12 850 (88.6) | 6950 (47.9) | 17.0 | 8.0 | 433 (223) | 700 (371) | 825 (441) |
| 2 | 3350 (23.1) | 1100 (7.6) | 14 900 (102.7) | 8700 (60.0) | 24.5 | 12.0 | 466 (241) | 669 (354) | 795 (424) |
| 3 | 5350 (36.9) | 1300 (9.0) | 17 600 (121.3) | 9900 (68.3) | 27.0 | 14.5 | 464 (240) | 792 (422) | 915 (491) |
| 7 | 2500 (17.2) | 1350 (9.3) | 15 650 (107.9) | 6150 (42.4) | 22.5 | 10.5 | 464 (240) | 514 (268) | 640 (338) |
| 8 | 2650 (18.3) | 1200 (8.3) | 15 600 (107.6) | 6150 (42.4) | 20.0 | 9.5 | 459 (237) | 522 (272) | 645 (341) |
| 15 | | | | | 21.0 | 13.0 | 479 (248) | 538 (281) | 662 (350) |

[a] The compression test specimens were cylinders 1.5 in (33 mm) in length and 0.5 in (13 mm) in diameter, machined from chill castings 2 in (51 mm) in length and 0.75 in (19 mm) in diameter. The Brinell tests were made on the bottom of parallel machined specimens cast in a mold 2 in (51 mm) in diameter and 0.625 in (16 mm) deep at room temperature.
[b] Data not available on alloy numbers 11 and 13.
[c] The specific gravity multiplied by 0.0361 equals the density in pounds per cubic inch.
[d] The values for yield point were taken from stress-strain curves at a deformation of 0.125% of gage length.
[e] Johnson's apparent elastic limit is taken as the unit stress at the point where the slope of the tangent to the curve is two-thirds times its slope at the origin.
[f] The ultimate strength values were taken as the unit load necessary to produce a deformation of 25% of the length of the specimen.
[g] These values are the average Brinell number of three impressions on each alloy using a 10-mm ball and a 500-kg load applied for 30 s.
SOURCE: ASTM.

**TABLE 8.14** Compositions for White Metal Bearing Alloys

| | Alloy number: Tin base: 1 | 2 | 3 | 11 | Lead base: 7 | 8 | 13 | 15 |
|---|---|---|---|---|---|---|---|---|
| Chemical composition, % | UNS—55191 | UNS—55193 | UNS—55189 | UNS—55188 | UNS—53581 | UNS—53565 | UNS—53346 | UNS—53620 |
| Tin | 90.0–92.0 | 88.0–90.0 | 83.0–85.0 | 86.0–89.0 | 9.3–10.7 | 4.5–5.5 | 5.5–6.5 | 0.8–1.2 |
| Antimony | 4.0–5.0 | 7.0–8.0 | 7.5–8.5 | 6.0–7.5 | 14.0–16.0 | 14.0–16.0 | 9.5–10.5 | 14.5–17.5 |
| Lead | 0.35 | 0.35 | 0.35 | 0.50 | Remainder[‡] | Remainder[‡] | Remainder[‡] | Remainder[‡] |
| Copper | 4.0–5.0 | 3.0–4.0 | 7.5–8.5 | 5.0–6.5 | 0.50 | 0.50 | 0.50 | 0.6 |
| Iron | 0.08 | 0.08 | 0.08 | 0.08 | 0.10 | 0.10 | 0.10 | 0.10 |
| Arsenic | 0.10 | 0.10 | 0.10 | 0.10 | 0.30–0.60 | 0.30–0.60 | 0.25 | 0.8–1.4 |
| Bismuth | 0.08 | 0.08 | 0.08 | 0.08 | 0.10 | 0.10 | 0.10 | 0.10 |
| Zinc | 0.005 | 0.005 | 0.005 | 0.005 | 0.005 | 0.005 | 0.005 | 0.005 |
| Aluminum | 0.005 | 0.005 | 0.005 | 0.005 | 0.005 | 0.005 | 0.005 | 0.005 |
| Cadmium | 0.05 | 0.05 | 0.05 | 0.05 | 0.05 | 0.05 | 0.05 | 0.05 |
| Total named elements, min | 99.80 | 99.80 | 99.80 | 99.80 | | | | |

* All values not given as ranges are maximum unless shown otherwise.
[†] Alloy Number 9 was discontinued in 1946 and numbers 4, 5, 6, 10, 11, 12, 16, and 19 were discontinued in 1959. A new number 11, similar to SAE Grade 11, was added in 1966.
[‡] To be determined by difference.
SOURCE: ASTM.

- *ANSI/AFBMA 16.1:* Airframe ball, roller, and needle roller bearings, metric design.
- *ANSI/AFBMA 16.2:* Airframe ball, roller, and needle roller bearings, inch design.
- *ANSI/AFBMA 18.1—1982:* Needle roller bearings, radial, metric design.
- *ANSI/AFBMA 18.2—1982:* Needle roller bearings, radial, inch design.
- *ANSI/AFBMA 21.1—1988:* Thrust needle roller and cage assemblies and thrust washers, metric design.
- *ANSI/AFBMA 21.2—1988:* Thrust needle roller and cage assemblies and thrust washers, inch design.

Figures 8.153 to 8.161 depict the various design configurations of the ball and roller bearing types that are currently being manufactured. The figures are reproduced with permission from the Anti-Friction Bearing Manufacturers' Association (extracts from ANSI/AFBMA Standard 1—1990).

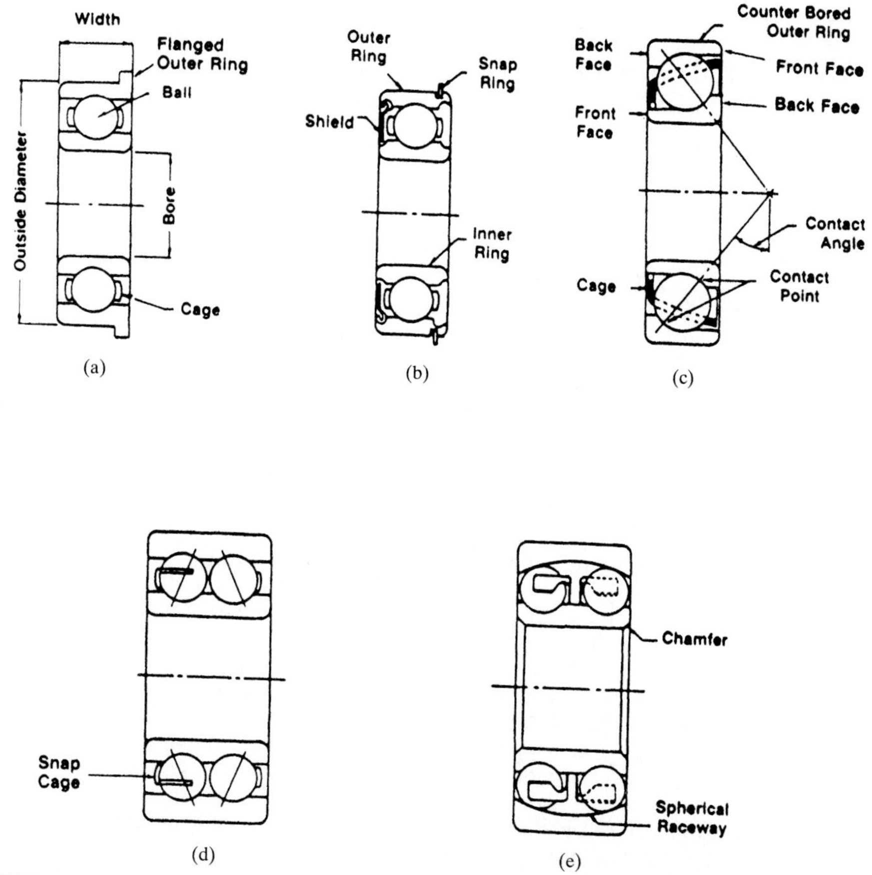


**FIGURE 8.153** Ball bearing. (*a*) single-row, deep-groove with flanged outer ring; (*b*) deep-groove, single-shielded, with snap ring; (*c*) angular contact; (*d*) double-row, deep-groove; (*e*) self-aligning, double-row. (*Copyright 1990 by the Anti-Friction Bearing Manufacturers Association, Inc. Reprinted with permission.*)

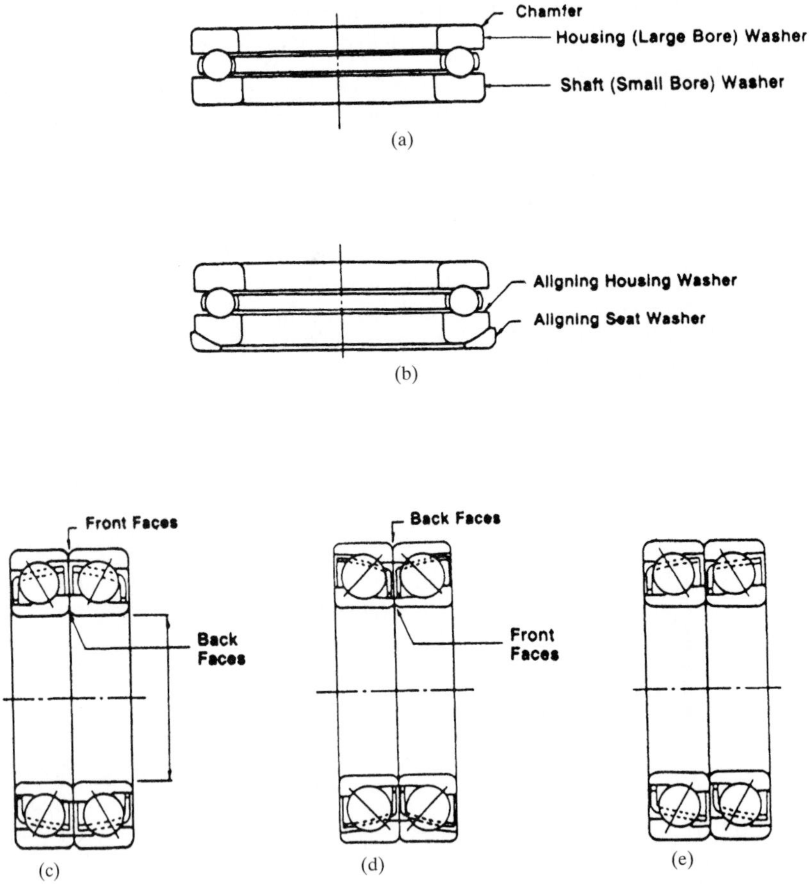


**FIGURE 8.154** (*a*) Ball thrust bearing; (*b*) ball thrust bearing with aligning washer; (*c*) face-to-face (front-to-front) duplex mounting; (*d*) back-to-back duplex mounting; (*e*) tandem mounting of two bearings. (*Copyright 1990 by the Anti-Friction Bearing Manufacturers Association, Inc. Reprinted with permission.*)

### 8.27.1 Load Ratings and Fatigue Life for Ball and Roller Bearings

The load rating and fatigue life for ball and roller bearing designs may be calculated using the procedures outlined in ANSI/AFBMA Standard 9—1990 and ANSI/AFBMA Standard 11—1990. The calculations are intricate and involved, requiring the use of many tables produced by the AFBMA and are thus not shown in this handbook. The calculations for tapered roller bearings are likewise intricate and involved, requiring the use of many tables, and these data are available from The Timken Company bearing catalog or the Torrington catalog. Other manufacturer catalogs are also available for the data and calculation procedures.

***Load and Speed Rules for Bearings.*** The following general rules which apply to bearings are approximations:

- Doubling the bearing load reduces bearing life to one-tenth.
- Reducing the load one-half increases bearing life approximately 10 times.

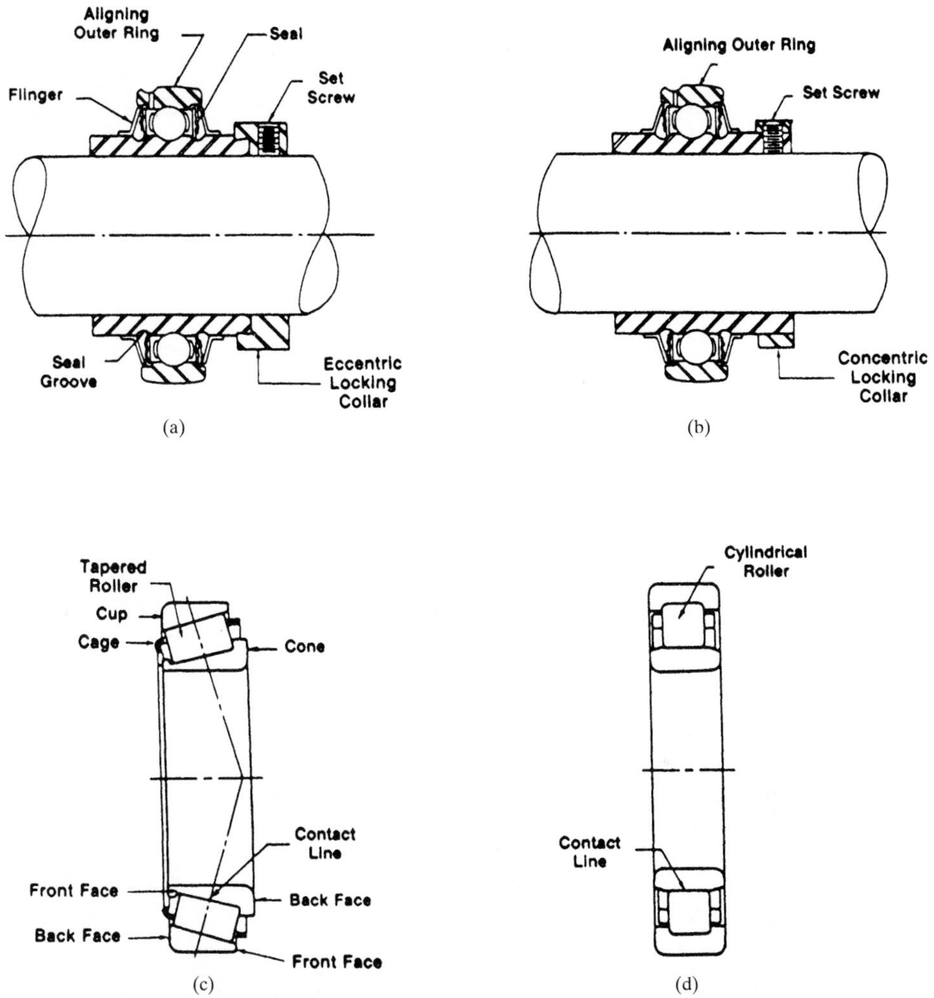


**FIGURE 8.155** (*a*) Extended inner ring bearing with accentric locking collar; (*b*) extended inner ring bearing with concentric locking collar; (*c*) radial tapered roller bearing; (*d*) radial cylindrical roller bearing. (*Copyright 1990 by the Anti-Friction Bearing Manufacturers Association, Inc. Reprinted with permission.*)

- Doubling the bearing speed reduces bearing life to one-half.
- Reducing the bearing speed by one-half doubles the bearing life.

### 8.27.2 Allowances for Fits (Applicable to Antifriction Bearings, etc.)

The fitting of a rolling-element bearing into its receiving housing or over its mounting shaft is a critical part of bearing applications and assembly. Figure 8.162 represents the different classes of fits and their respective dimensional limits for each class. Figure 8.162 is applicable not only to bearing fits but also to all classes of machined part fits (cylindrical parts) from run-

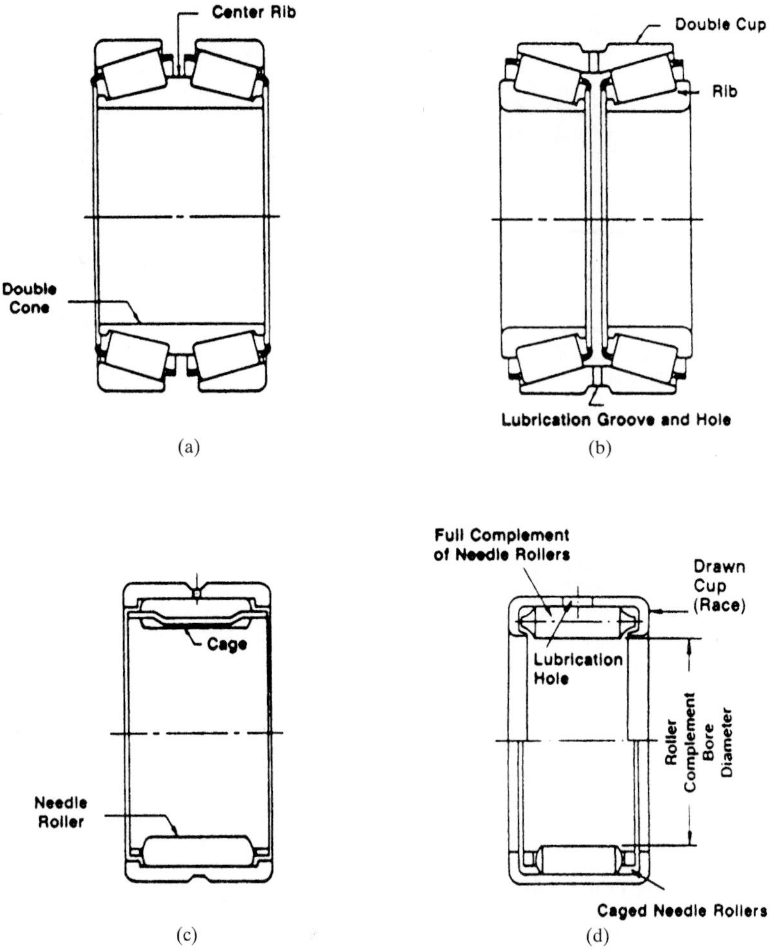


**FIGURE 8.156** (*a*) Double-row, double-cone tapered roller bearing; (*b*) double-row, double-cup tapered roller bearing; (*c*) needle roller bearing with cage, machined ring without inner ring; (*d*) needle roller bearing (drawn cup without inner ring), full-complement or caged needle rollers.

ning fits to forced and driving fits. These allowances for fits (shown in Fig. 8.162) will prove useful to the machinist, designer, and tool engineer in actual practice.

## 8.28 SELECTION OF BEARINGS: SOURCES AND PROCEDURES

### 8.28.1 Bearing Sources

All bearing manufacturers produce catalogs of their products, and many of them contain technical data and procedures for the use and design of bearing systems. Because many bearing manufacturers produce a very large array of types and designs, a central source of bearing selection would be advantageous and helpful to those involved in the selection of bearings.

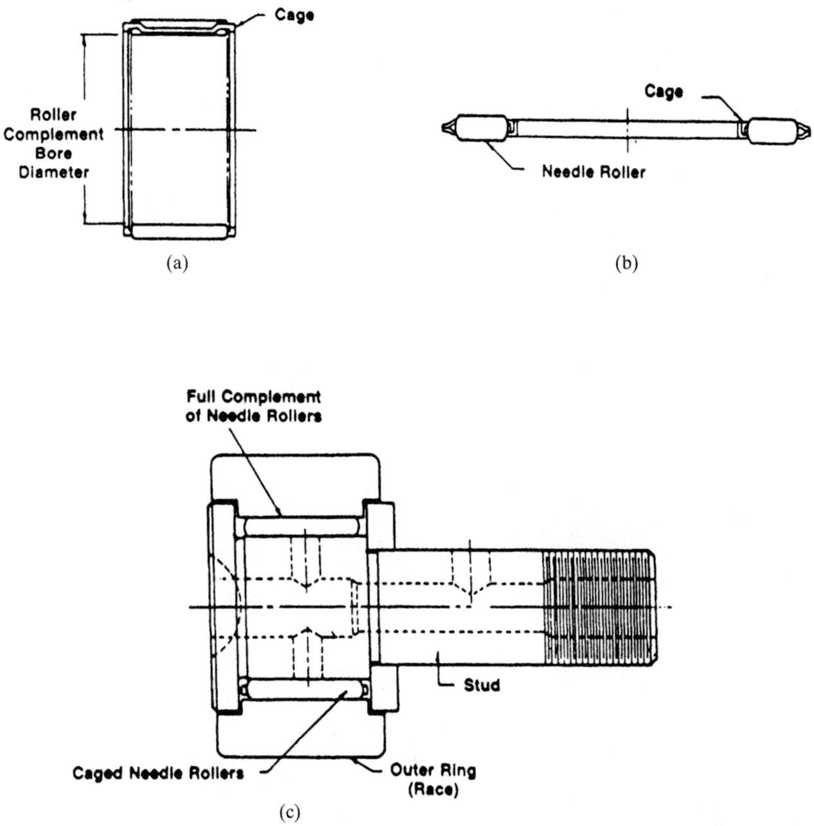


**FIGURE 8.157** (*a*) Radial needle roller and cage assembly, (*b*) thrust needle roller and cage assembly, (*c*) needle roller bearing—track roller, stud type—full complement or with cage.

Fortunately, there is a central source for the selection and cross-referencing of all the bearings manufactured in the United States, both inch series and metric series. The *Bearing Manual Cyclopedia* (registered trademark of Industrial Information Headquarters, Inc.) is the source for all bearing data, dimensions, and cross-references. This manual is produced every 2 to 3 years as a two-volume set and is available from

Industrial Information Headquarters Co., Inc.
2601 West 16th Street
Broadway, IL 60153
(708) 345-7944

Outside the United States, the manual may be obtained from

BHQ Export Services
3199 N. Shadeland Ave.
P.O. Box 26118
Indianapolis, IN 46226
(317) 545-2411

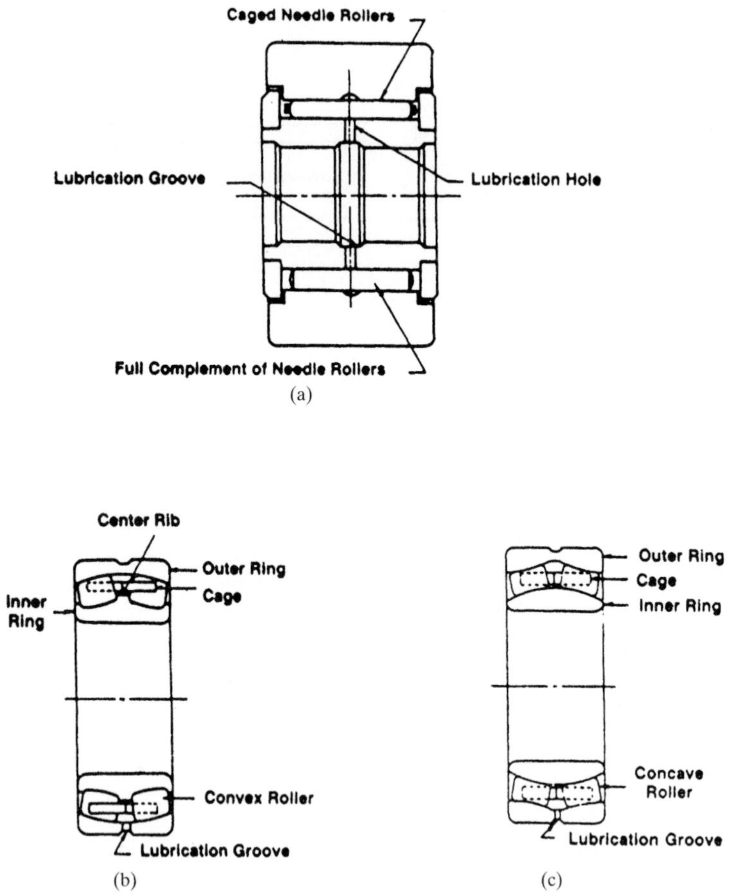


**FIGURE 8.158** (*a*) Needle roller bearing—track roller, yoke type—full complement or with cage; (*b*) spherical roller bearing—double row, convex roller—(raceway or outer ring, spherical); (*c*) spherical roller bearing—double-row, concave roller—(raceway or inner ring, spherical).

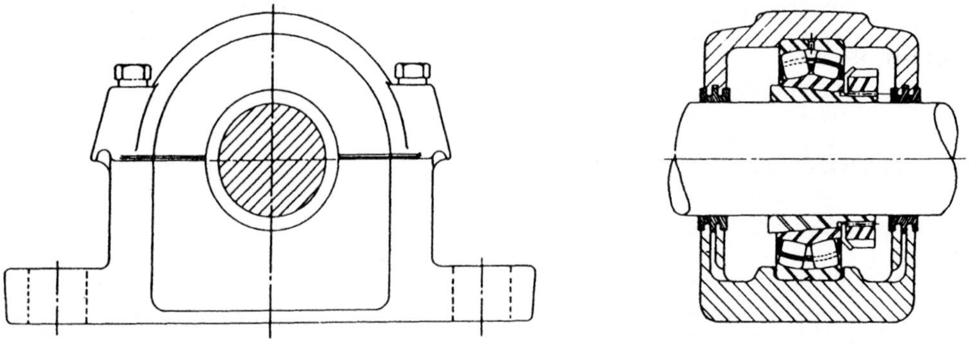

**FIGURE 8.159** Pillow block assembly.

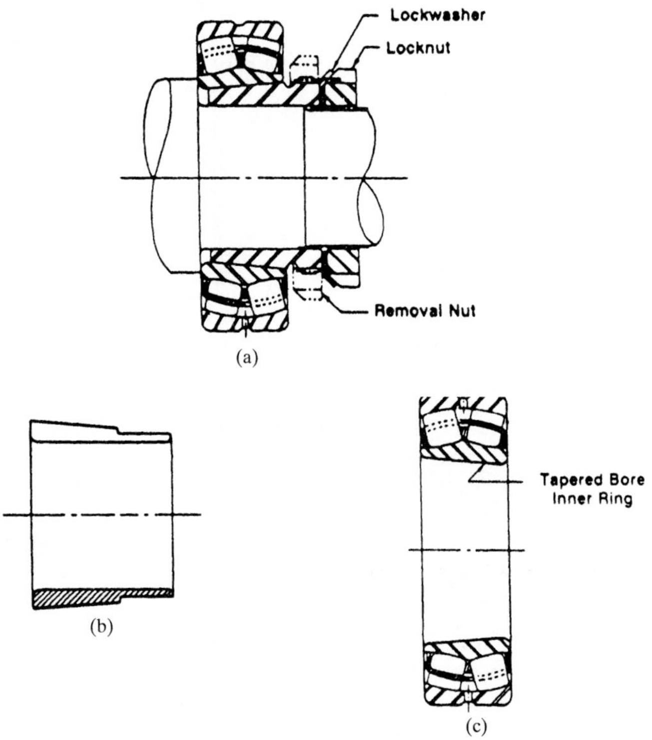


**FIGURE 8.160** (*a*) Withdrawal sleeve and bearing assembly, (*b*) adapter sleeve, (*c*) tapered bore bearing.

Figures 8.163 and 8.164 are reproductions taken from the *Bearing Manual* for illustrative purposes to show the type of information contained in the manual.

### 8.28.2 Bearing Loads in Mechanical Systems

The loads imposed on bearings used in gearing and belt-drive systems or other mechanical systems are calculated using the procedures shown in Sec. 8.10.1.

## *8.29 MECHANISMS AND LINKAGES: DESIGN, OPERATING PRINCIPLES, AND ANALYSIS*

The mechanisms and linkages described in this chapter have many applications for the product designer, tool engineer, and others involved in the design and manufacture of machinery, tooling, mechanical apparatus and electromechanical devices, products, and assemblies used in the industrial context. These mechanisms have been selected by the author, from a large assortment, as the more important, most commonly used types for a wide range of applica-

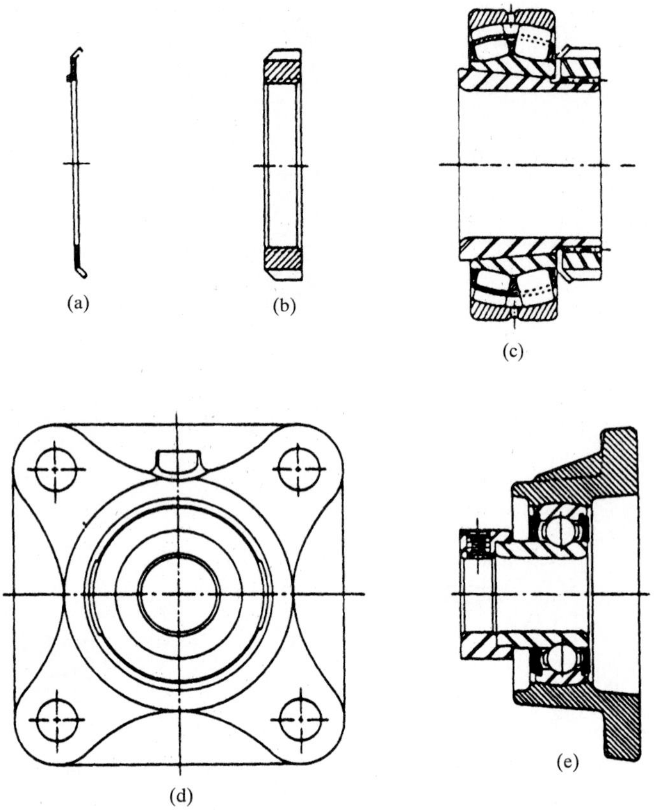


**FIGURE 8.161** (*a*) Lockwasher, (*b*) locknut, (*c*) adapter sleeve and bearing assembly, (*d*) glanged housing (4-bolt).

tions. A variety of important mechanical linkages are shown in Sec. 8.29.3, together with their mathematical solutions and principles of operation.

## 8.29.1 Basic and Compound Mechanisms

When you study the operating principles of these devices, you will be able to see the relationship to the basic simple machines such as the lever, wheel and axle, inclined plane or wedge, gear wheel, and so forth. There are six basic simple machines from which all machines and mechanisms may be constructed, either singly or in combination, not including the Rolomite mechanism. The hydraulic cylinder and gear wheel are also considered members of the basic simple machines.

Also, the gearing sections (Secs. 8.8 through 8.24) show many of the basic gear devices which have a multitude of uses in machine design and product design practice. Selected practical mechanisms are shown in Figs. 8.165 through 8.214, together with an explanation of the operation of the mechanism.

| Class | Nominal Diameters | Up to 0.500" | 0.5625-1" | 1.0625-2" | 2.0625-3" | 3.0625-4" | 4.0625-5" |
|---|---|---|---|---|---|---|---|
| | | | | Tolerances in Standard Holes ♦ | | | |
| A | High Limit | +0.00025 | +0.0005 | +0.00075 | +0.0010 | +0.0010 | +0.0010 |
| | Low Limit | -0.00025 | -0.00025 | -0.00025 | -0.0005 | -0.0005 | -0.0005 |
| | Tolerance | 0.0005 | 0.00075 | 0.0010 | 0.0015 | 0.0015 | 0.0015 |
| B | High Limit | +0.0005 | +0.00075 | +0.0010 | +0.00125 | +0.0015 | +0.00175 |
| | Low Limit | -0.0005 | -0.0005 | -0.0005 | -0.00075 | -0.00075 | -0.00075 |
| | Tolerance | 0.0010 | 0.00125 | 0.0015 | 0.0020 | 0.00225 | 0.0025 |
| | | | | Allowances for Forced Fits | | | |
| F | High Limit | +0.0010 | +0.0020 | +0.0040 | +0.0060 | +0.0080 | +0.0100 |
| | Low Limit | +0.0005 | +0.0015 | +0.0030 | +0.0045 | +0.0060 | +0.0080 |
| | Tolerance | 0.0005 | 0.0005 | 0.0010 | 0.0015 | 0.0020 | 0.0020 |
| | | | | Allowances for Driving Fits | | | |
| D | High Limit | +0.0005 | +0.0010 | +0.0015 | +0.0025 | +0.0030 | +0.0035 |
| | Low Limit | +0.00025 | +0.00075 | +0.0010 | +0.0015 | +0.0020 | +0.0025 |
| | Tolerance | 0.00025 | 0.00025 | 0.0005 | 0.0010 | 0.0010 | 0.0010 |
| | | | | Allowances for Push Fits | | | |
| P | High Limit | -0.00025 | -0.00025 | -0.00025 | -0.0005 | -0.0005 | -0.0005 |
| | Low Limit | -0.00075 | -0.00075 | -0.00075 | -0.0010 | -0.0010 | -0.0010 |
| | Tolerance | 0.0005 | 0.0005 | 0.0005 | 0.0005 | 0.0005 | 0.0005 |
| | | | | Allowances for Running Fits ■ | | | |
| X | High Limit | -0.0010 | -0.00125 | -0.00175 | -0.0020 | -0.0025 | -0.0030 |
| | Low Limit | -0.0020 | -0.00275 | -0.0035 | -0.00425 | -0.0050 | -0.00575 |
| | Tolerance | 0.0010 | 0.0015 | 0.00175 | 0.00225 | 0.0025 | 0.00275 |
| Y | High Limit | -0.00075 | -0.0010 | -0.00125 | -0.0015 | -0.0020 | -0.00225 |
| | Low Limit | -0.00125 | -0.0020 | -0.0025 | -0.0030 | -0.0035 | -0.0040 |
| | Tolerance | 0.0005 | 0.0010 | 0.00125 | 0.0015 | 0.0015 | 0.00175 |
| Z | High Limit | -0.0005 | -0.00075 | -0.00075 | -0.0010 | -0.0010 | -0.00125 |
| | Low Limit | -0.00075 | -0.00125 | -0.0015 | -0.0020 | -0.00225 | -0.0025 |
| | Tolerance | 0.00025 | 0.0005 | 0.00075 | 0.0010 | 0.00125 | 0.00125 |

(a)

| Class | High Limit | Low Limit |
|---|---|---|
| | $+ (D)^{0.5} \times 0.0006$ | $- (D)^{0.5} \times 0.0003$ |
| | $+ (D)^{0.5} \times 0.0008$ | $- (D)^{0.5} \times 0.0004$ |
| | $- (D)^{0.5} \times 0.0002$ | $- (D)^{0.5} \times 0.0006$ |
| | $- (D)^{0.5} \times 0.00125$ | $- (D)^{0.5} \times 0.0025$ |
| | $- (D)^{0.5} \times 0.001$ | $- (D)^{0.5} \times 0.0018$ |
| | $- (D)^{0.5} \times 0.0005$ | $- (D)^{0.5} \times 0.001$ |

NOTE: ♦ Tolerance is provided for holes which ordinary standard reamers can produce, in two grades, classes A and B, the selection of which is a question for the user's decision and dependent upon the quality of the work required. Some prefer to use class A as working limits and class B as inspection limits.

■ Running fits, which are the most commonly required, are divided into three grades; class X, for engine and other work where easy fits are desired; class Y, for high speeds and good average machine work; and class Z, for fine tooling work.

(b)

**FIGURE 8.162** (*a*) Table of allowances for different classes of fits of bearings and other cylindrical machined parts; (*b*) equations for determining allowances for classes of fits (see Fig. 8.162*a*).

Inch

# GROUND BEARINGS



## RADIAL - SINGLE ROW

Continued from the preceding page

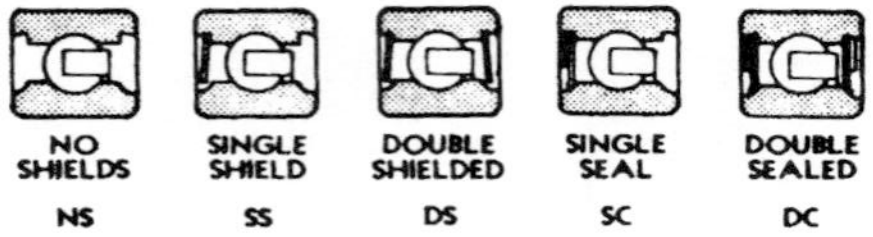


Courtesy Nice

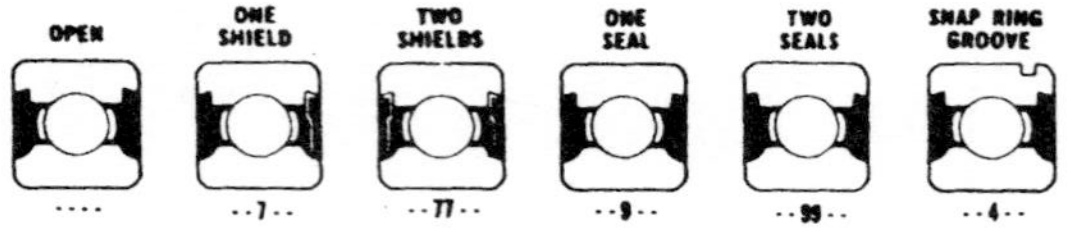


Courtesy Schatz

| DIMENSIONS | | | Nice | | | | | | | | | | Schatz*** | | | | |
|---|---|---|---|---|---|---|---|---|---|---|---|---|---|---|---|---|---|
| Bore | O. D. | Width | Open Type | | Single Shield | | Double Shield | | Single Seal | | Double Seal | | Open Type | Single Shield | Double Shield | Single Seal | Double Seal |
| Inches | Inches | Inches | Old No. | New No. | Old No. | New No. | Old No. | New No. | Old No. | New No. | Old No. | New No. | | | | | |
| 3/16 | 11/16 | 1/4 | 1601NS | 1601NSTN | 1601SS | 1601SSTN | 1601DS | 1601DSTN | | | | | BR-01* | BR-701* | BR-7701* | BR-901* | BR-9901* |
| 3/16 | 11/16 | 5/16 | | | | | | | 1601SC | 1601SCTN | 1601DC | 1601DCTN | | | | | |
| 1/4 | 11/16 | 1/4 | 1602NS | 1602NSTN | 1602SS | 1602SSTN | 1602DS | 1602DSTN | | | | | BR-02* | BR-702* | BR-7702* | BR-902* | BR-9902* |
| 1/4 | 11/16 | 5/16 | | | | | | | 1602SC | 1602SCTN | 1602DC | 1602DCTN | | | | | |
| 5/16 | 7/8 | 8/32 | 1603NS | 1603NSTN | 1603SS | 1603SSTN | 1603DS | 1603DSTN | | | | | BR-03** | BR-703** | BR-7703** | BR-903** | BR-9903** |
| 5/16 | 7/8 | 11/32 | | | | | | | 1603SC | 1603SCTN | 1603DC | 1603DCTN | | | | | |
| 5/16 | 29/32 | 9/32 | | | | | | | | | | | | | | | |
| 5/16 | 29/32 | 5/16 | 1605NS | 1605NSTN | 1605SS | 1605SSTN | 1605DS | 1605DSTN | 1605SC | 1605SCTN | 1605DC | 1605DCTN | BR-05 | BR-705 | BR-7705 | BR-905 | BR-9905 |
| 3/8 | 7/8 | 9/32 | 1604NS | 1604NSTN | 1604SS | 1604SSTN | 1604DS | 1604DSTN | | | | | BR-04** | BR-704** | BR-7704** | BR-904** | BR-9904** |
| 3/8 | 7/8 | 11/32 | | | | | | | 1604SC | 1604SCTN | 1604DC | 1604DCTN | | | | | |
| 3/8 | 29/32 | 5/16 | 1606NS | 1606NSTN | 1606SS | 1606SSTN | 1606DS | 1606DSTN | 1606SC | 1606SCTN | 1606DC | 1606DCTN | BR-06 | BR-706 | BR-7706 | BR-906 | BR-9906 |
| 3/8 | 29/32 | 19/32 | | | | | | | | | | | | | | | |
| 3/8 | 1- 1/8 | 3/8 | 1614NS | 1614NSTN | 1614SS | 1614SSTN | 1614DS | 1614DSTN | 1614SC | 1614SCTN | 1614DC | 1614DCTN | BR-14 | BR-714 | BR-7714 | BR-914 | BR-9914 |
| 7/16 | 29/32 | 5/16 | 1607NS | 1607NSTN | 1607SS | 1607SSTN | 1607DS | 1607DSTN | 1607SC | 1607SCTN | 1607DC | 1607DCTN | BR-07 | BR-707 | BR-7707 | BR-907 | BR-9907 |
| 7/16 | 1- 1/8 | 3/8 | 1615NS | 1615NSTN | 1615SS | 1615SSTN | 1615DS | 1615DSTN | 1615SC | 1615SCTN | 1615DC | 1615DCTN | BR-15 | BR-715 | BR-7715 | BR-915 | BR-9915 |
| 7/16 | 1- 3/8 | 7/16 | 1620NS | 1620NSTN | 1620SS | 1620SSTN | 1620DS | 1620DSTN | 1620SC | 1620SCTN | 1620DC | 1620DCTN | BR-20 | BR-720 | BR-7720 | BR-920 | BR-9920 |
| 1/2 | 1- 1/8 | 3/8 | 1616NS | 1616NSTN | 1616SS | 1616SSTN | 1616DS | 1616DSTN | 1616SC | 1616SCTN | 1616DC | 1616DCTN | BR-16 | BR-716 | BR-7716 | BR-916 | BR-9916 |
| 1/2 | 1- 3/8 | 7/16 | 1621NS | 1621NSTN | 1621SS | 1621SSTN | 1621DS | 1621DSTN | 1621SC | 1621SCTN | 1621DC | 1621DCTN | BR-21 | BR-721 | BR-7721 | BR-921 | BR-9921 |
| 9/16 | 1- 3/8 | 7/16 | 1622NS | 1622NSTN | 1622SS | 1622SSTN | 1622DS | 1622DSTN | 1622SC | 1622SCTN | 1622DC | 1622DCTN | BR-22 | BR-722 | BR-7722 | BR-922 | BR-9922 |
| 5/8 | 1- 3/8 | 7/16 | 1623NS | 1623NSTN | 1623SS | 1623SSTN | 1623DS | 1623DSTN | 1623SC | 1623SCTN | 1623DC | 1623DCTN | BR-23 | BR-723 | BR-7723 | BR-923 | BR-9923 |
| 5/8 | 1- 5/8 | 1/2 | 1628NS | 1628NSTN | 1628SS | 1628SSTN | 1628DS | 1628DSTN | 1628SC | 1628SCTN | 1628DS | 1628DCTN | BR-28 | BR-728 | BR-7728 | BR-928 | BR-9928 |
| 5/8 | 1- 3/4 | 1/2 | 1633NS | 1633NSTN | 1633SS | 1633SSTN | 1633DS | 1633DSTN | 1633SC | 1633SCTN | 1633DS | 1633DCTN | BR-33 | BR-733 | BR-7733 | BR-923 | BR-9933 |
| 11/16 | 1- 5/8 | 1/2 | | | | | | | | | | | BR-34 | BR-734 | BR-7734 | BR-934 | BR-9934 |
| 3/4 | 1- 5/8 | 1/2 | 1630NS | 1630NSTN | 1630SS | 1630SSTN | 1630DS | 1630DSTN | 1630SC | 1630SCTN | 1630DS | 1630DCTN | BR-30 | BR-730 | BR-7730 | BR-930 | BR-9930 |
| 3/4 | 1- 3/4 | 1/2 | 1635NS | 1635NSTN | 1635SS | 1635SSTN | 1635DS | 1635DSTN | 1635SC | 1635SCTN | 1635DS | 1635DCTN | BR-35 | BR-735 | BR-7735 | BR-935 | BR-9935 |
| 3/4 | 2 | 9/16 | 1638NS | 1638NSTN | 1638SS | 1638SSTN | 1638DS | 1638DSTN | 1638SC | 1638SCTN | 1638DS | 1638DCTN | BR-38 | BR-738 | BR-7738 | BR-938 | BR-9938 |
| 13/16 | 2 | 9/16 | | | | | | | | | | | BR-39 | NR-739 | BR-7739 | BR-939 | BR-9939 |
| 7/8 | 2 | 9/16 | 1640NS | 1640NSTN | 1640SS | 1640SSTN | 1640DS | 1640DSTN | 1640SC | 1640SCTN | 1640DS | 1640DCTN | BR-40 | BR-740 | BR-7740 | BR-940 | BR-9940 |
| 1 | 2 | 9/16 | 1641NS | 1641NSTN | 1641SS | 1641SSTN | 1641DS | 1641DSTN | 1641SC | 1641SCTN | 1641DS | 1641DCTN | BR-41 | BR-741 | BR-7741 | BR-941 | BR-9941 |
| 1- 1/8 | 2- 1/2 | 5/8 | 1652NS | 1652NSTN | 1652SS | 1652SSTN | 1652DS | 1652DSTN | 1652SC | 1652SCTN | 1652DS | 1652DCTN | BR-52 | BR-752 | BR-7752 | BR-952 | BR-9952 |
| 1- 1/4 | 2- 1/2 | 5/8 | 1654NS | 1654NSTN | 1654SS | 1654SSTN | 1654DS | 1654DSTN | 1654SC | 1654SCTN | 1654DS | 1654DCTN | BR-54 | BR-754 | BR-7754 | BR-954 | BR-9954 |
| 1- 1/4 | 2- 9/16 | 11/16 | 1657NS | 1657NSTN | 1657SS | 1657SSTN | 1657DS | 1657DSTN | 1657SC | 1657SCTN | 1657DS | 1657DCTN | | | | | |
| 1- 5/16 | 2- 9/16 | 11/16 | 1658NS | 1658NSTN | 1658SS | 1658SSTN | 1658DS | 1658DSTN | 1658SC | 1658SCTN | 1658DS | 1658DCTN | | | | | |

* Also available in 5/16 inches width.
** Also available in 11/32 inches width.
*** Schatz also makes this series with snap ring.

**FIGURE 8.163** Bearing manual sample sheet.

# TAPERED ROLLER BEARINGS

**Arranged by Bore Diameters**

## SINGLE ROW - STRAIGHT BORE

**Type TS**

| DIMENSIONS | | | | | BOWER | | TIMKEN | | TYSON | | Part Number |
|---|---|---|---|---|---|---|---|---|---|---|---|
| Cone Bore A | Cup O.D. B | Width C | Cone Length D | Cup Length E | | | | | | | |
| Inches | Inches | Inches | Inches | Inches | Cone | Cup | Cone | Cup | Cone | Cup | |
| * 0.7500 | 1.9380 | 0.9063 | 0.8480 | 0.6875 | 09074 | 09196 | 09074 | 09196 | 09074 | 09196 | |
| * 0.7500 | 2.2400 | 0.9063 | 0.8480 | 0.6875 | 09074 | 09194 | 09074 | 09194 | 09074 | 09194 | |
| 0.7500 | 2.2400 | 0.7625 | 0.7810 | 0.6250 | 1775 | 1729 | 1775 | 1729 | 1775 | 1729 | |
| 0.7500 | 2.2400 | 0.7625 | 0.7810 | 0.6250 | 1775 | 1729X | 1775 | 1729X | 1775 | 1729X | |
| 0.7870 | 1.8504 | 0.5662 | 0.5662 | 0.4375 | 05079 | 05185 | 05079 | 05185 | | | |
| 0.7874 | 1.9687 | 0.5313 | 0.5614 | 0.4750 | 07079 | 07196 | 07079 | 07196 | | | |
| 0.7874 | 2.0470 | 0.5910 | 0.5614 | 0.5000 | 07079 | 07204 | 07079 | 07204 | | | |
| 0.8125 | 1.9380 | 0.7813 | 0.7813 | 0.6250 | 12580 | 12520 | 12580 | 12520 | | | |
| 0.8125 | 2.4375 | 1.1250 | 1.1975 | 0.9375 | | | 3660 | 3620 | | | |
| 0.8125 | 2.5625 | 1.1250 | 1.1975 | 0.8750 | | | 3660 | 3623 | | | |
| 0.8437 | 1.9687 | 0.6900 | 0.7200 | 0.5500 | M12649 | M12610 | M12659 | M12610 | M12649 | M12610 | |
| 0.8750 | 1.6563 | 0.4400 | 0.4400 | 0.3400 | | | LL52549 | LL52510 | | | |
| 0.8750 | 1.9687 | 0.5313 | 0.5614 | 0.3750 | 07087 | 07196 | 07087 | 07196 | 07087 | 07196 | |
| 0.8750 | 1.9687 | 0.6900 | 0.7200 | 0.5500 | M12648 | M12610 | M12648 | M12610 | | | |
| 0.8750 | 2.0470 | 0.5910 | 0.5614 | 0.5000 | 07087 | 07204 | 07087 | 07204 | 07087 | 07204 | |
| 0.8750 | 2.0625 | 0.7625 | 0.7940 | 0.5625 | 1380 | 1328 | 1380 | 1328 | | | |
| 0.8750 | 2.1250 | 0.7625 | 0.7940 | 0.5625 | 1380 | 1329 | 1380 | 1329 | | | |
| 0.8750 | 2.2400 | 0.7625 | 0.7810 | 0.6250 | 1755 | 1729 | 1755 | 1729 | 1755 | 1729 | |
| * 0.8750 | 2.2400 | 0.7625 | 0.7810 | 0.6250 | 1755 | 1729X | 1755 | 1729X | 1755 | 1729X | |
| 0.8750 | 2.2500 | 0.8750 | 0.8750 | 0.6875 | | | 1280 | 1220 | | | |
| 0.8750 | 2.3125 | 0.7500 | 0.7620 | 0.5937 | 1975 | 1932 | 1975 | 1932 | | | |
| 0.8750 | 2.3750 | 0.7813 | 0.7620 | 0.6250 | 1975 | 1931 | 1975 | 1931 | | | |
| 0.8750 | 2.6150 | 0.9375 | 1.0013 | 0.7500 | 2684 | 2631 | 2684 | 2631 | | | |
| 0.9375 | 1.9687 | 0.5313 | 0.5614 | 0.3750 | 07093 | 07196 | 07093 | 07196 | | | |
| 0.9375 | 2.0470 | 0.5910 | 0.5614 | 0.5000 | 07093 | 07204 | 07093 | 07204 | | | |
| 0.9375 | 2.2400 | 0.7625 | 0.7810 | 0.6250 | 1779 | 1729 | 1779 | 1729 | 1779 | 1729 | |
| * 0.9375 | 2.2400 | 0.7625 | 0.7810 | 0.6250 | 1779 | 1729X | 1779 | 1729X | 1779 | 1729X | |
| 0.9375 | 2.4375 | 1.1250 | 1.1975 | 0.9375 | | | 3659 | 3620 | | | |
| 0.9375 | 2.5625 | 1.1250 | 1.1975 | 0.8750 | | | 3659 | 3623 | | | |
| 0.9375 | 2.6150 | 0.9375 | 1.0013 | 0.7500 | 2685 | 2631 | 2685 | 2631 | | | |
| 0.9375 | 2.7450 | 0.7500 | 0.7450 | 0.7500 | | | 26093 | 26274 | | | |
| 0.9375 | 2.8345 | 0.7480 | 0.7450 | 0.6250 | | | 26093 | 26283 | | | |
| 0.9375 | 3.0000 | 0.7480 | 0.7450 | 0.6250 | | | 26093 | 26300 | | | |
| 0.9835 | 1.9687 | 0.5313 | 0.5614 | 0.3750 | | | 07098 | 07196 | 07098 | 07196 | |
| 0.9835 | 2.0470 | 0.5910 | 0.5614 | 0.5000 | | | 07098 | 07204 | 07098 | 07204 | |
| 0.9835 | 2.4410 | 0.6300 | 0.6522 | 0.5625 | 17098 | 17244 | 17098 | 17244 | | | |
| 0.9842 | 1.9687 | 0.5313 | 0.5614 | 0.3750 | 07097 | 07196 | 07097 | 07196 | | | |
| 0.9842 | 2.0472 | 0.5910 | 0.5614 | 0.5000 | 07097 | 07204 | 07097 | 07204 | | | |
| 1.0000 | 1.9687 | 0.5313 | 0.5614 | 0.3750 | 07100 | 07196 | 07100 | 07196 | 07100 | 07196 | |
| 1.0000 | 1.9800 | 0.5600 | 0.5800 | 0.4200 | L44643 | L44610 | L44643 | L44610 | L44643 | L44610 | |

* Differs from preceding beraing number by cup or cone radius only.

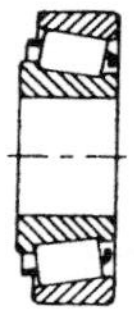

Courtesy Bower

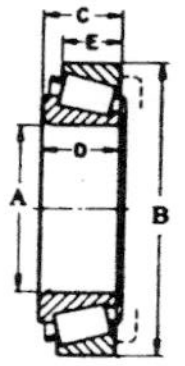


Courtesy Timken

Courtesy Tyson

**FIGURE 8.164** Bearing manual sample sheet.

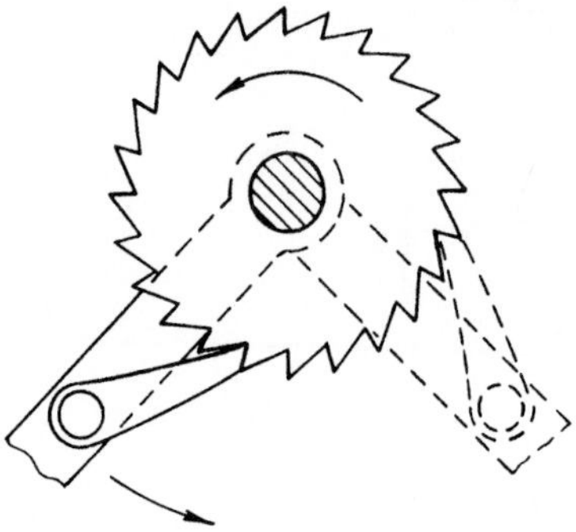

(a)

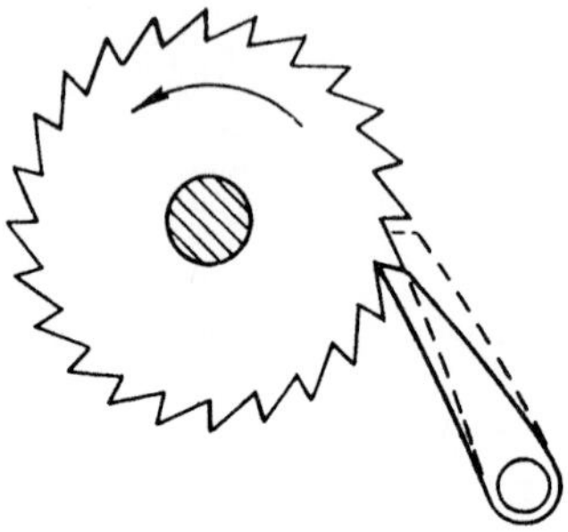

(b)

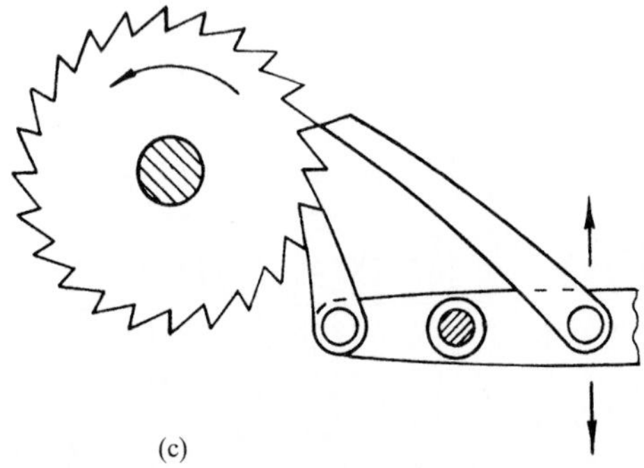

(c)

**FIGURE 8.165** Ratchet systems: (*a*) standard pawl and lever; (*b*) staggered pawls; (*c*) double pawls.

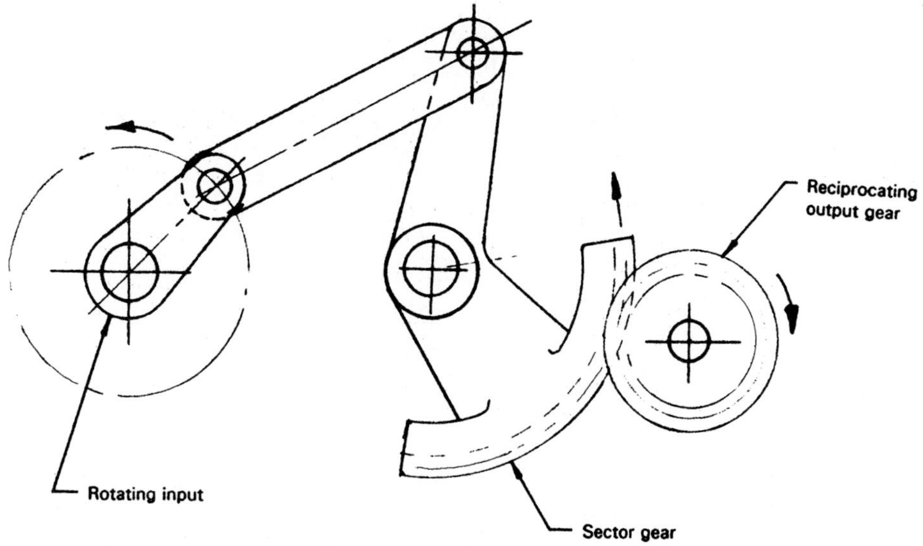


**FIGURE 8.166** Rotation at input produces reciprocating output via sector gear.

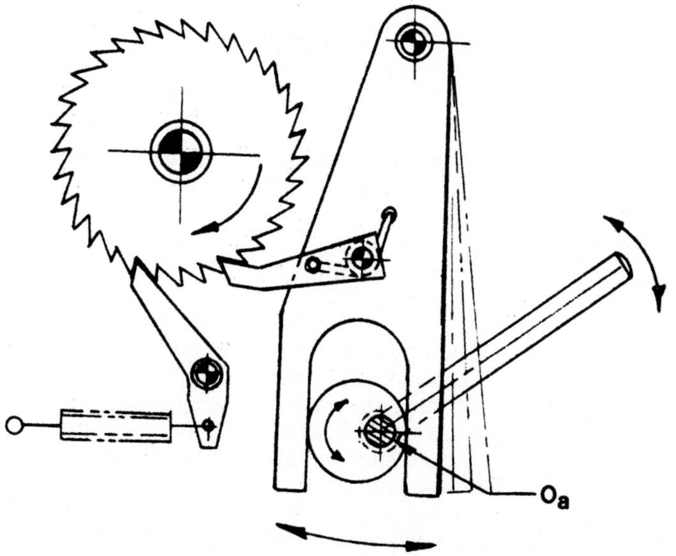


**FIGURE 8.167** Rotation at point $O_a$, either by lever or continuous, actuates the ratchet wheel in one direction (clockwise), while the holding pawl prevents the ratchet from reversing.

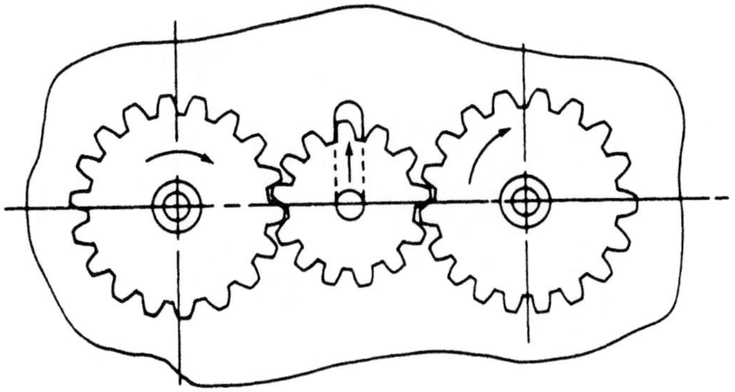

**FIGURE 8.168** A one-way gear drive. Reversing rotation of larger gears causes small center gear to rise in slot, thus disconnecting the train.

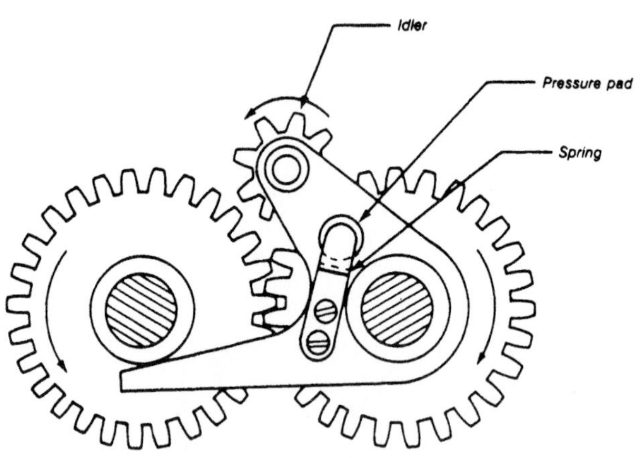


**FIGURE 8.169** Spring-loaded friction pads load the right gear. The idler meshes with and locks the gears when motion is reversed.

### 8.29.2 Space Mechanisms

There are potentially hundreds of space mechanisms, but only a few important types have been developed to date. A listing of the classification of all the kinematically possible pairs (joints) is shown in Fig. 8.215. Many modern industrial products utilize space mechanisms of one type or another, to perform functions and to operate in spaces which would be difficult using standard mechanisms. Figure 8.216 illustrates the nine most practical four-bar space mechanisms. These 9 four-bar space mechanisms have superior practicability because they contain only those joints which have area contact and are self-connecting. All these mechanisms can produce rotary or sliding output motion from a rotary input, the most common mechanical requirement for which linkage mechanisms are designed.

The type letters of the kinematic pairs shown in Fig. 8.216 are used to identify the mechanism by ordering the letter symbols consecutively around the closed kinematic chain. The first letter identifies the pair or joint connecting the input link and the fixed link; the last letter identifies the output link. Thus, a mechanism labeled R-S-C-R is a double-crank mechanism

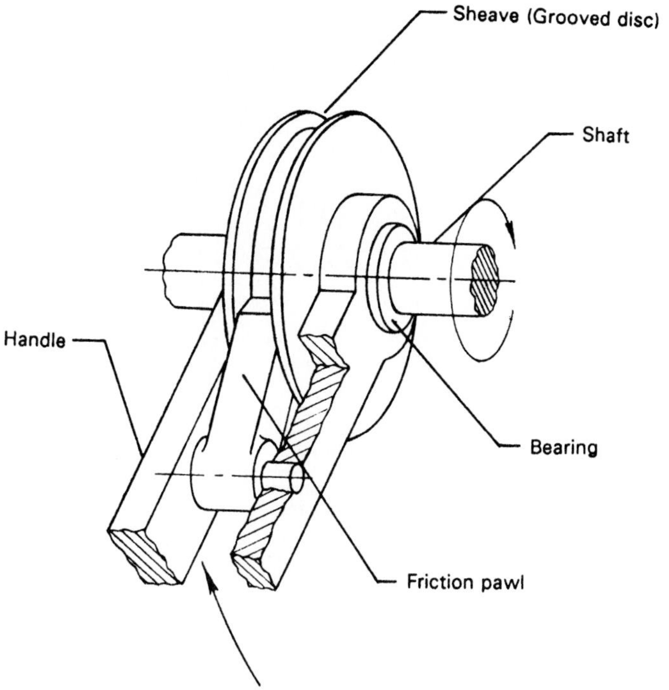


**FIGURE 8.170** A one-way friction action drive similar to a ratchet, but allowing various drive increments.

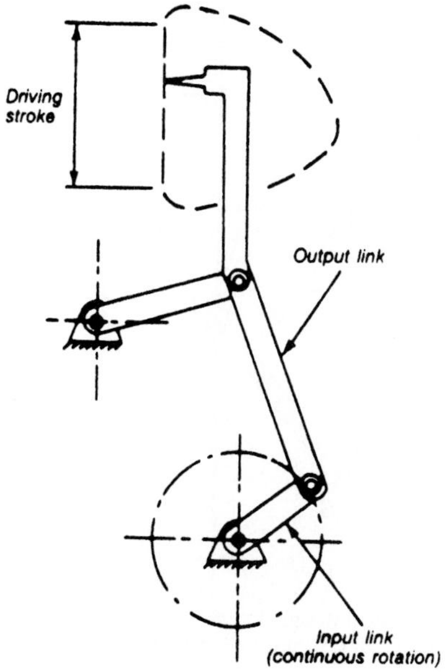


**FIGURE 8.171** In this straight-line linkage, the point on the output link describes a figure "D," a portion of which is straight-line motion. A very useful mechanism.

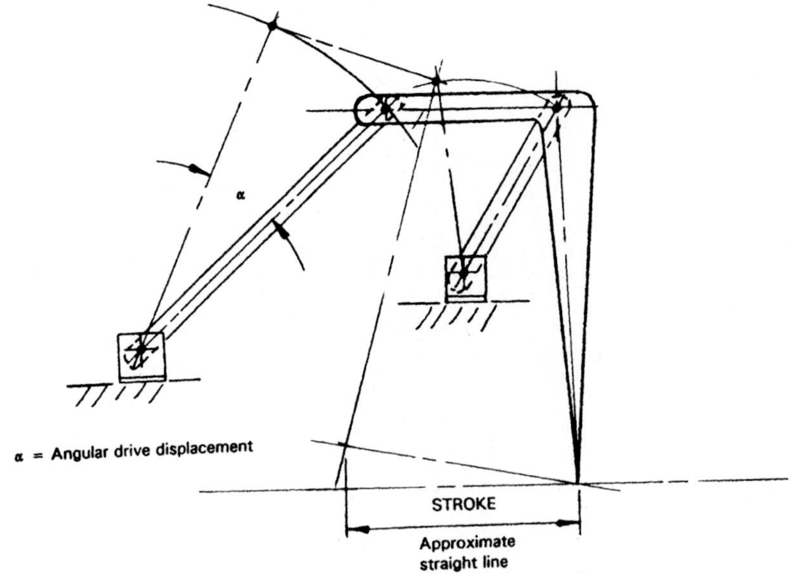


**FIGURE 8.172** A four-bar linkage which produces an approximate straight-line motion. A small angular displacement α produces a long, almost straight line.

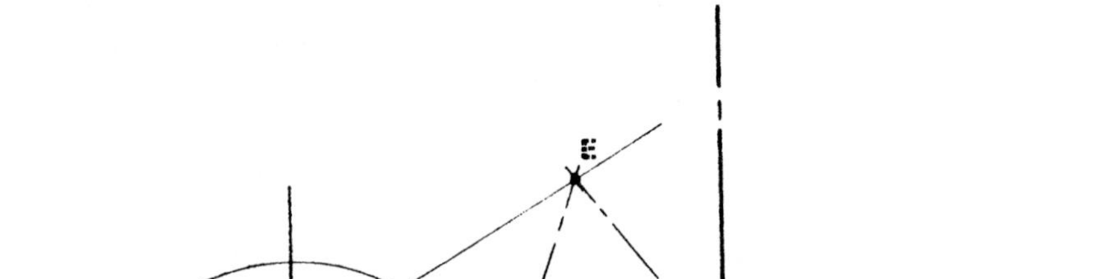

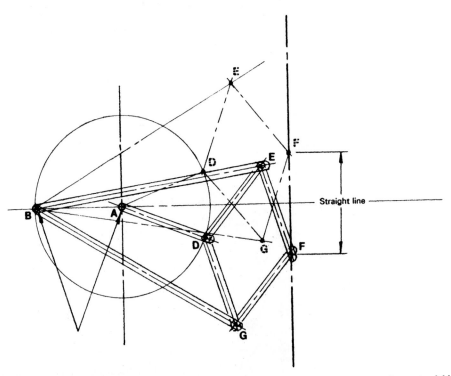


**FIGURE 8.173** The Peaucellier cell for generating a straight line with a linkage. The requirements of this linkage are $AB = BC$; $AD = AE$; and $CD$, $DF$, $FE$, and $EC$ must all be equal. The straight line is generated perpendicular to the line through points $A$ and $B$.

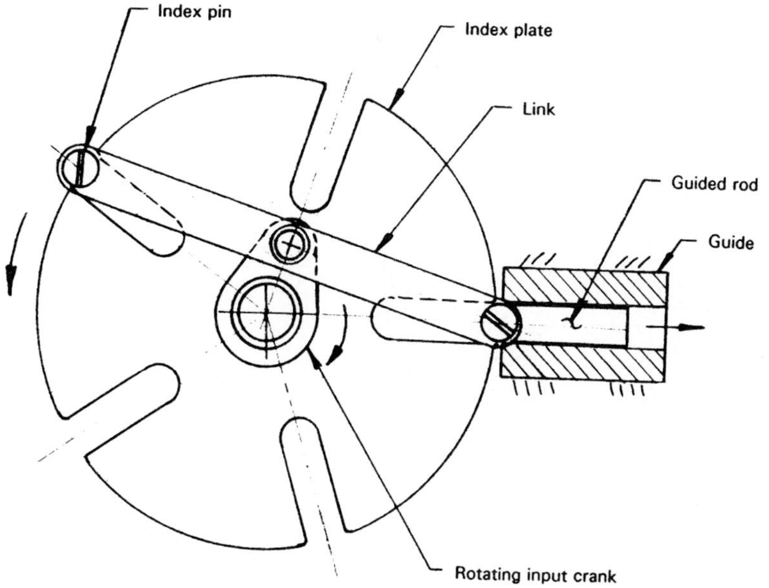


**FIGURE 8.174** The rotating input crank causes index pin to move the index plate, and the guided rod locks the index plate between indexing dwell periods.

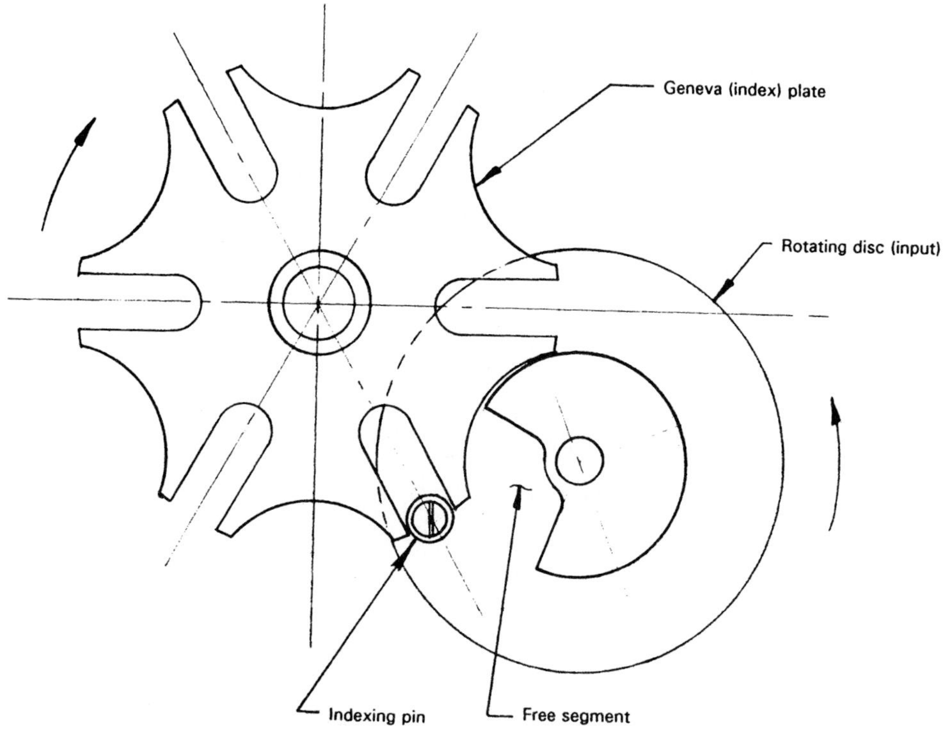


**FIGURE 8.175** An indexing mechanism which is locked by the segment on the rotating disk input during dwell periods.

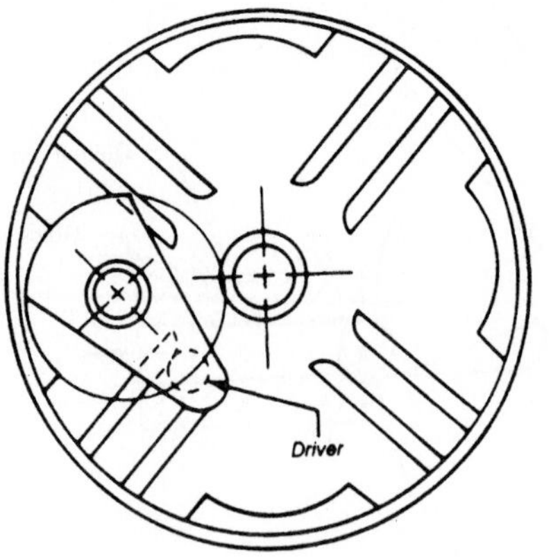


**FIGURE 8.176** An internal Geneva mechanism. The driver and driven wheel rotate in the same direction. Duration of the dwell is more than 180° of driver rotation.

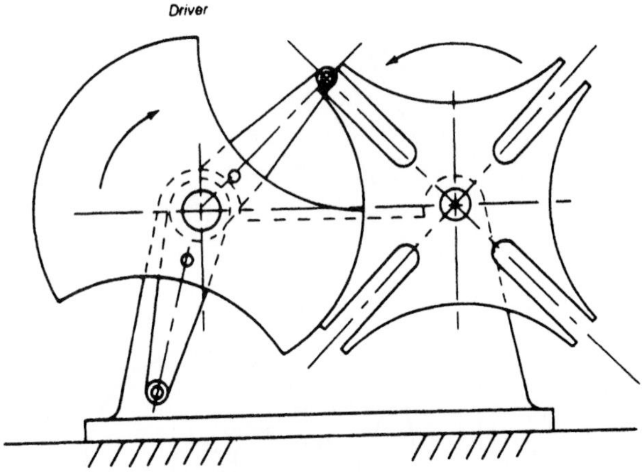


**FIGURE 8.177** In this Geneva mechanism, the duration of dwell time is changed by arranging the driving rollers unsymmetrically about their axes. This does not affect the duration of the motion periods.

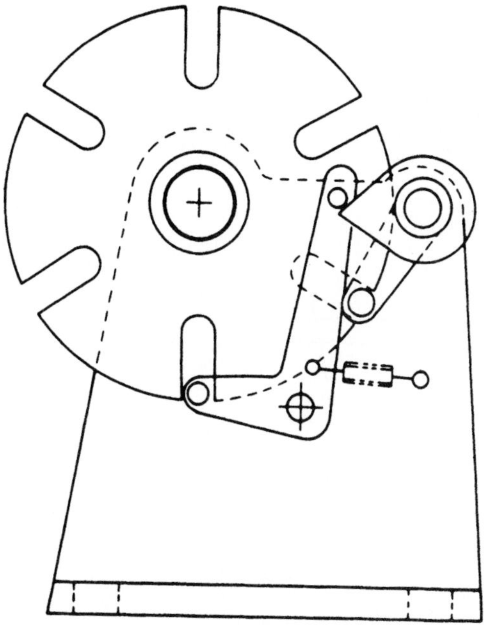

**FIGURE 8.178** As with most Geneva drives, the driver follower on the input crank enters a slot and rapidly indexes the output. On this mechanism, the roller in the locking arm (shown leaving the slot) enters a slot to prevent the index plate from rotating, when not indexing.

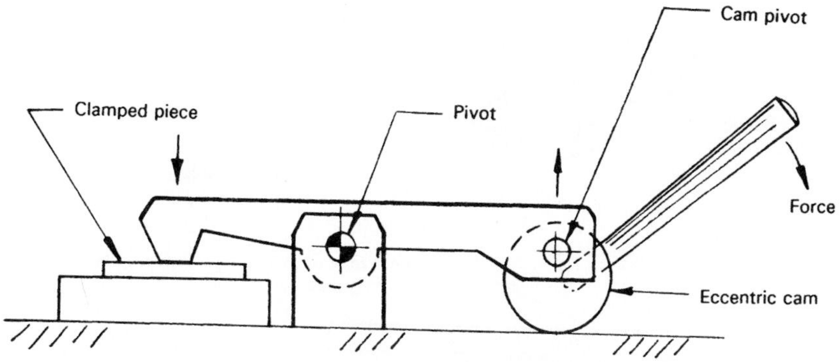


**FIGURE 8.179** A simple eccentric cam-action clamp mechanism.

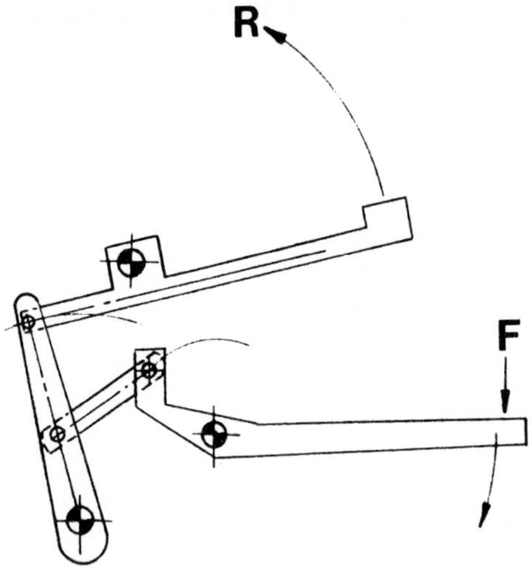


**FIGURE 8.180** A linkage for producing a long travel at *R* with a small displacement at *F*.

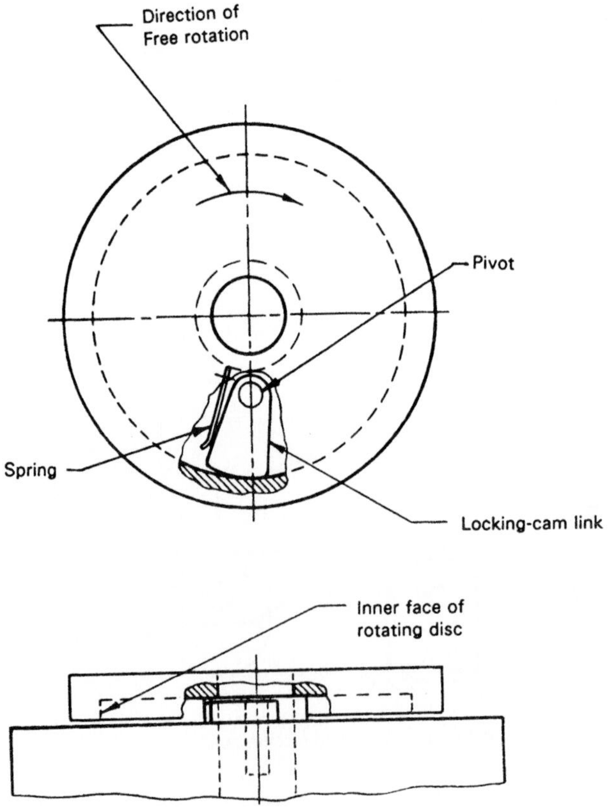


**FIGURE 8.181** A one-way rotating mechanism. The locking cam prevents counterclockwise rotation.

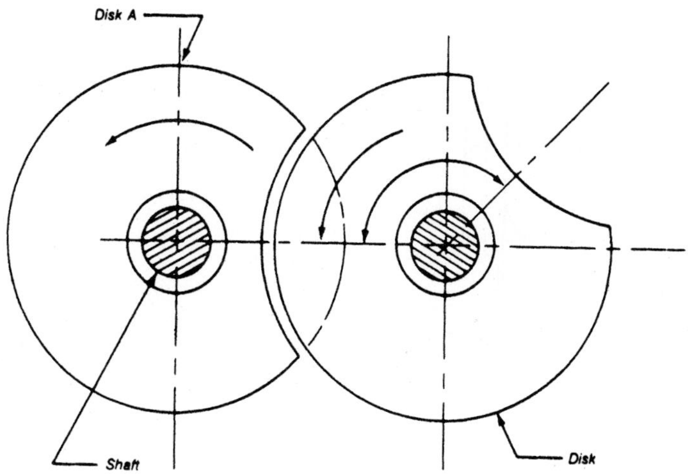


**FIGURE 8.182** Interlocking disk mechanism allows selective rotation of either shaft. As shown, disk *B* is free to rotate. Lining up the segment notches allows rotation of only one disk at a time.

**FIGURE 8.183** For each revolution of the input disk, the ratchet advances one tooth.

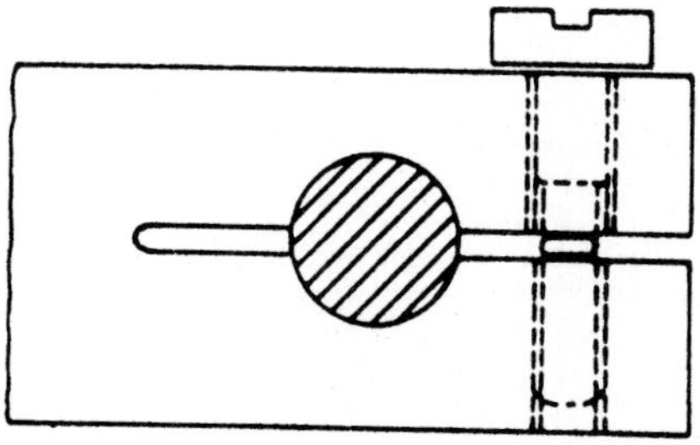

**FIGURE 8.184** A locking yoke clamp.

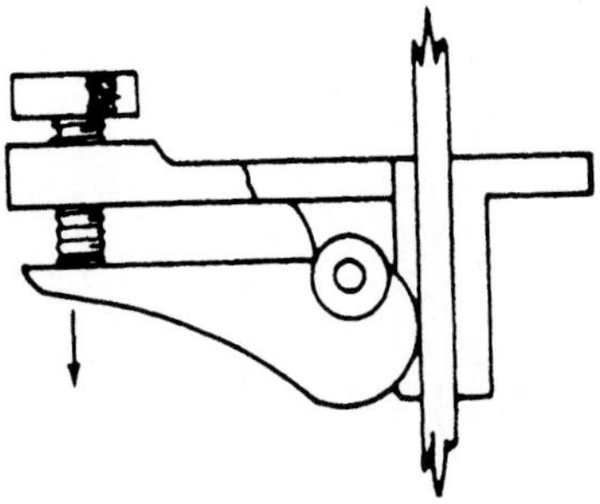

**FIGURE 8.185** A locking cam, actuated by a pressure screw.

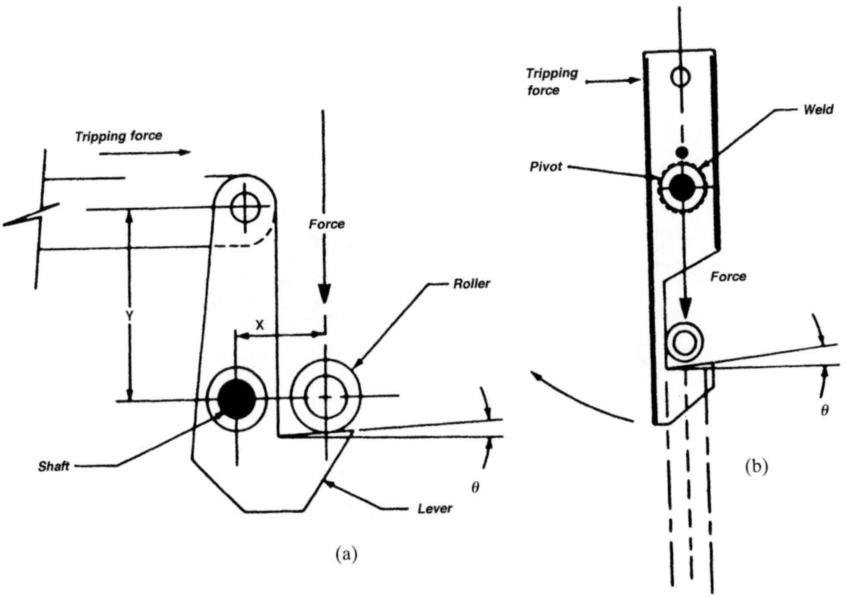


**FIGURE 8.186** Two sear tripping mechanisms for controlling a large force with small tripping pressures. The ratio of $X$ to $Y$ on the lever arms determines the required tripping pressure.

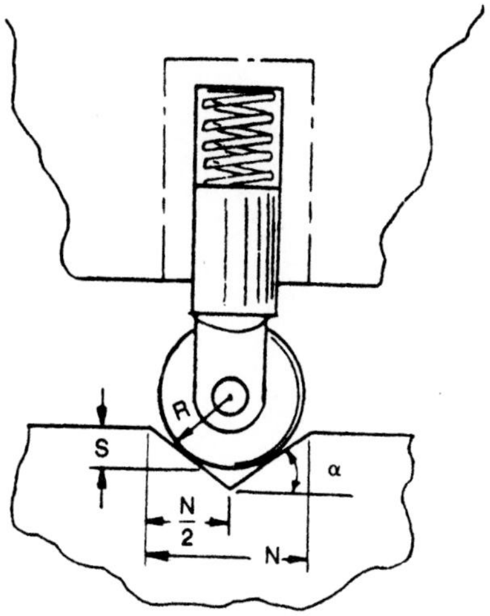


$$S, \text{ rise} = \frac{N \tan \alpha}{2} - R\left(\frac{1 - \cos \alpha}{\cos \alpha}\right)$$

$$R, \text{ radius} = \left(\frac{N \tan \alpha}{2} - S\right)\left(\frac{\cos \alpha}{1 - \cos \alpha}\right)$$

**FIGURE 8.187** A roller-wheel detent, spring-loaded. In this device, the rise and roller radius are determined by the indicated equations.

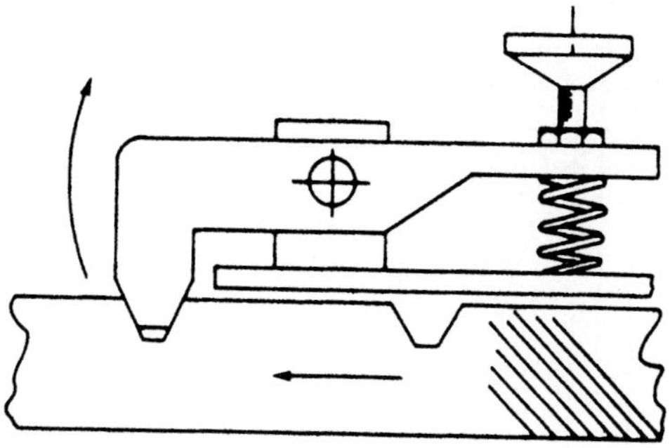

**FIGURE 8.188** A spring-loaded detent with release lever.

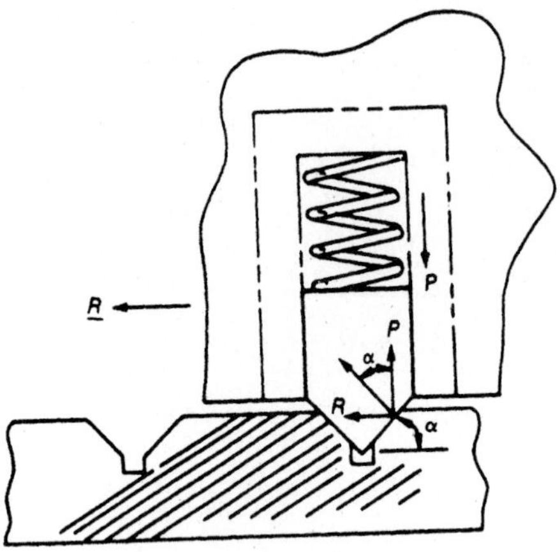


Holding power is R = P tan α;
for friction coefficient, F,
at contact surface
R = P (tan α + F)

**FIGURE 8.189** A spring-loaded detent for positioning and holding. Holding power is calculated from the equation shown.

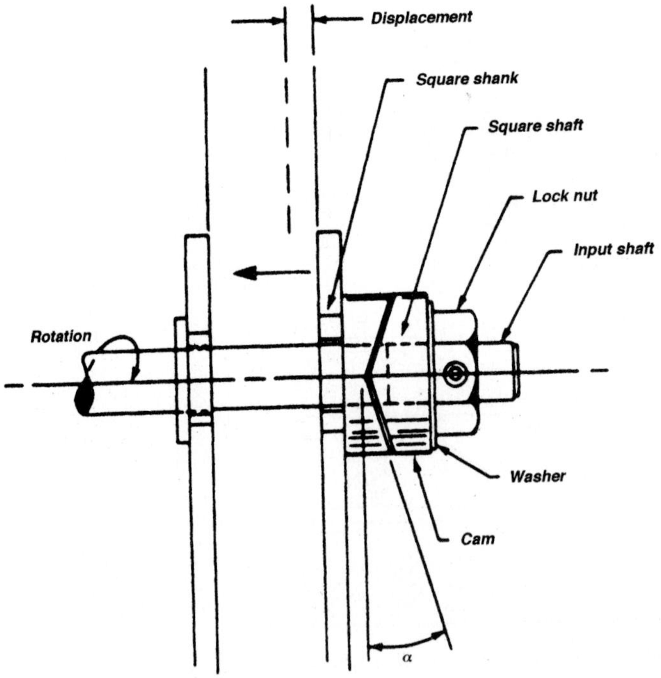


**FIGURE 8.190** A cam-actuated pressure clamping mechanism. Rotation of the input shaft displaces the profile cams and creates a clamping action. The clamping pressure of this mechanism is determined by the angle α in relation to the torque applied to the shaft. A photograph of the complete, actual mechanism is shown in Fig. 8.191, and the finished product is shown in Fig. 8.192.

**FIGURE 8.191** Clamping mechanism described in Fig. 8.190. Note the TIG welding on the central arm at the cam.

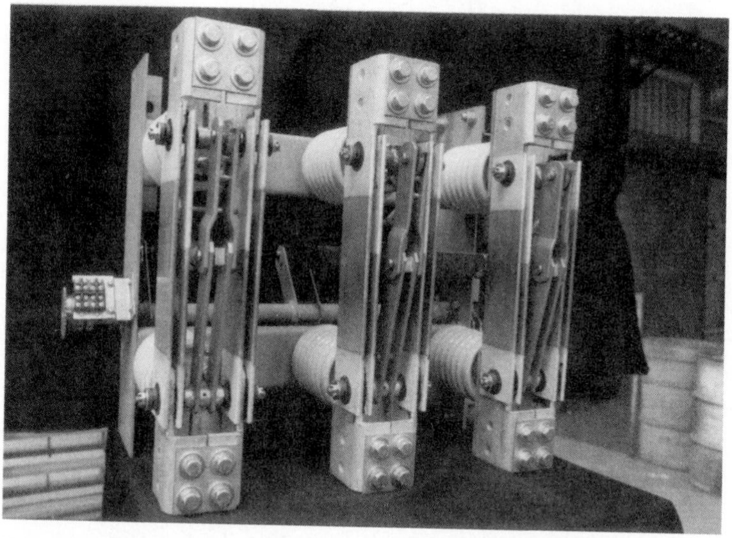

**FIGURE 8.192** The finished product detailed in Figs. 8.190 and 8.191.

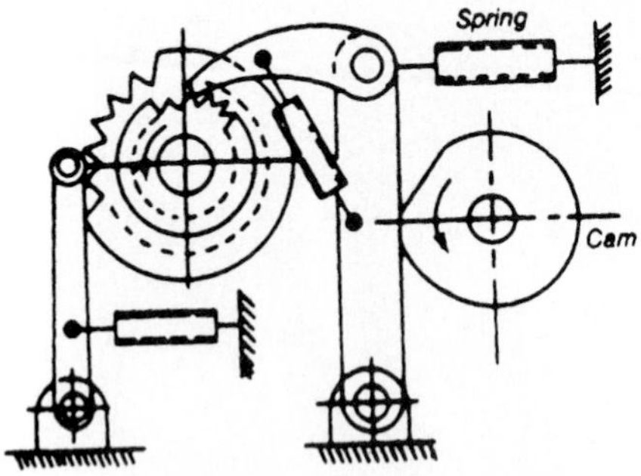


**FIGURE 8.193** A cam-driven ratchet. The ratchet advance is determined by the vertical location of the cam on the pawl's lever arm.

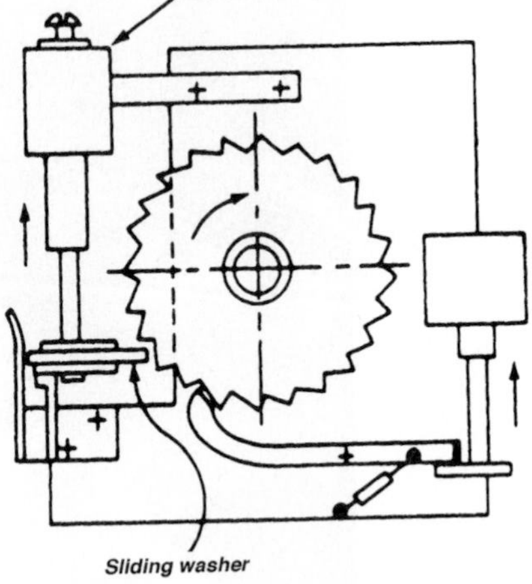


**FIGURE 8.194** A solenoid-operated ratchet with solenoid resetting mechanism. The teeth are engaged by the sliding washer.

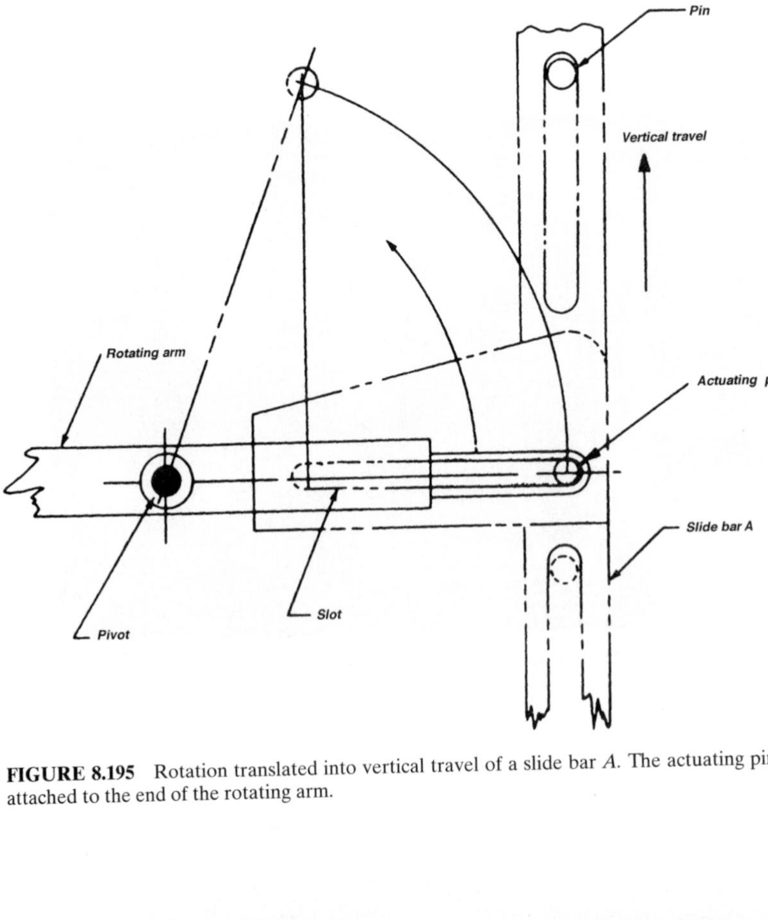


**FIGURE 8.195** Rotation translated into vertical travel of a slide bar *A*. The actuating pin is attached to the end of the rotating arm.

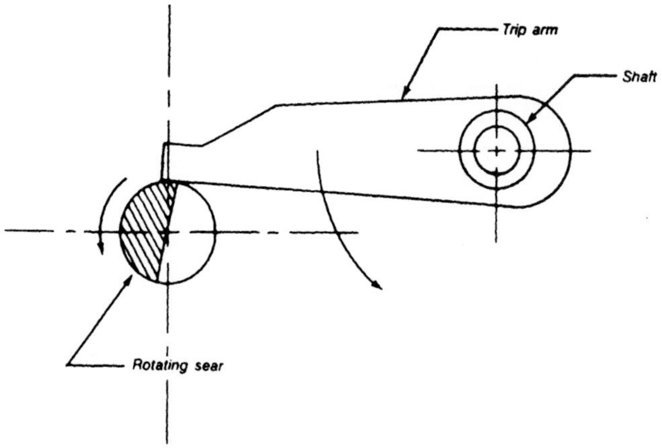


**FIGURE 8.196** A rotating sear mechanism used to actuate a large force with a small turning movement of the sear shaft. This type of sear is used to actuate the switching device shown in Figs. 6.71*a,* 6.71*b,* 6.72, and 8.225.

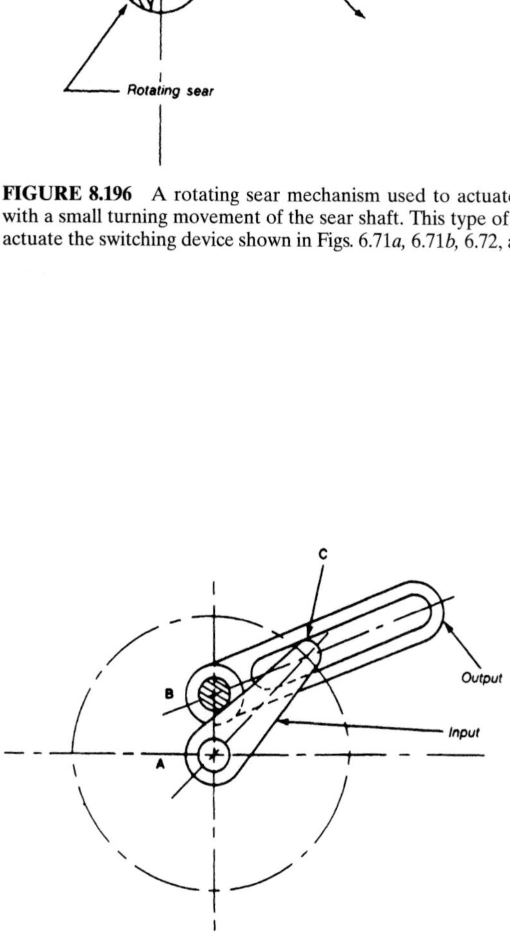


**FIGURE 8.197** The Whitworth "quick return," a simple method for varying the output motion of shaft *B*. The axes of shafts *A* and *B* are not collinear. The driving pin is at point C.

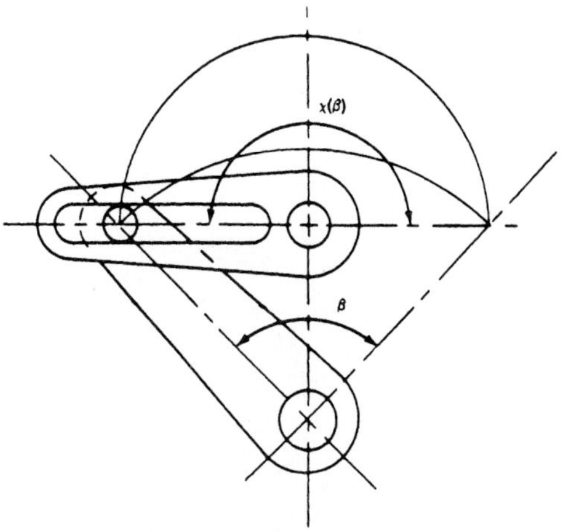


**FIGURE 8.198** A mechanism for enlarging the oscillating motion of a shaft. Rotating the input link through angle β will cause the follower arm to rotate $x(\beta)$, according to the center distance between the two shafts.

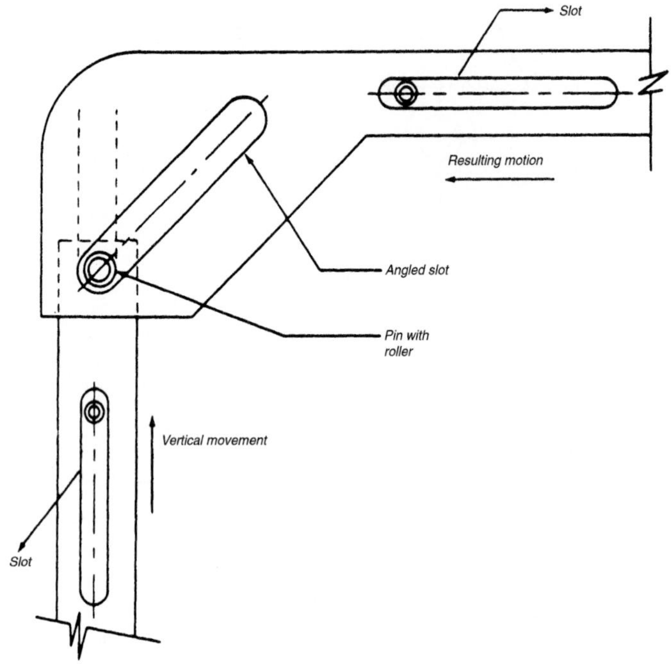


**FIGURE 8.199** A sliding mechanism for changing the direction of the linearly moving bars. When bar *A* is moved vertically, bar *B* moves to the left, as indicated. The angled slot can be repositioned for varying movements.

**FIGURE 8.200** A one-way internal ratcheting drive that transmits rotary motion in one direction only.

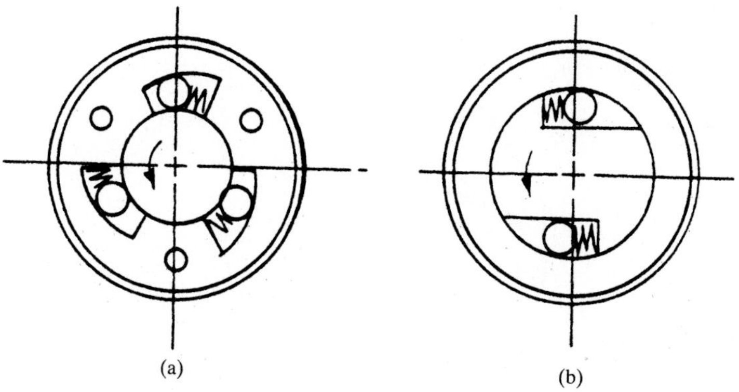


**FIGURE 8.201** (*a*) A one-way ball-lock clutch; (*b*) another variation of the one-way ball-lock clutch.

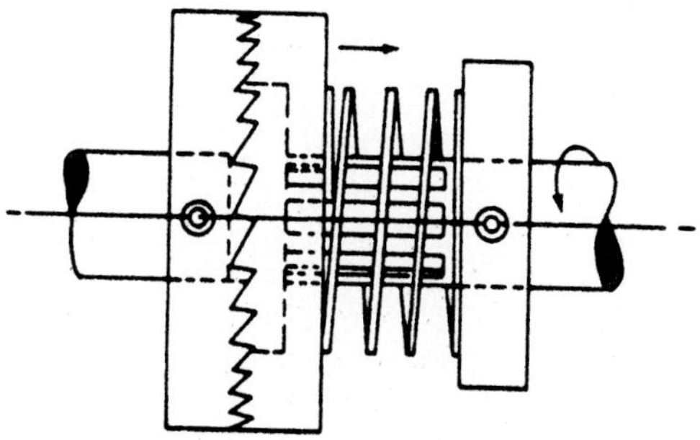

**FIGURE 8.202** A one-way ratcheting dog clutch with spring-loaded engagement.

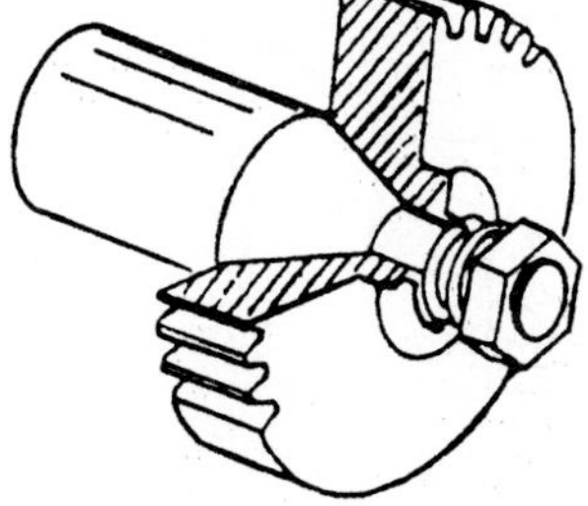

**FIGURE 8.203** A friction cone drive. The spring is put under pressure by tightening the nut, providing a means for allowing slippage in case of overload.

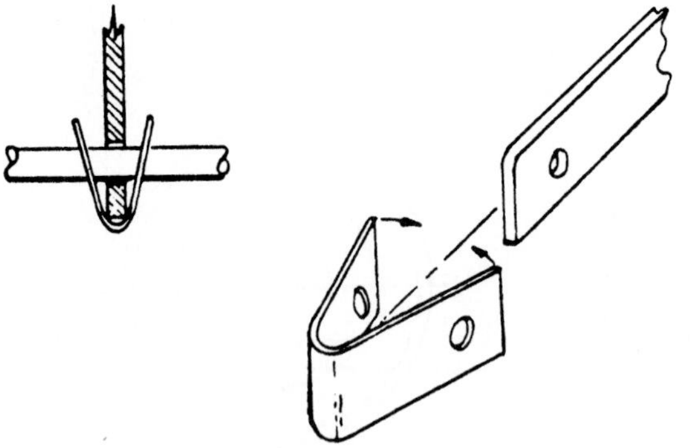

**FIGURE 8.204** A spring clip locking a shaft in position, with means for quick release.

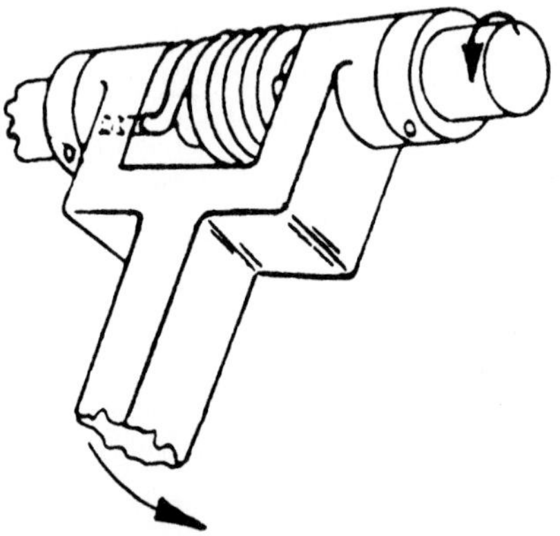

**FIGURE 8.205** A one-way drive using the tightening action of a torsion spring, allowing motion to be transmitted in one direction only.

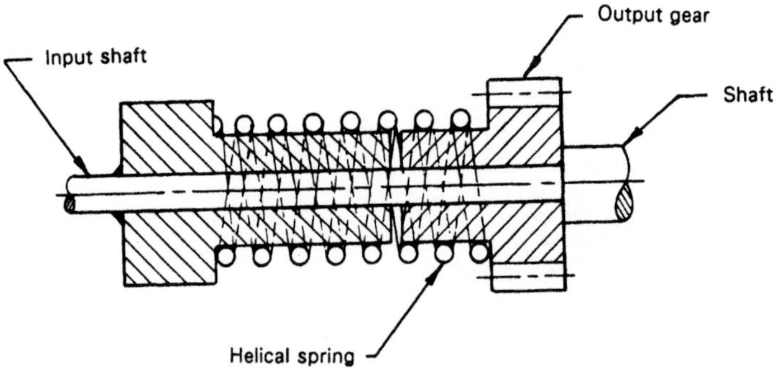


**FIGURE 8.206** A one-way spring clutch drive. In one direction, the spring tightens and transmits rotary motion; in the other direction, the spring unwinds and disconnects the shafts.

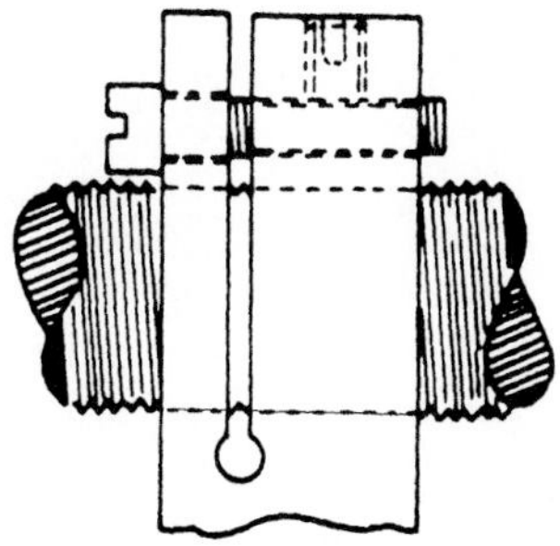

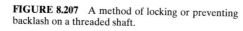

**FIGURE 8.207** A method of locking or preventing backlash on a threaded shaft.

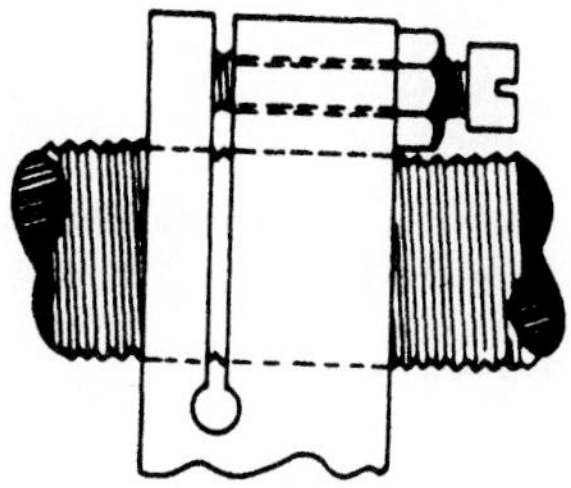

**FIGURE 8.208** Another method for controlling backlash.

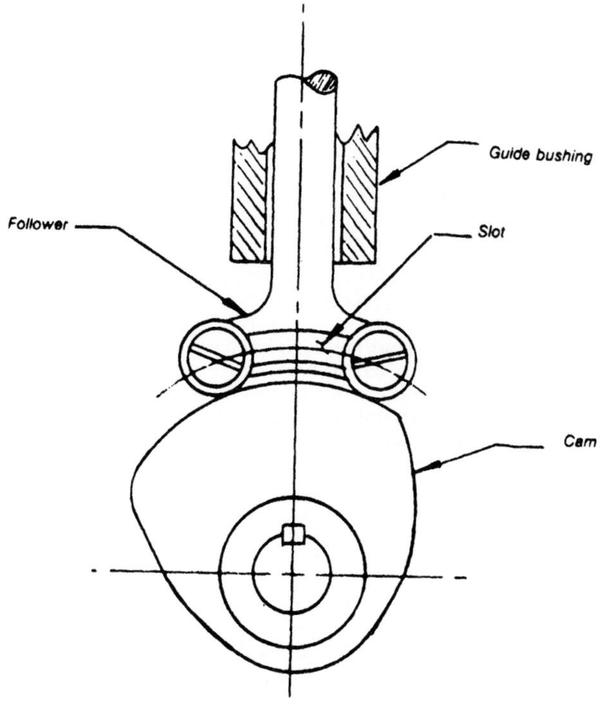


**FIGURE 8.209** An adjustable-dwell cam.

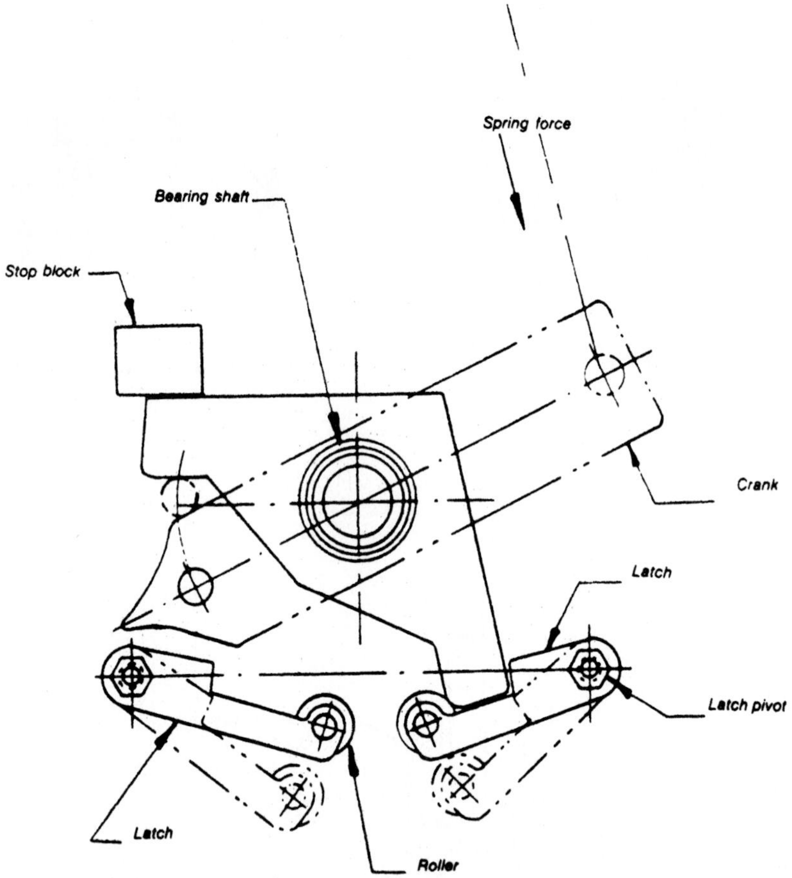


**FIGURE 8.210** A toggle-drive mechanism with automatic locking latches. In this compound mechanism, an input shaft rotates a spring-loaded crank until it passes top-dead-center. At this time, the crank tip unlocks the latch holding the A-shaped cam that is attached to the output shaft, driving the output shaft until the A cam strikes the stop block. At the end of its rotation, the A cam is again locked by the opposite latch. The latches are torsion spring-loaded. The entire mechanism is detailed in Figs. 8.211 to 8.214.

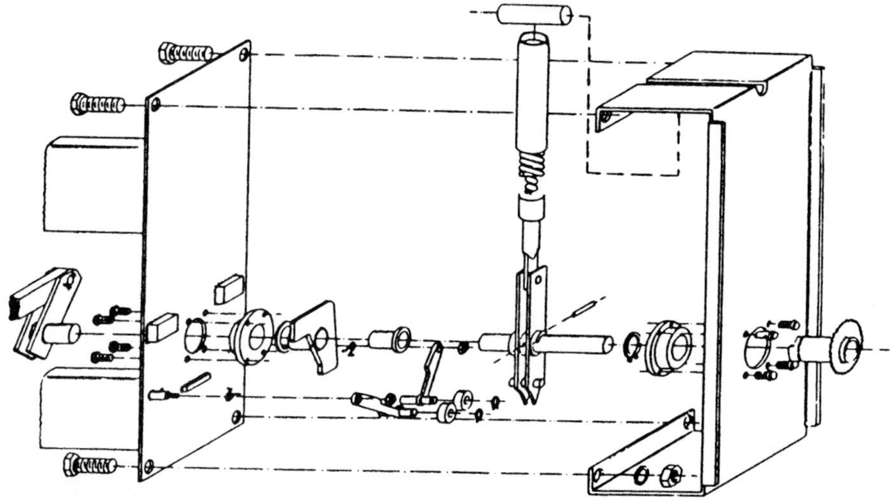

**FIGURE 8.211** Mechanism described in Fig. 8.210.

**FIGURE 8.212** Photograph of mechanism shown in Figs. 8.210 and 8.211.

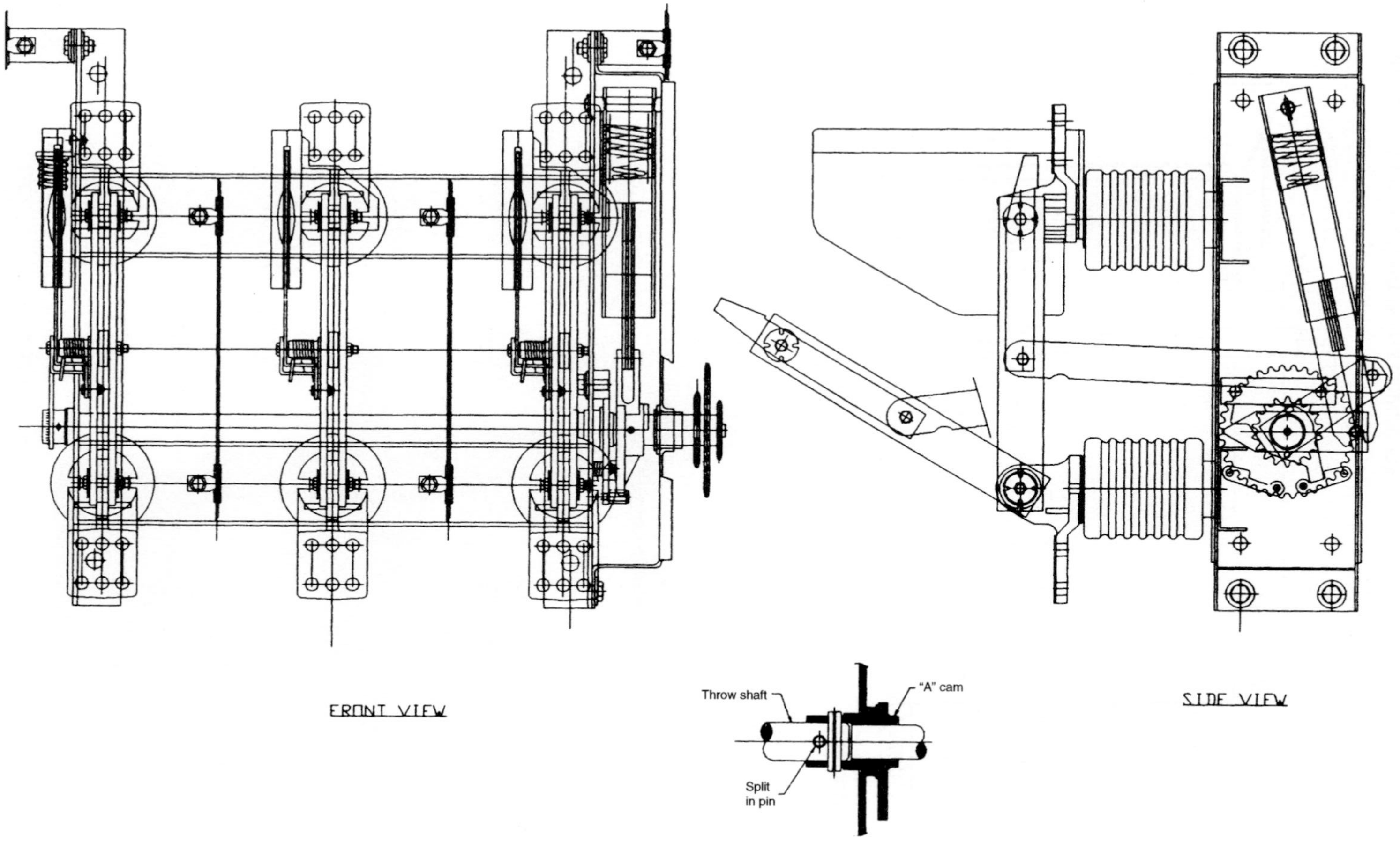


**FIGURE 8.213** AutoCad drawing of mechanism described in Fig. 8.210.

**FIGURE 8.214** The finished product of the switching mechanism described in Figs. 8.210 to 8.213.

| Degree of Freedom | Type Number ♦ | Type of Joint | |
|---|---|---|---|
| | | Symbol | Name |
| 1 | 100 | R | Revolute |
| | 010 | P | Prism |
| | 001 | H | Helix |
| 2 | 200 | T | Torus |
| | 110 | C | Cylinder |
| | 101 | $T_H$ | Torus-Helix |
| | 020 | ..... | ..... |
| | 011 | ..... | ..... |
| 3 | 300 | S | Sphere |
| | 210 | $S_S$ | Sphere-Slotted Cylinder |
| | 201 | $S_{SH}$ | Sphere-Slotted Helix |
| | 120 | $P_L$ | Plane |
| | 021 | ..... | ..... |
| | 111 | ..... | ..... |
| 4 | 310 | $S_G$ | Sphere-Groove |
| | 301 | $S_{GH}$ | Sphere-Grooved Helix |
| | 220 | $C_P$ | Cylinder-Plane |
| | 121 | ..... | ..... |
| | 211 | ..... | ..... |
| 5 | 320 | $S_P$ | Sphere-Plane |
| | 221 | ..... | ..... |
| | 311 | ..... | ..... |

Note: ♦ Number of freedoms, given in the order of $N_R$, $N_T$, $N_H$.
R = revolution joint, which permits rotation only; P = prism joint, which permits sliding motion only; H = helix or screw type of joint; C = cylinder joint, which permits both rotation and sliding (hence, has two degrees of freedom); S = sphere joint, which is the common ball joint permitting rotation in any direction, (three degrees of freedom).

**FIGURE 8.215** Classification of kinematic pairs: space mechanisms.

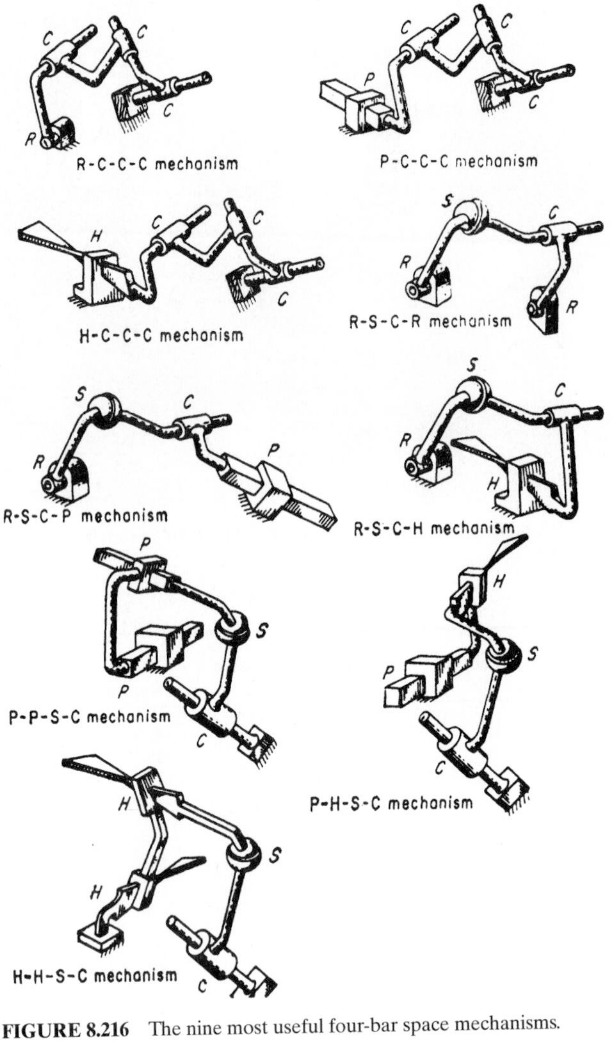


**FIGURE 8.216** The nine most useful four-bar space mechanisms.

with a spherical pair or joint between the input crank and the coupler, and a cylinder pair or joint between the coupler and the output link. The pair or joint designation letters are as follows (see also Fig. 8.215)

R Revolute joint, which permits rotation only.

P Prism joint, which permits sliding motion only.

H Helix- or screw-type joint.

C Cylinder joint, which permits both rotation and sliding (hence, the two degrees of freedom).

S Sphere joint, which is the common ball-joint permitting rotation in any direction (three degrees of freedom).

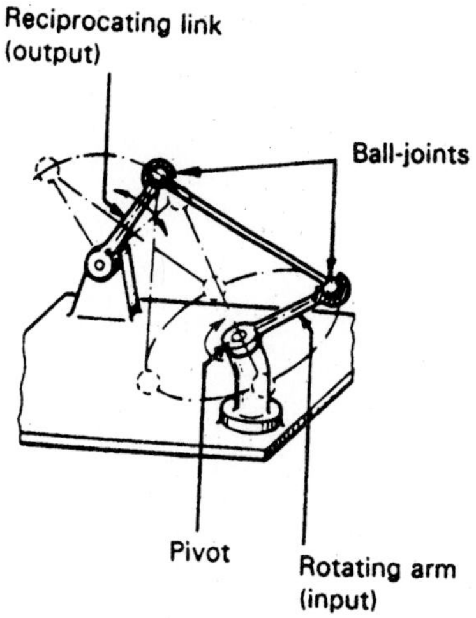


**FIGURE 8.217** A typical R-S-S-R space mechanism and linkage.

Figure 8.217 shows an R-S-S-R practicable space mechanism. This is one of the *maverick* space mechanisms, and one of the easiest and more practical to implement. By varying the length of the internal links, between the ball joints, various reciprocating outputs may be obtained. Mathematical analysis of space mechanisms is extremely complicated, but the advent of powerful CAD design stations using sophisticated motion analyses programs makes the work easier and quicker. The use of high-speed motion picture photography also allows the motions of these mechanisms to be seen and analyzed mechanically.

Three space mechanisms are known, which are classified as maverick space mechanisms:

- The Bennet R-R-R-R
- The R-S-S-R
- The R-C-C-R

These maverick mechanisms do not follow the kinematic rules shown in Fig. 8.215, and there may be undiscovered maverick types which would prove to be practicable.

### 8.29.3 Linkages

Linkages are an important element of machine design and are therefore detailed in this section, together with their mathematical solutions. Some of the more commonly used linkages are shown in Figs. 8.218 through 8.223. By applying these linkages to applications containing the simple machines, a wide assortment of workable mechanisms may be produced.

### 8.29.4 Linkage Analysis

#### 8.29.4.1 Toggle-Joint Linkages

Figure 8.218 shows the well-known and often used toggle mechanism. The mathematical relationships are shown in the figure. The famous Luger pistol bolt-lock action is based on the toggle-joint mechanism.

Figure 8.219 shows the application of a double-toggle joint, wherein the mechanical advantage may be multiplied. This mechanism is used in rock-crushing machinery, where an enormous force is required to crush rocks.

#### 8.29.4.2 The Four-Bar Linkage

Figure 8.220 shows the very important four-bar linkage, which is used in countless mechanisms. The linkage looks simple, but it was not until the 1950s that a mathematician named Freudenstein was able to find the mathematical relationship of this linkage and all its parts. The mathematical and geometric relationship of the four-bar linkage is therefore known as the *Freudenstein relationship,* after the mathematician, and is shown in Fig. 8.220. The geometry of the linkage may also be ascertained with the use of trigonometry, but the velocity ratios of points $p$ and $p'$ and the actions are complex, and with difficulty, can be solved using advanced mathematics.

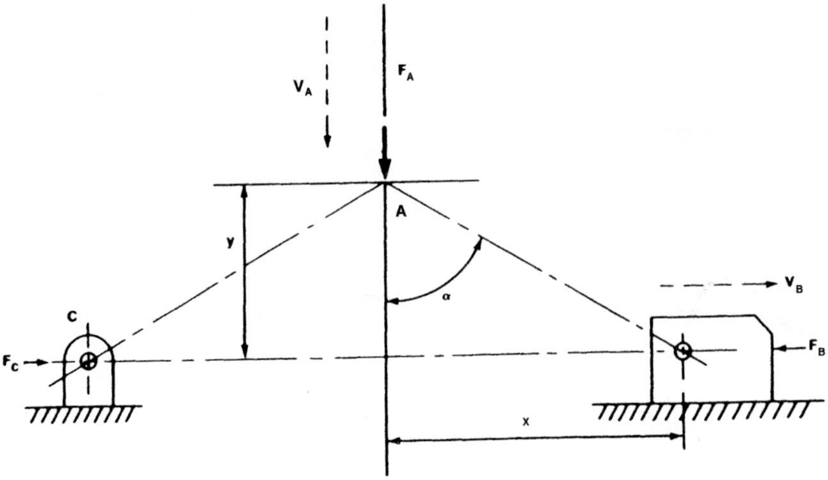


$$M_a = \frac{F_B}{F_A} = \frac{1}{2} \cdot \frac{x}{y} = \frac{1}{2} \tan \alpha = \frac{V_A}{V_B}.$$

As angle α approaches 90°, the links come into toggle, and the mechanical advantage and velocity ratio both approach infinity.

$M_a$ = Mechanical advantage (ratio)
$F_B$ = Force at point B
$F_A$ = Force at point A
$V_A$ = Velocity at point A
$V_B$ = Velocity at point B
X = Horizontal displacement
Y = Vertical displacement

**FIGURE 8.218** The standard "toggle" mechanism and linkage.

The use of high-speed photography on a four-bar mechanism makes the analysis possible without recourse to advanced mathematical methods, provided the mechanism can be photographed. When difficulty is encountered in the photographic attempt, a prototype mechanism may be made and this used for the photographic analysis. Certain four-bar mechanisms may be constructed in a way that creates numerous variables, making the mathematical solution, with any degree of accuracy, practically impossible.

### 8.29.4.3 Basic Linkages

In Fig. 8.221, the torque applied at point $T$ is known and we wish to find the force along link $F$. We proceed as follows. First, find the effective value of force $F_1$, which is

$$F_1 R = T \qquad F_1 = \frac{T}{R}$$

Then

$$\sin \phi = \frac{F_1}{F} \qquad F = \frac{F_1}{\sin \phi} \quad \text{or} \quad \frac{(T/R)}{\sin \phi}$$

[*Note:* $(T/R) = F_1$ = torque at $T$ divided by $R$ (radius).]

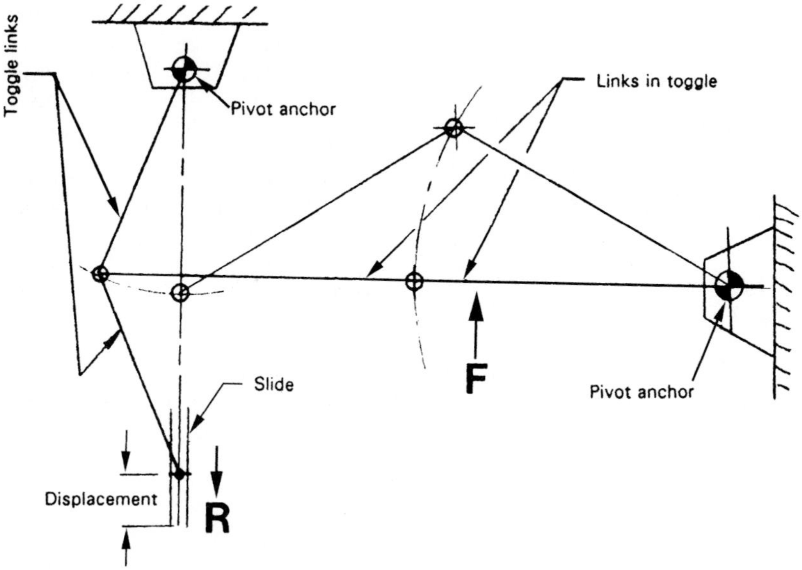


**FIGURE 8.219** The double-toggle-joint linkage.

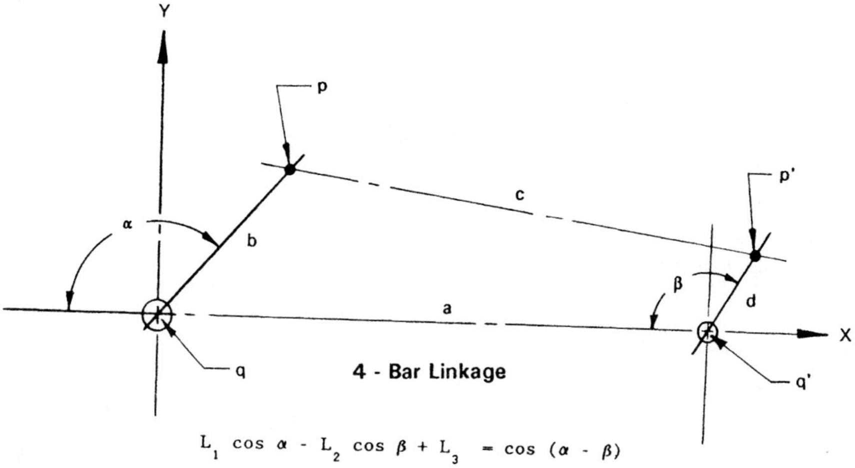


$$L_1 \cos \alpha - L_2 \cos \beta + L_3 = \cos(\alpha - \beta)$$

Freudenstein Relationship

where: $L_1 = a/d$

$L_2 = a/b$

and

$$L_3 = \frac{b^2 - c^2 + d^2 + a^2}{2\,bd}$$

**FIGURE 8.220** The geometry and mathematical relationship of the four-bar linkage.

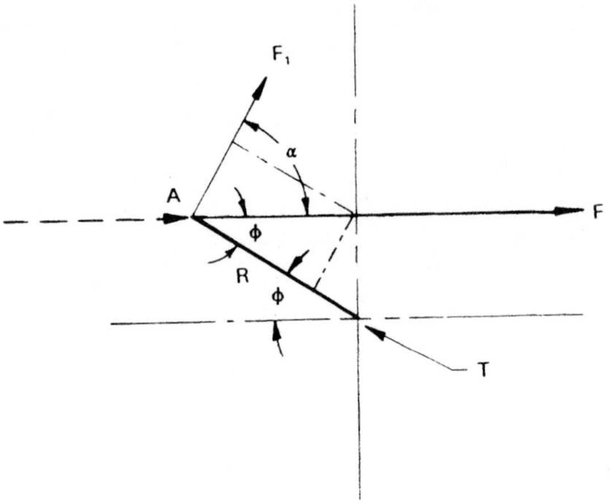


**FIGURE 8.221** A simple linkage.

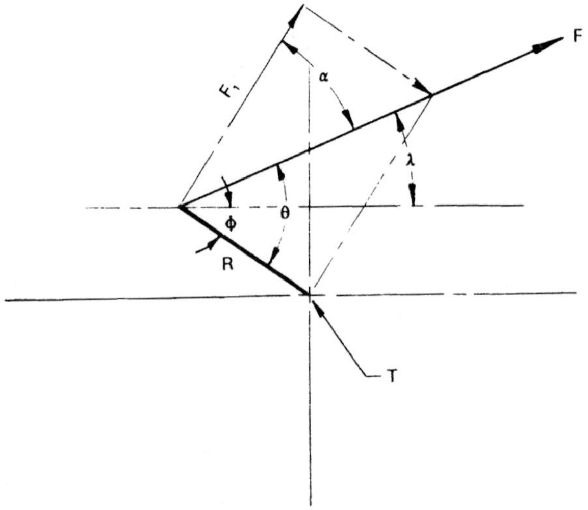


**FIGURE 8.222** A simple linkage.

In Fig. 8.222, the force $F$ acting at an angle $\theta$ is known, and we wish to find the torque at point $T$. First, we determine angle $\alpha$ from $\alpha = 90° - \theta$ and then proceed to find the vector component force $F_1$, which is

$$\cos\alpha = \frac{F_1}{F} \qquad \text{and} \qquad F_1 = F\cos\alpha$$

The torque at point $T$ is $F\cos\alpha R$, which is $F_1R$. (*Note:* $F_1$ is at 90° to $R$.)

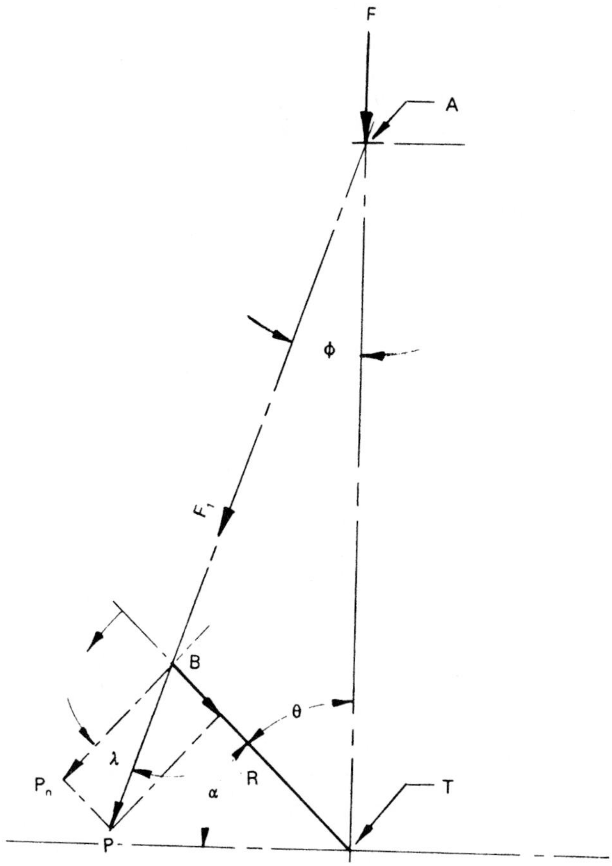


**FIGURE 8.223** The well-known engine crank linkage.

## 8.29.4.4 Crank Linkages

In Fig. 8.223, a downward force $F$ will produce a vector force $F_1$ in link $AB$. The instantaneous force at 90° to the radius arm $R$, which is $P_n$, will be:

$$F_1 \quad \text{or} \quad P = \frac{F}{\cos \alpha}$$

and

$$P_n = F_1 \quad \text{or} \quad P \cos \lambda \quad \text{or} \quad P_n = \frac{F}{\cos \phi} \sin (\phi - \theta)$$

The resulting torque at $T$ will be $T = P_n R$, where $R$ is the arm $BT$.

The preceding case is typical of a piston acting through a connecting rod to a crankshaft. This particular linkage is used often in machine design and the applications are numerous.

The preceding linkage solutions have their roots in engineering mechanics, further practical study of which may be made using Chap. 2 of this handbook and from other books listed in the bibliography.

## 8.30 MECHANISM AND LINKAGE APPLICATIONS

Figure 8.224 is a typical application for the four-bar linkage, showing a lever system of the four-bar arrangement applying torque to a shaft at $T_{Q1}$. The torque at the shaft will vary according to the force $F$ applied to the hand lever and also due to the changing angular relationships within the four-bar linkage. Applying standard engineering mechanics together with the Freudenstein relationship allows relatively simple calculations to be performed in the analysis of this linkage system.

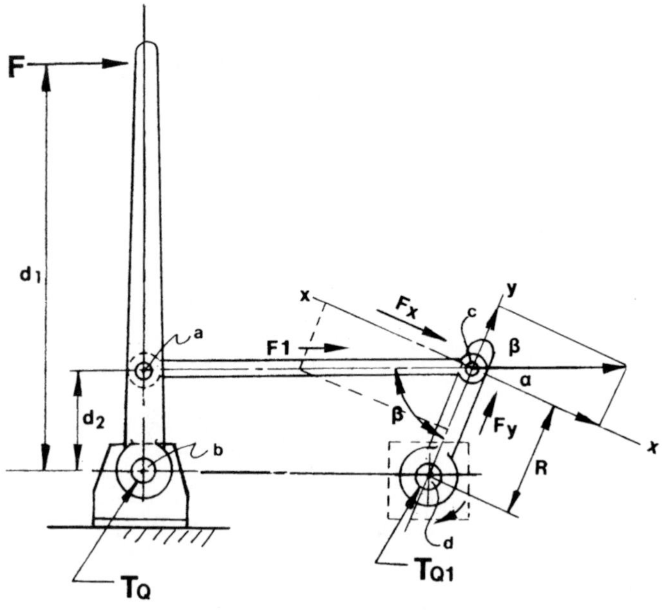


**FIGURE 8.224** A typical application of the four-bar linkage shown in Fig. 8.220.

Figure 8.225 is a typical industrial application of electromechanical design practice. Here, we have a complex mechanical device with electrical and electronic controls. The system on which these electromechanical controls are used can be seen in Figs. 6.71*a*, 6.71*b*, and 6.72 in Chap. 6. This switching system device and electrical and electronic controls were designed by the author for an industrial electric power distribution application.

Figure 8.226 is an AutoCAD drawing of the drive motor used to power the electromechanical device described in the preceding figures (Figs. 8.225, 6.71*a*, 6.71*b*, and 6.72). A brief study of the mechanisms involved here illustrates the various disciplines that the electromechanical design engineer uses in actual practice. Here, we see levers, springs, solenoids, electric motor, gears, linkages, a mechanical clutch, and other aspects of electromechanical design practice, including electrical and electronic components.

To illustrate the complexity and sophistication of modern electromechanical design, Fig. 8.227 illustrates the electromechanical shutter assembly of a modern single-lens-reflex camera. The shutter is made of titanium alloy and the timing control mechanism employs switch contacts and an electronic control printed-circuit board, all integrated to perform a com-

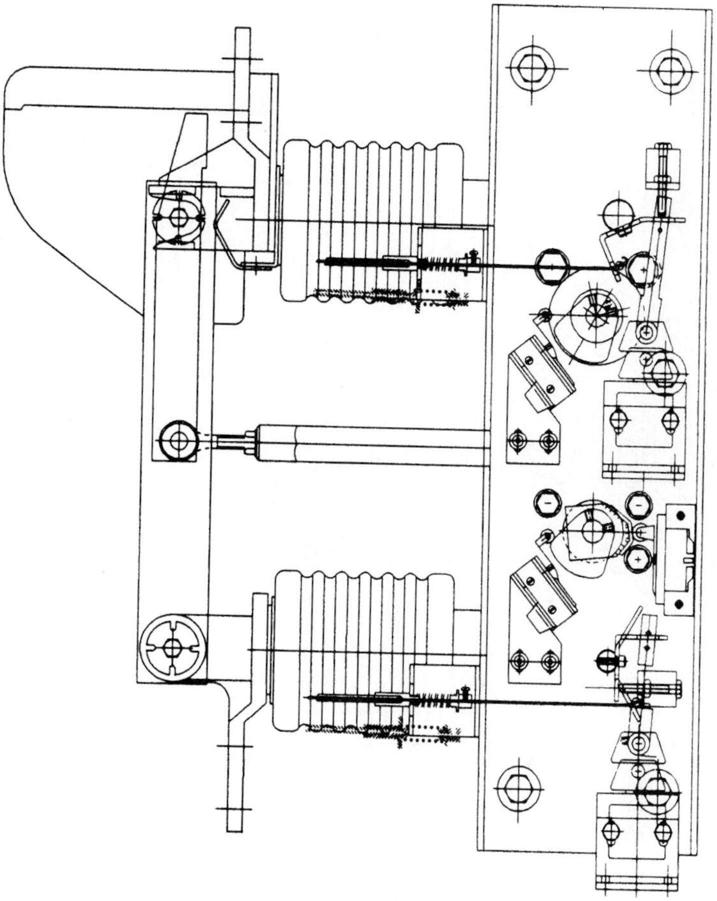

**FIGURE 8.225** A typical electromechanical product assembly drawing showing the control switches and tripping solenoids.

plex and highly accurate timing function, such as camera shutter mechanism and timing control.

Figure 8.228 is a typical schematic diagram of a mechanical system commonly referred to as a *kinematic diagram.* Here, the parts are not drawn to any scale or actual dimensional proportion. Rather, only the basic principles are shown to describe the function and operation of a particular mechanical or electromechanical system. Kinematic diagrams are an important part of the initial stages of design.

As a conclusion to this chapter, I cannot overemphasize the importance for modern design engineers to attempt to grasp all the basic disciplines involved in electromechanical design practice. The author has felt for many years that it is important for American industry to have as many people as possible who are trained in the combined disciplines of mechanical, electrical, and electronic design engineering practices and to know how to apply that training for the benefit of American industry. Industry needs more "general practitioners" and fewer "specialists" if the United States is to maintain substantial portions of the world market.

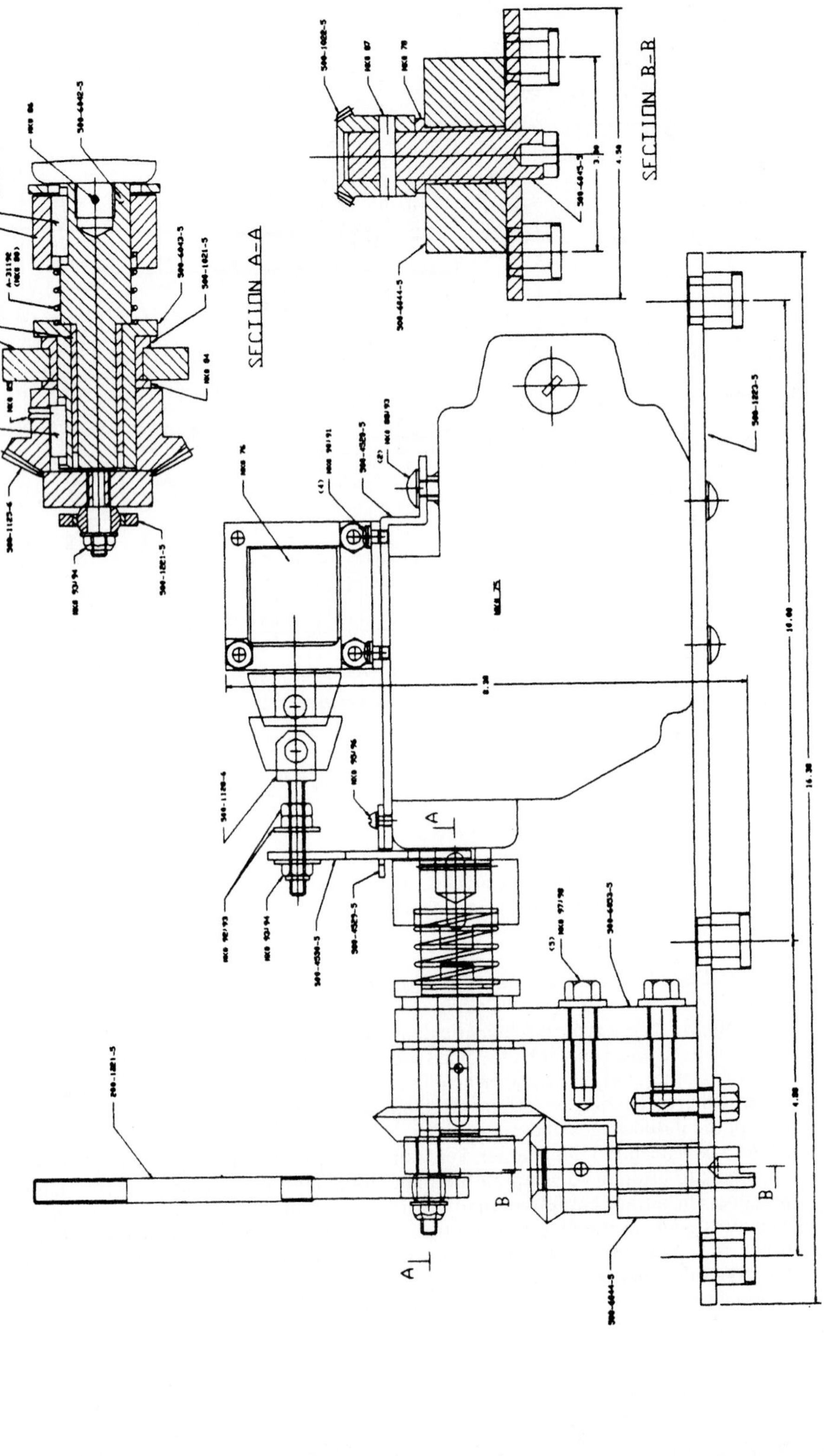


**FIGURE 8.226** A typical electromechanical mechanism used as the motor drive for the product shown in Figs. 8.225, 6.71*a*, 6.71*b*, and 6.72.

**FIGURE 8.227** A complex electromechanical camera shutter. Note the switches and printed-circuit board with components.

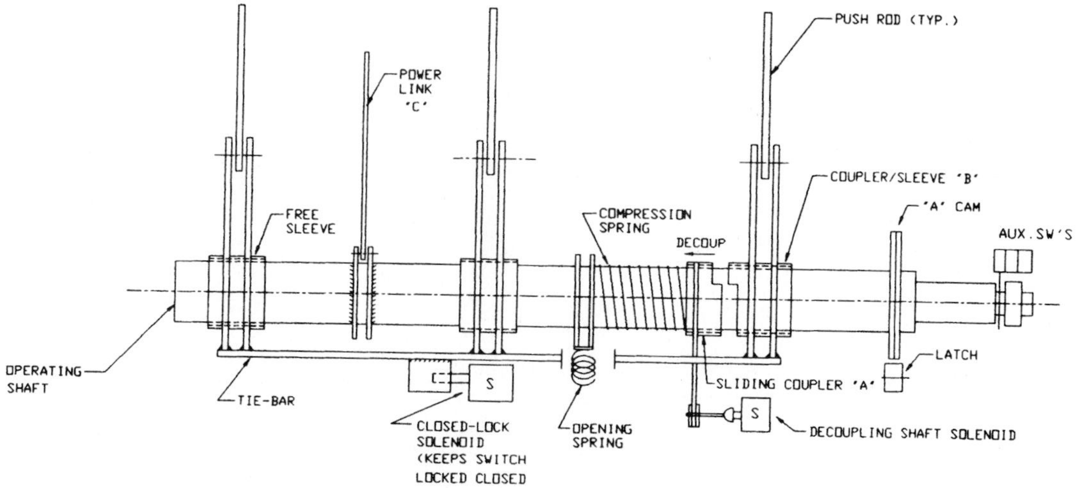


**FIGURE 8.228** A typical kinematic diagram used for preliminary design purposes.

## FURTHER READING


Chironis, N. P., 1965: *Mechanisms, Linkages and Mechanical Controls.* New York: McGraw-Hill.

Hall, A. S., A. R. Holowenko, and H. G. Laughlin, 1961: *Machine Design, Theory and Problems.* New York: McGraw-Hill.

Herkimer, H., 1952: *Engineering Illustrated Thesaurus.* Cleveland, Ohio: Chemical Rubber Publishing Company.

Hindhede, U., J. R. Zimmerman, B. R. Hopkins, J. R. Erisman, W. C. Hull, and J. D. Lang, 1983: *Machine Design Fundamentals.* New York: Wiley.

Nadjer, K. W., 1958: *Machine Designer's Guide,* 4th ed. Ann Arbor, Mich.: Edwards Bros.

Oberg, E., F. Jones, and H. Horton, 1986: *Machinery's Handbook,* 22d ed. New York: Industrial Press.

Shigley, J., and C. R. Mischke, 1986: *Standard Handbook of Machine Design,* New York: McGraw-Hill.

Spotts, M. F., 1953, "A practical method for designing gear trains," *Product Engineering,* February, p. 211.

# CHAPTER 9

# PNEUMATICS, HYDRAULICS, AIR HANDLING, AND HEAT

## 9.1 PNEUMATICS

*Pneumatics* is the study of air and gases and the relationship between volume, pressure, and temperature of the air or gas. To begin the discussion, let us define the normal condition of atmospheric air as follows:

At the reference conditions shown, the weight of atmospheric air is as follows:

| | |
|---|---|
| Temperature | 32°F (0°C) |
| Atmospheric pressure | 29.92 in of mercury (Hg), by barometer (14.7 psi at sea level); 760mm Hg |
| Weight | 0.08073 lb/ft$^3$; 1 lb of air occupies 12.387 ft$^3$ at the listed temperature and pressure |

Then, the weight of 1 ft$^3$ of air at any other pressure and temperature is

$$W = \frac{1.327 \times P_b}{T}$$

where $W$ = weight, lb/ft$^3$
$P_b$ = barometric pressure, in Hg
$T$ = absolute temperature, rankines (°R) (absolute 0 in rankine = –459.69°F; see Secs. 1.13 and 1.14 in Chap. 1)

### 9.1.1 Pressure, Volume, and Temperature of Air

The relationship between pressure, volume, and temperature may be expressed as

$$\frac{PV}{T} = 53.33$$

where $P$ = absolute pressure, lb/ft$^2$
$V$ = volume in ft$^3$ of 1 lb of air at the given temperature and pressure
$T$ = absolute temperature, °R

### 9.1.2 Adiabatic and Isothermal Compression or Expansion of Air

Adiabatic compression or expansion of air takes place without the transmission of heat to or from the air during the compression or expansion. The relations between $P$, $V$, and $T$ are given by the following series of equations:

$$\frac{P_2}{P_1} = \left(\frac{V_1}{V_2}\right)^{1.41} \qquad \frac{P_2}{P_1} = \left(\frac{T_2}{T_1}\right)^{3.46}$$

$$\frac{V_2}{V_1} = \left(\frac{P_1}{P_2}\right)^{0.71} \qquad \frac{V_2}{V_1} = \left(\frac{T_1}{T_2}\right)^{2.46}$$

$$\frac{T_2}{T_1} = \left(\frac{V_1}{V_2}\right)^{0.41} \qquad \frac{T_2}{T_1} = \left(\frac{P_2}{P_1}\right)^{0.29}$$

These equations are valid for both the U.S. customary and SI systems. Use the appropriate units as found in the conversion section.

The standard method of rearranging the equations for the unknown variable is illustrated by the following:

$$\frac{V_2}{V_1} = \left(\frac{P_1}{P_2}\right)^{0.71} \qquad V_2(P_2)^{0.71} = V_1(P_1)^{0.71} \qquad P_2 = \sqrt[0.71]{\frac{V_1(P_1)^{0.71}}{V_2}}$$

Isothermal compression or expansion takes place when a gas is compressed or expanded in conjunction with the addition or transmission of sufficient heat energy to maintain a constant temperature. Then, by Boyle's law for gases,

$$P_1V_1 = P_2V_2 = RT$$

where $P_1$ = initial absolute pressure, lb/ft$^2$
$P_2$ = absolute pressure after compression, lb/ft$^2$
$V_1$ = initial volume, ft$^3$
$V_2$ = volume of air after compression, ft$^3$
$R$ = 53.33 (universal gas constant)
$T$ = absolute temperature maintained during isothermal compression or expansion, °R

The formula $PV = RT$ is known as the *equation of state per mole* (mol) for air or gases and holds true for any given gas (atmospheric air included). The *mole* is defined as the gram-molecule. Then, for 1 g, this equation becomes

$$PV = \frac{RT}{m}$$

where $m$ = the molecular weight of the gas (or air). For $G$ grams, this becomes

$$PV = \frac{GRT}{m}$$

where $G$ = grams of gas or air.

### 9.1.3 Work or Energy Required in Compressing Air

Compressing air adiabatically, we obtain

$$E = 3.46\, P_1V_1\left[\left(\frac{P_2}{P_1}\right)^{0.29} - 1\right]$$

Compressing air isothermally, we have

$$E = P_1 V_1 \ln \frac{V_1}{V_2}$$

where $P_1$ = initial absolute pressure, lb/ft$^2$
$P_2$ = absolute pressure after compression, lb/ft$^2$
$V_1$ = initial volume, ft$^3$
$V_2$ = volume of air after compression, ft$^3$
$E$ = energy, ft · lb

Energy requirements for isothermal compression are considerably less than those for adiabatic compression. In practice, the actual energy requirements fall between the power required for each of the two methods.

### 9.1.4 Horsepower Requirements for Air Compression

For adiabatic compression, we obtain

$$\text{hp} = \frac{144nPVe}{33{,}000\,(e-1)} \left[ \left( \frac{P_2}{P} \right)^{(e-1)/ne} - 1 \right]$$

where $P$ = atmospheric pressure, lb/in$^2$
$P_2$ = absolute final pressure, lb/in$^2$
$n$ = number of stages of compression
$V$ = volume of air compressed per minute, ft$^3$
$e$ = 1.41 (constant for adiabatic compression)

For isothermal compression, we have

$$\text{hp} = \frac{144PV}{33{,}000} \left( \ln \frac{P_2}{P} \right)$$

### 9.1.5 Airflow in Pipes

Airflow in pipes is expressed as

$$v = \sqrt{\frac{25{,}000\,(\text{i.d.})\,P_1}{L}} \qquad \text{transposed for} \qquad P_1 = \frac{Lv^2}{25{,}000\,(\text{i.d.})}$$

where $v$ = air velocity, ft/sec
$P_1$ = pressure loss due to flow, oz/in$^2$
i.d. = inside diameter of pipe, in
$L$ = length of pipe, ft

The discharged air is therefore:

$$D = vA$$

where $v$ = air velocity, ft/sec (other symbols defined below) and

$$\text{hp} = \frac{v_1 P}{550}$$

where $D$ = discharged air, ft³/sec
$A$ = area of pipe, $ft^2$
$P$ = pressure, $lb/ft^2$
$v_1$ = volume of air moved in pipe, $ft^3/sec$
hp = horsepower requirement to move the air

### 9.1.6 Compressed Airflow in Pipes

When the pressure difference at the ends of the pipe is small, the volume of flow is then

$$V = 58\sqrt{\frac{P\,(\text{i.d.})^5}{WL}}$$

where $V$ = volume of air, $ft^3/min$.
$P$ = pressure difference, $lb/in^2$
i.d. = inside diameter of the pipe, in
$L$ = length of pipe, ft
$W$ = weight of entering air, $lb/ft^3$

### 9.1.7 Stresses in Pressurized Cylinders

The stress in the wall material of cylindrical containers may be calculated from the following equations. The ends of pressurized cylinders are normally made hemispherical or curved, so that the stress is equal to or less than the stress in the cylindrical wall. The internal gas pressure or hydraulic pressure is equal per unit area inside of a pressurized container. If the container is pressurized to 125 psi, every square inch of internal surface has 125 lbF exerted on it.

The following equations are used for low-pressure, high-pressure, and ultra-high-pressure cylinders. For low-pressure, thin-walled cylinders (see Fig. 9.1):

$$S = \frac{(\text{i.d.})\,P}{2t} \quad \text{(transpose for } t \text{ or } P\text{)}$$

For high-pressure, thick-walled cylinders (see Fig. 9.2):

$$S = \frac{(\text{o.d.})\,P}{2t} \quad \text{(transpose for } t \text{ or } P\text{)}$$

For very high pressure, Lamé's equation applies (see Fig. 9.3):

$$S = P\,\frac{R^2 + r^2}{R^2 - r^2} \qquad P = S\,\frac{R^2 - r^2}{R^2 + r^2}$$

$$R = r\sqrt{\frac{S+P}{S-P}} \qquad r = R\sqrt{\frac{S-P}{S+P}}$$

where $P$ = pressure, $lb/in^2$
$t$ = wall thickness, in
i.d. = inside diameter of cylinder, in
o.d. = outside diameter of cylinder, in
$S$ = stress, $lb/in^2$
$R$ = outside radius, in
$r$ = inside radius, in

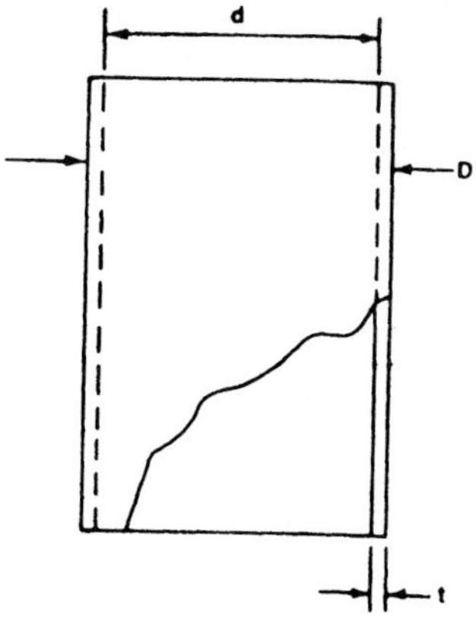


**FIGURE 9.1** Pressure cylinder.

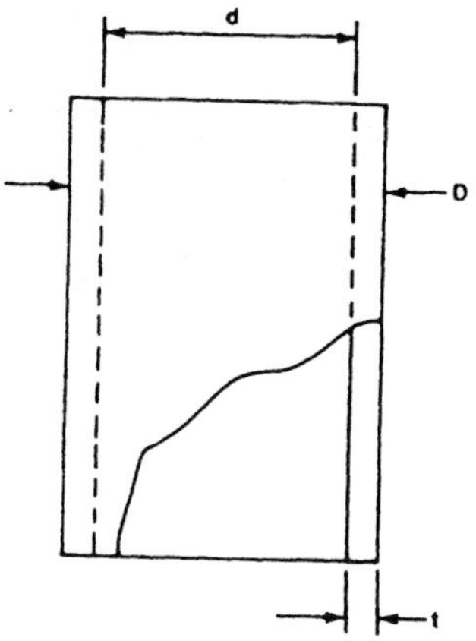


**FIGURE 9.2** High-pressure cylinder.

### 9.1.8 Gas Constants

Figure 9.4 lists the gas constants that are of value in calculating the various requirements encountered in basic pneumatic design procedures, using the preceding pneumatics equations.

## 9.2 BASIC HYDRAULICS

The power required to drive a hydraulic pump is

$$\text{hp} = \frac{PG_{pm}}{1714e}$$

where hp = horsepower requirement
$P$ = pressure of fluid, psi
$G_{pm}$ = gallons per minute to be pumped
$e$ = efficiency (e.g., 0.85 = 85 percent)

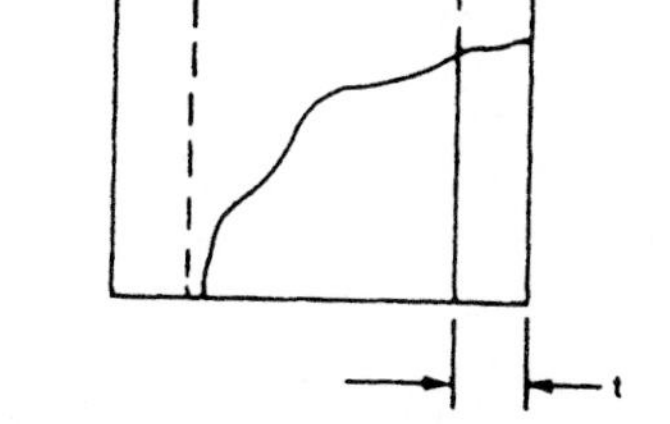

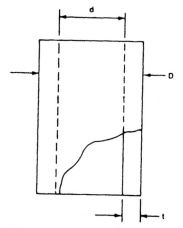

**FIGURE 9.3** Ultra-high-pressure cylinder.

The accuracy for any positive-displacement pump is ±5 percent.

Note regarding gauge and absolute pressures:

- psig = lb/in$^2$ gauge
- psia = lb/in$^2$ absolute
- psia = psig + atmospheric pressure [atmospheric pressure is taken as 14.7 lb/in$^2$ (psi) at sea level; barometric pressure as 29.92 in (760 mm) of mercury]

### 9.2.1 Fluid Flow, Pressure, and Volume Equivalents

For fluid flow equivalents, see Fig. 9.5; for pressure equivalents, see Fig. 9.6; for volume equivalents, see Fig. 9.7.

**Gas Constants - Molecular Weight, Density, Specific Gravity and Weight in Lbs/ft$^3$**

| Name of Gas | Molecular Weight | Density or Specific Gravity * | Weight: lb/ft$^3$ |
|---|---|---|---|
| Air (Normal) | ...... | 1.000 * | 0.08073 |
| Ammonia | ...... | 0.592 * | 0.04779 |
| Argon | 39.948 | 1.784 g/l | 0.11137 |
| Carbon dioxide | 44.010 | 1.977 g/l | 0.12342 |
| Carbon monoxide | ...... | 0.967 * | 0.07807 |
| Chlorine | ...... | 2.423 * | 0.19561 |
| Ethylene | ...... | 0.967 * | 0.07807 |
| Acetylene (Ethyne) | ...... | 0.920 * | 0.07427 |
| Helium | 4.0026 | 0.1785 g/l | 0.01114 |
| Hydrogen | 2.0159 | 0.0899 g/l | 0.00561 |
| Hydrogen Sulfide | 34.080 | 1.539 g/l | 0.09608 |
| Krypton | 83.800 | 3.736 g/l | 0.23323 |
| Natural gas (av.) | ...... | 0.47-0.48 * | 0.038-0.039 |
| Nitrous oxide | ...... | 1.527 * | 0.12327 |
| Nitrogen | ...... | 0.971 * | 0.07838 |
| Oxygen | 31.9988 | 1.429 g/l | 0.08921 |
| Ozone | 47.9982 | 2.144 g/l | 0.13385 |
| Radon | 222.000 | 9.730 g/l | 0.60742 |
| Sulfur dioxide | 64.060 | 2.927 g/l | 0.18273 |
| Sulfur hexafluoride | 146.05 | 6.602 g/l | 0.41215 |
| Xenon | 131.30 | 5.887 g/l | 0.36751 |

Note: * Specific gravity is in respect to air (1.000) at 32°F, 29.92" of mercury.
1 gram per liter (g/l) = 0.062428 lb/ft$^3$. 1 pound of air occupies 12.387 ft$^3$ at 32°F, 29.92" of mercury.

**FIGURE 9.4** Gas constants.

## 9.2.2 Basic Fluid Power Equations

The following simple equations will be of use in calculating some of the basic requirements in hydraulic design applications.

Torque:

$$T = \frac{5252\ \text{hp}}{\text{rpm}}$$

Hydraulic horsepower:

$$\text{hp} = \frac{PG_{pm}}{1714}$$

Velocity of oil flow in pipes:

$$v = \frac{0.3208G_{pm}}{A}$$

Hydraulic cylinder piston speed:

$$S_p = \frac{C_m}{A}$$

Thrust force of cylinder:

$$T_f = A_1 P$$

| gal/hr | gal/min | $ft^2$/hr | $ft^2$/min | L/hr | L/min | $cm^3$/min |
|---|---|---|---|---|---|---|
| 1 | 0.01667 | 0.1337 | $2.228 \times 10^{-3}$ | 3.7848 | 0.06308 | 63.08 |
| 60 | 1 | 8.022 | 0.1337 | 227.1 | 3.7848 | 3,784.8 |
| 7.48 | 0.1247 | 1 | 0.01667 | 28.32 | 0.472 | 472 |
| 448.8 | 7.48 | 60 | 1 | 0.47195 | 28.32 | $28.32 \times 10^{3}$ |
| 0.26418 | $4.403 \times 10^{3}$ | 0.03531 | $5.886 \times 10^{4}$ | 1 | 0.01667 | 16.67 |
| 15.8502 | $264.18 \times 10^{3}$ | 2.11887 | 0.03531 | 60 | 1 | 1,000 |
| $4.403 \times 10^{6}$ | $264.2 \times 10^{6}$ | $0.5886 \times 10^{6}$ | $35.3145 \times 10^{6}$ | $16.67 \times 10^{6}$ | 0.001 | 1 |

**FIGURE 9.5** Fluid power equivalents.

| atmos | bar | kPa/$cm^2$ | psi | in Hg | microns Hg | torr (mmHg) |
|---|---|---|---|---|---|---|
| 1 | 1.01325 | 1.0332 | 14.696 | 29.921 | $760 \times 10^3$ | 760 |
| 0.98692 | 1 | 1.01971 | 14.504 | 29.53 | $750.06 \times 10^3$ | 750.06 |
| 0.96784 | 0.98067 | 1 | 14.223 | 28.959 | $735.56 \times 10^3$ | 735.56 |
| 0.06805 | 0.06895 | 0.07031 | 1 | 2.036 | $51.72 \times 10^3$ | 51.72 |
| 0.03342 | 0.03364 | 0.03453 | 0.49116 | 1 | $2.54 \times 10^3$ | 25.4 |
| $1.3158 \times 10^{-6}$ | $1.3332 \times 10^{-6}$ | $1.3595 \times 10^{-6}$ | $19.337 \times 10^{-6}$ | $39.37 \times 10^{-6}$ | 1 | $1 \times 10^{3}$ |
| $1.3158 \times 10^{-3}$ | $1.3332 \times 10^{-3}$ | $1.3595 \times 10^{-3}$ | $19.337 \times 10^{-3}$ | $39.37 \times 10^{-3}$ | $1 \times 10^3$ | 1 |

**FIGURE 9.6** Pressure equivalents.

| $m^3$ | $ft^3$ | gal | L | qt | $in^3$ | $cm^3$ (ml) |
|---|---|---|---|---|---|---|
| 1 | 35.31 | 264.2 | 1,000 | 1,056.8 | $61.023 \times 10^3$ | $1 \times 10^6$ |
| $28.317 \times 10^{-3}$ | 1 | 7.4805 | 28.317 | 29.92 | 1,728 | $28.317 \times 10^3$ |
| $3.785 \times 10^{-3}$ | 0.1337 | 1 | 3.785 | 4 | 231 | 3,785 |
| $1 \times 10^{-3}$ | 0.03531 | 0.2642 | 1 | 1.057 | 61.023 | 1,000 |
| $9.463 \times 10^{-4}$ | 0.03342 | 0.25 | 0.9463 | 1 | 57.75 | 946.25 |
| $1.639 \times 10^{-5}$ | $5.787 \times 10^{-4}$ | $43.29 \times 10^{-4}$ | 0.01639 | 0.01732 | 1 | 16.387 |
| $1 \times 10^{-6}$ | $35.31 \times 10^{-6}$ | $2.642 \times 10^{-4}$ | $1 \times 10^{-3}$ | $10.568 \times 10^{-4}$ | 0.06102 | 1 |

**FIGURE 9.7** Volume equivalents.

Displacement and torque of a hydraulic motor:

$$T = \frac{DP_d}{24\pi}$$

$$D = \frac{24\pi T}{P_d}$$

where hp = horsepower
rpm = revolutions per minute
$T$ = torque, lb · ft
$P$ = gauge pressure, psig
$G_{pm}$ = oil flow, gal/min
$v$ = oil velocity, ft/sec
$A$ = inside pipe area, $in^2$
$S_p$ = piston travel speed, in/min

$C_m$ = oil flow into cylinder, $in^3/min$
$A_1$ = piston area, $in^2$
$D$ = displacement, $in^3/rev$
$P_d$ = pressure difference across motor, psi

Some approximate equivalents in hydraulics may be summarized as follows:

- For every unit horsepower of drive, the equivalent of 1 gal/min at 1500 psi can be obtained.
- Pump idling requirement is equal to approximately 5 percent of the pump's rated horsepower.
- Hydraulic oil volume is reduced approximately 0.5 percent for every 1000 psi of fluid pressure in the system.

### 9.2.3 Hydraulic Application Equations (Fluid Flow in Channels and Pipes)

The flow of water and other fluids in pipes may be calculated using variations of the Manning equation for open channels by substituting $d/4$ for the hydraulic radius $r$.

Manning's equation for water flow in open channels is given as

$$v = \frac{1.486}{n} r^{0.667} s^{0.500} \quad \text{(for channels)}$$

Manning's equation for water flow in closed pipes (when $d/4$ is substituted for the hydraulic radius $r$) therefore is expressed as

$$v = \frac{0.590}{n} d^{0.667} s^{0.500} \quad \text{(for pipes at full flow)}$$

The energy loss $h$ in pipes due to friction is expressed by

$$h = \frac{64\, lv^2}{R\, d\, 2g}$$

where $h$ = energy loss, ft · lb/lb;
$R$ = Reynolds number. The Reynolds number $R$ is the ratio of inertial and viscous forces in a fluid defined by the following relationship:

$$R = \frac{dv\rho}{\mu} = \frac{dv}{\upsilon}$$

In the preceding three equations and the following equations, the symbols are defined as follows:

$\mu$ = viscosity, slugs/ft · sec or lb · sec/ft$^2$ [also expressed in centipoise (cP); see later section]
$\upsilon = \mu/\rho$ = kinematic velocity, ft$^2$/sec
$\rho$ = density of fluid, slugs/ft$^3$ (also lb/ft$^3$)
$g$ = gravitational acceleration, ft/sec$^2$ (32.16 U.S. customary)
$v$ = velocity (average), ft/sec
$d$ = inside diameter of pipe, ft
$h$ = energy loss due to friction in $l$ feet of pipe, ft · lb/lb
$r$ = hydraulic radius = $a/p$ (where $a$ = area of cross section, ft$^2$; $p$ = wetted perimeter of pipe, ft)
$n$ = coefficient of roughness (see Table 9.1)
$d_1$ = inside diameter of pipe, in

$s = H_1/l$ = slope = drop in hydraulic gradient per foot
$H_1$ = loss of head due to friction
$l$ = length of pipe, ft
$Q = av$ = discharge, sec · ft

(*Note: Laminar* flow occurs when $R$ is less than 2000 and *turbulent* flow occurs when $R$ is greater than 4000.)

The Manning equation for *water flow* in pipes may be conveniently expressed in the following forms:

$$v = \frac{0.590}{n} d^{0.667} s^{0.500} \qquad Q = \frac{0.463}{n} d^{2.667} s^{0.500}$$

$$H_1 = 2.87\, n^2 \frac{lv^2}{d^{1.333}} \qquad d = \left( \frac{2.159\, Qn}{s^{0.500}} \right)^{0.375}$$

$$H_1 = 4.66\, n^2 \frac{lQ^2}{d^{5.333}} \qquad d_1 = \left( \frac{1.630\, Qn}{s^{0.500}} \right)^{0.375}$$

See the preceding symbol definitions.

### *Hydraulic Discharge in Pressurized Pipes*

$$Q = \frac{\pi (p_1 - p_2) d^4}{128\, \mu l}$$

where $p_1$ and $p_2$ = pressure difference at ends of flow. See preceding symbol definitions.

**TABLE 9.1** Values of $n$ (Surface Roughness Coefficients)

| Type of Pipe | Variation | | Use in Design | |
|---|---|---|---|---|
| | From | To | From | To |
| Clean uncoated cast iron | 0.011 | 0.015 | 0.013 | 0.015 |
| Clean coated cast iron | 0.010 | 0.014 | 0.012 | 0.014 |
| Rough cast iron | 0.015 | 0.035 | ...... | ...... |
| Riveted steel | 0.013 | 0.017 | 0.015 | 0.017 |
| Welded | 0.010 | 0.013 | 0.012 | 0.013 |
| Galvanized iron | 0.012 | 0.017 | 0.015 | 0.017 |
| Brass, copper and glass | 0.009 | 0.013 | ...... | ...... |
| Concrete, standard | 0.010 | 0.017 | ...... | ...... |
| Concrete with rough joints | ...... | ...... | 0.016 | 0.017 |
| Concrete, very smooth | ...... | ...... | 0.011 | 0.012 |
| Vitrified sewage | 0.010 | 0.017 | 0.013 | 0.015 |
| Clay drainage | 0.011 | 0.017 | 0.012 | 0.014 |

The average velocity of the flow is then

$$v = \frac{(p_1 - p_2)d^2}{32\,\mu l}$$

The relationship between pressure loss due to friction and pressure is therefore

$$\frac{(p_1 - p_2)}{\mu} = h = \frac{32\mu l v}{d^2 \rho g}$$

and introducing the Reynolds number $R$ yields

$$h = \frac{64\, l v^2}{R\, d 2g}$$

where $h$ = energy loss in $l$ feet of pipe, ft · lb/lb.

Obviously, if the rate of discharge of the water or fluid is known from calculation, and we know the pipe size and *head,* we can easily calculate the horsepower requirement for the pump motor for a particular application.

The Reynolds number $R$ may also be calculated from the following expression, where the viscosity of the fluid is given in centipoise (cP):

$$R = \frac{3160\, Q\, G_t}{D\,\mu}$$

where $R$ = Reynolds number (dimensionless)
$Q$ = flow rate of liquid, g/min
$G_t$ = specific gravity of liquid
$D$ = inside diameter of pipe, in
$\mu$ = viscosity of liquid, cP

### 9.2.4 Hydraulic and Air-Line Sizes and Thread Connections

Figure 9.8 shows the common hydraulic and air-line sizes available and the pipe thread sizes in common use for hydraulic and pneumatic applications.

## *9.3 AIR-HANDLING DATA*

The Air Movement and Control Association (AMCA) establishes standards, tests and certifies the performance of air-moving devices and equipment in the United States, and is a nationally recognized authority. The basics of air-moving or air-handling practice are detailed to a limited extent in this section. Air movement in shops and factory areas is an important aspect in manufacturing procedures, since it may materially affect manufacturing efficiencies.

### 9.3.1 Basic Laws for Air-Moving Equipment

***Air-Moving Terminology***

**dB** Decibel. A measure of the sound intensity produced by air-moving equipment.

**dB(A)** Sound-level reading on the A-weighted scale of a sound meter. The A weighting adjusts the response of the meter to that of the human ear.

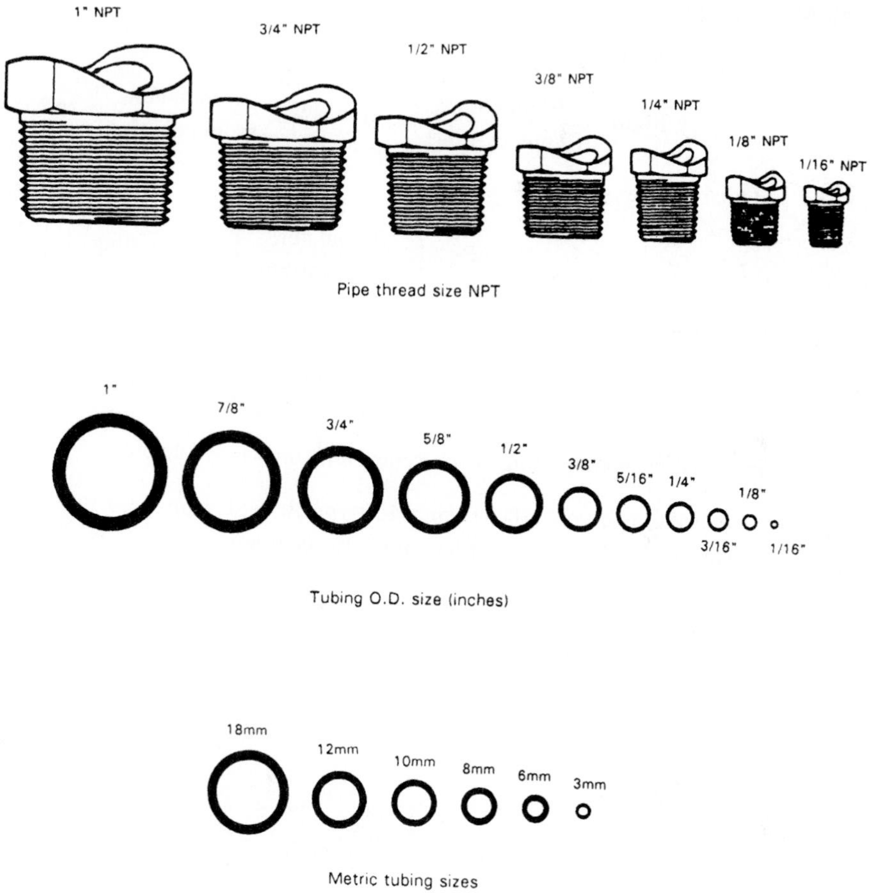


**FIGURE 9.8** Hydraulic and air-line sizes and threaded connections.

**ft³/min** Cubic feet per minute (also expressed CFM). A measure of volume flow rate or air-moving capability of an air-moving device. Volume of air moved past a fixed point per minute.

**Sone** An internationally recognized unit of sound intensity (loudness). One sone is equivalent to the sound made by an average refrigerator in a kitchen. A device which is rated at 6 sones sounds twice as loud as one operating at a level of 3 sones. Sones are thus a linear quantity, unlike the decibel, which is rated on a logarithmic scale.

**Sound power level or sound pressure level** The acoustic power radiating from a sound source, expressed in decibels (dB).

**SP** Static pressure. A measure of the resistance to movement of forced air through a system or installation caused by ductwork, inlets, louvers, etc. SP is measured in inches of water gauge (WG) or water column: the height in inches to which the pressure will lift a column of water. For any given system, the static pressure varies as the square of the flow rate.

### 9.3.2 Air-Moving Equations

Performance ratio:

$$\text{Ratio} = \frac{\text{ft}^3\text{/min new}}{\text{ft}^3\text{/min existing}} \qquad \text{rpm new} = \frac{\text{ft}^3\text{/min new}}{\text{ft}^3\text{/min existing}} \times \text{rpm existing}$$

Static pressure:

$$\text{SP new} = \left(\frac{\text{ft}^3\text{/min new}}{\text{ft}^3\text{/min existing}}\right)^2 \times \text{SP existing}$$

Horsepower:

$$\text{hp new} = \left(\frac{\text{ft}^3\text{/min new}}{\text{ft}^3\text{/min existing}}\right)^3 \times \text{hp existing}$$

A typical calculation using the preceding equations is as follows:

Existing conditions: 7000 $\text{ft}^3$/min; 1500 rpm; 0.75 SP, 1-hp motor.
Desired result: We now wish to move 10,000 $\text{ft}^3$/min.

$$\text{Ratio} = \frac{10{,}000}{7{,}000} = 1.429$$

Then the new rpm is

$$\text{rpm} = 1.429 \times 1500 = 2144 \text{ rpm}$$

Then the new SP is

$$\text{SP new} = \left(\frac{10{,}000}{7{,}000}\right)^2 \times 0.75 = 2.042 \times 0.75 = 1.53 \text{ SP}$$

Then the new horsepower required is

$$\text{hp new} = \left(\frac{10{,}000}{7{,}000}\right)^3 \times 1.0 = (1.429)^3 \times 1.0 = 2.92 \text{ or } 3 \text{ hp}$$

The preceding is an illustration in the use of air-moving equations.

***Air-Change Recommendations for Various Industrial Areas.*** Figure 9.9 lists the industry recommended air changes for various types of industrial work areas. To determine the required fan capacity to meet these recommendations, calculate the volume of the area to be ventilated in cubic feet and divide by the rate of air change shown in Fig. 9.9. The result will be the required $\text{ft}^3$/min, corresponding to fan ratings.

*Example.* Your foundry area contains 800,000 $\text{ft}^3$. The recommended air-change rate is 2 to 8 min per air change (take 5 as average). Then

$$\text{ft}^3\text{/min required} = \frac{800{,}000}{5} = 160{,}000 \text{ ft}^3\text{/min}$$

Check the fan ratings in $\text{ft}^3$/min in the fan catalogs and select one or more fans to do the job.

**Air Changes Recommended for Various Industrial Areas - Minutes per change**

| Ventilated Area | Minutes per Change |
|---|---|
| Assembly halls | 5-10 |
| Boiler rooms | 2-3 |
| Engine rooms | 3-5 |
| Factories | 5-10 |
| Foundries | 4-8 |
| Mills | 6-8 |
| Offices | 6-12 (smoking areas require higher range > 8) |
| Electroplating rooms | 3-5 |
| Rest rooms | 6-10 |
| Toilets | 4-6 |
| Transformer rooms | 2-5 |
| Warehouses | 5-10 |
| Painting facilities (spray) | 5-10 or more |

**FIGURE 9.9** Air-change requirements and recommendations.

## 9.4 TRANSMISSION OF HEAT

The total heat transmission through containment or building surfaces can be calculated from

$$H = AU\,(t_o - t_i)$$

where $H$ = sensible heat, Btu/h
$A$ = area of transmitting surface, ft$^2$
$U$ = coefficient of heat transmission of the transmitting surface
$t_o$ = outside temperature, °F
$t_i$ = inside temperature, °F

This equation can be used for calculating heat loss from a space or heat gain to a space. Coefficients of heat transmission and heat loss from personnel are shown in Fig. 9.10*a* and 9.10*b*.

Figure 9.11 shows heat transmission diagrammatically with respect to the preceding equation.

### 9.4.1 Ventilation

Infiltration air or air-change requirements were shown in Fig. 9.9.

***Sensible Heat of Ventilation or Infiltration.*** This quantity may be estimated from

$$H_s = 1.08\,V\,(t_o - t_i)$$

| Source | U Value |
|---|---|
| Single window | 1.13 |
| Double window | 0.50 |
| 12-in brick wall | 0.35 |
| 10-in concrete wall | 0.61 |
| 4-in concrete slab roof | 0.70 |
| Wood siding, ¾-in drywall w/3-in insulation | 0.07 |

(a)

| Design Condition: 1 man | Total Heat Loss, B.t.u/hr |
|---|---|
| At rest | 400 |
| Light work | 600 |
| Moderate work | 800 |

(b)

**FIGURE 9.10** (*a*) Coefficients of heat transmission; (*b*) heat loss from personnel.

where $H_s$ = sensible heat, Btu/h
$V$ = ft³/min of ventilation or infiltration air
$t_o$ = outside temperature, °F
$t_i$ = inside temperature, °F

Only infiltration air contributes to the internal heat load. Ventilation air, conditioned before entering the space, adds to the load of the conditioning or cooling equipment only.

***Air Quantity.*** The air quantity required to absorb a specific sensible-heat quantity is determined by

$$V = \frac{H_s}{60}\left(\frac{56}{t_a - t_d}\right)$$

where $V$ = quantity of air, ft³/min
$H_s$ = sensible heat, Btu/h
$t_a$ = space temperature, °F
$t_d$ = delivery temperature of air into space, °F

(*Note:* 1 Btu will raise 4.16 lb of air 1°F, or 56 ft³ of air 1°F.)

***Refrigerant Pressure versus Temperature.*** See Table 9.2 for air-conditioning refrigerant data.

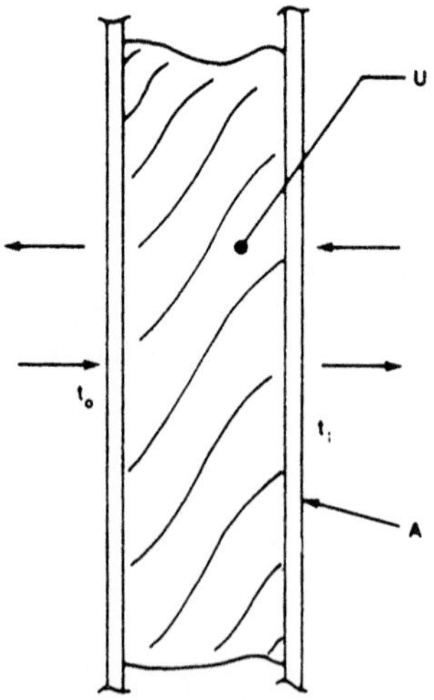


**FIGURE 9.11** Schematic of heat transmission.

## 9.5 SPECIFIC HEAT (THERMAL CAPACITY)

*Specific heat* is defined as the ratio of the heat required to raise the temperature of a unit weight of a substance 1° to the heat required to raise the temperature of a unit weight of pure water 1° at a specified temperature.

The units of measurement are calories per gram per °C in SI metric, and Btu per pound per °F in the U.S. customary system.

$$\text{Btu or calories} = Wct$$

**TABLE 9.2** Refrigerant Pressure versus Temperature-Pounds per Square Inch and Degrees Farenheit

| Temperature (degrees F) | R-12 | R-22 | Methyl Chloride R-40 | Ammonia R-717 |
|---|---|---|---|---|
| - 25 | 2.3● | 7.5 | 8.9● | 1.3 |
| - 20 | 0.6 | 10.3 | 6.1● | 3.6 |
| - 15 | 2.5 | 13.3 | 3.0● | 6.1 |
| - 10 | 4.5 | 16.6 | 0.3 | 9.0 |
| - 5 | 6.7 | 20.3 | 2.1 | 12.2 |
| 0 | 9.2 | 24.2 | 4.2 | 15.7 |
| 5 | 11.8 | 28.4 | 6.5 | 19.6 |
| 10 | 14.7 | 33.0 | 8.9 | 23.8 |
| 15 | 17.7 | 37.9 | 11.6 | 28.5 |
| 20 | 21.1 | 43.3 | 14.5 | 33.5 |
| 25 | 24.6 | 49.0 | 17.5 | 39.0 |
| 30 | 28.5 | 55.3 | 21.0 | 45.0 |
| 35 | 32.6 | 61.7 | 24.6 | 52.6 |
| 40 | 37.0 | 69.0 | 28.6 | 58.6 |
| 45 | 41.7 | 76.5 | 32.8 | 66.3 |
| 50 | 46.7 | 84.7 | 37.3 | 74.5 |
| 55 | 52.0 | 93.5 | 42.1 | 83.4 |
| 60 | 57.7 | 102.5 | 47.3 | 92.9 |
| 65 | 63.7 | 112.0 | 52.8 | 103.1 |
| 70 | 70.1 | 122.5 | 58.7 | 114.1 |
| 75 | 76.9 | 133.0 | 65.0 | 125.8 |
| 80 | 84.1 | 145.0 | 71.6 | 138.3 |
| 85 | 91.7 | 156.7 | 78.5 | 151.7 |
| 90 | 99.6 | 170.1 | 86.0 | 165.9 |
| 95 | 108.1 | 183.4 | 93.8 | 181.1 |
| 100 | 116.9 | 197.9 | 102.0 | 197.2 |
| 105 | 126.2 | 212.5 | 110.7 | 214.2 |
| 110 | 136.0 | 228.7 | 119.8 | 232.3 |
| 115 | 146.3 | 245.0 | 129.5 | 251.5 |
| 120 | 157.1 | 262.6 | 139.5 | 271.7 |
| 125 | 169.1 | 280.0 | ...... | 293.1 |
| 130 | 178.6 | 298.8 | ...... | ...... |
| 135 | 193.5 | 317.8 | ...... | ...... |
| 140 | 206.6 | 338.0 | ...... | ...... |

Note: Numbers marked (●) are inches vacuum. All other figures are gage pressure in psi.

where $W$ = weight of substance, lb or g
$c$ = specific heat of substance
$t$ = temperature rise, °F or °C

The specific heats for common materials are shown in Fig. 9.12.

### 9.5.1 Properties of Liquids

See Table 9.3.

## 9.6 TEMPERATURES OF MIXTURES AND HEAT REQUIREMENTS

When two volumes of liquids at different temperatures are mixed together, the final temperature of the mixture may be calculated from

$$T_m = \frac{cwt + c_1w_1t_1}{cw + c_1w_1}$$

| Substance | Specific Heat |
|---|---|
| Alcohol, | |
| Ethyl | 0.548 |
| Methyl | 0.601 |
| Aluminum | 0.214 |
| Benzene | 0.400 |
| Brass | 0.094 |
| Chromic acid (25%) | 0.825 |
| Copper | 0.094 |
| Ferric chloride (50%) | 0.750 |
| Glass, common | 0.199 |
| Graphite | 0.201 |
| Iron, cast | 0.130 |
| Iron, wrought | 0.110 |
| Lead | 0.031 |
| Mercury | 0.033 |
| Nickel | 0.109 |
| Platinum | 0.032 |
| Silver | 0.056 |
| Solder (60 Pb, 40 Sn) | 0.047 |
| Steel, low carbon | 0.116 |
| Sulfuric acid (50%) | 0.915 |
| Tin | 0.056 |
| Water (30°C) | 0.997 |
| Zinc | 0.095 |

Values apply between 32 to 212°F, except as noted.

**FIGURE 9.12** Specific heat of common materials.

where $w$, $w_1$ = weight of the two substances, lb or g
$t$, $t_1$ = temperatures of the two substances, °F or °C
$c$, $c_1$ = specific heat of the two substances
$T_m$ = final temperature of the mixture, °F or °C

## 9.6.1 Heat Required to Raise Temperature of Materials

You may calculate the amount of heat required to raise the temperature of a material from an initial starting temperature from the following equation:

$$\text{Btu's or calories} = ms\,(t_2 - t_1)$$

where $m$ = mass of the substance, lb or g
$s$ = specific heat of the substance
$t_2$ = final temperature of the substance, °F or °C
$t_1$ = initial temperature of the substance, °F or °C

## 9.6.2 Heat Equivalents

| | |
|---|---|
| 1 hp | 2546 Btu/h |
| 1 hp | 42.5 Btu/min |
| 1 ft · lb | 0.00129 Btu |
| 1 Btu | 778 ft · lb |
| 1 hp · h | 2546 Btu |
| 1 food cal | 1000 g · cal or 1 kg · cal |
| 1 g · cal | 0.00397 Btu |
| 1 g · cal | 41,855,000 erg (obsolete) |

**TABLE 9.3** Heat Properties of Common Liquids

| Liquid | Density Lbs/Ft³ | Average Specific Heat Btu.Lb.°F | Thermal Conductivity Btu.in/Hr.Ft³.°F | Heat of Vaporization Btu/Lb |
|---|---|---|---|---|
| Acetic acid 100% | 65.4 | 0.48 | 1.14 | 175 |
| Ammonia 100% | 47.9 | 1.10 | 3.48 | 589 |
| Amyl alcohol | 55.0 | 0.65 | ...... | 216 |
| Brine-sodium chloride | 74.1 | 0.786 | 2.88 | 730 |
| Butyl alcohol | 45.3 | 0.687 | ...... | 254 |
| Carbon tetrachloride | 98.5 | 0.210 | ...... | ...... |
| Ethyl alcohol | 50.4 | 0.600 | 1.30 | 370 |
| Ethyl chloride | 57.0 | 0.367 | ...... | 167 |
| Ethylene glycol | 70.1 | 0.555 | ...... | ...... |
| Formic acid | 69.2 | 0.525 | ...... | 216 |
| Freon 11 | 92.1 | 0.208 | 0.60 | ...... |
| Freon 12 | 81.8 | 0.232 | 0.492 | 62 |
| Freon 22 | 74.53 | 0.300 | 0.624 | ...... |
| Glycerine | 78.7 | 0.580 | 1.97 | ...... |
| Heptane | 38.2 | 0.490 | ...... | 137.1 |
| Hydrochloric acid 10% | 66.5 | 0.930 | ...... | ...... |
| Ice | 56.0 | 0.500 | 3.96 | ...... |
| Mercury | 845.0 | 0.033 | 59.64 | 117 |
| Methyl acetate | 54.8 | 0.470 | ...... | 176.5 |
| Methylene chloride | 82.6 | 0.288 | ...... | 142 |
| Nitric acid 7% | 64.7 | 0.920 | ...... | 918 |
| Nitric acid 95% | 93.5 | 0.500 | ...... | 207 |
| Gasoline | 41-43 | 0.530 | 0.936 | 116 |
| Toluene | 53.7 | 0.420 | 1.032 | ...... |
| Transformer oils | 56.3 | 0.420 | 0.900 | ...... |
| Phosphoric acid 10% | 65.4 | 0.930 | ...... | ...... |
| Propionic acid | 61.8 | 0.560 | ...... | 177.8 |
| Proply alcohol | 50.2 | 0.570 | ...... | 295.2 |
| Sea water | 64.2 | 0.940 | ...... | ...... |
| Sodium hydroxide 30% | 82.9 | 0.840 | ...... | ...... |
| Sulfuric acid 20% | 71.0 | 0.840 | ...... | ...... |
| Sulfuric acid 60% | 93.5 | 0.520 | 2.880 | 219 |
| Trichloroethylene | 91.3 | 0.230 | 0.840 | 103 |
| Vegetable oil | 57.5 | 0.430 | ...... | ...... |
| Water | 62.4 | 1.000 | 4.08 | 965 |
| Xylene | 53.8 | 0.411 | ...... | 149.2 |

## 9.7 HEAT LOSSES

System heat losses occur in three ways: conduction, convection, and radiation. Problems involving the addition or subtraction of heat from a system may be calculated for conduction, convection, and radiation using the following equations:

Conduction:

$$Q_{L1}, \text{Btu} = \frac{KA(T_2 - T_1)t_e}{L}$$

$$Q_{L1}, \text{Wh} = \frac{KA(T_2 - T_1)t_e}{3.412\ L}$$

Convection:

$$Q_{L2}, \text{Btu} = 3.412\ A_1 F_{vs} C_F t_e$$

$$Q_{L2}, \text{Wh} = A_1 F_{vs} C_F t_e$$

Radiation:

$$Q_{L3}, \text{Btu} = 3.412AF_{bb}et_e$$

$$Q_{L3}, \text{Wh} = AF_{bb}et_e$$

where $Q_{L1}$ = conduction heat losses, Btu or Wh
$K$ = thermal conductivity, Btu in/ft$^2 \cdot$ °F $\cdot$ h
$A$ = heat-transfer surface area, ft$^2$
$L$ = thickness of material, in
$T_2$, $T_1$ = temperatures across material ($T_2$ is the higher temperature)
$t_e$ = exposure time, h
$Q_{L2}$ = surface heat losses, Btu or Wh
$A_1$ = surface area, in$^2$

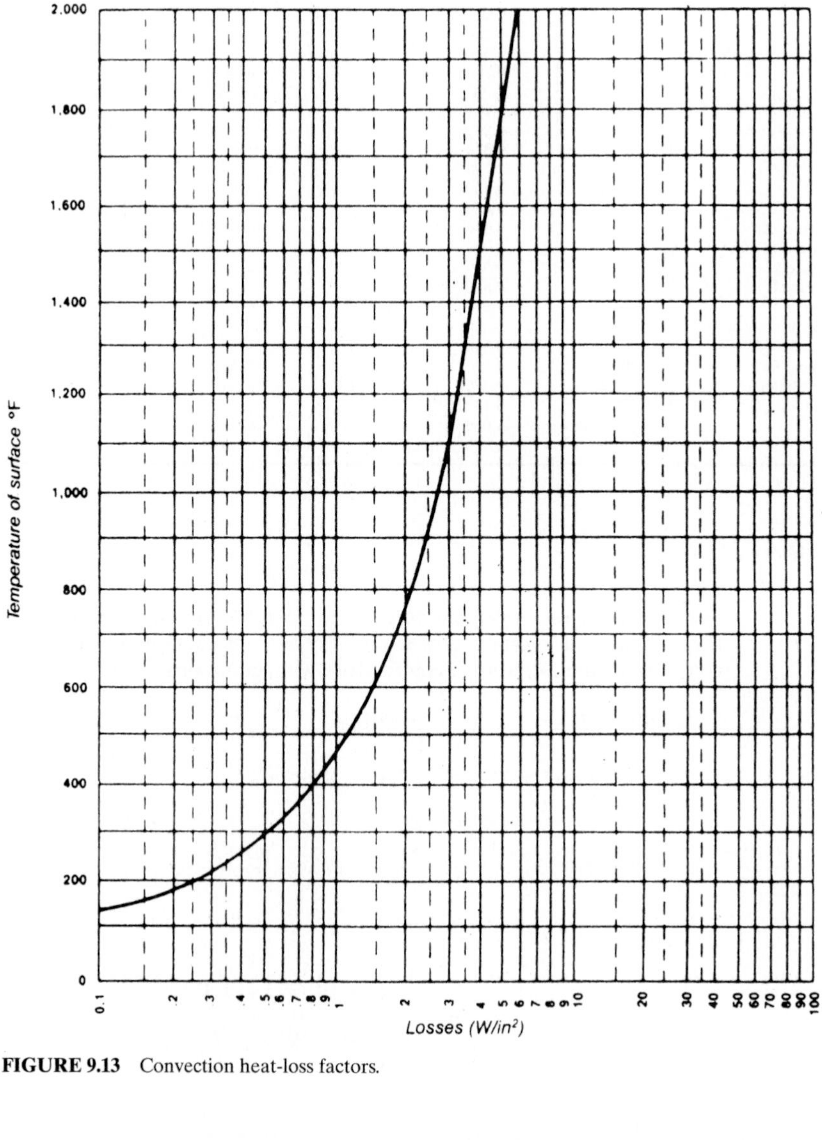


**FIGURE 9.13** Convection heat-loss factors.

$F_{vs}$ = vertical surface convection loss factor (see Fig. 9.13)
$C_F$ = surface orientation factor (whose typical values are horizontal top, 1.29; vertical, 1.00; horizontal bottom, 0.63)
$Q_{L3}$ = radiation heat losses, Btu or Wh
$F_{bb}$ = blackbody radiation-loss factor, W/in$^2$ (see Fig. 9.14)
$e$ = emmisivity correction factor of the material (also termed *emmisivity coefficient;* see Fig. 9.15)

*Example.* Calculate the radiation heat losses (Fig. 9.14). At 600°F, blackbody losses are 4 W/in$^2$. Copper, with a medium oxide surface ($e = 0.40$) would thus radiate $4 \times 0.04 = 0.16$ W/in$^2$ of surface area at 600°F.

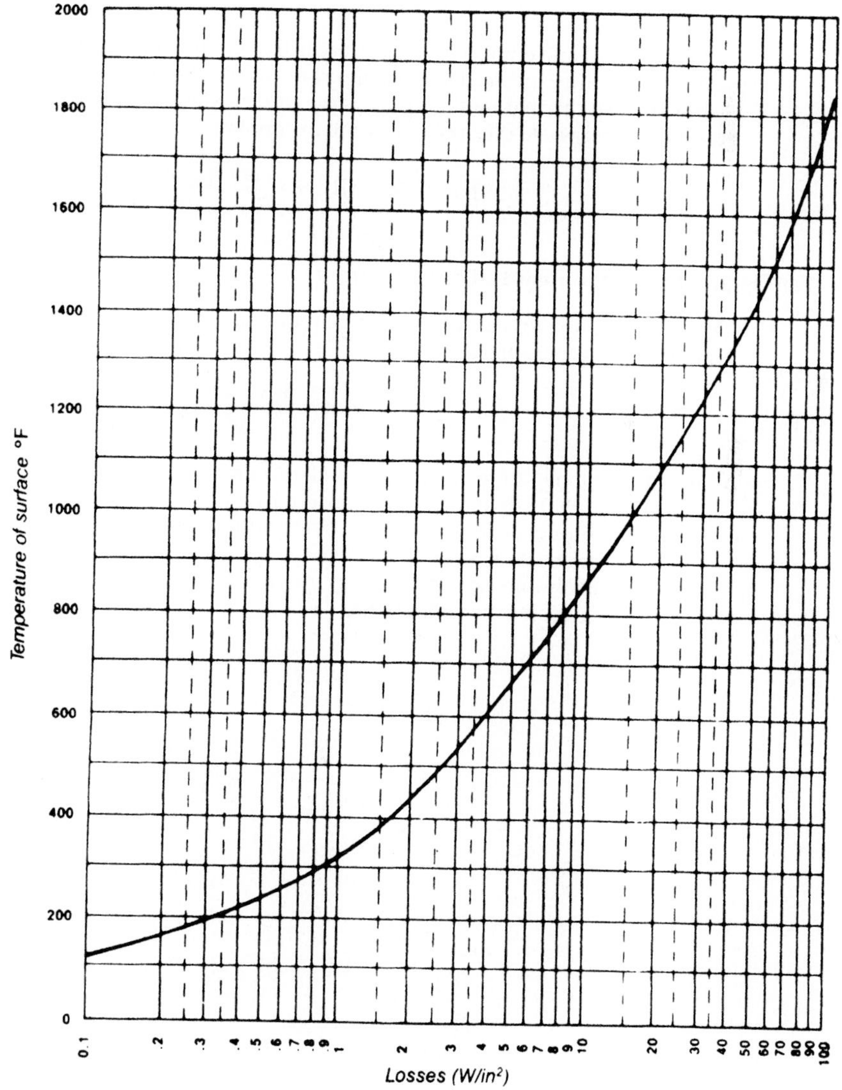


**FIGURE 9.14** Blackbody radiation heat-loss factors.

| Material | Specific Heat B.t.u/lb-°F | Emissivity Polished | Emissivity Med. Oxide | Emissivity Heavy Oxide |
|---|---|---|---|---|
| Aluminum | 0.24 | 0.09 | 0.11 | 0.22 |
| Copper | 0.10 | 0.04 | 0.40 | 0.65 |
| Cast iron | 0.12 | — | 0.80 | 0.85 |
| Steel | 0.12 | 0.10 | 0.75 | 0.85 |
| S/S 304 and 430 | 0.11 | 0.17 | 0.57 | 0.85 |
| Carbon | 0.20 | | | |
| Glass | 0.20 | (0.90 for most nonmetallics) | | |
| Plastic | 0.2 to 0.5 | | | |
| Rubber | 0.40 | | | |

**FIGURE 9.15** Specific heats and emissivity correction factors.

Specific heat and emmisivity values are listed for common materials in Fig. 9.15. Note that heater ratings are expressed in units of power (watts). Energy is expressed in watthours (Wh) or Btu (British thermal units). *Power* is the rate at which energy is being used (Btu/h or W). Power and energy are thus related by the expression

$$\text{Power, W or Btu/h} = \frac{\text{energy, Wh or Btu}}{\text{time, h}}$$

### 9.7.1 Fan Requirements for Enclosed Equipment

The basic equation for calculating the fan requirement in ft³/min for cooling an enclosure containing a heat-generating source is

$$\text{cfm} = \frac{3160\,(kW_e)}{\Delta T}$$

where cfm = rating of fan or fans required, $\text{ft}^3/\text{min}$

$kW_e$ = heat dissipated in the enclosure by a power source, kW

$\Delta T$ = difference in temperature of exhaust air from temperature of incoming or ambient air = $\Delta T = (t_f - t_a)$

$t_f$ = allowable exit air temperature, °F

$t_a$ = incoming air temperature, °F

# CHAPTER 10

# FASTENING AND JOINING TECHNIQUES AND DESIGN DATA

## 10.1 THREADED FASTENERS: BOLTS, SCREWS, AND NUTS

Bolts, screws, and nuts are the most commonly used types of fastening devices. The thread on a bolt or screw may be compared to an inclined plane wrapped around a cylinder, thus assuming the form of one of the basic machines. Many different thread form standards are used worldwide, but this section will present data pertaining only to the forms in common use such as the 60° V-thread form for the U.S. customary (inch) system and the metric (ISO) standard. The other important thread forms used worldwide and their geometry are shown in Sec. 10.2.

Threaded fasteners have countless applications in industry, and in many cases their performance is critical. Failure of a threaded fastener such as a heavily loaded bolt or bolts can cause severe injury, property damage, and death in critical applications. It is for these reasons that standards of performance have been devised and specified in the various American standards such as ANSI, ASTM, SAE, and Industrial Fasteners Institute (IFI). In recent years, it has been nationally discovered that counterfeit bolts are being imported into the United States. False head markings and inferior steels have been found, which have contributed to many failures in critical applications. Figure 10.1 shows a typical counterfeit bolt that failed during the installation process. This is a direct indication of improperly processed material or poor-quality material. Other cases have been cited where the failure was through the threaded section and also cracking due to improper heat treatment combined with faulty material. The author has personally seen at least three imported-bolt failures under normal operation within a time span of only one year.

Not only are the "counterfeit" fasteners causing problems, but improper material specifications have also been noted. The author recently had a sample of stainless steel analyzed that cracked during the welding process. The material was specified as type AISI 304 stainless steel, but when analyzed at a materials test laboratory, it did not conform to *any* known American stainless-steel series and also contained 10 times the maximum amount of sulfur allowed in the ASTM and SAE specifications for type 304 stainless steel. Defective materials are the *supplier's* responsibility.

As a precautionary procedure, hardware and materials that are used in critical applications should be certified and, if necessary, analyzed to ensure conformity to SAE, AISI, ASTM, or other applicable American standards. This will help avoid the problems associated with fastener and materials failures, and protect the American manufacturer against the possibility of lawsuits. See Chap. 15 concerning product liability.

Government legislation is in progress to control and, hopefully, eliminate the problems discussed above. Testing for conformance is forthcoming.

**FIGURE 10.1** Failure of the head of a counterfeit bolt (0.500–13) at installation.

### 10.1.1 Dimensions of Bolts, Screws, Nuts, and Washers

Basic design dimensions for American standard bolts most commonly used are shown in Fig. 10.2.

Basic design dimensions for American standard machine screws most commonly used are shown in Fig. 10.3.

The basic design dimensions for American standard miniature screws are shown in Fig. 10.4.

The dimensions shown in the figures are the most important ones needed for normal design purposes. The complete dimensional specifications and applicable tolerances for all

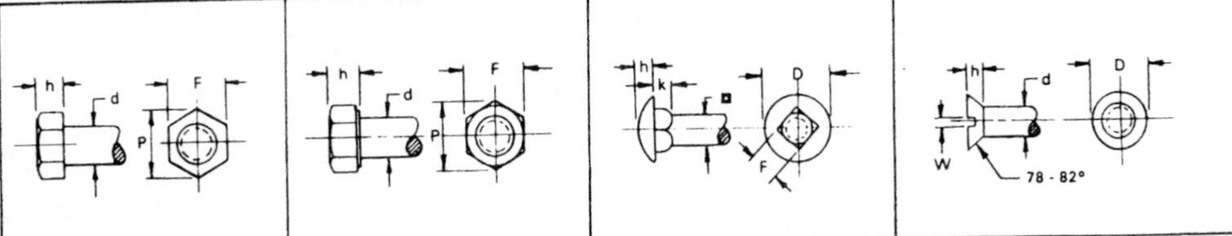


| | Hex Bolts | | | | Hex Cap Screw (Finished Hex Bolts) | | | | Round Head Square Neck Bolts | | | | Countersunk Bolts | | | |
|---|---|---|---|---|---|---|---|---|---|---|---|---|---|---|---|---|
| Nominal size | | | | | | | | | | | | | | | | |
| Basic Dia. | d | F | P | h | d | F | P | h | D | F | h | k | d | D | h | W ♦ |
| 0.2500 - 1/4 | 0.260 | 0.438 | 0.505 | 0.188 | 0.2500 | 0.438 | 0.505 | 0.163 | 0.594 | 0.260 | 0.145 | 0.156 | 0.260 | 0.493 | 0.150 | 0.064 |
| 0.3125 - 5/16 | 0.324 | 0.500 | 0.577 | 0.235 | 0.3125 | 0.500 | 0.577 | 0.211 | 0.719 | 0.324 | 0.176 | 0.187 | 0.324 | 0.618 | 0.189 | 0.072 |
| 0.3750 - 3/8 | 0.388 | 0.562 | 0.650 | 0.268 | 0.3750 | 0.562 | 0.650 | 0.243 | 0.844 | 0.388 | 0.208 | 0.219 | 0.388 | 0.740 | 0.225 | 0.081 |
| 0.4375 - 7/16 | 0.452 | 0.625 | 0.722 | 0.316 | 0.4375 | 0.625 | 0.722 | 0.291 | 0.969 | 0.452 | 0.239 | 0.250 | 0.452 | 0.803 | 0.226 | 0.081 |
| 0.5000 - 1/2 | 0.515 | 0.750 | 0.866 | 0.364 | 0.5000 | 0.750 | 0.866 | 0.323 | 1.094 | 0.515 | 0.270 | 0.281 | 0.515 | 0.935 | 0.269 | 0.091 |
| 0.5625 - 9/16 | ----- | ----- | ----- | ----- | 0.5625 | 0.812 | 0.938 | 0.371 | ----- | ----- | ----- | ----- | ----- | ----- | | |
| 0.6250 - 5/8 | 0.642 | 0.938 | 1.083 | 0.444 | 0.6250 | 0.938 | 1.083 | 0.403 | 1.344 | 0.642 | 0.344 | 0.344 | 0.642 | 1.169 | 0.336 | 0.116 |
| 0.7500 - 3/4 | 0.768 | 1.125 | 1.299 | 0.524 | 0.7500 | 1.125 | 1.299 | 0.483 | 1.594 | 0.768 | 0.406 | 0.406 | 0.768 | 1.402 | 0.403 | 0.131 |
| 0.8750 - 7/8 | 0.895 | 1.312 | 1.516 | 0.604 | 0.8750 | 1.312 | 1.516 | 0.563 | 1.844 | 0.895 | 0.469 | 0.469 | 0.895 | 1.637 | 0.470 | 0.147 |
| 1.0000 - 1 | 1.022 | 1.500 | 1.732 | 0.700 | 1.0000 | 1.500 | 1.732 | 0.627 | 2.094 | 1.022 | 0.531 | 0.531 | 1.022 | 1.869 | 0.537 | 0.166 |
| 1.1250 - 1-1/8 | 1.149 | 1.688 | 1.949 | 0.780 | 1.1250 | 1.688 | 1.949 | 0.718 | ----- | ----- | ----- | ----- | 1.149 | 2.104 | 0.604 | 0.178 |
| 1.2500 - 1-1/4 | 1.277 | 1.875 | 2.165 | 0.876 | 1.2500 | 1.875 | 2.165 | 0.813 | ----- | ----- | ----- | ----- | 1.277 | 2.337 | 0.671 | 0.193 |
| 1.3750 - 1-3/8 | 1.404 | 2.062 | 2.382 | 0.940 | 1.3750 | 2.062 | 2.382 | 0.878 | ----- | ----- | ----- | ----- | 1.404 | 2.571 | 0.738 | 0.208 |
| 1.5000 - 1-1/2 | 1.531 | 2.250 | 2.598 | 1.036 | 1.5000 | 2.250 | 2.598 | 0.974 | ----- | ----- | ----- | ----- | 1.531 | 2.804 | 0.805 | 0.240 |
| 1.7500 - 1-3/4 | 1.785 | 2.625 | 3.031 | 1.196 | 1.7500 | 2.625 | 3.031 | 1.134 | ----- | ----- | ----- | ----- | ----- | ----- | ----- | ----- |
| 2.0000 - 2 | 2.039 | 3.000 | 3.464 | 1.388 | 2.0000 | 3.000 | 3.464 | 1.263 | ----- | ----- | ----- | ----- | ----- | ----- | ----- | ----- |

Note: ♦ Minimum dimensions of slot widths. All other tabulated values are maximum dimensions. ◘ diameters same as hex bolts (d) of same basic size.

**FIGURE 10.2** Basic dimensions for American standard bolts.

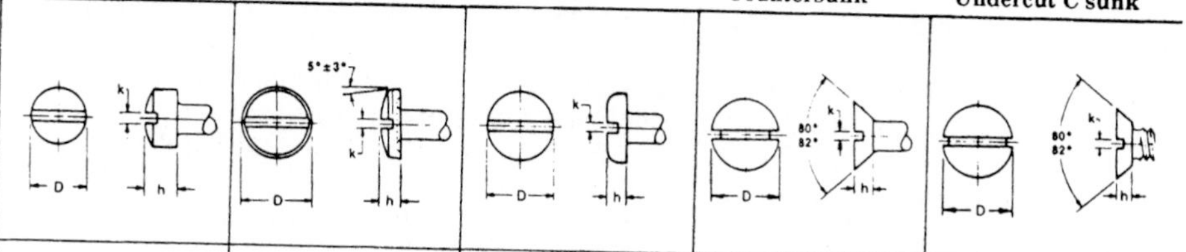


| Nominal Size | | Fillister Head | | | Binding Head | | | Pan Head | | | Countersunk | | | Undercut C'sunk | | |
|---|---|---|---|---|---|---|---|---|---|---|---|---|---|---|---|---|
| | Basic Dia. | D | h | k♦ | D | h | k♦ | D | h | k♦ | D | h | k♦ | D | h | k♦ |
| 0 - | 0.0600 | 0.096 | 0.055 | 0.016 | 0.126 | 0.032 | 0.016 | 0.116 | 0.039 | 0.016 | 0.119 | 0.035 | 0.016 | 0.119 | 0.025 | 0.016 |
| 1 - | 0.0730 | 0.118 | 0.066 | 0.019 | 0.153 | 0.041 | 0.019 | 0.142 | 0.046 | 0.019 | 0.146 | 0.043 | 0.019 | 0.146 | 0.031 | 0.019 |
| 2 - | 0.0860 | 0.140 | 0.083 | 0.023 | 0.181 | 0.050 | 0.023 | 0.167 | 0.053 | 0.023 | 0.172 | 0.051 | 0.023 | 0.172 | 0.036 | 0.023 |
| 3 - | 0.0990 | 0.161 | 0.095 | 0.027 | 0.208 | 0.059 | 0.027 | 0.193 | 0.060 | 0.027 | 0.199 | 0.059 | 0.027 | 0.199 | 0.042 | 0.027 |
| 4 - | 0.1120 | 0.183 | 0.107 | 0.031 | 0.235 | 0.068 | 0.031 | 0.219 | 0.068 | 0.031 | 0.225 | 0.067 | 0.031 | 0.225 | 0.047 | 0.031 |
| 5 - | 0.1250 | 0.205 | 0.120 | 0.035 | 0.263 | 0.078 | 0.035 | 0.245 | 0.075 | 0.035 | 0.252 | 0.075 | 0.035 | 0.252 | 0.053 | 0.035 |
| 6 - | 0.1380 | 0.226 | 0.132 | 0.039 | 0.290 | 0.087 | 0.039 | 0.270 | 0.082 | 0.039 | 0.279 | 0.083 | 0.039 | 0.279 | 0.059 | 0.039 |
| 8 - | 0.1640 | 0.270 | 0.156 | 0.045 | 0.344 | 0.105 | 0.045 | 0.322 | 0.096 | 0.045 | 0.332 | 0.100 | 0.045 | 0.332 | 0.070 | 0.045 |
| 10 - | 0.1900 | 0.313 | 0.180 | 0.050 | 0.399 | 0.123 | 0.050 | 0.373 | 0.110 | 0.050 | 0.385 | 0.116 | 0.050 | 0.385 | 0.081 | 0.050 |
| 12 - | 0.2160 | 0.357 | 0.205 | 0.056 | 0.454 | 0.141 | 0.056 | 0.425 | 0.125 | 0.056 | 0.438 | 0.132 | 0.056 | 0.438 | 0.092 | 0.056 |
| 1/4- | 0.2500 | 0.414 | 0.237 | 0.064 | 0.525 | 0.165 | 0.064 | 0.492 | 0.144 | 0.064 | 0.507 | 0.153 | 0.064 | 0.507 | 0.107 | 0.064 |
| 5/16- | 0.3125 | 0.518 | 0.295 | 0.072 | 0.656 | 0.209 | 0.072 | 0.615 | 0.178 | 0.072 | 0.635 | 0.191 | 0.072 | 0.635 | 0.134 | 0.072 |
| 3/8- | 0.3750 | 0.622 | 0.335 | 0.081 | 0.788 | 0.253 | 0.081 | 0.740 | 0.212 | 0.081 | 0.762 | 0.230 | 0.081 | 0.762 | 0.161 | 0.081 |
| 7/16- | 0.4375 | 0.625 | 0.368 | 0.081 | ---- | ---- | ---- | 0.863 | 0.247 | 0.081 | 0.812 | 0.223 | 0.081 | 0.812 | 0.156 | 0.081 |
| 1/2- | 0.5000 | 0.750 | 0.412 | 0.091 | ---- | ---- | ---- | 0.987 | 0.281 | 0.091 | 0.875 | 0.223 | 0.091 | 0.875 | 0.156 | 0.091 |

Note: ♦ Minimum dimensions of slot widths. All other tabulated values are maximum dimensions.

**FIGURE 10.3** Basic dimensions for American standard machine screws.

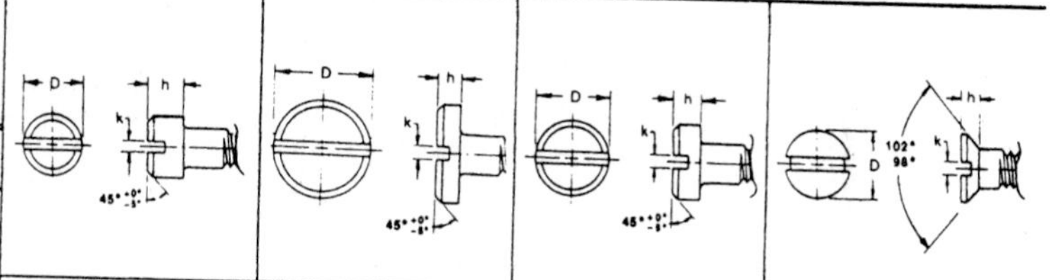


| Size | Thds/in | Basic dia. | Fillister | | | Binding | | | Pan | | | 100° Flat Head | | |
|---|---|---|---|---|---|---|---|---|---|---|---|---|---|---|
| | | | D | h | k♦ | D | h | k♦ | D | h | k♦ | D | h | k♦ |
| 30 UNM | 318 | 0.0118 | 0.021 | 0.012 | 0.003 | ---- | ---- | ---- | 0.025 | 0.010 | 0.003 | 0.023 | 0.007 | 0.003 |
| 40 UNM | 254 | 0.0157 | 0.025 | 0.016 | 0.003 | 0.041 | 0.010 | 0.004 | 0.033 | 0.012 | 0.004 | 0.029 | 0.008 | 0.003 |
| 50 UNM | 203 | 0.0197 | 0.033 | 0.020 | 0.004 | 0.051 | 0.012 | 0.005 | 0.041 | 0.016 | 0.005 | 0.037 | 0.011 | 0.004 |
| 60 UNM | 169 | 0.0236 | 0.041 | 0.025 | 0.005 | 0.062 | 0.016 | 0.007 | 0.051 | 0.020 | 0.007 | 0.045 | 0.013 | 0.005 |
| 80 UNM | 127 | 0.0315 | 0.051 | 0.032 | 0.007 | 0.082 | 0.020 | 0.008 | 0.062 | 0.025 | 0.008 | 0.056 | 0.016 | 0.007 |
| 100 UNM | 102 | 0.0394 | 0.062 | 0.040 | 0.008 | 0.103 | 0.025 | 0.012 | 0.082 | 0.032 | 0.012 | 0.072 | 0.019 | 0.008 |
| 120 UNM | 102 | 0.0472 | 0.082 | 0.050 | 0.012 | 0.124 | 0.032 | 0.015 | 0.103 | 0.040 | 0.015 | 0.092 | 0.025 | 0.012 |

Note: ♦ Minimum dimensions of slot widths. All other tabulated values are maximum dimensions.

**FIGURE 10.4** Basic dimensions for American miniature screws.

standard hardware or fasteners should be obtained from the fastener handbooks distributed by IFI (see Chap. 15).

The design specifications for strength of all types of threaded fasteners and the normal tightening torques and calculation procedures required for bolts and screws are shown later in this chapter.

Figure 10.5 shows the basic design dimensions for metric hexagonal (hex) cap screws.

Figure 10.6 shows samples of miniature metric screws with a machinists' scale for size comparison.

Figure 10.7 shows design dimensions of American standard hex nuts and jamb nuts.

Figure 10.8 shows design dimensions of metric standard nuts, type 1.

Figure 10.9 shows design dimensions of American standard flat washers.

Figure 10.10 shows design dimensions of metric standard flat washers.

Figure 10.11 shows design dimensions of American standard shoulder bolts and socket head cap screws, which are widely used in tooling applications as well as other mechanical design applications.

### *Standard Shoulder Screw Mechanical Data*

| | |
|---|---|
| Thread class | UNC-3A (ANSI/ASME B1.3) (no plating allowance is provided) |
| Material | Alloys of chrome, nickel, molybdenum, or vanadium |
| Hardness | C32 to C43 Rockwell hardness at the surface |
| Ultimate tensile strength | 140,000 psi based on minimum thread neck area |
| Shear strength | 84,000 psi in thread neck and shoulder areas |

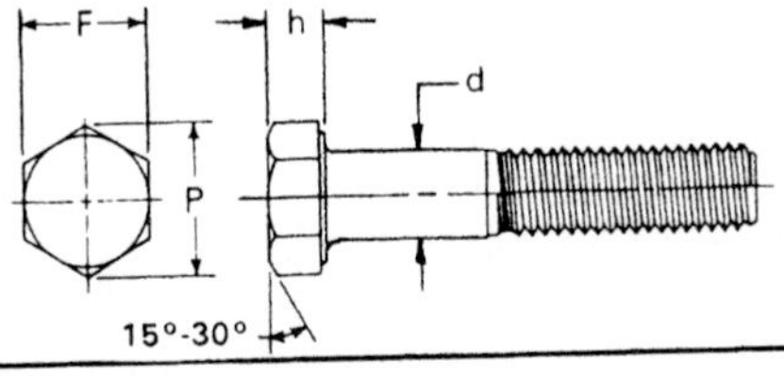


| Diameter & thread pitch | d | F | P | h |
|---|---|---|---|---|
| M5 - 0.8 | 5.00 | 8.00 | 9.24 | 3.65 |
| M6 - 1 | 6.00 | 10.00 | 11.55 | 4.15 |
| M8 - 1.25 | 8.00 | 13.00 | 15.01 | 5.50 |
| M10 - 1.50 | 10.00 | 16.00 | 18.48 | 6.63 |
| M12 - 1.75 | 12.00 | 18.00 | 20.78 | 7.76 |
| M14 - 2 | 14.00 | 21.00 | 24.25 | 9.09 |
| M16 - 2 | 16.00 | 24.00 | 27.71 | 10.32 |
| M20 - 2.5 | 20.00 | 30.00 | 34.64 | 12.88 |
| M24 - 3 | 24.00 | 36.00 | 41.57 | 15.44 |
| M30 - 3.5 | 30.00 | 46.00 | 53.12 | 19.48 |
| M36 - 4 | 36.00 | 55.00 | 63.51 | 23.38 |

Note: Tabulated dimensions are in millimeters and are maximum values.

**FIGURE 10.5** Basic dimensions for metric hex cap screws.

**FIGURE 10.6** An assortment of metric miniature screws.

| Size | Flats | Points | Thickness |
|---|---|---|---|
| #0 | 5/32 | 0.180 | 0.050 |
| #1 | 5/32 | 0.180 | 0.050 |
| #2 | 3/16 | 0.217 | 0.066 |
| #3 | 3/16 | 0.217 | 0.066 |
| #4 | 1/4 | 0.289 | 0.098 |
| #5 | 5/16 | 0.361 | 0.114 |
| #6 | 5/16 | 0.361 | 0.114 |
| #8 | 11/32 | 0.397 | 0.130 |
| #10 | 3/8 | 0.433 | 0.130 |
| 1/4 | 7/16 | 0.505 | 0.226<br>0.163-jamb nut |
| 5/16 | 1/2 | 0.577 | 0.273<br>0.195-jamb nut |
| 3/8 | 9/16 | 0.650 | 0.337<br>0.227-jamb nut |
| 7/16 | 11/16 | 0.794 | 0.385<br>0.260-jamb nut |
| 1/2 | 3/4 | 0.866 | 0.448<br>0.323-jamb nut |
| 5/8 | 15/16 | 1.083 | 0.559<br>0.387-jamb nut |
| 3/4 | 1 1/8 | 1.299 | 0.665<br>0.446-jamb nut |
| 7/8 | 1 5/16 | 1.516 | 0.776<br>0.510-jamb nut |
| 1 | 1 1/2 | 1.732 | 0.887<br>0.575-jamb nut |

**FIGURE 10.7** Standard American hex nuts and jamb nuts.

Figure 10.12 shows the head marking standard for SAE-graded bolts together with the tensile strength requirements and proof load ratings.

Figure 10.13 shows the head marking standard for ASTM-graded bolts together with the tensile strength requirements and proof load ratings.

| Size | Flats | Points | Thickness |
|---|---|---|---|
| M1.6x0.35 | 3.20 | 3.70 | 1.30 |
| M2x0.4 | 4.00 | 4.62 | 1.60 |
| M2.5x0.45 | 5.00 | 5.77 | 2.00 |
| M3x0.5 | 5.50 | 6.35 | 2.40 |
| M3.5x0.6 | 6.00 | 6.93 | 2.80 |
| M4x0.7 | 7.00 | 8.08 | 3.20 |
| M5x0.8 | 8.00 | 9.24 | 4.70 |
| M6x1 | 10.00 | 11.55 | 5.20 |
| M8x1.25 | 13.00 | 15.01 | 6.80 |
| M10x1.5 | 16.00 | 18.48 | 8.40 |
| M12x1.75 | 18.00 | 20.78 | 10.80 |
| M14x2 | 21.00 | 24.25 | 12.80 |
| M16x2 | 24.00 | 27.71 | 14.80 |
| M20x2.5 | 30.00 | 34.64 | 18.00 |
| M24x3 | 36.00 | 41.57 | 21.50 |
| M30x3.5 | 46.00 | 53.12 | 25.60 |
| M36x4 | 55.00 | 63.51 | 31.00 |

Note: Tabulated dimensions are in millimeters. 1 mm = 0.03937"

**FIGURE 10.8** Dimensions for metric standard nuts.

| Size | Outside Diameter | Thickness |
|---|---|---|
| #0 | 3/16 | 0.028 |
| #1 | 7/32 | 0.028 |
| #2 | 1/4 | 0.036 |
| #3 | 5/16 | 0.036 |
| #4 | 3/8 | 0.045 |
| #5 | 13/32 | 0.045 |
| #6 | 7/16 | 0.045 |
| #8 | 1/2 | 0.045 |
| #10 | 9/16 | 0.045 |
| 1/4 | 47/64 | 0.071 |
| 5/16 | 7/8 | 0.071 |
| 3/8 | 1 | 0.071 |
| 7/16 | 1 1/8 | 0.071 |
| 1/2 | 1 1/4 | 0.112 |
| 5/8 | 1 3/4 | 0.112 |
| 3/4 | 2 | 0.112 |
| 7/8 | 2 1/4 | 0.174 |
| 1 | 2 1/2 | 0.174 |

**FIGURE 10.9** Dimensions for American standard flat washers.

### 10.1.2 Grade Classification and Strength of Standard Bolts, Screws, and Nuts

Figure 10.14 shows the mechanical requirements and SAE identification markings for the different strength grades of bolts, screws, studs, SEMs, and U bolts.

Figure 10.15 shows the proof loads and tensile strength requirements for the different grades of threaded bolts and screws.

Figure 10.16 shows the strength requirements (allowable tensile loads) for machine screws only.

Figure 10.17 shows the proof loads and hardness requirements for the different grades of standard hex nuts.

| Size | Series | Outside Diameter | Thickness |
|---|---|---|---|
| 1.6 | Narrow | 4.00 | 0.70 |
| | Regular | 5.00 | 0.70 |
| | Wide | 6.00 | 0.90 |
| 2 | Narrow | 5.00 | 0.90 |
| | Regular | 6.00 | 0.90 |
| | Wide | 8.00 | 0.90 |
| 2.5 | Narrow | 6.00 | 0.90 |
| | Regular | 8.00 | 0.90 |
| | Wide | 10.00 | 1.20 |
| 3 | Narrow | 7.00 | 0.90 |
| | Regular | 10.00 | 1.20 |
| | Wide | 12.00 | 1.40 |
| 3.5 | Narrow | 9.00 | 1.20 |
| | Regular | 10.00 | 1.40 |
| | Wide | 15.00 | 1.75 |
| 4 | Narrow | 10.00 | 1.20 |
| | Regular | 12.00 | 1.40 |
| | Wide | 16.00 | 2.30 |
| 5 | Narrow | 11.00 | 1.40 |
| | Regular | 15.00 | 1.75 |
| | Wide | 20.00 | 2.30 |
| 6 | Narrow | 13.00 | 1.75 |
| | Regular | 18.80 | 1.75 |
| | Wide | 25.40 | 2.30 |
| 8 | Narrow | 18.80 | 2.30 |
| | Regular | 25.40 | 2.30 |
| | Wide | 32.00 | 2.80 |
| 10 | Narrow | 20.00 | 2.30 |
| | Regular | 28.00 | 2.80 |
| | Wide | 39.00 | 3.50 |
| 12 | Narrow | 25.40 | 2.80 |
| | Regular | 34.00 | 3.50 |
| | Wide | 44.00 | 3.50 |
| 14 | Narrow | 28.00 | 2.80 |
| | Regular | 39.00 | 3.50 |
| | Wide | 50.00 | 4.00 |
| 16 | Narrow | 32.00 | 3.50 |
| | Regular | 44.00 | 4.00 |
| | Wide | 56.00 | 4.60 |
| 20 | Narrow | 39.00 | 4.00 |
| | Regular | 50.00 | 4.60 |
| | Wide | 66.00 | 5.10 |
| 24 | Narrow | 44.00 | 4.60 |
| | Regular | 56.00 | 5.10 |
| | Wide | 72.00 | 5.60 |
| 30 | Narrow | 56.00 | 5.10 |
| | Regular | 72.00 | 5.60 |
| | Wide | 90.00 | 6.40 |
| 36 | Narrow | 66.00 | 5.60 |
| | Regular | 90.00 | 6.40 |
| | Wide | 110.00 | 8.50 |

Note: Tabulated dimensions are in millimeters.

**FIGURE 10.10** Dimensions for metric standard flat washers.

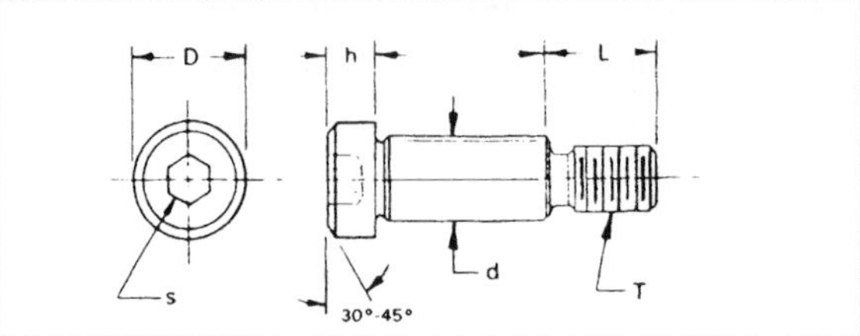


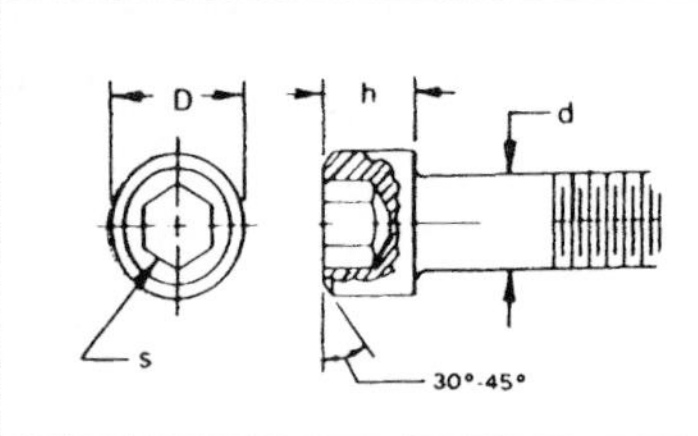


| Nominal size | | d | D | h | s | T | L | d | D | h | s |
|---|---|---|---|---|---|---|---|---|---|---|---|
| 4 | 0.1120 | ----- | ----- | ----- | ----- | ----- | ----- | 0.1120 | 0.183 | 0.112 | 0.094 |
| 5 | 0.1250 | ----- | ----- | ----- | ----- | ----- | ----- | 0.1250 | 0.205 | 0.125 | 0.094 |
| 6 | 0.1380 | ----- | ----- | ----- | ----- | ----- | ----- | 0.1380 | 0.226 | 0.138 | 0.109 |
| 8 | 0.1640 | ----- | ----- | ----- | ----- | ----- | ----- | 0.1640 | 0.270 | 0.164 | 0.141 |
| 10 | 0.1900 | ----- | ----- | ----- | ----- | ----- | ----- | 0.1900 | 0.312 | 0.190 | 0.156 |
| 1/4 | 0.2500 | 0.2480 | 0.375 | 0.188 | 0.125 | 0.190-24 | 0.375 | 0.2500 | 0.375 | 0.250 | 0.188 |
| 5/16 | 0.3125 | 0.3105 | 0.438 | 0.219 | 0.156 | 0.250-20 | 0.438 | 0.3125 | 0.469 | 0.312 | 0.250 |
| 3/8 | 0.3750 | 0.3730 | 0.562 | 0.250 | 0.188 | 0.312-18 | 0.500 | 0.3750 | 0.562 | 0.375 | 0.312 |
| 7/16 | 0.4375 | ----- | ----- | ----- | ----- | ----- | ----- | 0.4375 | 0.656 | 0.438 | 0.375 |
| 1/2 | 0.5000 | 0.4980 | 0.750 | 0.312 | 0.250 | 0.375-16 | 0.625 | 0.5000 | 0.750 | 0.500 | 0.375 |
| 5/8 | 0.6250 | 0.6230 | 0.875 | 0.375 | 0.312 | 0.500-13 | 0.750 | 0.6250 | 0.938 | 0.625 | 0.500 |
| 3/4 | 0.7500 | 0.7480 | 1.000 | 0.500 | 0.375 | 0.625-11 | 0.875 | 0.7500 | 1.125 | 0.750 | 0.625 |
| 7/8 | 0.8750 | ----- | ----- | ----- | ----- | ----- | ----- | 0.8750 | 1.312 | 0.875 | 0.750 |
| 1 | 1.0000 | 0.9980 | 1.312 | 0.625 | 0.500 | 0.750-10 | 1.000 | 1.0000 | 1.500 | 1.000 | 0.750 |
| 1-1/8 | 1.1250 | ----- | ----- | ----- | ----- | ----- | ----- | 1.1250 | 1.688 | 1.125 | 0.875 |
| 1-1/4 | 1.2500 | 1.2480 | 1.750 | 0.750 | 0.625 | 0.875-9 | 1.125 | 1.2500 | 1.875 | 1.250 | 0.875 |
| 1-3/8 | 1.3750 | ----- | ----- | ----- | ----- | ----- | ----- | 1.3750 | 2.062 | 1.375 | 1.000 |
| 1-1/2 | 1.5000 | 1.4980 | 2.125 | 1.000 | 0.875 | 1.125-7 | 1.500 | 1.5000 | 2.250 | 1.500 | 1.000 |
| 1-3/4 | 1.7500 | 1.7480 | 2.375 | 1.125 | 1.000 | 1.250-7 | 1.750 | 1.7500 | 2.625 | 1.750 | 1.250 |
| 2 | 2.0000 | 1.9980 | 2.750 | 1.250 | 1.250 | 1.500-6 | 2.000 | 2.0000 | 3.000 | 2.000 | 1.500 |

Note: All tabulated dimensions are maximum values. See text for materials, hardness, etc.

**FIGURE 10.11** Dimensions for American standard shoulder screws and socket head cap screws.

### 10.1.3 Tightening Torques and Clamp Loads of the Different Grades and Sizes of Machine Bolts

Figure 10.18 shows the tightening torques and clamp loads produced for class 2, 5, and 8 steel bolts. These three classes of bolts are the most commonly used in industry. Class 2 bolts are for noncritical commercial general applications, class 5 are extensively used where more tightening loads are required than allowable with class 2 bolts, and class 8 are used where critical, highly loaded bolt connections are required. Note that the American standard hex nuts are produced in classes 2, 5, and 8 per Fig. 10.17, which shows the proof load stress ratings for all three classes. The nut should be of the same strength class or higher class as the bolt used for each particular application. Locknuts are produced in the grades shown below, with their respective bolt grade:

| *Locknut grade* | *Bolt or screw grade* |
|---|---|
| A | 2 |
| B | 5 |
| C | 8 |

***Calculating Tightening Torque Values and Clamp Loads for Standard Bolts.*** The clamp load ranges shown in Fig. 10.18 are for applications developing 100 and 75 percent of the minimum yield strength for each respective bolt series. Torque values listed in this figure are given in pounds-feet.

| SAE bolt head markings | Diameter, in. | Tensile strength, psi | Proof load, psi | |
|---|---|---|---|---|
| 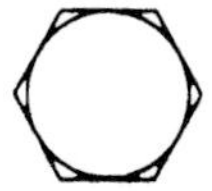 | 1/4 - 3/4<br>7/8 - 1 1/2 | 74,000 psi<br>60,000 | 55,000 psi<br>33,000 | Grade 2, *low or medium-carbon steel |
|  | 1/4 - 1<br>1 1/8 - 1 1/2 | 120,000<br>105,000 | 85,000<br>74,000 | Grade 5, *medium carbon steel (Q/T)† |
|  | - | 120,000 | 85,000 | Grade 5.1, screws |
|  | 1/4 - 1 | 120,000 | 85,000 | Grade 5.2, low-carbon martensitic steel (Q/T) |
| 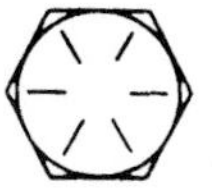 | 1/4 - 1 1/2 | 150,000 | 120,000 | Grade 8, medium-carbon alloy steel (Q/T) |
|  | 1/4 - 1 | 150,000 | 120,000 | Grade 8.2, low-carbon martensitic steel (Q/T) |

*Highest usage grades †(Q/T) designates quenched and tempered

**FIGURE 10.12** SAE bolt head markings and tensile and proof loads.

In Fig. 10.18, the maximum clamp load was determined using the minimum yield strength (psi) allowable for each bolt or screw grade (refer to Figs. 10.12 to 10.16). The minimum clamp load shown in Fig. 10.18 was determined by reducing the maximum clamp load by 25 percent.

*Example.* For a grade 5 bolt of ½ to 13 size, the minimum yield strength = 92,000 psi per Fig. 10.14. Stress area of a ½-to-13 bolt = 0.1419 in$^2$ (see figures in Sec. 10.2). Then, the clamp load is:

$$\begin{aligned}\text{Clamp load } L &= \text{minimum yield strength} \times \text{stress area of bolt}\\ &= 92{,}000 \times 0.1419\\ &= 13{,}055 \text{ lbf}\end{aligned}$$

The tightening torque is calculated from

$$T = KLD$$

where $T$ = tightening-torque, lb · in
$K$ = dynamic coefficient of friction, minimum = 0.15 (dry and zinc-plated)
$L$ = Clamp load, lbf
$D$ = nominal bolt diameter, in

| ASTM bolt head markings | Diameter, in. | Tensile strength, psi | Proof load, psi | |
|---|---|---|---|---|
| 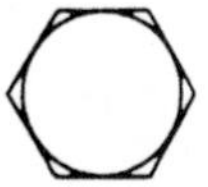 | 1/4 - 4 | A & B-60,000 min.<br>B-100,000 max. | | A307 carbon steel |
|  | 1/4 - 1<br>1 1/8 - 1 1/2<br>1 3/4 - 3 | 120,000<br>105,000<br>90,000 | 85,000<br>74,000<br>55,000 | A449 medium-carbon steel (Q/T) |
|   | 1/2 - 1<br>1 1/8 - 1 1/2 | 120,000<br>105,000 | 85,000<br>74,000 | A325 Type 1 medium carbon steel (Q/T) |
|   | 1/2 - 1<br>1 1/8 - 1 1/2 | 120,000<br>105,000 | 85,000<br>74,000 | A325 Type 2 low carbon martensitic steel (Q/T) |
|   | 1/2 - 1<br>1 1/8 - 1 1/2 | 120,000<br>105,000 | 85,000<br>74,000 | A325 Type 3 weathering steel (Q/T) |
|   | 1/2 - 1 1/2 | 150,000 | 120,000 | A490 alloy steel (Q/T) |
|  | 1/4 - 4 | 150,000 | 120,000 | A354 grade BD alloy steel (Q/T) |

*(Q/T) designates quenched and tempered*

**FIGURE 10.13** ASTM bolt head markings and tensile and proof loads.

Then $$T = 0.15 \times 13{,}055 \times 0.50 = 979 \text{ lb} \cdot \text{in or } 81.6 \text{ lb} \cdot \text{ft}$$

which agrees with the maximum torque value shown in Fig. 10.18 for this particular bolt (0.500 to 13, grade 5).

(*Note:* Do not specify torque loads as foot-pounds or inch-pounds, as these are measures of energy or work, not torque. Also, do not specify torque loads as lbs/ft as this is a force measurement per linear distance, and not a torque load. The metric equivalent of U.S. customary torque loads given in pounds-feet is newton-meters.)

The tightening torques and clamp loads for any machine bolt or screw may be effectively calculated using the above procedure. To find the minimum yield strength of the material from which the bolt or screw is made, see Chap. 5. Bolts or screws made from stainless steels or nonferrous alloys may thus be calculated. Because of the general nature of the torque equation and unknown coefficients of dynamic friction for various material combinations, it is advisable that the minimum yield strength of each material be reduced 25 to 30 percent prior to making the torque calculation from the calculated clamp load. Thus, if the minimum

| Grade designation | Products | Nominal size dia, in | Full-size bolts, screws, studs, sems: Proof load (stress), lb/in² | Full-size bolts, screws, studs, sems: Tensile strength (stress) min, lb/in² | Machine test specimens of bolts, screws, and studs: Yield[a] strength (stress) min, lb/in² | Machine test specimens: Tensile strength (stress) min, lb/in² | Machine test specimens: Elongation[f] min, % | Machine test specimens: Reduction of area min, % | Surface hardness: Rockwell 30N max | Core hardness: Rockwell Min | Core hardness: Rockwell Max | Grade identification marking |
|---|---|---|---|---|---|---|---|---|---|---|---|---|
| 1 | Bolts, screws, studs | ¼ to 1½ | 33,000[k] | 60,000 | 36,000[b] | 60,000 | 18 | 35 | — | B70 | B100 | None |
| 2 | Bolts, screws, studs | ¼ to ¾[c] | 55,000[k] | 74,000 | 57,000 | 74,000 | 18 | 35 | — | B80 | B100 | |
| | | Over ¾ to 1½ | 33,000 | 60,000 | 36,000[b] | 60,000 | 18 | 35 | — | B70 | B100 | None |
| 4 | Studs | ¼ to 1½ | 65,000 | 115,000 | 100,000 | 115,000 | 10 | 35 | — | C22 | C32 | None |
| 5 | Bolts, screws, studs | ¼ to 1 | 85,000 | 120,000 | 92,000 | 120,000 | 14 | 35 | 54 | C25 | C34 |  |
| | | Over 1 to 1½ | 74,000 | 105,000 | 81,000 | 105,000 | 14 | 35 | 50 | C19 | C30 | |
| 5.1[d] | Sems,[h] bolts, screws | No. 6 to ⅝ | 85,000 | 120,000 | — | — | — | — | 59.5[g] | C25 | C40[g] |  |
| | | No. 6 to ½ | | | | | | | | | | |
| 7[e] | Bolts, screws | ¼ to 1½ | 105,000 | 133,000 | 115,000 | 133,000 | 12 | 35 | 54 | C28 | C34 |  |
| 8 | Bolts, screws, studs | ¼ to 1½ | 120,000 | 150,000 | 130,000 | 150,000 | 12 | 35 | 58.6 | C33 | C39 |  |
| 8.1 | Studs | ¼ to 1½ | 120,000 | 150,000 | 130,000 | 150,000 | 10 | 35 | — | C32 | C38 | None |
| 8.2 | Bolts, screws | ¼ to 1 | 120,000 | 150,000 | 130,000 | 150,000 | 10 | 35 | 58.6 | C33 | C39 |  |

[a] Yield strength is stress at which a permanent set of 0.2% of gauge length occurs.
[b] Yield point shall apply instead of yield strength at 0.2% offset.
[c] Grade 2 requirements for sizes ¼ through ¾ in apply only to bolts and screws 6 in and shorter in length and to studs of all lengths. For bolts and screws longer than 6 in, grade 1 requirements shall apply.
[d] Grade 5 material heat treated before assembly with a hardened washer is an acceptable substitute.
[e] Grade 7 bolts and screws are roll threaded after heat treatment.
[f] Hex washer head and hex flange products without assembled washers shall have a core hardness not exceeding Rockwell C38 and a surface hardness not exceeding Rockwell 30N 57.5.
[g] Sems and similar products without washers.
[h] Not applicable to studs or slotted and cross-recess head products.
[i] Proof load test: Requirements in these grades only apply to stress relieved products.
SOURCE: Reprinted with permission, copyright 1992, Society of Automotive Engineers.

**FIGURE 10.14** Mechanical requirements for American bolts, screws, studs, SEMs, and U bolts.

## Proof Load and Tensile Strength Requirements*

| Nominal dia of product and threads per in | Stress area, in² | Grade 1 Proof load, lb | Grade 1 Tensile strength min, lb | Grade 2 Proof load, lb | Grade 2 Tensile strength min, lb | Grade 4 Proof load, lb | Grade 4 Tensile strength min, lb | Grades 5 and 5.2† Proof load, lb | Grades 5 and 5.2† Tensile strength min, lb | Grade 5.1 Proof load, lb | Grade 5.1 Tensile strength min, lb | Grade 7 Proof load, lb | Grade 7 Tensile strength min, lb | Grades 8, 8.1, and 8.2† Proof load, lb | Grades 8, 8.1, and 8.2† Tensile strength min, lb |
|---|---|---|---|---|---|---|---|---|---|---|---|---|---|---|---|
| Coarse-Thread Series—UNC | | | | | | | | | | | | | | | |
| No. 6–32 | 0.00909 | — | — | — | — | — | — | — | — | 750 | 1,100 | — | — | — | — |
| 8–32 | 0.0140 | — | — | — | — | — | — | — | — | 1,200 | 1,700 | — | — | — | — |
| 10–24 | 0.0175 | — | — | — | — | — | — | — | — | 1,500 | 2,100 | — | — | — | — |
| 12–24 | 0.0242 | — | — | — | — | — | — | — | — | 2,050 | 2,900 | — | — | — | — |
| 1/4–20 | 0.0318 | 1,050 | 1,900 | 1,750 | 2,350 | 2,050 | 3,650 | 2,700 | 3,800 | 2,700 | 3,800 | 3,350 | 4,250 | 3,800 | 4,750 |
| 5/16–18 | 0.0524 | 1,750 | 3,150 | 2,900 | 3,900 | 3,400 | 6,000 | 4,450 | 6,300 | 4,450 | 6,300 | 5,500 | 6,950 | 6,300 | 7,850 |
| 3/8–16 | 0.0775 | 2,550 | 4,650 | 4,250 | 5,750 | 5,050 | 8,400 | 6,600 | 9,300 | 6,600 | 9,300 | 8,150 | 10,300 | 9,300 | 11,600 |
| 7/16–14 | 0.1063 | 3,500 | 6,400 | 5,850 | 7,850 | 6,900 | 12,200 | 9,050 | 12,800 | 9,050 | 12,800 | 11,200 | 14,100 | 12,800 | 15,900 |
| 1/2–13 | 0.1419 | 4,700 | 8,500 | 7,800 | 10,500 | 9,200 | 16,300 | 12,100 | 17,000 | 12,100 | 17,000 | 14,900 | 18,900 | 17,000 | 21,300 |
| 9/16–12 | 0.182 | 6,000 | 10,900 | 10,000 | 13,500 | 11,800 | 20,900 | 15,500 | 21,800 | 15,500 | 21,800 | 19,100 | 24,200 | 21,800 | 27,300 |
| 5/8–11 | 0.226 | 7,450 | 13,600 | 12,400 | 16,700 | 14,700 | 25,400 | 19,200 | 27,100 | 19,200 | 27,100 | 23,700 | 30,100 | 27,100 | 33,900 |
| 3/4–10 | 0.334 | 11,000 | 20,000 | 18,400 | 24,700 | 21,700 | 38,400 | 28,400 | 40,100 | — | — | 35,100 | 44,400 | 40,100 | 50,100 |
| 7/8–9 | 0.462 | 15,200 | 27,700 | 15,200 | 27,700 | 30,000 | 53,100 | 39,300 | 55,400 | — | — | 48,500 | 61,400 | 55,400 | 69,300 |
| 1–8 | 0.606 | 20,000 | 36,400 | 20,000 | 36,400 | 39,400 | 69,700 | 51,500 | 72,700 | — | — | 63,600 | 80,600 | 72,700 | 90,900 |
| 1 1/8–7 | 0.763 | 25,200 | 45,800 | 25,200 | 45,800 | 49,600 | 87,700 | 56,500 | 80,100 | — | — | 80,100 | 101,500 | 91,600 | 114,400 |
| Fine-Thread Series—UNF | | | | | | | | | | | | | | | |
| No. 6–40 | 0.01015 | — | — | — | — | — | — | — | — | 850 | 1,200 | — | — | — | — |
| 8–36 | 0.01474 | — | — | — | — | — | — | — | — | 1,250 | 1,750 | — | — | — | — |
| 10–32 | 0.0200 | — | — | — | — | — | — | — | — | 1,700 | 2,400 | — | — | — | — |
| 12–28 | 0.0258 | — | — | — | — | — | — | — | — | 2,200 | 3,100 | — | — | — | — |
| 1/4–28 | 0.0364 | 1,200 | 2,200 | 2,000 | 2,700 | 2,350 | 4,200 | 3,100 | 4,350 | 3,100 | 4,350 | 3,800 | 4,850 | 4,350 | 5,450 |
| 5/16–24 | 0.0580 | 1,900 | 3,500 | 3,200 | 4,300 | 3,750 | 6,700 | 4,900 | 6,950 | 4,900 | 6,950 | 6,100 | 7,700 | 6,950 | 8,700 |
| 3/8–24 | 0.0878 | 2,900 | 5,250 | 4,800 | 6,500 | 5,700 | 10,100 | 7,450 | 10,500 | 7,450 | 10,500 | 9,200 | 11,700 | 10,500 | 13,200 |
| 7/16–20 | 0.1187 | 3,900 | 7,100 | 6,550 | 8,800 | 7,700 | 13,650 | 10,100 | 14,200 | 10,100 | 14,200 | 12,500 | 15,800 | 14,200 | 17,800 |
| 1/2–20 | 0.1599 | 5,300 | 9,600 | 8,800 | 11,800 | 10,400 | 18,400 | 13,600 | 19,200 | 13,600 | 19,200 | 16,800 | 21,300 | 19,200 | 24,000 |
| 9/16–18 | 0.203 | 6,700 | 12,200 | 11,200 | 15,000 | 13,200 | 23,300 | 17,300 | 24,400 | 17,300 | 24,400 | 21,300 | 27,000 | 24,400 | 30,400 |
| 5/8–18 | 0.256 | 8,450 | 15,400 | 14,100 | 18,900 | 16,600 | 29,400 | 21,800 | 30,700 | 21,800 | 30,700 | 26,900 | 34,000 | 30,700 | 38,400 |
| 3/4–16 | 0.373 | 12,300 | 22,400 | 20,500 | 27,600 | 24,200 | 42,900 | 31,700 | 44,800 | — | — | 39,200 | 49,600 | 44,800 | 56,000 |
| 7/8–14 | 0.509 | 16,800 | 30,500 | 16,800 | 30,500 | 33,100 | 58,500 | 43,300 | 61,100 | — | — | 53,400 | 67,700 | 61,100 | 76,400 |
| 1–12 | 0.663 | 21,900 | 39,800 | 21,900 | 39,800 | 43,100 | 76,200 | 56,400 | 79,600 | — | — | 69,600 | 88,200 | 79,600 | 99,400 |
| 1–14 uns | 0.679 | 22,400 | 40,700 | 22,400 | 40,700 | 44,100 | 78,100 | 57,700 | 81,500 | — | — | 71,300 | 90,300 | 81,500 | 101,900 |
| 1 1/8–12 | 0.856 | 28,200 | 51,400 | 28,200 | 51,400 | 55,600 | 98,400 | 63,300 | 89,900 | — | — | 89,900 | 113,800 | 102,700 | 128,400 |
| 1 1/4–12 | 1.073 | 35,400 | 64,400 | 35,400 | 64,400 | 69,700 | 123,400 | 79,400 | 112,700 | — | — | 112,700 | 142,700 | 128,800 | 161,000 |
| 1 3/8–12 | 1.315 | 43,400 | 78,900 | 43,400 | 78,900 | 85,500 | 151,200 | 97,300 | 138,100 | — | — | 138,100 | 174,900 | 157,800 | 197,200 |
| 1 1/2–12 | 1.581 | 52,200 | 94,900 | 52,200 | 94,900 | 102,800 | 181,800 | 117,000 | 166,000 | — | — | 166,000 | 210,300 | 189,700 | 237,200 |

* Proof loads and tensile strengths are computed by multiplying the proof load stresses and tensile strength stresses by the stress area of the thread.

The stress area of sizes and thread series may be computed from the formula: $A_s = 0.7854\left[D - \frac{0.9743}{n}\right]^2$, where $D$ equals nominal diameter in inches and $n$ equals threads per inch.

† Grades 5.2 and 8.2 applicable to sizes 1/4 through 1 in.

SOURCE: Reprinted with permission, copyright 1992, Society of Automotive Engineers.

**FIGURE 10.15**

| Nominal size or basic major dia of thread and threads per in | | Stress area, in$^2$ | Tensile strength,* lb, min | |
|---|---|---|---|---|
| | | | Grade 60M | Grade 120M |
| No. 4–40 | 0.112 | 0.00604 | 360 | 720 |
| 4–48 | 0.112 | 0.00661 | 390 | 780 |
| 5–40 | 0.125 | 0.00796 | 470 | 940 |
| 5–44 | 0.125 | 0.00830 | 490 | 980 |
| 6–32 | 0.138 | 0.00909 | 550 | 1100 |
| 6–40 | 0.138 | 0.01015 | 600 | 1200 |
| 8–32 | 0.164 | 0.0140 | 850 | 1700 |
| 8–36 | 0.164 | 0.01474 | 880 | 1750 |
| 10–24 | 0.190 | 0.0175 | 1050 | 2100 |
| 10–32 | 0.190 | 0.0200 | 1200 | 2400 |
| 12–24 | 0.216 | 0.0242 | 1450 | 2900 |
| 12–28 | 0.216 | 0.0258 | 1550 | 3100 |
| 1/4–20 | 0.250 | 0.0318 | 1900 | 3800 |
| 1/4–28 | 0.250 | 0.0364 | 2200 | 4350 |
| 5/16–18 | 0.312 | 0.0524 | 3,150 | 6,300 |
| 5/16–24 | 0.312 | 0.0580 | 3,500 | 6,950 |
| 3/8–16 | 0.375 | 0.0775 | 4,650 | 9,300 |
| 3/8–24 | 0.375 | 0.0878 | 5,250 | 10,500 |
| 7/16–14 | 0.438 | 0.1063 | 6,400 | 12,800 |
| 7/16–20 | 0.438 | 0.1187 | 7,100 | 14,200 |
| 1/2–13 | 0.500 | 0.1419 | 8,500 | 17,000 |
| 1/2–20 | 0.500 | 0.1599 | 9,600 | 19,200 |
| 9/16–12 | 0.562 | 0.182 | 10,900 | 21,800 |
| 9/16–18 | 0.562 | 0.203 | 12,200 | 24,400 |
| 5/8–11 | 0.625 | 0.226 | 13,600 | 27,100 |
| 5/8–18 | 0.625 | 0.256 | 15,400 | 30,700 |
| 3/4–10 | 0.750 | 0.334 | 20,100 | 40,100 |
| 3/4–16 | 0.750 | 0.373 | 22,400 | 44,800 |

* Tensile strength values for grade 60M and grade 120M are based on 60,000 and 120,000 lb/in$^2$, respectively.

SOURCE: Reprinted with permission, copyright 1992, Society of Automotive Engineers.

**FIGURE 10.16** Tensile load requirements for American machine screws.

yield strength of the material from which the bolt is made is 45,000 psi, reduce this value by 25 to 30 percent, calculate the allowable clamp load, and then calculate the torque required to produce this clamp load. (*Note:* The tightening torques and clamp loads for threaded fasteners may also be calculated using the equations for power screws shown in Sec. 8.6.)

***Controlling Preload on a Threaded Fastener.*** Using the turn of nut or head method, the bolt or screw is first torqued to 60 to 80 percent of its minimum yield strength (see preceding calculation method). Then, the nut or the head of the bolt (when a nut is not used, as in a tapped hole) is tightened an additional half-turn (180°). This amount of additional turn usually puts the bolt or screw into or beyond its yield strength point.

This method takes into account the recommendations of some authorities that the bolt should be tightened to its yield strength and is a reliable way to control preload on a threaded fastener. Structural steel joints have been preloaded as described by the construction and structures industries for more than 50 years. This method is quick and economical for the majority of applications found throughout industry. With a critical application, an accurate torque wrench should be used.

**Proof Load and Hardness Requirements for Nuts***

| Nut grade | Nut size dia, in | Proof load stress, lb/in²† Thread series UNC 8 UN | UNF, 12 UN and finer | Rockwell hardness |
|---|---|---|---|---|
| 2‡ | ¼ to 1½ | 90,000 | 90,000 | C32 max |
| 5 | ¼ to 1 | 120,000 | 109,000 | C32 max |
| | Over 1 to 1½ | 105,000 | 94,000 | C32 max |
| 8 | ¼ to ⅝ | 150,000 | 150,000 | C24–C32 |
| | Over ⅝ to 1 | | | C26–C34 |
| | Over 1 to 1½ | | | C26–C36 |

* Values listed are not normally applicable to jam, slotted, castle, heavy, or thick nuts.

† The proof load in pounds for a nut is computed by multiplying the proof load stress for the nut grade, size, and thread series, and the stress area for the applicable size and thread.

‡ Normally applicable to square nuts only. Also, square nuts normally available in Grade 2 only.

SOURCE: Reprinted with permission, copyright 1992, Society of Automotive Engineers.

**FIGURE 10.17** Proof load and hardness requirements for American standard nuts.

## 10.1.4 Setscrews, Self-Tapping, Thread-Forming, and Wood Screws

Figure 10.19 shows the American standard socket setscrews with dimensions and application data. The size of the Allen hex wrench required for each size is also shown in Fig. 10.23*c*.

Figure 10.20 shows a chart of the standard self-tapping and thread-forming screws. Types AB, A, B, BF, and C are thread-forming, while types D, F, G, T, BF, and BT are thread-cutting or self-tapping. Type U is a spiral screw type which is driven or press-fit into the appropriate-size hole. All thread forms shown are 60° V thread.

***Holes for Self-Tapping and Thread-Forming Screws.*** Extensive tables of recommended hole sizes for self-tapping and thread-forming screws are available from the manufacturer of the screws. Because of the great variety of materials and thickness ranges possible for applying these types of screws, no tables are given. In actual practice, you may measure the outside diameter of the screw thread and the root diameter, and then take the mean difference to find an approximate diameter for the particular screw and material combination. A few trial combinations will give you the exact drill size to use for your application.

Figure 10.21 shows American standard wood screws and their dimensions. Wood screws are used by pattern makers and in wooden form blocks for vacuum-forming equipment applications (vacuum-forming plastics).

Head styles of the various machine screws are shown in Fig. 10.22.

## 10.1.5 Wrench Clearances for Design Applications

An important point, which is sometimes forgotten in basic mechanical design work, is wrench clearances. Figure 10.23*a* and 10.23*b* shows an extensive listing of required clearances for open-end, box, and socket wrenches. Allowances should be made in initial design work to allow more clearance than shown in the figure when possible. Tabulated dimensions shown in the figure are minimum allowables. No allowance has been made for torquing devices, since these are extensive and variable per each manufacturer. Figure 10.23*c* shows the dimensions and mechanical properties of hex keys (Allen wrenches).

| Bolt size | SAE grade 2 Tightening torque range, lb·ft | SAE grade 2 Clamp load range, lb | SAE grade 5 Tightening torque range, lb·ft | SAE grade 5 Clamp load range, lb | SAE grade 8 Tightening torque range, lb·ft | SAE grade 8 Clamp load range, lb |
|---|---|---|---|---|---|---|
| ¼–20 | 5.7–4.3 | 1,813–1,360 | 9.1–6.9 | 2,926–2,195 | 12.9–9.7 | 4,134–3,101 |
| ¼–28 | 6.5–4.9 | 2,075–1,556 | 10.5–7.9 | 3,349–2,512 | 14.8–11.1 | 4,732–3,549 |
| 5⁄16–18 | 11.7–8.8 | 2,987–2,240 | 18.8–14.1 | 4,821–3,616 | 26.2–20.0 | 6,812–5,109 |
| 5⁄16–24 | 12.9–9.7 | 3,306–2,480 | 20.8–15.6 | 5,336–4,002 | 29.5–22.1 | 7,540–5,655 |
| ⅜–16 | 20.7–15.5 | 4,418–3,314 | 33.4–25.1 | 7,130–5,348 | 47.2–35.4 | 10,075–7,556 |
| ⅜–24 | 23.5–17.6 | 5,005–3,754 | 37.9–28.4 | 8,078–6,059 | 53.5–40.1 | 11,414–8,561 |
| 7⁄16–14 | 33.1–24.9 | 6,059–4,544 | 53.5–40.1 | 9,780–7,335 | 75.6–56.7 | 13,819–10,364 |
| 7⁄16–20 | 37.0–27.8 | 6,766–5,075 | 59.7–44.8 | 10,920–8,190 | 84.4–63.3 | 15,431–11,573 |
| ½–13 | 50.6–37.9 | 8,088–6,066 | 81.6–61.2 | 13,055–9,791 | 1115.3–86.5 | 18,447–13,835 |
| ½–20 | 57.0–42.7 | 9,114–5,835 | 91.9–69.0 | 14,711–11,033 | 130.0–97.4 | 20,787–15,590 |
| 9⁄16–12 | 73.0–54.7 | 10,374–7,780 | 117.7–88.1 | 16,744–12,558 | 166.4–124.8 | 23,660–17,745 |
| 9⁄16–18 | 81.4–61.0 | 11,571–8,678 | 131.3–98.1 | 18,676–14,007 | 185.6–139.2 | 26,390–19,793 |
| ⅝–11 | 100.6–75.5 | 12,882–9,662 | 162.4–121.8 | 20.792–15,594 | 229.5–172.1 | 29,380–22,035 |
| ⅝–18 | 114–85.5 | 14,592–10,944 | 184–138 | 23,552–17,664 | 260.0–195.0 | 33,280–24,960 |
| ¾–10 | 178.5–133.9 | 19,038–14,279 | 288–216 | 30,728–23,046 | 407.1–305.3 | 43,420–35,368 |
| ¾–16 | 199–149.5 | 21,261–15,946 | 321.7–241.3 | 34,316–25,737 | 454.6–341.0 | 48,490–45,045 |
| ⅞–9 | 288–216 | 26,334–19,751 | 464.9–348.7 | 42,504–31,878 | 656.9–492.7 | 60,060–45,045 |
| ⅞–14 | 317–238 | 29,013–19,751 | 512.2–384.1 | 46,828–35,121 | 723.7–542.8 | 66,170–49,628 |
| 1–8 | 432–324 | 34,542–25,907 | 696.9–522.7 | 55,752–41,814 | 984.8–738.6 | 78,780–59,085 |
| 1–12 | 472–354 | 37,791–28,343 | 761.1–571.8 | 60,996–45,747 | 1077–808 | 86,190–64,643 |

**FIGURE 10.18** Tightening torque requirements for American standard steel bolts.

## Socket Set Screws - Inch Series

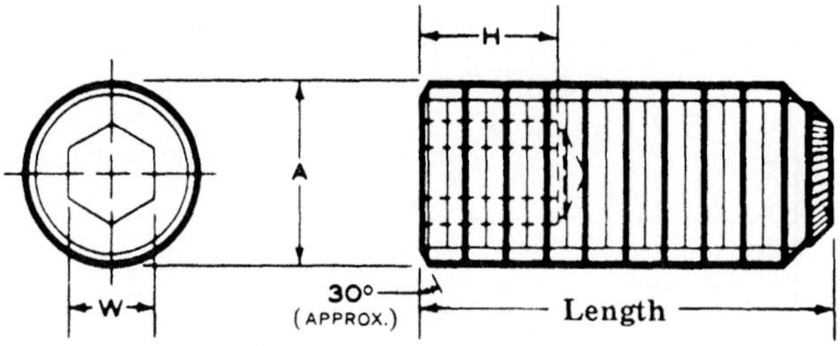


| | threads per inch | | Dimensions | | | | | | | | | | | |
|---|---|---|---|---|---|---|---|---|---|---|---|---|---|
| | | | A | | | C | | D | | F | | H | W |
| | | | max. | min. | | max. | min. | max. | min. | max. | min. | min. | nom |
| size | UNC | UNF | | UNC | UNF | | | | | | | | |
| #0 | .... | 80 | .0600 | .... | .0568 | .033 | .027 | .040 | .037 | .017 | .013 | .022 | .028 |
| #1 | 64 | 72 | .0730 | .0692 | .0695 | .040 | .033 | .049 | .045 | .021 | .017 | .028 | .035 |
| #2 | 56 | 64 | .0860 | .0819 | .0822 | .047 | .039 | .057 | .053 | .024 | .020 | .028 | .035 |
| #3 | 48 | 56 | .0990 | .0945 | .0949 | .054 | .045 | .066 | .062 | .027 | .023 | .040 | .050 |
| #4 | 40 | 48 | .1120 | .1069 | .1075 | .061 | .051 | .075 | .070 | .030 | .026 | .040 | .050 |
| #5 | 40 | 44 | .1250 | .1199 | .1202 | .067 | .057 | .083 | .078 | .033 | .027 | .050 | .0625 |
| #6 | 32 | 40 | .1380 | .1320 | .1329 | .074 | .064 | .092 | .087 | .038 | .032 | .050 | .0625 |
| #8 | 32 | 36 | .1640 | .1580 | .1585 | .087 | .076 | .109 | .103 | .043 | .037 | .062 | .0781 |
| #10 | 24 | 32 | .1900 | .1828 | .1840 | .102 | .088 | .127 | .120 | .049 | .041 | .075 | .0937 |
| 1/4 | 20 | 28 | .2500 | .2419 | .2435 | .132 | .118 | .156 | .149 | .0565 | .0585 | .100 | .125 |
| 5/16 | 18 | 24 | .3125 | .3038 | .3053 | .172 | .156 | .203 | .195 | .082 | .074 | .125 | .1562 |
| 3/8 | 16 | 24 | .3750 | .3656 | .3678 | .212 | .194 | .250 | .241 | .0987 | .0887 | .150 | .1875 |
| 7/16 | 14 | 20 | .4375 | .4272 | .4294 | .252 | .232 | .296 | .287 | .114 | .104 | .175 | .2187 |
| 1/2 | 13 | 20 | .5000 | .4891 | .4919 | .291 | .270 | .343 | .334 | .130 | .120 | .200 | .250 |
| 9/16 | 12 | 18 | .5625 | .5511 | .5538 | .332 | .309 | .390 | .379 | .1456 | .1356 | .200 | .250 |
| 5/8 | 11 | 18 | .6250 | .6129 | .6163 | .371 | .347 | .468 | .456 | .164 | .148 | .250 | .3125 |
| 3/4 | 10 | 16 | .7500 | .7371 | .7406 | .450 | .425 | .562 | .549 | .1955 | .1795 | .300 | .375 |
| 7/8 | 9 | 14 | .8750 | .8611 | .8647 | .530 | .502 | .656 | .642 | .2267 | .2107 | .400 | .500 |
| 1 | 8 | 12 | 1.0000 | .9850 | .9886 | .609 | .579 | .750 | .734 | .260 | .240 | .450 | .5625 |
| 1 1/8 | 7 | 12 | 1.1250 | 1.1086 | 1.1136 | .689 | .655 | .843 | .826 | .291 | .271 | .450 | .5625 |
| 1 1/4 | 7 | 12 | 1.2500 | 1.2336 | 1.2386 | .767 | .733 | .937 | .920 | .3225 | .3025 | .500 | .625 |
| 1 3/8 | 6 | 12 | 1.3750 | 1.3568 | 1.3636 | .848 | .808 | 1.031 | 1.011 | .3537 | .3337 | .500 | .625 |
| 1 1/2 | 6 | 12 | 1.5000 | 1.4818 | 1.4886 | .926 | .886 | 1.125 | 1.105 | .385 | .365 | .600 | .750 |

**FIGURE 10.19** American standard socket setscrews and styles of points: dimensions.

## Point Types

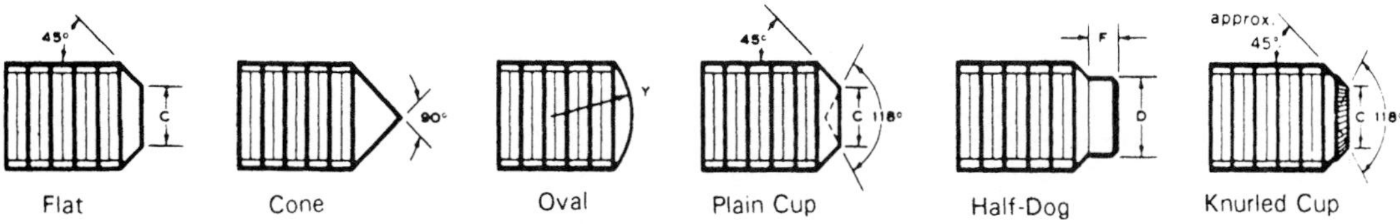


## Application Data

**recommended seating torques** Pound-inches

| Y rad. | tap drill size: size | tap drill size: UNC | tap drill size: UNF | regular and LOC-WEL screws: min. screw length | regular and LOC-WEL screws: alloy steel | regular and LOC-WEL screws: stainless | self-locking with NYLOK (1): min. screw length | self-locking with NYLOK (1): alloy steel | self-locking with NYLOK (1): stainless | self-locking with NYLOK (2): min. screw length | self-locking with NYLOK (2): alloy steel | self-locking with NYLOK (2): stainless |
|---|---|---|---|---|---|---|---|---|---|---|---|---|
| .047 | #0 | .... | 1.25mm | 3/32 | .5 | .4 | 3/32 | .4 | .3 | 1/8 | .5 | .4 |
| .055 | #1 | 1.5mm | 1.5mm | 3/32 | 1.5 | 1.2 | 1/8 | 1.2 | 1.0 | 5/32 | 1.5 | 1.2 |
| .062 | #2 | #50 | 1.85mm | 3/32 | 1.5 | 1.2 | 1/8 | 1.2 | 1.0 | 5/32 | 1.5 | 1.2 |
| .078 | #3 | #46 | 2.1mm | 3/32 | 5 | 4 | 5/32-3/16 | 4 | 3 | 7/32 | 5 | 4 |
| .084 | #4 | 2.3mm | #42 | 3/32 | 5 | 4 | 5/32-3/16 | 4 | 3 | 7/32 | 5 | 4 |
| .093 | #5 | #37 | #37 | 5/32 | 9 | 7 | 3/16 | 7 | 6 | 1/4 | 9 | 7 |
| .109 | #6 | #33 | #32 | 1/8 | 9 | 7 | 3/16 | 7 | 6 | 1/4 | 9 | 7 |
| .125 | #8 | #29 | 3.5mm | 1/8 | 20 | 16 | 1/4 | 16 | 13 | 5/16 | 20 | 16 |
| .141 | #10 | #24 | #20 | 5/32 | 33 | 26 | 1/4-5/16 | 28 | 22 | 3/8 | 33 | 26 |
| .188 | 1/4 | #6 | 5.5mm | 3/16 | 87 | 70 | 1/4-7/16 | 52 | 42 | 1/2 | 87 | 70 |
| .234 | 5/16 | G | I | 1/4 | 165 | 130 | 5/16-9/16 | 90 | 72 | 5/8 | 165 | 130 |
| .281 | 3/8 | O | 8.6mm | 5/16 | 290 | 230 | 3/8-5/8 | 200 | 160 | 3/4 | 290 | 230 |
| .328 | 7/16 | 9.4mm | 25/64 | 3/8 | 430 | 340 | 7/16-3/4 | 300 | 240 | 7/8 | 430 | 340 |
| .375 | 1/2 | 27/64 | 11.5mm | 1/2 | 620 | 500 | 1/2-3/4 | 500 | 400 | 7/8 | 620 | 500 |
| .422 | 9/16 | 31/64 | 1/2 | 1/2 | 620 | 500 | 1/2-3/4 | 500 | 400 | 7/8 | 620 | 500 |
| .468 | 5/8 | 17/32 | 14.5mm | 9/16 | 1,225 | 980 | 5/8-7/8 | 980 | 780 | 1 | 1,225 | 980 |
| .562 | 3/4 | 21/32 | 17.5mm | 5/8 | 2,125 | 1,700 | 3/4-1 1/8 | 1,700 | 1,360 | 1 1/4 | 2,125 | 1,700 |
| .656 | 7/8 | 49/64 | 20.5mm | 3/4 | 5,000 | 4,000 | 3/4-1 1/4 | 3,650 | 2,920 | 1 5/16 | 5,000 | 4,000 |
| .750 | 1 | 7/8 | 23.5 | 7/8 | 7,000 | 5,600 | 7/8-1 7/16 | 5,200 | 4,160 | 1 1/2 | 7,000 | 5,600 |
| .844 | 1 1/8 | 25mm | 1 3/64 | 1 | 7,000 | 5,600 | .... | .... | .... | .... | .... | .... |
| .938 | 1 1/4 | 1 7/64 | 1 11/64 | 1 1/8 | 9,600 | 7,700 | .... | .... | .... | .... | .... | .... |
| 1.032 | 1 3/8 | 1 7/32 | 1 19/64 | 1 1/4 | 9,600 | 7,700 | .... | .... | .... | .... | .... | .... |
| 1.125 | 1 1/2 | 34mm | 36mm | 1 1/4 | 11,320 | 9,100 | .... | .... | .... | .... | .... | .... |

Note: Materials: High grade alloy steel, austenitic stainless steel.
Hardness: Rc 45-53 for alloy steel grades. Thread class: 3A.

| Type | ANSI Standard |
|---|---|
| 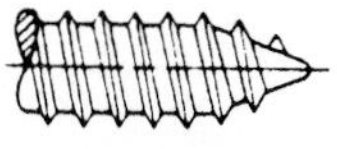 | AB |
| 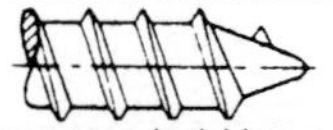 Not Recommended-Use type AB | A |
| 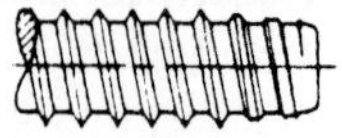 | B |
| 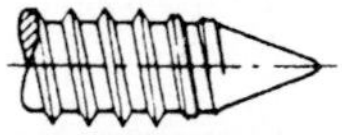 | BP |
| 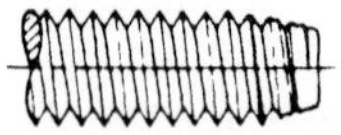 | C |
| 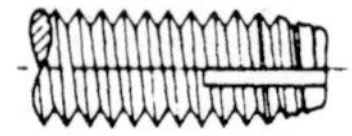 | D |
|  | F |
| 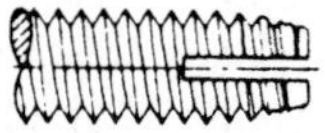 | G |
| 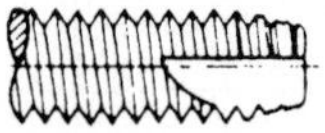 | T |
| 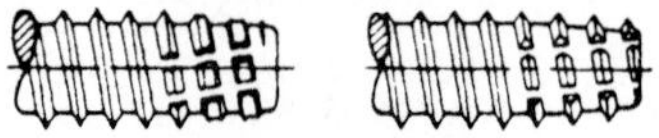 | BF |
| 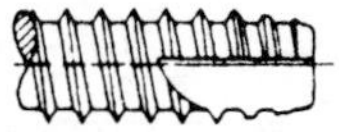 | BT |
|  | U |

**FIGURE 10.20** Self-tapping and thread-forming screw types, American standard.

## Flat Head

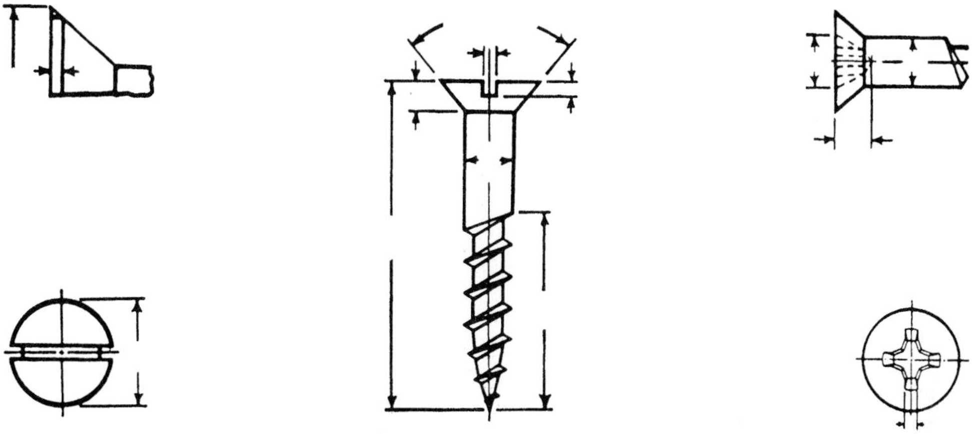

**SLOTTED** | **PHILLIPS**

| | D | A | | | S | H | | J | | T | | | D | M | | T | N | |
|---|---|---|---|---|---|---|---|---|---|---|---|---|---|---|---|---|---|---|
| Nominal Size | Basic Diameter of Screw | Head Diameter Max Sharp | Head Diameter Min Sharp | Head Diameter Absolute Min with Max S | Flat on Min Screw | Height of Head Max | Height of Head Min | Width of Slot Max | Width of Slot Min | Depth of Slot Max | Depth of Slot Min | Number Threads per Inch | Basic Diameter of Screw | Diameter of Recess Max | Diameter of Recess Min | Depth of Recess Max | Width of Recess Min | Driver Size |
| 0 | 0.060 | 0.119 | 0.105 | 0.099 | 0.002 | 0.035 | 0.026 | 0.023 | 0.016 | 0.015 | 0.010 | 32 | 0.060 | 0.069 | 0.056 | 0.043 | 0.014 | 0 |
| 1 | 0.073 | 0.146 | 0.130 | 0.123 | 0.003 | 0.043 | 0.033 | 0.026 | 0.019 | 0.019 | 0.012 | 28 | 0.073 | 0.077 | 0.064 | 0.051 | 0.015 | 0 |
| 2 | 0.086 | 0.172 | 0.156 | 0.147 | 0.003 | 0.051 | 0.040 | 0.031 | 0.023 | 0.023 | 0.015 | 26 | 0.086 | 0.102 | 0.089 | 0.063 | 0.017 | 1 |
| 3 | 0.099 | 0.199 | 0.181 | 0.171 | 0.004 | 0.059 | 0.048 | 0.035 | 0.027 | 0.027 | 0.017 | 24 | 0.099 | 0.107 | 0.094 | 0.068 | 0.018 | 1 |
| 4 | 0.112 | 0.225 | 0.207 | 0.195 | 0.004 | 0.067 | 0.055 | 0.039 | 0.031 | 0.030 | 0.020 | 22 | 0.112 | 0.128 | 0.115 | 0.089 | 0.018 | 1 |
| 5 | 0.125 | 0.252 | 0.232 | 0.220 | 0.005 | 0.075 | 0.062 | 0.043 | 0.035 | 0.034 | 0.022 | 20 | 0.125 | 0.154 | 0.141 | 0.086 | 0.027 | 2 |
| 6 | 0.138 | 0.279 | 0.257 | 0.244 | 0.005 | 0.083 | 0.069 | 0.048 | 0.039 | 0.038 | 0.024 | 18 | 0.138 | 0.174 | 0.161 | 0.106 | 0.029 | 2 |
| 7 | 0.151 | 0.305 | 0.283 | 0.268 | 0.005 | 0.091 | 0.076 | 0.048 | 0.039 | 0.041 | 0.027 | 16 | 0.151 | 0.189 | 0.176 | 0.121 | 0.031 | 2 |
| 8 | 0.164 | 0.332 | 0.308 | 0.292 | 0.006 | 0.100 | 0.084 | 0.054 | 0.045 | 0.045 | 0.029 | 15 | 0.164 | 0.204 | 0.191 | 0.136 | 0.032 | 2 |
| 9 | 0.177 | 0.358 | 0.334 | 0.316 | 0.006 | 0.108 | 0.091 | 0.054 | 0.045 | 0.049 | 0.032 | 14 | 0.177 | 0.214 | 0.201 | 0.146 | 0.033 | 2 |
| 10 | 0.190 | 0.385 | 0.359 | 0.340 | 0.007 | 0.116 | 0.098 | 0.060 | 0.050 | 0.053 | 0.034 | 13 | 0.190 | 0.258 | 0.245 | 0.146 | 0.034 | 3 |
| 12 | 0.216 | 0.438 | 0.410 | 0.389 | 0.008 | 0.132 | 0.112 | 0.067 | 0.056 | 0.060 | 0.039 | 11 | 0.216 | 0.283 | 0.270 | 0.171 | 0.036 | 3 |
| 14 | 0.242 | 0.491 | 0.461 | 0.437 | 0.009 | 0.148 | 0.127 | 0.075 | 0.064 | 0.068 | 0.044 | 10 | 0.242 | 0.303 | 0.290 | 0.191 | 0.039 | 3 |
| 16 | 0.268 | 0.544 | 0.512 | 0.485 | 0.010 | 0.164 | 0.141 | 0.075 | 0.064 | 0.075 | 0.049 | 9 | 0.268 | 0.327 | 0.314 | 0.216 | 0.045 | 3 |
| 18 | 0.294 | 0.597 | 0.563 | 0.534 | 0.011 | 0.180 | 0.155 | 0.084 | 0.072 | 0.083 | 0.054 | 8 | 0.294 | 0.378 | 0.365 | 0.230 | 0.062 | 4 |
| 20 | 0.320 | 0.650 | 0.614 | 0.582 | 0.012 | 0.196 | 0.170 | 0.084 | 0.072 | 0.090 | 0.059 | 8 | 0.320 | 0.393 | 0.380 | 0.245 | 0.065 | 4 |
| 24 | 0.372 | 0.756 | 0.716 | 0.679 | 0.013 | 0.228 | 0.198 | 0.094 | 0.081 | 0.105 | 0.069 | 7 | 0.372 | 0.424 | 0.411 | 0.276 | 0.069 | 4 |

**FIGURE 10.21** American standard wood screw dimensions.

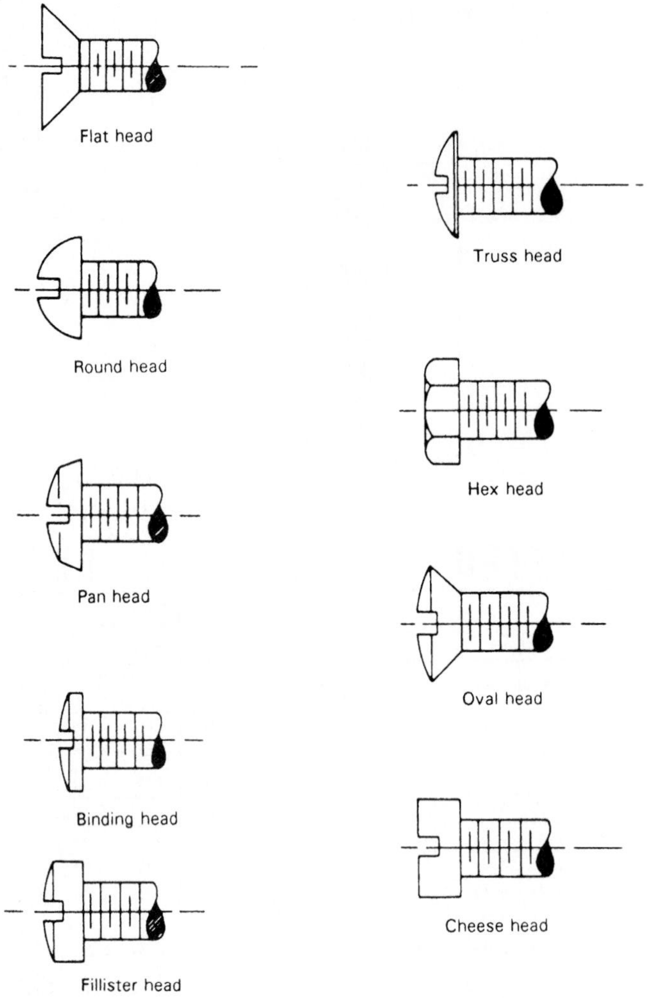


**FIGURE 10.22** Head styles of American standard machine screws.

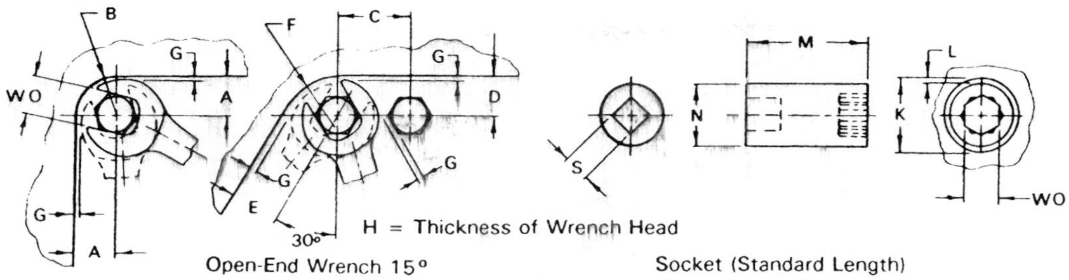


P = Torque that wrench will withstand, pound-inches

J = Torque that wrench will withstand, pound-inches

| W.O. Wrench Opening | Open End Wrench-15° | | | | | | | | | | | Socket Wrench (Standard Length) S = 0.25 | | | S = 0.375 | | | S = 0.500 | | | S = 0.750 | | |
|---|---|---|---|---|---|---|---|---|---|---|---|---|---|---|---|---|---|---|---|---|---|---|---|
| | A min. | B max. | C min. | D min. | E min. | F max. | G ref. | H max. | J min. | K min. | L ref. | M max. | N max. | P min. | M max. | N max. | P min. | M max. | N max. | P min. | M max. | N max. | P min. |
| 0.156 | 0.220 | 0.250 | 0.390 | 0.160 | 0.260 | 0.200 | 0.030 | 0.094 | 25 | .... | .... | .... | .... | .... | .... | .... | .... | .... | .... | .... | .... | .... | .... |
| 0.188 | 0.250 | 0.280 | 0.430 | 0.190 | 0.270 | 0.230 | 0.030 | 0.172 | 40 | 0.370 | 0.030 | 1.000 | 0.610 | 125 | .... | .... | .... | .... | .... | .... | .... | .... | .... |
| 0.250 | 0.280 | 0.340 | 0.530 | 0.270 | 0.310 | 0.310 | 0.030 | 0.172 | 60 | 0.470 | 0.030 | 1.000 | 0.610 | 200 | 1.250 | 0.690 | 250 | .... | .... | .... | .... | .... | .... |
| 0.312 | 0.380 | 0.470 | 0.660 | 0.280 | 0.390 | 0.390 | 0.060 | 0.203 | 125 | 0.550 | 0.030 | 1.000 | 0.610 | 300 | 1.250 | 0.690 | 400 | .... | .... | .... | .... | .... | .... |
| 0.344 | 0.420 | 0.500 | 0.750 | 0.340 | 0.450 | 0.450 | 0.060 | 0.203 | 175 | 0.580 | 0.030 | 1.000 | 0.519 | 450 | 1.250 | 0.690 | 675 | .... | .... | .... | .... | .... | .... |
| 0.375 | 0.420 | 0.500 | 0.780 | 0.360 | 0.450 | 0.520 | 0.060 | 0.219 | 250 | 0.620 | 0.030 | 1.000 | 0.580 | 550 | 1.250 | 0.690 | 900 | 1.500 | 0.880 | 1600 | .... | .... | .... |
| 0.438 | 0.470 | 0.590 | 0.890 | 0.420 | 0.520 | 0.640 | 0.060 | 0.250 | 375 | 0.750 | 0.030 | 1.000 | 0.683 | 550 | 1.250 | 0.880 | 1250 | 1.500 | 0.940 | 1700 | .... | .... | .... |
| 0.500 | 0.520 | 0.640 | 1.000 | 0.470 | 0.580 | 0.660 | 0.060 | 0.266 | 490 | 0.810 | 0.030 | 1.000 | 0.692 | 600 | 1.250 | 0.880 | 1450 | 1.500 | 0.940 | 2000 | .... | .... | .... |
| 0.562 | 0.590 | 0.770 | 1.130 | 0.520 | 0.660 | 0.700 | 0.060 | 0.297 | 700 | 0.870 | 0.030 | .... | .... | .... | 1.250 | 0.932 | 1600 | 1.500 | 0.940 | 2700 | .... | .... | .... |
| 0.594 | 0.640 | 0.830 | 1.210 | 0.530 | 0.700 | 0.700 | 0.060 | 0.344 | 800 | 0.920 | 0.030 | .... | .... | .... | 1.250 | 0.963 | 1750 | 1.562 | 0.970 | 3000 | .... | .... | .... |
| 0.625 | 0.640 | 0.830 | 1.230 | 0.560 | 0.700 | 0.700 | 0.060 | 0.344 | 935 | 0.960 | 0.030 | .... | .... | .... | 1.250 | 0.995 | 2000 | 1.562 | 1.000 | 3600 | .... | .... | .... |
| 0.688 | 0.770 | 0.920 | 1.470 | 0.660 | 0.880 | 0.800 | 0.060 | 0.375 | 1250 | 1.030 | 0.030 | .... | .... | .... | 1.250 | 1.068 | 2000 | 1.562 | 1.065 | 4300 | .... | .... | .... |
| 0.750 | 0.770 | 0.920 | 1.510 | 0.670 | 0.880 | 0.800 | 0.060 | 0.375 | 1500 | 1.120 | 0.030 | .... | .... | .... | 1.250 | 1.120 | 2000 | 1.562 | 1.130 | 5000 | .... | .... | .... |
| 0.781 | 0.830 | 0.960 | 1.550 | 0.690 | 0.890 | 0.840 | 0.060 | 0.375 | 1615 | 1.150 | 0.030 | .... | .... | .... | 1.250 | 1.126 | 2000 | 1.625 | 1.130 | 5000 | .... | .... | .... |
| 0.812 | 0.910 | 1.120 | 1.660 | 0.720 | 0.970 | 0.860 | 0.060 | 0.406 | 1710 | 1.200 | 0.030 | .... | .... | .... | 1.250 | 1.213 | 2000 | 1.625 | 1.222 | 5000 | .... | .... | .... |
| 0.875 | 0.970 | 1.150 | 1.810 | 0.800 | 1.060 | 0.910 | 0.060 | 0.438 | 2250 | 1.280 | 0.030 | .... | .... | .... | .... | .... | .... | 1.750 | 1.285 | 5000 | .... | .... | .... |
| 0.938 | 0.970 | 1.150 | 1.850 | 0.810 | 1.060 | 0.950 | 0.060 | 0.438 | 2750 | 1.370 | 0.030 | .... | .... | .... | .... | .... | .... | 1.750 | 1.410 | 5000 | .... | .... | .... |
| 1.000 | 1.050 | 1.230 | 2.00 | 0.880 | 1.160 | 1.060 | 0.060 | 0.500 | 3250 | 1.470 | 0.030 | .... | .... | .... | .... | .... | .... | 1.750 | 1.410 | 5000 | .... | .... | .... |
| 1.062 | 1.090 | 1.250 | 2.100 | 0.970 | 1.200 | 1.200 | 0.080 | 0.500 | 3500 | 1.550 | 0.030 | .... | .... | .... | .... | .... | .... | 1.844 | 1.506 | 5000 | .... | .... | .... |
| 1.125 | 1.140 | 1.370 | 2.210 | 1.000 | 1.270 | 1.230 | 0.080 | 0.500 | 4000 | 1.610 | 0.030 | .... | .... | .... | .... | .... | .... | 1.939 | 1.567 | 5000 | .... | .... | .... |
| 1.250 | 1.270 | 1.420 | 2.440 | 1.080 | 1.390 | 1.310 | 0.080 | 0.562 | 5250 | 1.890 | 0.030 | .... | .... | .... | .... | .... | .... | 2.000 | 1.723 | 5000 | 2.375 | 1.856 | 7250 |
| 1.312 | 1.390 | 1.690 | 2.630 | 1.170 | 1.520 | 1.340 | 0.080 | 0.562 | 6000 | 1.980 | 0.030 | .... | .... | .... | .... | .... | .... | .... | .... | .... | 2.600 | 1.920 | 8000 |
| 1.438 | 1.470 | 1.720 | 2.800 | 1.250 | 1.590 | 1.340 | 0.090 | 0.641 | 7500 | 2.140 | 0.030 | .... | .... | .... | .... | .... | .... | .... | .... | .... | 2.625 | 2.076 | 9550 |
| 1.500 | 1.470 | 1.720 | 2.840 | 1.270 | 1.590 | 1.450 | 0.090 | 0.641 | 8250 | 2.200 | 0.030 | .... | .... | .... | .... | .... | .... | .... | .... | .... | 2.625 | 2.170 | 10450 |
| 1.625 | 1.560 | 1.880 | 3.100 | 1.380 | 1.750 | 1.560 | 0.090 | 0.641 | 9000 | 2.390 | 0.030 | .... | .... | .... | .... | .... | .... | .... | .... | .... | 2.750 | 2.325 | 11750 |

Note: **P** = Torque that wrench will withstand, pound-inches. No allowances have been made for torque devices. **H** = wrench head thickness. **J** = Torque that wrench will withstand, pound-inches. Tabulated dimensions are in inches.

**FIGURE 10.23(*a*)** Wrench clearances for open-end wrenches and socket wrenches—American standard.

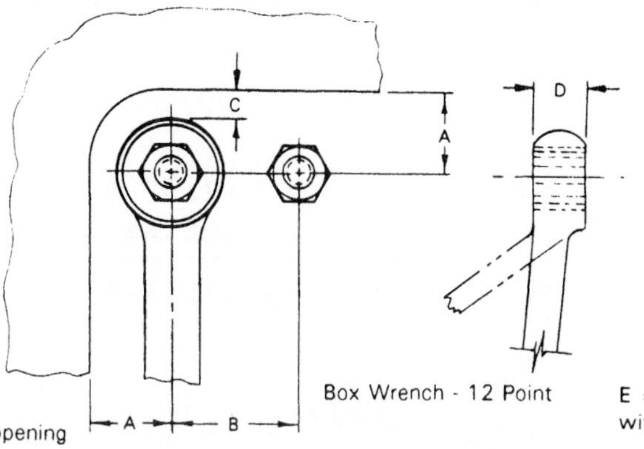


| W.O. | A min. | B min. | C ref. | D max. | E min. |
|---|---|---|---|---|---|
| 0.156 | 0.190 | 0.280 | 0.030 | 0.156 | 100 |
| 0.188 | 0.200 | 0.309 | 0.030 | 0.172 | 150 |
| 0.250 | 0.270 | 0.410 | 0.030 | 0.250 | 150 |
| 0.312 | 0.300 | 0.480 | 0.030 | 0.281 | 210 |
| 0.344 | 0.300 | 0.500 | 0.030 | 0.281 | 250 |
| 0.375 | 0.340 | 0.560 | 0.030 | 0.344 | 370 |
| 0.438 | 0.400 | 0.650 | 0.030 | 0.359 | 650 |
| 0.500 | 0.450 | 0.740 | 0.030 | 0.375 | 1020 |
| 0.562 | 0.500 | 0.830 | 0.030 | 0.406 | 1200 |
| 0.594 | 0.530 | 0.870 | 0.030 | 0.469 | 1200 |
| 0.625 | 0.560 | 0.920 | 0.030 | 0.469 | 2000 |
| 0.688 | 0.590 | 0.990 | 0.030 | 0.531 | 2300 |
| 0.750 | 0.660 | 1.090 | 0.030 | 0.594 | 2600 |
| 0.781 | 0.690 | 1.140 | 0.030 | 0.594 | 2600 |
| 0.812 | 0.720 | 1.190 | 0.030 | 0.594 | 3000 |
| 0.875 | 0.750 | 1.260 | 0.030 | 0.594 | 3300 |
| 0.938 | 0.780 | 1.320 | 0.030 | 0.656 | 4100 |
| 1.000 | 0.810 | 1.390 | 0.030 | 0.718 | 4900 |
| 1.062 | 0.840 | 1.450 | 0.030 | 0.781 | 5400 |
| 1.125 | 0.950 | 1.600 | 0.030 | 0.844 | 5900 |
| 1.250 | 0.980 | 1.700 | 0.030 | 0.875 | 7200 |
| 1.312 | 1.090 | 1.850 | 0.030 | 0.906 | 8000 |
| 1.438 | 1.220 | 2.050 | 0.030 | 1.000 | 8400 |
| 1.500 | 1.270 | 2.140 | 0.030 | 1.062 | 10450 |
| 1.625 | 1.340 | 2.280 | 0.030 | 1.156 | 11750 |

**FIGURE 10.23(*b*)** Wrench clearances for standard box wrenches—American standard.

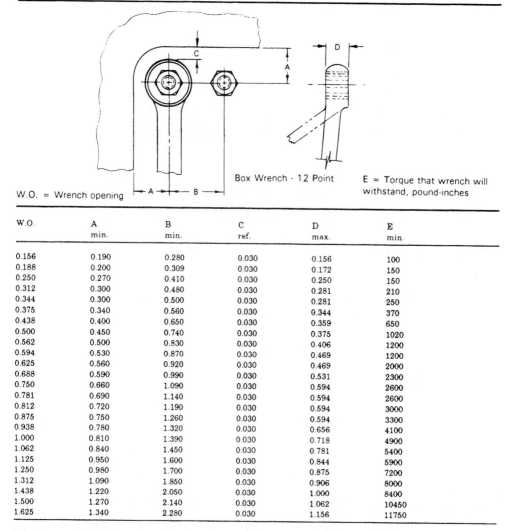

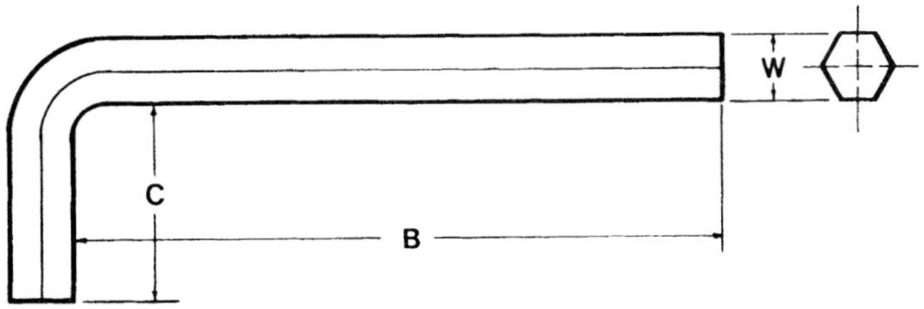


| | DIMENSIONS | | | | | MECHANICAL PROPERTIES | |
|---|---|---|---|---|---|---|---|
| | key size W | | B nom. | | C nom. | torsional shear strength Lb-ins min. | torsional yield Lb-ins min. |
| dash no. | max. | min. | short arm | long arm | | | |
| 1 | .028 | .0275 | 1.219 | 2.594 | .219 | 1.2 | 1.1 |
| 2 | .035 | .0345 | 1.219 | 2.672 | .344 | 2.4 | 2.1 |
| 3 | .050 | .049 | 1.656 | 2.844 | .531 | 7 | 6.0 |
| 4 | 1/16 | .0615 | 1.750 | 3.000 | .562 | 12 | 11 |
| 5 | 5/64 | .0771 | 1.875 | 3.188 | .609 | 26 | 23 |
| 6 | 3/32 | .0927 | 2.000 | 3.375 | .656 | 46 | 40 |
| 7 | 7/64 | .1077 | 2.125 | 3.562 | .703 | 73 | 63 |
| 8 | 1/8 | .1235 | 2.250 | 3.750 | .750 | 108 | 94 |
| 9 | 9/64 | .1391 | 2.375 | 3.960 | .796 | 154 | 134 |
| 10 | 5/32 | .1547 | 2.500 | 4.125 | .844 | 210 | 183 |
| 11 | 3/16 | .1860 | 2.750 | 4.500 | .938 | 364 | 317 |
| 12 | 7/32 | .2172 | 3.000 | 4.875 | 1.031 | 580 | 502 |
| 13 | 1/4 | .2480 | 3.250 | 5.250 | 1.125 | 860 | 750 |
| 14 | 5/16 | .3110 | 3.750 | 6.000 | 1.250 | 1,685 | 1,465 |
| 15 | 3/8 | .3730 | 4.250 | 6.750 | 1.375 | 2,900 | 2,520 |
| 16 | 7/16 | .4355 | 4.750 | 7.500 | 1.500 | 4,400 | 3,860 |
| 17 | 1/2 | .4975 | 5.250 | 8.250 | 1.625 | 6,600 | 5,750 |
| 18 | 9/16 | .5600 | 5.750 | 9.000 | 1.750 | 9,200 | 8,000 |
| 19 | 5/8 | .6225 | 6.250 | 9.750 | 1.875 | 12,650 | 11,000 |
| 20 | 3/4 | .7470 | 7.250 | 11.250 | 2.125 | 20,800 | 18,100 |
| 21 | 7/8 | .8720 | 8.250 | 12.750 | 2.375 | 29,200 | 25,400 |
| 22 | 1 | .9970 | 9.250 | 14.250 | 2.625 | 43,700 | 38,000 |
| 23 | 1 1/4 | 1.243 | 11.250 | .... | 3.000 | 71,900 | 62,500 |
| 24 | 1 1/2 | 1.493 | 13.250 | .... | 3.500 | 124,000 | 108,000 |
| 25 | 1 3/4 | 1.743 | 15.250 | .... | 4.000 | 198,000 | 172,000 |
| 26 | 2 | 1.993 | 17.250 | .... | 4.500 | 276,000 | 240,000 |

**FIGURE 10.23(*c*)** Dimensions and properties of hex keys—Allen wrenches (inch series).

Figure 10.24 shows the wrench opening for American standard nuts.

Figure 10.25 is a photograph of a sample of hardware components which shows some of the available screws, bolts, nuts, washers, pins, and special fasteners which are used throughout industry. The number and styles of standard hardware components are enormous, and new components are being developed constantly. The catalogs of the hardware manufacturers are available which show dimensions and mechanical properties of their particular fasteners or hardware components. The fastener handbooks which are available from IFI (see Chap. 15) are among the most complete and technically correct for both the inch series and metric components.

## 10.2 THREAD SYSTEMS: AMERICAN STANDARD AND METRIC (60° V)

The international standard screw threads consist of the unified inch series and the metric series. The metric series are standardized into the M and MJ profiles. The unified series profile is designated as *unified national* (UN). Another unified profile is designated as UNR,

| Maximum Width Across Flats of Nut | Wrench Opening Minimum | Wrench Opening Maximum |
|---|---|---|
| 5/32 | 0.158 | 0.163 |
| 3/16 | 0.190 | 0.195 |
| 7/32 | 0.220 | 0.225 |
| 1/4 | 0.252 | 0.257 |
| 9/32 | 0.283 | 0.288 |
| 5/16 | 0.316 | 0.322 |
| 11/32 | 0.347 | 0.353 |
| 3/8 | 0.378 | 0.384 |
| 7/16 | 0.440 | 0.446 |
| 1/2 | 0.504 | 0.510 |
| 9/16 | 0.566 | 0.573 |
| 5/8 | 0.629 | 0.636 |
| 11/16 | 0.692 | 0.699 |
| 3/4 | 0.755 | 0.763 |
| 13/16 | 0.818 | 0.826 |
| 7/8 | 0.880 | 0.888 |
| 15/16 | 0.944 | 0.953 |
| 1 | 1.006 | 1.015 |
| 1-1/16 | 1.068 | 1.077 |
| 1-1/8 | 1.132 | 1.142 |
| 1-1/4 | 1.257 | 1.267 |
| 1-5/16 | 1.320 | 1.331 |
| 1-3/8 | 1.383 | 1.394 |
| 1-7/16 | 1.446 | 1.457 |
| 1-1/2 | 1.508 | 1.520 |
| 1-5/8 | 1.634 | 1.646 |
| 1-11/16 | 1.696 | 1.708 |
| 1-13/16 | 1.822 | 1.835 |
| 1-7/8 | 1.885 | 1.898 |
| 2 | 2.011 | 2.025 |
| 2-1/16 | 2.074 | 2.088 |
| 2-3/16 | 2.200 | 2.215 |
| 2-1/4 | 2.262 | 2.277 |
| 2-3/8 | 2.388 | 2.404 |
| 2-7/16 | 2.450 | 2.466 |
| 2-9/16 | 2.576 | 2.593 |
| 2-5/8 | 2.639 | 2.656 |
| 2-3/4 | 2.766 | 2.783 |
| 2-13/16 | 2.827 | 2.845 |
| 2-15/16 | 2.954 | 2.973 |
| 3 | 3.016 | 3.035 |

Note: Wrenches are marked with the "Nominal size of wrench", which is equal to the basic or maximum width across the flats of the corresponding nut.

**FIGURE 10.24** Wrench openings for American standard nuts.

which has a rounded root on the external thread. The metric profile MJ also has a rounded root at the external thread and a larger minor diameter of both the internal and external thread. Both the UNR and MJ profiles are used for applications requiring high fatigue strength and are also employed in aerospace applications.

A constant-pitch unified series is also standardized and consists of 4, 6, 8, 12, 16, 20, 28, and 32 threads per inch. These are used for sizes over 1 inch diameter, and the 8UN, 12UN, and 16UN are the *preferred* pitches.

Figure 10.26*a* shows that both the unified and metric M series use the same profile geometry.

The other important screw thread systems are shown in Fig. 10.26*b* to 10.26*l* and include the acme, stub acme, Whitworth, and buttress. These figures show all the dimensional relationships and geometry of these thread systems for design purposes and reference.

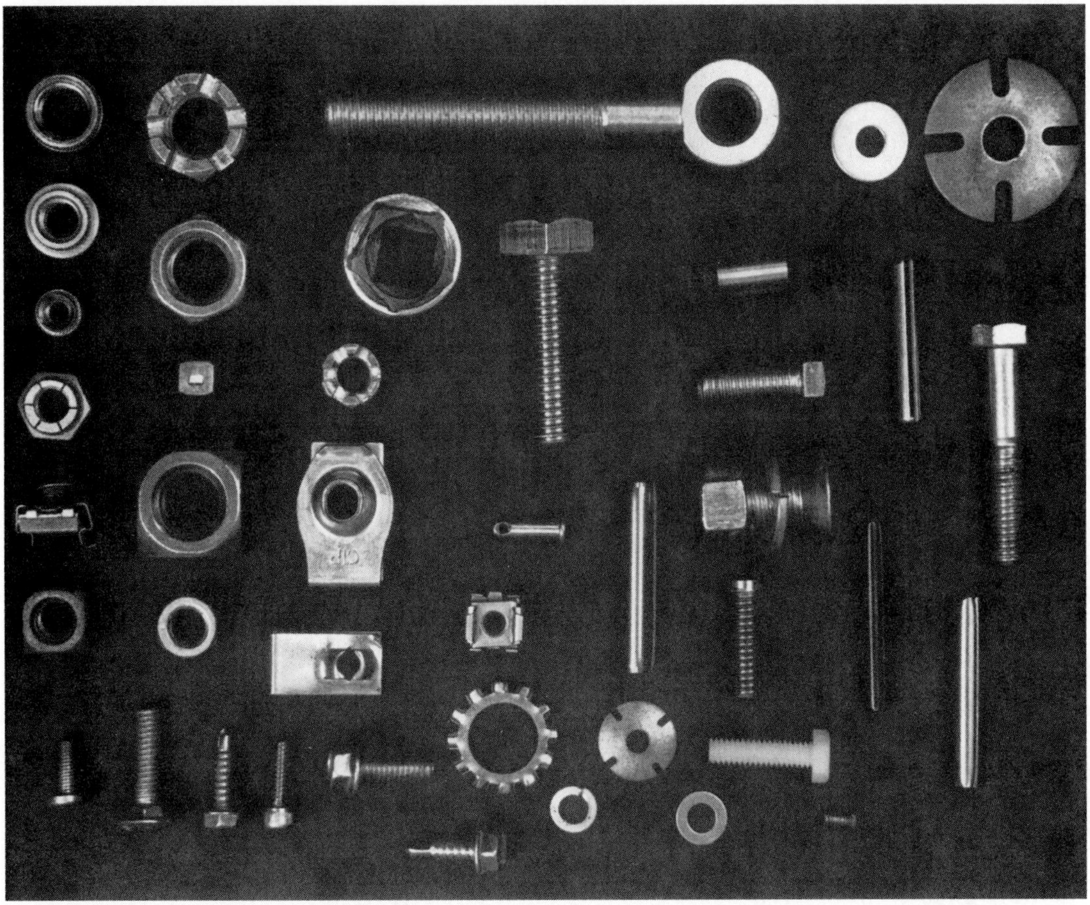

**FIGURE 10.25** Samples of standard hardware items.

The unified threads are further classified as UN A for external thread and UN B for internal threads. The UN series contains three fit classes, 1A, 2A, and 3A for external threads, and 1B, 2B, and 3B for internal threads. A typical engineering drawing "callout" for a coarse external class 2 thread in size ¼–20 would be shown as follows:

(0.250–20 UNC–2A)

A ¼–20 class 2 tapped hole would be shown as

(0.250–20 UNC–2B)

***Thread Fit Classes.*** Unified thread series and interference-fit threads are as follows:

Unified thread series:

Class:

1A External, loose fit for easy assembly and noncritical uses.

2A External, general applications where plating may be applied.

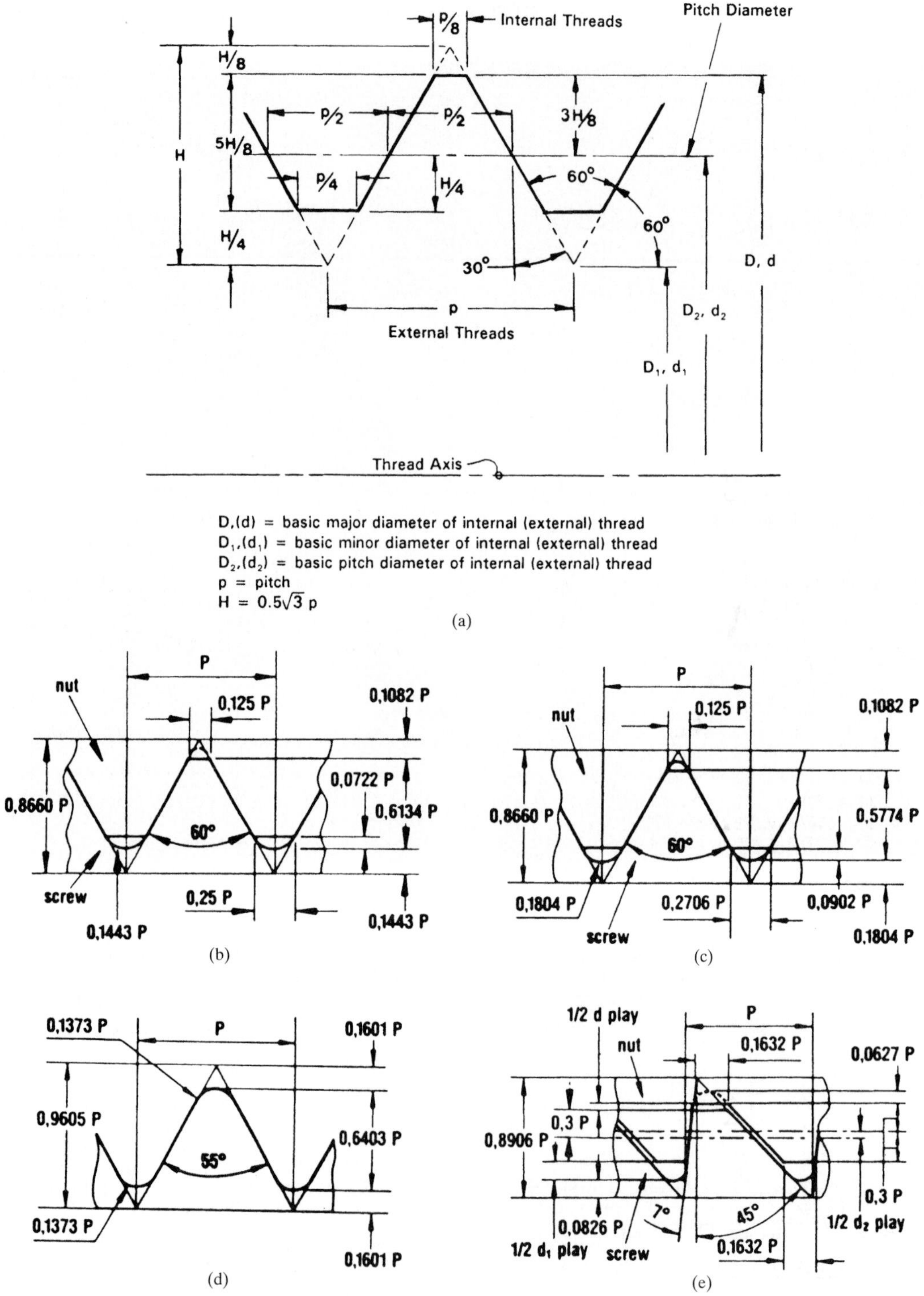


**FIGURE 10.26** (*a*) Basic thread profiles for unified national (UN) and metric (M) threads (ISO 68); thread form dimensions for (*b*) ISO M and UN; (*c*) UNJ controlled root radii; (*d*) Whitworth (BSW); (*e*) American buttress (7° face).

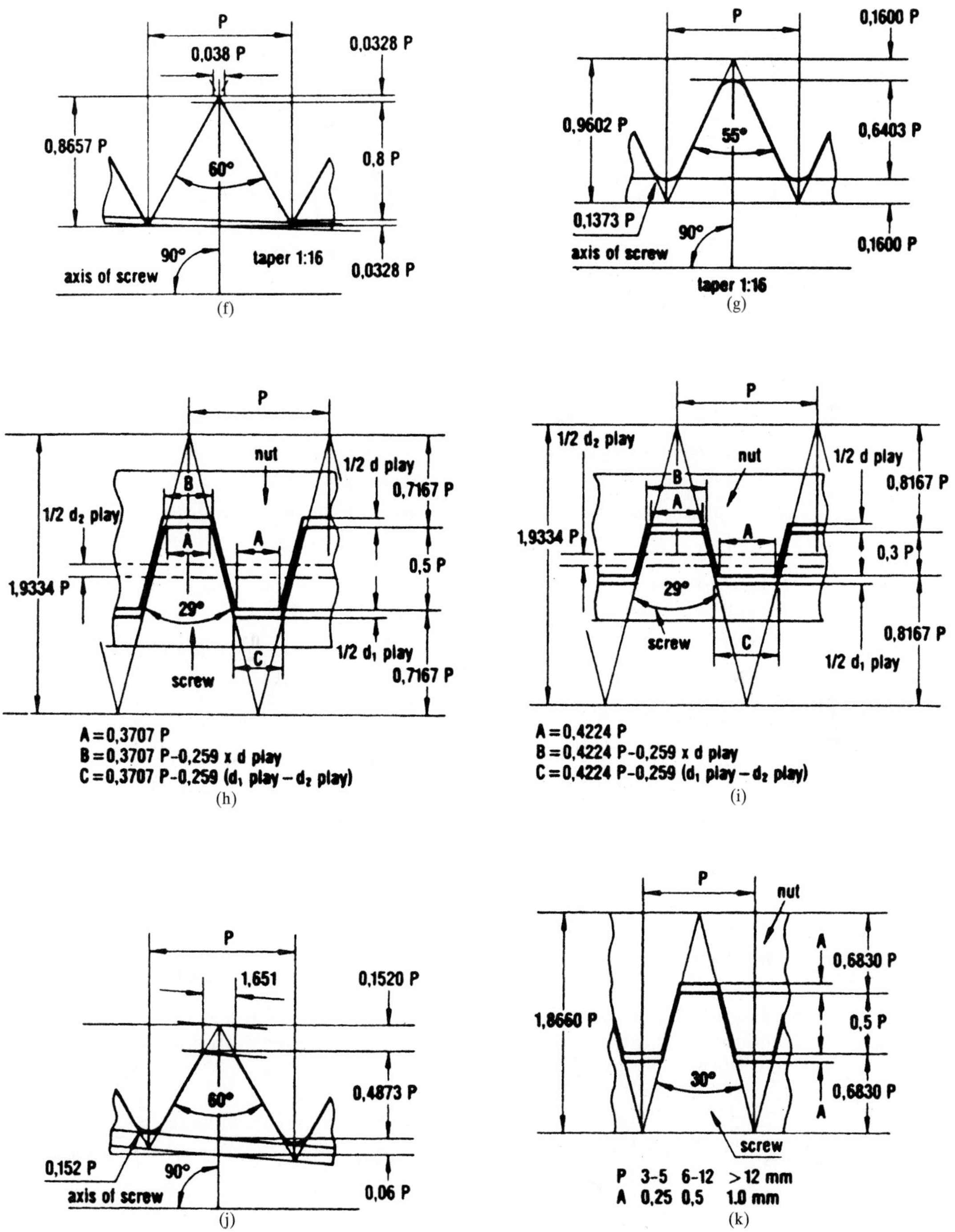


**FIGURE 10.26** Thread form dimensions for (*f*) NPT (American national pipe thread); (*g*) BSPT (British standard pipe thread); (*h*) acme (29°); (*i*) acme stub (29°); (*j*) API (taper 1:6 (V 0.38-in R); (*k*) TR DIN 103.

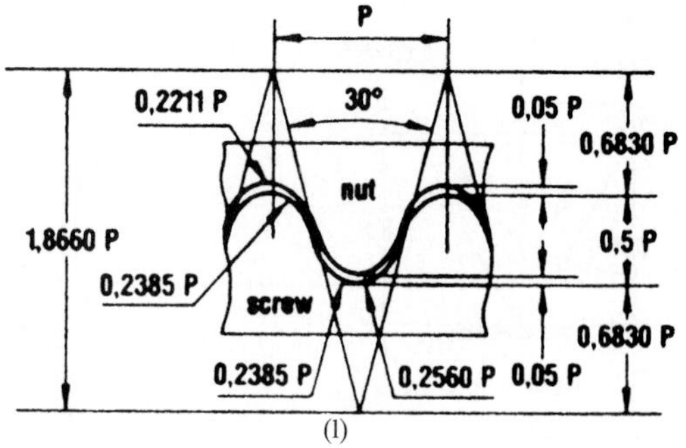


**FIGURE 10.26** Thread form dimensions for (*l*) RD DIN 405 (round).

3A External, tight fit used for great accuracy. No plating allowance is provided.
1B Internal, loose fit for easy assembly and noncritical uses.
2B Internal, general applications where plating may be applied.
3B Internal, tight fit used for great accuracy. No plating allowance is provided.

Class 5 Interference-fit threads:

Class 5 external:

NC5 HF For driving in hard ferrous materials >160 Bhn.
NC5 CSF For driving in copper alloys and soft ferrous materials <160 Bhn.
NC5 ONF For other nonferrous materials, any hardness.

Class 5 internal:

NC5 IF Entire ferrous material range.
NC5 INF Entire nonferrous material range.

Interference-fit threads are commonly used on threaded studs to assure a tight, vibration-resistant fit. The internal thread of a class 5 interference-fit application should be lubricated for best results when torquing the stud or externally threaded part into the class 5 fit-tapped hole.

## 10.2.1 Unified and Metric Thread Data

***Unified National Tap Drill Sizes.*** You may calculate the tap drill diameter for 75 and 100 percent thread for the unified and metric M profile series using the following equations. For unified national (UN) threads:

$$D_{td} = \begin{cases} D_m - \dfrac{0.947}{n} & \text{(tap drill diameter for 75\% thread)} \\ D_m - \dfrac{1.299}{n} & \text{(tap drill diameter for 100\% thread)} \end{cases}$$

where $D_{td}$ = diameter of tap drill, in
$D_m$ = major diameter of external thread or tap diameter, in
$n$ = number of threads per inch

For metric threads, M profile:

$$D_d = D_M - \frac{(\%T)(p)}{76.98} \qquad \%T = \frac{76.98}{p}(D_M - D_d)$$

where $D_d$ = drilled hole diameter, mm
$D_M$ = basic major diameter, mm
$\%T$ = percent of thread (70 percent = 0.70, etc.)
$p$ = pitch of thread, mm

(*Note:* Recommended percent of thread is usually taken as 72 to 75 for a general class 2 type fit, which allows for electroplatings such as zinc, cadmium, chrome, nickel, Parkerizing, tin, etc.)

Unified national coarse screw thread data are presented in Fig. 10.27.

Unified national fine screw thread data are presented in Fig. 10.28.

| Thread | Tap Drill | Decimal in | Stress Area in$^2$ | Basic Pitch Diameter, in |
|---|---|---|---|---|
| #1–64 | #53 | 0.0595 | 0.0026 | 0.0629 |
| #2–56 | #50 | 0.0700 | 0.0037 | 0.0744 |
| #3–48 | #47 | 0.0785 | 0.0048 | 0.0855 |
| #4–40 | #43 | 0.0890 | 0.0060 | 0.0958 |
| #5–40 | #38 | 0.1015 | 0.0080 | 0.1088 |
| #6–32 | #36 | 0.1065 | 0.0090 | 0.1177 |
| #8–32 | #29 | 0.1360 | 0.0140 | 0.1437 |
| #10–24 | #25 | 0.1495 | 0.0175 | 0.1629 |
| 1/4–20 | #7 | 0.2010 | 0.0318 | 0.2175 |
| 5/16–18 | F | 0.2570 | 0.0524 | 0.2764 |
| 3/8–16 | 5/16 | 0.3125 | 0.0775 | 0.3344 |
| 7/16–14 | T | 0.3580 | 0.1063 | 0.3911 |
| 1/2–13 | 27/64 | 0.4219 | 0.1419 | 0.4500 |
| 9/16–12 | 31/64 | 0.4844 | 0.1820 | 0.5084 |
| 5/8–11 | 17/32 | 0.5312 | 0.2260 | 0.5660 |
| 3/4–10 | 41/64 | 0.6406 | 0.3340 | 0.6850 |
| 7/8–9 | 49/64 | 0.7656 | 0.4620 | 0.8028 |
| 1–8 | 7/8 | 0.8750 | 0.6060 | 0.9188 |

**FIGURE 10.27** Screw thread data, UNC.

| Thread | Tap drill | Decimal in | Stress Area in$^2$ | Basic Pitch Diameter, in |
|---|---|---|---|---|
| #0–80 | 3/64 | 0.0469 | 0.0018 | 0.0519 |
| #1–72 | #53 | 0.0595 | 0.0027 | 0.0640 |
| #2–64 | #50 | 0.0700 | 0.0039 | 0.0759 |
| #3–56 | #45 | 0.0820 | 0.0052 | 0.0874 |
| #4–48 | #42 | 0.0935 | 0.0066 | 0.0985 |
| #5–44 | #37 | 0.1040 | 0.0083 | 0.1102 |
| #6–40 | #33 | 0.1130 | 0.0102 | 0.1218 |
| #8–36 | #29 | 0.1360 | 0.0147 | 0.1460 |
| #10–32 | #21 | 0.1590 | 0.0200 | 0.1697 |
| 1/4–28 | #3 | 0.2130 | 0.0364 | 0.2268 |
| 5/16–24 | I | 0.2720 | 0.0580 | 0.2854 |
| 3/8–24 | Q | 0.3320 | 0.0878 | 0.3479 |
| 7/16–20 | 25/64 | 0.3906 | 0.1187 | 0.4050 |
| 1/2–20 | 29/64 | 0.4531 | 0.1599 | 0.4675 |
| 9/16–18 | 33/64 | 0.5156 | 0.2030 | 0.5264 |
| 5/8–18 | 9/16 | 0.5625 | 0.2560 | 0.5889 |
| 3/4–16 | 11/16 | 0.6875 | 0.3730 | 0.7094 |
| 7/8–14 | 13/16 | 0.8125 | 0.5090 | 0.8286 |
| 1–12 | 29/32 | 0.9063 | 0.6630 | 0.9459 |

**FIGURE 10.28** Screw thread data, UNF.

Metric M profile thread data are presented in Fig. 10.29*a* and 10.29*b*.

American national drill sizes are shown in Fig. 10.31 (see p. 10.44). Both number and letter size drills are indicated with their decimal equivalents.

The standard limits of size for both American series and unified series screw threads are fully covered in Fig. 10.30. The tables shown in the figure are extracted from the National Bureau of Standards *Handbook H-28* and are in general agreement with the current ANSI standards. The tables cover the thread sizes 0–80 through 3–16UN.

***Engagement of Threads.*** The length of engagement of stud end or bolt end $E$ can be stated in terms of $D$, the major diameter of the thread. In general

- Steel stud in cast iron or steel: $E = 1.50D$
- Steel stud in hardened steel or high-strength bronze: $E = D$
- Steel stud in aluminum or magnesium alloys, shock loads: $E = 2.00D + 0.062$
- Steel stud as above, subjected to normal loads: $E = 1.50D + 0.062$

***Load to Break a Threaded Section.*** For screws or bolts

$$P_b = SA_{ts}$$

where $P_b$ = load to break the screw or bolt, lbf
$S$ = ultimate tensile strength of screw or bolt material, psi
$A_{ts}$ = tensile stress area of screw or bolt thread, in$^2$

| Thread Designation Dia. × Pitch, mm | Tap Drill, mm | Pitch Dia. 6H, Internal, mm | Pitch Dia. 6g, External, mm |
|---|---|---|---|
| M1.6 × 0.35 | 1.25 | 1.373 | 1.291 |
| M2 × 0.4 | 1.60 | 1.740 | 1.654 |
| M2.5 × 0.45 | 2.05 | 2.208 | 2.117 |
| M3 × 0.5 | 2.50 | 2.675 | 2.580 |
| M3.5 × 0.6 | 2.90 | 3.110 | 3.004 |
| M4 × 0.7 | 3.30 | 3.545 | 3.433 |
| M5 × 0.8 | 4.20 | 4.480 | 4.361 |
| M6 × 1 | 5.00 | 5.350 | 5.212 |
| M8 × 1.25 | 6.70 | 7.188 | 7.042 |
| M8 × 1 | 7.00 | 7.350 | 7.212 |
| M10 × 1.5 | 8.50 | 9.026 | 8.862 |
| M10 × 1.25 | 8.70 | 9.188 | 9.042 |
| M10 × 0.75 | — | 9.513 | 9.391 |
| M12 × 1.75 | 10.20 | 10.863 | 10.679 |
| M12 × 1.5 | — | 11.026 | 10.854 |
| M12 × 1.25 | 10.80 | 11.188 | 11.028 |
| M12 × 1 | — | 11.350 | 11.206 |
| M14 × 2 | 12.00 | 12.701 | 12.503 |
| M14 × 1.5 | 12.50 | 13.026 | 12.854 |
| M15 × 1 | — | 14.350 | 14.206 |
| M16 × 2 | 14.00 | 14.701 | 14.503 |
| M16 × 1.5 | 14.50 | 15.026 | 14.854 |
| M17 × 1 | — | 16.350 | 16.206 |
| M18 × 1.5 | 16.50 | 17.026 | 16.854 |
| M20 × 2.5 | 17.50 | 18.376 | 18.164 |
| M20 × 1.5 | 18.50 | 19.026 | 18.854 |
| M20 × 1 | — | 19.350 | 19.206 |
| M22 × 2.5 | 19.50 | 20.376 | 20.164 |
| M22 × 1.5 | 20.50 | 21.026 | 20.854 |
| M24 × 3 | 21.00 | 22.051 | 21.803 |
| M24 × 2 | 22.00 | 22.701 | 22.493 |
| M25 × 1.5 | — | 24.026 | 23.854 |

**FIGURE 10.29** (*a*) Metric thread data, M profile, internal and external.

| Nom Product Dia and Thread Pitch | Stress Area $mm^2$ | Class 4.6 | | | Class 4.8 | | | Class 5.8 | | | Classes 8.8 and 8.8.3 | | |
|---|---|---|---|---|---|---|---|---|---|---|---|---|---|
| | | Proof Load | | Tensile Strength, min | Proof Load | | Tensile Strength min | Proof Load | | Tensile Strength, min | Proof Load | | Tensile Strength, min |
| | | Method 1 | Method 2 | | Method 1 | Method 2 | | Method 1 | Method 2 | | Method 1 | Method 2 | |
| M1.6 × 0.35 | 1.27 | ... | ... | ... | 0.39 | 0.43 | 0.53 | ... | ... | ... | ... | ... | ... |
| M2 × 0.4 | 2.07 | ... | ... | ... | 0.64 | 0.70 | 0.87 | ... | ... | ... | ... | ... | ... |
| M2.5 × 0.45 | 3.39 | ... | ... | ... | 1.05 | 1.15 | 1.42 | ... | ... | ... | ... | ... | ... |
| M3 × 0.5 | 5.03 | ... | ... | ... | 1.56 | 1.71 | 2.11 | ... | ... | ... | ... | ... | ... |
| M3.5 × 0.6 | 6.78 | ... | ... | ... | 2.10 | 2.31 | 2.85 | ... | ... | ... | ... | ... | ... |
| M4 × 0.7 | 8.78 | ... | ... | ... | 2.72 | 2.99 | 3.69 | ... | ... | ... | ... | ... | ... |
| M5 × 0.8 | 14.2 | 3.20 | 3.41 | 5.68 | 4.40 | 4.83 | 5.96 | 5.40 | 5.96 | 7.38 | ... | ... | ... |
| M6 × 1 | 20.1 | 4.52 | 4.82 | 8.04 | 6.23 | 6.83 | 8.44 | 7.64 | 8.44 | 10.5 | ... | ... | ... |
| M8 × 1.25 | 36.6 | 8.24 | 8.78 | 14.6 | 11.3 | 12.4 | 15.4 | 13.9 | 15.4 | 19.0 | ... | ... | ... |
| M10 × 1.5 | 58.0 | 13.1 | 13.9 | 23.2 | 18.0 | 19.7 | 24.4 | 22.0 | 24.4 | 30.2 | ... | ... | ... |
| M12 × 1.75 | 84.3 | 19.0 | 20.2 | 33.7 | 26.1 | 28.7 | 35.4 | 32.0 | 35.4 | 43.8 | ... | ... | ... |
| M14 × 2 | 115 | 25.9 | 27.6 | 46.0 | 35.7 | 39.1 | 48.3 | 43.7 | 48.3 | 59.8 | ... | ... | ... |
| M16 × 2 | 157 | 35.3 | 37.7 | 62.8 | 48.7 | 53.4 | 65.9 | 59.7 | 65.9 | 81.6 | 94.2 | 104 | 130 |
| M20 × 2.5 | 245 | 55.1 | 58.8 | 98.0 | ... | ... | ... | 93.1 | 103 | 127 | 147 | 162 | 203 |
| M22 × 2.5 | 303 | ... | ... | ... | ... | ... | ... | ... | ... | ... | 182 | 200 | 251 |
| M24 × 3 | 353 | 79.4 | 84.7 | 141 | ... | ... | ... | 134 | 148 | 184 | 212 | 233 | 293 |
| M27 × 3 | 459 | ... | ... | ... | ... | ... | ... | ... | ... | ... | 275 | 303 | 381 |
| M30 × 3.5 | 561 | 126 | 135 | 224 | ... | ... | ... | ... | ... | ... | 337 | 370 | 466 |
| M36 × 4 | 817 | 184 | 196 | 327 | ... | ... | ... | ... | ... | ... | 490 | 539 | 678 |
| M42 × 4.5 | 1120 | 252 | 269 | 448 | ... | ... | ... | ... | ... | ... | ... | ... | ... |
| M48 × 5 | 1470 | 331 | 353 | 588 | ... | ... | ... | ... | ... | ... | ... | ... | ... |
| M56 × 5.5 | 2030 | 457 | 487 | 812 | ... | ... | ... | ... | ... | ... | ... | ... | ... |
| M64 × 6 | 2680 | 603 | 643 | 1070 | ... | ... | ... | ... | ... | ... | ... | ... | ... |
| M72 × 6 | 3460 | 779 | 830 | 1380 | ... | ... | ... | ... | ... | ... | ... | ... | ... |
| M80 × 6 | 4340 | 977 | 1040 | 1740 | ... | ... | ... | ... | ... | ... | ... | ... | ... |
| M90 × 6 | 5590 | 1260 | 1340 | 2240 | ... | ... | ... | ... | ... | ... | ... | ... | ... |
| M100 × 6 | 6990 | 1570 | 1680 | 2800 | ... | ... | ... | ... | ... | ... | ... | ... | ... |

Note: Table values are given in kilo-Newtons, (kN)

**FIGURE 10.29** (*b*) Proof loads and tensile strength requirements, metric ISO.

***Tensile Stress Area Calculation.*** The tensile stress area $A_{ts}$ of screws and bolts is derived from

$$A_{ts} = \frac{\pi}{4}\left(D - \frac{0.9743}{n}\right)^2 \quad \text{(for inch series threads)}$$

where $A_{ts}$ = tensile stress area, $in^2$
$D$ = basic major diameter of thread, in
$n$ = number of threads per inch

(*Note:* You may select the stress areas for unified bolts or screws by using Figs. 10.27 and 10.28, and you can derive the metric stress areas by converting millimeters to inches for each metric fastener and using the preceding equation.)

***Thread Engagement to Prevent Stripping.*** The calculation approach depends on materials selected.

**1.** *Same materials* chosen for both external threaded part and internal threaded part:

$$E_L = \frac{2A_{ts}}{\pi D_m\{\tfrac{1}{2} + [n(p_d - D_m)/\sqrt{3}]\}}$$

where $E_L$ = length of engagement of the thread, in
$D_m$ = maximum minor diameter of internal thread, in
$n$ = number of threads per in

## Standard Limits of Size: Unified and American Screw Threads

| Nominal size and threads per inch | Series designation | External | | | | | | | | | Internal | | | | | | |
|---|---|---|---|---|---|---|---|---|---|---|---|---|---|---|---|---|---|
| | | Class | Allowance, in | Major diameter limits, in | | | Pitch diameter limits, in | | | Minor diameter, in | Class | Minor diameter limits, in | | Pitch diameter limits, in | | | Major diameter, in |
| | | | | Max* | Min | Min' | Max* | Min | Tolerance | | | Min | Max | Min | Max | Tolerance | Min |
| No. 0–80 to ¼–13 | | | | | | | | | | | | | | | | | |
| 0–80 | NF | 2A | 0.0005 | 0.0595 | 0.0563 | | 0.0514 | 0.0496 | 0.0018 | 0.0442 | 2B | 0.0465 | 0.0514 | 0.0519 | 0.0542 | 0.0023 | 0.0600 |
| | | 3A | 0.0000 | 0.0600 | 0.0568 | | 0.0519 | 0.0506 | 0.0013 | 0.0447 | 3B | 0.0465 | 0.0514 | 0.0519 | 0.0536 | 0.0017 | 0.0600 |
| 1–64 | NC | 2A | 0.0006 | 0.0724 | 0.0686 | | 0.0623 | 0.0603 | 0.0020 | 0.0532 | 2B | 0.0561 | 0.0623 | 0.0629 | 0.0655 | 0.0026 | 0.0730 |
| | | 3A | 0.0000 | 0.0730 | 0.0692 | | 0.0629 | 0.0614 | 0.0015 | 0.0538 | 3B | 0.0561 | 0.0623 | 0.0629 | 0.0648 | 0.0019 | 0.0730 |
| 1–72 | NF | 2A | 0.0006 | 0.0724 | 0.0689 | | 0.0634 | 0.0615 | 0.0019 | 0.0554 | 2B | 0.0580 | 0.0635 | 0.0640 | 0.0665 | 0.0025 | 0.0730 |
| | | 3A | 0.0000 | 0.0730 | 0.0695 | | 0.0640 | 0.0626 | 0.0014 | 0.0560 | 3B | 0.0580 | 0.0635 | 0.0640 | 0.0659 | 0.0019 | 0.0730 |
| 2–56 | NC | 2A | 0.0006 | 0.0854 | 0.0813 | | 0.0738 | 0.0717 | 0.0021 | 0.0635 | 2B | 0.0667 | 0.0737 | 0.0744 | 0.0772 | 0.0028 | 0.0860 |
| | | 3A | 0.0000 | 0.0860 | 0.0819 | | 0.0744 | 0.0728 | 0.0016 | 0.0641 | 3B | 0.0667 | 0.0737 | 0.0744 | 0.0765 | 0.0021 | 0.0860 |
| 2–64 | NF | 2A | 0.0006 | 0.0854 | 0.0816 | | 0.0753 | 0.0733 | 0.0020 | 0.0662 | 2B | 0.0691 | 0.0753 | 0.0759 | 0.0786 | 0.0027 | 0.0860 |
| | | 3A | 0.0000 | 0.0860 | 0.0822 | | 0.0759 | 0.0744 | 0.0015 | 0.0668 | 3B | 0.0691 | 0.0753 | 0.0759 | 0.0779 | 0.0020 | 0.0860 |
| 3–48 | NC | 2A | 0.0007 | 0.0983 | 0.0938 | | 0.0848 | 0.0825 | 0.0023 | 0.0727 | 2B | 0.0764 | 0.0845 | 0.0855 | 0.0885 | 0.0030 | 0.0990 |
| | | 3A | 0.0000 | 0.0990 | 0.0945 | | 0.0855 | 0.0838 | 0.0017 | 0.0734 | 3B | 0.0764 | 0.0845 | 0.0855 | 0.0877 | 0.0022 | 0.0990 |
| 3–56 | NF | 2A | 0.0007 | 0.0983 | 0.0942 | | 0.0867 | 0.0845 | 0.0022 | 0.0764 | 2B | 0.0797 | 0.0865 | 0.0874 | 0.0902 | 0.0028 | 0.0990 |
| | | 3A | 0.0000 | 0.0990 | 0.0949 | | 0.0874 | 0.0858 | 0.0016 | 0.0771 | 3B | 0.0797 | 0.0865 | 0.0874 | 0.0895 | 0.0021 | 0.0990 |
| 4–40 | NC | 2A | 0.0008 | 0.1112 | 0.1061 | | 0.0950 | 0.0925 | 0.0025 | 0.0805 | 2B | 0.0849 | 0.0939 | 0.0958 | 0.0991 | 0.0033 | 0.1120 |
| | | 3A | 0.0000 | 0.1120 | 0.1069 | | 0.0958 | 0.0939 | 0.0019 | 0.0813 | 3B | 0.0849 | 0.0939 | 0.0958 | 0.0982 | 0.0024 | 0.1120 |
| 4–48 | NF | 2A | 0.0007 | 0.1113 | 0.1068 | | 0.0978 | 0.0954 | 0.0024 | 0.0857 | 2B | 0.0894 | 0.0968 | 0.0985 | 0.1016 | 0.0031 | 0.1120 |
| | | 3A | 0.0000 | 0.1120 | 0.1075 | | 0.0985 | 0.0967 | 0.0018 | 0.0864 | 3B | 0.0894 | 0.0968 | 0.0985 | 0.1008 | 0.0023 | 0.1120 |
| 5–40 | NC | 2A | 0.0008 | 0.1242 | 0.1191 | | 0.1080 | 0.1054 | 0.0026 | 0.0935 | 2B | 0.0979 | 0.1062 | 0.1088 | 0.1121 | 0.0033 | 0.1250 |
| | | 3A | 0.0000 | 0.1250 | 0.1199 | | 0.1088 | 0.1069 | 0.0019 | 0.0943 | 3B | 0.0979 | 0.1062 | 0.1088 | 0.1113 | 0.0025 | 0.1250 |
| 5–44 | NF | 2A | 0.0007 | 0.1243 | 0.1195 | | 0.1095 | 0.1070 | 0.0025 | 0.0964 | 2B | 0.1004 | 0.1079 | 0.1102 | 0.1134 | 0.0032 | 0.1250 |
| | | 3A | 0.0000 | 0.1250 | 0.1202 | | 0.1102 | 0.1083 | 0.0019 | 0.0971 | 3B | 0.1004 | 0.1079 | 0.1102 | 0.1126 | 0.0024 | 0.1250 |
| 6–32 | NC | 2A | 0.0008 | 0.1372 | 0.1312 | | 0.1169 | 0.1141 | 0.0028 | 0.0989 | 2B | 0.104 | 0.114 | 0.1177 | 0.1214 | 0.0037 | 0.1380 |
| | | 3A | 0.0000 | 0.1380 | 0.1320 | | 0.1177 | 0.1156 | 0.0021 | 0.0997 | 3B | 0.1040 | 0.1140 | 0.1177 | 0.1204 | 0.0027 | 0.1380 |
| 6–40 | NF | 2A | 0.0008 | 0.1372 | 0.1321 | | 0.1210 | 0.1184 | 0.0026 | 0.1065 | 2B | 0.111 | 0.119 | 0.1218 | 0.1252 | 0.0034 | 0.1380 |
| | | 3A | 0.0000 | 0.1380 | 0.1329 | | 0.1218 | 0.1198 | 0.0020 | 0.1073 | 3B | 0.1110 | 0.1186 | 0.1218 | 0.1243 | 0.0025 | 0.1380 |

**FIGURE 10.30** Standard limits of size (UN and American screw threads).

| | | | | | | | | | | | | | | | | | |
|---|---|---|---|---|---|---|---|---|---|---|---|---|---|---|---|---|---|
| 8–32 | NC | 2A | 0.0009 | 0.1631 | 0.1571 | | 0.1428 | 0.1399 | 0.0029 | 0.1248 | 2B | 0.130 | 0.139 | 0.1437 | 0.1475 | 0.0038 | 0.1640 |
| | | 3A | 0.0000 | 0.1640 | 0.1580 | | 0.1437 | 0.1415 | 0.0022 | 0.1257 | 3B | 0.1300 | 0.1389 | 0.1437 | 0.1465 | 0.0028 | 0.1640 |
| 8–36 | NF | 2A | 0.0008 | 0.1632 | 0.1577 | | 0.1452 | 0.1424 | 0.0028 | 0.1291 | 2B | 0.134 | 0.142 | 0.1460 | 0.1496 | 0.0036 | 0.1640 |
| | | 3A | 0.0000 | 0.1640 | 0.1585 | | 0.1460 | 0.1439 | 0.0021 | 0.1299 | 3B | 0.1340 | 0.1416 | 0.1460 | 0.1487 | 0.0027 | 0.1640 |
| 10–24 | NC | 2A | 0.0010 | 0.1890 | 0.1818 | | 0.1619 | 0.1586 | 0.0033 | 0.1379 | 2B | 0.145 | 0.156 | 0.1629 | 0.1672 | 0.0043 | 0.1900 |
| | | 3A | 0.0000 | 0.1900 | 0.1828 | | 0.1629 | 0.1604 | 0.0025 | 0.1389 | 3B | 0.1450 | 0.1555 | 0.1629 | 0.1661 | 0.0032 | 0.1900 |
| 10–32 | NF | 2A | 0.0009 | 0.1891 | 0.1831 | | 0.1688 | 0.1658 | 0.0030 | 0.1508 | 2B | 0.156 | 0.164 | 0.1697 | 0.1736 | 0.0039 | 0.1900 |
| | | 3A | 0.0000 | 0.1900 | 0.1840 | | 0.1697 | 0.1674 | 0.0023 | 0.1517 | 3B | 0.1560 | 0.1641 | 0.1697 | 0.1726 | 0.0029 | 0.1900 |
| 12–24 | NC | 2A | 0.0010 | 0.2150 | 0.2078 | | 0.1879 | 0.1845 | 0.0034 | 0.1639 | 2B | 0.171 | 0.181 | 0.1889 | 0.1933 | 0.0044 | 0.2160 |
| | | 3A | 0.0000 | 0.2160 | 0.2088 | | 0.1889 | 0.1863 | 0.0026 | 0.1649 | 3B | 0.1710 | 0.1807 | 0.1889 | 0.1922 | 0.0033 | 0.2160 |
| 12–28 | NF | 2A | 0.0010 | 0.2150 | 0.2085 | | 0.1918 | 0.1886 | 0.0032 | 0.1712 | 2B | 0.177 | 0.186 | 0.1928 | 0.1970 | 0.0042 | 0.2160 |
| | | 3A | 0.0000 | 0.2160 | 0.2095 | | 0.1928 | 0.1904 | 0.0024 | 0.1722 | 3B | 0.1770 | 0.1857 | 0.1928 | 0.1959 | 0.0031 | 0.2160 |
| 12–32 | NEF | 2A | 0.0009 | 0.2151 | 0.2091 | | 0.1948 | 0.1917 | 0.0031 | 0.1768 | 2B | 0.182 | 0.190 | 0.1957 | 0.1998 | 0.0041 | 0.2160 |
| | | 3A | 0.0000 | 0.2160 | 0.2100 | | 0.1957 | 0.1933 | 0.0024 | 0.1777 | 3B | 0.1820 | 0.1895 | 0.1957 | 0.1988 | 0.0031 | 0.2160 |
| ¼–20 | UNC | 1A | 0.0011 | 0.2489 | 0.2367 | | 0.2164 | 0.2108 | 0.0056 | 0.1876 | 1B | 0.196 | 0.207 | 0.2175 | 0.2248 | 0.0073 | 0.2500 |
| | | 2A | 0.0011 | 0.2489 | 0.2408 | 0.2367 | 0.2164 | 0.2127 | 0.0037 | 0.1876 | 2B | 0.196 | 0.207 | 0.2175 | 0.2223 | 0.0048 | 0.2500 |
| | | 3A | 0.0000 | 0.2500 | 0.2419 | | 0.2175 | 0.2147 | 0.0028 | 0.1887 | 3B | 0.1960 | 0.2067 | 0.2175 | 0.2211 | 0.0036 | 0.2500 |
| ¼–28 | UNF | 1A | 0.0010 | 0.2490 | 0.2392 | | 0.2258 | 0.2208 | 0.0050 | 0.2052 | 1B | 0.211 | 0.220 | 0.2268 | 0.2333 | 0.0065 | 0.2500 |
| | | 2A | 0.0010 | 0.2490 | 0.2425 | | 0.2258 | 0.2225 | 0.0033 | 0.2052 | 2B | 0.211 | 0.220 | 0.2268 | 0.2311 | 0.0043 | 0.2500 |
| | | 3A | 0.0000 | 0.2500 | 0.2435 | | 0.2268 | 0.2243 | 0.0025 | 0.2062 | 3B | 0.2110 | 0.2190 | 0.2268 | 0.2300 | 0.0032 | 0.2500 |
| ¼–32 | NEF | 2A | 0.0010 | 0.2490 | 0.2430 | | 0.2287 | 0.2255 | 0.0032 | 0.2107 | 2B | 0.216 | 0.224 | 0.2297 | 0.2339 | 0.0042 | 0.2500 |
| | | 3A | 0.0000 | 0.2500 | 0.2440 | | 0.2297 | 0.2273 | 0.0024 | 0.2117 | 3B | 0.2160 | 0.2229 | 0.2297 | 0.2328 | 0.0031 | 0.2500 |
| 5/16–18 | UNC | 1A | 0.0012 | 0.3113 | 0.2982 | | 0.2752 | 0.2691 | 0.0061 | 0.2431 | 1B | 0.252 | 0.265 | 0.2764 | 0.2843 | 0.0079 | 0.3125 |
| | | 2A | 0.0012 | 0.3113 | 0.3026 | 0.2982 | 0.2752 | 0.2712 | 0.0040 | 0.2431 | 2B | 0.252 | 0.265 | 0.2764 | 0.2817 | 0.0053 | 0.3125 |
| | | 3A | 0.0000 | 0.3125 | 0.3038 | | 0.2764 | 0.2734 | 0.0030 | 0.2443 | 3B | 0.2520 | 0.2630 | 0.2764 | 0.2803 | 0.0039 | 0.3125 |
| 5/16–24 | UNF | 1A | 0.0011 | 0.3114 | 0.3006 | | 0.2843 | 0.2788 | 0.0055 | 0.2603 | 1B | 0.267 | 0.277 | 0.2854 | 0.2925 | 0.0071 | 0.3125 |
| | | 2A | 0.0011 | 0.3114 | 0.3042 | | 0.2843 | 0.2806 | 0.0037 | 0.2603 | 2B | 0.267 | 0.277 | 0.2854 | 0.2902 | 0.0048 | 0.3125 |
| | | 3A | 0.0000 | 0.3125 | 0.3053 | | 0.2854 | 0.2827 | 0.0027 | 0.2614 | 3B | 0.2670 | 0.2754 | 0.2854 | 0.2890 | 0.0036 | 0.3125 |

**FIGURE 10.30** *(Continued)*

## Standard Limits of Size: Unified and American Screw Threads (*Continued*)

| Nominal size and threads per inch | Series designation | External | | | | | | | | | Internal | | | | | | |
|---|---|---|---|---|---|---|---|---|---|---|---|---|---|---|---|---|---|
| | | | | Major diameter limits, in | | | Pitch diameter limits, in | | | | | Minor diameter limits, in | | Pitch diameter limits, in | | | Major diameter, in |
| | | Class | Allowance, in | Max* | Min | Min | Max* | Min | Tolerance | Minor diameter, in | Class | Min | Max | Min | Max | Tolerance | Min |
| | | | | | | | No. 0–80 to ½–13 | | | | | | | | | | |
| 5/16–32 | NEF | 2A | 0.0010 | 0.3115 | 0.3055 | | 0.2912 | 0.2880 | 0.0032 | 0.2732 | 2B | 0.279 | 0.286 | 0.2922 | 0.2964 | 0.0042 | 0.3125 |
| | | 3A | 0.0000 | 0.3125 | 0.3065 | | 0.2922 | 0.2898 | 0.0024 | 0.2742 | 3B | 0.2790 | 0.2847 | 0.2922 | 0.2953 | 0.0031 | 0.3125 |
| 3/8–16 | UNC | 1A | 0.0013 | 0.3737 | 0.3595 | | 0.3331 | 0.3266 | 0.0065 | 0.2970 | 1B | 0.307 | 0.321 | 0.3344 | 0.3429 | 0.0085 | 0.3750 |
| | | 2A | 0.0013 | 0.3737 | 0.3643 | 0.3595 | 0.3331 | 0.3287 | 0.0044 | 0.2970 | 2B | 0.307 | 0.321 | 0.3344 | 0.3401 | 0.0057 | 0.3750 |
| | | 3A | 0.0000 | 0.3750 | 0.3656 | | 0.3344 | 0.3311 | 0.0033 | 0.2983 | 3B | 0.3070 | 0.3182 | 0.3344 | 0.3387 | 0.0043 | 0.3750 |
| 3/8–24 | UNF | 1A | 0.0011 | 0.3739 | 0.3631 | | 0.3468 | 0.3411 | 0.0057 | 0.3228 | 1B | 0.330 | 0.340 | 0.3479 | 0.2553 | 0.0074 | 0.3750 |
| | | 2A | 0.0011 | 0.3739 | 0.3667 | | 0.3468 | 0.3430 | 0.0038 | 0.3228 | 2B | 0.330 | 0.340 | 0.3479 | 0.3528 | 0.0049 | 0.3750 |
| | | 3A | 0.0000 | 0.3750 | 0.3678 | | 0.3479 | 0.3450 | 0.0029 | 0.3239 | 3B | 0.3300 | 0.3372 | 0.3479 | 0.3516 | 0.0037 | 0.3750 |
| 3/8–32 | NEF | 2A | 0.0010 | 0.3740 | 0.3680 | | 0.3537 | 0.3503 | 0.0034 | 0.3357 | 2B | 0.341 | 0.349 | 0.3547 | 0.3591 | 0.0044 | 0.3750 |
| | | 3A | 0.0000 | 0.3750 | 0.3690 | | 0.3547 | 0.3522 | 0.0025 | 0.3367 | 3B | 0.3410 | 0.3469 | 0.3547 | 0.3580 | 0.0033 | 0.3750 |
| 7/16–14 | UNC | 1A | 0.0014 | 0.4361 | 0.4206 | | 0.3897 | 0.3826 | 0.0071 | 0.3485 | 1B | 0.360 | 0.376 | 0.3911 | 0.4003 | 0.0092 | 0.4375 |
| | | 2A | 0.0014 | 0.4361 | 0.4258 | 0.4206 | 0.3897 | 0.3850 | 0.0047 | 0.3485 | 2B | 0.360 | 0.376 | 0.3911 | 0.3972 | 0.0061 | 0.4375 |
| | | 3A | 0.0000 | 0.4375 | 0.4272 | | 0.3911 | 0.3876 | 0.0035 | 0.3499 | 3B | 0.3600 | 0.3717 | 0.3911 | 0.3957 | 0.0046 | 0.4375 |
| 7/16–20 | UNF | 1A | 0.0013 | 0.4362 | 0.4240 | | 0.4037 | 0.3975 | 0.0062 | 0.3749 | 1B | 0.383 | 0.395 | 0.4050 | 0.4131 | 0.0081 | 0.4375 |
| | | 2A | 0.0013 | 0.4362 | 0.4281 | | 0.4037 | 0.3995 | 0.0042 | 0.3749 | 2B | 0.383 | 0.395 | 0.4050 | 0.4104 | 0.0054 | 0.4375 |
| | | 3A | 0.0000 | 0.4375 | 0.4294 | | 0.4050 | 0.4019 | 0.0031 | 0.3762 | 3B | 0.3830 | 0.3916 | 0.4050 | 0.4091 | 0.0041 | 0.4375 |
| 7/16–28 | UNEF | 2A | 0.0011 | 0.4364 | 0.4299 | | 0.4132 | 0.4096 | 0.0036 | 0.3926 | 2B | 0.399 | 0.407 | 0.4143 | 0.4189 | 0.0046 | 0.4375 |
| | | 3A | 0.0000 | 0.4375 | 0.4310 | | 0.4143 | 0.4116 | 0.0027 | 0.3937 | 3B | 0.3990 | 0.4051 | 0.4143 | 0.4178 | 0.0035 | 0.4375 |
| 1/2–12 | N | 2A | 0.0016 | 0.4984 | 0.4870 | | 0.4443 | 0.4389 | 0.0054 | 0.3962 | 2B | 0.410 | 0.428 | 0.4459 | 0.4529 | 0.0070 | 0.5000 |
| | | 3A | 0.0000 | 0.5000 | 0.4886 | | 0.4439 | 0.4419 | 0.0040 | 0.3978 | 3B | 0.4100 | 0.4223 | 0.4459 | 0.4511 | 0.0052 | 0.5000 |
| 1/2–13 | UNC | 1A | 0.0015 | 0.4985 | 0.4822 | | 0.4485 | 0.4411 | 0.0074 | 0.4041 | 1B | 0.417 | 0.434 | 0.4500 | 0.4597 | 0.0097 | 0.5000 |
| | | 2A | 0.0015 | 0.4985 | 0.4876 | 0.4822 | 0.4485 | 0.4435 | 0.0050 | 0.4041 | 2B | 0.417 | 0.434 | 0.4500 | 0.4565 | 0.0065 | 0.5000 |
| | | 3A | 0.0000 | 0.5000 | 0.4891 | | 0.4500 | 0.4463 | 0.0037 | 0.4056 | 3B | 0.4170 | 0.4284 | 0.4500 | 0.4548 | 0.0048 | 0.5000 |

**FIGURE 10.30** (*Continued*)

½–20 to 1½–12

| | | | | | | | | | | | | | | | | | |
|---|---|---|---|---|---|---|---|---|---|---|---|---|---|---|---|---|---|
| ½–20 | UNF | 1A | 0.0013 | 0.4987 | 0.4865 | | 0.4662 | 0.4598 | 0.0064 | 0.4374 | 1B | 0.446 | 0.457 | 0.4675 | 0.4759 | 0.0084 | 0.5000 |
| | | 2A | 0.0013 | 0.4987 | 0.4906 | | 0.4662 | 0.4619 | 0.0043 | 0.4374 | 2B | 0.446 | 0.457 | 0.4675 | 0.4731 | 0.0056 | 0.5000 |
| | | 3A | 0.0000 | 0.5000 | 0.4919 | | 0.4675 | 0.4643 | 0.0032 | 0.4387 | 3B | 0.4460 | 0.4537 | 0.4675 | 0.4717 | 0.0042 | 0.5000 |
| ½–28 | UNEF | 2A | 0.0011 | 0.4989 | 0.4924 | | 0.4757 | 0.4720 | 0.0037 | 0.4551 | 2B | 0.461 | 0.470 | 0.4768 | 0.4816 | 0.0048 | 0.5000 |
| | | 3A | 0.0000 | 0.5000 | 0.4935 | | 0.4768 | 0.4740 | 0.0028 | 0.4562 | 3B | 0.4610 | 0.4676 | 0.4768 | 0.4804 | 0.0036 | 0.5000 |
| 9/16–12 | UNC | 1A | 0.0016 | 0.5609 | 0.5437 | | 0.5068 | 0.4990 | 0.0078 | 0.4587 | 1B | 0.472 | 0.490 | 0.5084 | 0.5186 | 0.0102 | 0.5625 |
| | | 2A | 0.0016 | 0.5609 | 0.5495 | 0.5437 | 0.5068 | 0.5016 | 0.0052 | 0.4587 | 2B | 0.472 | 0.490 | 0.5084 | 0.5152 | 0.0068 | 0.5625 |
| | | 3A | 0.0000 | 0.5625 | 0.5511 | | 0.5084 | 0.5045 | 0.0039 | 0.4603 | 3B | 0.4720 | 0.4843 | 0.5084 | 0.5135 | 0.0051 | 0.5625 |
| 9/16–18 | UNF | 1A | 0.0014 | 0.5611 | 0.5480 | | 0.5250 | 0.5182 | 0.0068 | 0.4929 | 1B | 0.502 | 0.515 | 0.5264 | 0.5353 | 0.0089 | 0.5625 |
| | | 2A | 0.0014 | 0.5611 | 0.5524 | | 0.5250 | 0.5205 | 0.0045 | 0.4929 | 2B | 0.502 | 0.515 | 0.5264 | 0.5323 | 0.0059 | 0.5625 |
| | | 3A | 0.0000 | 0.5625 | 0.5538 | | 0.5264 | 0.5230 | 0.0034 | 0.4943 | 3B | 0.5020 | 0.5106 | 0.5264 | 0.5308 | 0.0044 | 0.5625 |
| 9/16–24 | NEF | 2A | 0.0012 | 0.5613 | 0.5541 | | 0.5342 | 0.5303 | 0.0039 | 0.5102 | 2B | 0.517 | 0.527 | 0.5354 | 0.5405 | 0.0051 | 0.5625 |
| | | 3A | 0.0000 | 0.5625 | 0.5553 | | 0.5354 | 0.5325 | 0.0029 | 0.5114 | 3B | 0.5170 | 0.5244 | 0.5354 | 0.5392 | 0.0038 | 0.5625 |
| 5/8–11 | UNC | 1A | 0.0016 | 0.6234 | 0.6052 | | 0.5644 | 0.5561 | 0.0083 | 0.5119 | 1B | 0.527 | 0.546 | 0.5660 | 0.5767 | 0.0107 | 0.6250 |
| | | 2A | 0.0016 | 0.6234 | 0.6113 | 0.6052 | 0.5644 | 0.5589 | 0.0055 | 0.5119 | 2B | 0.527 | 0.546 | 0.5660 | 0.5732 | 0.0072 | 0.6250 |
| | | 3A | 0.0000 | 0.6250 | 0.6129 | | 0.5660 | 0.5619 | 0.0041 | 0.5135 | 3B | 0.5270 | 0.5391 | 0.5660 | 0.5714 | 0.0054 | 0.6250 |
| 5/8–12 | N | 2A | 0.0016 | 0.6234 | 0.6120 | | 0.5693 | 0.5639 | 0.0054 | 0.5212 | 2B | 0.535 | 0.553 | 0.5709 | 0.5780 | 0.0071 | 0.6250 |
| | | 3A | 0.0000 | 0.6250 | 0.6136 | | 0.5709 | 0.5668 | 0.0041 | 0.5228 | 3B | 0.5350 | 0.5463 | 0.5709 | 0.5762 | 0.0053 | 0.6250 |
| 5/8–18 | UNF | 1A | 0.0014 | 0.6236 | 0.6105 | | 0.5875 | 0.5805 | 0.0070 | 0.5554 | 1B | 0.565 | 0.578 | 0.5889 | 0.5980 | 0.0091 | 0.6250 |
| | | 2A | 0.0014 | 0.6236 | 0.6149 | | 0.5875 | 0.5828 | 0.0047 | 0.5554 | 2B | 0.565 | 0.578 | 0.5889 | 0.5949 | 0.0060 | 0.6250 |
| | | 3A | 0.0000 | 0.6250 | 0.6163 | | 0.5889 | 0.5854 | 0.0035 | 0.5568 | 3B | 0.5650 | 0.5730 | 0.5889 | 0.5934 | 0.0045 | 0.6250 |
| 5/8–24 | NEF | 2A | 0.0012 | 0.6238 | 0.6166 | | 0.5967 | 0.5927 | 0.0040 | 0.5727 | 2B | 0.580 | 0.590 | 0.5979 | 0.6031 | 0.0052 | 0.6250 |
| | | 3A | 0.0000 | 0.6250 | 0.6178 | | 0.5979 | 0.5949 | 0.0030 | 0.5739 | 3B | 0.5800 | 0.5809 | 0.5979 | 0.6018 | 0.0039 | 0.6250 |
| 11/16–12 | N | 2A | 0.0016 | 0.6859 | 0.6745 | | 0.6318 | 0.6264 | 0.0054 | 0.5837 | 2B | 0.597 | 0.615 | 0.6334 | 0.6405 | 0.0071 | 0.6875 |
| | | 3A | 0.0000 | 0.6875 | 0.6761 | | 0.6334 | 0.6293 | 0.0041 | 0.5853 | 3B | 0.5970 | 0.6085 | 0.6334 | 0.6387 | 0.0053 | 0.6875 |
| 11/16–24 | NEF | 2A | 0.0012 | 0.6863 | 0.6791 | | 0.6592 | 0.6552 | 0.0040 | 0.6352 | 2B | 0.642 | 0.652 | 0.6604 | 0.6656 | 0.0052 | 0.6875 |
| | | 3A | 0.0000 | 0.6875 | 0.6803 | | 0.6604 | 0.6574 | 0.0030 | 0.6364 | 3B | 0.6420 | 0.6494 | 0.6604 | 0.6643 | 0.0039 | 0.6875 |
| 3/4–10 | UNC | 1A | 0.0018 | 0.7482 | 0.7288 | | 0.6832 | 0.6744 | 0.0088 | 0.6255 | 1B | 0.642 | 0.663 | 0.6850 | 0.6965 | 0.0115 | 0.7500 |
| | | 2A | 0.0018 | 0.7482 | 0.7353 | 0.7288 | 0.6832 | 0.6773 | 0.0059 | 0.6255 | 2B | 0.642 | 0.663 | 0.6850 | 0.6927 | 0.0077 | 0.7500 |
| | | 3A | 0.0000 | 0.7500 | 0.7371 | | 0.6850 | 0.6806 | 0.0044 | 0.6273 | 3B | 0.6420 | 0.6545 | 0.6850 | 0.6907 | 0.0057 | 0.7500 |
| 3/4–12 | N | 2A | 0.0017 | 0.7483 | 0.7369 | | 0.6942 | 0.6887 | 0.0055 | 0.6461 | 2B | 0.660 | 0.678 | 0.6959 | 0.7031 | 0.0072 | 0.7500 |
| | | 3A | 0.0000 | 0.7500 | 0.7386 | | 0.6959 | 0.6918 | 0.0041 | 0.6478 | 3B | 0.6600 | 0.6707 | 0.6959 | 0.7013 | 0.0054 | 0.7500 |

**FIGURE 10.30** *(Continued)*

**Standard Limits of Size: Unified and American Screw Threads (*Continued*)**

| Nominal size and threads per inch | Series desig- nation | External | | | | | | | | | Internal | | | | | | |
|---|---|---|---|---|---|---|---|---|---|---|---|---|---|---|---|---|---|
| | | Class | Allow- ance, in | Major diameter limits, in | | | Pitch diameter limits, in | | | Minor diam- eter, in | Class | Minor diam- eter limits, in | | Pitch diameter limits, in | | | Major diameter, in |
| | | | | Max* | Min | Min' | Max* | Min | Toler- ance | | | Min | Max | Min | Max | Toler- ance | Min |
| ½–20 to 1½–12 | | | | | | | | | | | | | | | | | |
| ¾–16 | UNF | 1A | 0.0015 | 0.7485 | 0.7343 | | 0.7079 | 0.7004 | 0.0075 | 0.6718 | 1B | 0.682 | 0.696 | 0.7094 | 0.7192 | 0.0098 | 0.7500 |
| | | 2A | 0.0015 | 0.7485 | 0.7391 | | 0.7079 | 0.7029 | 0.0050 | 0.6718 | 2B | 0.682 | 0.696 | 0.7094 | 0.7159 | 0.0065 | 0.7500 |
| | | 3A | 0.0000 | 0.7500 | 0.7406 | | 0.7094 | 0.7056 | 0.0038 | 0.6733 | 3B | 0.6820 | 0.6908 | 0.7094 | 0.7143 | 0.0049 | 0.7500 |
| ¾–20 | UNEF | 2A | 0.0013 | 0.7487 | 0.7406 | | 0.7162 | 0.7118 | 0.0044 | 0.6874 | 2B | 0.696 | 0.707 | 0.7175 | 0.7232 | 0.0057 | 0.7500 |
| | | 3A | 0.0000 | 0.7500 | 0.7419 | | 0.7175 | 0.7142 | 0.0033 | 0.6887 | 3B | 0.6960 | 0.7037 | 0.7175 | 0.7218 | 0.0043 | 0.7500 |
| 13/16–12 | N | 2A | 0.0017 | 0.8108 | 0.7994 | | 0.7567 | 0.7512 | 0.0055 | 0.7086 | 2B | 0.722 | 0.740 | 0.7584 | 0.7656 | 0.0072 | 0.8125 |
| | | 3A | 0.0000 | 0.8125 | 0.8011 | | 0.7584 | 0.7543 | 0.0041 | 0.7103 | 3B | 0.7220 | 0.7329 | 0.7584 | 0.7638 | 0.0054 | 0.8125 |
| 13/16–16 | UN | 2A | 0.0015 | 0.8110 | 0.8016 | | 0.7704 | 0.7655 | 0.0049 | 0.7343 | 2B | 0.745 | 0.759 | 0.7719 | 0.7782 | 0.0063 | 0.8125 |
| | | 3A | 0.0000 | 0.8125 | 0.8031 | | 0.7719 | 0.7683 | 0.0036 | 0.7358 | 3B | 0.7450 | 0.7533 | 0.7719 | 0.7766 | 0.0047 | 0.8125 |
| 13/16–20 | UNEF | 2A | 0.0013 | 0.8112 | 0.8031 | | 0.7787 | 0.7743 | 0.0044 | 0.7498 | 2B | 0.758 | 0.770 | 0.7800 | 0.7857 | 0.0057 | 0.8125 |
| | | 3A | 0.0000 | 0.8125 | 0.8044 | | 0.7800 | 0.7767 | 0.0033 | 0.7512 | 3B | 0.7580 | 0.7662 | 0.7800 | 0.7843 | 0.0043 | 0.8125 |
| ⅞–9 | UNC | 1A | 0.0019 | 0.8731 | 0.8523 | | 0.8009 | 0.7914 | 0.0095 | 0.7368 | 1B | 0.755 | 0.778 | 0.8028 | 0.8151 | 0.0123 | 0.8750 |
| | | 2A | 0.0019 | 0.8731 | 0.8592 | 0.8523 | 0.8009 | 0.7946 | 0.0063 | 0.7368 | 2B | 0.755 | 0.778 | 0.8028 | 0.8110 | 0.0082 | 0.8750 |
| | | 3A | 0.0000 | 0.8750 | 0.8611 | | 0.8028 | 0.7981 | 0.0047 | 0.7387 | 3B | 0.7550 | 0.7681 | 0.8028 | 0.8089 | 0.0061 | 0.8750 |
| ⅞–12 | N | 2A | 0.0017 | 0.8733 | 0.8619 | | 0.8192 | 0.8137 | 0.0055 | 0.7711 | 2B | 0.785 | 0.803 | 0.8209 | 0.8281 | 0.0072 | 0.8750 |
| | | 3A | 0.0000 | 0.8750 | 0.8636 | | 0.8209 | 0.8168 | 0.0041 | 0.7728 | 3B | 0.7850 | 0.7952 | 0.8209 | 0.8263 | 0.0054 | 0.8750 |
| ⅞–14 | UNF | 1A | 0.0016 | 0.8734 | 0.8579 | | 0.8270 | 0.8189 | 0.0081 | 0.7858 | 1B | 0.798 | 0.814 | 0.8286 | 0.8392 | 0.0106 | 0.8750 |
| | | 2A | 0.0016 | 0.8734 | 0.8631 | | 0.8270 | 0.8216 | 0.0054 | 0.7858 | 2B | 0.798 | 0.814 | 0.8286 | 0.8356 | 0.0070 | 0.8750 |
| | | 3A | 0.0000 | 0.8750 | 0.8647 | | 0.8286 | 0.8245 | 0.0041 | 0.7874 | 3B | 0.7980 | 0.8068 | 0.8286 | 0.8339 | 0.0053 | 0.8750 |
| ⅞–16 | UN | 2A | 0.0015 | 0.8735 | 0.8641 | | 0.8329 | 0.8280 | 0.0049 | 0.7968 | 2B | 0.807 | 0.821 | 0.8344 | 0.8407 | 0.0063 | 0.8750 |
| | | 3A | 0.0000 | 0.8750 | 0.8656 | | 0.8344 | 0.8308 | 0.0036 | 0.7983 | 3B | 0.8070 | 0.8158 | 0.8344 | 0.8391 | 0.0047 | 0.8750 |
| ⅞–20 | UNEF | 2A | 0.0013 | 0.8737 | 0.8656 | | 0.8412 | 0.8368 | 0.0044 | 0.8124 | 2B | 0.821 | 0.832 | 0.8425 | 0.8482 | 0.0057 | 0.8750 |
| | | 3A | 0.0000 | 0.8750 | 0.8669 | | 0.8425 | 0.8392 | 0.0033 | 0.8137 | 3B | 0.8210 | 0.8287 | 0.8425 | 0.8468 | 0.0043 | 0.8750 |
| 15/16–12 | UN | 2A | 0.0017 | 0.9358 | 0.9244 | | 0.8817 | 0.8760 | 0.0057 | 0.8336 | 2B | 0.847 | 0.865 | 0.8834 | 0.8908 | 0.0074 | 0.9375 |
| | | 3A | 0.0000 | 0.9375 | 0.9261 | | 0.8834 | 0.8793 | 0.0011 | 0.8353 | 3B | 0.8470 | 0.8575 | 0.8834 | 0.8889 | 0.0055 | 0.9375 |
| 15/16–16 | UN | 2A | 0.0015 | 0.9360 | 0.9266 | | 0.8954 | 0.8904 | 0.0050 | 0.8593 | 2B | 0.870 | 0.884 | 0.8969 | 0.9034 | 0.0065 | 0.9375 |
| | | 3A | 0.0000 | 0.9375 | 0.9281 | | 0.8969 | 0.8932 | 0.0037 | 0.8608 | 3B | 0.8700 | 0.8783 | 0.8969 | 0.9013 | 0.0049 | 0.9375 |

**FIGURE 10.30** (*Continued*)

| | | | | | | | | | | | | | | | | | |
|---|---|---|---|---|---|---|---|---|---|---|---|---|---|---|---|---|---|
| 15/16–20 | UNEF | 2A | 0.0014 | 0.9361 | 0.9280 | | 0.9036 | 0.8991 | 0.0045 | 0.8748 | 2B | 0.883 | 0.895 | 0.9050 | 0.9109 | 0.0059 | 0.9375 |
| | | 3A | 0.0000 | 0.9375 | 0.9294 | | 0.9050 | 0.9016 | 0.0034 | 0.8762 | 3B | 0.8830 | 0.8912 | 0.9050 | 0.9094 | 0.0044 | 0.9375 |
| 1–8 | UNC | 1A | 0.0020 | 0.9980 | 0.9755 | | 0.9168 | 0.9067 | 0.0101 | 0.8446 | 1B | 0.865 | 0.890 | 0.9188 | 0.9320 | 0.0132 | 1.0000 |
| | | 2A | 0.0020 | 0.9980 | 0.9830 | 0.9755 | 0.9168 | 0.9100 | 0.0068 | 0.8446 | 2B | 0.865 | 0.890 | 0.9188 | 0.9276 | 0.0088 | 1.0000 |
| | | 3A | 0.0000 | 1.0000 | 0.9850 | | 0.9188 | 0.9137 | 0.0051 | 0.8466 | 3B | 0.8650 | 0.8797 | 0.9188 | 0.9254 | 0.0066 | 1.0000 |
| 1–12 | UNF | 1A | 0.0018 | 0.9982 | 0.9810 | | 0.9441 | 0.9353 | 0.0088 | 0.8960 | 1B | 0.910 | 0.928 | 0.9459 | 0.9573 | 0.0114 | 1.0000 |
| | | 2A | 0.0018 | 0.9982 | 0.9868 | | 0.9441 | 0.9382 | 0.0059 | 0.8960 | 2B | 0.910 | 0.928 | 0.9459 | 0.9535 | 0.0076 | 1.0000 |
| | | 3A | 0.0000 | 1.0000 | 0.9886 | | 0.9459 | 0.9415 | 0.0044 | 0.8978 | 3B | 0.9100 | 0.9198 | 0.9459 | 0.9516 | 0.0057 | 1.0000 |
| 1–16 | UN | 2A | 0.0015 | 0.9985 | 0.9891 | | 0.9579 | 0.9529 | 0.0050 | 0.9218 | 2B | 0.932 | 0.946 | 0.9594 | 0.9659 | 0.0065 | 1.0000 |
| | | 3A | 0.0000 | 1.0000 | 0.9906 | | 0.9594 | 0.9557 | 0.0037 | 0.9233 | 3B | 0.9320 | 0.9408 | 0.9594 | 0.9643 | 0.0049 | 1.0000 |
| 1–20 | UNEF | 2A | 0.0014 | 0.9986 | 0.9905 | | 0.9661 | 0.9616 | 0.0045 | 0.9373 | 2B | 0.946 | 0.957 | 0.9675 | 0.9734 | 0.0059 | 1.0000 |
| | | 3A | 0.0000 | 1.0000 | 0.9919 | | 0.9675 | 0.9641 | 0.0034 | 0.9387 | 3B | 0.9460 | 0.9537 | 0.9675 | 0.9719 | 0.0044 | 1.0000 |
| 1 1/16–12 | UN | 2A | 0.0017 | 1.0608 | 1.0494 | | 1.0067 | 1.0010 | 0.0057 | 0.9586 | 2B | 0.972 | 0.990 | 1.0084 | 1.0158 | 0.0074 | 1.0625 |
| | | 3A | 0.0000 | 1.0625 | 1.0511 | | 1.0084 | 1.0042 | 0.0042 | 0.9603 | 3B | 0.9720 | 0.9823 | 1.0084 | 1.0139 | 0.0055 | 1.0625 |
| 1 1/16–16 | UN | 2A | 0.0015 | 1.0610 | 1.0516 | | 1.0204 | 1.0154 | 0.0050 | 0.9843 | 2B | 0.995 | 1.009 | 1.0219 | 1.0284 | 0.0065 | 1.0625 |
| | | 3A | 0.0000 | 1.0625 | 1.0531 | | 1.0219 | 1.0182 | 0.0037 | 0.9853 | 3B | 0.9950 | 1.0033 | 1.0219 | 1.0268 | 0.0049 | 1.0625 |
| 1 1/16–18 | NEF | 2A | 0.0014 | 1.0611 | 1.0524 | | 1.0250 | 1.0203 | 0.0047 | 0.9929 | 2B | 1.002 | 1.015 | 1.0264 | 1.0326 | 0.0062 | 1.0625 |
| | | 3A | 0.0000 | 1.0625 | 1.0538 | | 1.0264 | 1.0228 | 0.0036 | 0.9943 | 3B | 1.0020 | 1.0105 | 1.0264 | 1.0310 | 0.0046 | 1.0625 |
| 1 1/8–7 | UNC | 1A | 0.0022 | 1.1228 | 1.0982 | | 1.0300 | 1.0191 | 0.0109 | 0.9475 | 1B | 0.970 | 0.998 | 1.0322 | 1.0463 | 0.0141 | 1.1250 |
| | | 2A | 0.0022 | 1.1228 | 1.1064 | 1.0982 | 1.0300 | 1.0228 | 0.0072 | 0.9475 | 2B | 0.970 | 0.998 | 1.0322 | 1.0416 | 0.0094 | 1.1250 |
| | | 3A | 0.0000 | 1.1250 | 1.1086 | | 1.0322 | 1.0268 | 0.0054 | 0.9497 | 3B | 0.9700 | 0.9875 | 1.0322 | 1.0393 | 0.0071 | 1.1250 |
| 1 1/8–8 | N | 2A | 0.0021 | 1.1229 | 1.1079 | | 1.0417 | 1.0348 | 0.0069 | 0.9695 | 2B | 0.900 | 1.015 | 1.0438 | 1.0528 | 0.0090 | 1.1250 |
| | | 3A | 0.0000 | 1.1250 | 1.1100 | 1.1004 | 1.0438 | 1.0386 | 0.0052 | 0.9716 | 3B | 0.9900 | 1.0047 | 1.0438 | 1.0505 | 0.0067 | 1.1250 |
| 1 1/8–12 | UNF | 1A | 0.0018 | 1.1232 | 1.1060 | | 1.0691 | 1.0601 | 0.0090 | 1.0210 | 1B | 1.035 | 1.053 | 1.0709 | 1.0826 | 0.0117 | 1.1250 |
| | | 2A | 0.0018 | 1.1232 | 1.1118 | | 1.0691 | 1.0631 | 0.0060 | 1.0210 | 2B | 1.035 | 1.053 | 1.0709 | 1.0787 | 0.0078 | 1.1250 |
| | | 3A | 0.0000 | 1.1250 | 1.1136 | | 1.0709 | 1.0664 | 0.0045 | 1.0228 | 3B | 1.0350 | 1.0448 | 1.0709 | 1.0768 | 0.0059 | 1.1250 |
| 1 1/8–16 to 1 3/4–12 | | | | | | | | | | | | | | | | | |
| 1 1/8–16 | UN | 2A | 0.0015 | 1.1235 | 1.1141 | | 1.0829 | 1.0779 | 0.0050 | 1.0468 | 2B | 1.057 | 1.071 | 1.0844 | 1.0909 | 0.0065 | 1.1250 |
| | | 3A | 0.0000 | 1.1250 | 1.1156 | | 1.0844 | 1.0807 | 0.0037 | 1.0483 | 3B | 1.0570 | 1.0658 | 1.0844 | 1.0893 | 0.0049 | 1.1250 |
| 1 1/8–18 | NEF | 2A | 0.0014 | 1.1236 | 1.1149 | | 1.0875 | 1.0828 | 0.0047 | 1.0554 | 2B | 1.065 | 1.078 | 1.0889 | 1.0951 | 0.0062 | 1.1250 |
| | | 3A | 0.0000 | 1.1250 | 1.1163 | | 1.0889 | 1.0853 | 0.0036 | 1.0568 | 3B | 1.0650 | 1.0730 | 1.0889 | 1.0935 | 0.0046 | 1.1250 |
| 1 3/16–12 | UN | 2A | 0.0017 | 1.1858 | 1.1744 | | 1.1317 | 1.1259 | 0.0058 | 1.0836 | 2B | 1.097 | 1.115 | 1.1334 | 1.1409 | 0.0075 | 1.1875 |
| | | 3A | 0.0000 | 1.1875 | 1.1761 | | 1.1334 | 1.1291 | 0.0043 | 1.0853 | 3B | 1.0970 | 1.1073 | 1.1334 | 1.1390 | 0.0056 | 1.1875 |
| 1 3/16 | UN | 2A | 0.0015 | 1.1860 | 1.1766 | | 1.1454 | 1.1403 | 0.0051 | 1.1093 | 2B | 1.120 | 1.134 | 1.1469 | 1.1535 | 0.0066 | 1.1875 |
| | | 3A | 0.0000 | 1.1875 | 1.1781 | | 1.1469 | 1.1431 | 0.0038 | 1.1108 | 3B | 1.1200 | 1.1283 | 1.1469 | 1.1519 | 0.0050 | 1.1875 |
| 1 3/16–18 | NEF | 2A | 0.0015 | 1.1860 | 1.1773 | | 1.1499 | 1.1450 | 0.0049 | 1.1178 | 2B | 1.127 | 1.140 | 1.1514 | 1.1577 | 0.0063 | 1.1875 |
| | | 3A | 0.0000 | 1.1875 | 1.1788 | | 1.1514 | 1.1478 | 0.0036 | 1.1193 | 3B | 1.1270 | 1.1355 | 1.1514 | 1.1561 | 0.0047 | 1.1875 |

**FIGURE 10.30** *(Continued)*

## Standard Limits of Size: Unified and American Screw Threads (*Continued*)

| Nominal size and threads per inch | Series designation | External: Class | External: Allowance, in | External: Major diameter limits, in: Max* | External: Major diameter limits, in: Min | External: Major diameter limits, in: Min' | External: Pitch diameter limits, in: Max* | External: Pitch diameter limits, in: Min | External: Pitch diameter limits, in: Tolerance | External: Minor diameter, in | Internal: Class | Internal: Minor diameter limits, in: Min | Internal: Minor diameter limits, in: Max | Internal: Pitch diameter limits, in: Min | Internal: Pitch diameter limits, in: Max | Internal: Pitch diameter limits, in: Tolerance | Internal: Major diameter, in: Min |
|---|---|---|---|---|---|---|---|---|---|---|---|---|---|---|---|---|---|
| | | | | | | | | | 1¼–16 to 1⅜–12 | | | | | | | | |
| 1¼–7 | UNC | 1A | 0.0022 | 1.2478 | 1.2232 | | 1.1550 | 1.1439 | 0.0111 | 1.0725 | 1B | 1.095 | 1.123 | 1.1572 | 1.1716 | 0.0144 | 1.2500 |
| | | 2A | 0.0022 | 1.2478 | 1.2314 | 1.2232 | 1.1550 | 1.1476 | 0.0074 | 1.0725 | 2B | 1.095 | 1.123 | 1.1572 | 1.1668 | 0.0096 | 1.2500 |
| | | 3A | 0.0000 | 1.2500 | 1.2336 | | 1.1572 | 1.1517 | 0.0055 | 1.0747 | 3B | 1.0950 | 1.1125 | 1.1572 | 1.1644 | 0.0072 | 1.2500 |
| 1¼–8 | N | 2A | 0.0021 | 1.2479 | 1.2329 | 1.2254 | 1.1667 | 1.597 | 0.0070 | 1.0945 | 2B | 1.115 | 1.140 | 1.1688 | 1.1780 | 0.0092 | 1.2500 |
| | | 3A | 0.0000 | 1.2500 | 1.2350 | | 1.1688 | 1.1635 | 0.0053 | 1.0966 | 3B | 1.1150 | 1.1297 | 1.1688 | 1.1757 | 0.0069 | 1.2500 |
| 1¼–12 | UNF | 1A | 0.0018 | 1.2482 | 1.2310 | | 1.1941 | 1.1849 | 0.0092 | 1.1460 | 1B | 1.160 | 1.178 | 1.1959 | 1.2079 | 0.0120 | 1.2500 |
| | | 2A | 0.0018 | 1.2482 | 1.2368 | | 1.1941 | 1.1879 | 0.0062 | 1.1460 | 2B | 1.160 | 1.178 | 1.1959 | 1.2039 | 0.0080 | 1.2500 |
| | | 3A | 0.0000 | 1.2500 | 1.2386 | | 1.1959 | 1.1913 | 0.0046 | 1.1478 | 3B | 1.1600 | 1.1698 | 1.1959 | 1.2019 | 0.0060 | 1.2500 |
| 1¼–16 | UN | 2A | 0.0015 | 1.2485 | 1.2391 | | 1.2079 | 1.2028 | 0.0051 | 1.1718 | 2B | 1.182 | 1.196 | 1.2094 | 1.2160 | 0.0066 | 1.2500 |
| | | 3A | 0.0000 | 1.2500 | 1.2406 | | 1.2094 | 1.2056 | 0.0038 | 1.1733 | 3B | 1.1820 | 1.1908 | 1.2094 | 1.2144 | 0.0050 | 1.2500 |
| 1¼–18 | NEF | 2A | 0.0015 | 1.2485 | 1.2398 | | 1.2124 | 1.2075 | 0.0049 | 1.1803 | 2B | 1.190 | 1.203 | 1.2139 | 1.2202 | 0.0063 | 1.2500 |
| | | 3A | 0.0000 | 1.2500 | 1.2413 | | 1.2139 | 1.2103 | 0.0036 | 1.1818 | 3B | 1.1900 | 1.1980 | 1.2139 | 1.2186 | 0.0047 | 1.2500 |
| 1 5/16–12 | UN | 2A | 0.0017 | 1.3108 | 1.2994 | | 1.2567 | 1.2509 | 0.0058 | 1.2086 | 2B | 1.222 | 1.240 | 1.2584 | 1.2659 | 0.0075 | 1.3125 |
| | | 3A | 0.0000 | 1.3125 | 1.3011 | | 1.2584 | 1.2541 | 0.0043 | 1.2103 | 3B | 1.2220 | 1.2323 | 1.2584 | 1.2640 | 0.0056 | 1.3125 |
| 1 5/16–16 | UN | 2A | 0.0015 | 1.3110 | 1.3016 | | 1.2704 | 1.2653 | 0.0051 | 1.2343 | 2B | 1.245 | 1.259 | 1.2719 | 1.2785 | 0.0066 | 1.3125 |
| | | 5A | 0.0000 | 1.3125 | 1.3031 | | 1.2719 | 1.2681 | 0.0038 | 1.2358 | 3B | 1.2450 | 1.2533 | 1.2719 | 1.2769 | 0.0050 | 1.3125 |
| 1 5/16–18 | NEF | 2A | 0.0015 | 1.3110 | 1.3023 | | 1.2749 | 1.2700 | 0.0049 | 1.2428 | 2B | 1.252 | 1.265 | 1.2764 | 1.2827 | 0.0063 | 1.3125 |
| | | 3A | 0.0000 | 1.3125 | 1.3038 | | 1.2764 | 1.2728 | 0.0036 | 1.2443 | 3B | 1.2520 | 1.2605 | 1.2764 | 1.2811 | 0.0047 | 1.3125 |
| 1⅜–6 | UNC | 1A | 0.0024 | 1.3726 | 1.3453 | | 1.2643 | 1.2523 | 0.0120 | 1.1681 | 1B | 1.195 | 1.225 | 1.2667 | 1.2822 | 0.0155 | 1.3750 |
| | | 2A | 0.0024 | 1.3726 | 1.3544 | 1.3453 | 1.2643 | 1.2563 | 0.0080 | 1.1681 | 2B | 1.195 | 1.225 | 1.2667 | 1.2771 | 0.0104 | 1.3750 |
| | | 3A | 0.0900 | 1.3750 | 1.3568 | | 1.2667 | 1.2607 | 0.0060 | 1.1705 | 3B | 1.1950 | 1.2146 | 1.2667 | 1.2745 | 0.0078 | 1.3750 |
| 1⅜–8 | N | 2A | 0.0022 | 1.3728 | 1.3578 | 1.3503 | 1.2916 | 1.2844 | 0.0072 | 1.2194 | 2B | 1.240 | 1.265 | 1.2938 | 1.3031 | 0.0093 | 1.3750 |
| | | 3A | 0.0000 | 1.3750 | 1.3600 | | 1.2938 | 1.2884 | 0.0054 | 1.2216 | 3B | 1.2400 | 1.2547 | 1.2938 | 1.3008 | 0.0070 | 1.3750 |
| 1⅜–12 | UNF | 1A | 0.0019 | 1.3731 | 1.3559 | | 1.3190 | 1.3096 | 0.0094 | 1.2709 | 1B | 1.285 | 1.303 | 1.3209 | 1.3332 | 0.0123 | 1.3750 |
| | | 2A | 0.0019 | 1.3731 | 1.3617 | | 1.3190 | 1.3127 | 0.0063 | 1.2709 | 2B | 1.285 | 1.303 | 1.3209 | 1.3291 | 0.0082 | 1.3750 |
| | | 3A | 0.0000 | 1.3750 | 1.3636 | | 1.3209 | 1.3162 | 0.0047 | 1.2728 | 3B | 1.2850 | 1.2948 | 1.3209 | 1.3270 | 0.0061 | 1.3750 |

**FIGURE 10.30** (*Continued*)

| | | | | | | | | | | | | | | | | | |
|---|---|---|---|---|---|---|---|---|---|---|---|---|---|---|---|---|---|
| 1⅜–16 | UN | 2A | 0.0015 | 1.3735 | 1.3641 | | 1.3329 | 1.3278 | 0.0051 | 1.2968 | 2B | 1.307 | 1.321 | 1.3344 | 1.3410 | 0.0066 | 1.3750 |
| | | 3A | 0.0000 | 1.3750 | 1.3656 | | 1.3344 | 1.3306 | 0.0038 | 1.2983 | 3B | 1.3070 | 1.3158 | 1.3344 | 1.3394 | 0.0050 | 1.3750 |
| 1⅜–18 | NEF | 2A | 0.0015 | 1.3735 | 1.3648 | | 1.3374 | 1.3325 | 0.0049 | 1.3053 | 2B | 1.315 | 1.328 | 1.3389 | 1.3452 | 0.0063 | 1.3750 |
| | | 3A | 0.0000 | 1.3750 | 1.3663 | | 1.3389 | 1.3353 | 0.0036 | 1.3068 | 3B | 1.3150 | 1.3230 | 1.3389 | 1.3436 | 0.0047 | 1.3750 |
| 1$\frac{7}{16}$–12 | UN | 2A | 0.0018 | 1.4357 | 1.4243 | | 1.3816 | 1.3757 | 0.0059 | 1.3335 | 2B | 1.347 | 1.365 | 1.3834 | 1.3910 | 0.0076 | 1.4375 |
| | | 3A | 0.0000 | 1.4375 | 1.4261 | | 1.3834 | 1.3790 | 0.0044 | 1.3353 | 3B | 1.3470 | 1.3573 | 1.3834 | 1.3891 | .0057 | 1.4375 |
| 1$\frac{7}{16}$–16 | UN | 2A | 0.0016 | 1.4359 | 1.4265 | | 1.3953 | 1.3901 | 0.0052 | 1.3592 | 2B | 1.370 | 1.384 | 1.3969 | 1.4037 | 0.0068 | 1.4375 |
| | | 3A | 0.0000 | 1.4375 | 1.4281 | | 1.3969 | 1.3930 | 0.0039 | 1.3608 | 3B | 1.3700 | 1.3783 | 1.3969 | 1.4020 | 0.0051 | 1.4375 |
| 1$\frac{7}{16}$–18 | NEF | 2A | 0.0015 | 1.4360 | 1.4273 | | 1.3999 | 1.3949 | 0.0050 | 1.3678 | 2B | 1.377 | 1.390 | 1.4014 | 1.4079 | 0.0065 | 1.4375 |
| | | 3A | 0.0000 | 1.4375 | 1.4288 | | 1.4014 | 1.3977 | 0.0037 | 1.3693 | 3B | 1.3770 | 1.3855 | 1.4014 | 1.4062 | 0.0048 | 1.4375 |
| 1½–6 | UNC | 1A | 0.0024 | 1.4976 | 1.4703 | | 1.3893 | 1.3772 | 0.0121 | 1.2931 | 1B | 1.320 | 1.350 | 1.3917 | 1.4075 | 0.0158 | 1.5000 |
| | | 2A | 0.0024 | 1.4976 | 1.4794 | 1.4703 | 1.3893 | 1.3812 | 0.0081 | 1.2931 | 2B | 1.320 | 1.350 | 1.3917 | 1.4022 | 0.0105 | 1.5000 |
| | | 3A | 0.0000 | 1.5000 | 1.4818 | | 1.3917 | 1.3856 | 0.0061 | 1.2955 | 3B | 1.3200 | 1.3396 | 1.3917 | 1.3996 | 0.0079 | 1.5000 |
| 1½–8 | N | 2A | 0.0022 | 1.4978 | 1.4828 | 1.4753 | 1.4166 | 1.4093 | 0.0073 | 1.3444 | 2B | 1.365 | 1.390 | 1.4188 | 1.4283 | 0.0095 | 1.5000 |
| | | 3A | 0.0000 | 1.5000 | 1.4850 | | 1.4188 | 1.4133 | 0.0055 | 1.3466 | 3B | 1.3650 | 1.3797 | 1.4188 | 1.4259 | 0.0071 | 1.5000 |
| 1½–12 | UNF | 1A | 0.0019 | 1.4981 | 1.4809 | | 1.4440 | 1.4344 | 0.0096 | 1.3959 | 1B | 1.410 | 1.428 | 1.4459 | 1.4584 | 0.0125 | 1.5000 |
| | | 2A | 0.0019 | 1.4981 | 1.4867 | | 1.4440 | 1.4376 | 0.0064 | 1.3959 | 2B | 1.410 | 1.428 | 1.4459 | 1.4542 | 0.0083 | 1.5000 |
| | | 3A | 0.0000 | 1.5000 | 1.4886 | | 1.4459 | 1.4411 | 0.0048 | 1.3978 | 3B | 1.4100 | 1.4198 | 1.4459 | 1.4522 | 0.0063 | 1.5000 |
| 1½–16 | UN | 2A | 0.0016 | 1.4984 | 1.4890 | | 1.4578 | 1.4526 | 0.0052 | 1.4217 | 2B | 1.432 | 1.446 | 1.4594 | 1.4662 | 0.0068 | 1.5000 |
| | | 3A | 0.0000 | 1.5000 | 1.4906 | | 1.4594 | 1.4555 | 0.0039 | 1.4233 | 3B | 1.4320 | 1.4408 | 1.4594 | 1.4645 | 0.0051 | 1.5000 |
| 1½–18 | NEF | 2A | 0.0015 | 1.4985 | 1.4898 | | 1.4624 | 1.4574 | 0.0050 | 1.4303 | 2B | 1.440 | 1.452 | 1.4639 | 1.4704 | 0.0065 | 1.5000 |
| | | 3A | 0.0000 | 1.5000 | 1.4913 | | 1.4639 | 1.4602 | 0.0037 | 1.4318 | 3B | 1.4400 | 1.4480 | 1.4639 | 1.4687 | 0.0048 | 1.5000 |
| 1$\frac{9}{16}$–16 | N | 2A | 0.0016 | 1.5609 | 1.5515 | | 1.5203 | 1.5151 | 0.0052 | 1.4842 | 2B | 1.495 | 1.509 | 1.5219 | 1.5287 | 0.0068 | 1.5625 |
| | | 3A | 0.0000 | 1.5625 | 1.5531 | | 1.5219 | 1.5180 | 0.0039 | 1.4858 | 3B | 1.4950 | 1.5033 | 1.5219 | 1.5270 | 0.0051 | 1.5625 |
| 1$\frac{9}{16}$–18 | NEF | 2A | 0.0015 | 1.5610 | 1.5523 | | 1.5249 | 1.5199 | 0.0050 | 1.4928 | 2B | 1.502 | 1.515 | 1.5264 | 1.5329 | 0.0065 | 1.5625 |
| | | 3A | 0.0000 | 1.5625 | 1.5538 | | 1.5264 | 1.5227 | 0.0037 | 1.4943 | 3B | 1.5020 | 1.5105 | 1.5264 | 1.5312 | 0.0048 | 1.5625 |
| 1⅝–8 | N | 2A | 0.0022 | 1.6228 | 1.6078 | 1.6003 | 1.5416 | 1.5342 | 0.0074 | 1.4694 | 2B | 1.490 | 1.515 | 1.5438 | 1.5535 | 0.0097 | 1.6250 |
| | | 3A | 0.0000 | 1.6250 | 1.6100 | | 1.5438 | 1.5382 | 0.0056 | 1.4716 | 3B | 1.4900 | 1.5047 | 1.5438 | 1.5510 | 0.0072 | 1.6250 |
| 1⅝–12 | UN | 2A | 0.0018 | 1.6232 | 1.6118 | | 1.5691 | 1.5632 | 0.0059 | 1.5210 | 2B | 1.535 | 1.553 | 1.5709 | 1.5785 | 0.0076 | 1.6250 |
| | | 3A | 0.0000 | 1.6250 | 1.6136 | | 1.5709 | 1.5665 | 0.0044 | 1.5228 | 3B | 1.5350 | 1.5448 | 1.5709 | 1.5766 | 0.0057 | 1.6250 |

**FIGURE 10.30** (*Continued*)

## Standard Limits of Size: Unified and American Screw Threads (*Continued*)

| Nominal size and threads per inch | Series designation | External | | | | | | | | | Internal | | | | | | |
|---|---|---|---|---|---|---|---|---|---|---|---|---|---|---|---|---|---|
| | | | | Major diameter limits, in | | | Pitch diameter limits, in | | | | | Minor diameter limits, in | | Pitch diameter limits, in | | | Major diameter, in |
| | | Class | Allowance, in | Max* | Min | Min' | Max* | Min | Tolerance | Minor diameter, in | Class | Min | Max | Min | Max | Tolerance | Min |
| | | | | | | | 1⅝–16 to 1¾–12 | | | | | | | | | | |
| 1⅝–16 | UN | 2A | 0.0016 | 1.6234 | 1.6140 | | 1.5828 | 1.5776 | 0.0052 | 1.5467 | 2B | 1.557 | 1.571 | 1.5844 | 1.5912 | 0.0068 | 1.6250 |
| | | 3A | 0.0000 | 1.6250 | 1.6156 | | 1.5814 | 1.5805 | 0.0039 | 1.5483 | 3B | 1.5570 | 1.5658 | 1.5844 | 1.5895 | 0.0051 | 1.6250 |
| 1⅝–18 | NEF | 2A | 0.0015 | 1.6235 | 1.6148 | | 1.5874 | 1.5824 | 0.0050 | 1.5553 | 2B | 1.565 | 1.578 | 1.5889 | 1.5954 | 0.0065 | 1.6250 |
| | | 3A | 0.0000 | 1.6250 | 1.6163 | | 1.5889 | 1.5852 | 0.0037 | 1.5568 | 3B | 1.5650 | 1.5730 | 1.5889 | 1.5937 | 0.0048 | 1.6250 |
| 1$\frac{11}{16}$–16 | N | 2A | 0.0016 | 1.6859 | 1.6765 | | 1.6453 | 1.6400 | 0.0053 | 1.6092 | 2B | 1.620 | 1.634 | 1.6469 | 1.6538 | 0.0069 | 1.6875 |
| | | 3A | 0.0000 | 1.6875 | 1.6781 | | 1.6469 | 1.6429 | 0.0040 | 1.6108 | 3B | 1.6200 | 1.6283 | 1.6469 | 1.6521 | 0.0052 | 1.6875 |
| 1$\frac{11}{16}$–18 | NEF | 2A | 0.0015 | 1.6860 | 1.6773 | | 1.6499 | 1.6448 | 0.0051 | 1.6178 | 2B | 1.627 | 1.640 | 1.6514 | 1.6580 | 0.0066 | 1.6875 |
| | | 3A | 0.0000 | 1.6875 | 1.6788 | | 1.6514 | 1.6476 | 0.0038 | 1.6193 | 3B | 1.6270 | 1.6355 | 1.6514 | 1.6563 | 0.0049 | 1.6875 |
| 1¾–5 | UNC | 1A | 0.0027 | 1.7473 | 1.7165 | | 1.6174 | 1.6040 | 0.0134 | 1.5019 | 1B | 1.534 | 1.568 | 1.6201 | 1.6375 | 0.0174 | 1.7500 |
| | | 2A | 0.0027 | 1.7473 | 1.7268 | 1.7165 | 1.6174 | 1.6085 | 0.0089 | 1.5019 | 2B | 1.534 | 1.568 | 1.6201 | 1.6317 | 0.0116 | 1.7500 |
| | | 3A | 0.0000 | 1.7500 | 1.7295 | | 1.6201 | 1.6134 | 0.0067 | 1.5046 | 3B | 1.5340 | 1.5575 | 1.6201 | 1.6288 | 0.0087 | 1.7500 |
| 1¾–8 | N | 2A | 0.0023 | 1.7477 | 1.7327 | 1.7252 | 1.6665 | 1.6590 | 0.0075 | 1.5943 | 2B | 1.615 | 1.640 | 1.6688 | 1.6786 | 0.0098 | 1.7500 |
| | | 3A | 0.0000 | 1.7300 | 1.7350 | | 1.6688 | 1.6632 | 0.0056 | 1.5966 | 3B | 1.6150 | 1.6297 | 1.6688 | 1.6762 | 0.0074 | 1.7500 |
| 1¾–12 | UN | 2A | 0.0018 | 1.7482 | 1.7368 | | 1.6941 | 1.6881 | 0.0060 | 1.6460 | 2B | 1.660 | 1.678 | 1.6959 | 1.7037 | 0.0078 | 1.7500 |
| | | 3A | 0.0000 | 1.7500 | 1.7386 | | 1.6959 | 1.6914 | 0.0045 | 1.6478 | 3B | 1.6600 | 1.6698 | 1.6959 | 1.7017 | 0.0058 | 1.7500 |
| | | | | | | | 1¾–16 to 3–16 | | | | | | | | | | |
| 1¾–16 | UNEF | 2A | 0.0016 | 1.7484 | 1.7390 | | 1.7078 | 1.7025 | 0.0053 | 1.6717 | 2B | 1.682 | 1.696 | 1.7094 | 1.7163 | 0.0069 | 1.7500 |
| | | 3A | 0.0000 | 1.7500 | 1.7406 | | 1.7094 | 1.7054 | 0.0040 | 1.6733 | 3B | 1.6820 | 1.6906 | 1.7094 | 1.7146 | 0.0052 | 1.7500 |
| 1$\frac{13}{16}$–16 | N | 2A | 0.0016 | 1.8109 | 1.8015 | | 1.7703 | 1.7650 | 0.0053 | 1.7342 | 2B | 1.745 | 1.759 | 1.7719 | 1.7788 | 0.0069 | 1.8125 |
| | | 3A | 0.0000 | 1.8125 | 1.8031 | | 1.7719 | 1.7679 | 0.0040 | 1.7358 | 3B | 1.7450 | 1.7533 | 1.7719 | 1.7771 | 0.0052 | 1.8125 |
| 1⅞–8 | N | 2A | 0.0023 | 1.8727 | 1.8577 | 1.8502 | 1.7915 | 1.7838 | 0.0077 | 1.7193 | 2B | 1.740 | 1.765 | 1.7938 | 1.8038 | 0.0100 | 1.8750 |
| | | 3A | 0.0000 | 1.8750 | 1.8600 | | 1.7938 | 1.7881 | 0.0057 | 1.7216 | 3B | 1.7400 | 1.7547 | 1.7938 | 1.8013 | 0.0075 | 1.8750 |
| 1⅞–12 | UN | 2A | 0.0018 | 1.8732 | 1.8618 | | 1.8191 | 1.8131 | 0.0060 | 1.7710 | 2B | 1.785 | 1.803 | 1.8209 | 1.8287 | 0.0078 | 1.8750 |
| | | 3A | 0.0000 | 1.8750 | 1.8636 | | 1.8209 | 1.8164 | 0.0045 | 1.7728 | 3B | 1.7850 | 1.7948 | 1.8209 | 1.8267 | 0.0058 | 1.8750 |
| 1⅞–16 | UN | 2A | 0.0016 | 1.8734 | 1.8640 | | 1.8328 | 1.8275 | 0.0053 | 1.7967 | 2B | 1.807 | 1.821 | 1.8344 | 1.8413 | 0.0069 | 1.8750 |
| | | 3A | 0.0000 | 1.8750 | 1.8656 | | 1.8344 | 1.8304 | 0.0040 | 1.7983 | 3B | 1.8070 | 1.8158 | 1.8344 | 1.8396 | 0.0052 | 1.8750 |

**FIGURE 10.30** (*Continued*)

| | | | | | | | | | | | | | | | | | |
|---|---|---|---|---|---|---|---|---|---|---|---|---|---|---|---|---|---|
| $1^{15}/_{16}$–16 | N | 2A | 0.0016 | 1.9359 | 1.9265 | | 1.8953 | 1.8899 | 0.0054 | 1.8592 | 2B | 1.870 | 1.884 | 1.8969 | 1.9039 | 0.0070 | 1.9375 |
| | | 3A | 0.0000 | 1.9375 | 1.9281 | | 1.8969 | 1.8929 | 0.0040 | 1.8608 | 3B | 1.8700 | 1.8783 | 1.8969 | 1.9021 | .0052 | 1.9375 |
| 2–4½ | UNC | 1A | 0.0029 | 1.9971 | 1.9641 | | 1.8528 | 1.8385 | 0.0143 | 1.7245 | 1B | 1.759 | 1.795 | 1.8557 | 1.8743 | 0.0186 | 2.0000 |
| | | 2A | 0.0029 | 1.9971 | 1.9751 | 1.9641 | 1.8528 | 1.8433 | 0.0095 | 1.7245 | 2B | 1.759 | 1.795 | 1.8557 | 1.8681 | 0.0124 | 2.0000 |
| | | 3A | 0.0000 | 2.0000 | 1.9780 | | 1.8557 | 1.8486 | 0.0071 | 1.7274 | 3B | 1.7590 | 1.7861 | 1.8557 | 1.8650 | 0.0093 | 2.0000 |
| 2–8 | N | 2A | 0.0023 | 1.9977 | 1.9827 | 1.9752 | 1.9165 | 1.9087 | 0.0078 | 1.8443 | 2B | 1.865 | 1.890 | 1.9188 | 1.9289 | 0.0101 | 2.0000 |
| | | 3A | 0.0000 | 2.0000 | 1.9850 | | 1.9188 | 1.9130 | 0.0058 | 1.8466 | 3B | 1.8650 | 1.8797 | 1.9188 | 1.9264 | 0.0076 | 2.0000 |
| 2–12 | UN | 2A | 0.0018 | 1.9982 | 1.9868 | | 1.9441 | 1.9380 | 0.0061 | 1.8960 | 2B | 1.910 | 1.928 | 1.9459 | 1.9538 | 0.0079 | 2.0000 |
| | | 3A | 0.0000 | 2.0000 | 1.9886 | | 1.9459 | 1.9414 | 0.0045 | 1.8978 | 3B | 1.9100 | 1.9198 | 1.9459 | 1.9518 | 0.0059 | 2.0000 |
| 2–16 | UNEF | 2A | 0.0016 | 1.9984 | 1.9890 | | 1.9578 | 1.9524 | 0.0054 | 1.9217 | 2B | 1.932 | 1.946 | 1.9594 | 1.0664 | 0.0070 | 2.0000 |
| | | 3A | 0.0000 | 2.0000 | 1.9906 | | 1.9594 | 1.9554 | 0.0040 | 1.9233 | 3B | 1.9320 | 1.9408 | 1.9594 | 1.9646 | 0.0052 | 2.0000 |
| $2^{1}/_{16}$–16 | N | 2A | 0.0016 | 2.0609 | 2.0515 | | 2.0203 | 2.0149 | 0.0054 | 1.9842 | 2B | 1.995 | 2.009 | 2.0219 | 2.0289 | 0.0070 | 2.0625 |
| | | 3A | 0.0000 | 2.0625 | 2.0531 | | 2.0219 | 2.0179 | 0.0040 | 1.9858 | 3B | 1.9950 | 2.0033 | 2.0219 | 2.0271 | 0.0052 | 2.0625 |
| $2^{1}/_{8}$–8 | N | 2A | 0.0024 | 2.1226 | 2.1076 | 2.1001 | 2.0414 | 2.0335 | 0.0079 | 1.9692 | 2B | 1.990 | 2.015 | 2.0438 | 2.0540 | 0.0102 | 2.1250 |
| | | 3A | 0.0000 | 2.1250 | 2.1100 | | 2.0438 | 2.0379 | 0.0059 | 1.9716 | 3B | 1.9900 | 2.0047 | 2.0438 | 2.0515 | 0.0077 | 2.1250 |
| $2^{1}/_{8}$–12 | UN | 2A | 0.0018 | 2.1232 | 2.1118 | | 2.0691 | 2.0630 | 0.0061 | 2.0210 | 2B | 2.035 | 2.053 | 2.0709 | 2.0788 | 0.0079 | 2.1250 |
| | | 3A | 0.0000 | 2.1250 | 2.1136 | | 2.0709 | 2.0664 | 0.0045 | 2.0228 | 3B | 2.0350 | 2.0448 | 2.0709 | 2.0768 | 0.0059 | 2.1250 |
| $2^{1}/_{8}$–16 | UN | 2A | 0.0016 | 2.1234 | 2.1140 | | 2.0828 | 2.0774 | 0.0054 | 2.0467 | 2B | 2.057 | 2.071 | 2.0844 | 2.0914 | 0.0070 | 2.1250 |
| | | 3A | 0.0000 | 2.1250 | 2.1156 | | 2.0844 | 2.0803 | 0.0041 | 2.0483 | 3B | 2.0570 | 2.0658 | 2.0844 | 2.0896 | 0.0052 | 2.1250 |
| $2^{3}/_{16}$–16 | N | 2A | 0.0016 | 2.1859 | 2.1765 | | 2.1453 | 2.1399 | 0.0054 | 2.1092 | 2B | 2.120 | 2.134 | 2.1469 | 2.1539 | 0.0070 | 2.1875 |
| | | 3A | 0.0000 | 2.1875 | 2.1781 | | 2.1469 | 2.1428 | 0.0041 | 2.1108 | 3B | 2.1200 | 2.1283 | 2.1469 | 2.1521 | 0.0052 | 2.1875 |
| $2^{1}/_{4}$–4½ | UNC | 1A | 0.0029 | 2.2471 | 2.2141 | | 2.1028 | 2.0882 | 0.0146 | 1.9745 | 1B | 2.009 | 2.045 | 2.1057 | 2.1247 | 0.0190 | 2.2500 |
| | | 2A | 0.0029 | 2.2471 | 2.2251 | 2.2141 | 2.1028 | 2.0931 | 0.0097 | 1.9745 | 2B | 2.009 | 2.045 | 2.1057 | 2.1183 | 0.0126 | 2.2500 |
| | | 3A | 0.0000 | 2.2500 | 2.2280 | | 2.1057 | 2.0984 | 0.0073 | 1.9774 | 3B | 2.0090 | 2.0361 | 2.1057 | 2.1152 | 0.0095 | 2.2500 |
| $2^{1}/_{4}$–8 | N | 2A | 0.0024 | 2.2476 | 2.2326 | 2.2251 | 2.1664 | 2.1584 | 0.0080 | 2.0942 | 2B | 2.115 | 2.140 | 2.1688 | 2.1792 | 0.0104 | 2.2500 |
| | | 3A | 0.0000 | 2.2500 | 2.2350 | | 2.1688 | 2.1628 | 0.0060 | 2.0966 | 3B | 2.1150 | 2.1297 | 2.1688 | 2.1766 | 0.0078 | 2.2500 |
| $2^{1}/_{4}$–12 | UN | 2A | 0.0018 | 2.2482 | 2.2368 | | 2.1941 | 2.1880 | 0.0061 | 2.1460 | 2B | 2.160 | 2.178 | 2.1959 | 2.2038 | 0.0079 | 2.2500 |
| | | 3A | 0.0000 | 2.2500 | 2.2386 | | 2.1959 | 2.1914 | 0.0045 | 2.1478 | 3B | 2.1600 | 2.1698 | 2.1959 | 2.2018 | 0.0059 | 2.2500 |
| $2^{1}/_{4}$–16 | UN | 2A | 0.0016 | 2.2484 | 2.2390 | | 2.2078 | 2.2024 | 0.0054 | 2.1717 | 2B | 2.182 | 2.196 | 2.2094 | 2.2164 | 0.0070 | 2.2500 |
| | | 3A | 0.0000 | 2.2500 | 2.2406 | | 2.2094 | 2.2053 | 0.0041 | 2.1733 | 3B | 2.1820 | 2.1908 | 2.2094 | 2.2146 | 0.0052 | 2.2500 |
| $2^{5}/_{16}$–16 | N | 2A | 0.0017 | 2.3108 | 2.3014 | | 2.2702 | 2.2647 | 0.0055 | 2.2341 | 2B | 2.245 | 2.259 | 2.2719 | 2.2791 | 0.0072 | 2.3125 |
| | | 3A | 0.0000 | 2.3125 | 2.3031 | | 2.2719 | 2.2678 | 0.0041 | 2.2358 | 3B | 2.2450 | 2.2533 | 2.2719 | 2.2773 | 0.0054 | 2.3125 |

**FIGURE 10.30** *(Continued)*

## Standard Limits of Size: Unified and American Screw Threads (*Continued*)

| Nominal size and threads per inch | Series designation | External | | | | | | | | | | Internal | | | | | | |
|---|---|---|---|---|---|---|---|---|---|---|---|---|---|---|---|---|---|---|
| | | Class | Allowance, in | Major diameter limits, in | | | Pitch diameter limits, in | | | Minor diameter, in | | Class | Minor diameter limits, in | | Pitch diameter limits, in | | | Major diameter, in |
| | | | | Max* | Min | Min' | Max* | Min | Tolerance | | | | Min | Max | Min | Max | Tolerance | Min |
| **1¾–16 to 3–16** | | | | | | | | | | | | | | | | | | |
| 2⅜–12 | UN | 2A | 0.0019 | 2.3731 | 2.3617 | | 2.3190 | 2.3128 | 0.0062 | 2.2709 | | 2B | 2.285 | 2.303 | 2.3209 | 2.3290 | 0.0081 | 2.3750 |
| | | 3A | 0.0000 | 2.3750 | 2.3636 | | 2.3209 | 2.3163 | 0.0046 | 2.2728 | | 3B | 2.2850 | 2.2948 | 2.3209 | 2.3269 | 0.0060 | 2.3750 |
| 2⅜–16 | UN | 2A | 0.0017 | 2.3733 | 2.3639 | | 2.3327 | 2.3272 | 0.0055 | 2.2966 | | 2B | 2.307 | 2.321 | 2.3344 | 2.3416 | 0.0072 | 2.3750 |
| | | 3A | 0.0000 | 2.3750 | 2.3656 | | 2.3344 | 2.3303 | 0.0041 | 2.2983 | | 3B | 2.3070 | 2.3158 | 2.3344 | 2.3398 | 0.0054 | 2.3750 |
| 2 7/16–16 | N | 2A | 0.0017 | 2.4358 | 2.4264 | | 2.3953 | 2.3897 | 0.0055 | 2.3591 | | 2B | 2.370 | 2.384 | 2.3969 | 2.4041 | 0.0072 | 2.4375 |
| | | 3A | 0.0000 | 2.4375 | 2.4281 | | 2.3969 | 2.3928 | 0.0041 | 2.3608 | | 3B | 2.3700 | 2.3783 | 2.3969 | 2.4023 | 0.0054 | 2.4375 |
| 2½–4 | UNC | 1A | 0.0031 | 2.4969 | 2.4612 | | 2.3345 | 2.3190 | 0.0155 | 2.1902 | | 1B | 2.229 | 2.267 | 2.3376 | 2.3578 | 0.0202 | 2.5000 |
| | | 2A | 0.0031 | 2.4969 | 2.4731 | 2.4612 | 2.3345 | 2.3241 | 0.0104 | 2.1902 | | 2B | 2.229 | 2.267 | 2.3376 | 2.3511 | 0.0135 | 2.5000 |
| | | 3A | 0.0000 | 2.5000 | 2.4762 | | 2.3376 | 2.3298 | 0.0078 | 2.1933 | | 3B | 2.2290 | 2.2594 | 2.3376 | 2.3477 | 0.0101 | 2.5000 |
| 2½–8 | N | 2A | 0.0024 | 2.4976 | 2.4826 | 2.4751 | 2.4164 | 2.4082 | 0.0082 | 2.3442 | | 2B | 2.365 | 2.390 | 2.4188 | 2.4294 | 0.0106 | 2.5000 |
| | | 3A | 0.0000 | 2.5000 | 2.4850 | | 2.4188 | 2.4127 | 0.0061 | 2.3466 | | 3B | 2.3650 | 2.3797 | 2.4188 | 2.4268 | 0.0080 | 2.5000 |
| 2½–12 | UN | 2A | 0.0019 | 2.4981 | 2.4867 | | 2.4440 | 2.4378 | 0.0062 | 2.3959 | | 2B | 2.410 | 2.428 | 2.4459 | 2.4540 | 0.0081 | 2.5000 |
| | | 3A | 0.0000 | 2.5000 | 2.4886 | | 2.4459 | 2.4413 | 0.0046 | 2.3978 | | 3B | 2.4100 | 2.4198 | 2.4459 | 2.4519 | 0.0060 | 2.5000 |
| 2½–16 | UN | 2A | 0.0017 | 2.4983 | 2.4889 | | 2.4577 | 2.4522 | 0.0055 | 2.4216 | | 2B | 2.432 | 2.446 | 2.4594 | 2.4666 | 0.0072 | 2.5000 |
| | | 3A | 0.0000 | 2.5000 | 2.4906 | | 2.4594 | 2.4553 | 0.0041 | 2.4233 | | 3B | 2.4320 | 2.4408 | 2.4594 | 2.4648 | 0.0054 | 2.5000 |
| 2⅝–12 | UN | 2A | 0.0019 | 2.6231 | 2.6117 | | 2.5690 | 2.5628 | 0.0062 | 2.5209 | | 2B | 2.535 | 2.553 | 2.5709 | 2.5790 | 0.0081 | 2.6250 |
| | | 3A | 0.0000 | 2.6250 | 2.6136 | | 2.5709 | 2.5663 | 0.0046 | 2.5228 | | 3B | 2.5350 | 2.5448 | 2.5709 | 2.5769 | 0.0060 | 2.6250 |
| 2⅝–16 | UN | 2A | 0.0017 | 2.6233 | 2.6139 | | 2.5827 | 2.5772 | 0.0055 | 2.5466 | | 2B | 2.557 | 2.571 | 2.5844 | 2.5916 | 0.0072 | 2.6250 |
| | | 3A | 0.0000 | 2.6250 | 2.6156 | | 2.5844 | 2.5808 | 0.0041 | 2.5483 | | 3B | 2.5570 | 2.5658 | 2.5844 | 2.5898 | 0.0054 | 2.6250 |
| 2¾–4 | UNC | 1A | 0.0032 | 2.7468 | 2.7111 | | 2.5844 | 2.5686 | 0.0158 | 2.4401 | | 1B | 2.479 | 2.517 | 2.5876 | 2.6082 | 0.0206 | 2.7500 |
| | | 2A | 0.0032 | 2.7468 | 2.7230 | 2.7111 | 2.5844 | 2.5739 | 0.0105 | 2.4401 | | 2B | 2.479 | 2.517 | 2.5876 | 2.6013 | 0.0137 | 2.7500 |
| | | 3A | 0.0000 | 2.7500 | 2.7262 | | 2.5876 | 2.5797 | 0.0079 | 2.4433 | | 3B | 2.4790 | 2.5094 | 2.5876 | 2.5979 | 0.0103 | 2.7500 |
| 2¾–8 | N | 2A | 0.0025 | 2.7475 | 2.7325 | 2.7250 | 2.6663 | 2.0580 | 0.0083 | 2.5941 | | 2B | 2.615 | 2.640 | 2.6688 | 2.6796 | 0.0108 | 2.7500 |
| | | 3A | 0.0000 | 2.7500 | 2.7350 | | 2.6688 | 2.6625 | 0.0063 | 2.5966 | | 3B | 2.6150 | 2.6297 | 2.6688 | 2.6769 | 0.0081 | 2.7500 |
| 2¾–12 | UN | 2A | 0.0019 | 2.7481 | 2.7367 | | 2.6940 | 2.6878 | 0.0062 | 2.6459 | | 2B | 2.660 | 2.678 | 2.6959 | 2.7040 | 0.0081 | 2.7500 |
| | | 3A | 0.0000 | 2.7500 | 2.7386 | | 2.6959 | 2.6913 | 0.0046 | 2.6478 | | 3B | 2.6600 | 2.6698 | 2.6959 | 2.7019 | 0.0060 | 2.7500 |

**FIGURE 10.30** (*Continued*)

| | | | | | | | | | | | | | | | | |
|---|---|---|---|---|---|---|---|---|---|---|---|---|---|---|---|---|
| 2¾–16 | UN | 2A | 0.0017 | 2.7483 | 2.7389 | | 2.7077 | 2.7022 | 0.0055 | 2.6716 | 2B | 2.682 | 2.696 | 2.7094 | 2.7166 | 0.0072 | 2.7500 |
| | | 3A | 0.0000 | 2.7500 | 2.7406 | | 2.7094 | 2.7053 | 0.0041 | 2.6733 | 3B | 2.6820 | 2.6908 | 2.7094 | 2.7148 | 0.0054 | 2.7500 |
| 2⅞–12 | UN | 2A | 0.0019 | 2.8731 | 2.8617 | | 2.8190 | 2.8127 | 0.0063 | 2.7709 | 2B | 2.785 | 2.803 | 2.8209 | 2.8291 | 0.0082 | 2.8750 |
| | | 3A | 0.0000 | 2.8750 | 2.8636 | | 2.8209 | 2.8162 | 0.0047 | 2.7728 | 3B | 2.7850 | 2.7948 | 2.8209 | 2.8271 | 0.0062 | 2.8750 |
| 2⅞–16 | UN | 2A | 0.0017 | 2.8733 | 2.8639 | | 2.8327 | 2.8271 | 0.0056 | 2.7966 | 2B | 2.807 | 2.821 | 2.8344 | 2.8417 | 0.0073 | 2.8750 |
| | | 3A | 0.0000 | 2.8750 | 2.8656 | | 2.8344 | 2.8302 | 0.0042 | 2.7983 | 3B | 2.8070 | 2.8158 | 2.8344 | 2.8399 | 0.0055 | 2.8750 |
| 3–4 | UNC | 1A | 0.0032 | 2.9968 | 2.9611 | | 2.8344 | 2.8183 | 0.0161 | 2.6901 | 1B | 2.729 | 2.767 | 2.8376 | 2.8585 | 0.0209 | 3.0000 |
| | | 2A | 0.0032 | 2.9968 | 2.9730 | 2.9611 | 2.8344 | 2.8237 | 0.0107 | 2.6901 | 2B | 2.729 | 2.767 | 2.8376 | 2.8515 | 0.0139 | 3.0000 |
| | | 3A | 0.0000 | 3.0000 | 2.9762 | | 2.8376 | 2.8296 | 0.0080 | 2.6933 | 3B | 2.7290 | 2.7594 | 2.8376 | 2.8480 | 0.0104 | 3.0000 |
| 3–8 | N | 2A | 0.0026 | 2.9974 | 2.9824 | 2.9749 | 2.9162 | 2.9077 | 0.0085 | 2.8440 | 2B | 2.865 | 2.890 | 2.9188 | 2.9299 | 0.0111 | 3.0000 |
| | | 3A | 0.0000 | 3.0000 | 2.9850 | | 2.9188 | 2.9124 | 0.0064 | 2.8466 | 3B | 2.8650 | 2.8797 | 2.9188 | 2.9271 | 0.0083 | 3.0000 |
| 3–12 | UN | 2A | 0.0019 | 2.9981 | 2.9867 | | 2.9440 | 2.9377 | 0.0063 | 2.8959 | 2B | 2.910 | 2.928 | 2.9459 | 2.9541 | 0.0082 | 3.0000 |
| | | 3A | 0.0000 | 3.0000 | 2.9886 | | 2.9459 | 2.9412 | 0.0047 | 2.5978 | 3B | 2.9100 | 2.9198 | 2.9459 | 2.9521 | 0.0062 | 3.0000 |
| 3–16 | UN | 2A | 0.0017 | 2.9983 | 2.9889 | | 2.9577 | 2.9521 | 0.0056 | 2.9216 | 2B | 2.932 | 2.946 | 2.9594 | 2.9667 | 0.0073 | 3.0000 |
| | | 3A | 0.0000 | 3.0000 | 2.9906 | | 2.9594 | 2.9552 | 0.0042 | 2.9233 | 3B | 2.9320 | 2.9408 | 2.9594 | 2.9649 | 0.0055 | 3.0000 |

* For class 2A threads having an additive finish, the maximum is increased to the basic size, the value being the same as for class 3A shown in this column.

† For unfinished hot-rolled material.

SOURCE: Extracted from National Bureau of Standards Handbook H28 (1967), Part I (Screw-Thread Standards for Federal Services), which is in general agreement with *American Standard Unified Screw Threads* (ASA B1.1-1960).

**FIGURE 10.30** (*Continued*)

| Drill No. | Decimal | Drill No. | Decimal | Drill No. | Decimal |
|---|---|---|---|---|---|
| 97 | 0.0059 | 56 | 0.0465 | 15 | 0.180 |
| 96 | 0.0063 | 55 | 0.052 | 14 | 0.182 |
| 95 | 0.0067 | 54 | 0.055 | 13 | 0.185 |
| 94 | 0.0071 | 53 | 0.0595 | 12 | 0.189 |
| 93 | 0.0075 | 52 | 0.0635 | 11 | 0.191 |
| 92 | 0.0079 | 51 | 0.067 | 10 | 0.1935 |
| 91 | 0.0083 | 50 | 0.070 | 9 | 0.196 |
| 90 | 0.0087 | 49 | 0.073 | 8 | 0.199 |
| 89 | 0.0091 | 48 | 0.076 | 7 | 0.201 |
| 88 | 0.0095 | 47 | 0.0785 | 6 | 0.204 |
| 87 | 0.010 | 46 | 0.076 | 5 | 0.2055 |
| 86 | 0.0105 | 45 | 0.082 | 4 | 0.209 |
| 85 | 0.011 | 44 | 0.086 | 3 | 0.213 |
| 84 | 0.0115 | 43 | 0.089 | 2 | 0.221 |
| 83 | 0.012 | 42 | 0.0935 | 1 | 0.228 |
| 82 | 0.0125 | 41 | 0.096 | A | 0.234 |
| 81 | 0.013 | 40 | 0.098 | B | 0.238 |
| 80 | 0.0135 | 39 | 0.0995 | C | 0.242 |
| 79 | 0.0145 | 38 | 0.1015 | D | 0.246 |
| 78 | 0.016 | 37 | 0.104 | E | 0.250 |
| 77 | 0.018 | 36 | 0.1065 | F | 0.257 |
| 76 | 0.020 | 35 | 0.110 | G | 0.261 |
| 75 | 0.021 | 34 | 0.111 | H | 0.266 |
| 74 | 0.0225 | 33 | 0.113 | I | 0.272 |
| 73 | 0.024 | 32 | 0.116 | J | 0.277 |
| 72 | 0.025 | 31 | 0.120 | K | 0.281 |
| 71 | 0.026 | 30 | 0.1285 | L | 0.290 |
| 70 | 0.028 | 29 | 0.136 | M | 0.295 |
| 69 | 0.0292 | 28 | 0.1405 | N | 0.302 |
| 68 | 0.033 | 27 | 0.144 | O | 0.316 |
| 67 | 0.032 | 26 | 0.147 | P | 0.323 |
| 66 | 0.035 | 25 | 0.1495 | Q | 0.332 |
| 65 | 0.035 | 24 | 0.152 | R | 0.339 |
| 64 | 0.035 | 23 | 0.154 | S | 0.348 |
| 63 | 0.037 | 22 | 0.157 | T | 0.358 |
| 62 | 0.038 | 21 | 0.159 | U | 0.368 |
| 61 | 0.039 | 20 | 0.161 | V | 0.377 |
| 60 | 0.040 | 19 | 0.166 | W | 0.386 |
| 59 | 0.041 | 18 | 0.1695 | X | 0.397 |
| 58 | 0.042 | 17 | 0.173 | Y | 0.404 |
| 57 | 0.043 | 16 | 0.177 | Z | 0.413 |

**FIGURE 10.31** American standard drill sizes.

$A_{ts}$ = tensile stress area of screw thread as given by the equation in item 2 (below), in
$p_d$ = minimum pitch diameter of external thread, in

**2.** *Different materials*; i.e., internal threaded part of lower strength than external threaded part:

***a.*** Determine relative strength of external thread and internal thread from

$$R = \frac{A_{se}(S_e)}{A_{si}(S_i)}$$

where $R$ = relative strength factor
$A_{se}$ = shear area of external thread, in$^2$
$A_{si}$ = shear area of internal thread, in$^2$
$S_e$ = tensile strength of external thread material, psi
$S_i$ = tensile strength of internal thread material, psi

***b.*** If $R$ is ≤1, the length of engagement as determined by the equation in item 1 (above) is adequate to prevent stripping of the internal thread. If $R$ is >1, the length of engagement $G$ to prevent internal thread strip is

$$G = E_L R$$

In the immediately preceding equation, $A_{se}$ and $A_{si}$ are the shear areas and are calculated as follows:

$$A_{se} = \pi n E_L D_m \left[ \frac{1}{2n} + \frac{(p_d - D_m)}{\sqrt{3}} \right] \qquad A_{si} = \pi n E_L D_M \left[ \frac{1}{2n} + \frac{(D_M - D_p)}{\sqrt{3}} \right]$$

where $D_p$ = maximum pitch diameter of internal thread, in; $D_M$ = minimum major diameter of external thread, in. (Other symbols have been defined previously.)

## 10.3 RIVETS

Rivets form a large class of fastening devices and are manufactured in many types and varieties and various materials. Rivets are made from carbon steels, aluminum alloys, brass, copper, bronze, or other materials agreed on by the manufacturer and purchaser.

Head forms for standard rivets include button, truss, brazier, coopers, oval, and flush. The flush head rivets are provided in the following countersunk head angles: 60°, 78°, 90°, 100°, 120°, 144°, and 150°. The most common countersunk forms are the 90° and 100° types. To employ these types of rivets, a countersinking tool is used to cut the recess for the flush head. When the material is thin, a process known as *dimpling* or *double dimpling* is employed. Figure 10.32*a* and 10.32*b* shows the dimpling methods.

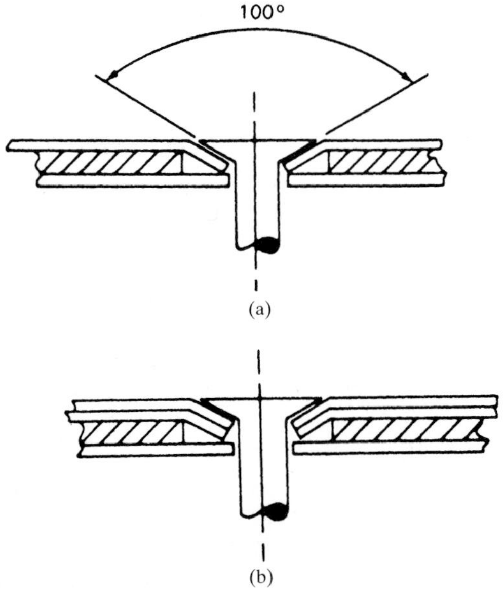


**FIGURE 10.32** (*a*) Dimpling; (*b*) double dimpling.

A dimpling tool is used to produce the countersunk recess for the rivet, and this method is used on aerospace vehicles to attach the "skins" or outer metal layers on the craft.

Modern methods of adhesive bonding are being used extensively in conjunction with spot welding to manufacture large sections of aerospace vehicles in lieu of riveting. But riveting is still used extensively for many fastening applications throughout industry and the construction trades.

Rivets are made for "blind hole" applications in the form of pull-stem pop rivets, drive-stem rivets, and explosive rivets. Other rivet forms include solid-stem, tubular, and semi-tubular.

***Rivet Edge Distance.*** The position or location of rivets from the edge of a part is normally 2*d*, the rivet shank diameter, with an absolute minimum of 1.5*d*. The lateral spacing between rivets, known as *pitch*, is determined by the load requirements of the riveted joint. See Fig. 10.33 for illustration of edge distance.

Rivet symbols on engineering drawings take the forms shown in Fig. 10.34. These symbols are frequently encountered on aerospace vehicle assembly drawings of the craft's structure.

## 10.3.1 Basic Stresses in Riveted Joints

Calculate the three loads as shown.

**1.** Single-shear tensile load $F_s$ (see Fig. 10.35) is

$$F_s = nAS_s$$

where $n$ = number of rivets
$A$ = cross-sectional area of one rivet, in$^2$
$S_s$ = allowable shearing stress of the rivet, psi

**2.** Bearing-stress tensile load $F_b$ (see Fig. 10.36) is

$$F_b = nA_1S_b$$

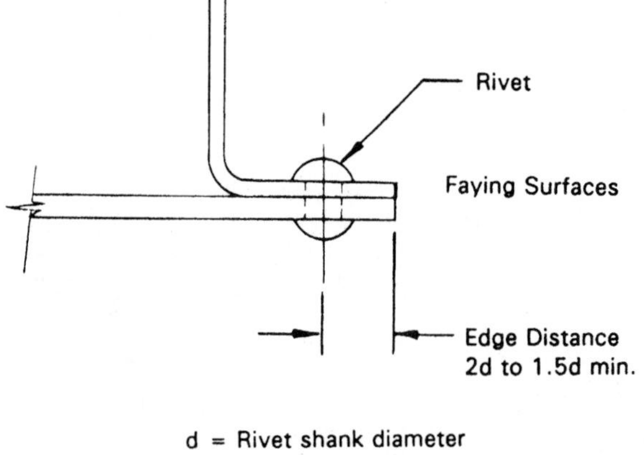


**FIGURE 10.33** Standard edge distance for rivets.

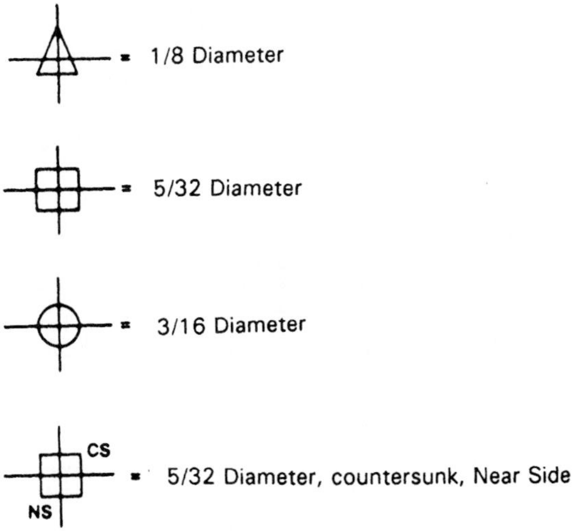


**FIGURE 10.34** Standard rivet symbols.

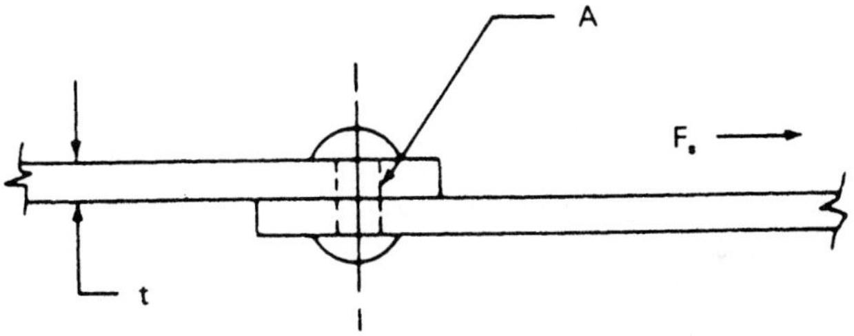


**FIGURE 10.35** Single-shear area.

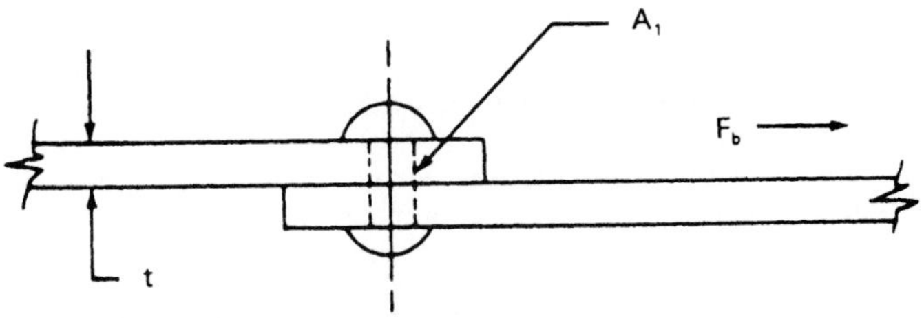


**FIGURE 10.36** Pitch and bearing-stress area.

where $n$ = number of rivets
$A_1$ = projected bearing area (diameter of rivet × $t$), in$^2$
$S_b$ = allowable bearing stress of the material, psi

**3.** Safe load $F_t$ based on tensile stress (see Fig. 10.37) is

$$F_t = nA_2S_t$$

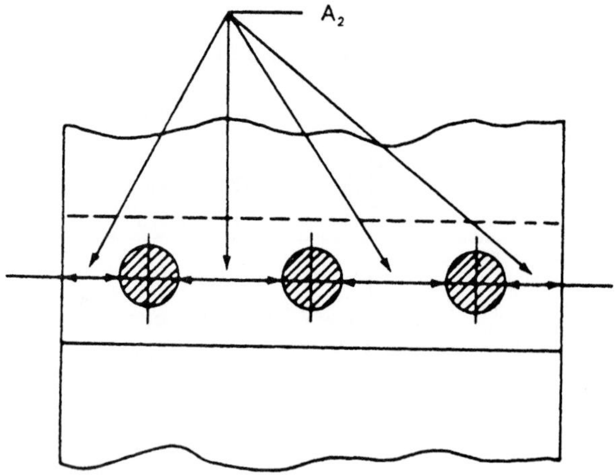


**FIGURE 10.37** Pitch and bend areas.

where $n$ = number of rivets
$A_2$ = area of plate between rivets, in$^2$
$S_t$ = allowable tensile strength of the material, psi

The *least* of the three calculated loads will be the safe tensile load for the riveted joint.

In machine design, commonly used allowable design stresses for structural steel and plain steel rivets used for riveted joints are

- 11,000 psi, tensile
- 8800 psi, shear
- 19,000 psi, compressive or bearing

These values are well below the ultimate stress allowables listed in the ASME boiler code and are therefore conservative.

Rivet placement, pitch, and sizing requirements on complex, heavily loaded dynamic structures are usually determined by the structural designer in collaboration with stress analysts and dynamics engineers.

Other manufacturing techniques such as welding, spot welding, and structural adhesive bonding have replaced riveting in many applications. Although this be the case, riveting is still employed in a great number of industrial applications. The types and varieties of rivets are numerous, and a complete description and dimensional data as well as material specifications are contained in the IFI handbooks on fasteners, in both the U.S. customary and metric systems.

### 10.3.2 General Sizing of Rivets

For rivet sizing (general, noncritical applications), the following approximations may be used:

$$d = 1.25\sqrt{t} \quad \text{to} \quad 1.45\sqrt{t}$$

where $d$ = diameter of rivet shank, in or mm (use next-larger size) and $t$ = material thickness, in.

## *10.4 PINS*

Pins of various types are used throughout industry. Pins are available in different sizes, styles, and materials. This section will detail the most common usage and popular pins used in all types of design and assembly applications. The common-usage pins include

- Clevis pins
- Cotter pins
- Spring pins (roll pins)
- Spiral spring pins (see Fig. 10.41, later)
- Taper pins
- Dowel pins
- Grooved pins
- Quick-release pins

### 10.4.1 Clevis Pins

Clevis pins are used where a quick detachable pin is of benefit from a design and manufacturing standpoint. Figure 10.38 shows the normal sizes and dimensions of standard clevis pins. The clevis pin may be made in various materials such as carbon steel, stainless steel, and aluminum alloy.

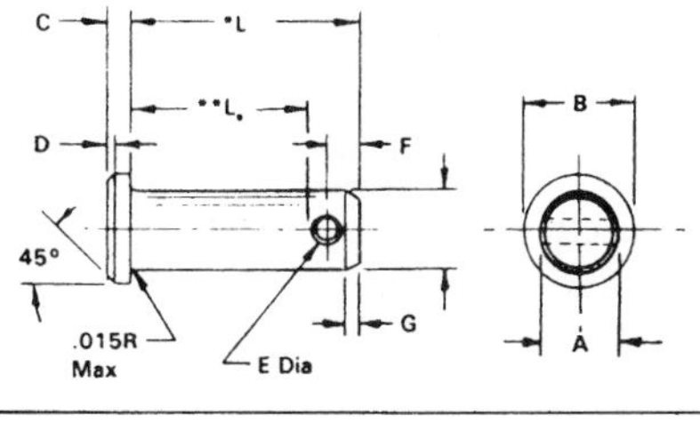


| | A | | B | C | D | E | | | | |
|---|---|---|---|---|---|---|---|---|---|---|
| Basic Pin Diameter | Shank Diameter Max | Min | Head Diameter Max | Head Height Max | Head Chamfer ± 0.01 | Hole Diameter Max | Min | F | G | Cotter Pin Size |
| 0.188 | 0.186 | 0.181 | 0.32 | 0.07 | 0.02 | 0.088 | 0.073 | 0.09 | 0.055 | 1/16 |
| 0.250 | 0.248 | 0.243 | 0.38 | 0.10 | 0.03 | 0.088 | 0.073 | 0.09 | 0.055 | 1/16 |
| 0.312 | 0.311 | 0.306 | 0.44 | 0.10 | 0.03 | 0.119 | 0.104 | 0.12 | 0.071 | 3/32 |
| 0.375 | 0.373 | 0.368 | 0.51 | 0.13 | 0.03 | 0.119 | 0.104 | 0.12 | 0.071 | 3/32 |
| 0.438 | 0.436 | 0.431 | 0.57 | 0.16 | 0.04 | 0.119 | 0.104 | 0.12 | 0.071 | 3/32 |
| 0.500 | 0.496 | 0.491 | 0.63 | 0.16 | 0.04 | 0.151 | 0.136 | 0.15 | 0.089 | 1/8 |
| 0.625 | 0.621 | 0.616 | 0.82 | 0.21 | 0.06 | 0.151 | 0.136 | 0.15 | 0.089 | 1/8 |
| 0.750 | 0.746 | 0.741 | 0.94 | 0.26 | 0.07 | 0.182 | 0.167 | 0.18 | 0.110 | 5/32 |
| 0.875 | 0.871 | 0.866 | 1.04 | 0.32 | 0.09 | 0.182 | 0.167 | 0.18 | 0.110 | 5/32 |
| 1.000 | 0.996 | 0.991 | 1.19 | 0.35 | 0.10 | 0.182 | 0.167 | 0.18 | 0.110 | 5/32 |

*L = Total length under head
**$L_e$ = Length of effective grip
Effective grip lengths must be selected from manufacturer's catalogs.

**FIGURE 10.38** Clevis pin dimensions.

### 10.4.2 Cotter Pins

Cotter pins are a very common and economical form of fastening device which see countless uses and applications throughout industry. The two common forms of cotter pins with sizes and dimensions are shown in Fig. 10.39. The cotter pin is normally used with the standard clevis pin.

### 10.4.3 Spring Pins (Roll Pins)

The slotted type of spring pin (sometimes called a *roll pin*) is used in applications where economy is important. This type of pin has been used to replace the tapered pin, dowel pin, and grooved pin in many applications because of its low cost and ease of preparation and assembly into holes with loose tolerances. Spring temper carbon steel is the normal material used for these pins, but they are also available in other materials. Figure 10.40 shows the sizes, dimensions, and recommended hole diameters for their application plus the double-shear values for design reference. Placement of the slot in relation to the shock loads in this type of pin can affect its performance and shock-absorbing qualities.

### 10.4.4 Spiral Spring Pins (Coiled Spring Pins)

The spiral spring pin was developed after the standard slotted spring pin in order to provide more shock resistance and a tighter fit in the drilled hole for the pin. These pins are made for

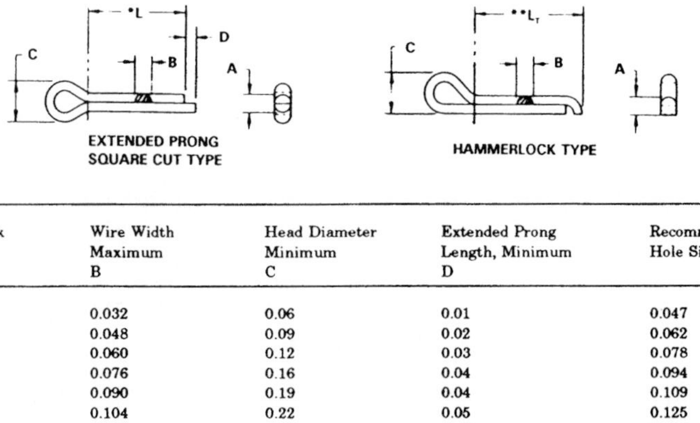


| Basic Pin Diameter | Total Shank Diameter A | Wire Width Maximum B | Head Diameter Minimum C | Extended Prong Length, Minimum D | Recommended Hole Size |
|---|---|---|---|---|---|
| 0.031 | 0.032 | 0.032 | 0.06 | 0.01 | 0.047 |
| 0.047 | 0.048 | 0.048 | 0.09 | 0.02 | 0.062 |
| 0.062 | 0.060 | 0.060 | 0.12 | 0.03 | 0.078 |
| 0.078 | 0.076 | 0.076 | 0.16 | 0.04 | 0.094 |
| 0.094 | 0.090 | 0.090 | 0.19 | 0.04 | 0.109 |
| 0.109 | 0.104 | 0.104 | 0.22 | 0.05 | 0.125 |
| 0.125 | 0.120 | 0.120 | 0.25 | 0.06 | 0.141 |
| 0.141 | 0.134 | 0.134 | 0.28 | 0.06 | 0.156 |
| 0.156 | 0.150 | 0.150 | 0.31 | 0.07 | 0.172 |
| 0.188 | 0.176 | 0.176 | 0.38 | 0.09 | 0.203 |
| 0.219 | 0.207 | 0.207 | 0.44 | 0.10 | 0.234 |
| 0.250 | 0.225 | 0.225 | 0.50 | 0.11 | 0.266 |
| 0.312 | 0.280 | 0.280 | 0.62 | 0.14 | 0.312 |
| 0.375 | 0.335 | 0.335 | 0.75 | 0.16 | 0.375 |
| 0.438 | 0.406 | 0.406 | 0.88 | 0.20 | 0.438 |
| 0.500 | 0.473 | 0.473 | 1.00 | 0.23 | 0.500 |
| 0.625 | 0.598 | 0.598 | 1.25 | 0.30 | 0.625 |
| 0.750 | 0.723 | 0.723 | 1.50 | 0.36 | 0.750 |

* L = Length
** $L_T$ = Total length
Allow extra length for spreading and securing
Available lengths to be selected from manufacturer's catalogs.

**FIGURE 10.39** Cotter pin dimensions.

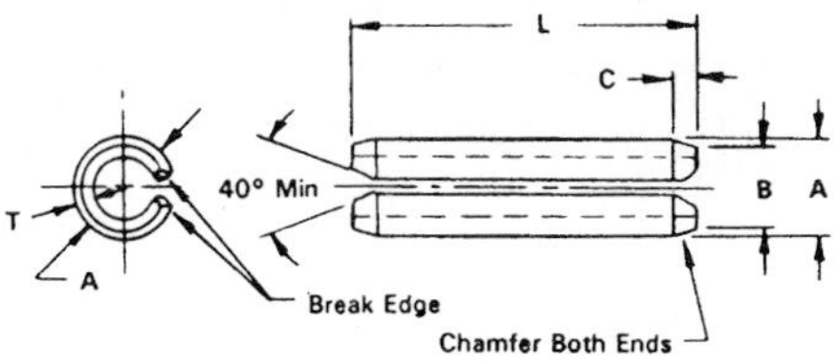


| | A | | B | C | T | | | | | |
|---|---|---|---|---|---|---|---|---|---|---|
| Basic Pin Diameter | Pin Diameter Max | Min | Chamfer Dia. | Chamfer Lth. | Stock Thickness | Hole Diameter Recommended Max | Min | Double Shear Load, Lb. ♦ AISI 1070 1095 & 420 | AISI 302 | Beryllium Copper |
| 0.062 | 0.069 | 0.066 | 0.059 | 0.028 | 0.012 | 0.065 | 0.062 | 425 | 350 | 270 |
| 0.078 | 0.086 | 0.083 | 0.075 | 0.032 | 0.018 | 0.081 | 0.078 | 650 | 550 | 400 |
| 0.094 | 0.103 | 0.099 | 0.091 | 0.038 | 0.022 | 0.097 | 0.094 | 1,000 | 800 | 660 |
| 0.125 | 0.135 | 0.131 | 0.122 | 0.044 | 0.028 | 0.129 | 0.125 | 2,100 | 1,500 | 1,200 |
| 0.141 | 0.149 | 0.145 | 0.137 | 0.044 | 0.028 | 0.144 | 0.140 | 2,200 | 1,600 | 1,400 |
| 1.156 | 0.167 | 0.162 | 0.151 | 0.048 | 0.032 | 0.160 | 0.156 | 3,000 | 2,000 | 1,800 |
| 0.188 | 0.199 | 0.194 | 0.182 | 0.055 | 0.040 | 0.192 | 0.187 | 4,400 | 2,800 | 2,600 |
| 0.219 | 0.232 | 0.226 | 0.214 | 0.065 | 0.048 | 0.224 | 0.219 | 5,700 | 3,550 | 3,700 |
| 0.250 | 0.264 | 0.258 | 0.245 | 0.065 | 0.048 | 0.256 | 0.250 | 7,700 | 4,600 | 4,500 |
| 0.312 | 0.328 | 0.321 | 0.306 | 0.080 | 0.062 | 0.318 | 0.312 | 11,500 | 7,100 | 6,800 |
| 0.375 | 0.392 | 0.385 | 0.368 | 0.095 | 0.077 | 0.382 | 0.375 | 17,600 | 10,000 | 10,100 |
| 0.438 | 0.456 | 0.448 | 0.430 | 0.095 | 0.077 | 0.445 | 0.437 | 20,000 | 12,000 | 12,200 |
| 0.500 | 0.521 | 0.513 | 0.485 | 0.110 | 0.094 | 0.510 | 0.500 | 25,800 | 15,500 | 16,800 |
| 0.625 | 0.650 | 0.640 | 0.608 | 0.125 | 0.125 | 0.636 | 0.625 | 46,000 | 18,800 | ..... |
| 0.750 | 0.780 | 0.769 | 0.730 | 0.150 | 0.150 | 0.764 | 0.750 | 66,000 | 23,200 | ..... |

Length L, is selected from manufacturer's catalogs. ♦ Other materials may be available.

**FIGURE 10.40** Spring pin dimensions.

standard, light-duty, and heavy-duty applications and are produced in various materials to suit the application. The author has conducted studies of slotted spring pins and spiral spring pins and has found that the heavy-duty slotted spring pin will withstand more shock loading cycles than the standard spiral spring pin. Figure 10.41 shows the sizes, dimensions, recommended hole sizes, and double-shear values of these pins. Figure 10.42 shows spiral and spring pins that have failed in service because of excessive shock loads. Slotted spring washers are also shown in which cyclic overloading was the mode of failure.

## 10.4.5 Taper Pins

The standard taper pin was widely used before the advent of the spring pin series and grooved pins. This pin type is still used in some industrial applications. Figure 10.43 shows the standard sizes and dimensions of taper pins.

## 10.4.6 Dowel Pins (Hardened and Ground Machine Type)

The hardened and ground dowel pin is widely used in tooling applications and a wide range of indexing applications where great accuracy is required. Figure 10.44 shows the available standard sizes and dimensions of this important class of pin.

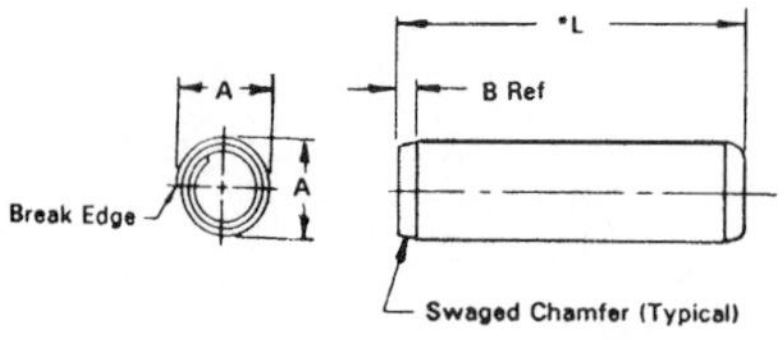


| Basic Pin Diameter | A: Pin Diameter, Std Duty Max | A: Pin Diameter, Heavy Duty Max | A: Pin Diameter, Light Duty Max | B: Chamfer Length Ref. | Hole Size Max | Hole Size Min | Double Shear Load, Min., Lb ♦ Std Duty 1070, 1095 and 420 | Std Duty 302 | Heavy Duty 1070, 1095 and 420 | Heavy Duty 302 | Light Duty 1070, 1095 and 420 | Light Duty 302 |
|---|---|---|---|---|---|---|---|---|---|---|---|---|
| 0.031 | 0.035 | ..... | ..... | 0.024 | 0.032 | 0.031 | 75 | 60 | ..... | ..... | ..... | ..... |
| 0.047 | 0.052 | ..... | ..... | 0.024 | 0.048 | 0.046 | 170 | 140 | ..... | ..... | ..... | ..... |
| 0.062 | 0.072 | 0.070 | 0.073 | 0.028 | 0.065 | 0.061 | 300 | 250 | 450 | 350 | ..... | 135 |
| 0.078 | 0.088 | 0.086 | 0.089 | 0.032 | 0.081 | 0.077 | 475 | 400 | 700 | 550 | ..... | 225 |
| 0.094 | 0.105 | 0.103 | 0.106 | 0.038 | 0.097 | 0.093 | 700 | 550 | 1,000 | 800 | 375 | 300 |
| 0.109 | 0.120 | 0.118 | 0.121 | 0.038 | 0.112 | 0.108 | 950 | 750 | 1,400 | 1,125 | 525 | 425 |
| 0.125 | 0.138 | 0.136 | 0.139 | 0.044 | 0.129 | 0.124 | 1,250 | 1,000 | 2,100 | 1,700 | 675 | 550 |
| 0.156 | 0.171 | 0.168 | 0.172 | 0.048 | 0.160 | 0.155 | 1,925 | 1,550 | 3,000 | 2,400 | 1,100 | 875 |
| 0.188 | 0.205 | 0.202 | 0.207 | 0.055 | 0.192 | 0.185 | 2,800 | 2,250 | 4,400 | 3,500 | 1,500 | 1,200 |
| 0.219 | 0.238 | 0.235 | 0.240 | 0.065 | 0.224 | 0.217 | 3,800 | 3,000 | 5,700 | 4,600 | 2,100 | 1,700 |
| 0.250 | 0.271 | 0.268 | 0.273 | 0.065 | 0.256 | 0.247 | 5,000 | 4,000 | 7,700 | 6,200 | 2,700 | 2,200 |
| 0.312 | 0.337 | 0.334 | 0.339 | 0.080 | 0.319 | 0.308 | 7,700 | 6,200 | 11,500 | 9,200 | 4,440 | 3,500 |
| 0.375 | 0.403 | 0.400 | 0.405 | 0.095 | 0.383 | 0.370 | 11,200 | 9,000 | 17,600 | 14,000 | 6,000 | 5,000 |
| 0.438 | 0.469 | 0.466 | 0.471 | 0.095 | 0.446 | 0.431 | 15,200 | 13,000 | 22,500 | 18,000 | 8,400 | 6,700 |
| 0.500 | 0.535 | 0.532 | 0.537 | 0.110 | 0.510 | 0.493 | 20,000 | 16,000 | 30,000 | 24,000 | 11,000 | 8,800 |
| 0.625 | 0.661 | 0.658 | ..... | 0.125 | 0.635 | 0.618 | 31,000 | 25,000 | 46,000 | 37,000 | ..... | ..... |
| 0.750 | 0.787 | 0.784 | ..... | 0.150 | 0.760 | 0.743 | 45,000 | 36,000 | 66,000 | 53,000 | ..... | ..... |

* Length L is selected from manufacturer's catalogs. ♦ Other materials may be available.

**FIGURE 10.41** Spiral spring pin dimensions.

## 10.4.7 Grooved Pins

The solid groove pin is one of the most popular fastening devices in use today. Its applications are limitless, its use is economical and practical, and it is easy to implement. The hole diameter required for its application is not as critical as that of other solid pins. Seven standard styles or types are recognized as American national standards. Figure 10.45*a* shows the different types, sizes, and dimensions; Fig. 10.45*b* shows the standard available lengths; Fig. 10.45*c* shows the recommended hole sizes for their application; and Fig. 10.45*d* gives the standard minimum double-shear loads for these pins.

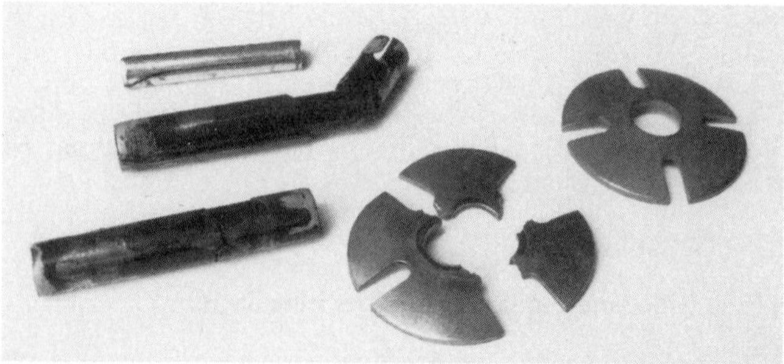

**FIGURE 10.42** Spring pin, spiral spring pin, and slotted washer fatigue failure samples.

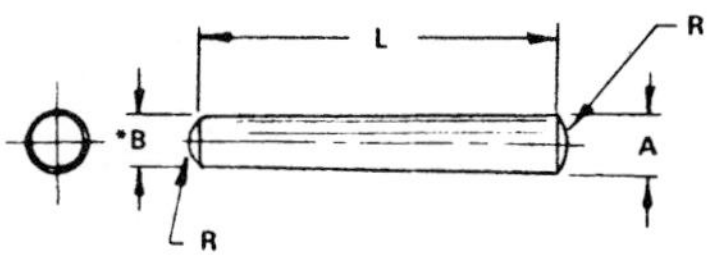


| Pin Size Number and Basic Pin Dia | | A Major Diameter (Large end) Commercial Class Max | Min | Precision Class Max | Min | R End Crown Radius Maximum |
|---|---|---|---|---|---|---|
| 7/0 | 0.0625 | 0.0638 | 0.0618 | 0.0635 | 0.0625 | 0.072 |
| 6/0 | 0.0780 | 0.0793 | 0.0773 | 0.0790 | 0.0780 | 0.088 |
| 5/0 | 0.0940 | 0.0953 | 0.0933 | 0.0950 | 0.0940 | 0.104 |
| 4/0 | 0.1090 | 0.1103 | 0.1083 | 0.1100 | 0.1090 | 0.119 |
| 3/0 | 0.1250 | 0.1263 | 0.1243 | 0.1260 | 0.1250 | 0.135 |
| 2/0 | 0.1410 | 0.1423 | 0.1403 | 0.1420 | 0.1410 | 0.151 |
| 0 | 0.1560 | 0.1573 | 0.1553 | 0.1570 | 0.1560 | 0.166 |
| 1 | 0.1720 | 0.1733 | 0.1713 | 0.1730 | 0.1720 | 0.182 |
| 2 | 0.1930 | 0.1943 | 0.1923 | 0.1940 | 0.1930 | 0.203 |
| 3 | 0.2190 | 0.2203 | 0.2183 | 0.2200 | 0.2190 | 0.229 |
| 4 | 0.2500 | 0.2513 | 0.2493 | 0.2510 | 0.2500 | 0.260 |
| 5 | 0.2890 | 0.2903 | 0.2883 | 0.2900 | 0.2890 | 0.299 |
| 6 | 0.3410 | 0.3423 | 0.3403 | 0.3420 | 0.3410 | 0.351 |
| 7 | 0.4090 | 0.4103 | 0.4083 | 0.4100 | 0.4090 | 0.419 |
| 8 | 0.4920 | 0.4933 | 0.4913 | 0.4930 | 0.4920 | 0.502 |
| 9 | 0.5910 | 0.5923 | 0.5903 | 0.5920 | 0.5910 | 0.601 |
| 10 | 0.7060 | 0.7073 | 0.7053 | 0.7070 | 0.7060 | 0.716 |
| 11 | 0.8600 | 0.8613 | 0.8593 | ..... | ..... | 0.870 |
| 12 | 1.0320 | 1.0333 | 1.0313 | ..... | ..... | 1.042 |
| 13 | 1.2410 | 1.2423 | 1.2403 | ..... | ..... | 1.251 |
| 14 | 1.5210 | 1.5223 | 1.5203 | ..... | ..... | 1.531 |

* B dimension varies per length. Length L to be selected from manufacturer's catalogs.

**FIGURE 10.43** Taper pin dimensions.

### 10.4.8 Quick-Release Pins

The quick-release pin is available in many types and finds use in design applications where a quick-release action of a fastened joint or part is required. Most of these types of pins contain a pushbutton release action which allows them to be removed by a straight pulling action, after the release button is pressed. The quick-release pin series is available in carbon steel, alloy steels, and stainless steels. Applications include tool engineering, aerospace vehicles, and many other applications where a strong, quick-release fastener is required. The quick-release pin is normally used for shear loading applications only.

## 10.5 RETAINING RINGS

Retaining rings have many uses in the design and maintenance of modern equipment, including

- Shaft retention
- Bearing retention

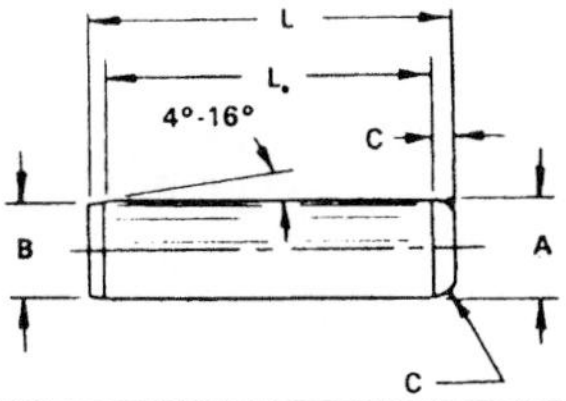


| Nominal | Pin Diameter - A | | | | | | B | | C | | |
|---|---|---|---|---|---|---|---|---|---|---|---|
| | Standard Series | | | Oversize Series | | | Point Dia | | Crown Radius | | Double Shear Load, Lb |
| Diameter | Basic | Max | Min | Basic | Max | Min | Max | Min | Max | Min | Carbon or Alloy Steel |
| 0.0625 | 0.0627 | 0.0628 | 0.0626 | 0.0635 | 0.0636 | 0.0634 | 0.058 | 0.048 | 0.020 | 0.008 | 800 |
| 0.0781 | 0.0783 | 0.0784 | 0.0782 | 0.0791 | 0.0792 | 0.0790 | 0.074 | 0.064 | 0.026 | 0.010 | 1,240 |
| 0.0938 | 0.0940 | 0.0941 | 0.0939 | 0.0948 | 0.0949 | 0.0947 | 0.089 | 0.079 | 0.031 | 0.012 | 1,800 |
| 0.1250 | 0.1252 | 0.1253 | 0.1251 | 0.1260 | 0.1261 | 0.1259 | 0.120 | 0.110 | 0.041 | 0.016 | 3,200 |
| 0.1562 | 0.1564 | 0.1565 | 0.1563 | 0.1572 | 0.1573 | 0.1571 | 0.150 | 0.140 | 0.052 | 0.020 | 5,000 |
| 0.1875 | 0.1877 | 0.1878 | 0.1876 | 0.1885 | 0.1886 | 0.1884 | 0.180 | 0.170 | 0.062 | 0.023 | 7,200 |
| 0.2500 | 0.2502 | 0.2503 | 0.2501 | 0.2510 | 0.2511 | 0.2509 | 0.240 | 0.230 | 0.083 | 0.031 | 12,800 |
| 0.3125 | 0.3127 | 0.3128 | 0.3126 | 0.3135 | 0.3136 | 0.3134 | 0.302 | 0.290 | 0.104 | 0.039 | 20,000 |
| 0.3750 | 0.3752 | 0.3753 | 0.3751 | 0.3760 | 0.3761 | 0.3759 | 0.365 | 0.350 | 0.125 | 0.047 | 28,700 |
| 0.4375 | 0.4377 | 0.4378 | 0.4376 | 0.4385 | 0.4386 | 0.4384 | 0.424 | 0.409 | 0.146 | 0.055 | 39,100 |
| 0.5000 | 0.5002 | 0.5003 | 0.5001 | 0.5010 | 0.5011 | 0.5009 | 0.486 | 0.471 | 0.167 | 0.063 | 51,000 |
| 0.6250 | 0.6252 | 0.6253 | 0.6251 | 0.6260 | 0.6261 | 0.6259 | 0.611 | 0.595 | 0.208 | 0.078 | 79,800 |
| 0.7500 | 0.7502 | 0.7503 | 0.7501 | 0.7510 | 0.7511 | 0.7509 | 0.735 | 0.715 | 0.250 | 0.094 | 114,000 |
| 0.8750 | 0.8752 | 0.8753 | 0.8751 | 0.8760 | 0.8761 | 0.8759 | 0.860 | 0.840 | 0.293 | 0.109 | 156,000 |
| 1.0000 | 1.0002 | 1.0003 | 1.0001 | 1.0010 | 1.0011 | 1.0009 | 0.980 | 0.960 | 0.333 | 0.125 | 204,000 |

Note: Sizes 0.0781 and 0.1562 diameter not recommended for new design.
L = Total pin length; $L_e$ = Length of engagement.
Dowel pins listed are available in nominal lengths from 0.1875 to 6.000 inches.
Consult the manufacturer's catalogs for available lengths.

**FIGURE 10.44** Dowel pin dimensions.

- Retention of parts on shafts
- Spring retention
- Vertical and horizontal shaft support

The standard retaining ring is normally made of spring steel, although other materials such as beryllium copper alloys and stainless steels are sometimes used. Figure 10.46 shows some of the main retaining types.

The allowable thrust loads for each particular type of retaining ring must be obtained from the retaining ring manufacturers' handbooks. The calculation procedures for various applications are also shown in these handbooks. There are at least 25 to 30 different types or styles of retaining rings. The critical dimensions for machining the grooves which hold the retaining ring in place are also obtained from the manufacturers' handbooks. Figure 10.47 shows a typical page from the Waldes Truarc retaining ring handbook. This sample page is for the 5100 series external type ring, and the typical machining dimensions for the retaining groove may be seen under "Groove Dimensions" in the table. Dimensions and application data are also shown.

Note that electroplating thickness can interfere with the proper functioning of the ring in the groove. Overplating can cause the ring to come out of the retaining groove during operation of the mechanism on which the ring is used. Therefore, specify a maximum plating thickness of 0.0002 in (0.2 mil) or allow additional clearance for the ring in the groove when the part is to be electroplated.

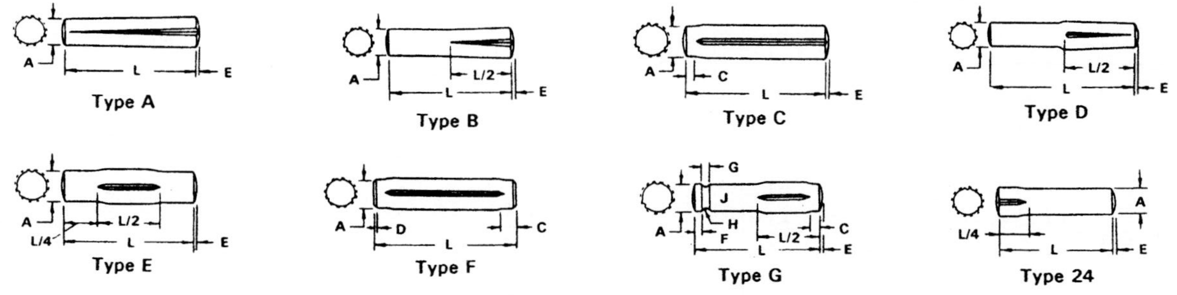


| Basic Pin Diameter | A Max | C Ref. | D Min. | E Max | F Min. | G Max. | H Ref. | J Max |
|---|---|---|---|---|---|---|---|---|
| 0.0312 | 0.0312 | 0.015 | ..... | ..... | ..... | ..... | ..... | ..... |
| 0.0469 | 0.0469 | 0.031 | ..... | ..... | ..... | ..... | ..... | ..... |
| 0.0625 | 0.0625 | 0.031 | 0.016 | 0.0115 | ..... | ..... | ..... | ..... |
| 0.0781 | 0.0781 | 0.031 | 0.016 | 0.0137 | ..... | ..... | ..... | ..... |
| 0.0938 | 0.0938 | 0.031 | 0.016 | 0.0141 | 0.028 | 0.041 | 0.016 | 0.067 |
| 0.1094 | 0.1094 | 0.031 | 0.016 | 0.0160 | 0.028 | 0.041 | 0.016 | 0.082 |
| 0.1250 | 0.1250 | 0.031 | 0.016 | 0.0180 | 0.059 | 0.041 | 0.031 | 0.088 |
| 0.1563 | 0.1563 | 0.062 | 0.031 | 0.0220 | 0.059 | 0.057 | 0.031 | 0.109 |
| 0.1875 | 0.1875 | 0.062 | 0.031 | 0.0230 | 0.059 | 0.057 | 0.031 | 0.130 |
| 0.2188 | 0.2188 | 0.062 | 0.031 | 0.0270 | 0.091 | 0.072 | 0.047 | 0.151 |
| 0.2500 | 0.2500 | 0.062 | 0.031 | 0.0310 | 0.091 | 0.072 | 0.047 | 0.172 |
| 0.3125 | 0.3125 | 0.094 | 0.047 | 0.0390 | 0.122 | 0.104 | 0.062 | 0.214 |
| 0.3750 | 0.3750 | 0.094 | 0.047 | 0.0440 | 0.122 | 0.135 | 0.062 | 0.255 |
| 0.4375 | 0.4375 | 0.094 | 0.047 | 0.0520 | 0.185 | 0.135 | 0.094 | 0.298 |
| 0.5000 | 0.5000 | 0.094 | 0.047 | 0.0570 | 0.185 | 0.135 | 0.094 | 0.317 |

**FIGURE 10.45** (*a*) Grooved pin types and dimensions.

| Nominal Length | Nominal Size | | | | | | | | | | | | | | |
|---|---|---|---|---|---|---|---|---|---|---|---|---|---|---|---|
| | 1/32 | 3/64 | 1/16 | 5/64 | 3/32 | 7/64 | 1/8 | 5/32 | 3/16 | 7/32 | 1/4 | 5/16 | 3/8 | 7/16 | 1/2 |
| 1/8 | Y | Y | Y | | | | | | | | | | | | |
| 1/4 | Y | Y | Y | Y | Y | Y | Y | | | | | | | | |
| 3/8 | Y | Y | Y | Y | X | X | X | X | X | | | | | | |
| 1/2 | Y | Y | Y | Y | X | X | X | X | X | X | X | | | | |
| 5/8 | | Y | Y | Y | X | X | X | X | X | X | X | X | | | |
| 3/4 | | | Y | Y | X | X | X | X | X | X | X | X | X | | |
| 7/8 | | | Y | Y | X | X | X | X | X | X | X | X | X | X | |
| 1 | | | Y | Y | X | X | X | X | X | X | X | X | X | X | X |
| 1-1/4 | | | | | X | X | X | X | X | X | X | X | X | X | X |
| 1-1/2 | | | | | | | X | X | X | X | X | X | X | X | X |
| 1-3/4 | | | | | | | | X | X | X | X | X | X | X | X |
| 2 | | | | | | | | X | X | X | X | X | X | X | X |
| 2-1/4 | | | | | | | | | X | X | X | X | X | X | X |
| 2-1/2 | | | | | | | | | | X | X | X | X | X | X |
| 2-3/4 | | | | | | | | | | X | X | X | X | X | X |
| 3 | | | | | | | | | | X | X | X | X | X | X |
| 3-1/4 | | | | | | | | | | | X | X | X | X | X |
| 3-1/2 | | | | | | | | | | | | X | X | X | X |
| 3-3/4 | | | | | | | | | | | | | X | X | X |
| 4 | | | | | | | | | | | | | X | X | X |
| 4-1/4 | | | | | | | | | | | | | X | X | X |
| 4-1/2 | | | | | | | | | | | | | | X | X |

Note: Carbon steel pins are normally available in the marked sizes by X and Y. X designates all types of pins; Y designates all types except type G. Other lengths may be available from different manufacturers.

**FIGURE 10.45** (*b*) Sizes and lengths of grooved pins.

| Recommended Hole Sizes For Grooved Pins | | | |
|---|---|---|---|
| Nominal Pin Size | Drill Size | Hole Diameter | |
| | | Max | Min |
| 1/32 | 1/32 | 0.0324 | 0.0312 |
| 3/64 | 3/64 | 0.0482 | 0.0469 |
| 1/16 | 1/16 | 0.0640 | 0.0625 |
| 5/64 | 5/64 | 0.0798 | 0.0781 |
| 3/32 | 3/32 | 0.0956 | 0.0938 |
| 7/64 | 7/64 | 0.1113 | 0.1094 |
| 1/8 | 1/8 | 0.1271 | 0.1250 |
| 5/32 | 5/32 | 0.1587 | 0.1563 |
| 3/16 | 3/16 | 0.1903 | 0.1875 |
| 7/32 | 7/32 | 0.2219 | 0.2188 |
| 1/4 | 1/4 | 0.2534 | 0.2500 |
| 5/16 | 5/16 | 0.3166 | 0.3125 |
| 3/8 | 3/8 | 0.3797 | 0.3750 |
| 7/16 | 7/16 | 0.4428 | 0.4375 |
| 1/2 | 1/2 | 0.5060 | 0.5000 |

**FIGURE 10.45** (*c*) Hole sizes for grooved pins.

| Nominal Pin Size | Double Shear Load, Min, lb | | | |
|---|---|---|---|---|
| | Material | | | |
| | Low Carbon Steel | Alloy Steel (Rockwell C45 to 50) | Corrosion Resistant Steel | Brass |
| 1/32 | 104 | 202 | 143 | 64 |
| 3/64 | 220 | 430 | 300 | 136 |
| 1/16 | 402 | 785 | 540 | 250 |
| 5/64 | 624 | 1,215 | 860 | 386 |
| 3/32 | 896 | 1,750 | 1,240 | 555 |
| 7/64 | 1,222 | 2,380 | 1,685 | 757 |
| 1/8 | 1,600 | 3,115 | 2,200 | 990 |
| 5/32 | 2,494 | 4,860 | 3,440 | 1,540 |
| 3/16 | 3,588 | 6,990 | 4,960 | 2,220 |
| 7/32 | 4,884 | 9,520 | 6,760 | 3,020 |
| 1/4 | 6,380 | 12,430 | 8,840 | 3,950 |
| 5/16 | 9,970 | 19,420 | 13,750 | 6,170 |
| 3/8 | 11,620 | 27,950 | 19,800 | 9,050 |
| 7/16 | 15,820 | 38,060 | 27,000 | 12,100 |
| 1/2 | 20,600 | 49,700 | 35,200 | 15,800 |

**FIGURE 10.45** (*d*) Double-shear loads for grooved pins.

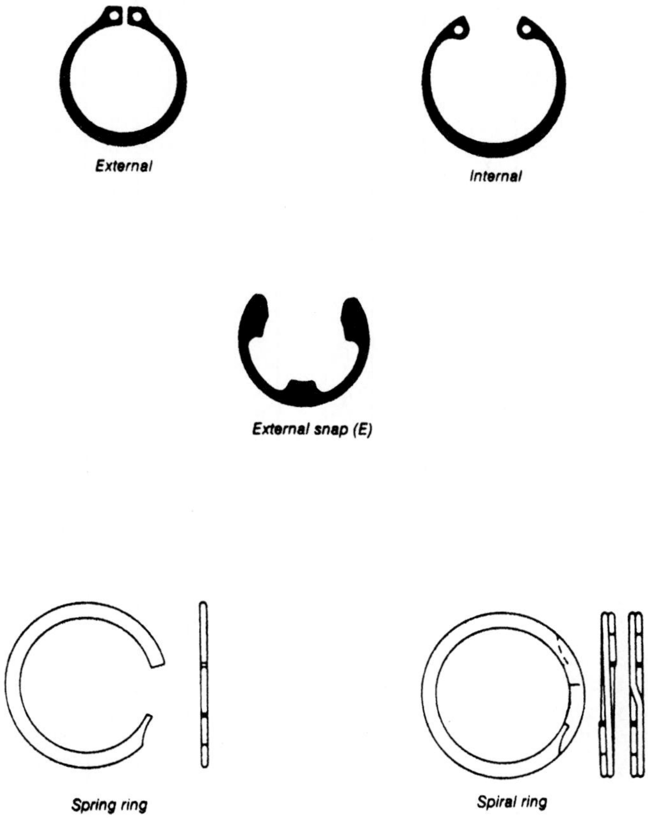


**FIGURE 10.46** Sample types of retaining rings.

BASIC
*external series*
**5100**

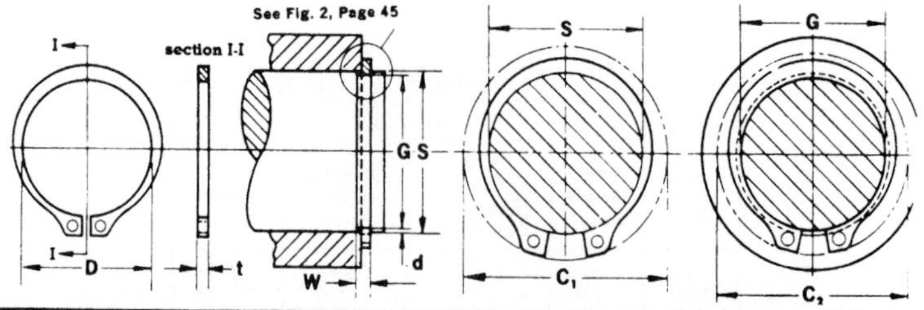


MIL-R-21248
MS 16624

Thickness **t** applies only to unplated rings. For plated and stainless steel (Type H) rings, add .002" to the listed maximum thickness. Maximum ring thickness will be at least .0002" less than the listed minimum groove width (**W**).

T.I.R. (total indicator reading) is the maximum allowable deviation of concentricity between groove and shaft.

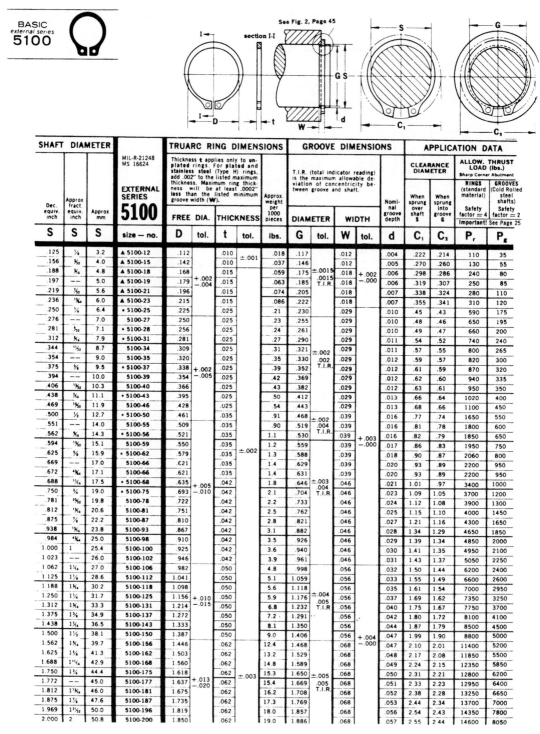

| SHAFT DIAMETER: Dec. equiv. inch S | SHAFT DIAMETER: Approx fract. equiv. inch S | SHAFT DIAMETER: Approx mm S | EXTERNAL SERIES 5100 size — no. | TRUARC RING DIMENSIONS: FREE DIA. D | FREE DIA. tol. | THICKNESS t | THICKNESS tol. | Approx. weight per 1000 pieces lbs. | GROOVE DIMENSIONS: DIAMETER G | DIAMETER tol. | WIDTH W | WIDTH tol. | Nominal groove depth d | APPLICATION DATA: CLEARANCE DIAMETER: When sprung over shaft S $C_1$ | CLEARANCE DIAMETER: When sprung into groove G $C_2$ | ALLOW. THRUST LOAD (lbs.) Sharp Corner Abutment: RINGS (standard material) Safety factor = 4 $P_r$ | GROOVES (Cold Rolled steel shafts) Safety factor = 2 $P_g$ |
|---|---|---|---|---|---|---|---|---|---|---|---|---|---|---|---|---|---|
| .125 | 1/8 | 3.2 | ▲ 5100-12 | .112 | | .010 | ±.001 | .018 | .117 | | .012 | | .004 | .222 | .214 | 110 | 35 |
| .156 | 5/32 | 4.0 | ▲ 5100-15 | .142 | | .010 | | .037 | .146 | | .012 | | .005 | .270 | .260 | 130 | 55 |
| .188 | 3/16 | 4.8 | ▲ 5100-18 | .168 | | .015 | | .059 | .175 | ±.0015 .0015 T.I.R. | .018 | +.002 −.000 | .006 | .298 | .286 | 240 | 80 |
| .197 | — — | 5.0 | ▲ 5100-19 | .179 | +.002 −.004 | .015 | | .063 | .185 | | .018 | | .006 | .319 | .307 | 250 | 85 |
| .219 | 7/32 | 5.6 | ▲ 5100-21 | .196 | | .015 | | .074 | .205 | | .018 | | .007 | .338 | .324 | 280 | 110 |
| .236 | 15/64 | 6.0 | ▲ 5100-23 | .215 | | .015 | | .086 | .222 | | .018 | | .007 | .355 | .341 | 310 | 120 |
| .250 | 1/4 | 6.4 | • 5100-25 | .225 | | .025 | | .21 | .230 | | .029 | | .010 | .45 | .43 | 590 | 175 |
| .276 | — — | 7.0 | 5100-27 | .250 | | .025 | | .23 | .255 | | .029 | | .010 | .48 | .46 | 650 | 195 |
| .281 | 9/32 | 7.1 | • 5100-28 | .256 | | .025 | | .24 | .261 | | .029 | | .010 | .49 | .47 | 660 | 200 |
| .312 | 5/16 | 7.9 | • 5100-31 | .281 | | .025 | | .27 | .290 | | .029 | | .011 | .54 | .52 | 740 | 240 |
| .344 | 11/32 | 8.7 | 5100-34 | .309 | | .025 | | .31 | .321 | | .029 | | .011 | .57 | .55 | 800 | 265 |
| .354 | — — | 9.0 | 5100-35 | .320 | | .025 | | .35 | .330 | ±.002 .002 T.I.R. | .029 | | .012 | .59 | .57 | 820 | 300 |
| .375 | 3/8 | 9.5 | • 5100-37 | .338 | +.002 −.005 | .025 | | .39 | .352 | | .029 | | .012 | .61 | .59 | 870 | 320 |
| 394 | — — | 10.0 | 5100-39 | .354 | | .025 | | .42 | .369 | | .029 | | .012 | .62 | .60 | 940 | 335 |
| .406 | 13/32 | 10.3 | 5100-40 | .366 | | .025 | | .43 | .382 | | .029 | | .012 | .63 | .61 | 950 | 350 |
| .438 | 7/16 | 11.1 | • 5100-43 | .395 | | .025 | | .50 | .412 | | .029 | | .013 | .66 | .64 | 1020 | 400 |
| .469 | 15/32 | 11.9 | 5100-46 | .428 | | .025 | | .54 | .443 | | .029 | | .013 | .68 | .66 | 1100 | 450 |
| .500 | 1/2 | 12.7 | • 5100-50 | .461 | | .035 | | .91 | .468 | | .039 | | .016 | .77 | .74 | 1650 | 550 |
| .551 | — — | 14.0 | 5100-55 | .509 | | .035 | | .90 | .519 | ±.002 .004 T.I.R. | .039 | | .016 | .81 | .78 | 1800 | 600 |
| .562 | 9/16 | 14.3 | • 5100-56 | .521 | | .035 | | 1.1 | .530 | | .039 | | .016 | .82 | .79 | 1850 | 650 |
| .594 | 19/32 | 15.1 | 5100-59 | .550 | | .035 | | 1.2 | .559 | | .039 | +.003 −.000 | .017 | .86 | .83 | 1950 | 750 |
| .625 | 5/8 | 15.9 | • 5100-62 | .579 | | .035 | ±.002 | 1.3 | .588 | | .039 | | .018 | .90 | .87 | 2060 | 800 |
| .669 | — — | 17.0 | 5100-66 | .621 | | .035 | | 1.4 | .629 | | .039 | | .020 | .93 | .89 | 2200 | 950 |
| .672 | 43/64 | 17.1 | 5100-66 | .621 | | .035 | | 1.4 | .631 | | .039 | | .020 | .93 | .89 | 2200 | 950 |
| .688 | 11/16 | 17.5 | • 5100-68 | .635 | +.005 −.010 | .042 | | 1.8 | .646 | ±.003 .004 T.I.R. | .046 | | .021 | 1.01 | .97 | 3400 | 1000 |
| .750 | 3/4 | 19.0 | • 5100-75 | .693 | | .042 | | 2.1 | .704 | | .046 | | .023 | 1.09 | 1.05 | 3700 | 1200 |
| .781 | 25/32 | 19.8 | 5100-78 | .722 | | .042 | | 2.2 | .733 | | .046 | | .024 | 1.12 | 1.08 | 3900 | 1300 |
| .812 | 13/16 | 20.6 | 5100-81 | .751 | | .042 | | 2.5 | .762 | | .046 | | .025 | 1.15 | 1.10 | 4000 | 1450 |
| .875 | 7/8 | 22.2 | 5100-87 | .810 | | .042 | | 2.8 | .821 | | .046 | | .027 | 1.21 | 1.16 | 4300 | 1650 |
| .938 | 15/16 | 23.8 | 5100-93 | .867 | | .042 | | 3.1 | .882 | | .046 | | .028 | 1.34 | 1.29 | 4650 | 1850 |
| .984 | 63/64 | 25.0 | 5100-98 | .910 | | .042 | | 3.5 | .926 | | .046 | | .029 | 1.39 | 1.34 | 4850 | 2000 |
| 1.000 | 1 | 25.4 | 5100-100 | .925 | | .042 | | 3.6 | .940 | | .046 | | .030 | 1.41 | 1.35 | 4950 | 2100 |
| 1.023 | — — | 26.0 | 5100-102 | .946 | | .042 | | 3.9 | .961 | | .046 | | .031 | 1.43 | 1.37 | 5050 | 2250 |
| 1.062 | 1 1/16 | 27.0 | 5100-106 | .982 | | .050 | | 4.8 | .998 | | .056 | | .032 | 1.50 | 1.44 | 6200 | 2400 |
| 1.125 | 1 1/8 | 28.6 | 5100-112 | 1.041 | | .050 | | 5.1 | 1.059 | | .056 | | .033 | 1.55 | 1.49 | 6600 | 2600 |
| 1.188 | 1 3/16 | 30.2 | 5100-118 | 1.098 | | .050 | | 5.6 | 1.118 | | .056 | | .035 | 1.61 | 1.54 | 7000 | 2950 |
| 1.250 | 1 1/4 | 31.7 | 5100-125 | 1.156 | +.010 −.015 | .050 | | 5.9 | 1.176 | ±.004 .005 T.I.R | .056 | | .037 | 1.69 | 1.62 | 7350 | 3250 |
| 1.312 | 1 5/16 | 33.3 | 5100-131 | 1.214 | | .050 | | 6.8 | 1.232 | | .056 | | .040 | 1.75 | 1.67 | 7750 | 3700 |
| 1.375 | 1 3/8 | 34.9 | 5100-137 | 1.272 | | .050 | | 7.2 | 1.291 | | .056 | | .042 | 1.80 | 1.72 | 8100 | 4100 |
| 1.438 | 1 7/16 | 36.5 | 5100-143 | 1.333 | | .050 | | 8.1 | 1.350 | | .056 | | .044 | 1.87 | 1.79 | 8500 | 4500 |
| 1.500 | 1 1/2 | 38.1 | 5100-150 | 1.387 | | .050 | | 9.0 | 1.406 | | .056 | +.004 −.000 | .047 | 1.99 | 1.90 | 8800 | 5000 |
| 1.562 | 1 9/16 | 39.7 | 5100-156 | 1.446 | | .062 | | 12.4 | 1.468 | | .068 | | .047 | 2.10 | 2.01 | 11400 | 5200 |
| 1.625 | 1 5/8 | 41.3 | 5100-162 | 1.503 | | .062 | | 13.2 | 1.529 | | .068 | | .048 | 2.17 | 2.08 | 11850 | 5500 |
| 1.688 | 1 11/16 | 42.9 | 5100-168 | 1.560 | | .062 | | 14.8 | 1.589 | | .068 | | .049 | 2.24 | 2.15 | 12350 | 5850 |
| 1.750 | 1 3/4 | 44.4 | 5100-175 | 1.618 | | .062 | ±.003 | 15.3 | 1.650 | | .068 | | .050 | 2.31 | 2.21 | 12800 | 6200 |
| 1.772 | — — | 45.0 | 5100-177 | 1.637 | +.013 −.020 | .062 | | 15.4 | 1.669 | ±.005 .005 T.I.R. | .068 | | .051 | 2.33 | 2.23 | 12950 | 6400 |
| 1.812 | 1 13/16 | 46.0 | 5100-181 | 1.675 | | .062 | | 16.2 | 1.708 | | .068 | | .052 | 2.38 | 2.28 | 13250 | 6650 |
| 1.875 | 1 7/8 | 47.6 | 5100-187 | 1.735 | | .062 | | 17.3 | 1.769 | | .068 | | .053 | 2.44 | 2.34 | 13700 | 7000 |
| 1.969 | 1 31/32 | 50.0 | 5100-196 | 1.819 | | .062 | | 18.0 | 1.857 | | .068 | | .056 | 2.54 | 2.43 | 14350 | 7800 |
| 2.000 | 2 | 50.8 | 5100-200 | 1.850 | | .062 | | 19.0 | 1.886 | | .068 | | .057 | 2.55 | 2.44 | 14600 | 8050 |

Important! See Page 25

**FIGURE 10.47** Sample Truarc retaining ring data sheet. *(Source: Waldes Truarc.)*

Figure 10.48 shows a retaining ring interchangeability chart for various manufacturers and also lists the appropriate military standard number.

### 10.5.1 X Washers (Split Washers)

An X washer is a form of retaining ring that is unique. It is used in applications similar to the standard spring retaining ring except that it may be installed with a pair of common pliers as shown in Fig. 10.49. The dimensions and sizes of the presently available X-washer series are also shown in Fig. 10.49. The X washer is sometimes called a *split washer* and may be listed as such in fastener catalogs, although that is the incorrect name.

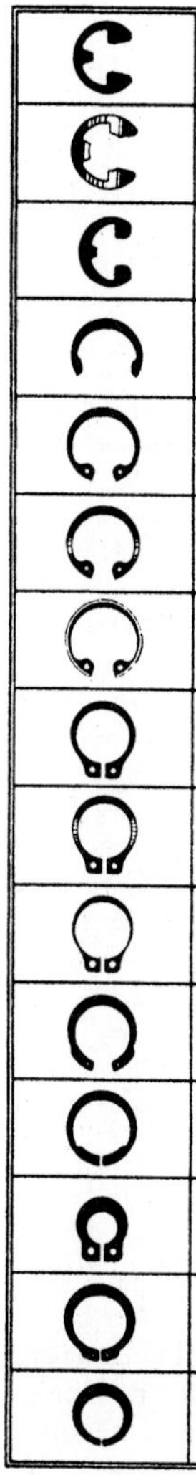

| Roto Clip | Waldes | I.R.R. | Anderton | Mil Standard |
|---|---|---|---|---|
| E | 5133 | 1000 | 1500 | 16633 |
| BE | 5131 | 1001 | 1501 | 16634 |
| RE | 5144 | 1200 | 1540 | 3215 |
| C | 5103 | 2000 | 1800 | 16632 |
| HO | 5000 | 3000 | 1300 | 16625 |
| BHO | 5001 | 3001 | 1301 | 16629 |
| VHO | 5002 | --- | 1302 | 16631 |
| SH | 5100 | 3100 | 1400 | 16624 |
| BSH | 5101 | 3101 | 1401 | 16628 |
| VSH | 5102 | --- | 1402 | 16630 |
| HOI | 5008 | 4000 | 1308 | 16627 |
| SHI | 5108 | 4100 | 1408 | 16626 |
| SHF | 5555 | 7100 | 1440 | 90707 |
| SHR | 5160 | --- | 1460 | 3217 |
| SHM | 5560 | --- | --- | --- |

**FIGURE 10.48** Retaining ring interchangeability table and index.

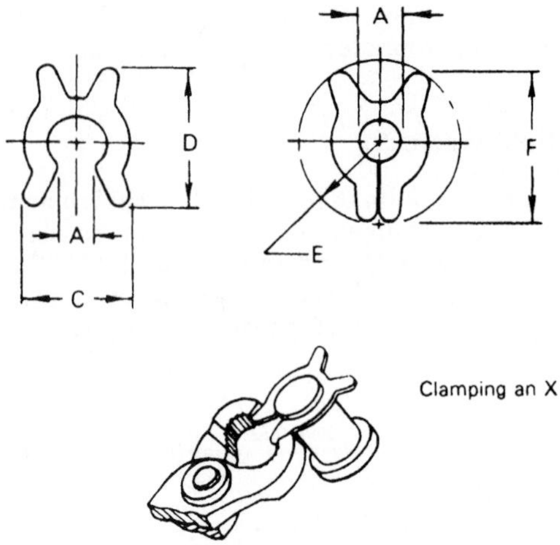


X - Washer Dimensions

| A | B* | C | D | E | F |
|---|---|---|---|---|---|
| .086 | .025 | .320 | .406 | .210 | .406 |
| .098 | .055 | .364 | .490 | .297 | .475 |
| .130 | .055 | .430 | .575 | .359 | .556 |
| .164 | .065 | .523 | .687 | .422 | .665 |
| .190 | .065 | .593 | .745 | .437 | .730 |
| .222 | .075 | .622 | .776 | .469 | .775 |
| .256 | .075 | .698 | .905 | .500 | .890 |
| .285 | .075 | .822 | .986 | .563 | .984 |
| .317 | .089 | .872 | 1.100 | .609 | 1.078 |
| .347 | .089 | .948 | 1.190 | .688 | 1.188 |
| .381 | .089 | 1.060 | 1.297 | .797 | 1.281 |

Note: B = thickness (inches)

**FIGURE 10.49** X washer sizes and dimensions.

An X washer is normally used for a one-time application only. Once the washer is installed on a shaft and removed with pliers, it should not be reused. Although only applicable for one-time installation, these devices are suitable and economical for many applications in product design and manufacturing. The fact that these washers or rings may be installed and removed with common pliers, and not special tools, is an asset in their application.

## 10.6 SET, CLAMP, AND SPLIT COLLARS

These devices have many uses in product design and manufacturing. They are used to retain shafts, withstand high thrust loads, and space parts and are often welded to plate cams, plate sprockets, and indexing plates to form a hub with which the part is retained on a shaft. The standard set collar is simply a machined ring with setscrews installed to hold the ring on a shaft or cylindrical end of a part. Figure 10.50 shows the standard set-collar sizes and dimensions. These collars are normally made from low-carbon steel with zinc plating as a finish.

| DIMENSIONS | | | |
|---|---|---|---|
| BORE | O. D. | WIDTH | SET SCREW |
| 3/16 | 7/16 | 1/4 | 8-32 |
| 1/4 | 1/2 | 9/32 | 10-32 |
| 5/16 | 5/8 | 11/32 | 10-32 |
| 3/8 | 3/4 | 3/8 | 1/4-20 |
| 7/16 | 7/8 | 7/16 | 1/4-20 |
| 1/2 | 1" | 7/16 | 1/4-20 |
| 9/16 | 1" | 7/16 | 1/4-20 |
| 5/8 | 1-1/8 | 1/2 | 5/16-18 |
| 11/16 | 1-1/4 | 9/16 | 5/16-18 |
| 3/4 | 1-1/4 | 9/16 | 5/16-18 |
| 13/16 | 1-5/16 | 9/16 | 5/16-18 |
| 7/8 | 1-1/2 | 9/16 | 5/16-18 |
| 15/16 | 1-5/8 | 9/16 | 5/16-18 |
| 1" | 1-5/8 | 5/8 | 5/16-18 |
| 1-1/16 | 1-3/4 | 5/8 | 5/16-18 |
| 1-1/8 | 1-3/4 | 5/8 | 5/16-18 |
| 1-3/16 | 2" | 11/16 | 3/8-16 |
| 1-1/4 | 2" | 11/16 | 3/8-16 |
| 1-5/16 | 2-1/8 | 11/16 | 3/8-16 |
| 1-3/8 | 2-1/8 | 3/4 | 3/8-16 |
| 1-7/16 | 2-1/4 | 3/4 | 3/8-16 |
| 1-1/2 | 2-1/4 | 3/4 | 3/8-16 |
| 1-9/16 | 2-1/2 | 13/16 | 3/8-16 |
| 1-5/8 | 2-1/2 | 13/16 | 3/8-16 |
| 1-11/16 | 2-1/2 | 13/16 | 3/8-16 |
| 1-3/4 | 2-3/4 | 7/8 | 1/2-13 |
| 1-13/16 | 2-3/4 | 7/8 | 1/2-13 |
| 1-7/8 | 2-3/4 | 7/8 | 1/2-13 |
| 1-15/16 | 3" | 7/8 | 1/2-13 |
| 2" | 3" | 7/8 | 1/2-13 |
| 2-1/16 | 3" | 7/8 | 1/2-13 |
| 2-1/8 | 3" | 7/8 | 1/2-13 |
| 2-3/16 | 3-1/4 | 15/16 | 1/2-13 |
| 2-1/4 | 3-1/4 | 15/16 | 1/2-13 |
| 2-5/16 | 3-1/4 | 15/16 | 1/2-13 |
| 2-3/8 | 3-1/4 | 15/16 | 1/2-13 |
| 2-7/16 | 3-1/2 | 1" | 1/2-13 |
| 2-1/2 | 3-1/2 | 1" | 1/2-13 |
| 2-9/16 | 3-3/4 | 1-1/8 | 1/2-13 |
| 2-5/8 | 3-3/4 | 1-1/8 | 1/2-13 |
| 2-11/16 | 4" | 1-1/8 | 1/2-13 |
| 2-3/4 | 4" | 1-1/8 | 1/2-13 |
| 2-13/16 | 4-1/4 | 1-1/8 | 1/2-13 |
| 2-7/8 | 4-1/4 | 1-1/8 | 1/2-13 |
| 2-15/16 | 4-1/4 | 1-1/8 | 1/2-13 |
| 3" | 4-1/4 | 1-1/8 | 1/2-13 |

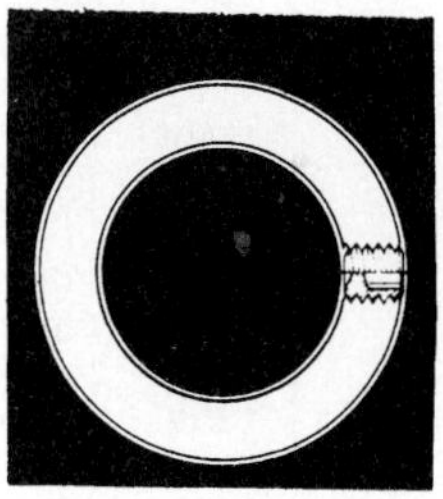

**FIGURE 10.50** Set collars *(Source: Ruland Catalog of Collars and Couplings, Ruland Manufacturing Co., Inc., Watertown, Mass.)*

| DIMENSIONS | | | |
|---|---|---|---|
| BORE | O.D. | WIDTH | CLAMP SCREW |
| 1/8 | 1/2 | .235 | 4-40 |
| 3/16 | 9/16 | .235 | 4-40 |
| 1/4 | 5/8 | .281 | 4-40 |
| 5/16 | 11/16 | .281 | 4-40 |
| 3/8 | 7/8 | .343 | 6-32 |
| 7/16 | 15/16 | .343 | 6-32 |
| 1/2 | 1-1/8 | .406 | 8-32 |
| 9/16 | 1-1/4 | .437 | 10-32 |
| 5/8 | 1-5/16 | .437 | 10-32 |
| 11/16 | 1-3/8 | .437 | 10-32 |
| 3/4 | 1-1/2 | 1/2 | 1/4-28 |
| 13/16 | 1-5/8 | 1/2 | 1/4-28 |
| 7/8 | 1-5/8 | 1/2 | 1/4-28 |
| 15/16 | 1-3/4 | 1/2 | 1/4-28 |
| 1 | 1-3/4 | 1/2 | 1/4-28 |
| 1-1/16 | 1-7/8 | 1/2 | 1/4-28 |
| 1-1/8 | 1-7/8 | 1/2 | 1/4-28 |
| 1-3/16 | 2-1/16 | 1/2 | 1/4-28 |
| 1-1/4 | 2-1/16 | 1/2 | 1/4-28 |
| 1-5/16 | 2-1/8 | 9/16 | 1/4-28 |
| 1-3/8 | 2-1/4 | 9/16 | 1/4-28 |
| 1-7/16 | 2-1/4 | 9/16 | 1/4-28 |
| 1-1/2 | 2-3/8 | 9/16 | 1/4-28 |
| 1-9/16 | 2-3/8 | 9/16 | 1/4-28 |
| 1-5/8 | 2-5/8 | 11/16 | 5/16-24 |
| 1-11/16 | 2-3/4 | 11/16 | 5/16-24 |
| 1-3/4 | 2-3/4 | 11/16 | 5/16-24 |
| 1-13/16 | 2-7/8 | 11/16 | 5/16-24 |
| 1-7/8 | 2-7/8 | 11/16 | 5/16-24 |
| 1-15/16 | 3 | 11/16 | 5/16-24 |
| 2 | 3 | 11/16 | 5/16-24 |
| 2-1/16 | 3-1/8 | 3/4 | 5/16-24 |
| 2-1/8 | 3-1/4 | 3/4 | 5/16-24 |
| 2-3/16 | 3-1/4 | 3/4 | 5/16-24 |
| 2-1/4 | 3-1/4 | 3/4 | 5/16-24 |
| 2-5/16 | 3-3/8 | 3/4 | 5/16-24 |
| 2-3/8 | 3-1/2 | 3/4 | 5/16-24 |
| 2-7/16 | 3-1/2 | 3/4 | 5/16-24 |
| 2-1/2 | 3-3/4 | 7/8 | 3/8-24 |
| 2-9/16 | 3-7/8 | 7/8 | 3/8-24 |
| 2-5/8 | 3-7/8 | 7/8 | 3/8-24 |
| 2-11/16 | 4 | 7/8 | 3/8-24 |
| 2-3/4 | 4 | 7/8 | 3/8-24 |
| 2-13/16 | 4-1/4 | 7/8 | 3/8-24 |
| 2-7/8 | 4-1/4 | 7/8 | 3/8-24 |
| 2-15/16 | 4-1/4 | 7/8 | 3/8-24 |
| 3 | 4-1/4 | 7/8 | 3/8-24 |

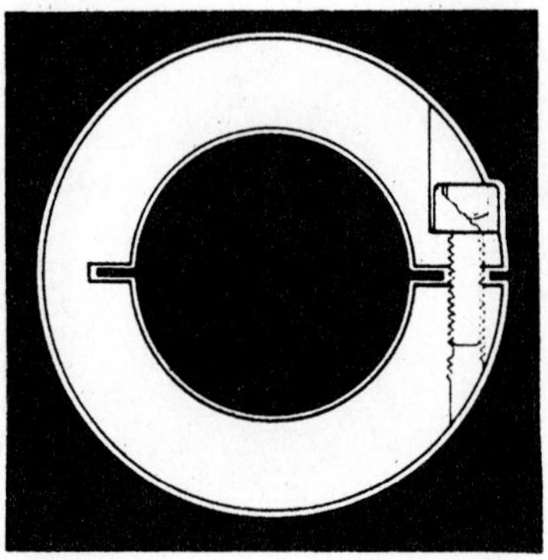

**FIGURE 10.51** Clamp collars. *(Source: Ruland Catalog of Collars and Couplings, Ruland Manufacturing Co., Inc., Watertown, Mass.)*

Clamp collars are similar to set collars, except that they have a slit through one wall and are clamped on a shaft by tightening the clamp screw provided on the collar. These types of collars are also made with internal threads for adjustment and a more positive clamping force. Clamp collars are normally made of carbon steels, aluminum alloys, and stainless steels. Figure 10.51 shows the sizes and dimensions of standard clamp collars, and Fig. 10.52 shows the data for the internally threaded clamp-collar series.

| DIMENSIONS | | | |
|---|---|---|---|
| THREADED BORE | O.D. | WIDTH | CLAMP SCREW |
| 8-32 | 1/2 | .235 | 4-40 |
| 10-24 | 9/16 | .235 | 4-40 |
| 10-32 | 9/16 | .235 | 4-40 |
| 1/4-20 | 5/8 | .281 | 4-40 |
| 1/4-28 | 5/8 | .281 | 4-40 |
| 5/16-18 | 11/16 | .281 | 4-40 |
| 5/16-24 | 11/16 | .281 | 4-40 |
| 3/8-16 | 7/8 | .343 | 6-32 |
| 3/8-24 | 7/8 | .343 | 6-32 |
| 7/16-14 | 15/16 | .343 | 6-32 |
| 7/16-20 | 15/16 | .343 | 6-32 |
| 1/2-13 | 1-1/8 | .406 | 8-32 |
| 1/2-20 | 1-1/8 | .406 | 8-32 |
| 5/8-11 | 1-5/16 | .437 | 10-32 |
| 5/8-18 | 1-5/16 | .437 | 10-32 |
| 3/4-10 | 1-1/2 | 1/2 | 1/4-28 |
| 3/4-16 | 1-1/2 | 1/2 | 1/4-28 |
| 7/8-9 | 1-5/8 | 1/2 | 1/4-28 |
| 7/8-14 | 1-5/8 | 1/2 | 1/4-28 |
| 1-8 | 1-3/4 | 1/2 | 1/4-28 |
| 1-12 | 1-3/4 | 1/2 | 1/4-28 |
| 1-14 | 1-3/4 | 1/2 | 1/4-28 |
| 1-1/8-7 | 1-7/8 | 1/2 | 1/4-28 |
| 1-1/8-12 | 1-7/8 | 1/2 | 1/4-28 |
| 1-1/4-7 | 2-1/16 | 1/2 | 1/4-28 |
| 1-1/4-12 | 2-1/16 | 1/2 | 1/4-28 |
| 1-3/8-6 | 2-1/4 | 9/16 | 1/4-28 |
| 1-3/8-12 | 2-1/4 | 9/16 | 1/4-28 |
| 1-1/2-6 | 2-3/8 | 9/16 | 1/4-28 |
| 1-1/2-12 | 2-3/8 | 9/16 | 1/4-28 |
| 1-3/4-16 | 2-3/4 | 11/16 | 5/16-24 |
| 2"-12 | 3 | 11/16 | 5/16-24 |

**FIGURE 10.52** Threaded clamp collars. *(Source: Ruland Catalog of Collars and Couplings, Ruland Manufacturing Co., Inc., Watertown, Mass.)*

Split collars and threaded split collars are also available and widely used in machine design. The data for these types of clamping collars are shown in Figs. 10.53 and 10.54.

## 10.7 MACHINERY BUSHINGS, SHIMS, AND ARBOR SPACERS

***Machinery Bushings.*** *Machinery bushings* are a special form of flat washer commonly made of low-carbon mild steel. They are used as spacers between gears, pulleys, and sprockets and also as filler spacers for parts mounted on shafting. These bushings are manufactured in the following gauges and diameters:

- 18-gauge (0.048-in)
- 14-gauge (0.075-in)
- 10-gauge (0.134-in)
- $^3/_{16}$-gauge (0.1875-in)

Inside diameters range from 0.500 to 3.00 in.

| DIMENSIONS | | | |
|---|---|---|---|
| BORE | O.D. | WIDTH | CLAMP SCREW |
| 1/8 | 1/2 | .235 | 4-40 |
| 3/16 | 9/16 | .235 | 4-40 |
| 1/4 | 5/8 | .281 | 4-40 |
| 5/16 | 11/16 | .281 | 4-40 |
| 3/8 | 7/8 | .343 | 6-32 |
| 7/16 | 15/16 | .343 | 6-32 |
| 1/2 | 1-1/8 | .406 | 8-32 |
| 9/16 | 1-1/4 | .437 | 10-32 |
| 5/8 | 1-5/16 | .437 | 10-32 |
| 11/16 | 1-3/8 | .437 | 10-32 |
| 3/4 | 1-1/2 | 1/2 | 1/4-28 |
| 13/16 | 1-5/8 | 1/2 | 1/4-28 |
| 7/8 | 1-5/8 | 1/2 | 1/4-28 |
| 15/16 | 1-3/4 | 1/2 | 1/4-28 |
| 1 | 1-3/4 | 1/2 | 1/4-28 |
| 1-1/16 | 1-7/8 | 1/2 | 1/4-28 |
| 1-1/8 | 1-7/8 | 1/2 | 1/4-28 |
| 1-3/16 | 2-1/16 | 1/2 | 1/4-28 |
| 1-1/4 | 2-1/16 | 1/2 | 1/4-28 |
| 1-5/16 | 2-1/8 | 9/16 | 1/4-28 |
| 1-3/8 | 2-1/4 | 9/16 | 1/4-28 |
| 1-7/16 | 2-1/4 | 9/16 | 1/4-28 |
| 1-1/2 | 2-3/8 | 9/16 | 1/4-28 |
| 1-9/16 | 2-3/8 | 9/16 | 1/4-28 |
| 1-5/8 | 2-5/8 | 11/16 | 5/16-24 |
| 1-11/16 | 2-3/4 | 11/16 | 5/16-24 |
| 1-3/4 | 2-3/4 | 11/16 | 5/16-24 |
| 1-13/16 | 2-7/8 | 11/16 | 5/16-24 |
| 1-7/8 | 2-7/8 | 11/16 | 5/16-24 |
| 1-15/16 | 3 | 11/16 | 5/16-24 |
| 2 | 3 | 11/16 | 5/16-24 |
| 2-1/16 | 3-1/8 | 3/4 | 5/16-24 |
| 2-1/8 | 3-1/4 | 3/4 | 5/16-24 |
| 2-3/16 | 3-1/4 | 3/4 | 5/16-24 |
| 2-1/4 | 3-1/4 | 3/4 | 5/16-24 |
| 2-5/16 | 3-3/8 | 3/4 | 5/16-24 |
| 2-3/8 | 3-1/2 | 3/4 | 5/16-24 |
| 2-7/16 | 3-1/2 | 3/4 | 5/16-24 |
| 2-1/2 | 3-3/4 | 7/8 | 3/8-24 |
| 2-9/16 | 3-7/8 | 7/8 | 3/8-24 |
| 2-5/8 | 3-7/8 | 7/8 | 3/8-24 |
| 2-11/16 | 4 | 7/8 | 3/8-24 |
| 2-3/4 | 4 | 7/8 | 3/8-24 |
| 2-13/16 | 4-1/4 | 7/8 | 3/8-24 |
| 2-7/8 | 4-1/4 | 7/8 | 3/8-24 |
| 2-15/16 | 4-1/4 | 7/8 | 3/8-24 |
| 3 | 4-1/4 | 7/8 | 3/8-24 |

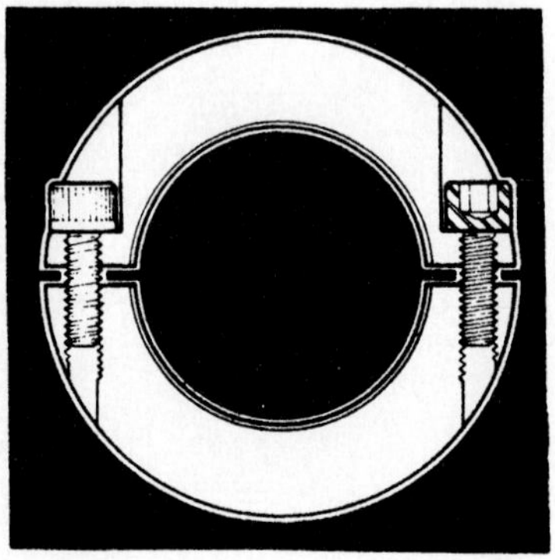

**FIGURE 10.53** Split collars. *(Source: Ruland Catalog of Collars and Couplings, Ruland Manufacturing Co., Inc., Watertown, Mass.)*

***Steel Shims.*** *Steel shims* are thin steel rings with a plain center hole and are used for building up gears and bearings and to provide proper clearance between mating parts. Figure 10.55 lists the sizes and thicknesses normally available for steel shims.

***Steel Arbor Spacers.*** *Steel arbor spacers* are thin steel rings with a keyway center hole and are used for accurately spacing milling cutters, slitter knives, and gang saws on keyway arbors.

| DIMENSIONS | | | |
|---|---|---|---|
| BORE | O.D. | WIDTH | CLAMP SCREW |
| 8-32 | 1/2 | .235 | 4-40 |
| 10-24 | 9/16 | .235 | 4-40 |
| 10-32 | 9/16 | .235 | 4-40 |
| 1/4-20 | 5/8 | .281 | 4-40 |
| 1/4-28 | 5/8 | .281 | 4-40 |
| 5/16-18 | 11/16 | .281 | 4-40 |
| 5/16-24 | 11/16 | .281 | 4-40 |
| 3/8-16 | 7/8 | .343 | 6-32 |
| 3/8-24 | 7/8 | .343 | 6-32 |
| 7/16-14 | 15/16 | .343 | 6-32 |
| 7/16-20 | 15/16 | .343 | 6-32 |
| 1/2-13 | 1-1/8 | .406 | 8-32 |
| 1/2-20 | 1-1/8 | .406 | 8-32 |
| 5/8-11 | 1-5/16 | .437 | 10-32 |
| 5/8-18 | 1-5/16 | .437 | 10-32 |
| 3/4-10 | 1-1/2 | 1/2 | 1/4-28 |
| 3/4-16 | 1-1/2 | 1/2 | 1/4-28 |
| 7/8-9 | 1-5/8 | 1/2 | 1/4-28 |
| 7/8-14 | 1-5/8 | 1/2 | 1/4-28 |
| 1-8 | 1-3/4 | 1/2 | 1/4-28 |
| 1-12 | 1-3/4 | 1/2 | 1/4-28 |
| 1-14 | 1-3/4 | 1/2 | 1/4-28 |
| 1-1/8-7 | 1-7/8 | 1/2 | 1/4-28 |
| 1-1/8-12 | 1-7/8 | 1/2 | 1/4-28 |
| 1-1/4-7 | 2-1/16 | 1/2 | 1/4-28 |
| 1-1/4-12 | 2-1/16 | 1/2 | 1/4-28 |
| 1-3/8-6 | 2-1/4 | 9/16 | 1/4-28 |
| 1-3/8-12 | 2-1/4 | 9/16 | 1/4-28 |
| 1-1/2-6 | 2-3/8 | 9/16 | 1/4-28 |
| 1-1/2-12 | 2-3/8 | 9/16 | 1/4-28 |
| 1-3/4-16 | 2-3/4 | 11/16 | 5/16-24 |
| 2"-12 | 3 | 11/16 | 5/16-24 |

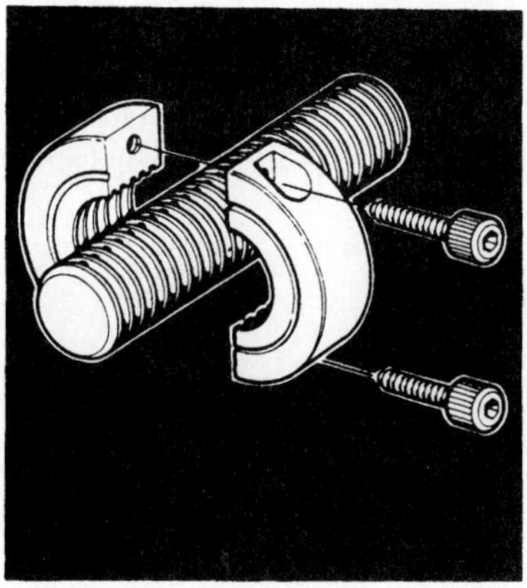

**FIGURE 10.54** Split threaded clamp collars. *(Source: Ruland Catalog of Collars and Couplings, Ruland Manufacturing Co., Inc., Watertown, Mass.)*

Steel shims and steel arbor spacers are made of AISI 1010, fully hardened, cold-rolled low-carbon steel. Figure 10.55 also lists the sizes and thicknesses of steel arbor spacers.

## 10.8 SPECIALTY FASTENERS

The specialty fastener component lines available today are great in numbers and types. In this section, we will detail and list only those specialty fasteners that have become common and which are widely used in new product design and manufacturing.

Figure 10.25 shows a variety of standard and specialty fasteners. At the bottom center of Fig. 10.25 is an eyebolt, and immediately above this is a pentahead bolt and its socket wrench. This five-sided-head bolt is used to make equipment tamperproof, as any standard wrench will not grip the head for removal. The lower right of the figure shows a variety of swage nuts. The lower left of the photograph shows a slotted spring washer which is used to maintain pressure on joints which have a central pivot point. Almost half of the fasteners shown in Fig. 10.25 can be considered as specialty fasteners. A partial listing of some of the common specialty fasteners would include

**Sizes and Thicknesses of Steel Shims - (Inches)**

| ID | OD | Thickness Ranges |
|---|---|---|
| 0.375 | 0.625 | 0.001 0.002 0.003 0.004 0.005 0.006 0.007 0.008 0.010 0.012 0.015 0.020 0.025 0.031 0.047 |
| 0.500 | 0.750 | 0.0062 0.093 0.125 (These thicknesses available in all ID and OD sizes) |
| 0.625 | 1.000 | |
| 0.750 | 1.125 | |
| 0.875 | 1.375 | |
| 1.000 | 1.500 | |
| 1.125 | 1.625 | |
| 1.250 | 1.750 | |
| 1.375 | 1.875 | |
| 1.500 | 2.125 | |
| 1.750 | 2.750 | |
| 2.000 | 2.750 | |

**Sizes and Thicknesses of Steel Arbor Spacers - (Inches)**

| ID | OD | Thickness Ranges |
|---|---|---|
| 0.500 | 0.750 | 0.001 0.002 0.003 0.004 0.005 0.006 (These thicknesses available for diameters listed) |
| 0.625 | 1.000 | |
| 0.750 | 1.125 | |
| 0.875 | 1.375 | |
| 1.000 | 1.500 | |
| 1.250 | 1.750 | |
| 1.500 | 2.125 | |
| 2.000 | 2.250 | |
| 0.750 | 1.125 | 0.007 0.008 0.010 0.012 0.015 0.020 (These thicknesses available for diameters listed) |
| 0.875 | 1.375 | |
| 1.000 | 1.500 | |
| 1.250 | 1.750 | |
| 1.500 | 2.125 | |
| 2.000 | 2.750 | |
| 1.000 | 1.500 | 0.025 0.031 0.047 0.062 0.093 0.125 (These thicknesses available for diameters listed) |
| 1.250 | 1.750 | |
| 1.500 | 2.125 | |
| 2.000 | 2.750 | |

**FIGURE 10.55** Dimensions and gauges of shims and arbor spacers.

- Acorn nuts
- Floating nuts
- Plastic bolts
- Split-lock nuts
- SEMs
- Weld nuts
- Various plastic washers
- Sealing washers
- T-slot nuts and bolts
- Push nuts (Pal nuts)
- Various types of weld studs
- Sheet-metal nuts
- Nylok bolts and others.

***Specialty Fasteners in Common Use.*** Figure 10.56 shows the different types of SEMs (screw and captive washer assemblies) available today. Note that on the SEM, the screw is either

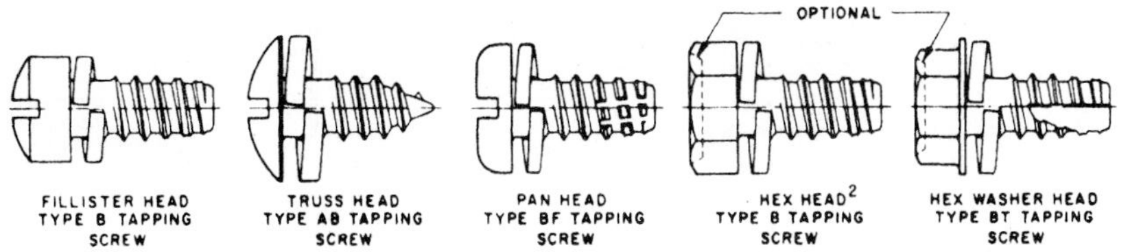


REPRESENTATIVE EXAMPLES OF HELICAL SPRING LOCK WASHER SEMS

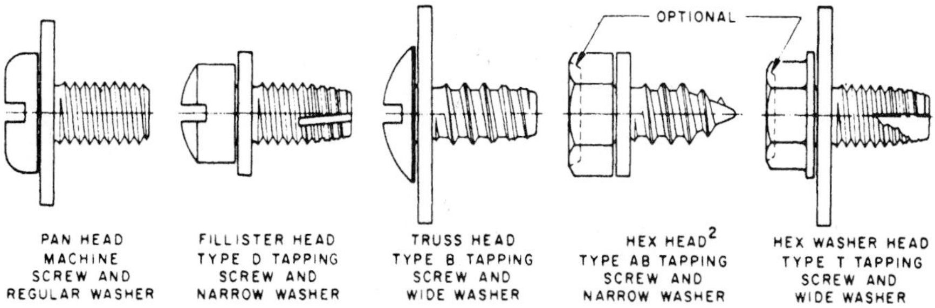


REPRESENTATIVE EXAMPLES OF PLAIN WASHER SEMS

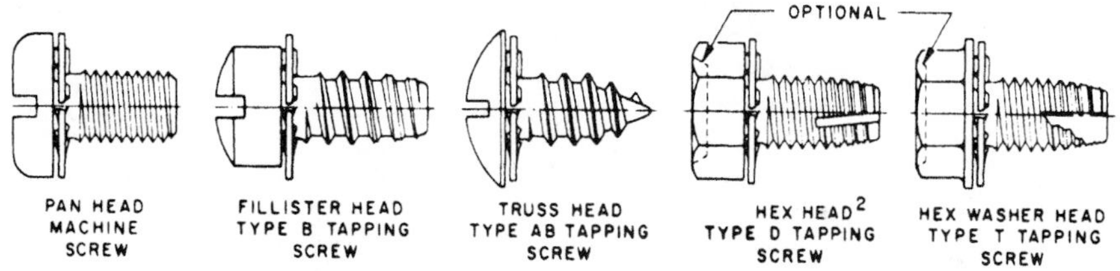


REPRESENTATIVE EXAMPLES OF INTERNAL TOOTH LOCK WASHER SEMS

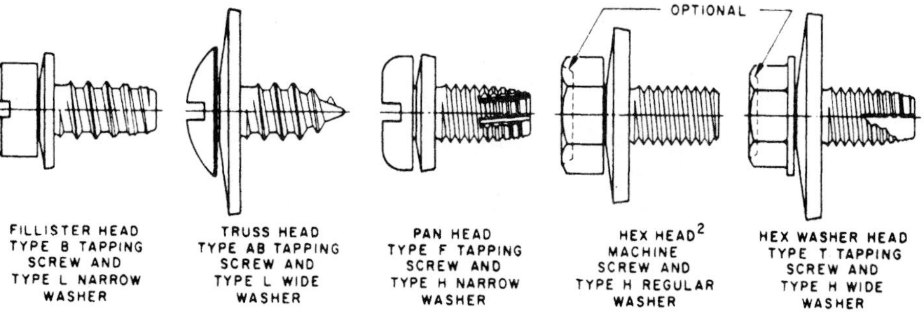


REPRESENTATIVE EXAMPLES OF CONICAL SPRING WASHER SEMS

**FIGURE 10.56** Types of SEMs.

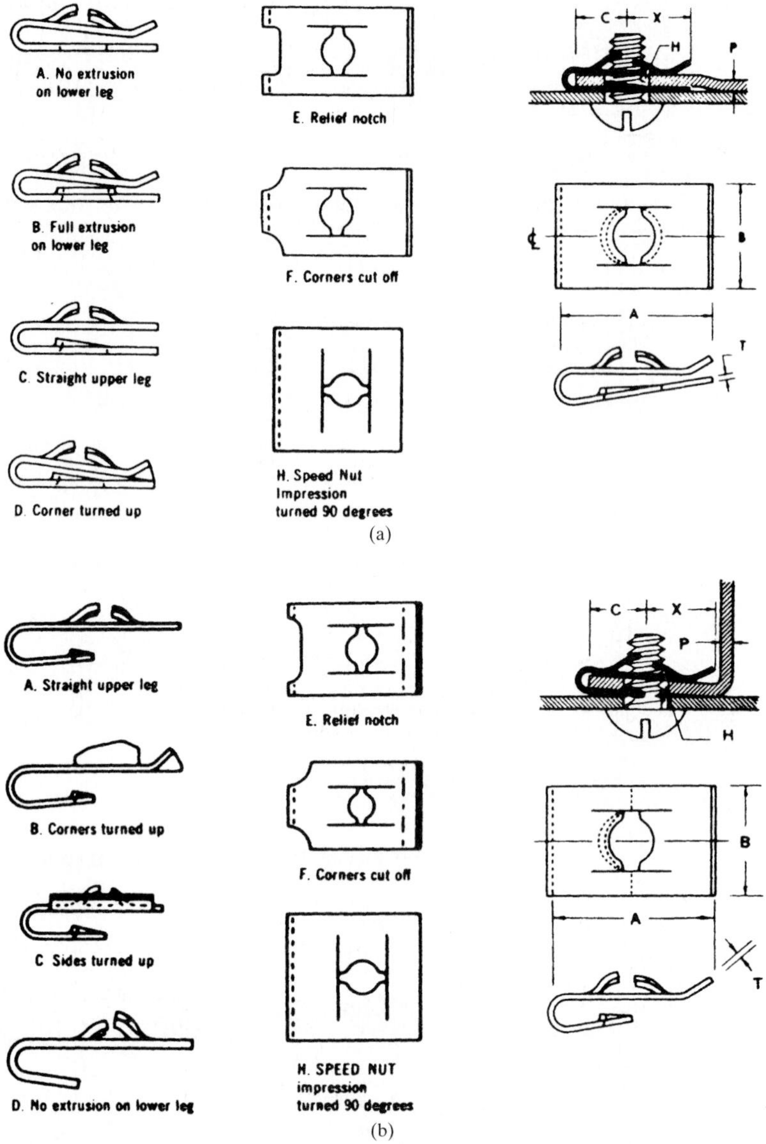


**FIGURE 10.57** Tinnerman speed nuts: (*a*) "U" type, (*b*) "J" type.

thread-forming or thread-tapping. This makes this class of fastener useful and economical in rapid assembly applications such as automotive equipment manufacturing. SEMs are specified in the American national standards ANSI/ASME B18.13.

Figure 10.57 shows some of the widely used Tinnerman types of speed nuts, which are made of high-carbon, spring-tempered steel. These types of speed nuts are produced in sizes from 6 to 32 through 5⁄16 to 18 or larger in special cases. The Tinnerman type U and J nuts are

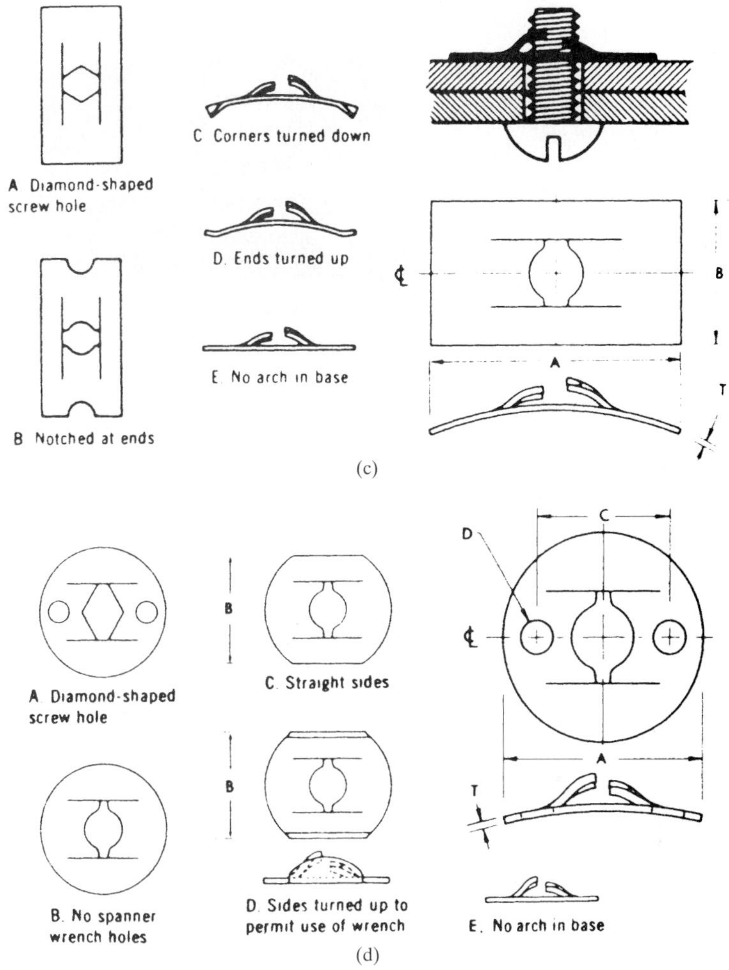


**FIGURE 10.57** (*c*) flat, (*d*) round.

widely used to fasten sheet-metal screw covers onto sheet-metal enclosures. The flat and round types are used on through-bolt sheet-metal applications, such as automotive equipment and electronic chassis work. These are economical, efficient fasteners whose applications are limitless.

Another specialty type of fastener that is widely used is the swage nut. The swage nut is produced in different styles or types, one of which is shown in Fig. 10.58*A*. The swage nut is highly useful in applications where the thread cannot be efficiently or effectively produced in the parent metal which must be fastened to another part. Swage nuts are used in switchgear equipment where copper bus bars are fastened together and it is not practical to tap the soft copper bars for the bolting application. These nuts are also used on thin sheet-metal parts where a strong joint is required and not enough material thickness is available for tapping the sheet metal. The swage nut is normally made from carbon steel with zinc or cadmium plating, stainless steels, and aluminum alloys. Figure 10.58*A* shows a typical PEM-type nut. The

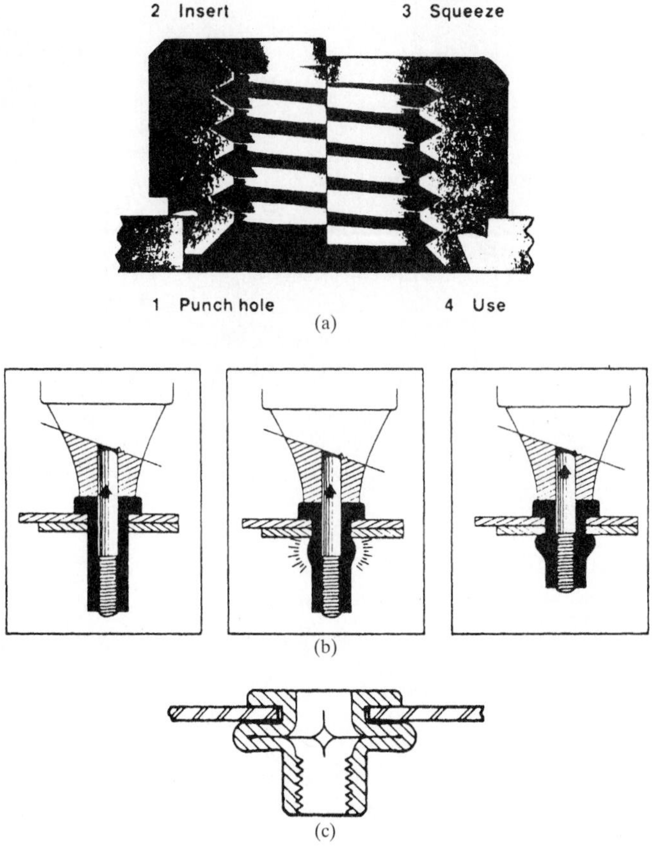


**FIGURE 10.58** (*a*) Clinch or swage nut; (*b*) Rivnut installation procedure; (*c*) installed Plusnut.

Rivnut and Plusnut, which are produced by BF Goodrich Company, are shown in Fig. 10.58*b* (Rivnut) and 10.58*c* (Plusnut). These types of "blind" fasteners have countless applications in industry and are also produced with sealed ends for liquid-proofing applications. The Rivnut is widely used in the aerospace industry.

***Electroplating Fasteners.*** High-quality fasteners such as the Unbrako series of socket head cap screws and shoulder bolts, which use the UNR thread profile, may be precision plated per the table shown in Fig. 10.59. Other types of fasteners may also use the plating specifications shown in Fig. 13.4. For more complete information on electroplating, see Chap. 13.

## 10.9 WELDING, BRAZING, AND SOLDERING

Welding, brazing, and soldering are all important methods of joining and fastening metals and alloys. In this section, we will detail the various methods or processes and materials used in these three branches of joining techniques.

**Precision Plating Specifications for Fasteners**

| Type | Thickness (minimum), in | | | Pre- or postplate treatments or instructions | Typical specifications |
|---|---|---|---|---|---|
| | A | B | C | | |
| Cadmium | 0.0002 | 0.0003 | 0.0005 | Clear postplate dip | AMS 2400 and QQ-P-416, type I |
| Cadmium | 0.0002 | 0.0003 | 0.0005 | Olive drab chromate | QQ-P-416, type II |
| Cadmium | 0.0002 | 0.0003 | 0.0005 | Iridescent dichromate | QQ-P-416, type II |
| Zinc | 0.0002 | 0.0003 | 0.0005 | Clear bright | ASTM B633, type III |
| Zinc | 0.0002 | 0.0003 | 0.0005 | Olive drab chromate | ASTM B633, type II |
| Zinc | 0.0002 | 0.0003 | 0.0005 | Iridescent dichromate | ASTM B633, type II |
| Zinc | 0.0002 | 0.0003 | 0.0005 | Supplementary phosphate | ASTM B633, type IV |
| Silver | 0.0002 | 0.0003 | 0.0005 | Nickel strike | AMS 2410, AMS 2411 |
| Black oxide | Alloy or carbon steel | 18-8 stainless | | | AMS 2485, Mil-C-13924 |
| Dull nickel | 0.0002 | 0.0003 | 0.0005 | | AMS 2403 |
| Copper | 0.0002 | 0.0003 | 0.0005 | | AMS 2418 |
| Tin | 0.0002 | 0.0003 | 0.0005 | | AMS 2408, Mil-T-10727, type I |
| Phosphate (class A) (Parker-Lubrite) | Dry | Nondrying oil | | Manganese phosphate | AMS 2481, DOD-P-16232 |
| Phosphate (class B) (Parkerizing) | Dry | Drying oil | | Zinc phosphate | AMS 2480, DOD-P-16232 |
| Cadmium | 0.0002 | 0.0003 | 0.0005 | Black dye over olive drab chromate | QQ-P-416 type II except color |
| Cadmium | 0.0002 | 0.0003 | 0.0005 | Fluoborate bath, bake at 375° for 23 h, iridescent dichromate | NAS 672 |
| Silver | 0.0002 | 0.0003 | 0.0005 | Copper strike | AMS 2412 |
| Nickel | 0.0002 NI | | | Thermal treat 630° | AMS 2416 |
| Cadmium | 0.0001 CD | | | | |
| Vacuum cadmium | 0.0002 | 0.0003 | 0.0005 | | Mil-C-8837, type I |
| Cadmium | 0.0002 | 0.0003 | 0.0005 | Supplementary phosphate | QQ-P-416, type III |
| Vacuum cadmium | 0.0002 | 0.0003 | 0.0005 | Iridescent dichromate "Cronak" or equivalent | Mil-C-8837, type II |
| Molydisulfide coating | | | | Available with a variety of carriers, concentrations, and treatments to customer requirements | |
| Passivation | | | | For austenitic series stainless steel | QQ-P-35 |
| Passivation | | | | For 400 series stainless steel | |
| Cadmium | 0.0002 | 0.0003 | 0.0005 | Clear postplate dip | AMS 2401 |
| Sermetel "Metric" blue dye | | | | | AMS 2506 |
| IVD aluminum | 0.0010 | 0.0005 | 0.0003 | Supplementary chromate | Mil-C-83488, type II |

**FIGURE 10.59** Precision plating specifications for threaded fasteners.

## 10.9.1 Welding

*Welding* is a fusion process for joining metals. The heat of application causes mixing of the joint metals or of the filler metal and the joint metal. The resulting joint is as strong as the parent metal, provided the weld is correctly made.

There are numerous welding methods or techniques in common use for a vast array of applications. The modern welding methods or techniques are catagorized in Fig. 10.60, which shows the process and the American Welding Society (AWS) designation. Both the modern welding and cutting processes of metals are shown in Fig. 10.60.

Although the list of processes shown in Fig. 10.60 is extensive, the majority of welding involves the following methods:

- Stick welding (fluxed rod)—SMAW
- Metal inert-gas (MIG) welding—GMAW
- Tungsten inert-gas (TIG) welding—GTAW
- Stud arc welding—SW

**Welding Processes and Designations**

| Process | Designation |
|---|---|
| **Arc welding (AW)** | |
| Atomic-hydrogen welding | AHW |
| Bare-metal arc welding | BMAW |
| Carbon arc welding | CAW |
| -gas | CAW-G |
| -shielded | CAW-S |
| -twin | CAW-T |
| Flux-cored arc welding | FCAW |
| -electrogas | FCAW-EG |
| Gas-metal arc welding | GMAW |
| -electrogas | GMAW-EG |
| -pulsed arc | GMAW-P |
| -short circuiting arc | GMAW-S |
| Gas-tungsten arc welding | GTAW |
| -pulsed arc | GTAW-P |
| Plasma arc welding | PAW |
| Plasma gas-metal arc welding | PAW-GMAW |
| Shielded-metal arc welding | SMAW |
| Stud arc welding | SW |
| Submerged arc welding | SAW |
| -series | SAW-S |
| **Solid-State Welding (SSW)** | |
| Cold welding | CW |
| Diffusion welding | DFW |
| Explosive welding | EXW |
| Forge welding | FOW |
| Friction welding | FRW |
| Hot press welding | HPW |
| Roll welding | ROW |
| Ultrasonic welding | USW |
| **Other Welding** | |
| Electron-beam welding | EBW |
| Electroslag welding | ESW |
| Flow welding | FLOW |
| Induction welding | IW |
| Laser beam welding | LBW |
| Thermite welding | TW |
| **Oxyfuel Gas Welding (OFW)** | |
| Air acetylene welding | AAW |
| Oxyacetylene welding | OAW |
| Oxyhydrogen welding | OHW |
| Pressure gas welding | PGW |
| **Resistance Welding (RW)** | |
| Flash welding | FW |
| High frequency resistance welding | HFRW |
| Percussion welding | PEW |
| Projection welding | RPW |
| Resistance seam welding | RSEW |
| Resistance spot welding | RSW |
| Upset welding | UW |

(a)

**Metal Cutting Processes and Designations**

| Process | Designation |
|---|---|
| **Thermal Oxygen Cutting (OC)** | |
| Chemical flux cutting | FOC |
| Metal powder cutting | POC |
| Oxyfuel gas cutting | OFC |
| -oxyacetylene | OFC-A |
| -oxyhydrogen | OFC-H |
| -oxynatural gas | OFC-N |
| -oxypropane | OFC-P |
| Oxygen arc cutting | AOC |
| Oxygen lance cutting | LOC |
| **Arc cutting (AC)** | |
| Air-carbon arc cutting | AAC |
| Carbon arc cutting | CAC |
| Gas-metal arc cutting | GMAC |
| Gas-tungsten arc cutting | GTAC |
| Metal arc cutting | MAC |
| Plasma arc cutting | PAC |
| Shielded-metal arc cutting | SMAC |
| **Other cutting** | |
| Electron-beam cutting | EBC |
| Laser beam cutting | LBC |

(b)

**FIGURE 10.60** (*a*) Welding processes and designations (AWS); (*b*) metal-cutting processes and designations (AWS).

Whether the welding process is gas or arc, the welder may proceed to do the weld in the *forward* or *backward* direction. The direction of welding is left to the judgment of the welder and the object being welded. When the welder is looking down at the welded joint, this is normally called *in-position* welding. When the welder is looking up at the joint, this is normally called *out-of-position* welding. Any orientation not looking "down" at the weld can be considered as "out-of-position" welding.

The various welding processes produce physical characteristics which may normally be determined by direct eye inspection to identify the process. In Fig. 10.61, we see a welded assembly produced using the MIG process (GMAW). Here, the weld is rather rough looking and was difficult to clean, mechanically. The welding heat setting and the diameter of the weld wire play important roles in producing a neat, clean weld which is also mechanically sound. Figure 10.62 is the same type of joint (using the same parts), which has been welded using the

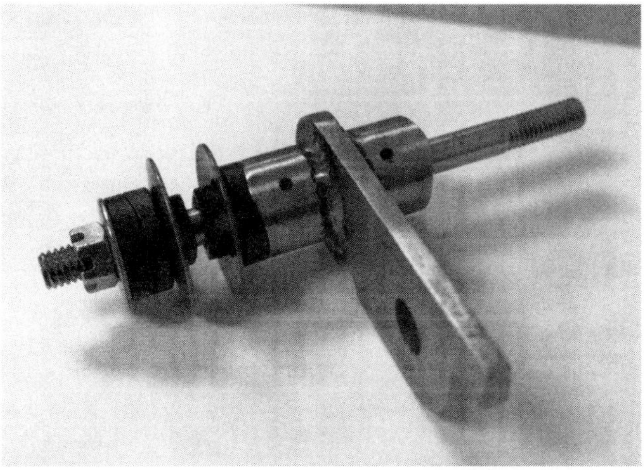

**FIGURE 10.61** MIG-welded assembly (stainless steel).

**FIGURE 10.62** TIG-welded assembly (stainless steel, abrasive-cleaned).

TIG process (GTAW). It is immediately apparent that the TIG process in this application is more advantageous in terms of both strength and cosmetic appearance. This illustrated joint is of parts manufactured from AISI 304-type stainless steel. In Fig. 10.62, the welded assembly was sand-blasted after welding. The TIG weld in this application was efficiently and correctly made by a skilled welder. Welding is an art as well as a science, with a large amount of practice and experience required to produce a skilled welder.

In a welded assembly of numerous piece parts, the welder's skill and experience play an important role in producing an acceptable final welded assembly. Many welded assemblies require additional machining after the welding process, because of the strains produced when the welded joints cool to room temperature. Allowance must be made in the design of the assembly to accommodate these welding distortions, which are sometimes unavoidable.

***Weld Strength: Related Equations and Tables.*** Here, we will present the equations for the approximate strength of welded joints. Any equation involving a process with many variables, such as welding, can be only an approximation. With this in mind, we allow a factor of safety when designing and calculating the strengths of welded joints.

***Fillet Welds.*** Refer to Fig. 10.63*a*. The basic welding equations for the fillet weld are as follows:

$$h = 0.707(l) \qquad hL = \frac{P}{S_i} \qquad A = \frac{P}{S_i} \qquad L = \frac{P}{S_i h}$$

where $l$ = leg dimension of fillet weld, in
$L$ = length of fillet weld, in
$h$ = weld throat height, in
$P$ = load, lb
$S_i$ = induced stress, psi
$A$ = throat area, in$^2$

***Butt Welds (Primary).*** The tensile stress in a butt weld induced by a tensile load $P$ (Fig. 10.63*b*) is

$$S_t = \frac{P}{td}$$

where $S_t$ = tensile stress, psi
$P$ = tensile load, lb
$t$ = material thickness, in
$d$ = width of butt-welded joint, in

***Minimum Leg Size for Fillet Welds.*** Refer to Fig. 10.64*b*.

***Fillet Welds and Partial-Penetration Groove Welds (Design Strengths).*** Design allowable strengths and shear forces for these types of welds are listed in Fig. 10.64*a*.

*Example.* The allowable unit force for a fillet weld with a 0.25-in leg, using 80,000-psi weld rod or wire, is 16,500 × 0.25 = 4125 lb per linear inch. Thus, if the weld joint is 3 in long, the force allowable is 4125 × 3 = 12,375 lb.

***Plug Welds.*** Plug welds (Fig. 10.65*a* and 10.65*b*) are useful in sheet-metal and structural design applications. Plug welds are primarily used for shear loads, although not limited to this type of load. A plug weld may be subjected to a combination of shear and tensile loads. The typical sizes of plugs welds are shown in Fig. 10.64*c* for various applications or combinations of material thicknesses.

Figure 10.65*a* and 10.65*b* illustrates the typical plug weld.

*Note:* The preceding weld strength figures and tables, allowables, and examples are for *static* loads only. When the welded members are dynamically or cyclically loaded, a factor of safety should be applied. A safety factor of 3 should be applied for general dynamic condi-

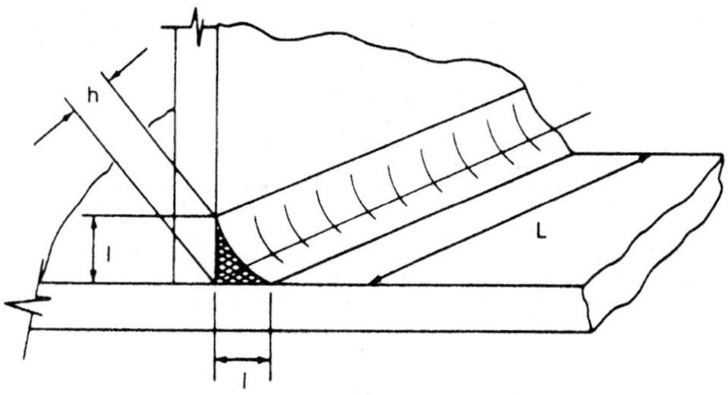


(a)

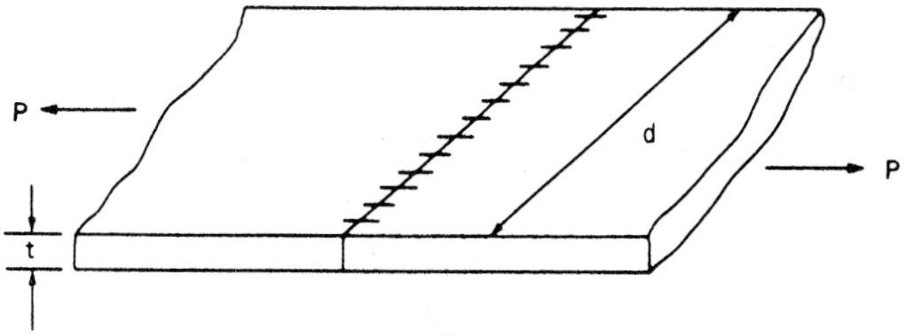


(b)

**FIGURE 10.63** (*a*) Fillet weld; (*b*) butt weld.

tions. In other words, if the weld joint was calculated to withstand a load of 3000 lbf for dynamic conditions, this load should be reduced to a 1000 lbf maximum allowable. (Divide the calculated load by 3 to arrive at the allowable load, with the factor of safety applied.)

***Circular Fillet Weld Strength***

Axial load, circular member (see Fig. 10.66*a*)

$$F = \frac{P}{\pi d}$$

Torsional load, circular member (see Fig. 10.66*b*)

$$F = \frac{2T}{\pi d^2}$$

Bending load, circular member (see Fig. 10.66*c*)

$$S_b = \frac{32M\,(d + 2h)}{\pi[(d + 2h)^4 - d^4]}$$

where $F$ = unit force, lb/linear in
$P$ = load, lb
$d$ = inside diameter of tube, in
$T$ = torque, lb-in
$h$ = throat of weld, in
$M$ = bending moment, lb-in
$S_b$ = bending stress, psi

| Strength of Weld* Rod or Wire Metal, psi | Allowable Shear Stress on Throat (*h*), psi | Allowable Unit Force, lb/Linear in. |
|---|---|---|
| 60,000 | 17,000 | 12,500 × *l* |
| 80,000 | 23,000 | 16,500 × *l* |
| 100,000 | 29,000 | 21,000 × *l* |
| 120,000 | 35,000 | 25,000 × *l* |

*For intermediate weld-rod strengths, interpolation may be used. In the above table *h* = weld throat, *l* = length of leg of the fillet (see Fig.

(a)

| Thickness of Thicker Plate (*t*), in | Minimum* Leg Size (*l*), in |
|---|---|
| Up to 1/4 | 1/8 |
| > 1/4 to 1/2 | 3/16 |
| > 1/2 to 3/4 | 1/4 |
| > 3/4 to 1 1/2 | 5/16 |
| > 1 1/2 to 2 1/2 | 3/8 |

*Also minimum throat (*h*) of partial penetration groove weld. Leg of weld (*l*) should not exceed thickness of thinner plate.

(b)

| Gauge or Thickness of Thinner Member, in | Plug-weld Hole Diameter, in |
|---|---|
| 1/16 or #16 gauge | 1/4 to 3/8 |
| 3/32 or #13 gauge | 1/2 |
| 1/8 or #11 gauge | 5/8 |
| 3/16 or #7 gauge | 5/8 |
| 1/4 | 3/4 |
| 3/8 | 1 |
| 1/2 | 1 1/4 |

(c)

**FIGURE 10.64** (*a*) Weld strength allowables; (*b*) minimum leg sizes of fillet welds; (*c*) plug weld sizes.

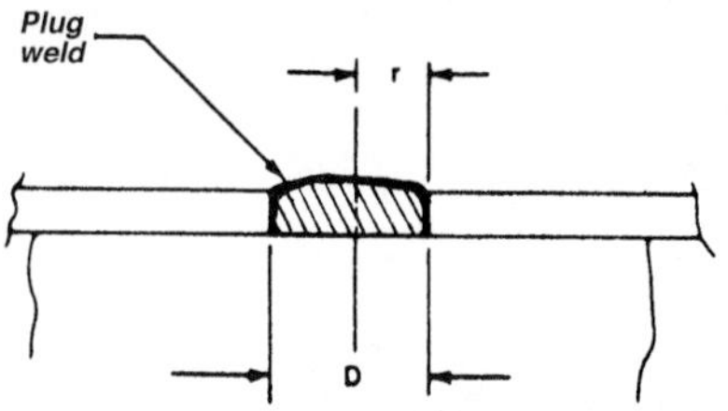


(a)

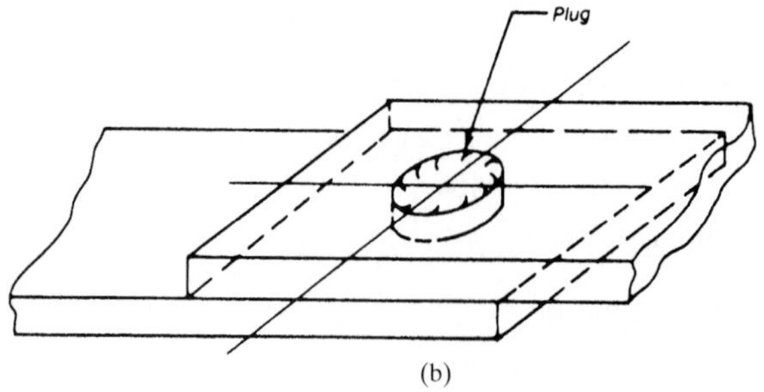


(b)

**FIGURE 10.65** (*a*) Section of plug weld; (*b*) plug-welded plates.

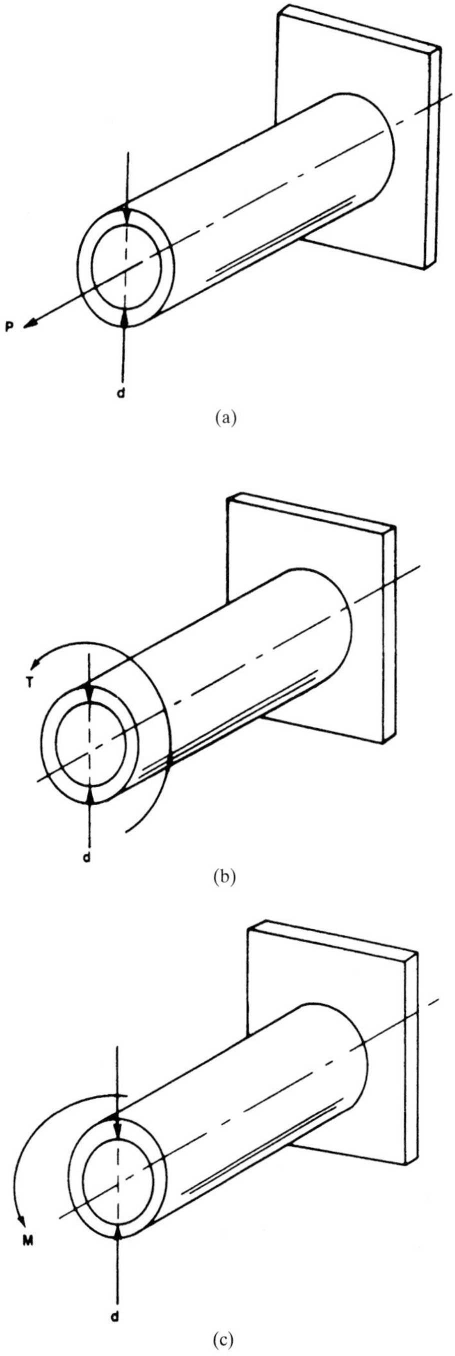


**FIGURE 10.66** (*a*) Axial load on welded tube; (*b*) torsional load on welded tube; (*c*) bending load on welded tube.

***Specifying Welds.*** The type of welding, weld-rod strength and type, fillet or bead size, and location and length of welds all must be specified on the welding drawings of a part or assembly. Standard weld symbols recognized by the AWS should be used on the engineering drawings (see weld symbols).

Thin-section parts or any part or assembly that may pose a weld-distortion problem should be reviewed in coordination with the welding department or welder prior to final design or beginning the work. The experienced welder usually knows or can determine welding sequences to prevent distortion or keep it to a minimum. Welding sequence instructions may be required on the welding drawing.

Secondary machining operations are usually performed on a welded part or assembly, after the welding operation. This is to correct unavoidable distortion or dimensional changes that take place on a welded part or assembly.

To reduce cost and save welding time, the amount of welding on a part or assembly should be kept to a minimum, in accordance with the strength requirements of the design or sealing requirements.

***Standard Weld Symbols.*** The basic weld symbols shown in Fig. 10.67 should be used on all welding drawings, especially if the welded part is sent to an outside vendor or subcontractor. If in-house symbols are used, these should be noted on the welding drawings, so that outside vendors or subcontractors know their exact meaning. The symbols shown in Fig. 10.67 depict the symbols which are recognized by AWS, American Iron and Steel Institute, AISC, ASME, SAE, and other authorities and specification agencies.

***Elements of the Welding Symbol.*** When a weld is specified on an American standard engineering drawing, it should conform to certain characteristics, which are shown in Fig. 10.68. In this way, uniformity and complete understanding are maintained between the welder and the design engineer.

Typical welding drawing *callouts* or symbols are shown in Fig. 10.69, with an explanation of their meaning.

***Types of Weld Joints.*** There are many types of weld joints or designs; the basic ones are shown in Fig. 10.70. The various joints have been designed for different applications and strengths. Other characteristics are designed into the weld joint, such as minimal outgassing, dynamic strength, deep penetration, and pressure vessel applications. The weld joints which require special preparation, such as machining, filing, or grinding, are more expensive to produce and are thus used for special applications. The majority of industrial welding consists of the simple fillet and butt-welded joints, followed by the single-V and double-V joints.

***Welding Application Data.*** Welding is one of the most common and important means of fastening. Its applications

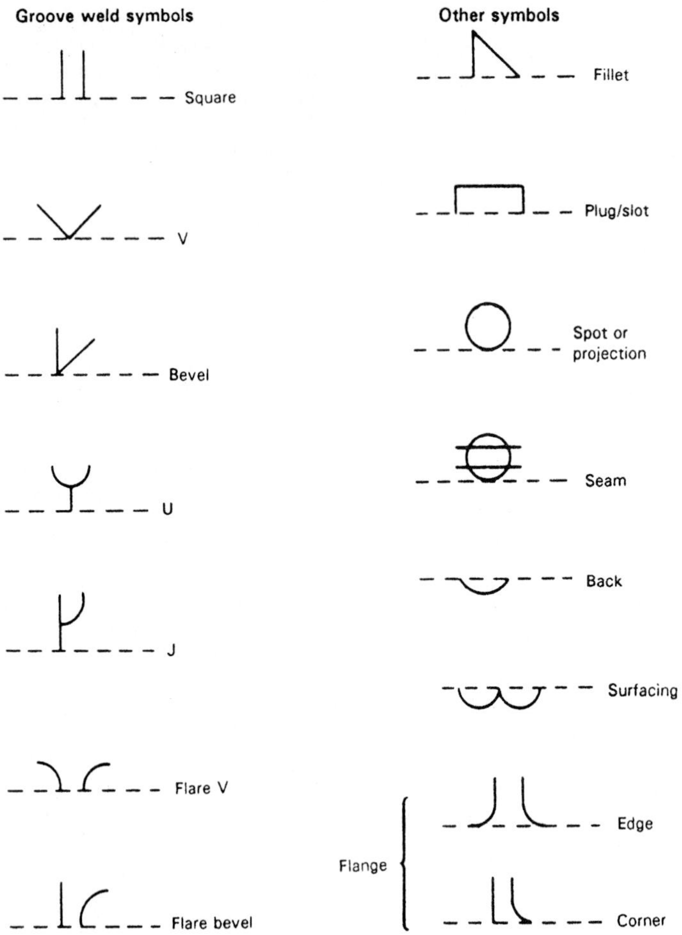


**FIGURE 10.67** Industry standard weld-type symbols.

are limitless, and the technology is constantly changing. Given here is a listing of various welding process applications which will be helpful to the design engineer in considering weld applications.

*Thin-gauge metal welding:* Small welding flames or small arcs are required for welding thin-gauge metals. The TIG process is especially useful for producing small, accurate welds on thin materials, under 11-gauge (0.1196-in). To prevent buckling of large-area, thin materials, the application of *heat sinks* is useful. Heat sinks may take the form of wet burlap bags or large blocks of metal clamped to the welded parts or sections. Applying tack welds in a specified sequence may also help prevent buckling of large, thin sections, prior to beginning the final seams.

*Preheating:* Large sections or masses of metal usually require a preheat stage, where the parts are heated a few hundred degrees Fahrenheit prior to beginning the welding process. This prevents thermal shock and minimizes distortion and possible cracking of the welds.

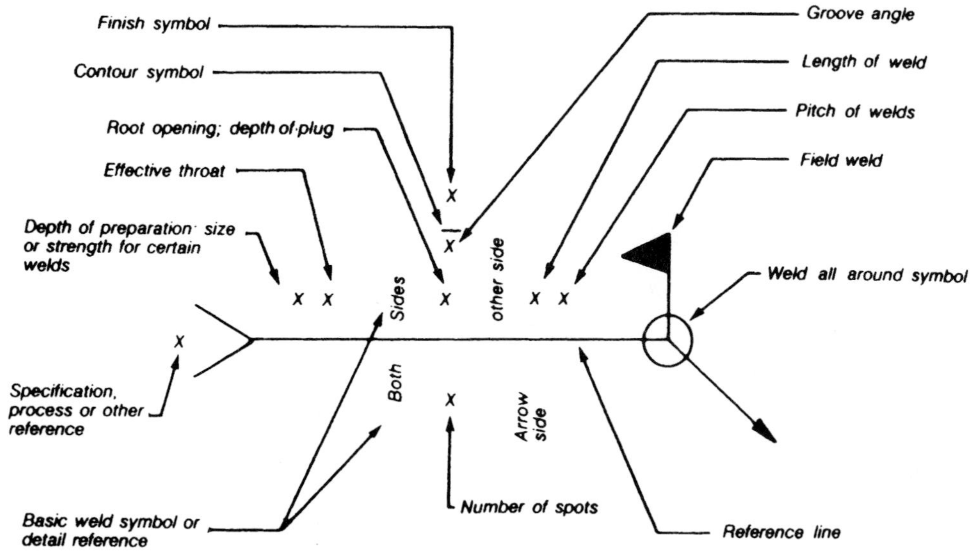


**FIGURE 10.68** Elements of the welding symbol.

*Air cooling:* Welded parts or assemblies are normally allowed to cool to ambient temperature after the welding process is completed. Do not water quench welded parts immediately after welding. Because of the high temperatures generated in the welding process, cracking or distortion may occur. Changes in the grain structure of the metal may occur if the hot, welded part is suddenly cooled by water quenching.

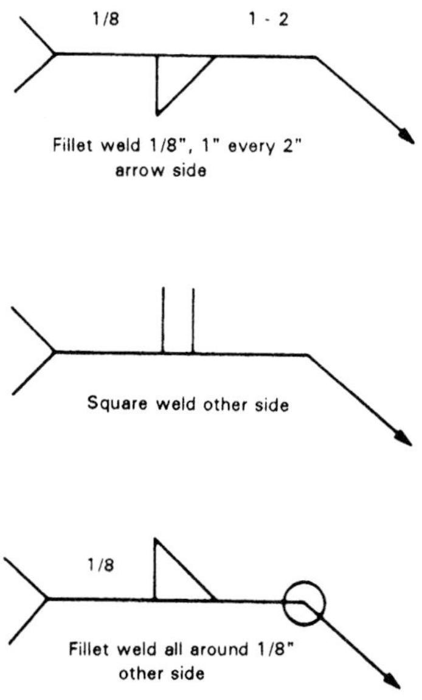


**FIGURE 10.69** Typical weld callouts for engineering drawings.

*Welding bases or platforms:* A flat, level area is required for welding large assemblies. This is usually provided by structural beams embedded in the weld shop floor. The beams must be straight and leveled with a transit or leveling instrument. For smaller welded parts and assemblies, the popular welding table is used. Figure 10.71 shows a typical steel grid welding table, which is used in many welding departments. This type of table is level and has square openings, where different types of clamping and squaring tools may be attached to hold the welded assembly prior to the welding operation.

Notice the screening and plastic shielding which are located around the welding table area. This shielding prevents the intense ultraviolet radiation generated by the welding arc from reaching the eyes of other personnel. The intensity of the welding arc radiation is sufficiently high to damage the membrane covering the human cornea and eyeball. It is not necessary to look directly at the welding arc for damage to occur to the eye. The arc rays can penetrate the side of the eye indirectly and cause damage. The usual effects of looking at a welding arc are the feeling that sand has entered the eye. This usually begins to show some hours after exposure to the arc, either directly or indirectly, and is the result of scar tissue formation in the damaged eye tissues.

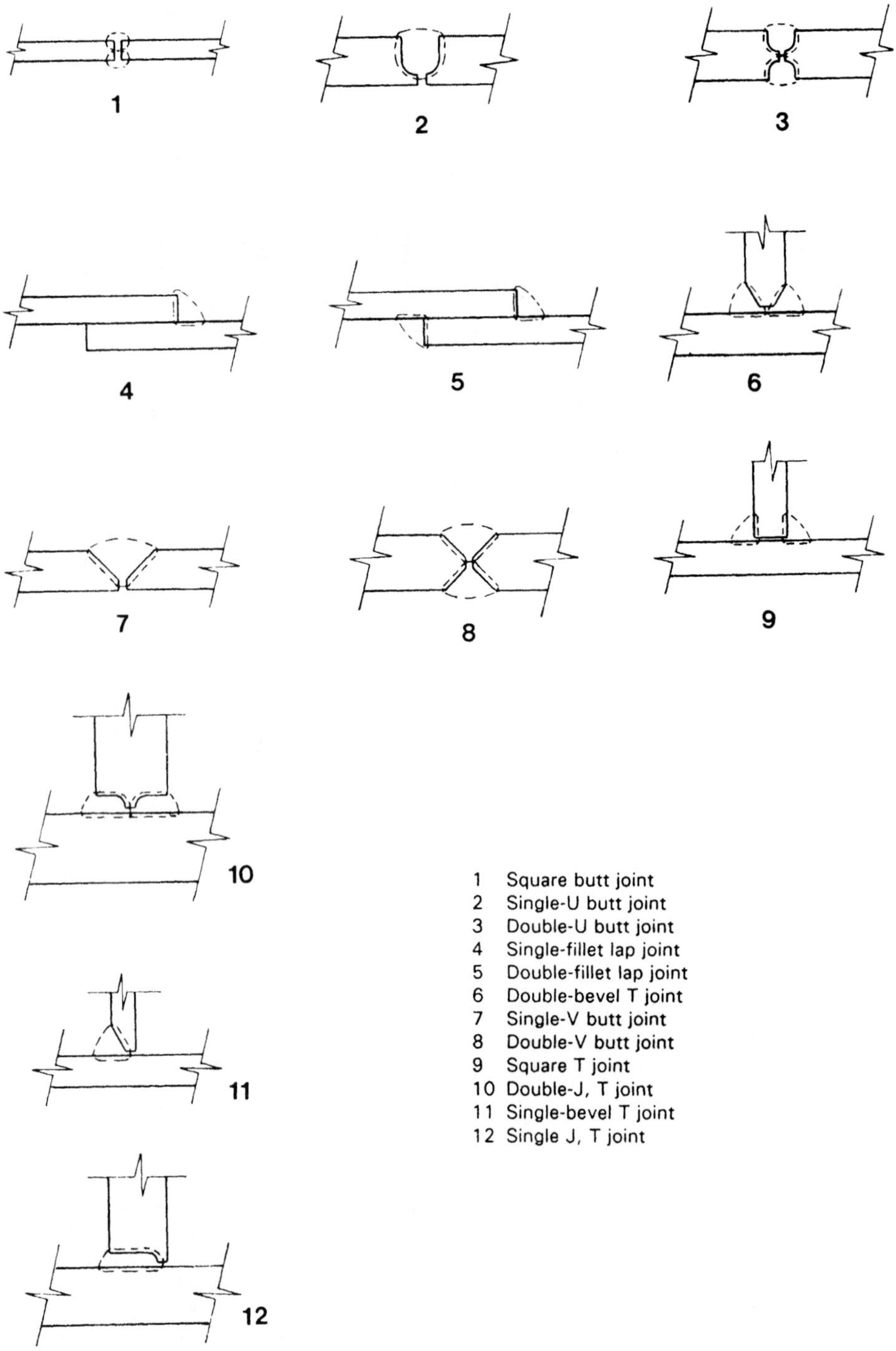


**FIGURE 10.70** Types of welded joints.

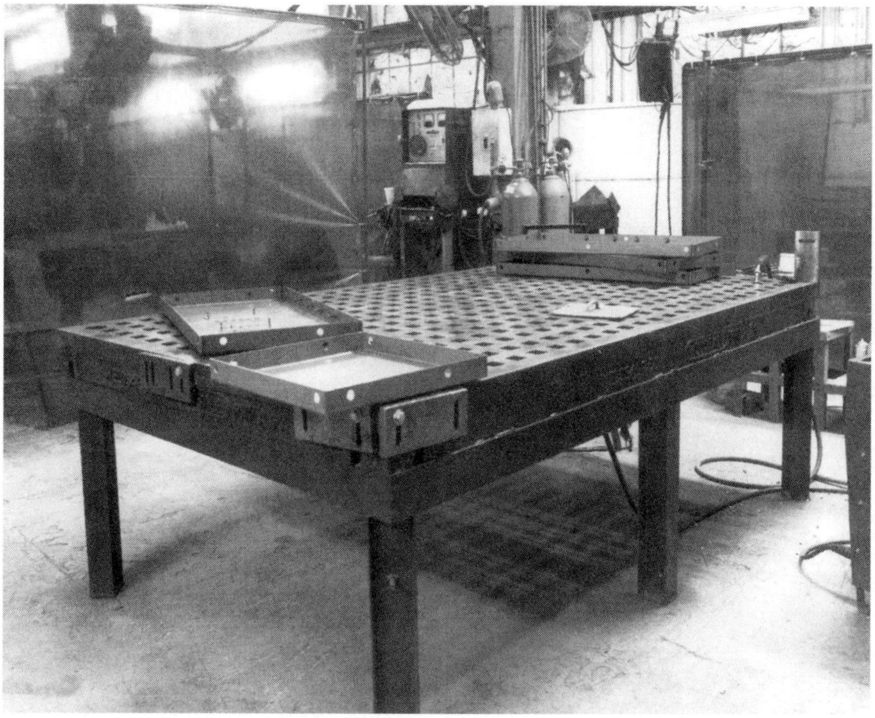

**FIGURE 10.71** A cast-iron grid welding table.

*Welding stainless steels:* The electrode used should also be stainless steel, matching the application. Types AISI 300 through 303 should not be welded, as these are machining grades. Type AISI 304 is a preferred stainless steel for welding applications, with AISI type 304L a special low-carbon grade for critical applications such as those used on aerospace vehicles. Too much carbon or sulfur in a stainless steel produces cracking during the welding process. AISI 308, 309S, 310S, 316, and 316L are also suitable for welding applications, being low in carbon content. AISI Type 316 and 316L stainless steels are widely used in highly corrosive environments such as chemical and food processing applications. The 300 series of stainless steels are austenitic (nonmagnetic) and cannot be hardened by heat-treatment procedures. See Chap. 4 for more information on stainless steels.

**FIGURE 10.72** A laser-cut steel part (C1018 cold-rolled steel).

*Welding carbon and alloy steels:* The low-carbon grades from AISI 1010 through 1020 are readily weldable as they are low in carbon content. Some of the medium-carbon grades are weldable with caution, while the high-carbon grades are not recommended for welding (AISI 1045 to 1095). The higher carbon content in the steel causes cracking during the welding process. Low-alloy and alloy-grade steels are weldable according to type and welding process employed. See Chap. 4 for more data on the low-alloy and alloy steels.

*Cutting metals:* Oxyacetylene cutting (OFC-A) is the most common welding process for cutting ferrous metals. Most ferrous metals cut cleanly using this method except the stainless steels, high-manganese steels, and special alloy steels. To cut these difficult materials, special techniques are required wherein the molten metal can be blown away from the flame or arc during the cutting process. Laser-beam cutting (LBC) is a modern method of cutting difficult materials and extremely thin materials. Figure 10.72 shows a small part made of carbon steel which was cut using the laser beam method. This part is approximately 2 in long and 0.25 in thick. Part quantity required did not justify building a stamping die for this part, and milling the part from stock was expensive; therefore, LBC was selected for the method of manufacture.

Many branches of industry rely on the welding processes for economically producing their equipment and machinery. Figure 10.73 shows a typical electrical power distribution industry lineup of switchgear, which relies heavily on the welding processes to fasten and join the many sheet metal parts and structures required in the equipment (see also Figs. 6.85, 6.86, and 6.88).

### 10.9.2 Brazing

Brazing employs a nonferrous filler metal, usually in wire or paste form, to join metal parts at a temperature above approximately 800°F, but below that of the melting point of the base metals being joined.

**FIGURE 10.73** Industrial electrical power distribution equipment of welded sheet-metal construction.

Various fluxes are used in the brazing process to remove oxides on the base metal parts, so that the filler metal can adhere to them strongly. Sal ammoniac is a good general-purpose flux for brazing copper, phosphor bronze, and stainless steels.

The brazing of tungsten alloys to various base metals is accomplished by pretinning the brazing surfaces of the tungsten alloy part in a nitrogen-atmosphere furnace before joining the tungsten alloy part with the base metal part. This is done because it is extremely difficult to remove oxide layers which would form on the tungsten part, unless it were pretinned and protected from the atmosphere.

Many of the various designs of electrical contact points are brazed to their base metal mountings, although riveting is sometimes employed.

***Brazing Heat.*** Heat for brazing parts may be supplied by the following methods:

- Torch brazing
- Furnace brazing
- Resistance brazing
- Dip brazing
- Vacuum-furnace brazing
- Induction brazing

When there is an unusual condition in the brazing method or process, a brazing specialist should be consulted. There are so many brazing materials and fluxes available that only an expert in the field may be able to solve your particular brazing problem.

In the normal brazing process, a suitable brazing wire or paste is selected and the parts to be brazed are set up or clamped into position (brazing fixtures are sometimes employed) after the joint has been fluxed with the proper flux for the application. Heat is applied to the parts to be joined and when hot enough, the filler wire is fed into the brazed joint, where a wetting action takes place, and the joint is filled with the melted brazing filler. The joint is allowed to air-cool to room temperature. The method of fluxing and applying the filler metal is critical for a well-brazed joint. It requires considerable skill and practice to braze a difficult joint or complicated setup. Excess brazing filler should not be used on any brazed joint, as it wastes material and produces an inefficient joint.

***Brazing Pastes and Alloys.*** Brazing and soldering alloys are available as wire, strip, preformed shapes, and pastes. Brazing pastes are applied to the joint prior to heating and are therefore applicable for manual as well as automated processing. Flux is also compounded in some of the brazing pastes, making their use easily applicable to production operations.

Lucas-Milhaupt (Handy and Harman Company) produces an extensive array of pastes for soldering and brazing in their Handy-Flo series 110, 120, 210, 310, 320, and 330.

With some of the new brazing products it is now possible to join ceramics, diamonds, and other difficult-to-wet materials.

### 10.9.3 Soldering

The use of lead or tin base alloys having melting points below 800°F for joining metal parts is known as *soft soldering.* Such joints seldom have great mechanical strength, although they may be required to carry electric currents.

The three basic standard soldering alloys include

- 60% tin–40% lead: melting point 375°F
- 50% tin–50% lead: melting point 420°F
- 40% tin–60% lead: melting point 460°F

Many other alloys are available, and these may be referenced through the ASTM standards or the *McGraw-Hill Machining and Metalworking Handbook*. The standard 60-40 solder is useful for most electrical and electronic soldering and a small amount of antimony may be added for improving quality and strength. The electronic wire solders usually are made with a rosin flux core. The flux is noncorrosive and nonconducting and may be removed after the soldering operation with naphtha, Varsol, or enamel paint thinners. Lacquer-type paint thinners should not be used, because they may damage electrionic components. Rosin-cored solders can be stored for extensive periods (years) without degradation of the rosin flux core.

Preforms are also produced in soldering alloys. The preform is a stamped shape, made of solder alloy, which will fit a particular device or application. Some of the delicate soldering operations would be difficult or impossible without preforms.

Figure 10.74*a* and 10.74*b* shows a typical printed-circuit (PC) board which had been automatically soldered in a "wave" soldering operation. This board was taken from a piece of computer equipment (plug-in module). All the copper foil lines on the PC board are covered with a masking coating, and only the component termination points are neatly soldered.

***Soldering Aluminum.*** Aluminum soldering employs a soldering alloy of 60 to 75% tin; the remainder is zinc. Special aluminum fluxes are available, and the process is carried out at a temperature of 550 to 775°F.

Commercial and high-purity aluminum are easiest to solder, with wrought alloys that contain no more than 1% magnesium or manganese the next easiest. The heat-treatable alloys are more difficult. Forged and cast aluminum alloys are not recommended for soldering,

**FIGURE 10.74** (*a*) Wave-soldered printed-circuit-board assembly (rear).

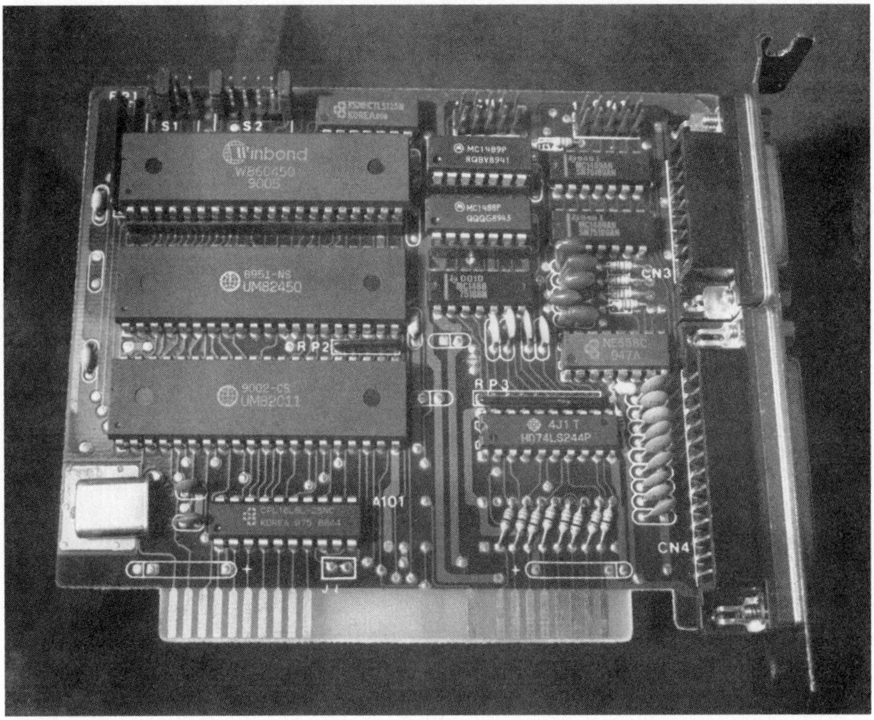

**FIGURE 10.74** (*b*) Front side of PC board shown in (*a*).

although it may be possible to a limited extent. The low-temperature brazing alloys may do the job more efficiently, with a higher-strength joint possible.

***Ultrasonic Soldering.*** The following metals and alloys may be soldered ultrasonically: aluminum, brass, copper, germanium, magnesium, silicon, and silver.

## 10.10 ADHESIVE BONDING

Adhesive bonding is used very frequently in modern manufacturing operations. The methods are generally fast, strong, and economical when used in the proper manner on appropriate articles. Entire fuselage sections are bonded or cemented on modern aircraft to replace riveting. The method not only affords lighter weight but is also stronger and requires no maintenance. The bonding strength of cyanoacrylate adhesives is on the order of thousands of pounds in tension for one square inch of bonded surface area.

***Structural Adhesives.*** Following is a list of the various types of adhesives used in industrial applications:

- *Hot-melt adhesives:* thermoplastic resins—100% solids
- *Dispersion or solution adhesives:* thermoplastic resins—20 to 50% solids

- *Silicone adhesives:* thermoset—100% solids; rubber-like with high impact and peel strengths
- *Anaerobic adhesives:* thermoset—100% solids; liquid; thread-locking uses such as Loctite series adhesives
- *Phenolic or urea adhesives:* thermoset resins—100% solids or solution
- *Epoxy adhesives:* thermoset—100% solids; liquid
- *Polyurethane adhesives:* thermoset—100% solids; liquid
- *Cyanoacrylate adhesives:* thermoset—100% solids; liquid; hazard of bonding skin on contact; almost instant set, 30-sec to 5-min cure; high strength
- *Modified acrylic adhesives:* thermoset—100% solids; liquid or paste; fast cure, 3 to 60 sec

***Weld Bonding.*** A combination of spot welding and adhesive bonding, this process is used on aerospace assemblies to cut costs, reduce weight, and increase strength. The method affords lighter weight than riveting, is stronger, yields air- and fueltight joints, and produces a smooth exterior surface that reduces wind resistance. The structural adhesives are applied prior to the spot-welding operation; then the bonded and welded assembly is oven-cured at an elevated temperature for a predetermined time interval.

# CHAPTER 11

# SHEET-METAL DESIGN, LAYOUT, AND FABRICATION PRACTICES

The branch of metalworking known as *sheet metal* represents a large and important element. Sheet-metal parts are used in countless commercial and military products and are found on almost every product produced by the metalworking industries throughout the world.

Sheet metal gauges run from under 0.001 to 0.500 in. Hot-rolled steel products can run from ½ in thick to 18-gauge (0.0478 in) and still be considered as *sheet,* although many authorities consider any sheet-metal thickness of 0.250 in and greater as plate. Cold-rolled steel sheets are generally available from stock in sizes from 10-gauge (0.1345-in) down to 28-gauge (0.0148-in). Other sheet thicknesses are available as special-order *mill-run* products, when the basic order is large enough to justify a mill-run order. Large manufacturers who use vast tonnages of steel products such as the automobile manufacturers, appliance industry, switchgear producers, and other sheet-metal fabricators may order their steel to their own company specifications (composition, gauges, and physical properties).

The steel sheets are supplied in flat form or rolled into coils. Flat-form sheets are made to specific standard sizes, unless ordered to special nonstandard dimensions. Common American steel mill parlance denotes 0.250 in-thick steel plate as *ten-pound-two* (10-lb-2). This designation is derived from the fact that one square foot of normal mild carbon steel ¼ in thick weighs 10.2 lb. Thus a 1-$ft^2$ piece of ⅛-in-thick steel would weigh about 5.1 lb and is called *five-pound-one* (5-lb-1).

## 11.1 CARBON AND LOW-ALLOY STEEL SHEETS

Carbon steel sheet and coil are produced in the following grades or classes:

**1.** Hot-rolled:
- ***a.*** Low-carbon (commercial quality)
- ***b.*** Pickled and oiled
- ***c.*** 0.40/0.50 carbon
- ***d.*** Abrasion-resisting
- ***e.*** Hi-Form (A715)—high-strength, low-alloy (grade 50; grade 80)
- ***f.*** A607 specification—high-strength, low-alloy (INX 45; INX 50; Ex-Ten 50)
- ***g.*** A606 specification—high-strength, low-alloy (Cor-Ten)

**2.** Cold-rolled:
- ***a.*** Low-carbon (commercial quality)
- ***b.*** Special killed (drawing quality)

*c.* Auto prototype (special killed—drawing quality)
*d.* Vitreous enameling
*e.* Plating quality
*f.* Stretcher leveled

The code numbers for the hot-rolled group indicate the yield strength of the high-strength, low-alloy steels; e.g., INX 45 = 45,000 psi yield; grade 80 = 80,000 psi yield. Applications for these sheet steels are as follows.

### *Hot-Rolled Applications*

*Low-carbon* (*commercial quality*): Conforms to ASTM A569 and is used for tanks, barrels, farm implements, and other applications where surface quality is not critical or important.

*Pickeled and oiled:* Conforms to ASTM A569 and is used for automotive parts, switchgear, appliances, toys, and other applications where a better surface quality is required and paint and enamel adhere well. Carbon content is 0.10 maximum, and the stock may be easily formed and welded the same as low-carbon commercial quality.

*0.40/0.50 carbon:* Have 50 percent more yield strength and abrasion resistance than low-carbon sheets. May be heat-treated for more strength and hardness. Used for scrapers, blades, tools, and other applications requiring a strong, moderate-cost steel sheet.

*Abrasion-resistant:* Medium carbon content and higher manganese greatly improves resistance to abrasion. Brinell hardness = 210 minimum. Uses include scrapers, liners, chutes, conveyors, and other applications requiring a strong, abrasion-resistant steel sheet. Formability is moderate.

*High-strength, low-alloy sheets*

*A607 specification:* Lowest-cost low-alloy steel sheet. Low carbon content assures good formability. Excellent weldability. Typical uses include utility poles, transmission towers, automotive parts, truck trailers, and other applications requiring a low-cost, high-strength alloy steel sheet.

*A606 specification:* Five times more resistant to atmospheric corrosion than low-carbon steel. Excellent weldability and formability.

*A715 specification:* Fine-grained columbium bearing series of high-strength steel. Enhanced bending and forming properties. Tough and fatigue-resistant, with excellent weldability using all welding processes. Yield point levels range from 40,000 to 80,000 psi.

### *Cold-Rolled Applications*

*Low-carbon* (*commercial quality*): Produced with a high degree of gauge accuracy, with uniform physical characteristics. Excellent surface for painting (enamel or lacquer). Good for stamping and moderate drawing applications. Improved welding and forming characteristics with uses such as household appliances, truck bodies, signs, panels, and many other applications.

*Special killed* (*drawing quality*): Used for severe forming and drawing applications. Freedom from age hardening and fluting. Conforms to ASTM A365 specification.

*Auto prototype* (*special killed—drawing quality*): Used for prototype work and other deep drawing applications. Closely controlled gauge thickness with better tolerances compared to commercial-quality grades.

*Vitreous enameling:* Cold-rolled from commercially pure iron ingots for porcelain-enameled products. Textured surface and suitable for forming and moderate drawing applications and flatwork. Conforms to ASTM A424, grade A.

*Plating quality:* Two finishes are provided which are suitable for most plating applications: commercial bright and extralight matte.

*Stretcher leveled:* Uniform, high-quality matte sheets, further processed by stretching to provide superior flatness. Furnished resquared or not resquared. Resquared sheets have the stretching gripper marks removed. Used in the manufacture of table tops, cabinets, truck body panels, partitions, templates, and many other applications. Conforms to ASTM A336 specification.

***Galvanized Sheet and Coil.*** The galvanic coating is zinc and is applied to the standard steel sheets or coils in two basic methods: hot-dipped galvanized and electrogalvanized. The hot-dipped galvanized processes are known as *Ti-Co* (titanium-cobalt) *galvanized,* galvanized bonderized, galvannealed, galvannealed A, and hot-dipped galvanized. Galvanizing specifications are found in ASTM A526 and A527. Some of the hot-dipped sheets may have 1.25 oz of zinc per square foot of surface area and others, a lighter deposit.

Electrogalvanized sheets are cold-rolled steel sheets zinc-coated by electrolytic deposition and conform to ASTM A591. These sheets should be painted if exposed to outdoor conditions. They can be formed, rolled, or stamped without flaking, peeling, or cracking of the zinc coating. These galvanized sheets have the same gauge thickness as cold-rolled sheets. Applications include cabinets, signs, and light fixtures where an excellent finish is required. Coating weight is typically 0.1 oz/ft$^2$, or each side is 0.00008 in thick. Trade names include Paint-Lok, Bethzin, Gripcoat, Lifecote 1, and Weirzin Bonderized.

***Aluminized and Long-Terne Sheets.*** Sheet steel is also aluminized and produced in long-terne sheet. Aluminized steel sheet is hot-dip-coated on both sides with aluminum silicon alloy by the continuous method. Strong and corrosion-resistant, aluminized sheet is also low-cost. The aluminum coating is typically 0.001 in thick on both sides or 0.40 oz/ft$^2$. Aluminized sheet conforms to ASTM A463 specification. Applications include dry kiln fan walls, dryers, incinerators, mufflers, and oven and space heater components.

Long-terne sheet is a soft steel, coated with an 85% lead and 15% tin alloy for maximum ease in soldering. These long-terne sheets conform to ASTM A308 specification and are used for soldered tanks, automotive accessories, hood and radiator work, and many other stamped and formed products.

As can be seen from the preceding description of sheet steels that are commercially available, the selection of a particular steel for a particular sheet-metal application is relatively easy. Not only are there a great number of different sheet-metal stocks available, but special sheet steels may be ordered to your specifications when quantities are large enough to justify their production by the American steelmakers.

## 11.2 NONFERROUS SHEET METAL

The nonferrous sheet metals include aluminum and aluminum alloys, copper and copper alloys, magnesium alloys, titanium alloys, and other special alloys. See Chap. 5 for data and specifications on the nonferrous and also the ferrous materials which are specified by ASTM and SAE. Supplier catalogs are also available from companies such as Ryerson, Vincent, Atlantic, Alcoa, Reynolds, Anaconda, and Chase which may be selected from the industrial supplier master indexes such as the *Thomas Register of American Manufacturers.* The *McGraw-Hill Machining and Metalworking Handbook* has an extensive list of standards specifications for SAE and ASTM ferrous and nonferrous metals and alloys and their stock tolerances (for sheets, plates, bars, rods, tubes, pipes, hexagons, and other shapes).

## 11.3 MACHINERY FOR SHEET-METAL FABRICATION

Some of the typical machinery found in a large manufacturing plant for processing and producing sheet-metal parts would include

- Shears, hydraulic and squaring
- Press brakes
- Leaf brakes
- Roll-forming machines
- Automatic (CNC), multistation punch presses
- Single-die punch presses—Strippit and Unipunch setups
- Slitting machines
- Stretching-bending machines
- Hydropresses (Marforming presses—Martin-Marietta Corp.)
- Pin routers
- Yoder hammers
- Spin-forming machines
- Tumbling and deburring machines
- Sandblasting equipment
- Explosive forming facilities
- Ironworkers (for structural shapes) and many others

The electromechanical designer and tool engineer should be familiar with all machinery used to manufacture parts in the factory. These specialists must know the limitations of the machinery that will produce the parts as designed and tooled. Coordination of design with the tooling and manufacturing departments within a company is essential to the quality and economics of the products which are manufactured. Our modern machinery has been designed and is constantly being improved to allow us to manufacture a quality product at an affordable price to the consumer. Medium- to large-size companies can no longer afford to manufacture products whose quality standards do not meet the demands and requirements of the end user.

### 11.3.1 Modern Sheet-Metal Manufacturing Machinery

The processing of sheet metal begins with the hydraulic shear, where the material is squared and cut to size for the next operation. Figure 11.1 shows a typical hydraulic shear with a capacity of 120-in cut in up to 7-gauge steel sheet. These types of machines are the workhorses in the typical sheet-metal department as all operations on sheet-metal parts start at the shears or slitting machines.

Figure 11.2 shows a Wiedemann Optishear, which shears and squares the sheet metal to a high degree of accuracy. Blanks which are used in blanking, punching, and forming dies are produced on this machine, as are other flat and accurate pieces, which proceed to the next stage of manufacture.

The flat, sheared sheet-metal parts may then be routed to the punch presses, where holes of various sizes and patterns are produced. Figure 11.3 shows a large CNC-controlled multistation turret punch press, which is both highly accurate and very high-speed. The centerlines of punched holes can be maintained within 0.003 to 0.005 in over relatively long spans, with closer tolerances on smaller centerline distances.

**FIGURE 11.1** Large hydraulic shears. Capacity 60 × 120-in 7-gauge sheet steel.

**FIGURE 11.2** The Weidemann Optishear, squaring shears, CNC-controlled.

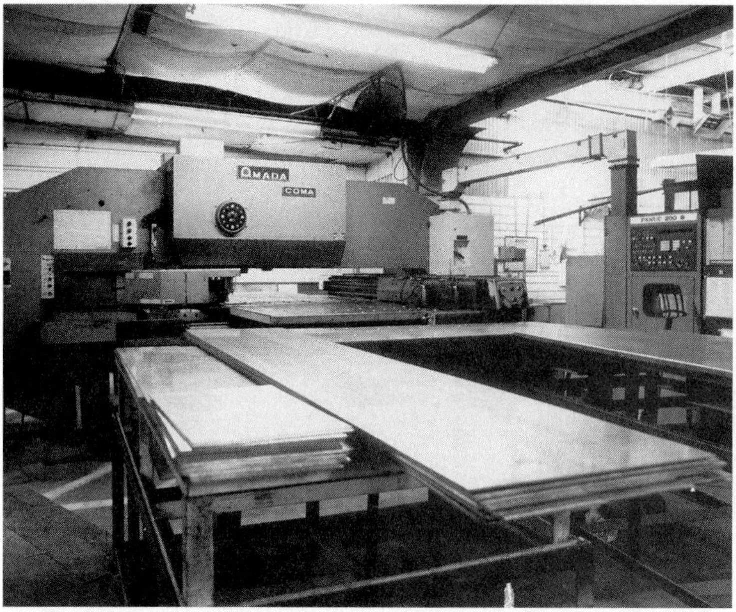


**FIGURE 11.3** Large multistation punch press, CNC-controlled, high-speed.

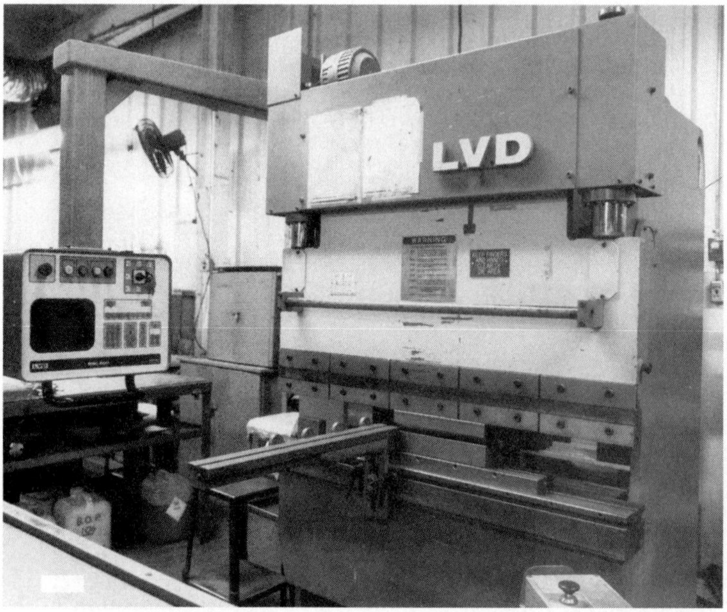


**FIGURE 11.4** Small, digital readout press brake for bending small sheet-metal parts.

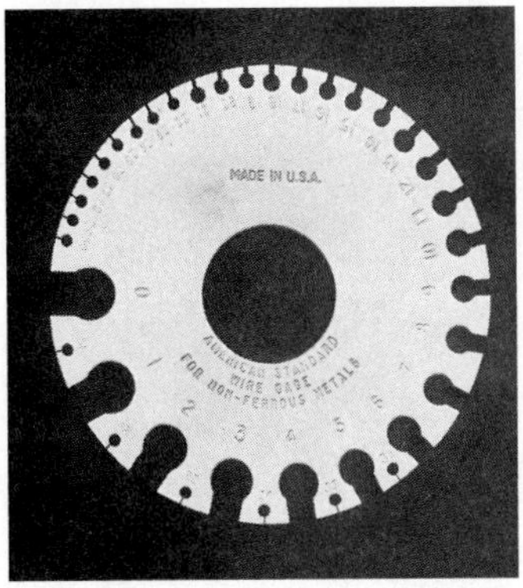


**FIGURE 11.5** A sheet-metal measuring gauge for nonferrous sheet stock.

After a sheet-metal part is sheared and punched, it may require a press brake operation to form flanges or produce hemmed edges. Figure 11.4 shows a small, digital readout press brake, which has a digital readout for the press depth of stroke and back-gauge dimension. A small press brake such as that shown in Fig. 11.4 may be used for bending small sheet-metal parts, flat bar stock, and copper and aluminum bus bars.

To quickly check a sheet-metal gauge, either ferrous or nonferrous, a tool similar to that shown in Fig. 11.5 is frequently used in the sheet-metal department. The tool gauge shown in the figure is for nonferrous metals such as stainless-steel sheets and nonferrous wires.

## 11.4 GAUGING SYSTEMS

To specify the thickness of different metal products such as steel sheets, wire, strip and tubing, and music wire, a host of gauging systems have been developed over the course of many years. Shown in Fig. 11.6 are the common gauging systems used for commercial steel sheets, strip and tubing, brass, and steel wire. The steel sheets column of Fig. 11.6 are the gauges and equivalent thicknesses which are used by American steel sheet manufacturers and steelmakers. This gauging system can be immediately recognized by its 11-gauge equivalent of 0.1196 in, which is standard today for this very common and high-usage gauge of sheet steel.

Figure 11.7 is a table of gauging systems which were widely used in the past, although some are still in use today, including the American or Brown and Sharpe system. The Brown and Sharpe system is also shown in Fig. 11.6, but there it is indicated in only four-place decimals.

### 11.4.1 Aluminum Sheet-Metal Standard Thicknesses

Aluminum is widely used in the aerospace industry and over the years, the gauges of aluminum sheets have developed on their own and are generally available in the thicknesses shown in Fig. 11.8. The fact that the final weight of an aerospace vehicle is very critical to its performance has played an important role in the development of the standard aluminum gauges, wherein the strength:weight ratio is critical and a few thousandths of an inch extra on an aluminum sheet will mean more final weight of the aerospace vehicle.

The gauges or thicknesses of other metals and alloys, together with the stock tolerances for all the common flat materials and shapes, are shown in the *McGraw-Hill Machining and Metalworking Handbook* (1994). Effective detail part design *cannot* be accomplished without knowledge of these mill tolerances on stock materials. Both the sheet-metal designer and the tool engineer must know mill product stock tolerances accurately. The tolerances on metal sheets are important for blanking and progressive die-bending operations. The tolerances on sheets, bars, rods, tubing, pipe, and other shapes must be known so that parts may be fitted without machining where possible. As an example, the use of a cold-drawn steel round rod would usually not be suitable for shafting applications. This is because the outside diameter of the rod usually has too much tolerance spread to fit shaft bearings without machining (reducing the diameter), or it will be too loose. Special *shafting grades* of steels and steel alloys are produced for this simple reason. This also applies to rods and tubes that must fit inside one another without recourse to a machining operation.

**Standard gauge decimals.**

| Gauge no. | Brass (Brown & Sharpe) | Steel sheets* | Strip and Tubing | Steel wire ga.† |
|---|---|---|---|---|
| 6-0 | 0.5800 | — | — | 0.4615 |
| 5-0 | 0.5165 | — | 0.500 | 0.4305 |
| 4-0 | 0.4600 | — | 0.454 | 0.3938 |
| 3-0 | 0.4096 | — | 0.425 | 0.3625 |
| 2-0 | 0.3648 | — | 0.380 | 0.3310 |
| 0 | 0.3249 | — | 0.340 | 0.3065 |
| 1 | 0.2893 | — | 0.300 | 0.2830 |
| 2 | 0.2576 | — | 0.284 | 0.2625 |
| 3 | 0.2294 | 0.2391 | 0.259 | 0.2437 |
| 4 | 0.2043 | 0.2242 | 0.238 | 0.2253 |
| 5 | 0.1819 | 0.2092 | 0.220 | 0.2070 |
| 6 | 0.1620 | 0.1943 | 0.203 | 0.1920 |
| 7 | 0.1443 | 0.1793 | 0.180 | 0.1770 |
| 8 | 0.1285 | 0.1644 | 0.165 | 0.1620 |
| 9 | 0.1144 | 0.1495 | 0.148 | 0.1483 |
| 10 | 0.1019 | 0.1345 | 0.134 | 0.1350 |
| 11 | 0.0907 | 0.1196 | 0.120 | 0.1205 |
| 12 | 0.0808 | 0.1046 | 0.109 | 0.1055 |
| 13 | 0.0720 | 0.0897 | 0.095 | 0.0915 |
| 14 | 0.0641 | 0.0747 | 0.083 | 0.0800 |
| 15 | 0.0571 | 0.0673 | 0.072 | 0.0720 |
| 16 | 0.0508 | 0.0598 | 0.065 | 0.0625 |
| 17 | 0.0453 | 0.0538 | 0.058 | 0.0540 |
| 18 | 0.0403 | 0.0478 | 0.049 | 0.0475 |
| 19 | 0.0359 | 0.0418 | 0.042 | 0.0410 |
| 20 | 0.0320 | 0.0359 | 0.035 | 0.0348 |
| 21 | 0.0285 | 0.0329 | 0.032 | 0.0317 |
| 22 | 0.0253 | 0.0299 | 0.028 | 0.0286 |
| 23 | 0.0226 | 0.0269 | 0.025 | 0.0258 |
| 24 | 0.0201 | 0.0239 | 0.022 | 0.0230 |
| 25 | 0.0179 | 0.0209 | 0.020 | 0.0204 |
| 26 | 0.0159 | 0.0179 | 0.018 | 0.0181 |
| 27 | 0.0142 | 0.0164 | 0.016 | 0.0173 |
| 28 | 0.0126 | 0.0149 | 0.014 | 0.0162 |
| 29 | 0.0113 | 0.0135 | 0.013 | 0.0150 |
| 30 | 0.0100 | 0.0120 | 0.012 | 0.0140 |
| 31 | 0.0089 | 0.0105 | 0.010 | 0.0132 |
| 32 | 0.0080 | 0.0097 | 0.009 | 0.0128 |
| 33 | 0.0071 | 0.0090 | 0.008 | 0.0118 |
| 34 | 0.0063 | 0.0082 | 0.007 | 0.0104 |
| 35 | 0.0056 | 0.0075 | 0.005 | 0.0095 |
| 36 | 0.0050 | 0.0067 | 0.004 | 0.0090 |
| 37 | 0.0045 | 0.0064 | — | 0.0085 |
| 38 | 0.0040 | 0.0060 | — | 0.0080 |

*Common commercial standard.
†Reference only.

**FIGURE 11.6** Common gauging systems.

## 11.5 SHEET-METAL FABRICATION METHODS

Many methods have been developed for working or fabricating sheet-metal parts. The methods include those employed to cut, punch, and form sheet-metal parts and are summarized below:

| Number of wire gauge | American or Brown & Sharpe | Birmingham or Stubs' Iron wire | Washburn & Moen, Worcester, Mass. | W. & M. steel music wire | American S. & W. Co's. music wire gauge | Stubs' steel wire | U.S. Standard gauge for sheet and plate iron and steel | Number of wire gauge |
|---|---|---|---|---|---|---|---|---|
| 00000000 | | | | 0.0083 | | | | 00000000 |
| 0000000 | | | | 0.0087 | | | | 0000000 |
| 000000 | | | | 0.0095 | 0.004 | | 0.46875 | 000000 |
| 00000 | | | | 0.010 | 0.005 | | 0.4375 | 00000 |
| 0000 | 0.460 | 0.454 | 0.3938 | 0.011 | 0.006 | | 0.40625 | 0000 |
| 000 | 0.40964 | 0.425 | 0.3625 | 0.012 | 0.007 | | 0.375 | 000 |
| 00 | 0.3648 | 0.380 | 0.3310 | 0.0133 | 0.008 | | 0.34375 | 00 |
| 0 | 0.32486 | 0.340 | 0.3065 | 0.0144 | 0.009 | | 0.3125 | 0 |
| 1 | 0.2893 | 0.300 | 0.2830 | 0.0156 | 0.010 | 0.227 | 0.28125 | 1 |
| 2 | 0.025763 | 0.284 | 0.2625 | 0.0166 | 0.011 | 0.219 | 0.265625 | 2 |
| 3 | 0.22942 | 0.259 | 0.2437 | 0.0178 | 0.012 | 0.212 | 0.250 | 3 |
| 4 | 0.20431 | 0.238 | 0.2253 | 0.0188 | 0.013 | 0.207 | 0.234375 | 4 |
| 5 | 0.18194 | 0.220 | 0.2070 | 0.0202 | 0.014 | 0.204 | 0.21875 | 5 |
| 6 | 0.16202 | 0.203 | 0.1920 | 0.0215 | 0.016 | 0.201 | 0.203125 | 6 |
| 7 | 0.14428 | 0.180 | 0.1770 | 0.023 | 0.018 | 0.199 | 0.1875 | 7 |
| 8 | 0.12849 | 0.165 | 0.1620 | 0.0243 | 0.020 | 0.197 | 0.171875 | 8 |
| 9 | 0.11443 | 0.148 | 0.1483 | 0.0256 | 0.022 | 0.194 | 0.15625 | 9 |
| 10 | 0.10189 | 0.134 | 0.1350 | 0.027 | 0.024 | 0.191 | 0.140625 | 10 |
| 11 | 0.090742 | 0.120 | 0.1205 | 0.0284 | 0.026 | 0.188 | 0.125 | 11 |
| 12 | 0.080808 | 0.109 | 0.1055 | 0.0296 | 0.029 | 0.185 | 0.109375 | 12 |
| 13 | 0.071961 | 0.095 | 0.0915 | 0.0314 | 0.031 | 0.182 | 0.09375 | 13 |
| 14 | 0.064084 | 0.083 | 0.0800 | 0.0326 | 0.033 | 0.180 | 0.078125 | 14 |
| 15 | 0.057068 | 0.072 | 0.0720 | 0.0345 | 0.035 | 0.178 | 0.0703125 | 15 |
| 16 | 0.05082 | 0.065 | 0.0625 | 0.036 | 0.037 | 0.175 | 0.0625 | 16 |
| 17 | 0.045257 | 0.058 | 0.0540 | 0.0377 | 0.039 | 0.172 | 0.05625 | 17 |
| 18 | 0.040303 | 0.049 | 0.0475 | 0.0395 | 0.041 | 0.168 | 0.050 | 18 |
| 19 | 0.03589 | 0.042 | 0.0410 | 0.0414 | 0.043 | 0.164 | 0.04375 | 19 |
| 20 | 0.031961 | 0.035 | 0.0348 | 0.0434 | 0.045 | 0.161 | 0.0375 | 20 |
| 21 | 0.028462 | 0.032 | 0.03175 | 0.046 | 0.047 | 0.157 | 0.034375 | 21 |
| 22 | 0.025347 | 0.028 | 0.0286 | 0.0483 | 0.049 | 0.155 | 0.03125 | 22 |
| 23 | 0.022571 | 0.025 | 0.0258 | 0.051 | 0.051 | 0.153 | 0.028125 | 23 |
| 24 | 0.0201 | 0.022 | 0.0230 | 0.055 | 0.055 | 0.151 | 0.025 | 24 |
| 25 | 0.0179 | 0.020 | 0.0204 | 0.0586 | 0.059 | 0.148 | 0.021875 | 25 |
| 26 | 0.01594 | 0.018 | 0.0181 | 0.0626 | 0.063 | 0.146 | 0.01875 | 26 |
| 27 | 0.014195 | 0.016 | 0.0173 | 0.0658 | 0.067 | 0.143 | 0.0171875 | 27 |
| 28 | 0.012641 | 0.014 | 0.0162 | 0.072 | 0.071 | 0.139 | 0.015625 | 28 |
| 29 | 0.011257 | 0.013 | 0.0150 | 0.076 | 0.075 | 0.134 | 0.0140625 | 29 |
| 30 | 0.010025 | 0.012 | 0.0140 | 0.080 | 0.080 | 0.127 | 0.0125 | 30 |
| 31 | 0.008928 | 0.010 | 0.0132 | | 0.085 | 0.120 | 0.0109375 | 31 |
| 32 | 0.00795 | 0.009 | 0.0128 | | 0.090 | 0.115 | 0.01015625 | 32 |
| 33 | 0.00708 | 0.008 | 0.0118 | | 0.095 | 0.112 | 0.009375 | 33 |
| 34 | 0.006304 | 0.007 | 0.0104 | | | 0.110 | 0.00859375 | 34 |
| 35 | 0.005614 | 0.005 | 0.0095 | | | 0.108 | 0.0078125 | 35 |
| 36 | 0.005 | 0.004 | 0.0090 | | | 0.106 | 0.00703125 | 36 |
| 37 | 0.004453 | | | | | 0.103 | 0.006640625 | 37 |
| 38 | 0.003965 | | | | | 0.101 | 0.00625 | 38 |
| 39 | 0.003531 | | | | | 0.099 | | 39 |
| 40 | 0.003144 | | | | | 0.097 | | 40 |

**FIGURE 11.7** Older gauging systems; some still in use in the United States.

### 11.5.1 Sheet-Metal Cutting Methods

- *Shearing:* The sheet-metal stock sheet is cut on a hydraulic-powered shear to its appropriate flat pattern or blank size.
- *Slitting:* The sheet-metal stock sheet is run through a slitting machine, where accurate widths are slit with slitting knives to various lengths. The edges are sharp and accurate, with excellent straightness along the edges.

**Aluminum sheet metal thicknesses and weights.**

| Standard thickness, in | Weight, lb/ft$^2$ |
|---|---|
| 0.010 | 0.141 |
| 0.016 | 0.226 |
| 0.020 | 0.282 |
| 0.025 | 0.353 |
| 0.032 | 0.452 |
| 0.040 | 0.564 |
| 0.050 | 0.706 |
| 0.063 | 0.889 |
| 0.071 | 1.002 |
| 0.080 | 1.129 |
| 0.090 | 1.270 |
| 0.100 | 1.411 |
| 0.125 | 1.764 |
| 0.160 | 2.258 |
| 0.190 | 2.681 |
| 0.250 | 3.528 |

Weight based on an average aluminum weight of 0.098 lb/in$^3$.

**FIGURE 11.8** Aluminum standard sheet-metal gauges.

- *Welding process cutting:* The sheet-metal stock sheet may be oxyacetylene-torch-cut, oxy-hydrogen-cut, oxygen-lance-cut, plasma-arc-cut, arc-cut, laser-beam-cut, or water-jet-cut. All these methods are used today to cut not only sheet metal but also bar stock, plates, structures and extruded shapes, and other materials, both metallic and nonmetallic.

## 11.5.2 Sheet-Metal Punching Methods

- *Turret punch press:* The sheet-metal stock sheet or blank is punched with various holes of different shapes by the punching dies contained on the punch press. Punch presses usually have a revolving turret which contains punches of different sizes and shapes, and these are interchangeable. The modern multistation punch presses are often CNC-controlled and high-speed.
- *Die punching:* The sheet-metal blank is placed in a punching die, where a pattern of holes is punched simultaneously with one stroke of the punching press. The high-tonnage brake press is often used to provide the power stroke required on the punching die block.
- *Strippit punching:* The sheet-metal blank is punched one hole at a time on a Strippit punch press. The punching dies may be quickly changed for punching different sizes or shapes of holes, such as round, square, rectangular, oval, or obround or other special shapes for which the dies are designed.

## 11.5.3 Sheet-Metal Forming Methods

***Press Brake.*** This type of machine tool is found in every sheet-metal department and is used to bend flanges, hems, and other special shapes. Figure 11.9 shows the various bending abilities of the press brake when equipped with the proper tooling.

***Die Bending, Forming, and Molding.*** Hard dies are produced for making bends and molds, and forming and drawing sheet-metal parts. The forming or drawing dies may be all metal or a combination of metal and neoprene pads, which force the metal against the die

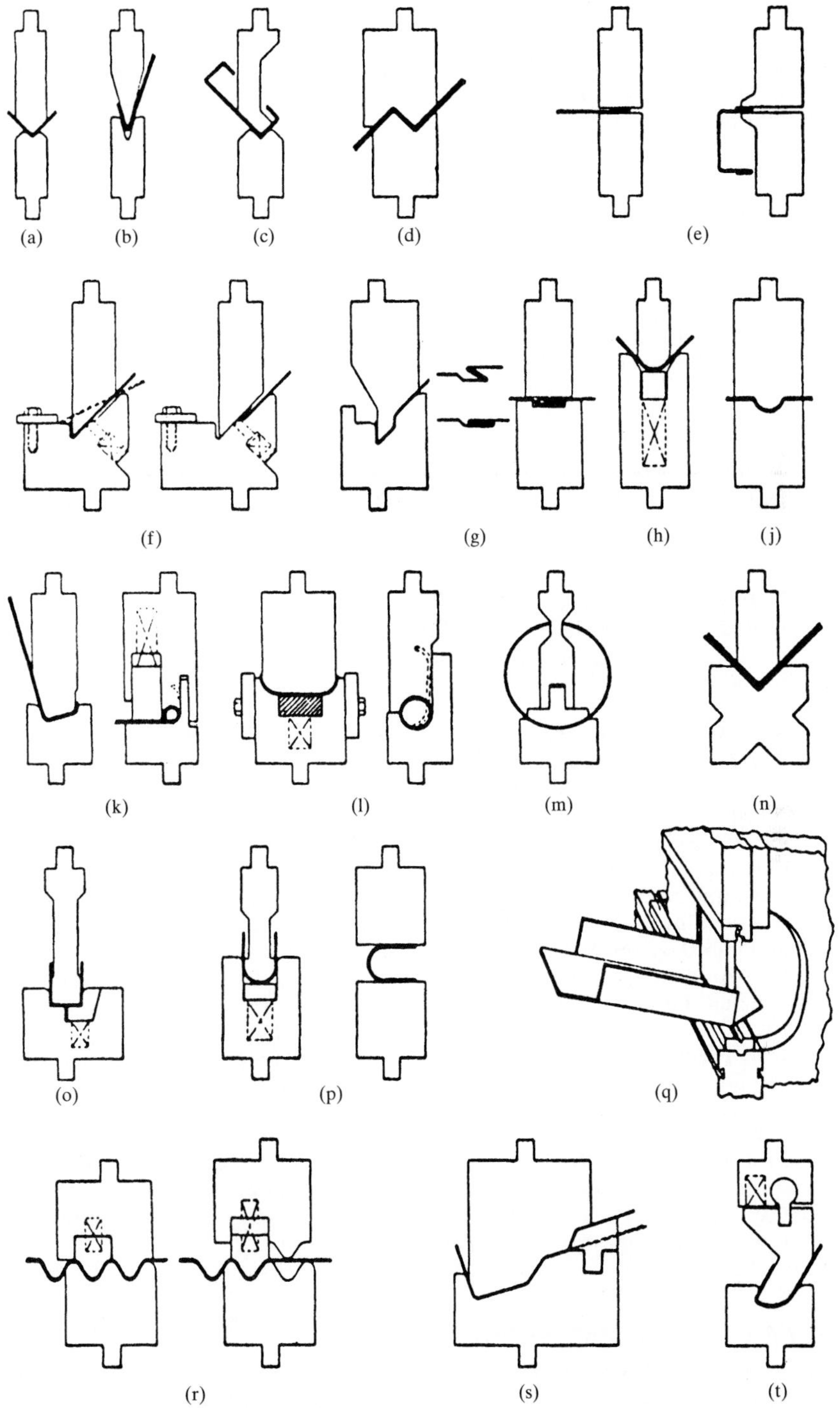


**FIGURE 11.9** Bending capabilities of the press brake when used with proper tooling accessories.

block or male form. This process is widely used in the aerospace industry, where aluminum sheet-metal parts are formed on hydropresses (*Marforming* is a hydropressing process originated at the Martin Company, Middle River, Maryland). Large lead alloy form blocks are also used in the aircraft industry to form large, compound curved surfaces in sheet metal, generally aluminum and magnesium alloys.

***Yoder Hammering.*** This is a specialized metal-forming operation, where a rapidly moving set of vertical forming hammers of various shapes are used to form special surfaces on sheet-metal sections in a hammering process. This is a manual operation which requires operator skill and practice and is relatively rare today.

***Spin Forming.*** In this sheet-metal forming process, a sheet-metal disk is rotated in a special type of lathe tool, while a special forming tool is pressed against the rotating disk of metal in a rotary swaging operation, thus forming or spin drawing the part to the required shape against its forming pattern. The process is limited to metal sections which are of a symmetrical, rotated section, such as bell shapes, cones, parabolic sections, and cylinders.

***Explosive Forming.*** In this process of metal forming, a shaped charge of explosive is suspended above a sheet-metal flat pattern, both of which are submerged under water. When the explosive charge is detonated, the shock waves from the explosion exert a hydrostatic pressure against the sheet metal, forcing it against a forming die block almost instantaneously. The sheet-metal part conforms to the shape of the die block. Very complex sheet-metal parts may be formed with this process. Some companies in the United States are devoted entirely to this specialized form of sheet-metal fabrication. This process dates back to the 1950s, when it was used by the Glen L. Martin Company, Middle River, Maryland, in their aircraft manufacturing facilities.

***Stretcher Bending.*** This process for sheet-metal forming uses a machine known as a *stretcher bender,* wherein a sheet-metal-formed section such as an angle, channel, or Z (zee) section is pulled against a radiused die block to accurately form structural frame sections to a specified radius of curvature. The frame structures of rockets and aerospace vehicles which have a single radius of curvature are formed on the stretcher-bending machine. This is the only cost-effective method known for producing such sheet-metal frame parts quickly and accurately on a relatively simple machine. In this operation, the straight frame section which was press-brake-bent is held in a set of grippers at each end of the part. The part is stretched and pulled against the radiused die block simultaneously by the gripper arms, thus forming the part to the specified radius. Allowance is made for springback of the formed part by overbending and then allowing the metal to spring back or return to the correct form.

***Roll Forming.*** In the roll-forming process, a flat strip of sheet metal is fed into the roll-forming machine, which has a series of rolling dies whose shape gradually changes as the metal is being fed past each stage of rolls, until the final roll-formed section is completed. The number of different cross sections of roll-formed sheet-metal parts is limitless. The roll-formed part is usually made to a specific length or stock length, or may be produced to any special length required. Figure 11.10*a* and 11.10*b* shows sample pages of roll-formed sections taken from the Dahlstrom catalog of molded and rolled sections, Dahlstrom Manufacturing Corporation, Jamestown, New York.

## 11.6 SHEET-METAL FLAT PATTERNS

The correct determination of the flat-pattern dimensions of a sheet-metal part which is formed or bent is of prime importance to sheet-metal workers, designers, and design drafters.

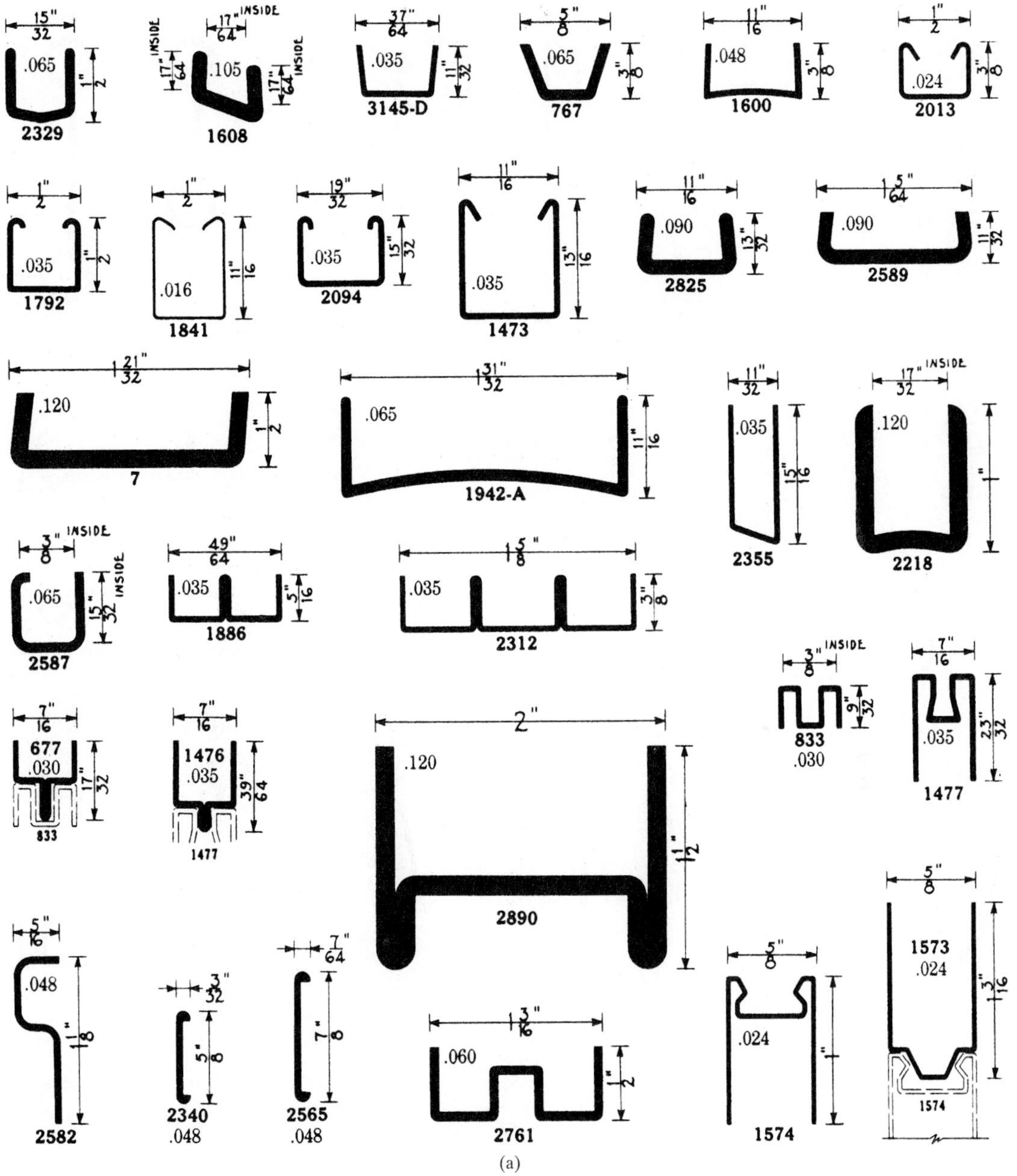


(a)

**FIGURE 11.10** (*a*) Roll-formed shapes sample catalog sheet.

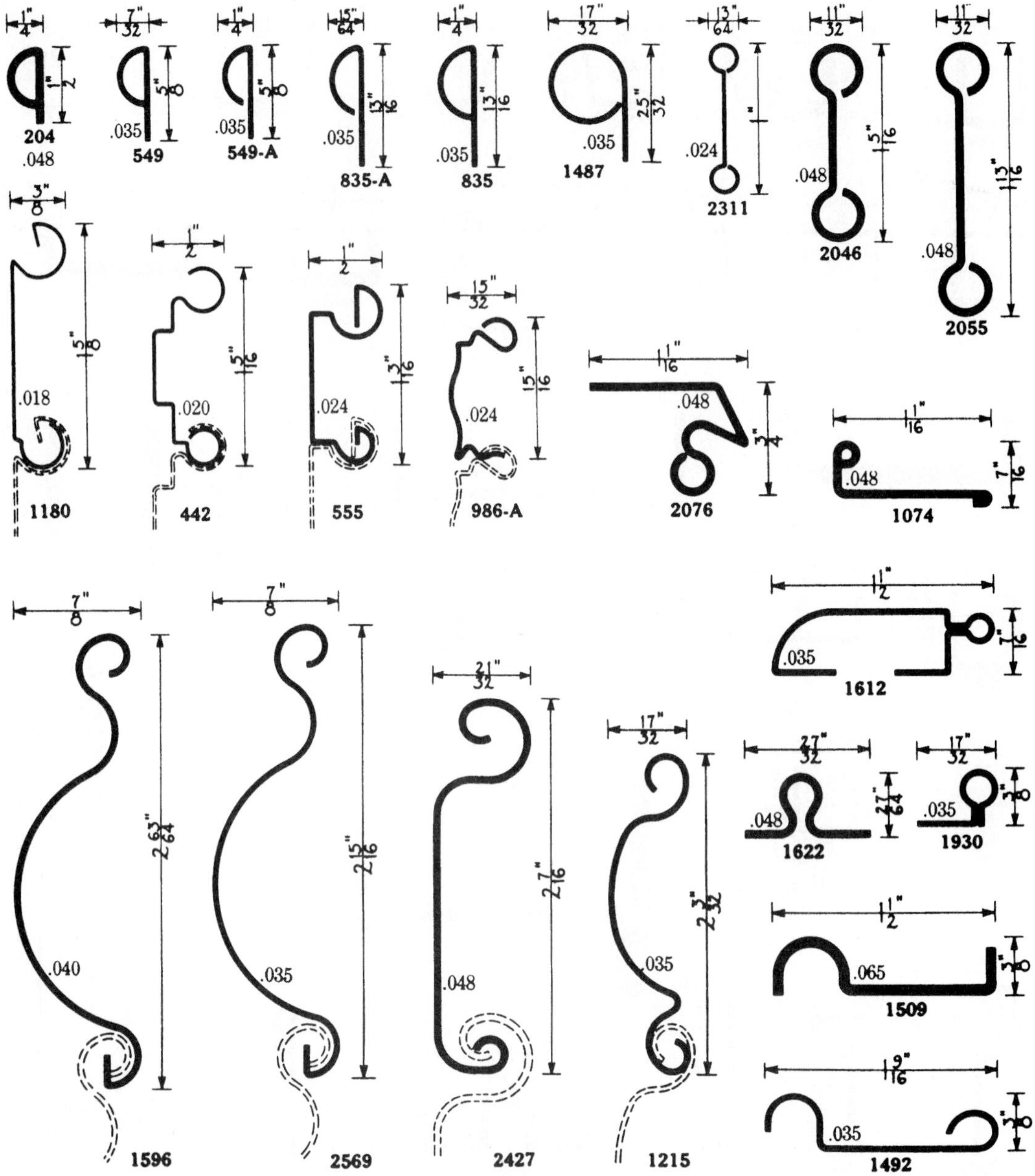


(b)

**FIGURE 11.10** (*b*) Other available roll-formed shapes.

There are three methods for performing the calculations to determine flat patterns which are considered as normal practice (the method chosen can also determine the accuracy of the results):

- By bend deduction (BD) or setback
- By bend allowance (BA)
- By inside dimensions [inside mold line (IML)] for sharply bent parts only

There are also other methods in use for calculating the flat-pattern length of sheet-metal parts. Some take into consideration the ductility of the material, and others are based on extensive experimental data for determining BA. The methods included in this section are accurate when the bend radius has been properly selected for each particular gauge and condition of the material. When the proper bend radius is selected, there is no stretching of the *neutral axis* within the part (the neutral axis is generally accepted as being located 0.445 × material thickness away from the IML). See Fig. 11.12 (below).

***Methods of Determining Flat Patterns.*** Refer to Fig. 11.11.

*Method 1:* by bend deduction or setback

$$L = a + b - (\text{setback})$$

*Method 2:* by bend allowance

$$L = a' + b' + c$$

where $c$ = bend allowance or length along neutral axis.

*Method 3:* by inside dimensions or IML

$$L = (a - T) + (b - T)$$

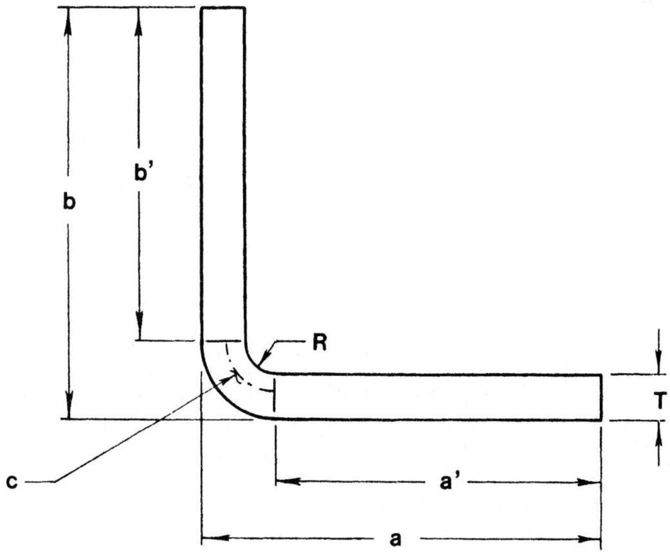


**FIGURE 11.11** Sheet-metal angle; determining the flat pattern.

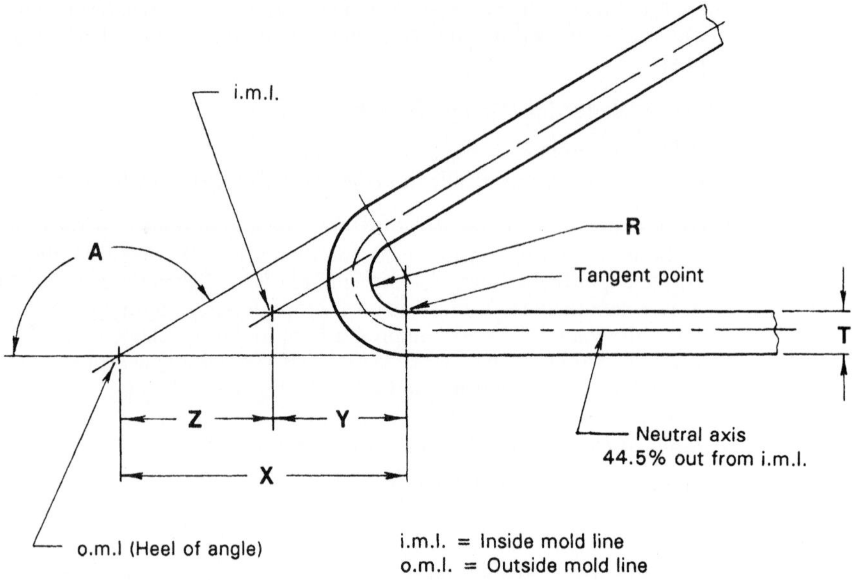


**FIGURE 11.12** Geometry of a sheet-metal angle for flat pattern calculations.

The calculation of bend allowance (BA) and bend deduction (BD) (setback) is keyed to Fig. 11.12 and is as follows:

$$\text{BA} = A(0.01745R + 0.00778T)$$

$$\text{BD} = \left(2\tan\frac{1}{2}A\right)(R+T) - (\text{BA})$$

$$X = \left(\tan\frac{1}{2}A\right)(R+T) \qquad Z = T\left(\tan\frac{1}{2}A\right)$$

$$Y = X - Z \quad \text{or} \quad R\left(\tan\frac{1}{2}A\right)$$

On "open" angles that are bent less than 90° (see Fig. 11.13):

$$X = \left(\tan\frac{1}{2}A\right)(R+T)$$

The method used to calculate the sheet-metal flat pattern may be determined by designer option, company standards, order of accuracy, and method of manufacture. For soft steel (1010, etc.) the inside dimension method (method 3, described previously) is used whenever the material may be bent with a sharp or minimal inside bend radius (0.062 in or less). The inside bend method is accurate enough for all gauges up to and including 0.375-in-thick stock where the tolerance of bent parts is ±0.032 in. On stock thicknesses from 7-gauge to 0.375 in, 0.062 in is added to the sum of the inside bend dimensions and divided across the bend, with 0.032 in going into each leg or flange inside dimension. The 0.062-in allowance is added for each 90° bend in the part. This method is popular in industries such as the electrical switchgear and appliance industries, where great accuracy is not required.

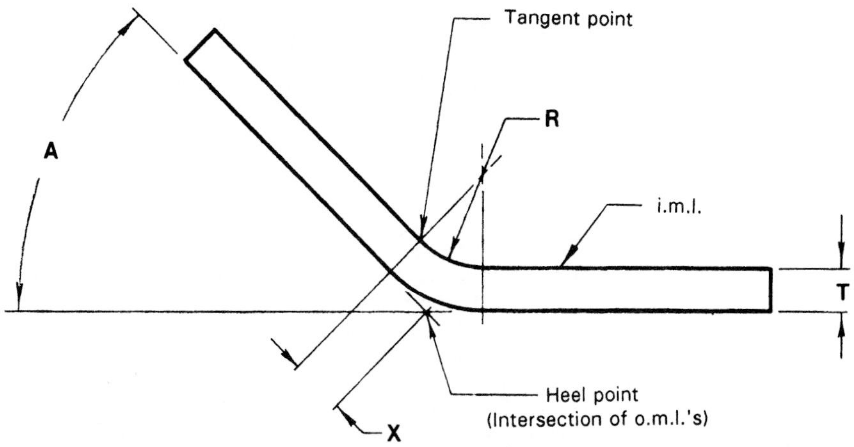


**FIGURE 11.13** An open sheet-metal angle and its geometry.

For recommended bend radii on various materials and gauges, see Sec. 11.6.2. Also, your company may have the recommended bend radii listed in the design manuals used in the design and tool engineering departments. All aerospace companies and automakers have this information as part of the company design standards. Many companies develop their own bend radii through testing and experimentation.

For very accurate flat pattern dimensions, intended for aerospace vehicles, automotive work, appliances, and other consumer products, the bend deduction or setback method is used in lofting procedures and standard sheet-metal tooling drawings. The tooling department is usually responsible for generating the flat pattern drawings of parts produced with stamping dies, punching dies, and drawing dies. The engineering department is usually responsible for generating the lofting drawings for flat patterns of regular and irregular shapes. In *lofting,* the part is drawn very accurately in flat pattern on flat metal sheets with specially prepared surfaces or directly onto heavy Mylar drafting film. The loft or drawing is then photographed and the pattern transferred to another metal sheet in full scale.

The part is then accurately cut out and becomes a master pattern. A stack of sheets can then be pin-routed or tracer-milled using the master template as a guide or jig. The cutout parts are then sent to the forming dies or the brakes for the final bending or forming operations. With modern CNC equipment, the part outline may be programmed and then automatically cut on the appropriate machine prior to the bending or forming operations.

When the sheet-metal part is to be press-brake-bent only, on radiused bending dies, the bend allowance method can be used to calculate the flat pattern whenever accuracy and a specific inside bend radius are required. Bend deduction or setback may also be used in this case.

### 11.6.1 Setback or J Chart for Determining Bend Deductions

Figure 11.14*a* shows a form of BD or setback chart known as a *J chart.* You may use this chart to determine BD or setback when the angle of bend, material thickness, and inside bend radius are known. The chart shown in the figure shows a sample line running from the top to the bottom and drawn through the $\frac{3}{16}$-in radius and the material thickness of 0.075 in. For a 90° bend, read across from the right to where the line intersects the closest curved line in the body of the chart. In this case, it can be seen that the line intersects the curve whose value is 0.18. This value is then the required setback or BD for a bend of 90° in a part whose thickness is 0.075 in, with an inside bend radius of $\frac{3}{16}$ in. If we check this setback or bend deduction

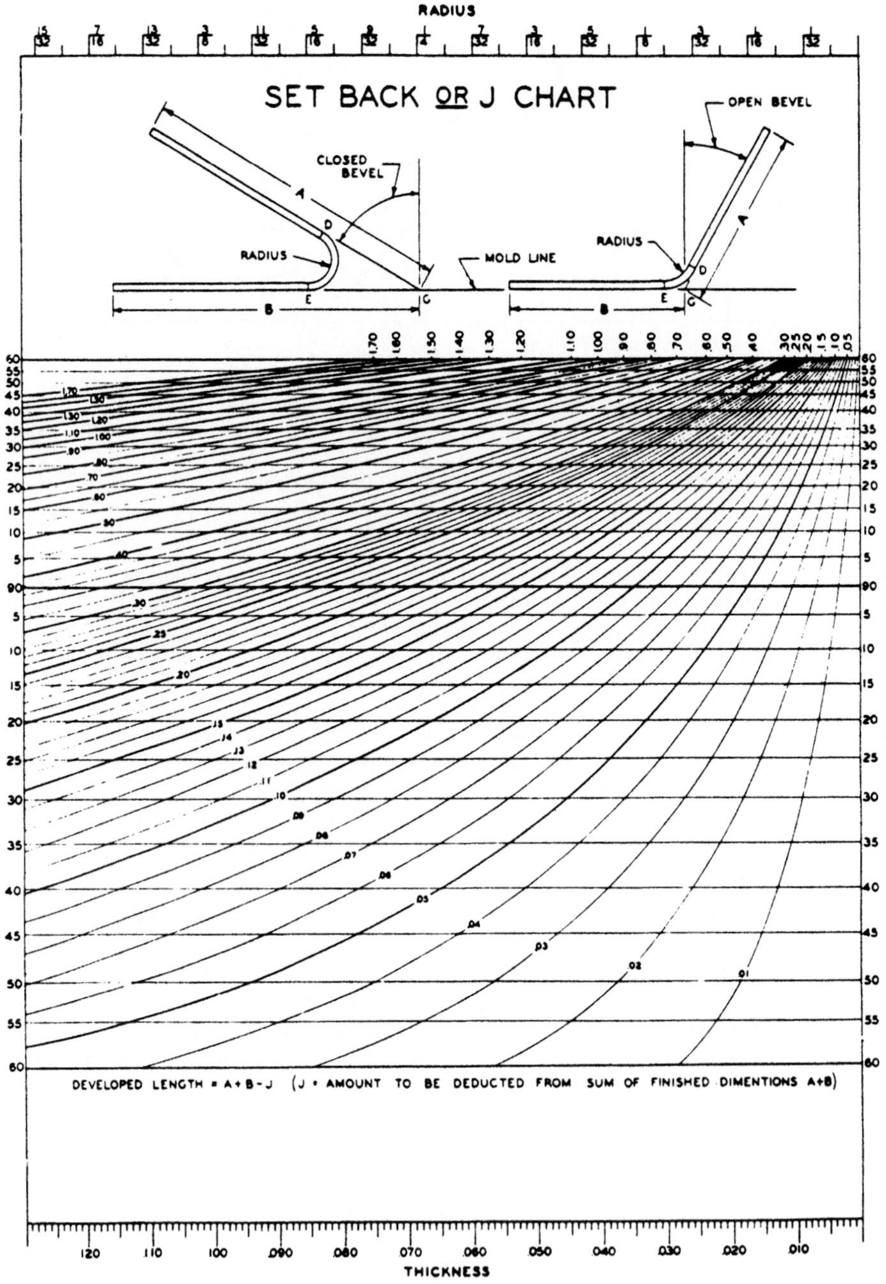


(a)

**FIGURE 11.14** (*a*) Setback or J chart for determining bend deduction or setback.

value using the appropriate equations shown previously, we can check the value given by the J chart.

*Checking.* Bend deduction (BD) or setback is given as

$$\text{BD or setback} = \left(2 \tan \frac{1}{2} A\right)(R + T) - \text{BA}$$

We must first find the bend allowance from

$$\text{BA} = A\ (0.01745R + 0.00778T)$$
$$= 90\ (0.01745 \times 0.1875 + 0.00778 \times 0.075)$$
$$= 90\ (0.003855)$$
$$= 0.34695$$

Now, substituting the bend allowance of 0.34695 into the BD equation yields

$$\text{BD or setback} = \left(2 \tan \frac{1}{2}\ (90)\right)(0.1875 + 0.075) - 0.34695$$
$$= (2 \times 1)\ (0.2625) - 0.34695$$
$$= 0.525 - 0.34695$$
$$\text{BD} = 0.178 \text{ or } 0.18 \text{ as shown in Fig. 11.14}$$

The J chart of Fig. 11.14 is then an important tool for determining the bend deduction or setback of sheet-metal flat patterns without recourse to tedious calculations. The accuracy of this chart has been shown to be of a high order. This chart and the equations were developed after extensive experimentation and practical working experience in the aerospace industry.

## 11.6.2 Bend Radii for Aluminum Alloy and Steel Sheet (Average)

Figures 11.14*b* and 11.14*c* show average bend radii for various aluminum alloys and steel sheets. For other bend radii in different materials and gauges, refer to Sec. 9.5.2 of the *McGraw-Hill Machining and Metalworking Handbook.*

| | Steel designation | |
|---|---|---|
| Material Gauge | 1020 | 302-303-304 |
| 0.010 | 1/32 | 1/32 |
| 0.020 | 1/32 | 1/32 |
| 0.030 | 1/32 | 1/32 |
| 0.040 | 1/32 | 1/32 |
| 0.050 | 1/32 | 1/32 |
| 0.060 | 1/32 | 1/16 |
| 0.070 | 1/32 | 1/16 |
| 0.080 | 1/32 | 1/16 |
| 0.090 | 1/16 | 1/16 |
| 0.120 | 1/16 | 1/8 |
| 0.190 | 1/8 | 1/4 |
| 0.250 | 1/8 | 1/4 |

(b)

**FIGURE 11.14** (*b*) Bend radii for steel sheets.

| Material Gauge | Aluminum designation<br>6061-T6<br>5052-H36<br>1100-H18 | 5052-H22<br>3003-H14 | 2024-T3 |
|---|---|---|---|
| 0.010 | 1/16 | 1/32 | 1/16 |
| 0.020 | 1/16 | 1/32 | 1/16 |
| 0.030 | 1/16 | 1/32 | 1/8 |
| 0.040 | 1/8 | 1/32 | 1/4 |
| 0.050 | 1/8 | 1/32 | 1/4 |
| 0.060 | 1/8 | 1/32 | 1/4 |
| 0.070 | 1/4 | 1/16 | 1/4 |
| 0.080 | 1/4 | 1/32 | 3/8 |
| 0.090 | 3/8 | 1/8 | 3/8 |
| 0.120 | 3/8 | 1/8 | 1/2 |
| 0.190 | 3/4 | 1/4 | 3/4 |
| 0.250 | 1 | 1/2 | 1 |

(c)

**FIGURE 11.14** (*c*) Bend radii for aluminum sheets.

## 11.7 SHEET-METAL DEVELOPMENTS AND TRANSITIONS

The layout of sheet metal as required in "development and transition" parts is an important phase of sheet-metal design and practice. The methods included here will prove useful in many design and working applications. These methods have application in ductwork, aerospace vehicles, automotive equipment, and other areas of product design and development requiring the use of transitions and developments.

When sheet metal is to be formed into a curved section, it may be laid out, or *developed* with reasonable accuracy by *triangulation,* if it forms a simple curved surface, without compound curves or curves in multiple directions. Sheet-metal curved sections are found on many products, and if a straightedge can be placed flat against *elements* of the curved section, accurate layout or development is possible using the methods shown in this section.

On double-curved surfaces such as those found on automobile and truck bodies and aircraft, forming dies are created from a full-scale model in order to duplicate these compound curved surfaces in sheet metal. The full-scale models used in aerospace vehicle manufacturing facilities are commonly called *mockups,* and the models used to transfer the compound curved surfaces are made by toolmakers in the tooling department.

### 11.7.1 Skin Development

Skin development on aerospace vehicles or other applications may be accomplished by triangulation when the surface is not double-curved. In Fig. 11.15 we see a side view of the nose section of a simple aircraft.

If we wish to develop the outer skin or sheet metal between stations 20.00 and 50.00, the general procedure is as follows.

The "master lines" of the curves at stations 20.00 and 50.00 must be determined. In actual practice, the curves are developed by the master-lines engineering group of the company, or you may know or develop your own curves. The procedure for layout of the flat pattern is as follows (see also Fig. 11.16):

1. Divide curve *A* into a number of equally spaced points. Use the spline lengths (arc distances), *not* chordal distances.
2. Lay an accurate triangle tangent to one point on curve *A,* and by parallel action, transfer the edge of the triangle back to curve *B* and mark a point where the edge of the triangle is

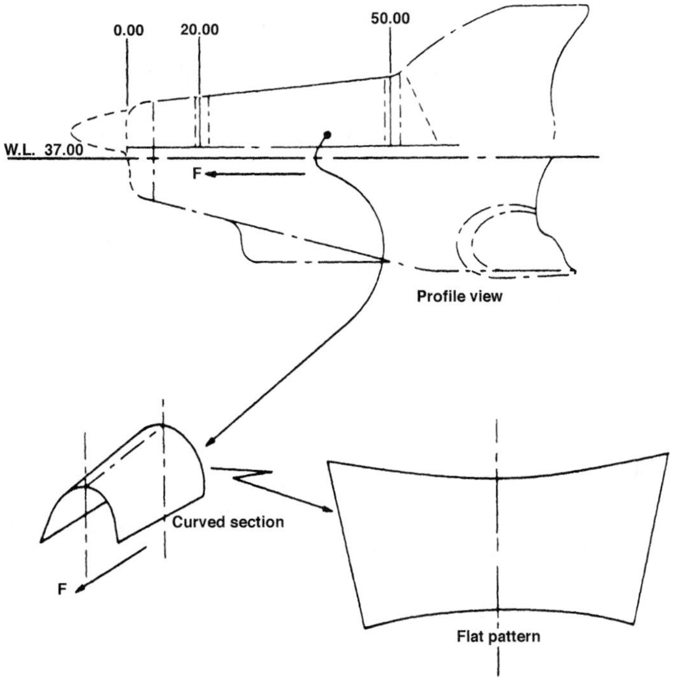


**FIGURE 11.15** Outer-skin development on an aircraft.

tangent to curve *B* (e.g., point *b* on curve *A* back to point *h* on curve *B;* see Fig. 11.16). Then parallel-transfer all points on curve *A* back to curve *B* and label all points for identification. Draw the element lines and diagonals on the frontal view: 1*A*, 2*B*, 3*C*, etc. (Fig. 11.16*b*).

3. Construct a true-length diagram as shown in Fig. 11.16*a*, where all the element and diagonal true lengths can be found (elements are 1, 2, 3, 4, etc; diagonals are *A, B, C, D,* etc.). The true distance between the two curves is 30.00 (50.00 – 20.00 from Fig. 11.15).
4. Transfer the element and diagonal true-length lines to the triangulation flat-pattern layout as shown in Fig. 11.16*c*. The triangulated flat pattern is completed by transferring all elements and diagonals to the flat-pattern layout.

### 11.7.2 Canted-Station Skin Development (Bulkheads at an Angle to Axis)

When the planes of curves *A* and *B* (Fig. 11.17) are *not* perpendicular to the axis of the curved section, layout procedures to determine the true lengths of the element and diagonal lines are as shown in Fig. 11.17. The remainder of the procedure is as explained in Sec. 11.7.1 to develop the triangulated flat pattern.

In aerospace terminology, the locations of points on the craft are determined by station, waterline, and buttline. These terms are defined as follows:

*Station:* The numbered locations from the front to the rear of the craft.

*Waterline:* The vertical locations from the lowest point to the highest point of the craft.

*Buttline:* The lateral locations from the centerline of the axis of the craft to the right and to the left of the axis of the craft. There are right buttlines and left buttlines.

With these three axes, any exact point on the craft may be described or dimensioned.

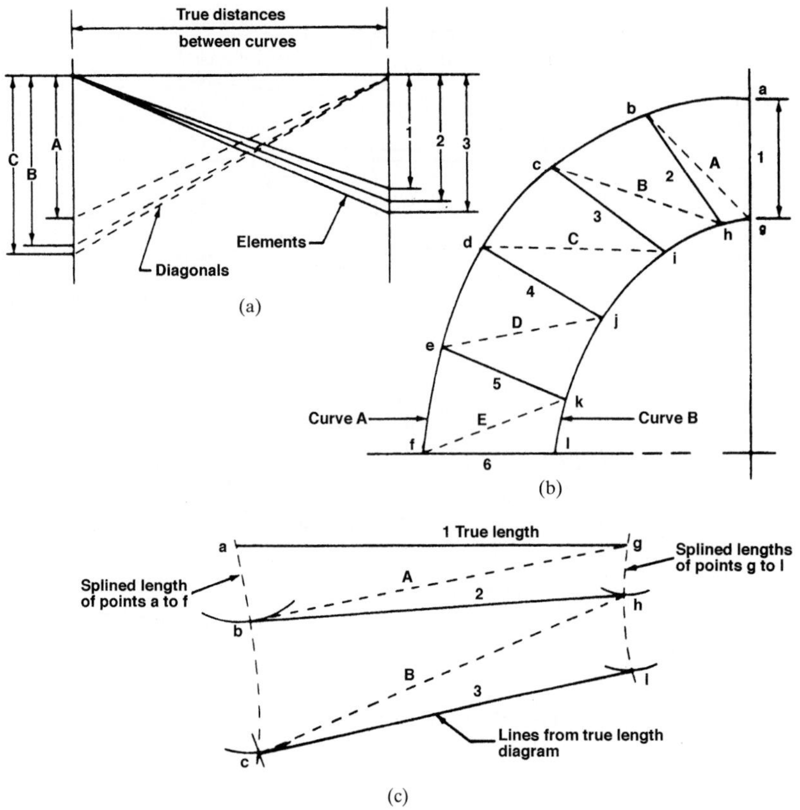


**FIGURE 11.16** Layout of the flat pattern for the aircraft skin shown in Fig. 11.15: (*a*) true-length diagram; (*b*) frontal view; (*c*) triangulation of skin in flat pattern.

## 11.8 DEVELOPING FLAT PATTERNS

Developing flat patterns can be done by bend deduction or setback. Figure 11.18 shows a type of sheet-metal part that may be bent on the press brake. The flat pattern part is bent on the brake, with the center of bend line (CBL) held on the bending die centerline. The machine's back gauge is set by the operator in order to form the part. If you study the figure closely, you can see how the dimensions progress, the bend deduction is drawn in, and the next dimension taken from the end of the first bend deduction. The next dimension is then measured, the bend deduction drawn in for that bend, and then the next dimension taken from the end of the second bend deduction, and so on. Notice that the second bend deduction is larger due to the larger radius of the second bend (0.16*R*).

## 11.9 STIFFENING SHEET-METAL PARTS

On many sheet-metal parts which have large areas, stiffening can be achieved by creasing the metal in an X configuration by means of brake bending. On certain parts where great stiffness

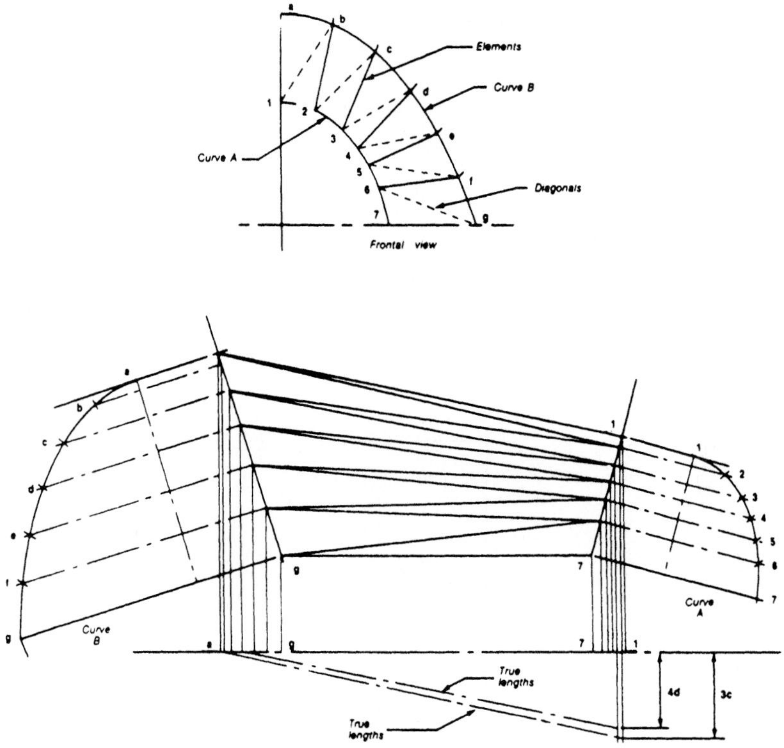


**FIGURE 11.17** Canted-station skin development.

and rigidity are required, the method called *beading* is employed. The beading is carried out at the same time the part is being hydropressed, Marformed, or hard-die-formed (see Fig. 11.19*b*).

Another method for stiffening the edge of a long sheet-metal part is to hem or *Dutch-bend* the edge as shown in Fig. 11.19*a*. In aerospace and automotive sheet-metal parts, flanged *lightening holes* are used as shown in Fig. 11.19*c*. The lightening hole makes the part not only lighter in weight but also more rigid. This method is commonly used in wing ribs, airframes, and gussets or brackets. The lightening hole need not be circular, but can take any convenient shape as required by the application.

## 11.10 SHEET-METAL FAYING SURFACES

*Faying surfaces* are surfaces where two sheet-metal parts come into contact with one another, such as the joining of a flange with a web section, or two flanges. All faying surfaces should be primer-painted or finished in some manner to prevent corrosion. A standard finish for many applications is zinc chromate primer.

Other finishes include zinc plating, nickel plating, cadmium plating, and chrome plating. As cadmium plating and plating solutions are toxic, zinc plating is used in its place, where the application allows this. New designs should specify zinc or nickel in place of cadmium, unless there is a specific technical reason for using cadmium.

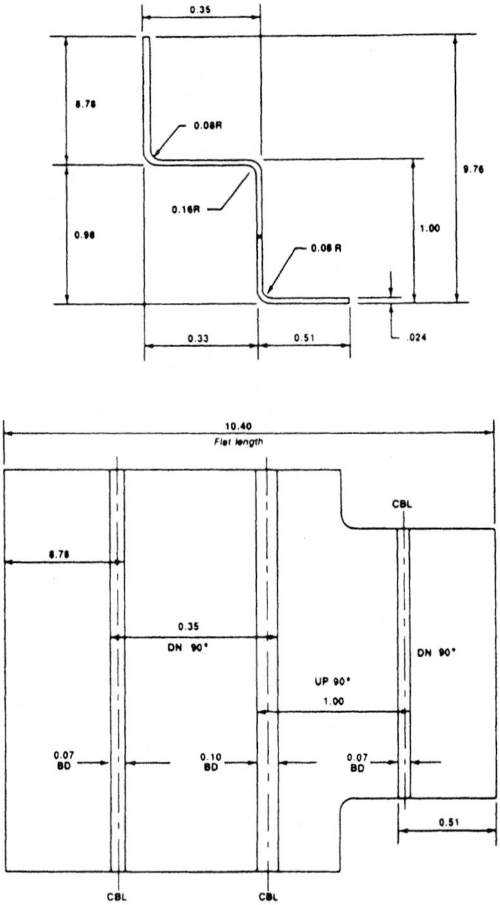


**FIGURE 11.18** Development of a sheet-metal flat pattern for accurate bending.

## 11.11 DESIGN POINTS FOR SHEET-METAL PARTS

- Do not specify flanges that are too narrow for the type of bending operation. That is, do not design a part with a 0.375-in flange width, unless you have the equipment or dies to produce such a short flange on the press brake or other type of machine. Bottoming dies can enable the bending of such a flange in very light gauges up to 11.
- Do not specify gauges that are heavier than necessary for the function of the part. An exception to this rule is code specifications such as for electrical switchgear equipment, where a minimum of 11-gauge steel is specified for certain sections of the equipment, whether it is structurally required or not. The heavy-gauge limit in this application is to prevent the spread of fires between adjacent units if there is an electrical fault.
- Use proper edge distances for hardware components, keeping the flanges as narrow as practical in respect to their required rigidity or strength.

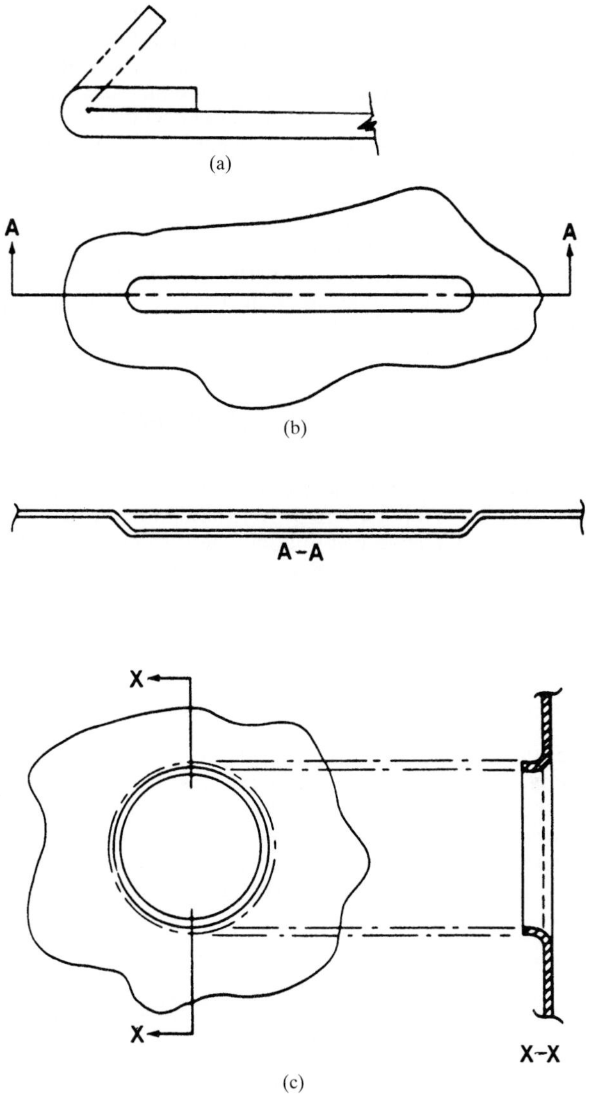


**FIGURE 11.19** (*a*) Sheet-metal Dutch bend or hem; (*b*) stiffening beads for sheet metal; (*c*) lightening holes.

- Do not design a part which is impractical to bend on the type of machinery with which your operation is equipped.
- Keep brake-formed parts in a size range where the parts can be handled manually by the brake operators, unless your operation is equipped with automatic machinery or other special handling equipment.
- When using hot-rolled steel sheets, use tolerances on the parts that are functionally related to the equipment used in your operation. That is, do not expect to hold dimensions to $\pm\frac{1}{64}$ in when you are producing 11-gauge hot-rolled sheet-metal parts on a press brake. Normal

shop tolerance on general sheet-metal parts is usually ±1⁄32 in for parts under 3 ft in length in both directions. One reason why hot-rolled steel sheet-metal parts between 16 and 7 gauge need a generous tolerance is that the steel sheet-metal thickness varies from sheet to sheet. This variation can reach ±½ a gauge step on a batch of steel of the same gauge. The variation of the gauge causes the flanges to be bent to different outside dimensions even though the back-gauge dimension is the same and is very apparent when more than two bends are made on the same part.

In high speed, mass-production operations there is no remedy for this condition of variable thickness within the same gauge hot-rolled steel sheet material, except a favorable tolerance spread on the dimensions of the parts. Some of the larger companies specify the gauge variation limits when they order steel sheet metal, but the mill order must be large in order to do this. These problems do not occur when using gauge-accurate cold-rolled steel sheets.

## 11.12 TYPICAL TRANSITIONS AND DEVELOPMENTS

The following transitions and developments are the most common types, and learning or using them for reference will prove useful in many industrial applications. Using the principles shown will enable you to apply these to many different variations or geometric forms. The principles shown and described in Secs. 11.7 (introduction), 11.7.1, and 11.7.2 will prove helpful in trying to understand and put to practice the transitions and developments shown in this section. The construction of the true-length diagrams is of particular importance.

### 11.12.1 Development of a Truncated Right Pyramid

Refer to Fig. 11.20. Draw the projections of the pyramid that show a normal view of (1) the base or right section and (2) the axis. Lay out the pattern for the pyramid, and then superimpose the pattern to the truncation.

Since this is a portion of a right regular pyramid, the lateral edges are all of equal length. The lateral edges *OA* and *OD* are parallel to the frontal plane and consequently show in their true length on the front view. With the center at $O_1$, taken at any convenient place, and a

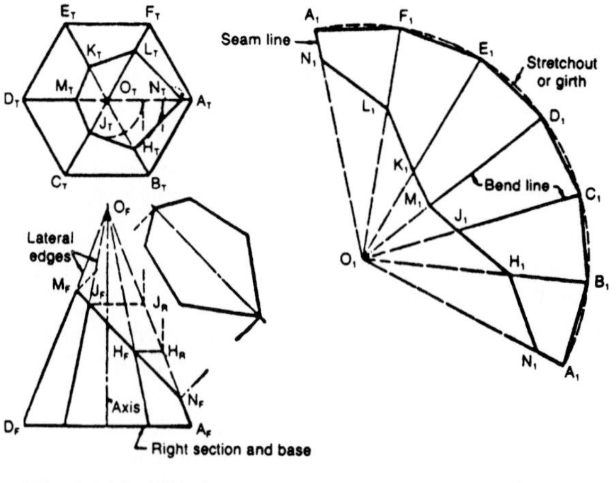


**FIGURE 11.20** Development of a truncated right pyramid.

radius $O_FA_F$, draw an arc that is the stretchout of the pattern. On it, step off the six equal sides of the hexagonal base, obtained from the top view, and connect these points successively with each other and with the vertex $O_1$, thus forming the pattern for the pyramid.

The intersection of the cutting plane and lateral surfaces is developed by laying off the true length of the intercept of each lateral edge on the corresponding line of the development. The true length of each of these intercepts, such as *OH, OJ,* and so on, is found by rotating it about the axis of the pyramid until it coincides with $O_FA_F$ as previously explained. The path of any point, as *H,* will be projected on the front view as a horizontal line. To obtain the development of the entire surface of the truncated pyramid, attach the base; also find the true size of the cut face and attach it on a common line.

### 11.12.2 To Develop an Oblique Pyramid

Refer to Fig. 11.21. Since the lateral edges are unequal in length, the true length of each must be found separately by rotating it parallel to the frontal plane. With $O_1$ taken at any convenient place, lay off the seam line $O_1A_1$ equal to $O_FA_R$. With $A_1$ as center and radius $O_1B_1$ equal to $O_FB_R$, describe a second arc intersecting the first in vertex $B_1$. Connect the vertices $O_1$, $A_1$, $B_1$, thus forming the pattern for the lateral surface *OAB.* Similarly, lay out the pattern for the remaining three lateral surfaces, joining them on their common edges. The stretchout is equal to the summation of the base edges. If the complete development is required, attach the base on a common line.

### 11.12.3 To Develop a Truncated Right Cylinder

Refer to Fig. 11.22. The development of a cylinder is similar to that of a prism. Draw two projections of the cylinder:

1. A normal view of a right section
2. A normal view of the elements

In rolling the cylinder out on a tangent plane, the base or right section, which is perpendicular to the axis, will develop into a straight line. For convenience in drawing, divide the normal

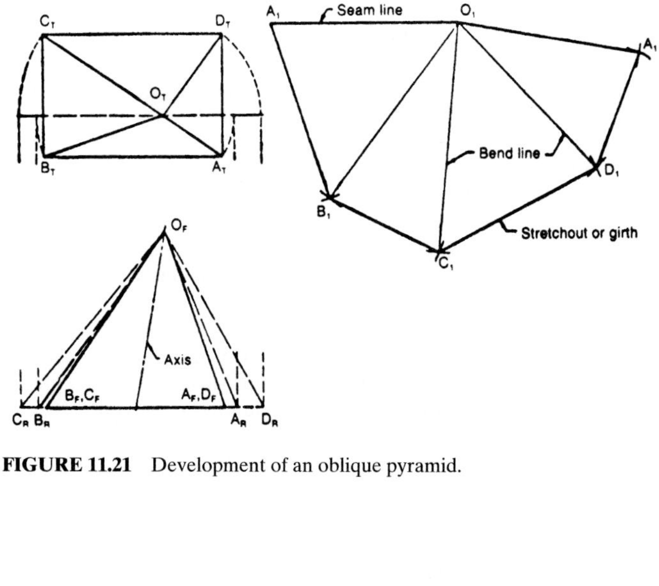


**FIGURE 11.21** Development of an oblique pyramid.

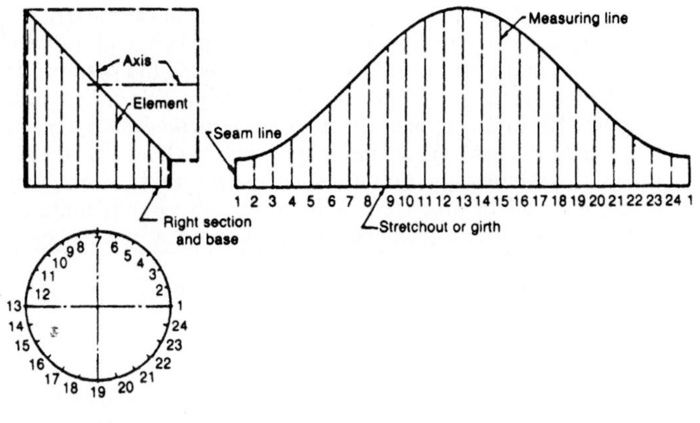


**FIGURE 11.22** Development of a truncated right cylinder.

view of the base, shown in Fig. 11.22 in the bottom view, into a number of equal parts by points that represent elements. These divisions should be spaced so that the chordal distances approximate the arc closely enough to make the stretchout practically equal to the periphery of the base or right section.

Project these elements to the front view. Draw the stretchout and measuring lines; the cylinder is now treated as a multiside prism. Transfer the lengths of the elements in order, either by projection or by using dividers, and join the points thus found by a smooth curve. Sketch the curve in very lightly, freehand, before fitting the french curve or ship's curve to it. This development might be the pattern for one-half of a two-piece elbow.

Three-piece, four-piece, or five-piece elbows may be drawn similarly, as illustrated in Fig. 11.23. As the base is symmetrical, only one-half of it need be drawn. In these cases, the intermediate pieces as *B*, *C*, and *D* are developed on a stretchout line formed by laying off the perimeter of a right section. If the right section is taken through the middle of the piece, the stretchout line becomes the center of the development.

Evidently, any elbow could be cut from a single sheet without waste, if the seams were made alternately on the long and short sides.

### 11.12.4 To Develop a Truncated Right Circular Cone

Refer to Fig. 11.24. Draw the projection of the cone that will show a normal view of (1) the base or right section and (2) the axis. First, develop the surface of the complete cone and then superimpose the pattern for the truncation.

Divide the top view of the base into a sufficient number of equal parts, so that the sum of the resulting chordal distances will closely approximate the periphery of the base. Project these points to the front view and draw front views of the elements through them. With center $A_1$ and a radius equal to the slant height $A_F I_F$, which is the true length of all the elements, draw an arc, which is the stretchout. Lay off on it the chordal divisions of the base, obtained from the top view. Connect these points 2, 3, 4, 5, and so forth with $A_1$, thus forming the pattern for the cone.

Find the true length of each element from vertex to cutting plane by rotating it to coincide with the contour element $A_1$, and lay off this distance on the corresponding line of the development. Draw a smooth curve through these points. The pattern for the cut surface is obtained from the auxiliary view.

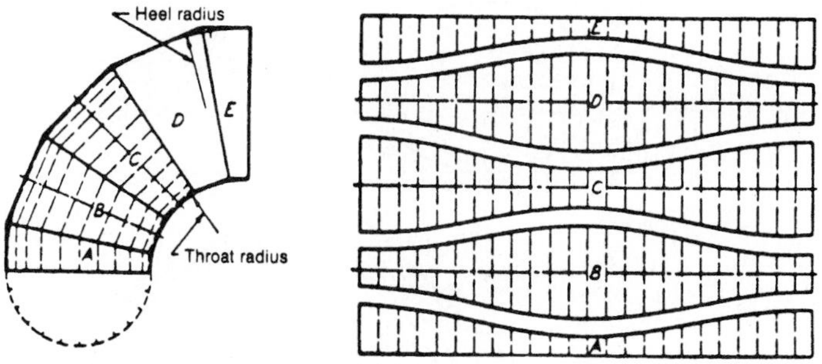


**FIGURE 11.23** Development of elbows.

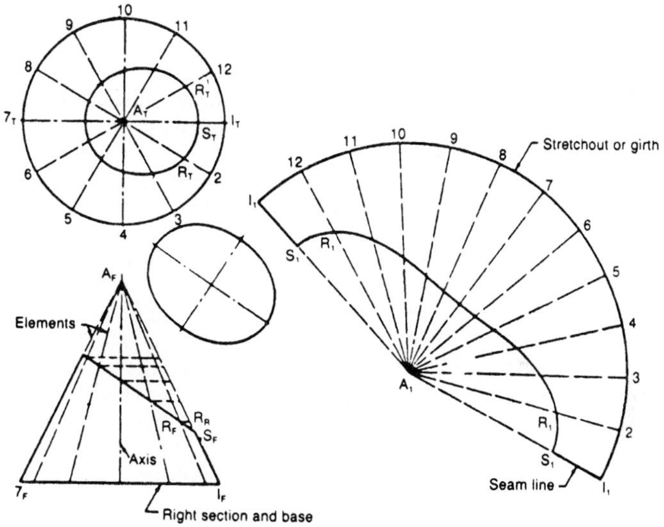


**FIGURE 11.24** Development of a truncated right circular cone.

***Triangulation.*** Nondevelopable surfaces are developed approximately by assuming them to be made of narrow sections of developable surfaces. The commonest and best method for approximate development is triangulation; that is, the surface is assumed to be made up of a large number of triangular strips or plane triangles with very short bases. This method is used for all warped surfaces and also for oblique cones. Oblique cones are single-curved surfaces and are capable of true theoretical development, but can be developed much more easily and accurately by triangulation.

## 11.12.5 To Develop an Oblique Cone

Refer to Fig. 11.25. An oblique cone differs from a cone of revolution in that the elements are all of different lengths. The development of a right circular cone is made up of a number of

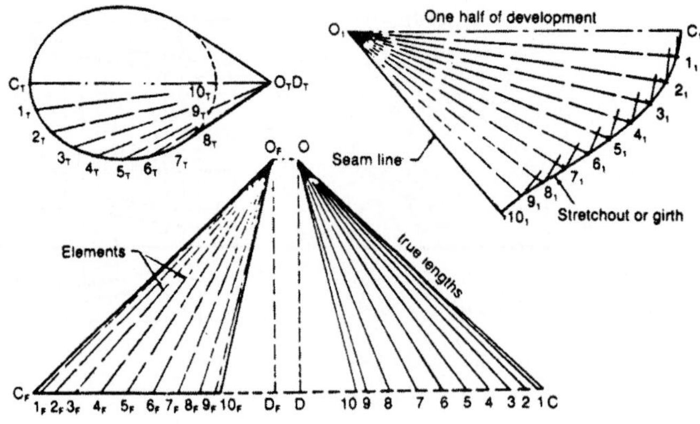


**FIGURE 11.25** Development of an oblique cone.

equal triangles meeting at the vertex, whose sides are elements and whose bases are the chords of short arcs of the base of the cone. In the oblique cone, each triangle must be found separately.

Draw two views of the cone showing a normal view of (1) the base and (2) the altitude. Divide the true size of the base, shown in Fig. 11.25 in the top view, into a number of equal parts, sufficient that the sum of the chordal distances will closely approximate the length of the base curve. Project these points to the front view of the base. Through these points and the vertex, draw the elements in each view.

Since the cone is symmetrical about a frontal plane through the vertex, the elements are shown on only the front half of it. Also, only one-half of the development is drawn. With the seam on the shortest element, the element $OC$ will be the centerline of the development and may be drawn directly at $O_1C_1$, as its true length is given by $O_FC_F$.

Find the true length of the elements by rotating them until they are parallel to the frontal plane, or by constructing a *true-length diagram.* The true length of any element will be the hypotenuse of a triangle, one leg the length of the projected element as seen in the top view, and the other leg equal to the altitude of the cone. Thus, to make the diagram, draw the leg $OD$ coinciding with or parallel to $O_FD_F$. At $D$ and perpendicular to $OD$, draw the other leg, and lay off on it the lengths $D_1$, $D_2$, etc., equal to $D_T1_T$, $D_T2_T$, etc., respectively. Distances from $O$ (origin) to points on the base of the diagram are the true lengths of the elements.

Construct the pattern for the front half of the cone as follows. With $O_1$ as the center and radius $O1$, draw an arc. With $C_1$ as center and the radius $C_T1_T$, draw a second arc intersecting the first at $1_1$. Then $O_11_1$ will be the developed position of the element $O1$. With $1_1$ as the center and radius $1_T2_T$ draw an arc intersecting a second arc with $O_1$ as center and radius $O2$, thus locating $2_1$. Continue this procedure until all the elements have been transfered to the development. Connect the points $C_1$, $1_1$, $2_1$, etc., with a smooth curve, the stretchout line, to complete the development.

### 11.12.6 Conical Connection between Two Cylindrical Pipes

Refer to Fig. 11.26. The method used in drawing the pattern is the application of the development of an oblique cone. One-half of the elliptical base is shown in true size in an auxiliary view (attached to the front view in Fig. 11.26). Find the true size of the base from its major and minor axes; divide it into a number of equal parts, so that the sum of these chordal distances closely approximates the periphery of the curve. Project these points to the front and top

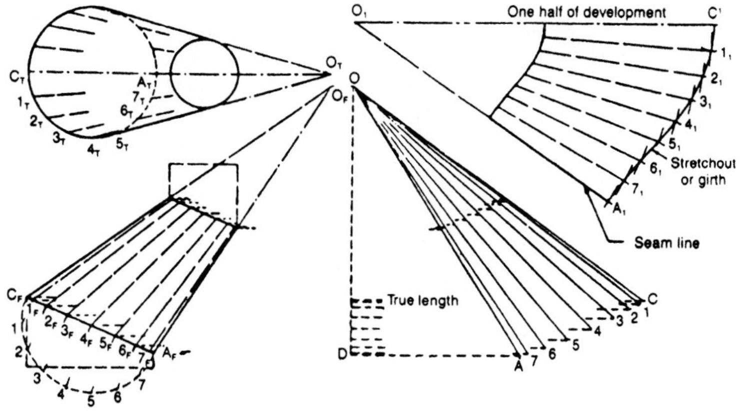


**FIGURE 11.26** Development of a conical connection between two cylindrical pipes.

views. Draw the elements in each view through these points and find the vertex $O$ by extending the contour elements until they intersect.

The true length of each element is found by using the vertical distance between its ends as the vertical leg of the diagram and its horizontal projection as the other leg. As each true length from vertex to base is found, project the upper end of the intercept horizontally across from the front view to the true length of the corresponding element to find the true length of the intercept. The development is drawn by laying out each triangle in turn, from vertex to base as in Sec. 11.12.5, starting on the centerline $O_1C_1$, and then measuring on each element its intercept length. Draw smooth curves through these points to complete the pattern.

## 11.12.7 To Develop Transition Pieces

Refer to Figs. 11.27 and 11.28. Transitions are used to connect pipes or openings of different shapes or cross sections.

Figure 11.27, showing a transition piece for connecting a round pipe and a rectangular pipe, is typical. These pieces are always developed by triangulation. The piece shown in Fig. 11.27 is, evidently, made up of four triangular planes whose bases are the sides of the rectangle, and four parts of oblique cones whose common bases are arcs of the circle and whose vertices are at the corners of the rectangle. To develop the piece, make a true-length diagram as shown in Sec. 11.12.6. After the true length of $O1$ is found, all the sides of triangle $A$ will be known. Attach the developments of cones $B$ and $B^1$, then those of triangle $C$ and $C^1$, and so on.

Figure 11.28 is another transition piece joining a rectangle to a circular pipe whose axes are not parallel. By using a partial right-side view of the round opening, you can find the divisions of the bases of the oblique cones. (As the object is symmetrical, only one-half of the opening need be divided.) The true lengths of the elements are obtained as shown in Fig. 11.27.

## 11.12.8 Triangulation of Warped Surfaces

The approximate development of a warped surface is made by dividing it into a number of narrow quadrilaterals and then splitting each of these into two triangles by a diagonal line, which is assumed to be a straight line, although it is really a curve.

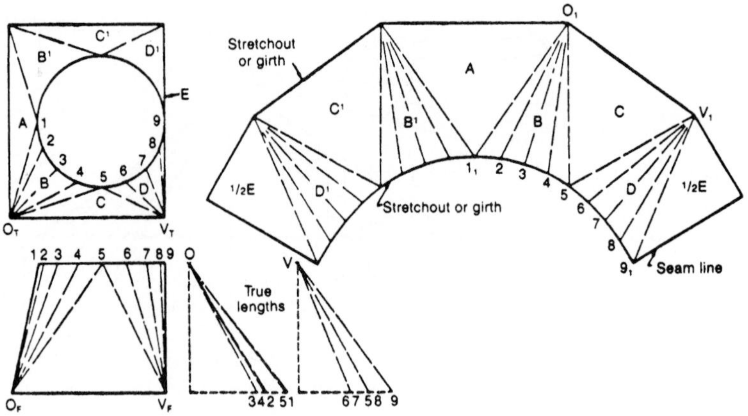


**FIGURE 11.27** Development for connecting a round to rectangular pipe.

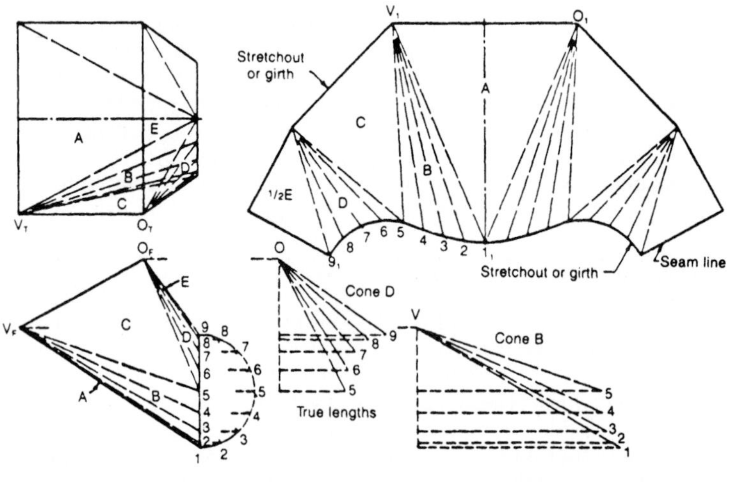


**FIGURE 11.28** Development for joining a circular pipe to a rectangle whose axes are not parallel.

Figure 11.29 shows a warped transition piece that connects an ovular (upper) pipe with a right circular cylindrical pipe (lower). Find the true size of one-half the elliptical base by rotating it until horizontal about an axis through 1, when its true shape will be seen. The major axis is $1–7_R$, and the minor axis through $4_R$ will be equal to the diameter of the lower pipe. Divide the semiellipse into a sufficient number of equal parts, and project these to the top and front views. Divide the top semicircle into the same number of equal parts and connect similar points on each end, thus dividing the surface into approximate quadrilaterals. Cut each into two triangles by a diagonal. On true-length diagrams, find the lengths of the elements and the diagonals, and draw the development by constructing the true sizes of the triangles in regular order.

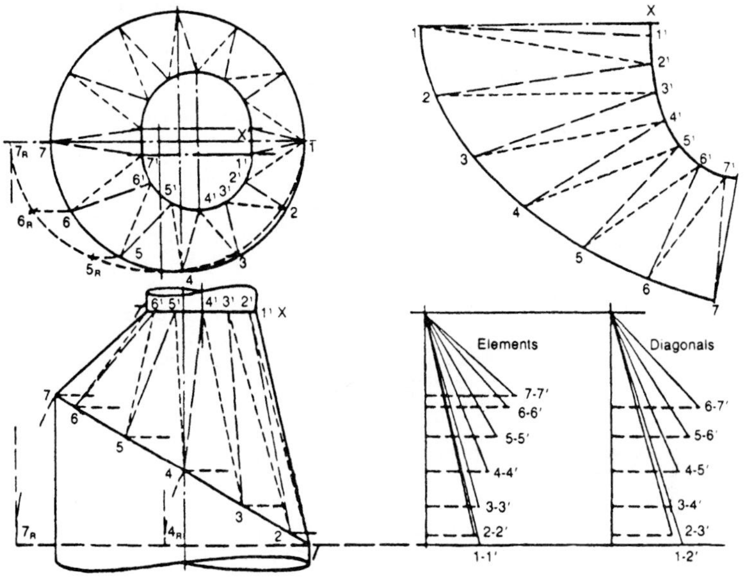

**FIGURE 11.29** Development of a warped transition piece.

## 11.13 SHEET-METAL FABRICATION PRACTICES

Figures 11.30 through 11.37 illustrate some of the methods used in sheet-metal design and fabrication practices. A detailed description of the figures is summarized here:

*Figure 11.30:* This figure shows the accepted methods of relieving the corner stresses and deformation which occur when sheet-metal flanges meet at a 90° corner. Circular punch, oblong punch, and sawcuts are illustrated.

*Figure 11.31:* A sheet-metal angle may be offset (joggled) to fit over another sheet-metal part (common practice in the aerospace industries). Corners are formed and fastened by welding directly or with gusset plates.

*Figure 11.32:* A partially closed sheet-metal part which would normally be roll-formed can be produced on the press brake using the sequence shown in this illustration. The outer flanges are bent; then the center is bent to form a W section; finally, the section is completed with a flattening die.

*Figure 11.33a, 11.33b:* A return flange on a sheet-metal part made on a press brake is possible only when a *gooseneck* die is used in the operation. See Fig. 11.33*a*. Using the modern leaf brake, return flanges can easily be bent while the sheet stock remains in the flat position during bending. Figure 11.33*b* shows a modern CNC-controlled leaf brake

*Figure 11.34:* A sheet-metal *stake* is shown in illustration (*a*). Stakes are used as stops and also for spring anchors. Illustration (*b*) shows how an integral flange is produced in the web of a sheet-metal part. The sheet metal must first be die punched prior to the bending operation.

*Figure 11.35:* The *punch tap* is an economical method for producing tapped holes in sheet-metal parts. The hole is first punched and extruded by the same die and then tapped with the appropriate thread. The maximum practical size of the punch-tapped hole is generally 0.375-16 in 11-gauge sheet steel.

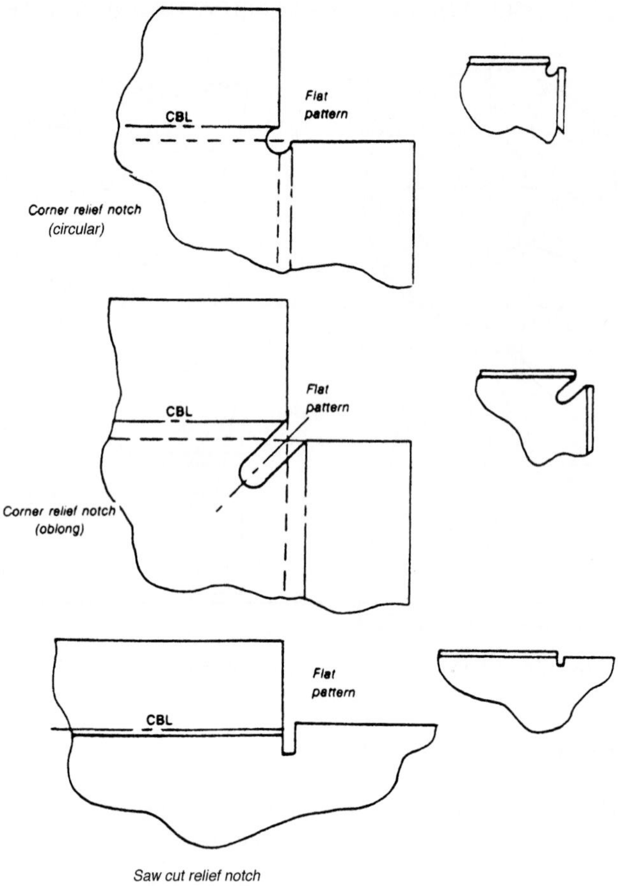


**FIGURE 11.30** Methods for relieving corners of 90° bent flanges.

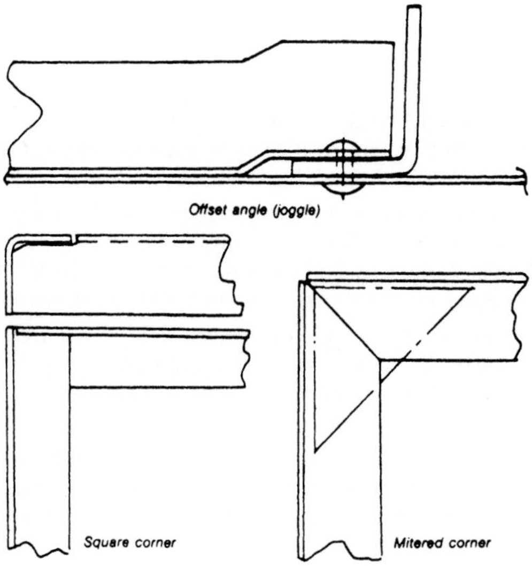


**FIGURE 11.31** Joggling angles and corners for sheet-metal members.

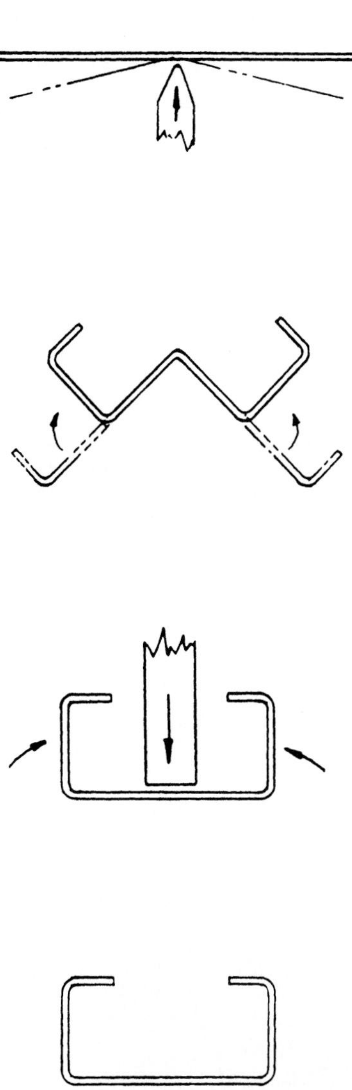

**FIGURE 11.32** Method of forming a partially closed section using the gooseneck bending die.

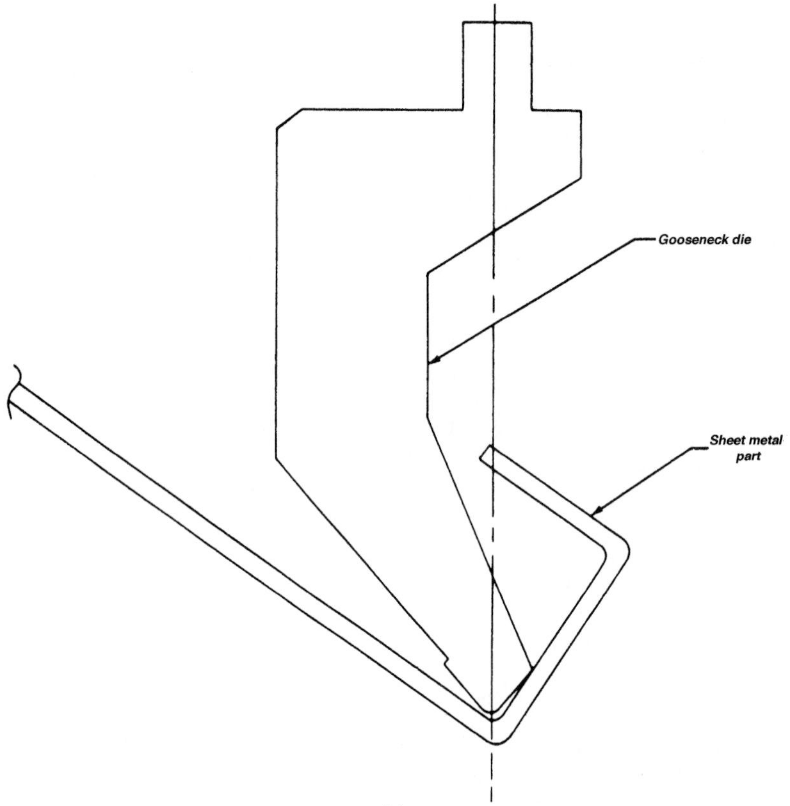


(a)

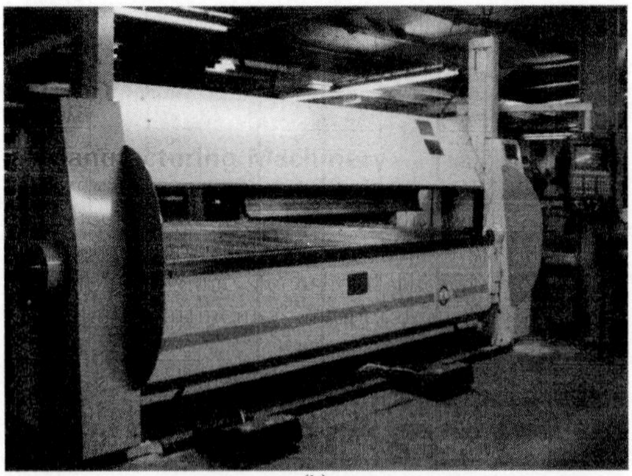

(b)

**FIGURE 11.33** (*a*) A return flange produced on the press brake; (*b*) a modern CNC-controlled leaf brake.

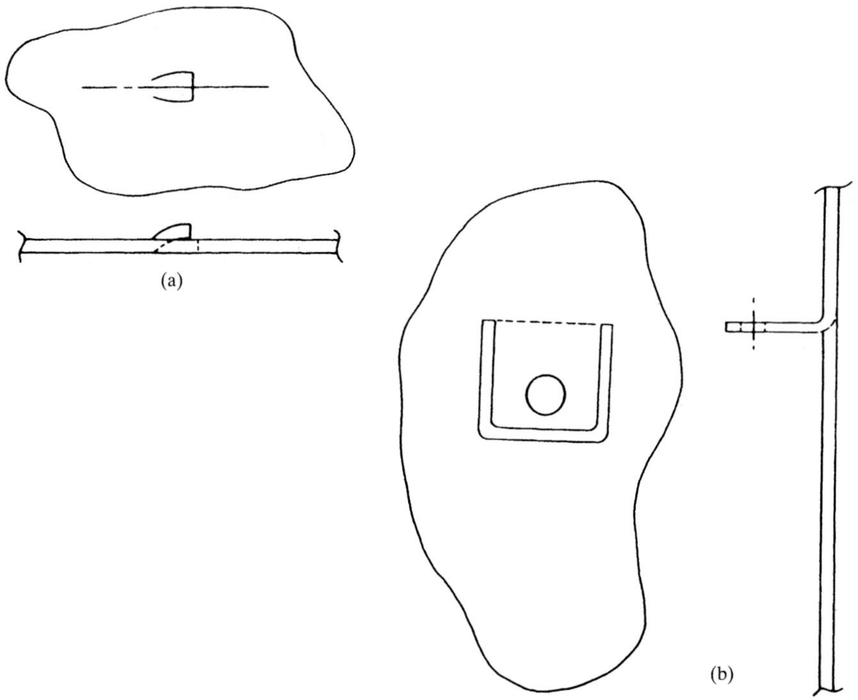


**FIGURE 11.34** (*a*) A sheet-metal stake; (*b*) an integral flange formed from the part web.

*Figure 11.36:* Illustration (*a*) shows the common *corner break* or stiffening notch which is formed in the heel section of a sheet-metal angle. These simple metal deformation methods add a great amount of strength and stiffness to light-gauge sheet-metal parts.

*Figure 11.37:* This figure illustrates some of the common methods of applying gaskets to sheet metal doors and the sheet-metal configurations required to effect these gasketing methods.

## *11.14 LIGHT-GAUGE SHEET-METAL STRUCTURAL FORMS: DIMENSIONS AND STRENGTHS*

Figures 11.38 through 11.48* show the complete dimensions and properties of sheet-metal structural shapes which have a wide range of uses in industrial applications. Light-gauge cold-formed steel sheet-metal shapes have a high strength:weight ratio and are used in countless applications.

* Extracted from American Iron and Steel Institute (AISI): *Light Gage Cold-Formed Steel Design Manual.* The latest edition of this engineering design manual may be obtained from the AISI (see Chap. 15, Sec. 15.1, for the address).

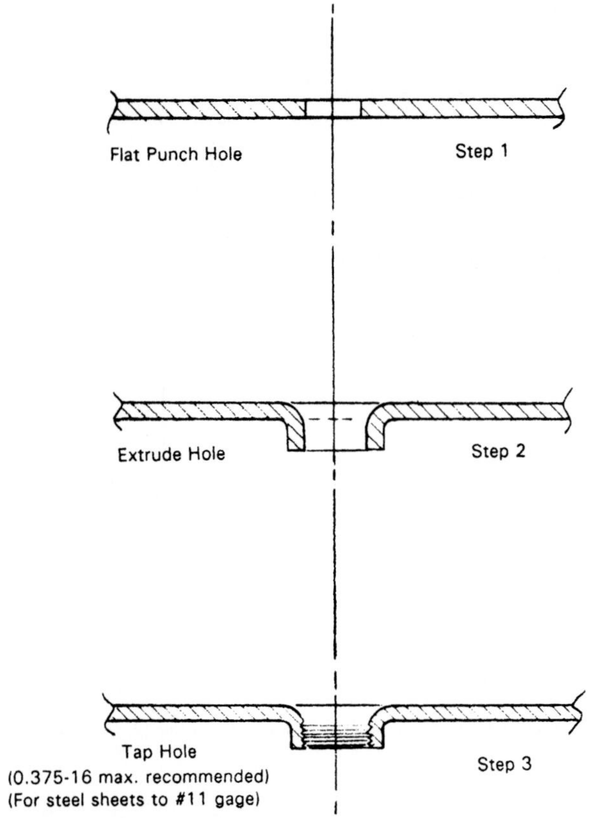


**FIGURE 11.35** Method and stages in producing the punch-tapped hole in sheet-metal parts.

In the following figures, the properties about the $X$–$X$ and $Y$–$Y$ axes are most important from a design standpoint for calculating the simple strength capabilities of each particular shape. The symbols in the figures are defined as follows:

$I_{x,y}$ = moment of inertia, $in^4$ (about the $X$–$X$ or $Y$–$Y$ axes)
$S_{x,y}$ = section modulus, $in^3$ (about the $X$–$X$ or $Y$–$Y$ axes)
$r_{x,y}$ = radius of gyration, in (about the $X$–$X$ or $Y$–$Y$ axes)
$x$ = location of centroid, in (center of gravity of the section)

The equations for calculating the deflection under a given load and the maximum stress imposed on the member for various conditions of loading as well as column bending calculations may be found in other sections of the handbook. (See Index.)

## 11.15 THE EFFECTS OF COLD WORKING STEEL

It has long been known that any cold working, such as cold stretching, bending, and twisting, affects the mechanical properties of steel. Generally, such operations produce strain harden-

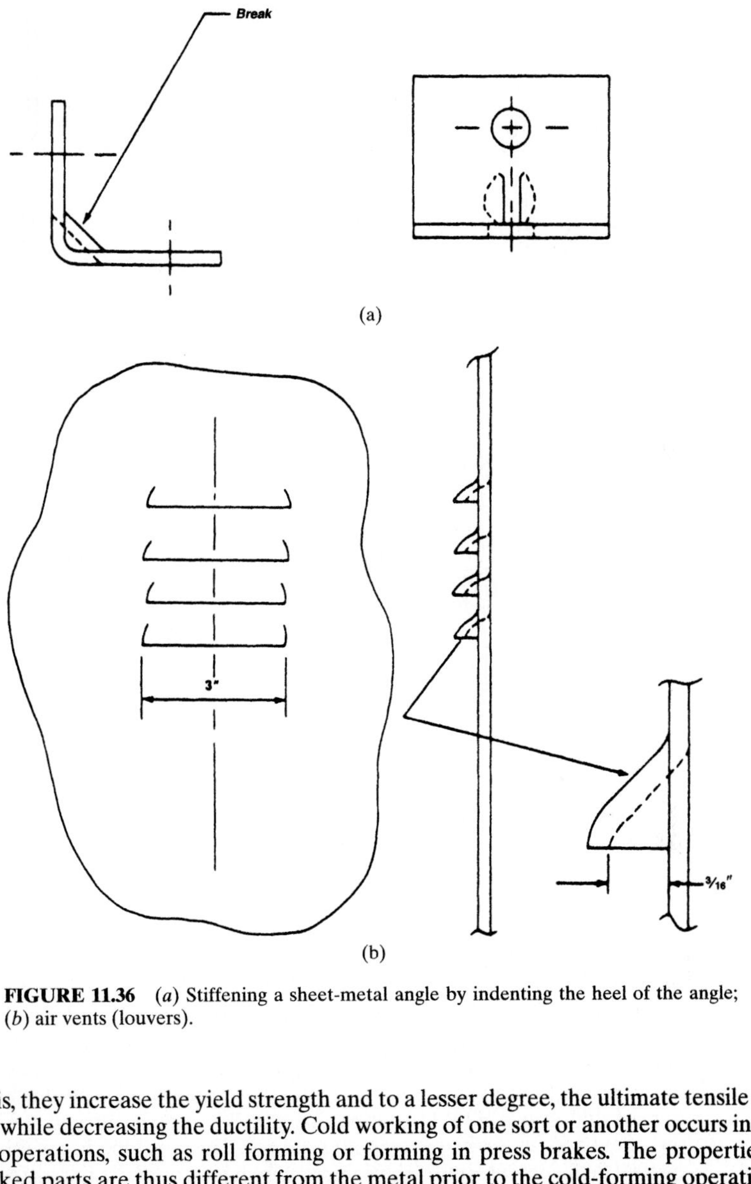


**FIGURE 11.36** (*a*) Stiffening a sheet-metal angle by indenting the heel of the angle; (*b*) air vents (louvers).

ing; that is, they increase the yield strength and to a lesser degree, the ultimate tensile strength of steels, while decreasing the ductility. Cold working of one sort or another occurs in all cold-forming operations, such as roll forming or forming in press brakes. The properties of the cold-worked parts are thus different from the metal prior to the cold-forming operations. The effects of cold forming depend strongly on the details of the particular cold-forming process. The effects of cold forming are also much more evident in the bent corners than in the flat sections. Metallurgically different kinds of structural carbon steels react differently to the same cold-forming process. The actual effect of any cold-forming operation on sheet steels is, of course, of an extremely complex nature. Unusual or excessive cold working of structural sheet steel may render the formed section unsuitable for a particular application. In other words, the cold-worked section may be weakened by excessive cold-forming operations.

Bending and buckling of sheet-metal sections used as columns and bracing or support beams may be calculated by referring to Chap. 5 and Figs. 11.38 through 11.48. (See also Index.)

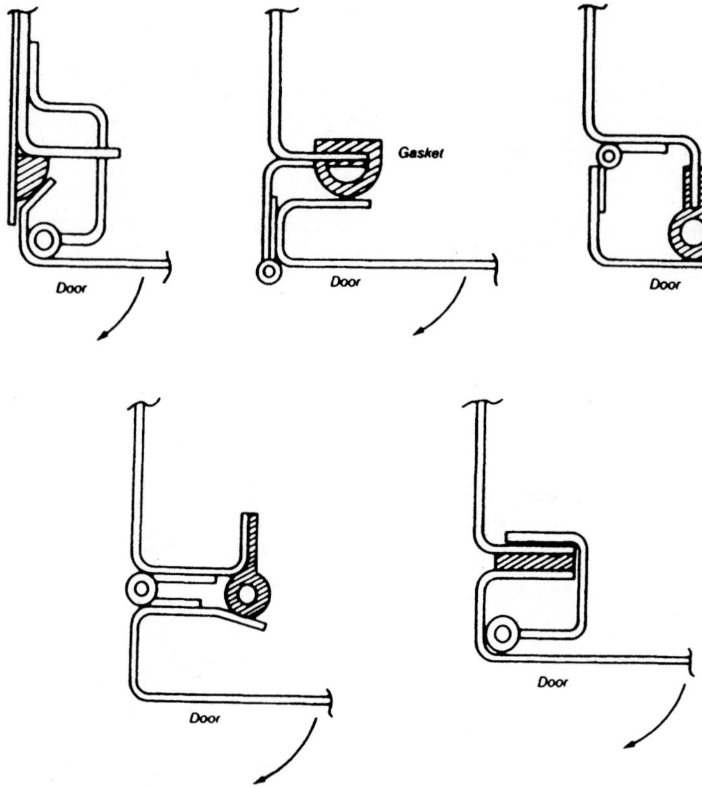


**FIGURE 11.37** Methods of designing doors for applying sealing gaskets.

An interesting aspect of the cold-working applications may be seen in Fig. 11.49. Here, we see a copper bus bar that has been axially twisted into a 90° spiral bend. This practice is common in industry where the direction of the flats on the bar must be changed in a limited amount of space, while at the same time eliminating a bolted joint. This copper bus bar was bent on a specially designed machine, where various-size bars may be bent axially to different angles. The machine used to bend this particular bar was designed by the author, and has been in operation for 15 years, with no maintenance required. The machine referred to will bend bars of cold-drawn ETP 110 copper from 0.125 × 1.00 in through 0.625 × 8 in, and a 90° bending operation requires approximately 15 sec. This machine is capable of developing in excess of 100,000 lb · in of torque. The gripper jaws which hold the bar are spaced an appropriate distance (according to bar size), to allow an almost perfect 90° axial, spiral bend (unsupported).

Figure 11.50 shows a common machine which is used to cut and notch structural steel shapes, such as angles, channels, and Zs (zee's) or sheet-metal roll-formed shapes, efficiently and accurately.

In structural welding work, the structural shapes must often be notched, mitered, or cut at an angle prior to the welding operation. This machine allows accurate notches to be made for corners and other structural intersections.

Figure 11.51 shows a lineup of electrical switchgear which is made almost entirely of sheet-metal cold-formed parts. The common gauges of sheet steel used on this type of industrial equipment include 7-, 11-, 13-, and 16-gauge. Most of the sheet steel used in this type of application is hot-rolled, pickeled, and oiled, commercial quality.

The sheet-metal flat pattern calculation methods for producing the detail parts for this type of large equipment consist mainly of the inside bend method, as outlined in a previous section, where the sheet metal is sharply bent on the press brake. Bend allowance or bend deduction methods are thus eliminated, since the detail parts are allowed to have large tolerances. In comparison, the aircraft and aerospace industries use exact procedures for producing the detail sheet-metal parts of aircraft and aerospace vehicles, where the part tolerances are very close. For this type of accurate work, the bend deduction and bend allowance meth ods must be used and the flat patterns of the parts accurately calculated and drawn in the *lofting* procedures. Blanking and forming dies may also be employed when the part quantities are sufficient.

Figure 11.52 shows a detail sheet-metal part that was blanked on blanking dies and bent on a progressive bending die set. This part is made of 11-gauge sheet steel and has closely controlled dimensions after bending.

Figure 11.53 is an AutoCAD drawing of a complex sheet-metal part shown in flat pattern. A part such as this may be programmed for fabrication on a high-speed multistation punch press that is CNC-controlled. Chapter 14 shows some of the modern sheet-metal fabrication equipment that is commonly used today in modern manufacturing plants worldwide. If the part shown in Fig. 11.53 has a requirement for quantities of a 100,000 or more, the design and construction of a progressive blanking and bending die set may be economically feasible to produce the parts.

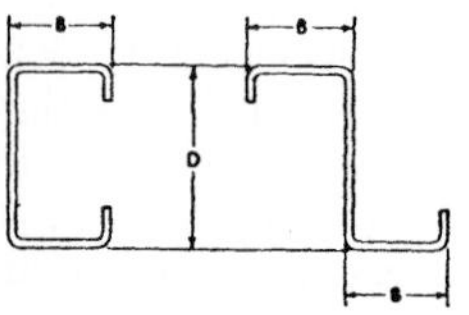

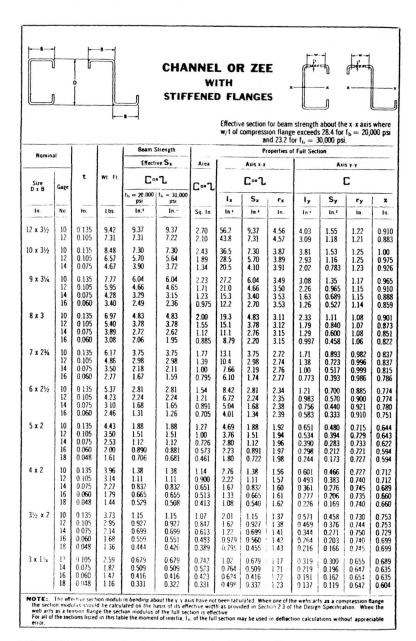

**CHANNEL OR ZEE WITH STIFFENED FLANGES**

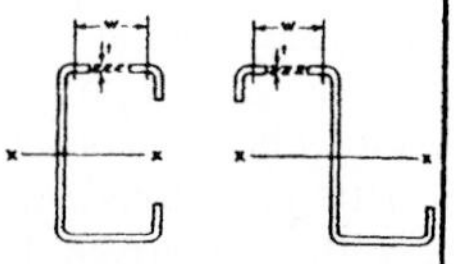


Effective section for beam strength about the x-x axis where w/t of compression flange exceeds 28.4 for $f_b = 20{,}000$ psi and 23.2 for $f_b = 30{,}000$ psi.

| Nominal | | | | Beam Strength | | Properties of Full Section | | | | | | | |
|---|---|---|---|---|---|---|---|---|---|---|---|---|---|
| | | | | Effective $S_x$ | | Area | Axis x-x | | | Axis y-y | | | |
| | | | | C or Z | | C or Z | C or Z | | | C | | | |
| Size D x B | Gage | t | Wt. Ft. | $f_b$ = 20,000 psi | $f_b$ = 30,000 psi | | $I_x$ | $S_x$ | $r_x$ | $I_y$ | $S_y$ | $r_y$ | x |
| In. | No | In. | Lbs. | In.³ | In.³ | Sq. In. | In.⁴ | In.³ | In. | In.⁴ | In.³ | In. | In. |
| 12 x 3½ | 10 | 0.135 | 9.42 | 9.37 | 9.37 | 2.70 | 56.2 | 9.37 | 4.56 | 4.03 | 1.55 | 1.22 | 0.910 |
| | 12 | 0.105 | 7.31 | 7.31 | 7.22 | 2.10 | 43.8 | 7.31 | 4.57 | 3.09 | 1.18 | 1.21 | 0.883 |
| 10 x 3½ | 10 | 0.135 | 8.48 | 7.30 | 7.30 | 2.43 | 36.5 | 7.30 | 3.87 | 3.81 | 1.53 | 1.25 | 1.00 |
| | 12 | 0.105 | 6.57 | 5.70 | 5.64 | 1.89 | 28.5 | 5.70 | 3.89 | 2.93 | 1.16 | 1.25 | 0.975 |
| | 14 | 0.075 | 4.67 | 3.90 | 3.72 | 1.34 | 20.5 | 4.10 | 3.91 | 2.02 | 0.783 | 1.23 | 0.926 |
| 9 x 3¼ | 10 | 0.135 | 7.77 | 6.04 | 6.04 | 2.23 | 27.2 | 6.04 | 3.49 | 3.08 | 1.35 | 1.17 | 0.965 |
| | 12 | 0.105 | 5.95 | 4.66 | 4.65 | 1.71 | 21.0 | 4.66 | 3.50 | 2.26 | 0.965 | 1.15 | 0.910 |
| | 14 | 0.075 | 4.28 | 3.29 | 3.15 | 1.23 | 15.3 | 3.40 | 3.53 | 1.63 | 0.689 | 1.15 | 0.888 |
| | 16 | 0.060 | 3.40 | 2.49 | 2.36 | 0.975 | 12.2 | 2.70 | 3.53 | 1.26 | 0.527 | 1.14 | 0.859 |
| 8 x 3 | 10 | 0.135 | 6.97 | 4.83 | 4.83 | 2.00 | 19.3 | 4.83 | 3.11 | 2.33 | 1.11 | 1.08 | 0.901 |
| | 12 | 0.105 | 5.40 | 3.78 | 3.78 | 1.55 | 15.1 | 3.78 | 3.12 | 1.79 | 0.840 | 1.07 | 0.873 |
| | 14 | 0.075 | 3.89 | 2.72 | 2.62 | 1.12 | 11.1 | 2.76 | 3.15 | 1.29 | 0.600 | 1.08 | 0.851 |
| | 16 | 0.060 | 3.08 | 2.06 | 1.95 | 0.885 | 8.79 | 2.20 | 3.15 | 0.997 | 0.458 | 1.06 | 0.822 |
| 7 x 2¾ | 10 | 0.135 | 6.17 | 3.75 | 3.75 | 1.77 | 13.1 | 3.75 | 2.72 | 1.71 | 0.893 | 0.982 | 0.837 |
| | 12 | 0.105 | 4.86 | 2.98 | 2.98 | 1.39 | 10.4 | 2.98 | 2.74 | 1.38 | 0.723 | 0.996 | 0.837 |
| | 14 | 0.075 | 3.50 | 2.18 | 2.11 | 1.00 | 7.66 | 2.19 | 2.76 | 1.00 | 0.517 | 0.999 | 0.815 |
| | 16 | 0.060 | 2.77 | 1.67 | 1.59 | 0.795 | 6.10 | 1.74 | 2.77 | 0.773 | 0.393 | 0.986 | 0.786 |
| 6 x 2½ | 10 | 0.135 | 5.37 | 2.81 | 2.81 | 1.54 | 8.42 | 2.81 | 2.34 | 1.21 | 0.700 | 0.885 | 0.774 |
| | 12 | 0.105 | 4.23 | 2.24 | 2.24 | 1.21 | 6.72 | 2.24 | 2.35 | 0.983 | 0.570 | 0.900 | 0.774 |
| | 14 | 0.075 | 3.10 | 1.68 | 1.65 | 0.891 | 5.04 | 1.68 | 2.38 | 0.756 | 0.440 | 0.921 | 0.780 |
| | 16 | 0.060 | 2.46 | 1.31 | 1.26 | 0.705 | 4.01 | 1.34 | 2.39 | 0.583 | 0.333 | 0.910 | 0.751 |
| 5 x 2 | 10 | 0.135 | 4.43 | 1.88 | 1.88 | 1.27 | 4.69 | 1.88 | 1.92 | 0.651 | 0.480 | 0.715 | 0.644 |
| | 12 | 0.105 | 3.50 | 1.51 | 1.51 | 1.00 | 3.76 | 1.51 | 1.94 | 0.534 | 0.394 | 0.729 | 0.643 |
| | 14 | 0.075 | 2.53 | 1.12 | 1.12 | 0.726 | 2.80 | 1.12 | 1.96 | 0.390 | 0.283 | 0.733 | 0.622 |
| | 16 | 0.060 | 2.00 | 0.890 | 0.881 | 0.573 | 2.23 | 0.891 | 1.97 | 0.298 | 0.212 | 0.721 | 0.594 |
| | 18 | 0.048 | 1.61 | 0.706 | 0.681 | 0.461 | 1.80 | 0.722 | 1.98 | 0.244 | 0.173 | 0.727 | 0.594 |
| 4 x 2 | 10 | 0.135 | 3.96 | 1.38 | 1.38 | 1.14 | 2.76 | 1.38 | 1.56 | 0.601 | 0.466 | 0.727 | 0.712 |
| | 12 | 0.105 | 3.14 | 1.11 | 1.11 | 0.900 | 2.22 | 1.11 | 1.57 | 0.493 | 0.383 | 0.740 | 0.712 |
| | 14 | 0.075 | 2.27 | 0.832 | 0.832 | 0.651 | 1.67 | 0.832 | 1.60 | 0.361 | 0.276 | 0.745 | 0.689 |
| | 16 | 0.060 | 1.79 | 0.665 | 0.655 | 0.513 | 1.33 | 0.665 | 1.61 | 0.277 | 0.206 | 0.735 | 0.660 |
| | 18 | 0.048 | 1.44 | 0.529 | 0.508 | 0.413 | 1.08 | 0.540 | 1.62 | 0.226 | 0.169 | 0.740 | 0.660 |
| 3½ x 2 | 10 | 0.135 | 3.73 | 1.15 | 1.15 | 1.07 | 2.01 | 1.15 | 1.37 | 0.571 | 0.458 | 0.730 | 0.753 |
| | 12 | 0.105 | 2.95 | 0.927 | 0.927 | 0.847 | 1.62 | 0.927 | 1.38 | 0.469 | 0.376 | 0.744 | 0.753 |
| | 14 | 0.075 | 2.14 | 0.699 | 0.699 | 0.613 | 1.22 | 0.699 | 1.41 | 0.344 | 0.271 | 0.750 | 0.729 |
| | 16 | 0.060 | 1.68 | 0.559 | 0.551 | 0.483 | 0.979 | 0.560 | 1.42 | 0.264 | 0.203 | 0.740 | 0.699 |
| | 18 | 0.048 | 1.36 | 0.444 | 0.426 | 0.389 | 0.795 | 0.455 | 1.43 | 0.216 | 0.166 | 0.745 | 0.699 |
| 3 x 1¾ | 12 | 0.105 | 2.59 | 0.679 | 0.679 | 0.742 | 1.02 | 0.679 | 1.17 | 0.319 | 0.300 | 0.655 | 0.689 |
| | 14 | 0.075 | 1.82 | 0.509 | 0.509 | 0.523 | 0.764 | 0.509 | 1.21 | 0.219 | 0.196 | 0.647 | 0.635 |
| | 16 | 0.060 | 1.47 | 0.416 | 0.416 | 0.423 | 0.624 | 0.416 | 1.22 | 0.181 | 0.162 | 0.654 | 0.635 |
| | 18 | 0.048 | 1.16 | 0.331 | 0.322 | 0.331 | 0.498 | 0.332 | 1.23 | 0.137 | 0.119 | 0.642 | 0.604 |

**NOTE:** The effective section moduli in bending about the y-y axis have not been tabulated. When one of the webs acts as a compression flange the section modulus should be calculated on the basis of its effective width as provided in Section 2.3 of the Design Specification. When the web acts as a tension flange the section modulus of the full section is effective.
For all of the sections listed in this table the moment of inertia, $I_x$, of the full section may be used in deflection calculations without appreciable error.

**FIGURE 11.38** Properties of channels or zee's with stiffened legs.

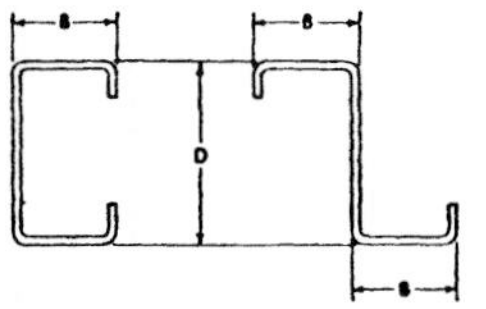

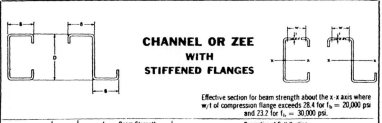

## CHANNEL OR ZEE WITH STIFFENED FLANGES

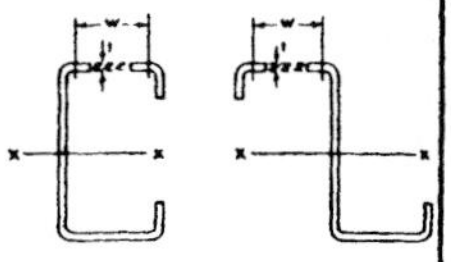


Effective section for beam strength about the x-x axis where w/t of compression flange exceeds 28.4 for $f_b$ = 20,000 psi and 23.2 for $f_b$ = 30,000 psi.

| Nominal Size D x B | Nominal Gage | t | Wt. Ft. | Beam Strength: Effective $S_x$ (C or Z) $f_b$ = 20,000 psi | Beam Strength: Effective $S_x$ (C or Z) $f_b$ = 30,000 psi | Properties of Full Section: Area (C or Z) | Axis x-x (C or Z) $I_x$ | Axis x-x $S_x$ | Axis x-x $r_x$ | Axis y-y (C) $I_y$ | Axis y-y $S_y$ | Axis y-y $r_y$ | Axis y-y x |
|---|---|---|---|---|---|---|---|---|---|---|---|---|---|
| In. | No. | In. | Lbs. | In.³ | In.³ | Sq. In. | In.⁴ | In.³ | In. | In.⁴ | In.³ | In. | In. |
| 12 x 3½ | 10 | 0.135 | 9.42 | 9.37 | 9.37 | 2.70 | 56.2 | 9.37 | 4.56 | 4.03 | 1.55 | 1.22 | 0.910 |
| | 12 | 0.105 | 7.31 | 7.31 | 7.22 | 2.10 | 43.8 | 7.31 | 4.57 | 3.09 | 1.18 | 1.21 | 0.883 |
| 10 x 3½ | 10 | 0.135 | 8.48 | 7.30 | 7.30 | 2.43 | 36.5 | 7.30 | 3.87 | 3.81 | 1.53 | 1.25 | 1.00 |
| | 12 | 0.105 | 6.57 | 5.70 | 5.64 | 1.89 | 28.5 | 5.70 | 3.89 | 2.93 | 1.16 | 1.25 | 0.975 |
| | 14 | 0.075 | 4.67 | 3.90 | 3.72 | 1.34 | 20.5 | 4.10 | 3.91 | 2.02 | 0.783 | 1.23 | 0.926 |
| 9 x 3¼ | 10 | 0.135 | 7.77 | 6.04 | 6.04 | 2.23 | 27.2 | 6.04 | 3.49 | 3.08 | 1.35 | 1.17 | 0.965 |
| | 12 | 0.105 | 5.95 | 4.66 | 4.65 | 1.71 | 21.0 | 4.66 | 3.50 | 2.26 | 0.965 | 1.15 | 0.910 |
| | 14 | 0.075 | 4.28 | 3.29 | 3.15 | 1.23 | 15.3 | 3.40 | 3.53 | 1.63 | 0.689 | 1.15 | 0.888 |
| | 16 | 0.060 | 3.40 | 2.49 | 2.36 | 0.975 | 12.2 | 2.70 | 3.53 | 1.26 | 0.527 | 1.14 | 0.859 |
| 8 x 3 | 10 | 0.135 | 6.97 | 4.83 | 4.83 | 2.00 | 19.3 | 4.83 | 3.11 | 2.33 | 1.11 | 1.08 | 0.901 |
| | 12 | 0.105 | 5.40 | 3.78 | 3.78 | 1.55 | 15.1 | 3.78 | 3.12 | 1.79 | 0.840 | 1.07 | 0.873 |
| | 14 | 0.075 | 3.89 | 2.72 | 2.62 | 1.12 | 11.1 | 2.76 | 3.15 | 1.29 | 0.600 | 1.08 | 0.851 |
| | 16 | 0.060 | 3.08 | 2.06 | 1.95 | 0.885 | 8.79 | 2.20 | 3.15 | 0.997 | 0.458 | 1.06 | 0.822 |
| 7 x 2¾ | 10 | 0.135 | 6.17 | 3.75 | 3.75 | 1.77 | 13.1 | 3.75 | 2.72 | 1.71 | 0.893 | 0.982 | 0.837 |
| | 12 | 0.105 | 4.86 | 2.98 | 2.98 | 1.39 | 10.4 | 2.98 | 2.74 | 1.38 | 0.723 | 0.996 | 0.837 |
| | 14 | 0.075 | 3.50 | 2.18 | 2.11 | 1.00 | 7.66 | 2.19 | 2.76 | 1.00 | 0.517 | 0.999 | 0.815 |
| | 16 | 0.060 | 2.77 | 1.67 | 1.59 | 0.795 | 6.10 | 1.74 | 2.77 | 0.773 | 0.393 | 0.986 | 0.786 |
| 6 x 2½ | 10 | 0.135 | 5.37 | 2.81 | 2.81 | 1.54 | 8.42 | 2.81 | 2.34 | 1.21 | 0.700 | 0.885 | 0.774 |
| | 12 | 0.105 | 4.23 | 2.24 | 2.24 | 1.21 | 6.72 | 2.24 | 2.35 | 0.983 | 0.570 | 0.900 | 0.774 |
| | 14 | 0.075 | 3.10 | 1.68 | 1.65 | 0.891 | 5.04 | 1.68 | 2.38 | 0.756 | 0.440 | 0.921 | 0.780 |
| | 16 | 0.060 | 2.46 | 1.31 | 1.26 | 0.705 | 4.01 | 1.34 | 2.39 | 0.583 | 0.333 | 0.910 | 0.751 |
| 5 x 2 | 10 | 0.135 | 4.43 | 1.88 | 1.88 | 1.27 | 4.69 | 1.88 | 1.92 | 0.651 | 0.480 | 0.715 | 0.644 |
| | 12 | 0.105 | 3.50 | 1.51 | 1.51 | 1.00 | 3.76 | 1.51 | 1.94 | 0.534 | 0.394 | 0.729 | 0.643 |
| | 14 | 0.075 | 2.53 | 1.12 | 1.12 | 0.726 | 2.80 | 1.12 | 1.96 | 0.390 | 0.283 | 0.733 | 0.622 |
| | 16 | 0.060 | 2.00 | 0.890 | 0.881 | 0.573 | 2.23 | 0.891 | 1.97 | 0.298 | 0.212 | 0.721 | 0.594 |
| | 18 | 0.048 | 1.61 | 0.706 | 0.681 | 0.461 | 1.80 | 0.722 | 1.98 | 0.244 | 0.173 | 0.727 | 0.594 |
| 4 x 2 | 10 | 0.135 | 3.96 | 1.38 | 1.38 | 1.14 | 2.76 | 1.38 | 1.56 | 0.601 | 0.466 | 0.727 | 0.712 |
| | 12 | 0.105 | 3.14 | 1.11 | 1.11 | 0.900 | 2.22 | 1.11 | 1.57 | 0.493 | 0.383 | 0.740 | 0.712 |
| | 14 | 0.075 | 2.27 | 0.832 | 0.832 | 0.651 | 1.67 | 0.832 | 1.60 | 0.361 | 0.276 | 0.745 | 0.689 |
| | 16 | 0.060 | 1.79 | 0.665 | 0.655 | 0.513 | 1.33 | 0.665 | 1.61 | 0.277 | 0.206 | 0.735 | 0.660 |
| | 18 | 0.048 | 1.44 | 0.529 | 0.508 | 0.413 | 1.08 | 0.540 | 1.62 | 0.226 | 0.169 | 0.740 | 0.660 |
| 3½ x 2 | 10 | 0.135 | 3.73 | 1.15 | 1.15 | 1.07 | 2.01 | 1.15 | 1.37 | 0.571 | 0.458 | 0.730 | 0.753 |
| | 12 | 0.105 | 2.95 | 0.927 | 0.927 | 0.847 | 1.62 | 0.927 | 1.38 | 0.469 | 0.376 | 0.744 | 0.753 |
| | 14 | 0.075 | 2.14 | 0.699 | 0.699 | 0.613 | 1.22 | 0.699 | 1.41 | 0.344 | 0.271 | 0.750 | 0.729 |
| | 16 | 0.060 | 1.68 | 0.559 | 0.551 | 0.483 | 0.979 | 0.560 | 1.42 | 0.264 | 0.203 | 0.740 | 0.699 |
| | 18 | 0.048 | 1.36 | 0.444 | 0.426 | 0.389 | 0.795 | 0.455 | 1.43 | 0.216 | 0.166 | 0.745 | 0.699 |
| 3 x 1¼ | 12 | 0.105 | 2.59 | 0.679 | 0.679 | 0.742 | 1.02 | 0.679 | 1.17 | 0.319 | 0.300 | 0.655 | 0.689 |
| | 14 | 0.075 | 1.82 | 0.509 | 0.509 | 0.523 | 0.764 | 0.509 | 1.21 | 0.219 | 0.196 | 0.647 | 0.635 |
| | 16 | 0.060 | 1.47 | 0.416 | 0.416 | 0.423 | 0.624 | 0.416 | 1.22 | 0.181 | 0.162 | 0.654 | 0.635 |
| | 18 | 0.048 | 1.16 | 0.331 | 0.322 | 0.331 | 0.498 | 0.332 | 1.23 | 0.137 | 0.119 | 0.642 | 0.604 |

**NOTE:** The effective section moduli in bending about the y-y axis have not been tabulated. When one of the webs acts as a compression flange the section modulus should be calculated on the basis of its effective width as provided in Section 2.3 of the Design Specification. When the web acts as a tension flange the section modulus of the full section is effective.
For all of the sections listed in this table the moment of inertia, $I_x$, of the full section may be used in deflection calculations without appreciable error.

**FIGURE 11.38** (*Continued*)

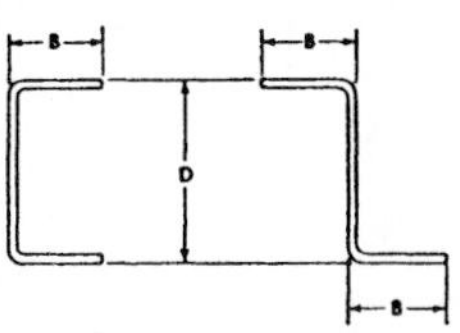

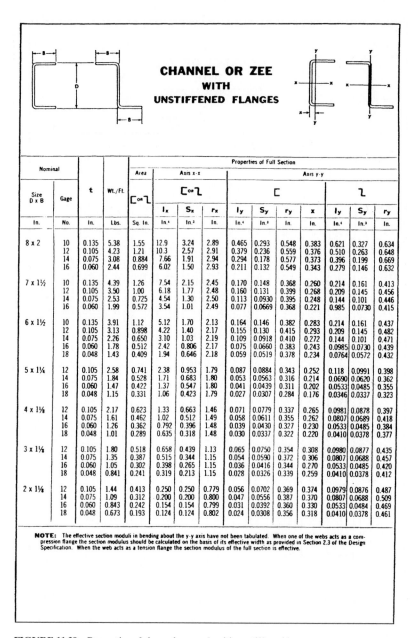

## CHANNEL OR ZEE WITH UNSTIFFENED FLANGES

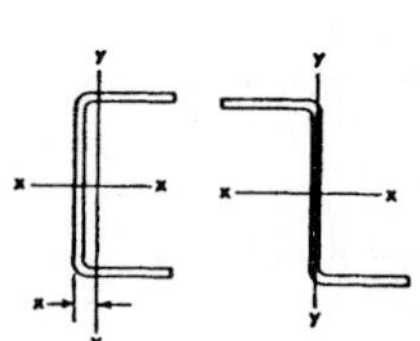


| Nominal | | t | Wt./Ft. | Properties of Full Section | | | | | | | | | | |
|---|---|---|---|---|---|---|---|---|---|---|---|---|---|---|
| | | | | Area | Axis x-x | | | Axis y-y | | | | | | |
| | | | | [ or Z | [ or Z | | | [ | | | | Z | | |
| Size D x B | Gage | t | Wt./Ft. | | $I_x$ | $S_x$ | $r_x$ | $I_y$ | $S_y$ | $r_y$ | x | $I_y$ | $S_y$ | $r_y$ |
| In. | No. | In. | Lbs. | Sq. In. | In.$^4$ | In.$^3$ | In. | In.$^4$ | In.$^3$ | In. | In. | In.$^4$ | In.$^3$ | In. |
| 8 x 2 | 10 | 0.135 | 5.38 | 1.55 | 12.9 | 3.24 | 2.89 | 0.465 | 0.293 | 0.548 | 0.383 | 0.621 | 0.327 | 0.634 |
| | 12 | 0.105 | 4.23 | 1.21 | 10.3 | 2.57 | 2.91 | 0.379 | 0.236 | 0.559 | 0.376 | 0.510 | 0.263 | 0.648 |
| | 14 | 0.075 | 3.08 | 0.884 | 7.66 | 1.91 | 2.94 | 0.294 | 0.178 | 0.577 | 0.373 | 0.396 | 0.199 | 0.669 |
| | 16 | 0.060 | 2.44 | 0.699 | 6.02 | 1.50 | 2.93 | 0.211 | 0.132 | 0.549 | 0.343 | 0.279 | 0.146 | 0.632 |
| 7 x 1½ | 10 | 0.135 | 4.39 | 1.26 | 7.54 | 2.15 | 2.45 | 0.170 | 0.148 | 0.368 | 0.260 | 0.214 | 0.161 | 0.413 |
| | 12 | 0.105 | 3.50 | 1.00 | 6.18 | 1.77 | 2.48 | 0.160 | 0.131 | 0.399 | 0.268 | 0.209 | 0.145 | 0.456 |
| | 14 | 0.075 | 2.53 | 0.725 | 4.54 | 1.30 | 2.50 | 0.113 | 0.0930 | 0.395 | 0.248 | 0.144 | 0.101 | 0.446 |
| | 16 | 0.060 | 1.99 | 0.572 | 3.54 | 1.01 | 2.49 | 0.077 | 0.0669 | 0.368 | 0.221 | 0.985 | 0.0730 | 0.415 |
| 6 x 1½ | 10 | 0.135 | 3.91 | 1.12 | 5.12 | 1.70 | 2.13 | 0.164 | 0.146 | 0.382 | 0.283 | 0.214 | 0.161 | 0.437 |
| | 12 | 0.105 | 3.13 | 0.898 | 4.22 | 1.40 | 2.17 | 0.155 | 0.130 | 0.415 | 0.293 | 0.209 | 0.145 | 0.482 |
| | 14 | 0.075 | 2.26 | 0.650 | 3.10 | 1.03 | 2.19 | 0.109 | 0.0918 | 0.410 | 0.272 | 0.144 | 0.101 | 0.471 |
| | 16 | 0.060 | 1.78 | 0.512 | 2.42 | 0.806 | 2.17 | 0.075 | 0.0660 | 0.383 | 0.243 | 0.0985 | 0.0730 | 0.439 |
| | 18 | 0.048 | 1.43 | 0.409 | 1.94 | 0.646 | 2.18 | 0.059 | 0.0519 | 0.378 | 0.234 | 0.0764 | 0.0572 | 0.432 |
| 5 x 1¼ | 12 | 0.105 | 2.58 | 0.741 | 2.38 | 0.953 | 1.79 | 0.087 | 0.0884 | 0.343 | 0.252 | 0.118 | 0.0991 | 0.398 |
| | 14 | 0.075 | 1.84 | 0.528 | 1.71 | 0.683 | 1.80 | 0.053 | 0.0563 | 0.316 | 0.214 | 0.0690 | 0.0620 | 0.362 |
| | 16 | 0.060 | 1.47 | 0.422 | 1.37 | 0.547 | 1.80 | 0.041 | 0.0439 | 0.311 | 0.202 | 0.0533 | 0.0485 | 0.355 |
| | 18 | 0.048 | 1.15 | 0.331 | 1.06 | 0.423 | 1.79 | 0.027 | 0.0307 | 0.284 | 0.176 | 0.0346 | 0.0337 | 0.323 |
| 4 x 1⅛ | 12 | 0.105 | 2.17 | 0.623 | 1.33 | 0.663 | 1.46 | 0.071 | 0.0779 | 0.337 | 0.265 | 0.0981 | 0.0878 | 0.397 |
| | 14 | 0.075 | 1.61 | 0.462 | 1.02 | 0.512 | 1.49 | 0.058 | 0.0611 | 0.355 | 0.262 | 0.0807 | 0.0689 | 0.418 |
| | 16 | 0.060 | 1.26 | 0.362 | 0.792 | 0.396 | 1.48 | 0.039 | 0.0430 | 0.327 | 0.230 | 0.0533 | 0.0485 | 0.384 |
| | 18 | 0.048 | 1.01 | 0.289 | 0.635 | 0.318 | 1.48 | 0.030 | 0.0337 | 0.322 | 0.220 | 0.0410 | 0.0378 | 0.377 |
| 3 x 1⅛ | 12 | 0.105 | 1.80 | 0.518 | 0.658 | 0.439 | 1.13 | 0.065 | 0.0750 | 0.354 | 0.308 | 0.0980 | 0.0877 | 0.435 |
| | 14 | 0.075 | 1.35 | 0.387 | 0.515 | 0.344 | 1.15 | 0.054 | 0.0590 | 0.372 | 0.306 | 0.0807 | 0.0688 | 0.457 |
| | 16 | 0.060 | 1.05 | 0.302 | 0.398 | 0.265 | 1.15 | 0.036 | 0.0416 | 0.344 | 0.270 | 0.0533 | 0.0485 | 0.420 |
| | 18 | 0.048 | 0.841 | 0.241 | 0.319 | 0.213 | 1.15 | 0.028 | 0.0326 | 0.339 | 0.259 | 0.0410 | 0.0378 | 0.412 |
| 2 x 1⅛ | 12 | 0.105 | 1.44 | 0.413 | 0.250 | 0.250 | 0.779 | 0.056 | 0.0702 | 0.369 | 0.374 | 0.0979 | 0.0876 | 0.487 |
| | 14 | 0.075 | 1.09 | 0.312 | 0.200 | 0.200 | 0.800 | 0.047 | 0.0556 | 0.387 | 0.370 | 0.0807 | 0.0688 | 0.509 |
| | 16 | 0.060 | 0.843 | 0.242 | 0.154 | 0.154 | 0.799 | 0.031 | 0.0392 | 0.360 | 0.330 | 0.0533 | 0.0484 | 0.469 |
| | 18 | 0.048 | 0.673 | 0.193 | 0.124 | 0.124 | 0.802 | 0.024 | 0.0308 | 0.356 | 0.318 | 0.0410 | 0.0378 | 0.461 |

**NOTE:** The effective section moduli in bending about the y-y axis have not been tabulated. When one of the webs acts as a compression flange the section modulus should be calculated on the basis of its effective width as provided in Section 2.3 of the Design Specification. When the web acts as a tension flange the section modulus of the full section is effective.

**FIGURE 11.39** Properties of channels or zee's with unstiffened legs.

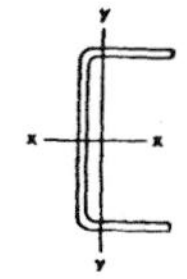

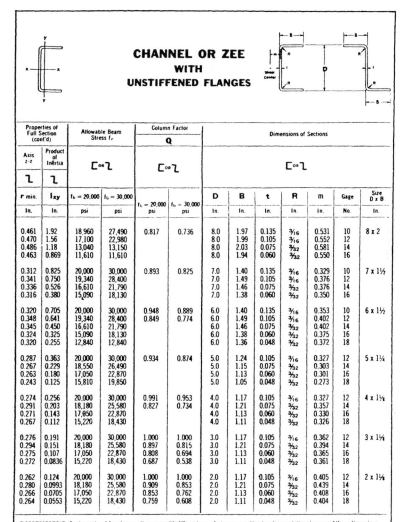

## CHANNEL OR ZEE WITH UNSTIFFENED FLANGES

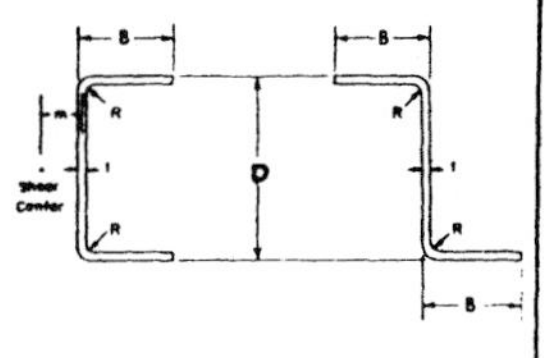


| Properties of Full Section (cont'd) | | Allowable Beam Stress $f_r$ | | Column Factor Q | | Dimensions of Sections | | | | | | |
|---|---|---|---|---|---|---|---|---|---|---|---|---|
| Axis Z-Z | Product of Inertia | [ or Z | | [ or Z | | [ or Z | | | | | | |
| r min. | $I_{xy}$ | $f_b$ = 20,000 | $f_b$ = 30,000 | $f_b$ = 20,000 | $f_b$ = 30,000 | D | B | t | R | m | Gage | Size D x B |
| In. | In. | psi | psi | psi | psi | In. | In. | In. | In. | In. | No. | In. |
| 0.461 | 1.92 | 18,960 | 27,490 | 0.817 | 0.736 | 8.0 | 1.97 | 0.135 | 3/16 | 0.531 | 10 | 8 x 2 |
| 0.470 | 1.56 | 17,100 | 22,980 | | | 8.0 | 1.99 | 0.105 | 3/16 | 0.552 | 12 | |
| 0.486 | 1.18 | 13,040 | 13,150 | | | 8.0 | 2.03 | 0.075 | 3/32 | 0.581 | 14 | |
| 0.463 | 0.869 | 11,610 | 11,610 | | | 8.0 | 1.94 | 0.060 | 3/32 | 0.550 | 16 | |
| 0.312 | 0.825 | 20,000 | 30,000 | 0.893 | 0.825 | 7.0 | 1.40 | 0.135 | 3/16 | 0.329 | 10 | 7 x 1½ |
| 0.341 | 0.750 | 19,340 | 28,400 | | | 7.0 | 1.49 | 0.105 | 3/16 | 0.376 | 12 | |
| 0.336 | 0.526 | 16,610 | 21,790 | | | 7.0 | 1.46 | 0.075 | 3/32 | 0.376 | 14 | |
| 0.316 | 0.380 | 15,090 | 18,130 | | | 7.0 | 1.38 | 0.060 | 3/32 | 0.350 | 16 | |
| 0.320 | 0.705 | 20,000 | 30,000 | 0.948 | 0.889 | 6.0 | 1.40 | 0.135 | 3/16 | 0.353 | 10 | 6 x 1½ |
| 0.348 | 0.641 | 19,340 | 28,400 | 0.849 | 0.774 | 6.0 | 1.49 | 0.105 | 3/16 | 0.402 | 12 | |
| 0.345 | 0.450 | 16,610 | 21,790 | | | 6.0 | 1.46 | 0.075 | 3/32 | 0.402 | 14 | |
| 0.324 | 0.325 | 15,090 | 18,130 | | | 6.0 | 1.38 | 0.060 | 3/32 | 0.375 | 16 | |
| 0.320 | 0.255 | 12,840 | 12,840 | | | 6.0 | 1.36 | 0.048 | 3/32 | 0.372 | 18 | |
| 0.287 | 0.363 | 20,000 | 30,000 | 0.934 | 0.874 | 5.0 | 1.24 | 0.105 | 3/16 | 0.327 | 12 | 5 x 1¼ |
| 0.267 | 0.229 | 18,550 | 26,490 | | | 5.0 | 1.15 | 0.075 | 3/32 | 0.303 | 14 | |
| 0.263 | 0.180 | 17,050 | 22,870 | | | 5.0 | 1.13 | 0.060 | 3/32 | 0.301 | 16 | |
| 0.243 | 0.125 | 15,810 | 19,850 | | | 5.0 | 1.05 | 0.048 | 3/32 | 0.273 | 18 | |
| 0.274 | 0.256 | 20,000 | 30,000 | 0.991 | 0.953 | 4.0 | 1.17 | 0.105 | 3/16 | 0.327 | 12 | 4 x 1⅛ |
| 0.291 | 0.203 | 18,180 | 25,580 | 0.827 | 0.734 | 4.0 | 1.21 | 0.075 | 3/32 | 0.357 | 14 | |
| 0.271 | 0.143 | 17,050 | 22,870 | | | 4.0 | 1.13 | 0.060 | 3/32 | 0.330 | 16 | |
| 0.267 | 0.112 | 15,220 | 18,430 | | | 4.0 | 1.11 | 0.048 | 3/32 | 0.326 | 18 | |
| 0.276 | 0.191 | 20,000 | 30,000 | 1.000 | 1.000 | 3.0 | 1.17 | 0.105 | 3/16 | 0.362 | 12 | 3 x 1⅛ |
| 0.294 | 0.151 | 18,180 | 25,580 | 0.897 | 0.815 | 3.0 | 1.21 | 0.075 | 3/32 | 0.394 | 14 | |
| 0.275 | 0.107 | 17,050 | 22,870 | 0.808 | 0.694 | 3.0 | 1.13 | 0.060 | 3/32 | 0.365 | 16 | |
| 0.272 | 0.0836 | 15,220 | 18,430 | 0.687 | 0.538 | 3.0 | 1.11 | 0.048 | 3/32 | 0.361 | 18 | |
| 0.262 | 0.124 | 20,000 | 30,000 | 1.000 | 1.000 | 2.0 | 1.17 | 0.105 | 3/16 | 0.405 | 12 | 2 x 1⅛ |
| 0.280 | 0.0993 | 18,180 | 25,580 | 0.909 | 0.853 | 2.0 | 1.21 | 0.075 | 3/32 | 0.439 | 14 | |
| 0.266 | 0.0705 | 17,050 | 22,870 | 0.853 | 0.762 | 2.0 | 1.13 | 0.060 | 3/32 | 0.408 | 16 | |
| 0.264 | 0.0553 | 15,220 | 18,430 | 0.759 | 0.608 | 2.0 | 1.11 | 0.048 | 3/32 | 0.404 | 18 | |

**DIMENSIONS:** Equipment and forming practices vary with different manufacturers, resulting in minor variations in some of these dimensions. These minor variations do not affect the published properties. Consult the manufacturer for actual weight per foot and actual dimensions.

Column form factors, Q, for members having webs with w/t-ratios in excess of 60 are not shown. See limitations of Section 2.3.3(a) of the Specification applicable to element stiffened by simple lip.

**FIGURE 11.39** *(Continued)*

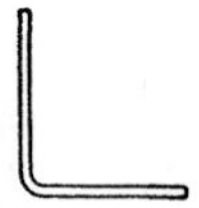

## EQUAL LEG ANGLE
## WITH
## UNSTIFFENED LEGS

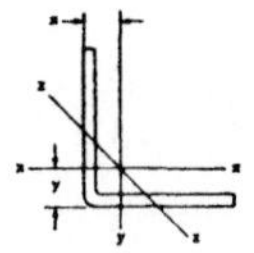

| Nominal | | t | Wt./Ft. | Properties of Full Section | | | | | | | Beam Strength $f_b$ = 20,000 psi | | |
|---|---|---|---|---|---|---|---|---|---|---|---|---|---|
| | | | | Area | Axis x-x and Axis y-y | | | | Axis z-z | | M max. | | M max. |
| Size | Gage | | | | I | S | r | x = y | I | r | $f_c$ | Comp. x-L-x Tension | Comp. x-Γ-x Tension |
| In. | No. | In. | Lbs. | Sq. In. | $In.^4$ | $In.^3$ | In. | In. | $In.^4$ | In. | psi | In.-Lb. | In.-Lb. |
| 4 x 4 | 10 | 0.135 | 3.66 | 1.05 | 1.715 | 0.582 | 1.28 | 1.07 | 0.662 | 0.794 | 12,300 | 7,160 | 11,640 |
| 3 x 3 | 10 | 0.135 | 2.72 | 0.781 | 0.712 | 0.324 | 0.955 | 0.819 | 0.271 | 0.589 | 15,340 | 4,970 | 6,480 |
| | 12 | 0.105 | 2.16 | 0.620 | 0.586 | 0.262 | 0.972 | 0.817 | 0.224 | 0.601 | 12,600 | 3,300 | 5,240 |
| 2½ x 2½ | 10 | 0.135 | 2.25 | 0.646 | 0.407 | 0.223 | 0.793 | 0.694 | 0.153 | 0.487 | 17,080 | 3,810 | 4,460 |
| | 12 | 0.105 | 1.79 | 0.515 | 0.338 | 0.182 | 0.811 | 0.692 | 0.128 | 0.499 | 14,600 | 2,660 | 3,640 |
| 2 x 2 | 10 | 0.135 | 1.78 | 0.511 | 0.204 | 0.141 | 0.632 | 0.569 | 0.0756 | 0.385 | 18,830 | 2,650 | 2,820 |
| | 12 | 0.105 | 1.43 | 0.410 | 0.173 | 0.116 | 0.649 | 0.567 | 0.0643 | 0.396 | 16,830 | 1,950 | 2,320 |
| | 14 | 0.075 | 1.08 | 0.311 | 0.144 | 0.092 | 0.680 | 0.570 | 0.0555 | 0.423 | 12,590 | 1,160 | 1.840 |
| | 16 | 0.060 | 0.840 | 0.241 | 0.104 | 0.069 | 0.658 | 0.545 | 0.0404 | 0.409 | 11,040 | 760 | 1,380 |

The allowable bending moments shown in this Table apply only when the sections are adequztely braced laterally.

Where the vertical legs of the angles are in compression, $M_{max}$ is based on the values of $f_c$ (see 3.2 of Design Specification) indicated; where the vertical legs of the angles are in tension $M_{max}$ is based on $f_b$ (tension) since the compression stress is always less than $f_c$ for the sections listed.

Because it is virtually impossible to load single angle struts concentrically, the design of any such strut should take the eccentricity into account.

**FIGURE 11.40** Properties of equal leg angles with unstiffened legs.

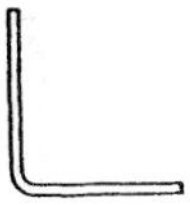

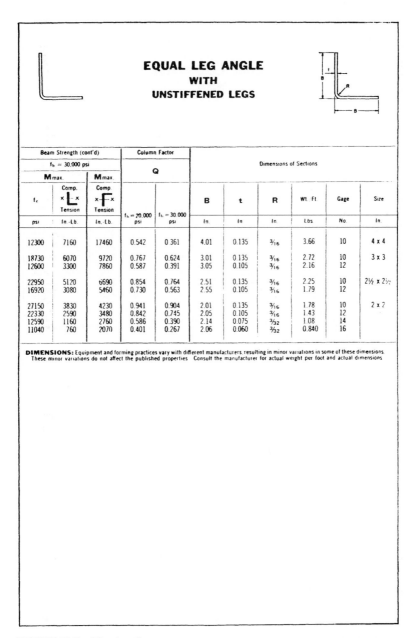

## EQUAL LEG ANGLE WITH UNSTIFFENED LEGS

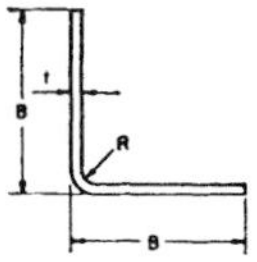


| Beam Strength (cont'd) $f_b$ = 30,000 psi | | | Column Factor Q | | Dimensions of Sections | | | | | |
|---|---|---|---|---|---|---|---|---|---|---|
| $f_c$ | $M_{max}$ Comp. L Tension | $M_{max}$ Comp. F Tension | $f_b$ = 20,000 psi | $f_b$ = 30,000 psi | B | t | R | Wt. Ft | Gage | Size |
| psi | In.-Lb. | In.-Lb. | | | In. | In | In | Lbs | No. | In. |
| 12300 | 7160 | 17460 | 0.542 | 0.361 | 4.01 | 0.135 | 3/16 | 3.66 | 10 | 4 x 4 |
| 18730 | 6070 | 9720 | 0.767 | 0.624 | 3.01 | 0.135 | 3/16 | 2.72 | 10 | 3 x 3 |
| 12600 | 3300 | 7860 | 0.587 | 0.391 | 3.05 | 0.105 | 3/16 | 2.16 | 12 | |
| 22950 | 5120 | 6690 | 0.854 | 0.764 | 2.51 | 0.135 | 3/16 | 2.25 | 10 | 2½ x 2½ |
| 16920 | 3080 | 5460 | 0.730 | 0.563 | 2.55 | 0.105 | 3/16 | 1.79 | 12 | |
| 27150 | 3830 | 4230 | 0.941 | 0.904 | 2.01 | 0.135 | 3/16 | 1.78 | 10 | 2 x 2 |
| 22330 | 2590 | 3480 | 0.842 | 0.745 | 2.05 | 0.105 | 3/16 | 1.43 | 12 | |
| 12590 | 1160 | 2760 | 0.586 | 0.390 | 2.14 | 0.075 | 3/32 | 1.08 | 14 | |
| 11040 | 760 | 2070 | 0.401 | 0.267 | 2.06 | 0.060 | 3/32 | 0.840 | 16 | |

**DIMENSIONS:** Equipment and forming practices vary with different manufacturers, resulting in minor variations in some of these dimensions. These minor variations do not affect the published properties. Consult the manufacturer for actual weight per foot and actual dimensions.

**FIGURE 11.40** (*Continued*)

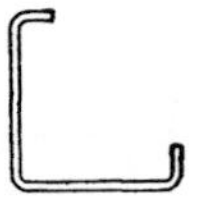

## EQUAL LEG ANGLE WITH STIFFENED LEGS

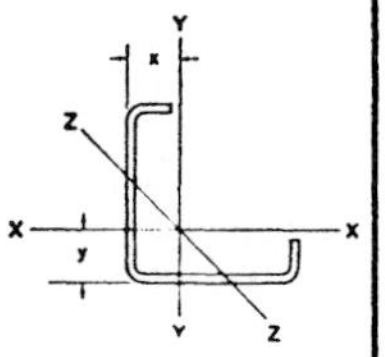


| Nominal | | t | Wt./Ft. | Area | Properties of Full Section | | | | | | |
|---|---|---|---|---|---|---|---|---|---|---|---|
| | | | | | Axis x-x and Axis y-y | | | | Axis z-z | |
| Size | Gage | | | | I | S | r | x = y | I | r |
| In. | No. | In. | Lbs. | Sq. In. | In.$^4$ | In.$^3$ | In. | In. | In.$^4$ | In. |
| 4 x 4 | 10 | 0.135 | 4.46 | 1.28 | 2.62 | 0.962 | 1.43 | 1.29 | 1.25 | 0.988 |
| 3 x 3 | 10 | 0.135 | 3.34 | 0.957 | 1.08 | 0.536 | 1.06 | 0.993 | 0.531 | 0.745 |
| | 12 | 0.105 | 2.59 | 0.743 | 0.864 | 0.416 | 1.08 | 0.973 | 0.407 | 0.740 |
| 2½ x 2½ | 10 | 0.135 | 2.68 | 0.768 | 0.579 | 0.342 | 0.868 | 0.818 | 0.277 | 0.600 |
| | 12 | 0.105 | 2.15 | 0.617 | 0.494 | 0.286 | 0.895 | 0.824 | 0.236 | 0.618 |
| 2 x 2 | 10 | 0.135 | 2.21 | 0.633 | 0.306 | 0.233 | 0.695 | 0.695 | 0.159 | 0.501 |
| | 12 | 0.105 | 1.78 | 0.512 | 0.267 | 0.198 | 0.722 | 0.701 | 0.137 | 0.517 |
| | 14 | 0.075 | 1.33 | 0.381 | 0.222 | 0.154 | 0.763 | 0.693 | 0.107 | 0.530 |
| | 16 | 0.060 | 1.00 | 0.287 | 0.153 | 0.108 | 0.731 | 0.646 | 0.071 | 0.497 |

**NOTE:** The properties listed in this Table apply only when the sections are adequately braced laterally.

Unless lipped angle compression struts are checked for torsional buckling, these Q-values apply only to situations where such torsional buckling is prevented, as for instance when two angles are connected back to back.

Because it is virtually impossible to load single angle struts concentrically, the design of any such strut should take the eccentricity into account.

**FIGURE 11.41** Properties of equal leg angles with stiffened legs.

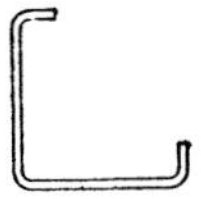

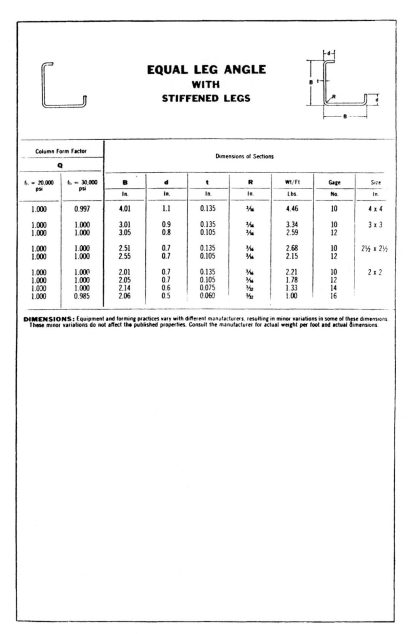

## EQUAL LEG ANGLE WITH STIFFENED LEGS

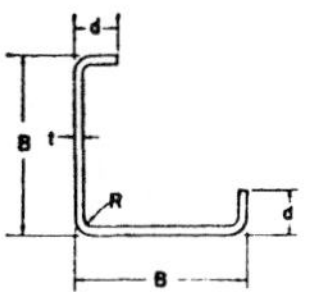


| Column Form Factor Q | | Dimensions of Sections | | | | | | |
|---|---|---|---|---|---|---|---|---|
| $f_b$ = 20,000 psi | $f_b$ = 30,000 psi | B In. | d In. | t In. | R In. | Wt/Ft Lbs. | Gage No. | Size In. |
| 1.000 | 0.997 | 4.01 | 1.1 | 0.135 | 3/16 | 4.46 | 10 | 4 x 4 |
| 1.000 | 1.000 | 3.01 | 0.9 | 0.135 | 3/16 | 3.34 | 10 | 3 x 3 |
| 1.000 | 1.000 | 3.05 | 0.8 | 0.105 | 3/16 | 2.59 | 12 | |
| 1.000 | 1.000 | 2.51 | 0.7 | 0.135 | 3/16 | 2.68 | 10 | 2½ x 2½ |
| 1.000 | 1.000 | 2.55 | 0.7 | 0.105 | 3/16 | 2.15 | 12 | |
| 1.000 | 1.000 | 2.01 | 0.7 | 0.135 | 3/16 | 2.21 | 10 | 2 x 2 |
| 1.000 | 1.000 | 2.05 | 0.7 | 0.105 | 3/16 | 1.78 | 12 | |
| 1.000 | 1.000 | 2.14 | 0.6 | 0.075 | 3/32 | 1.33 | 14 | |
| 1.000 | 0.985 | 2.06 | 0.5 | 0.060 | 3/32 | 1.00 | 16 | |

**DIMENSIONS:** Equipment and forming practices vary with different manufacturers, resulting in minor variations in some of these dimensions. These minor variations do not affect the published properties. Consult the manufacturer for actual weight per foot and actual dimensions.

**FIGURE 11.41** (*Continued*)

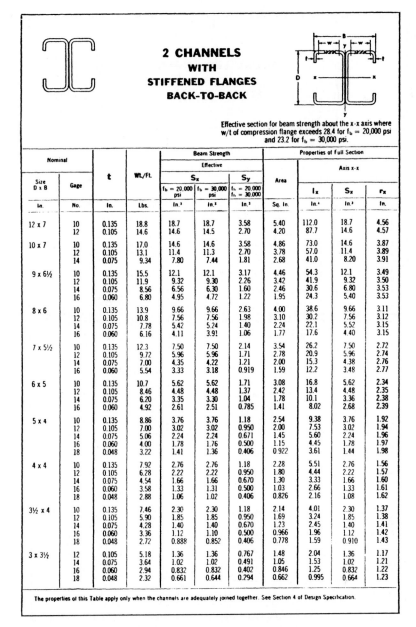

## 2 CHANNELS WITH STIFFENED FLANGES BACK-TO-BACK

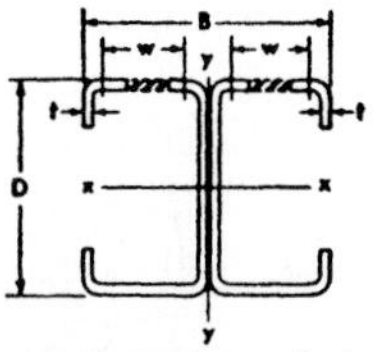


Effective section for beam strength about the x-x axis where w/t of compression flange exceeds 28.4 for $f_b$ = 20,000 psi and 23.2 for $f_b$ = 30,000 psi.

| Nominal | | t | Wt./Ft. | Beam Strength, Effective | | | Properties of Full Section | | | |
|---|---|---|---|---|---|---|---|---|---|---|
| | | | | $S_x$ | | $S_y$ | Area | Axis x-x | | |
| Size D x B | Gage | | | $f_b$ = 20,000 psi | $f_b$ = 30,000 psi | $f_b$ = 20,000 $f_b$ = 30,000 | | $I_x$ | $S_x$ | $r_x$ |
| In. | No. | In. | Lbs. | In.$^3$ | In.$^3$ | In.$^3$ | Sq. In. | In.$^4$ | In.$^3$ | In. |
| 12 x 7 | 10 | 0.135 | 18.8 | 18.7 | 18.7 | 3.58 | 5.40 | 112.0 | 18.7 | 4.56 |
| | 12 | 0.105 | 14.6 | 14.6 | 14.5 | 2.70 | 4.20 | 87.7 | 14.6 | 4.57 |
| 10 x 7 | 10 | 0.135 | 17.0 | 14.6 | 14.6 | 3.58 | 4.86 | 73.0 | 14.6 | 3.87 |
| | 12 | 0.105 | 13.1 | 11.4 | 11.3 | 2.70 | 3.78 | 57.0 | 11.4 | 3.89 |
| | 14 | 0.075 | 9.34 | 7.80 | 7.44 | 1.81 | 2.68 | 41.0 | 8.20 | 3.91 |
| 9 x 6½ | 10 | 0.135 | 15.5 | 12.1 | 12.1 | 3.17 | 4.46 | 54.3 | 12.1 | 3.49 |
| | 12 | 0.105 | 11.9 | 9.32 | 9.30 | 2.26 | 3.42 | 41.9 | 9.32 | 3.50 |
| | 14 | 0.075 | 8.56 | 6.56 | 6.30 | 1.60 | 2.46 | 30.6 | 6.80 | 3.53 |
| | 16 | 0.060 | 6.80 | 4.95 | 4.72 | 1.22 | 1.95 | 24.3 | 5.40 | 3.53 |
| 8 x 6 | 10 | 0.135 | 13.9 | 9.66 | 9.66 | 2.63 | 4.00 | 38.6 | 9.66 | 3.11 |
| | 12 | 0.105 | 10.8 | 7.56 | 7.56 | 1.98 | 3.10 | 30.2 | 7.56 | 3.12 |
| | 14 | 0.075 | 7.78 | 5.42 | 5.24 | 1.40 | 2.24 | 22.1 | 5.52 | 3.15 |
| | 16 | 0.060 | 6.16 | 4.11 | 3.91 | 1.06 | 1.77 | 17.6 | 4.40 | 3.15 |
| 7 x 5½ | 10 | 0.135 | 12.3 | 7.50 | 7.50 | 2.14 | 3.54 | 26.2 | 7.50 | 2.72 |
| | 12 | 0.105 | 9.72 | 5.96 | 5.96 | 1.71 | 2.78 | 20.9 | 5.96 | 2.74 |
| | 14 | 0.075 | 7.00 | 4.35 | 4.22 | 1.21 | 2.00 | 15.3 | 4.38 | 2.76 |
| | 16 | 0.060 | 5.54 | 3.33 | 3.18 | 0.919 | 1.59 | 12.2 | 3.48 | 2.77 |
| 6 x 5 | 10 | 0.135 | 10.7 | 5.62 | 5.62 | 1.71 | 3.08 | 16.8 | 5.62 | 2.34 |
| | 12 | 0.105 | 8.46 | 4.48 | 4.48 | 1.37 | 2.42 | 13.4 | 4.48 | 2.35 |
| | 14 | 0.075 | 6.20 | 3.35 | 3.30 | 1.04 | 1.78 | 10.1 | 3.36 | 2.38 |
| | 16 | 0.060 | 4.92 | 2.61 | 2.51 | 0.785 | 1.41 | 8.02 | 2.68 | 2.39 |
| 5 x 4 | 10 | 0.135 | 8.86 | 3.76 | 3.76 | 1.18 | 2.54 | 9.38 | 3.76 | 1.92 |
| | 12 | 0.105 | 7.00 | 3.02 | 3.02 | 0.950 | 2.00 | 7.53 | 3.02 | 1.94 |
| | 14 | 0.075 | 5.06 | 2.24 | 2.24 | 0.671 | 1.45 | 5.60 | 2.24 | 1.96 |
| | 16 | 0.060 | 4.00 | 1.78 | 1.76 | 0.500 | 1.15 | 4.45 | 1.78 | 1.97 |
| | 18 | 0.048 | 3.22 | 1.41 | 1.36 | 0.406 | 0.922 | 3.61 | 1.44 | 1.98 |
| 4 x 4 | 10 | 0.135 | 7.92 | 2.76 | 2.76 | 1.18 | 2.28 | 5.51 | 2.76 | 1.56 |
| | 12 | 0.105 | 6.28 | 2.22 | 2.22 | 0.950 | 1.80 | 4.44 | 2.22 | 1.57 |
| | 14 | 0.075 | 4.54 | 1.66 | 1.66 | 0.670 | 1.30 | 3.33 | 1.66 | 1.60 |
| | 16 | 0.060 | 3.58 | 1.33 | 1.31 | 0.500 | 1.03 | 2.66 | 1.33 | 1.61 |
| | 18 | 0.048 | 2.88 | 1.06 | 1.02 | 0.406 | 0.826 | 2.16 | 1.08 | 1.62 |
| 3½ x 4 | 10 | 0.135 | 7.46 | 2.30 | 2.30 | 1.18 | 2.14 | 4.01 | 2.30 | 1.37 |
| | 12 | 0.105 | 5.90 | 1.85 | 1.85 | 0.950 | 1.69 | 3.24 | 1.85 | 1.38 |
| | 14 | 0.075 | 4.28 | 1.40 | 1.40 | 0.670 | 1.23 | 2.45 | 1.40 | 1.41 |
| | 16 | 0.060 | 3.36 | 1.12 | 1.10 | 0.500 | 0.966 | 1.96 | 1.12 | 1.42 |
| | 18 | 0.048 | 2.72 | 0.888 | 0.852 | 0.406 | 0.778 | 1.59 | 0.910 | 1.43 |
| 3 x 3½ | 12 | 0.105 | 5.18 | 1.36 | 1.36 | 0.767 | 1.48 | 2.04 | 1.36 | 1.17 |
| | 14 | 0.075 | 3.64 | 1.02 | 1.02 | 0.491 | 1.05 | 1.53 | 1.02 | 1.21 |
| | 16 | 0.060 | 2.94 | 0.832 | 0.832 | 0.402 | 0.846 | 1.25 | 0.832 | 1.22 |
| | 18 | 0.048 | 2.32 | 0.661 | 0.644 | 0.294 | 0.662 | 0.995 | 0.664 | 1.23 |

The properties of this Table apply only when the channels are adequately joined together. See Section 4 of Design Specification.

**FIGURE 11.42** Properties of two channels with stiffened legs.

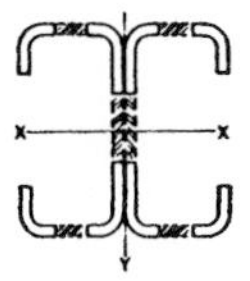

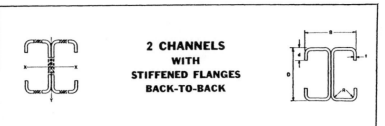

## 2 CHANNELS WITH STIFFENED FLANGES BACK-TO-BACK

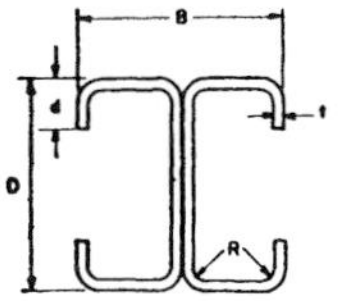

| Properties of Full Section (cont'd) Axis y-y | | | Column Factor Q | | Dimensions of Sections | | | | | (x-x box section) | | Nominal | |
|---|---|---|---|---|---|---|---|---|---|---|---|---|---|
| $I_y$ | $S_y$ | $r_y$ | $f_b$ = 20,000 psi | $f_t$ = 30,000 psi | D | B | d | t | R | $I_y$ | $r_y$ | Gage | Size D x B |
| In.⁴ | In.³ | In. | | | In. | In. | In. | In. | In. | In.⁴ | In. | No. | In. |
| 12.52 | 3.58 | 1.52 | 0.751 | 0.703 | 12.0 | 7.0 | 1.0 | 0.135 | 3/16 | 44.28 | 2.86 | 10 | 12 x 7 |
| 9.45 | 2.70 | 1.50 | 0.689 | 0.640 | 12.0 | 7.0 | 0.9 | 0.105 | 3/16 | 34.95 | 2.88 | 12 | |
| 12.52 | 3.58 | 1.60 | 0.819 | 0.770 | 10.0 | 7.0 | 1.0 | 0.135 | 3/16 | 38.00 | 2.80 | 10 | 10 x 7 |
| 9.45 | 2.70 | 1.58 | 0.756 | 0.705 | 10.0 | 7.0 | 0.9 | 0.105 | 3/16 | 29.96 | 2.81 | 12 | |
| 6.33 | 1.81 | 1.54 | 0.634 | 0.567 | 10.0 | 7.0 | 0.7 | 0.075 | 3/32 | 21.80 | 2.85 | 14 | |
| 10.31 | 3.17 | 1.52 | 0.852 | 0.803 | 9.0 | 6.5 | 1.0 | 0.135 | 3/16 | 29.45 | 2.57 | 10 | 9 x 6½ |
| 7.34 | 2.26 | 1.47 | 0.785 | 0.737 | 9.0 | 6.5 | 0.8 | 0.105 | 3/16 | 23.25 | 2.60 | 12 | |
| 5.19 | 1.60 | 1.45 | 0.674 | 0.607 | 9.0 | 6.5 | 0.7 | 0.075 | 3/32 | 16.98 | 2.62 | 14 | |
| 3.96 | 1.22 | 1.42 | 0.590 | 0.522 | 9.0 | 6.5 | 0.6 | 0.060 | 3/32 | 13.67 | 2.65 | 16 | |
| 7.90 | 2.63 | 1.41 | 0.887 | 0.837 | 8.0 | 6.0 | 0.9 | 0.135 | 3/16 | 22.28 | 2.36 | 10 | 8 x 6 |
| 5.94 | 1.98 | 1.38 | 0.821 | 0.774 | 8.0 | 6.0 | 0.8 | 0.105 | 3/16 | 17.60 | 2.38 | 12 | |
| 4.20 | 1.40 | 1.37 | 0.719 | 0.653 | 8.0 | 6.0 | 0.7 | 0.075 | 3/32 | 12.92 | 2.40 | 14 | |
| 3.19 | 1.06 | 1.34 | 0.635 | 0.565 | 8.0 | 6.0 | 0.6 | 0.060 | 3/32 | 10.39 | 2.42 | 16 | |
| 5.90 | 2.14 | 1.29 | 0.924 | 0.876 | 7.0 | 5.5 | 0.8 | 0.135 | 3/16 | 16.38 | 2.15 | 10 | 7 x 5½ |
| 4.72 | 1.71 | 1.30 | 0.861 | 0.814 | 7.0 | 5.5 | 0.8 | 0.105 | 3/16 | 12.93 | 2.16 | 12 | |
| 3.33 | 1.21 | 1.29 | 0.764 | 0.701 | 7.0 | 5.5 | 0.7 | 0.075 | 3/32 | 9.49 | 2.18 | 14 | |
| 2.53 | 0.919 | 1.26 | 0.686 | 0.615 | 7.0 | 5.5 | 0.6 | 0.060 | 3/32 | 7.68 | 2.20 | 16 | |
| 4.26 | 1.71 | 1.18 | 0.962 | 0.919 | 6.0 | 5.0 | 0.7 | 0.135 | 3/16 | 11.60 | 1.94 | 10 | 6 x 5 |
| 3.42 | 1.37 | 1.19 | 0.905 | 0.858 | 6.0 | 5.0 | 0.7 | 0.105 | 3/16 | 9.18 | 1.95 | 12 | |
| 2.60 | 1.04 | 1.21 | 0.815 | 0.759 | 6.0 | 5.0 | 0.7 | 0.075 | 3/32 | 6.78 | 1.95 | 14 | |
| 1.96 | 0.785 | 1.18 | 0.744 | 0.674 | 6.0 | 5.0 | 0.6 | 0.060 | 3/32 | 5.48 | 1.97 | 16 | |
| 2.36 | 1.18 | 0.962 | 0.994 | 0.964 | 5.0 | 4.0 | 0.7 | 0.135 | 3/16 | 5.97 | 1.53 | 10 | 5 x 4 |
| 1.90 | 0.950 | 0.973 | 0.951 | 0.907 | 5.0 | 4.0 | 0.7 | 0.105 | 3/16 | 4.75 | 1.54 | 12 | |
| 1.34 | 0.671 | 0.961 | 0.858 | 0.811 | 5.0 | 4.0 | 0.6 | 0.075 | 3/32 | 3.54 | 1.56 | 14 | |
| 1.00 | 0.500 | 0.934 | 0.801 | 0.745 | 5.0 | 4.0 | 0.5 | 0.060 | 3/32 | 2.86 | 1.58 | 16 | |
| 0.813 | 0.406 | 0.939 | 0.737 | 0.670 | 5.0 | 4.0 | 0.5 | 0.048 | 3/32 | 2.31 | 1.58 | 18 | |
| 2.35 | 1.18 | 1.02 | 1.000 | 0.998 | 4.0 | 4.0 | 0.7 | 0.135 | 3/16 | 4.98 | 1.48 | 10 | 4 x 4 |
| 1.90 | 0.950 | 1.03 | 0.994 | 0.968 | 4.0 | 4.0 | 0.7 | 0.105 | 3/16 | 3.97 | 1.49 | 12 | |
| 1.34 | 0.670 | 1.02 | 0.926 | 0.885 | 4.0 | 4.0 | 0.6 | 0.075 | 3/32 | 2.96 | 1.51 | 14 | |
| 1.00 | 0.500 | 0.987 | 0.875 | 0.821 | 4.0 | 4.0 | 0.5 | 0.060 | 3/32 | 2.40 | 1.53 | 16 | |
| 0.812 | 0.406 | 0.992 | 0.810 | 0.738 | 4.0 | 4.0 | 0.5 | 0.048 | 3/32 | 1.94 | 1.53 | 18 | |
| 2.35 | 1.18 | 1.05 | 1.000 | 1.000 | 3.5 | 4.0 | 0.7 | 0.135 | 3/16 | 4.47 | 1.45 | 10 | 3½ x 4 |
| 1.90 | 0.950 | 1.06 | 1.000 | 0.991 | 3.5 | 4.0 | 0.7 | 0.105 | 3/16 | 3.57 | 1.45 | 12 | |
| 1.34 | 0.670 | 1.05 | 0.959 | 0.922 | 3.5 | 4.0 | 0.6 | 0.075 | 3/32 | 2.67 | 1.48 | 14 | |
| 1.00 | 0.500 | 1.02 | 0.914 | 0.861 | 3.5 | 4.0 | 0.5 | 0.060 | 3/32 | 2.16 | 1.50 | 16 | |
| 0.812 | 0.406 | 1.02 | 0.851 | 0.779 | 3.5 | 4.0 | 0.5 | 0.048 | 3/32 | 1.75 | 1.50 | 18 | |
| 1.34 | 0.767 | 0.951 | 1.000 | 1.000 | 3.0 | 3.5 | 0.7 | 0.105 | 3/16 | 2.31 | 1.25 | 12 | 3 x 3½ |
| 0.860 | 0.491 | 0.906 | 0.985 | 0.954 | 3.0 | 3.5 | 0.5 | 0.075 | 3/32 | 1.74 | 1.29 | 14 | |
| 0.703 | 0.402 | 0.912 | 0.949 | 0.910 | 3.0 | 3.5 | 0.5 | 0.060 | 3/32 | 1.41 | 1.29 | 16 | |
| 0.515 | 0.294 | 0.881 | 0.900 | 0.840 | 3.0 | 3.5 | 0.4 | 0.048 | 3/32 | 1.14 | 1.31 | 18 | |

**DIMENSIONS:** Equipment and forming practices vary with different manufacturers, resulting in minor variations in some of these dimensions. These minor variations do not affect the published properties. Consult the manufacturer for actual weight per foot and actual dimensions.

**FIGURE 11.42** (*Continued*)

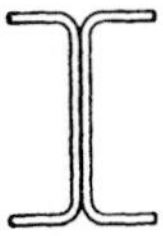

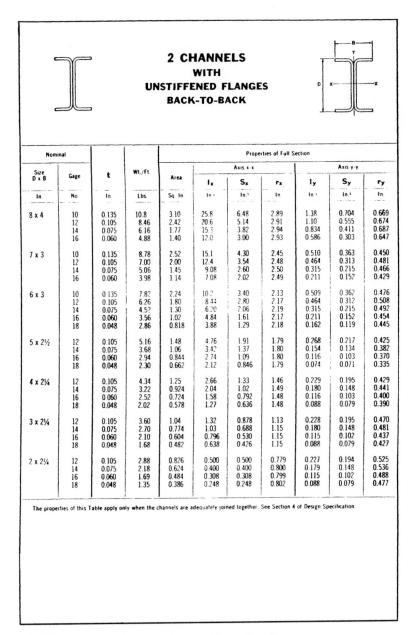

## 2 CHANNELS WITH UNSTIFFENED FLANGES BACK-TO-BACK

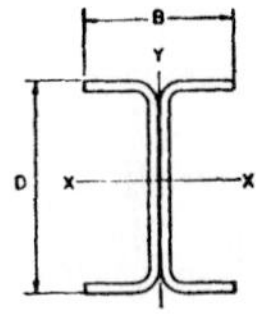


| Nominal | | | | Properties of Full Section | | | | | | | |
|---|---|---|---|---|---|---|---|---|---|---|---|
| | | | | | Axis x-x | | | Axis y-y | | | |
| Size D x B | Gage | t | Wt./Ft. | Area | $I_x$ | $S_x$ | $r_x$ | $I_y$ | $S_y$ | $r_y$ |
| In. | No | In | Lbs. | Sq. In | $In.^4$ | $In.^3$ | In | $In.^4$ | $In.^3$ | In. |
| 8 x 4 | 10 | 0.135 | 10.8 | 3.10 | 25.8 | 6.48 | 2.89 | 1.38 | 0.704 | 0.669 |
| | 12 | 0.105 | 8.46 | 2.42 | 20.6 | 5.14 | 2.91 | 1.10 | 0.555 | 0.674 |
| | 14 | 0.075 | 6.16 | 1.77 | 15.3 | 3.82 | 2.94 | 0.834 | 0.411 | 0.687 |
| | 16 | 0.060 | 4.88 | 1.40 | 12.0 | 3.00 | 2.93 | 0.586 | 0.303 | 0.647 |
| 7 x 3 | 10 | 0.135 | 8.78 | 2.52 | 15.1 | 4.30 | 2.45 | 0.510 | 0.363 | 0.450 |
| | 12 | 0.105 | 7.00 | 2.00 | 12.4 | 3.54 | 2.48 | 0.464 | 0.313 | 0.481 |
| | 14 | 0.075 | 5.06 | 1.45 | 9.08 | 2.60 | 2.50 | 0.315 | 0.215 | 0.466 |
| | 16 | 0.060 | 3.98 | 1.14 | 7.08 | 2.02 | 2.49 | 0.211 | 0.152 | 0.429 |
| 6 x 3 | 10 | 0.135 | 7.82 | 2.24 | 10.2 | 3.40 | 2.13 | 0.509 | 0.362 | 0.476 |
| | 12 | 0.105 | 6.26 | 1.80 | 8.44 | 2.80 | 2.17 | 0.464 | 0.312 | 0.508 |
| | 14 | 0.075 | 4.52 | 1.30 | 6.20 | 2.06 | 2.19 | 0.315 | 0.215 | 0.492 |
| | 16 | 0.060 | 3.56 | 1.02 | 4.84 | 1.61 | 2.17 | 0.211 | 0.152 | 0.454 |
| | 18 | 0.048 | 2.86 | 0.818 | 3.88 | 1.29 | 2.18 | 0.162 | 0.119 | 0.445 |
| 5 x 2½ | 12 | 0.105 | 5.16 | 1.48 | 4.76 | 1.91 | 1.79 | 0.268 | 0.217 | 0.425 |
| | 14 | 0.075 | 3.68 | 1.06 | 3.42 | 1.37 | 1.80 | 0.154 | 0.134 | 0.382 |
| | 16 | 0.060 | 2.94 | 0.844 | 2.74 | 1.09 | 1.80 | 0.116 | 0.103 | 0.370 |
| | 18 | 0.048 | 2.30 | 0.662 | 2.12 | 0.846 | 1.79 | 0.074 | 0.071 | 0.335 |
| 4 x 2¼ | 12 | 0.105 | 4.34 | 1.25 | 2.66 | 1.33 | 1.46 | 0.229 | 0.195 | 0.429 |
| | 14 | 0.075 | 3.22 | 0.924 | 2.04 | 1.02 | 1.49 | 0.180 | 0.148 | 0.441 |
| | 16 | 0.060 | 2.52 | 0.724 | 1.58 | 0.792 | 1.48 | 0.116 | 0.103 | 0.400 |
| | 18 | 0.048 | 2.02 | 0.578 | 1.27 | 0.636 | 1.48 | 0.088 | 0.079 | 0.390 |
| 3 x 2¼ | 12 | 0.105 | 3.60 | 1.04 | 1.32 | 0.878 | 1.13 | 0.228 | 0.195 | 0.470 |
| | 14 | 0.075 | 2.70 | 0.774 | 1.03 | 0.688 | 1.15 | 0.180 | 0.148 | 0.481 |
| | 16 | 0.060 | 2.10 | 0.604 | 0.796 | 0.530 | 1.15 | 0.115 | 0.102 | 0.437 |
| | 18 | 0.048 | 1.68 | 0.482 | 0.638 | 0.426 | 1.15 | 0.088 | 0.079 | 0.427 |
| 2 x 2¼ | 12 | 0.105 | 2.88 | 0.826 | 0.500 | 0.500 | 0.779 | 0.227 | 0.194 | 0.525 |
| | 14 | 0.075 | 2.18 | 0.624 | 0.400 | 0.400 | 0.800 | 0.179 | 0.148 | 0.536 |
| | 16 | 0.060 | 1.69 | 0.484 | 0.308 | 0.308 | 0.799 | 0.115 | 0.102 | 0.488 |
| | 18 | 0.048 | 1.35 | 0.386 | 0.248 | 0.248 | 0.802 | 0.088 | 0.079 | 0.477 |

The properties of this Table apply only when the channels are adequately joined together. See Section 4 of Design Specification.

**FIGURE 11.43** Properties of two channels with unstiffened legs.

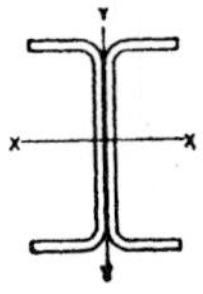


## 2 CHANNELS WITH UNSTIFFENED FLANGES BACK-TO-BACK

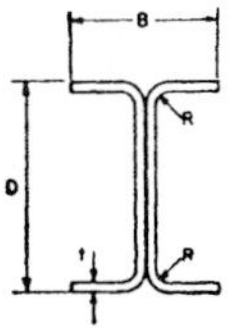


| Allowable Beam Stress $f_c$ psi: $f_b = 20{,}000$ psi | Allowable Beam Stress $f_c$ psi: $f_b = 30{,}000$ psi | Column Factor Q: $f_b = 20{,}000$ psi | Column Factor Q: $f_b = 30{,}000$ psi | D In. | B In. | t In. | R In. | Wt./Ft. Lbs. | Gage No. | Size D x B In. |
|---|---|---|---|---|---|---|---|---|---|---|
| 18960 | 27490 | 0.817 | 0.736 | 8.0 | 3.934 | 0.135 | 3/16 | 10.8 | 10 | 8 x 4 |
| 17100 | 22980 | | | 8.0 | 3.972 | 0.105 | 3/16 | 8.46 | 12 | |
| 13040 | 13150 | | | 8.0 | 4.052 | 0.075 | 3/32 | 6.16 | 14 | |
| 11610 | 11610 | | | 8.0 | 3.882 | 0.060 | 3/32 | 4.88 | 16 | |
| 20000 | 30000 | 0.893 | 0.825 | 7.0 | 2.810 | 0.135 | 3/16 | 8.78 | 10 | 7 x 3 |
| 19340 | 28400 | | | 7.0 | 2.972 | 0.105 | 3/16 | 7.00 | 12 | |
| 16610 | 21790 | | | 7.0 | 2.928 | 0.075 | 3/32 | 5.06 | 14 | |
| 15090 | 18130 | | | 7.0 | 2.758 | 0.060 | 3/32 | 3.98 | 16 | |
| 20000 | 30000 | 0.948 | 0.889 | 6.0 | 2.810 | 0.135 | 3/16 | 7.82 | 10 | 6 x 3 |
| 19340 | 28400 | 0.849 | 0.774 | 6.0 | 2.972 | 0.105 | 3/16 | 6.26 | 12 | |
| 16610 | 21790 | | | 6.0 | 2.928 | 0.075 | 3/32 | 4.52 | 14 | |
| 15090 | 18130 | | | 6.0 | 2.758 | 0.060 | 3/32 | 3.56 | 16 | |
| 12840 | 12840 | | | 6.0 | 2.722 | 0.048 | 3/32 | 2.86 | 18 | |
| 20000 | 30000 | 0.934 | 0.874 | 5.0 | 2.472 | 0.105 | 3/16 | 5.16 | 12 | 5 x 2½ |
| 18550 | 26490 | | | 5.0 | 2.302 | 0.075 | 3/32 | 3.68 | 14 | |
| 17050 | 22870 | | | 5.0 | 2.258 | 0.060 | 3/32 | 2.94 | 16 | |
| 15810 | 19850 | | | 5.0 | 2.098 | 0.048 | 3/32 | 2.30 | 18 | |
| 20000 | 30000 | 0.991 | 0.953 | 4.0 | 2.346 | 0.105 | 3/16 | 4.34 | 12 | 4 x 2¼ |
| 18180 | 25580 | 0.827 | 0.734 | 4.0 | 2.428 | 0.075 | 3/32 | 3.22 | 14 | |
| 17050 | 22870 | | | 4.0 | 2.258 | 0.060 | 3/32 | 2.52 | 16 | |
| 15220 | 18430 | | | 4.0 | 2.222 | 0.048 | 3/32 | 2.02 | 18 | |
| 20000 | 30000 | 1.000 | 1.000 | 3.0 | 2.346 | 0.105 | 3/16 | 3.60 | 12 | 3 x 2¼ |
| 18180 | 25580 | 0.897 | 0.815 | 3.0 | 2.428 | 0.075 | 3/32 | 2.70 | 14 | |
| 17050 | 22870 | 0.808 | 0.694 | 3.0 | 2.258 | 0.060 | 3/32 | 2.10 | 16 | |
| 15220 | 18430 | 0.687 | 0.538 | 3.0 | 2.222 | 0.048 | 3/32 | 1.68 | 18 | |
| 20000 | 30000 | 1.000 | 1.000 | 2.0 | 2.346 | 0.105 | 3/16 | 2.88 | 12 | 2 x 2¼ |
| 18180 | 25580 | 0.909 | 0.853 | 2.0 | 2.428 | 0.075 | 3/32 | 2.18 | 14 | |
| 17050 | 22870 | 0.853 | 0.762 | 2.0 | 2.258 | 0.060 | 3/32 | 1.69 | 16 | |
| 15220 | 18430 | 0.759 | 0.608 | 2.0 | 2.222 | 0.048 | 3/32 | 1.35 | 18 | |

**DIMENSIONS:** Equipment and forming practices vary with different manufacturers, resulting in minor variations in some of these dimensions These minor variations do not affect the published properties. Consult the manufacturer for actual weight per foot and actual dimensions.

Column form factors, Q, for members having webs with w/t-ratios in excess of 60 are not shown. See limitations of Section 2.3.3(a) of the Specification applicable to element stiffened by simple lip.

**FIGURE 11.43** *(Continued)*

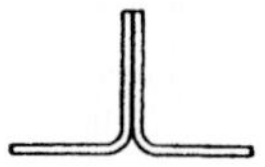

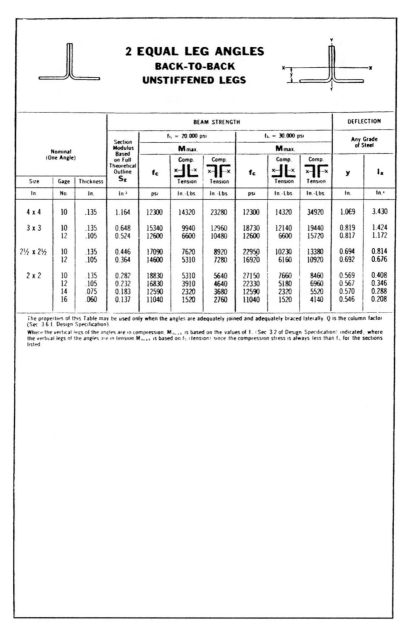

## 2 EQUAL LEG ANGLES
### BACK-TO-BACK
### UNSTIFFENED LEGS

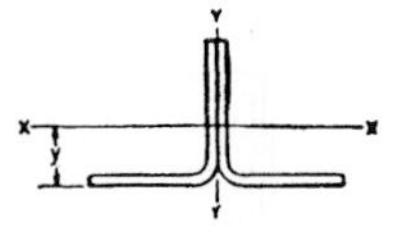


| Nominal (One Angle) | | | BEAM STRENGTH | | | | | | | DEFLECTION | |
|---|---|---|---|---|---|---|---|---|---|---|---|
| | | | Section Modulus Based on Full Theoretical Outline | $f_b$ = 20,000 psi | | | $f_b$ = 30,000 psi | | | Any Grade of Steel | |
| | | | | | $M_{max}$ | | | $M_{max}$ | | | |
| Size | Gage | Thickness | $S_x$ | $f_c$ | Comp. / Tension | Comp. / Tension | $f_c$ | Comp. / Tension | Comp. / Tension | y | $I_x$ |
| In. | No. | In. | In.$^3$ | psi | In.-Lbs. | In.-Lbs. | psi | In.-Lbs. | In.-Lbs. | In. | In.$^4$ |
| 4 x 4 | 10 | .135 | 1.164 | 12300 | 14320 | 23280 | 12300 | 14320 | 34920 | 1.069 | 3.430 |
| 3 x 3 | 10 | .135 | 0.648 | 15340 | 9940 | 12960 | 18730 | 12140 | 19440 | 0.819 | 1.424 |
| | 12 | .105 | 0.524 | 12600 | 6600 | 10480 | 12600 | 6600 | 15720 | 0.817 | 1.172 |
| 2½ x 2½ | 10 | .135 | 0.446 | 17090 | 7620 | 8920 | 22950 | 10230 | 13380 | 0.694 | 0.814 |
| | 12 | .105 | 0.364 | 14600 | 5310 | 7280 | 16920 | 6160 | 10920 | 0.692 | 0.676 |
| 2 x 2 | 10 | .135 | 0.282 | 18830 | 5310 | 5640 | 27150 | 7660 | 8460 | 0.569 | 0.408 |
| | 12 | .105 | 0.232 | 16830 | 3910 | 4640 | 22330 | 5180 | 6960 | 0.567 | 0.346 |
| | 14 | .075 | 0.183 | 12590 | 2320 | 3680 | 12590 | 2320 | 5520 | 0.570 | 0.288 |
| | 16 | .060 | 0.137 | 11040 | 1520 | 2760 | 11040 | 1520 | 4140 | 0.546 | 0.208 |

The properties of this Table may be used only when the angles are adequately joined and adequately braced laterally. Q is the column factor (Sec. 3.6.1, Design Specification).

Where the vertical legs of the angles are in compression, $M_{max}$ is based on the values of $f_c$ (Sec. 3.2 of Design Specification) indicated; where the vertical legs of the angles are in tension $M_{max}$ is based on $f_b$ (tension) since the compression stress is always less than $f_c$ for the sections listed.

**FIGURE 11.44** Properties of two equal leg angles with unstiffened legs.

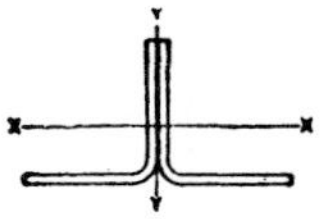


## 2 EQUAL LEG ANGLES
### BACK-TO-BACK
### UNSTIFFENED LEGS

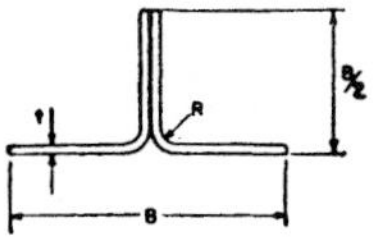


| COLUMN PROPERTIES | | | | | DIMENSIONS | | | | |
|---|---|---|---|---|---|---|---|---|---|
| Full Theoretical Outline | | | Q | | B | Thickness t | Radius R | Weight per Ft. | Nominal Size |
| Area | $r_x$ | $r_y$ | $f_b$ = 20,000 psi | $f_b$ = 30,000 psi | | | | | |
| Sq. In. | In. | In. | | | In. | In. | In. | Lbs. | In. |
| 2.10 | 1.28 | 1.67 | 0.542 | 0.361 | 8.030 | .135 | 3/16 | 7.32 | 4 x 4 |
| 1.56 | 0.955 | 1.26 | 0.767 | 0.624 | 6.030 | .135 | 3/16 | 5.44 | 3 x 3 |
| 1.24 | 0.972 | 1.27 | 0.587 | 0.391 | 6.110 | .105 | 3/16 | 4.32 | |
| 1.29 | 0.793 | 1.05 | 0.854 | 0.764 | 5.030 | .135 | 3/16 | 4.50 | 2½ x 2½ |
| 1.03 | 0.811 | 1.07 | 0.730 | 0.563 | 5.110 | .105 | 3/16 | 3.58 | |
| 1.02 | 0.632 | 0.850 | 0.941 | 0.904 | 4.030 | .135 | 3/16 | 3.56 | 2 x 2 |
| 0.820 | 0.649 | 0.862 | 0.842 | 0.745 | 4.110 | .105 | 3/16 | 2.86 | |
| 0.622 | 0.680 | 0.887 | 0.586 | 0.390 | 4.276 | .075 | 3/32 | 2.16 | |
| 0.482 | 0.658 | 0.855 | 0.401 | 0.267 | 4.128 | .060 | 3/32 | 1.68 | |

**DIMENSIONS:** Equipment and forming practices vary with different manufacturers, resulting in minor variations in some of these dimensions. These minor variations do not affect the published properties. Consult the manufacturer for actual weight per foot and actual dimensions.

**FIGURE 11.44** (*Continued*)

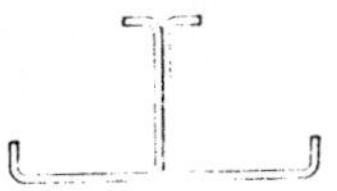

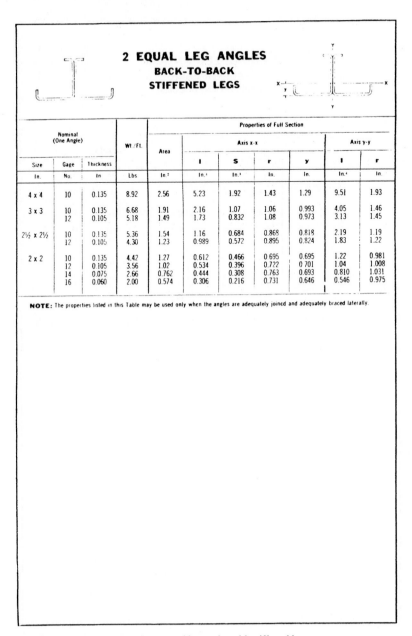

## 2 EQUAL LEG ANGLES
### BACK-TO-BACK
### STIFFENED LEGS

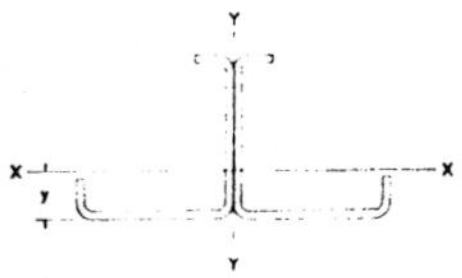


| Nominal (One Angle) | | | Wt./Ft. | Properties of Full Section | | | | | | | |
|---|---|---|---|---|---|---|---|---|---|---|---|
| | | | | Area | Axis x-x | | | | | Axis y-y | |
| Size | Gage | Thickness | | | I | S | r | y | | I | r |
| In. | No. | In | Lbs | In.² | In.⁴ | In.³ | In. | In. | | In.⁴ | In. |
| 4 x 4 | 10 | 0.135 | 8.92 | 2.56 | 5.23 | 1.92 | 1.43 | 1.29 | | 9.51 | 1.93 |
| 3 x 3 | 10 | 0.135 | 6.68 | 1.91 | 2.16 | 1.07 | 1.06 | 0.993 | | 4.05 | 1.46 |
| | 12 | 0.105 | 5.18 | 1.49 | 1.73 | 0.832 | 1.08 | 0.973 | | 3.13 | 1.45 |
| 2½ x 2½ | 10 | 0.135 | 5.36 | 1.54 | 1.16 | 0.684 | 0.868 | 0.818 | | 2.19 | 1.19 |
| | 12 | 0.105 | 4.30 | 1.23 | 0.989 | 0.572 | 0.895 | 0.824 | | 1.83 | 1.22 |
| 2 x 2 | 10 | 0.135 | 4.42 | 1.27 | 0.612 | 0.466 | 0.695 | 0.695 | | 1.22 | 0.981 |
| | 12 | 0.105 | 3.56 | 1.02 | 0.534 | 0.396 | 0.722 | 0.701 | | 1.04 | 1.008 |
| | 14 | 0.075 | 2.66 | 0.762 | 0.444 | 0.308 | 0.763 | 0.693 | | 0.810 | 1.031 |
| | 16 | 0.060 | 2.00 | 0.574 | 0.306 | 0.216 | 0.731 | 0.646 | | 0.546 | 0.975 |

**NOTE:** The properties listed in this Table may be used only when the angles are adequately joined and adequately braced laterally.

**FIGURE 11.45** Properties of two equal leg angles with stiffened legs.

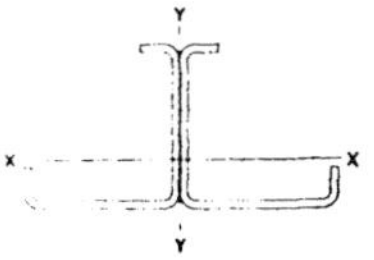

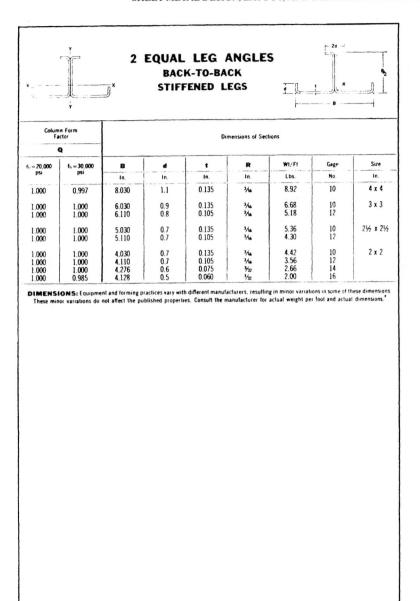

## 2 EQUAL LEG ANGLES
### BACK-TO-BACK
### STIFFENED LEGS

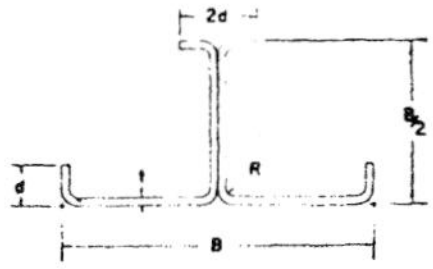


| Column Form Factor Q | | Dimensions of Sections | | | | | | |
|---|---|---|---|---|---|---|---|---|
| $f_{b} = 20{,}000$ psi | $f_{b} = 30{,}000$ psi | B In. | d In. | t In. | R In. | Wt/Ft Lbs. | Gage No. | Size In. |
| 1.000 | 0.997 | 8.030 | 1.1 | 0.135 | 3/16 | 8.92 | 10 | 4 x 4 |
| 1.000 | 1.000 | 6.030 | 0.9 | 0.135 | 3/16 | 6.68 | 10 | 3 x 3 |
| 1.000 | 1.000 | 6.110 | 0.8 | 0.105 | 3/16 | 5.18 | 12 | |
| 1.000 | 1.000 | 5.030 | 0.7 | 0.135 | 3/16 | 5.36 | 10 | 2½ x 2½ |
| 1.000 | 1.000 | 5.110 | 0.7 | 0.105 | 3/16 | 4.30 | 12 | |
| 1.000 | 1.000 | 4.030 | 0.7 | 0.135 | 3/16 | 4.42 | 10 | 2 x 2 |
| 1.000 | 1.000 | 4.110 | 0.7 | 0.105 | 3/16 | 3.56 | 12 | |
| 1.000 | 1.000 | 4.276 | 0.6 | 0.075 | 3/32 | 2.66 | 14 | |
| 1.000 | 0.985 | 4.128 | 0.5 | 0.060 | 3/32 | 2.00 | 16 | |

**DIMENSIONS:** Equipment and forming practices vary with different manufacturers, resulting in minor variations in some of these dimensions. These minor variations do not affect the published properties. Consult the manufacturer for actual weight per foot and actual dimensions.*

**FIGURE 11.45** *(Continued)*

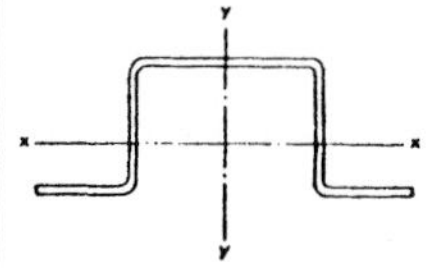


## HAT SECTIONS

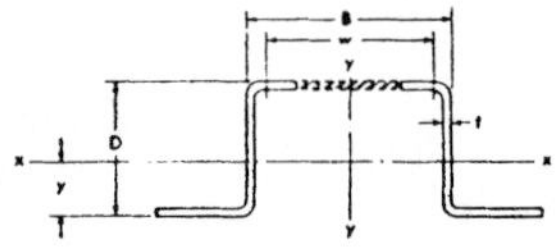

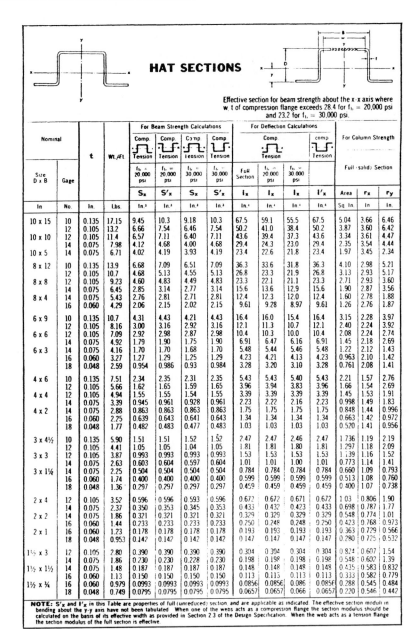

Effective section for beam strength about the x-x axis where w/t of compression flange exceeds 28.4 for $f_b$ = 20,000 psi and 23.2 for $f_b$ = 30,000 psi.

| Nominal | | t | Wt./Ft. | For Beam Strength Calculations | | | | For Deflection Calculations | | | | For Column Strength | | |
|---|---|---|---|---|---|---|---|---|---|---|---|---|---|---|
| | | | | Comp. / Tension | Comp. / Tension | Comp. / Tension | Comp. / Tension | | Comp. / Tension | | comp / Tension | Full (solid) Section | | |
| Size D x B | Gage | | | $f_b$ = 20,000 psi | $f_b$ = 20,000 psi | $f_b$ = 30,000 psi | $f_b$ = 30,000 psi | Full Section | $f_b$ = 20,000 psi | $f_b$ = 30,000 psi | | | | |
| | | | | $S_x$ | $S'_x$ | $S_x$ | $S'_x$ | $I_x$ | $I_x$ | $I_x$ | $I'_x$ | Area | $r_x$ | $r_y$ |
| In | No. | In. | Lbs. | In.$^3$ | In.$^3$ | In.$^3$ | In.$^3$ | In.$^4$ | In.$^4$ | In.$^4$ | In.$^4$ | Sq In. | In | In. |
| 10 x 15 | 10 | 0.135 | 17.15 | 9.45 | 10.3 | 9.18 | 10.3 | 67.5 | 59.1 | 55.5 | 67.5 | 5.04 | 3.66 | 6.46 |
| | 12 | 0.105 | 13.2 | 6.66 | 7.54 | 6.46 | 7.54 | 50.2 | 41.0 | 38.4 | 50.2 | 3.87 | 3.60 | 6.42 |
| 10 x 10 | 12 | 0.105 | 11.4 | 6.57 | 7.11 | 6.40 | 7.11 | 43.6 | 39.4 | 37.3 | 43.6 | 3.34 | 3.61 | 4.47 |
| | 14 | 0.075 | 7.98 | 4.12 | 4.68 | 4.00 | 4.68 | 29.4 | 24.3 | 23.0 | 29.4 | 2.35 | 3.54 | 4.44 |
| 10 x 5 | 14 | 0.075 | 6.71 | 4.02 | 4.19 | 3.93 | 4.19 | 23.4 | 22.6 | 21.8 | 23.4 | 1.97 | 3.45 | 2.34 |
| 8 x 12 | 10 | 0.135 | 13.9 | 6.68 | 7.09 | 6.51 | 7.09 | 36.3 | 33.6 | 31.8 | 36.3 | 4.10 | 2.98 | 5.21 |
| | 12 | 0.105 | 10.7 | 4.68 | 5.13 | 4.55 | 5.13 | 26.8 | 23.3 | 21.9 | 26.8 | 3.13 | 2.93 | 5.17 |
| 8 x 8 | 12 | 0.105 | 9.23 | 4.60 | 4.83 | 4.49 | 4.83 | 23.3 | 22.1 | 21.1 | 23.3 | 2.71 | 2.93 | 3.60 |
| | 14 | 0.075 | 6.45 | 2.85 | 3.14 | 2.77 | 3.14 | 15.6 | 13.6 | 12.9 | 15.6 | 1.90 | 2.87 | 3.56 |
| 8 x 4 | 14 | 0.075 | 5.43 | 2.76 | 2.81 | 2.71 | 2.81 | 12.4 | 12.3 | 12.0 | 12.4 | 1.60 | 2.78 | 1.88 |
| | 16 | 0.060 | 4.29 | 2.06 | 2.15 | 2.02 | 2.15 | 9.61 | 9.28 | 8.97 | 9.61 | 1.26 | 2.76 | 1.87 |
| 6 x 9 | 10 | 0.135 | 10.7 | 4.31 | 4.43 | 4.21 | 4.43 | 16.4 | 16.0 | 15.4 | 16.4 | 3.15 | 2.28 | 3.97 |
| | 12 | 0.105 | 8.16 | 3.00 | 3.16 | 2.92 | 3.16 | 12.1 | 11.3 | 10.7 | 12.1 | 2.40 | 2.24 | 3.92 |
| 6 x 6 | 12 | 0.105 | 7.09 | 2.92 | 2.98 | 2.87 | 2.98 | 10.4 | 10.3 | 10.0 | 10.4 | 2.08 | 2.24 | 2.74 |
| | 14 | 0.075 | 4.92 | 1.79 | 1.90 | 1.75 | 1.90 | 6.91 | 6.47 | 6.16 | 6.91 | 1.45 | 2.18 | 2.69 |
| 6 x 3 | 14 | 0.075 | 4.16 | 1.70 | 1.70 | 1.68 | 1.70 | 5.48 | 5.44 | 5.46 | 5.48 | 1.22 | 2.12 | 1.43 |
| | 16 | 0.060 | 3.27 | 1.27 | 1.29 | 1.25 | 1.29 | 4.23 | 4.21 | 4.13 | 4.23 | 0.963 | 2.10 | 1.42 |
| | 18 | 0.048 | 2.59 | 0.954 | 0.986 | 0.93 | 0.984 | 3.28 | 3.20 | 3.10 | 3.28 | 0.761 | 2.08 | 1.41 |
| 4 x 6 | 10 | 0.135 | 7.51 | 2.34 | 2.35 | 2.31 | 2.35 | 5.43 | 5.43 | 5.40 | 5.43 | 2.21 | 1.57 | 2.76 |
| | 12 | 0.105 | 5.66 | 1.62 | 1.65 | 1.59 | 1.65 | 3.96 | 3.94 | 3.83 | 3.96 | 1.66 | 1.54 | 2.69 |
| 4 x 4 | 12 | 0.105 | 4.94 | 1.55 | 1.55 | 1.54 | 1.55 | 3.39 | 3.39 | 3.39 | 3.39 | 1.45 | 1.53 | 1.91 |
| | 14 | 0.075 | 3.39 | 0.945 | 0.961 | 0.928 | 0.961 | 2.23 | 2.22 | 2.16 | 2.23 | 0.998 | 1.49 | 1.83 |
| 4 x 2 | 14 | 0.075 | 2.88 | 0.863 | 0.863 | 0.863 | 0.863 | 1.75 | 1.75 | 1.75 | 1.75 | 0.848 | 1.44 | 0.996 |
| | 16 | 0.060 | 2.25 | 0.639 | 0.643 | 0.641 | 0.643 | 1.34 | 1.34 | 1.34 | 1.34 | 0.663 | 1.42 | 0.972 |
| | 18 | 0.048 | 1.77 | 0.482 | 0.483 | 0.477 | 0.483 | 1.03 | 1.03 | 1.03 | 1.03 | 0.520 | 1.41 | 0.956 |
| 3 x 4½ | 10 | 0.135 | 5.90 | 1.51 | 1.51 | 1.52 | 1.52 | 2.47 | 2.47 | 2.46 | 2.47 | 1.736 | 1.19 | 2.19 |
| | 12 | 0.105 | 4.41 | 1.05 | 1.05 | 1.04 | 1.05 | 1.81 | 1.81 | 1.80 | 1.81 | 1.297 | 1.18 | 2.09 |
| 3 x 3 | 12 | 0.105 | 3.87 | 0.993 | 0.993 | 0.993 | 0.993 | 1.53 | 1.53 | 1.53 | 1.53 | 1.139 | 1.16 | 1.52 |
| | 14 | 0.075 | 2.63 | 0.603 | 0.604 | 0.597 | 0.604 | 1.01 | 1.01 | 1.00 | 1.01 | 0.773 | 1.14 | 1.41 |
| 3 x 1½ | 14 | 0.075 | 2.25 | 0.504 | 0.504 | 0.504 | 0.504 | 0.784 | 0.784 | 0.784 | 0.784 | 0.660 | 1.09 | 0.793 |
| | 16 | 0.060 | 1.74 | 0.400 | 0.400 | 0.400 | 0.400 | 0.599 | 0.599 | 0.599 | 0.599 | 0.513 | 1.08 | 0.760 |
| | 18 | 0.048 | 1.36 | 0.297 | 0.297 | 0.297 | 0.297 | 0.459 | 0.459 | 0.459 | 0.459 | 0.400 | 1.07 | 0.738 |
| 2 x 4 | 12 | 0.105 | 3.52 | 0.596 | 0.596 | 0.593 | 0.596 | 0.672 | 0.672 | 0.671 | 0.672 | 1.03 | 0.806 | 1.90 |
| | 14 | 0.075 | 2.37 | 0.350 | 0.353 | 0.345 | 0.353 | 0.433 | 0.432 | 0.423 | 0.433 | 0.698 | 0.787 | 1.77 |
| 2 x 2 | 14 | 0.075 | 1.86 | 0.321 | 0.321 | 0.321 | 0.321 | 0.329 | 0.329 | 0.329 | 0.329 | 0.548 | 0.774 | 1.01 |
| | 16 | 0.060 | 1.44 | 0.233 | 0.233 | 0.233 | 0.233 | 0.250 | 0.248 | 0.248 | 0.250 | 0.423 | 0.768 | 0.973 |
| 2 x 1 | 16 | 0.060 | 1.23 | 0.178 | 0.178 | 0.178 | 0.178 | 0.193 | 0.193 | 0.193 | 0.193 | 0.363 | 0.729 | 0.566 |
| | 18 | 0.048 | 0.953 | 0.142 | 0.142 | 0.142 | 0.142 | 0.147 | 0.147 | 0.147 | 0.147 | 0.280 | 0.725 | 0.532 |
| 1½ x 3 | 12 | 0.105 | 2.80 | 0.390 | 0.390 | 0.390 | 0.390 | 0.304 | 0.304 | 0.304 | 0.304 | 0.824 | 0.607 | 1.54 |
| | 14 | 0.075 | 1.86 | 0.230 | 0.230 | 0.228 | 0.230 | 0.198 | 0.198 | 0.198 | 0.198 | 0.548 | 0.602 | 1.39 |
| 1½ x 1½ | 14 | 0.075 | 1.48 | 0.187 | 0.187 | 0.187 | 0.187 | 0.148 | 0.148 | 0.148 | 0.148 | 0.435 | 0.583 | 0.832 |
| | 16 | 0.060 | 1.13 | 0.150 | 0.150 | 0.150 | 0.150 | 0.113 | 0.113 | 0.113 | 0.113 | 0.333 | 0.582 | 0.779 |
| 1½ x ¾ | 16 | 0.060 | 0.979 | 0.0993 | 0.0993 | 0.0993 | 0.0993 | 0.0856 | 0.0856 | 0.086 | 0.0856 | 0.288 | 0.545 | 0.484 |
| | 18 | 0.048 | 0.749 | 0.0795 | 0.0795 | 0.0795 | 0.0795 | 0.0657 | 0.0657 | 0.066 | 0.0657 | 0.220 | 0.546 | 0.442 |

**NOTE:** $S'_x$ **and** $I'_x$ **in this Table are** properties of *full* (unreduced) section, and are applicable as indicated. The effective section moduli in bending about the y-y axis have not been tabulated. When one of the webs acts as a compression flange the section modulus should be calculated on the basis of its effective width as provided in Section 2.3 of the Design Specification. When the web acts as a tension flange the section modulus of the full section is effective.

**FIGURE 11.46** Properties of hat sections.

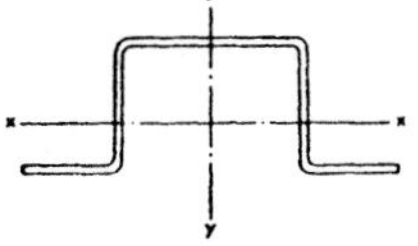

## HAT SECTIONS

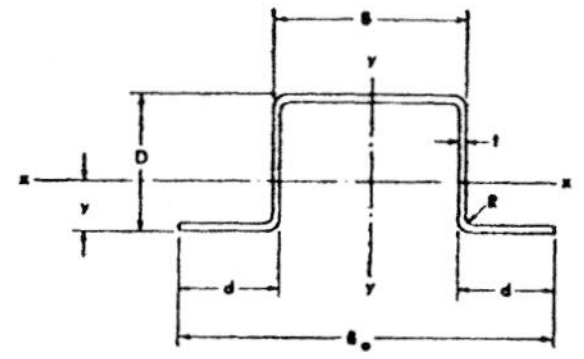

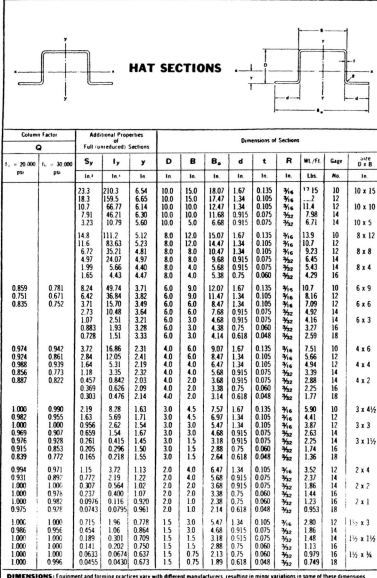

| Column Factor Q: $f_b$ = 20,000 psi | Column Factor Q: $f_b$ = 30,000 psi | Additional Properties of Full (unreduced) Sections: $S_y$ (In.³) | $I_y$ (In.⁴) | y (In.) | Dimensions of Sections: D (In.) | B (In.) | $B_o$ (In.) | d (In.) | t (In.) | R (In.) | Wt./Ft. (Lbs.) | Gage (No.) | Size D x B (In.) |
|---|---|---|---|---|---|---|---|---|---|---|---|---|---|
| | | 23.3 | 210.3 | 6.54 | 10.0 | 15.0 | 18.07 | 1.67 | 0.135 | 3/16 | 17 15 | 10 | 10 x 15 |
| | | 18.3 | 159.5 | 6.65 | 10.0 | 15.0 | 17.47 | 1.34 | 0.105 | 3/16 | [illegible] | 12 | |
| | | 10.7 | 66.77 | 6.14 | 10.0 | 10.0 | 12.47 | 1.34 | 0.105 | 3/16 | 11.4 | 12 | 10 x 10 |
| | | 7.91 | 46.21 | 6.30 | 10.0 | 10.0 | 11.68 | 0.915 | 0.075 | 3/32 | 7.98 | 14 | |
| | | 3.23 | 10.79 | 5.60 | 10.0 | 5.0 | 6.68 | 0.915 | 0.075 | 3/32 | 6.71 | 14 | 10 x 5 |
| | | 14.8 | 111.2 | 5.12 | 8.0 | 12.0 | 15.07 | 1.67 | 0.135 | 3/16 | 13.9 | 10 | 8 x 12 |
| | | 11.6 | 83.63 | 5.23 | 8.0 | 12.0 | 14.47 | 1.34 | 0.105 | 3/16 | 10.7 | 12 | |
| | | 6.72 | 35.21 | 4.81 | 8.0 | 8.0 | 10.47 | 1.34 | 0.105 | 3/16 | 9.23 | 12 | 8 x 8 |
| | | 4.97 | 24.07 | 4.97 | 8.0 | 8.0 | 9.68 | 0.915 | 0.075 | 3/32 | 6.45 | 14 | |
| | | 1.99 | 5.66 | 4.40 | 8.0 | 4.0 | 5.68 | 0.915 | 0.075 | 3/32 | 5.43 | 14 | 8 x 4 |
| | | 1.65 | 4.43 | 4.47 | 8.0 | 4.0 | 5.38 | 0.75 | 0.060 | 3/32 | 4.29 | 16 | |
| 0.859 | 0.781 | 8.24 | 49.74 | 3.71 | 6.0 | 9.0 | 12.07 | 1.67 | 0.135 | 3/16 | 10.7 | 10 | 6 x 9 |
| 0.751 | 0.671 | 6.42 | 36.84 | 3.82 | 6.0 | 9.0 | 11.47 | 1.34 | 0.105 | 3/16 | 8.16 | 12 | |
| 0.835 | 0.752 | 3.71 | 15.70 | 3.49 | 6.0 | 6.0 | 8.47 | 1.34 | 0.105 | 3/16 | 7.09 | 12 | 6 x 6 |
| | | 2.73 | 10.48 | 3.64 | 6.0 | 6.0 | 7.68 | 0.915 | 0.075 | 3/32 | 4.92 | 14 | |
| | | 1.07 | 2.51 | 3.21 | 6.0 | 3.0 | 4.68 | 0.915 | 0.075 | 3/32 | 4.16 | 14 | 6 x 3 |
| | | 0.883 | 1.93 | 3.28 | 6.0 | 3.0 | 4.38 | 0.75 | 0.060 | 3/32 | 3.27 | 16 | |
| | | 0.728 | 1.51 | 3.33 | 6.0 | 3.0 | 4.14 | 0.618 | 0.048 | 3/32 | 2.59 | 18 | |
| 0.974 | 0.942 | 3.72 | 16.86 | 2.31 | 4.0 | 6.0 | 9.07 | 1.67 | 0.135 | 3/16 | 7.51 | 10 | 4 x 6 |
| 0.924 | 0.861 | 2.84 | 12.05 | 2.41 | 4.0 | 6.0 | 8.47 | 1.34 | 0.105 | 3/16 | 5.66 | 12 | |
| 0.988 | 0.939 | 1.64 | 5.31 | 2.19 | 4.0 | 4.0 | 6.47 | 1.34 | 0.105 | 3/16 | 4.94 | 12 | 4 x 4 |
| 0.856 | 0.773 | 1.18 | 3.35 | 2.32 | 4.0 | 4.0 | 5.68 | 0.915 | 0.075 | 3/32 | 3.39 | 14 | |
| 0.887 | 0.822 | 0.457 | 0.842 | 2.03 | 4.0 | 2.0 | 3.68 | 0.915 | 0.075 | 3/32 | 2.88 | 14 | 4 x 2 |
| | | 0.369 | 0.626 | 2.09 | 4.0 | 2.0 | 3.38 | 0.75 | 0.060 | 3/32 | 2.25 | 16 | |
| | | 0.303 | 0.476 | 2.14 | 4.0 | 2.0 | 3.14 | 0.618 | 0.048 | 3/32 | 1.77 | 18 | |
| 1.000 | 0.990 | 2.19 | 8.28 | 1.63 | 3.0 | 4.5 | 7.57 | 1.67 | 0.135 | 3/16 | 5.90 | 10 | 3 x 4½ |
| 0.982 | 0.955 | 1.63 | 5.69 | 1.71 | 3.0 | 4.5 | 6.97 | 1.34 | 0.105 | 3/16 | 4.41 | 12 | |
| 1.000 | 1.000 | 0.956 | 2.62 | 1.54 | 3.0 | 3.0 | 5.47 | 1.34 | 0.105 | 3/16 | 3.87 | 12 | 3 x 3 |
| 0.969 | 0.907 | 0.659 | 1.54 | 1.67 | 3.0 | 3.0 | 4.68 | 0.915 | 0.075 | 3/32 | 2.63 | 14 | |
| 0.976 | 0.928 | 0.261 | 0.415 | 1.45 | 3.0 | 1.5 | 3.18 | 0.915 | 0.075 | 3/32 | 2.25 | 14 | 3 x 1½ |
| 0.915 | 0.853 | 0.205 | 0.296 | 1.50 | 3.0 | 1.5 | 2.88 | 0.75 | 0.060 | 3/32 | 1.74 | 16 | |
| 0.839 | 0.772 | 0.165 | 0.218 | 1.55 | 3.0 | 1.5 | 2.64 | 0.618 | 0.048 | 3/32 | 1.36 | 18 | |
| 0.994 | 0.971 | 1.15 | 3.72 | 1.13 | 2.0 | 4.0 | 6.47 | 1.34 | 0.105 | 3/16 | 3.52 | 12 | 2 x 4 |
| 0.931 | 0.892 | 0.772 | 2.19 | 1.22 | 2.0 | 4.0 | 5.68 | 0.915 | 0.075 | 3/32 | 2.37 | 14 | |
| 1.000 | 1.000 | 0.307 | 0.564 | 1.02 | 2.0 | 2.0 | 3.68 | 0.915 | 0.075 | 3/32 | 1.86 | 14 | 2 x 2 |
| 1.000 | 0.978 | 0.237 | 0.400 | 1.07 | 2.0 | 2.0 | 3.38 | 0.75 | 0.060 | 3/32 | 1.44 | 16 | |
| 1.000 | 0.982 | 0.0976 | 0.116 | 0.920 | 2.0 | 1.0 | 2.38 | 0.75 | 0.060 | 3/32 | 1.23 | 16 | 2 x 1 |
| 0.975 | 0.928 | 0.0743 | 0.0795 | 0.961 | 2.0 | 1.0 | 2.14 | 0.618 | 0.048 | 3/32 | 0.953 | 18 | |
| 1.000 | 1.000 | 0.715 | 1.96 | 0.778 | 1.5 | 3.0 | 5.47 | 1.34 | 0.105 | 3/16 | 2.80 | 12 | 1½ x 3 |
| 0.986 | 0.956 | 0.454 | 1.06 | 0.864 | 1.5 | 3.0 | 4.68 | 0.915 | 0.075 | 3/32 | 1.86 | 14 | |
| 1.000 | 1.000 | 0.189 | 0.301 | 0.709 | 1.5 | 1.5 | 3.18 | 0.915 | 0.075 | 3/32 | 1.48 | 14 | 1½ x 1½ |
| 1.000 | 1.000 | 0.141 | 0.202 | 0.750 | 1.5 | 1.5 | 2.88 | 0.75 | 0.060 | 3/32 | 1.13 | 16 | |
| 1.000 | 1.000 | 0.0633 | 0.0674 | 0.637 | 1.5 | 0.75 | 2.13 | 0.75 | 0.060 | 3/32 | 0.979 | 16 | 1½ x ¾ |
| 1.000 | 0.996 | 0.0455 | 0.0430 | 0.673 | 1.5 | 0.75 | 1.89 | 0.618 | 0.048 | 3/32 | 0.749 | 18 | |

**DIMENSIONS:** Equipment and forming practices vary with different manufacturers, resulting in minor variations in some of these dimensions. These minor variations do not affect the published properties. Consult the manufacturer for actual weight per foot and actual dimensions.

Column form factors, Q, for members having webs with w/t-ratios in excess of 60 are not shown. See limitations of Section 2.3.3(a) of the Specification applicable to element stiffened by a simple lip.

**FIGURE 11.46** (*Continued*)

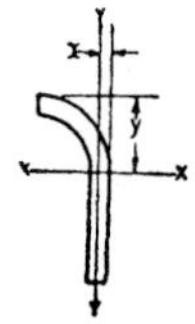


## ONE FLANGE STIFFENER

(Which Includes One 90° Corner)

## PROPERTIES AND DIMENSIONS

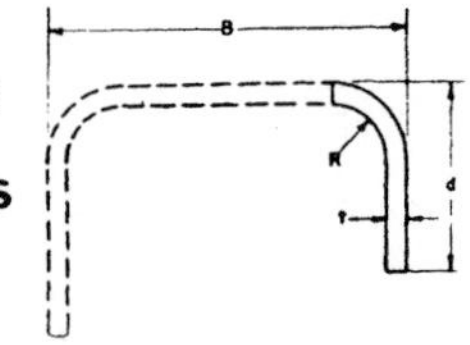


Stock Width of Blank Taken at t/3 Distance From Inner Surface.

| Nominal Gage | Thickness t | Depth d | PROPERTIES X-X AXIS $I_x$ | X-X AXIS y | Y-Y AXIS $I_y$ | Y-Y AXIS x | Area | DIMENSIONS Max Flange B | Blank Width | Thickness t | Radius R | Depth d | Nominal Gage |
|---|---|---|---|---|---|---|---|---|---|---|---|---|---|
| No. | In. | In. | In.⁴ | In. | In.⁴ | In. | Sq. In. | In. | In. | In. | In. | In. | No. |
| | | 1.0 | .01255 | .4737 | .000796 | .1005 | .14554 | 3.87 | 1.043 | .135 | 3/16 | 1.0 | |
| | | .9 | .00916 | .4250 | .000759 | .1039 | .13204 | 3.25 | 0.943 | .135 | 3/16 | .9 | |
| | | .8 | .00643 | .3766 | .000719 | .1080 | .11854 | 2.81 | 0.843 | .135 | 3/16 | .8 | |
| 10 | .135 | .7 | .00430 | .3286 | .000674 | .1133 | .10504 | 2.53 | 0.743 | .135 | 3/16 | .7 | 10 |
| | | .6 | .00269 | .2811 | .000621 | .1200 | .09154 | 2.37 | 0.643 | .135 | 3/16 | .6 | |
| | | .5 | .00153 | .2346 | .000556 | .1291 | .07804 | 2.30 | 0.543 | .135 | 3/16 | .5 | |
| | | .4 | .00076 | .1896 | .000474 | .1420 | .06454 | 2.27 | 0.443 | .135 | 3/16 | .4 | |
| | | .9 | .00735 | .4205 | .000475 | .0850 | .10337 | 4.28 | 0.957 | .105 | 3/16 | .9 | |
| | | .8 | .00518 | .3719 | .000453 | .0886 | .09287 | 3.34 | 0.857 | .105 | 3/16 | .8 | |
| | | .7 | .00348 | .3237 | .000428 | .0932 | .08237 | 2.66 | 0.757 | .105 | 3/16 | .7 | |
| 12 | .105 | .6 | .00219 | .2761 | .000398 | .0992 | .07187 | 2.22 | 0.657 | .105 | 3/16 | .6 | 12 |
| | | .5 | .00126 | .2292 | .000362 | .1072 | .06137 | 1.98 | 0.557 | .105 | 3/16 | .5 | |
| | | .4 | .00064 | .1836 | .000314 | .1185 | .05087 | 1.88 | 0.457 | .105 | 3/16 | .4 | |
| | | .9 | .00493 | .4351 | .000081 | .0475 | .07031 | 6.79 | 0.918 | .075 | 3/32 | .9 | |
| | | .8 | .00348 | .3855 | .000076 | .0487 | .06281 | 4.94 | 0.818 | .075 | 3/32 | .8 | |
| 14 | .075 | .7 | .00234 | .3361 | .000072 | .0502 | .05531 | 3.50 | 0.718 | .075 | 3/32 | .7 | 14 |
| | | .6 | .00148 | .2869 | .000067 | .0522 | .04781 | 2.46 | 0.618 | .075 | 3/32 | .6 | |
| | | .5 | .00086 | .2379 | .000062 | .0549 | .04031 | 1.77 | 0.518 | .075 | 3/32 | .5 | |
| | | .4 | .00044 | .1894 | .000055 | .0589 | .03281 | 1.40 | 0.418 | .075 | 3/32 | .4 | |
| | | .8 | .00283 | .3836 | .000048 | .0400 | .05044 | 7.50 | 0.825 | .060 | 3/32 | .8 | |
| | | .7 | .00191 | .3341 | .000045 | .0414 | .04444 | 5.19 | 0.725 | .060 | 3/32 | .7 | |
| 16 | .060 | .6 | .00121 | .2848 | .000043 | .0432 | .03844 | 3.46 | 0.625 | .060 | 3/32 | .6 | 16 |
| | | .5 | .00071 | .2357 | .000040 | .0456 | .03244 | 2.23 | 0.525 | .060 | 3/32 | .5 | |
| | | .4 | .00036 | .1871 | .000036 | .0492 | .02644 | 1.48 | 0.425 | .060 | 3/32 | .4 | |
| | | .7 | .00155 | .3325 | .000030 | .0344 | .03567 | 7.96 | 0.731 | .048 | 3/32 | .7 | |
| 18 | .048 | .6 | .00099 | .2831 | .000028 | .0360 | .03087 | 5.19 | 0.631 | .048 | 3/32 | .6 | 18 |
| | | .5 | .00058 | .2340 | .000026 | .0382 | .02607 | 3.20 | 0.531 | .048 | 3/32 | .5 | |
| | | .4 | .00030 | .1853 | .000024 | .0414 | .02127 | 1.87 | 0.431 | .048 | 3/32 | .4 | |

**FIGURE 11.47** Properties of one-flange stiffener.

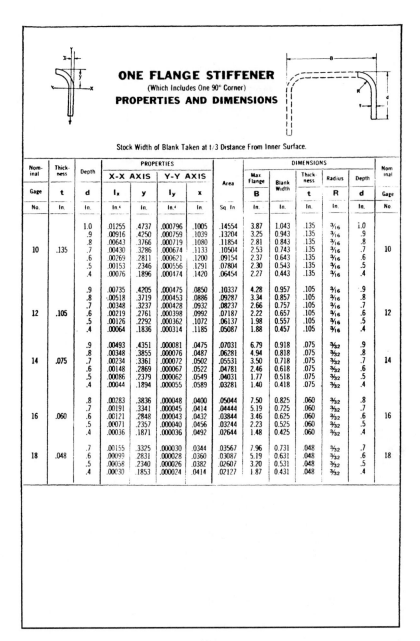

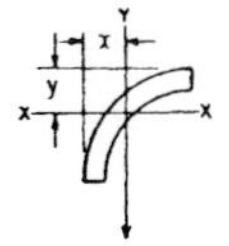

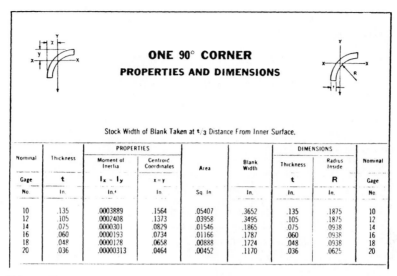

**ONE 90° CORNER**
**PROPERTIES AND DIMENSIONS**

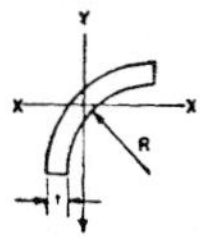


Stock Width of Blank Taken at t/3 Distance From Inner Surface.

| Nominal | Thickness | PROPERTIES: Moment of Inertia | PROPERTIES: Centroid Coordinates | Area | Blank Width | DIMENSIONS: Thickness | DIMENSIONS: Radius Inside | Nominal |
|---|---|---|---|---|---|---|---|---|
| Gage | t | $I_x = I_y$ | x = y | | | t | R | Gage |
| No. | In. | $In.^4$ | In. | Sq. In. | In. | In. | In. | No. |
| 10 | .135 | .0003889 | .1564 | .05407 | .3652 | .135 | .1875 | 10 |
| 12 | .105 | .0002408 | .1373 | .03958 | .3495 | .105 | .1875 | 12 |
| 14 | .075 | .0000301 | .0829 | .01546 | .1865 | .075 | .0938 | 14 |
| 16 | .060 | .0000193 | .0734 | .01166 | .1787 | .060 | .0938 | 16 |
| 18 | .048 | .0000128 | .0658 | .00888 | .1724 | .048 | .0938 | 18 |
| 20 | .036 | .00000313 | .0464 | .00452 | .1170 | .036 | .0625 | 20 |

**FIGURE 11.48** Properties of one 90° corner.

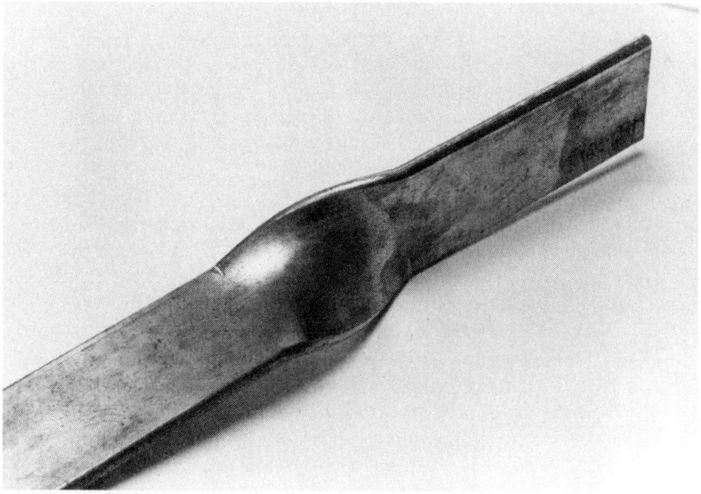

**FIGURE 11.49** A copper bus bar with a 90° axial twist.

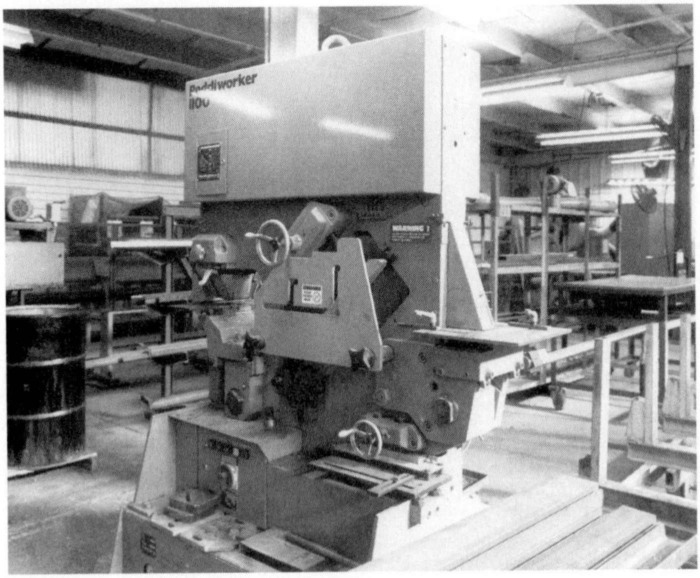

**FIGURE 11.50** A structural shape notching and cutting machine.

**FIGURE 11.51** A typical switchgear lineup made of bent, welded, and bolted sheet steel.

**FIGURE 11.52** A typical die-punched and progressive die-bent sheet-metal part of high accuracy.

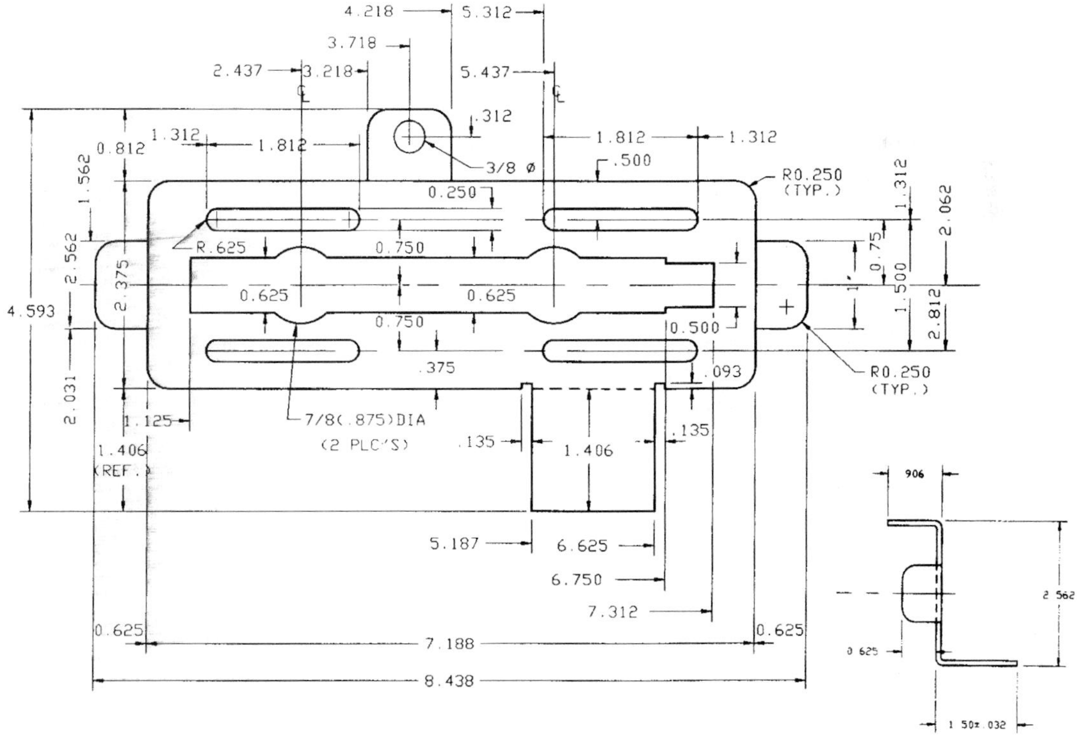


**FIGURE 11.53** Typical AutoCad drawing of a complex sheet-metal part.

# CHAPTER 12

# CASTINGS, MOLDINGS, EXTRUSIONS, AND POWDER-METAL TECHNOLOGY

## 12.1 CASTINGS

The casting of metals and alloys is an important branch of the metalworking industry and of prime importance to the electromechanical design engineer. Castings allow parts to be made at a rapid pace, with controlled accuracy, using a large variety of metals and alloys. Castings replace machined parts which would otherwise be difficult or impossible to machine and very costly to manufacture. At the same time, it should be noted that castings cannot replace many types of machined parts due to material, configuration, stress requirements, dynamics, and other physical considerations. There are many types or processes used in the casting or foundry industries to produce cast parts in different materials and for various dimensional accuracy requirements. The casting processes or methods that are in use today include

1. Sand casting
2. Shell casting
3. Carbon dioxide casting
4. Fluid sand casting
5. Composite mold casting
6. Plaster mold casting
7. Slush casting
8. Evaporative pattern casting (EPC)
9. Die casting
10. Permanent mold casting
11. Ceramic mold casting
12. Investment casting (lost-wax process)

The casting method or process chosen by the design engineer and the foundry is determined by the following factors:

- Type of metal to be cast
- Size of part to be cast

- Required cast accuracy of the part
- Economics
- Required secondary operations such as machining, hardening, welding, and plating

### 12.1.1 Sand Casting

In the sand-casting process, a wooden, plastic, or metal pattern is packed in a special sand, which is dampened with water, and then removed, leaving a hollow space having the shape of the part. The pattern is purposely made larger than the size of the cast part to allow for shrinkage of the casting as it cools.

The mold consists of two steel frames, which are called the *cope* (top half) and the *drag* (bottom half). See Fig. 12.1 for an illustration of a typical sand-casting setup.

Sand cores may be placed in the cavity to produce holes in the part where required. Once the cope and drag are clamped together, the molten metal is poured into the "gate" of the mold. Vent holes placed appropriately in the mold allow hot gases to escape from the mold cavity during pouring. The pouring temperature of the metal is always made a few hundred degrees higher than the melting point, so that the metal has good fluidity during the pour and so that it does not cool prematurely, causing voids in the part.

Sand casting is the least expensive of all the casting processes on a part-to-part basis, but a need for secondary machining operations may indicate the use of one of the other casting processes. Figure 12.2 shows a sand-cast part made of a copper alloy. This part would be difficult to machine from solid stock.

Figure 12.3 shows another sand-cast part of intricate design. This part is made of 356-T6 aluminum alloy (heat-treated). Note that at the center section of the part, there is a sprocket for number 40 ANSI roller chain. The sprocket in this application need not have a high degree of accuracy, as the sprocket and chain action is only intermittent. A part of such design would be extremely difficult and costly to manufacture using the machining processes or methods.

Figure 12.4 illustrations *a* through *h* show some of the gating and venting methods employed in sand casting and similar casting processes.

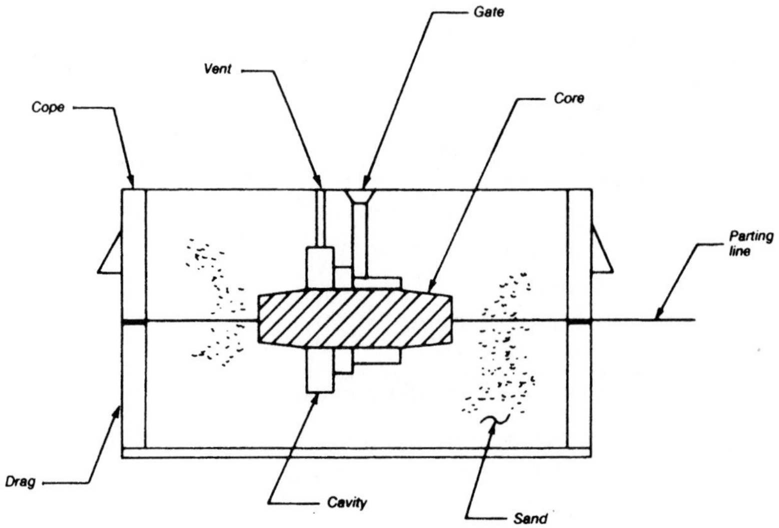


**FIGURE 12.1** A typical sand-casting mold setup.

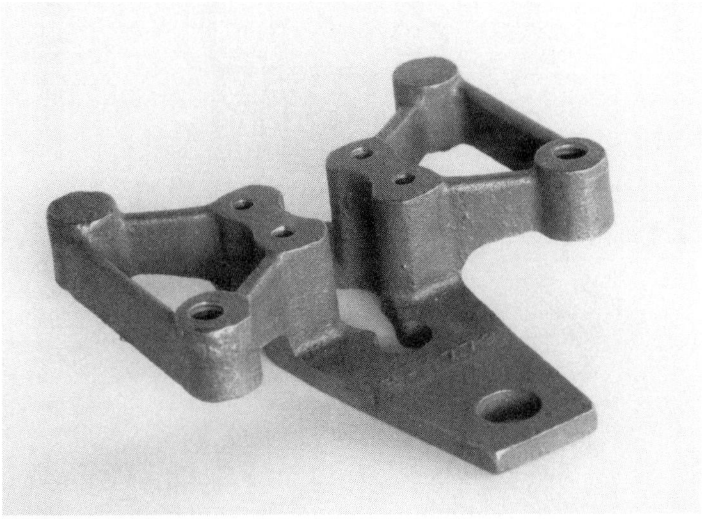

**FIGURE 12.2** An intricate sand-cast copper alloy part.

**FIGURE 12.3** An intricate sand-cast aluminum alloy part (356-T6).

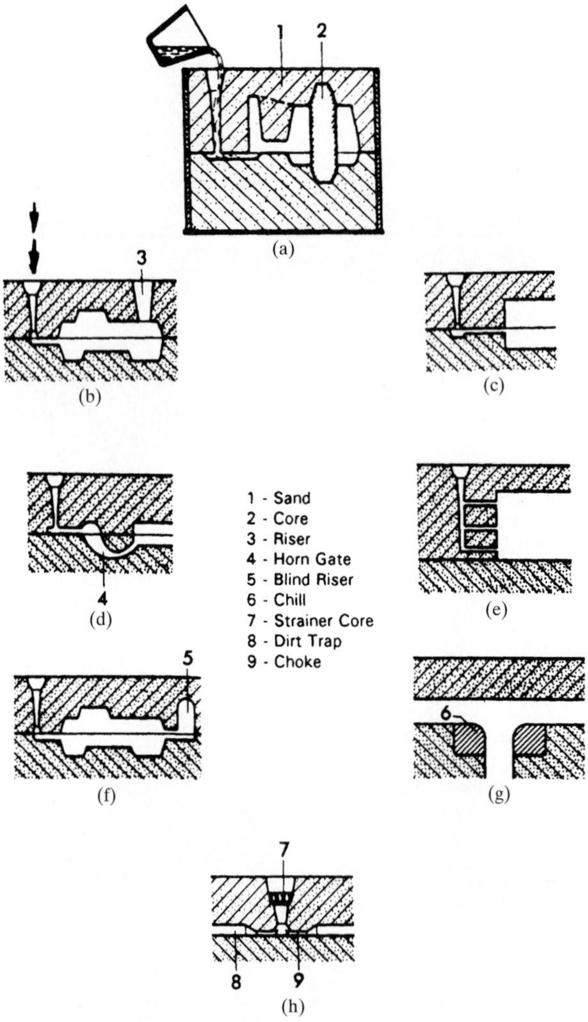


**FIGURE 12.4** Molds and gating methods.

Figure 12.5 shows a typical combination casting–machining engineering drawing. This drawing provides enough information for the patternmaker, foundry, and machine shop to be able to produce the part.

## 12.1.2 Shell Casting

This process entails forming a mold from a mixture of sand and a thermosetting resin binder. The sand-thermoset mix is placed against a heated pattern, causing the resin to bind the sand particles and form a strong shell. After the shell is cured and stripped from the pattern, cores may be set. The cope and drag are secured together and placed in a flask with added backup material. The mold is then ready for pouring the molten metal.

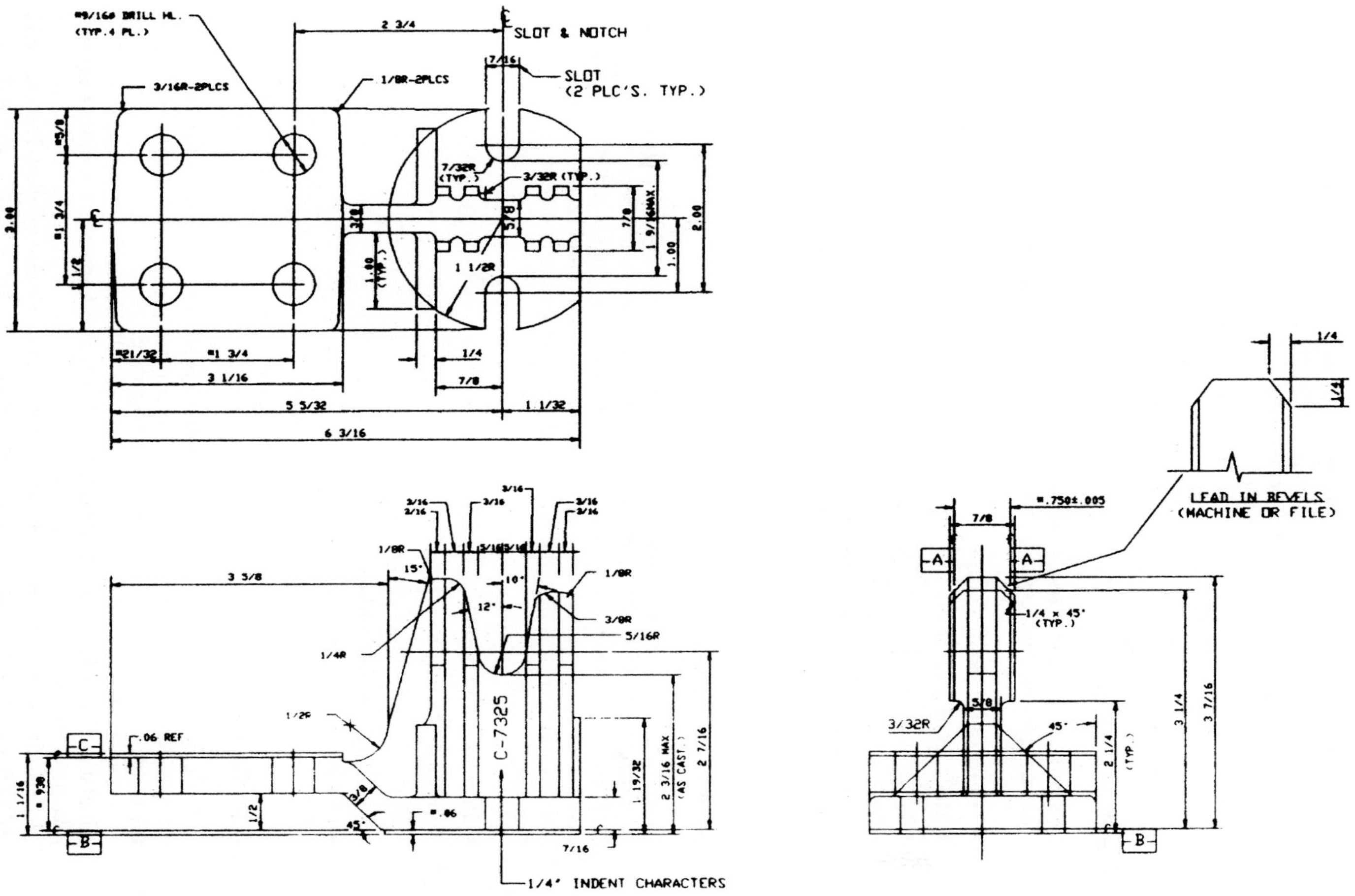


**FIGURE 12.5** A combination pattern, casting, and machining drawing.

### 12.1.3 Carbon Dioxide Casting

The carbon dioxide process uses sodium silicate (water glass) binders instead of the clay binders employed in conventional sand casting. Treated with carbon dioxide gas, the sodium silicate–sand mixture is dried and strengthened.

Ready-for-use cores or molds can be made using this process in a few minutes, with no baking required. Excellent dimensional accuracy is obtained even while making cores rapidly. This process is used more often to make cores than to make molds.

### 12.1.4 Plaster Mold Casting

This process is used when nonferrous metals must be cast more accurately than is possible with conventional sand casting. The four recognized plaster mold processes are

- Conventional mold
- Match-plate pattern
- The Antioch method
- The foamed plaster method

Castings produced by the plaster mold process have smoother surfaces, better accuracy, and finer detail than those made by conventional sand casting, but they are also more expensive.

### 12.1.5 Composite Mold Casting

Composite mold casting utilizes different sections of the mold and cores made by different methods so that the greatest advantage is obtained from each process in the appropriate section. This process is usually chosen for aluminum parts and is sometimes called *premium-quality casting* or *engineered casting.* The use of plaster mold sections affords accuracy and stability wherever it is essential on the cast part. Composite molds are used for the following reasons:

- Decreased cost of mold material
- Increased casting accuracy
- Decreased amount of gassing
- Improved finish or surface
- Quicker processing time

### 12.1.6 Investment Casting

In this casting process, an expendable pattern is coated with a refractory slurry that sets at ambient temperature. The expendable pattern (wax or plastic) is then melted out of the refractory shell. Ceramic cores are used as required.

There are two distinct processes or methods followed in investment casting: shell investment and solid investment. Investment casting is also known as the "lost wax" process and *precision casting.*

Cast parts can be produced in almost any pourable metal or alloy. The finished parts are dimensionally accurate and are generally used as-cast. The process is used for high accuracy, mass-produced parts, and making jewelry.

The economics of this process must be weighed against the complexity of the part. Simple parts are generally not economical to produce with this process. The process is advantageous for applicable parts not in excess of 10 lb weight, although heavier parts are frequently investment cast.

### 12.1.7 Ceramic Mold Casting

Ceramic molding techniques are proprietary processes. They utilize permanent patterns and fine-grain zircon and calcined, high-alumina Mullite slurries for molding.

As in other processes, the molds are constructed as a *cope* and a *drag.* Fine detail may be produced with high-dimensional accuracy. The refractory mold allows the pouring of high-melting-point metals and alloys. The ferrous alloys and metals are more commonly cast with this process. Aluminum, beryllium copper, titanium, ductile iron, carbon, and low-alloy steels and tool steels are cast using ceramic-mold processes.

Two of the proprietary processes for ceramic-mold casting are the Shaw method and the unicast method.

### 12.1.8 Permanent Mold Casting

In this process, a metal mold consisting of two or more parts produces the cast parts. Metal, sand, or plaster cores are also used, in which case the process is known as *semipermanent mold casting.*

Intricate castings can be produced, but mold cost is high. Not all metals and alloys can be cast, and some shapes cannot be made because of the parting line, or the difficulty of removing the part from the mold. Suitable casting metals include

| | |
|---|---|
| Aluminum alloys | ≤30 lb |
| Magnesium alloys | ≤15 lb |
| Copper alloys | ≤20 lb |
| Zinc alloys | ≤20 lb |
| Gray iron (hypereutectic) | ≤30 lb |

A variation of this process chills the cast metal rapidly, producing enhanced properties with regard to grain configuration and size. Surface qualities of permanent mold casting are better than conventional sand castings, but the mold cost must always be evaluated with respect to quantity of parts produced and secondary operations required (machining) to produce a finished part. Many parts cast with this process can be used as-cast when close tolerances are not required. Figure 12.6 shows a relatively complex part which was cast by the permanent mold chill-cast process in an aluminum bronze, high-strength alloy.

The type of metals which can be permanent-mold-cast are limited to those whose melting points are not above the copper-base alloys. Pouring higher temperature alloys causes permanent damage to the steel molds.

### 12.1.9 Die Casting

Die castings are made by forcing molten metal under high pressure into permanent molds called *dies.* The advantages of die casting include

- Complex shapes possible
- Thin-walled sections possible
- High production rates

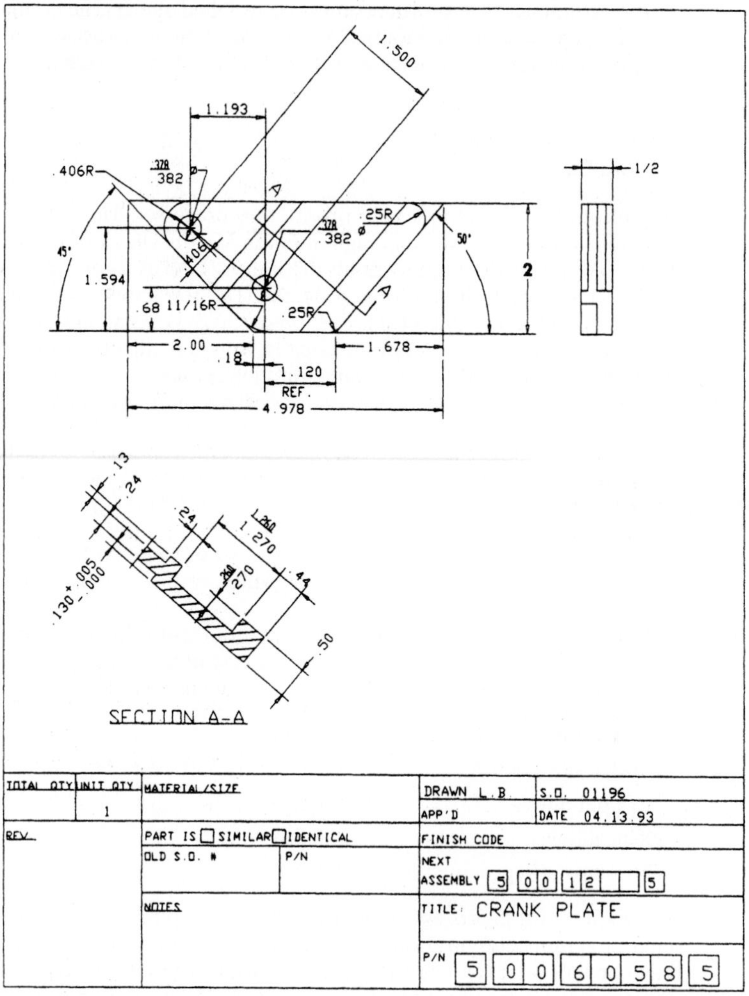


**FIGURE 12.6** Intricate part made by the chill-cast permanent-mold method (aluminum bronze).

- High dimensional accuracy
- High volume of parts with little change in the dies
- Minimum surface preparation required for plating

The disadvantages of die casting include

- Casting size limited: 50 lb seldom exceeded
- Air entrapment and porosity difficulties in complex shapes
- Expensive machinery and dies
- Limited to metals having melting points no higher than copper-base alloys

Figure 12.7 shows some typical die-cast parts made of zinc alloy. The intricate designs shown are easily produced using the die-casting process.

### 12.1.10 Evaporative Pattern Casting (EPC)

Although this process has been known and patents issued as early as 1958, it has not developed until very recently. In this process, a plastic expendable pattern (usually Styrofoam) is coated with a refractory slurry and cured. The composition of the coating is critical to the casting process, as it must allow outgassing of the vaporized Styrofoam during the pouring of the molten metal. When the molten metal pours into the mold, the Styrofoam vaporizes and gasses out of the mold and through the sand, leaving the molten metal in the void left by the Styrofoam. The process allows the coated Styrofoam pattern to be packed in sand as in the conventional sand-casting process, except that a cope and drag are not needed since the pattern is not mechanically removed, but vaporizes.

This process permits the casting of any pourable metal or alloy. Complexity of parts is generally not a problem, and cores may be utilized as in the investment process. Complex shapes are devised by gluing sections of the Styrofoam patterns together. Risers and gates form part of the pattern. The entire refractory coated assembly is packed in sand and then cast. This process is highly applicable to some of the automated systems and robotization.

### 12.1.11 Slush Casting

This process is limited to hollow castings. Zinc or lead-base alloys are generally utilized. Products made by this process include lamp bases and parts and consumer novelty items.

In this process, the molten metal is poured into a split bronze mold and allowed to set a specified time, after which the mold is inverted and the remaining liquid metal poured out. Remaining is a thin shell that has hardened at the mold surface, thus producing a hollow cast shape.

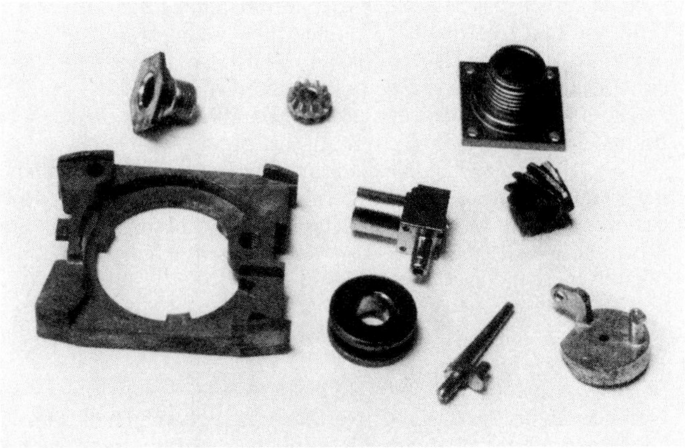

**FIGURE 12.7** Samples of zinc alloy die-cast parts.

## 12.2 FERROUS METAL ALLOYS USED IN CASTING

The general ferrous metals and alloys used to produce castings include white iron, gray iron, malleable iron (ferritic and pearlitic), carbon steels, alloy steels, and stainless steels.

### 12.2.1 Gray Iron Castings

Carbon content in this iron ranges from 2.8 to 4%. Gray iron is poured at the lowest casting temperature and has the best castability and least shrinkage of all the ferrous cast metal alloys.

Gray iron may be heat-treated in softening for better machinability or hardening for wear resistance. Pouring temperatures range from 2500 to 2700°F. Gray iron castings may be repaired by welding (shielded metal arc or oxyacetylene gas).

### 12.2.2 Ductile Iron Castings

The composition and handling of ductile irons are very similar to those of gray irons. The difference between the two irons is that in ductile iron, solidified graphite is spherical, whereas it is in flake form in gray iron. Metallurgy of the two irons is similar. Pouring temperatures range from 2500 to 2700°F. Ductile iron, as the name implies, has high ductility with resulting high impact strength for shock-load applications.

### 12.2.3 Malleable Iron Castings

This iron is produced from base metal having the following general composition:

| | |
|---|---|
| Carbon | 2 to 3% |
| Silicon | 1 to 1.8% |
| Manganese | 0.2 to 0.5% |

There are traces of sulfur, phosphorus, boron, and aluminum, and the remainder is iron.

Pouring range is 2700 to 2900°F. Ferritic malleable iron and pearlitic malleable iron receive different heat treatments to induce their metallurgical differences. Ferritic malleable iron must have a carbon-free matrix, and pearlitic malleable iron must have a matrix containing a controlled amount of carbon in the combined form.

Liquid-quenched and tempered malleable iron is made by two processes, both of which produce high-quality, high-strength cast irons.

Note that white cast iron is frequently formed intentionally at high-wear points of a casting by incorporating chills in the mold. The *chill* is an insert that causes rapid cooling of the cast iron, with the consequent formation of white iron at the points selected. White cast iron is brittle but extremely hard and wear-resistant. Figure 12.4, illustration *g*, shows the position of chills in a sand-casting mold (item 6).

### 12.2.4 Steel and Alloy Steel Castings

Green sand, dry sand, shell, investment, and ceramic molds are all used to produce steel and alloy steel castings. Plaster molds cannot be used because the high pouring temperature of steels destroys plaster molds. Green sand is the most widely used method for producing steel castings. Pouring temperature of steels for casting will reach as high as 3200°F, which is common.

Large steel castings are commonly produced in dry sand molds and may range from 1 ton to >100 tons in weight. Steel castings are welded for repairs and for joining two or more castings into a structural assembly.

## 12.3 REPRESENTATIVE CASTING METALS AND ALLOYS

Selected data on commonly used engineering alloys for casting processes are shown in Figs. 12.8 through 12.13. These represent only a small number of the materials that are available, but include many favorite engineering alloys that have been proved for many design applications. Chapter 4 lists some of the more common materials used in industrial applications.

Figure 12.8 shows the physical properties of the common cast irons.

Figure 12.9 shows the mechanical properties of cast structural steels.

Figure 12.10 shows the designations and properties of cast stainless steels.

Figure 12.11 shows the designations, types of processes, and physical properties of common cast aluminum alloys.

Figure 12.12 shows the designations and mechanical properties of cast magnesium alloys.

Figure 12.13 shows the Copper Development Association (CDA) designations and properties and application data for the common cast copper alloys.

| | Tensile Strength psi | Yield Strength psi | Elongation % in 2 in. | Impact ft-lbs | Young's Modulus psi |
|---|---|---|---|---|---|
| Ductile iron ASTM A536-77Δ | 90,000 / 150,000 | 60,000 / 125,000 | 15 / 40 | 2 / 10 | $22 \times 10^6$ / $25 \times 10^6$ |
| Gray iron* ASTM A48-76 | 20,000 / 65,000 | 12,000 / 40,000 | * | 0.5 / 1.0 | $12 \times 10^6$ / $20 \times 10^6$ |
| Malleable iron (ferritic) | 50,000 / 55,000 | 30,000 / 35,000 | 70 / 90 | 10 / 18 | $25 \times 10^6$ |
| Malleable iron (pearlitic) | 60,000 / 100,000 | 40,000 / 90,000 | 20 / 35 | 1 / 10 | $26 \times 10^6$ / $28 \times 10^6$ |
| White iron | 20,000 / 50,000 | — | 3 / 10 | — | — |

*Gray iron is not generally used for impact applications.
ΔDuctile iron is also Austempered (ADI), which approximately doubles its strength.

**FIGURE 12.8** Physical properties of cast irons.

| Grade | Tensile Strength, psi | Yield Strength, psi | Elongation % in 2 in | Impact ft-lbs | Endurance Limit, psi |
|---|---|---|---|---|---|
| 60,000 | 63,000 | 35,000 | 30 | 12 | 30,000 |
| 65,000 | 68,000 | 38,000 | 28 | 35 | 30,000 |
| 70,000 | 75,000 | 42,000 | 27 | 30 | 35,000 |

Modulus of elasticity, typical: $29 \times 10^6$ to $30 \times 10^6$.

**FIGURE 12.9** Mechanical properties of cast carbon steels.

| Designation | Tensile Strength, psi | Yield Strength, psi | Elongation % in 2 in | Impact Charpy | Young's Modulus, psi |
|---|---|---|---|---|---|
| CA-15 | 115,000 | 100,000 | 30 | 35 | $29 \times 10^6$ |
| CC-50 *(HC) | 70,000 / 110,000 | 65,000 | 18 | 45 | $29 \times 10^6$ |
| CE-30 *(HE) | 87,000 / 92,000 | 63,000 | 18 | 10 | $25 \times 10^6$ |
| CH-20 *(HH) | 80,000 / 88,000 | 50,000 | 38 | 15 | $28 \times 10^6$ |

**Typical physical properties of cast alloy steels**

| | |
|---|---|
| Melting temperature | 2,700° to 2,800°F |
| Thermal conductivity (B.t.u-ft/hr-ft$^2$-°F) | 18 to 27 |
| Density, lb/in$^3$ | 0.283 to 0.284 |
| Electrical resistivity (micro-ohm cm) | 227 |

**ASTM and SAE specifications—cast alloy steels**

| | |
|---|---|
| General mechanical | A27, A148 |
| Low temperature applications | A352, A757 |
| Weldability | A216 |
| Automotive applications | SAE J435 |

*Heat-resistant grades. †ACI (Alloy Casting Institute) designations.

**FIGURE 12.10** Cast stainless steels.

## *12.4 ASTM-LISTED CAST IRONS AND STEELS*

The ASTM specification cast irons and steels have physical and chemical characteristics which make them suitable for the types of applications and services listed below:

- High-temperature applications
- Low-temperature applications
- Impact resistance
- Structural applications
- Pressure vessel service
- High ductility
- Corrosion resistance

Lists of these irons and steels may be obtained from the ASTM materials manuals, although many of these are shown in the *McGraw-Hill Machining and Metalworking Handbook.*

### 12.4.1 ASTM—Cast Irons and Cast Steels

*ASTM A159-83 (R 1988):* Standard specification for automotive gray iron castings. The grades of gray cast iron consist of the following:

| Alloy Designation | Type | Uses and Typical Strengths |
|---|---|---|
| UNS A02010 (ANSI 201.0) | S PM | Very high strength, high impact, High ductility, high cost US = 60 YS = 50 El = 5.0 |
| UNS A02060 (ANSI 206.0) | S PM | High tensile and yield, structural parts, automotive and aerospace US = 40 YS = 24 El = 8.0 |
| UNS A02080 (ANSI 208.0) | S PM | Manifolds, valve bodies, pressure-tightness applications US = 20 YS = 12 El = 1.5 |
| UNS A02220 (ANSI 222.0) | S PM | Pistons and air-cooled cylinder heads US = 30/40 |
| UNS A03190 (ANSI 319.0) | S | Low cost, general purpose alloy US = 31 YS = 20 El = 1.5 |
| UNS A03540 (ANSI 354.0) | PM | High strength premium alloy US = 48 YS = 37 El = 3.0 |
| UNS A03560 (ANSI 356.0) | S PM | For intricate work, good strength and ductility US = 33 YS = 22 El = 3.0 |
| UNS A03600 (ANSI 360.0) | D | Very good casting and strength US = 44 YS = 25 El = 2.5 |
| UNS A13600 (ANSI A360) | D | Excellent casting, corrosion resistance, thin walls, intricate parts US = 46 YS = 24 El = 3.5 |
| UNS A03840 (ANSI 384.0) | D | General purpose alloy, thin sect. US = 48 YS = 24 El = 2.5 |
| UNS A03900 (ANSI 390.0) | D | High wear resistance, cylinder heads, pistons, engine crankcases US = 41 YS = 35 El = 1.0 |

In the preceding table: *US* = ultimate strength, *YS* = yield strength in kpsi, *El* = elongation, % in 2 in, *S* = sand casting alloy, *PM* = permanent mold casting alloy, *D* = die casting alloy.

**FIGURE 12.11** Aluminum casting alloys.

| | |
|---|---|
| G1800 | ferritic-pearlitic |
| G2500 | pearlitic-ferritic |
| G3000 | pearlitic |
| G3500 | pearlitic |
| G4000 | pearlitic |

*ASTM A148/A 148M-89a:* Standard specification for steel castings, high strength, for structural applications.

*ASTM A297/A297M-89:* Standard specification for steel castings, iron chromium, iron chromium nickel; heat-resistant; for general applications.

*ASTM A352/A352M-89:* Standard specification for steel castings, ferritic and martensitic, for pressure-containing parts, suitable for low-temperature service.

*ASTM A436-84:* Standard specification for austenitic gray iron castings.

*ASTM A439-83:* Standard specification for austenitic ductile iron castings.

*ASTM A487/A487M-89a:* Standard specification for steel castings suitable for pressure service.

| Designation | Temper | US | YS | Elongation % in 2 in |
|---|---|---|---|---|
| ΔM10100 | F | 20,000 | — | — |
| (ASTM-AM 100A) | T4 | 34,000 | — | 6 |
| SAE 502 | T6 | 35,000 | 17,000 | — |
| ΔM11630 | F | 26,000 | 11,000 | 4 |
| (ASTM-AZ 63A) | T4 | 34,000 | 11,000 | 7 |
| SAE 50 | T6 | 34,000 | 16,000 | 3 |
| *M11910 | — | 34,000 | 23,000 | 3 |
| (ASTM-AZ 91A) | | | | |
| SAE 501 | | | | |
| *M11912 | — | 34,000 | 23,000 | 3 |
| (ASTM-AZ 91B) | | | | |
| SAE 501A | | | | |
| ΔM11920 | F | 23,000 | 11,000 | — |
| (ASTM-AZ 92A) | T4 | 34,000 | 11,000 | 6 |
| | T6 | 34,000 | 18,000 | 1 |
| ΔM16630 | T6 | 40,000 | 27,000 | 5 |
| (ASTM-ZE 63A) | | | | |

*Automotive die-casting alloys
ΔSand castings alloys

**FIGURE 12.12** Magnesium casting alloys.

*ASTM A743/743M-89:* Standard specification for castings, iron chromium, iron chromium nickel; corrosion-resistant; for general applications.

***Brinell Hardness Measurements.*** For castings and other applications, calculation of the Brinell hardness number (Bhn) can be performed using the equation shown in Sec. 4.2 of Chap. 4. (See also Index.)

## 12.5 PLASTIC MOLDINGS

There are two classifications of plastics and their moldings: thermoplastics and thermoset plastics. Thermoplastics are basically the same chemically after molding as they were in the raw form. This means that once molded, they may be reused, in most cases, by chopping the parts into small pieces and remelting with a percentage of new material. Thermoset plastics, once molded, cannot be remolded or reprocessed since they have a one-way chemistry that alters their as-molded characteristics from their raw constituents.

Types of thermoplastics include

- Acrylonitrile-butadiene-styrene (ABS)
- Acetal (Delrin)
- Acrylic (Lucite, Plexiglas)
- Cellulosics (acetates)
- Fluoroplastics (PTFE, FEP, PFA, CTFE, ETFE, PVDF)
- Nylon
- Phenylene oxide
- Polycarbonate (Lexan)

**Alloy No. C80100**
Composition: 99.95% copper, 0.05 trace elements
Conductivity: 100% IACS
Tensile strength: 19,000 psi min. to 25,000 psi typical
Yield strength: 6,500 psi min. to 9,000 psi typical (0.2% offset)
Typical uses: Electrical and thermal conductors, corrosion resistance applications.
Similar alloys: C80300, C80500, C80700, C80900
All are difficult to cast, with low casting yields.

**Alloy No. C81500 (chrome-copper)**
Composition: 1% chromium, balance copper
Conductivity: 82% IACS
Tensile strength: 45,000 psi min. to 51,000 psi typical (heat treated)
Yield strength: 35,000 psi. to 40,000 psi typical (0.5% exten. under load)
Typical uses: Electrical and thermal conductors where high strength and hardness are required. A premium quality alloy.

**Alloy No. C81700**
Composition: 0.4% beryllium, 0.9 cobalt, 0.9 nickel, 1.0 silver
Conductivity: 48% IACS
Tensile strength: 85,000 psi min. to 92,000 psi typical
Yield strength: 62,000 psi min. to 68,000 psi typical (0.2% offset)
Typical uses: Electrical and thermal conductors where high strength and hardness are required.

**Alloy No. C82400 (former name: 165C)**
Composition: 1.7% beryllium, 0.25 cobalt, remainder copper
Conductivity: 25% IACS
Tensile strength: 145,000 psi min. to 150,000 psi typical (heat treated)
Yield strength: 135,000 psi min. to 140,000 psi typical (0.2% offset)
Typical uses: Molds, cams, bushings, gears, bearings
Similar alloys: C82800, C82700, C82600

**Alloy No. C83600 (115, leaded red brass, composition bronze)**
Composition: 85% copper, 5.0 lead, 5.0 tin, 5.0 zinc
Conductivity: 15% IACS
Tensile strength: 30,000 psi min. to 37,000 psi typical (as cast)
Yield strength: 14,000 psi min. to 17,000 psi typical (0.5% exten. under load)
Typical uses: Pipe fittings, pump castings, impellers, small gears
Similar alloys: C83800, C84200, C84400, C84500

**Alloy No. C85200 (400, leaded yellow brass)**
Composition: 72% copper, 3.0 lead, 1.0 tin, 24.0 zinc
Conductivity: 18% IACS
Tensile strength: 35,000 psi min. to 38,000 psi typical (as cast)
Yield strength: 12,000 psi min. to 13,000 psi typical (0.5% exten. under load)
Typical uses: Ferrules, valves, hardware, plumbing fittings
Similar alloys: C85300, C85400, C85500 *(Continued)*

**FIGURE 12.13** Cast copper alloys: properties and applications.

- Polyester (e.g., Mylar)
- Polyethylene
- Polyimide
- Polyphenylene sulfide
- Polypropylene
- Polystyrene
- Polysulfone
- Polyurethane
- Polyvinyl chloride

**Alloy No. C86100 (high-strength yellow brass, 90 kpsi manganese bronze)**
Composition: 5.0 aluminum, 67.0 copper, 3.0 iron, 4.0 manganese, 21.0 zinc
Conductivity: 7.5% IACS
Tensile strength: 90,000 psi min. to 95,000 psi typical
Yield strength: 45,000 psi min. to 50,000 psi typical (0.2% offset)
Typical uses: Marine castings, gears, bushings, bearings
Similar alloys: C86200, C86300, C86400, C86500

**Alloy No. C87800 (die-cast silicon brass)**
Composition: 82.0 copper, 4.0 silicon, 14.0 zinc
Conductivity: 6.7% IACS
Tensile strength: 85,000 psi typical
Yield strength: 50,000 psi typical (0.2% offset)
Typical uses: High-strength die castings, lever arms, brackets, hex nuts, clamps
Similar alloys: C87900

**Alloy No. C94400 (phosphor bronze, 312)**
Composition: 81.0 copper, 11.0 lead, .35 phosphorus, 8.0 tin
Conductivity: 10% IACS
Tensile strength: 32,000 psi typical
Yield strength: 16,000 psi typical (0.5% exten. under load)
Typical uses: Bushings, bearings, electrical items

**Alloy No. C95400 (aluminum-bronze 9C, 415)**
Composition: 11.0 aluminum, 85.0 copper, 4.0 iron
Conductivity: 13% IACS
Tensile strength: 75,000 psi min. to 85,000 psi typical
Yield strength: 30,000 psi min. to 35,000 psi typical (0.5% exten. under load)
Typical uses: Bearings, worms, gears, bushings, valve guides
Similar alloys: C95200, C95300, C95500

**Alloy No. C95700 (manganese aluminum bronze)**
Composition: 8.0 aluminum, 75.0 copper, 3.0 iron, 12.0 manganese, 2.0 nickel
Conductivity: 3.1% IACS
Tensile strength: 90,000 psi min. to 95,000 psi typical
Yield strength: 40,000 psi min. to 45,000 psi typical (0.5% exten. under load)
Typical uses: Impellers, propellers, safety tools, valves, pump castings

Note: The preceding alloy numbers refer to the UNS numbering system for copper cast alloys. For a complete listing of copper cast alloys, refer to the CDA handbook of cast copper alloys available from the Copper Development Association. Refer to Chapter 15 of this handbook.

**FIGURE 12.13** (*Continued*)

Types of thermoset plastics include

- Alkyd
- Allyl (diallyl phthalate)
- Amino (urea, melamine)
- Epoxy (including cycloaliphatic)
- Phenolic
- Polyester
- Polyurethane
- Silicone

Polyesters and polyurethanes include thermosets that are also thermoplastics. Thermoset grades are usually filled with reinforcing materials such as glass fibers, carbon, and mineral fibers for added strength. Thermoset plastics are usually more dimensionally stable, are more heat resistant, and have better electrical properties than thermoplastics.

Complex molded shapes of the plastics are analyzed today using the advanced finite element analysis (FEA) techniques available for the personal computer (PC) and engineering design stations.

Figure 12.14 shows a typical intricate plastic part, which must be dimensionally accurate as well as chemical-resistant.

### 12.5.1 Prototypes of the Plastics

Building a prototype plastic part is a compromise, since the part is usually machined from plastic blocks and slabs and will not, in most cases, duplicate the exact performance of the finished part that is made in a mold.

The closest duplicate of a plastic production part is made by molding a prototype in mild-steel molds made specifically for the prototype. This method is expensive but is the closest method to use if you wish to avoid expensive rework or changes to a finished production mold. This approach will give the most accurate test results prior to building the actual production mold.

### 12.5.2 Properties and Characteristics of Modern Plastics

Widely used plastics are discussed in Chap. 4. The typical applications of the modern plastics are given there.

### 12.5.3 Design of Molded Plastic Parts

Final selection of plastic type and part configuration or design should be reviewed and coordinated with the moldmaker and plastic part manufacturer before the final design drawings are made. Moldmakers and molded-part manufacturers can alert the designer to the many problem areas that are prevalent on many preliminary designs for plastic parts.

Plastic part design handbooks are available from all the leading producers of plastic materials such as DuPont, General Electric, and Monsanto. These manuals or handbooks cover detail design, appropriate calculation techniques, and complete chemical, physical, and elec-

**FIGURE 12.14** A thermoplastic molded part (chemical-resistant).

trical properties of the materials. The design handbooks may be secured by writing directly to the plastics sections of the large suppliers or through their distributors.

### 12.5.4 Plastics Molding Machinery and Molds

A great variety of machinery is made for molding plastic parts, most of which are expensive and require specialized techniques for their operation. Figures 12.15 and 12.16 show a typical group of machinery that is required for producing parts made of cycloaliphatic epoxy. This class of thermoset plastic has gained wide recognition in the electric power distribution industry for parts that support or brace high-voltage current-carrying busses and parts of switching devices such as breakers and switches of all classes, up to 34.5 kV. Figure 12.15 shows the mixing and dispensing machine, and Fig. 12.16 shows the complete group of equipment, with the mold-clamping machine at the front of the photograph.

Because of the high injection pressures developed in this plastic molding process, the clamping machine must withstand a high separating load on the mold to prevent it from opening during the injection process. In this process, the hot, molten plastic is injected under pressure into the mold, where it is held for the "curing" time or setting time prior to being released from the mold. In the molding of cycloaliphatic epoxy, after the proper cure, the plastic part is removed from the mold and placed in an oven set to a preselected temperature and is baked for a predetermined time interval. Cycloaliphatic epoxy is being used in the electrical industries as a substitute for wet-process porcelain, which for many years was one of the few materials available for this type of service. Glass and polyester-glass thermosets are also used in electrical applications, as are other electrical-grade epoxies and other types of thermoset plastics. Chapter 4 shows some sample material specification sheets for polyester-glass, polycarbonate, Valox, and acrylic.

**FIGURE 12.15** Mixing station and control panel for cycloaliphatic epoxy casting process.

**FIGURE 12.16** Complete cycloaliphatic epoxy casting equipment (clamping machine in front).

The tonnage and size of the clamping machine for injection-molding plastics is determined by the volume of plastic to be injected at one time (a single shot). A machine that handles parts up to 20 $in^3$ would be considered of moderate size, while a machine required to injection-mold parts with a volume of 60 $in^3$ would be considered large. Clamping machines are made in sizes up to hundreds of tons' clamping capacity, and these machines are usually found at the larger plastic molding manufacturers.

Figure 12.16*a* is an AutoCAD drawing of a part made from urea formaldehyde using the compression-molding process. The stock thermoset material is molded under extreme pressure and heat in a clamping-type mold. When set, the part is expelled from the mold using pushpins or extractors which push the finished part out of the mold cavity. Excellent dimensional stability is to be expected using the compression-molding process with thermoset plastics. Figure 12.16*b* shows a finished plastics mold that is used in the molding of cycloaliphatic resins (epoxies). This master mold is used to cast duplicate molding plastic molds to produce the finished parts. Many duplicate molds can be made from the master mold for an increased production rate.

## 12.6 EXTRUSIONS

An *extruded* shape is one in which the material being extruded is pushed, under high pressure and high temperature, through a set of extrusion dies that have the shape of the cross section of the part. Extrusion is similar to drawing, except that in drawing, the drawn part is pulled out of or through the dies or series of dies of progressive sizes. The material is cold-worked more effectively when it is drawn, rather than extruded.

Thermoset plastics are also *pultruded* to various design cross sections. In this process, the thermoset, which is reinforced with glass or other fibers, is pulled through a set of pultruding dies. Pultruded, reinforced thermoset plastic parts are *extremely* strong and show tremendous flexural strength, even in small cross-sectional-area pultruded shapes.

Aluminum alloys are particularly well suited for the extrusion processes. Very large, high-tonnage extrusion presses are required for the extrusion of aluminum and its alloys.

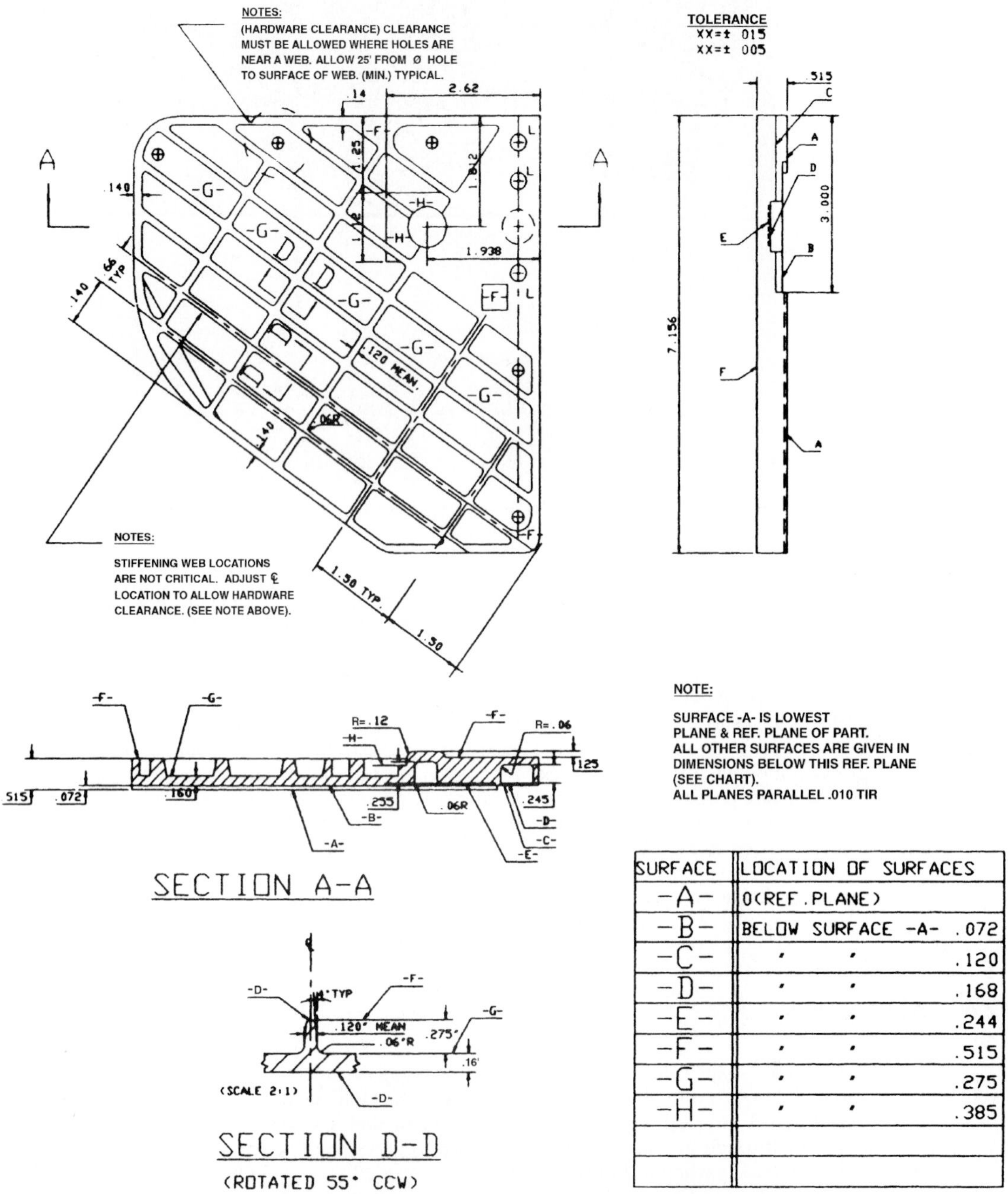


| SURFACE | LOCATION OF SURFACES | |
|---|---|---|
| -A- | 0(REF. PLANE) | |
| -B- | BELOW SURFACE -A- | .072 |
| -C- | " " | .120 |
| -D- | " " | .168 |
| -E- | " " | .244 |
| -F- | " " | .515 |
| -G- | " " | .275 |
| -H- | " " | .385 |
| | | |
| | | |

**FIGURE 12.16*a*** AutoCAD drawing for a part to be made from urea formaldehyde thermoset plastic.

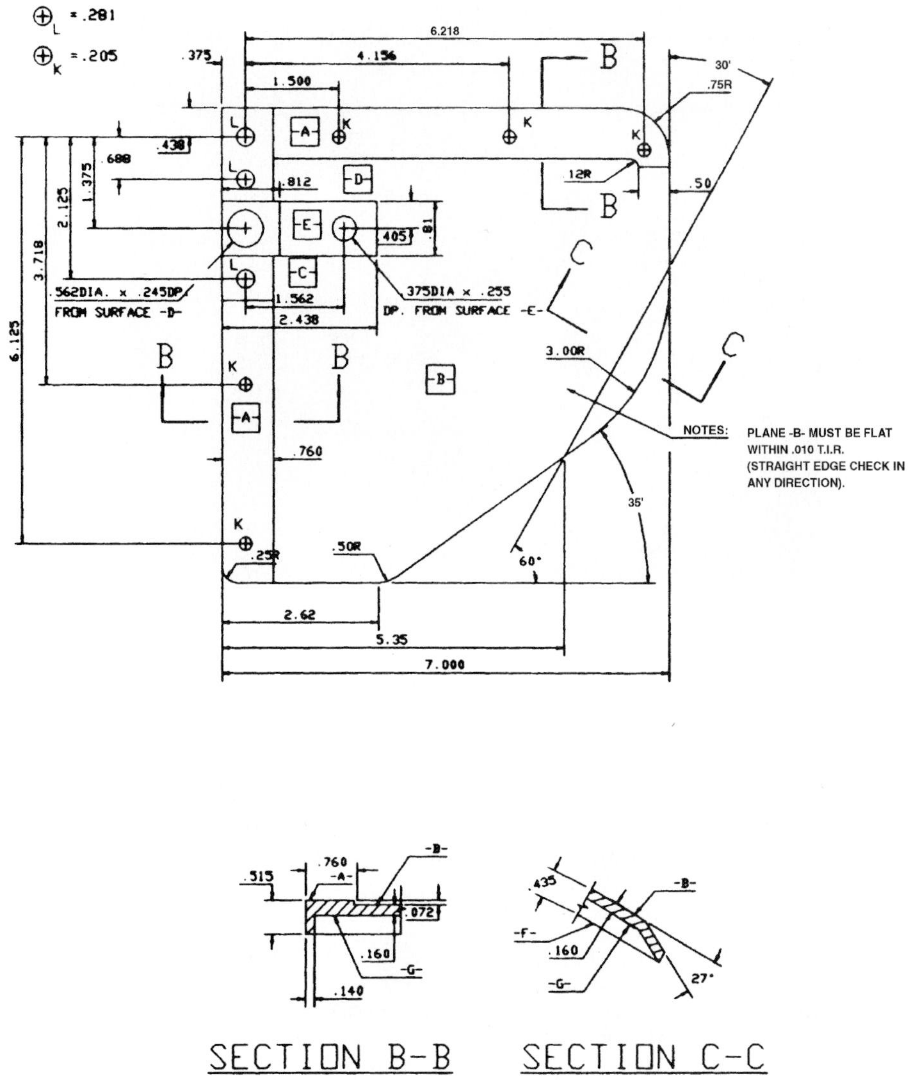


NOTES: DIMENSIONS REFER TO

1—SIZE OF FINISHED PART
2—MATERIAL: UREA FORMALDEHYDE
'BEETLE' #8023
3—COLOR: NATURAL
4—SPECIFICATION: A-30606.

**FIGURE 12.16*a*** (*Continued*)

**FIGURE 12.16*b*** A finished tool steel mold.

There are generally two widely available classes or grades of aluminum extrusions: structural aluminum and architectural aluminum shapes. These two classes of extrusions are available in the following alloys and tempers:

| Class | Shape | Alloy, temper, and specification |
|---|---|---|
| Structural | Angles | 6061-T6, ASTM B308 |
| | Channels | ″ |
| | Tee's | ″ |
| | Zee's | ″ |
| | H beams and I beams | ″ |
| | Wide-flange beams | ″ |
| Architectural | Angles | 6063-T52, ASTM B221 |
| | Channels | ″ |
| | Tee's | ″ |
| | Zee's | ″ |

Aluminum alloy extrusions are also available in the following standard shapes:

- Round rod
- Square bar
- Rectangular bar
- Hexagonal bar
- Round tube
- Square tube
- Rectangular tube

The alloy designations of these shapes may be either 2011, 2017, 2024, 6061, or 6063.

An aluminum alloy extrusion may be designed to any practical cross-sectional shape that the extrusion dies and available extrusion presses will accommodate, and is usually limited only by the designers' imagination and ingenuity and the size of the part.

Some of the standard extruded shapes that are available from various sources nationwide are shown in Figs. 12.17 and 12.18. The temper designations and lengths available are shown in these figures.

***Plastic Extrusions.*** The number of available plastic extruded shapes available is extremely large. These plastic extruded shapes are available from the plastic manufacturers whose catalogs may be obtained through the trade magazines such as *The American Machinist, Machine Design, Product Design and Development,* and *Modern Machine Shop.* If you work with or need plastic extruded shapes, keep a series of these catalogs in your reference files. The plastic extruded shapes not only make the design job easier but also enhance the appearance of the final product and facilitate assembly in many cases.

## 12.7 POWDER-METAL TECHNOLOGY

Powder-metal parts are made by compressing a highly purified metallic powder in a set of dies under extremely high pressure and then fusing the particles in an oven under controlled high

**STANDARD ANGLES** ①

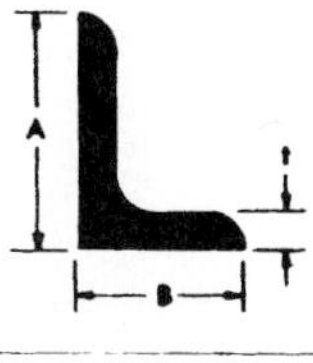


**ANGLES—UNEQUAL LEGS, EXTRUDED**
**Alloy 6061-T6 • Length—25 ft**

| Dimensions, in. A | B | t | Est wt per ft, lb |
|---|---|---|---|
| N 1-1/4 | 3/4 | 3/32 | .21 |
| N 1-1/4 | 1 | 1/8 | .31 |
| N 1-1/2 | 3/4 | 1/8 | .31 |
| 1-1/2 | 3/4 | N 3/16 | .46 |
| N 1-1/2 | 1 | 5/32 | .43 |
| 1-1/2 | 1 | N 1/4 | .66 |
| 1-1/2 | 1-1/4 | 1/8 | .38 |
| 1-1/2 | 1-1/4 | 3/16 | .57 |
| 1-1/2 | 1-1/4 | 1/4 | .74 |
| 1-3/4 | 1-1/4 | 1/8 | .42 |
| 1-3/4 | 1-1/4 | 3/16 | .62 |
| 1-3/4 | 1-1/4 | 1/4 | .81 |
| 2 | 1-1/2 | 1/8 | .50 |
| 2 | 1-1/2 | 3/16 | .73 |
| 2 | 1-1/2 | 1/4 | .96 |
| 2 | 1-1/2 | 3/8 | 1.38 |
| N 2-1/2 | 1-1/2 | 3/16 | .85 |
| 2-1/2 | 1-1/2 | 1/4 | 1.11 |
| 2-1/2 | 1-1/2 | N 5/16 | 1.36 |
| 2-1/2 | 2 | 1/8 | .65 |
| 2-1/2 | 2 | 3/16 | .96 |
| 2-1/2 | 2 | 1/4 | 1.26 |

| Dimensions, in. A | B | t | Est wt per ft, lb |
|---|---|---|---|
| 2-1/2 | 2 | 5/16 | 1.55 |
| 2-1/2 | 2 | 3/8 | 1.83 |
| 3 | 2 | 3/16 | 1.07 |
| 3 | 2 | 1/4 | 1.40 |
| 3 | 2 | 5/16 | 1.73 |
| 3 | 2 | 3/8 | 2.05 |
| 3 | 2 | N 7/16 | 2.35 |
| 3 | 2-1/2 | 1/4 | 1.54 |
| 3 | 2-1/2 | 5/16 | 1.90 |
| 3 | 2-1/2 | 3/8 | 2.25 |
| 3-1/2 | 2-1/2 | 1/4 | 1.68 |
| 3-1/2 | 2-1/2 | 5/16 | 2.08 |
| 3-1/2 | 2-1/2 | 3/8 | 2.47 |
| 3-1/2 | 2-1/2 | N 1/2 | 3.23 |
| 3-1/2 | 3 | 1/4 | 1.84 |
| 3-1/2 | 3 | 5/16 | 2.28 |
| 3-1/2 | 3 | N 3/8 | 2.70 |
| 4 | 3 | 1/4 | 1.99 |
| 4 | 3 | 5/16 | 2.46 |
| 4 | 3 | 3/8 | 2.93 |

| Dimensions, in. A | B | t | Est wt per ft, lb |
|---|---|---|---|
| 4 | 3 | 1/2 | 3.83 |
| 4 | 3 | 5/8 | 4.69 |
| N 4 | 3-1/2 | 3/8 | 3.13 |
| 4 | 3-1/2 | N 1/2 | 4.10 |
| 5 | 3 | 3/8 | 3.35 |
| 5 | 3 | 1/2 | 4.40 |
| 5 | 3-1/2 | 5/16 | 3.01 |
| 5 | 3-1/2 | 3/8 | 3.58 |
| 5 | 3-1/2 | N 7/16 | 4.15 |
| 5 | 3-1/2 | 1/2 | 4.70 |
| 5 | 3-1/2 | 5/8 | 5.79 |
| 6 | 3-1/2 | 5/16 | 3.39 |
| 6 | 3-1/2 | N 3/8 | 4.04 |
| 6 | 3-1/2 | 1/2 | 5.31 |
| 6 | 4 | 3/8 | 4.24 |
| 6 | 4 | N 7/16 | 4.91 |
| 6 | 4 | 1/2 | 5.58 |
| 6 | 4 | 5/8 | 6.88 |
| 6 | 4 | 3/4 | 8.48 |
| N 8 | 6 | 5/8 | 9.84 |
| 8 | 6 | N 11/16 | 10.76 |
| 8 | 6 | 3/4 | 11.68 |

① Angle sections listed as "standard" are approximations of American Standard sections. For elements of sections and detailed dimensions, consult *Alcoa Structural Handbook.*
N Not stocked at plant.

**FIGURE 12.17** Aluminum standard extruded angles.

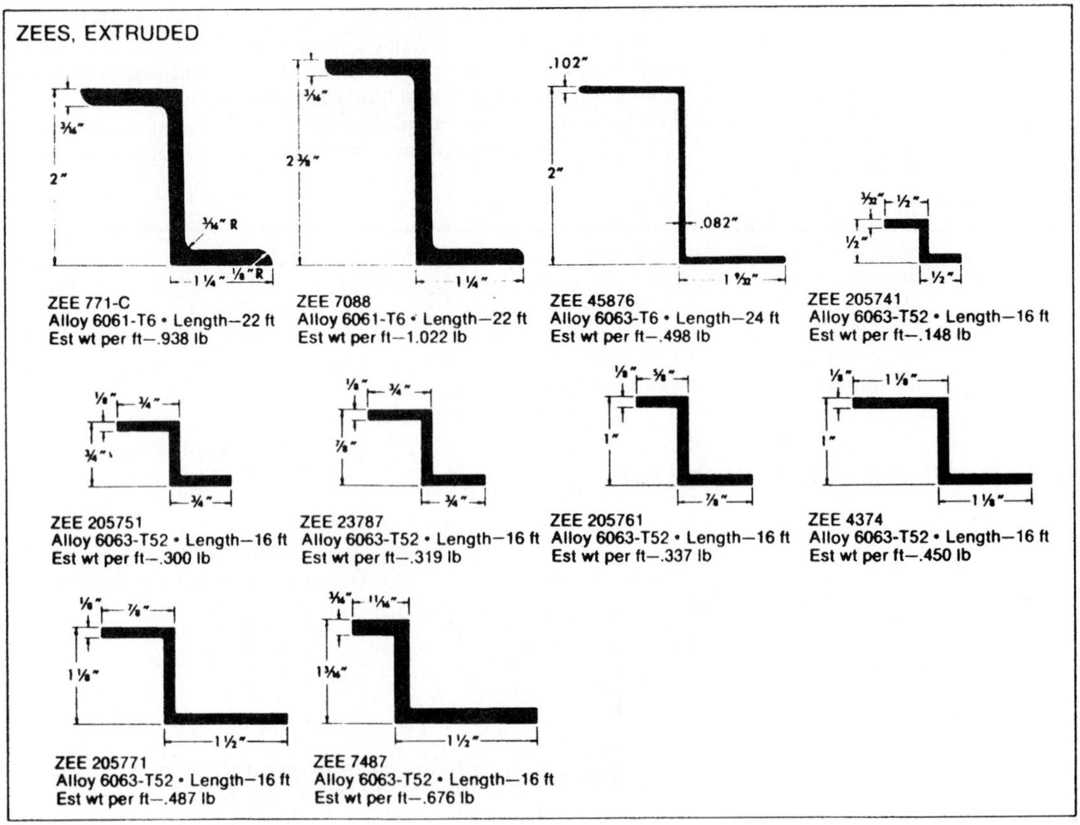


**FIGURE 12.18** Aluminum extruded zee's.

temperature. This compression process compacts the powder metal until the part is approximately 90 percent the density of the solid metal or alloy. The density of the part can be controlled to an extent that the open pores in the powder-metal part, after fusion, may be impregnated with lubricants or filler resins and binders. The well-known sintered-bronze journal bearings with impregnated lubrication are prime examples of powder-metal technology.

Powder-metal parts may be made of aluminum alloys, copper alloys, steels, and stainless steels. Other metals and alloys are also possible for use in powder-metal processing.

The modern powder-metal part may be impregnated with resins and binders, which make possible the application of electroplated finishes such as copper, zinc, nickel, and chromium. If the part is made of one of the corrosion-resistant stainless steels, the plating process may be eliminated.

Parts may be produced using powder metal that normally would be difficult to machine. Most of the powder-metal parts shown in Fig. 12.19, for example, would be difficult to make using other processes. The parts shown could be made using the investment casting process or machining techniques, but would be costly to produce and time-consuming. Figure 12.20 is a closer view of the three small three-lobed parts, which show a minute hole in the center of the parts. Each division on the scale shown in Fig. 12.20 is equal to 0.020 in (one-fiftieth of an inch). All the parts shown in both figures are easy and economical to produce using powder-metal technology.

**FIGURE 12.19** A sample of powder-metal parts.

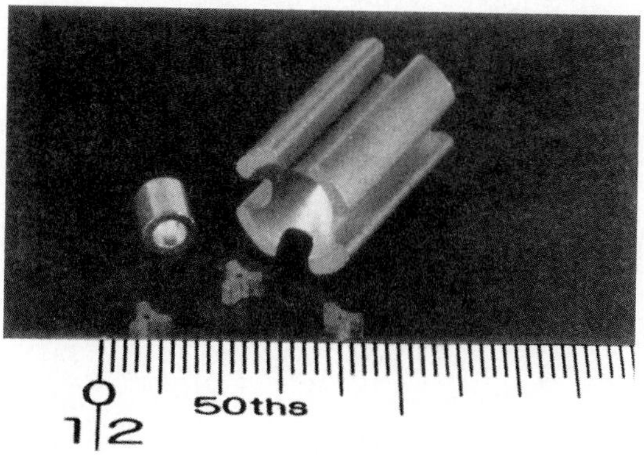


**FIGURE 12.20** Closeup view of powder-metal parts showing small size.

### 12.7.1 Design of Powder-Metal Parts

The design and manufacturing of powder-metal parts is not difficult if the basic rules of powder-metal part design procedures are followed and if the powder-metal part manufacturer is consulted during the design stages. The part manufacturer will advise you if the part can be produced as designed and what remedial actions to take or what design changes are necessary to produce the parts. The manufacturer will also advise of the availability of the different metals and alloys for producing the part.

Basic rules for powder-metal part design are

- Keep the outline elements of the part on lines parallel to the part axis or direction of compression of the male die.
- Do not design parts with reverse angles along the axis (the part cannot be extracted from the dies if this occurs).

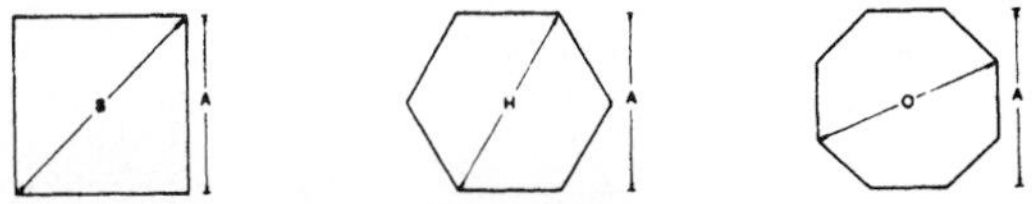


S=1.414A H=1.155A O=1.082A

| A Size in inches | Distance across corners in inches S Square | H Hexagon | O Octagon | A Size in inches | Distance across corners in inches S Square | H Hexagon | O Octagon |
|---|---|---|---|---|---|---|---|
| 1/8 | .177 | .144 | .135 | 2 1/8 | 3.005 | 2.454 | 2.300 |
| 3/16 | .265 | .217 | .203 | 2 3/16 | 3.094 | 2.526 | 2.368 |
| 1/4 | .354 | .289 | .271 | 2 1/4 | 3.182 | 2.598 | 2.435 |
| 5/16 | .442 | .361 | .338 | 2 5/16 | 3.270 | 2.670 | 2.503 |
| 3/8 | .530 | .433 | .406 | 2 3/8 | 3.359 | 2.742 | 2.571 |
| 7/16 | .619 | .505 | .474 | 2 7/16 | 3.447 | 2.815 | 2.638 |
| 1/2 | .707 | .577 | .541 | 2 1/2 | 3.536 | 2.887 | 2.706 |
| 9/16 | .795 | .650 | .609 | 2 9/16 | 3.624 | 2.959 | 2.774 |
| 5/8 | .884 | .722 | .677 | 2 5/8 | 3.712 | 3.031 | 2.841 |
| 11/16 | .972 | .794 | .744 | 2 11/16 | 3.801 | 3.103 | 2.909 |
| 3/4 | 1.061 | .866 | .812 | 2 3/4 | 3.889 | 3.175 | 2.977 |
| 13/16 | 1.149 | .938 | .879 | 2 13/16 | 3.977 | 3.248 | 3.044 |
| 7/8 | 1.237 | 1.010 | .947 | 2 7/8 | 4.066 | 3.320 | 3.112 |
| 15/16 | 1.326 | 1.083 | 1.015 | 2 15/16 | 4.154 | 3.392 | 3.180 |
| 1 | 1.414 | 1.155 | 1.082 | 3 | 4.243 | 3.464 | 3.247 |
| 1 1/16 | 1.503 | 1.227 | 1.150 | 3 1/8 | 4.419 | 3.608 | 3.383 |
| 1 1/8 | 1.591 | 1.299 | 1.218 | 3 1/4 | 4.596 | 3.753 | 3.518 |
| 1 13/16 | 1.679 | 1.371 | 1.285 | 3 3/8 | 4.773 | 3.897 | 3.653 |
| 1 1/4 | 1.768 | 1.443 | 1.353 | 3 1/2 | 4.950 | 4.041 | 3.788 |
| 1 5/16 | 1.856 | 1.516 | 1.421 | 3 5/8 | 5.126 | 4.186 | 3.924 |
| 1 3/8 | 1.945 | 1.588 | 1.488 | 3 3/4 | 5.303 | 4.330 | 4.059 |
| 1 7/16 | 2.033 | 1.660 | 1.556 | 3 7/8 | 5.480 | 4.474 | 4.194 |
| 1 1/2 | 2.121 | 1.732 | 1.624 | 4 | 5.657 | 4.619 | 4.330 |
| 1 9/16 | 2.210 | 1.804 | 1.691 | 4 1/4 | 6.010 | 4.907 | 4.600 |
| 1 5/8 | 2.298 | 1.876 | 1.759 | 4 1/2 | 6.364 | 5.196 | 4.871 |
| 1 11/16 | 2.386 | 1.949 | 1.827 | 4 3/4 | 6.717 | 5.485 | 5.141 |
| 1 3/4 | 2.475 | 2.021 | 1.894 | 5 | 7.071 | 5.774 | 5.412 |
| 1 13/16 | 2.563 | 2.093 | 1.962 | 5 1/4 | 7.425 | 6.062 | 5.683 |
| 1 7/8 | 2.652 | 2.165 | 2.031 | 5 1/2 | 7.778 | 6.351 | 5.953 |
| 1 15/16 | 2.740 | 2.237 | 2.097 | 5 3/4 | 8.132 | 6.640 | 6.224 |
| 2 | 2.828 | 2.309 | 2.165 | 6 | 8.485 | 6.928 | 6.494 |
| 2 1/16 | 2.917 | 2.382 | 2.232 | | | | |

**FIGURE 12.21** Table for dimensions of squares, hexagons, and octagons.

- Keep wall sections or webs as thick as possible.
- Single-tapered holes in the part are possible when the direction of taper allows extraction from the dies after compression of the powder metal.
- Limit the size or volume of the part to that which may be produced with available machinery.
- Do not specify electroplated finishes unless necessary (the part may be produced in a corrosion-resistant alloy if necessary).
- Holes in the part may be controlled to close tolerances.
- Check with the part manufacturer to ascertain whether the part can sustain the imposed stress loads anticipated.
- The outline accuracy of parts can be closely controlled.

Some of the powder-metal part producers can provide design manuals or brochures to the design engineer which outline in more detail the design procedures for powder-metal parts.

Many parts that are normally or routinely machined or cast may be made of powder metal when the stress or load requirements are satisfied by the powder-metal part. The production of complex small metal parts is often very easily and economically accomplished using powder-metal technology. Secondary operations on machined parts, such as drilling, deburring, reaming, and broaching, may be eliminated using powder-metal technology.

Small-part design should always be reviewed to see if the design requirements can be met using powder-metal technology. Powder-metal technology is much further advanced today than in the past, where it was used mainly to produce plain and flanged bronze bearings with impregnated lubrication.

## 12.8 TABLE FOR SQUARES, HEXAGONS, AND OCTAGONS

Figure 12.21 shows the distances across the corners of squares, hexagons, and octagons of different standard stock sizes. This table can save a lot of valuable time when these dimensions, of the shapes shown, are required for design purposes.

## FURTHER READING

American Society for Metals, 1992: *Metals Handbook,* 8th ed., *Properties and Selection,* vol. 1; *Forging and Casting,* vol. 5; *Atlas of Microstructures,* vol. 7. Metals Park, Ohio.

Society of Automotive Engineers, 1992: *SAE Handbook, Materials,* vol. 1.

# CHAPTER 13

# ENGINEERING FINISHES, PLATING PRACTICES, AND SPECIFICATIONS

Commercial products and equipment of all types and classifications require finishes of one type or other. These finishes range from basic oxide coatings to the various paints and plastics to electrodeposited and hot-dipped metals. The common electroplating metals include zinc, copper, chromium, nickel, silver, gold, palladium, rhodium, tin, lead, and cadmium. The hot-dip metals include tin, lead, zinc, and aluminum.

The finish or plating used on any particular part should contribute to the engineering qualities of the finished product, and also to the cosmetic appearance, when required. The desired effects of the finish could include weather protection, resistance to corrosive chemicals, heat resistance, physical appearance, electrical conductivity, wear resistance, resistance to galvanic corrosion, and improved lubrication qualities.

It is the design engineer's responsibility to specify the finish characteristics and specifications on a part or assembly. Designers should be aware of the types of finishes and plating processes that are commercially available and how to specify them on the design and detail part drawings or specifications. Arbitrary selection of a finish or metallic plating and its thickness range can lead to many design problems relating to corrosion, dimensional interferences, and cost.

This chapter of the handbook is intended to familiarize the electromechanical design engineer, and other engineering support personnel, with common finishing processes, procedures for specifying thicknesses of plating, and the appropriate industrial standard specifications which control these finishes.

## 13.1 FINISHES

The finishing processes and methods in common use for engineering applications are outlined here.

**1.** Mechanical finishes
   *a.* Sanding or grinding
   *b.* Brushing (scratch and satin)
   *c.* Sandblasting (abrasive cleaning and texturing of surfaces)
   *d.* Shot peening and tumbling (metal or ceramic ball)
   *e.* Burnishing
   *f.* Mechanical powder plating
   *g.* Polishing

2. Chemical finishing
   - *a.* Etching
   - *b.* Bonderite
   - *c.* Alodine
   - *d.* Iridite
   - *e.* Phosphatizing
   - *f.* Passivating (stainless steels)
   - *g.* Black oxide
   - *h.* Blueing
   - *i.* Parkerizing (a phosphatizing finish)
   - *j.* Teflon coating and surface impregnation
   - *k.* Other plastic coatings
3. Electrolytic oxides
   - *a.* Sulfuric acid anodize (Alumilite—Alcoa process)
   - *b.* Chromic acid anodize (aerospace and other applications)
   - *c.* Martin hardcoat anodize (Martin-Marietta process)
   - *d.* Electropolishing
   - *e.* Various other hardcoat processes of anodizing
     (*Note:* Anodizing may be performed on aluminum, magnesium, titanium, and zinc.)
4. Electroplating
   - *a.* Copper plate
   - *b.* Cadmium plate
   - *c.* Chromium plate (bright, hard, and black)
   - *d.* Nickel plate (bright, black, and chromium nickel)
   - *e.* Gold plate
   - *f.* Silver plate
   - *g.* Tin plate (tin cadmium, tin lead, and tin nickel)
   - *h.* Indium plate
   - *i.* Rhodium plate
   - *j.* Palladium plate
   - *k.* Zinc plate (nickel zinc, zinc with chromate)
   - *l.* Lead plate
5. Electroless plating
   - *a.* Boron nickel alloy
   - *b.* Chromium
   - *c.* Cobalt
   - *d.* Gold
   - *e.* Iron
   - *f.* Nickel
6. Plasma, vapor phase, and vacuum-deposition coatings
   - *a.* Fluorides
   - *b.* Chlorides
   - *c.* Silver
   - *d.* Aluminum
   - *e.* Other metals and compounds
7. Hot-dip plating
   - *a.* Hot-dip tin
   - *b.* Hot-dip zinc
   - *c.* Hot-dip aluminum
   - *d.* Hot-dip lead

The base metals and their alloys that are commonly electroplated are iron, steel, stainless steel (more commonly passivated), aluminum, copper, brass, bronze, titanium, and magnesium.

## 13.2 CORROSION OF METALS: PRINCIPLES

The design engineer has four choices when planning for corrosion resistance of a metallic part:

1. Use a corrosion-resistant alloy steel
2. Use a nonferrous, noncorroding metal or metallic alloy
3. Use plated steel or anodized alloys
4. Use a chemically coated, painted, or plastic-coated metal part (ferrous or nonferrous)

The third choice is usually the most economical, although it may not be the best choice. Plated steel parts are the most common and suffice for most applications, although the choice of a more expensive alloy may be mandatory because of required antimagnetic properties, low electrical resistivity, restrictions on the weight of the part (such as in manned aircraft and aerospace vehicles), and other considerations.

***Basic Principles of Corrosion of Metals and Metallic Alloys.*** All metals and alloys have a specific relative electrical potential. When metals of different electrical potential, such as steel and copper, are in contact in the presence of moisture (electrolyte), a low-energy electric current flows from the metal having the higher potential to the one having the lower potential. This is called *galvanic action.* One result is that corrosion of the metal having the higher potential (steel, in the example here) is accelerated.

The mechanism is an anode reaction, a cathode reaction, the conduction of electrons through the metal from anode to cathode, and the conduction of ions through the electrolyte solution. Corrosion occurs in the anode area, while the cathode area is protected.

It is important to know from which of two metals current will flow. Figure 13.1 shows the galvanic or electromotive series for common engineering materials that gives potential differences (in volts) with respect to the hydrogen electrode. The chart of Fig. 13.1 will immediately show the difference in potential for these materials. The greater the potential difference, the greater the galvanic action, or the faster the corrosion will progress. Per convention, the anode is considered (+) and the cathode (−). Gold is the most noble metal and magnesium is one of the least noble metals. The galvanic action is shown in Fig. 13.1*a*.

If you study Fig. 13.1 closely, you will see why the different electrochemical batteries were developed. The potential differences between nickel, cadmium, zinc, iron, lead, and silver are such that electrochemical battery systems were able to evolve around these elements, when used with the proper electrolyte solutions.

Corrosion has been estimated as causing over $125 billion per year in damages in the United States alone. Corrosion is at work 24 h per day, 365 days a year, and is a never-ending problem. Authorities on corrosion problems believe that much of this damage and loss can be prevented if design engineers pay closer attention to potential corrosion problems during the design process. Thinking about corrosion problems while selecting the design materials is the first step a product designer should undertake.

As an example, the selection of high-strength steels which will go into an environment containing sulfides or raw hydrogen is a grave design mistake. The high-strength steels will fail in these environments, although the authorities are not quite sure why this takes place, except that hydrogen embrittlement will occur.

The use of dissimilar metals is another often cited cause of corrosion problems. The welding of stainless steels which are not low in carbon content is another cause of stress-induced corrosion. Austenitic stainless steels, with high nickel content and low carbon, are recommended for welding applications (AISI 304, AISI 304L, AISI 316, etc.).

One of the best technical books for electromechanical design engineers to use as a reference for corrosion problems is that by Pludek (see bibliography at end of chapter).

| Engineering Material | Potential |
|---|---|
| Magnesium & magnesium alloys | -2.37v |
| Aluminum (1100 series) | -1.66 |
| Zinc | -0.76 |
| Chromium | -0.74 |
| Type 304 stainless steel (active) | ------ |
| Type 316 stainless steel (active) | ------ |
| Steel, iron, cast iron | -0.41 |
| Aluminum (2024) | ------ |
| Cadmium | -0.40 |
| Nickel | -0.23 |
| Tin | -0.14 |
| Lead | -0.13 |
| Copper | +0.34 |
| Silver solder | ------ |
| Nickel (passive) | ------ |
| Chromium-iron (passive) | ------ |
| Type 304 & 316 Stainless steel (passive) | ------ |
| Silver | +0.80 |
| Titanium | ------ |
| Platinum | +1.20 |
| Gold | +1.50 |

**FIGURE 13.1** Galvanic-electromotive series.

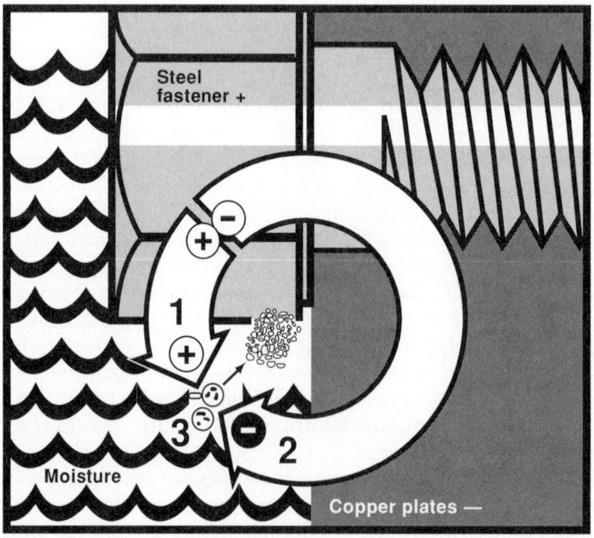


**FIGURE 13.1*a*** Diagram of the galvanic action.

***Electrochemical Equivalents for Various Metals.*** The electrochemical equivalents and other related data for the various plating metals and alloys are shown in Fig. 13.2. If you study the data for chromium, you will see why this metal is a difficult metal to plate, because it takes a large quantity of electrical energy for thin coatings. Cadmium, on the other hand, is an easy metal to plate, but it is a toxic element, as are its salts. Cadmium and mercury are almost equally poisonous heavy-metal elements.

***Electroplating Practice.*** The table of Fig. 13.3 lists the common plating metals and alloys and their possible plating baths, uses, and plating requirements. Procedures for actually performing different plating operations are not shown in this handbook. Many of the plating processes and procedures are proprietary; their procedures are closely guarded trade secrets. Platings such as bright chromium, black chromium, gold, and nickel chromium use proprietary procedures, which are not published in any book. Plating has been called a "black art," and you will understand why when you try to obtain information on the exact processes and procedures used in some of the industrial plating procedures and practices. Plating procedures are difficult and must be closely controlled to exacting standards if the finished article is to be of high quality.

## 13.3 ELECTROPLATING DATA AND SPECIFICATIONS

### 13.3.1 Electroplating and Oxide Layer Thickness Ranges

Many plating thickness specifications call for plating thicknesses in micrometers (μm), but others may be specified in mils (0.001 in = 1 mil)

$$1\ \mu\text{m} = 0.00004\ \text{in or}\ \frac{4}{100}\ \text{mils (0.04 mil)}$$

$$25.4\ \mu\text{m} = 0.001\ \text{in} = 1\ \text{mil}$$

$$5\ \mu\text{m} = 0.0002\ \text{in} = 0.2\ \text{mil} = \frac{2}{10}\ \text{mil}$$

Both measurement units are given in the following listings, which summarize, in outline form, thickness specifications for various platings and coatings on a number of base-metal types.

### 13.3.2 Anodic Coating Thickness (Anodized Parts)

| Type and/or description | Thickness, mil (μm) |
|---|---|
| Aluminum ASTM B580-79 | |
| A, hard coat | 2.0 (50) |
| B, architectural | 0.7 (18) |
| D, automobile exterior | 0.3 (8) |
| E, interior—*a* | 0.2 (5) |
| F, interior—*b* | 0.1 (3) |
| G, chromic acid | 0.04 (1) |
| Aluminum Mil-A-8625C | |
| Chromic acid | 0.05–0.3 (1.3–8) |
| Sulfuric acid | 0.1–1.0 (3–30) |
| Hard coat | 0.5–4.5 (13–114) |

| Metal | mg/coulomb | gm/amp hr | oz/amp hr | Specific Gravity | oz/sq ft for 0.001 in | amp hr to deposit 0.001"/sq ft ♦ | Symbol |
|---|---|---|---|---|---|---|---|
| Cadmium | 0.58 | 2.10 | 0.074 | 8.64 | 0.72 | 9.73 | Cd |
| Chromium | 0.09 | 0.32 | 0.011 | 7.10 | 0.59 | 51.8 | Cr |
| Copper | 0.33 | 1.19 | 0.042 | 8.92 | 0.74 | 17.7 | Cu |
| Gold | 0.68 | 2.45 | 0.087 | 19.3 | 1.61 | 18.6 | Au |
| Iron | 0.29 | 1.04 | 0.037 | 7.90 | 0.66 | 17.9 | Fe |
| Lead | 1.07 | 3.87 | 0.136 | 11.3 | 0.94 | 6.91 | Pb |
| Nickel | 0.30 | 1.10 | 0.039 | 8.90 | 0.74 | 19.0 | Ni |
| Palladium | 0.28 | 0.10 | 0.035 | 12.0 | 0.10 | 28.6 | Pd |
| Platinum | 0.51 | 1.82 | 0.065 | 21.4 | 1.78 | 27.6 | Pt |
| Rhodium | 0.27 | 0.10 | 0.034 | 12.5 | 1.04 | 30.8 | Rh |
| Silver | 1.12 | 4.02 | 0.142 | 10.5 | 0.88 | 6.20 | Ag |
| Tin | 0.31 | 1.11 | 0.039 | 7.30 | 0.61 | 15.6 | Sn |
| Zinc | 0.34 | 1.22 | 0.43 | 7.10 | 0.59 | 13.7 | Zn |

**NOTE: ♦ Equals specific gravity x 0.08323**

**FIGURE 13.2** Electrochemical equivalents.

| Type and/or description | Thickness, mil (μm) |
|---|---|
| Magnesium AMS 2478B | |
| Acid, full coat | 0.9–1.6 (23–41) |
| Magnesium Mil-M-45202C | |
| Light HAE | 0.1–0.3 (3–8) |
| Heavy HAE | 1.3–1.7 (33–43) |

### 13.3.3 Electroplating Thicknesses and Specifications

*Cadmium plate*

Range: 0.2 to 1.0 mil (5 to 25 μm)

Specification reference: ASTM A165-80

*Note:* Cadmium metal is toxic, as are its salts. Use may be restricted.

*Chromium plate*

Range: 0.01 to 6.0 mil (0.3 to 150 μm)

Decorative: 0.01 mil (0.3 μm)

Engineering: 2 mil (50 μm)

Specify hard-chrome on the drawing for corrosion resistance

Heavy chrome plating range (4 to 6 mil) will have porosity

Specification reference: Fed QQ-C-320B

*Copper plate*

Range: 0.1 to 5 mil (3 to 125 μm)

Specification reference: Mil-C-14550A (used as a base for other platings)

*Gold plate*

Range: 0.02 to 1.5 mil (0.5 to 38 μm) in eight thickness classes

Specification reference: Mil-G-45204B

*Nickel plate*

Range: 0.2 to 7.9 mil (5 to 200 μm); engineering coatings

Specification reference: ASTM B689-81 (may be soft- or hard-deposited within other specifications)

*Nickel with chromium plate on steel*

Range: 0.4 to 1.6 mil nickel; 0.01 mil chromium

Kinds of coatings include bright, dull, and layered

Specification reference: ASTM B456-79

Also covers nickel with chromium on copper and copper alloys

*Palladium plate*

Range: 0.05 to 0.2 mil (1.3 to 5 μm)

Specification reference: ASTM B679-80

*Rhodium plate*

Range: 0.008 to 0.25 mil (0.2 to 6.4 μm); engineering coatings

Specification reference: ASTM B634-78

*Silver plate*

Range: 0.04 to 1.6 mil (1 to 40 μm)

Types include mat finish, bright finish, tarnish-resistant (chromate-treated)

| Plating Metal | Typical Uses | Plating Solution Type | Plating Bath Temp. °F | Current Density Amps/ft$^2$ | Volts DC | Throwing Power | Time to Deposit 0.001" |
|---|---|---|---|---|---|---|---|
| Cadmium | Protection | Cyanide | 70-95 | 15-45 | 1-4 | Good | 20 min. |
| Chromium | Decorative, Engineering (Hard), Cylinder liners | Chromic acid | 120 | 200-250 | 6-8 | Poor | 2 hours |
| Copper | Electroforming, Undercoat Stop-off for casehardening | Acid<br>Cyanide<br>Rochelle | 75-120<br>75-100<br>140-160 | 15-40<br>5-15<br>20-60 | 1-2<br>1.5-3<br>2-3 | Fair<br>Good<br>Good | 35 min<br>90 min<br>45 min |
| Gold | Decorative, Electronics, PC boards | Cyanide<br>Proprietary | 120-160<br>...... | 5-15<br>...... | 2-6<br>...... | Good<br>...... | ......<br>...... |
| Indium | Bearing surfaces | Cyanide<br>Sulfate<br>Fluoborate | 70-75<br>70-75<br>70-90 | 10-150<br>20<br>50-100 | ......<br>......<br>...... | Good<br>Poor<br>Good | ......<br>......<br>...... |
| Iron | Electroforming, Repair | Chloride<br>Sulfate | 190<br>70-75 | 60<br>20 | ......<br>...... | ......<br>...... | 20 min<br>1 hour |
| Lead | Protection Bearing surfaces | Fluoborate | 70-75 | 10-80 | 0.5 | Good | 40 min |
| Nickel | Protection, Decorative, Base coat for chromium | Sulfate-chloride<br>Sulfamate<br>Fluoborate | 75-100 | Varies | 0.5-3 | Fair | 30 min |
| Rhodium | Decorative, Optical | Sulfate<br>Phosphate | 110-120 | 10-80 | 2.5-5 | ...... | ...... |
| Silver | Protective, Decorative, Electrical contacts | Cyanide | 80 | 5-15 | 1 | Good | ...... |
| Tin | Protective, | Sulfate | 70-75 | 40 | 1-3 | Fair | 15 min |

| | | | | | | | |
|---|---|---|---|---|---|---|---|
| | Food & dairy, | Fluoborate | 75-100 | 50 | ...... | Good | 10 min |
| | Bearings, | Stannate | 150-190 | 40 | 4-8 | Excellent | 30 min |
| Zinc | Protective | Sulfate | 75-100 | 15-400 | ...... | Fair | 10 min |
| | | Cyanide | 100 | 10-50 | ...... | Good | 40 min |
| | | | **Alloys** | | | | |
| Brass | Rubber bonding, Decorative | Cyanide | 75-100 | 3-10 | 2-3 | Good | ...... |
| Bronze | Decorative, Base for chromium, Stop-off for steel | Cyanide-stannate | 155 | 20-100 | 3-6 | Excellent | 30 min |
| Lead-tin | Bearings Solderability, Electrotyping | Fluoborate | 70-75 | 60 | 1-2 | Good | ...... |
| Tin-zinc | Solderability | Cyanide-stannate | 150 | 10-75 | 4-5 | Excellent | 30 min |
| Tin-nickel | Printed circuits | Chloride-fluoride | 150 | 25 | 1-2 | Excellent | 30 min |

**FIGURE 13.3** Electroplating practices: average operating conditions.

Specification reference: ASTM B700-81

*Note:* Silver plating of electrical current carrying parts such as switch blades, breaker contacts, and other sliding parts is generally given as 0.2 mil (5 μm), except where industry specifications may call for 3 mil (76 μm). The 3-mil coating is more effective on surfaces subject to heavy friction, such as large, sliding electrical contacts such as those found on electrical power distribution equipment.

*Tin plate*

Range: 0.2 to 1.2 mil (5 to 30 μm)

Specification reference: ASTM B545-72

*Note:* On electrical switchgear and other industrial equipment, 3 mil thickness is usually specified for service at paper mills and other environments where hydrogen sulfide in the atmosphere corrodes silver and other platings.

*Zinc plating on steel*

Range: 0.2 to 1 mil (5 to 25 μm) service conditions 1 through 4

The standard general thickness on most parts is 0.2 mil (0.0002 in). For severe service, the plating thickness should be 13 to 15 μm or 0.00057 in, or $^{6}/_{10}$ mil (0.0006 in). Hot-dip coatings of 1 mil (0.001 in) are given to parts that are subject to extremely severe service.

Specification reference: ASTM B633-78

*Note:* Bright zinc plating on steel parts (usually sheet metal and machined parts) is often given a yellow chromate conversion coating after the zinc plate, imparting a gold, iridescent appearance. Many commercial parts are given this treatment for corrosion resistance in lieu of painting. Conversion coatings are also available for silver, copper and its alloys, tin, aluminum, magnesium, zinc-base die-casting alloys, and electroplated chromium. During the conversion process, a complex chromium metal gel forms on the part surface, which contains hexavalent and trivalent chromium. When the soft gel coating dries, it becomes hard and somewhat abrasion-resistant. This then forms what is the preferred plating and coating for many engineering applications.

### 13.3.4 Plating Metals: Characteristics and Properties

*Cadmium*

Melting point: 320.9°C (609.6°F)

Specific gravity: 8.65 at 20°C

Soft, bluish-white metal, similar to zinc in appearance. Cadmium is toxic, as are its compounds. Being replaced by zinc in many plating applications because of its toxicity. Used to plate mechanical fasteners, etc. but should not be specified unless required by special application.

*Chromium*

Melting point: 1890°C (3434°F)

Specific gravity: 7.19 at 20°C

Steel gray, lustrous, and very hard metal. On a scale of 1 to 10, it would be 9, with only diamond at 10. Hexavalent chromium compounds are toxic. Excellent corrosion resistance for many applications. Attacked by hydrochloric acid, in which it dissolves. Heavily plated deposits become porous on the outer layers. Used to plate hand tools, surgical instruments, decorative trim, cutting tools and drills, and wear-resistant surfaces.

*Copper*

Melting point: 1083°C (1981°F)

Specific gravity: 8.96 at 20°C

Reddish-colored, malleable, and ductile metal. Excellent conductor of heat and electric current. Good corrosion resistance. Used as a plating base on irons and steels, and for other plating metals such as chromium, tin, and nickel.

*Gold*

Melting point: 1063°C (1945°F)

Specific gravity: 19.32 at 20°C

Yellow, soft metal which is extremely malleable and ductile. Excellent conductor of heat and electric current. Excellent corrosion resistance. The metal alloys are measured in karats; 24 karats is pure gold (18 karats = $^{18}/_{24}$ = 0.75 or 75% pure gold; 14 karats = $^{14}/_{24}$ = 0.583 or 58.3% pure gold, etc.). Used to plate electrical contacts, printed circuits (PCs), watches, and other jewelry.

*Indium*

Melting point: 156.6°C (313.9°F)

Specific gravity: 7.31 at 20°C

Very soft, silvery-white metal. May be alloyed to produce very low-melting-point alloys. The metal is toxic.

*Nickel*

Melting point: 1453°C (2647°F)

Specific gravity: 8.90 at 20°C

Nickel is hard, malleable, ductile, slightly ferromagnetic, and a fair conductor of heat and electric current. Silvery-white colored metal that will take a high polish. Excellent corrosion resistance. Used to plate handguns, hand tools, decorative trim, and corrosion-resistant containers.

*Palladium*

Melting point: 1552°C (2825.6°F)

Specific gravity: 12.02 at 20°C

Steel-white metal which is soft and ductile when annealed. Very hard when cold-worked. Does not tarnish in air. Uses include engineering plating and plating of electrical contacts.

*Rhodium*

Melting point: 1966°C (3570°F)

Specific gravity: 12.41 at 20°C

Silvery-white metal with low electrical resistance. Highly resistant to corrosion. Plated rhodium is extremely hard. The base metal is also ductile. Used to plate electrical contacts.

*Silver*

Melting point: 960.8°C (1761.4°F)

Specific gravity: 10.50 at 20°C

Brilliant white, lustrous metal. Harder than gold but still very malleable and ductile. Highest thermal and electrical conductivity of all metals. Stable in normal air, but tarnishes when exposed to ozone, hydrogen sulfide, and air containing sulfur. Used to plate electrical contacts, conducting surfaces such as bus bars, eating utensils, and jewelry. Silver was previously used to plate PC boards, but exhibited a detrimental property known as *silver migration.* Thin "whiskers" of silver would form on PC-board conductor foil in the presence of an electric current, and short-circuit the circuits on the PC board by bridging the spaces between the conductor patterns. This property of silver was not discovered until some time after PC boards were put into service on many military and commercial products. The result was a loss of many millions of dollars in damages, correctable only by replacing all the defective PC boards using gold- or tin-plated conductor foil.

*Tin*

Melting point: 231.9°C (449.4°F)

Specific gravity: 7.31 (white tin); 5.75 (gray tin) at 20°C

Silvery-white metal which is malleable and slightly ductile. Resists sea water and tap water but is attacked by strong acids, alkalies, and acid salts. Used to plate electrical contacts, bus bars, and corrosion-resistant containers and tubing.

*Zinc*

Melting point: 419.4°C (786.9°F)

Specific gravity: 7.13 at 20°C

Bluish-white lustrous, brittle metal. Fair conductor of electricity. Will burn in air at a red heat, producing clouds of white zinc oxide. Good corrosion resistance and used widely to plate irons and steels.

### 13.3.5 Summary of Plating and Finishing

The tabular list in Fig. 13.4 covers most of the presently available and widely used plating processes and protective finishes, excluding paints and plastic coatings. These finishes have wide range usage in aerospace, commercial aircraft, automotive industry, consumer products, and other industrial applications. The name, applications, and appearance of the finish are all listed in the Fig. 13.4. For further information concerning the latest military specifications and American standard specifications for these finishes, consult the plating companies or finishers.

The effective corrosion resistance of selective platings and finishes is shown in Fig. 13.5. The thicknesses or ranges are shown in inches. As can be seen from the data in the table of Fig. 13.5, hard chromium is one of the best engineering finishes for metal products, although it is more expensive than the other common platings.

## 13.4 COLORING PROCESSES FOR METALS AND ALLOYS

Following is a listing of some of the coloring processes used to color metals and alloys. Coloring of metals may be for appearance and also for additional protection against corrosion and rusting. These processes have evolved over a span of many years and are widely used in the metal finishing industries. Most of them involve the use of chemical action together with the application of heat.

It should be noted that the finished appearance of the metal being colored is affected to a great extent by how well the base metal itself is finished and polished. As an example, in the blueing process for handguns and rifles, the base metal is very carefully finished and highly polished prior to blueing. If the metal parts are not finished and highly polished, the blueing process will not hide the defects in the base metal.

## 13.5 COLORING METALS (FERROUS AND NONFERROUS)

***Black Oxide.*** Irons and steels may be given a black finish or black oxide finish by either of two processes:

1. Heat the ferrous metal part to approximately 500 to 700°F and plunge it into a container of good-quality machine oil. Reheat and plunge the part into the oil a number of times until the desired depth of black is obtained. Use precautions on hardened and tempered parts.

2. A thin black oxide coating may also be applied to iron and steel parts by immersing the parts in a boiling solution of sodium hydroxide and mixtures of nitrates and nitrites (sodium nitrate, potassium nitrate, etc.).

***Blueing Irons and Steels.*** Many ferrous metal parts receive a blueing operation to enhance their appearance and also as a rust-preventive coating. Handguns, rifles, and shotguns normally receive a blueing operation, unless chromium or satin nickel are specified.

The following blueing process imparts a fine blue to blue-black finish. Clean the metal with a potassium bichromate–sulfuric acid solution, then wash with ammonium hydroxide solution and wipe dry with a clean, lint-free cloth. Apply ammonium polysulfide until the desired depth of blueing is obtained. The finish may be made nearly black by repeating the process of applying the ammonium polysulfide. A light machine oil or silicone cloth wipe will then additionally provide the surface with excellent corrosion and rust resistance.

***Phosphatizing Irons and Steels.*** Three types of phosphate coatings (conversion coatings) are given to ferrous metals: zinc phosphate (Parkerizing), iron phosphate, and manganese phosphate. Zinc phosphate coatings vary in color from light to dark gray; iron phosphate, dark gray; and manganese phosphate, dark gray, becoming black with service. The phosphate coatings are used for paint bases, to facilitate cold working of the metal, and for rust prevention. The zinc phosphate or Parkerized finish is used on many military small arms, knives, and bayonets.

The phosphatizing solution contains approximately 3 to 5% of phosphoric acid, by volume. The part to be phosphatized is first thoroughly chemically cleaned and dipped into a solution of 2% by volume of hydrochloric acid for approximately 15 sec and then water-rinsed and dried. The part is then immersed into the phosphatizing bath for the time required to impart the desired coating. Zinc plates and zinc parts are also phosphatized in the same manner. Stainless steels and certain alloy steels cannot be phosphatized. Also, hardened and case-hardened steel parts are difficult to phosphatize in some cases.

***Passivation of Copper and Copper Alloys.*** A passivation coating may be applied to copper and its alloys, which imparts a blue-green color or patina. This coating is corrosion-resistant. The passivation solution may be made using the following solution composition:

1. 6 lb of ammonium sulfate
2. 3 oz of copper sulfate
3. 1.4 fluid oz of technical ammonia [specific gravity (sp.gr.) = 0.09]
4. 6.5 gal of water

Apply the solution with a spray to the chemically cleaned copper or copper alloy parts in six applications, drying between applications. The patina appears after approximately 6 h and continues as the part weathers.

***Copper and Copper Alloy Blackening (Alloys Containing >85% Copper).*** To color copper and its alloys black, use the following procedure. Make a solution consisting of 4 oz of arsenious oxide, 8 fluid oz of hydrochloric acid (sp.gr. = 1.16), and 1 gal of water. Then

1. Heat the solution to 175 to 200°F.
2. Immerse the copper or copper alloy parts until a uniform black color is obtained.
3. Brush the parts while wet, dry the parts, and apply a protective clear finish such as polyurethane varnish.

| Finish | Application | Appearance |
|---|---|---|
| Brass | Used on high strength fasteners where a decorative finish is required | Gold iridescent |
| Cadmium | General purpose plating. High degree of corrosion resistance especially in salt atmospheres. Soft finish not for threaded applications requiring frequent tightening cycles. Cadmium is toxic. Zinc recommended as replacement. | Silver-gray |
| Cadmium and clear chromate | Suited for applications for hand contact. Corrosion resistant | Clear |
| Cadmium and black chromate | Recommended where black color is desired. Note: chromate can be applied in red, green, blue or other color as suits the application. | Black or color |
| Cadmium and bronze chromate | Normally used in military applications for identification of plating type. | Gold iridescent |
| Cadmium and yellow iridescent chromate | Similar to bronze | Yellow iridescent |
| Cadmium and olive drab chromate | Good for paint bonding applications. | Dull olive |
| Cadmium and phosphate | Similar to cadmium and olive drab chromate | Black |
| Copper | Used on Allen threaded products. Also for high heat applications. | Copper-matte |
| Chromium, decorative | Decorative and corrosion applications. Plating is usually very thin. | Bright or satin |
| Chromium, hard | A most durable plating which is extremely hard. Expensive. Engineering aplications. Cutting tools and drills. | Satin |
| Lead | For soldering applications. | Matte |
| Nickel | Decorative appearance, excellent wear and corrosion resistance. | Matte or bright |
| Silver | Electrical properties, utensils, appearance. | Matte or bright |
| Tin | Anti-seize properties, corrosion resistance, non-toxic. | Grayish-white |
| Zinc | Popular plating for many applications; hardware, steel sheets, machined parts. | Bluish-white |
| Zinc and clear chromate | Similar to cadmium and clear chromate. Corrosion resistance, economical. | Clear |
| Zinc and black chromate | Similar to cadmium and black chromate. Where black color is desired. | Black |

| | | |
|---|---|---|
| Zinc and bronze chromate | Similar to cadmium and bronze chromate. Color applications. | Gold iridescent |
| Zinc and yellow iridescent chromate | Similar to cadmium and yellow iridescent chromate. Color applications. Popular engineering finish. Economical and attractive for industrial applications. | |
| Zinc and olive drab chromate | Similar to cadmium and olive drab chromate. Good paint adhesion properties. | Dull olive |
| Zinc and phosphate | Similar to cadmium and phosphate. Black for color applications. | Black |
| Anodic | Finishes for aluminum, magnesium, and titanium. Many types and thicknesses available. | Many colors |
| Black oxide | For threaded fasteners such as set-screws. Only mildly corrosion and rust resistant. Normally applied to irons and steels. Often oiled. | Black |
| Iron phosphate | For hardware which is severely handled. Decorative gray finish. | Oily gray-black (oil finish) |
| Manganese phosphate | For thread lubrication applications. For frictional contact applications. Supplied oiled or waxed for additional protection | Gray-black (oil finish) |
| Black nitrate | For stainless steel fasteners which require uniform black finish. | Dull or luster black |
| Passivation | For stainless steel fasteners and parts to prevent corrosion. Acid treatment. | Bright or matte |
| Zinc phosphate | For exposed fasteners requiring good paint adhesion properties. Dry or lubricated. | Dull gray (oil coating) |
| Teflon coating | Corrosion resistant and high lubricity. For non-galling anti-seizing applications. Used for bearings. New processes incorporate anodizing with teflon impregnation. | Gray to black |

**FIGURE 13.4** Summary of plating and finishing types and applications.

| Plating or Surface Treatment | Specification | Type | Class | Minimum Thickness | Salt Spray Test (minimum hours) | |
|---|---|---|---|---|---|---|
| | | | | | White Corrosion | Rust |
| Anodize | Mil-C-5541A | ..... | ..... | ..... | 168 | ..... |
| | Mil-A-8625B | I, II | ..... | ..... | 240 | ..... |
| Cadmium and Chromate | AMS-2400L1 | ..... | ..... | 0.0002-0.0003 | 100 | ..... |
| | AMS-2400L2 | ..... | ..... | 0.0002-0.0004 | 100 | ..... |
| | AMS-2400L3 | ..... | ..... | 0.0003-0.0005 | 150 | ..... |
| Cadmium and Chromate | QQ-P-416a | II | 1 | 0.0005 | 96 | ..... |
| | QQ-P-416a | II | 2 | 0.0003 | 96 | ..... |
| | QQ-P-416a | II | 3 | 0.0002 | 96 | ..... |
| | Mil-C-8837 | II | 1 | 0.0005 | 96 | ..... |
| | Mil-C-8837 | II | 2 | 0.0003 | 96 | ..... |
| | Mil-C-8837 | II | 3 | 0.0002 | 96 | ..... |
| Chromium, hard | Mil-C-11436 | ..... | ..... | 0.0012 | ..... | 100 |
| Lead | Mil-L-13808 | I, II | 3 | 0.00025 | ..... | 24 |
| | Mil-L-13808 | I, II | 2 | 0.0005 | ..... | 48 |
| Manganese Phosphate, dry | Mil-P-16232C | M | 3 | 0.0002 | ..... | 1.5 |
| Manganese Phosphate, oil | Mil-P-16232C | M | 2 | 0.0002 | ..... | 24 |
| Tin | Mil-T-10727A | I | ..... | 0.0002 | ..... | 24 |
| Zinc | Mil-Z-325a | I | 2 | 0.0005 | ..... | 96 |
| | Mil-Z-325a | I | 3 | 0.0002 | ..... | 36 |
| Zinc Phosphate, dry | Mil-P-16232C | Z | 3 | 0.0002 | ..... | 2 |
| | Mil-P-16232C | Z | 2 | 0.0002 | ..... | 48 |
| Zinc and Chromate | QQ-Z-325a | II | 2 | 0.0005 | 96 | ..... |
| | AMS-2402E | ..... | ..... | 0.0002-0.0003 | 100 | ..... |
| | QQ-Z-325a | II | 3 | 0.0002 | 96 | ..... |
| | AMS-2402E | ..... | ..... | 0.0003-0.0005 | 150 | ..... |
| Zinc and Phosphate | QQ-Z-325a | III | 2 | 0.0005 | ..... | 96 |
| | QQ-Z-325a | III | 3 | 0.0002 | ..... | 36 |

**FIGURE 13.5** Effective corrosion resistance of selective platings.

***Coloring Brasses.*** Brass may be given a green color per the following procedure. Mix a solution containing 1 oz of ferric nitrate, 6 oz of sodium thiosulfate, and 1 gal of water. Then

1. Heat the solution to 160 to 180°F.
2. Immerse the brass parts in the solution until the desired color is obtained.
3. Dry and coat with a clear protective finish such as clear lacquer or polyurethane varnish.

## 13.6 ETCHING METALS

The solutions used to etch various metals are often called *mordants,* and some of the solutions used in the etching processes are described as follows.

***Etching Irons, Steels, and Zinc Plate.*** A popular solution for etching irons, steels, and zinc plates consists of a solution of 2 oz of 50% nitric acid mixed into 15 oz of water. This is a $\frac{1}{16}$ solution (1 part nitric acid to 15 parts water, by volume). Always pour the acid into the water; never pour water into the acid. Pouring water into acid may cause a mixing reaction which could cause the acid to splash from the mixing container. The solution is relatively slow-acting unless used with a splash-type etching machine. For hand etching, a stronger solution can be used, but the etching cut becomes rough at the edges of the etching action. Solutions as strong as $\frac{1}{8}$ can be used; 1 part concentrated nitric acid to 7 parts water, by volume.

***Etching Copper and Copper Alloys.*** *Ferric chloride* is often used to etch copper and its alloys. A 40°Bé solution of ferric chloride used at 75 to 80°F etches copper cleanly and not too rapidly so as to produce a rough etched edge. A 40°Bé solution is made by mixing 20 oz of ferric chloride (anhydrous) to water to make a final volume of 1 liter (1000 ml). The specific gravity of this solution can vary from 1.37 to 1.38. The designation *°Bé* (Baumé; pronounced "Bow-may") is defined in respect to specific gravity by the following equation:

$$\text{Specific gravity} = \frac{145}{145 - {}^\circ\text{Bé}}$$

With this equation, you may determine °Bé if you know what specific gravity you want or have, or you may determine the specific gravity required of the solution if you know the °Bé you wish to produce. (*Note:* The reaction of ferric chloride produces a great amount of heat when it is dissolved in water. The water temperature should be between 50 and 75°F prior to mixing the ferric chloride solution.)

Solutions of ferric chloride as low as 30°Bé are used for fast etching of copper and its alloys, while the 40 to 42°Bé solutions are used for etching intaglio printing plates or photogravure work. These solutions have also been used to etch stainless steels. (*Caution:* The ferric chloride bath should be contained in a glass, plastic, or wooden tank, as the solution is highly corrosive. Do not dispose of these solutions directly into standard drainage systems. The solutions should be diluted in water and neutralized with a base chemical such as sodium bicarbonate. Do not allow the solutions to come into contact with the skin; rubber gloves should be worn at all times during use of these solutions.)

Printed-circuit boards for electronic applications have been produced using ferric chloride solutions for many years. Using the lower °Bé solutions produces a fast and relatively accurate etching action or cut, and the ferric chloride is economical and long-lasting. Perchlorate chemicals are also used for etching PC boards. Chapter 6 details the engineering procedures and manufacturing processes used to fabricate PC boards for prototypes and small production quantities.

Titanium heaters are used to heat the ferric chloride solutions at large PC fabrication facilities.

***Etching Aluminum and Aluminum Alloys.*** Most aluminum alloys can be etched using sodium hydroxide solutions of varying strengths. Sodium hydroxide is commonly called *caustic soda* or *lye* and is a low-cost chemical. Sodium hydroxide is very corrosive or caustic and is poisonous. When aluminums are etched using sodium hydroxide solutions, one of the reaction products is hydrogen gas, which is highly explosive when mixed with normal air. Proper venting must be employed when using this process for etching aluminum alloys.

The etching action is relatively rapid, especially if the sodium hydroxide solution is above $\frac{1}{16}$ (1 part sodium hydroxide to 15 parts water). Strong solutions produce a violent foaming reaction with the production of substantial amounts of hydrogen gas.

## 13.7 ANODIZING

*Anodizing* is a process of oxidation, produced in an electrolytic bath. Aluminum is the metal and alloys which are most commonly anodized, although the anodizing process may be performed on magnesium, zinc, and titanium. This section will cover the anodize coatings and processes for aluminum and its alloys only.

The anodic coating, which is aluminum oxide, is extremely hard and abrasive, especially aluminum *hardcoat* anodized surfaces. The classifications for aluminum and aluminum alloy anodic coatings are as follows:

*Type I:* Chromic acid anodizing, conventional coatings produced from chromic acid baths. Shall not be applied to alloys containing >5% copper or >7% silicon or when alloying elements exceed 7.5%.

*Type IB:* Chromic acid anodizing, low-voltage (20-V) process. Heat-treatable alloys (T4, T6, etc.) should be tempered prior to anodizing.

*Type II:* Sulfuric acid anodizing, conventional coatings produced from sulfuric acid baths. Heat-treatable alloys (T4, T6, etc.) should be tempered prior to anodizing.

*Type III:* Hard anodic coatings (hardcoat). Shall not be applied to alloys containing >5% copper, or >8% silicon, unless agreed on by the supplier. Heat treatable alloys (T4, T6, etc.) should be tempered.

Classes are as follows:

*Class 1:* nondyed, natural, including dichromate sealing

*Class 2:* dyed

Standard specifications for anodized aluminum and aluminum alloys are

| | |
|---|---|
| ASTM B244 | Thickness of anodic coatings, measurement of |
| ANSI/ASTM B 137 | Weight of coatings on anodized aluminum, measurement of |
| ASTM B 117 | Method of salt-spray (fog) testing |

***Sealing Anodized Aluminum and Aluminum Alloys.*** All types of anodizing must be sealed using any of the following methods after the electrolytic anodized coating is applied:

- Immersion in an aqueous solution of 5% sodium dichromate (15 min at 90 to 100°C)
- Immersion in deionized water (15 min at 100°C)
- Immersion in an aqueous solution of nickel or cobalt acetate (15 min at 100°C)
- Teflon impregnation processes (for sealing and lubricity)

***Anodic Coating Design Data.*** Radii of curvature on anodized parts are as follows:

| Nominal coating thickness, in | Radius of curvature (outside/inside), in |
|---|---|
| 0.001 | ~0.032 |
| 0.002 | ~0.062 |
| 0.003 | ~0.093 |
| 0.004 | ~0.125 |

Thickness ranges of anodic coatings on aluminum and aluminum alloys are as follows:

| Coating type | Thickness range, in |
|---|---|
| I and IB | 0.00002–0.0003 |
| II | 0.00007–0.0010 |
| III | 0.0005–0.0045 |

See Fig. 13.6 for minimum typical anodic coating thicknesses on various aluminum alloys, per type.

| Alloy Designation | Thickness of Coating, inches | |
|---|---|---|
| | Type I and IB | Type II |
| 1100 | 0.000029 | 0.000093 |
| 2024-T4 | ...... | 0.000125 |
| 2024-T6 | 0.000044 | ...... |
| 3003 | 0.000035 | 0.000103 |
| 5052 | 0.000033 | 0.000098 |
| 5056 | 0.000021 | ...... |
| 6061-T6 | 0.000034 | 0.000099 |
| 7075-T6 | 0.000040 | ...... |
| Alclad 2014-T6 | 0.000045 | ...... |
| Alclad 7075-T6 | 0.000041 | ...... |
| 295-T6 | ...... | 0.000107 |
| 356-T6 | ...... | 0.000102 |
| 514 | ...... | 0.000086 |

**Note: Anodic coating types I, IB and II are normally applied as thin coats, while type III (Hardcoat) is normally applied as the thicker coatings. See table of anodic thicknesses following.**

**FIGURE 13.6** Anodic coating thicknesses.

See Fig. 13.7 for maximum attainable thicknesses of anodic coatings on selected aluminum alloys.

### *Design Notes for Anodized Parts*

- A 2-mil (0.002-in) hardcoat anodic coating will penetrate the part 0.001 in and protrude from the part 0.001 in. A 1.000-in-diameter part which is anodized 2 mils (0.002 in) will have a finished diameter of 1.002 in. Half of the coating is inside the part, and half is on the outside.
- Avoid blind holes in parts.
- Avoid hollow weldments (drill 0.250-in-diameter weep holes in the part).
- Avoid steel inserts.
- Avoid sharp corners (see radii chart preceding; section on anodic design data).
- Avoid heavy to thin cross sections on the part.
- Allow for the anodic coating in your design tolerances on the part.

### *Anodic Coating Specifications*

| Name | Hardcoat anodize | Chromic anodize | Sulfuric anodize |
|---|---|---|---|
| Army-Navy | Mil-A-8625, Ty-3 | Mil-A-8625, Ty-1 | Mil-A-8625, Ty-2 |
| General Electric | AMS-2468D | AMS-2470H | AMS-2471D |
| Boeing | Code-302 | Code-300 | Code-301 |
| IBM | 41-207 | 41-204 | 41-203 |
| Grumman | G-9031 | 9030B | G-9032 |

Hardcoat anodize processes are as follows:

- Martin
- Alumilite
- Alpha
- Mae
- Sanford
- Boeing
- Scionic
- Hardas
- Imperv-X

Hardcoat-Teflon processes are

- Amphodize
- Hardtef
- Analon
- Tufram
- Polylube
- Lukon
- Nituf
- Kalon
- Ptfe

| Alloy Designation | Anodic Coating Thickness - inches | Color |
|---|---|---|
| 380 | 0.0006 | Gray |
| 360 | 0.001 | Gray |
| 319 | 0.0014 | Gray |
| 1100 | 0.0017 | Bronze |
| 2011 | 0.0021 | Light gray |
| 2014 | 0.0025 | Gray |
| 2019 | 0.0028 | Bronze |
| 6262 | 0.0031 | Black |
| 6061 | 0.0035 | Black |
| 2024 | 0.0040 | Bronze |
| 2017 | 0.0045 | Bronze |
| 6063 | 0.0050 | Black |
| 5052 | 0.0053 | Black |
| 2618 | 0.0055 | Gray |
| 2219 | 0.0058 | Gray |
| 218 | 0.0062 | Gray |
| 7079 | 0.0065 | Dark gray |
| 355 | 0.0075 | Gray |
| 7075 | 0.0082 | Bronze |
| Almag 35 | 0.0100 | Gray |
| 356 | 0.0120 | Light gray |

Note: All above coatings may be dyed black. Coatings over 0.003 inches thick tend to chip and become milky in color and should be used only in the salvage of parts.
Source: Anodic, Inc., Stevenson, CT 06491.

**FIGURE 13.7** Maximum anodic coating thicknesses.

- Smoothcoat
- Sanfordize
- Hardlube

Hardcoat anodize physical data are

| | |
|---|---|
| Hardness | 65 to 70 Rockwell C scale (harder than hard chrome) |
| Color | Dark gray to black |
| Dielectric strength | 800 V/mil |
| Machining | Grinding, lapping, polishing, and honing |
| Sealing | Dichromate, nickel, or cobalt acetate; hot-water or Teflon impregnation |
| Resistivity | $10^8$ to $10^{12}$ $\Omega \cdot$ cm |

(*Note:* Parts which have been anodized cannot be reanodized, nor can the anodic coating be made thicker.)

## 13.8 PAINT FINISHING

Paints are the most used of all finishes for metal products. The types and different varieties and colors available are limitless. Paint technology improves and changes constantly. Attesting to the quality and long-lasting ability of modern paints for metals are automotive vehicles and aircraft. Modern industrial paints not only must withstand the weather and corrosive elements, but many are chemically protected from the harmful effects of ultraviolet radiation from the sun.

Common industrial paints include

- Enamels (baked and air-dried)
- Lacquers
- Epoxies (1- and 2-part systems)
- Varnishes (glyptols, polyurethanes, and special chemical-resistant agents)
- Latex- and water-based paints

Pigments such as zinc and titanium oxide, carbon black, zinc chromate, Prussian blue, cobalt blue, and many others can be added to the paint for coloring and to improve durability.

Products that are built to customer or military specifications frequently have the type of finish listed in the specifications, together with the method of base-metal preparation, such as phosphatizing, zinc chromate primer, and zinc primer. The color and the required minimum dry-film paint thickness are also frequently set out in the equipment specifications.

Some of the high-quality equipment manufacturers in the United States apply multiple coatings to their products, producing a finish which is not only attractive but also long-lasting and durable. Large equipment, primarily made of sheet steel, as emphasized in the switchgear manufacturing industry, will often be multiple-coated. Some of the manufacturers use a phosphatizing process on the base metal, followed by a zinc-based primer, which is then epoxy coated and then finally given a high-quality polyurethane paint finish. Equipment of this caliber can be expected to withstand outdoor weathering for many years without finish problems.

When equipment is improperly base-metal-prepared and then given a few coats of low-quality finish, corrosion usually begins soon after the equipment is exposed to the elements. Finish problems on large, fabricated metal equipment and structures are often a serious defect, and it becomes difficult to prevent further corrosion, even though repainting is under-

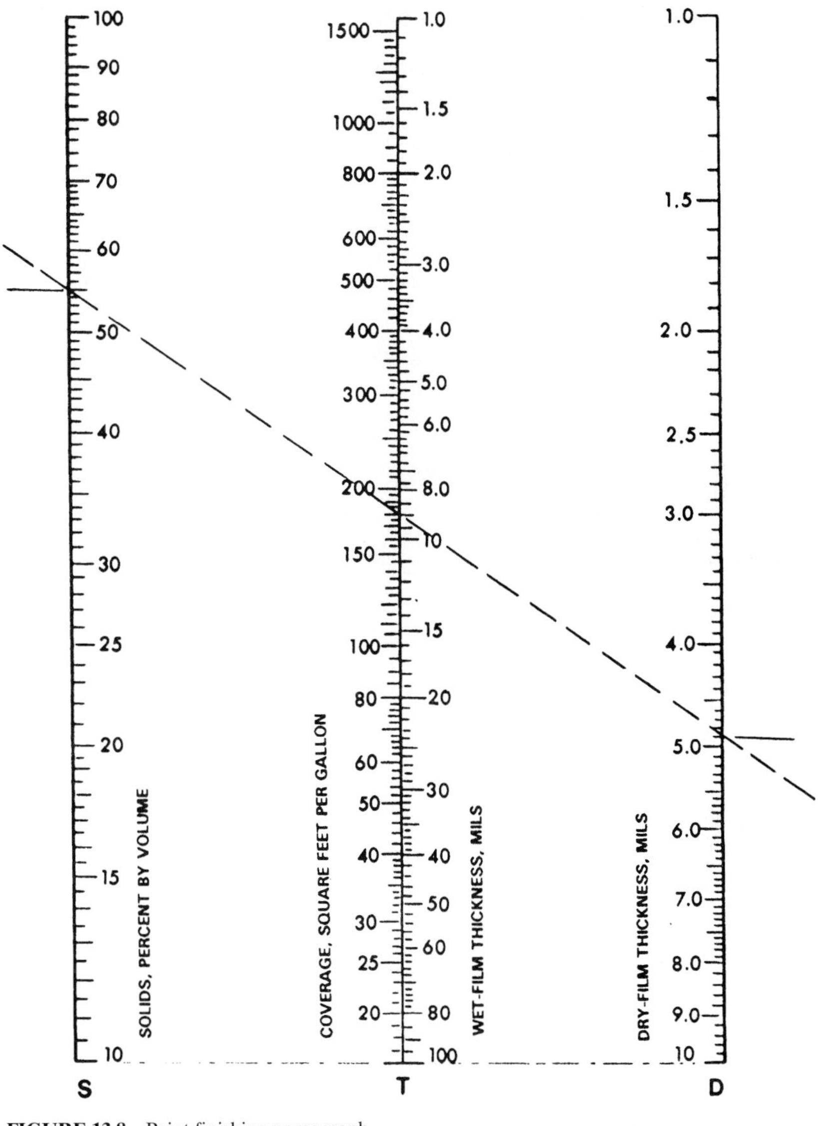


**FIGURE 13.8** Paint finishing nomograph.

taken. If the base-metal preparation and primers are not selected or applied correctly, corrosion problems will *always* be prevelant.

Specific design questions and requirements for the paint finish must be directed to the proper paint manufacturer if high-quality results are to be expected. High-quality paints and correct base-metal preparation techniques are expensive and require the proper equipment and facilities. There are many government regulations in force today that stipulate the procedures to be followed for environmental protection relative to the use of paint finishes and other toxic chemicals. These regulations are formulated and enforced by the Environmental Protection Agency (EPA) and Occupational Safety and Health Administration (OSHA).

### 13.8.1 Estimating Paint Film Thickness and Coverage

Estimating the quantity of materials for a painting project is not as simple as expected. The surface area coverage of any particular paint is usually listed on the paint container. But this coverage is for average wet-film area and does not take into consideration the dry-film thickness required on the product.

To determine the coverage for a 1-mil (0.001-in) dry-film thickness for a paint containing less than 100% solids, multiply the percentage of nonvolatile solids by 1604 and divide by 100. The percent of nonvolatile solids is found on the paint container label or from the paint manufacturers' literature. The wet-film thickness required for a specific dry-film thickness is found by dividing the desired dry-film thickness by the percent solids by volume of the coating to be used. (Wet-film thickness is measured during application by a wet-film thickness gauge.)

A fast and accurate estimate of wet-film thickness required to obtain a specified dry-film thickness and theoretical coverage in square feet of area per gallon can be made using the nomograph shown in Fig. 13.8. To use the nomograph of Fig. 13.8, connect the dry-film thickness on scale $D$ with the percent solids on scale $S$ and read the wet-film thickness required on scale $T$ together with the square feet of coverage per gallon.

## FURTHER READING

Durney, L. J., 1984: *Electroplating Engineering Handbook,* 4th ed. New York: Van Nostrand Reinhold.

*Machine Design,* 1986: "Fasteners Edition." Cleveland, Ohio: Penton.

Pludek, V. R., 1977: *Design and Corrosion Control.* New York: Macmillan.

Weast, R. C., 1969: *Handbook of Chemistry and Physics,* 50th ed. Cleveland, Ohio: The Chemical Rubber Company.

# CHAPTER 14

# MANUFACTURING MACHINERY AND DIMENSIONING AND TOLERANCING PRACTICES

A practical understanding and knowledge of the capabilities of the equipment, machinery, and tools found in a modern manufacturing facility are of prime importance to the electro-mechanical or product design engineer. Most aspects of part design are directly related to the abilities of the equipment, tools, and machinery used to manufacture each particular part. Inexperienced product designers often design parts that simply cannot be produced practically or economically, using the equipment available at their manufacturing facilities. Therefore, a thorough understanding of the capabilities and limits of the manufacturing machinery is essential to successful design engineering practice. Much of this experience is gained through the interaction of the designer or design engineer with the manufacturing personnel who operate the various equipments and machinery in the manufacturing facility, and more importantly, with the manufacturing and tool engineering departments.

## 14.1 MANUFACTURING MACHINERY AND MACHINE TOOLS

There are enormous varieties of modern manufacturing equipment, tools, and machinery available today. A large assortment of different machinery, tools, and equipment may be found in the medium-size manufacturing facility, and many more are found in the larger facilities. Certain machinery and equipment are indigenous to various industries such as foundries, forges, plastics molders, platers, optical facilities, extrusion facilities, and heat treaters.

The general, large manufacturing facility for mechanical type products would contain machinery and equipment similar to the following: lathes, milling machines, CNC turning centers, CNC vertical machining centers, hydraulic shears, squaring shears, stock cutting machines, boring mills, drill presses, shapers, planers, surface grinders, cylindrical grinders, EDM machines (CNC), screw machines, water-cutting machines, laser cutters, plasma cutters, welding machines, press brakes, leaf brakes, roll-forming machines, CNC turret punch presses, slitting machines, hydropresses, stretcher benders, spin-forming machines, friction welders, spot welders, seam welders, and many more.

The designer or design engineer, to be effective, should be familiar with all machinery used to manufacture parts in the factory. One need not be an expert in the theory and operation of these machines, but one must know their capabilities and limitations. Coordination of design engineering with the manufacturing and tooling departments within a company is *essential* to the sound and economical production of parts and equipment. This coordination is vital to the fulfillment of the design engineering function. An experienced design engineer always acknowledges and considers shop or factory suggestions on any phase of detail part or finished product design that will affect manufacturing, by improving the product or decreasing cost.

### 14.1.1 Typical Manufacturing Machinery, Equipment, and Tools

To familiarize those who are not aware of some of the more modern manufacturing equipment, tools, and machinery, the following figures will be of help.

One of the most important pieces of machinery for sheet-metal fabrication is the punch press. In Chap. 11 (Fig. 11.3), one may see the modern type of CNC-controlled high-speed multistation punch press that is used to produce the various holes and patterns found on the flat-pattern sheet-metal part. The advent of this type of machine has drastically reduced the time and thus the cost of producing complex sheet-metal parts. The accuracy of the parts (hole pattern relationships) has also been greatly improved over older equipment.

Figure 11.2 (in Chap. 11) shows the modern squaring shear which is used to accurately cut flat steel and aluminum sheets or other metal sheets to precise squareness and dimension. This machine is more accurate than a standard hydraulic shear, is CNC-controlled for optimum use of stock material, and is also high-speed.

The normal sheet-metal flat sheet (flat pattern) is then usually sent to the press brake, where the various flanges are bent. Figure 14.1 shows a large, high-tonnage press brake in which the back-gauge and press stroke are controlled by a digital panel, shown to the far right in the photograph. Figure 14.2 shows a small press brake where small sheet-metal parts are bent and formed to their final configuration. Large press brakes are also used as the actuating force to operate blanking dies and progressive dies. Figure 14.3 shows a modern leaf brake. In the leaf brake, the sheet stock remains flat, while the rotating front gate pivots up to produce the bend in the sheet-metal part. Leaf brakes can produce bends in sheet metal that cannot be produced on the press brake unless it is equipped with special bending dies. The leaf brake shown in Fig. 14.3 is CNC-controlled and produces accurate bends at a high rate of speed compared to the press brake.

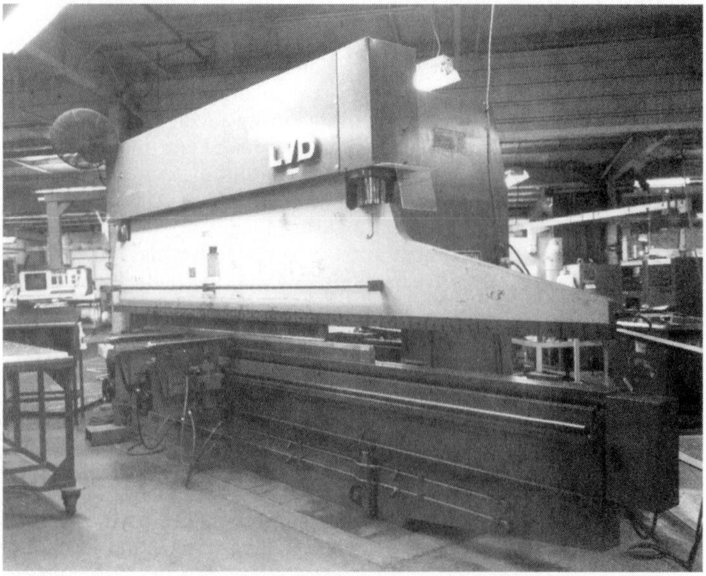


**FIGURE 14.1** A high-tonnage press brake with digital controls.

**FIGURE 14.2** A small press brake with digital controls.

**FIGURE 14.3** A modern CNC leaf brake with controller.

Moving on to machinery used in the machine shop, Fig. 14.4 shows the workhorse of machining operations, the geared-head engine lathe. This particular machine has a digital readout panel which is used to control the cross-feed and carriage travel. Figure 14.5 is a closer view of the same machine, looking into the head stock, showing the cross-feed controls and three-jaw chuck used to hold the workpiece during the turning operation.

For milling operations, the universal manual milling machine is frequently used in the model shop and tooling shop. The milling machine shown in Fig. 14.6 is the popular Bridge-

FIGURE 14.4 Engine lathe with digital controls.

FIGURE 14.5 Closeup view looking into spindle of engine lathe at the cross-feed.

port universal mill. This machine is preferred by many experienced master machinists and tool and die makers because of its extreme rigidity and controllable accuracy, together with its versatility for performing any milling operation. Figure 14.7 is a closeup view of a ball-end milling operation being performed on an aluminum alloy part.

The preceding figures for the lathe and milling machines showed the manual type whose actions are directly controlled by the machinist or machine operator. Figure 14.8 shows a modern CNC-controlled vertical machining center that is capable of very high speed and accurate work, changes its own tools, and is controlled by an onboard computer and a computer program developed in the tool engineering department. Many of these modern CNC

**FIGURE 14.6** The Bridgeport universal milling machine with digital controls.

**FIGURE 14.7** Closeup view of ball-end milling operation on an aluminum alloy part.

machines are equipped for three-, four-, or five-axis machining. Although this machine is a basic milling machine, additional control axes and tooling accessories allow threads, gears, and other highly complex parts to made be easily, accurately, and quickly. Figure 14.9 is a detail view of the CNC control panel of the machine shown in Fig. 14.8. This control panel contains the microprocessor and memory that deliver the machining instructions to the machine. The instructions for the program for the controller are sent electronically, directly from the tool engineering department. Figure 14.10 shows a gang milling operation being performed on this machine on a palletizing system. The pallet contains the separate parts which are all machined in one continuous operation. Various automatic tool changes are made during the operation to correspond to the machining operation required, such as drilling, reaming, boring, tapping, and milling.

For turning operations, the automatic CNC turning center is employed. Figure 14.11 shows a typical CNC-controlled turning center. The microprocessor controller is shown to the right in the photograph. Figure 14.12 shows another CNC turning center of the double-spindle type. This type of machine is currently being replaced by the newer two-independent-spindle machines with 32-bit microprocessors. These newer machines are not only faster, but are much more versatile because of the interaction of two independent spindles. Parts may be produced on these newer machines which were impossible to produce on the older models.

Figure 14.13 shows the versatile and valuable electrostatic discharge machine (EDM). With the EDM, it is possible to produce dies and other difficult parts very efficiently and with extreme accuracy. The machine shown in Fig. 14.13 is a wire EDM. A vertically moving wire with an electrical charge does the actual cutting operation. A 4-in-thick block of hardened tool steel may be accurately cut with machines of this type. The newer EDM machines are faster and also have two independent axes by which the wire may be accurately controlled. A cone-shaped hole may be accurately cut in thick, tough tool steel with the newer machines. The hole is not limited to a cylindrical or conical shape; any type of sloping compound outline may be cut on varying angles and pitches. The EDM machines have taken out much of the tedious, demanding work required in producing dies.

**FIGURE 14.8** A modern CNC vertical machining center with controller panel to the right.

**FIGURE 14.9** Closeup view of CNC controller unit of the machining center shown in Fig. 14.8.

**FIGURE 14.10** A palletizing gang milling operation on a vertical machining center.

FIGURE 14.11 A CNC turning center with controller to the right.

The EDM machines are also produced in random-access-memory (RAM) EDM form, where a preshaped electrode of positive form is lowered into a block of tool steel to produce an impression of great accuracy and fine finish for tooling molds that are used for injection and compression molding of plastic parts. Die-casting molds may also be produced on the RAM EDM machines.

(a)

FIGURE 14.12 A double-spindle CNC turning center with controller (two views).

(b)

**FIGURE 14.12** (*Continued*)

**FIGURE 14.13** A CNC-controlled EDM machine.

Figure 14.14 is a closeup view of the wire and wireguides on a wire EDM machine. Figure 14.15 shows in detail the type of cut that is produced by the wire EDM machine. The figure is that of a steel ratchet, the teeth of which were cut on this type of machine. Note the extremely fine and sharp cut produced by this process. Figure 14.16*a* shows a sample die set that was produced by EDM from a piece of hardened tool steel. The male and female parts of this die set were produced by different programs. Figure 14.16*b* shows the two pieces of the die set assembled into one block. The side clearance between the mating parts is 0.001 in. This is a good example of the type of accuracy that is attainable with the EDM processes.

With the advances made with machine tools in the past 40 years or so, advances have also been made with the tools used to measure the parts produced by the new machines. Figure 14.17 shows the newer type of digital micrometer that is now being widely used in industry. Tools of this caliber can accurately read to five decimal places. Figure 14.18 shows a typical, standard type of material hardness tester that is used to verify material hardness. Because there are numerous types of hardness measuring systems, there are also numerous tools or instruments for measuring them.

Figure 14.19 shows a finished blanking die set. This is a log-run, high-quality die set typical of the many that are continuously being used throughout industry. The blanking or stamping die is a critical element in the economical mass production of parts. Die sets are produced for blanking, punching, forming, and bending and in combinations known as *progressive dies.*

Figure 14.20 shows a finished machined part that is to be used for producing a mold for the manufacture of plastic parts. The high degree of finish on the part shown was not produced as a secondary operation; this is how the part was finish machined using the modern, high-speed CNC machine tools and newer types of cutting inserts available today.

**FIGURE 14.14** Closeup view of wire and guides on an EDM machine.

**FIGURE 14.15** Closeup view of a steel ratchet wheel cut on the EDM machine.

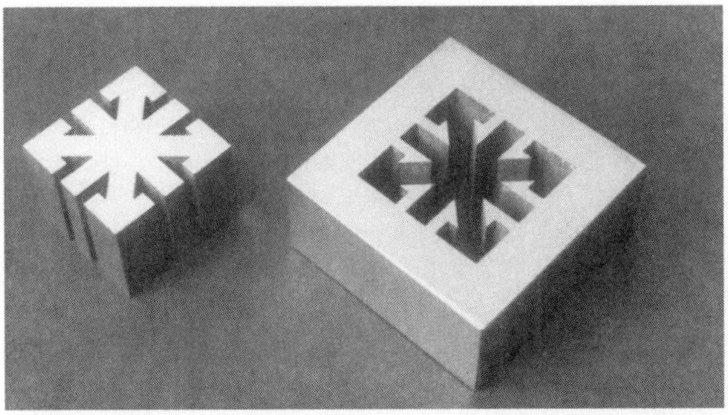

(a)

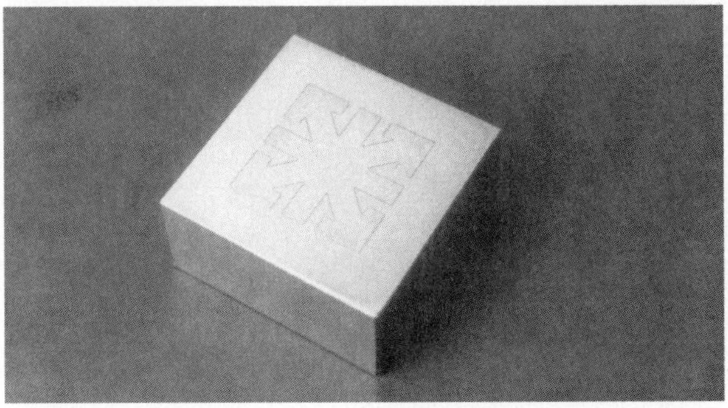

(b)

**FIGURE 14.16** (*a*) EDM cut hardened tool steel die set; (*b*) the die set of Fig. 14.16*a* with parts assembled together to show the close fit.

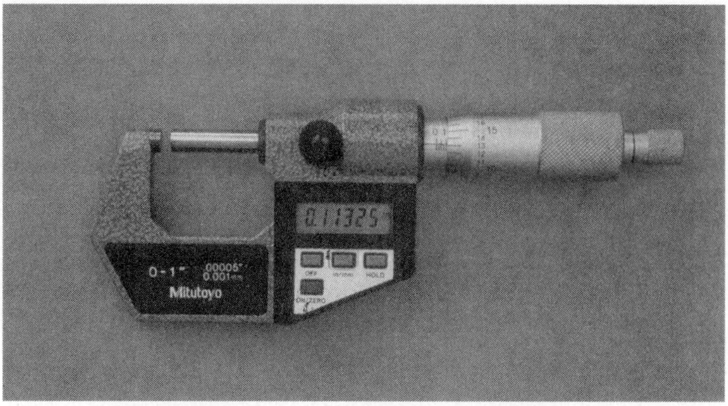


FIGURE 14.17 A modern digital micrometer reading to 0.00005 in.

FIGURE 14.18 A typical material hardness testing machine.

**FIGURE 14.19** A high-quality, high-production blanking die set.

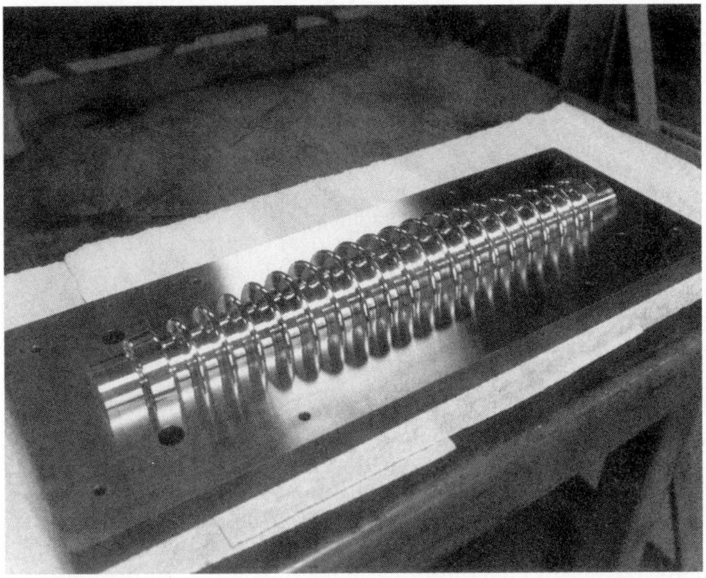

**FIGURE 14.20** A master mold pattern cut from aluminum alloy on a CNC turning center.

## 14.2 BASIC DIMENSIONING AND TOLERANCING PROCEDURES

### 14.2.1 General Dimensioning and Tolerancing Practices

Our definition of *tolerance* is the amount of dimensional variation allowed on a part or assembly of parts. Tolerance is equal to the difference between maximum and minimum limits of the specified dimension. For example

$$5.000 \pm 0.010 \text{ has a tolerance of } 0.020$$

$$\frac{5.000}{4.995} \text{ has a tolerance of } 0.005$$

$$\frac{6.129}{6.121} \text{ has a tolerance of } 0.008$$

#### 14.2.1.1 Unilateral and Bilateral Tolerances

1. Unilateral tolerance is used to relate the total tolerance to a basic dimension in one direction only. For example

$$5.000 \begin{smallmatrix} +0.002 \\ -0.000 \end{smallmatrix} \quad \text{(positive unilateral)}$$

$$5.000 \begin{smallmatrix} +0.000 \\ -0.002 \end{smallmatrix} \quad \text{(negative unilateral)}$$

2. Bilateral tolerance is used to relate the total tolerance to a basic dimension in both plus and minus directions. For example

$$5.000 \pm 0.005 \quad \text{(tolerance} = 0.010)$$

$$5.000 \begin{smallmatrix} +0.002 \\ -0.006 \end{smallmatrix} \quad \text{(tolerance} = 0.008)$$

3. Showing toleranced dimensions in an over-and-under form is as illustrated:

For shafts: $\frac{5.005}{4.995}$ (larger dimension "over"), $T = 0.010$

For holes: $\frac{5.002}{5.007}$ (smaller dimension "over"), $T = 0.005$

***Tolerancing Practices.*** Tolerances are applied to show the permissible variation in the direction that is the least critical. If variation in either direction is equally critical, the bilateral tolerance is used. If a variation in one direction is more critical than a variation in another direction, the unilateral tolerance should be given in the least critical direction.

For example, if you want the diameter of a round rod not to exceed 3.000 but will allow it to go under to 2.995, this is the same as stating

$$3.000 \begin{smallmatrix} +0.000 \\ -0.005 \end{smallmatrix}$$

Tolerances on the centerline distances between holes in a part are usually bilateral.

For nonmating surfaces, unilateral or bilateral tolerances can be used, but on mating surfaces the tolerances should be unilateral in all but a few cases.

***Locating Toleranced Dimensions.*** A common baseline or point in each plane of a part should be used to establish dimensions.

### 14.2.1.2 Tolerance Studies

At many companies, the design and checking groups perform on assembly drawings what is commonly referred to as a *tolerance study*. This is an in-depth analysis of all dimensions and tolerances on the detail parts to confirm the interchangeability of mass-produced parts going into the final product or subassembly.

The method generally followed to do the study involves comparing in combination all matching surfaces and hole-pattern dimensional tolerances at maximum and minimum conditions. If interference or mismatch occurs, the tolerances must be balanced to afford interchangeability of all parts and subassemblies.

This study is essential and critical when producing great quantities of parts to eliminate special fitting, which would destroy the ability to interchange parts.

A thorough designer will perform a tolerance study on the mechanism or assembly of parts and balance his/her tolerances on all dimensions to create interchangeability. For standard tolerances and fits, both English and metric, refer to ANSI B4.1-1967, R 1979 and ANSI B4.2-1978.

Parts must also be toleranced to allow for electrodeposited finishes such as zinc plating, hard-coat anodizing, and Teflon coating. (See Chap. 13 for plating thicknesses.)

General tolerancing procedures, force fits, sliding fits, etc., are described in detail in handbooks listed in the bibliography sections of previous chapters; alternatively, the information may be obtained from ANSI or SAE standards (see also Chap. 15, Sec. 15.1).

The following are standards references:

| | |
|---|---|
| Dimensioning and tolerancing | ANSI Y14.5M-1982 |
| Standard limits and fits | ANSI B4.1-1967, R 1979 |
| Basic metric fits | ANSI B4.2-1978 |
| Geometric tolerancing | ANSI Y14.5M-1982 |
| Surface texture | ANSI B46.1-1978 |
| Drawing practices (surface texture) | ANSI Y14.36-1978 |

### 14.2.2 Tolerance Accumulation

The tolerance values resulting from the three methods of dimensioning parts or assemblies as shown in Fig. 14.21*a*–14.21*c* are as follows. We will then have the following conditions:

1. *Chain dimensioning:* The maximum variation between two features is equal to the sum of the tolerances on the intermediate distances. This results in the greatest tolerance accumulation. See Fig. 14.21*a*.
2. *Baseline dimensioning:* The maximum variation between two features is equal to the sum of the tolerances on the two dimensions from their origin to the features. This results in a reduction of the tolerance accumulation. See Fig. 14.21*b*.
3. *Direct dimensioning:* The maximum variation between two features is controlled by the tolerance on the dimension between the features. This results in the least tolerance. See Fig. 14.21*c*.

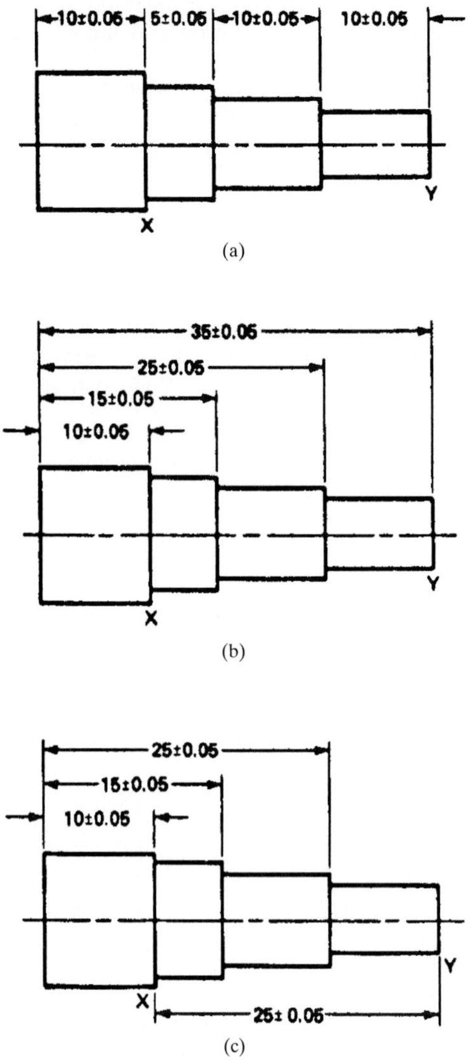


**FIGURE 14.21** (*a*) Chain dimensioning—greatest tolerance accumulation between *X* and *Y*; (*b*) baseline dimensioning—lesser tolerance accumulation between *X* and *Y*; (*c*) direct dimensioning—least tolerance accumulation between *X* and *Y*.

## 14.2.3 Dimensioning per ANSI Y14.5M-1982

ANSI Y14.5M-1982 covers dimensioning, tolerancing, and related practices for use on engineering drawings and related documents. The International System of Units (SI) is featured in this standard because SI units are expected to supercede United States (U.S.) customary units specified on engineering drawings. Nevertheless, it should be understood that U.S. customary units are equally applicable to this standard. Many United States companies and manufacturing facilities of all types still employ the U.S. customary or inch system.

Decimal dimensioning shall be used on engineering drawings except where certain commercial commodities are identified, such as pipe size, lumber size, and wire sizes.

### *Millimeter Dimensioning*

- When the dimension is less than one millimeter, a zero shall precede the decimal point.
- Whole-number dimensions use neither a decimal point nor a zero following.
- On dimensions which exceed a whole number, the last digit to the right of the decimal point shall not be followed by a zero.

(*Note:* The exception to the last rule is for bilateral tolerances or limits, where a zero is used by itself, as part of the tolerance or limit.)

***Decimal-Inch Dimensioning.*** The decimal-inch system is explained in ANSI B87.1, and the following shall be observed when specifying decimal-inch dimensions on engineering drawings:

- A zero is not used *before* the decimal point for dimensional values less than one inch.
- A dimension is expressed to the same number of decimal places as its tolerance. Zeros are added to the right of the decimal point where necessary.
- Decimal points must be uniform and large enough to be clearly visible and meet the requirements of ANSI Y14.2M.
- The conversion and rounding of U.S. customary units are covered in ANSI Z210.1.

### *Glossary of Terms—Dimensioning and Tolerancing*

The following glossary of terms will be useful to the design engineer, in that the use of or recognition of these terms related to dimensioning and tolerancing will prevent misunderstanding between individuals and companies in relation to the meaning of these terms.

**Actual size** The measured size.

**Basic dimension** A numerical value used to describe the theoretically exact size, profile, orientation, or location of a feature or datum target. It is the basis from which permissible variations are established by tolerances on other dimensions, in notes, or in feature control frames.

**Bilateral tolerance** A tolerance in which variation is permitted in both directions from the specified dimension.

**Datum** A theoretically exact point, axis, or plane derived from the true geometric counterpart of a specified datum feature. A datum is the origin from which the location or geometric characteristics of features of a part are established.

**Datum feature** An actual feature of a part that is used to establish a datum.

**Datum target** A specified point, line, or area on a part used to establish a datum.

**Dimension** A numerical value expressed in appropriate units of measure and indicated on a drawing and in other documents along with lines, symbols, and notes to define the size or geometric characteristic, or both, of a part or part feature.

**Feature** The general term applied to a physical portion of a part, such as a surface, hole, or slot.

**Feature of size** One cylindrical or spherical surface, or a set of two plane parallel surfaces, each of which is associated with a size dimension.

**Full indicator movement (FIM)** The total movement of an indicator when appropriately applied to a surface to measure its variations.

**Geometric tolerance** The general term applied to the category of tolerances used to control form, profile, orientation, location, and runout.

**Least-material condition (LMC)** The condition in which a feature of size contains the least amount of material within the stated limits of size—for example, maximum hole diameter, minimum shaft diameter.

**Limits of size** The specified maximum and minimum sizes.

**Maximum material condition (MMC)** The condition in which a feature of size contains the maximum amount of material within the stated limits of size—for example, minimum hole diameter, maximum shaft diameter.

**Reference dimension** A dimension, usually without tolerance, used for information purposes only. It is considered auxiliary information and does not govern production or inspection operations. A reference dimension is a repeat of a dimension or is derived from other values shown on the drawing or on related drawings.

**Regardless of feature size (RFS)** The term used to indicate that a geometric tolerance or datum reference applies at any increment of size of the feature within its size tolerance.

**Tolerance** The total amount by which a specific dimension is permitted to vary. The tolerance is the difference between the maximum and minimum limits.

**True position** The theoretically exact location of a feature established by basic dimensions.

**Unilateral tolerance** A tolerance in which variation is permitted in one direction from the specified dimension.

**Virtual condition** The boundary generated by the collective effects of the specified MMC limit of size of a feature and any applicable geometric tolerances.

***Applications of Dimensions.*** The following figures show the applications of dimensions and dimension lines:

| | |
|---|---|
| Fig. 14.22 | Angular units |
| Fig. 14.23 | Millimeter dimensions |
| Fig. 14.24 | Decimal-inch dimensions |
| Fig. 14.25 | Application of dimensions |
| Fig. 14.26 | Grouping of dimensions |
| Fig. 14.27 | Spacing of dimensions |

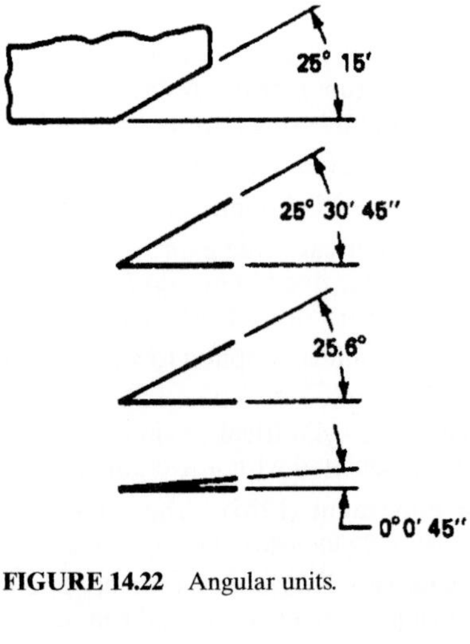


**FIGURE 14.22** Angular units.

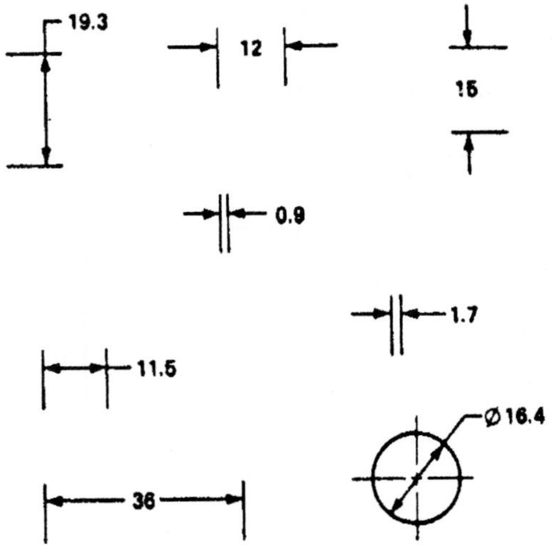


**FIGURE 14.23** Millimeter dimensions.

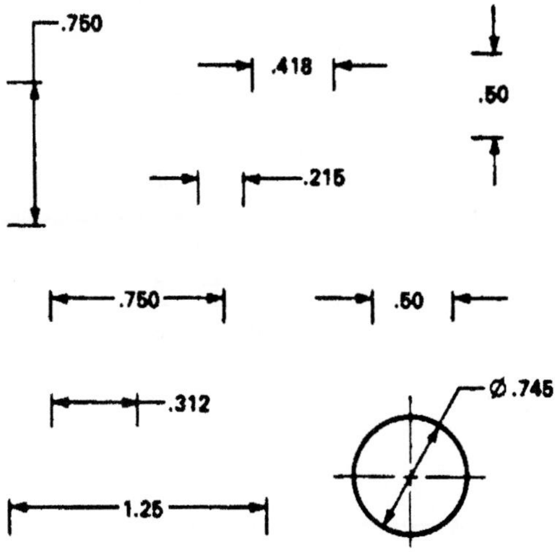


**FIGURE 14.24** Decimal-inch dimensions.

| | |
|---|---|
| Fig. 14.28 | Staggered dimensions |
| Fig. 14.29 | Oblique extension lines |
| Fig. 14.30 | Breaks in extension lines |
| Fig. 14.31 | Point locations |
| Fig. 14.32 | Leaders |
| Fig. 14.33 | Reading directions |
| Fig. 14.34 | Intermediate reference dimension |

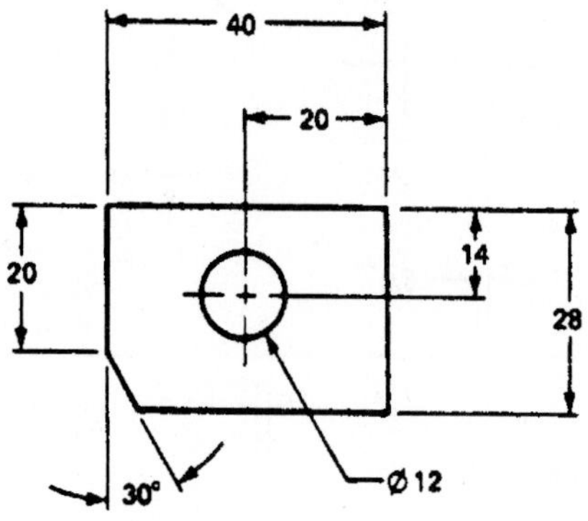


**FIGURE 14.25** Application of dimensions.

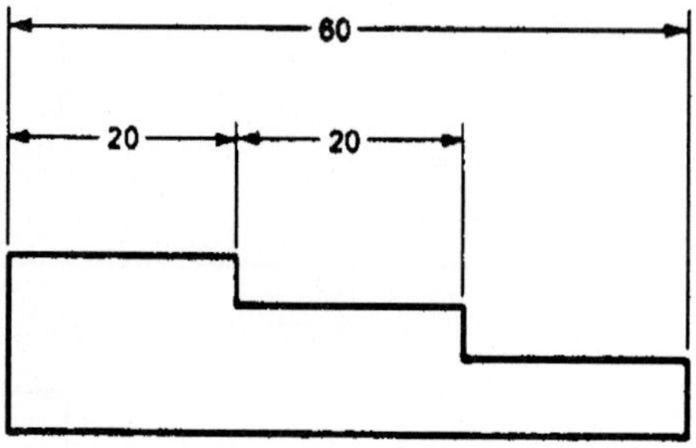


**FIGURE 14.26** Grouping of dimensions.

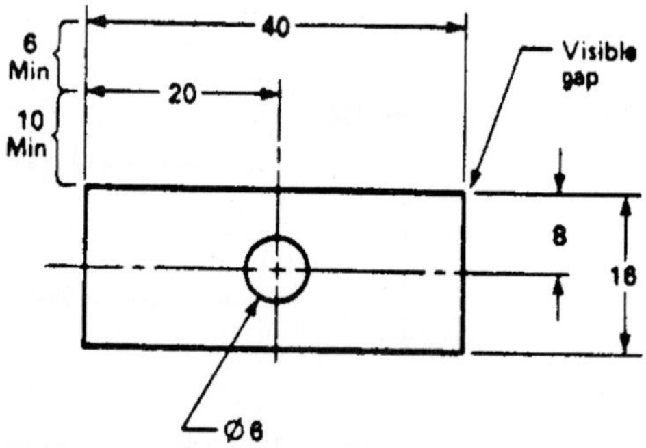


**FIGURE 14.27** Spacing of dimensions.

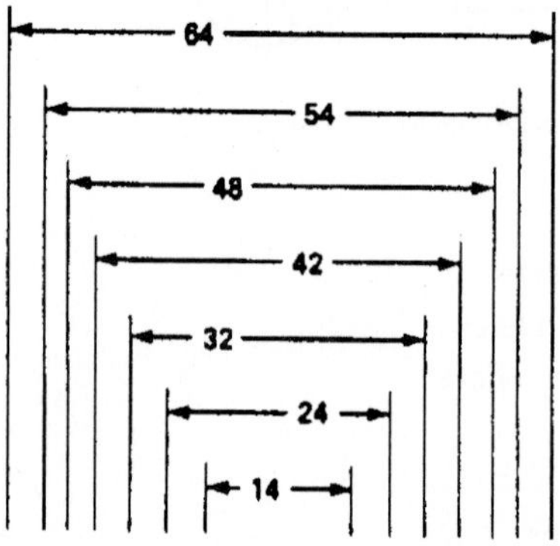


**FIGURE 14.28** Staggered dimensions.

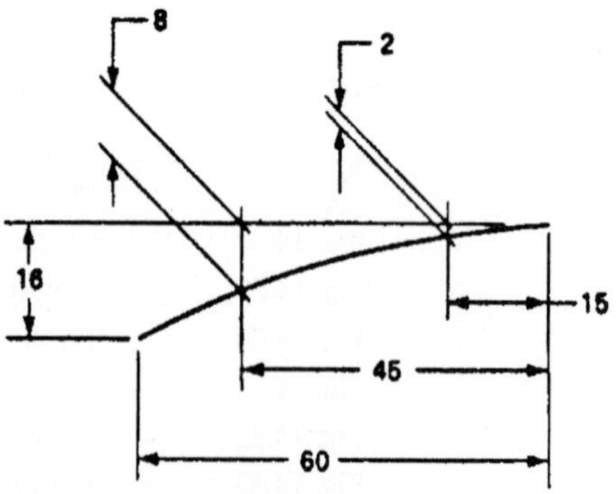


**FIGURE 14.29** Oblique extension lines.

| | |
|---|---|
| Fig. 14.35 | Diameters |
| Fig. 14.36 | Radii |
| Fig. 14.37 | Chords, arcs, and angles |
| Fig. 14.38 | Tabulated outline |
| Fig. 14.39 | Symmetrical outlines |
| Fig. 14.40 | Coordinate or offset outline |
| Fig. 14.41 | Round holes |
| Fig. 14.42 | Rectangular coordinate dimensioning |
| Fig. 14.43 | Polar coordinate dimensioning |
| Fig. 14.44 | Repetitive features and dimensions |

### 14.2.4 ANSI Y14.5M-1982 Tolerancing Practices

Per ANSI Y14.5M-1982, tolerances may be expressed as follows:

- As direct limits or as tolerance values applied directly to a dimension.
- As a geometric tolerance.

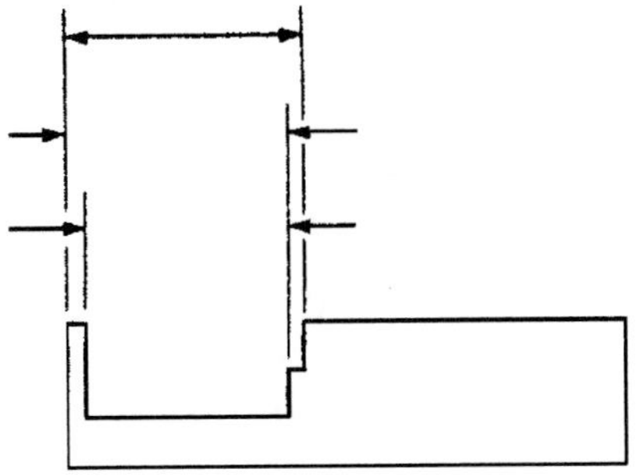

**FIGURE 14.30** Breaks in dimension lines.

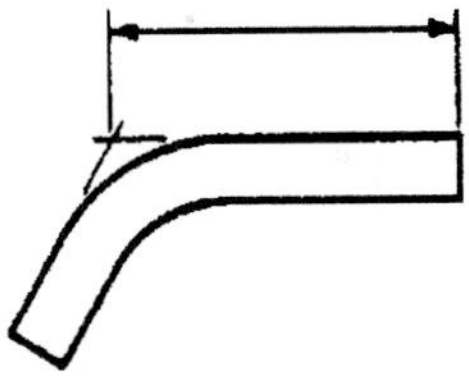

**FIGURE 14.31** Point locations.

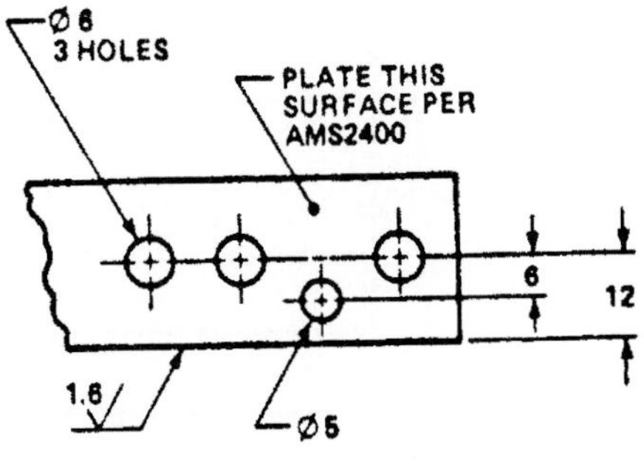


**FIGURE 14.32** Leaders.

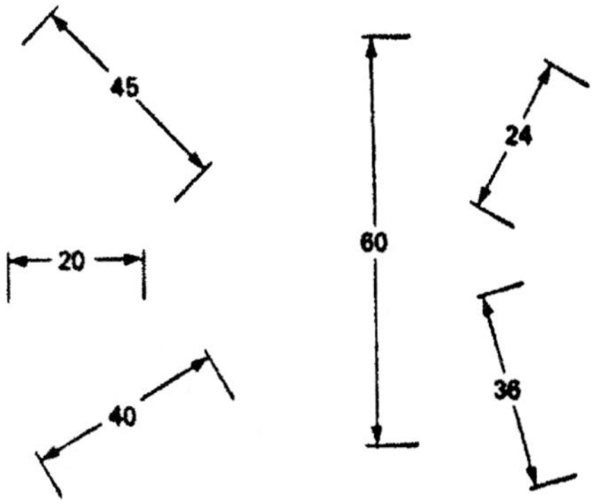


**FIGURE 14.33** Reading directions of dimensions.

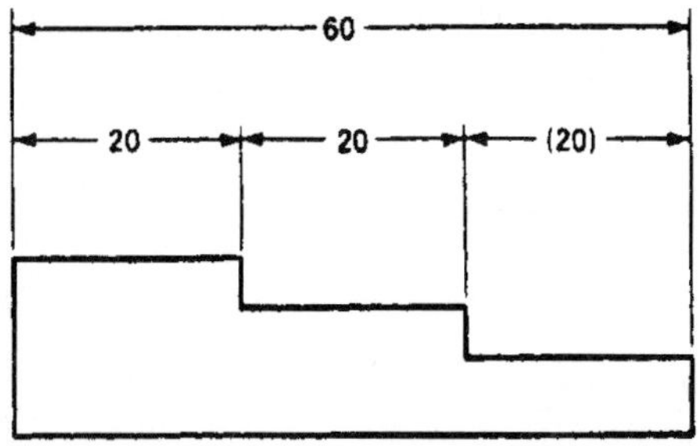


FIGURE 14.34 The intermediate reference dimension.

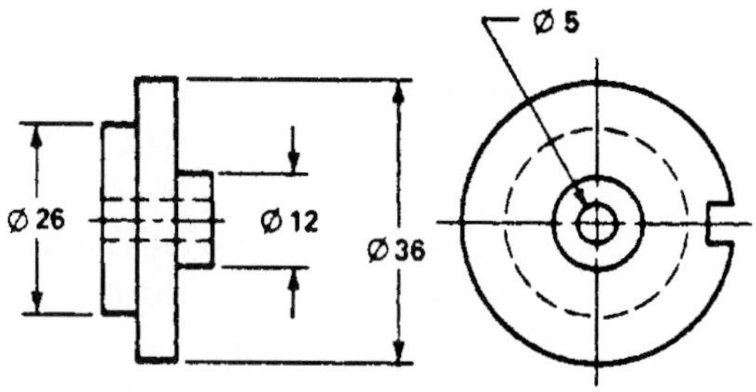


FIGURE 14.35 Dimensioning diameters.

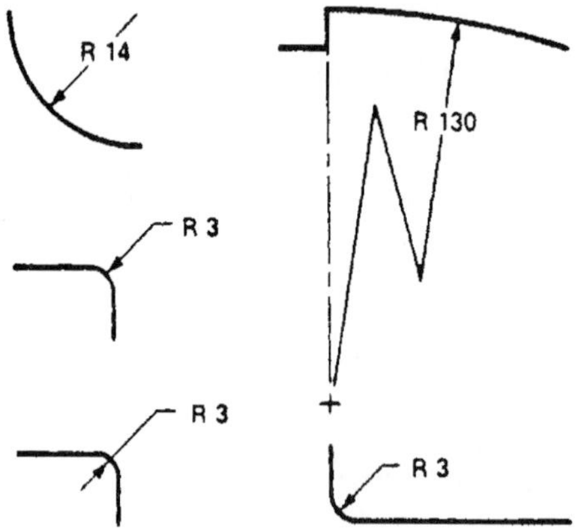


FIGURE 14.36 Dimensioning radii.

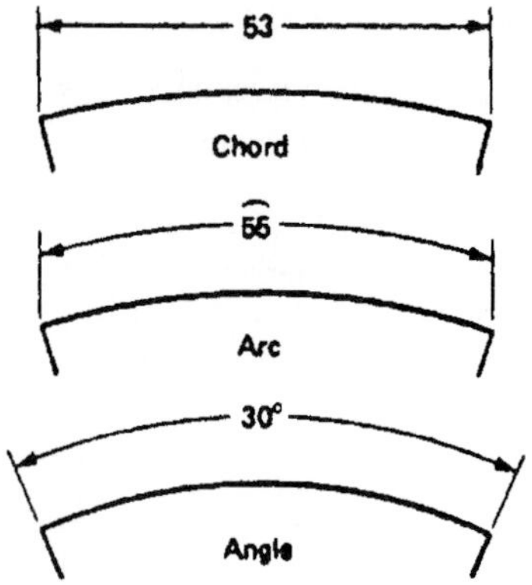


FIGURE 14.37 Dimensioning chords, arcs, and angles.

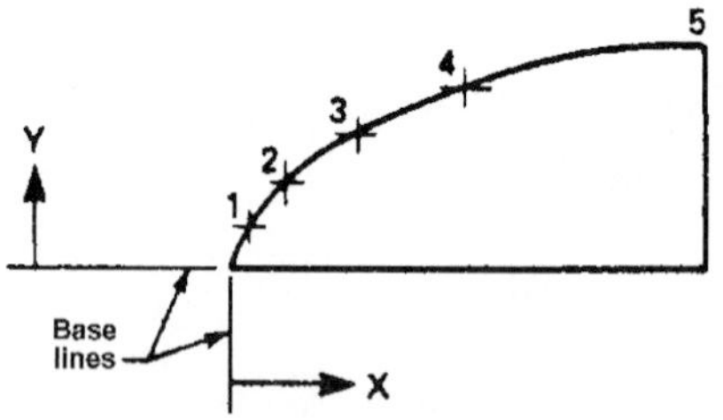

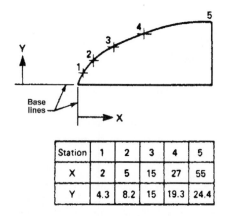

| Station | 1 | 2 | 3 | 4 | 5 |
|---|---|---|---|---|---|
| X | 2 | 5 | 15 | 27 | 55 |
| Y | 4.3 | 8.2 | 15 | 19.3 | 24.4 |

FIGURE 14.38 Tabulated outline dimensions.

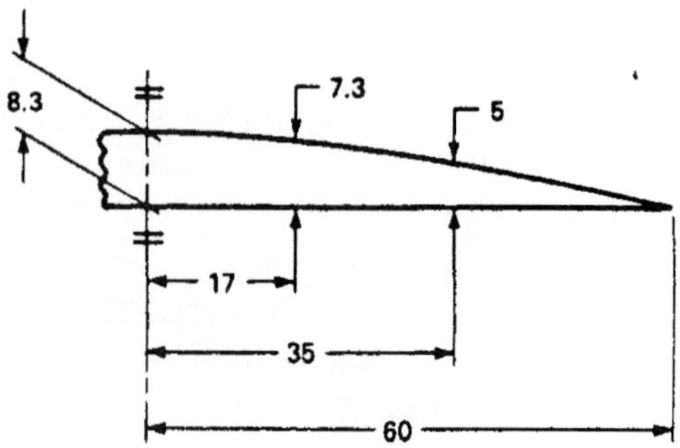


FIGURE 14.39 Symmetrical outlines.

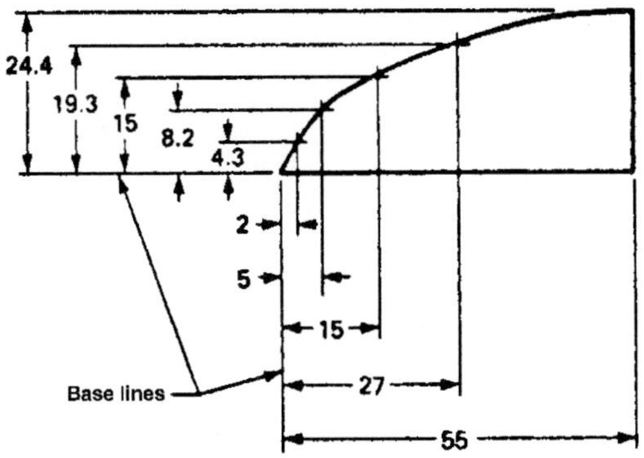


**FIGURE 14.40** Coordinate or offset outline dimensions.

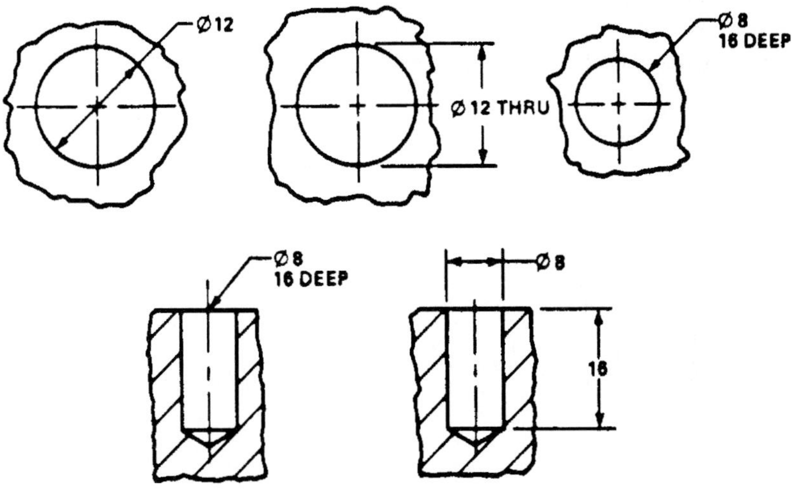


**FIGURE 14.41** Dimensioning round holes.

- In a note referring to specific dimensions.
- As specified in other documents referenced on the engineering drawing for specific features or processes.
- In a general tolerance block referring to all dimensions on the engineering drawing, unless specified otherwise.

Tolerances on dimensions that locate features of size may be applied directly to the locating dimensions or specified by the positional tolerancing method.

Unless otherwise specified, where a general tolerance note on the drawing includes angular tolerances, it applies to features shown at specified angles and at *implied* 90° angles, i.e., intersections of centerlines, corners of parts (internal and external), or other obvious areas not specifically shown to have angles other than 90°.

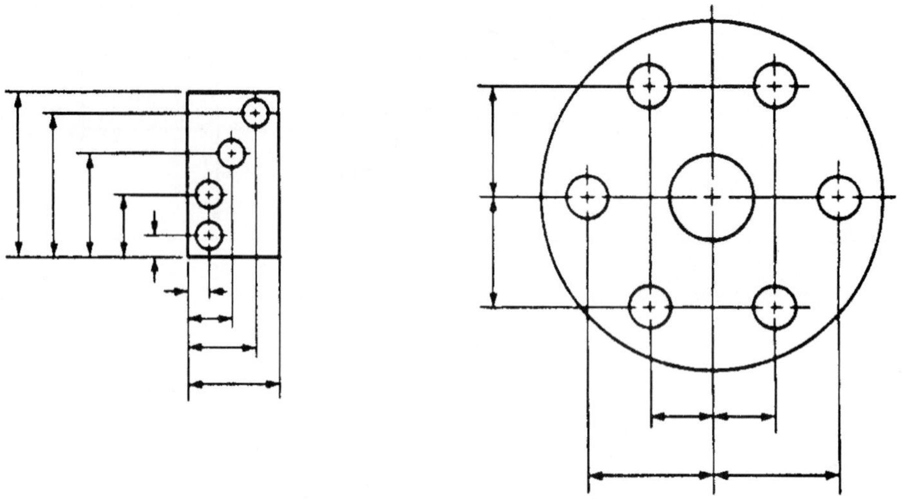

**FIGURE 14.42** Rectangular coordinate dimensioning.

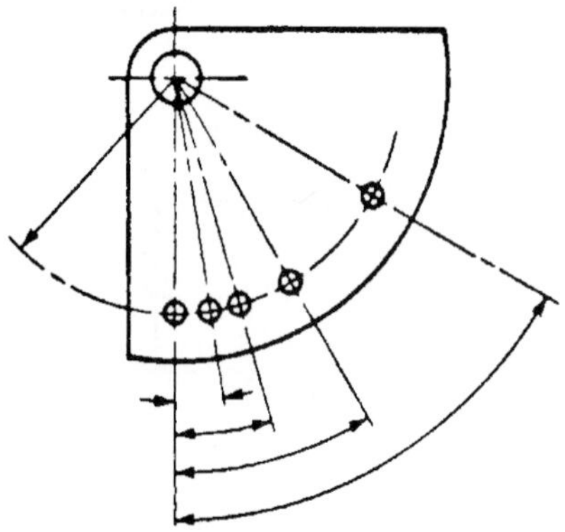

**FIGURE 14.43** Polar coordinate dimensioning.

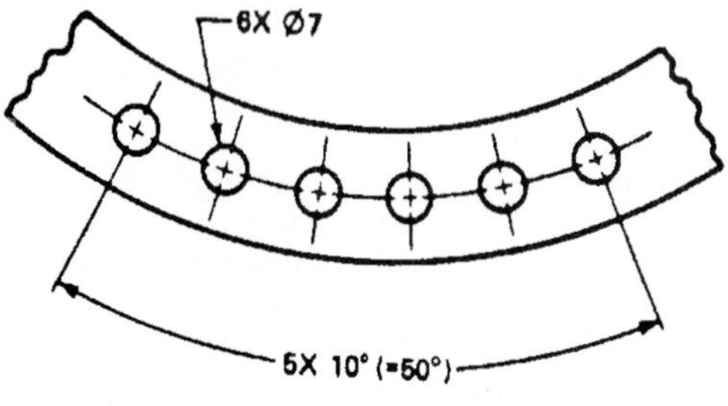


**FIGURE 14.44** Dimensioning repetitive features.

### 14.2.5 Direct Tolerancing Methods

Limits and directly applied tolerance values are specified as follows:

- *Limit dimensions:* The high limit or maximum value is placed above the low limit or minimum value. As a single-line callout, the low limit precedes the high limit with a dash separating the values.
- *Plus and minus tolerancing:* The basic dimension is given first, followed by a plus-and-minus expression of tolerance.
- *Tolerance limits:* All tolerance limits are *absolute.*
- *Dimensional limits before or after plating:* For plated or coated parts, the engineering drawing or referenced document shall state whether the dimensions are *before* or *after* plating; i.e., "Dimensional limits apply before plating" or "Dimensional limits apply after plating."

### 14.2.6 Positional Tolerancing

*Positional* or *location tolerancing* defines a zone within which the center, axis, or center plane of a feature of size is permitted to vary from the *true* or exact position.

*Geometric tolerancing* is the general term applied to the category of tolerances used to control form, profile, orientation, location, and runout.

### 14.2.7 Examples of ANSI Y14.5M-1982 Dimensioning and Tolerancing Practices

Figure 14.45 shows a typical engineering drawing using the ANSI form for dimensioning, tolerancing, and positioning.

For a complete description and operational instructions for the use of ANSI standard dimensioning and tolerancing practices, see ANSI Y14.5M-1982, which may be obtained directly from ANSI, Inc. See Chap. 15 for addresses and acronyms of American standards organizations, specification authorities, societies, and institutes.

### 14.2.8 Design Notes on Dimensioning and Tolerancing

From an electromechanical design standpoint, the dimensioning and tolerancing practices used on engineering drawings for the design and manufacturing of parts, subassemblies, and assemblies should take the following points into consideration:

- Close tolerances add cost to a finished product.
- The tolerance should be balanced to the function of the part.
- Arbitrary selection of a *general* tolerance can cause design and fit problems on the finished product or create unnecessary work.
- Use care in the selection of bilateral and unilateral tolerances.
- Remember that modern CNC-controlled turning centers, machining centers, EDM machines, CNC-controlled punch presses, and other CNC-controlled equipment are capable of producing parts with closer tolerances than was possible in the past. Spindle accuracies are higher and CNC movement controls are very accurate on modern machine tools and equipment.
- Use tables of preferred limits and fits only when applicable.
- Select plating thickness limits (range) carefully so as to prevent dimensional interference between mating parts.

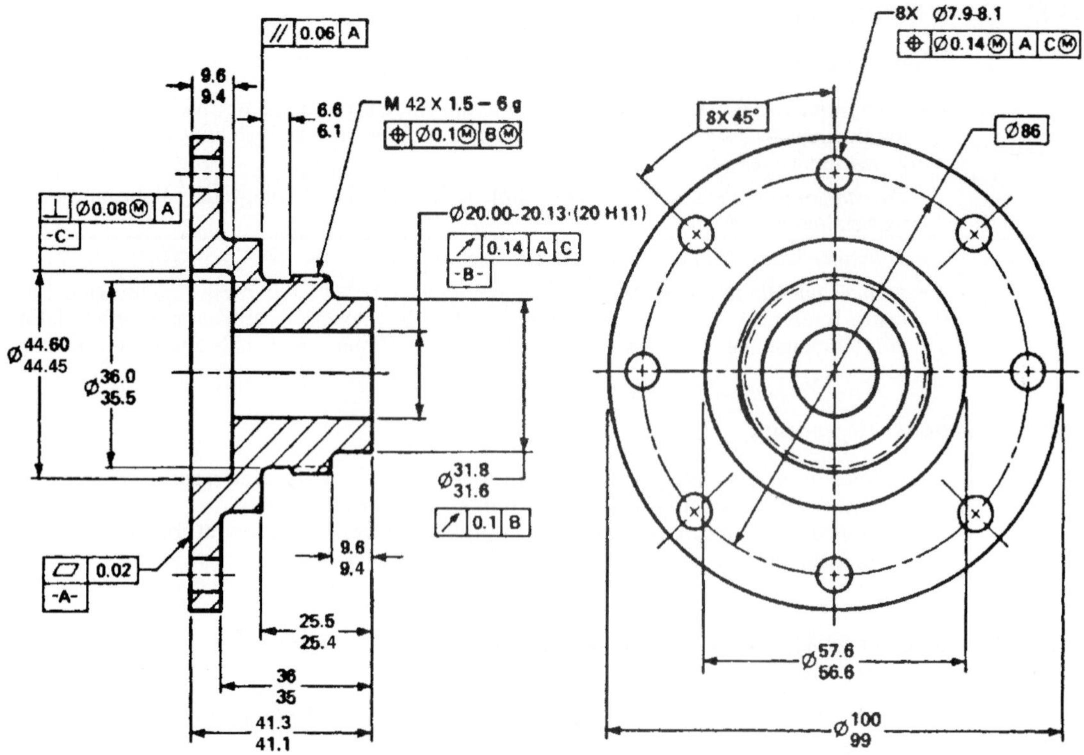


**FIGURE 14.45** Feature control frame placement dimensions per ANSI Y14.5M-1982.

- To control a large tolerance spread due to many parts in a dynamic assembly, design an adjustment means in the mechanism at one or more critical positions.
- Tooling fixtures and tooled parts help control tolerance ranges to a great extent on assemblies and complex mechanisms.
- Machined finishes are surface textures and therefore can be considered to have a tolerance [root-mean-square (rms) value]. Therefore, specify only a machined or tooled surface finish that is functional to the part.

The experienced electromechanical designer or product design engineer must be proficient in dimensioning and tolerancing practices to be most effective and successful at the design function. Many design and manufacturing problems occur when the engineering drawings are not dimensioned and toleranced properly and effectively. It is therefore recommended by the author that a thorough study be made of dimensioning and tolerancing practices through the use of ANSI Y14.5M-1982 (reaffirmed 1988) or later revision.

### 14.2.9 Symbols Used in ANSI Y14.5M-1982 and ISO Dimensioning and Tolerancing

See Fig. 14.46 for the ANSI and ISO symbols currently used for dimensioning and tolerancing.

| SYMBOL FOR: | ANSI Y14.5 | ISO |
|---|---|---|
| STRAIGHTNESS | ⏤ | ⏤ |
| FLATNESS | ⏥ | ⏥ |
| CIRCULARITY | ○ | ○ |
| CYLINDRICITY | ⌭ | ⌭ |
| PROFILE OF A LINE | ⌒ | ⌒ |
| PROFILE OF A SURFACE | ⌓ | ⌓ |
| ALL AROUND PROFILE | ←○— | NONE |
| ANGULARITY | ∠ | ∠ |
| PERPENDICULARITY | ⊥ | ⊥ |
| PARALLELISM | // | // |
| POSITION | ⌖ | ⌖ |
| CONCENTRICITY/COAXIALITY | ◎ | ◎ |
| SYMMETRY | NONE | ⌯ |
| CIRCULAR RUNOUT | ↗ | ↗ |
| TOTAL RUNOUT | ⌰ | ⌰ |
| AT MAXIMUM MATERIAL CONDITION | Ⓜ | Ⓜ |
| AT LEAST MATERIAL CONDITION | Ⓛ | NONE |
| REGARDLESS OF FEATURE SIZE | Ⓢ | NONE |
| PROJECTED TOLERANCE ZONE | Ⓟ | Ⓟ |
| DIAMETER | Ø | Ø |
| BASIC DIMENSION | [50] | [50] |
| REFERENCE DIMENSION | (50) | (50) |
| DATUM FEATURE | [-A-] | ▲ OR ▲—[A] |
| DATUM TARGET | (Ø6 / A1) | (Ø6 / A1) |
| TARGET POINT | X | X |
| DIMENSION ORIGIN | ⌀→ | NONE |
| FEATURE CONTROL FRAME | [⌖ \| Ø0.5Ⓜ \| A \| B \| C] | [⌖ \| Ø0.5Ⓜ \| A \| B \| C] |
| CONICAL TAPER | ⌲ | ⌲ |
| SLOPE | ⌳ | ⌳ |
| COUNTERBORE/SPOTFACE | ⌴ | NONE |
| COUNTERSINK | ⌵ | NONE |
| DEPTH/DEEP | ↧ | NONE |
| SQUARE (SHAPE) | □ | □ |
| DIMENSION NOT TO SCALE | $\underline{15}$ | $\underline{15}$ |
| NUMBER OF TIMES/PLACES | 8X | 8X |
| ARC LENGTH | $\overset{\frown}{105}$ | NONE |
| RADIUS | R | R |
| SPHERICAL RADIUS | SR | NONE |
| SPHERICAL DIAMETER | SØ | NONE |

**FIGURE 14.46** ANSI and ISO dimensioning and tolerancing symbols.

### 14.2.10 Typical Industrial Design Engineering Drawings

There are certain exceptions to the recommended dimensioning and tolerancing practices described in this chapter.

- Engineering drawings that are used exclusively within the confines of any particular company or organization may use drawing styles and dimensioning and tolerancing practices that differ from those shown in the preceding sections of this chapter. Some companies are still using the English fractional system of dimensioning, even though the fractional system was discontinued from general industrial use in the early 1950s.
- Different symbolisms, for the different manufacturing processes such as welding, brazing, assembly, and so forth, are also used by different companies.
- Tooling drawings such as that shown in Fig. 14.47 also may differ from the ANSI and ISO standards. Figure 14.47 shows a part which is to be die-stamped, and there are *no* tolerances on the drawing, with the exception of the hole location, which is produced at a different stage of manufacturing as a separate operation. The EDM machine that is used to cut the dies for this part will allow the part to be stamped exactly to the dimensions shown, i.e., to the precise three-place decimals as indicated on the drawing.
- Companies worldwide, 50 years ago, used different dimensioning and tolerancing systems, yet produced some of the finest mass-produced equipment available at any time period. The Ford Motor Company at one time used their own measuring system, for which the appropriate scales were produced. The Martin Company, from the 1930s to the 1950s, also used scales and drawing equipment made to their specifications for producing trimetric

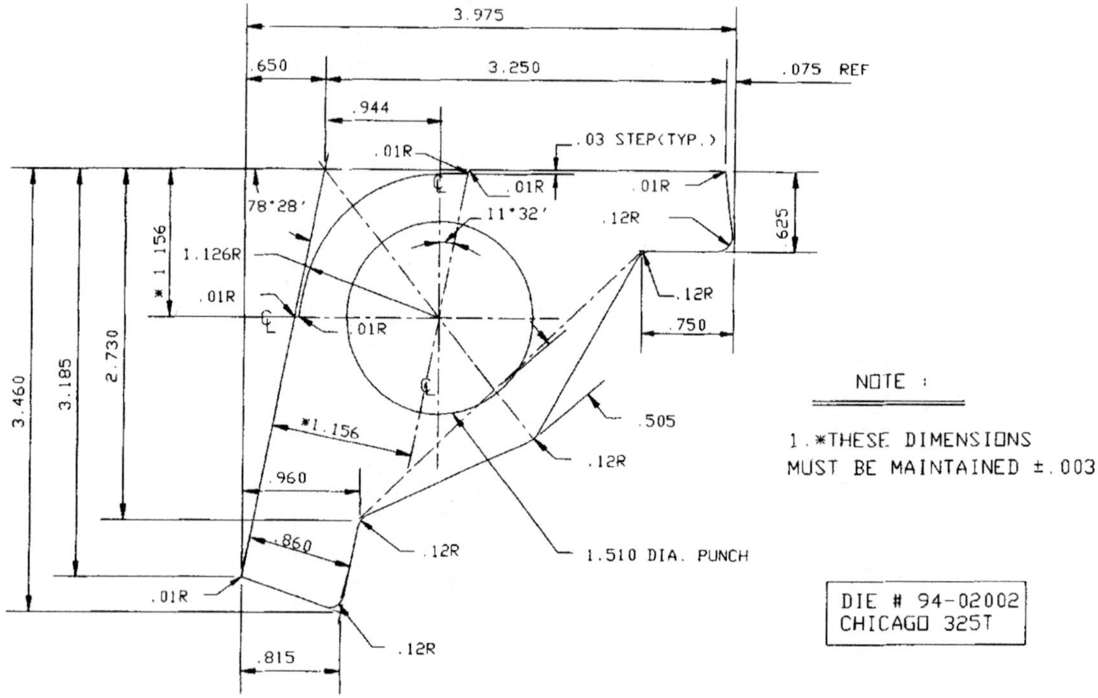


**FIGURE 14.47** AutoCad drawing showing typical industrial practice on drawing for tooled part.

drawings from full-scale lofts. (*Lofts* are precise scale drawings of flat-pattern sheet-metal parts done on specially prepared metal sheets or very heavy and dimensionally stable Mylar, a polyester film.)

Although there are exceptions to the modern dimensioning and tolerancing systems as indicated previously, the advent of ANSI Y14.5M, SI, ISO 9000, and other national and international standards will make it difficult for American companies to compete in the world marketplace unless they adhere to these worldwide standards. One requirement of these systems is the drawing, dimensioning, and tolerancing practices and engineering documentation control. Also, it is to be noted that when any particular company's engineering drawings are sent to outside vendors or subcontractors, these drawings should always be prepared to the national and international standards; otherwise, misinterpretations of the drawings may occur, causing losses, rejections, and possible lawsuits.

National and international quality-control standards for manufactured products will be discussed in Chap. 15. There, we will look at the new systems of product quality control such as total quality management (TQM) and ISO 9000 and also product liability and lawsuits resulting from product faults.

# CHAPTER 15
# SUBJECTS OF IMPORTANCE TO THE DESIGN ENGINEER

## 15.1 SOCIETIES, ASSOCIATIONS, INSTITUTES, AND SPECIFICATION AUTHORITIES

Following is a listing of recognized specification authorities, societies, and institutes from which the design engineer can obtain specifications and standards covering many areas of engineering design and manufacturing.

Many of the standards are revised periodically by these authorities to keep pace with changing technology. It is therefore suggested by the author that copies of the American standards which you require to perform your job function be obtained directly from the standards organizations which generate the particular standard. In this way, you are assured that you will be using the most recent standard and its revisions. Update notifications to the standards listings are distributed periodically to those on the organization mailing lists, or annual standards listing manuals may be ordered directly from the standards organizations such as ASTM, SAE, ANSI, AGMA, and AISC.

Many of the standards organizations publish handbooks and design manuals related to their particular fields which are technically excellent. Indexes to the standards are also available for pinpointing your area of interest in the form of annual standards catalogs.

### *Standards Organizations and Acronyms*

Aeronautical Materials Specifications (AMS)
Society of Automotive Engineers, Inc.
400 Commonwealth Drive
Warrendale, PA 15096

The Aluminum Association
818 Connecticut Avenue, NW
Washington, DC 20006

American Chain Association (ACA)
1000 Vermont Avenue, NW
Washington, DC 20005

American Engineering Model Society (AEMS)
Box 2066
Aiken, SC 29801

American Gear Manufacturers Association (AGMA)
1901 North Fort Meyer Drive, Suite 1000
Arlington, VA 22209

American Institute of Steel Construction (AISC)
400 North Michigan Avenue
Chicago, IL 60611

American Iron and Steel Institute (AISI)
1000 16th Street, NW
Washington, DC

American National Standards Institute, Inc. (ANSI)
11 West 42nd Street
New York, NY 10036

American Petroleum Institute (API)
300 Corrigan Tower Building
Dallas, TX 75201

American Society for Metals (ASM)
Metals Park, OH 44073

American Society for Testing and Materials (ASTM)
1916 Race Street
Philadelphia, PA 19103

American Society of Lubrication Engineers (ASLE)
838 Busse Highway
Park Ridge, IL 60068

American Society of Mechanical Engineers (ASME)
345 East 47th Street
New York, NY 10017

American Supply and Machinery Manufacturers Association (ASMMA)
1230 Keith Building
Cleveland, OH 44115

American Welding Society (AWS)
550 Northwest Lejeune Road
PO Box 351040
Miami, FL 33135

Anti-Friction Bearing Manufacturers Association (AFBMA)
2341 Jefferson Davis Highway, Suite 1015
Arlington, VA 22202

Association of American Railroads (AAR)
59 East Van Buren Street
Chicago, IL 60605

Bearing Specialists Association (BSA)
221 North LaSalle Street, Suite 2026
Chicago, IL 60601

Copper Development Association, Inc. (CDA)
405 Lexington Avenue
New York, NY 10017

Edison Electric Institute (EEI)
750 Third Avenue
New York, NY 10017

Farm Equipment Manufacturers Association (FEMA)
230 South Bemiston, Suite 809
St. Louis, MO 63105

Federal Specifications
U.S. Navy Supply Depot
5801 Tabor Avenue
Philadelphia, PA 19120

Industrial Fasteners Institute (IFI)
1505 East Ohio Building
1717 East 9th Street
Cleveland, OH 44114

Ingot Number
Brass and Bronze Ingot Institute
33 North LaSalle St., Room 3500
Chicago, IL 60602

Institute of Electrical and Electronics Engineers (IEEE)
345 East 47th Street
New York, NY 10017

Institute of Industrial Engineers (IIE)
25 Technology Park
Atlanta, GA 30092

Instrument Society of America (ISA)
PO Box 12277
Research Triangle Park, NC 27709

Mechanical Power Transmission Association (MPTA)
3451 West Church Street
Evanston, IL 60203

Military Specifications
U.S. Naval Supply Depot
5801 Tabor Avenue
Phiadelphia, PA 19120

National Aerospace Standards Committee (NASC)
1321 Fourteenth Street
Washington, DC 20005

National Bureau of Standards (NBS)
Washington, DC 20234

National Electrical Manufacturers Association (NEMA)
2101 L Street NW, Suite 300
Washington, DC 20037

National Fire Protection Association (NFPA—National Electrical Code)
Batterymarch Park
Quincy, MA 02269

National Industrial Belting Association (NIBA)
1900 Arch Street
Philadelphia, PA 19103

National Lubricating Grease Institute (NLGI)
4635 Wyandotte Street
Kansas City, MO 64112

National Safety Council (NSC)
444 North Michigan Avenue
Chicago, IL 60611

Robot Institute of America (RIA)
PO Box 1366
Dearborn, MI 48121

Rubber Manufacturers Association (RMA)
1901 Pennsylvania Avenue, NW
Washington, DC 20006

Society of Automotive Engineers (SAE)
400 Commonwealth Drive
Warrendale, PA 15096

Society of Manufacturing Engineers (SME)
One SME Drive
PO Box 930
Dearborn, MI 48128

Spring Manufacturers Institute (SMI)
380 West Palatine Road
Wheeling, IL 60090

The Welding Institute (TWI)
PO Box 5268
Hilton Head Island, SC 29928

Underwriters Laboratories, Inc. (UL)
333T Pfingsten Road
Northbrook, IL 60062

## 15.2 ANSI STANDARDS APPLICABLE TO PRODUCT DESIGN ENGINEERING AND MANUFACTURING PROCESSES AND INSPECTION

The American National Standards Institute (ANSI) issues standards in the form of published pamphlets which define the geometry, dimensions and tolerances, inspection limits, test procedures, and other control data and specifications important to the design and manufacture of thousands of items. Components, materials, and specifications such as screw thread systems, bolts, screws, washers, nuts, splines, pins, cutting tools, steels, nonferrous alloys, plastics, mechanical devices, various electromechanical equipment and machinery, and a host of other items of importance to the machining, metalworking, and electromechanical industries are all defined and specified in the various American national standards, including ANSI, SAE, ASTM, AGMA, AISC, AISI, NEMA, NEC, AFBMA, SMI, IEEE, and many others.

The purpose of these standards is to define the various physical, chemical, electrical, and mechanical characteristics of materials, components, systems, assemblies of equipment, and inspection and testing procedures. The American national standards provide a means for obtaining order and conformity among American manufactured products.

On August 24, 1969, the American Standards Association (ASA) was restructured as the United States of America Standards Institute, and standards which were approved as American standards were designated as *USA Standards.* On October 6, 1969, the name was then changed to *American National Standards Institute* (ANSI). The present standards designation is *ANSI* instead of *ASA* or *USAS.*

The American National Standards Institute works in collaboration with other national organizations such as ASTM, SAE, AWS, ASME, AGMA, AISI, and IEEE in an effort to consolidate the American standards data and publications generated by these other national specifications and standards organizations.

The electromechanical industries and other industries nationwide depend on the combined national standards of all the various American societies, institutes, and associations in order to have guidelines, specifications, and design and test procedures for designing and manufacturing their products.

Many products and materials are required by purchasing specifications to conform to the various American standards and will not be accepted by the purchaser unless they do conform to the specified standards. For example, when a design engineer specifies a material on the design drawing of a spring, the material as listed on the drawing may be as such:

0.156-in-diameter music wire per ASTM A228

When the material is delivered to the spring manufacturer, it should conform to the specifications of this standard designation (ASTM A228), both physically and chemically. If this specified material were to be tested and analyzed at a materials test laboratory, it would be required to conform to the ASTM A228 specification. Failure of the material to conform to the specification could result in part failure in service. In American industry today, this problem of failure to conform to American standard specifications is becoming more prevalent and is evident in the hardware failures found in counterfeit bolts, screws, and nuts and in the material failures found in nonconforming ferrous and nonferrous alloys which are being imported into the United States in ever-increasing quantities. A photograph of a counterfeit bolt that broke during installation can be seen in Fig. 10.1 in Chap. 10. The failure of imported and sometimes nationally made parts or hardware which contain various types of screw threads that are not made to the ANSI standards or NBS standards also causes problems to product manufacturers in the form of excessively loose or binding fits. Imported roller chain has also been cited as not conforming to ANSI standard specifications, with breakages noted well below the standard tensile strength allowables.

The designations of the various ANSI standards that are of prime importance to the machining, metalworking, and electromechanical industries are listed in this section by subject category.

It was the author's and publisher's decision to omit extracts of the various ANSI and other American national standards in the handbook for the following reasons:

- The basic standards that are applicable to the electromechanical industries are too extensive to republish.
- The standards are being constantly revised to keep pace with changing technologies.
- Whenever a standard is revised, in effect, the handbook containing these standards would be out of date and contain obsolete data.

Companies that rely on the data contained in the national standards published by ANSI and the other national standards organizations should keep copies of the standards which apply to their work in the standards or engineering departments of their organizations. The ANSI standards listed in this section of the handbook are considered by the author to be the main, basic standards required for the machining, mechanical, and metalworking industries.

Most of the national associations, societies, and institutes that generate American standards applicable to the metalworking and electromechanical industries are listed in this chapter of the handbook, together with the current addresses. Standards may be purchased directly from these organizations, and most of the listed organizations have catalogs available which specify the various technical publications they produce.

***Listing of ANSI Standards by Category***

Category 1 Thread systems
Category 2 Fastening and joining devices
Category 3 Machining practices
Category 4 Tools and tooling
Category 5 Mechanical components
Category 6 Welding

| | |
|---|---|
| Category 7 | Heat treatment |
| Category 8 | Tolerances and fits |
| Category 9 | Drawing symbols and formats |
| Category 10 | Gauging and inspection |

Category 1—thread systems:

| | |
|---|---|
| ANSI B1.9-1973 | Buttress inch screw threads |
| ANSI B1.10-1958 (R 1988) | Unified miniature screw threads |
| ANSI B1.11-1958 (R 1989) | Microscope objective thread |
| ANSI B1.18M-1982 (R 1987) | Metric screw threads for commercial mechanical fasteners—boundry profile defined |
| ANSI B1.20.3-1976 (R 1982) | Dryseal pipe threads (inch) |
| ANSI B1.20.4-1976 (R 1982) | Dryseal pipe threads (metric translation of B1.20.3-1976) |
| ANSI/ASME B1.1-1989 | Unified inch screw threads (UN and UNR thread form) |
| ANSI/ASME B1.5-1988 | Acme screw threads |
| ANSI/ASME B1.7M-1984 | Screw threads, definitions and letter symbols for |
| ANSI/ASME B1.8-1988 | Stub acme screw threads |
| ANSI/ASME B1.12-1987 | Screw threads—class 5 interference-fit thread |
| ANSI/ASME B1.13M-1983 (R 1989) | Metric screw threads—M profile |
| ANSI/ASME B1.20.1-1983 | Pipe threads (general-purpose), inch |
| ANSI/ASME B1.20.7-1966 (R 1983) | Hose coupling screw threads |

Category 2—fastening and joining devices:

| | |
|---|---|
| ANSI B18.1.1-1972 (R 1989) | Small solid rivets (≤0.4375 in diameter) |
| ANSI B18.1.2-1972 (R 1989) | Large rivets (≥0.500 in diameter) |
| ANSI B18.2.1-1981 | Square and hexagonal (hex) bolts and screws, inch series |
| ANSI B18.2.3.1M-1979 (R 1989) | Screws, metric hex cap |
| ANSI B18.2.3.2M-1979 (R 1989) | Screws, metric formed hex |
| ANSI B18.2.3.3M-1979 (R 1989) | Screws, metric heavy hex |
| ANSI B18.2.3.5M-1979 (R 1989) | Bolts, metric hex |
| ANSI B18.2.3.6M-1979 (R 1989) | Bolts, metric heavy hex |
| ANSI B18.2.3.7M-1979 (R 1989) | Bolts, metric heavy hex structural |
| ANSI B18.2.3.8M-1981 | Screws, metric hex lag |
| ANSI B18.2.4.1M-1979 (R 1989) | Hex nuts, style 1, metric |
| ANSI B18.2.4.2M-1979 (R 1989) | Hex nuts, style 2, metric |
| ANSI B18.2.4.3M-1979 (R 1989) | Hex nuts, slotted, metric |
| ANSI B18.2.4.4M-1982 | Nuts, metric hex flange |
| ANSI B18.2.4.5M-1979 (R 1990) | Hex jam nuts, metric |
| ANSI B18.2.4.6M-1979 (R 1990) | Hex nuts, heavy, metric |
| ANSI B18.5.2.1M-1981 | Bolts, metric round head, short square neck |

| | |
|---|---|
| ANSI B18.6.1-1981 | Wood screws, inch series |
| ANSI B18.6.2-1972 (R 1983) | Slotted head cap screws, square head setscrews, and slotted headless setscrews |
| ANSI B18.6.3-1972 (R 1983) | Slotted and recessed head machine screws and machine screw nuts |
| ANSI B18.6.4-1981 | Screws, tapping and metallic drive, inch series, thread forming, and cutting |
| ANSI B18.7-1972 (R 1980) | Semi-tubular rivets, full tubular rivets, split rivets and rivet caps, general-purpose |
| ANSI B18.8.1-1972 (R 1983) | Clevis pins and cotter pins |
| ANSI B18.8.2-1978 (R 1989) | Taper pins, dowel pins, straight pins, grooved pins, and spring pins, inch series |
| ANSI B18.9-1958 (R 1989) | Plow bolts |
| ANSI B18.11-1961 (R 1983) | Miniature screws |
| ANSI B18.17-1968 (R 1983) | Wing nuts, thumb screws, and wing screws |
| ANSI B18.22M-1981 | Washers, metric plain |
| ANSI B18.22.1-1965 (R 1981) | Plain washers |
| ANSI/ASME B18.1.3M-1983 (R 1989) | Metric small solid rivets |
| ANSI/ASME B18.2.2-1987 | Square and hex nuts, inch series |
| ANSI/ASME B18.2.3.4M-1984 | Screws, metric hex flange |
| ANSI/ASME B18.2.3.9M-1984 | Metric heavy hex flange screws |
| ANSI/ASME B18.3-1986 | Socket cap, shoulder and setscrews, inch series |
| ANSI/ASME B18.3.1M-1986 | Screws, socket head cap, metric series |
| ANSI/ASME B18.3.3M-1986 | Hex socket head shoulder screws, metric series |
| ANSI/ASME B18.3.4M-1986 | Screws, hex socket button head cap, metric |
| ANSI/ASME B18.3.5M-1986 | Hexagon socket flat countersunk head cap screws, metric series |
| ANSI/ASME B18.3.6M-1986 | Screws, hexagon socket set, metric series |
| ANSI/ASME B18.5.2.2M-1982 | Bolts, metric round head square neck |
| ANSI/ASME B18.6.5M-1986 | Metric thread forming and thread cutting tapping screws |
| ANSI/ASME B18.6.7M-1985 | Metric machine screws |
| ANSI/ASME B18.13-1987 | Screw and washer assemblies—SEMs (inch series) |
| ANSI/ASME B18.15-1985 | Forged eyebolts |
| ANSI/ASME B18.21.1-1990 | Lock washers |
| ANSI/ASME B18.21.2M-1990 | Lock washers, metric series |

Category 3—machining:

| | |
|---|---|
| ANSI B5.8-1972 (R 1988) | Chucks and chuck jaws |
| ANSI B5.10-1981 (R 1987) | Machine tapers |
| ANSI B5.16-1952 (R 1986) | Accuracy of engine and tool-room lathes |

| | |
|---|---|
| ANSI B17.1-1967 (R 1989) | Keys and keyseats |
| ANSI B17.2-1967 (R 1978) | Woodruff keys and keyseats |
| ANSI B74.2-1982 | Shapes and sizes of grinding wheels and shapes, sizes and identification of mounted wheels, specifications for |
| ANSI B74.13-1990 | Markings for identifying grinding wheels and other bonded abrasives |
| ANSI B94.2-1983 (R 1988) | Reamers |
| ANSI B94.3-1965 (R 1984) | Straight cutoff blades for lathes and screw machines |
| ANSI B94.7-1980 (R 1987) | Hobs |
| ANSI B94.8-1967 (R 1987) | Inserted blade milling cutter bodies |
| ANSI B94.11M-1979 (R 1987) | Twist drills, straight shank and taper shank combined drills and countersinks |
| ANSI B94.21-1986 (R 1987) | Gear shaper cutters |
| ANSI B94.49-1975 (R 1986) | Spade drill blades and spade drill holders |
| ANSI/ASME B5.1M-1985 | T-slots—their bolts, nuts and tongues |
| ANSI/ASME B94.6-1984 | Knurling |
| ANSI/ASME B94.9-1987 | Taps, cut and ground threads |
| ANSI/ASME B94.19-1985 | Milling cutters and end mills |

Category 4—tools and tooling accessories:

| | |
|---|---|
| ANSI B5.25-1978 (R 1986) | Punch-and-die sets (inch) |
| ANSI B5.25M-1980 (R 1986) | Punch-and-die sets (metric) |
| ANSI B94.14-1968 (R 1987) | Punches—basic head type |
| ANSI B94.14.1-1977 (R 1984) | Punches—basic head type (metric) |
| ANSI B94.33-1974 (R 1986) | Jig bushings |
| ANSI B107.6-1978 (R 1987) | Box, open-end, combination and flare nut wrenches (inch series) |
| ANSI B107.9-1978 (R 1987) | Box, open-end, combination and flare nut wrenches (metric series) |
| ANSI B107.10M-1982 (R 1988) | Socket wrenches, handles and attachments for hand (inch and metric series) |
| ANSI/ASME B107.5M-1987 | Socket wrenches, hand (metric) |
| ANSI/ASME B107.8M-1984 | Adjustable wrenches |
| ANSI/ASME B107.19-1987 | Pliers, retaining ring |

Category 5—mechanical components:

| | |
|---|---|
| ANSI B29.2M-1982 (R 1987) | Inverted-tooth (silent) chains and sprockets |
| ANSI B29.6M-1983 (R 1988) | Steel detachable link chains and sprockets |
| ANSI B29.10M-1981 (R 1987) | Heavy-duty offset sidebar transmission roller chains and sprocket teeth |
| ANSI B29.15-1973 (R 1987) | Heavy-duty roller-type conveyor chains and sprocket teeth |

| | |
|---|---|
| ANSI B29.19-1976 (R 1987) | A and CA550 and 620 roller chains, attachments, and sprockets |
| ANSI B92.1-1970 (R 1982) | Involute splines and inspection, inch version |
| ANSI B92.2M-1981 (R 1989) | Involute splines, metric module |
| ANSI/ASME B29.1M-1986 | Precision power transmission roller chains, attachments, and sprockets |
| Category 6—welding: | |
| ANSI/AWS D9.1-90 | Sheet-metal welding code |
| ANSI/AWS D10.12-89 | Recommended practices and procedures for welding low-carbon steels |
| ANSI/AWS D11.2-89 | Guide for welding iron castings |
| ANSI/AWS D14.2-86 | Machine tool weldments, specification for metal cutting |
| Category 7—heat treatment: | |
| ANSI/SAE AMS 2728 | Heat treatment of wrought copper beryllium alloy parts |
| ANSI/SAE AMS 2756 | Gas nitriding of steel parts |
| ANSI/SAE AMS 2757 | Gaseous nitrocarburizing |
| ANSI/SAE AMS 2759 | Heat treatment of steel parts, general requirements |
| ANSI/SAE AMS 2759/3 | Heat treatment of precipitation corrosion-resisting and maraging steel parts |
| ANSI/SAE AMS 2759/4 | Heat treatment of austenitic corrosion-resistant steel parts |
| ANSI/SAE AMS 2759/5 | Heat treatment of martensitic corrosion-resistant steel parts |
| ANSI/SAE AMS 2759/6 | Heat treatment and gas nitriding of low-alloy steel parts |
| ANSI/SAE AMS 2760A | Heat treatment—carbon, low-alloy, and specialty steels |
| ANSI/SAE AMS 2770D | Heat treatment of aluminum alloy parts |
| ANSI/SAE AMS 2775A | Case hardening of titanium and titanium alloys |
| Category 8—tolerances and fits: | |
| ANSI B4.1-1967 (R 1987) | Preferred limits and fits for cylindrical parts |
| ANSI B4.2-1978 (R 1984) | Preferred metric limits and fits |
| ANSI B4.3-1978 (R 1984) | General tolerances for metric-dimensioned products |
| ANSI B89.3.1-1972 (R 1988) | Out-of-roundness, measurement of |
| ANSI B89.6.2-1973 (R 1988) | Temperature and humidity environment for dimensional measurement |
| ANSI Y14.5M-1982 (R 1988) | Dimensioning and tolerancing |
| ANSI/ASME B1.22M-1985 | Gauges and gauging practice for MJ series metric screw threads |
| ANSI/ASME B107.17M-1985 | Gauges, wrench openings, reference |

Category 9—drawing symbols and formats:

| | |
|---|---|
| ANSI Y10.20-1975 (R 1988) | Mathematic signs and symbols for use in physical sciences and technology |
| ANSI Y14.1-1980 (R 1987) | Drawing sheet size and format |
| ANSI Y14.7.1-1971 (R 1988) | Gear drawing standards—part 1, for spur helical, double-helical, and rack |
| ANSI Y14.7.2-1978 (R 1989) | Gear and spline drawing standards—part 2, bevel and hypoid gears |
| ANSI Y14.17-1966 (R 1987) | Fluid power diagrams |
| ANSI Y14.36-1978 (R 1987) | Surface texture symbols |
| ANSI Y32.10-1967 (R 1987) | Graphic symbols for fluid power diagrams |

Category 10—gauging and inspection:

| | |
|---|---|
| ANSI B4.4M-1981 (R 1987) | Inspection of workpieces |
| ANSI B89.3.1-1972 (R 1988) | Out-of-roundness, measurement of |
| ANSI/ASME B1.2-1983 | Gauges and gauging for unified screw threads |
| ANSI/ASME B1.3M-1986 | Gauging systems for dimensional acceptability, inch and metric screw threads (UN, UNR, UNJ, M, and MJ) |
| ANSI/ASME B1.16M-1984 | Gauges and gauging for metric M screw threads |
| ANSI/ASME B1.19M-1984 | Gauges for metric screw threads for commercial mechanical fasteners—boundry profile defined |
| ANSI/ASME B107.17M-1985 | Gauges, wrench openings, reference |

***Standards and Approval Agencies and Acronyms***

Air Movement and Control Association (AMCA)
American Gas Association (AGA)
American Gear Manufacturers Association (AGMA)
American National Standards Institute (ANSI)
American Refrigeration Association (ARI)
American Society of Heating, Refrigeration, and Air-conditioning Engineers (ASHRAE)
American Society of Mechanical Engineers (ASME)
American Society of Sanitary Engineering (ASSE)
American Standards Association (ASA) (now ANSI)
Applied Research Laboratories (ARL)
Army-Navy Standard (AN)
Association of Home Appliance Manufacturers (AHAM)
California Energy Commission (CEC)
Canadian Gas Association (CGA)
Canadian Standards Association (CSA)

Department of Transportation (U.S.) (DOT)
Deutchland Ingineering Normalization (German Engineering/Industrial Standard) (DIN)
ETL Testing Laboratories (ETL)
Factory Mutual (FM)
Florida Solar Energy Center (FSEC)
Gas Appliance Manufacturers' Association (GAMA)
Home Ventilating Institute (HVI)
International Association of Plumbing and Mechanical Officials (IAPMO)
International Electrotechnical Commission (IEC)
International Standards Organization (ISO)
Joint Army-Navy Standard (JAN)
Military Standard (MS)
National Aeronautics and Space Administration (NASA)
National Electrical Code (NEC)
National Electrical Manufacturers' Association (NEMA)
National Fire Protection Association (NFPA)
National Sanitation Foundation (NSF)
Occupational Safety and Health Administration (OSHA)
Outdoor Power Equipment Institute (OPEI)
Solar Rating and Certification Corporation (SRCC)
Underwriters' Laboratories (UL)
United States Department of Agriculture (USDA)
USA Standard (USAS) (now ANSI)

## 15.3 APPROVAL ASSOCIATIONS AND THEIR TRADEMARKS

Refer to Fig. 15.1.

## 15.4 FREQUENTLY USED MECHANICAL AND ELECTRICAL STANDARDS

ANSI mechanical standards:

| | |
|---|---|
| B1.1 | Unified inch screw threads |
| B1.5 | Acme screw threads |
| B1.8 | Stub acme screw threads |
| B1.9 | Buttress inch screw threads |
| B1.10 | Unified miniature screw threads |

| | |
|---|---|
| **UL Listed** | Products or systems are evaluated by Underwriters Laboratories with respect to hazards to life and property. Listing signifies that production samples of the product have been found to comply with the established requirements. |
| **UL Recognized** | An evaluation by Underwriters Laboratories of component parts which will be used later in a complete product or system. These would be factory-installed components in UL listed equipment. |
| **CSA Certified** | Product or system has been evaluated by Canadian Standards Association through examination, testing, and inspection, and complies with applicable standards of safety and/or performance. |
| **Static Control** | This refers to products and materials that discourage the formation of static electricity or are designed to drain static charges to the ground. |
| **OSHA** | The product is designed to meet the requirements of the Occupational Safety and Health Administration, which sets standards necessary or appropriate to provide safe or healthful employment and places of employment. |
| **FM** | Factory Mutual System is a group of mutual insurance companies that provides insurance from fire, explosion, accidents and other hazards. Services include fire prevention, inspection, research and consultation. |
| **ANSI** | Product meets the requirements of the American National Standards Institute, an organization which coordinates safety, engineering, and industrial standards. |
| **OSHA/ANSI** | Product meets requirements of Occupational Safety and Health Administration and American National Standards Institute. |
| **Explosionproof** | Product may be used in at least one of the hazardous locations defined by the National Electrical Code. These are classified according to the nature of the hazard. |
| **Federal Specification** | These specifications describe essential and technical requirements for items, materials, or services bought by the U S Government. |
| **Military Specification** | These specifications describe requirements for products bought by the U S armed forces. |

**FIGURE 15.1** Trademarks of approval associations.

| | |
|---|---|
| B2.1 | Pipe threads |
| B4.1 | Preferred limits and fits for cylindrical parts |
| B4.2 | Preferred metric limits and fits |
| B17.1 | Keys and keyseats |
| B17.2 | Woodruff keys and keyseats |
| B18.2.1 | Square and hex bolts and screws |
| B18.2.2 | Square and hex nuts |
| B18.3 | Socket cap, shoulder, and setscrews |
| B18.5 | Round head bolts |
| B18.6.3 | Slotted and recessed-head machine screws and nuts |
| B18.8.1 | Clevis pins and cotter pins |
| B18.21.1 | Lock washers |
| B18.22.1 | Plain washers |
| Y14.5 | Dimensioning and tolerancing |
| Y14.36 | Surface texture symbols |
| ANSI H35.1-1982 | Alloy and temper designations for aluminum |
| ANSI/NFPA 65-1980 | Processing and finishing of aluminum |
| ANSI H35.2-1982 | Aluminum mill products, tolerances, dimensions |
| ANSI/ASTM B224-80 | Classification of coppers |
| ANSI/NFPA 48-1982 | Magnesium |
| ANSI/IEEE 268-1982 | Metric practice |
| ANSI/NFPA 481-1982 | Titanium and alloys |
| ANSI/AWS C5.6-79 | Arc welding, gas metal |

Electrical standards (ANSI and others):

| | |
|---|---|
| ANSI Y14.15a-1971 | Electrical and electronics diagrams, interconnection diagrams |
| Mil-std-15-1, 1961 | Graphical drawing symbols |
| Y32.2 and 32.2a-1964 | USA Standard for graphical symbols for electrical diagrams |
| C6.1-1956 or latest revision | USA Standard terminal markings for electrical apparatus |
| ANSI/NFPA 70-1984 | National Electrical Code |
| ANSI/EIA | Components for electronic equipment, value RS-385-1970 (R 1983) |
| ANSI/IPC D-310A-1977 | Artwork generation and measurement techniques, printed circuit |
| ANSI/IEEE 315-1975 | Symbols for electrical and electronic diagrams |
| ANSI/IEEE 200-1975 | Electrical and electronic parts reference designations |

## 15.5 UNITED STATES PATENTS

All designers, at one time or another, have thought about patenting their ideas or inventions.

The United States Patent Office issues two types of patents that are of interest to the designer. One is termed a *letters patent;* the other, a *design patent.* The letters patent, when granted, remains in force for 17 years from the date of issue and is the type of patent that covers new ideas and devices. The design patent is issued to cover graphic designs or pictorial representations of forms and is generally of less use to designers who wish to protect their ideas and devices.

We will therefore limit this discussion to the process of the letters patent.

***The Process of Letters Patent.*** Anyone anticipating obtaining a letters patent for an original idea or invention should do the following:

- Keep an accurate record of your idea or invention in a record-type notebook that is bound so that pages cannot be inserted or removed without detection. Use a separate book for each different invention.
- Date the beginning of every daily entry and sign or initial every page.
- Cross out changes or errors and initial them. Do not erase.
- When the notebook is complete, describing your invention in detail, have it notarized and witnessed by two parties who understand the basic concepts of the invention.
- Store all separate, related papers and drawings with the notebook. Have these notarized, also.
- Prepare a preliminary write-up for a description of your invention, stating what it does, how it functions, why it is new, how it is better than similar inventions (or if it is unique), and other pertinent facts and data. Your patent attorney will work from your written description, drawings, and verbal statements.
- Check as many sources as possible to convince yourself that your idea is indeed new and patentable.

At this stage, it is wise to decide on your patent attorney. I do not recommend that anyone attempt to obtain a letters patent without the aid and guidance of a patent attorney—that is, unless you yourself are trained in patent law and procedures.

The patent attorney you select will need to review your invention disclosure, as well as all applicable records and drawings, and will need copies of all your data relating to the invention.

The first stage in seeking a letters patent is to make a search in the Patent Office to see if there are any other inventions that have anticipated your invention or ideas. Patents from foreign countries that have patent agreements with the United States must also be searched. The search is conducted by your attorney or a professional searcher assigned by your attorney.

If the search does not disclose another patented invention that anticipates your invention, the patent attorney will inform you of such, and procedures may continue.

During this stage, your attorney will write the preliminary disclosure, including the *critical* claims, which are the heart of the letters patent. The necessary patent drawings are made, describing the invention pictorially. After you check the disclosure and approve it, the document is registered at the Patent Office (i.e., as a patent application). In a few weeks, you will receive a patent registration receipt from the Patent Office, which renders the invention *patent pending*.

In time, your disclosure will be reviewed by a patent examiner, who is an official of the Patent Office. Examiners specialize in different disciplines, and yours will usually be trained in the field of your invention. Your disclosure and its claims will be ruled on by the examiner, duly noted, and eventually returned to your patent attorney for the required action.

If the ruling is favorable and the claims allowed, you will be informed that the letters patent will be granted. After the necessary fees are paid, the letters patent will be issued to you. Of course, it takes some months to complete the procedure. The United States Patent Office reviews thousands of patent applications simultaneously.

A United States Letters Patent (Fig. 15.2) is a very formal, impressive document, and anyone who is granted one should be proud.

**FIGURE 15.2** The United States letters patent.

The procedures involved in obtaining a letters patent can be long and difficult, according to the complexity of the invention and the action taken by the examiner. A good deal of money and effort is required of one who seeks to obtain a letters patent. At this point I wish to set forth some facts about patents that I have gained through experience.

**1.** If you signed a patent agreement form with your employer, the patent must be assigned to the employer if it involves the type of work performed by your employer, and if the employer desires the assignment.

**2.** If two or more people are issued the patent, the patent rights belong equally to all whose names appear as the inventors. There is no such condition as owning one-half, one-third, or one-fourth of a patent; all the parties own 100 percent of the patent and may proceed on that basis. Auxiliary agreements that are legally binding are often made by the inventors.

**3.** The issuance of the letters patent to you is only the beginning of the procedures required to market your invention. As the inventor, you must

   ***a.*** Find a developer or manufacturer for your invention.

   ***b.*** Negotiate and sign royalty and assignment contracts.

   ***c.*** See that your invention is developed and marketed.

   These tasks must be done before there can be remuneration for your skill, time, expense, and effort.

**4.** You or those to whom you assigned the patent rights must enforce those rights. In other words, you must keep your invention from being infringed on by others. Enforcement can require time, effort, and legal expenses on your part.

**5.** Your invention may be considered as "close to" or infringing on another invention, even though you were issued a patent. In this case, you could be sued by another inventor or company. If you go to court because of this and lose the contestment, you are liable to fines, court costs, and the revoking of your patent and/or related patents.

6. You must make sure your patent assignment contract has a "no shelving" clause. This assures that the invention will not be neglected, with no action taken to develop and market it.
7. If possible, your patent attorney should be present during contract assignments or other legal proceedings in order to review and comment on the contract or proceedings.

This, then, is a brief description of patent procedures and recommendations. Your patent attorney will inform and guide you on all the requirements of pursuing a letters patent, and you should follow this advice.

***Patent Drawing Requirements.*** Patent drawings that form part of your disclosure are usually prepared by an experienced patent draftsman for fees varying according to the complexity of the drawings. Many designers are expert drafters themselves and are allowed to prepare their own patent drawings, provided the patent drawing preparation requirements are adhered to.

All patent drawings are drawn in ink on a specified-size, white Bristol board. Black India ink is usually employed, in various line weights and configurations, with a specified minimum space between lines. The patent drawing must be clean and clear, and depict the invention in various views to describe properly its functions or salient features. Third-angle orthographic projection, trimetric views, isometric views, and perspective views are all permitted, as are auxiliary views, section cuts, detail views, and exploded views.

One of the simplest, most effective ways to produce the finished, inked patent drawing is to first prepare a penciled drawing showing all the important features required. Then, placing the Bristol board patent-drawing form over the penciled drawing, trace through the drawing using a light table. This method allows a clean, sharp finished drawing to be produced with a minimum of effort. The alternate approach of inking over penciled lines directly onto the Bristol board is both tedious and prone to errors.

A small light table can be easily constructed using $\frac{1}{2} \times 4$-in lumber as a frame. Fluorescent lights can be added into the frame, and a $\frac{1}{4}$-in plate-glass top positioned over the lights, secured by a notched area in the edge of the top of the frame. Fluorescent lights are recommended so that excess heat is not transferred through the glass working surface; incandescent lights produce too much heat for this application.

The level of illumination should be sufficient to show fine detail on the penciled drawing through the Bristol board. If the penciled drawing is too light, reproduce a copy on a photocopy machine, such as are used in many offices. The recommended size of the light table is 16 in wide × 24 in high. The cost of the table will be recovered in producing only a few finished drawings. In the case of a complex patent drawing, it will be recovered in the production of the first drawing.

Modern, tubular-fed ink pens are recommended for inking the finished patent drawing. These pens are available in various line widths from very fine to thick. The point of the pen should be *jeweled* to produce the best results and also to prevent point wear and the gouging of the Bristol board surface. Undercut, inking triangles should be employed for line work to prevent ink from running under the triangle and ruining the drawing. Prepunched drafting templates should be used wherever possible for accuracy and consistency in drawing.

Patent-drawing samples, line requirements, views, and symbols are detailed in a booklet prepared by the Government Printing Office. The booklet may also be obtained from your patent attorney.

An official patent drawing is shown in Fig. 15.3*a*–15.3*c* for reference.

## 15.6 PRODUCT LIABILITY

Safety in product design should be of prime importance to all designers.

# United States Patent

[11] **3,596,139**

[72] Inventor **Ronald A. Walsh**
**191 Plymouth Lane, Apt. A., Glen Burnie, Md. 21061**
[21] Appl No **870,518**
[22] Filed **Oct. 22, 1969**
[45] Patented **July 27, 1971**

[54] **IMPROVED ELECTRONIC COMPONENT ASSEMBLY CYLINDRICAL SHELL HOUSING WITH INNER PERIPHERAL RADIATING FIN CIRCUIT BOARD FASTENER MEANS**
**10 Claims, 6 Drawing Figs.**

[52] **U.S. Cl.** .......... **317/101 R,** 174/68.5
[51] **Int. Cl.** .......... **H05k 5/06**
[50] **Field of Search** .......... 317/101, 99, 101 DH, 100, 101 D; 174/68.5, 52, 72, 82; 339/17, 119, 128, 129; 200/168 G

[56] **References Cited**

UNITED STATES PATENTS

| | | | |
|---|---|---|---|
| 2,633,526 | 3/1953 | Snyder | 174/68.5 X |
| 2,820,866 | 1/1958 | Graybill et al. | 200/168 (G) X |
| 2,833,966 | 5/1938 | Goodier et al. | 317/101 |
| 2,976,806 | 3/1961 | Risk et al. | 317/101 |
| 3,087,095 | 4/1963 | McConkey, Jr et al. | 317/99 |
| 3,219,886 | 11/1965 | Katzin | 317/101 (D) UX |
| 3,234,433 | 2/1966 | Braunagel | 174/68.5 X |
| 3,257,585 | 6/1966 | Ransom et al. | 317/101 |
| 3,414,806 | 12/1968 | Carr | 174/68.5 X |
| 3,500,131 | 3/1970 | Seeley et al. | 317/101 (DH) X |

*Primary Examiner*—J. R. Scott
*Attorney*—John F. McClellan, Sr.

**ABSTRACT:** Demountable assembly and packaging means and method securing rectangular circuit boards in axially symmetrical tubular configuration in a cylindrical housing without the use of conventional attachment hardware, are described, including longitudinal grooves inside the housing with mountings fixed radially inwardly from the grooves, the inner ends of the mountings protruding between and receiving the edges of adjacent component boards, and the component boards and mountings having integral lay-in wireways permitting unfolding the assembly on removal from the housing without necessity for unwiring; the end closure of the housing is provided with inwardly detachable connectors receiving leads from the component boards to facilitate mounting and demounting; an axial spider having radial legs tightening the assembly of mountings and component boards is optionally provided.

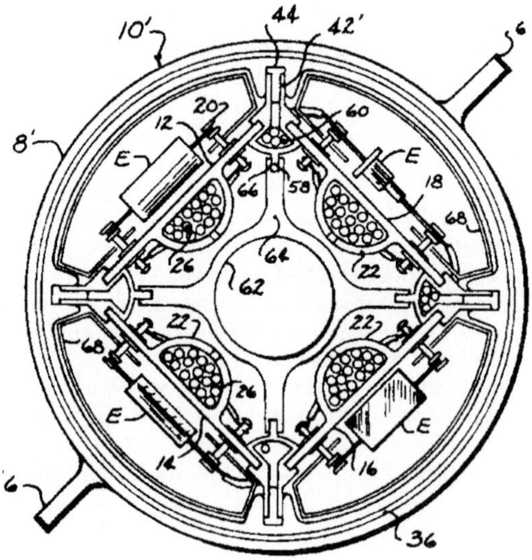


**FIGURE 15.3** (*a*) Cover sheet of U.S. letters patent.

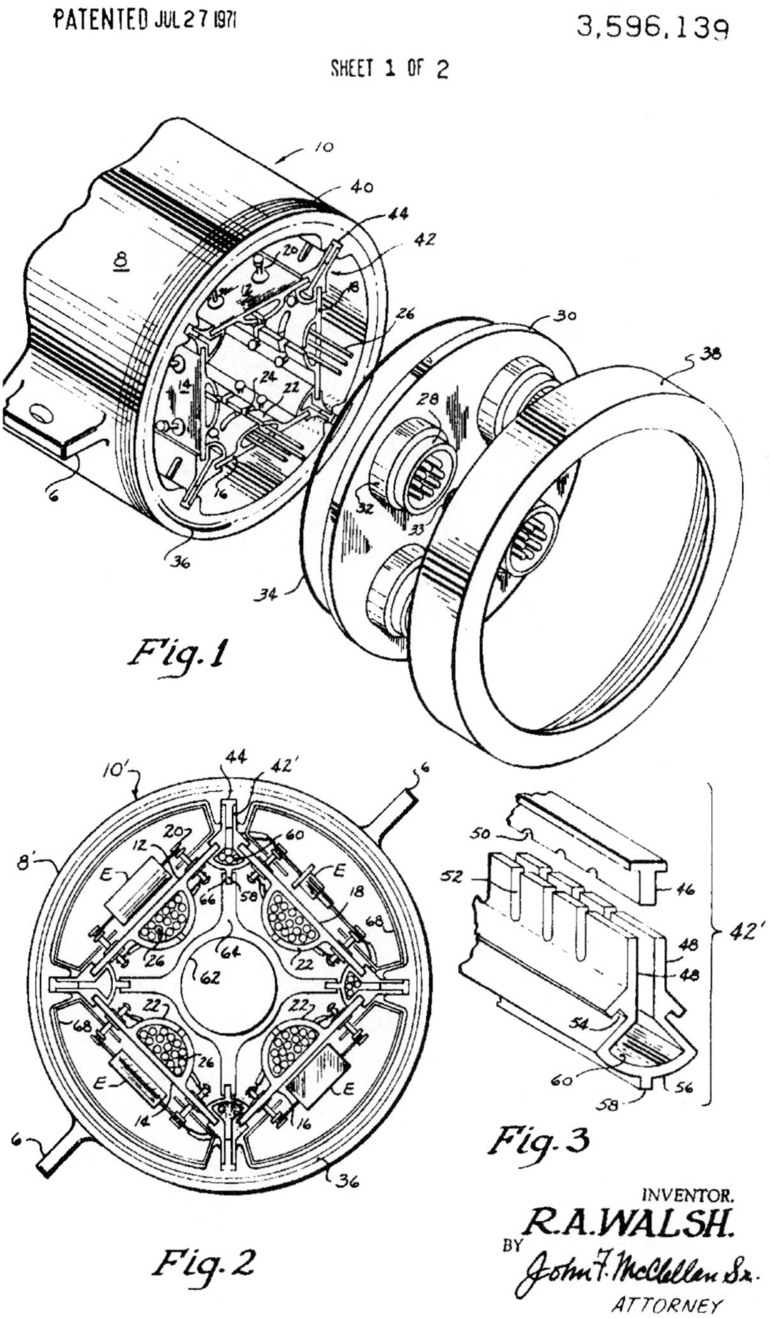


**FIGURE 15.3** (*b*) Patent drawing sample.

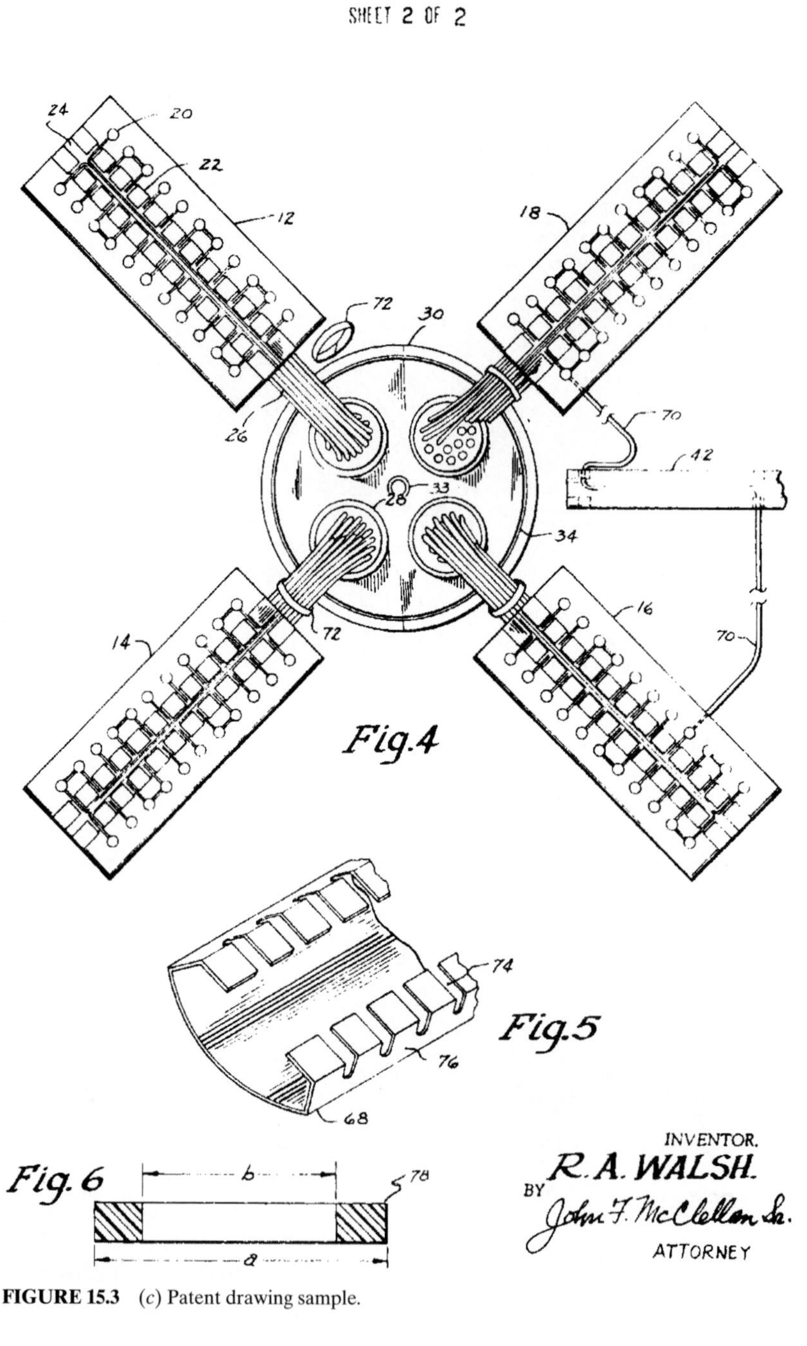


**FIGURE 15.3** (*c*) Patent drawing sample.

There have been almost five hundred awards of a million-dollars-plus granted since 1962 in the United States arising from product liability lawsuits brought against manufacturers. It is of prime concern then that designers never sacrifice safety for profit. Some of the criteria to consider and make note of during the design of a product are

1. All possible hazards in using or misusing the product
2. The environment in which the product is used
3. The typical user of the product, and the typical nonuser
4. All instructions and warnings that are to be presented with the product

Many products are inherently dangerous and cannot be rendered "undangerous." Examples of dangerous products include

- Automotive equipment (vehicles)
- Aeronautical equipment
- Power tools
- Many chemicals
- Most machinery
- Electrical equipment
- Firearms and munitions

The list is, of course, only partial. When you stop and look closely, almost every implement manufactured and used by humans can be dangerous if misapplied. This leaves the product designer with much to think about when designing a product.

Some of the strongest safeguards available to product designers and manufacturers against product liability lawsuits are

1. The use of prototypes and fully functional models to demonstrate a product's characteristics. Also important are quality control tests of random production-run samples.
2. The accurate recording of qualitative test results, including high-speed film records and laboratory test equipment readings.
3. The application of warning labels and the supplying of accurate instructions with the product. Warning labels must attract the attention of the user, be clear and understandable, and, finally, convey the nature and extent of the probable harm resulting from misuse or failure to follow the warnings and instructions.

Following the safeguards listed above, the product designer and manufacturer will be in a strong position if a product liability lawsuit is brought against the manufacturer in relation to one of their products.

## 15.7 PROTOTYPES OF PRODUCTS

The prototype or working model of a product serves many purposes, including

- To prove the functionality and practicality of a design
- To provide test results for product evaluation
- To provide an indication of the feasibility of cost reduction and design-improvement measures prior to final design

Many products would be difficult to develop without the use of a prototype. It allows designers to see if their ideas are functional, and safe, meet specifications or standards requirements, and are practical to manufacture.

Some products and engineering projects cannot have a prototype phase. These types of projects must rely on design calculations, proven specifications, design standards, and experience.

Before a prototype can have any engineering value, it must be produced in the following manner:

1. It must be constructed *exactly* as described by the engineering drawings and instructions. Otherwise, design and detail errors could go undetected.
2. It must contain the same materials and processes as shown on the engineering drawings, wherever possible. Exceptions must be noted and evaluated.
3. It must be tested in accordance with its functional usage and requirements, using standard test procedures as specified by ASTM, ANSI, NEMA, SAE, AGMA, or other applicable organizations.

Only through these procedures can an accurate evaluation be made as to functionality, practicality, reliability, safety, and cost.

Detailed records should be kept of prototype tests and evaluations, together with the appropriate engineering calculations used in the design.

A product may require many prototype stages in order to render it suitable for manufacture.

## 15.8 COMPUTERS IN DESIGN

The use of computers to aid in the design of products is now relatively common practice. Many types of computations can be performed quickly and accurately using mathematics programs. The use of finite element analysis (FEA) and statistics allows the evaluation of complex parts and assemblies.

Computer-aided design (CAD) and computer-aided engineering (CAE) programs are readily available, enabling designs to be constructed and viewed on the computer's display screen. Parts can be represented in three dimensions as a trimetric view and can be rotated to various viewing angles. The computer can take orthographic projections and automatically present a three-view pictorial representation of the part. Images on the viewing screen can be sent electronically to a plotter, which will draw whatever is sent from the screen.

Complex mechanical and electrical or electronic drawings and diagrams can be constructed on the screen using graphics programs. The drawings can be produced in various layers, in different colors for each layer. The layers can be shown individually, or combined to show the total picture.

A big advantage of using graphics programs is that major and minor changes can be made to a drawing before it is put on paper. And the drawings are easily filed on floppy disks as a backup storage medium, in case the finished drawings are lost or destroyed. Also, the plotters produce a top-quality drawing that is neat and easy to read. Some plotters have multiple-colored pens that make it possible for drawings to be produced in colors for easier reading.

Computers allow the designer to produce first-class technical writing by means of scientific word processors. These technical word processors allow the designer to display any type of equation along with a written description, making it easier to prepare engineering-analysis reports, tables, and test reports. When a graphics program and scientific word processor are used together, it is possible to prepare instruction books or technical books for a product.

The computational power of some of the mathematics programs is formidable, allowing the designer to perform many complex calculations in a short time and with high accuracy. Some of these programs provide a printout of the screen, allowing accurate records to be produced for reference and future use.

**FIGURE 15.4** The personal computer using AutoCAD.

The use of computers in engineering is a necessity today for certain applications, and constant progress in computer technology will give the designer even more computational power in the future.

Figure 15.4 shows a modern AutoCAD station using a fast 486, 50-MHz personal computer (PC) to produce engineering drawings, layouts, and geometric studies. AutoCAD version 12 is being used here to produce an assembly drawing of an electric motor drive unit for a high-power switching device, which is a good example of an electromechanical product.

Figure 15.5*a* and 15.5*b* are photographs of an actual electromechanical switching device, developed by the author, on the assembly line. Figure 15.5*b* is a closeup view of the high-pressure clamping mechanism that is shown in Sec. 8.30 in Chap. 8.

Figure 15.6 is a three-dimensional exploded view drawing of one of the poles of the three-phase switching device shown in Fig. 15.5*a* and 15.5*b* as it is being prepared for inclusion into the instruction book for the finished product.

Powerful computer designer stations, as used in the larger or more advanced companies, are capable of generating the design and motion-study drawings required by the design engineer during the design stages of product development. Geometric relations, dimensional interference indications, and stress calculations are also performed on the more powerful computer designer stations.

**FIGURE 15.5** (*a*) High-voltage, high-current switching device.

**FIGURE 15.5** (*b*) Detail view of clamping mechanism of device shown in Fig. 15.5.

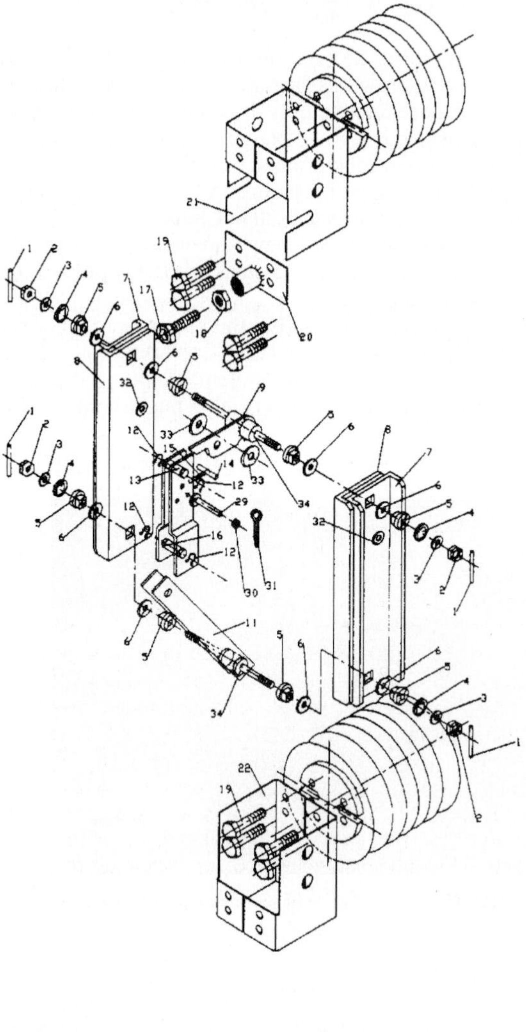

**FIGURE 15.6** An AutoCAD drawing for a product instruction book.

Computers are valuable assets to any design engineering department, even in the smaller companies, and they are indispensable for many engineering and manufacturing applications and functions in modern industry.

## 15.9 ELECTROMECHANICAL SYSTEM DESIGN

Figure 15.7 is a good example to illustrate the many functions of modern electromechanical design engineering practice. This figure is of a complete electric motor control system that was developed in order to improve the serviceability and performance and drastically reduce the cost of a standard electrical motor control system for an electric power distribution switching device unit.

In this system, the previously used costly reversing contactors and electromechanical relay delay timers were replaced by standard electronic and integrated-circuit (IC) components. If you look at the full-wave bridge rectifying circuit section, you will see the classic voltage-divider circuit explained in Chap. 6 of the handbook. Resistor $R_1$ forms the voltage-divider portion of the circuit, followed by a filter capacitor and a voltage-clamping zener diode to maintain the voltage in this branch of the circuit at 24 V dc.

The area outline marked by the letter A is the relay delay timer portion of the circuit, utilizing the 555 timer IC. Diode $D_1$ may be omitted, since this circuit is not a high-speed circuit for driving TTL ICs. The 1-MΩ potentiometer $R_5$ may be changed to 5 MΩ if a delay range of 0 to 100 sec is desired. This circuit has a delay time range of 0 to 10 sec, which is adequate for this application. This motor control circuit was developed to electrically operate the switching devices shown in Figs. 6.67 to 6.79. The PC foil pattern for the circuit is shown in Fig. 6.153, full scale.

The 555 IC timer section of the circuit of Fig. 15.7 was used again in another relay delay timer, the circuit of which is shown in Fig. 15.8. This circuit was used on the breaker-action type interrupter switch shown in Figs. 6.71*a*, 6.71*b*, and 6.72. This relay delay timer (Fig. 15.8) was incorporated into the opening or trip solenoid circuit of the interrupter switch, to prevent the switch from being tripped open until a time lapse of a minimum of 3 sec after it has been closed. This time delay is required to give the downstream fault-clearing devices time to clear a power system fault, since the load-interrupter type of switch is not capable of interrupting fault currents that can occur in high-power electrical distribution systems. An interrupter switch is capable of clearing or interrupting a fault only if it is equipped with high-power fault-interrupting fuses (expulsion type). Interrupter switches are normally designed to close into system fault currents, but not to open during system fault currents.

Figure 15.9*a* and 15.9*b* shows the completed relay delay timer whose schematic diagram was shown in Fig. 15.8. Figure 15.10*a* and 15.10*b* shows the PC foil pattern of the circuit. Figure 15.10*a* is the component side, and 15.10*b* is the bottom side, looking through the PC board.

It is important for electromechanical designers to know that company-designed electromechanical devices, such as that shown for this relay delay timer, are easy to design, and are easy and economical to manufacture as substitutes for expensive commercial relay delay timers and other electromechanical and electronic devices. The 1994 cost of this relay delay timer module is approximately $12.50, and that of a commercially available relay delay timer that performs the same function is approximately $50.00. This small electromechanical device can control relays rated to 40 A dc without an increase in the size of the PC board shown in Fig. 15.9. With the IC circuits shown in Chap. 6 of this handbook and in the circuit design manuals and brochures available from the linear IC manufacturers, a great variety of useful and cost-effective electromechanical devices and control systems can be designed and implemented by the modern, trained electromechanical design engineer, as part of the complete, integrated mechanical, electrical, and electronic product.

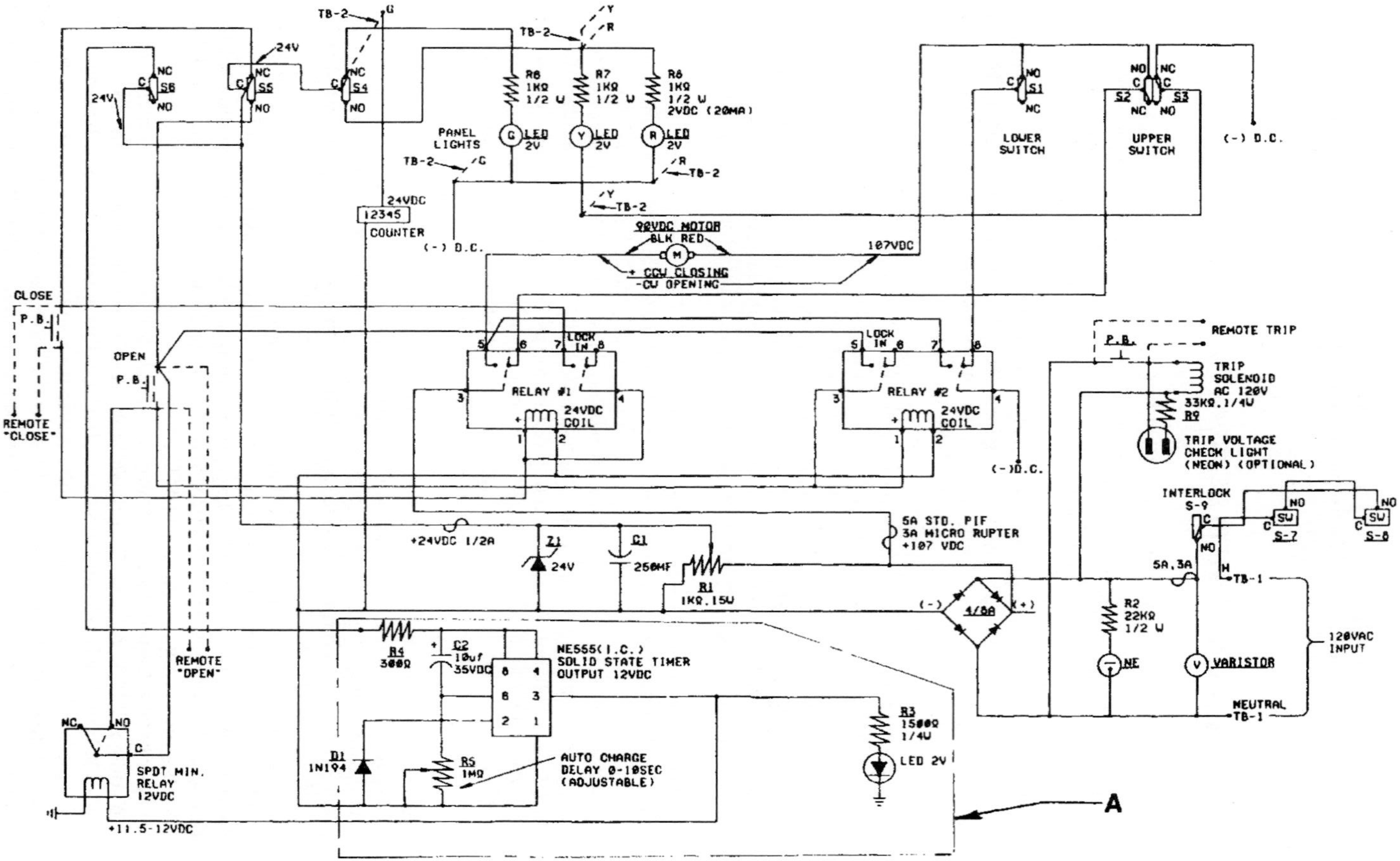


**FIGURE 15.7** A complex motor control system schematic diagram.

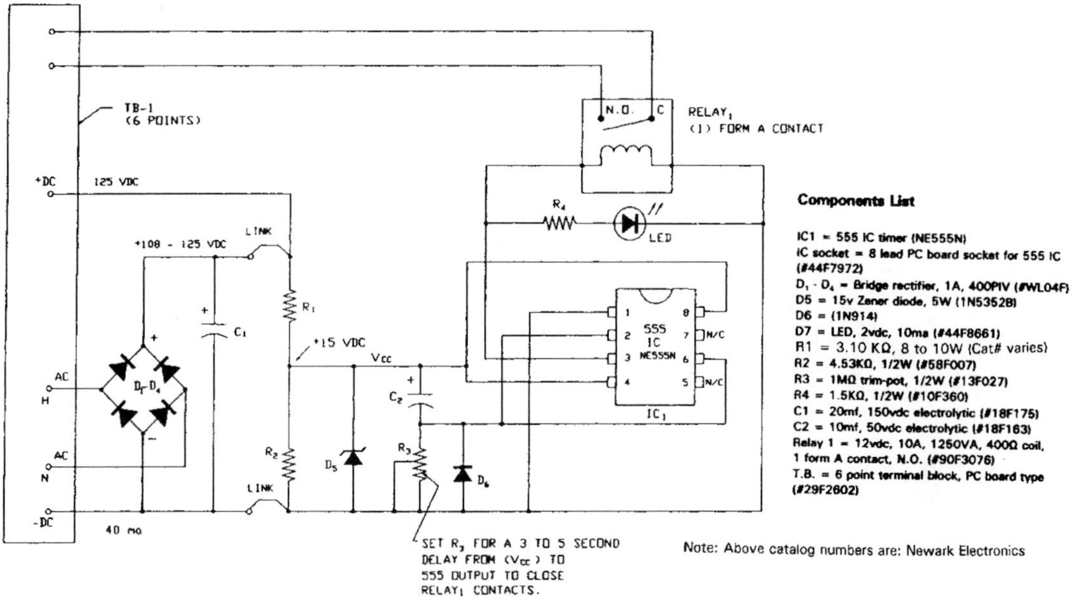


**FIGURE 15.8** Schematic diagram of a relay delay timer using ICs.

In developing a control and operating system like the one described, the electromechanical designer must review and analyze many factors:

- Circuit design
- PC board design
- Electrical and/or electronic packaging
- Motor selection
- Drive-train configurations
- Power requirements
- System function and safety interlocks
- Economics of the design
- Production of the required engineering documentation

Further, there are a host of other design factors that are strictly mechanical. As an example, Fig. 8.220 shows the geometry of a typical, four-bar drive linkage that in this system actuates a spring-charged, electrical-power-interrupting device. Many other mechanical parts were required for the complete system.

The coordination of all these factors should culminate in a prototype of the system for evaluation and incorporation of changes necessary to ready the system for production. It is during the prototype stage that final design-change suggestions and cost-cutting measures should be implemented.

In large companies bringing together complex electromechanical systems, many different departments must coordinate their efforts to arrive at a finished product. By way of example, the engineering and production effort required to produce an aerospace vehicle would generally entail the following activities:

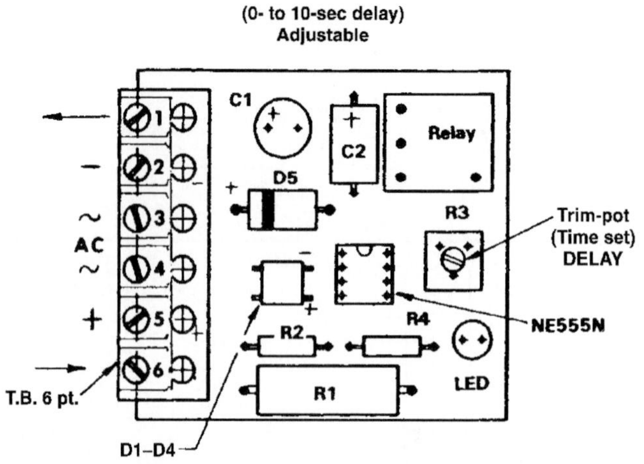


(a)

(b)

**FIGURE 15.9** (*a, b*) Full-scale assembly of the relay delay timer shown in Figure 15.8.

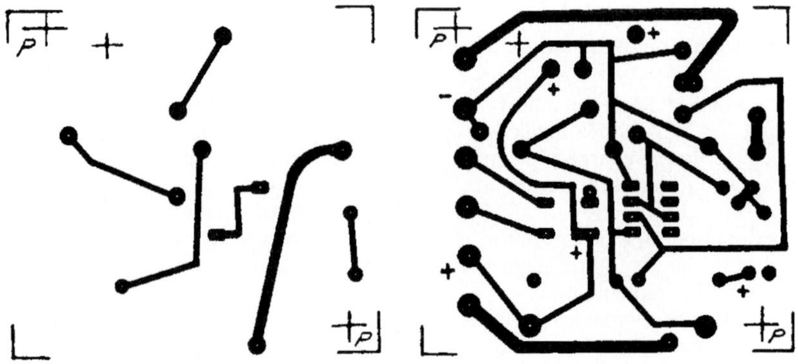

(a) (b)

**FIGURE 15.10** (*a*) Component side foil pattern for relay delay timer of Fig. 15.9; (*b*) bottom foil pattern, looking through the PC board.

*I.* Engineering effort
- *A.* Conceptual engineering (advanced design department)
- *B.* Design engineering groups
  - *1.* Structural design group
  - *2.* Master lines and lofting group
  - *3.* Power-plant engineering group
  - *4.* Pneumatics-hydraulics engineering group
  - *5.* Electrical-electronics engineering group
  - *6.* Design integration group
  - *7.* Flight-test instrumentation engineering group
- *C.* Stress analysis and dynamics engineering groups
- *D.* Weights engineering group
- *E.* Liaison engineering group
- *F.* Tool engineering group

*II.* Support effort
- *A.* Materials tests and process development laboratory
- *B.* Standards group
- *C.* Production illustration group
- *D.* Tooling department

*III.* Purchasing department

*IV.* Sales department

*V.* Manufacturing effort
- *A.* Production planning and coordination department
- *B.* Inspection and quality control departments
- *C.* Manufacturing departments
  - *1.* Machine shop
  - *2.* Sheet metal
  - *3.* Welding
  - *4.* Subassembly
  - *5.* Final assembly
  - *6.* Printed circuits
  - *7.* Electronic assembly
  - *8.* Wiring department
  - *9.* Test department
  - *10.* Plating and finishing
  - *11.* Heat-treating department
  - *12.* Model shop and mockup
  - *13.* Tool and die manufacturing department

## 15.10 QUALITY-CONTROL SYSTEMS

Various standards for controlling the quality of industrial and consumer products have been implemented both nationally and internationally. These standards are important to both manufacturers and product design engineers and engineering departments. In the United States, a system known as *total quality management* (TQM) has been in effect for some years. Other quality-control systems have been implemented nationally and on a company-level basis. These systems help the consumer as well as the manufacturer.

Internationally, the quality-control or quality-assurance system of most importance to American manufacturers is the ISO 9000 system. On January 1, 1993, the European Community (EC), consisting of 12 countries, officially formed a unified market. This began or started the adoption of the ISO 9000 system. All United States companies wishing to sell products to the European market will use the ISO 9000 system as their quality standard.

ISO 9000 establishes a *minimum* quality standard that may be expanded on, based on future requirements. The U.S. Department of Defense may replace standards MIL-I-45208A and MIL-Q-9858A with the ISO 9000 system. After January 1, 1993, U.S. companies may be required to be in compliance with ISO 9000 in order to continue to do business with the EC nations. Note that the American TQM system will exceed the stated ISO 9000 guidelines. Sixty countries have adopted ISO 9000 as their national quality standard.

The EC Consumer Council Ministers have adopted a general product safety directive. This directive imposes general safety requirements on *all* consumer products sold in the EC market. U.S. exporters to the EC must familiarize themselves with the EC product liability regulations. Use of the ISO 9000 standard may be mandatory in the near future in order for U.S. companies to sell their products to the EC and other nations who have adopted the ISO 9000 standard.

***ISO 9000 System.*** The ISO 9000 system is divided into a series of standards, including

- ISO 9001
- ISO 9002
- ISO 9003
- ISO 9004

ISO 9001, 9002, and 9003 are audited quality-control standards, while ISO 9004 contains the guidelines to quality management elements and is not audited. The ISO standard which your company may require depends on your manufacturing and product setup. The breakdown of the ISO 9000 system is as follows:

| | |
|---|---|
| ISO 9001 | Model for quality assurance in design and development, production, installation, and servicing |
| ISO 9002 | Model for quality assurance in production and installation |
| ISO 9003 | Model for quality assurance in final inspection and test |
| ISO 9004 | Guidelines to quality management elements |

Additional information on the entire ISO 9000 system may be obtained from the following organizations:

National Technical Information Service (NTIS)
5285 Port Royal Road
Springfield, VA 22161
Tel. (703) 487-4650
Fax (703) 321-8547

Registration Accreditation Board (RAB)
611 East Wisconsin Avenue
Milwaukee, WI 53202
Tel. (414) 272-8575

U.S. Government Printing Office (GPO)
Superintendent of Documents
Washington, DC 20402
Tel. (202) 783-3238
Fax (202) 275-2529

## 15.11 TEST LABORATORIES

Product testing and evaluation is required in many branches of industry, in order to legally sell your products. The testing is a *legal* aspect of your company's products because some of the purchasing contracts signed by the manufacturer may stipulate that the product is tested and in conformance to certain American standards such as ANSI, NEMA, ASTM, SAE, and IEEE. Also, the product manufacturer may be required by the purchaser to submit proof of product test conformance and acceptability, by way of certified laboratory test reports.

This requirement holds true for electrical power distribution equipment, consumer products with electrical and electronic components (e.g., UL approval), materials suppliers, welding certifications, manned aircraft, aerospace vehicles, and many others.

A list of all American and some foreign test laboratories is contained in the ASTM book *Directory of Testing Laboratories,* 1992. The directory is available from the American Society for Testing and Materials (see Sec. 15.1 in this chapter for ASTM address).

Sample sheets from the *ASTM Directory of Testing Laboratories* are shown in Figs. 15.11 and 15.12.

Photographs of equipment being tested are shown in Figs. 15.13 and 15.14. Figure 15.13 shows a test being conducted on a reactor core for a three-phase current-limiting reactor assembly. The reactor core under test is dissipating 27 kW of heat energy during the current test. This reactor core is rated 15 kV, 3000 A, continuous-current, 95-kV BIL.

Figure 15.14 is a photograph that was taken in an electrical test laboratory during an insulator test. The photograph shows a 300,000-V impulse flashover to ground. The arc traveled down the insulator and across an insulated bus bar to the ground connection. Both Figs. 15.13 and 15.14 were taken at the Powercon high-power test laboratory in Severn, Maryland. The camera shutter must remain open long enough to catch this ultra-fast-impulse test procedure and then close to complete the exposure.

## 15.12 ENGINEERING DRAWING SIZES AND FORMATS

Figure 15.15 shows the U.S. customary and SI engineering drawing sizes that are normally used for engineering documentation.

# ASTM Directory of TESTING LABORATORIES

## THE CODING SYSTEM

In addition to the two descriptive sections, each laboratory listing contains two coded sections: FIELDS OF TESTING/LAB SERVICES and MATERIALS AND PRODUCTS. These codes are explained below.

In the FIELDS OF TESTING/LAB SERVICES section, one or more two-character abbreviations are shown. The laboratory may perform some or all of the tests or services represented by a particular abbreviation. Therefore you should contact the laboratory to ensure that it can perform the specific test or service required.

In the MATERIALS AND PRODUCTS section, again one or more abbreviations are shown. An abbreviation appears either by itself or with a number or number string after it in parentheses. If only the abbreviation appears, then that laboratory handles *all* of the materials and products within that category. For example, a laboratory that lists AGRI (without a number or number string) can handle all the commodities within that category—i.e., 1-food, 2-agricultural products, etc.). On the other hand, if a number or number string follows the abbreviation—for example, ELAS (1) or NONMET (3-8), then that laboratory can handle *only* the commodities delineated by those numbers. For example, a listing of AGRI (4) means that the laboratory handles only forestry products.

### FIELDS OF TESTING

AC **Acoustic/Vibration**—(including noise, shock resistance)

BI **Biological Testing**—(including biochemical, toxicological, pharmacological, bacteriological)

CH **Chemical Testing**—(including wet chemistry, associated physical tests, viscosity, density, particle size, volumetric, gravimetric, corrosion, electrochemistry)

CR **Chromatography**—(including gas, liquid, ion exchange, paper, column, thin layer, gel permeation, gas chromatography/mass spectrometry, liquid chromatography/mass spectrometry)

EE **Electrical/Electronic Testing**—(including electromagnetic, magnetic)

GE **Geotechnical Testing**—(including soil and rock tests, geophysical, seismographic)

ME **Mechanical Testing**—(including tensile, compression, impact, hardness, torsion, fracture, fatigue, shear ductility, metallography, calibration and testing of mechanical equipment, hydraulics)

MT **Metrology**—(measurement of mass, length, and time and their immediate derivatives such as angle, volume, and pressure; calibration)

ND **Nondestructive Evaluation**—(including neutron and X-ray, radiography, ultrasonics, penetrants, magnetic particle, eddy current, acoustic emission, leak testing)

OP **Optics/Photometry**—(including appearance, color, reflectance, gloss, transmittance, luminance)

RD **Radiation, Ionizing**—(including radioactivity, radiochemistry)

SN **Sensory Evaluation**—(psychophysical testing including odor, taste, texture)

SP **Spectroscopy**—(including atomic absorption, emission, fluorescence, infrared, Fourier transform infrared, ultraviolet, visible, X-ray diffraction, nuclear magnetic resonance, Raman, inductively coupled plasma, mass spectrometry)

SA **Surface Analysis/Microscopy**—(including Auger, ion scattering, secondary ion mass spectrometry, X-ray or ultraviolet induced electron emission spectrometry, ion microscopy, transmission electron microscopy, scanning electron microscopy)

TA **Thermal Analysis**—(including differential thermal analysis, differential calorimetric analysis, calorimetry, thermogravimetric analysis, thermomechanical analysis)

TH **Thermal/Fire Testing**—(including heat, temperature, thermal conductivity, fire, flammability, smoke/toxicity)

### LAB SERVICES

BW **Building and Welding Inspection**—(including quality assurance, site monitoring, field surveys, sampling, certification)

CE **Construction Materials Engineering**

EN **Environmental Impact**—(including chemical and biological techniques for measuring impacts)

ES **Environmental Simulation**—(including weathering, accelerated testing, natural exposure)

FR **Forensic**—(including failure analysis, expert testimony)

PP **Product Performance**—(including labeling, reliability, serviceability, certification)

SF **Safety**—(including industrial hygiene)

**FIGURE 15.11** Testing laboratory directory code system.

## MATERIALS AND PRODUCTS

**AGRI—Animal and Plant Products**
1 Food
2 Agricultural products (excluding food)
3 Animal and fishery products (excluding food), including leather, furs
4 Forestry products

**FIBER—Textiles and Fibrous Materials and Products**
1 Textile mill products
2 Textile products, finished products and apparel
3 Geotextiles
4 Lumber, wood, cordage, furniture
5 Paper, packaging and related products

**ELAS—Elastomers and Protective Coatings**
1 Plastics, rubbers, resins
2 Adhesives (organic resins), glues
3 Paints, varnishes, lacquers, printing inks and related products

**COMP—Composite Materials**

**NONMET—Nonmetallic Minerals and Products**
1 Bituminous and other organic materials, coal, tar
2 Cement, asbestos products, concrete, lime, gypsum
3 Soil and rock, building stones, aggregates, sand
4 Ceramics; clay and clay products
5 Glass and glass products
6 Semiconductor materials and devices
7 Petroleum refinery products, including asphaltic materials; petrochemicals; lubricants
8 Petroleum crudes, natural gas
9 Oil shale and tar sands

**METAL—Metallic Materials and Products**
1 Metallic ores, powders
2 Ferrous alloys, steels
3 Nonferrous metals, alloys
4 Stainless alloys including nickel, chromium
5 Primary metals: bar, sheet, pigs, ingots
6 Reactive and refractory metals
7 Metallic coatings
8 Metallic fasteners
9 Metallic products, semifabricated; extrusions; rolled sections
10 Metallic components, cast, forged, welded, pressed

**CONSTR—Constructions**
1 Building constructions including foundations
2 Plumbing, carpentry, tile assemblies, roofing, glazing, and other trade work
3 Highways, bridges, tunnels, and other related constructions
4 Pipelines

**MACH—Machinery**
1 Electrical machinery, equipment, supplies, appliances
2 Machinery, miscellaneous, implements, engines, turbines
3 Boilers, pressure vessels, pipework
4 Transportation vehicles and equipment including tires

**CHEM—Chemicals and Chemical Products**
1 Chemical compounds and allied products (excluding human medicinals)
2 Pharmaceuticals
3 Soaps, detergents, water treatments
4 Fertilizers, feeds, pesticides
5 Ordnance, ammunition, explosives

**LAB—Laboratory, Scientific, Medical Equipment**
1 Scientific instruments
2 Laboratory apparatus, supplies
3 Medical devices
4 Computers

**MARK—Marketplace Products-Consumer and Business**
1 Office, printing, lithographic
2 Educational products
3 Sports equipment, toys, musical instruments
4 Consumer goods, miscellaneous

**BIOMAT—Environmental, Biological Materials**
1 Air; indoor and outdoor atmospheres; stack emissions, noise levels
2 Water; ground water, wastewater, high purity, industrial effluent, saline, recycled, rain, surface, process
3 Soil and rock for environmental uses
4 Hazardous waste, solid (nuclear and chemical)
5 Leachates
6 Animal, human tissues

ASTM

This card is used in conjunction with ASTM's *Directory of Testing Laboratories*, available from:

**ASTM ■ 1916 Race Street ■ Philadelphia, PA 19103-1187**
215-299-5400 ■ TWX 710-670-1037 ■ FAX: 215-977-9679

**FIGURE 15.11** (*Continued*)

**MARYLAND** Baltimore

0481
**INSECT CONTROL & RESEARCH, INC.**
1330 Dillon Heights Ave., Baltimore, MD 21228
(301) 747-4500 Dr. Robin G. Todd, Dir.

**SPECIALTY:** Screening and efficacy testing of pesticides.

**TESTING/LAB SERVICES:** BI/EN, PP

**MATERIALS & PRODUCTS:** CHEM (4)

**EQUIPMENT; TESTING CAPABILITIES; APPLICATIONS:** Screening with common lab insects; maintain colonies of flies, mosquitoes, cockroaches, fleas, lice, bedbugs, stored product pests, etc. Wind tunnel tests for LD50, topical application, residual, aerosol testing, lab and field. Package penetration and repellent studies.

**STAFF:** 10 inc. 3 PHD/MD, 5 tech.

0482
**INSPECTORATE - DANIEL C. GRIFFITH**
(Samplers & Analysts, Inc.) U.S.A. Inc., 1540 Caton Center Dr., Suite D, Baltimore, MD 21227
(301) 536-0506 Fax: (301) 536-0509
James H. Rosemond, Jr., Div. Mgr., Coal

**SPECIALTY:** Coal sampling, inspection, testing, and analysis including Hardgrove, Arnu, Giesler, Ash Fusion, and proximate analysis.

**TESTING/LAB SERVICES:** CH, CR, OP, SP, TA

**MATERIALS & PRODUCTS:** NONMET (1-5); METAL (1-10); CHEM (1); LAB (1-2); BIOMAT (2-3, 5)

**EQUIPMENT; TESTING CAPABILITIES; APPLICATIONS:** Cargo sampling. Crushing and blending equipment for reducing gross samples to laboratory size. ES, AAS, UV, TA, inorganic chemical analysis. Strong emphasis on wet chemistry. Classical methods. Extensive ferro alloy experience. Establishment of standards. Particle size. Water analysis.

**STAFF:** 14 inc. 5 engineers & scientists, 3 tech.

0483
**MARTEL LABORATORY SERVICES, INC.**
1025 Cromwell Bridge Rd., Baltimore, MD 21204
(301) 825-7790 Fax: (301) 821-1054 Paul R. Jackson, V.P.

**SPECIALTY:** Testing of water, wastewater, hazardous wastes, petroleum products, alloys, paint.

**TESTING/LAB SERVICES:** BI, CH, CR, GE, SP/EN, SF

**MATERIALS & PRODUCTS:** AGRI (1); ELAS (3); NONMET (7); METAL (2-4); BIOMAT (1-5)

**EQUIPMENT; TESTING CAPABILITIES; APPLICATIONS:** Environmental analysis (for RCRA, NPDES, EPA Superfund, etc.) by GC, GC/MS, AA, ICP, wet chemistry. Specification testing of new petroleum products. Spectrometric analysis of used oil for aviation, automotive industries. Microbiological testing of water, food. Analysis of alloys, paint, air. Sampling and geophysical services.

**STAFF:** 49 inc. 1 PHD/MD, 22 engineers & scientists, 20 tech.

0484
**MET ELECTRICAL TESTING CO., INC.**
916 W. Patapsco Ave., Baltimore, MD 21230
(301) 354-2200 Fax: (301) 354-1624 Leonard Frier, Pres.

**SPECIALTY:** Independent product testing for safety certification, listing and field testing of equipment and installations.

**TESTING/LAB SERVICES:** AC, CR, EE, ME, SP/BW, ES, PP, SF

**MATERIALS & PRODUCTS:** MACH (1-2); LAB; MARK (4)

**EQUIPMENT; TESTING CAPABILITIES; APPLICATIONS:** Specialist in testing of electrical transmission and distribution equipment. RFI, EMI and environmental testing to military standards. FCC Part 68 and Part 15 certification testing. DOC Canadian testing, safety testing; listing, labeling, and performance testing of electrical products.

**STAFF:** 100 inc. 24 engineers & scientists, 50 tech.

**BRANCHES:** MD, PA

0485
**PENNIMAN & BROWNE, INC.**
PO Box 65309, 6252 Falls Rd., Baltimore, MD 21209
(301) 825-4131 Fax: (301) 321-7384
Herbert E. Wilgis, Jr., Pres.

**SPECIALTY:** Environmental, chemical, fuel, soils, water, air, engineering, metallurgical, marine.

**TESTING/LAB SERVICES:** CH, CR, ME, OP, SP, TH/BW, CE, EN, FR, PP, SF

**MATERIALS & PRODUCTS:** ELAS; NONMET (1-3, 7); METAL (1-5, 7-10); CONSTR (1, 3); CHEM (1, 3-4); LAB (1-2); BIOMAT (1-5)

**EQUIPMENT; TESTING CAPABILITIES; APPLICATIONS:** GC/MS, GC, AA, IR, XRF (for lead), instrumented, wet chemical analysis of water, soils, GC/MS, UV, plastics, coal, petroleum products. CMT laboratory, field inspection of portland cement concrete, bituminous concrete, soils/masonry. Marine chemistry. Asbestos. Metals/alloys: chemical, physical, microstructural analysis. Environmental audits and consulting.

**STAFF:** 30 inc. 1 PHD/MD, 13 engineers & scientists, 13 tech.

0486
**POWELL LABS, LTD.**
1915 Aliceanna St., Baltimore, MD 21231
(301) 732-1606 Strati Yorgiadis, Partner

**SPECIALTY:** High-purity water and metallurgical analyses, wastewater analysis.

**TESTING/LAB SERVICES:** CH, CR, SP

**MATERIALS & PRODUCTS:** CHEM (3); BIOMAT (2)

**EQUIPMENT; TESTING CAPABILITIES; APPLICATIONS:** High-purity water, ion chromatography, atomic absorption, graphite furnace, tube deposit analyses, metal overheating and corrosion failure, wastewater, chemical cleaning, solvent analyses, and selection.

**STAFF:** 3 inc. 3 engineers & scientists.

**1992 ASTM DIRECTORY OF TESTING LABORATORIES**

**FIGURE 15.12** Sample page from *Directory of Testing Laboratories,* published by ASTM.

**FIGURE 15.13** A reactor core under test at a testing laboratory.

**FIGURE 15.14** An electrical impulse test of an insulator system at a testing laboratory.

| Drawing Size, Inches | | Drawing Size, Metric (SI) | |
|---|---|---|---|
| Designation | Size, inches | Designation | Size, mm |
| A | 8.5 x 11 | A0 | 841 x 1149 |
| B | 11 x 17 | A1 | 594 x 841 |
| C | 17 x 22 | A2 | 420 x 594 |
| D | 22 x 34 | A3 | 297 x 420 |
| E | 34 x 44 | A4 | 210 x 297 |
| F | 28 x 40 | ..... | ..... |

**FIGURE 15.15** Table of standard engineering drawing sizes.

# INDEX

## ABOUT THE AUTHOR

Ronald A. Walsh is director of research and development at the Powercon Corporation in Severn, Maryland. An industrial product designer for almost 40 years with such companies as Bendix Radio, American Machine and Foundry, and Martin Marietta, he holds three U.S. patents for electronic packaging systems and mechanical devices and five copyrights. Ronald Walsh worked at Cape Canaveral, Florida, as a liaison engineer on the Titan and Gemini aerospace vehicles and participated in the design of the Titan, Gemini, and Apollo aerospace vehicles and numerous manned military and commercial aircraft and military guided missiles. He has worked with some of the pioneers of advanced aircraft and aerospace vehicle design, atomic energy processing equipment, and military and commercial electrical and electronic equipment. He is the author of the *McGraw-Hill Machining and Metalworking Handbook.*